Catalysis and Reaction Engineering Division 2014

AA003017

Core Programming Area at the 2014 AIChE Annual Meeting

Atlanta, Georgia, USA
16-21 November 2014

Volume 1 of 2

ISBN: 978-1-5108-1252-9

Printed from e-media with permission by:

Curran Associates, Inc.
57 Morehouse Lane
Red Hook, NY 12571

Some format issues inherent in the e-media version may also appear in this print version.

Printed by Curran Associates, Inc. (2015)

AIChE
120 Wall Street, FL 23
New York, NY 10005-4020

Phone: (800) 242-4363
Fax: (203) 775-5177

www.aiche.org

Additional copies of this publication are available from:

Curran Associates, Inc.
57 Morehouse Lane
Red Hook, NY 12571 USA
Phone: 845-758-0400
Fax: 845-758-2634
Email: curran@proceedings.com
Web: www.proceedings.com

TABLE OF CONTENTS

VOLUME 1

VOLUME 2

UNIVERSITEIT GENT

FACULTY OF ENGINEERING AND ARCHITECTURE

Acid Strength Effects on Acid-Base Cooperative Aldol Condensation

J. Lauwaert[1], E.G. Moschetta[2], P. Van Der Voort[3], J.W. Thybaut[1], C.W. Jones[2], G.B. Marin[1]

[1]*Laboratory for Chemical Technology, Ghent University*
[2]*School of Chemical & Biomolecular Engineering, Georgia Institute of Technology*
[3]*Center for Ordered Materials, Organometallics & Catalysis, Ghent University*

AIChE Annual Meeting, Atlanta, USA, November 16th 2014

1

AIChE Annual Meeting, Atlanta, USA, November 16th 2014

Aldol condensation

- Reaction between two components with carbonyl groups (ketone/aldehyde)
- Important reaction to create new C-C bonds
  ➔ yield larger and more complex molecules

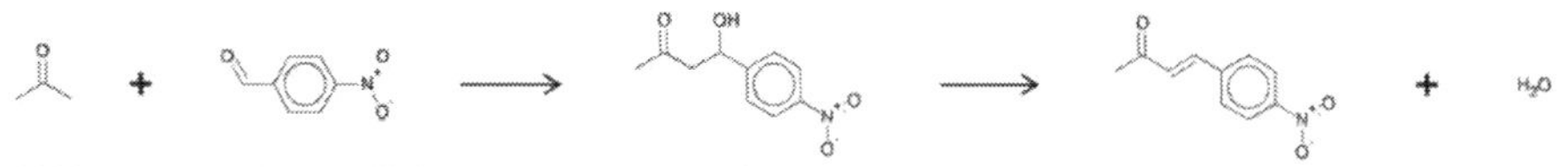

- Applications in the production of:
  - Vitamins and pharmaceutical compounds
  - Solvents
  - Food preservatives
  - Dyes, flavors and fragrances

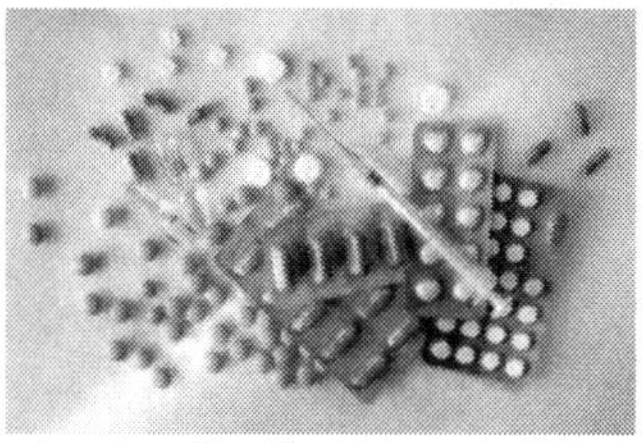

- In industry, catalyzed using homogeneous strong base catalysts such as NaOH

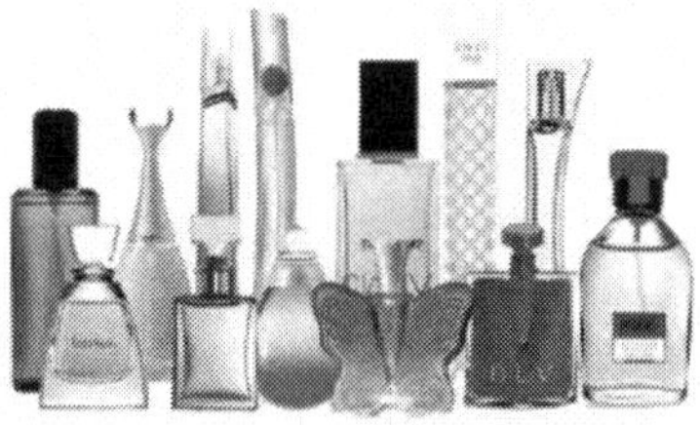

2

Aminated silica materials

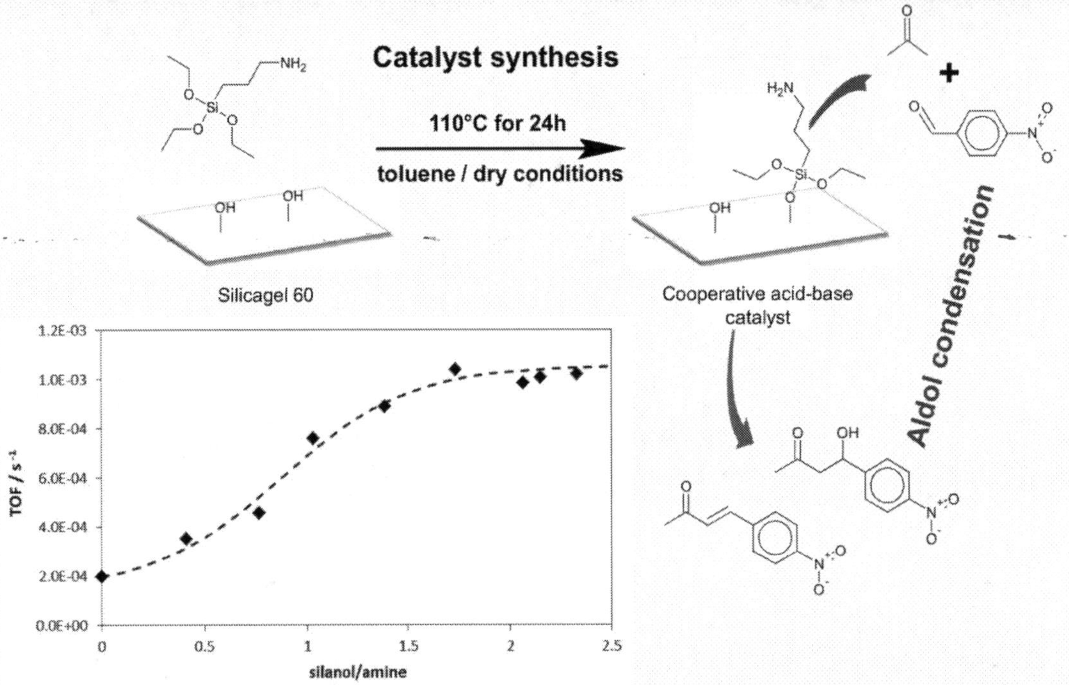

J. Lauwaert, E. De Canck, D. Esquivel, J. W. Thybaut, P. Van Der Voort, G. B. Marin, ChemCatChem, **2014**, 6, 255-264
N. A. Brunelli, C. W. Jones, J. Catal., **2013**, 308, 60-72

Promoting effect of the silanols

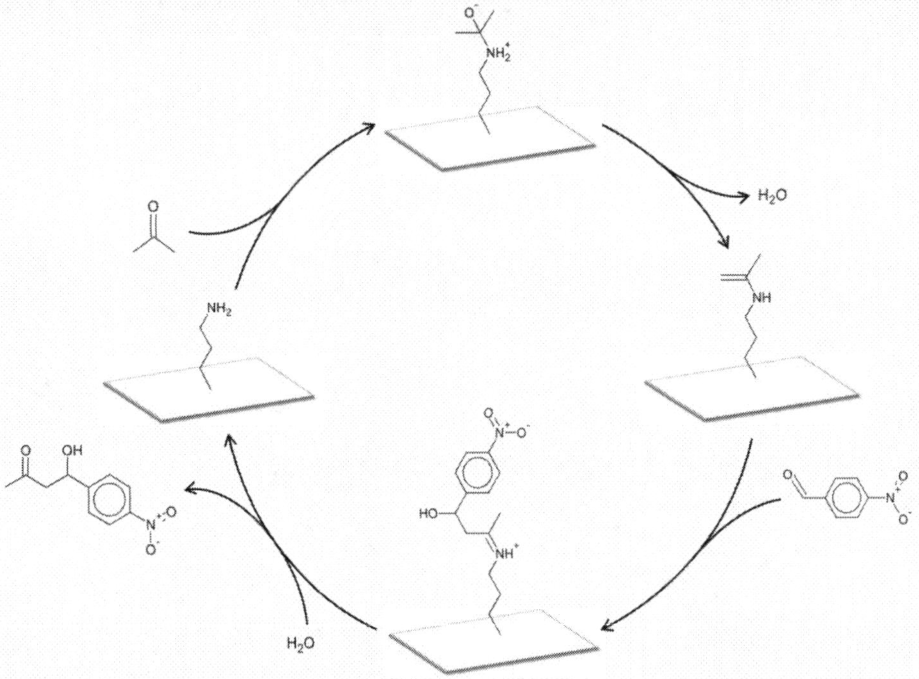

N. A. Brunelli, C. W. Jones, J. Catal., **2013**, 308, 60-72
J. Lauwaert, E. De Canck, D. Esquivel, J. W. Thybaut, P. Van Der Voort, G. B. Marin, ChemCatChem, **2014**, 6, 255-264

Promoting effect of the silanols

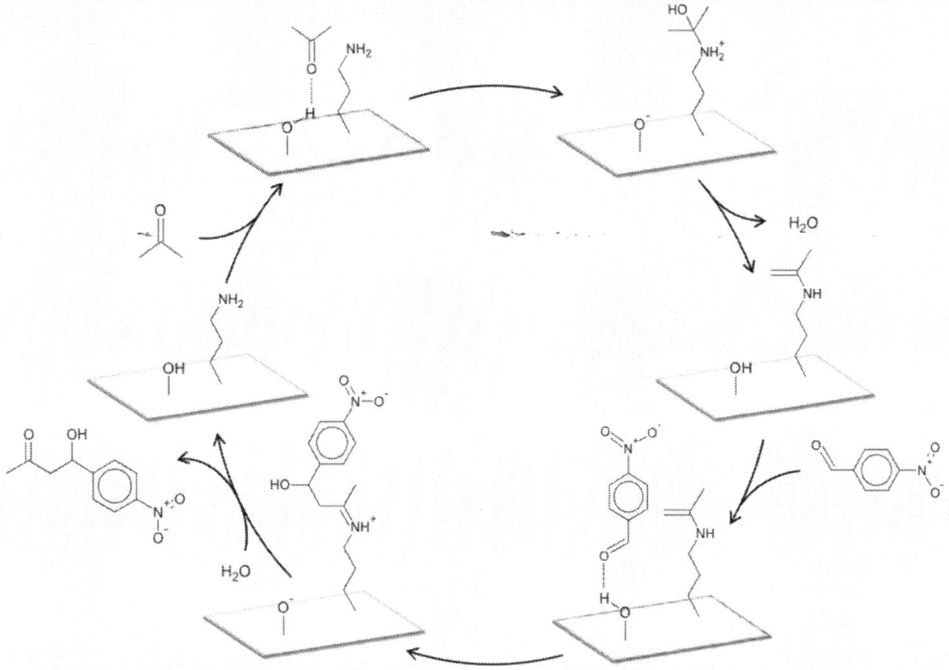

N. A. Brunelli, C. W. Jones, J. Catal., **2013**, 308, 60-72
J. Lauwaert, E. De Canck, D. Esquivel, J. W. Thybaut, P. Van Der Voort, G. B. Marin, ChemCatChem, **2014**, 6, 255-264

Arrangement of the amines

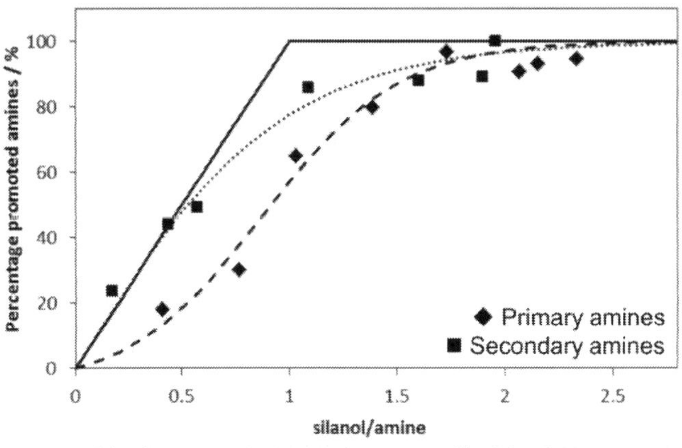

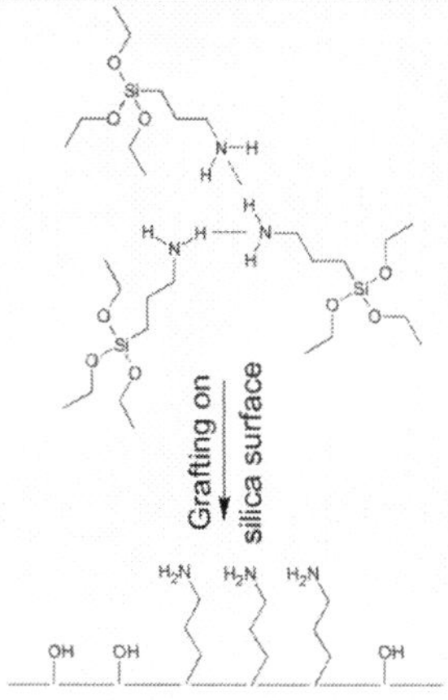

—— Ideal arrangement ········ Random arrangement - - - - Clustered arrangement

  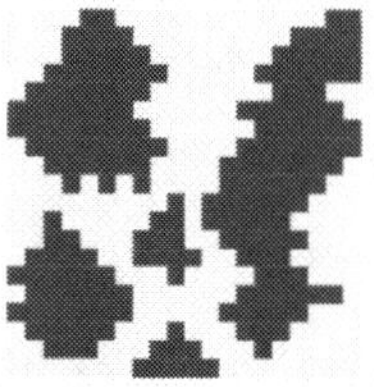

J. Lauwaert, E. De Canck, D. Esquivel, J. W. Thybaut, P. Van der Voort, G. B. Marin, ChemCatChem, **2014**, 6, 255-264
J. Lauwaert, E. De Canck, D. Esquivel, P. Van der Voort, J. W. Thybaut, G. B. Marin, Catal. Today, **2014**, **(Accepted)**
J. C. Hicks, R. Dabestani, A. C. Buchanan, C. W. Jones, Chem Mater, **2006**, 18, 5022-5032

Effects of the amine group

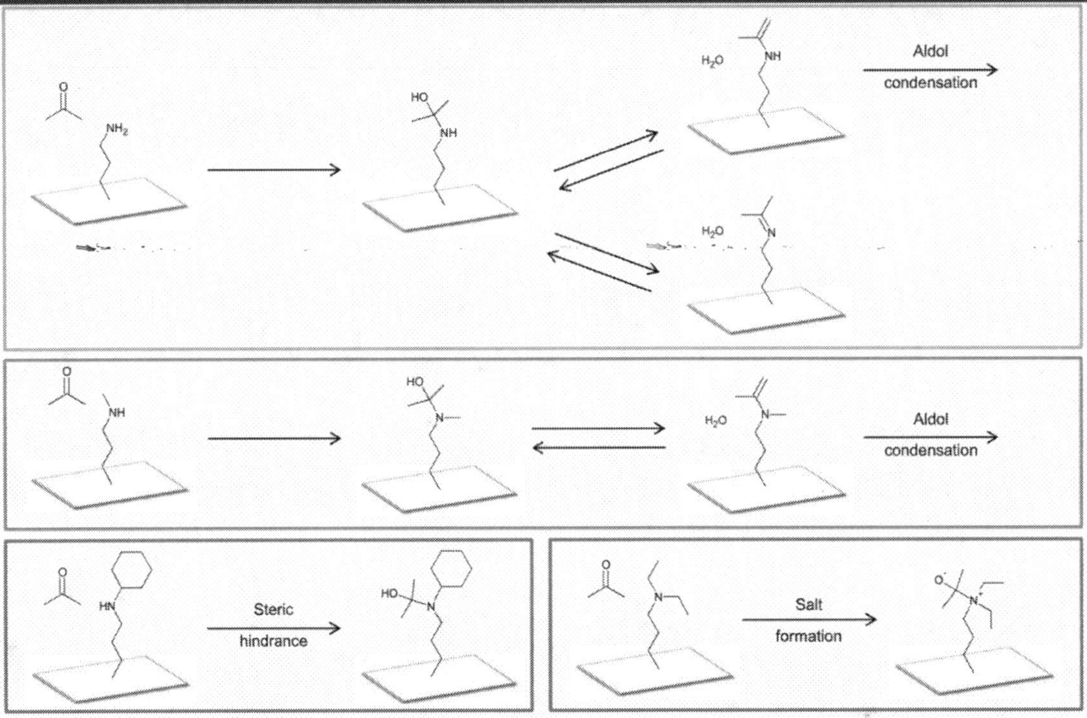

J. Lauwaert, E. De Canck, D. Esquivel, P. Van der Voort, J. W. Thybaut, G. B. Marin, Catal. Today, **2014**, **(Accepted)**

Effect of the acid strength

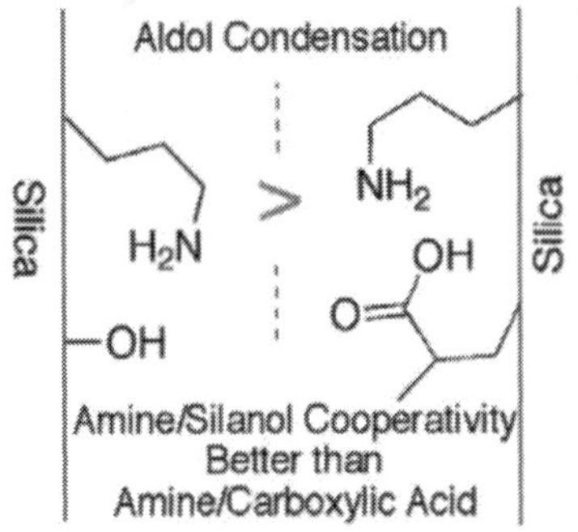

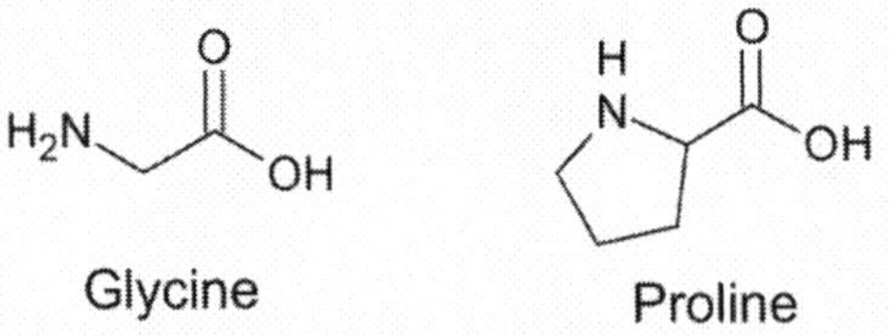

Glycine and proline are used as homogeneous aldol condensation catalysts.

Goal of this work: investigate the effect of the acid strength of the promoting site in a more systematic manner and with a more optimal arrangement of the sites!

R. K. Zeidan, M. E. Davis, J. Catal., **2007**, 247, 379-382
N. A. Brunelli, K. Venkatasubbaiah, C. W. Jones, Chem. Mater., **2012**, 24, 2433-2442
N. A. Brunelli, C. W. Jones, J. Catal., **2013**, 308, 60-72
F. Tanaka, R. Thayumanavan, N. Mase, C. F. Barbas, Tetrahedron Lett., **2004**, 45, 325-328
B. List, R. A. Lerner, C. F. Barbas, J. Am. Chem. Soc., **2000**, 122, 2395-2396

Outline

- Introduction
  - Promoting effect and accessibility of surface silanols
  - Effects of the amine group
  - Scope of the work
- Catalyst synthesis
- Catalyst characterization
- Performance testing
  - Reference catalyst
  - Acid strength effects
  - The promoting effect of silanols
- Conclusions

Catalyst synthesis: Functionalization

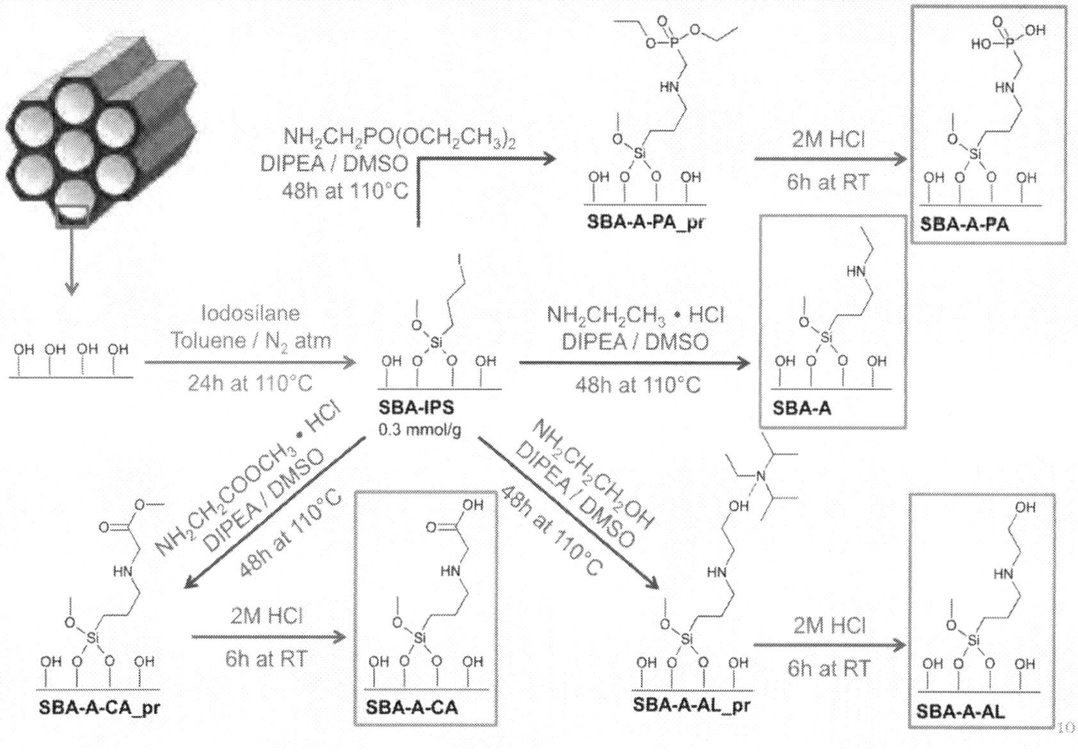

Catalyst synthesis: Functionalization

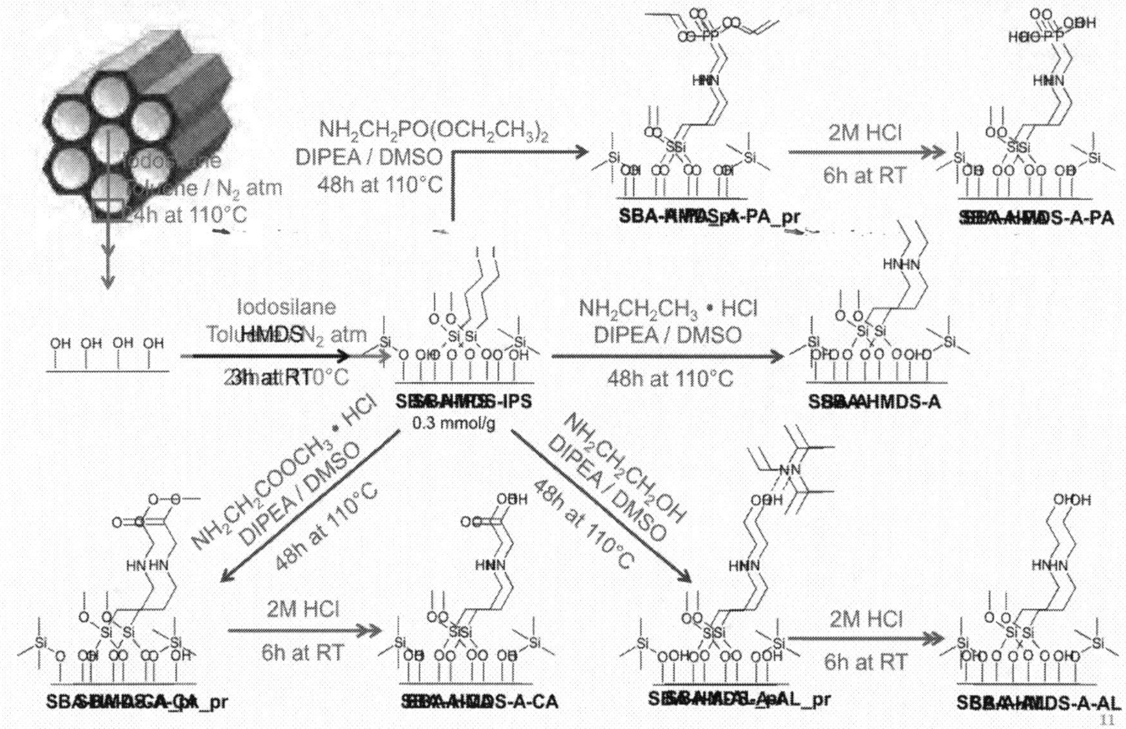

Outline

- Introduction
  - Promoting effect and accessibility of surface silanols
  - Effects of the amine group
  - Scope of the work
- Catalyst synthesis
- Catalyst characterization
- Performance testing
  - Reference catalyst
  - Acid strength effects
  - The promoting effect of silanols
- Conclusions

Catalyst characterization

Catalyst	BET surface area (m^2/g)	Average pore size (Å)	Total pore volume (cm^3/g)	Amine loading (mmol/g)
SBA-15 support	912	80	1.10	0.00
SBA-A	704	81	0.96	0.27
SBA-A-AL	644	81	0.96	0.31
SBA-A-CA	693	79	0.93	0.31
SBA-A-PA	662	76	0.89	0.24
SBA-HMDS-A	506	70	0.81	0.21
SBA-HMDS-A-AL	608	70	0.93	0.23
SBA-HMDS-A-CA	603	70	0.90	0.15
SBA-HMDS-A-PA	612	70	0.91	0.15

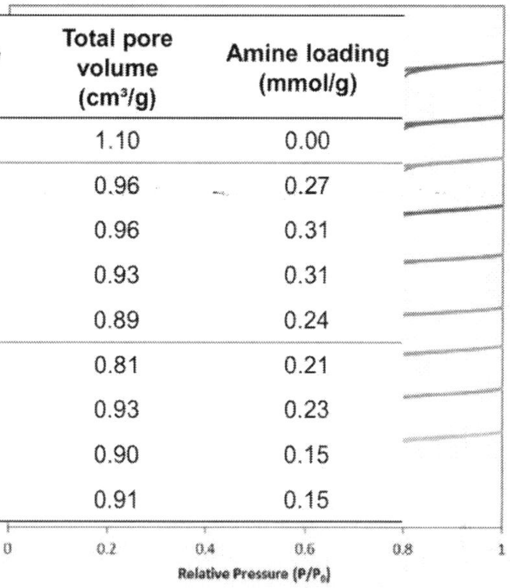

Catalyst characterization

CP MAS ^{13}C NMR

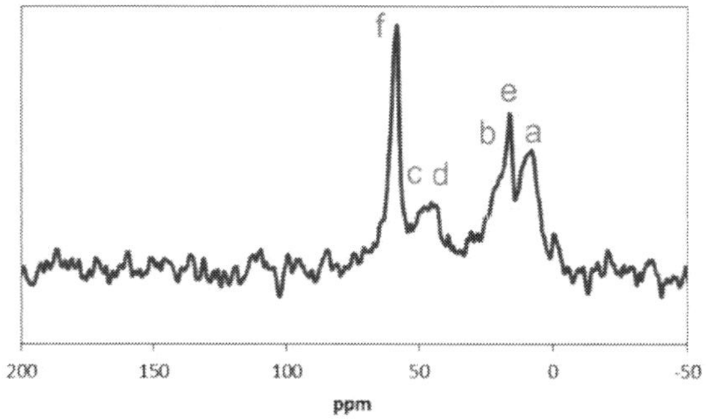

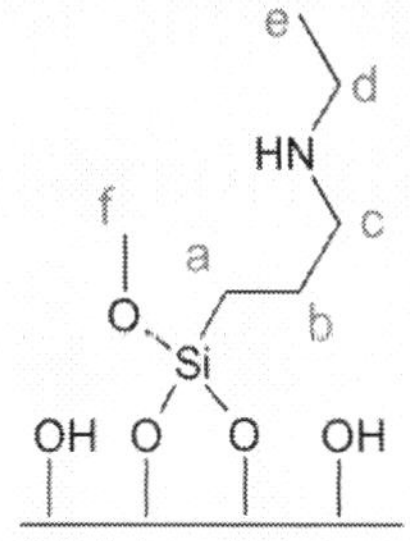

SBA-A

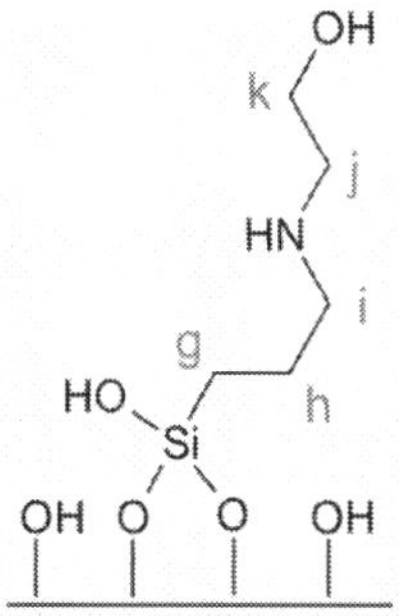

SBA-A-AL

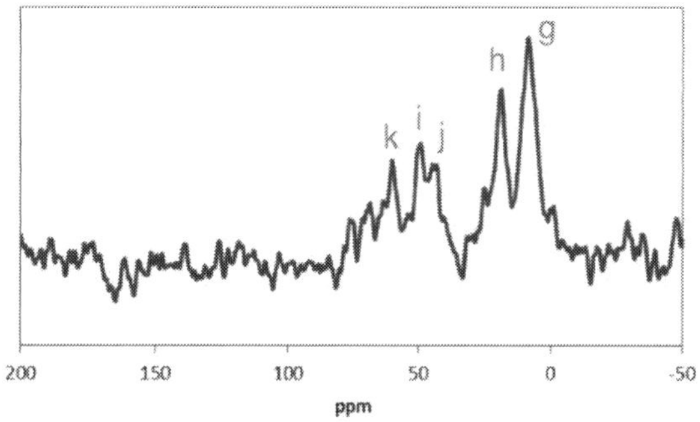

Catalyst characterization

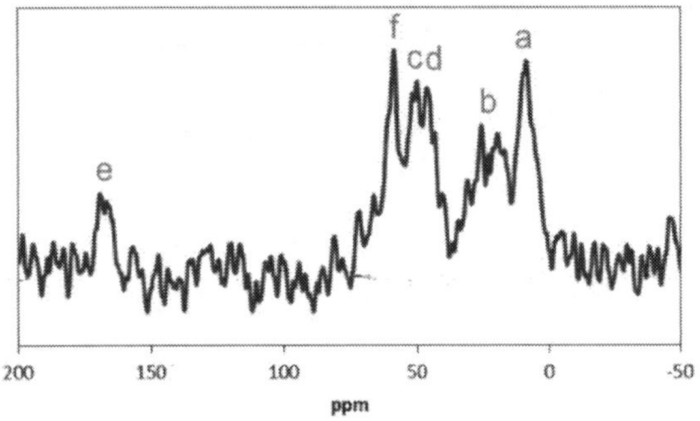

CP MAS ^{13}C NMR

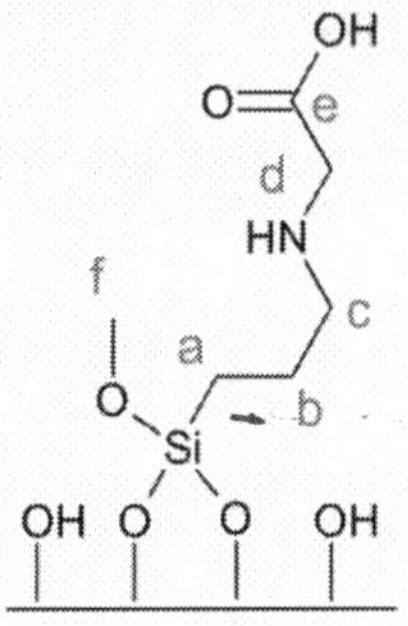

SBA-A-CA

FTIR spectroscopy

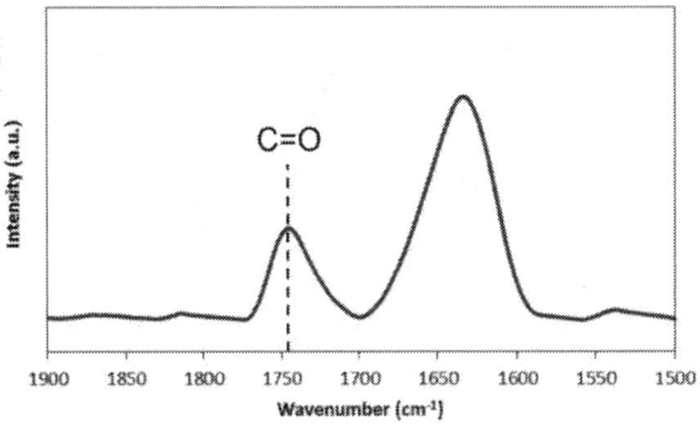

AIChE Annual Meeting, Atlanta, USA, November 16th 2014

Catalyst characterization

CP MAS ^{13}C NMR

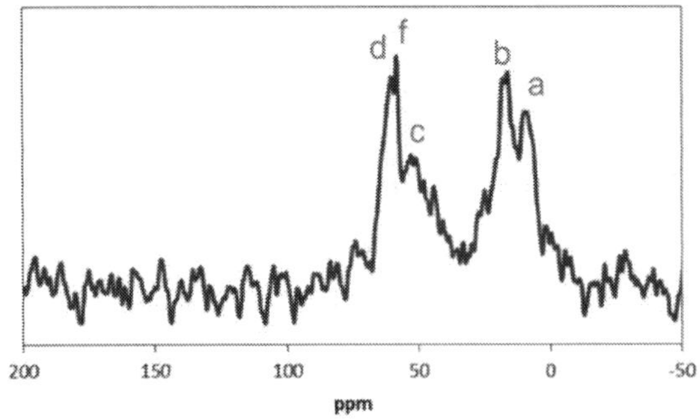

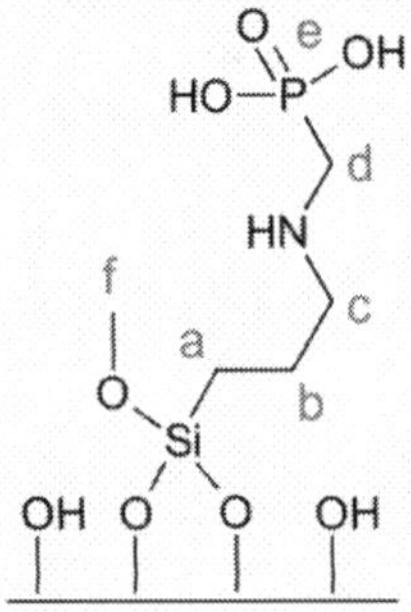

SBA-A-PA

CP MAS ^{31}P NMR

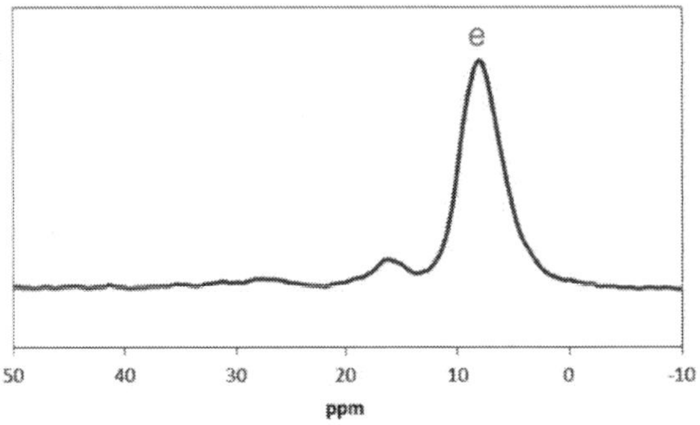

Outline

- Introduction
    - Promoting effect and accessibility of surface silanols
    - Effects of the amine group
    - Scope of the work
- Catalyst synthesis
- Catalyst characterization
- Performance testing
    - Reference catalyst
    - Acid strength effects
    - The promoting effect of silanols
- Conclusions

Reference catalyst

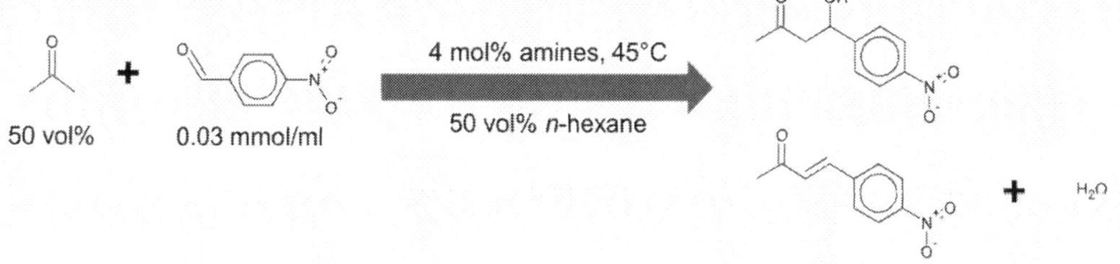

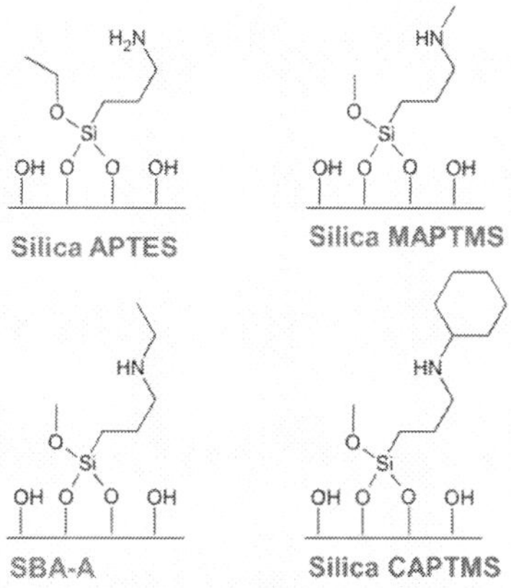

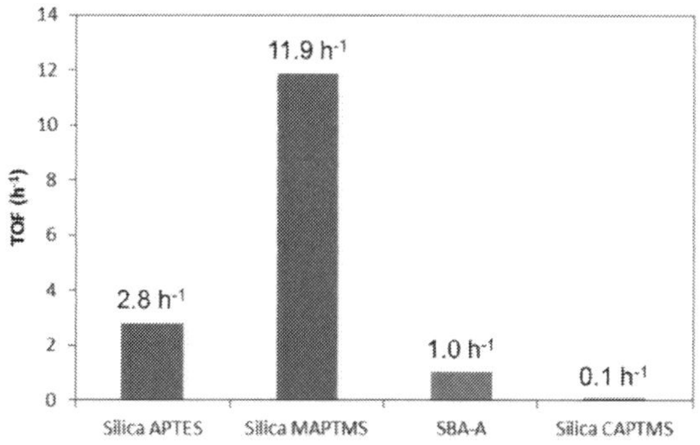

J. Lauwaert, E. De Canck, D. Esquivel, P. Van der Voort, J. W. Thybaut, G. B. Marin, Catal. Today, **2014**, **(Accepted)**

Acid strength effects

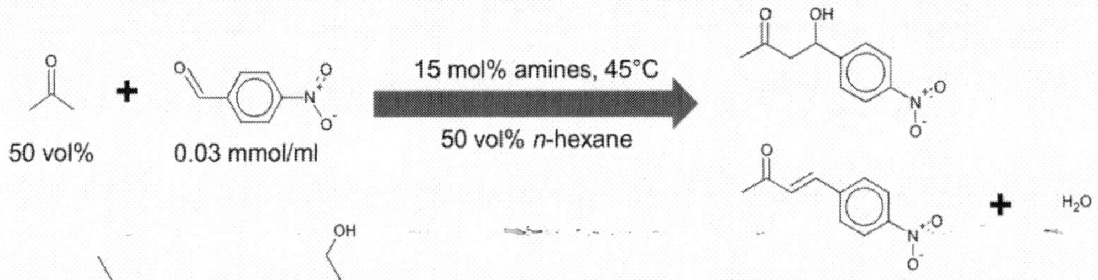

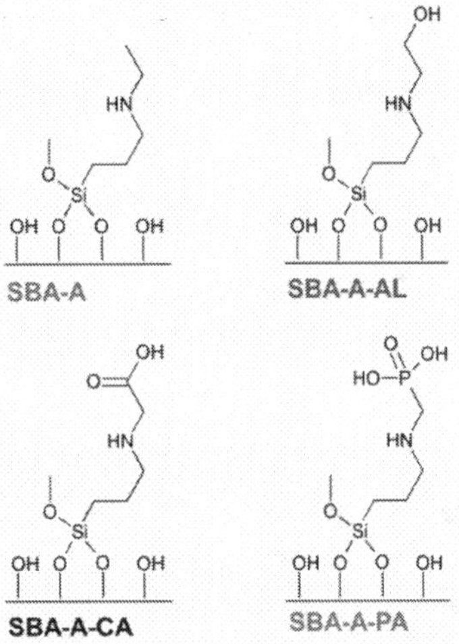

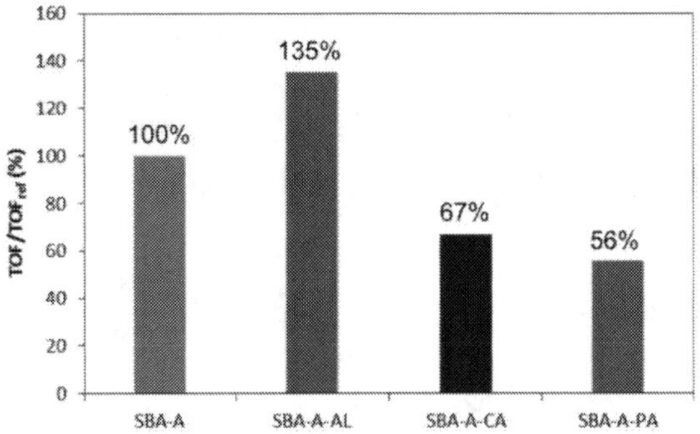

19

The promoting effects of silanols

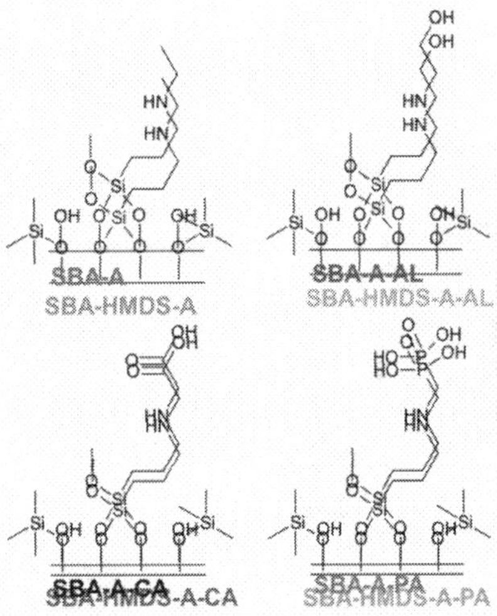

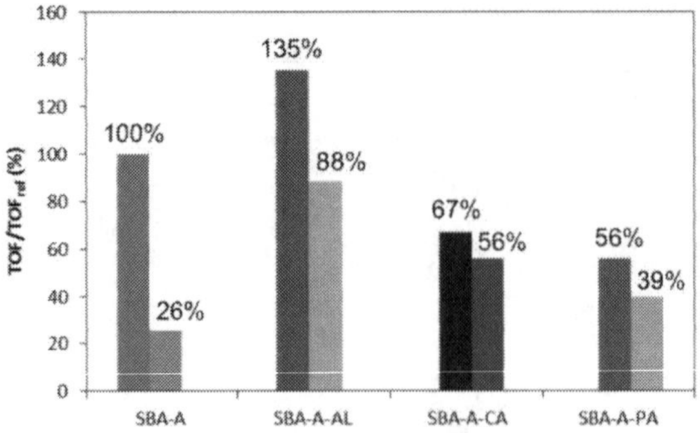

20

Outline

- Introduction
  - Promoting effect and accessibility of surface silanols
  - Effects of the amine group
  - Scope of the work
- Catalyst synthesis
- Catalyst characterization
- Performance testing
  - Reference catalyst
  - Acid strength effects
  - The promoting effect of silanols
- Conclusions

Conclusions

- A SBA-15 support was successfully functionalized with a series of new acid-base functions
- Changing the substituent of a secondary amine from a methyl to an ethyl group increases the steric hindrance drastically
- It has been confirmed that amine-silanol cooperativity is better than amine-carboxylic acid cooperativity or amine-phosphoric acid cooperativity
- The activity of the secondary amine could be increased by bringing the promoting alcohol group closer to the amine
- After removing the silanols from the surface, the catalysts comprising both an amine and an acid site retained most of their activities
- These results suggests that a silica support functionalized with a methyl substituted secondary amine and an alcohol group on the β-carbon might be a good aldol condensation catalyst

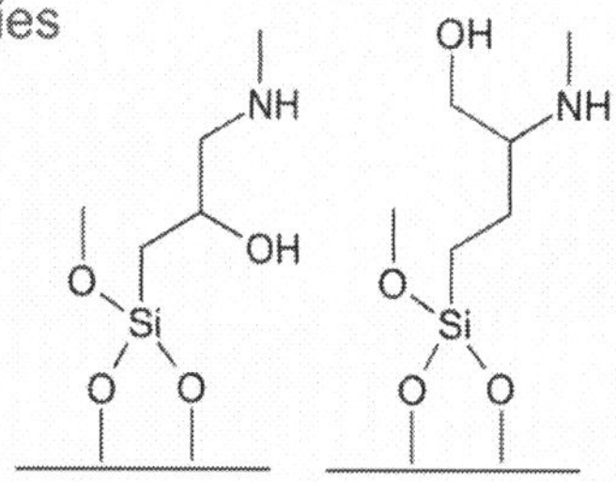

Questions and Answers

Thank you for your kind attention !

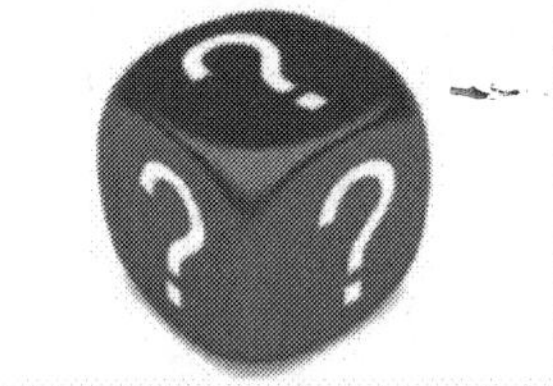

This work was supported by:

The Long Term Structural Methusalem Funding by the Flemish Goverment

fwo
Opening new horizons
Fonds Wetenschappelijk Onderzoek – Vlaanderen
Research Foundation – Flanders

U.S. Department of Energy
Office of Basic Energy Sciences
through Catalysis Contract No. DEFG02-03ER15459.

Tuning the accessibility of the silanols

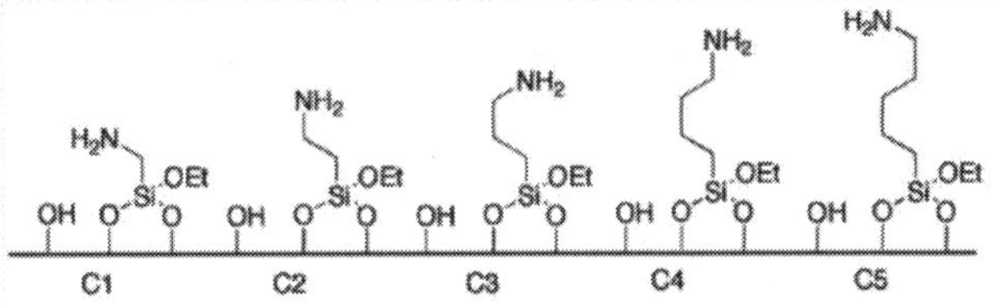

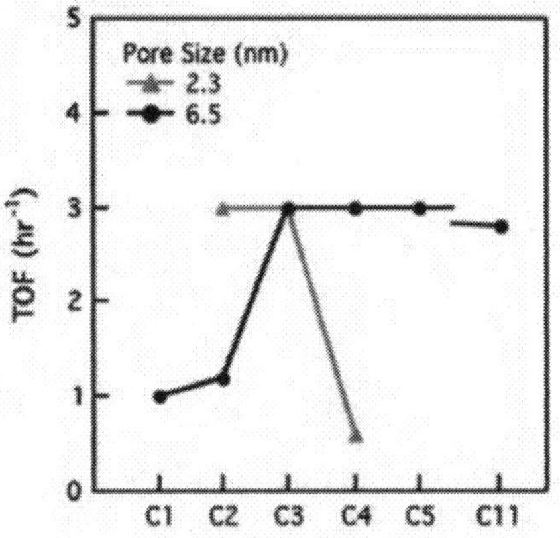

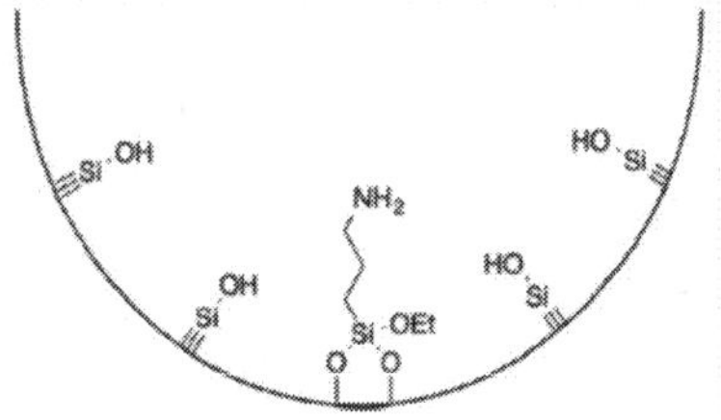

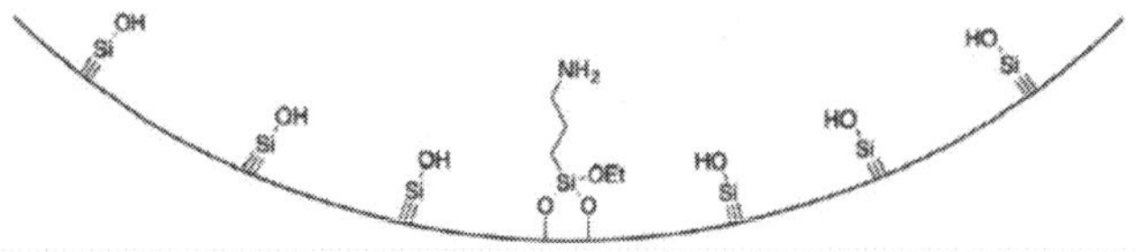

N. A. Brunelli, S. A. Didas, K. Venkatasubbaiah, C. W. Jones, J. Am. Chem. Soc., **2012**, 134, 13950-13953
N. A. Brunelli, C. W. Jones, J. Catal., **2013**, 308, 60-72

Catalyst synthesis: SBA-15 support

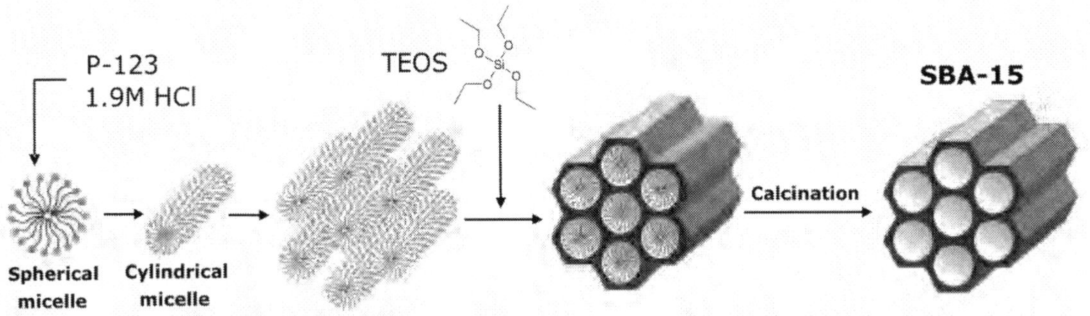

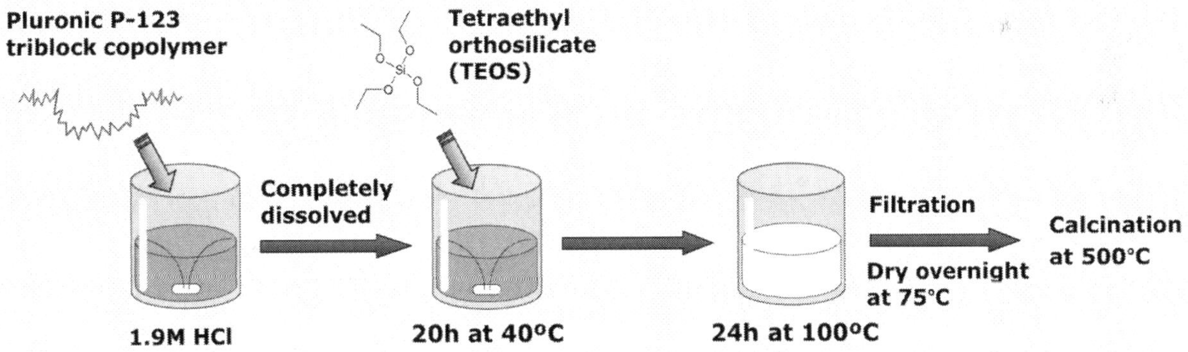

25

AICHE Fall 2014
Dates: Nov 16-21 (Atlanta)
Submission Deadline: May 12

Consequences of Solvation and Microporous Channel Environment during Alkene Oligomerization on Solid Acid Catalysts

Michele L. Sarazen[1], Eric Doskocil[2] and Enrique Iglesia[1*]

[1]*Department of Chemical and Biomolecular Engineering, University of California, Berkeley, Berkeley, California 94720, United States*
[2]*BP Products North America Inc., 150 West Warrenville Rd., Naperville, Illinois 60563, United States*
*iglesia@berkeley.edu

Oxygenates, such as ethanol, isopropanol and isobutanol, can be produced from renewable resources and readily dehydrated to alkenes on acid catalysts. The conversion of these light alkenes on solid acids to larger hydrocarbons that are useful as fuels involves concurrent oligomerization, isomerization, β-scission, hydrogen transfer and cyclization reactions [1]. Here, we examine the effects of confinement and connectivity on the observed product distribution during propene oligomerization as well as the consequences of the void size on turnover rates (per accessible proton) for solid acid catalysts with a wide range of acid strength and void environments, such as zeolites, mesoporous aluminosilicates and supported polyoxometalates (POM).

Oligomerization turnover rates are proportional to propene pressure on microporous zeolites (H-TON, H-MFI, H-MOR, H-BEA and H-FAU) and mesoporous solid acids (amorphous silica-alumina (ASA), MCM-41), consistent with similar C-C formation steps and with infrared evidence for isopropoxide most abundant surface intermediates (MASI). Turnover rates depend on an effective rate constant, k_{eff}, which reflects free energy differences between the cationic C-C bond formation transition state and a neutral alkoxide with a gas phase propene:

$$r_{C_3} = k_{eff}\,(C_3) = (k_{oligo}K_{z,3})(C_3) \quad (1)$$

where $K_{z,3}$ is the equilibrium constant for propene physisorption and k_{oligo} is the rate constant for C-C bond formation from an isopropoxide and a physisorbed propene. The value of k_{eff} depends sensitively on aluminosilicate void size, because transition states are solvated preferentially over the relevant precursors as voids become smaller and the number of van der Waals contacts with the void environment increases. H-TON (10-MR; 1-D zeolite) gives 10-fold larger rates than protons of similar acid strength within ASA mesopores. Turnover rates on H-MFI and H-BEA lie between those on H-TON and H-ASA, but are similar to each other, consistent with the similar size of the channel intersections in H-MFI and 12-membered ring channels in H-BEA. The predominant presence of protons in MFI intersections is consistent with increasing C_9/C_6 ratio with increasing Al content (Si/Al=14-140), suggestive of sites being located where larger chain growth can occur, while turnover rates (per proton) measured on these H-MFI samples remain the same, indicative of all samples similarly solvating the relevant transition state. Similar conclusions concerning the location of protons in MFI have been drawn using methanol dehydration as a probe reaction where dimethyl ether formation rate constants for samples with different Al content (Si/Al=22-118) are the same, again reflecting similar solvation of the relevant transition state with respect to the relevant precursor [2].

Different solid acids generate diverse product distributions during alkene oligomerization, not according to the ability of the confining voids to solvate C-C bond formation transitions states, but according to the connectivity of these confining voids, which affects relative rates of oligomerization, isomerization, hydride transfer and β-scission. The relevant selectivity for a given solid acid catalyst is a true oligomer selectivity (χ), which is defined as the fraction of products that egress as true oligomers (C_{3n} for propene) as opposed to smaller alkenes formed from β-scission as hydride transfer is negligible and isomerization does not change chain length, but does reach pseudo-equilibrium with the gas phase. One-dimensional samples, both zeolites (H-TON, H-MOR) and mesoporous acids (H-ASA, Al-MCM-41, POM), exhibit high χ values (0.9-1), irrespective of channel size. In contrast, three-dimensional exhibit lower χ values than their 1-D counterparts with χ = 0.64, 0.35, and 0.31 for H-BEA, H-MFI and H-FAU, respectively at 0.2 bar. The three-dimensional zeolites contain voids larger than the smallest aperture where chain growth forms products that must undergo β-scission to egress through the crystal as smaller species. The relative ratio of the size of these voids to their subsequent channels dictates the selectivity of oligomerization to β-scission products; lack of this undulation leads to species remaining intact, as observed in the 1-D micro- and mesoporous samples. These results and their conceptual interpretations allow for optimizing chain growth of light alkenes to useful products by utilizing zeolite frameworks of varying size and connectivity as well as mesoporous solid acids to affect catalytic rates and product selectivities.

[1] Garwood, W. E., ACS Symp. Ser., 218, 383 (1983).
[2] Jones, A.J., Carr, R.T., Zones, S.I., Iglesia, E. J. Catal. 312, 58 (2014).

Financial support from BP through the XC^2 Program and National Science Foundation Graduate Research Fellowship Programs are gratefully acknowledged.

389553 In Situ Spectroscopic Identification of Active Components in the "Hydrocarbon Pool" during Methanol-to-Hydrocarbons Conversion

Abstract:

Methanol-to-hydrocarbons conversion is believed to proceed under participation of a "hydrocarbon pool" present inside the pores of a zeolite or zeotype catalyst. Many constituents of the pool have been identified, with alkylbenzenes being the most frequently reported species. However, it is debated which components participate in the product formation and how the products are actually formed. The goal of this investigation was to identify active components in the pool. HZSM-5 and H-beta were used as catalysts at temperatures of 275 to 310 °C and a methanol pressure of 10 kPa. The primary tool for species identification was in situ diffuse reflectance UV-vis-NIR spectroscopy.

Spectra of both catalysts during methanol conversion revealed the presence of charged hydrocarbon species, mostly cations of alkylbenzenes and alkyl-substituted cyclopentadienes; on H-beta, cations of larger polycyclic aromatic compounds were formed in addition. The number of alkyl groups on the rings was estimated from the spectroscopic data and found to be larger in the case of H-beta, consistent with the larger pores of this zeolite in comparison to HZSM-5. Correlations between the pool composition and catalyst performance were sought by imposing changes in conditions, monitoring the response, and seeking parallel trends in the catalytic and spectroscopic data. From this analysis, cyclopentenyl species emerge as most promising key intermediates.

376767 A Mechanistic Basis for the Effects of Crystallite Size on Light Olefin Selectivity in Methanol to Hydrocarbons Conversion on MFI

Abstract:

The change in selectivity of methanol to hydrocarbons conversion (MTH) with change in MFI crystallite size has been mechanistically linked to the transport restrictions within the crystallite. Three MFI samples with varying diffusion lengths were investigated: SPP MFI (~1 nm), 3DOm-i MFI (~20 nm), and 17-micron MFI (~8500 nm). The light olefin (ethene + propene) selectivity, for the reaction of dimethyl ether at 623 K, increased from 21.8 C% in SPP MFI to 45.8 C% in the 17-micron MFI at 51-59% dimethyl ether conversion. We postulate that as the crystallite size of MFI increases, the intra-crystalline residence time of bulkier hydrocarbons like polymethylbenzenes increases, enabling these molecules to undergo multiple methylation/dealkylation cycles before exiting the crystal. Transport restrictions were also introduced externally in a conventional ZSM-5 sample by silylation using tetraethyl orthosilicate. Silylation treatment partially or completely blocks the pore openings of MFI and increases the residence time of bulkier hydrocarbons inside the crystal, similar to a large crystallite MFI. Three silylated MFI samples were obtained by repeating the treatment 1-3 times. Transport restrictions in these silylated MFI samples were described by adsorption uptake measurements of 2,2-dimethylbutane at 293 K and 20 kPa 2,2-dimethylbutane pressure. The light olefin selectivity, for the reaction of dimethyl ether at 623 K and at 50-56% dimethyl ether conversion, increased from 27.5 C% in conventional MFI, which had lower transport restrictions, to 49.7 C% in the sample that was silylated three times and therefore had higher transport restrictions.

389739 Van Der Waals Corrected First-Principles Study of the Methanol-to-DME Reaction Mechanism on H-ZSM-5

Abstract:

Conversion of methanol to olefins (MTO) or hydrocarbons (MTH) over acidic catalysts, in particular H-ZSM-5 (a 3-D channel zeolite with MFI framework), is a feasible route for the production of commodity chemicals (*e.g.* propylene) and liquid fuels, and has been a popular subject of catalytic studies. Although the overall process is very complicated, the first step is accepted to be methanol dehydration to dimethyl ether (DME). The formation of DME from methanol may proceed through two competing pathways. The associative pathway is characterized by the initial adsorption of a methanol dimer followed by water elimination, while in the dissociative pathway water formation occurs from a single methanol molecule followed by the DME formation step after a second methanol molecule adsorbs.

The active sites of H-ZSM-5 are Brønsted acid sites that can be found in 12 distinguishable locations, which may be grouped as belonging to the sinusoidal channel, the straight channel, or the intersection of each. We have recently demonstrated that the geometric and chemical properties of these 12 sites span a wide range of values, and that these variations may favor different reaction pathways [1]. Hence, H-ZSM-5 is ideally suited as a model zeolite to gain a thorough understanding of the mechanism of methanol dehydration, the effect of pore confinement, and the role of acidity. This information can be used to derive structure-function relationships for improved catalyst design.

A theoretical study of the reaction mechanism for DME formation at the DFT-GGA level has been conducted recently over H-ZSM-22 (a 1-D channel zeolite with TON framework), where the dominant mechanism under industrially relevant conditions was determined to be the dissociative pathway. The effect of active site acidity on reaction barriers was also reported [2]. Here, we use periodic van der Waals density functional theory (vdW-DF) calculations to study the reaction mechanism of methanol dehydration over H-ZSM-5, a zeolite catalyst with known activity for this reaction that is frequently used for MTO/MTH. We show for representative active sites in the straight and sinusoidal channels, and at the channel intersection that the associative pathway possesses a lower reaction barrier at 0 K than the two steps of the dissociative pathway. Entropy calculations are carried out to compute the Gibbs free energy change of transition states with respect to the reference state (i.e. zeolite + gaseous reactants), from which we determine the dominant mechanism at realistic reaction temperatures.

The use of the vdW-DF exchange-correlation functional enables us to investigate the significance of transition state stabilization through pore confinement effects. Large differences in reaction barriers are obtained when transition states are calculated in the absence and presence of vdW interactions. Moreover, vdW interactions prove to have a significant influence on the dominant reaction mechanism. Similar to the results of the study by Moses and Nørskov [2], exclusion of vdW interactions suggests dominance of the dissociative pathway for all practical temperature conditions. However, when vdW interactions are included the transition from the associative pathway at low temperatures to the dissociative pathway occurs at higher temperatures. These results imply that for practical conditions both pathways are feasible. The significance of geometry-dependent confinement effects on the reaction mechanism is investigated by calculating reaction barriers at T10, T11, and T12 sites representing sinusoidal, straight, and intersection positions, respectively.

To investigate the sensitivity to the strength and number of acid sites, we first tuned the active site acidity at specific locations in the MFI unit cell by using three different heteroatom substitutions, *i.e.* aluminum, gallium, and indium [2]. The resulting activation barriers are used as the site acidity descriptor and the relationship with other site properties is investigated. The number of acid sites is in practice altered by varying the Si/Al ratio. At lower Si/Al ratios the probability of finding two nearby acid sites increases, and we estimated the acidity and the respective kinetic parameters for DME formation for a Brønsted acid site pair. Characterization of the site pair acidity using the calculated adsorption energies of probe molecules (CH_3OH, NH_3) revealed that the average acidity of two neighboring acid sites in H-ZSM-5 is similar to that of a single site model. However, the stabilization of a methanol dimer and the transition state of the associative pathway shows distinct differences when a site pair is present. In summary, the results of our detailed and accurate DFT simulations can illuminate the underlying factors influencing the mechanism of the methanol-to-DME reaction over H-ZSM-5 and provide valuable insight into the design of improved catalysts.

[1] Arian Ghorbanpour, Jeffrey D. Rimer, Lars C. Grabow, Periodic, vdW-corrected density functional theory investigation of the effect of Al siting in H-ZSM-5 on chemisorption properties and site-specific acidity, *Catalysis Communications* (2014), doi: 10.1016/j.catcom.2014.04.005.
[2] Poul Georg Moses, Jens K. Nørskov, Methanol to dimethyl ether over ZSM-22: a periodic density functional theory study, *ACS Catalysis* **3** (2013) 735-745.

380587 Understanding Local Environments of Aluminum Heteroatoms in Bulk and Mesostructured Zeolite Catalysts

Abstract:

The chemical reaction activities and/or selectivities of porous catalysts depend strongly on the local compositions and structures near active site moieties, as well as the mass transport of reactant and product molecules to and from such sites. In zeolite catalysts, the solid acidity depends on the locations, local structural environments, and distributions of heteroatoms (*e.g.*, Al, B) in the silicate framework. Recently, a variety of approaches have been pursued for synthesizing zeolitic materials with a combination of meso- and nanoscale porosities, which have resulted in improved transport and catalytic performance, especially for large molecules. However, the specific locations of heteroatoms and their correlations with macroscopic Brønsted acidity are challenging to understand and control, in part because of subtle differences in local bonding geometries and distributions of nearest-neighbor environments near heteroatom sites. Despite the lack of long-range atomic ordering of heteroatoms in zeolite frameworks, their local structural features can be understood by combining spectroscopic, scattering, and macroscopic property analyses that provide complementary insights across multiple length and time scales. In particular, solid-state 2D $^{27}Al\{^{29}Si\}$ nuclear magnetic resonance (NMR) measurements are sensitive to through-bond scalar (J) couplings, as manifested by correlated signals from covalently bonded framework moieties. Using these methods, the connectivities of ^{27}Al heteroatoms with nearby ^{29}Si framework sites are established in conventional bulk and surfactant-directed mesostructured zeolites with BEA and MFI topologies. These analyses reveal that Al heteroatoms are incorporated into distinct framework sites that manifest differences in their respective Brønsted acidities, as measured by the ^{31}P isotropic chemical shifts of adsorbed phosphine oxide probe molecules. Such molecular-level insights about the local compositions and structures near the catalytically active sites are expected to aid the development of new zeolite catalysts with improved macroscopic reaction, adsorption, and transport properties for diverse engineering applications.

The Effects of Alkyl Linker Length and Pore Size on the Bifunctional Catalytic Activity of Amine-Functionalized Silica Materials for C-C Bond Formation

Eric G. Moschetta, Nicholas A. Brunelli, Virginia E. Collier, George I. Lindy, and Christopher W. Jones*

School of Chemical & Biomolecular Engineering

Georgia Institute of Technology, 311 Ferst Drive NW, Atlanta, GA 30332

*cjones@chbe.gatech.edu

Bifunctional catalysts employ two catalytic species that interact cooperatively to increase the catalytic activity of the material such that the observed activity is greater than the sum of the activities of the separate catalytic species. These bifunctional, or cooperative, catalysts are designed to separate the catalytic moieties spatially on the heterogeneous support to allow them to interact with each other, yet without deactivation of either entity. The interactions of the catalytic species with themselves and/or the reactants enhance the catalytic activity in a manner that may be unattainable otherwise. Additionally, the heterogeneous nature of the catalytic material allows for facile separation from the reactive mixture. Mesoporous silica is a popular choice as a support for bifunctional catalysts due to its chemically modifiable surface silanols, which can also act as weak Brønsted acids in an acid-base bifunctional material. Trialkoxysilanes bearing functional groups such as amines, thiols, and carboxylic acids are the most common choices for chemical modification of the silica surface. The chemical nature of the catalytic units and the structure of the heterogeneous support dictate the catalytic activity of the synthesized bifunctional material.

We present a study of the effects of the alkyl linker length of the aminosilane and the pore diameter of the mesoporous silica on the cooperative catalytic activity of amino-silanol acid-base catalysts. Our model reactions are the aldol and nitroaldol condensations, two common methods for C-C bond formation in organic synthesis. We prepared catalytic materials using MCM-41 and SBA-15 silicas of various pore diameters and functionalized them with aminosilanes of different alkyl linker lengths. Additionally, we characterized these materials using XRD, TGA, SEM, nitrogen physisorption, and elemental analysis to ensure structural integrity and to determine the amine content of each sample. We then evaluated each catalyst for its activity in the aldol and nitrolaldol condensations and calculated turnover frequencies (TOFs) during the first hour of reaction. Our results demonstrate that the optimal linker length depends on the pore diameter of the silica, meaning that the alkyl linker of the amine must be long and flexible enough in a given pore diameter to interact with the silanols on the surface of the silica.[1]

[1] N. A. Brunelli, C. W. Jones, *J. Catal.* **2013**, *308*, 60–72.

THE NATURE OF THE ACTIVE Cu SPECIES IN Cu/CHA MATERIALS FOR THE SELECTIVE CATALYTIC REDUCTION OF NOx WITH NH_3

Feng Gao, Marton Kollar, Yilin Wang,
Janos Szanyi, and Charles H.F. Peden

Institute for Integrated Catalysis, Pacific Northwest National Laboratory, Richland, WA 99354
chuck.peden@pnnl.gov

Introduction

Metal-exchanged molecular sieves with a Chabazite (CHA) structure are of considerable current research interest due to the very recent commercial success of their application for the selective catalytic reduction (SCR) of NOx with ammonia on diesel vehicles [1]. Catalysts with outstanding SCR performance (activity, hydrothermal stability, etc.) include the aluminosilicate zeolite, SSZ-13, and the silico-alumino-phosphate, SAPO-34, with ion-exchanged Cu. Apart from the commercial success, detailed catalyst structures, reaction mechanisms, and structure-activity relationships are still lacking. In this presentation, we will describe results of more recent studies of Cu/SSZ-13 that specifically focus on the nature of the active Cu species under a variety of reaction conditions. Catalysts with varying Si/Al ratios and Cu loadings have been prepared and studied. We use a number of characterization methods including FTIR, x-ray absorption (XAS), and electron paramagnetic resonance (EPR) spectroscopies. Temperature-programmed reduction (TPR) experiments in the presence or absence of water are also highlighted. Besides these catalyst characterization measurements, we will also provide NH_3-SCR and NH_3 oxidation kinetics over these catalysts at high space velocity conditions for the development of structure-activity relationships.

Experimental

Na/SSZ-13 with a Si/Al ratio of ~6 was synthesized using a procedure detailed in reference [2]. For the two higher Si/Al ratio samples, synthesis was conducted hydrothermally using a method modified slightly from a protocol developed recently by Deka, et al. [3] with reactant ratios varied to provide differing Si/Al ratios. After exchange of Na^+ for NH_4^+, aqueous $CuSO_4$ solutions with varying concentrations were used to prepare Cu/SSZ-13 catalysts with Cu loadings. Reaction tests were performed using a plug-flow reactor and products were measured using an online FTIR detector. Catalysts were characterized with ICP, surface area/pore volume measurements, temperature programmed reduction (TPR), and x-ray absorption, electron paramagnetic resonance (EPR), nuclear magnetic resonance (NMR), and FTIR spectroscopies.

Results and Discussion

A portion of the results to be discussed are presented here. Of particular interest is the temperature at which Cu reduces from the +2 to +1 oxidation states during TPR experiments. This occurs at moderate (150-500 °C) temperatures and often in distinct temperature ranges as a function of Cu loading [1]. For somewhat higher Cu-loaded (Cu/Al > 0.1) SSZ-13 samples with Si/Al = 6, H_2-TPR results for fully hydrated and fully dehydrated samples have been reported previously [4]. In more recent studies [5], experiments under identical conditions were conducted on Si/Al = 12 and 35 samples with various Cu loadings, and the results are plotted in Fig. 1. These results suggest that a number of factors determine the $Cu^{+2} \rightarrow Cu^{+1}$ reduction temperature: 1) Cu-ion location in the SSZ-13 structure (as indicted in our prior studies [4]); 2) the extent of Cu-ion 'bonding' to the zeolite (i.e., whether Cu ions interact with 1 or 2 ion exchange sites); and 3) whether reduction is carried out in the presence or absence of water, with water likely hydrolyzing Cu-zeolite O bonds and resulting in a mobile Cu-ion species.

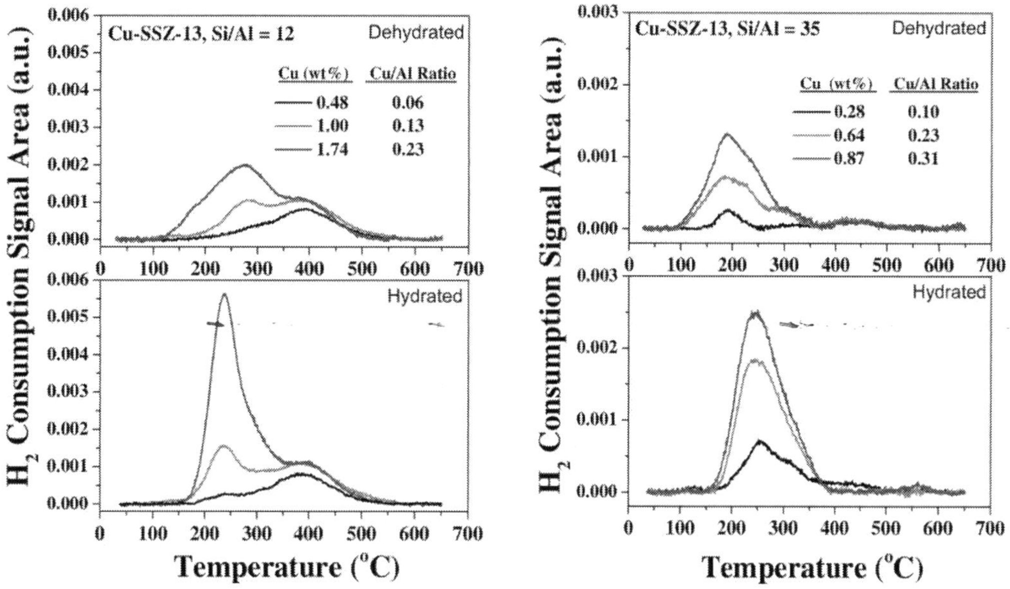

(a) (b)

Figure 1: Temperature-programmed reduction (TPR) data for Cu-SSZ-13 samples with various Cu loadings for (a) Si/Al = 12 and (b) Si/Al = 35. The upper panels show TPR for dehydrated samples while the lower panels show TPR for fully hydrated samples.

On the basis of FTIR studies of the region of the spectra where T-O-T zeolite vibrations are observed, we recently noted that the presence of water eliminates newly formed (perturbed) T-O-T vibrations that occur in Cu/SSZ-13 catalysts [6]. These results also strongly suggest that the Cu/zeolite-O bonds have been hydrolyzed resulting in mobile Cu-ion species. Significantly, the perturbed T-O-T vibrations reappear when the catalyst is heated to approximately 250 °C where water is largely removed from the zeolite. This latter result is especially interesting because this is the same temperature range at which an unusual dip in performance (the so-called "seagull" shape) for the "standard" (i.e., NOx = NO) SCR reaction is observed for Cu/SSZ-13 catalysts at lower Cu loadings [5]. We propose that a hydrolyzed, readily reducible and mobile Cu species is responsible for high standard SCR rates at lower temperatures, while Cu ions, strongly interacting with the zeolite upon dehydration, are more difficult to reduce and are only active at higher reaction temperatures.

Acknowledgement. This work was supported by the U.S. DOE/EERE/Vehicle Technologies Office.

References

(1) F. Gao, J.H. Kwak, J. Szanyi and C.H.F. Peden, *Topics in Catal.*, **2013**, 56, 1441, and references therein.
(2) D.W. Fickel and R.F. Lobo, *J. Phys. Chem. C*, **2010**, 114, 1633.
(3) U. Deka, A. Juhin, E.A. Eilertsen, H. Emerich, M.A. Green, S.T. Korhonen, B.M. Weckhuysen and A. M. Beale, *J. Phys. Chem. C* **2012**, 116, 4809.
(4) F. Gao, E.D. Walter, E.M. Karp, J.Y. Luo, R.G. Tonkyn, J.H. Kwak, J. Szanyi and C.H.F. Peden, *J. Catal.*, **2013**, 300, 20.
(5) F. Gao, E.D. Walter, M. Kollar, Y. Wang, J. Szanyi and C.H.F. Peden, *J. Catal.*, **2014**, submitted for publication.
(6) J.H. Kwak, T. Varga, C.H.F. Peden, F. Gao, J.C. Hanson and J. Szanyi *J. Catal.*, **2014**, in press.

Characterizing the Exchange Sites in Cu-SSZ-13 and Cu-SAPO-34 During the Selective Catalytic Reduction of NO_x with NH_3: A Detailed Study from First Principles

Nitrogen oxides (NO_x) are one of the main air pollutants present in the exhaust from diesel engines, which are popular for vehicle transportation due to their efficiency and durability [1]. However, NO_x emission control is a challenge in these "lean-burn" engines [2]. As is well known, the selective catalytic reduction (SCR) of NO_x with NH_3 is a reaction between the NO, NO_2, and O_2 oxidants and the NH_3 reductant to form N_2 and H_2O. Copper-exchanged small-pore zeolites, such as Cu-SSZ-13 and Cu-SAPO-34, display excellent catalytic activity and hydrothermal stability in SCR of NO_x, as has been shown in a number of recent studies [3, 4]. 6 membered rings, 4 membered rings and 8 membered rings build these zeolites, as shown in Figure 1.

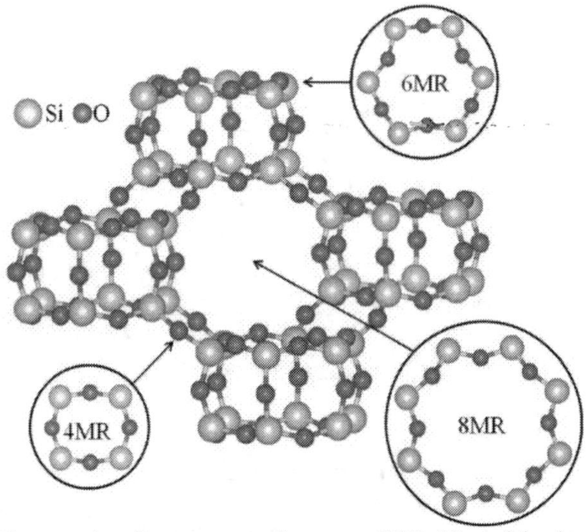

Figure 1. Structure of pure SSZ-13 with its corresponding 4 membered rings (4MR), 6 membered rings (6MR) and 8 membered rings (8MR).

Currently, a number of issues still need to be addressed for CHA-based SCR catalysts including the effect of reactants on the active Cu-species location. By using first-principles calculations based on density functional theory (DFT), several possible adsorption positions for Cu in the Cu-SSZ-13, along with the effects of extra lattice species on the Cu location were studied. We found that the 6 membered ring site is the most energetically favorable while the 8 membered ring sites are less favorable with energy differences about 0.5 eV with respect to the 6 membered ring site, which is in agreement with prior studies [5, 6]. Upon the molecular adsorption of extra-lattice species, the energy differences between Cu in the 8 membered ring and 6 membered ring sites decreases and almost disappears [7]. Similar conclusions are obtained when examining the influence of extra-lattice species on the location of Cu in Cu-SAPO-34.

X-ray absorption spectroscopy (XAS) is a powerful technique for characterizing heterogeneous catalysts. For example, the appearance of a pre-edge signal in the X-ray absorption near edge spectrum (XANES) has been associated with a reduced Cu(I) ion through a comparison to a reference sample [5]. To further interpret these results, supercell DFT can be useful. In particular, the origin of the pre-edge feature as observed for Cu-SSZ-13 under standard SCR conditions was examined here. We found that the pre-edge peak in K-edge XANES of a Cu(I) ion in a 8 membered ring site is induced by a 1s to 4s transition while there is no pre-edge peak when the Cu(I) ion is in the 6 membered ring site. As a result, the appearance of the pre-edge feature in the XANES spectra does not seem to necessarily correlate with a reduced Cu(I) ion. On the other hand, the molecular adsorption of extra-lattice species on Cu atoms located in 6 membered rings and 8 membered rings results in the same K-edge XANES spectrum. We will present and compare these results to those obtained on Cu-SAPO-34. These results are expected to greatly aid our understanding of the properties of Cu-SSZ-13 and Cu-SAPO-34 as SCR catalysts.

References

1. Deka, U., I. Lezcano-Gonzalez, B.M. Weckhuysen, and A.M. Beale, *Local Environment and Nature of Cu Active Sites in Zeolite-Based Catalysts for the Selective Catalytic Reduction of NOx.* ACS Catalysis, 2013. **3**: p. 413-427.
2. Matsumoto, S.i., *Catalytic Reduction of Nitrogen Oxides in Automotive Exhaust Containing Excess Oxygen by NOx Storage-Reduction Catalyst.* CATTECH, 2000. **4**(2): p. 102-109.
3. Kwak, J.H., R.G. Tonkyn, D.H. Kim, J. Szanyi, and C.H. Peden, *Excellent activity and selectivity of Cu-SSZ-13 in the selective catalytic reduction of NOx with NH3.* Journal of Catalysis, 2010: p. 1-4.
4. Ma, L., Y. Cheng, G. Cavataio, R.W. McCabe, L. Fu, and J. Li, *Characterization of commercial Cu-SSZ-13 and Cu-SAPO-34 catalysts with hydrothermal treatment for NH3-SCR of NOx in diesel exhaust.* Chemical Engineering Journal, 2013. **225**: p. 323-330.
5. McEwen, J.-S., T. Anggara, W.F. Schneider, V.F. Kispersky, J.T. Miller, W.N. Delgass, and F.H. Ribeiro, *Integrated operando X-ray absorption and DFT characterization of Cu–SSZ-13 exchange sites during the selective catalytic reduction of NOx with NH3.* Catal. Today, 2012. **184**: p. 129-144.
6. Korhonen, S., D.W. Fickel, R.F. Lobo, B. Weckhuysen, and A. Beale, *Isolated Cu2+ ions: active sites for selective catalytic reduction of NO.* Chem. Commun., 2011. **47**(2): p. 800.
7. Zhang, R., J.-S. McEwen, M. Kollar, F. Gao, Y. Wang, J. Szanyi, and C. Peden, *NO Chemisorption on Cu/SSZ-13: A Comparative Study from Infrared Spectroscopy and DFT Calculations.* ACS Catalysis (submitted), 2014.

378123 Titration of Brønsted Acid Sites in Cu-Zeolites and Their Mechanistic Roles in Standard SCR

Abstract:

Small-pore, bifunctional, Cu-exchanged zeolites with the chabazite (CHA) topology are used in commercial deNO$_X$ applications using selective catalytic reduction (SCR) with NH_3. Here, we discuss methods that enable using NH_3 as a stoichiometric chemical titrant for H^+ sites in ZSM-5 (MFI) and SSZ-13 (CHA) zeolites in their H- and partially-Cu-exchanged forms, and discuss how these methods provide new insight into the mechanistic roles of H^+ sites in standard SCR.

In H-ZSM-5 and Cu-ZSM-5 samples with varying aluminum content (Si/Al = 17-90), H^+ sites were titrated selectively and counted consistently using three different NH_3 titration procedures and one reactive *n*-propylamine titration method. The number of H^+ sites on ZSM-5 zeolites measured by direct chemical titration was often less than the number of framework Al (Al_f) atoms determined from ^{27}Al MAS NMR spectra (H^+:Al_f = 0.7–1.0), providing yet another demonstration that Al_f structures can be imprecise surrogates for active H^+ sites on zeolites. In contrast with ZSM-5, the number of H^+ sites accessible to NH_3 on H-SSZ-13 (Si:Al = 4.5) and Cu-SSZ-13 (Cu:Al = 0-0.20) was higher by at least a factor of four than the number of H^+ sites accessible to *n*-propylamine under the saturation conditions studied. The number of H^+ sites on these Cu-SSZ-13 samples decreased with increasing Cu loading with a 2:1 stoichiometry, consistent with the exchange of isolated Cu^{2+} ions for two H^+ sites associated with proximal Al_f atoms. We conclude that NH_3 is more appropriately suited to titrate all accessible H^+ sites in small-pore Cu-zeolites, because residual OH groups in SSZ-13 were not detected in infrared spectra after NH_3 exposure, and because NH_3 is the reductant in NO_X SCR reactions [1].

Standard SCR rates were measured under differential conditions and normalized by the number of isolated Cu^{2+} sites, quantified by combining elemental analysis with ambient UV-Visible and X-ray absorption spectral data [2]. SCR turnover rates (473 K, per Cu^{2+}) show a zero-order dependence on the concentration of residual H^+ sites that remain after Cu^{2+} exchange on Cu-SSZ-13 [1]. The primary function of these residual H^+ sites appears to be for NH_3 storage under the conditions studied. We also show how NH_3 titrations performed under different gaseous environments can probe the nature of active H^+ sites that participate in standard SCR turnovers.

References:

[1] Bates, S. A.; Delgass, W. N.; Ribeiro, F. H.; Miller, J. T.; Gounder, R. *J. Catal.* **2014**, *312*, 26.

[2] Bates, S. A.; Verma, A. A.; Paolucci, C. P.; Parekh, A. A.; Anggara, T.; Yezerets, A.; Schneider, W. F.; Miller, J. T.; Delgass, W. N.; Ribeiro, F. H. *J. Catal.* **2014**, *312*, 87.

Nitrogen Oxides (NOx) are by-product pollutants of combustion reactions in air. Their selective reduction in oxygen rich environments is a key challenge in realizing more stringent emission regulations. NOx Selective Catalytic Reduction (SCR) is the catalyzed reaction through which NOx species are selectively reduced in an excess oxygen environment. Metallic SCR catalysts have problems with poisoning and selectivity, so recent focus has shifted to the use of zeolite catalysts, which are porous aluminosilicate materials. The metal-exchanged zeolite Cu-SSZ-13 in particular has shown excellent activity, selectivity, and hydrothermal stability and has already seen industrial usage. The fundamental reasons why Cu-SSZ-13 exhibits these traits are not well understood. **A primary obstacle for further improvement of SCR catalysts is a lack of molecular-level understanding of SCR activity in current catalysts**.

Our previous work has identified single Cu species in the six membered ring of Cu-SSZ-13 as an active site of SCR. Here we investigate the mechanistic chemistry through which the reaction proceeds at those sites utilizing hybrid (HSE06) Density Functional Theory (DFT) planewave calculations. To accomplish this we:

1) Probe the interaction between the SCR gas species and the active site using DFT and *ab-initio* molecular dynamics. We then validate the calculation results through comparison to *operando* X-ray adsorption experiments performed by our collaborators at Purdue and Argonne National Laboratory

2) Construct a thermodynamic phase diagram to compare the relative stability of the adsorbed species as well as other possible reaction intermediates under SCR conditions.

3) Elucidate a reaction mechanism for SCR that is consistent with the findings above.

We conclude that the SCR mechanism involves cyclic redox of the isolated Cu species between its 2+ and 1+ oxidation states. We show that no single gas species will both spontaneously adsorb and reduce/oxidize the Cu and thus the redox require formation of intermediates. The reduction half cycle occurs via an intermediate formed selectively in the presence of O_2 from NH_3 and NO. This formation generates an excess acid site, which has been quantified experimentally. The oxidation half cycle involves an intermediate formed from NO and O_2 to form nitrite, which reacts with the NH_4^+ formed from the generated acid site.

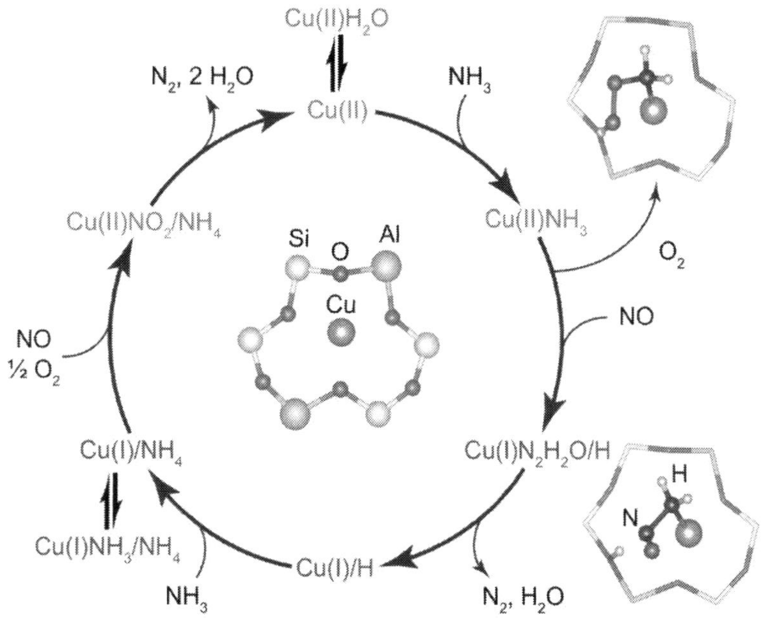

Promoting effect of WO_3 on deNO$_x$ activity of Ag/Al_2O_3 for ethanol/SCR

Pyung Soon Kim[1,2], Yelin Kim[2], Byong K. Cho[2], In-Sik Nam[2,*]

[1]Exhaust Emission Engineering Team, Research & Development Division, Hyundai Motor Group, 772-1, Jangduk, Hwaseong, 445-706, Korea
[2]School of Environmental Science and Engineering / Department of Chemical Engineering, Pohang University of Science and Technology (POSTECH), san 31 Hyoja, Pohang, 790-784, Korea
[*]Corresponding author; Tel: +82-54-279-2264, Fax: +82-54-279-8299, E-mail: isnam@postech.ac.kr

Ethanol/SCR over Ag/Al_2O_3 has been attracting a growing attention from both industry and academia in recent years due in part to the recent U.S. mandates for the use of biofuels in gasoline fuel[1] along with the increasing availability of E85 fuel (containing 85% ethanol in gasoline) in the U.S., and in part to its superior deNOx efficiency compared with other HC/SCR, especially in the low temperature region.[2-5] However, its effective temperature window (ETW) still remains to be a major barrier for its commercial application in the energy efficient modern vehicles[6,7], prompting extensive research efforts to widen its ETW for an enhanced deNOx performance. For example, the Ag metal loadings of Ag/Al_2O_3 and/or the operating conditions such as the C_1/NO_x and E/SD (ethanol/simulated diesel) feed ratios have been optimized to widen the ETW of Ag/Al_2O_3.[2,8-10] Also, a catalyst promoter[10] or a bifunctional reductant[11] has been used for the same purpose with varying degrees of success.

Regarding the use of W as a catalyst promoter, Blanchard et al.[12] reported a detrimental effect of W on the NO_x removal activity of $Ag/W/Al_2O_3$ by the n-C_{10}/SCR, as the W on Al_2O_3 blocked and hindered the interaction of the active Ag species with the Al_2O_3 support. On the other hand, promoting effects of W were reported for a variety of the SCR reactions over PGM-based catalysts such as the reduction of NO by CO on Pt/WO_3–SiO_2[13], the CH_4/SCR on Pd/WO_x–ZrO_2,[14] and the H_2/SCR on Pt/WO_x–ZrO_2.[15]

In this study, we report the deNO$_x$ activity of Ag/Al_2O_3 modified by W for ethanol/SCR, demonstrating a dramatic increase of the NO_x reduction efficiency compared to that by the Ag/Al_2O_3 proper. The $Ag/W/Al_2O_3$ catalysts with varying W loadings were prepared by the incipient wetness method, in which the γ-Al_2O_3 support was impregnated sequentially with an aqueous solution of $(NH_4)_6H_2W_{12}O_{40}\cdot xH_2O$ followed by the Ag impregnation using an aqueous solution of $AgNO_3$. For a comparative study, both Ag/Al_2O_3

and Ag_2WO_4-Al_2O_3 (a physical mixture of Ag_2WO_4 and Al_2O_3) were also prepared as model catalysts in elucidating the role of the interaction between Ag and WO_3 in the deNOx activity of the Ag/W/Al_2O_3 catalyst. Note that the nanoparticles of Ag_2WO_4 were synthesized at room temperature according to the precipitation method in the literature.[16] The prepared catalysts were oven-dried overnight at 110 °C and calcined in a muffle furnace at 550 °C for 5 h. A conventional packed-bed flow reactor system was used for the catalytic activity test.[2,10] The standard feed stream contained 400 ppm NO, 800 ppm C_2H_5OH, 6% O_2, 2.5% H_2O and He balance (GHSV: 60,000 h^{-1}). A variety of the catalyst characterization techniques such as XPS, H_2-TPR, and UV-vis has been employed to elucidate the effect of W on the physicochemical properties and the deNOx activity of Ag/W/Al_2O_3.

Fig. 1 compares the deNO$_x$ performance of Ag(4.0)/Al_2O_3, Ag_2WO_4(4.0 wt.% of Ag)-Al_2O_3, Ag(4.0)/W(6.0)/Al_2O_3, and W(6.0)/Al_2O_3 in terms of the NO$_x$-to-N_2 conversion via ethanol/SCR at the C_1/NO$_x$ feed ratio of 4 with a constant NO feed concentration of 400 ppm. It clearly demonstrates that the addition of W to the Ag/Al_2O_3 catalyst leads to an enhanced NO$_x$ reduction to N_2 over a wide temperature window under the present reaction condition. Results of XPS studies have indicated the formation of the oxygen-bridged Ag and W structure (Ag-O-W) on the surface of Ag/W/Al_2O_3. Also important to note in Fig. 1 is that the deNOx performance of Ag(4.0)/W(6.0)/Al_2O_3 is higher than the sum of the individual deNOx performance of Ag(4.0)/Al_2O_3 and W(6.0)/Al_2O_3 over the entire temperature range tested, indicative of a synergistic interaction between Ag and W on Ag(4.0)/W(6.0)/Al_2O_3 for NO reduction.

The deNOx performance of Ag_2WO_4-Al_2O_3 is superior to that of Ag/Al_2O_3, especially at low temperatures below 350 °C, while it is slightly lower than that of Ag(4.0)/W(6.0)/Al_2O_3 at high temperatures above 350 °C, probably due to the lower W content included in Ag_2WO_4-Al_2O_3 (Ag: 4.0 wt.%, W: 3.0 wt.%). Since the Ag_2WO_4-Al_2O_3 catalyst is mainly comprised of the oxygen-bridged silver and tungsten (Ag-O-W) structure, this observation indicates that W plays a key role in widening the ETW of the Ag-based ethanol/SCR catalysts via its synergistic interaction with Ag through the formation of the Ag-O-W structure on its surface.

The oxidation of ethanol by oxygen in the absence of NO over Ag(4.0)/Al_2O_3 and Ag(4.0)/W(6.0)/Al_2O_3 was also investigated to further elucidate the role of W in the ethanol/SCR. As presented in Fig. 2, Ag(4.0)/W(6.0)/Al_2O_3 produces much more CH_3CHO than Ag(4.0)/Al_2O_3 does (Fig. 2a), while producing much less CO_2 (Fig. 2b). This is a clear indication that the addition of W to the Ag/Al_2O_3 catalyst promotes the partial oxidation of HC forming more reactive intermediate reductant such as aldehydes, while suppressing the complete oxidation thereof to CO_2, whereby resulting in the enhancement of the NO reduction activity. Included also in the presentation will be the details of catalyst characterizations by H_2-TPR, UV-vis and XPS showing the change in the state of Ag and W as a function of W loadings.

References

1. Environmental Protection Agency, Federal Register 78:158 (2013) 49794.
2. M. K. Kim, P. S. Kim, J. H. Baik, I. S. Nam, B. K. Cho, S. H. Oh, Appl. Catal. B: Environ. 105 (2011) 1.
3. P. S. Kim, M. K. Kim, B. K. Cho, I. -S. Nam, S. H. Oh, J. Catal. 301 (2013) 65.
4. Y. Yu, J. Zhao, Y. Yan, X. Han, H. He, Appl. Catal. B: Environ. 136 (2013) 103.
5. J. A. Pihl, T. J. Toops, G. B. Fisher, B. H. West, Catal. Today 231 (2014) 46.
6. M. Iwamoto, H. Yabiro, S. Shundo, Y. Yu-u, N. Mizno, Appl. Catal. 69 (1991) L15.
7. R. Burch, J.P. Breen, F.C. Meunier, Appl. Catal. B: Environ. 39 (2002) 283.
8. Y.B. Yu, H. He, Q.C. Feng, H.W. Gao, X. Yang, Appl. Catal. B: Environ. 49 (2004) 159.
9. C. Zhang, X. Shi, H. He, Chinese J. Catal. 26 (2005) 645.
10. P. S. Kim, B. K. Cho, I.-S. Nam, J. W. Choung, ChemCatChem. (2014) doi: 10.1002/cctc.201402085
11. P. S. Kim, M. K. Kim, B. K. Cho, I. -S. Nam, J. Catal. 292 (2012) 44.
12. J. Blanchard, R.P. Doherty, H.-Y. Law, C. Méthivier, C. Thomas, Top. Catal. 56 (2013) 134.
13. J.R. Regalbuto, E.E. Wolf, J. Catal. 109 (1988) 12.
14. Y.-H. Chin, W.A. Alvarez, D.E. Resasco, Catal. Today 62 (2000) 159.
15. F.J.P. Schott, P. Balle, J. Adler, S. Kureti, Appl. Catal. B: Environ. 87 (2009) 18.
16. T. George, S. Joseph, S. Mathew. Pramana J. Phys. 65 (2005) 793.

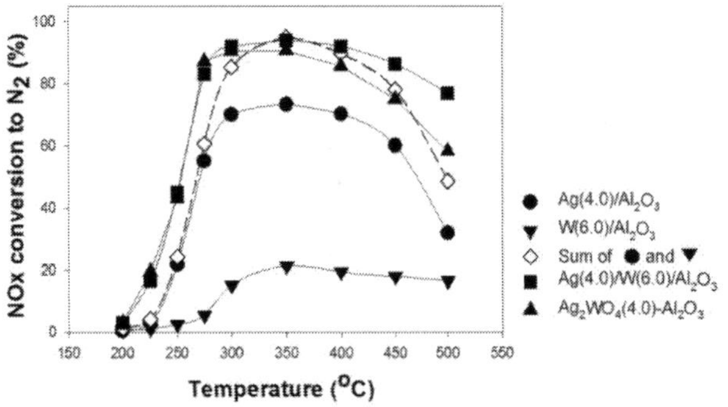

Fig. 1. Comparison of NO_x-to-N_2 conversion by ethanol/SCR over Ag-based Al_2O_3 catalysts.

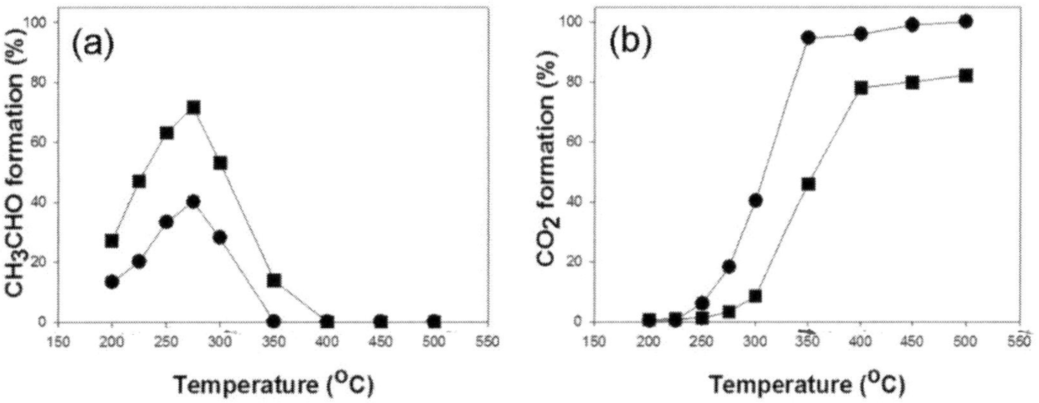

Fig. 2. Comparison of products from ethanol oxidation by oxygen in the absence of NO: $Ag(4.0)/Al_2O_3$ (●) vs. $Ag(4.0)/W(6.0)/Al_2O_3$(■)

376107 Highly Dispersed Fe-Oxide Catalysts for Reduction of NOx By CO: Synthesis, Evaluation, & Operando XAS

Abstract:

The reduction of oxides of nitrogen (NOx) without the use of platinum group metals (PGM) remains a challenge in automotive exhaust catalysis. Current strategies have employed selective catalytic reduction (SCR) over Cu- or Fe-exchanged zeolites.[1] The structure of the transition metal-oxide active sites in the zeolite framework has yet to be clarified, but it is generally accepted that isolated ions or other low nuclearity species are responsible.[2-4] Dispersion of the metal oxide on a high surface area support represents a potential route to the identification of the NO reduction active site by creating sites that mimic those found in the ion-exchanged zeolites. However, the surface oxide species created by typical synthetic routes to supported metal oxides are largely governed by thermodynamics, and it is difficult to obtain narrow distributions of metal-oxide surface species. Therefore, the current study utilizes a unique synthetic route to highly dispersed Fe-oxides on a non-zeolitic CeO_2support for reduction of NO by CO.

The Fe-oxide catalysts were impregnated onto CeO_2 using Fe coordinated to a bulky organic ligand EDTA and a Na cation (denoted as NaFeEDTA). Surface densities of 0-1.5 Fe/nm^2 were prepared. These catalysts are compared to the traditional $Fe(NO_3)_3$ precursor impregnated on Na-promoted CeO_2 over the same surface density range. Through comprehensive characterization by UV-vis, XRD, Raman, XAS, Mössbauer, and TPR, it was discovered that the NaFeEDTA route was preferred over $Fe(NO_3)_3$ to produce a narrow distribution of smaller, non-crystalline, surface Fe-oxide species with excellent redox cyclability ($Fe^{3+}\leftrightarrow Fe^{2+}$sites).

Evaluation of the steady-state activity for NO reduction by CO demonstrated that the NaFeEDTA catalysts exhibited the highest activity over the studied surface density range. Additionally, a maximum activity was established at ~1.0 Fe/nm^2 before plateauing at higher surface densities. The plateau in activity is explained by quantifying the redox cyclable Fe sites ($Fe^{3+}\leftrightarrow Fe^{2+}$) using TPR/TPO cycles, which correlated with both structure and activity.

This reaction was further probed using *operando* XANES. The results indicate that removal of NO from the reaction stream begins concurrent with reduction of Fe and Ce. This observation suggests that reduction occurs at the Fe-O-Ce interface and this event is necessary to initiate activity. Further investigation of the role of this site in the NO reduction mechanism will be investigated by *in situ* FTIR spectroscopy during NO adsorption and temperature programmed surface reaction under flowing CO gas. The current study is expected to expand the understanding of the necessary redox properties and active site structures of Fe-oxide active sites for NO reduction during automotive exhaust catalysis.

References

1. Kröcher, Stud. Surf. Sci. Catal, vol. 171 (2007) 261.
2. M.S. Kumar, M. Schwidder, W. Grünert, U. Bentrp, A. Brückner J. Catal. 239 (2006) 173.
3. J.-S. McEwen, T. Anggara, W.F. Schneider, V.F. Kispersky, J.T. Miller, W.N. Delgass, F.H. Ribeiro, Catal. Today 184 (2012) 129.
4. J. Szanyi, J.H. Kwak, H. Zhu, C.H.F. Peden, Phys. Chem. Chem. Phys. 15 (2013) 2368.

387365 Effect of Aging Conditions on Model Pd Three-Way Catalysts

Abstract:

Palladium plays a critical role in automotive three-way catalysts (TWC). Three-way catalysts deactivate with time, and this paper explores the effect of different aging environments on the properties and performance of model Pd catalysts supported on alumina or ceria-zirconia (CZO). A systematic study is carried out to determine how different aging environments affect the oxidation state and dispersion of Pd. In addition, the influence of the support and the Pd particle size are discussed with special emphasis on its relationship to the activity in various oxidation reactions. The catalysts are extensively characterized by hydrogen and carbon monoxide chemisorption, temperature programmed reduction and oxidation, ex-situ and in-situ XRD, and XPS. As probe reactions, the oxidation of CO, methane, and propylene are used, along with the water shift reaction. The effect of aging high temperatures, either lean or rich, can have differing effects on metal particle size in this family of catalysts.

In-Situ ATR-IR Study on Formation of Intermediates of APR Reactions at Room and Elevated Temperature over a Pt/γ-Al_2O_3 Catalyst

Jungseob So, John Copeland, Guo Shiou Foo, Yoona Chung
David Sholl, Carsten Sievers

November 16, 2014
Atlanta, GA

Georgia Institute of Technology
School of Chemical & Biomolecular Engineering

Hydrogen production

- Hydrogen is a long-term solution as sustainable clean energy
- Hydrogen is in great demand in many areas such as production of ammonia, glycols liquid fuels, and application in fuel cells
- Problems of commercial process for producing hydrogen
    - Availability of natural gas and coal is finite.
    - The emission of CO_2 is leading to global warming.

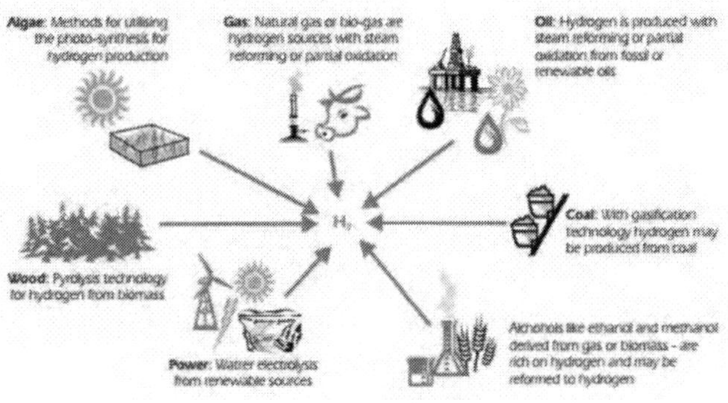

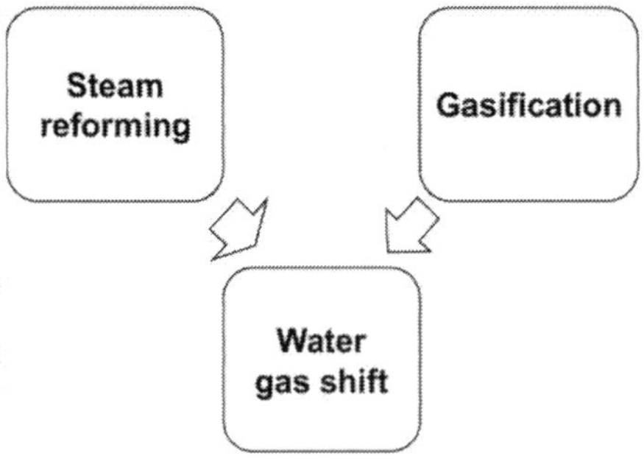

Georgia Institute of Technology
School of Chemical & Biomolecular Engineering

Hydrogen from biomass

- Hydrogen derived from biomass-derived carbon-hydrates is a natural way for sustainable energy production
- What are advantages?
  - Carbon-neutral processing.
  - Reforming of oxygenated hydrocarbons for hydrogen production is thermodynamically favorable at significantly lower temperatures than those required for alkanes with similar number of carbon atoms

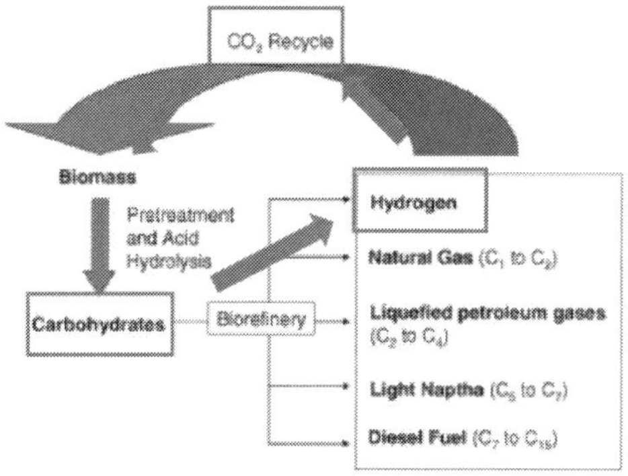

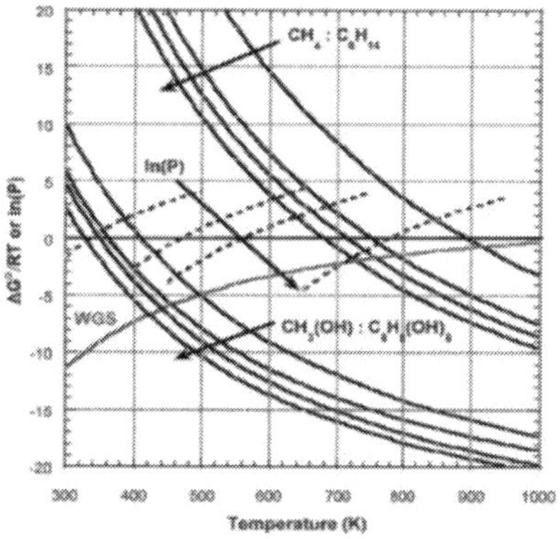

Georgia Institute of Technology
School of Chemical & Biomolecular Engineering

R.R. Davda, J.W. Shabaker, G.W. Huber, R.D. Cortright, J.A. Dumesic, Appl. Catal, 56 (2005) 171
G.W. Huber, J.A. Dumesic, Catal. Today, 111 (2006) 119.

Aqueous Phase Reforming

Hydrogen Route

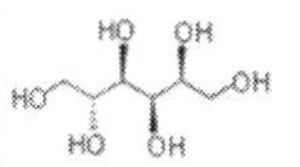

Sorbitol $+ 6H_2O \xrightarrow{225 - 265°C} 13H_2 + 6CO_2$

Raney Nickel or Pt/Al_2O_3

Alkane Route

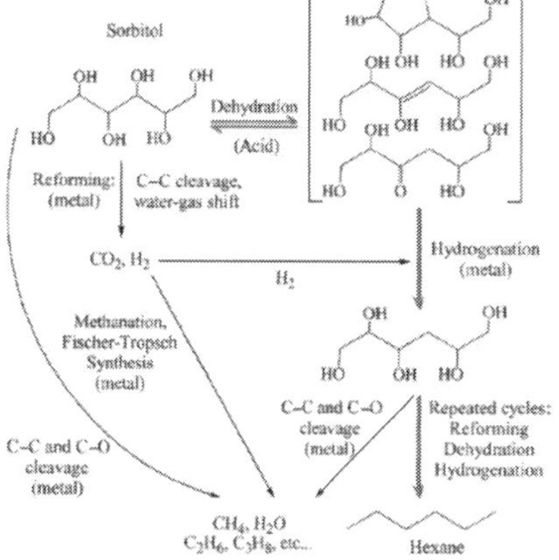

Catalyst: Pt or Pd on Al_2O_3 or SiO_2-Al_2O_3

- Aqueous Phase Reforming(APR) of polyols and sugars is a promsing altenative for hydrogen proudction
- APR is used to convert sugar alcohol to hydrogen and/or alkanes.
- High hydrogen yields are obtained over catalysts with a high activity for C-C bond cleavage (e.g. Pt, Ni).
- Alkanes are produced by a combination of acid catalyzed dehydration and metal catalyzed hydrogenation.

Georgia Institute of Technology
School of Chemical & Biomolecular Engineering

R.D. Cortright, R.R. Davda, J.A. Dumesic, Nature 418 (2002) 964.
G.W. Huber, J.W. Shabaker, J.A. Dumesic, Science 300 (2003) 2075.
G.W. Huber, R.D. Cortright, J.A. Dumesic, Angew. Chem.-Int. Ed. 43 (2004) 1549.

Aqueous Phase Reforming: Mechanism

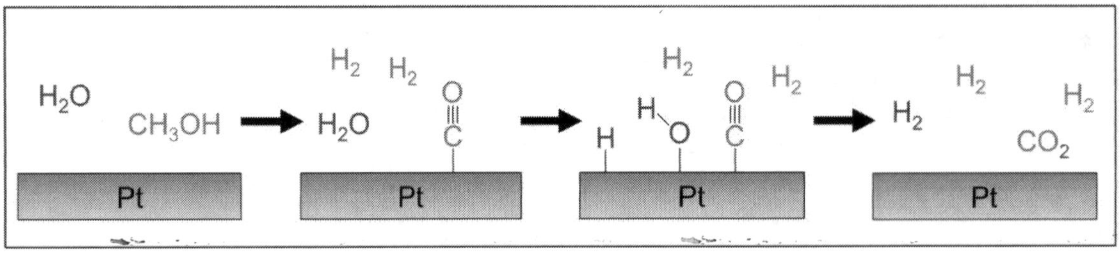

Glycerol: $C_3H_8O_3 + 3H_2O \rightarrow 3CO_2 + 7H_2$

- Aqueous phase reforming consists of decomposition of oxygenates to form CO and hydrogen followed by water-gas shift.
- How does the composition and size of metal particles affect the rate of formation and conversion of CO?
- What influence do co-adsorbed oxygenates (reactants or products from side reactions) have?
- Why is APR of larger oxygenates less efficient?

Georgia Institute of Technology
School of Chemical & Biomolecular Engineering

R. He, R.R. Davda, and J.A. Dumesic, J. Phys. Chem. B 109 (2005) 2810.

Attenuated Total Reflection Spectroscopy

- IR spectroscopy is a powerful tool to study detailed interactions of reactants and reaction-intermediates with catalyst surface
- We direct the IR beam into a crystal of relatively higher refractive index.
- If the angle of incidence of the IR bean is greater than critical angle, the IR beam reflects from the internal surface and also create the evanescent waves.
- We analyses the evanescent waves which normally penetrate up to 5 micrometer to analyze the surface of catalyst.

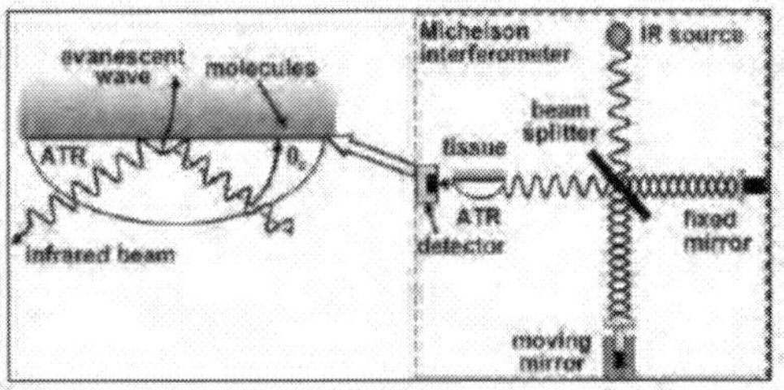

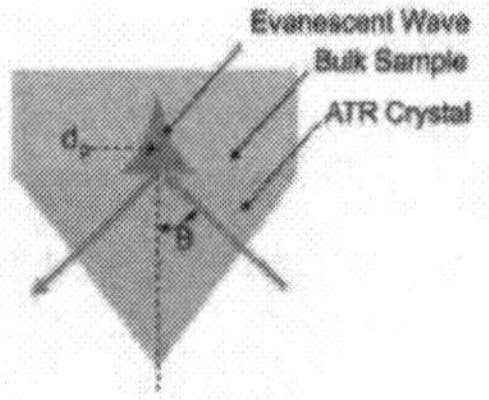

$$\theta_c = \sin^{-1}\left(\frac{n_2}{n_1}\right) \qquad d_p = \frac{\lambda}{2\pi(n_1^2 \sin^2\theta - n_2^2)^{\frac{1}{2}}}$$

Georgia Institute of Technology
School of Chemical & Biomolecular Engineering

Attenuated Total Reflection Spectroscopy

6

- With this approach, surface species on a catalyst can be probed effectively while keeping the contribution from the liquid solvent marginal to preserve many important regions of IR spectrum

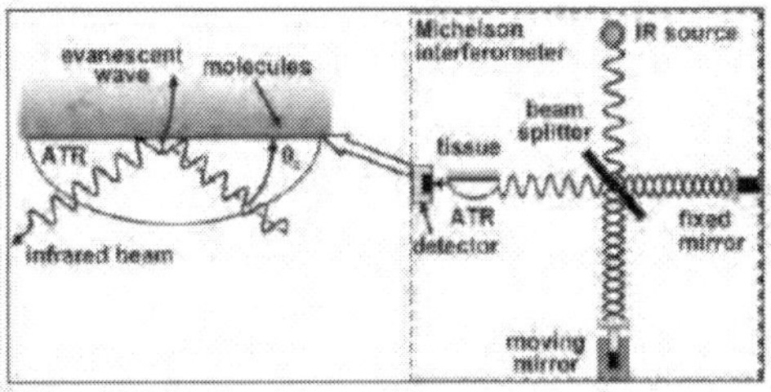

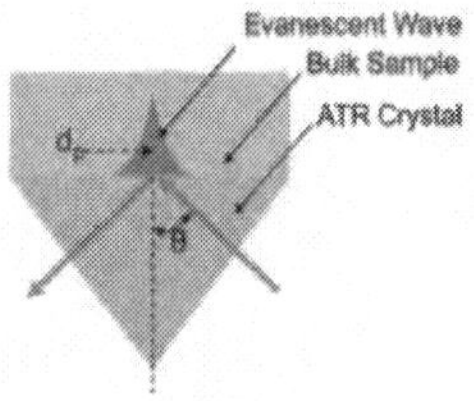

$$\theta_c = \sin^{-1}\left(\frac{n_2}{n_1}\right) \qquad d_p = \frac{\lambda}{2\pi(n_1^2 \sin^2\theta - n_2^2)^{\frac{1}{2}}}$$

Georgia Institute of Technology
School of Chemical & Biomolecular Engineering

Experimental Setup, ATR-IR

7

- Experimental Method
  - A slurry of Pt/γ-Al_2O_3 in water is deposited on the ZnSe crystal and the water is allowed to evaporate, 5 times

Pt/Al_2O_3
0.1 M glycerol or glucose in water
Flow rate: 5 ml/min

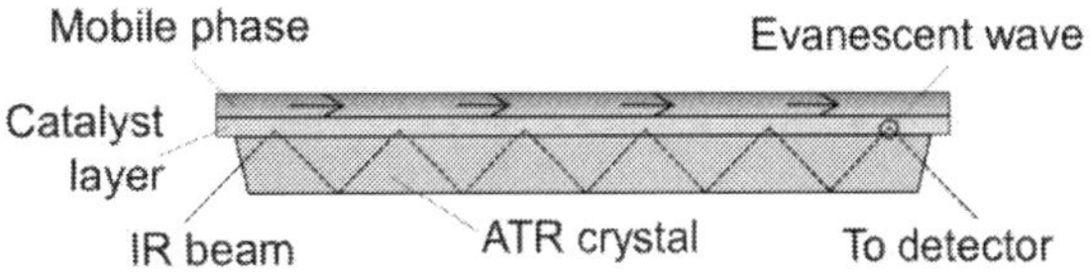

- Several solutions were used as mobile phase
  - Degassed water and aqueous glycerol and glucose solution
  - Water with Dissolved O_2, H_2

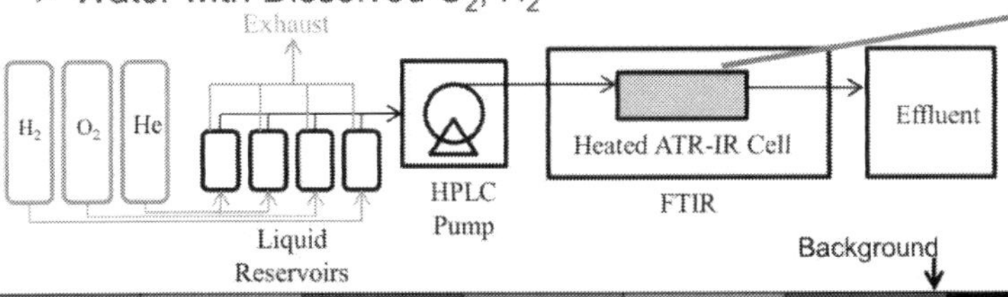

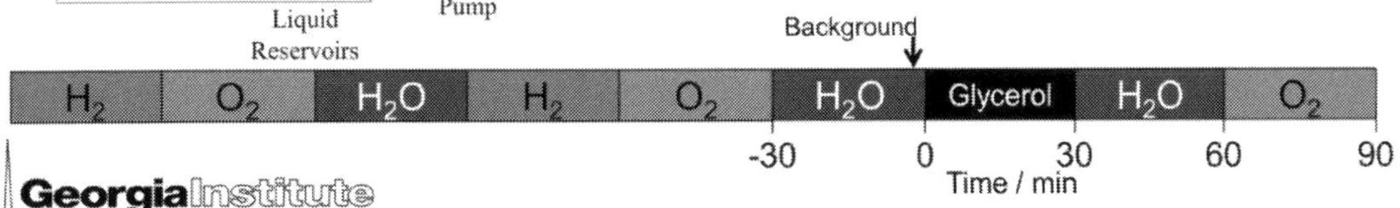

Georgia Institute of Technology
School of Chemical & Biomolecular Engineering

J.R. Copeland, G.S. Foo, L.A. Harrison, C. Sievers, Catal. Today, 205 (2013) 49.

CO Formation on Pt/Al_2O_3

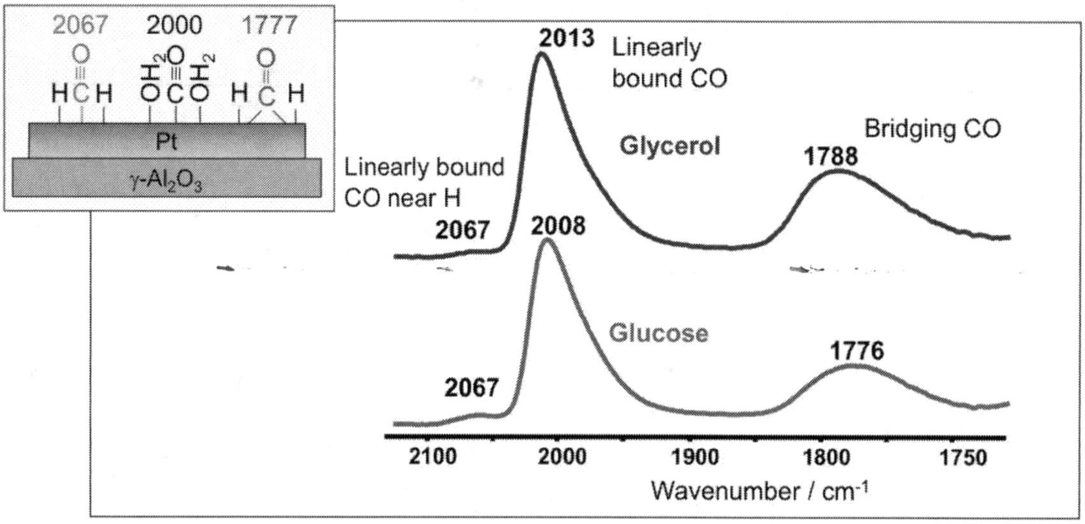

- Glycerol and glucose is converted over Pt/Al_2O_3 to linear and bridge CO as well as surface bound hydrogen.
- Three types of CO surface species formed, linear CO preferentially mixed with surface hydrogen atom, linear CO preferentially mixed with H_2O, and bridging CO.

Georgia Institute of Technology
School of Chemical & Biomolecular Engineering

J.R. Copeland, G.S. Foo, L.A. Harrison, C. Sievers, Catal. Today, 205 (2013) 49.

Peak position and integrated area

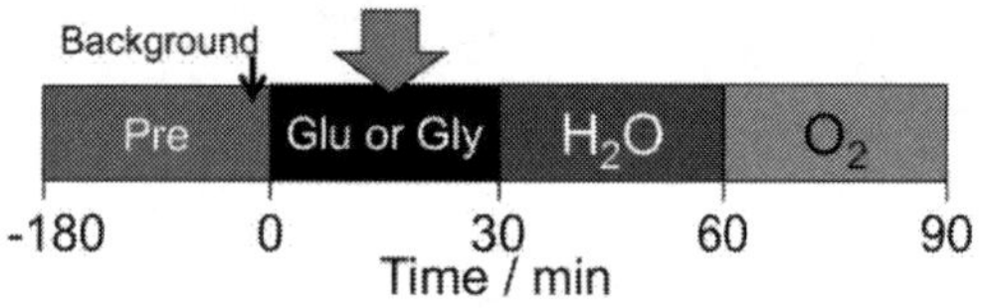

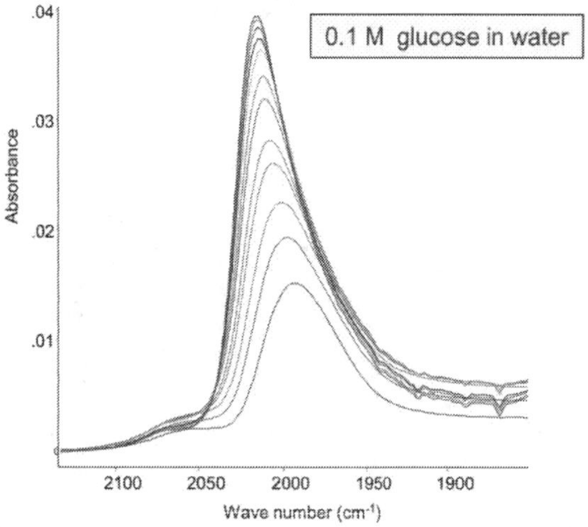

- Dipole-dipole interactions become stronger as coverage increases.

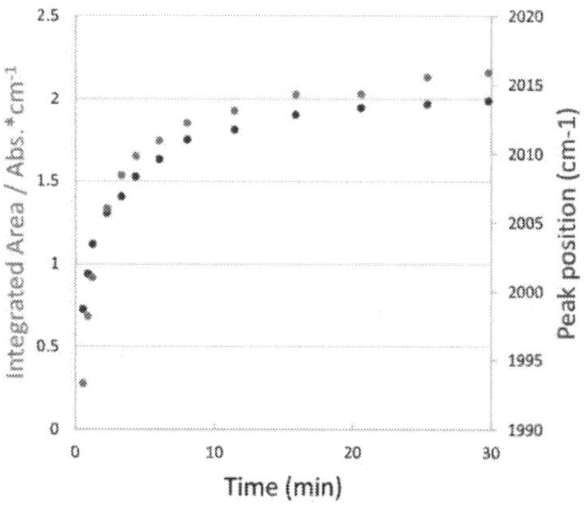

Georgia Institute of Technology
School of Chemical & Biomolecular Engineering

Glycerol – Water Gas Shift at 50 °C

10

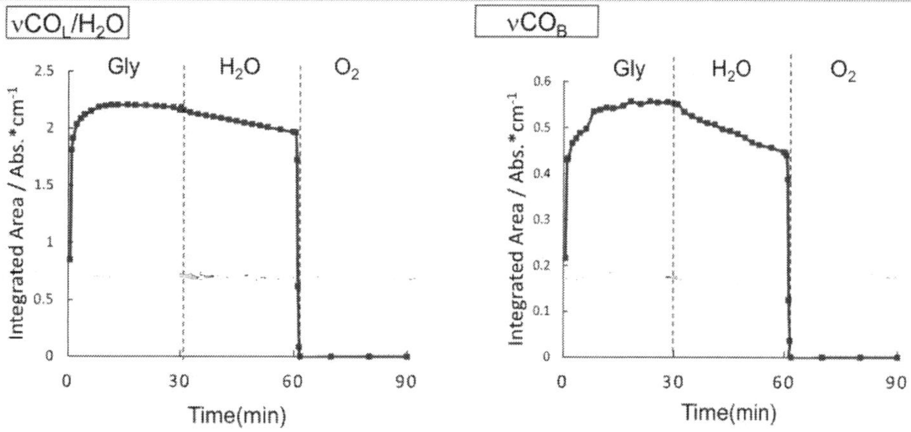

- The maximum of νCO_L is attributed to competitive adsorption of hydrogen.
- Certain linear and bridging CO species can be converted by water-gas-shift at 50 °C.
- Bridging CO is more reactive than linear CO.

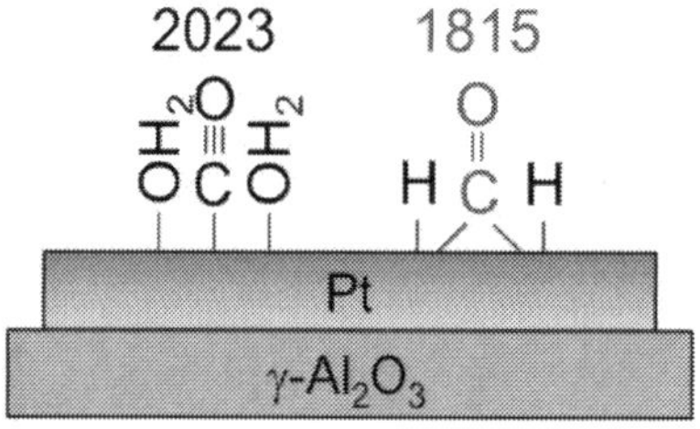

Georgia Institute of Technology
School of Chemical & Biomolecular Engineering

* B. Hammer, O. H. Nielsen, J. K. Norskov, Catal. Lett, 46 (1997) 31.
J.R. Copeland, G.S. Foo, L.A. Harrison, C. Sievers, Catal. Today, 205 (2013) 49.

Water-Gas Shift

11

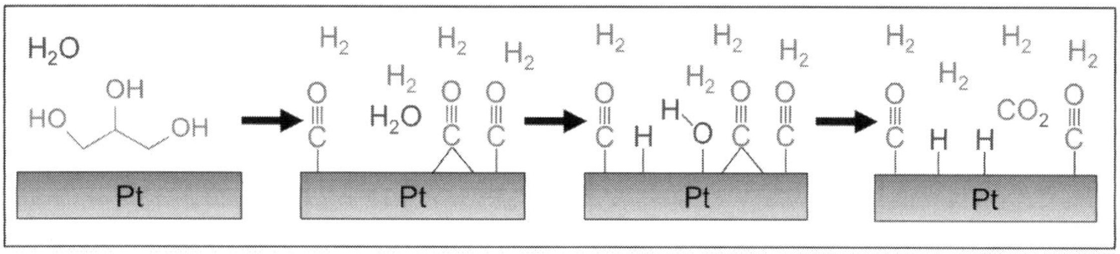

- Sites that bind CO both in linear and bridging coordination appear to be active for water-gas shift at 50 °C.
- Adsorbed hydrogen remains preventing reoccupation of some of these sites by CO species.

Georgia Institute of Technology
School of Chemical & Biomolecular Engineering

J.R. Copeland, G.S. Foo, L.A. Harrison, C. Sievers, Catal. Today, 205 (2013) 49.

Glycerol at Room Temperature

13

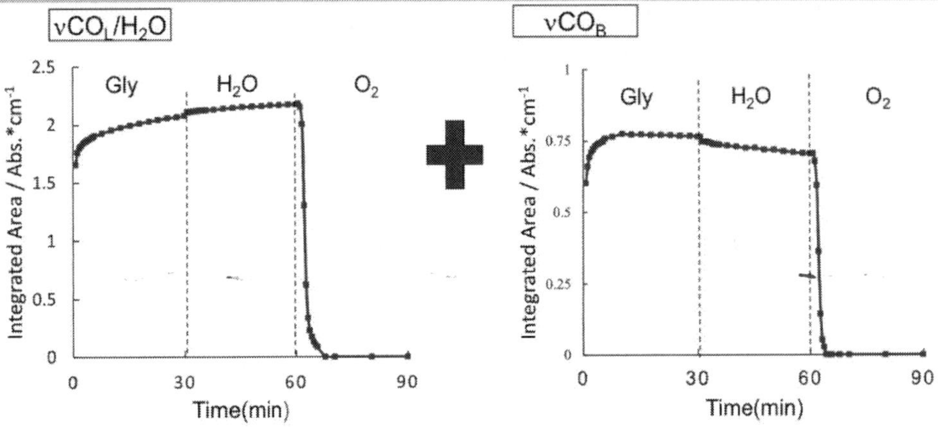

- Linearly bound CO keeps increasing up to the water flow.
- Once bridge CO reaches a maximum, it keeps gradually decreasing up to the water flow.

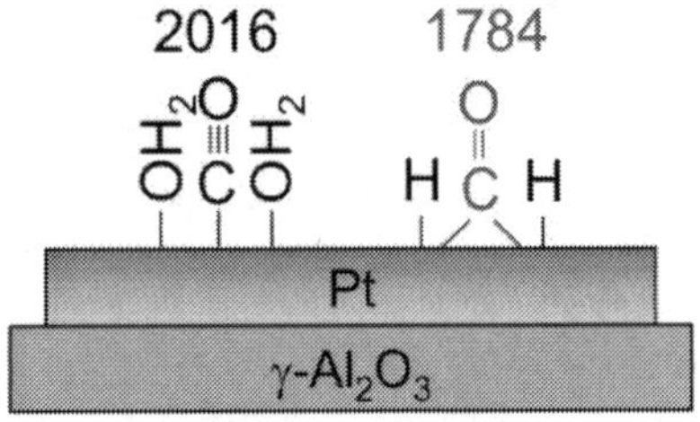

Georgia Institute of Technology
School of Chemical & Biomolecular Engineering

Glycerol at Room Temperature

14

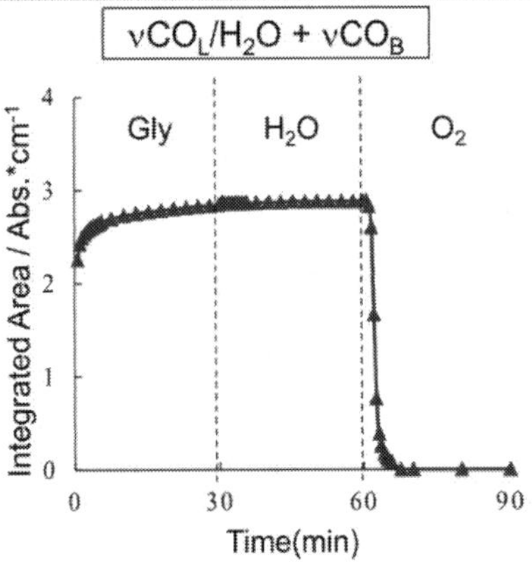

- Sum of the integrated area linear and bridging CO is constant with no further supply of glycerol reactants.
- Once the linear and bridging CO is formed on the surface, CO site conversion occurs at room temperature.

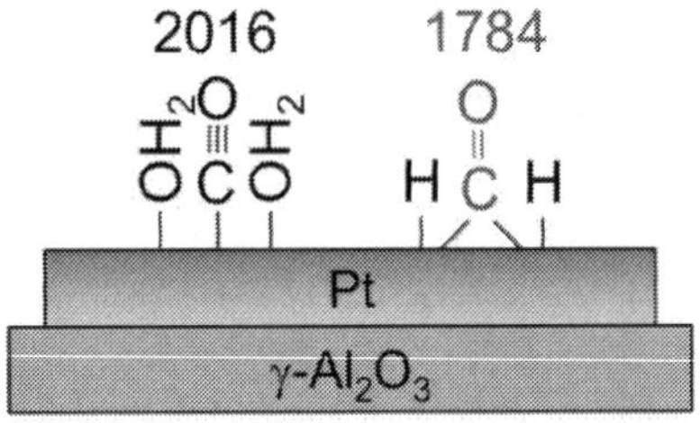

Georgia Institute of Technology
School of Chemical & Biomolecular Engineering

Glycerol at Different Temperatures

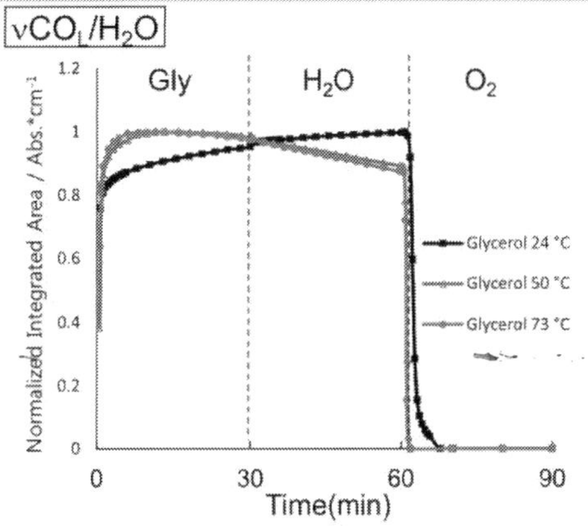

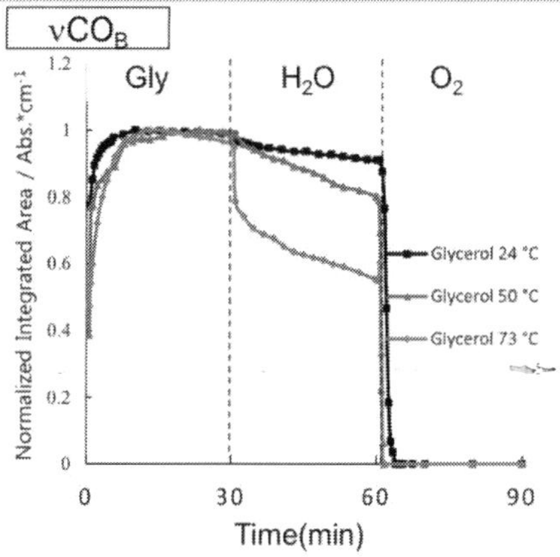

- Linear CO which is inactive at the low temperature become active at the elevated temperature, indicating that apparent activation energy was overcome.
- Disappearance rates of linear CO are similar at the both elevated temperatures.
- Some changes in peak areas may be due to migration of CO.
- As the temperature increases, the disappearance rate of bridging CO increases.

Georgia Institute of Technology
School of Chemical & Biomolecular Engineering

Comparison of Glycerol and Glucose

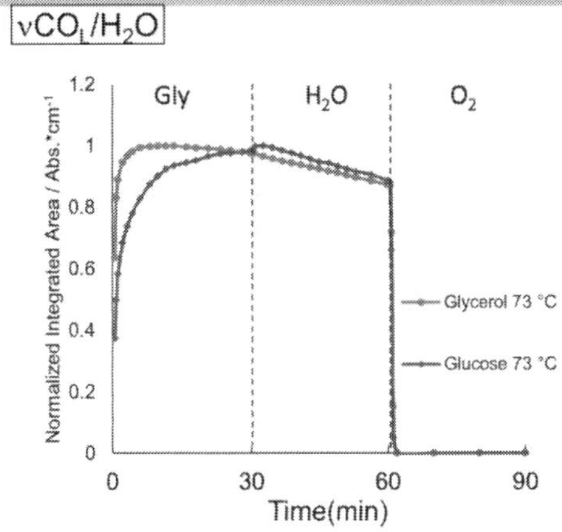

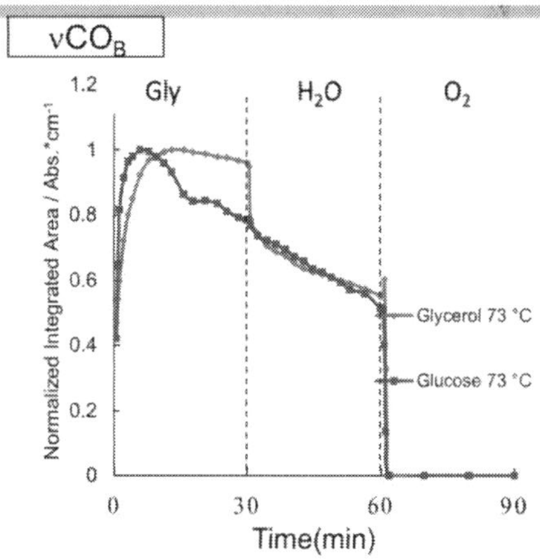

- Bridging and linear CO derived from glycerol and glucose went through the water gas shift reaction at 73 °C.
- Linear CO derived from glucose slowly formed compared to glycerol, which might be due to the presence of heavy diverse intermediates.
- The formation rate of bridging CO from glucose is faster than from glycerol.
- The disappearance rate during water flow is not affected by the nature of the feed.

Georgia Institute of Technology
School of Chemical & Biomolecular Engineering

Conclusions

17

- Glycerol and glucose are converted to surface bound CO over Pt/Al_2O_3 at room and elevated temperature.
- CO can migrate between sites at the room temperature.
- At slightly elevated temperatures, CO surface species are oxidized by the WGS reaction.
- Bridging CO is more reactive than linear CO.
- Glucose intermediates from the decomposition inhibit the formation of linear CO surface species more than glycerol intermediates.

Georgia Institute of Technology
School of Chemical & Biomolecular Engineering

Acknowledgements

18

The Sievers group
Gautami Newalkar
Akil Syed
Jessica Ewbank
Guo Shiou Foo
Chukwuemeka Okolie
Sireesha Aluri
Yulia Demidova

The Jones group
Shuai Tan

Funding:

GT startup funds

Paper Science and Engineering Fellow ship from Renewable Bioproducts Institute

Georgia Institute of Technology
School of Chemical & Biomolecular Engineering

364849 Catalytic Dehydration of Methyl Lactate

Abstract:

Catalytic dehydration of methyl lactate to acrylic acid and methyl acrylate on NaY is investigated via in-situ transmission FTIR spectroscopy. Ion exchange, Bronsted acid site-mediated dehydration and displacement reaction are proposed to be the key mechanistic steps in the surface mediated chemical transformation. Design principles for bifunctional catalysts have been developed based on the mechanistic insights will be discussed.

382902 Nature of the Active Site for Ammonia Standard SCR and NO Oxidation Reactions

Abstract:

Copper chabazite (CHA) zeolites are known for their stability under harsh hydrothermal conditions and performance in reducing NO_x emissions. This study aims at understanding the mechanistic details and the types of active sites required for both NO oxidation and NH_3 standard selective catalytic reduction (SCR) reactions by combining experimental techniques along with first-principles density functional theory (DFT) calculations.

Aluminosilicate CHA zeolites (SSZ-13, Si/Al = 4.5) with varying amounts of exchanged Cu ions were prepared by liquid phase ion exchange using $Cu(NO_3)_2$ as the Cu precursor. For the standard SCR reaction (473 K), the reaction order with respect to NH_3 ranged between -0.2 to 1, that for O_2 ranged from 0.2 to 0.5, while the order with respect to NO ranged between 0.7 and 0.9. The reaction rate per gram catalyst increased linearly with Cu loading up to Cu/Al = 0.2, where the maximum rate was 3.8 x 10^{-6} mol NO (g cat)$^{-1}$ s^{-1}. Further increases in the Cu loading above Cu/Al = 0.2 led to decreases in the standard SCR reaction rate (per g). UV-Vis and *operando* X-Ray absorption spectroscopy (XAS) revealed the presence of only isolated Cu^{2+} ions up to Cu/Al = 0.2, confirming that isolated Cu^{2+} is the active site for standard SCR. Statistical analysis, assuming a random distribution of Al in the SSZ-13 framework, estimated that the maximum number of framework Al (Al_f) pairs needed to stabilize isolated Cu^{2+} ions was $Cu^{2+}/Al_f = 0.18$ for Si/Al = 4.5. The role of Brønsted acid sites was investigated by quantifying their number using an NH_3 adsorption technique developed in-house to ensure that only residual Brønsted acid sites after Cu exchange were titrated. No correlation between the standard SCR reaction rate and the number of residual Brønsted acid sites was observed, indicating that the reaction was not limited by the number of Brønsted acid sites under the conditions studied (473 K, SSZ-13 Si/Al = 4.5, Cu/Al < 0.2). Only for samples with Cu/Al > 0.2, Cu clustering was observed by XAS and NO oxidation rates in the absence of water were detected, suggesting that Cu clusters are the active sites for NO oxidation. *In situ* XANES indeed showed a linear correlation between the NO oxidation reaction rate (per mol Cu, 573 K) and the fraction of Cu present as Cu_xO_y clusters. DFT calculations showed favorable free energies for forming Cu-O-Cu dimer species in the 8-membered rings of the CHA structure once the isolated Cu^{2+} ion density in the 6-membered rings reached saturation. Thus, we conclude that isolated Cu^{2+} ions in the 6-membered rings are the active sites for NH_3 standard SCR reaction, while Cu_xO_y clusters formed in the 8-membered rings of the CHA cages are the active sites for the NO oxidation reaction.

In situ Surface Chemistries and Catalytic Performance of 3D Ordered Macroporous Cu-Fe Catalysts for Higher Alcohols Synthesis from Syngas

Yongwu Lu,[a] Fei Yu,[a,*] Junjun Shan,[b] Franklin (Feng) Tao [b]

[a] Department of Agricultural and Biological Engineering, Mississippi State University, Mississippi State, MS 39762, United States

[b] Department of Chemical & Petroleum Engineering and Department of Chemistry, University of Kansas, Lawrence, KS 66047, United States

2014 AICHE Annual Meeting
Atlanta, GA
Nov 16–21, 2014

Outline

- 1. Introduction (Biomass-to-Liquids;3DOM)
- 2. Objective
- 3. Experimental Materials
- 4. Results and Discussion
- 5. Conclusions
- 6. Future work
- 7. Acknowledgment

Outline

- 1. Introduction (Biomass-to-Liquids;3DOM)
- 2. Objective
- 3. Experimental Materials
- 4. Results and Discussion
- 5. Conclusions
- 6. Future work
- 7. Acknowledgment

Biofuels produced from renewable resources

— as an alternative to fossil oil.

— mitigate global warming.

— minimize the fossil fuel burning and CO_2 production.

Biomass to liquid fuels (BTL) technology is one of the most promising technologies to produce useful and high-value biofuels.

Biomass-to-Liquid fuels (BTL)

- Biomass gasification to yield producer gas (a type of syngas consists of CO, H_2, CO_2, N_2, CH_4 etc.),
- Cleaning of producer gas to remove contaminants,
- Catalytic conversion of clean producer gas to liquid fuels in the fixed-bed reactor system.

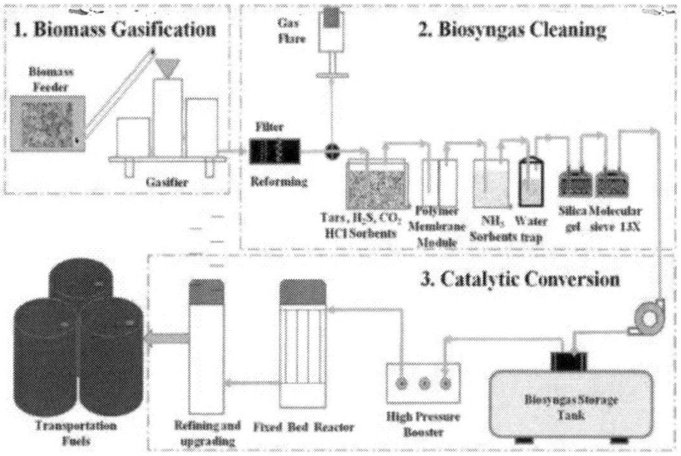

BTL at Mississippi State (Lab Scale)

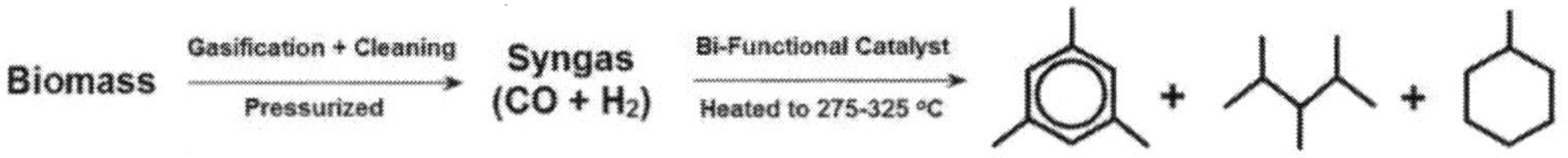

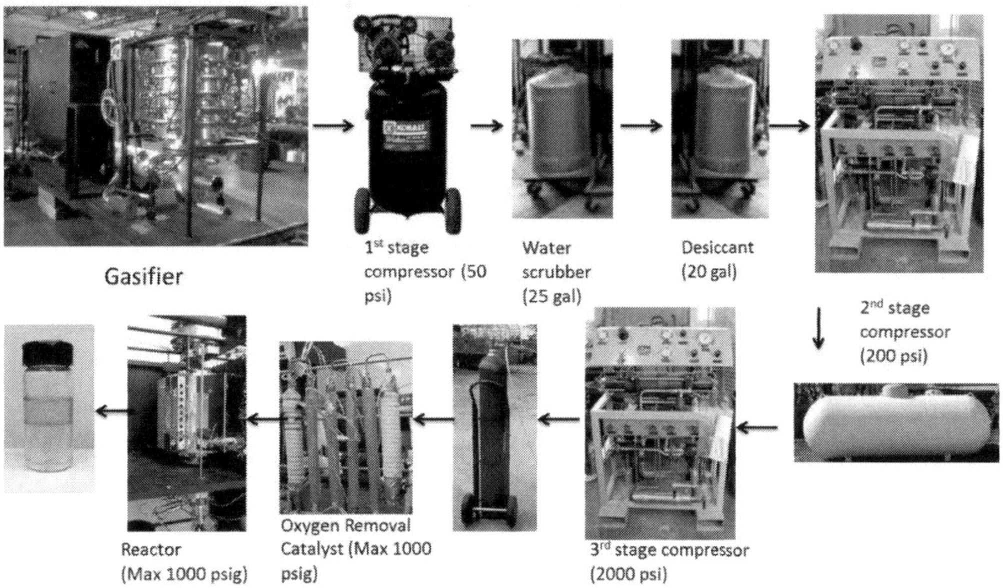

BTL at Mississippi State (Pilot scale)

Gasifier

Water Scrubber, Desiccant, Oxygen Removal Catalyst, More Desiccant

1st, 2nd, 3rd stage compressor (2000 psi)

Quantity of **Six,** 2 in. diameter, 5 ft. long catalyst bed reactors

Catalytic conversion of syngas to liquid fuels

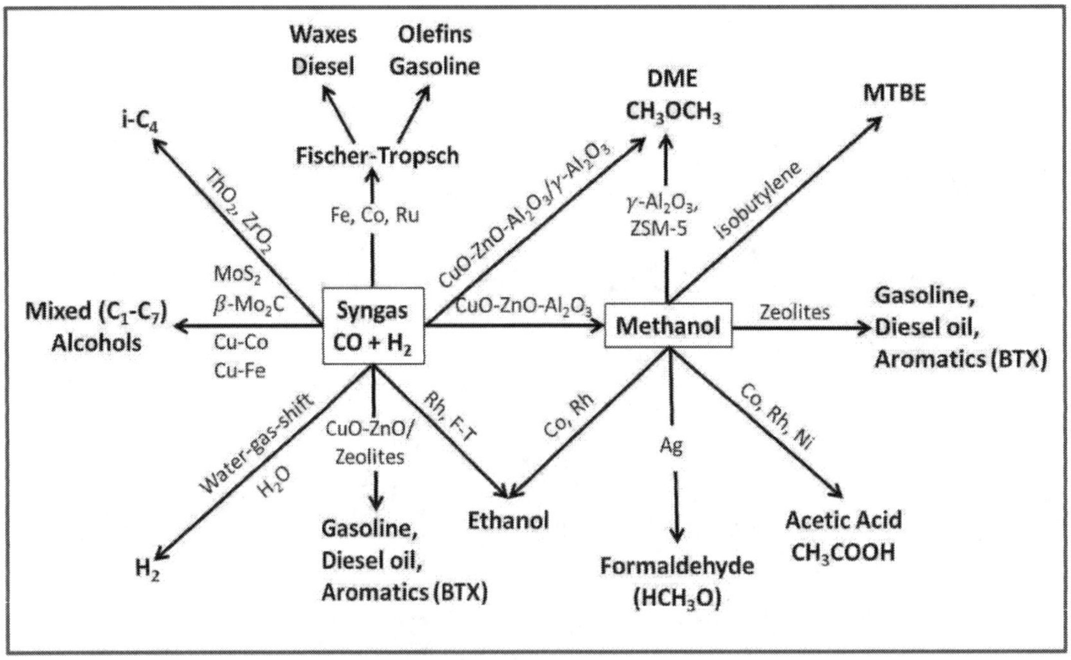

Velu Subramani et al, *Energy & Fuel*, **2008**, *22*, 814.

Catalytic conversion of syngas to liquid fuels

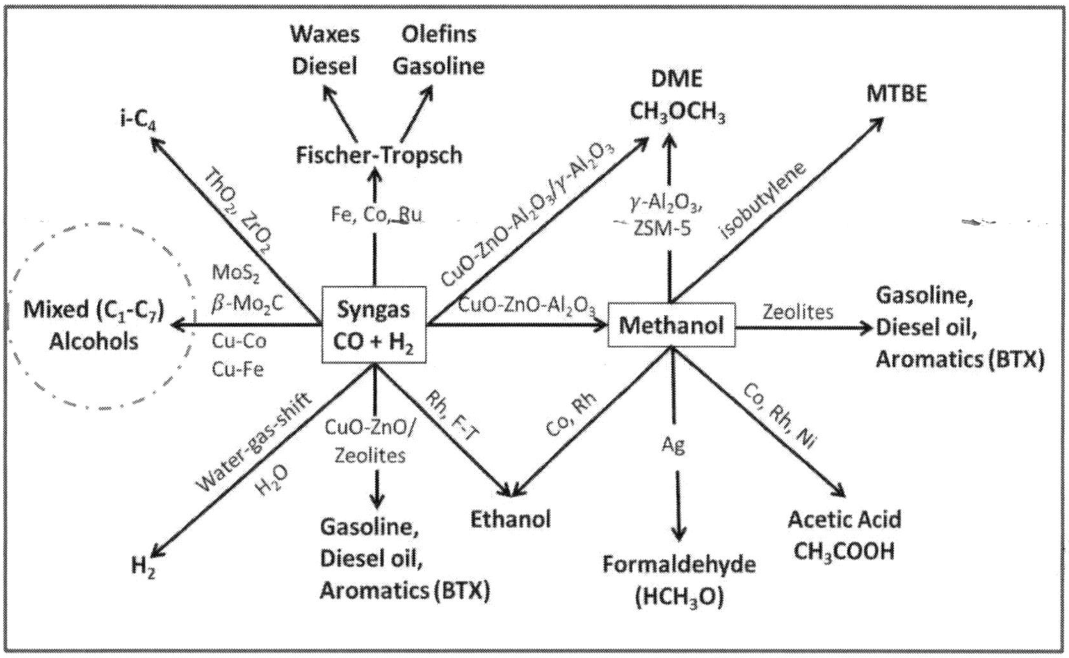

Velu Subramani et al, *Energy & Fuel*, **2008**, *22*, 814.

Why Higher alcohols?

Short-chain higher alcohols (mainly C_2—C_5) can be used as oxygenate fuel and fuel additives.

- The difference between gasoline, diesel hydrocarbon and alcohols is that alcohols contain oxygen.
- Adding alcohols to petroleum oil allows the fuel to combust more completely because of the presence of oxygen, which increases the combustion efficiency and reduces air pollution.

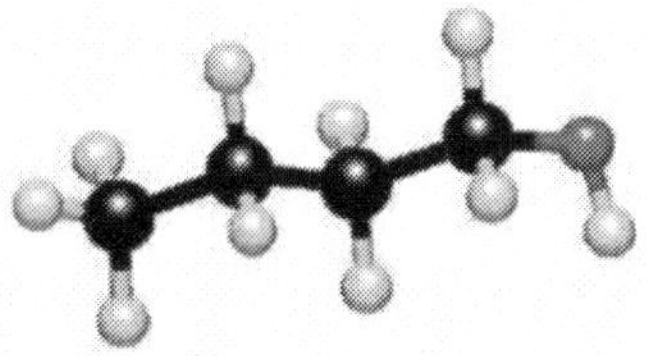

Motivation: Why Higher alcohols?

Short-chain higher alcohols (mainly C_2—C_5) can be used as oxygenate fuel and fuel additives.

- Reduction of greenhouse gas emissions.
- Reduction of toxic exhaust emissions.
- Enhancement of overall energy efficiency.

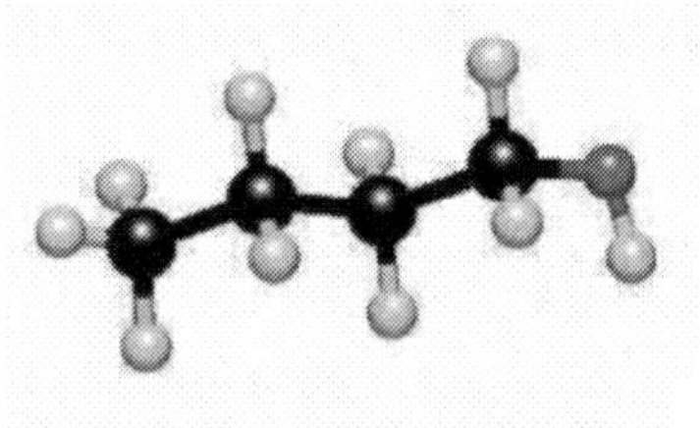

Typical catalyst systems for HAS from syngas

1. **Modified methanol synthesis catalysts.**
   (methanol is the main product, C_{2+} alcohols selectivity is limited by the kinetics of chain growth. The C_1 to C_2 step is the rate limiting step for higher alcohols synthesis)

2. **Noble metal catalysts, Rh etc.**
   (limited in availability, high cost, lower activity, unattractive for commercial application)

3. **Modified Molybdenum based catalysts: Carbides and Sulfides**
   (high temperature, high pressure)

4. **Modified FT catalysts: Cu-Co, Cu-Fe** (higher activity, C_{2+} selectivity, WGS)
   (One of the most promising candidates, it has moderate activity and selectivity)

J. J. Spivey, A. Egbebi, **Chem. Soc. Rev. 2007**, 36, 1514.

Challenges of the Catalyst Design

- **Catalysts with high selectivity towards higher alcohols.**
- **Sufficiently active catalysts.**
- **Cheap catalysts, i. e. based on abundant elements.**
- **Chemical stability in synthesis gas.**

The main problem lies in the catalyst

1. Be efficient in breaking the C–O bond to produce long chain intermediates C_nH_z* (alkyl species), but retain a large coverage of intact CO molecules to form C_nH_zCO* (acyl species).
2. Be efficient in hydrogenating C_nH_zCO* intermediates into alcohols, but not too good, as high hydrogenation activity leads to termination as C_xH_y.

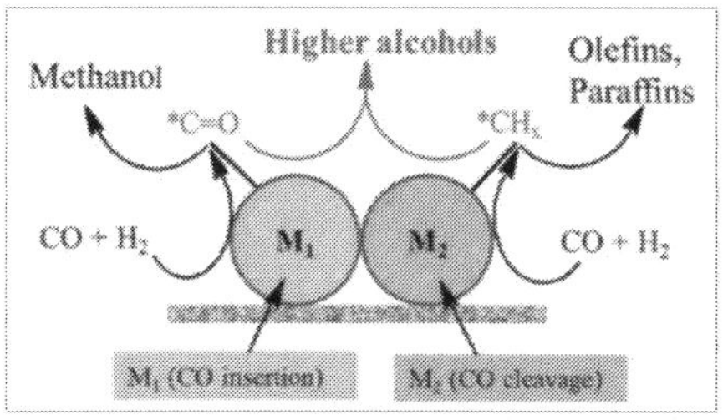

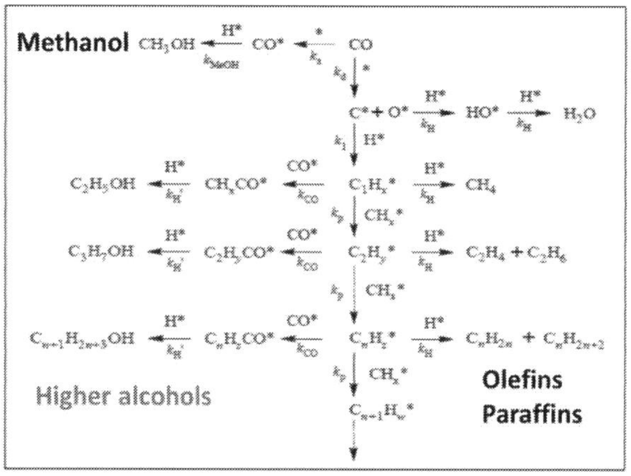

X. Xu, E. B. M. Doesburg, J. J. F. Scholten, *Catal. Today*. **1987**, *2*, 125.

Benefits of 3DOM material in catalysis

- High throughput permits short contact time.
- Benefit the diffusion of high molecular weight product.
- No significant pressure build-up.
- Possibility of combining multiple catalytic components in one material.

A. Stein, F. Li, N. R. Denny, *Chem. Mater.*, **2008**, *20*, 649

Outline

- 1. Introduction (Biomass-to-Liquids;3DOM)
- **2. Objective**
- 3. Experimental Materials
- 4. Results and Discussion
- 5. Conclusions
- 6. Future work
- 7. Acknowledgment

Objective

- Synthesize the 3DOM Cu-Fe based catalyst for higher alcohol synthesis from syngas with high selectivity and high yield compared with traditional non-porous catalysts.

Objective

- Synthesize the 3DOM Cu-Fe based catalyst for higher alcohol synthesis from syngas with high selectivity and high yield compared with traditional non-porous catalysts.

- In-situ AP-XPS to identify the active site of the 3DOM Cu-Fe catalyst for higher alcohol synthesis from syngas.

Outline

- 1. Introduction (Biomass-to-Liquids;3DOM)
- 2. Objective
- **3. Experimental Materials**
- 4. Results and Discussion
- 5. Conclusions
- 6. Future work
- 7. Acknowledgment

Experimental materials

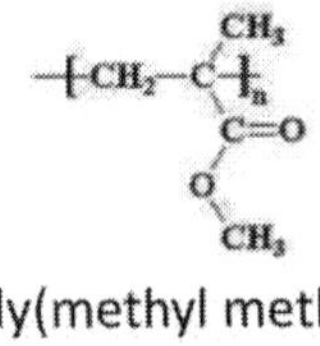

Poly(methyl methacrylate)
Emulsion

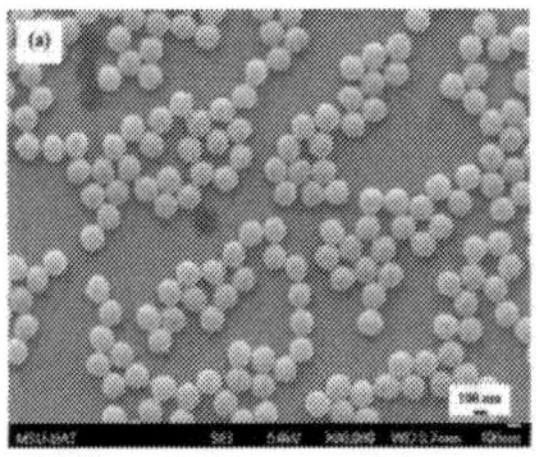

SEM image of PMMA

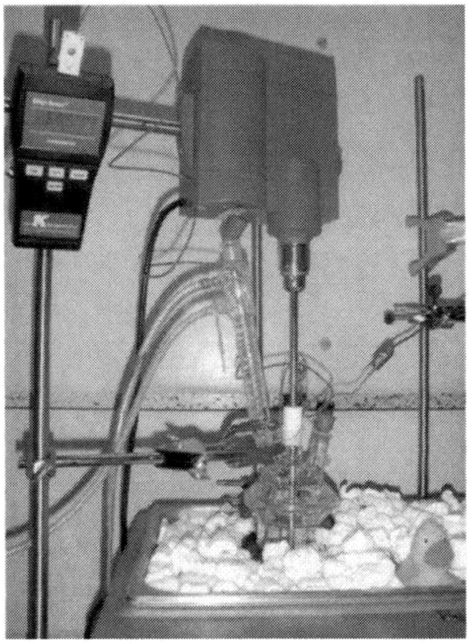

Reactor for PMMA synthesis

Experimental materials

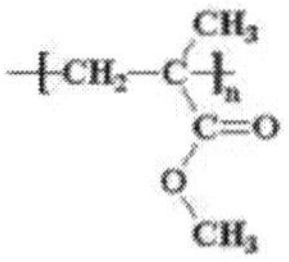

Poly(methyl methacrylate)
Emulsion

PMMA

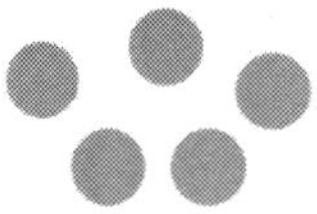

Ethylene glycol (EG)-methanol mixed solution of copper and iron nitrate.

Precursors

Experimental materials

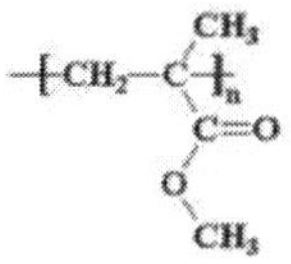

Poly(methyl methacrylate)
Emulsion

PMMA

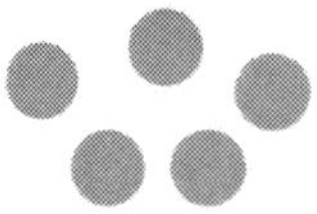

Ethylene glycol (EG)-methanol mixed solution of copper and iron nitrate.

Precursors

Centrifuged

Colloidal Crystal Template (CCT)

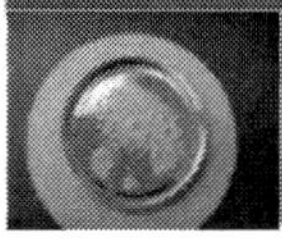

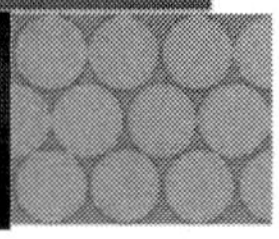

SEM image of PMMA Colloidal Crystal Template

Experimental materials

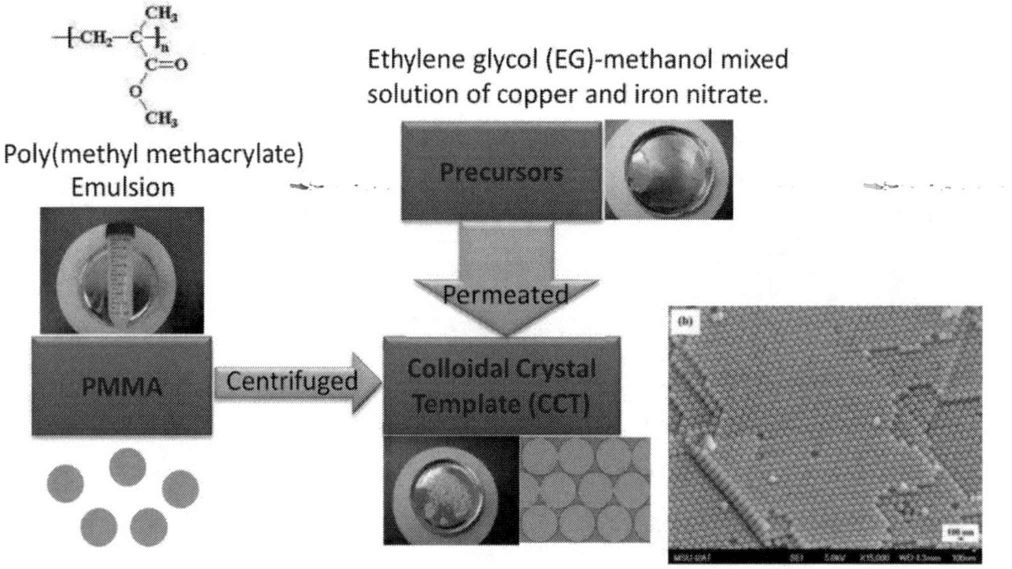

SEM image of PMMA
Colloidal Crystal Template

Experimental materials

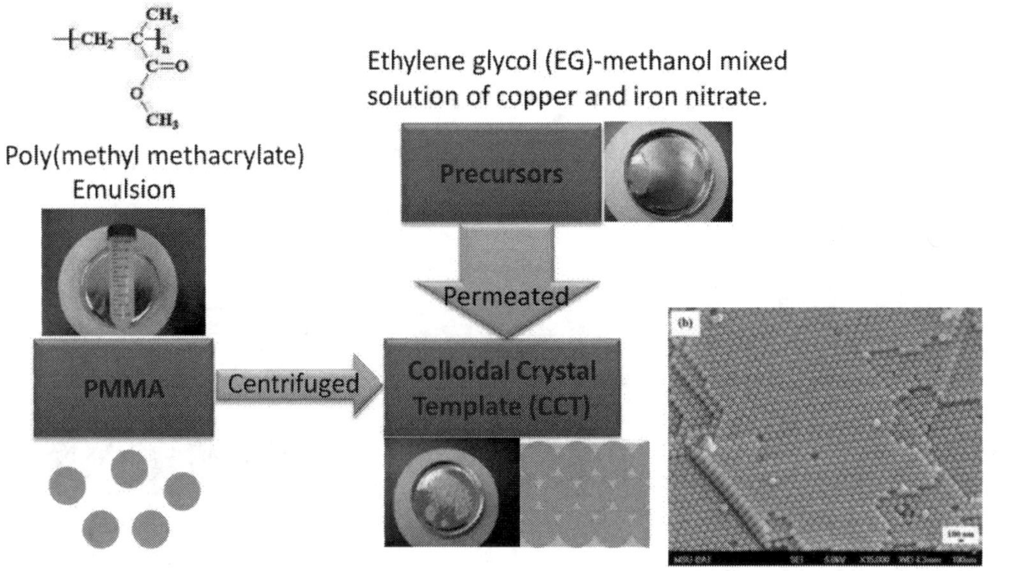

SEM image of PMMA
Colloidal Crystal Template

Experimental materials

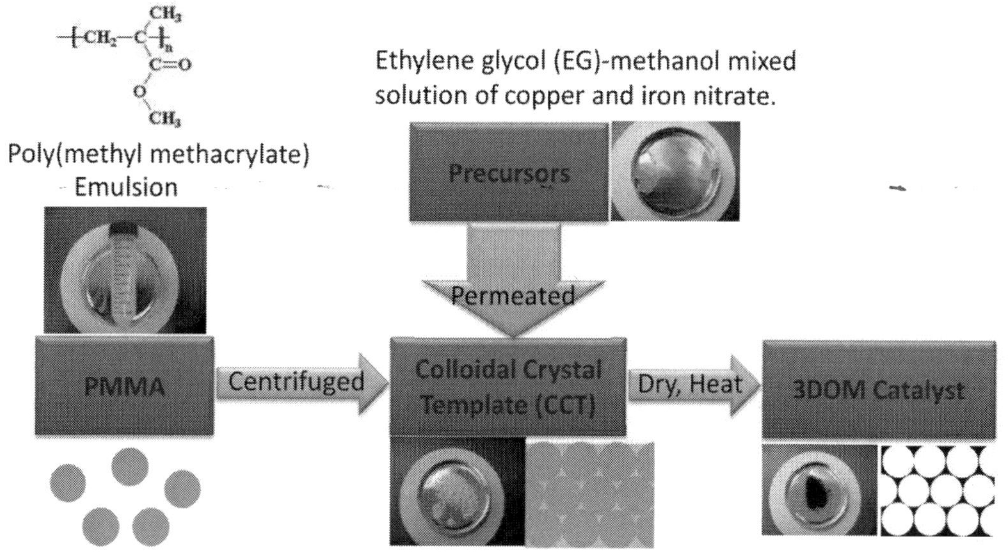

Glyoxylate route

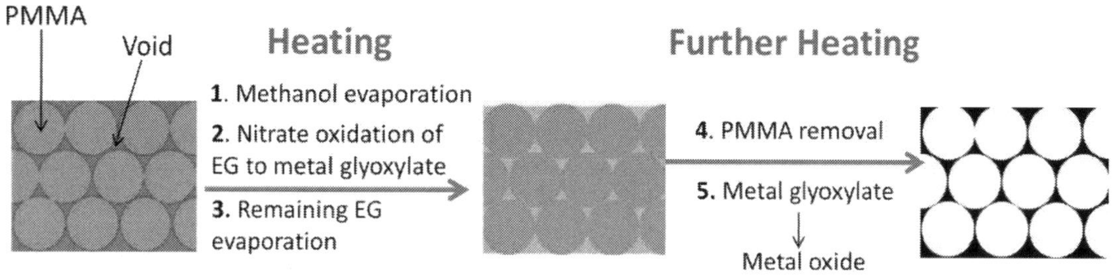

EG-methanol solution of metal nitrate in PMMA colloidal crystal template

Metal glyoxylate derivatives

3DOM catalyst mixed CuFe-oxide

EG: Ethylene glycol (CH_2OH-CH_2OH)

Glyoxylate: $[HC(OH)O–COO]^{2-}$, $[C_2H_2O_4]^{2-}$

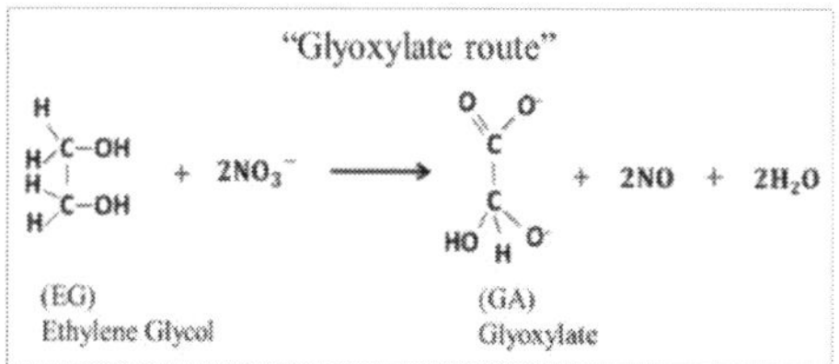

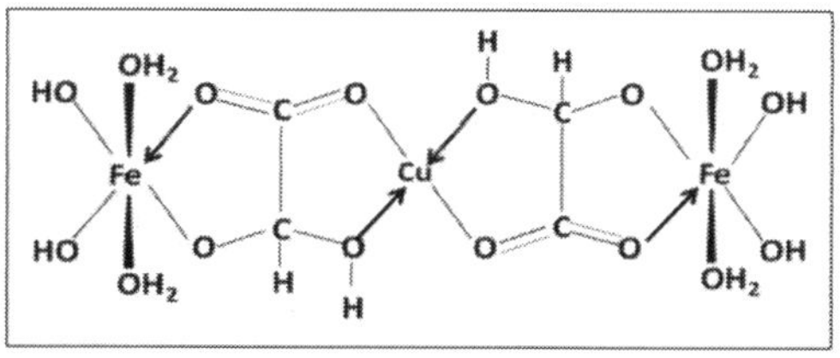

Outline

- 1. Introduction (Biomass-to-Liquids;3DOM)
- 2. Objective
- 3. Experimental Materials
- 4. Results and Discussion
- 5. Conclusions
- 6. Future work
- 7. Acknowledgment

The morphology of 3DOM Cu-Fe Catalysts

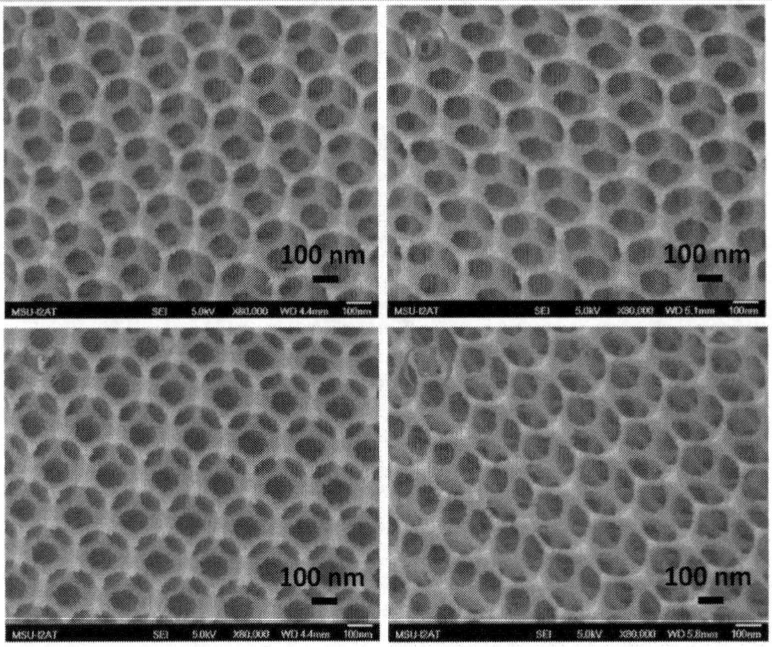

SEM images of fresh 3DOM Cu-Fe based catalyst, (a) Cu/Fe=1:1; (b) Cu/Fe=2:1; (c) Cu/Fe=3:1; (d) Cu/Fe=1:2.

Characterization of fresh 3DOM catalyst (SEM, TEM)

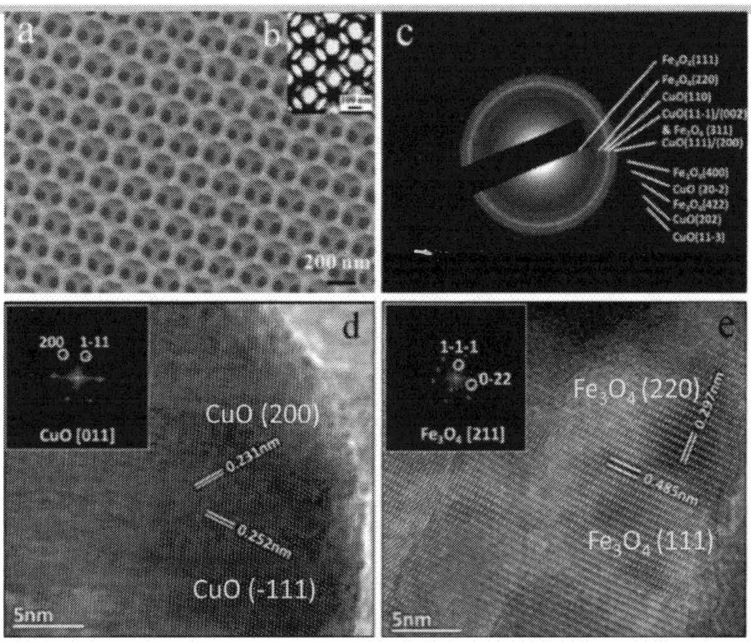

Fresh 3DOM Cu_2Fe_1 catalyst. a) SEM, and b) TEM images. c) The indexed SAED pattern of (b). d) High resolution TEM image of CuO, inset is FFT. e) HRTEM image of Fe_3O_4, inset is FFT.

Y. Lu et al. *ChemCatChem*, **2014**, *6*, 473-478.

Microstructure of fresh 3DOM Cu-Fe catalysts

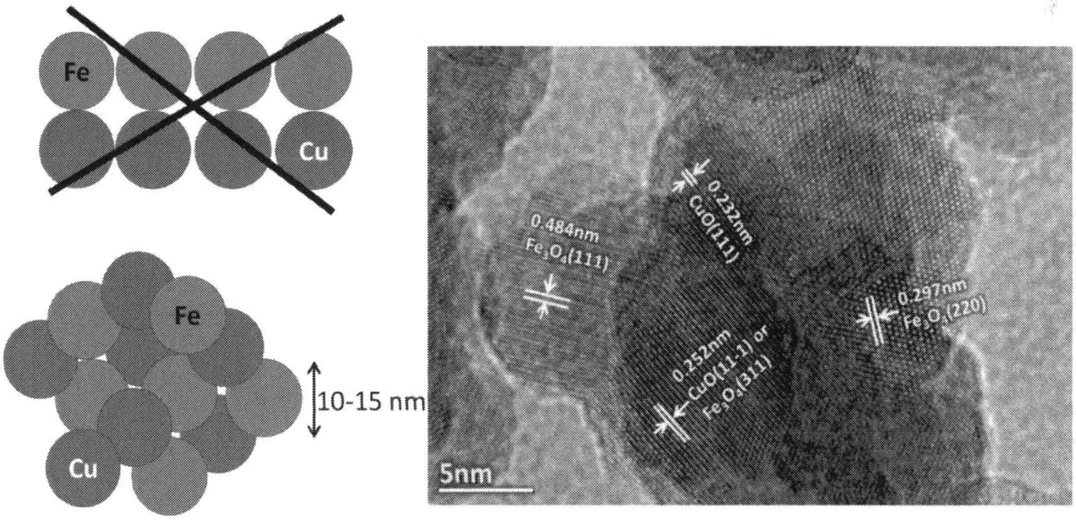

HRTEM images of fresh 3DOM Cu_2Fe_1 catalyst, indicating homogeneous distribution of nanoparticles of 3DOM Cu-Fe catalysts .

The crystallite structure of fresh 3DOM Cu-Fe catalyst

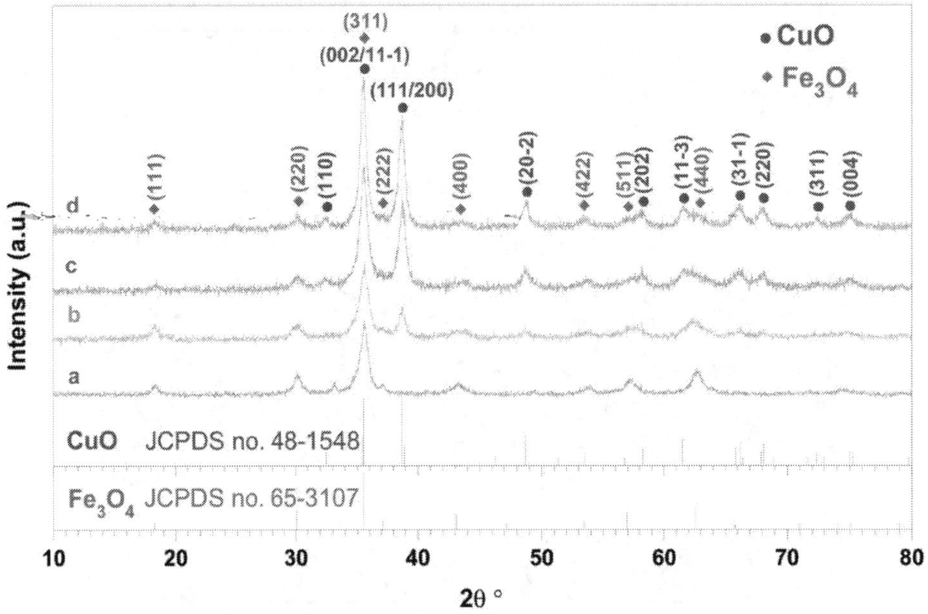

XRD pattern of fresh 3DOM Cu-Fe based catalyst (a) 3DOM Cu_1Fe_2. (b) 3DOM Cu_1Fe_1. (c) 3DOM Cu_2Fe_1. (d) 3DOM Cu_3Fe_1.

Characterization of fresh 3DOM Cu-Fe catalyst (XPS)

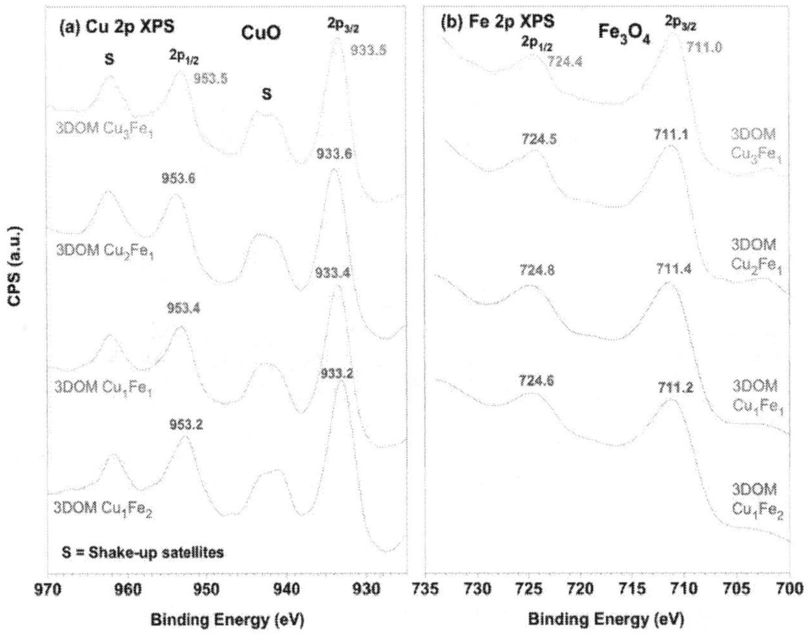

High resolution scan XPS spectra for Cu 2p, Fe 2p of 3DOM Cu-Fe catalyst.

Catalytic performance (3DOM vs. Non ordered porous)

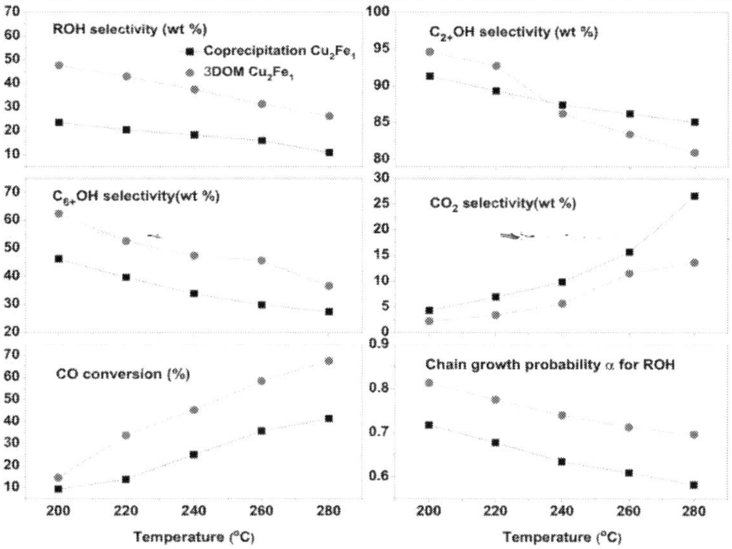

Catalytic tests were carried out at P = 700 psi, T = 260 °C, GHSV = 2000 h^{-1}, H_2/CO =1.0. Chain growth probability α are calculated via the equation of $W_n = n(1-\alpha)^2\alpha^{n-1}$, where W_n stands for the mass fraction of products containing n carbon atoms.

Y. Lu et al. *ChemCatChem*, **2014**, *6*, 473-478.

What is the Active Site of Cu-Fe Catalyst for HAS from Syngas?

What is the Active Site of Cu-Fe Catalyst for HAS from Syngas?

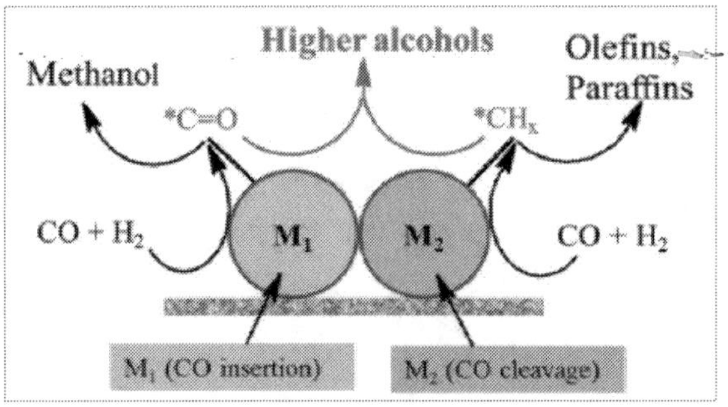

Oxygenation Site　　Carbon Chain Grown Site

In situ AP-XPS at Dr. Franklin Tao's Group

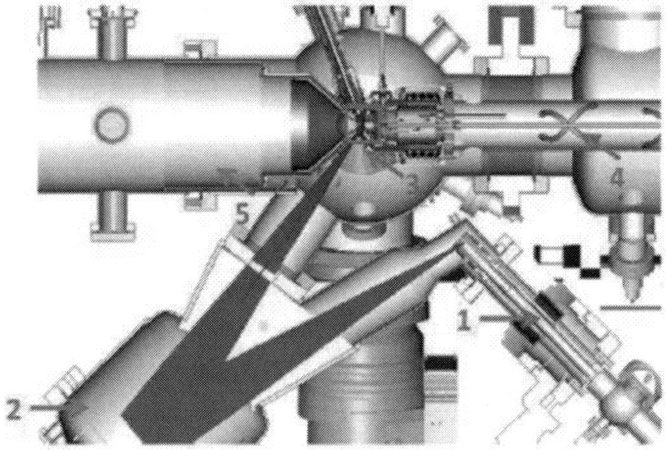

In-house X-ray source, Al Kα for the excitation of photoelectrons from catalysts in the gaseous environment of the reaction cell. 1: electron gun; 2: monochromator of Al Kα ; 3: reaction cell; 4: manipulator of the reaction cell; 5: differential pumping stages connected to the energy analyzer.

Franklin Tao, *Chem. Commun.*, **2012**, *48*, 3812-3814.

In situ AP-XPS Sample Preparation

Au thin film (0.4 mm thick, 99.99%, VWR) was used as a substrate to load a catalyst.

Au foil was deliberately roughened using a SiC knife to increase adhesion.

A certain amount of 3DOM Cu_2Fe_1 catalyst was suspended in ethanol (100%) and deposited on pre-cleaned Au foil.

The deposition was repeated until the Au foil was completed covered in sample.

Ethanol left in the sample on the Au foil is vaporized by placing Au foil in a vacuum oven at 60 °C. The dry sample is ready for XPS studies.

High-resolution spectra (Cu 2p, Cu (LMM), Fe 2p, O 1s, C 1s) were gathered using an average of 35 scans with pass energy of 23.5 eV and a step size of 0.1 eV.

All spectra are calibrated to their corresponding $Au4f_{7/2}$ which is 84.0 eV.

In situ AP-XPS Study

Unlike in vacuum studies, reactant gases are introduced to flow through the catalyst at a certain temperature in the catalysis reactor of AP-XPS while data acquisition is going on.

Flow rate in the reactor was measured through a mass flow meter installed between each gas source and the entrance of the flow reaction cell.

The total pressure of the mixture gas of reactor is measured with a capacitance gauge installed at the entrance. The pressure at the exit is measured with another capacitance gauge. An average of the pressures at entrance and exit is defined to be the pressure above the catalyst in the catalytic reactor.

The catalyst is heated through heating the vacuum side of a sample stage using an e-beam heater installed in the vacuum section between the external wall of the catalytic reactor and the internal wall of the UHV chamber. The gaseous side is the internal wall of the reactor.

In situ AP-XPS Studies of 3DOM Cu_2Fe_1 catalyst

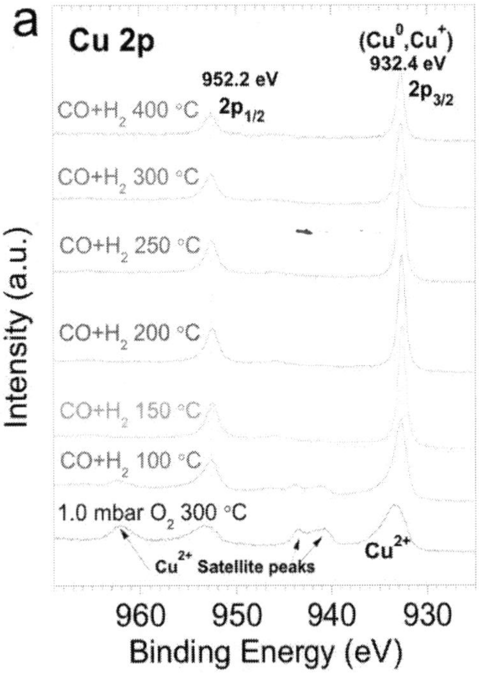

Using 0.5 mbar H_2 + 0.5 mbar CO at various temperatures

In situ AP-XPS Studies of 3DOM Cu_2Fe_1 catalyst

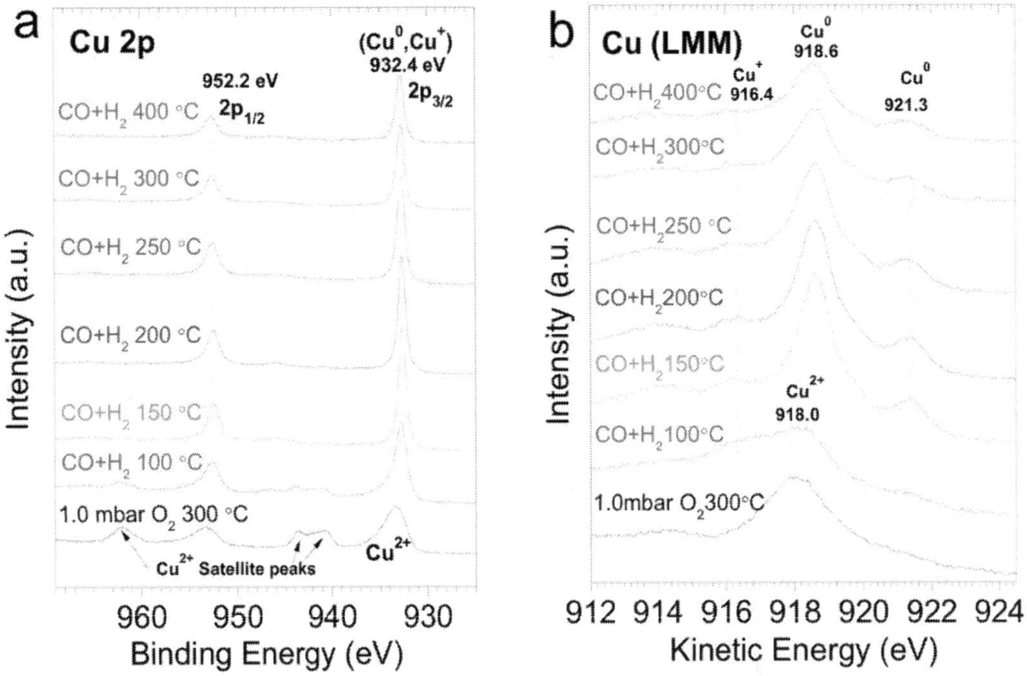

Using 0.5 mbar H_2 + 0.5 mbar CO at various temperatures

In situ AP-XPS Studies of 3DOM Cu_2Fe_1 catalyst

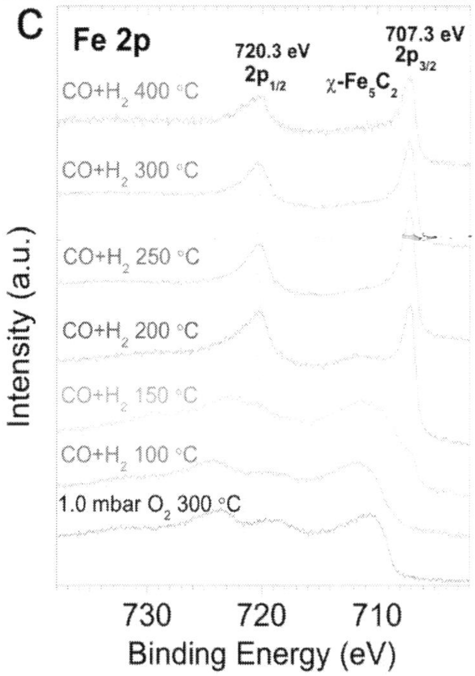

Using 0.5 mbar H_2 + 0.5 mbar CO at various temperatures.

In situ AP-XPS Studies of 3DOM Cu_2Fe_1 catalyst

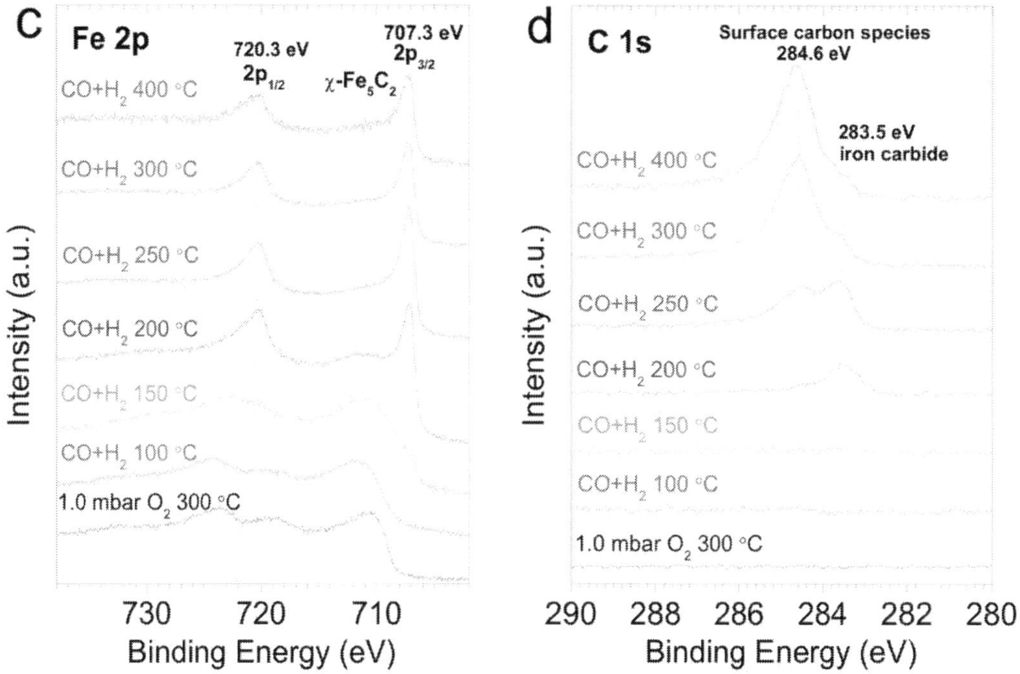

Using 0.5 mbar H_2 + 0.5 mbar CO at various temperatures.

In situ AP-XPS Studies of 3DOM Cu_2Fe_1 catalyst

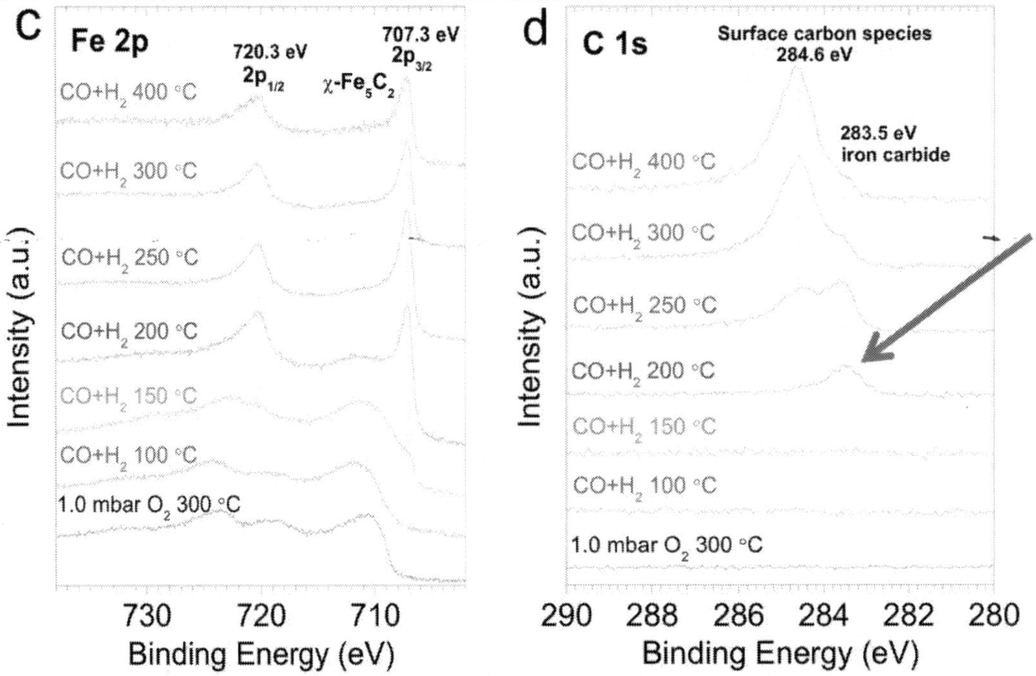

Using 0.5 mbar H_2 + 0.5 mbar CO at various temperatures.

Characterization of reduced catalysts (ex situ XRD)

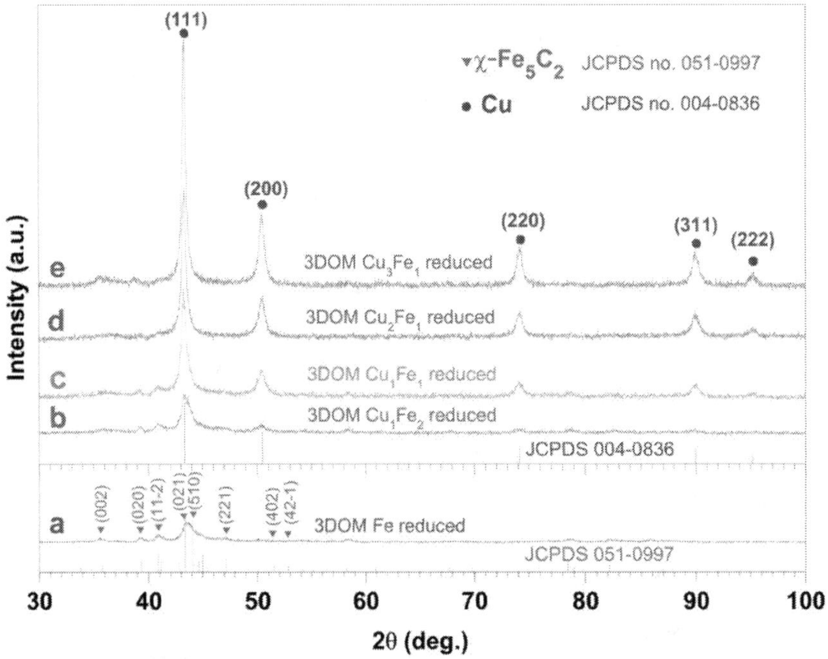

XRD of 3DOM catalyst after reduction via syngas (H_2/CO = 1.0).

The crystallite structure of reduced 3DOM Cu-Fe catalysts (ex situ XRD)

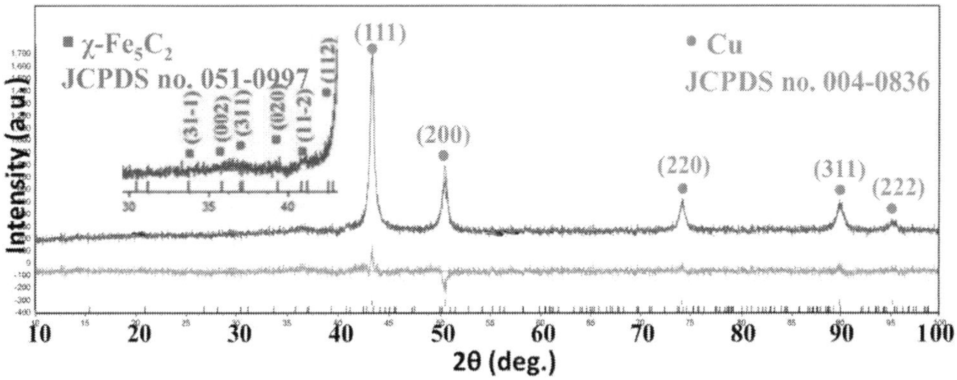

3DOM Cu_2Fe_1 catalyst after reduction via syngas (H_2/CO = 1.0). Rietveld refinements of XRD patterns using TOPAS 5.0 software.

The characteristic under-estimation of the intensity near the maximum of the 200 peak at ca. 50.4° 2ϑ in the XRD pattern can be explained with the presence of twin boundaries and stacking faults resulting in a broadening of this reflection, which usually causes a shift of the 111 and 200 peaks towards each other.

Y. Lu et al. *ChemCatChem*, **2014**, *6*, 473-478.

Reduced 3DOM Cu_2Fe_1 catalysts (Mössbauer spectra)

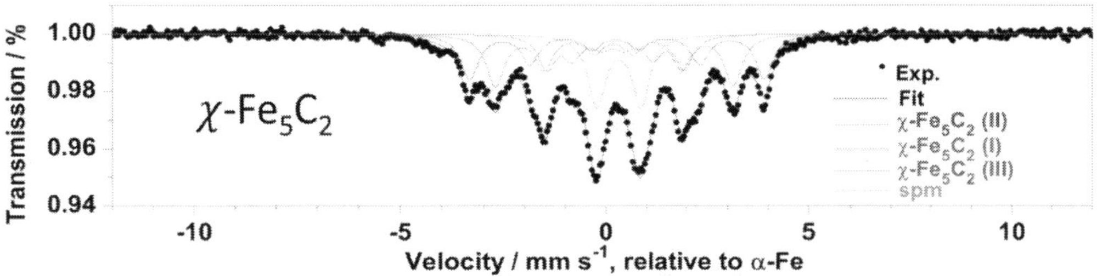

The fitted Mössbauer spectrum of the reduced 3DOM Cu_2Fe_1 catalysts with three sextets confirms the existence of χ-Fe_5C_2.

Generally speaking, the iron carbide active site is critical for the carbon chain growth in FT reaction.

It has been found that the iron carbide particle size significantly affect the carbon chain growth in FT synthesis.

The long-chain product selectivity depends on the nature of iron carbide site.

Characterization of reduced catalysts (Mössbauer)

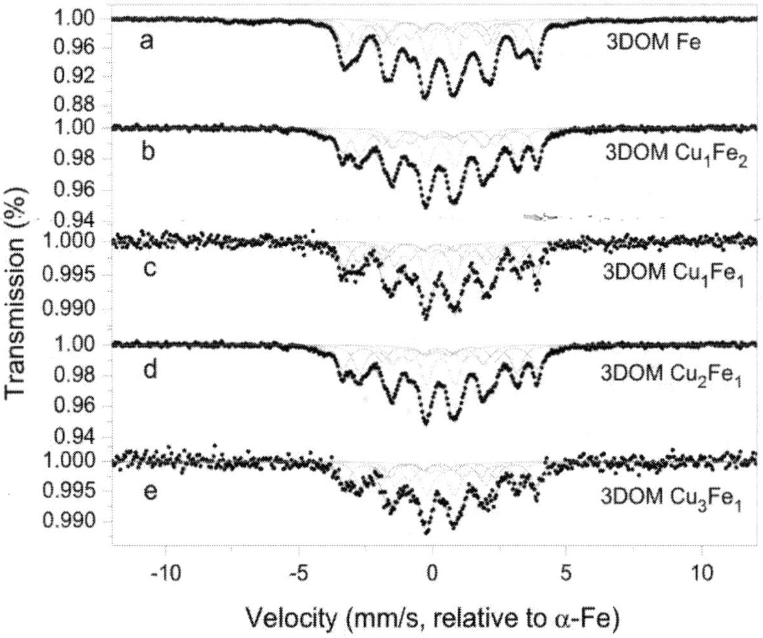

χ-Fe_5C_2

Three fitted sextet

Mössbauer spectra of the reduced 3DOM Cu-Fe catalysts measured at 27 °C.

Insight into the Cu^0 and χ-Fe_5C_2 dual site

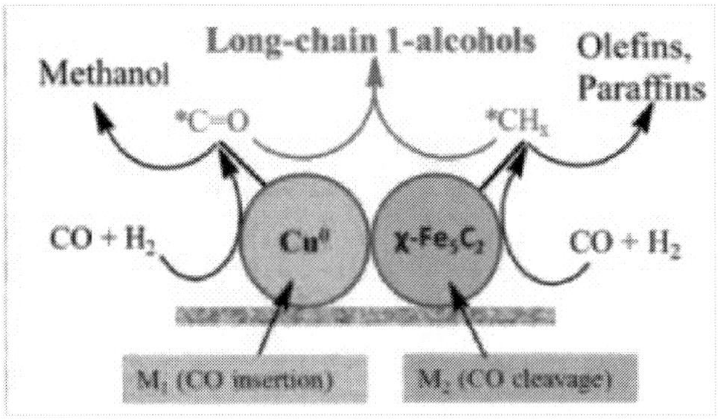

Oxygenation Site

Carbon Chain Grown Site

Insight into the Cu^0 and χ-Fe_5C_2 dual site (HRTEM)

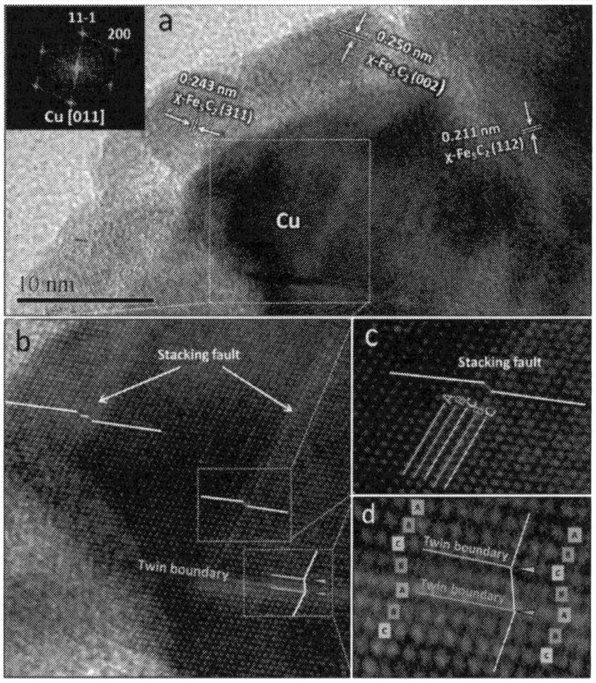

3DOM Cu_2Fe_1 catalyst after reduction via syngas ($H_2/CO = 1.0$). a) HRTEM, inset is FFT. b) is a close-up in (a). c) and d) are close-ups of in (b) showing stacking fault and twin boundary.

The highly active Cu is a defective form of nanoparticle with planar defects.

HRTEM observations provide evidence that this type of planar defects are stacking faults and twin boundaries.

Y. Lu et al. *ChemCatChem*, **2014**, *6*, 473-478.

Insight into the Cu^0 and χ-Fe_5C_2 dual site

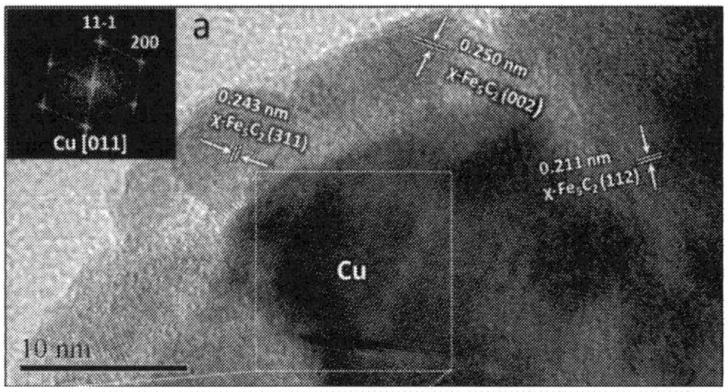

The **Cu nanoparticle** containing stacking fault and twin boundaries is intimately surrounded by hägg carbide **χ-Fe_5C_2 nanoparticles**.

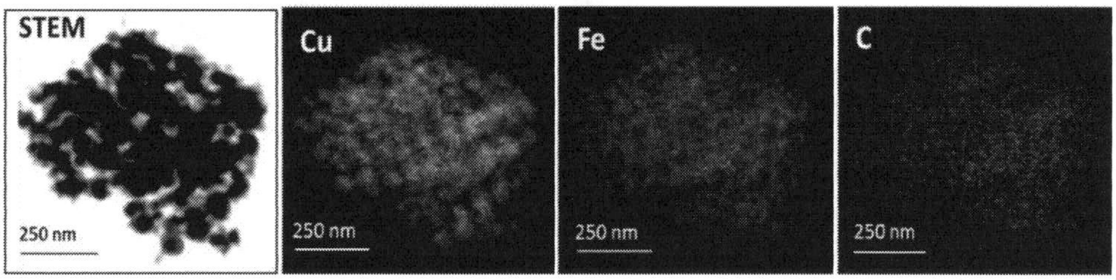

STEM-EDS mapping further evidence that the reduced 3DOM Cu_2Fe_1 catalysts possessing a high density of uniform and stable distributions of active Cu and Fe_xC_y nanoparticles.

Atomic Cu steps

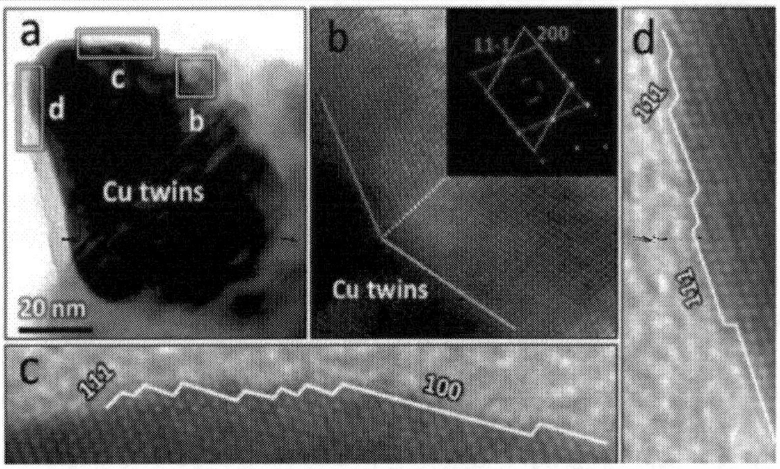

3DOM Cu_2Fe_1 catalyst after reduction via syngas (H_2/CO = 1.0). a) HRTEM image shows Cu nanoparticle involving an abundance of twin boundaries. b) typical Cu twins. c) and d) formed atomic Cu steps, these steps will serve as high-activity sites for CO association.

The presence of steps at the Cu surface, stabilized by stacking fault or twin boundaries terminating at the surface, is one of the reasons for generating high activity for alcohol synthesis.

Y. Lu et al. *ChemCatChem*, **2014**, *6*, 473-478.

Catalytic performance (Total 1-alcohols yield)

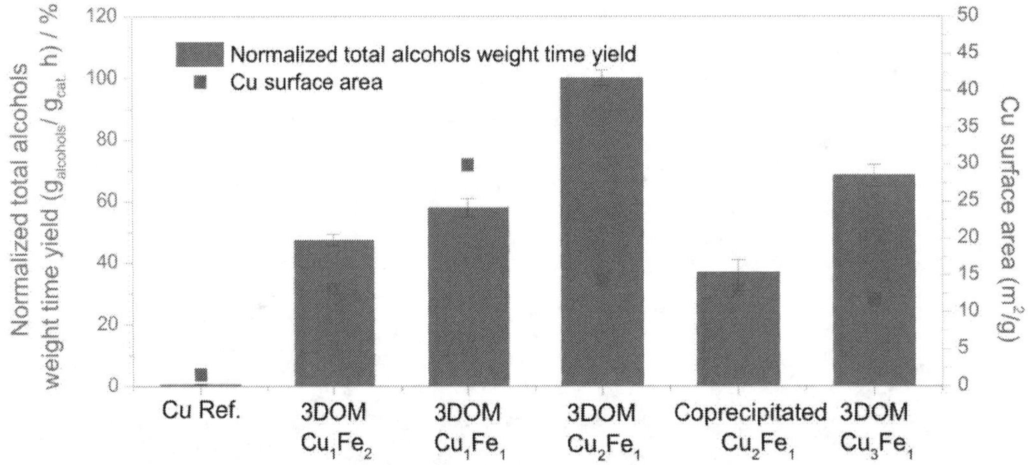

Catalytic activities and N_2O chemisorption Cu surface area (P =700 psi, T = 260 °C, GHSV = 2000 h^{-1}, H_2/CO =1.0, time-on-stream 120 h, normalized to the most active sample).

Y. Lu et al. *ChemCatChem*, **2014**, *6*, 473-478.

Intrinsic Activities

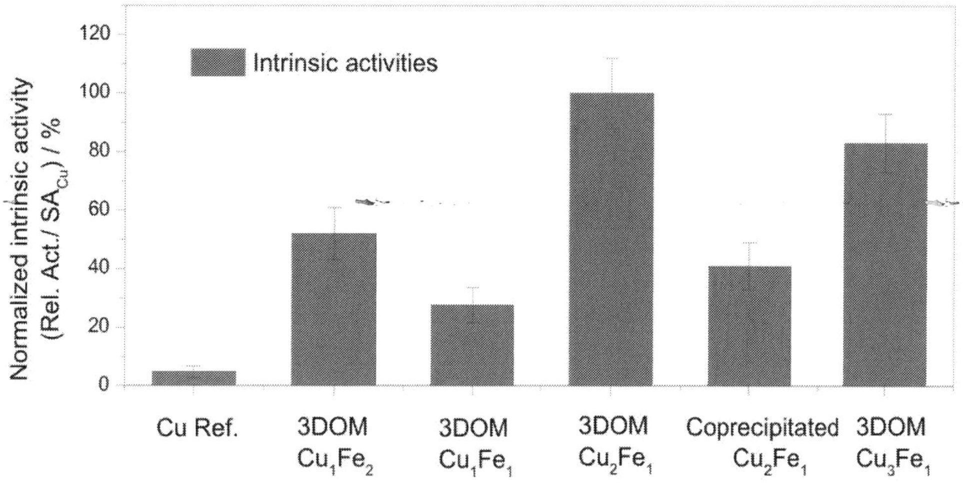

Intrinsic activities per N_2O Chemisorption Cu surface area, normalized to the most active sample.

Y. Lu et al. *ChemCatChem*, **2014**, *6*, 473-478.

Defect analysis of X-ray diffraction data

M. S. Paterson, *J. Appl. Phys.* **1952**, *23*, 805: (h+k+l = 3N±1) broadened and shifted; (h+k+l = 3N) not affected

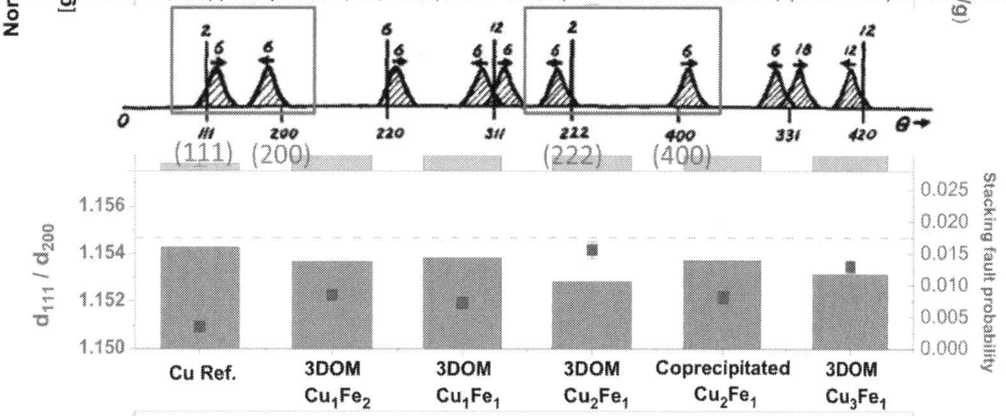

Deviation of d_{111}/d_{200} observed in XRD and the resulting Cu stacking fault probability.

The ratio of d_{111}/d_{200} (a ratio of the distances between the (111) and (200) lattice planes), is constant at $2/\sqrt{3}$ =1.1547 for an ideal defect-free Cu fcc structure.

The stacking fault probability δ is calculated according to $\delta = 8.5 \times [(2/\sqrt{3}) - (d_{111}/d_{200})]$.

M. Behrens et al. *Science* **2012**, *336*, 893.

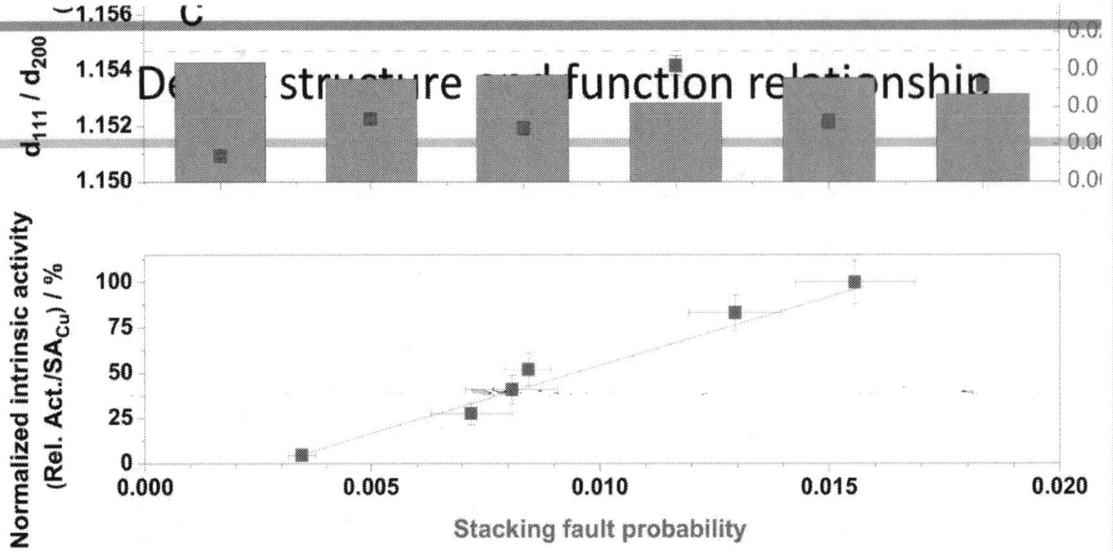

Cu intrinsic activity has a linear relationship with the stacking fault probability.

The stacking fault probability δ is calculated according to $\delta = 8.3 \times [(2/\sqrt{3}) - (d_{111}/d_{200})]$.

Y. Lu et al. *ChemCatChem*, **2014**, *6*, 473-478.

Outline

- 1. Introduction (Biomass-to-Liquids;3DOM)
- 2. Objective
- 3. Experimental Materials
- 4. Results and Discussion
- 5. Conclusions
- 6. Future work
- 7. Acknowledgment

Conclusions

(1) In situ AP-XPS evidences that Cu^0 and hägg carbide χ-Fe_5C_2 are the active dual sites for higher alcohols synthesis from syngas.

(2) Atomic steps on Cu surface, induced by lattice plane transitions or planar defects, serve as high-activity oxygenation site for CO association; and active χ-Fe_5C_2 intimately surrounds the Cu surface for CO dissociation, resulting in a synergetic effect between the actively stable Cu–Fe_xC_y dual sites for long-chain 1-alcohols synthesis from syngas.

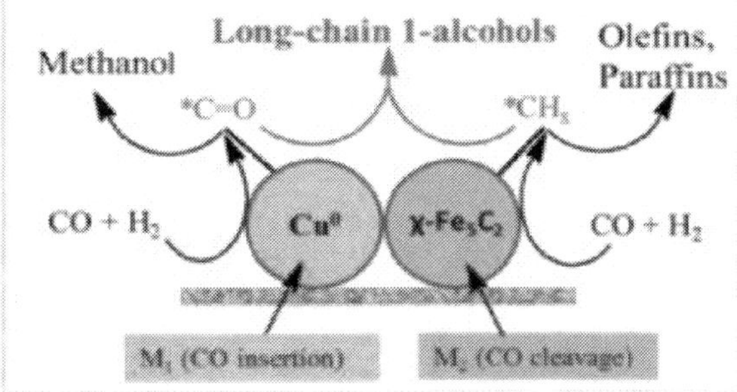

Future Work

(1) In-situ characterization: XRD, X-ray Absorption Spectroscopy to investigate the active site of the catalyst under reaction condition or during catalysis.

(2) DFT calculation to investigate the reaction mechanism for higher alcohols synthesis from syngas over Cu-Fe catalysts.

Acknowledgments

Collaboration with **Dr. Franklin Tao** (In situ AP-XPS) at **University of Kansas** .

We thank Ms. Amanda Lawrence (TEM), Mr. Erick Vasquez and Mr. Ersan Eyiler (XPS), Dr. Baobao Cao (HRTEM), Dr. Xiaohao Liu (Mössbauer) for technical support.

Financial support was given by the **US Department of Energy** under Award Number (DE-FG3606GO86025) and **US Department of Agriculture** under Award Number (AB567370MSU).

An *In Situ* Molecular Spectroscopic Study of Ethylene Polymerization by Supported CrO_x/SiO_2 Catalysts

Anisha Chakrabarti and Israel E. Wachs

Operando Molecular Spectroscopy & Catalysis Laboratory,

Department of Chemical Engineering, Lehigh University, Bethlehem, PA 18015 USA

The discovery that ethylene could be converted to high-density polyethylene (HDPE) by supported CrO_x/SiO_2 catalysts was made by J.P. Hogan and R.L. Banks of Phillips Petroleum Company in the 1950s.[1] This catalyst is now responsible for over half of the production of HDPE sold globally.[2] Although supported CrO_x/SiO_2 ethylene polymerization catalysts have been extensively studied over the years, the same fundamental structural and mechanistic questions are still being debated today with researchers being no closer to a common view. For example, the initial molecular structure of the surface $Cr^{+6}O_x$ site has been proposed to be present as either isolated surface dioxo CrO_4, isolated surface mono-oxo CrO_5, and surface dimers of Cr_2O_7, and reduction of the surface chromia oxidation state to either Cr^{+2} or Cr^{+3} during ethylene polymerization.

To address the molecular details of ethylene polymerization by supported CrO_x/SiO_2 catalysts, modern *in situ* molecular spectroscopic techniques were applied before and during reaction conditions. The initial supported CrO_x/SiO_2 catalyst was found to only contain isolated chromia sites (very high UV-vis edge energy value), a significant fraction of CrO_4 coordinated sites (pronounced XANES pre-edge feature) and two distinct surface chromia sites (dioxo $(O=)_2CrO_2$ and mono-oxo $O=CrO_4$)[3]. *In situ* Raman spectroscopy during ethylene polymerization revealed that the surface dioxo CrO_4 site reduces and becomes activated more easily than the surface mono-oxo CrO_5 site. The corresponding *in situ* UV-vis experiments verified the reduction of Cr^{+6} to Cr^{+3} for both the dioxo and mono-oxo surface chromia sites during ethylene polymerization. The condensed polyethylene product formed on the catalysts during the polymerization reaction was monitored with *in situ* IR spectroscopy and key surface reaction intermediates were detected. This presentation will address the catalytic roles of the two distinct surface CrO_x species on SiO_2 in the ethylene polymerization reaction.

(1) Hogan, J. Paul. *App. Ind. Cat.* **1983,** 1, 149.
(2) McDaniel, M.P. *Adv. Catal.* **2010,** 53.
(3) Lee, Edward L. and Wachs, Israel E. *J. Phys. Chem. C.* **2008,** 112, 6487.

384703 Illuminating Surface Atoms in Nanoparticles By Differential X-Ray Absorption Spectroscopy

Abstract:

The adsorption of Ar on Pd nanoparticles (1 nm dia, silica-supported) at 77 K results in restructuring of surface Pd atoms, as evidenced by differential extended x-ray absorption spectroscopy (EXAFS). EXAFS analysis shows that 9 ± 2 nearest-neighbor (NN) Pd-Pd bonds expand by 0.104 ± 0.005 Å as a result of Ar adsorption. This is compared to the total number of NN Pd-Pd bonds in the Pd nanoparticles of ~140. We discuss how the differential EXAFS technique is not limited to a particular set of materials, but instead it can be applied to any system in which modulation causes small changes in local structure. To support our experimental observation of Ar-induced surface restructuring and to differentiate between proposed EXAFS models, we conducted a series of ReaxFF-MD simulations in which model Pd clusters were exposed to both vacuum and an Ar gas phase at 77 K. The simulations demonstrate that Ar may drive surface reconstruction of under-coordinated atoms, thus leading to an increased Pd-Pd bond length of surface Pd atoms, in agreement with the experimental observations. Based on observations of the atomistic simulations, it is likely that under-coordinated atoms are trapped in metastable states at 77 K and Ar provides the kinetic energy needed to overcome the barrier for surface restructuring. We note the possibility of using Ar to probe the number of under-coordinated sites of catalyst nanoparticles which could be compared to selectivity or activity.

380074 Effect of Deviation from Ideality of Different Pore Geometries for Zero Order Electrochemical Reactions

Abstract:

The study of realistic catalytic pore domains in electrochemical systems is vital in determining the region of parameters that are acceptable for describing, accurately, chemical processes taking place within modelling domains. Models are useful in predictions of system behaviors and assessing experimental results relevant to electrochemical applications such as batteries, fuel cells. A "deviation" from ideal geometry of porous domains for both rectangular and cylindrical domains are analyzed and compared with their ideal geometries. These systems are studied by employing the microscopic scale species mass balancing conservation equation with diffusion and zero order electrochemical reaction present. The equation is scaled up to macroscopic level by employing area averaging-based approach to obtain a system level information. The results obtained from the macroscopic level identified potential deviations in the concentration profile. This study shows the impact of irregular geometry (which is more realistic in analyzing porous systems) on using the often assumed ideal geometry and quantifies useful guidelines for modelling these complex systems.

385166 Design of a Novel Polymeric Acid Catalyst for Biomass Hydrolysis

Abstract:

A novel polymeric solid acid catalyst consisting of dual functional chains for biomass hydrolysis was successfully synthesized. A poly (styrene sulfonic acid) (PSSA) polymer chain is immobilized on a substrate and used to catalyze biomass hydrolysis. A neighboring poly (vinyl imidazolium chloride) ionic liquid (PIL) polymer chain is grafted to help solubilize lignocellulosic biomass and enhance the catalytic activity. The immobilized polymeric catalyst demonstrates over 97% total reducing sugar (TRS) yield in ionic liquid. This solid acid catalyst is highly efficient, stable, reusable and environmentally friendly. In order to understand its catalytic activity and further optimize its performance, the acidity of PSSA chains as well as the interactions among the PSSA, PIL polymer chains and cellulose were investigated using combined quantum mechanical (QM) and classical molecular dynamics (MD) simulations. In addition, the interaction free energies were determined using metadynamics (MTD) simulations. Our results demonstrate that polymer chain length and chain density play a crucial role in the catalytic activities of this polymer catalyst. Additional improvement for catalytic activity by substitution or copolymerization was suggested. This solid acid catalyst can be tuned and potentially be used for many other catalytic conversions.

Modeling the Production of Aromatics from Biomass in Solid Acid Zeolites

Ryan Patet, Stavros Caratzoulas, Dionisios G. Vlachos

Center for Catalysis and Energy Innovation (CCEI), Department of Chemical and Biomolecular Engineering, University of Delaware, 221 Academy Street, Newark, DE 19716

The production of value-added fuels and chemicals from lignocellulosic biomass is an active area of research motivated by a supply of unused biomass material, desire for independence from fossil fuels, and reduction in environmental impact from the production and utilization of these products.[1] Aromatic chemicals such as benzene, toluene, and xylene, which are currently manufactured by the catalytic reforming of naphtha, are produced in high volumes for use, for example, in the polymer industry.[2] Production of these high volume chemicals from biomass would help to meet the aforementioned.

In one specific example, production of p-xylene from biomass-derived 2,5-dimethylfuran (DMF) and ethylene has been achieved with high selectivity to the desired aromatic product using solid acid zeolite catalysts through a two-step reaction process.[3,4] The DMF and ethylene initially undergo a Diels-Alder cycloaddition, followed by a dehydration of the resulting oxanorbornene intermediate to form the final aromatic product.[4] The ability of solid acid zeolites to catalyze this reaction sequence has been shown to be due to the ability of Brønsted acids to protonate the intermediate and lower the reaction barrier of the dehydration step.[4]

The production of p-xylene from DMF and ethylene has been successful, but applying this approach to benzene and toluene from furan and 2-methylfuran, respectively, has resulted in lower yields due to the formation of undesired polymer byproducts.[5] By tailoring the acidic environment of the solid-acid zeolites, selectivity towards the desired aromatic products over the undesired polymer byproducts could be achieved. Being able to model this acidic environment in order to understand the chemistry and make predictions about new catalytic environments becomes difficult computationally, due to the number of atoms needed to define the pore of the system. The size constraints of the system can be managed using the ONIOM method, in which a small model portion of the system is treated using expensive quantum mechanical density functional theory (DFT) calculations, while the remainder of the system is treated with a less-expensive molecular mechanics model.[6,7]

In this study, ONIOM models have been developed which can accurately describe the acidic environment found within the pores of different zeolite structures. Comparison to a large dataset of experimental data shows that our model has predictive ability. With these extensively assessed models, the effect of different zeolite frameworks (e.g. H-Beta vs. H-Y), as well as the nature of the acid sites have been investigated for their effect in the chemistry of the production of aromatics. We will discuss how the reactants, intermediates, and products are stabilized within the pore environments of zeolites and use this understanding toward the design of catalysts capable of improving the selectivity to the desired aromatic products.

1. Climent MJ, Corma A, Iborra S. Converting Carbohydrates to Bulk Chemicals and Fine Chemicals Over Heterogeneous Catalysts. *Green Chem.* 2011;13:520-540.
2. Zukauskas V, Perego C. Aromatics. In: Amadei C, ed. *Encyclopaedia of Hydrocarbons.* Vol II. Rome: Istituto della Enciclopedia Italiana; 2006:591-614.
3. Williams CL, Chang c-C, Do P, et al. Cycloaddition of Biomass-Derived Furans for Catalytic Production of Renewable p-Xylene. *ACS Catal.* 2012;2:935-939.
4. Nikbin N, Do PT, Caratzoulas S, Lobo RF, Dauenhauer PJ, Vlachos DG. A DFT Study of the Acid-Catalyzed Conversion of 2,5-Dimethylfuran and Ethylene to p-Xylene. *J. Catal.* 2013;297:35-43.
5. Wang D, Osmundsen CM, Taarning E, Dumesic JA. Selective Production of Aromatics from Alkylfurans over Solid Acid Catalysts. *ChemCatChem.* 2013;5:1-8.
6. Svensson M, Humbel S, Froese RDJ, Matsubara T, Sieber S, Morokuma K. ONIOM: A Multilayered Integrated MO+MM Method for Geometry Optimizations and Single Point Energy Predictions: A Test for Diels-Alder Reactions and Pt(P(t-Bu)(3))(2)+H-2 Oxidative Additiion. *J. Phys. Chem.* 1996;100(50):19357-19363.
7. Dapprich S, Komaromi I, Byun KS, Morokuma K, Frisch MJ. A New ONIOM Implementation in Gaussian98 - Part 1: The Calculation of Energies, Gradients, Vibrational Frequencies and Electric Field Derivatives. *J. Mol. Struct. (Theochem).* 1999;461:1-21.

Mechanistic Insights into Fructose dehydration to 5-(hyroxymethyl)furfural

Vinit Choudhary, Liu Yang, Georgios Tsilomelekis, Stavros Caratzoulas, Vladimiros Nikolakis, Stanley I. Sandler, and Dionisios G. Vlachos
Catalysis Center for Energy Innovation, Department of Chemical and Biomolecular Engineering, University of Delaware, Newark, DE 19716 USA

Development of improved catalysts to design efficient and economic processes for biomass conversion to fuels and chemicals entail mechanistic insights into the reaction pathways. The objective of this study is to gain fundamental insights into the fructose dehydration to 5-(hydroxymethyl) furfural (HMF). Recent studies showed that closed-ring reaction pathway dominates in fructose dehydration to HMF. However, many different closed-ring pathways have been proposed. Additionally, fundamental questions such as the effect of solvent on the dominant reaction pathway remain unanswered. Here we investigate the fructose dehydration to HMF using isotopic-labeling experiments and NMR spectroscopy along with ab initio calculations with implicit and explicit water. A pronounced kinetic-isotope effect (KIE) is observed for HCl catalyzed fructose-D1 conversion in aqueous media, whereas no KIE is observed in fructose-D5 indicating that hydrogen transfer from C1 to C2 is the rate limiting step, consistent with earlier quantum mechanics/molecular mechanics calculations published by some of the authors. A plausible reaction intermediate in the fructose dehydration has been identified, and deuterium atoms from fructose have been mapped to various products using NMR spectroscopy that provide insights into the dominant reaction pathway. Finally, the effect of solvent on the mechanism will be discussed.

A CFD Pressure Drop Model for Microfibrous Entrapped Catalyst Filters using Micro Scale Imaging

Qiang Gu and Bruce J. Tatarchuk
Center for Microfibrous Material Manufacturing (CM3), Department of Chemical Engineering, 212 Ross Hall, Auburn University, AL 36849

Abstract

A new structure of matter called microfibrous entrapped catalyst (MFEC) was developed to improve overall performance over conventional monolith reactors. This structure is manufactured by wet-lay process of metal fibers. Catalyst support particles (γ-Al2O3 150-250 μm) were entrapped by the sinter-lock structure of the metal fibers. Pressure drop across the reactor is a crucial factor when designing the reactor. This work presents experimental and simulation method used to evaluate pressure drop across this novel material.

Because of the dimension difference in fiber and filters, pressure drop modeling for filter structures is usually difficult if the details like fiber dimension and entrapped particle shapes are considered. This issue is solved by studying the fiber media in both micro and macro scale scope. Micro scale Computational Fluid Dynamics (CFD) pressure drop model for microfibrous material entrapped catalyst (MFEC) is first established using SEM imaging. The micro scale CFD model is based on theoretical calculation of velocity profile within the fiber material, which shows the plug flow pattern. SEM imaging for top and side view of the fiber media is obtained to sketch the geometric structure of the simulation area. Fiber and particle contribution to the pressure drop are studied separately. Shape factor and fiber compressibility are also considered for model accuracy. Micro scale pressure drop simulation results are then transferred into macro scale pleated filter structures for pressure drop modeling. Two kinds of filter fairing are studied for pressure drop reduction performance. All simulation results are compared with experiment data.

Keywords: Pressure Drop, CFD, SEM, MFEC

1. Methods and Experiments

1.1 Velocity profile in flat microfibrous material

Theoretical calculation of flow in microfibrous media is done by solving equation 4 (Figure 1), which is volume average Navier-Stoke equation for porous media developed by Vafai and Tien

— (4)

1.2 Numerical process

All velocity schemes are discretized using central difference for r and z directions with Euler explicit method. The form for iteration is arranged as

— — — — — —

— — — — — — (8)

1.3 Flat sheet pressure drop measurement

The quadratic form pressure drop was measured using a 1 inch diameter steel pipe. Then 80 psi of house air was supplied as velocity inlet for microfibrous material.

1.4 SEM imaging for model detail

A sample SEM image of the microfibrous media and its corresponding sketch are shown in Figure 2 and Figure 3. Since only the top few layers are visible and easy to distinguish from one another, sketching of the fiber media structure by SEM was based on these layers. As to the inner layers, the top layers are translated because all layers are randomly dispersed and assumed to be similar in performance.

2. Results and Discussion

2.1 Fiber only

The pressure drop across a blank 8µm nickel fiber (Figure 4) was first simulated. The fiber material used is 1.2% in volume loading. The distance between fibers is determined by SEM imaging method. The simulation applied for four layer of unit cell which is 1.1 mm Experimental tests are based on 4 mm thick fiber sheet and these results will be compared on a pressure drop per millimeter basis.

2.2 Fiber with round particles

Since volume loading of particles in entrapped fiber sheets is much higher than fiber itself, the particles contribute to a large portion of pressure drop to the total pressure drop of the sheet. Typically volume loading of particles ranges from 20% - 30% for metal fiber sheet. In this application, particle volume loading is set at 20%. Particles used are ground and sieved to 150-250 m. In this part, the particles are assumed to be ideal in shape (sphere) and average in size (200 m). In the simulation, the particles are sketched as spheres with surfaces defined as wall conditions. Other simulation criteria are the same as fiber only case. shows SEM imaging of particle-entrapped fiber media, which is sketched in as simulation unit cell.

2.3 Fiber with elliptic particles

Shown by SEM imaging, shape of entrapped particles is not perfectly round. In this simulation, shape factor also known as sphericity has been investigated for contribution to the total pressure drop of the fiber material. Karwa *et al.* has empirically obtained shaped factors using Blake Kozeny equation for microfibrous material (Table 1). In this research, 180-210 µm particles were used and shape factor of 0.8 was chose as simulation criteria. Pressure drop data were shown in Figure 5.

2.4 Fiber compressibility

To improve CFD model accuracy, fiber and particle spacing has been adjusted according to the thickness tested (Figure 6). Due to the volume loading increment of fibers and entrapped particles, the pressure model accuracy has been improved dramatically especially at higher face velocities because of the larger compressions at these velocities (Figure 7).

2.5 Pressure drop for pleated MFEC using micro scale data

In high volumetric tests, the microfibrous media is engineered into pleat structure to reduce pressure drop. These structure help with reduction of amount of air going into each of the pleats. The pressure drop and velocity profile for these complicated structures are simulated using previous flat sheet simulation results. In addition, the pleat structure has brought in pleat tips which increased the total pressure drops. Fairing added to the pleat tip has been tested experimentally and by simulation using FLUENT CFD to explain the benefit of those fairings (Figure 8,9,10).

3. Summary

Pleat structure pressure drop of the microfibrous media was simulated based on the simulation of flat media. Fairings added to reduce pressure drop were investigated, which showed triangle fairings help with gradual compression of the air and the total pressure drop. Optimum pleat number was also obtained based on simulation results.

Reference

Angirasa D., Forced convective heat transfer in metallic fibrous materials, Journal of Heat Transfer, Vol. 124, pp 739-745, 2002

Bird, R., W. Stewart, E. Lightfoot, Transport Phenomena, 2nd ed. New York, John Wiley & Son, Inc., 2001

Caesar, T., Schroth T., The influence of pleat geometry on the pressure drop in deep-pleated cassette filters, Filtration and Separation, Vol. 39(9), pp 48-54, 2002

Cahela, D.R. Tatarchuk, B.J., Permeability of sintered microfibrous composites for heterogeneous catalysis and other chemical processing opportunities, Catalysis Today Vol. 69, pp 33-39, 2001

Chen, D.R., Pui, D.Y.H., Tang Y.M., Filter pleating design for cabin air filtration, Technical paper 960944, Annual index/abstract of SAE technical papers, Society of Automotive engineers, Warredale, PA., 1996

Dhandapani B., Oyama S.T., Gas phase ozone decomposition catalyst, Applied Catalysis B: Environmental Vol. 11, pp 120-166, 1997

Harris D.K., Cahela D.R., Tatarchuk B.J., Wet layup and sintering of metal-containing microfibrous composites for chemical processing opportunities, Composites: Part A Vol. 32, pp 1117-1126, 2001

Hill R.J., Koch D.L., The transition from steady to weakly turbulent flow in a close-packed ordered array of spheres, Journal of Fluid Mechanics, Vol. 465 pp 59-97, 2002

Kalluri R.R., Cahela D.R., Tatarchuk B.J., Comparative heterogeneous contacting efficiency in fixed bed reactor: Opportunities for new microstructured systems, Applied Catalysis B: Environmental Vol. 90, pp 507-515, 2009

Kalluri R. R., Duggirala R. K., Cahela D. R., Roy C. J., and Tatarchuk B. J., CFD analysis of mass transfer enhancement due to inert structures present in fluid flow paths in heterogeneous contacting systems, manuscript submitted to computers and chemical engineering, 2008

Kameya T., Urano K., Catalytic Decomposition of ozone gas by a Pd impregnated MnO2 catalyst, Journal of environmental engineering, pp 286-292, 2002

Karwa A.N., Tatarchuk B.J., Pressure drop and aerosol filtration efficiency of microfibrous entrapped catalyst and sorbent media: Semi-empirical models, Separation and Purification Technology Vol. 86 pp 55-63, 2012

Rivers, D.R., Murphy D.J., Air filter performance under variable air volume conditions, ASHRAE Transactions Vol. 106(2), pp 131-144, 2000

Sothen R.A., Tatarchuk B.J., A semi-empirical pressure drop model: part I- pleated filters, HVAC&R Research, Vol. 14(6), 2008

Vafai, K., Tedn, C. L., Boundary and inertia effect on flow and heat transfer in porous media, International J. Heat Mass Transfer, pp 1183-1190, 1982

Wang X., Kim K., Lee C., and Kim J., Prediction of air filter efficiency and pressure drop in air filtration media using a stochastic simulation, Fiber and Polymers, Vol. 9(1), pp 34-38, 2008

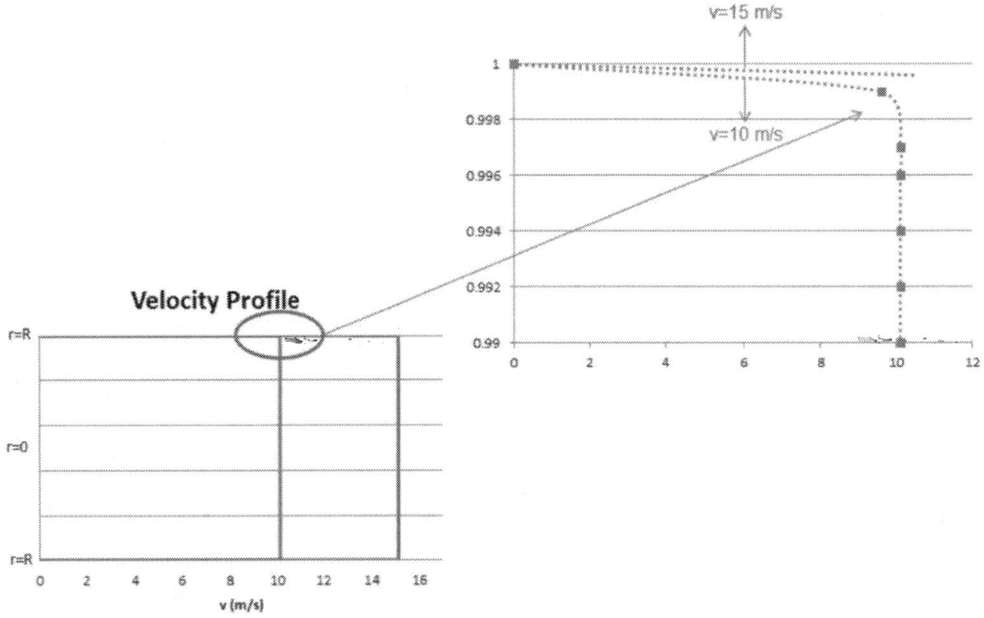

Figure 1 Velocity Profile of microfiber material at two different face velocities with high Reynolds numbers

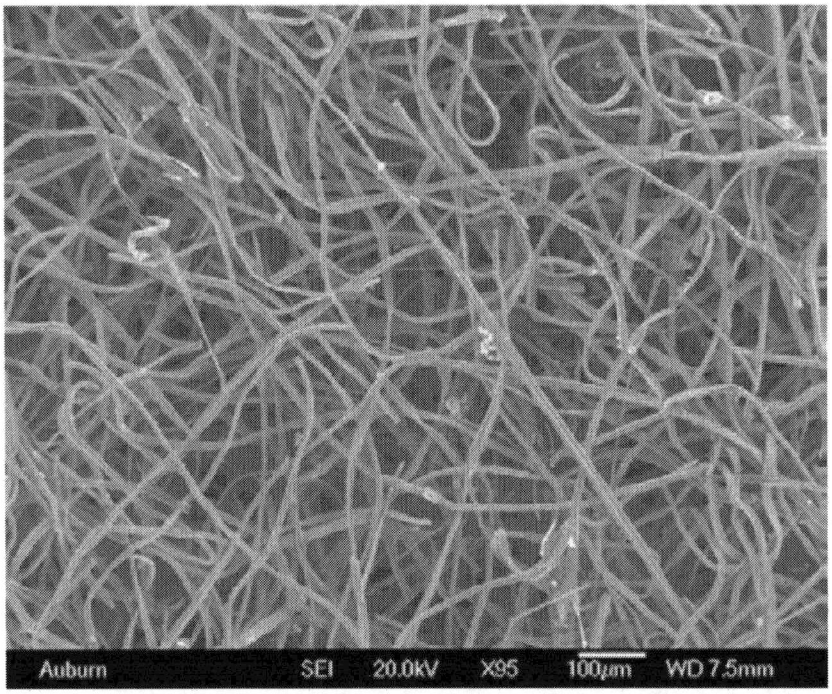

Figure 2 Sample SEM imaging of sintered microfibrous material surface

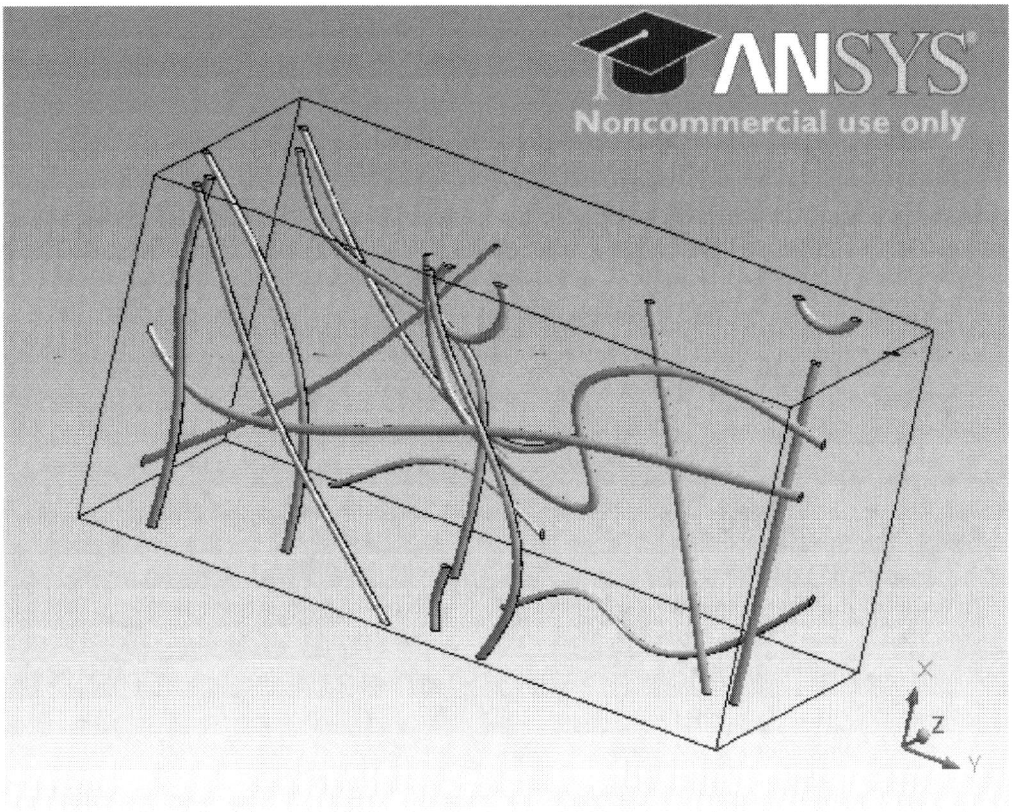

Figure 3 Sketch of selected area of microfibrous material for simulation

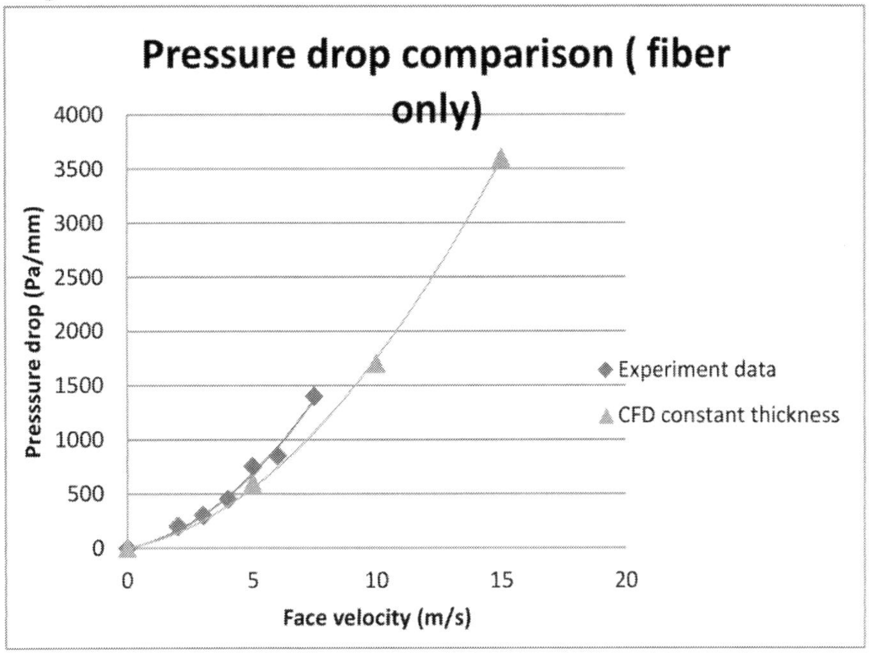

Figure 4 Pressure drop comparison between the model and experiment data

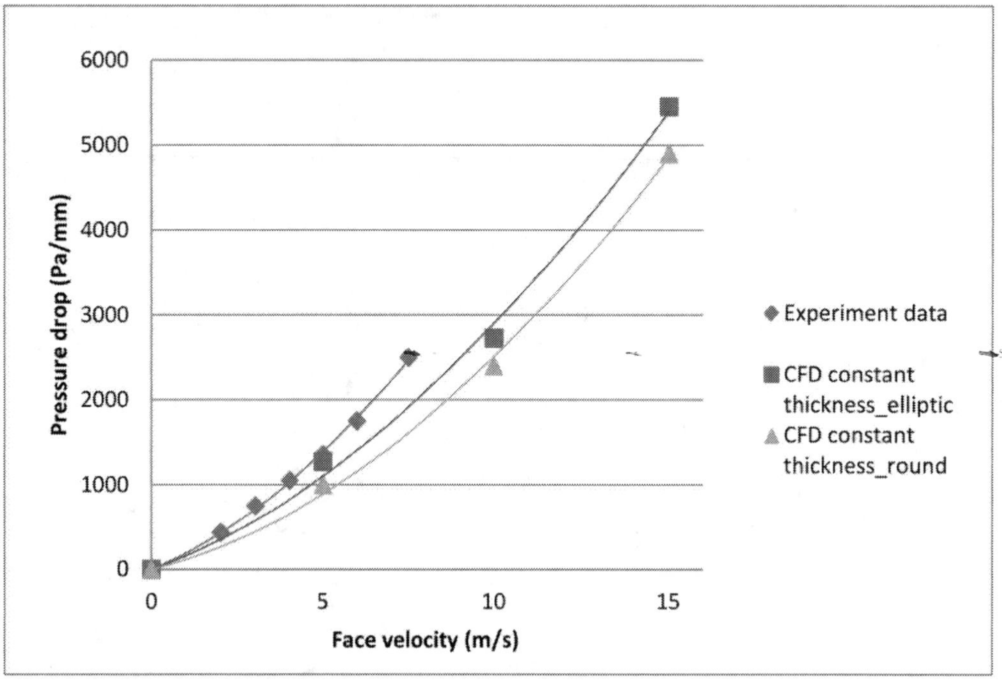

Figure 5 Pressure drop comparison of experiment, round particle and elliptic particle media by CFD simulation

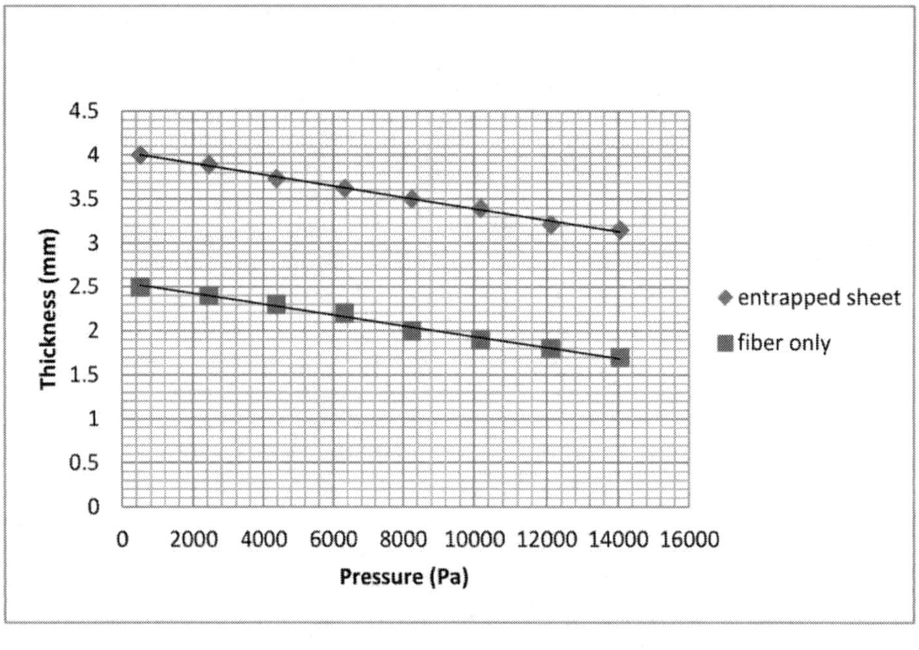

Figure 6 Microfiber media thickness at different pressures; initial volume loading, fiber 1.5%, particles 20%; initial thickness 4 mm

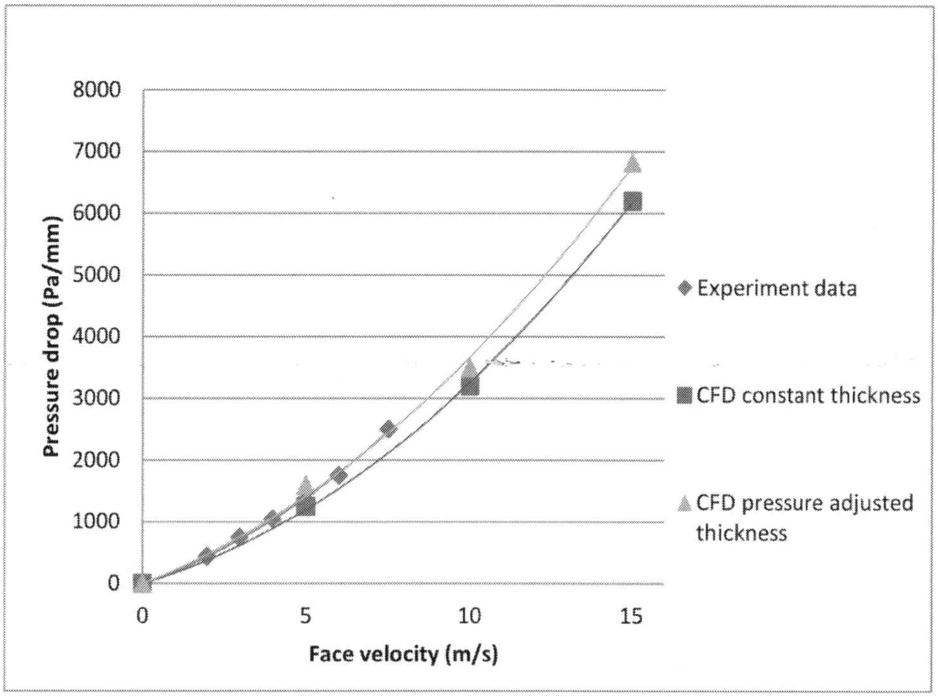

Figure 7 Pressure drop for fiber with elliptic particle case; CFD simulation in both uncompressed and compressed consideration; experiment result collected up to 7.5 m/s

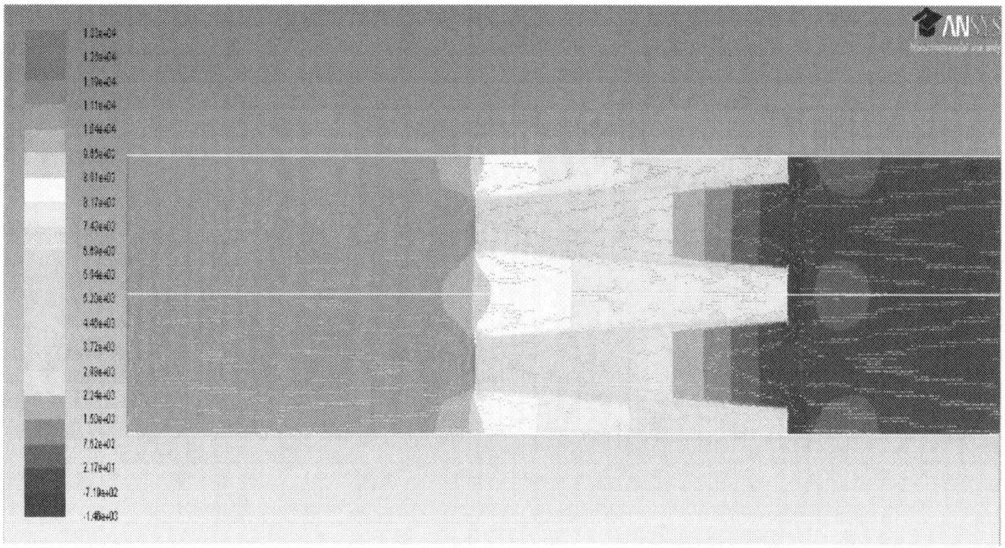

Figure 8 Pressure drop simulation for pleated microfibrous material filter

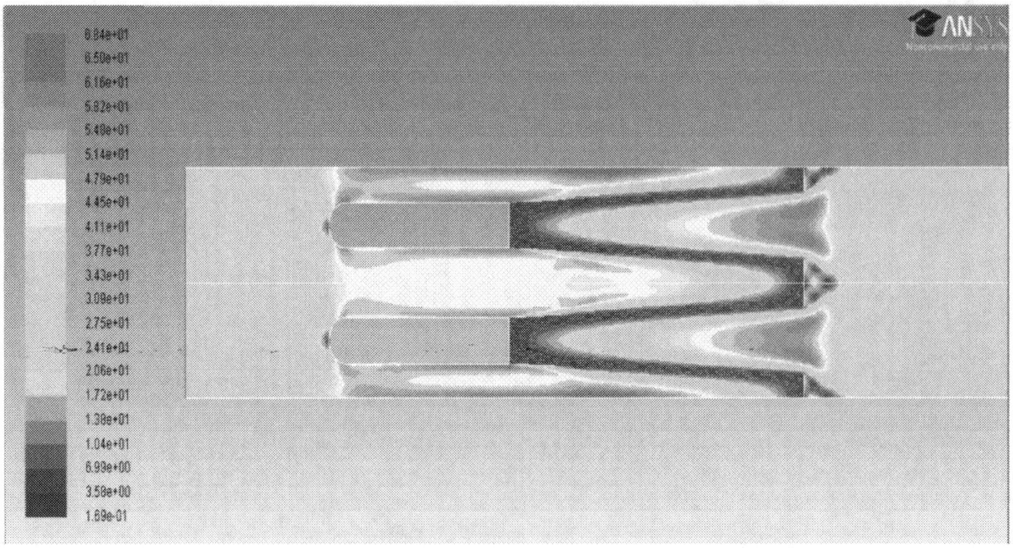

Figure 9 Velocity profile for pleated microfibrous material filter with round-top fairings; eddy flow removed by added fairings

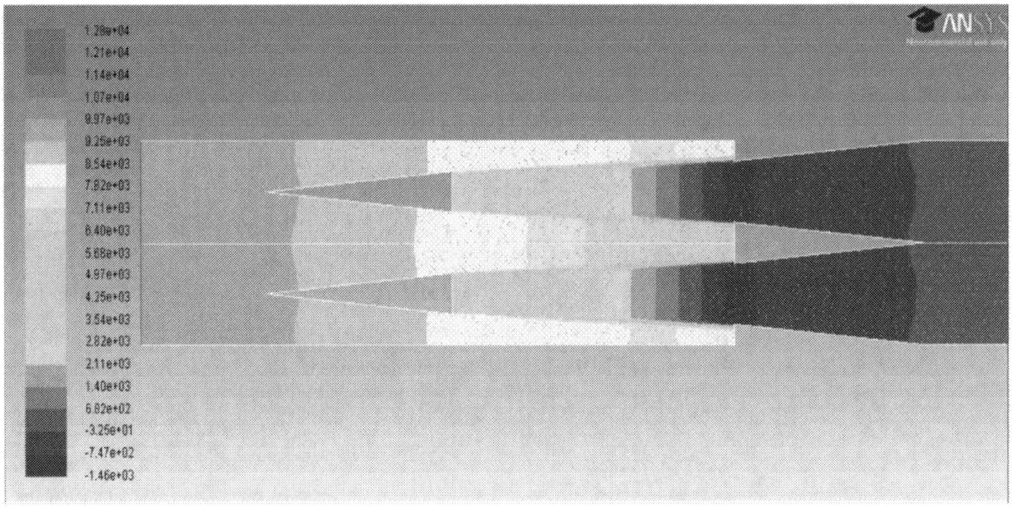

Figure 10 Pressure drop simulation for pleated microfibrous material filters with triangle fairings at both upstream and downstream

Kinetics of Jatropha oil methanolysis

Mario Andrés Noriega Valencia[1], Paulo César Narváez Rincón[1], Christoph Heinz[2], Nevardo Bello Yaya[1], Juan Guillermo Cadavid Estrada[1]

1.Grupo de Procesos Químicos y Bioquímicos, Departamento de Ingeniería Química. Facultad de Ingeniería, Universidad Nacional de Colombia, Bogotá, Colombia.
2 Technische Universität München, Chemistry Department, Lichtenbergstraße 4, D-85748 Garching b. München, Germany.

*pcnarvaezr@unal.edu.co

Abstract

Biodiesel is currently produced by methanolysis of vegetable oils such as rapeseed, palm and soybean oils. However, Jatropha oil is considered one of the most promising sources for producing biodiesel in Asia, Europe and Africa, The knowledge of the kinetics of his methanolysis reaction is extremely important because it enables the design of reactors with suitable dimensions and the operation in optimal conditions. In this work the kinetics of Jatropha oil methanolysis was studied. The effect of temperature, catalyst concentration and methanol to oil molar ratio on Jatropha oil methanolysis were studied in the range from 40 to 60 °C, 0.2 to 1wt% (NaOH) and 3 to 12, respectively. A 100% conversion of Jatropha oil with a 97% yield to fatty acid methyl esters (FAME) were obtained after 60 minutes of reaction at 60°C, with 0.6wt% of NaOH, and methanol to oil molar ratio of 6:1. Catalyst concentration increments between 0.2wt% and 0.6wt% had a positive effect on the reaction rate and on the yield to FAME. However, the 0.2wt% catalyst concentration did not show activity at all reaction temperatures. A second order kinetic model that includes the effect of temperature and catalyst concentration on reaction rate describes adequately Jatropha oil methanolysis, as validated by the Fisher–Snedecor test of unbiased variances. Kinetic model parameters of Jatropha oil methanolysis, Palm oil methanolysis and Palm oil ethanolysis were compared and showed similar values and behavior.

Keywords: Jatropha oil, Palm oil, Methanolysis, Ethanolysis Biodiesel, Kinetic model, Mathematic model

Catalyst-Scale Modeling of Fischer-Tropsch Synthesis from Syngas Using COMSOL Multiphysics

Arvind. Nanduri[1] and Patrick L. Mills[*1]
[1]Texas A&M University-Kingsville, Department of Chemical & Natural Gas Engineering
*Corresponding author: Texas A&M University-Kingsville, Department of Chemical & Natural Gas Engineering, EC 303D, MSC 193, Kingsville, TX-78363-8202, USA. Email: Patrick.Mills@tamuk.edu

Abstract: A 2-D particle model for different catalyst shapes (solid cylinder, hollow cylinder, 4-hole cylinder, modified 4-hole cylinder and 7-hole cylinder) was simulated to compare the particle-level performance for the Fischer-Tropsch Synthesis (FTS). A Fe-based micro-kinetic olefin re-adsorption model developed by Wang *et al.* (2008) was coupled with the Soave-Redlich-Kwong (SRK) equation of state to describe the particle-scale transport-kinetic interactions and phase behavior for the gas-phase FTS [4]. The intra-particle effectiveness factor, liquid-to-vapor ratio, CO conversion and intra-particle volume-averaged concentration of diesel were analyzed at different process conditions to compare the performance of different catalyst particle shapes.

Keywords: catalyst particle shape, Fischer-Tropsch Synthesis, vapor-liquid-equilibrium (VLE), gas phase micro-kinetics.

1. Introduction

The Fischer-Tropsch Synthesis (FTS) is a highly exothermic condensation polymerization reaction of syngas ($CO+H_2$) in the presence of Fe/Co/Ru-based catalysts to produce a wide range of paraffins, olefins and oxygenates, the latter of which is often called *syncrude*. Multi-Tubular Fixed Bed Reactors (MTFBR) and Slurry Bubble Column reactors (SBCR) are widely employed for FTS processes [1]. To understand the reactor-scale catalyst performance, it is important to first analyze particle-scale transport-kinetic interactions.

The FTS reaction network produces hydrocarbons with carbon numbers typically ranging from 1 to 100 so the catalyst pores in this process can be potentially filled with liquid wax (C_{20}+) leading to high diffusional limitations. To model such a reaction network and account for all species, micro-kinetic rate expressions for each individual species must be coupled with the intraparticle solubility of gases in liquid wax and transport equations for the various reaction species in the porous catalyst. This paper describes the initial results obtained from such an effort using COMSOL Multiphysics™ as the numerical engine for solving the model equations.

2. Catalyst Pellet Model

The isothermal 2-D catalyst pellet model used to describe transport-kinetic interactions is presented in this section.

A total number of 20 paraffins (C_1 to C_{20}), 19 olefins (C_2 to C_{20}) and 4 key components (H_2, CO, CO_2, and H_2O) are considered in the reaction network, which leads to 43 second-order nonlinear differential equations for the specie mass balances. The micro-kinetic and thermodynamic expressions used in the model can be found elsewhere [3-5]. Five different particle shapes based on a solid cylinder, sphere, and hollow cylinders having the same effective diffusion length ($L_e = V_p/S_p$) are simulated in this paper [2]. The catalyst shapes are illustrated in Figure 1.

2.1 Specie mass balance equations

The specie mass balance equations for each particle shape are summarized below.

General specie balance

$$\nabla.(-D_{ei}\nabla C_i) = \rho_p \sum_{j=1}^{44}\sum_{i=1}^{43} \alpha_{ij}R_{ij}$$

Specie balance in cylindrical coordinates

$$\frac{1}{r}\frac{\partial}{\partial r}\left(D_{ei} r \frac{\partial C_i}{\partial r}\right) = -\rho_p \sum_{j=1}^{44}\sum_{i=1}^{43} \alpha_{ij}R_{ij}$$

The specie mass balance for all the particle shapes considered in this work are solved in primitive variables where *r* is the radial coordinate. Particle end effects are neglected so specie concentrations are a function of the radial coordinate only.

Numerical solutions showed that the intra-particle temperature gradient was on the order of a few degrees owing to the high effective thermal conductivity of the particle. For this reason, the energy balance was not included in any subsequent simulations.

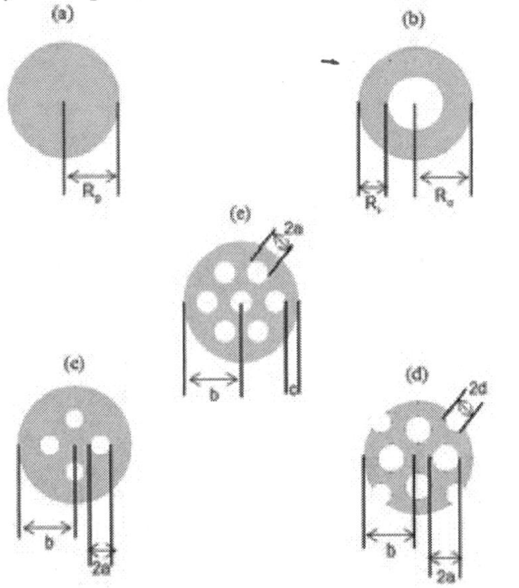

Figure 1. Catalyst particle shapes. (a) cylinder, (b) hollow cylinder (H-Cylinder), (c) 4-hole cylinder (4-H Ring), (d) modified 4-hole cylinder (M-4-H Ring), and (e) 7-hole cylinder (7-H Ring).

2.2 Boundary conditions

The boundary conditions correspond to specified concentrations at the pellet surface (Dirichlet conditions). The boundary conditions are summarized in Table-1.

Table 1. Boundary Conditions.

Cylinder	At $r = R_p$ (surface), $C_i = C_{i,bulk}$
H-Cylinder	At $r = R_o$ (outer surface) and $r = R_i$ (inside surface), $C_i = C_{i,bulk}$
4-H Ring	$C_i = C_{i,bulk}$ on all boundaries exposed to bulk phase
M-4-H Ring	$C_i = C_{i,bulk}$ on all boundaries exposed to bulk phase
7-H Ring	$C_i = C_{i,bulk}$ on all boundaries exposed to bulk phase

2.3 Vapor-Liquid-Equilibrium calculations

The vapor-liquid equilibrium for the FTS reaction mixture is described here using the Soave-Redlich- Kwong equation of state [3]. The flash calculations are based upon the Rachford-Rice objective function:

$$F(\alpha_g) = \sum_i \frac{z_i(K_i - 1)}{\left(1 + \alpha_g(K_i - 1)\right)} = 0$$

In this application, the Rachford-Rice objective function is a polynomial of degree 43 (43 distinct roots). Since the fraction α_g = V/(V+L) is a positive number less than 1, only the roots that make physical sense are considered.

The initial estimate of the distribution coefficient is obtained using Wilson's correlation [3]:

$$K_i^{\text{guess value}} = \frac{P_{ic}}{P} \exp\left(5.37(1 + \omega_i)\left(1 - \frac{T_{ic}}{T}\right)\right)$$

Here, the distribution coefficient is defined as the ratio of the vapor to liquid mixture fugacity coefficients:

$$K_i = \frac{\phi_i^{V}}{\phi_i^{L}}$$

3. Modeling Strategy

The 2-D diffusion-reaction system of equations was solved using the transport of diluted species module in COMSOL Multiphysics™ assuming steady-state conditions. The Rachford-Rice objective function (VLE calculations) was solved using the coefficient form of the PDE solver. The catalyst properties and operating conditions used in the models are listed in Tables 2 - 4. The solution for the specie concentration profiles were based upon a convergence factor of 10^{-3}.

To avoid convergence issues, the reaction rates for individual species were multiplied by a perturbation factor of 10^{-4} and this factor was slowly increased to 1. Since the FTS is a relatively fast diffusion-reaction system, numerical instabilities were encountered in the interior region of the particle as the CO and CO_2 concentrations approached zero. The convergence issues were solved by using built-in logical operators, such as CO = if (CO ≤ 0, eps, CO), and CO_2 = if (CO_2 ≤ 0, eps, CO_2) which resulted in positive values for the specie concentrations. Mesh refinement was manually performed until the concentration profiles were relatively constant and satisfied the convergence criterion.

Table 2. Dimensions of different catalyst particle shapes with equal characteristic diffusion length.

Cylinder	L = 3 mm & R = 1 mm
H-Cylinder	L = 3 mm, R_o = 2 mm & R_i = 1 mm
4-H Ring	L = 3 mm, b = 2 mm, a = 0.36 mm & c = 0.54 mm
M-4-H Ring	L = 6 mm, b = 2 mm, a = 0.5 mm, c = 0.5 mm & d = 0.415 mm
7-H Ring	L = 6 mm, b = 3 mm, a = 0.6 mm & c = 0.6 mm

Table 3. Catalyst properties.

Density of pellet, ρ_p	1950 (kg/m^3)
Porosity of pellet, ε	0.51
Tortuosity, τ	2.6

Table 4. Operating conditions.

Temperature, K	493, 523 & 533
Pressure, bar	25 & 30
H_2/CO	2

4. Results and Discussion

4.1 Concentration profiles of key components, fuel gas and diesel

The 2-D isothermal intra-particle concentration profiles of the key components (H_2, CO, CO_2 and H_2O), fuel gas (C1-C2) and diesel (C13-C17) for different catalyst particle shapes are shown in Figure 2 - 7. The volume-averaged concentration of each component is used as the basis for comparison. The average concentration profiles for all operating conditions are shown in Figure 8-19. The profiles show that the CO concentration approaches zero for all shapes due to the presence of significant diffusional limitations. The CO_2 concentration reaches a maximum and then decreases due to the reverse Water-Gas-Shift (WGS) reaction. The reverse WGS reaction produces CO, which is then consumed in the subsequent hydrocarbon producing FT reactions. Figure 12 and 13 show that the reverse WGS reaction does not occur in the M-4-H ring.

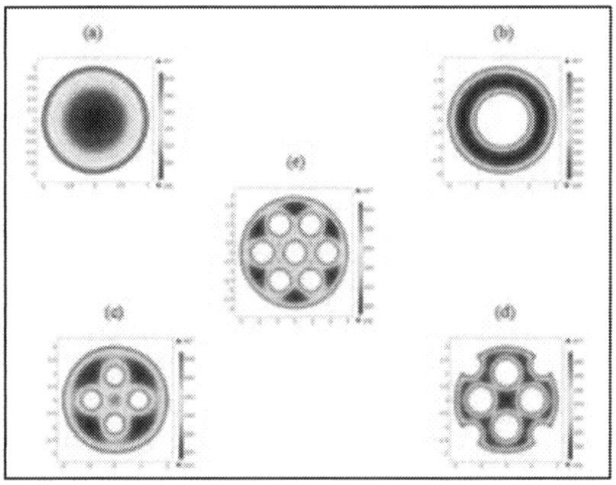

Figure 2. Intra-particle concentration profile of H_2 (mol/m^3) at T = 493 K and P = 25 bar. (a) cylinder, (b) hollow cylinder (H-Cylinder), (c) 4-hole cylinder (4-H Ring), (d) modified 4-hole cylinder (M-4-H Ring), and (e) 7-hole cylinder (7-H Ring).

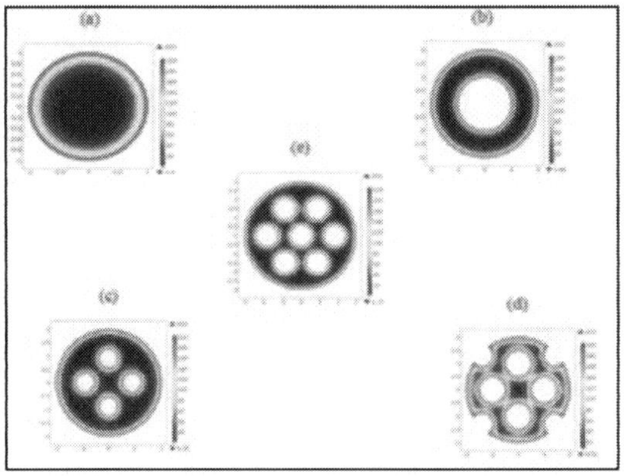

Figure 3. Intra-particle concentration profile of CO (mol/m^3) at T = 493 K and P = 25 bar. (a) cylinder, (b) hollow cylinder (H-Cylinder), (c) 4-hole cylinder (4-H Ring), (d) modified 4-hole cylinder (M-4-H Ring), and (e) 7-hole cylinder (7-H Ring).

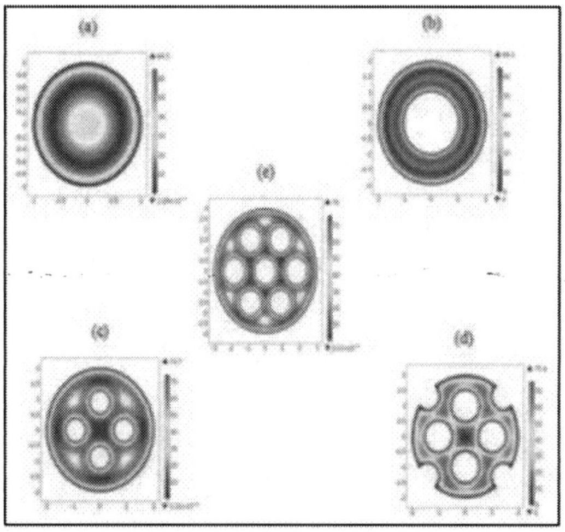

Figure 4. Intra-particle concentration profile of CO_2 (mol/m^3) at T = 493 K and P = 25 bar. (a) cylinder, (b) hollow cylinder (H-Cylinder), (c) 4-hole cylinder (4-H Ring), (d) modified 4-hole cylinder (M-4-H Ring), and (e) 7-hole cylinder (7-H Ring).

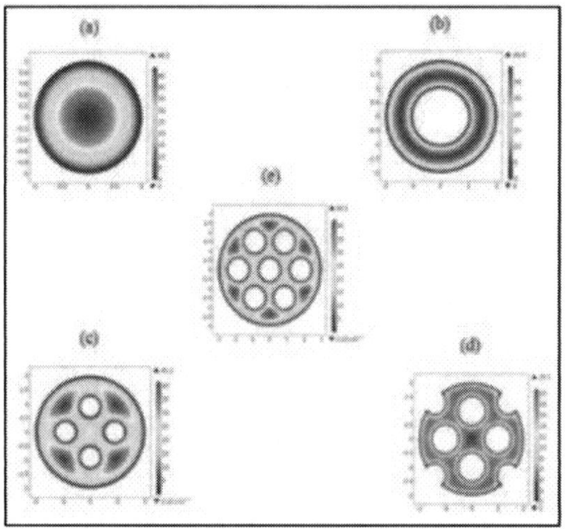

Figure 5. Intra-particle concentration profile of H_2O (mol/m^3) at T = 493 K and P = 25 bar. (a) cylinder, (b) hollow cylinder (H-Cylinder), (c) 4-hole cylinder (4-H Ring), (d) modified 4-hole cylinder (M-4-H Ring), and (e) 7-hole cylinder (7-H Ring).

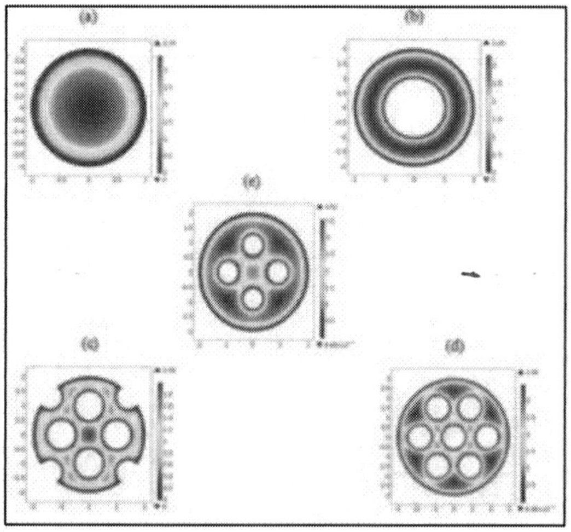

Figure 6. Intra-particle concentration profile of fuel gas (mol/m^3) at T = 493 K and P = 25 bar. (a) cylinder, (b) hollow cylinder (H-Cylinder), (c) 4-hole cylinder (4-H Ring), (d) modified 4-hole cylinder (M-4-H Ring), and (e) 7-hole cylinder (7-H Ring).

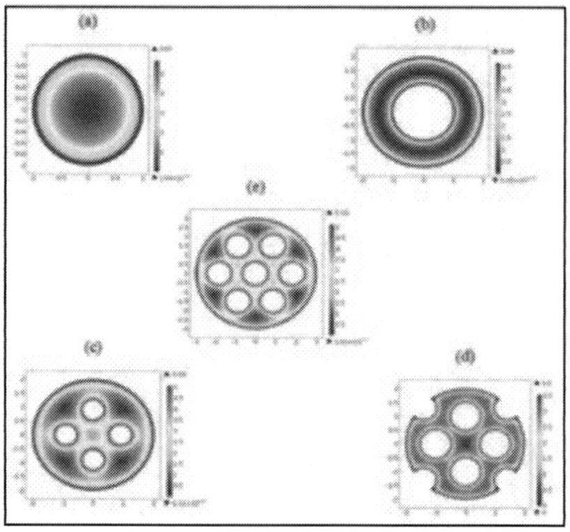

Figure 7. Intra-particle concentration profile of diesel (mol/m^3) at T = 493 K and P = 25 bar. (a) cylinder, (b) hollow cylinder (H-Cylinder), (c) 4-hole cylinder (4-H Ring), (d) modified 4-hole cylinder (M-4-H Ring), and (e) 7-hole cylinder (7-H Ring).

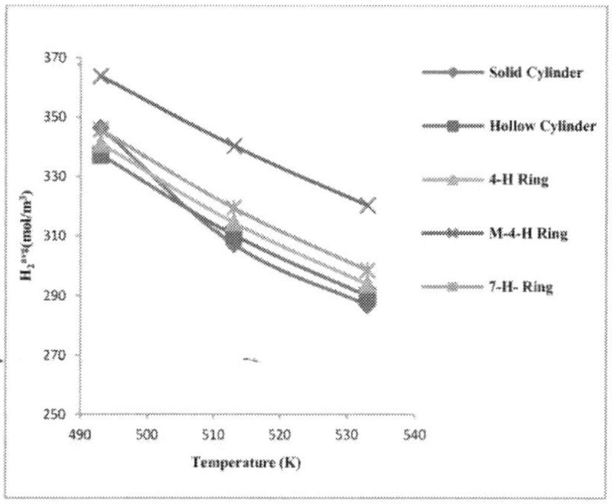

Figure 8. Average concentration of H_2 (mol/m^3) for T = 493 K, 513 K and 533 K and P = 25 bar.

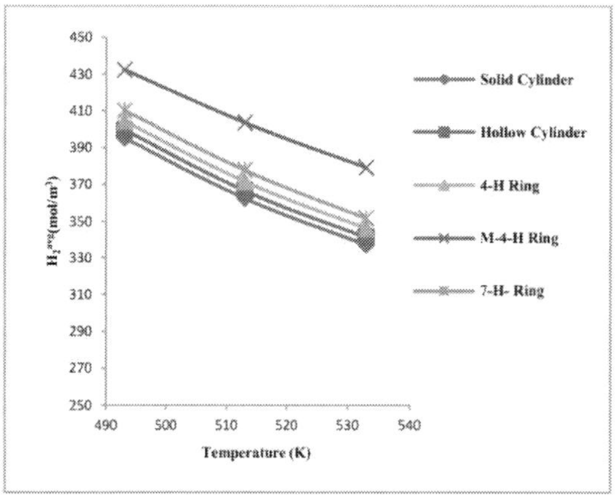

Figure 9. Average concentration of H_2 (mol/m^3) for T = 493 K, 513 K and 533 K and P = 30 bar.

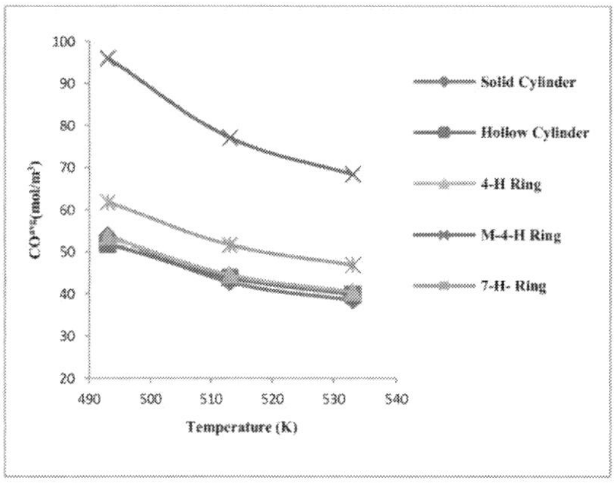

Figure 10. Average concentration of CO (mol/m^3) for T = 493 K, 513 K and 533 K and P = 25 bar.

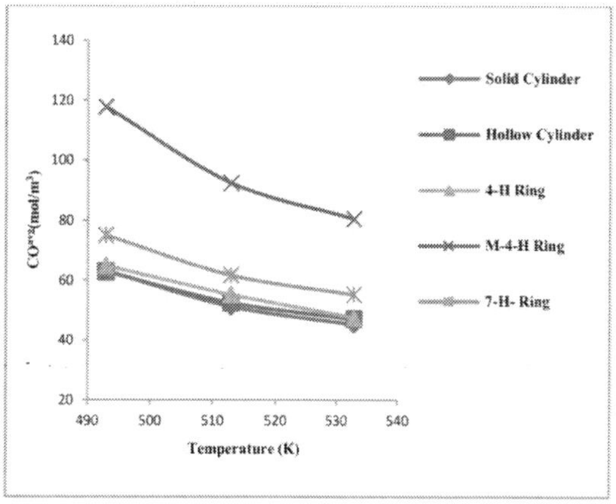

Figure 11. Average concentration of CO (mol/m^3) for T = 493 K, 513 K and 533 K and P = 30 bar.

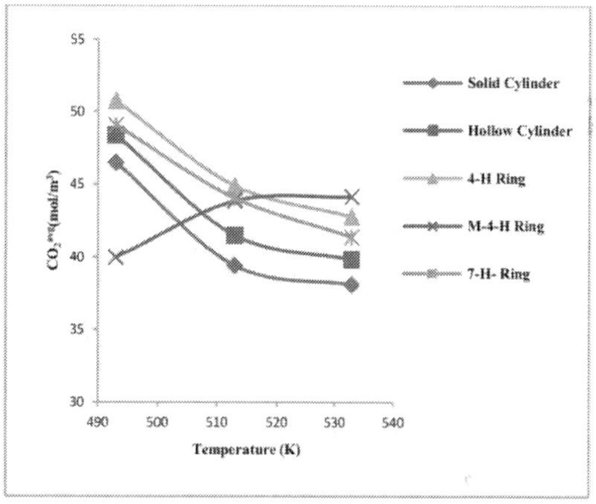

Figure 12. Average concentration of CO_2 (mol/m^3) for T = 493 K, 513 K and 533 K and P = 25 bar.

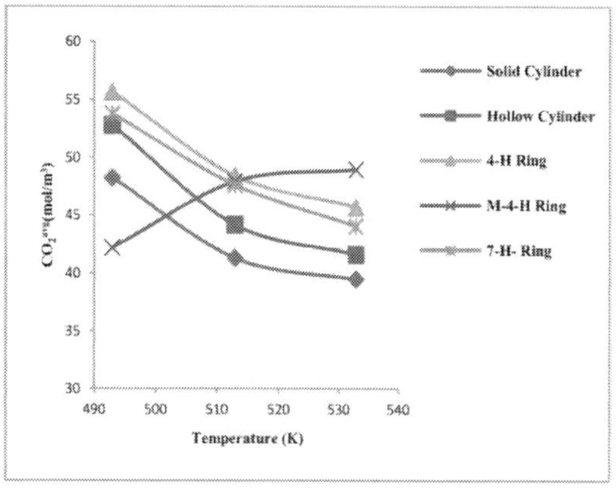

Figure 13. Average concentration of CO_2 (mol/m^3) for T = 493 K, 513 K and 533 K and P = 30 bar.

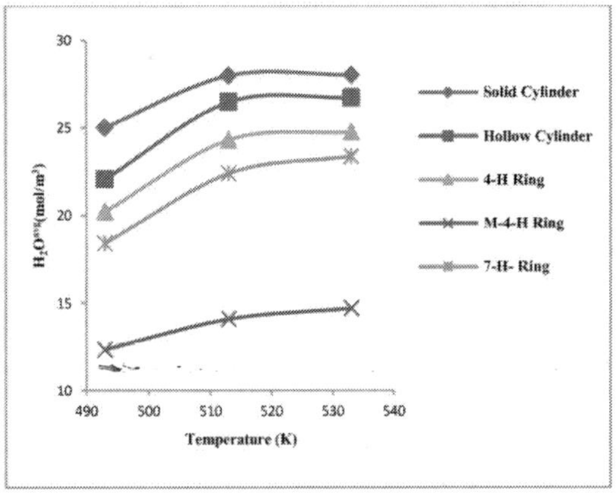

Figure 14. Average concentration of H_2O (mol/m^3) for T = 493 K, 513 K and 533 K and P = 25 bar.

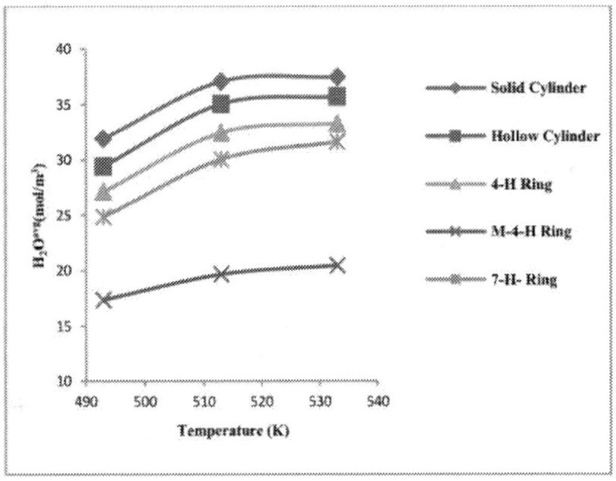

Figure 15. Average concentration of H_2O (mol/m^3) for T = 493 K, 513 K and 533 K and P = 30 bar.

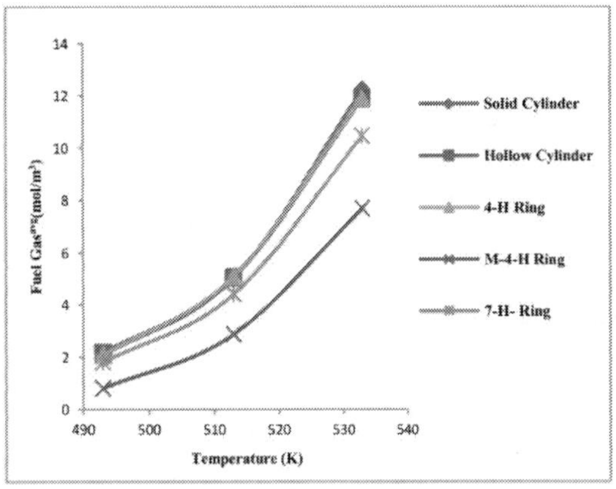

Figure 16. Average concentration of fuel gas (mol/m^3) for T = 493 K, 513 K and 533 K and P = 25 bar.

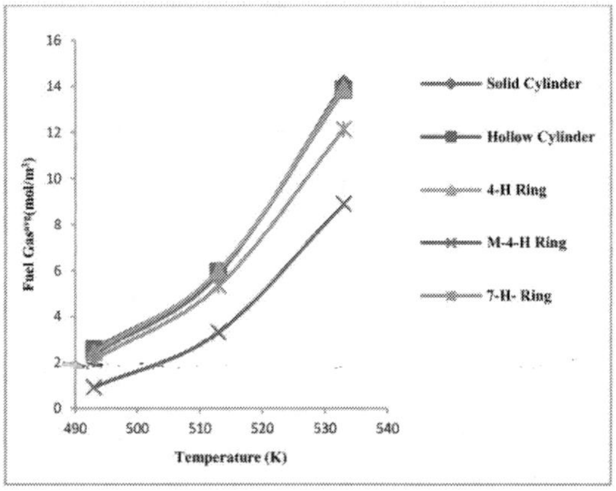

Figure 17. Average concentration of fuel gas (mol/m^3) for T = 493 K, 513 K and533 K and P = 30 bar.

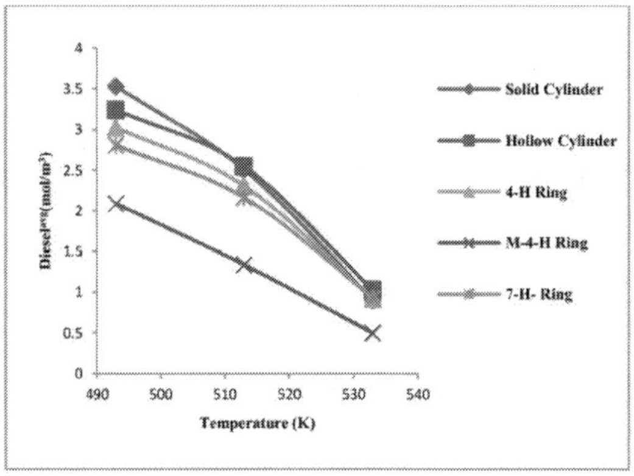

Figure 18. Average concentration of diesel (mol/m^3) for T = of 493 K, 513 K and 533 K and P = 25 bar.

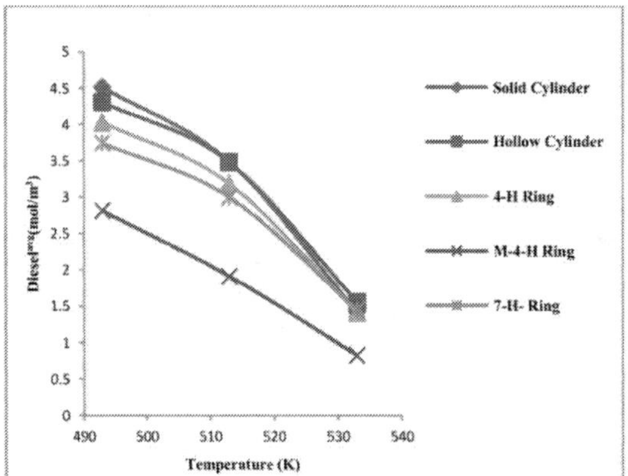

Figure 19. Average concentration of diesel (mol/m^3) for T = 493 K, 513 K and 533 K and P = 30 bar.

4.2 Overall effectiveness factor

The overall effectiveness factor (η) for different particle shapes are shown in Figure 20 - 21. The effectiveness factor has a decreasing trend with temperature and as the center of the particle is approached. It can be observed that the M-4-H ring has the highest effectiveness factor, but the reactor-scale performance and diesel range selectivity for all catalyst particle shapes has to be analyzed in more detail to identify the preferred catalyst shape. Other factors must also be considered, such as mechanical strength of the catalyst as well as pressure drop.

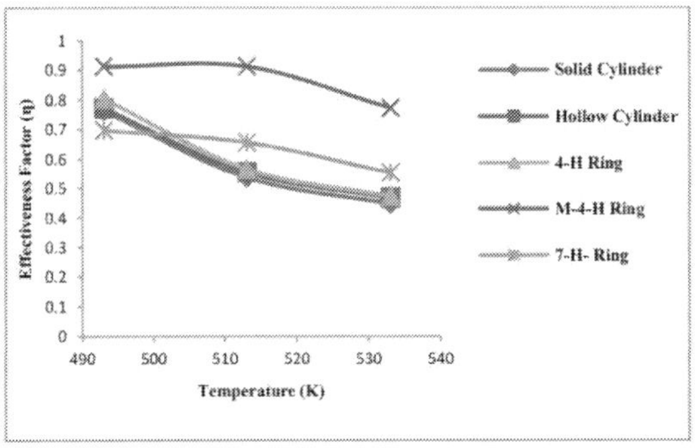

Figure 20. Effectiveness factor (η) for T = 493 K, 513 K and 533 K and P = 25 bar.

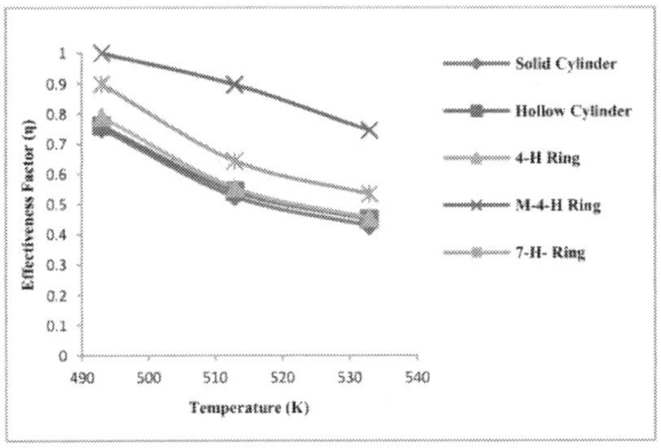

Figure 21. Effectiveness factor (η) for T = of 493 K, 513 K and 533 K and P = 30 bar.

4.3 Intra-particle and average liquid-to-vapor ratio

The intra-particle and volume average liquid-to-vapor molar ratio (L/V) for different particle shapes is shown in Figures 22 - 24. These results show that the M-4-H ring and 7-H ring particles have the smallest liquid fraction when compared to the solid cylinder, the H-cylinder, and the 4-H ring. This result is also supported by comparing concentration profiles of key components (see Figures 2 - 7) where the later particle shapes exhibit the most significant diffusional limitation. These results also show the importance of understanding the phase behavior of FT products in the catalyst pores since the reaction kinetics for liquid phase and gas phase would differ. However, this lies outside the scope of the current work.

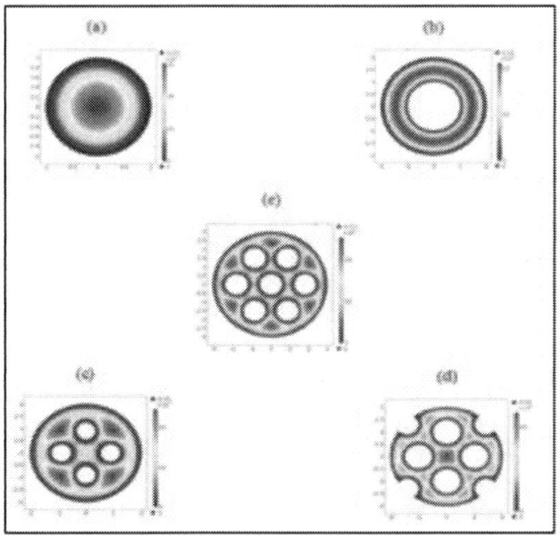

Figure 22. Intra-particle liquid-to-vapor ratio at T = 493 K and P = 25 bar. (a) cylinder, (b) hollow cylinder (H-Cylinder), (c) 4-hole cylinder (4-H Ring), (d) modified 4-hole cylinder (M-4-H Ring), and (e) 7-hole cylinder (7-H Ring).

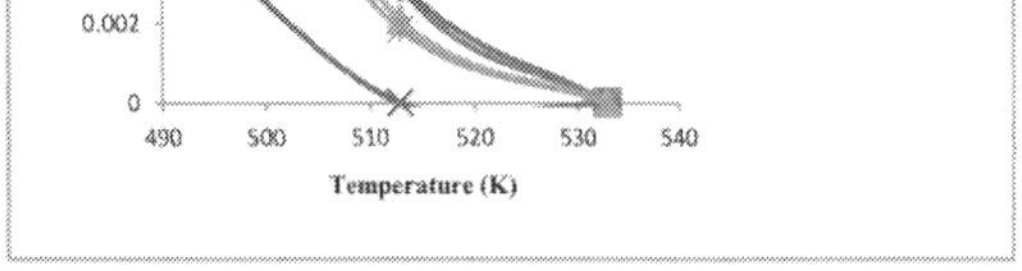

Figure 23. Liquid-to-vapor ratio for T = 493 K, 513 K and 533 K and P = 25 bar.

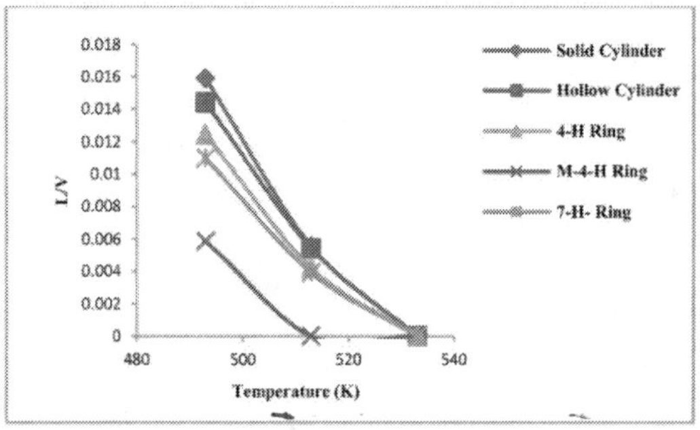

Figure 24. Liquid-to-vapor ratio for T = 493 K, 513 K and 533 K and P = 30 bar.

5. Summary and Conclusions

The transport-kinetic interactions for FTS in a 2-D catalyst pellet model using a detailed micro-kinetic model were successfully analyzed for the first time using COMSOL Multiphysics™. This study is an important extension to the previous work by Wang *et al.* (2008) who used the same micro-kinetic model to study the F-T reaction-diffusion mechanism in a spherical catalyst, but did not study the effect of catalyst shape nor report the inter-particle average diesel concentration. Also, the numerical technique used by Wang *et al.* (2008) was based upon a custom computer code that implemented the orthogonal collocation method to solve the two-point boundary value system of ODEs. The results in the current work will provide the basis for modeling the catalyst bed using the packed bed reactor model in COMSOL Multiphysics 5.0 as well other extensions that account for additional particle-scale transport effects.

6. Nomenclature

C_i	Concentration of species 'i' (mol/m^3)
$C_{i,bulk}$	Concentration of species 'i' in bulk phase (mol/m^3)
C_i^{avg}	$\frac{\int C_i dv}{V_p}$ (mol/m^3)
D_{ei}	Effective diffusivity of species *i* (m^2/s)
K_i	Equilibrium ratio
L	Number of moles in the liquid phase (moles)
L_p	Length of catalyst pellet (mm)
P	Pressure (bar)
P_{ic}	Critical pressure of component *i* (bar)
r	Radial coordinate (mm)
R_p	Radius of catalyst pellet (mm)
R_i	Internal radius of hollow cylinder (mm)
R_o	External radius of hollow cylinder (mm)
R_{ij}	Rate of component *i* in jth reaction (mol kg^{-1} s^{-1})
T	Temperature (K)
T_{ic}	Critical temperature of component *i* (K)
V	Number of moles in vapor phase
z_i	Number of moles of component i

Greek Letters

α_g	Intra-particle vapor fraction
α_{ij}	Stoichiometric coefficient of component *i* in reaction *j*
η	Overall effectiveness factor $\left(\frac{\int \Sigma_{i,j} \alpha_{ij} R_{ij} dv}{(\Sigma_{i,j} \alpha_{ij} R_{ij})_{max} V_p}\right)$
ρ_p	Density of catalyst pellet (kg/m^3)
ε	Catalyst pellet porosity
τ	Catalyst tortuosity

7. References

1. D. A. Wood, C. Nwaoha, and B. F. Towler, Gas-To-Liquids (GTL): A Review of an Industry Offering Several Routes for Monetizing Natural Gas, *Journal of Natural Gas Science and Engineering*, **9**, 196-208 (2012)

2. N. J. Mariani, C. Mocciaro, O. M. Martinez, and G. F. Barreto, Estimation of Effectiveness Factor for Arbitrary Particle Shape and Non-Linear Kinetics, *Industrial & Engineering Chemistry Research*, **48**, 1172-1177 (2009)

3. Yi-Ning Wang, Yuan-Yuan Xu, Hong-Wei Xiang, Yong-Wang Li, and Bi-Jiang Zhang, Modeling of Catalyst Pellets for Fischer-Tropsch Synthesis, *Industrial & Engineering Chemistry Research*, **40**, 4324-4335 (2001)

4. Y. N. Wang, Y. W. Li, L. Bai, Y. L. Zhao, and B. J. Zhang, Correlation for Gas-Liquid Equilibrium Prediction in Fischer-Tropsch Synthesis, *Fuel*, **78**, 911-917 (1999)

5. Yi-Ning Wang *et al.*, Kinetic Modelling of Fischer-Tropsch Synthesis Over an Industrial Fe-Cu-K Catalyst, *Fuel*, **82**, 195-213 (2003)

362621 Decontamination of Opaque Fluids Using Heterogeneous Photo Catalysis: Critical Role of Mobile Electron Holes on Semiconductor Surfaces

Abstract:

Heterogeneous photocatalysis is a technique to destroy pollutants in water solutions with a big promise, using a semiconductor and near UV-irradiation (1). However, decontamination of opaque fluids using photocatalysts and near UV irradiation involves major technical challenges, mainly because the resistance for the light transmission in a slurry, and then it is necessary to use the catalyst attached to a surface. This study considers a thin TiO_2 layer placed inside and in the bottom of a CREC-Photoreactor cell containing the opaque fluid. In this new setup, an external UV-lamp was placed below the unit, facing the interface support-TiO_2 film and the TiO_2 film received irradiation before photons could reach the opaque fluid. This new photoreactor cell is employed for the photocatalytic degradation of malic and malonic acids, typical apple juice components (2). The reactor setup was run for 30 minutes in the dark in order to reach absorption equilibrium on the TiO_2layer surface. Once adsorption equilibrium was reached, the lamp was turned on and the photocatalytic degradation was initiated. Samples were taken every hour for 15 hours to analyze the change in malic and malonic acid concentrations during the photocatalytic conversion, and were analyzed using HPLC and TOC.

Conversion of organic species can only proceed through the "dark side" of the TiO_2 layer, which is in direct contact with the fluid. Under the selected operating conditions both external mass transfer limitations and photolysis are found to be negligible. It was found that the relatively high fluid circulation in the near thin film region, estimated at 24 and 47 cm/s respectively, promotes high mixing and negligible external mass transfer on the overall photoconversion rate. Macroscopic radiation balances shows that 92% of near UV radiation is absorbed by the 'back side" of the TiO_2-film. Photocatalytic degradation experiments with 10, 20, 30 and 40 ppm malic acid initial concentrations, show that malonic acid is a main intermediate. Complete malic acid conversion occurs after 5-8 hours of irradiation. Kinetic modeling of malic and malonic acid photodegradation with kinetic parameter estimation is performed using both an "in series" and an "in series-parallel" reaction networks, represented by a set of differential equations of the Langmuir – Hinshelwood type (3). The "in series-parallel" reaction network displays better ability for predicting CO_2 formation, showing maximum quantum yields of 14.2%. According to literature, the required time for holes and electrons to reach the 1.3 10^{-6} m agglomerate surface as in the present study, is significantly $1.3x10^{-10}$s and $6.5x10^{-10}$s respectively. This is significantly smaller than the $8.9x10^{-5}$s expected charges lifetime as quoted by (4) Wilke, K. and Breuer, H.D (1999).

Malic acid photoconversion allows one to demonstrate that Langmuir-Hinshelwood kinetics in conjunction with adequate reaction pathways such as the "in series-parallel" (RN2) network are required for photocatalytic kinetic modeling. However, "in series-parallel" provides a better description of the CO_2formed.

Given that in the CREC-Photoreactor cell with a thin TiO_2 film, photocatalysis can only proceed via the transfer of mobile "h^+" sites from the irradiated side to the "dark side', this study demonstrates the significance of this step on the overall photocatalysis mechanism.

References.

(1) de Lasa H, Serrano B, Salaices M. Photocatalytic Reaction Engineering. Springer New York. 2005.

(2) Duffy S, Churey J, Worobo R, Schaffner DW. Analysis and modeling of the variability associated with UV inactivation of Escherichia coli in apple cider. J. Food Prot. 2000, 63, 1587-590.

(3) Moreira J, Serrano B, Ortiz A, de Lasa H. A unified kinetic model for phenol photocatalytic degradation over TiO2 photocatalysts. Chem. Eng. Sci. 2012, 78, 186-203.

(4) Wilke K, Breuer H D, The influence of transition metal doping on the physica and photocatalytic properties of titania, J. of Photochemistry and photobiology, A, Chemistry, 1999, 121, 49 – 53.

Photo-assisted CO and NO redox reaction over CuO/CeO_2-TiO_2 Catalysts

Johannes W. Schwank[1], Nan Liu[1,2], Xiaoyin Chen[1], Jinli Zhang[2]
[1]*Department of Chemical Engineering, University of Michigan, Ann Arbor, MI 48109, USA*
[2]*Department of Chemical Engineering, Tianjin University, Tianjin 300072, China*
E-mail of corresponding author: schwank@umich.edu

Reduction of NO_x with CO represents an effective and economical technique for simultaneous removal of NO_x and CO. The thermal catalytic NO and CO redox reaction requires relatively high temperatures up to 450 °C. Here, we investigate the photo-assisted catalytic NO and CO redox reaction over CuO/CeO_2-TiO_2 catalyst at lower temperature via in-situ DRIFT (Diffuse Reflectance Infrared Fourier Transform spectroscopy). Enhanced CO conversion (64.4%) under UV irradiation was achieved at 65 °C, compared with 11.5% in absence of UV light. Thermal catalysis became dominating at higher temperatures (200 °C). A promoting effect of ceria on NO reduction with CO has been revealed and rationalized on the basis of physical and chemical characterization results. XPS and TPR results showed that the electronic state interaction between Ce and the other two components (Ti and Cu) in the catalysts induced more efficient oxygen transport from ceria to copper species and resulted in higher oxygen vacancy concentration that leads to higher catalytic activity for the NO and CO reaction.

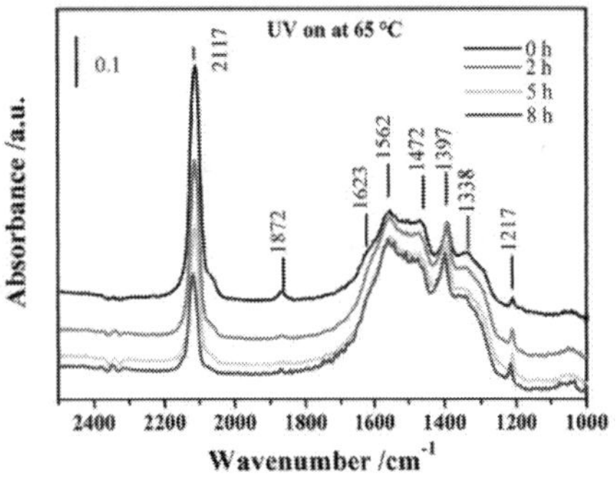

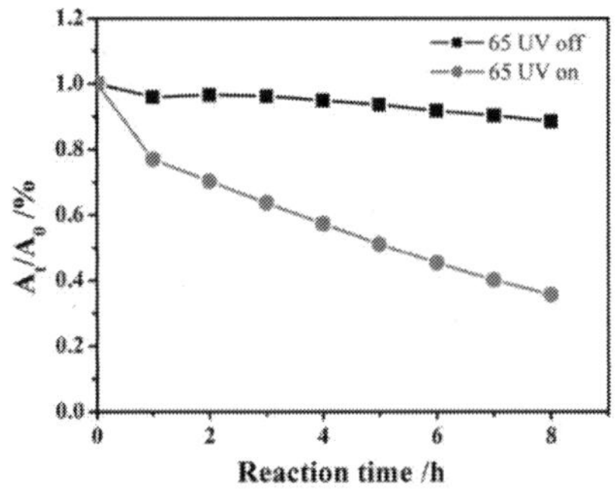

378905 Hydrogen Via Heterogeneous Photocatalysis with Binary Promoted TiO2

Abstract:

The need for renewable energy focuses attention on hydrogen obtained by the sustainable and green methods. The new family of the binary promoted photocatalysts N-M-TiO_2 (N = nitrogen, M = none, Cr, Co, Ni, Cu) have been synthesized via sol-gel method followed by drying and calcination. The binary promoted photocatalysts have been characterized by Raman spectroscopy, BET, UV-Vis Diffuse Reflectance Spectroscopy (UV-Vis DRS), and photoluminescence spectroscopy. The synthesis of the N-M-TiO_2 photocatalysts via hydrolysis of Ti isopropoxide at pH>11 in the presence of nitrogen precursor yields the non-crystalline solids that do not produce any hydrogen, while hydrolysis at the neutral pH yields the photocatalytically active materials containing nanocrystalline anatase. The N-M-TiO_2 photocatalysts show absorption bands at 370-450 nm (due to N promoter) and 500-800 nm (due to M promoter). The photocatalysts have been tested for the photocatalytic production of hydrogen from water with and without sustainable sacrificial donor glycerol, under ambient conditions, and near-UV/visible light. The N-Ni-TiO_2 is 5-fold more effective than the benchmark P25 TiO_2, while N-Cu-TiO_2 is 44-fold more active. The photocatalytic activity of the N-M-TiO2 increases from Cr to Co to Ni and Cu, while the photoluminescence decreases; the change in activity is due to the modulation of charge recombination. Certain N-M-TiO2 can be prepared in-situ by photochemical reduction of the precursors, and the activity of the in-situ synthesized N-M-TiO2 was determined as the function of experimental conditions and the amount of the deposited promoter element. Acknowledgments: A.S. thanks Research Corporation for Science Advancement (RCSA) for his Cottrell College Science grant, and Rutgers University for his Research Council Award.

Photocatalytic application of SiO_2/TiO_2 Thin Film on Glass

Benjawan Moongraksathum and Yu-Wen Chen,*

Department of Chemical engineering, National Central University, Jhong-Li, 32001, Taiwan

**Fax number: +886 3 4252296, E-mail: ywchen@ncu.edu.tw*

Abstract: SiO_2-TiO_2 neutral sols were prepared by peroxo sol-gel method. They were coated on the glass substrate. The characteristics of the SiO_2/TiO_2 hydrosol and film were characterized by SEM, TEM, HRTEM, XPS and UV-Vis. SiO_2/TiO_2 film with Si/Ti atomic ratio of 5 had the highest activity based on the same amount of TiO_2 and high antireflection property.

Keywords: Titania, Silica, Photocatalyst, Antireflection.

1. Introduction

Titania has been extensively used as a photocatalyst. However, when TiO_2 is coated on glass, the light transmission is not high enough. This study aimed to investigate the synthesis of SiO_2/TiO_2 hydrosols with the different compositions of SiO_2, which possess high light transmittance and high photocatalytic activity simultaneously. The presence of SiO_2 in the films could inhibit the diffusion of sodium cations from the glass substrate during heat treatment process[1]. It also has antireflection and antiglare properties.

2. Experimental

$TiCl_4$ was slowly added into 1 M HCl aqueous solution under stirring, and kept in ice bath. NH_4OH was then added slowly to form $Ti(OH)_4$. The pH value of the solution was adjusted to ~9. After aging and stirring for a period of time, $Ti(OH)_4$ gel was filtered and washed with distilled water several times until no chloride ion was detected. The gel was dispersed in distilled water. 30 wt% H_2O_2 was added to the solution to oxidize the gel to form titanium peroxide. The resultant solution was heated at 95 °C for 4 h. SiO_2 hydrosol was added into the TiO_2 sol during heating. The SiO_2/TiO_2 sol was then dip-coated on glass substrate several times. The effect of SiO_2 migration barrier on the sodium migration from soda-lime substrate was also investigated. The sample was denoted as ST(x) where x is Si/Ti atomic ratio. For comparison, the conventional multilayer coating of SiO_2 and TiO_2 was also prepared, and denoted as S/T film. The samples were characterized by XRD, SEM, TEM, HR-TEM, UV-vis and XPS. Methylene blue destruction was also tested under UVC and UVA light irradiation.

3. Results and discussion

The TiO_2 and SiO_2/TiO_2 hydrosols were prepared by peroxo sol-gel method[2] by using $TiCl_4$ as precursor. It was found that the as-prepared TiO_2 sol was stable light-yellow transparent sol which contained the rhombus shape of TiO_2 nanoparticles dispersed in the sol. In addition, the as-prepared SiO_2/TiO_2 sols with SiO_2:TiO_2 mole ratio of 3:1 and 5:1, denoted as ST(3) and ST(5), respectively, could be observed that the TiO_2 nanoparticles have been found to embed deeply in the spherical SiO_2 particles. This type of combination possesses some better advantages than the others due to the efficient reversible absorption of target organic substances and short traveling distance from the adsorbed site to the active[3]. For antireflection study, the films were prepared by dip-coating in order to obtain the thickness of about 1 μm, and It was found that the transmittance increased along with the presence of silicon dioxide and could enhance the light transmittance up to 4% in case of ST(5) film when compared to the bare glass substrate, whereas the TiO_2 film decreased at least 5% of transmittance (see Fig. 1). Prior to study the effect of the sodium migration from substrate into the film layer, the as-prepared films were treated at 500 °C for 1 h, it could observe that ST(5) film could be detected a small amount of 0.04 atomic % sodium content on the surface, whereas the multilayer coating showed the content of sodium up to 6.57 atomic % but less that on the bare TiO_2. For the photocatalytic degradation of methylene blue aqueous solution, the TiO_2 film represented the highest activity (see Fig. 2). However, based on the same amount of TiO_2 species existed in the films, the presence of SiO_2 in the film could enhance the photocatalytic activity. The ST(5) film was also active to destruct methylene blue under UVA illumination (365 nm).

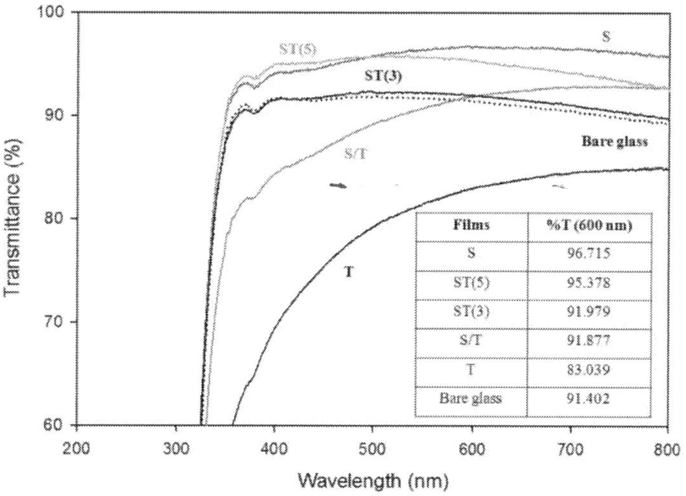

Films	%T (600 nm)
S	96.715
ST(5)	95.378
ST(3)	91.979
S/T	91.877
T	83.039
Bare glass	91.402

Figure 1. Transmittance (%) of the films.

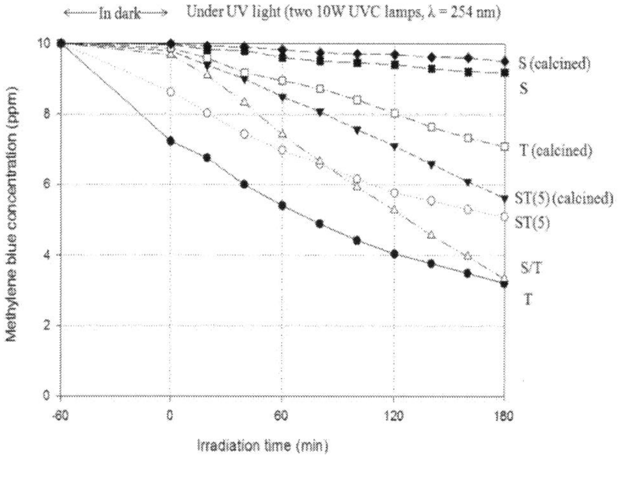

Figure 2. Photocatalytic degradation of methylene blue of various films.

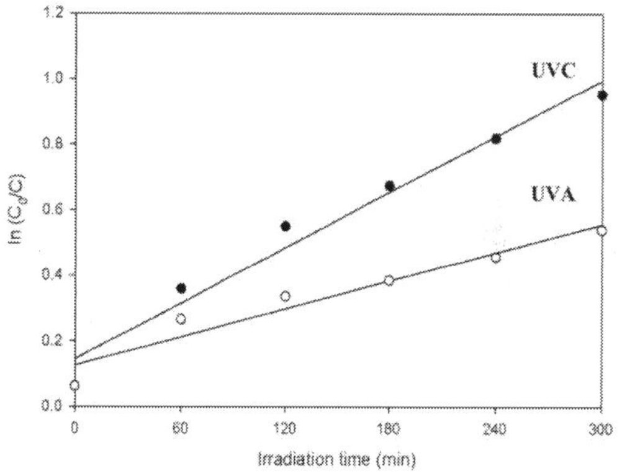

Figure 3. Photocatalytic activity of ST(5) film under different UV sources.

4. Conclusions

Bi-functional TiO_2-SiO_2 material of antireflection and degradation of organic substance were prepared by using peroxo sol-gel method with the preparation condition of prolonging heating the sol at 95 ^{0}C for 4 h. This can be the alternative way to produce the film coating on substrate which possesses high light transmittance, prevents the sodium diffusion from substrate to film and preserves high photocatalytic activity simultaneously.

References

1. H.J. Nam, T. Amemiya, M. Murabayashi, K. Itoh, J. Phys. Chem. B 108 (2004) 8254.
2. N. Sasirekha, B. Rajesh, Y.W. Chen, Thin Solid Films 518 (2009) 43.
3. G. Degan, S. Sampath, O. Lev, Chem. Mater. 7 (1995) 446.

Photocatalysis, where solar energy provides activation energy can in principle contribute to offsetting some of the demand on fossil fuels compared to conventional catalysis. Major efficiency improvements are still needed, however, before photocatalysis becomes a practical and economically viable technology (1). Previous work has shown that photoactivity occurs and even increases at high temperatures (2). This work investigates two new photocatalysis phenomena that potentially occur at high temperature with the deposition of Au nanoparticles on the surface of TiO_2. Both phenomena involve the migration of photo-generated electrons to the unfilled bands of deposited Au nanoparticles according to a well-known mechanism (3). First, this decreases the rate of electron-hole recombination in TiO_2, which could increase its photocatalytic activity at high temperature. Second, this could cause an electrochemical promotion of the Au catalytic activity. The photoactivities of TiO_2 and Au/TiO_2 catalysts prepared by flame-spray pyrolysis and photodeposition were evaluated by C_2H_4 oxidation experiments in a novel quartz plate reactor and in in-situ DRIFTS experiments. Results showed no photocatalytic promotion of Au, likely due to its relatively large average particle size of 11.6 nm, measured by STEM. However, a new photocatalytic reaction pathway was discovered on TiO_2 alone which had a maximum rate near 400°C, and which was enhanced by Au nanoparticles in the presence of H_2O. This new pathway may represent an exciting new type of photocatalysis, accessible at high temperature, and inspires future research to understand what other reactions may be catalyzed by its chemistry.

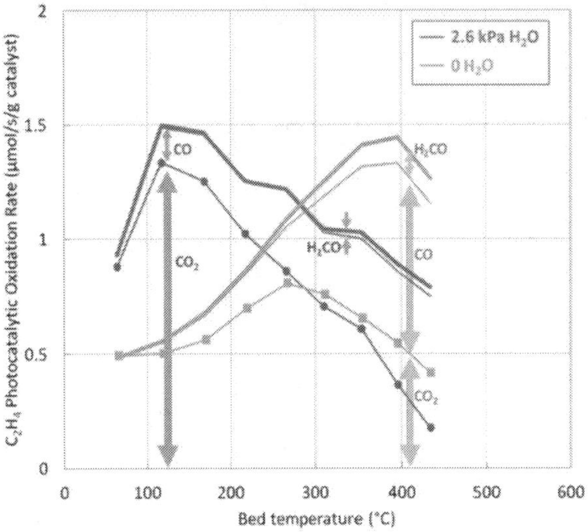

REFERENCES
1. H.H. Mohamed and D.W. Bahnemann, *Appl. Catal., B* 128 (2012) 91.
2. T.A. Westrich, K.A. Dahlberg, M. Kaviany, and J.W. Schwank, *J. Phys. Chem. C* 115 (2011) 16537.
3. T. Hirakawa and P.V. Kamat, *J. Am. Chem. Soc.* 127 (2005) 3928.

369976 Acid-Base Bi-Functional Catalysis for Deoxygenation during Cellulose Fast Pyrolysis

Abstract:

Catalytic fast pyrolysis of cellulose was performed in a micropyrolyzer coupled with online product analysis by GC-MS/FID. Acid, base, transitional metal oxide and acid-base bi-functional catalyst were evaluated relative to their deoxygenation performance. The bi-functional catalyst was prepared by acid treatment of a natural mixed metal oxide, serpentine. Temperature programmed desorption measurement was used to evaluate the acid and base properties of the catalytic materials. The natural serpentine was basic, while medium and strong acid sites were introduced through acid treatment. Using the same number of acid/base sites, all of the catalytic materials were used in pyrolysis for deoxygenation comparison. The results showed the bi-functional catalyst achieved the best balance between deoxygenation and bio-oil yield among the tested materials. For the best catalytic material, the oxygen content of the condensed product from cellulose was decreased by 27 % and its high heating value increased from 15.41 MJ kg^{-1} to 22.78 MJ kg^{-1}, with 60 % of the initial energy content remaining in the condensed product. It appeared that during catalysis, the acidic sites primarily promoted dehydration while the basic sites mainly promoted fragmentation reactions. Compared to a physical mixture of acid and base materials, the bi-functional catalyst appeared to promote deoxygenation reactions more effectively presumably due to the presence of adjacent acid and base sites.

Operando Molecular Spectroscopy during Catalytic Biomass Pyrolysis

Christopher J. Keturakis and Israel E. Wachs*

Operando Molecular Spectroscopy and Catalysis Laboratory, Chemical Engineering Department, Lehigh University, Bethlehem, PA 18015 USA

**iew0@lehigh.edu*

This work focuses on the molecular events taking place on the surface of solid acid catalysts during biomass pyrolysis. Although biomass pyrolysis is a topic of great interest such studies have only investigated pyrolysis product yield and bio-oil composition, little fundamental information has been reported about the molecular events taking place during the catalytic biomass pyrolysis reaction over solid acid mixed oxide catalysts: catalytic active sites, their activation and deactivation as well as promotion, surface reaction intermediates and surface reaction pathways. Popular HZSM-5 catalysts with different Si/Al ratios, representing different amounts of acidity, were studied for biomass pyrolysis. Additionally, model supported Al_2O_3/SiO_2 catalysts were synthesized, characterized (*in situ* Raman, *in situ* IR, and *ex situ* ^{27}Al NMR spectroscopy), and studied for biomass pyrolysis because the structure and location of the AlO_x sites are known and easily controllable, allowing for a comparison between HZSM-5 sites and well-defined AlO_x sites.

In order to elucidate the molecular events taking place on the catalyst surface, cutting edge *in situ* and *operando* spectroscopic characterization studies (Raman and IR) during the catalytic biomass pyrolysis reaction were undertaken with representative biomass monosaccharides (glucose, xylose, and arabinose), polysaccharides (cellulose and hemicelluloses xylan, arabinoxylan, and xyloglucan), and lignin. Time resolved spectroscopy and gas chromatography-mass spectrometry (GC/MS or MS) *operando* style experiments during catalytic biomass pyrolysis provide a direct correlation between catalyst structure, surface reaction intermediates, and pyrolysis products. Insights into the pyrolysis mechanism, catalytic active sites, structure-activity/selectivity relationships, and catalyst deactivation will be discussed based on these fundamental studies during pyrolysis with well-defined solid acid catalysts and model biomass compounds.

369569 Investigation of Primary and Secondary Reactions during the Pyrolysis of Different Celluloses

Abstract:

A micropyrolyzer/GC-MS/FID system was used to investigate the primary and secondary reactions of cellulose pyrolysis when sample weight, particle size, length scale, feedstock, crystallinity, and chain length were varied. Initial work involved estimating the heating rate of the sample to reaction temperature in the system. The sample heating rate was found to be an order of magnitude lower than pyrolysis reaction time. It was also found that with larger samples mass transfer limitations existed, leading to secondary reactions including the formation of secondary char and gas, as well as the dehydration/decomposition of levoglucosan to low molecular weight products, furans, dehydrated pyranose, and char. The char from cellulose pyrolysis was found to catalyze the secondary reactions. Using a system that minimized transport limitations, celluloses of differing crystallinity, chain length, and feedstock source were pyrolyzed. Similar product distributions were found for each sample, indicating that the formation of primary products was not influenced by crystallinity, chain length, or feedstock.

Spatiotemporally Resolved Diffuse Reflectance in situ Spectroscopy (DRiSP) of Pyrolyzing Wood Particles

Alex D. Paulsen, Blake R. Hough, C. Luke Williams, Andrew R. Teixeira, Daniel T. Schwartz, Jim Pfaendtner, and Paul J. Dauenhauer

Fast pyrolysis of biomass has potential for placing gasoline, diesel, and chemicals with renewable liquids derived from renewable sources [1]. However, pyrolysis of biomass requires significant optimization to achieve commercial viability, and technical details of the pyrolysis process remain unknown [2,3] before pyrolysis can contribute to the existing oil-based transportation system. One method of process development utilized detailed kinetic reaction models for predicting the distribution of chemicals derived from biomass fast pyrolysis, which provide the foundation for design of chemical reactors to targeted products. In this work, the goal is development of a detailed reaction-transport model to improve upon current descriptions of fast pyrolysis. A new technique is proposed [4], called SpatioTemporally-Resolved - Diffuse Reflectance in situ Spectroscopy of Particles (STR-DRiSP), which can measure the carbohydrate composition of reacting biomass particles with high spatial (10 micron) and temporal (1 millisecond) resolution. Compositional profiles of reacting particles were utilized in development of a comprehensive two-dimensional single particle model, comprised of a multi-step, semi-global reaction mechanism, prescribed particle shrinkage, and thermophysical properties dependent on temperature, composition and particle orientation.

[1] M.S. Mettler, D.G. Vlachos, P.J. Dauenhauer, "Top Ten Fundamental Challenges of Biomass Pyrolysis,"*Energy & Environmental Science* **2012**, 5, 7797-7809.
[2] M.S. Metter, A.D. Paulsen, D.G. Vlachos, P.J. Dauenhauer, "The Chain Length Effect in Pyrolysis: Bridging the Gap between Glucose and Cellulose," 2012. *Green Chemistry* **2012**, 14, 1284-1288.
[3] M.S. Mettler, A.D. Paulsen, D.G. Vlachos, P.J. Dauenhauer, "Pyrolytic Conversion of Cellulose to Fuels: Levoglucosan Deoxygenation via Elimination and Cyclization within Molten Biomass," *Energy & Environmental Science* **2012**, 5, 7864-7868.
[4] Alex D. Paulsen, Blake R. Hough, C. Luke Williams, Andrew R. Teixeira, Daniel T. Schwartz, Jim Pfaendtner, Paul J. Dauenhauer, "Fast Pyrolysis of Wood for Biofuels: Spatiotemporally Resolved Diffuse Reflectance in situ Spectroscopy of Particles," ChemSusChem **2014**, 7(3), 765-776.

391095 Biomass decomposition reaction sequence analysis using simultaneous gas sampling and temperature measurements

Abstract:

Understanding the reaction sequence and associated kinetics are imperative for industrial biomass conversion applications. One of the lesser known aspects is the actual energy change happening during the conversion process within the biomass sample. Many literature studies reveal different temperature profiles developing during the process mainly without measuring the gaseous quantification of volatiles released inside the biomass sample. The use of gas component detection methods in concert with thermal analysis can help in better understanding the reaction sequence and mechanisms governing the decomposition behavior. The aim of this paper is to present an innovative intra-particle gas sampling technique that provides insight on the reaction sequence and energy changes inside biomass particle. To our knowledge, this is the first intra-particle gas sampling measurement performed.

The experimental tests have been performed on wooden spheres ranging in diameter from 19.0 mm to 31.8 mm at operating temperatures of 500°C, in a tubular reactor under non-oxidant controlled conditions. Simultaneous measurements were obtained for product gas evolution and temperature during the process. In addition the behavior of biomass samples during devolatilization was determined by thermal analysis: thermogravimetric (TG) and differential scanning calorimetry (DSC).

The TG analysis revealed a rapid mass loss in the temperature range 290°C to 410°C and a total mass loss of 75% - 76%. The DSC analysis showed exothermic behavior above 305°C and two maximum peaks at 350°C and 413°C. Evolved gases such as H_2, CO_x and hydrocarbons were extracted through a probe using a micro sample system. The measurements showed a simultaneous release of CO_2 and CO with a maximum occurring around 325°C, for larger samples. The CO release, while lower in absolute concentration, is much more stable during the entire heating of the particle. The measurements showed a pronounced transition between hydrogen release and subsequent hydrocarbons release. The light hydrocarbons such as methane and ethane were observed in high quantities, 19% and 0.9% respectively. It is anticipated this data will help modeling thermochemical processes ultimately leading to predictive capability.

386818 Catalytic Pyrolysis of Char and Coke Precursors: An Investigation of the Formation Mechanisms of Char and Coke

Abstract:

Catalytic pyrolysis of char and coke precursors: an investigation of the formation mechanisms of char and coke

Shoucheng Du, David P. Gamliel, Julia A. Valla, George M. Bollas

Department of Chemical & Biomolecular Engineering, University of Connecticut, Storrs, CT

Abstract

Fast catalytic pyrolysis of lignocellulosic biomass, in the presence of proper pore structure zeolite catalysts, such as ZSM-5, has proven to provide enhanced bio-oil selectivity to aromatic hydrocarbons. [1-3] However, not all carbon content in biomass is converted to bio-oil, due to the production of carbon oxides (CO and CO_2), olefins and solid residue (char and coke). The production of solid carbonaceous residue not only results in the loss of carbon from the product bio-oil, but also causes catalyst deactivation, which largely affects the product distribution and bio-oil selectivity. Control of coke and char formation during pyrolysis could be possible through innovative catalyst and process designs, in which the fundamental understanding of the formation mechanisms of char and coke should be viewed as a prerequisite. The latter requires insights into coke and char structural characteristics and formation chemistry. Brewer et al. [4] proposed a structure of char formed by slow and fast thermal pyrolysis of switch grass using ^{13}C-NMR analysis. They showed that aromatic clusters of 7-8 rings terminated by carbonyl and hydroxyl groups are representative of the composition of thermal char. Valle et al. [5] studied catalytic upgrading of bio-oil with methanol over ZSM-5. They showed two origins of coke, thermal and catalytic, by performing temperature programmed oxidation (TPO). They also observed that catalytic coke is deposited mainly inside the zeolite crystal channels; whereas thermal coke is deposited mainly outside the zeolite crystals. Cheng and Huber [6] studied catalytic pyrolysis of furan over ZSM-5. They found the soluble coke mostly consists of aromatic rings and carbonyl groups. They also identified the molecular weight distribution of soluble coke by using gel permeation chromatography (GPC) and concluded that the maximum molecular weight of soluble coke is beyond the limitation of a GPC column (M_w >10^4).

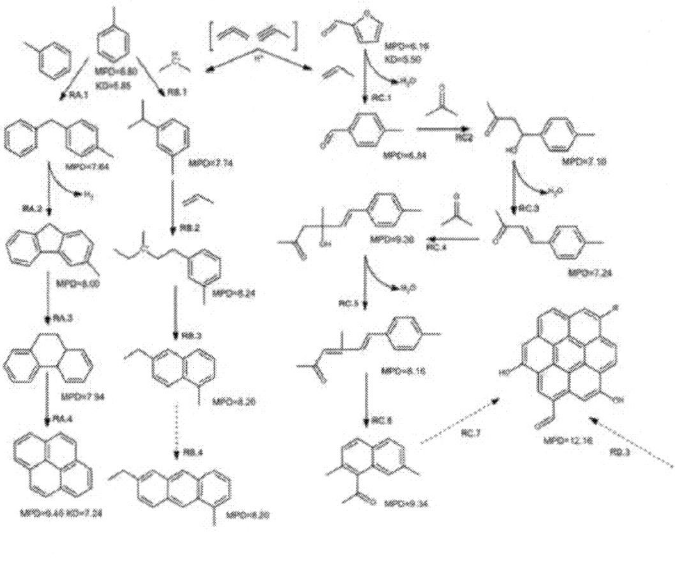

Figure 1 Some reaction pathways for coke and char formation. (a) RA ¨C toluene self-alkylation via RA.1 (alkylation), RA.2 (dehydrogenative coupling), RA.3 (iso- merization), and RA.4 (hydrogen transfer and repetition of RA.1¨CRA.5); (b) RB ¨C coke formation via RB.1 (alkylation on the nucleus with carbenium ions), RB.2 (side alkylation and isomerization), RB.3 (cyclization), and RB.4 (repetition of RB.2, RB.3); (c) RC ¨C char formation from furfural via RC.1 (Diels¨CAlder with propylene), and aldol condensations (RC.2¨CRC.7)

Research efforts have focused on understanding the reaction mechanisms of biomass catalytic pyrolysis, and those of coke and char formation. Figure 1 illustrates some pathways (RA, RB and RC) for the formation of char and coke as identified previously [7]. For instance, when toluene is chosen as the precursor, coke forms under self-alkylation reactions [8] and/or side alkylation with propylene [9]. Starting from furfural, formation of char may undergo Diels-Alder reactions to form tolualdehyde, followed by a series of Aldol condensation steps.

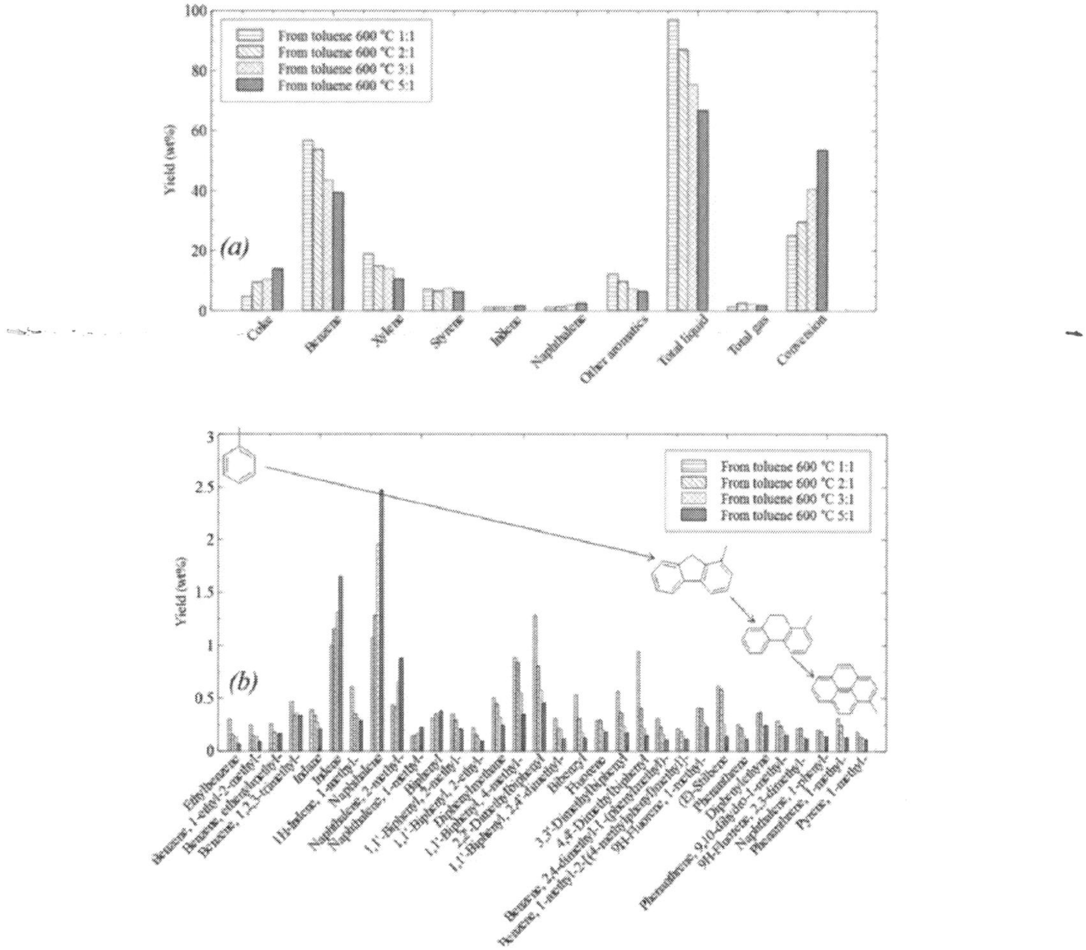

Figure 2 Major (a) and secondary (b) product distributions from catalytic pyrolysis of toluene at 600 °C, 1-5 catalyst to toluene ratios

An effective way to study reaction mechanisms is the employment of model compounds [6,10"C13]. In this study, toluene and tolualdehyde are used as model compounds for the identification of reaction pathways, responsible for coke and char formation, and the role of the catalyst. Figure 2 shows the major product distribution (a) and more detailed liquid product selectivity (b) from the catalytic pyrolysis of toluene at 600 °C and varying catalyst to feed ratio. Specifically, as the catalyst to toluene ratio increases, the coke yield increases, whereas the yields to benzene, xylene and other aromatics, such as 1-methyl-9H-fluorene and 9, 10-dihydro-1-methyl-phenanthrene, decrease. This suggests that the aforementioned liquid products act as intermediates for the formation of coke during catalytic pyrolysis, which is consistent with the previously proposed mechanisms (RA). This also shows that the production of aromatics is competitive to coke and char formation. Moreover, the yield of 1-methyl-pyrene decreases with increasing coke yield, which implies that 1-methyl-pyrene is the largest coke precursor detectable by GC-MS. Furthermore, elemental analysis of the coke/char from catalytic pyrolysis of tolualdehyde with acetone under different catalyst to feed ratios show that as the catalyst to feed ratio increases, catalytic pyrolysis of tolualdehyde with acetone produces coke/char of higher H/C ratios. This translates to less fused aromatics with more side chains [4,7], or generally less condensed aromatic structures in the coke/char.

As indicated, liquid product analysis reflects the reaction intermediates in coke/char formation and the elemental analysis implies the structure of final coke/char products. Combining these analyses with other techniques, such as Raman spectroscopy and solid-state NMR, more detailed information on the reaction pathways towards coke and char is revealed. This allows for a comprehensive identification of the most significant pathways to coke and char in biomass catalytic pyrolysis. It also confirms that aromatics selectivity in biomass catalytic pyrolysis is directly competitive to coke and char formation.

Acknowledgement

The study is funded by the National Science Foundation Award CBET-1236738.

Reference

[1] J. Jae, G.A. Tompsett, A.J. Foster, K.D. Hammond, S.M. Auerbach, R.F. Lobo, G.W. Huber, J. Catal. 279 (2011) 257.

[2] A. Aho, N. Kumar, K. Eranen, T. Salmi, M. Hupa, D. Murzin, Fuel 87 (2008) 2493.

[3] S.D. Stefanidis, K.G. Kalogiannis, E.F. Iliopoulou, a a Lappas, P. a Pilavachi, Bioresour. Technol. 102 (2011) 8261.

[4] C.E. Brewer, K. Schmidt-rohr, J.A. Satrio, R.C. Brown, 28 (2009).

[5] B. Valle, P. Castaño, M. Olazar, J. Bilbao, A.G. Gayubo, J. Catal. 285 (2012) 304.

[6] Y.-T. Cheng, G.W. Huber, ACS Catal. 1 (2011) 611.

[7] S. Du, J.A. Valla, G.M. Bollas, Green Chem. 15 (2013) 3214.

[8] H.S. Cerqueira, G. Caeiro, L. Costa, F. Ramôa Ribeiro, J. Mol. Catal. A Chem. 292 (2008) 1.
[9] R. Quintana-Sol¨®rzano, J.W. Thybaut, G.B. Marin, R. Lødeng, A. Holmen, Catal. Today 107-108 (2005) 619.
[10] L. Hora, V. Kelbichov¨¢, O. Kikhtyanin, O. Bortnovskiy, D. Kubička, Catal. Today 223 (2014) 138.
[11] S. Mani, J.R. Kastner, A. Juneja, Fuel Process. Technol. 114 (2013) 118.
[12] F. Lannuzel, R. Bounaceur, R. Michels, G. Scacchi, P.-M. Marquaire, J. Anal. Appl. Pyrolysis 87 (2010) 236.
[13] T. Odedairo, R.J. Balasamy, S. Al-Khattaf, J. Mol. Catal. A Chem. 345 (2011) 21.

389282 A Partial-Combustion Pyrolysis Model for an Energy + Biochar Reactor Design

Abstract:

The simultaneous production of biochar and thermal energy through slow pyrolysis is a promising process for efficient biomass utilization. Partial combustion allows the reactor to be energy self-sufficient. The addition of oxygen (air) to the pyrolyzer, however, alters the reactor conditions, thus affecting the pyrolysis reaction kinetics and, ultimately, the biochar properties. The goals of this study are to develop a partial combustion reaction model to allow the design of a continuous, energy self-sufficient slow pyrolysis system that produces appreciable amounts of biochar for soil application, and thermal and electrical energy to operate a water desalination unit. This water desalination unit will use multiple effect distillation (MED), a thermal desalination method, to treat brackish groundwater. The overall system will be used in rural to convert agricultural residues to water suitable for irrigation or human consumption, and biochar to improve agricultural soil fertility and water holding capacity.

An auger reactor will be used for the pyrolysis and partial combustion of biomass. Often, such reactors are characterized by (a) biomass residence times that may be two to three orders of magnitude longer than the residence times of gas, and (b) very high solid to gas mass ratios. These conditions allow for significant simplification of the equations governing the transient operation of the reactor. A network of parallel and sequential reactions was used to describe the kinetics of biomass pyrolysis. Since the feed gas to the reactor will also contain some oxygen, partial combustion of the produced char and volatiles was also considered to provide the energy needed for the desalination process. The kinetic constants of all reactions were estimated using thermogravimetry and differential scanning calorimetry. Assuming plug flow for the solid and gas phases, transient mass balances were developed for all the gaseous and solid species, using one-dimensional partial differential equations for the solids and two-dimensional equations with convection and dispersion terms for the gases. Finally, two-dimensional PDEs were used to describe the transient energy balances for the solid and gas phases, with heat exchange through the reactor wall. The system of coupled equations describing the operation of the reactor was solved using finite differences to determine the key parameters for optimal operation of our system: biomass feed rate and residence time, maximum pyrolysis temperature and fraction of air in the gas feed. A parametric study was also carried out to determine how feedstock properties (composition, particle size, moisture content etc.) affect the net energy ratio of our reactor, as well as the amount and the properties of produced biochar. The net energy ratio of the reactor was defined as: (Enthalpy of exit gases + Energy content of produced biochar) / (Energy content of biomass + Energy used to heat the reactor).

The selection of polymorphic metal oxides as candidate materials for applications in heterogeneous catalysis depends strongly on the accurate prediction of polymorph phase stability under particular synthesis reaction conditions. By changing the reaction conditions under which polymorphic materials are synthesized, the relative energetic stabilities of polymorph structures can be altered, causing metastable material phases to become thermodynamically stable. In cases where metastable phases are energetically similar to stable phases, experimental procedures such as epitaxial thin film stabilization or moderately changing synthesis temperature can be employed to access these phases without the costly application of high pressure. When subjected to high temperatures or pressures in compression and decompression experimental procedures, metastable phases within energetic windows of epitaxial stabilization can observe phase transitions that differ from those observed by stable phases either thermodynamically or kinetically.

Prediction of phase stability and transition pressures is particularly important in the context of photocatalysis, as the anatase polymorph of TiO_2 is the most reactive active catalytic material of commonly studied TiO_2 polymorphs that exists at ambient pressure in most experimental setups. However, TiO_2 polymorphs that are accessible only at higher pressures, such as α-PbO_2 and fluorite, can be suitable active photocatalyst candidates in mixed-metal oxides, while polymorphs that are accessible under different synthesis procedures performed at ambient conditions, such as rutile and brookite, can serve as supports for metal catalysts that determine metal surface orientation. In the case of a single polymorph composition placed under limited reaction conditions, phase transitions can be affordably characterized by solely experimental research, leading to the discovery of the least costly conditions under which desired high pressure polymorphs can be synthesized. However, when considering a large set of BO_2 polymorph compositions (B = Ti, V, Ru, and Ir) over a wide range of temperature and pressure conditions, the experimental discovery of all possible pressure-induced phase transitions for the stabilized phases of these BO_2 is intractable. Thus, a first-principles methodology capable of accurately and efficiently predicting the reaction condition dependent relative energetic stabilities of and phase transition pressures between polymorphic materials must be developed. Given that the energetics of 3*d* metal containing oxides can be computed inaccurately by standard DFT GGA functionals, the DFT+U method and self-consistent implementation of the linear response method for determining Hubbard U values are employed.

In this work, the effects of incrementally changing the U value on the 3*d* orbitals of Ti cations in TiO_2 polymorphs is initially inspected, revealing that the prediction of relative energetic stability ordering of the Rutile, Anatase, and Columbite polymorphs is consistent with experiment in the approximate range U = 2.8 – 4.2 eV. Previous first principles studies have resolved self-consistent linear response U values for the Rutile and Anatase polymorph Ti cations of 3.23 and 3.25 eV, respectively. Applying the average of these two theoretically calculated U values to the Rutile, Anatase, and Columbite polymorphs successfully predicts their experimentally observed energetic stability ordering. Imposing a U parameter of 3.24 eV on these three polymorphs also leads to the almost exact prediction of two of the three phase transition pressures involving them, namely the ambient pressure transition from Anatase to Rutile and the transition from Rutile to Columbite at 10.0 GPa. The transition from Anatase to Columbite, which occurs at 2.6 GPa in experiment, is moderately underestimated by first principles calculations at 1.4 GPa. Given the reproducibility of experimental results in bulk TiO_2 polymorphs, relative energetic stability and phase transition calculations are applied to bulk VO_2, RuO_2, and IrO_2 polymorph sets, while calculations are repeated on these polymorph sets using different pseudopotentials to ensure that predictions are not significantly affected by pseudopotential selection. Additional Hubbard U values applied to the resolution of energetic results calculated via self-consistent linear response theory are performed for comparative purposes.

378272 A DFT+U and Ab Initio Thermodynamics Approach to Hydrocarbon Oxidation Mechanism Determination on Oxide Catalysts: Propane Oxidation on Zr-Doped CeO2

Abstract:

Elucidation of the elementary reaction processes involved in hydrocarbon oxidation on oxide catalysts can help guide active site optimization. This talk will discuss the difficulties and approaches developed for considering the complex elementary mechanism of propane oxidation over metal oxide catalysts. We use density functional theory (DFT+U) methods to examine the reforming of propane over the Zr-doped CeO_2 (1 1 1) surface. Numerous modeling and mechanistic questions arise in modeling this multistep reaction on oxides. The surface redox and coverage state in the reaction environment impacts energetics and must be considered. Phase diagrams of Zr-doped and pure CeO_2 (1 1 1) are created based upon the partial pressure of oxygen and hydrogen, demonstrating that the surface will be reduced under operating conditions. All elementary energetics along the preferential path for propane reforming were identified, and differences in path with oxygen pressure variance were identified. The reaction path varies depending on the oxygen chemical potential, as this alters at which step in the mechanism the surface will re-oxidize. The use of the stoichiometric oxide surface as the reactive surface for propane oxidation is found to misrepresent the reaction energetics and path, as the stable surface under propane reforming conditions is highly reduced.

379749 Benchmarking Modern Range Separated DFT Functionals and *Ab Initio* Wavefunction Theory-in-DFT Embedding for Computational Catalysis Applications

Abstract:

Computational quantum chemistry theories facilitate scientific discovery by enabling first-principles predictions of physical phenomena. Of course, the quality of these predictions hinges on the theory capturing the system's underlying physics. Conventional GGA- and (to a lesser extent) hybrid exchange-correlation (XC) functionals remain the workhorse approaches for elucidating catalytic properties thanks to their favorable balance of accuracy and computational cost. While commonly used XC functionals have been developed about two or more decades ago, recent developments in quantum chemistry have afforded new methods that may be of interest to the catalysis community but have not yet been widely adapted. We present a comparative benchmark study to show how different methods treat energy profiles of common reaction steps in catalysis (beta-hydride eliminations, reductive coupling, etc.) This presentation shows the relative importance of different treatments of dispersion and self-interaction error found in XC functionals. We also illustrate the utility of Manby and Miller's *ab initio* Wavefunction Theory-in-DFT Embedding calculation schemes for benchmarking DFT methods and future outlook for first-principles modeling of homogeneous and heterogeneous catalysis.

381823 Contributions of Dispersion Forces to R-3-Methylcyclohexanone Physisorption on Low and High Miller Index Cu Surfaces

Abstract:

There has been a growing trend of modeling surface-adsorbate interaction with density functional theory (DFT); however, earlier calculations with nonhybrid functional failed to account for the van der Waals forces. This had led to a significant underestimation of total adsorption energy for both chemisorbed and physisorbed system. The development of dispersion corrected methods had attempted to address this deficiency and closed the gap between simulation and experiment. In this work, we examined the physisorption of R-3-Methycyclohexanone (R3MCHO) on low and high Miller index Cu surfaces with D2, vdW-optB86b, and vdW-optB88 dispersion corrected methods. The dispersion corrected adsorption energies were found to be more comparable to results obtained through temperature programmed desorption (TPD) than those calculated using nonhybrid functional. However, the corrected methods had shown a tendency to overbind the adsorbate. Finally, the calculations had indicated an unexpected potential adsorbate induced surface reconstruction on the Cu(110).

367533 What More Can U Do: Alternative Applications of DFT+U

Abstract:

DFT+U and related approaches remain a popular way to fix electron delocalization problems in practical exchange-correlation functional implementations of density functional theory. In DFT+U, standard XC functionals are augmented with a Hubbard model term that can successfully localize electrons in the ground state in molecules and open gaps in extended systems, producing insulators where common XC functionals might incorrectly predict metallic behavior. Nevertheless, ongoing challenges for DFT+U's ascendancy include the fact that the approach is often perceived and employed as an ad hoc, empirical correction. Here, we explain some alternative approaches wherein we not only calculate U from linear response of the system we are studying but we also motivate chemically-based renormalizations of linear-response calculations to obtain more broadly applicable values of U. We also demonstrate how the U can be viewed as a tool for tuning eigenvalues generally, even in cases with only 2p electrons present. In this case, we use a +U term to correct eigenvalues in closely spaced tautomeric states of molecules, thereby also correctin energetic spacing. Fundamentally, there's more that U can do and much of the efficacy of this approach lies in its ability to tune properties in a chemically-derived manner.

372653 Assessing the Reliability of Calculated Catalytic Ammonia Synthesis Rates with Beef-Vdw

Abstract:

Using a general method for estimating the exchange-correlation uncertainty in density functional theory calculations we estimate the uncertainty on catalytic rate of ammonia synthesis over a range of transition metal catalysts. In general the estimated error is significantly reduced due to correlations in the error between various intermediates and surfaces, highlighting the need for systematic methods of error estimation. Uncertainties are shown to be a complex function of reaction condition and catalyst material, and we demonstrate that relative rates between different catalysts are better described than the absolute rates. A new approach is introduced where catalysts can be screened based on the probability that their rate will be an improvement over a given standard, rather than on the basis of absolute predicted rates. The implications and challenges of analyzing sources of uncertainty beyound exchange-correltaion energy are discussed, along with the potential opportunities for application of the probabilistic approach to other areas of computational materials design.

Single-Step Delamination of a MWW Borosilicate Layered Zeolite Precursor under Mild Conditions without Surfactant and Sonication

*Xiaoying Ouyang [1], Son-Jong Hwang [2], Ron C. Runnebaum [1], Dan Xie [3], Ying-Jen Wanglee [1], Thomas Rea [3], Stacey I. Zones * [1, 3], and Alexander Katz * [1]*

[1] *Department of Chemical and Biomolecular Engineering, University of California at Berkeley, Berkeley, California 94720*

[2] *Department of Chemical Engineering, California Institute of Technology, Pasadena, California 91125*

[3] *Chevron Energy Technology Company, Richmond, California 94802*

Abstract

This presentation focuses on synthesis of delaminated Layered borosilicate zeolite precursor ERB-1P (Si/B = 11). Our approach lies on isomorphous substitution of Al for B using a simple aqueous $Al(NO_3)_3$ treatment. TEM demonstrates transformation of rectilinear layers in the precursor to single and curved layers in the delaminated material, and characterization by PXRD shows loss of long-range order. N_2 physisorption and base titration confirm the expected decrease of micropore volume and increase in external surface area for delaminated materials relative to their calcined 3D zeolite counterpart. FTIR and multinuclear NMR spectroscopies demonstrate synthesis of Brønsted acid sites upon delamination. Comparative synthetic studies demonstrate that this new delamination method requires (i) a borosilicate layered zeolite precursor, in which boron atoms can be isomorphously substituted by aluminum, (ii) neutral amine pore fillers instead of rigid and large quaternary amine SDAs, and (iii) careful temperature control, with the preferred temperature window being around 135 °C for ERB-1P delamination. Acylation of 2-methoxynaphthalene was used as a model reaction which is only catalyzed by the acid sites located near the external surface, instead of the acid sites located within the 10-MR micropores to investigate the catalytic benefits of delamination.[1,2] A partially dealuminated delaminated material (e.g. ERB-1-del-135) displays a 2.3-fold enhancement in its initial rate of catalysis relative to the 3D calcined material. This is nearly equal to its 2.5-fold measured increase in external surface area. Our simple, surfactant- and sonication-free, mild delamination method is expected to find broad synthetic utility.

(1) Andy, P.; Garcia-Martinez, J.; Lee, G.; Gonzalez, H.; Jones, C. W.; Davis, M. E. *J. Catal.* **2000**, *192*, 215-223.

(2) Ouyang, X.; Hwang, S.-J.; Runnebaum, R. C.; Xie, D.; Wanglee, Y.-J.; Rea, T.; Zones, S. I.; Katz, A. *J. Am. Chem. Soc.* **2013**, 136, 1449-1461.

388734 Strategies for the Encapsulation of Metal Clusters within Zeolites and Consequences for Catalytic Stability, Reactivity and Selectivity

Abstract:

The encapsulation of metal clusters within zeolite voids can protect such clusters from coalescence and growth and also from contact with feed impurities, while allowing reactions to benefit from the product and reactant shape selectivity conferred by the molecular size of zeolite windows and voids [1-3]. We have developed general protocols for the synthesis of metal clusters encapsulated within small and medium-pore zeolites and for the understanding of catalytic consequences of encapsulation [4-6]. The small apertures within these materials preclude post-synthetic encapsulation protocols, which involve solvated metal-oxo oligomers that cannot diffuse into these microporous solids. Therefore, we have developed encapsulation methods that use ligand-stabilized metal precursors that do not precipitate as colloidal oxyhydroxides at the high pH and temperatures required for hydrothermal zeolite crystallization. For zeolites requiring temperatures too high for stable ligated precursors, we have achieved encapsulation by first placing clusters within zeolites that form at milder conditions and then retaining encapsulation during their crystalline interconversion into the zeolite framework of interest. Specifically, these combined strategies have led to the successful encapsulation of a wide range of metal clusters (Pt, Pd, Ru, Rh, Ir, Re, and Ag) within small-pore (SOD, LTA, GIS, and ANA) and medium-pore (MFI) zeolites.

The success of encapsulation, the size uniformity and surface cleanliness of metal clusters were demonstrated by combining electron microscopy, chemisorptive titrations, and catalytic reactions of small and large molecules. Transmission electron microscopy showed that clusters were small (1.1-1.9 nm) and uniform in size, consistent with measurements by H_2 or O_2 chemisorptions (1.2-1.8 nm), indicating that clusters surfaces were accessible and clean and therefore suitable for catalysis on their surfaces. The encapsulated clusters did not change in size during oxidative thermal treatments at temperatures (573–873 K) that led to significant growth for clusters of comparable size and composition dispersed instead on mesoporous SiO_2. The preferential encapsulation of metal clusters within zeolite voids was confirmed by measuring the ratio of oxidative dehydrogenation (ODH) and hydrogenation rates for small (methanol, ethene and toluene) and large (isobutanol, isobutene and trimethyl benzene) reactants on unconstrained clusters dispersed on SiO_2 ($X_{SiO2} = r_{small\ reactant}/r_{large\ reactant}$) and on clusters of the same metal in zeolites ($\chi_{zeolite}$). The ratio of these relative reactivities denotes the encapsulation selectivity parameter ($\Phi = \chi_{zeolite}/\chi_{SiO2}$), indicative of the ratio of active surfaces within zeolite crystals to total exposed metal surface area. This encapsulation selectivity parameter would approach unity for clusters with unimpeded access to reactants, such as those at external zeolite surfaces. The encapsulation selectivity (Φ) values were 7-83 for all samples indicating the predominant presence of active sites within zeolite structures. The selective confinement of these clusters within microporous regions also protected cluster surfaces from contact with catalytic poisons, such as thiophene and H_2S, which cannot access such regions, and allowed ethene hydrogenation and H_2-D_2 exchange rates to proceed at rates unperturbed by titrants that fully suppressed reactivity on similar clusters dispersed on mesoporous SiO_2.

1. Weisz, P.; Frilette, V.; Maatman, R.; Mower, E., J. Catal. 1962, 1 (4), 307.

2. Wu, J. C. S.; Goodwin, J. G.; Davis, M., J. Catal. 1990, 125, 488.

3. Zhan, B.-Z.; Iglesia, E., Angew. Chem. Int. Ed. 2007, 46, 3697.

4. Choi, M.; Wu, Z.; Iglesia, E., J. Am. Chem. Soc. 2010, 132 (26), 9129.

5. Goel, S.; Wu, Z.; Zones, S. I.; Iglesia, E., J. Am. Chem. Soc. 2012, 134 (42), 17688.

6. Wu, Z.; Goel, S.; Choi, M.; Iglesia, E., J. Catal. 2014, 311, 458.

The authors gratefully acknowledge financial support for these studies from Chevron Energy Technology Company.

389929 Catalyst Development for Ultra-Selective Production of p-Xylene from Cycloaddition of Dimethylfuran

Abstract:

The need for sustainable production of everyday materials in addition to market volatility of petroleum-based feedstocks has motivated research into the production of renewable aromatic chemicals from biomass. Specific chemicals of interest include *p*-xylene, the feedstock for polyethylene terephthalate (PET). We have proposed a renewable method of producing renewable *p*-xylene by cycloaddition of biomass-derived dimethylfuran (DMF) and ethylene, which serves as the last step in a complete process for producing p-xylene from cellulose[1]. The reaction occurs by symmetry-allowed [4 + 2] Diels-Alder cycloaddition of ethylene and DMF and subsequent aromatization by acid-catalyzed dehydration.

In this work we highlight enhanced selectivity to *p*-xylene can be achieved by using microporous catalysts with large pore and tunable acidity. High selectivity to *p*-xylene (~90%) can be achieved when the conversion of DMF reaches >99%. It is shown that 2,5-hexanedione, formed by the hydrolysis of DMF in the initial stage of the reaction (conversion of DMF <60%), can reform DMF to maintain the equilibrium between DMF/water and 2,5-hexanedione, which significantly improves overall selecitivty to *p*-xylene[2]. This study reveals that the superior performance of large-pore zeolites for selective production of *p*-xylene can be attributed to: (a) their large pore structure and tunable partilce size for fast mass tranport, (b) their resistance to deactivation, allowing for high conversion of DMF and (c) its ability to catalyze dehydration of the Diels-Alder cycloadduct without catalyzing other side reactions. In addition, it was also found that *p*-xylene, once formed from DMF, did not readily isomerize to *o*- or *m*-xylene at the considered reaction conditions, which provides a significant process advantage by eliminating the expensive separation of xylene isomers. The discovery that large pore zeolites with tunable acidity exhibits high selectivity with high catalytic reaction rate significantly improve the potential for lignocellulosic biomass to serve as a renewable feedstock for aromatic chemicals.

References:

1. Williams, C. L.; Chang, C.-C.; Do, P.; Nikbin, N.; Caratzoulas, S.; Vlachos, D. G.; Lobo, R. F.; Fan, W.; Dauenhauer, P. J., "Cycloaddition of Biomass-Derived Furans for Catalytic Production of Renewable p-Xylene". *ACS Catal.* **2012,** *2*(6), 935-939.

2. Chang, C.-C.; Green, S. K.; Williams, C. L.; Dauenhauer, P. J.; Fan, W., "Ultra-selective cycloaddition of dimethylfuran for renewable p-xylene with H-BEA". *Green Chem.* **2014,** *16* (2), 585-588.

388744 Tailoring the Physicochemical Properties of Zeolite Catalysts through Molecular Design

Abstract:

There are growing opportunities for the use of zeolite catalysts in a wide range of applications. Despite their extensive use in commercial processes, an understanding of their growth mechanism(s) remains elusive. The rational design of zeolite catalysts calls for more versatile synthetic approaches capable of *a priori* tailoring crystal properties, such as crystal size, morphology, and surface architecture. Here we will discuss the design of industrially relevant zeolites, notably LTL, MOR and CHA. Focusing on this class of porous materials presents an opportunity to explore the impact of crystal engineering on targeted catalyst performance (e.g., activity and lifetime). To this end, we will discuss our studies of zeolite crystallization in the presence of zeolite growth modifiers (ZGMs), which are molecules that selectively bind to specific surfaces of zeolite crystals and mediate anisotropic growth rates [1]. We systematically examined a library of modifiers ranging in their structure and functional moieties to assess the physicochemical properties that regulate their efficacy and specificity as zeolite crystal growth modifiers. As we will demonstrate, the judicious selection of ZGMs can markedly alter zeolite crystal morphology from high aspect ratio (needle-like) crystals and micron-sized spheroidal crystals to thin discs and small spheres, respectively, with concomitant reduction in the internal diffusion pathlength and external surface area. Our studies reveal that ZGMs permit the control of zeolite crystal size spanning more than three orders of magnitude. This work also revealed that ZGM hydrophobicity and the spatial sequencing of binding moieties are effective molecular descriptors of its efficacy [2].

Collectively, these studies seek to establish a commercially viable approach to optimize catalyst performance and to provide general heuristics for designing zeolites. We will present these fundamental studies and also discuss how techniques, such as atomic force microscopy [3], can be used to probe the mechanisms of ZGM action. Moreover, we will present time elapsed studies of crystallization in order to elucidate the impact of modifiers on zeolite nucleation and growth.

References:

[1] Lupulescu, A.I. and Rimer, J.D., Tailoring Silicalite-1 Crystal Morphology with Molecular Modifiers, *Angew. Chem. Int. Ed.* 51 (2012) 3345-3349.

[2] Lupulescu, A.I., Kumar, M., Rimer, J.D., A Facile Strategy to Design Zeolite L Crystals with Tunable Morphology and Surface Architecture, *J. Amer. Chem. Soc.* 135 (2013) 6608-6617

[3] Lupulescu, A.I. and Rimer, J.D., In Situ Imaging of Silicalite-1 Surface Growth Reveals the Mechanism of Crystallization, *Science* 2014 (In Press)

380355 Textural and Catalytic Properties of Mo Loaded Meso-/Microporous Layered Zeolites for Direct Methane Conversion

Abstract:

The hierarchical meso-/microporous lamellar zeolites consist of a zeolitic layer structure, with microporosity and mesoporosity within the layers and between the layers, respectively. The consequences of lamellar meso-/microporous structures on distribution and activity of active sites (Brønsted acid and metal sites) in molybdenum (Mo) impregnated MFI were investigated by textural and acidity property characterizations and direct methane aromatization (DMA) reactions. In comparison with their microporous analogies, which are Mo impregnated conventional microporous MFI with different crystallite sizes, the Mo/lamellar MFI zeolites showed lower number of free Brønsted acid sites and lower number of near surface Mo species. The presence of mesoporosity in the Mo/lamellar zeolites facilitates the dispersion of Mo in zeolite micro-channels to form interacting Mo-Brønsted acid site species. The catalytic DMA reactions showed that the Mo loaded meso-/microporous lamellar zeolites enabled efficient methane conversion, aromatic product formation, and coke accumulation in the initial stage of reaction, while their performances are similar to those of Mo loaded conventional microporous zeolite catalysts in the long term run. A plausible explanation to this scenario is that the active sites surrounded by mesopores account for initial fast reaction rate and deactivation rates of Mo/ lamellar zeolite catalysts in DMA reactions; while the active sites in microporous channels are responsible for the long term catalyst activity due to the shape selectivity of the zeolite micropores.

Preparation of isolated iron species in ZSM-5 for high selectivity alcohol formation from lower alkanes under mild conditions

Catalysis with Microporous and Mesoporous Materials II

AIChE Meeting November 2014

Dr. Michael M Forde

University of the West Indies

Trinidad & Tobago

Graham J. Hutchings, Nikolaos Dimitratos, Robert Armstrong, Qian He and Peter Wells

Work conducted at Cardiff Catalysis Institute, Cardiff United Kingdom

Outline

- Overview of previous work
- Materials Preparation
- Catalysis
- Materials Characterisation
- Key findings

Many approaches to C-H activation

General Objectives

- ✓ Lower alkane oxidation to alcohols (and acids)
- ✓ Highly Selective
- ✓ Appreciable Conversion
- ✓ Benign conditions (temperature, pressure, solvents etc.)

New but not so new chemistry

- ➢ Nature has some well designed materials for the process
- ➢ A lot of previous work in this area

C-H activation is the "holy grail" of catalysis

Examples of Previous Approaches

Homogeneous domain – strong acid conditions, methyl ester made

Periana and co-workers: Angew. Chem. Int. Ed. 43 (2004) 4626

Sen and co-workers: J. Am. Chem. Soc. 113 (1991) 700

X. Bao and co-workers: J. Am. Chem. Soc. 128 (2006) 16028

Y. Wang and co-workers: Adv. Synth. Catal. 349 (2007) 1199

Heterogeneous domain- meso/microporous materials + metals

Mizuno and co-workers, Appl. Catal. A 158 (1997) L47 ;

J. Phys.Chem. B (104) 2000 5940; J. Catal. 184 (1999)550

G. I. Panov and co-workers Appl. Catal. 83 (1992) 31

H. Groothaert and co-workers J. Am. Chem. Soc 127 (2005) 1394

Our Reaction network - methane

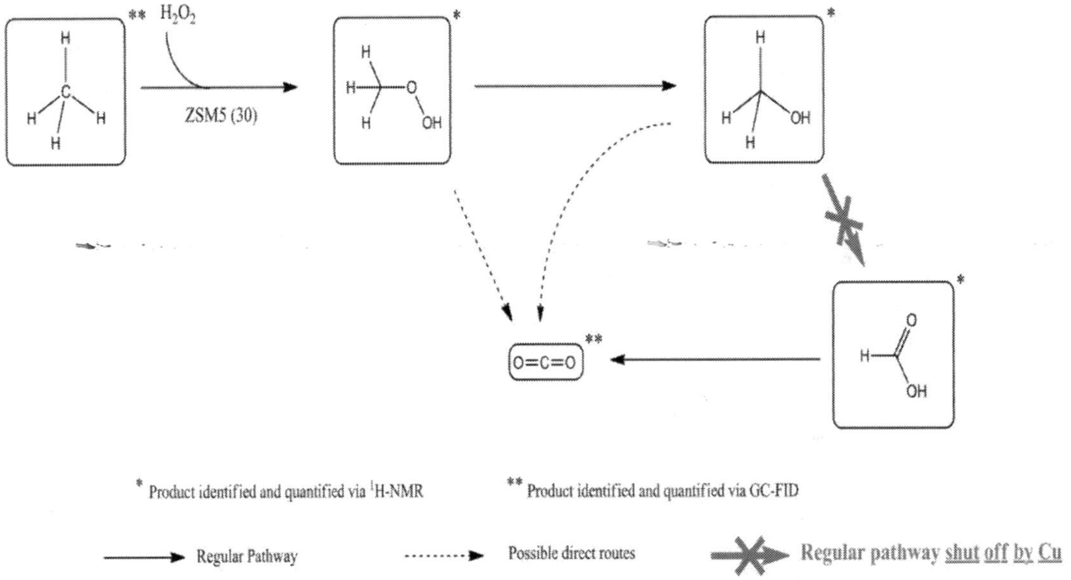

Angew. Chem. Int. Ed., **2012**, *51*, 5129–5133; Chem. Eur. J. (2012) 18, 15735 – 15745
ACS Catal., **2013**, *3*, 1835-1844; *ACS Catal*, **2013**, *3*, 689-699

Our Reaction Network – ethane

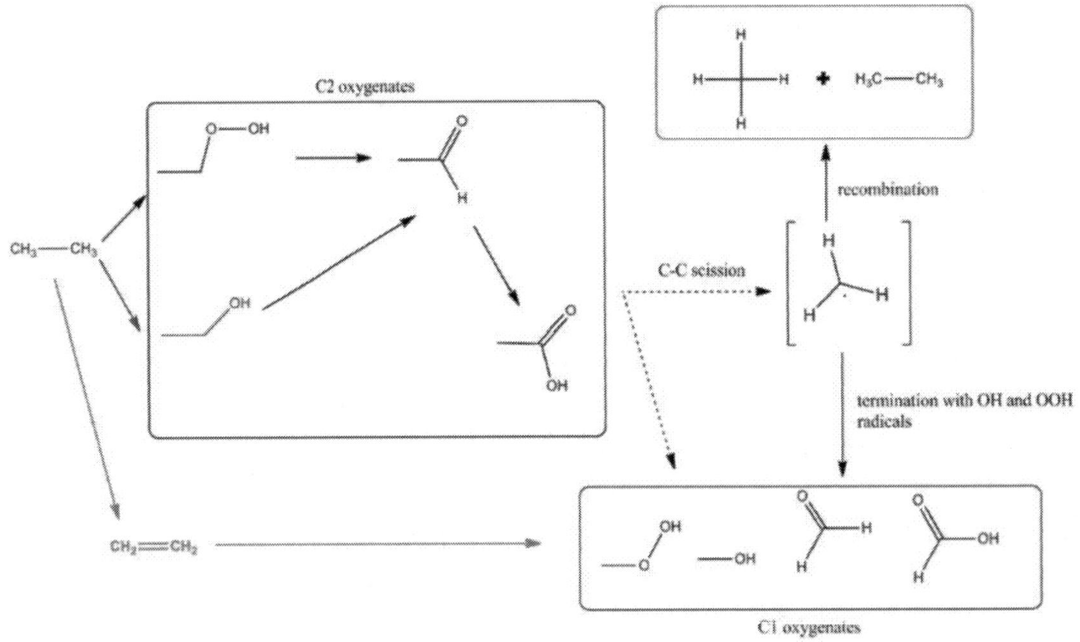

J. Am. Chem. Soc., **2013**, *135*, 11087–11099.

Key highlights- ethane oxidation

✓ Low COx selectivity always observed

✓ High conversion (50% in batch) easily achieved

✓ Ethylene was a major product when Cu was incorporated into the catalyst

✓ Little or no control of ethanol selectivity (too unstable) but direct route to ethanol observed

Interest in how to prepare the ZSM-5 to have higher ethanol selectivity

Catalyst Preparation

Chemical Vapour Impregnation

➢ Mix Metal Precursor (acetylacetonate salt) + H-ZSM-5(30)
➢ Heat under vacuum with stirring
➢ Cool, removed and heat treat

Technique Characteristics

✓ Can be done in any lab with standard equipment
✓ Multi-metallics and layered deposition possible
✓ Works will many powdered supports (not carbons)

Catalyst preparation

Heat treatment

Conventional - calcine in air or calcine in air then helium

Our methods

- ➢ Calcine in air
- ➢ Heat treat only in H_2/Ar or He (no air)

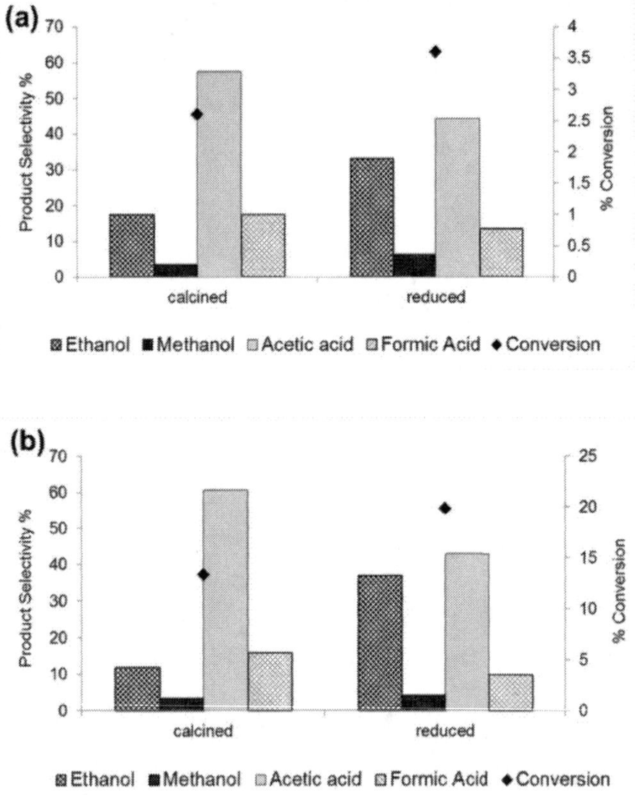

Catalysis

➢ **Improved ethanol selectivity even at high conversion**

Catalysis- Methane

Catalyst treatment	Total Productivity[a][b] mol kg(cat)$^{-1}$h^{-1}	MeOH Productivity [c] mol kg(cat)$^{-1}$h^{-1}	MeOH Selectivity (%) [e]
Air, 550°C	15.8	1.9	12
H_2/Ar, 550°C	16.0	5.1	32.1
ZSM-5 Pre-treated in air then 550°C, 3h H_2/Ar	16.7	6.0	36.0

Alcohol Selectivity Effect also observed with Methane oxidation

Characterisation-UV

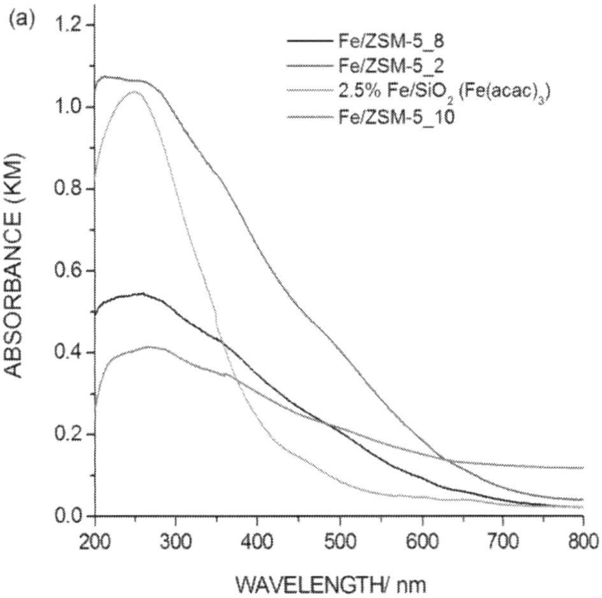

Air treated - Typical of what is in the literature

Characterisation- UV

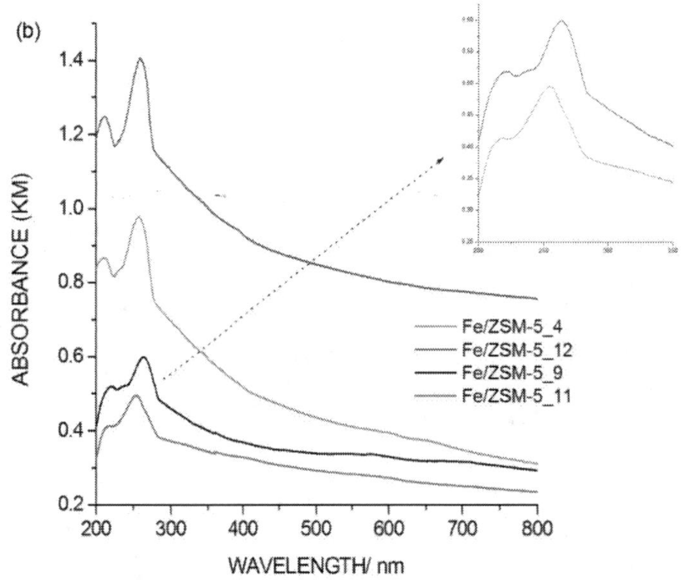

H_2 Treated- well defined "isolated" iron species

XAS Work

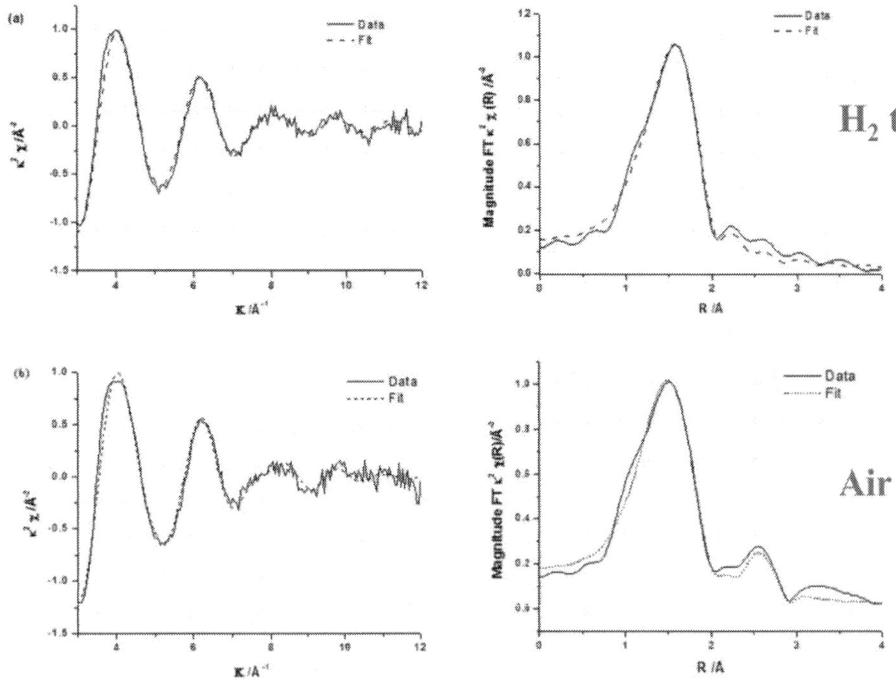

H_2 treated

Air treated

XAS Work

Catalyst	Treatment	Abs Sc	N	R / Å	$2\sigma^2$ / Å^2	E_f / eV	R_{factor}
		Fe-O	2.5 (9)	1.95 (3)	0.005 (3)		
Fe /ZSM-5_#8[1]	550°C, Air	Fe-O	3.5 (9)	2.08 (2)	0.005 (3)	-1 (1)	0.003
		Fe -Fe	1.1 (3)	3.04 (3)	0.009 (4)		
		Fe-O	1.7 (7)	1.95 (1)	0.005 (4)		
Fe /ZSM-5_#9[2]	550°C, H_2/Ar					0 (1)	0.003
		Fe-O	3.9 (9)	2.09 (1)	0.004 (3)		

Primary coordination sphere of Fe in reduced sample is 6 „O"
No firm evidence of Fe-Fe interactions

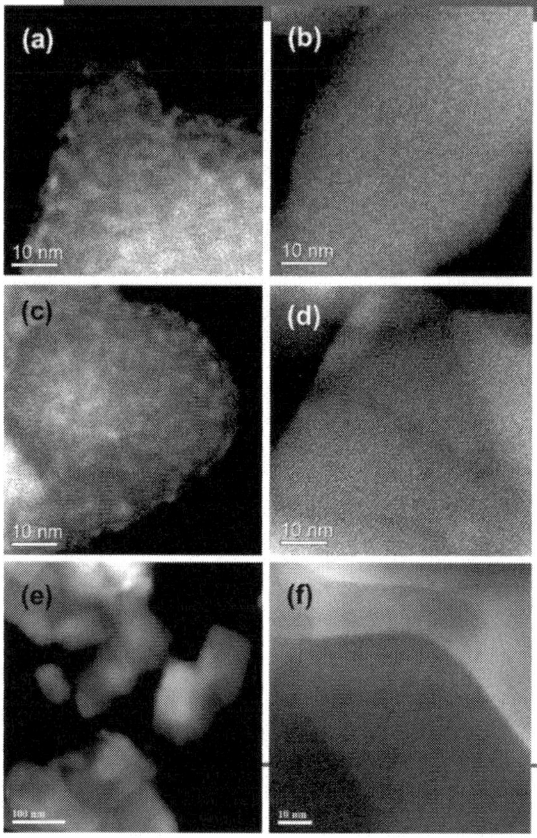

HAADF-TEM

A,B,C - Air treated Fe/ZSM-5
D- bare ZSM-5 air treated
F- H2/Ar treated Fe/ZSM-5

No evidence of surface Fe in "reduced" material

Reusability

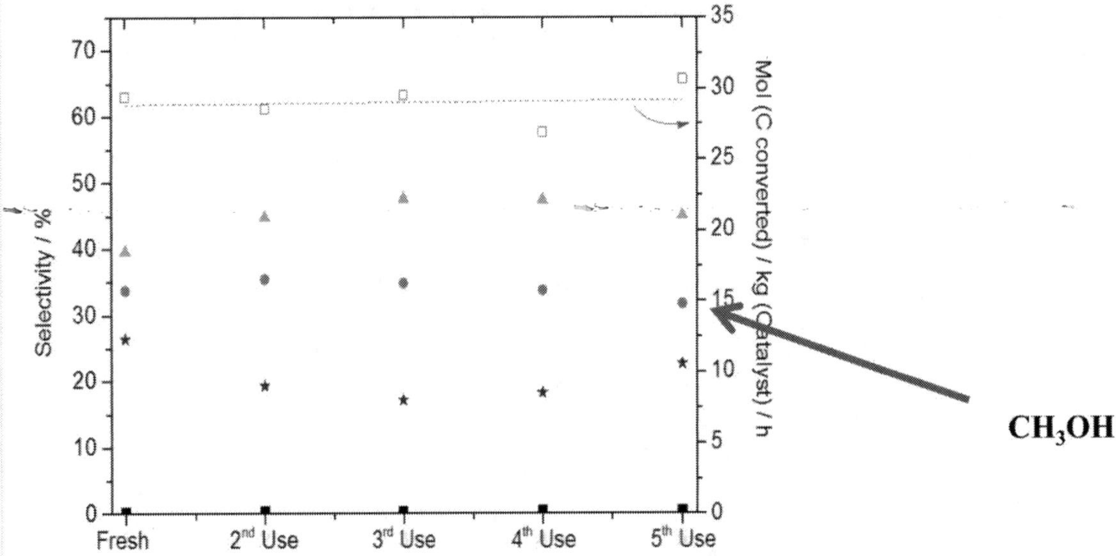

Effects not transient with methane oxidation

Reusability

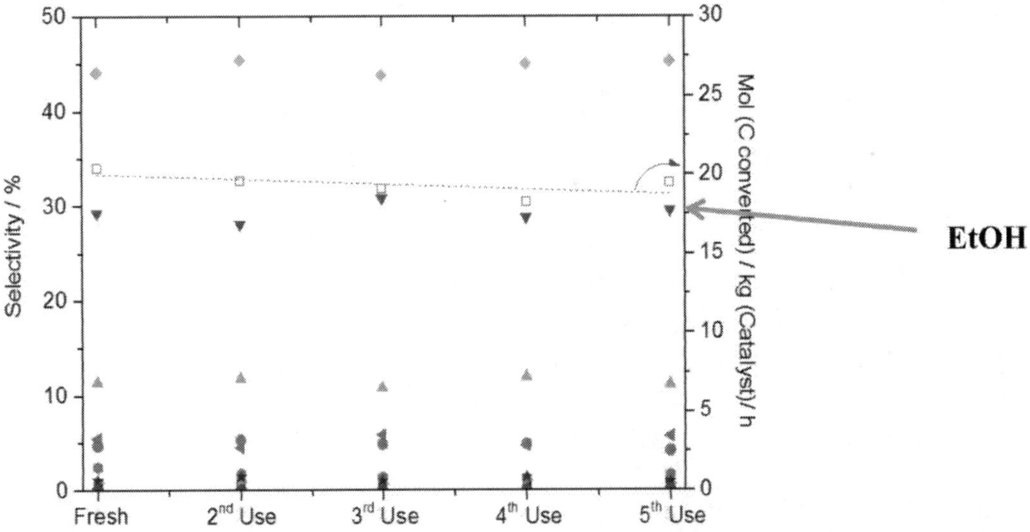

Effects not transient in ethane oxidation

Summary

"New type" of Fe species observed

- Isolated, Oxygen only coordination sphere
- Beneficial effect on alcohol selectivity for both ethane and methane
- Reusable, thus non-transient effect
- Enhanced productivity in direct route to ethanol from ethane implies

possibility direct route for methanol, involving O_2 possibly

Summary

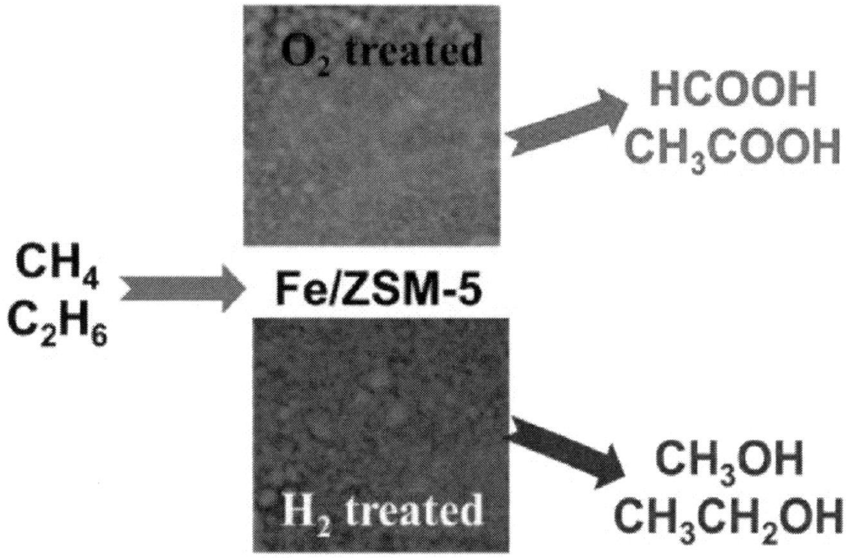

Chemical Science 2014, 5, 3603-3616

Thanks for listening
Thanks to the sponsors

CardiffCatalysisInstitute

EPSRC for funding (EP/I019693/1, EP/K014714/1) and Diamond Light Source for provision of beam time (SP8071-1)
The Research Campus Harwell are also acknowledged for use of facilities and support of their staff.

Niobium-exchanged mesoporous silicates as catalysts for ethylene epoxidation

Wenjuan Yan,[a, b] Anand Ramanathan,[a] Bala Subramaniam*[a, b]
(a) Center for Environmentally Beneficial Catalysis, Lawrence, KS, (b) Department of Chemical and Petroleum Engineering, University of Kansas, Lawrence, KS

With the increased availability of ethane as a collateral product from shale gas recovery, alternate ethylene oxide (EO) technologies that eliminate CO_2 as byproduct has gained renewed importance. This talk will present investigations on niobium (Nb)-incorporated TUD-1 (amorphous mesoporous silicate) as potential epoxidation catalysts. Nb was incorporated in varying amounts into TUD-1 support, synthesized by a sol-gel technique where niobium chloride ($NbCl_5$) was used as the Nb precursor. The resulting materials, termed as Nb-TUD-1, showed Type IV adsorption isotherm typical of TUD-1 type mesoporous structures. The BET surface area of Nb-TUD-1 decreased from approximately 628 m^2/g to 358 m^2/g with an increase in Nb content from 1.6 to 12.2 wt.%. Average pore diameter varied from 4.4 nm at the highest Nb loading to 7.3 nm at the lowest loading. It was noticed from diffuse reflectance UV-Vis spectra of Nb-TUD-1 samples that NbO_4 tetrahedral units and oligomeric NbO_4 units co-exist at all the loadings. In all materials, no band characteristic of bulk crystalline Nb_2O_5 was evident. The synthesized materials showed ethylene epoxidation activity at mild temperatures (35 °C) and pressure (50 bar), using H_2O_2 as oxidant and methanol as solvent. No detectable CO_2 was observed as byproduct. The measured epoxidation activity is in the range of (341~1,000 mg EO/h/g Nb), similar in magnitude to those observed with Ag-based catalyst used in conventional process (700-4,400 mg EO/h/g Ag) and homogeneous Re-based catalyst (1,610~4,970 mg EO/h/g Re).[1] However, Nb leaching and H_2O_2 decomposition are major drawbacks. Catalyst modification by silylation and similar pretreatment dramatically decrease Nb leaching and H_2O_2 utilization. These results will be presented and discussed.

1. Ghanta, M., et al., Highly Selective Homogeneous Ethylene Epoxidation in Gas (Ethylene)-Expanded Liquid: Transport and Kinetic Studies. AIChE Journal, 2013. 59(1): p. 180-187.

376445 Production of Sugars, Lignin and Lignin Monomers from Biomass Using Gamma-Valerolactone As a Solvent

Abstract:

Widespread production of biomass-derived fuels and chemicals requires cost-effective processes for breaking down cellulose, hemicellulose and lignin into their constituent monomers. Soluble carbohydrates can be produced at high yields (70 to 90%) from corn stover, hardwood and softwood in a solvent mixture of gamma-valerolactone (GVL), water, and dilute acid (0.05 wt% H_2SO_4). When used as a solvent, GVL promotes thermocatalytic saccharification of biomass by complete solubilization of the biomass, including the lignin fraction, and can be produced at high-yields from biomass itself (about 60% yield from cellulose). Rates of acid-catalyzed reactions in GVL appear to be linked to the increased catalytic activity of the proton due to solvent effects.

The carbohydrates can be recovered and concentrated (up to 127 g/L) in an aqueous solution by addition of CO_2 to form a CO_2-expanded GVL phase no longer miscible with water. This strategy is well suited for catalytic upgrading to furans or fermentative upgrading to ethanol at high titers and near theoretical yield. In parallel, the solubilized lignin stream can be recovered at high yields (>70% of the original lignin) by precipitation in water. Analysis of the isolated lignin by NMR demonstrated that its chemical properties remained remarkably similar to native lignin or enzymatically-isolated lignin despite the high temperatures and acid used during processing. Because of this, we show that this stream can be converted catalytically converted to lignin monomers at similar yields to those obtained from unprocessed biomass.

389390 Synthesis of Sulfonated 3DOm Carbon Catalysts for Hydrolysis of β(1-4) Glucan

Abstract:

Selective hydrolysis of cellulose into glucose is the key process for the utilization of biomass. It is the first step in the reaction network for converting biomass into fuels and high-value chemicals[1-3]. Conventional processes for the hydrolysis of cellulose using mineral acids, cellulase enzymes, and supercritical water have several drawbacks, such as the high cost for separation of products and catalysts, corrosion hazards, severe controls of enzymes, low selectivity and waste fluids. The increasing need for environmentally sustainable chemical processes has stimulated the use of recyclable heterogeneous acid catalysts as replacements for the unrecyclable homogeneous acid catalysts and enzyme-based techniques. In particular, sulfonated carbon catalysts ($C\text{-}SO_3H$) have shown promising activity for the selective hydrolysis of cellulose in water under moderate conditions. However, due to the limited understanding of the interaction between the solid catalyst and water-insoluble cellulose, the catalytic property of carbon catalysts for this reaction is still elusive.

The aim of this study is to design a bifunctional heterogeneous catalyst with both acid and adsorption sites for the selective hydrolysis of β(1-4) glucan to glucose. Three dimensionally ordered mesoporous (3DOm) carbon with a cage size of 35 nm was successfully functionalized with sulfonic acid through covalent attachment of sulfonic acid-containing aryl radicals produced by homogeneous reduction of diazonium salt with hypophosphorous acid. It is found that the adsorption of saccharides on this carbon catalyst could be dominated by the hydrophobic interaction between the π-domains on the carbon surface and the CH groups on the saccharides.

The carbon catalyst was used to hydrolyze acidulated ball-milled cellulose (ABMC) to produce glucose. ABMC was prepared by incipient wetness of microcrystalline cellulose with dilute sulfuric acid and ball-milled, producing short β(1-4) chains (DP $<$ 10) with α(1-6) branching. This ABMC has excellent solubility (10% in water) compared to ordinary β(1-4) glucan. The turn over frequency (TOF, (mole glucose)(mole $H^+)^{-1}h^{-1}$) for ABMC reacted with the carbon catalyst and HCl was 1.5 and 0.18, respectively. A max glucose yield of 65% was obtained by reacting ABMC with the carbon catalyst. NMR studies revealed that β(1-4) glycosidic bonds break down rapidly while the α(1-6) bonds remain mostly unreacted using the carbon catalyst. The high reaction rate of the β(1-4) glycosidic bonds in ABMC when using the carbon catalyst is likely due to the interaction between α(1-6) branched β(1-4) glucan chains and adsorption on the carbon surface.

References:

1. Dornath, P.; Fan, W., "Dehydration of fructose into furans over zeolite catalyst using carbon black as adsorbent". *Microporous Mesoporous Mater.* **2014,** *191* (0), 10-17.

2. Chang, C.-C.; Green, S. K.; Williams, C. L.; Dauenhauer, P. J.; Fan, W., "Ultra-selective cycloaddition of dimethylfuran for renewable p-xylene with H-BEA". *Green Chem.* **2014,** *16* (2), 585-588.

3. Williams, C. L.; Chang, C.-C.; Do, P.; Nikbin, N.; Caratzoulas, S.; Vlachos, D. G.; Lobo, R. F.; Fan, W.; Dauenhauer, P. J., "Cycloaddition of Biomass-Derived Furans for Catalytic Production of Renewable p-Xylene". *ACS Catal.* **2012,** *2* (6), 935-939.

378810 Synergistic Effect Between Defect Sites and Functional Groups in the Hydrolysis of Cellulose Using Activated Carbon

Abstract:

Carbon with Brønsted acid sites can function as a stable and effective solid acid catalyst for reactions in hot liquid water. It is able to hydrolyze cellulose, an abundant and renewable biomass, into glucose, which is a versatile platform molecule. However, there are limited reports on the effect of different chemical treatments and the nature of active sites. In this study, the effect of chemical oxidation by H_2O_2 and H_2SO_4 on activated carbon was investigated, and the materials were used in the hydrolysis of cellulose for comparison. The oxidation procedure was characterized by XRD, FTIR, ^{13}C DP MAS NMR, Boehm titration, N_2 physisorption, CO_2 physisorption and Raman spectroscopy. Oxidation by H_2O_2 imparts acidic functional groups such as phenol, lactone and carboxylic acid, while treatment with H_2SO_4 imparts all of those oxygen functional groups along with sulfonic acid groups. Characterization reveals that H_2O_2 is capable of generating and enlarging defect sites, while H_2SO_4 creates more edges in the carbon material, and this effect is more pronounced with higher temperature.

Adsorption isotherms demonstrated that sulfonic groups enhance the adsorption of glucose monomers more than phenols, lactones and carboxylic acids by hydrogen bonding, and the influence of this enhanced-adsorption effect decreases as the length of glucan chain increases. This can be explained by the increasing influence of van der Waals interactions between the CH groups and the polyaromatic rings. All of the chemically treated materials exhibit catalytic activities in the hydrolysis of cellulose despite the presence of weakly acidic functional groups. Carbon treated by H_2SO_4 showed higher reactivity even though it has less functional groups and lower pore volume distribution. This can be attributed to the synergistic effect between the edge/defect sites and functional groups in the carbon material, where the glucan chains are immobilized and the exposed glycosidic bonds are forced to interact with the in-plane functional groups.

Fossil fuels provide approximately three quarters of today's fuel energy, but current fossil fuel world reserves will not meet future energy demands. The burning of these energy sources is one of the main causes of the increase of CO_2 concentration in the atmosphere that has been associated with Climate Change. Consequently a transition from nonrenewable (fossil fuels) energy sources to renewable ones is desirable. Biomass is the main renewable source that can substitute fossil fuels, but to achieve this goal some technical challenges must be overcome. One of them is the design of catalysts with good thermo solvent stability, and this is especially true for aqueous liquid phase processing as is the case of fructose dehydration to 5-hydroxymethyl-furfural (HMF). In this work we report the continuous production of HMF from fructose using mesoporous CMK-3 carbon materials functionalized with methyl-, ethyl- and phenyl- sulfonic acid groups. These materials were characterized using scanning electron microscope (SEM), nitrogen adsorption - desorption isotherms, X-Ray diffraction (XRD), elemental analysis and acid titration. The mesoporous carbon materials have the desired long-range order of mesopores, textural properties and have been grafted with the desired sulfonic functional groups. We will discuss the catalytic performance of the sulfonic acid functionalized mesoporous carbons using kinetic data obtained in batch and flow reactors. Preliminary results indicate that the phenylsulfonic catalysts are the most active and selective towards HMF production with 100% conversion of fructose and 44% yield towards HMF in 30 min at 145 °C. We will also present a study of the catalysts stability under reaction conditions in a packed bed reactor. Finally, possible deactivation modes will be presented.

386444 Unraveling Solvation Effects on 5-Hydroxylmethylfurfural Degradation: Insights from Catalytic and Spectroscopic Studies

Abstract:

The prospective use of lignocellulosic biomass as a feedstock of value-added chemicals and fuels requires the effective and synergistic utilization of catalysts and solvents. The necessity to develop efficient and economic processes for the production of biomass derived platform chemicals, such as 5- hydroxymethylfurfural (HMF) has motivated research for the development of carbohydrate dehydration processes using heterogeneous[1, 2] and/or homogeneous[3, 4] catalysts. The use of organic co-solvents is one of the methods used to enhance HMF yields by reducing the rates of its rehydration and humin-formation reactions. Although many solvent mixtures have been tested in this respect, the reasons of enhanced HMF stability in aqueous aprotic co-solvent mixtures are still not clearly understood.

In the present work, vibrational spectroscopy is coupled with catalytic experiments in an effort to elucidate the origins of HMF stability and to provide "solvent-solute interactions" – activity relationships. Attenuated total reflection FTIR spectra of HMF in a wide composition range of aqueous aprotic (i.e. DMSO, g-valerolactone, THF etc.) and protic (i.e. methanol, ethanol etc.) solvent mixtures were measured in order to unravel the effect of the co-solvent on the structural and vibrational properties of dissolved HMF. Deconvolution of pertinent wavenumber regions was carried out in an attempt to discriminate HMF-solvent interactions at different mixed solvent compositions. Different intermolecular interaction have been ascribed that are responsible for the non-conventional change of the HMF C=O and O-H vibrational modes under different solvent composition[5]. Our spectroscopic analysis has revealed the first experimental evidence of preferential interaction of HMF by aprotic solvents that that also depends on the co-solvent nature and amount.

HMF rehydration experiments were carried out in several mixed-solvent mixtures under different conditions (i.e. temperature, and acid concentration), and the effect of co-solvent type and concentration on the HMF stability is quantified. In addition to the possible effects of co-solvent on the intrinsic reaction rate constant, the rehydration and humin formation reaction rates can also be affected by the water activity, and proton concentration. The realative contribution of each of these factors on HMF stability/ reactivity is determined and will be analyzed in an effort to provide solvation/activity relationships based on our spectroscopic analysis. Our findings will also be discussed with respect to the possible HMF rehydration and humin formation reaction mechanisms.

References

[1] V.V. Ordomsky, J. van der Schaaf, J.C. Schouten, T.A. Nijhuis, 5 (2012) 1812-1819.

[2] Y. Zhang, J.J. Wang, J.W. Ren, X.H. Liu, X.C. Li, Y.J. Xia, G.Z. Lu, Y.Q. Wang, Catal Sci Technol, 2 (2012) 2485-2491.

[3] Y. Roman-Leshkov, J.A. Dumesic, Top Catal, 52 (2009) 297-303.

[4] J.N. Chheda, Y. Roman-Leshkov, J.A. Dumesic, Green Chemistry, 9 (2007) 342-350.

[5] G. Tsilomelekis, T.R. Josephson, V. Nikolakis, S. Caratzoulas, Chemsuschem, 7 (2014) 117-126.

378449 Kinetics and Insights into Glucose Dehydration Catalyzed By Homogeneous Lewis Acidic Metal Salts in Water

Abstract:

Dehydration of glucose catalyzed by various Lewis acidic metal salts was studied with water as reaction media. The glucose conversion kinetic profile was used to reflect the Lewis acid character of different metal ions. It was found that the pH value of the aqueous solution played an important role in controlling the Lewis activities. For the lanthanide chlorides, their Lewis acidity was comparable under the pH values studied (from 2.5 to 5.5). However, the Lewis acidity strength of other metal salts, such as aluminum chloride, showed dependence on the pH value of the solution. Activation energies with various Lewis acids were also calculated with both glucose and fructose to obtain more insight about the strength of the catalyst-substrate interaction as well as the dehydration reaction. The kinetic isotope effect with labeled glucose molecules was also studied to explore a more mechanistic understanding of the dehydration, which likely involves the 2-H atom of the glucose molecule in the transition state.

371604 Solvent-Induced Frequency Shifts of 5-Hydroxymethylfurfural and Their Role in Its Stability

Abstract:

5-Hydroxymethylfurfural (HMF) is a promising platform-chemical for sustainable fuels and plastics. A major issue in the production of HMF is the side reactions that lead to significant loss of sugars and high cost. Use of polar aprotic solvents improves HMF yields from sugar dehydration by preventing HMF side reactions, but the influence of solvent on the structure of HMF is not fundamentally understood. In an effort to understand solvent effects on HMF stability and yield, solvent-induced frequency shifts (SIFS) of the carbonyl stretching vibration ν(C=O) of 5-hydroxymethylfurfural were measured in seven protic, polar aprotic, and non-polar solvents. The Gutmann Acceptor Number (AN), an empirical measure of Lewis acidity, was found to correlate with the measured frequency shifts. The SIFS were then investigated using *ab initio* electronic structure calculations, treating the solvent implicitly and with an explicit solvent ligand interacting with the carbonyl. The conductor polarizable continuum model (CPCM) solvation model predicted that ν(C=O) shifted as a function of the dielectric constant, in agreement with the analytical predictions of the Kirkwood-Bauer-Magat (KBM) theory for a dipole in a dielectric continuum, but in disagreement with experimental ν(C=O). Experimental ν(C=O) were best predicted using gas-phase complexes of HMF and explicit solvent ligand. NBO analysis and Bader's Atoms in Molecules theory were used to investigate the electronic structure and hydrogen bonding in these complexes. SIFS arose from H-bonding interactions, as observed in delocalization of carbonyl lone-pair electrons to H-bonding solvent σ*(X-H) orbitals, and in increased charge density and decrease in local potential energy at the H-bond (3,-1) critical point in the charge density. Consequently, by predicting the experimental SIFS and examining the electronic structure, we find the first theoretical evidence for treating Gutmann's solvent AN as a measure of solvent Lewis acidity. Application of these findings in understanding the stability of HMF in various solvents will also be discussed.

Abstract Submission to Be Considered as an Oral Presentation at ***AICHE 2014 Annual Meeting*** in *T4A06 Biochemical Conversion Processes in Forest/Plant Biomass Biorefineries II section*

Chemical Conversion of Biorefinery Lignin to Value Added Platform Chemicals.

Ruoshui Ma[1], Vince Hebert[1], Jinwen Zhang[1], Tom Candy[2], Wei Wang[2], Xiao Zhang[1]

1. Washington State University, Richland, WA, 99354
2. Solvay Chemicals Inc., Houston, TX, 77098

Lignin is the largest nature source of renewable material with an aromatic skeleton. Depolymerizing lignin to low molecular weight aromatic and phenolic compounds offers attractive opportunities to produce a range of high value products/chemicals. Lignin can be depolymerized into low molecular weight phenolic compounds (LMWPC) by a number of chemical and/or biological means. Among these pathways, oxidation strategy has shown its advantages to selectively increase functionality of the final products.

In this presentation, we will report recently work from our group on applying peroxygen chemicals to depolymerize and convert biorefinery lignin into platform low molecular weight phenolic and quinone compounds.

Presenting author: Ruoshui Ma, Chemical Engineering, Washington State University, Richland, WA, 99354

Contact: ruoshui.ma@email.wsu.edu

xiaozhang@tricity.wsu.edu

Abstract Submission to Be Considered as an Oral Presentation at AICHE 2014 Annual Meeting in T4A03 Chemical Conversion Processes in Forest/Plant Biorefineries I session

Metal catalyzed conversion of Biomass to Maximize Hemicellulose Sugar Conversion to Valuable Organic Acids

Carlos Alvarez-Vasco, Xiao Zhang

Hemicellulose usually represents between 22 to 30% of biomass dry weight been along with cellulose and lignin the main biomass constituents. However, due to its lower degree of polymerization and crystallinity compared to cellulose, hemicellulose is more sensitive to thermal and chemical degradation during biomass pretreatment. With current pretreatment technologies, hemicellulose degradation compounds do not add value to the overall process. In fact, they may even be detrimental for subsequent fermentation or purification steps.

This study investigated the potential of metallic ions addition during pretreatment in order to maximize hemicellulose conversion to valuable chemicals and cellulose recovery to fermentable sugars. Main benefices of metallic ions addition include its potential higher selectivity toward specific products and the possibility of been recovery and use in several runs. In addition, the natural occurrence of metallic ions in biomass structure (K and Ca on ash content) or in forest residues (Al and Fe due to soil contamination) are a potential source of this ions.

Biomass metallic ion catalysis takes advantage of the natural occurrence of these metallic ions to catalyze the conversion of Hemicellulose to organic acids. Major conversion pathways involved in carbohydrate metal catalyzed conversion of biomass and hemicellulose model compounds (i.e. glucomannan and xylan) are elucidated. Considerable amounts of carboxylic acids including lactic acid, glycolic acid, succinic acid and formic acid are produced along with pretreated biomass. The addition of metallic ions during pretreatment increased the formation of products for both biomass and tested carbohydrate polymers model compounds. In addition, the role of metallic ion catalyzed pretreatment conditions on resulting enzymatic cellulose-to-glucose yields of pretreated biomass are study. This study demonstrated an integrated metallic ion treatment of biomass with the opportunity to produce valuable chemicals from hemicellulose in order to maximize the value of biomass carbohydrates.

Presenting author: Carlos Alvarez-Vasco

Bioproducts Science Engineering Laboratory, Washington State University, 2710 Crimson Way, Voiland School of Chemical Engineering and Bioengineering, Richland, WA 99354

Contact: c.alvarezvasco@email.wsu.edu, xiaozhang@tricity.wsu.edu

Guangzhou Institute of Energy Conversion， Chinese Academy of Sciences

Catalytic Hydroprocessing of Fatty Acid Methyl Esters to Jet Fuels

Pengmei Lv

Marriott Marquis Atlanta,

November 17, 2014,

Chemical Conversion Processes in Forest/Plant Biorefineries

AIChE
The Global Home of Chemical Engineers

Table of contents

1. Background

Problems related to biodiesel

- ☐ A higher cold filter plugging point;
- ☐ Containing oxygen, less stable;
- ☐ Lower heating value

Catalytic hydrotreated and isomerized to bio-aviation fuel

State of the art of the catalysts

➤**Microporous Si-Al molecular sieve:** Relative strong acidity and good hydrothermal stability, whereas its micro pore size impedes the diffusional mass transfer of reactants and products.

➤**This case:** The molecular size of FAME reaches over 4.4 nm.

Demand: A kind of molecular sieve with larger pore size, namely mesoporous molecular sieve is in need.

2.Molecular sieve synthesis and characterization

(1) Introduction

(2) Synthesis

Principle, catalyst design, catalyst preparation

(3) Characterization

SEM, XRD, NMR, surface area, pore volume, diameter and acidity

(1) INTRODUCTION

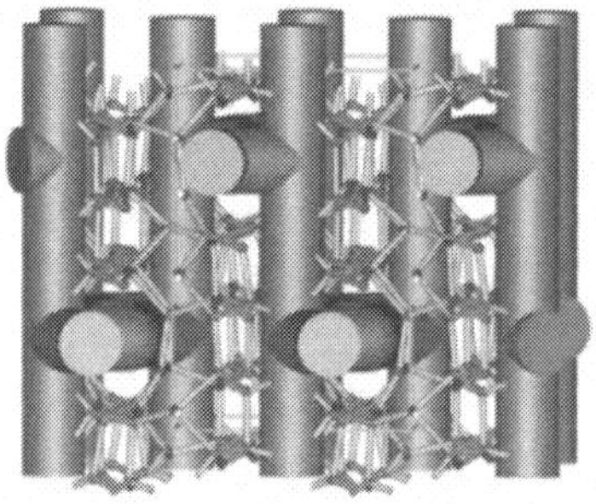

- **Definition:** As defined by IUPAC (International Union of Pure and Applied Chemistry), mesoporous materials means their pore size ranges between 2-50 nm.
- **Molecular sieve:** A class of materials with uniform and ordered porous structure .
- **Typical spatial structure:** Cage-like ordered arrangement.
- **The first ordered mesoporous SiO_2 material :** Synthesized by Mobil company in 1992.
- **Generally, two categories:** Silicon-based mesoporous material and non-silicon based mesoporous material.

(2) SYNTHESIS

➢ Synthesis principle

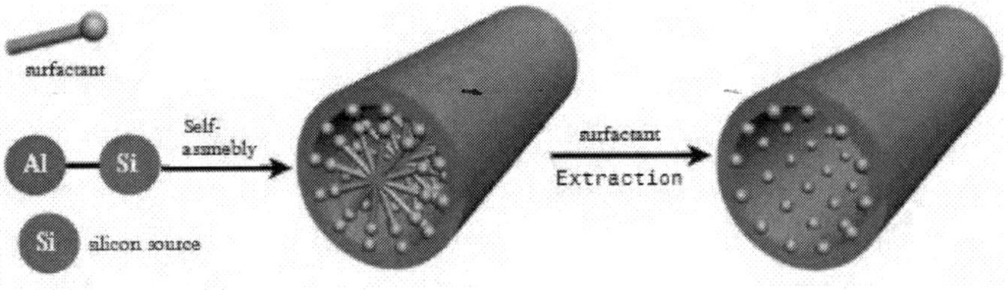

Supramolecular assembly through cooperative effect of hybrid organic - inorganic species.

➢ Catalyst design

(1) **Carrier** – Mesoporous molecular sieve, dependent upon the size scale of FAME molecuar. Surfactant selection and hydrothermal crystallization process investigation。

(2) Main active metal component and complex component- Ni, the main active component for hydrotreating reactions, M selected to be transition metal, e.g. Co, Zr, etc.

(3) **Appropriate Bronsted and Lewis acid sites distribution** – Involved reactions include FAME hydrotreatment, isomerization and crack, etc, achieved through the Si/Al ratio regulation and preparation process.

(3) CHARACTERIZATION

➤ SEM characterization

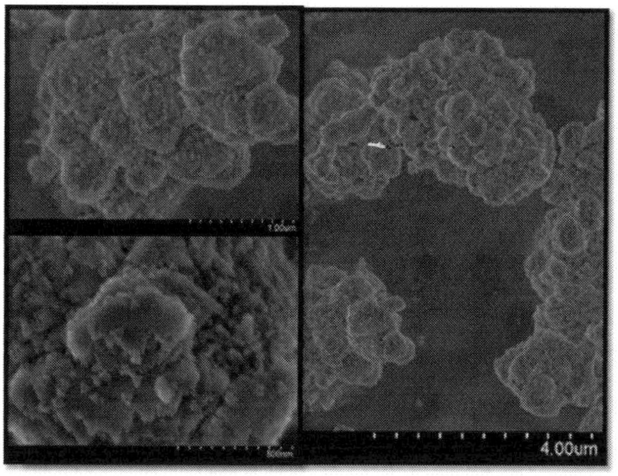

A serrated surface of 10wt%Ni/Hβ

➤ XRD characterization

2θ/(°)	Interplanar spacing, D/nm	relative intensity
7.8	1.14±0.02	MS
12.0	0.74±0.02	W
13.6	0.67±0.02	W
21.0	0.425±0.01	W
22.4	0.397±0.01	VS
29.4	0.30±0.01	W
41.0	0.22±0.01	W

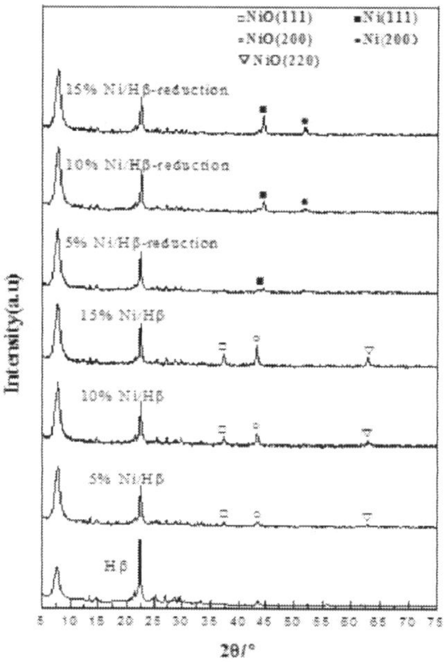

- After Ni loaded, although the intensity changes, its crystal structure was maintained well;
- With the increase of Ni loading, NiO diffraction peak becomes more and more strong.

➢ NMR characterization

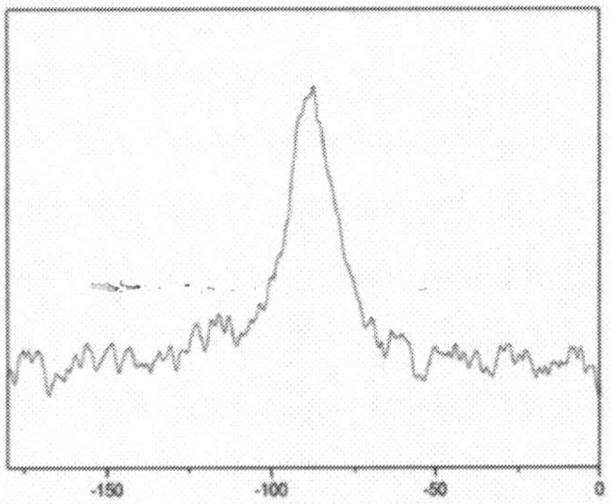

Al/SBA-15 ^{29}Si MAS NMR

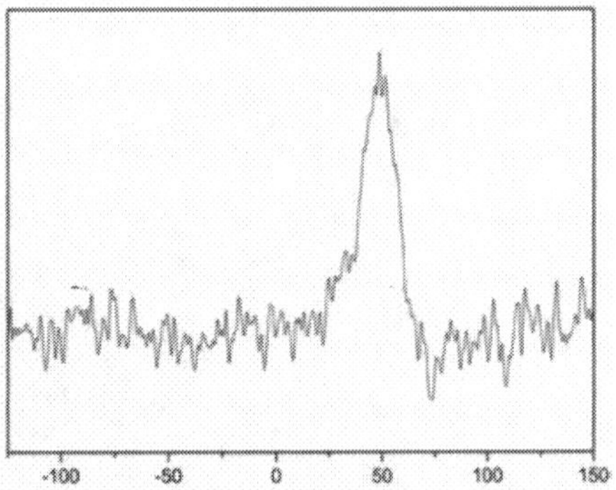

Al/SBA-15 ^{27}Al MAS NMR

➢ NMR shows that Al appears a four-coordinated form in the framework of molecular sieve to generate Lewis acid sites.

➢ Surface area, pore volume, diameter and acidity

			Physical structures			Acid density (mmol·g^{-1})		
Catalyst	Ni loading (wt%)	Si/Al	BET surface area (m^2·g^{-1})	Pore volume (cm^3·g^{-1})	Pore diameter (nm) (HK method)	Weak acid sites	Strong acid sites	Total amount
Hβ	0	25	554.4	0.20	0.66	0.31	0.30	0.61
Ni/Hβ	5	25	511.7	0.19	0.63	0.39	0.33	0.72
Ni/Hβ	10	25	483.4	0.18	0.62	1.05	0.34	1.39
Ni/Hβ	15	25	458.1	0.17	0.60	0.85	0.18	1.03
Ni/Hβ	10	50	462.2	0.17	0.62	0.92	0.22	1.14
Ni/Hβ	10	150	445.8	0.17	0.62	0.83	0.13	0.96
SiO_2	0	/	356.5	0.84	9.6(BJH method)	0	0	0
Ni/ SiO_2	10	/	315.9	0.72	8.5(BJH method)	0.11	0.09	0.20

- NiO particles located on the surface of Hβ create a weakly acidic environment after H_2 reduction forming the H^+ as Brønsted acid ($Ni^{2+} + H_2 = Ni + 2H^+$).
- On the other hand, the particles of Ni may block the Hβ microchannels, thus decreasing the accessibility of NH_3.

3. Catalytic performance

(1) Different Ni loading and Si/Al ratio

(2) Different temperatures

(3) Different time and reusability test

The composition of FAMEs

FAMEs	Fatty acid compositions[a] of FAMEs structure	Formula	Compositions (wt%)
Methyl Myristic	$C_{14:0}$[b]	$C_{15}H_{30}O_2$	0.35
Methyl Hexadecanoate	$C_{16:0}$	$C_{17}H_{34}O_2$	21.22
Methyl Octadecanoate	$C_{18:0}$	$C_{19}H_{38}O_2$	33.74
Methyl Oleate	$C_{18:1}$	$C_{19}H_{36}O_2$	39.52
Methyl Linoleate	$C_{18:2}$	$C_{19}H_{34}O_2$	0.19
Methyl Linolenate	$C_{18:3}$	$C_{19}H_{32}O_2$	4.23
Methyl Arachidate	$C_{20:0}$	$C_{21}H_{42}O_2$	0.75

[a] Fatty acid composition was assigned on the basis of alkyl chain length.
[b] Nomenclature refers to the number of carbon atoms and number of C=C double bonds.

Hydrotreatment of FAMEs was carried out in a 100 mL stainless slurry reactor equipped with a polytetrafluoroethylene (PTFE) liner and mechanical stirrer operating in batch mode.

(1) Different Ni loading and Si/Al ratio

Catalyst	Ni loading (wt%)	Si/Al	Conversion (%)	Selectivity (%)						
				C_{15}	iso-C_{16}	C_{16}	C_{17}	iso-C_{18}	C_{18}	C_{15}–C_{18}
Hβ	0	25	3.2	1.2	0	23.9	5.5	0	65.8	96.4
Ni/Hβ	5	25	95.3	4.0	0.8	14.1	14.6	2.7	52.7	88.9
Ni/Hβ	10	25	100	4.7	1.2	14.1	17.1	3.6	52.5	93.2
Ni/Hβ	15	25	98.7	7.6	0.9	9.2	27.2	2.6	30.3	77.8
Ni/Hβ	10	50	97.2	2.2	0.5	17.5	10.1	1.8	61.3	93.4
Ni/Hβ	10	150	96.8	0.7	0.1	19.0	7.6	0.5	66.1	94.0
Ni	Pure Ni	/	37.2	26.9	0	0.7	66.3	0	0.2	94.1
Ni/SiO_2	10	/	91.2	19.8	0	1.6	70.6	0	2.7	94.7

Reaction conditions: FAMEs (1.0 g); dodecane (4.0 g); catalyst (0.1 g); 270 ° C; 1 MPa H_2; 8 h.

- A pure Hβ support without Ni loading , very low conversion of 3.2%.
- Comparatively, for pure Ni, a higher conversion of 37.2%. This proves that Ni alone can catalyze hydrotreatment of FAMEs to alkanes without support, including hydrogenation, hydrogenolysis, hydrodecarboxylation reactions, et al.
- Hβ can provide a larger catalytic specific surface area and more contact sites to the reactant than pure Ni metal.
- For Ni/Hβ, compared with other products, the content of C15 and C17 is relative low. This proves that the catalytic hydrotreatment occurs mainly via HDO without carbon loss.

Catalyst	Ni loading (wt%)	Si/Al	Conversion (%)	Selectivity (%)						
				C_{15}	iso-C_{16}	C_{16}	C_{17}	iso-C_{18}	C_{18}	C_{15}–C_{18}
Hβ	0	25	3.2	1.2	0	23.9	5.5	0	65.8	96.4
Ni/Hβ	5	25	95.3	4.0	0.8	14.1	14.6	2.7	52.7	88.9
Ni/Hβ	10	25	100	4.7	1.2	14.1	17.1	3.6	52.5	93.2
Ni/Hβ	15	25	98.7	7.6	0.9	9.2	27.2	2.6	30.3	77.8
Ni/Hβ	10	50	97.2	2.2	0.5	17.5	10.1	1.8	61.3	93.4
Ni/Hβ	10	150	96.8	0.7	0.1	19.0	7.6	0.5	66.1	94.0
Ni	Pure Ni	/	37.2	26.9	0	0.7	66.3	0	0.2	94.1
Ni/SiO_2	10	/	91.2	19.8	0	1.6	70.6	0	2.7	94.7

For pure Ni or supported on SiO_2 ,

- A relative higher content C_{15} and C_{17} is obtained. This indicates the catalyst with no or little acidity may favor hydrodecarboxylation or hydrodecarbonylation reaction.
- The iso-alkanes content is 0. On the account of no or little acidity of these 2 catalysts, this proves that the acidic sites are the active centers for isomerization reaction.

(2) Different temperatures

Temperature (°C)	Conversion (%)	Selectivity (%) C_{15}	iso-C_{16}	C_{16}	C_{17}	iso-C_{18}	C_{18}	C_{15}–C_{18}
230	49.4	0.8	0	18.3	7.2	0	72.4	98.7
240	62.4	1.3	0.4	17.4	8.3	0.7	70.6	98.7
250	70.5	1.7	0.7	16	10.5	1.6	68	98.5
260	97.4	2.9	1	15.6	13.1	3	60.5	96.1
270	100	4.7	1.2	14.1	17.1	3.6	52.5	93.2
280	100	4	0.8	12.8	13.6	3.2	43.7	78.1

Reaction conditions: FAMEs (1.0 g); dodecane (4.0 g); catalysts (0.1 g); 1 MPa H_2; 8 h

- A higher reaction temperature promotes not only the hydrotreatment of FAMEs – especially the hydrodecarbonylation or hydrodecarboxylation pathway – but also the enhanced cracking of long-chain alkane products.

(3) Different time and reusability test

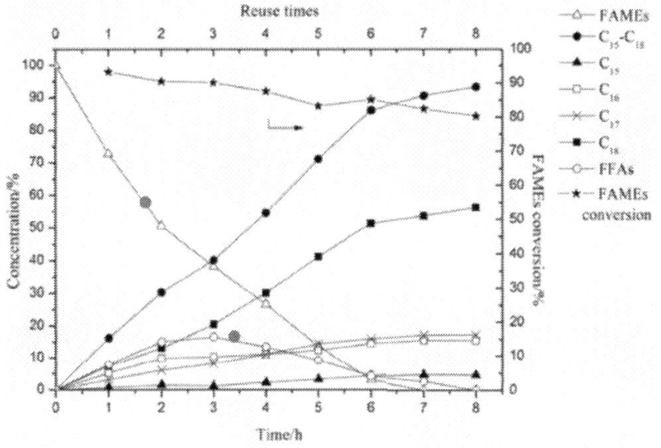

FAME hydrotreatment catalyzed by 10%Ni/Hβ

- The conversion of FAMEs in 1H decreased 27.2%. This illustrates the hydrogenation reduction of unsaturated bond is a rapid process.
- The intermediate product of FFAs reached the highest concentration of 16.5% at the 3rd hour. In the first 3H, FFAs accumulation rate was higher than that of its conversion rate. After 3h, FFAs was converted into alkanes.

4. Conclusion

Hydrotreatment and isomerization pathway

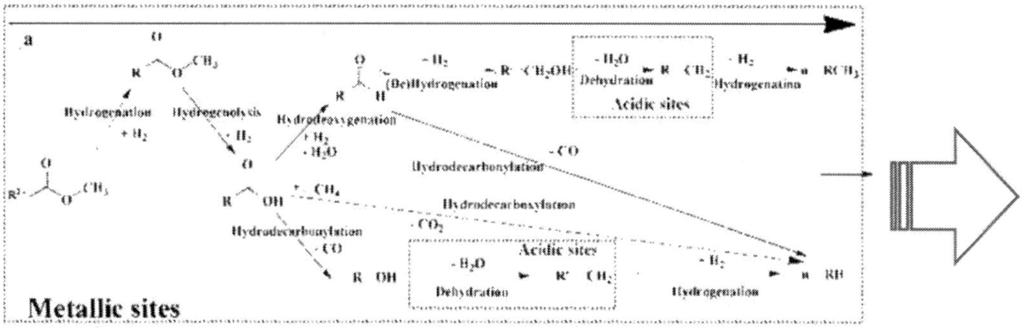

(1) Metal active center catalyzed hydrogenation of unsaturated bonds in the alkyl chain.

(2) Hydrogenolysis of the formed saturated FAMEs to produce carboxylic acids, e.g., hexadecanoic and octadecanoic acids.

a. subsequent hydrodeoxygenation of carboxylic acids to aldehydes then to alcohols, followed by dehydration via acidic functionalities in the zeolite catalyst to alkenes and subsequent hydrogenation to alkanes.

b. carboxylic acids directly change to alkanes via decarboxylation with carbon loss.

c. carboxylic acids undergo decarbonylation to alkanes.

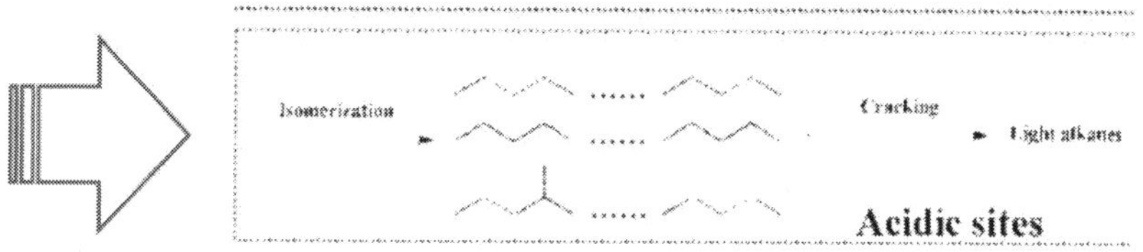

(3) All long-chain alkanes could be subjected to hydroisomerization and hydrocracking at the function of acidic sites on the catalyst.

(4) Carbon dioxide and carbon monoxide may react with H_2 to produce methane and water.

① The amount of acid sites increase with Al content. The acid sites promote the isomerization reaction.

② The catalyst with no or little acidity may favor hydrodecarboxylation or hydrodecarbonylation reaction.

③ Metallic sites are the active center for hydrodeoxygenation. A higher reaction temperature promotes the cracking of long-chain alkane products.

④ With certain range of metal loading and acid amounts, the catalyst will show an optimal catalytic performance on conversion and selectivity.

⑤ 10 wt% Ni/Hβ with Si/Al = 25 shows the best performance both for the deoxygenation reaction and isomerization reaction.

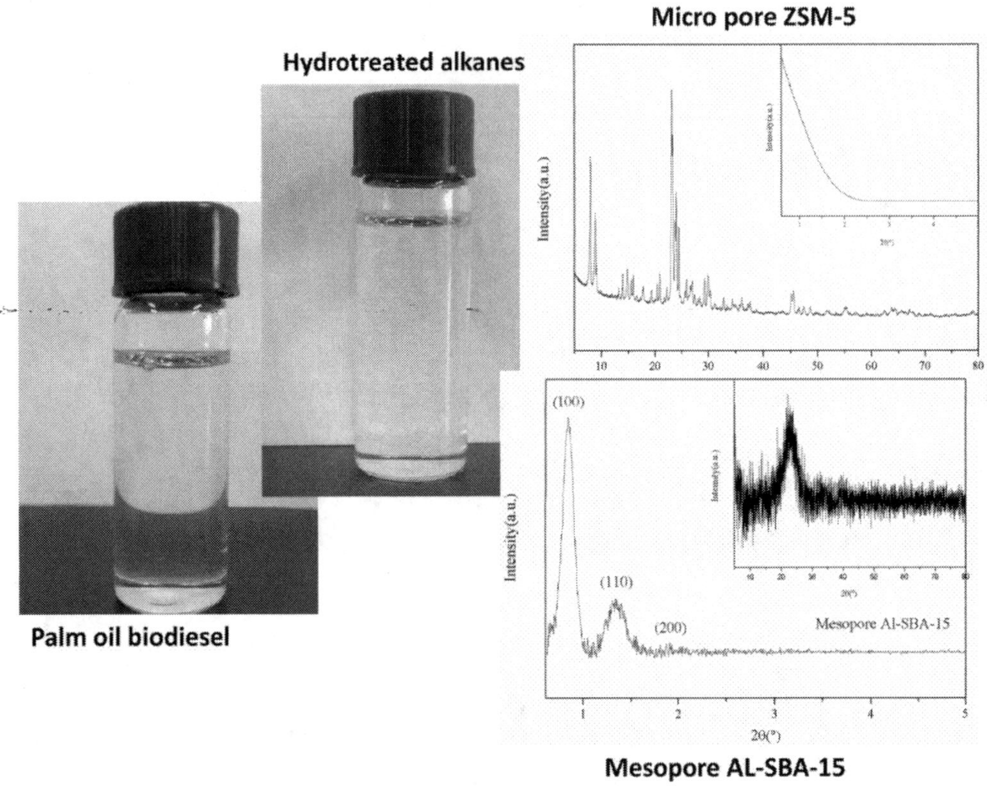

Thanks for your attention!

389277 Saccharification and Lignin Fractionation of Lignocellulosic Biomass in Molten Salt Hydrate System

Abstract:

Molten salt hydrate (MSH) has unique properties, such as high boiling point and depressed freezing point, very low water vapor pressure, enhanced acidity, and high enthalpy of mixing. Some MSHs are able to swell and solubilize cellulose through breaking inter- and intra-molecule hydrogen bonds of cellulose. Compared to ionic liquids, MSHs are inexpensive and easier to prepare and recover. Recently, our investigation indicated that lignocellulosic biomass (such as corn stover, switchgrass, hardwood and softwood) could be quickly and completely saccharified without any pretreatment under moderate conditions (110-130 °C for 10-60 min) in molten salt hydrate systems. Both cellulose and hemicellulose in lignocellulosic biomass were completely hydrolyzed into fermentable sugars with high yield and selectivity (>90%). Lignin in the biomass was remained as insoluble residue and could be easily separated for co-products development. In this presentation, we report the results of process optimization of biomass saccharification in molten salt hydrate, sugar-salt separation after the saccharification, lignin characterization and co-product potential, and techno-economic analysis of the system.

372234 Conversion of Lignin to Hydrocarbon Fuels Via Catalytic Depolymerization and Hydrogenation

Abstract:

Three different solvent systems (sodium hydroxide solution, ethanol aqueous solution and aqueous solution of phosphoric acid) were chosen to study the conversion of lignin to hydrocarbons via catalytic depolymerization and hydrogenation. A two-step method, selective depolymerization of lignin to liquid compounds rich in phenolics followed by hydrogenation of the obtained liquid, was developed. It was found that sodium hydroxide solution showed the best performance among the three, as the yield of phenolic compounds could reach 60%, with a selectivity of 85% when used for lignin degradation. What's more, the selectivity for hydrocarbons could reach 97% when the derived phenolic compounds were hydrogenated. The preliminary mechanism study using model compounds indicated that the conversion of lignin to hydrocarbons is a multi-step process consisting of alkaline hydrolysis, demethylation, demethoxylation and transalkylation.

373657 Catalytic Conversion of Xylose-Rich Hydrolyzate from Autohydrolysis into Furfural

Abstract:

Catalytic conversion of xylose-rich hydrolyzate from autohydrolysis into furfural

Lu Liu*, Jing Du, Hou-min Chang, Hasan Jameel, Sunkyu Park

Department of Forest Biomaterials, North Carolina State University, Raleigh, NC 27695, USA

* Presenting author: llui@ncsu.edu

Furfural, the dehydration product of xylose, can be served as a sustainable substitute for petroleum-based building blocks used in production of fine chemicals and plastics. Aiming at the improvement of current method to produce furfural from xylose, our study is focused on an efficient and environmentally friendly method to produce furfural using the autolydrolysis hydrolyzate from lignocellulosic biomass. The optimization process of temperature, pressure, and other experimental conditions in lab-scale are described based on xylose conversion and furfural yield in both systems, pure xylose solution and biomass hydrolyzate. Different types of solid catalyst and solvent system were applied to compare with non-catalytic scenario to achieve high yield of furfural production. In model system of pure xylose solution, the furfural yield was not significantly improved by adding catalyst, while the xylose conversion yield was significantly increased. About 40% furfural yield was achieved in 2% xylose solution after 3 hours of reaction in non-catalytic scenario with about 30% xylose left. With biomass hydrolyzate from autohydrolysis, an enhanced furfural yield and xylose conversion yield was achieved. To maximize conversion yield of xylose, a reactor was modified to allow the produced furfural and steam to be released and collected. The challenges and opportunities of hydrolyzate conversion into furfural will be discussed in details, considering hydrolyzate impurities, conversion mechanism, and scale-up feasibility.

362955 Optimize Activity and Durability of Multimetallic Electrocatalysts By Tuning the Nanostructure

Abstract:

Chemical-electrical energy conversion and storage are greatly attractive for the development of sustainable energy. Among the chemical-electrical energy conversion devices, polymer electrolyte membrane fuel cell (PEMFC) is one promising technology for portable applications, from as small as portable electronics, to as large as automobiles. One of the key components for PEMFCs is the electrocatalyst. Development of high-performance electrocatalyst nanomaterials relies on tuning material structures at nanoscale. This is in particular manifested in the design of electrocatalysts demanding both high activity and durability. Through solution-phase-synthesis, high quality nanocrystals are prepared at various morphologies (shape-controlled) and compositions (composition-controlled). The nanostructure (i.e. selective exposing specific low-index surface, controlling core-shell structure, and etc.) of nanocrystals are also carefully tuned. Using well-controlled nanomaterials, we explore the correlation between nanostructure and activity (and/or durability), and thus optimize the balance of activity and durability. Herein, a group of high performance electrocatalysts are developed for various electrocatalytic processes, including oxygen reduction reaction (ORR) and hydrogen evolution reaction (HER). It is well-demonstrated that the careful manipulation of nanostructure of electrocatalysts is crucial to improve the activity and durability.

383164 Development of Novel Pt Monolayer Electrocatalysts for the Oxygen Reduction Reaction in PEM Fuel Cells

Abstract:

Platinum exhibits superior oxygen reduction reaction (ORR) activity compared with other elements, making it an attractive cathode material for hydrogen fuel cell applications. The main problems with pure platinum electro-catalysts are large activation losses, limited stability, and high material cost. These deficiencies have greatly inhibited the development of efficient, reliable, and inexpensive proton exchange membrane fuel cells (PEM FCs). Previous experimental and computational studies have shown that it is possible to improve the performance of platinum cathodes by selectively tuning the chemical reactivity of catalytic sites through geometric and electronic effects induced by alloying. This has led to the development of a number of Pt-based alloy electrocatalysts for this reaction.

Our group has previously developed a first principles model that can predict the ORR activity for various Pt alloys. I will discuss how we used this model in an attempt to develop novel electrocatalysts with superior performance compared to pure platinum. I will particularly focus on the design, synthesis, and testing of so-called second generation Pt alloys which contain an alloy core, Pt shell exposed to the reactants, and a protective Au buffer between the alloy core and the Pt shell. Extensive in situ and ex situ characterization verify the atomic and electronic structure of these core-shell Pt electrocatalysts. The oxygen reduction activity is determined through electrochemical testing of supported catalysts on a rotating disc electrode (RDE) in a three-electrode cell. All electrochemical measurements were performed at room temperature in a 0.1M $HClO_4$ electrolyte.

Structural and Electrochemical Properties of Pt Supported on Sn-doped Indium Oxide

S. Zhao(Department of Chemical and Biomolecular Engineering at the University of Connecticut), Y. Liu, A. Wangstrom and W. E. Mustain (University of Connecticut, Department of Chemical and Biomolecular Engineering)

Abstract Text:

Proton exchange membrane fuel cells (PEMFCs) have drawn significant attention as an alternative to environmentally unfriendly power sources. However in order to realize its commercialization, many methods have been proposed to develop new electrocatalysts with improved performance such as catalytic activity and stability. One considerable improvement to reduce Pt loading and cost for the oxygen reduction reaction on the cathode is to incorporate elements with higher stability into existing active electrocatalysts in the PEMFC systems.

We are focusing our work on studying the activity and stability of Sn-doped indium oxide (ITO) as the supporting material for platinum to catalyze the oxygen reduction reaction (ORR) in both acidic and alkaline media [1]. We have found that the catalytic activity of Pt supported on ITO electrocatalysts for ORR can be enhanced drastically through synergistic effects between surface Sn and Pt, including preferential Pt faceting, which maximized the surface area of the Pt {111} facet that is highly active toward ORR. We also found that for our Pt/ITO prepared by lower Pt loading galvanic displacement yields a single type of particle with preferential {111} faceting, which gives yields its characteristic high activity. Other than higher activity or less Pt loading and better stability proves ITO to be a promising supporting material. [2]

In this work, we present a comparison of the structural and electrochemical properties of Pt supported on ITO in both acidic and alkaline media. Galvanic displacement deposition method was adopted to deposit Pt onto our freshly prepared ITO. Transmission electron microscopy images were taken before and after the stability test of the electrocatalyst to study the structural difference. In order to investigate the composition change on the material surface, X-ray photoelectron spectroscopy was used to characterize the result. For electrochemical properties, cyclic voltammetry and polarization curves were taken to calculate the electrocatalytic surface area, the specific activity and mass activity. [3] Koutechy-Levich plot and Tafel plot were also drawn to determine the mechanism of the electrochemical reactions. These results show Pt supported on ITO in both acidic and alkaline media are promising for the integration of devices.

This talk will focus on the experimental method, physical and electrochemical characterizations of our electrocatalysts in both acidic and alkaline environments.

References.

1. Y. Liu and W.E. Mustain, "High stability, high activity Pt/ITO oxygen reduction electrocatalysts", *J. Am. Chem. Soc.*, **135** (2013) 530
2. Y. Liu and W.E. Mustain, "Stability limitations for Pt/Sn–In_2O_3 and Pt/In–SnO_2 in acidic electrochemical systems", *Electrochimica Acta*, **115** (2014) 116-125
3. Y. Liu and W.E. Mustain, "High Stability, High Activity Pt/ITO Oxygen Reduction Electrocatalysts". Abstract 86, 223rd Meeting of the Electrochemical Society, San Francisco, CA.

Symposium Selection:
20040 Electrocatalysis for PEM Fuel Cells in the Catalysis and Reaction Engineering (CRE) Division

Submitter's E-mail Address:
shuai.zhao@uconn.edu

Preferred Presentation Format:

Oral -- *withdraw if not accepted for preferred format*

First Author (Corresponding)
Shuai Zhao
Affiliation(s): University of Connecticut, Department of Chemical and Biomolecular Engineering
Phone Number: 8609040267
E-mail Address: shuai.zhao@uconn.edu

Second Author
Ms. Ying Liu
Affiliation(s): University of Connecticut, Department of Chemical and Biomolecular Engineering
Phone Number:
E-mail Address: yil09001@engr.uconn.edu

Third Author
Abbey Wangstrom
Affiliation(s): University of Connecticut, Department of Chemical and Biomolecular Engineering
Phone Number: 8604864020
E-mail Address: abbey.wangstrom@uconn.edu

Fourth Author
Dr. William E Mustain
Affiliation(s): Department of Chemical and Biomolecular Engineering at the University of Connecticut
Address:
191 Auditorium Road, Unit 3222
Storrs, CT, 06269
USA
Phone Number: 630-338-9777
E-mail Address: mustain@engr.uconn.edu

Abstract

Development of renewable energy technologies requires advanced catalysts for efficient electrical-chemical energy conversion reactions. For practical implementation of fuel cells, one of the main challenges is the need for substantially improved electrocatalysts versus platinum (Pt) for the oxygen reduction reaction (ORR). Here we report Pt-Re alloy as a novel ORR catalyst. An organic solvothermal approach is developed to synthesize monodisperse and homogeneous Pt_3Re alloy nanoparticles. Electrochemical studies show that these nanoparticles exhibit an improvement factor of 4 for the ORR compared to commercial Pt catalysts of similar particle size. A bifunctional mechanism has been interpreted by density function theory (DFT) calculations for the alloy catalyst, involving the surface modifications of oxygen coverage and electronic structures.

Dimethyl Ether Electro-Oxidation on Platinum Surfaces

Luke T. Roling, Jeffrey A. Herron, Winny Budiman, Peter Ferrin, and Manos Mavrikakis

Dimethyl ether (DME) has attracted significant interest as an alternative fuel, particularly in direct low-temperature fuel cell applications.[1] Pt catalysts have demonstrated good activity for DME electro-oxidation, but require a high overpotential due to poisoning by CO. This motivates a deeper understanding of the reaction mechanism for the design of improved catalysts. This reaction demonstrates substantial structure sensitivity, which has been studied experimentally on Pt(111)[2], Pt(100)[3], and stepped[4] Pt surfaces. Of these, Pt(100) demonstrates the highest DME electro-oxidation activity. Using periodic, self-consistent density functional theory (DFT), we perform a detailed evaluation of DME electro-oxidation kinetics and thermodynamics on Pt(100), Pt(111), and Pt(211) surfaces. We use these data to postulate probable reaction mechanisms on each surface as a function of overpotential. By comparing these mechanisms and key elementary steps involved in electro-oxidation, we uncover the origin of the observed structure sensitivity of DME electro-oxidation on Pt surfaces.

1. Kerangueven, G.; Coutanceau, C.; Sibert, E.; Leger, J. M.; Lamy, C., Journal of Power Sources **2006**, 157, 318.

2. Lu, L. L.; Yin, G. P.; Tong, Y. J.; Zhang, Y.; Gao, Y. Z.; Osawa, M.; Ye, S., Journal of Electroanalytical Chemistry **2008**, 619, 143.

3. Lu, L. L.; Yin, G. P.; Tong, Y. J.; Zhang, Y.; Gao, Y. Z.; Osawa, M.; Ye, S., Journal of Electroanalytical Chemistry **2010**, 642, 82.

4. Li, H.; Calle-Vallejo, F.; Kolb, M. J.; Kwon, Y.; Li, Y.; Koper, M. T. M., Journal of the American Chemical Society **2013**, 135, 38.

PtRu nanotubes and PtRu coated CuNWs as catalysts for methanol oxidation reaction in direct methanol fuel cells

Jie Zheng and Yushan Yan*

Department of Chemical and Biomolecular Engineering
University of Delaware
150 Academy Street, Newark, DE 19716, USA
*yanys@udel.edu

Direct methanol fuel cells (DMFCs) are an attractive alternative to hydrogen-fueled proton exchange fuel cells for powering portable electronic devices because of their high energy density and ease of fuel transportation and storage. While nearly zero overpotential is observed for hydrogen oxidation reaction in acid, large overpotential exists even on the state-of-the-art PtRu/C MOR catalyst.[1] Therefore, developing catalysts with high MOR activity is of great importance.

MOR is known to be a structure sensitive reaction with Pt(110) facet showing the highest specific activity among low index surfaces.[2] One-dimensional (1D) nanostructures with 2D extended surfaces such as nanowires, nanotubes and nanorods often have the preferential exposure of certain facets which will enhance the MOR activity.[3] Studies have shown that platinum nanotubes (PtNTs) have higher MOR specific activity than platinum nanoparticles supported on carbon (Pt/C)[4], which might be attributed to the exposure of (110) facets of PtNTs. However, it remains unclear whether 1D PtRu nanostructures have higher MOR activity.

In our work, we synthesized PtRu nanotubes (PtRuNTs) and PtRu coated Cu nanowires (PtRu/CuNWs) via complete or partial galvanic displacement of CuNWs by Pt and Ru precursors. By varying the Pt and Ru precursor ratio, PtRuNTs and PtRu/CuNWs with different Pt:Ru ratio were obtained. Their MOR activities were evaluated by cyclic voltammetry using rotating disk electrodes and compared with the benchmark commercial catalysts PtRu/C (HiSPEC® 12100).

Our results show that the optimal Pt:Ru raito is about 4 for both PtRuNTs and PtRu/CuNWs. Enhanced specific MOR activities were observed on PtRuNTs and PtRu/CuNWs compared with the benchmark PtRu/C, with specific activity in the order of PtRu/CuNWs > PtRuNTs > PtRu/C. However, PtRuNTs showed smaller mass activity than PtRu/C due to their small electrochemical surface areas (ECSAs). PtRu/CuNWs showed comparable mass activity compared with PtRu/C, attributed to the improved specific activity as well as increased ECSA. X-ray photoelectron spectroscopy (XPS) reveals a stronger electron transfer from Ru to Pt on PtRu/CuNWs, which may have led to a stronger back-donation of electron from Pt to CO (the poisoning intermediate), and consequently a weaker Pt-CO bond and a higher MOR activity on PtRu/CuNWs.

References

1. A. Hamnett, in *Handbook of Fuel Cells*, John Wiley & Sons, Ltd (2010).
2. E. Herrero, K. Franaszczuk and A. Wieckowski, *The Journal of Physical Chemistry*, **98**, 5074 (1994).
3. C. Koenigsmann and S. S. Wong, *Energy & Environmental Science*, **4**, 1161 (2011).
4. S. M. Alia, G. Zhang, D. Kisailus, D. Li, S. Gu, K. Jensen and Y. Yan, *Advanced Functional Materials*, **20**, 3742 (2010).

387331 Performance of Ordered Mesosporous Carbon (OMC) Supported Pt-Electrocatalysts in PEMFC

Abstract:

A series of platinum nanoparticles supported on OMC electrocatalysts with different Pt loading have been prepared for polymer electrolyte membrane fuel cell (PEMFC) applications. In this work, we prepared OMC and modified-OMC using SBA-15 and Ni-SBA-15, respectively, as a hard template with sucrose as source of carbon. The catalysts are characterized by using N_2 adsorption-desorption isotherm (BET), transmission electron microscopy (TEM), small angle x-ray diffraction (XRD), thermo-gravimetric and differential calorimetric analysis (TGA-DSC), temperature programmed reduction (TPR), and Fourier transform infrared spectroscopy (FTIR). The BET results showed that the synthesized OMC possesses surface area in excess of 1000 m^2/g. TEM and XRD results confirm the ordered mesoporous structure. The FTIR spectra showed the complete removal of the silica template. The Pt-catalyst was tested for its performance and polarization on a 5 cm^2 membrane electrode assembly (MEA) under controlled experimental conditions. The polarization test and cyclic voltammetry analysis of the modified-OMC exhibited improved performance compared to the conventional OMC. The electrochemically active surface area (ECSA) evaluated for modified-OMC was observed to be significantly higher than that of the OMC and commercial carbon black supported catalyst.

Effects of CO, H_2 and C_3H_6 on the NH_3-SCR reactions over a Cu-chabazite SCR catalyst

Yang Zheng, Michael P. Harold*, Dan Luss*

Department of Chemical & Biomolecular Engineering
University of Houston
Houston, TX 77204

Abstract

The effects of reducing agents of CO, H_2 and C_3H_6 on the performance of a Cu-chabazite catalyst were investigated, including NO_2 decomposition, NH_3-SCR reactions, NH_3 storage, as well as N_2O production. Different reductants affect NH_3-SCR reactions to different extents by means of different pathways. Propylene is most effective in promoting NO_2 decomposition to NO via organic intermediates. CO effectively reduces nitrates to nitrites that can then react with NO_2 to release NO, while H_2 has a slightly promoting effect when NO_2 becomes thermodynamically unstable at high temperatures. As for SCR reactions, propylene competes with NH_3 for adsorbed NO_2 or nitrates to release NO and thus increase NO/NOx ratio. This leads to the dominance of either fast (NH_3 + NO + NO_2) or standard SCR (NH_3 + NO + O_2) reaction under slow SCR (NH_3 + NO_2)condition if C_3H_6 present. The competition of CO with NO for nitrates reduction to nitrites, has a negligible effect on both standard and fast SCR reactions but a promoting effect on slow SCR reaction. In contrast to steady-state operation, lean/rich cycling enhances cycle-averaged NOx conversion of all NH_3-SCR reactions when adding either C_3H_6 or CO + H_2 in rich phase. Both high NO_2/NOx ratio and long lean-phase purge can mitigate propene poisoning, enabling the combined contribution of both HC- and NH_3-SCR pathways. In addition, decreased N_2O emission from slow SCR reaction is observed when CO or C_3H_6 present partly due to each reaction with ammonium nitrates. All three reductants do not compete with NH_3 for storage sites and only C_3H_6 inhibits NH_3 oxidation. This study underscores the importance of accounting for the impact of different reducing agents on conventional NH_3-SCR reaction mechanism during the rich regeneration for integrated systems (LNT + SCR, SCR on filter) or cold-start period.

Development of SSZ-13 based sulfur resistant NH_3-SCR catalysts

Chao Wang, Erdem Sasmaz, Benjamin Galloway, Bihter Padak, Jochen Lauterbach*
Department of Chemical Engineering
University of South Carolina
Columbia, SC, 29208, U.S.A.
*lauteraj@cec.sc.edu

The ammonia-selective catalytic reduction (NH_3-SCR) of NO_x is a widely used technology for the reduction of NO_x emissions in stationary sources (gas-, oil-, and coal-power plants) and mobile sources (e.g., diesel engines). Previous studies have shown that the activity of SCR catalysts depends on a wide parameter space including the catalyst type (i.e., V_2O_5/TiO_2 vs. zeolites), metal type, synthesis method, and reaction parameters (flue gas composition, reaction temperature, and space velocity). Moreover, the presence of excess O_2, SO_2 and H_2O vapor in combustion exhaust affects both catalyst selectivity and stability at high and low temperatures.

Many studies have shown that zeolites are promising catalysts to be operated under coal combustion flue gas conditions. Previous work indicates that SSZ-13 based zeolite catalyst own higher thermal stability and NO_x conversion than other zeolite catalysts under ideal flue gas condition (NO, NH_3, O_2 and N_2) [1]. In this work, SSZ-13 catalysts loaded with Cu, Fe, Ce are synthesized and characterized to test NH_3-SCR activity under SO_2 and H_2O feed conditions. All catalysts are found to deactivate at 250°C with the addition of SO_2 and H_2O vapor. Among them, Cu-SSZ-13 shows the best poisoning resistance and keeps more than 90% NO conversion at high temperature. At low temperature, catalysts deactivate more than at high temperature, which might be due to the formation of $(NH_4)_2SO_4$ on the catalyst [2]. XPS results support the possible formation of $(NH_4)_2SO_4$, indicating that both NH_4^+ and SO_4^{2-} present on the surface of the catalysts. Detailed deactivation mechanism for different metal loading SSZ-13 catalysts will be explored.

References

1. D. W. Fickel, E. D. Addioa, J. A. Lauterbach, R. F. Lobo, Appl. Catal. B Environmental 102, 441 (2011)
2. G. S. Qi, R. T, Yang, R. Chang, Appl. Catal. B Environmental, 51, 93 (2004).

387796 Understanding of the Transition Metal Ion Properties and Their Impact on Various Catalytic Functions of Zeolite-Based SCR Catalysts

Abstract:

Copper and iron-zeolite SCR catalysts are widely used in US and European diesel after-treatment system to achieve drastic reduction in NOx emissions. The type of transition metal ion in zeolite has a significant impact on their key performance and durability characteristics including the response to sulfur containing exhaust gas. For example, it has been shown that at low temperatures, under standard SCR conditions, Cu-zeolites show higher steady state NOx conversion than Fe-zeolites and reverse is true at high temperatures. These differences have been reported in literature, but the underlying mechanisms were not very well understood mainly due to lack of comparable reference catalysts, which potentially introduces confounding effects for various reasons. In this article, we made an attempt to address the impact of only the transition metal ion on various SCR catalytic functions, by using a single source of zeolite BEA in the preparation of H-, Fe- and Cu-BEA samples.

Similar molar amounts of copper and iron were ion-exchanged into H-BEA and washcoated on the cordierite support by Johnson Matthey Inc. NOx conversion with NH_3 as a reductant and NO and NH_3 oxidation, and their storage were among the catalytic properties evaluated. The above catalytic properties were measured before and after sulfation of the SCR catalysts. Sulfation of the catalysts is achieved by exposing them to a mixture of SO_2 and SO_3 containing gas feed. The cumulative amount of sulfur stored on the catalyst and the rate of SOx release from sulfated SCR catalysts were determined using temperature programmed desorption under oxidizing conditions. Detailed experimental set up and conditions were described from our previous publications on SCR topic [1].

Below 300C, Fe-BEA showed lower NOx conversion than Cu-BEA despite the former catalyst more active for NO oxidation to NO_2, a potential reaction intermediate. Upon sulfation, the NOx reduction activity of Cu-BEA degraded much more than Fe-BEA SCR catalysts. Adsorption and desorption of various probe molecules such as NO_2, SOx, and NH_3 indicated that the Fe-BEA is relatively more acidic than Cu-BEA therefore, the former catalyst is prone to strong NH_3 adsorption on Fe active species inhibiting the NOx conversion at lower temperatures. On the other hand, more acidic Fe-BEA shows lower affinity to acidic species such as sulfur oxides. This study reconciles a number of observed differences in catalytic properties of Fe- and Cu-zeolite SCR catalysts, mainly based on their acid-base characteristics. A systematic and comprehensive analysis, which yielded practically relevant insights into the catalyst performance and sulfation and desulfation behaviors, will be presented.

[1] Ashok Kumar, Michael A. Smith, Krishna Kamasamudram, Neal W. Currier, Hongmei An, Aleksey Yezerets, "Impact of different forms of feed sulfur on small-pore Cu-zeolite SCR catalyst", Catalysis Today, 231 (2014) 75-82

374792 Controlled Synthesis of Novel Interweaved Titanium Oxide Nanotubes Confined Metal Oxide Catalytic Formulations: Effect of Morphology on Selective Catalytic Reduction (SCR) of NOx By NH3

Abstract:

The textural properties of various metal oxide confined titania nanotubes and their influence on surface active sites for NH_3−SCR were studied. Remarkably, the surface texture and tubular morphology of Me/TNT-H catalysts greatly promoted the NO_x conversions and temperature window compared to their nanoparticles. It can be ascribed to the preferentially located active species in the tubular framework. The impregnation of MnO_x over TNT-SA catalyst occur bare minimum because of the low specific surface area and low porevolume compared to other titania natotube supports, and this leads to poor deNO_x activity and narrow temperature window. Raman and X-ray diffraction patterns indicated that tubular structure in Mn/TNT samples significantly increases the concentration of oxygen vacancies and thus facilitate NO oxidation to NO_2. The relative atomic percentage value of Mn^{4+}/Mn^{3+} characterized by deconvoluted XPS spectra was considerably high for the Mn(0.25)/TNT-H catalyst (Mn^{4+}/Mn^{3+} = 2.15) whereas, surface Mn^{3+} species are in equivalence with MnO_2 phase in other catalysts (Mn^{4+}/Mn^{3+} = 0.36-1.34). In H_2-TPR, the disappearance of high-temperature reduction peak at 482 °C suggests that the Mn^{4+} species are dominant in Mn/TNT-H sample compared to Mn/TiO_2 nanoparticles. Our Raman results illustrate the redshift of 14 cm^{-1} from 640 cm^{-1} (TiO_2-Hombikat) to 626 cm^{-1} in Mn/TNTs-H sample which indicates the formation of a new phase or structural difference. In comparison with the Mn supported on TiO_2 (Hombikat) nanoparticles, the Mn species on the developed tubular Mn/TNT-H were less pronounced to coagulate under the reaction temperatures (100-300 °C) and demonstrated an impressive promotion in the deNO_x performance and SCR activity window. The intrinsic activity of the prepared catalysts measured under differential reaction conditions and the effect of morphological features will be discussed in the presentation.

Bimetallic Pt-Pd Catalyst kinetics for CO and C_3H_6 Oxidation in Diesel Oxidation Catalysts (DOC)

Melanie J. Hazlett and William S. Epling
University of Houston, Houston, TX (USA)
*wsepling@uh.edu

Introduction

Today's diesel engines are more fuel efficient than gasoline engines, but even so increases in fuel economy are needed. With this in mind and due to increasingly stringent environmental policies on NO_X, hydrocarbons (HC), CO, CO_2 and particulate matter (PM) emissions from diesel engines, low temperature combustion (LTC) engines have been developed. The emissions from these LTC engines have significantly lower concentrations of emitted NO_X (by a factor of 6) and PM (by a factor of 8), however the HC and CO concentrations are much higher, by factors of 20 and 7 respectively. Also the exhaust temperatures are 40-70°C lower. With these changes, significantly more emphasis needs to be put on the activity of the oxidation catalyst function and low temperature NO_X reduction performance.

Diesel oxidation catalysts (DOCs) are designed to oxidize CO, NO, and HCs, and typically contain Pt and Pd. It has been shown that adding Pd to Pt based DOCs leads to less Pt sintering relative to monometallic Pt catalysts [1]. Pt is a much better NO oxidation catalyst than Pd, and Pd is less sensitive to CO poisoning than Pt [2]. Non-uniformly distributed oxidation catalysts have been studied [3], where the premise for DOCs is axially distributing active sites will minimize the amount of metal used to meet conversion targets. A non-linear relationship between oxidation performance and Pt or Pd content has been observed, with different minimum Pt:Pd ratios achieving low light-off temperatures for different compounds [4]. With these features, there is a unique possibility that DOCs can be designed with gradients in Pt and Pd amounts, as well as ratios, that could lead to improved low temperature performance.

Methodology

Pt, Pd, and different ratios of Pt-Pd bimetallic catalysts, supported on Al_2O_3, were prepared via incipient wetness impregnation. For the data to be described, all catalysts are based on 1 wt% Pt catalyst loading, with a molar equivalent of Pd considered (0.55 wt%). Catalysts were dried overnight and calcined for 4 hours at 550°C. Catalysts were pressed and sieved to 40-60 mesh, and placed in a quartz tube packed bed reactor with SiO_2 pellets of same size to minimize temperature gradients. Temperature programmed oxidation (TPO) experiments were conducted with CO (1000-4000ppm), and C_3H_6 (500-2000 ppm) with excess O_2 (6-10 vol%) from 100-300°C. Outlet concentrations were measured with a MKS MultiGas2030 FTIR gas analyzer. Results were modeled as a PFR, with low conversion (1-5%) data used to calculate kinetic parameters (activation energies and frequency factors).

Results and Discussion

The CO and C_3H_6 kinetics for monometallic Pt and Pd have been tested and modeled, and we are currently characterizing the kinetics for various bimetallic Pt-Pd catalysts with varying Pt:Pd ratios. We started with the Voltz kinetics [5] as a basis. Using these, CO oxidation kinetics calculated for the Pt catalyst using LHHW kinetics ($r = kC_{CO}C_{O2}^{1/2}/(1+KC_{CO})^2$) show good agreement with experimental data, although we did modify the mechanism to include dissociative oxygen adsorption instead of molecular adsorption. Oxidation kinetics for Pd were also modified, since PdO is the assumed active state rather than metallic Pd. This also led to good fits between experimental and modeling results. At this stage we are characterizing kinetic and performance data from the bimetallic catalysts, and these will be included in the presentation, with insight into designs for improved CO and HC oxidation included.

In terms of performance, a 1:1 ratio of Pt:Pd typically proved to have the best T_{50} value (temperature where 50% conversion was achieved). This was true for CO oxidation, C_3H_6 oxidation and the oxidation of both when added together. However, different ratios led to the best T_{90} values. Again, this suggests a possible optimized design to attain the best performance with minimum precious metal content.

Overall, the goal of this research is designing a better DOC (reduction in precious metal content or improved performance), by taking advantage of the effects of different Pt-Pd ratios. Kinetic parameters calculated can be used in a monolith reactor model to determine optimal axial distribution of PGM.

REFERENCES

1. J.C. Summers, and L.L. Hegedus, *Journal of Catalysis* 51(1978)185.
2. R. Chen, Z. Chen, B. Ma, X. Hao, N. Kapur, J. Hyun, K. Cho, and B. Shan, *Computational and Theoretical Chemistry* 987(2012)77.
3. Y.D. Kim, S.J. Jeong, and W.S. Kim, *Chemical Engineering Science* 64(2009)1373.
4. M. Skoglundh, L.O. Lowendahl, and J.E. Ottersted, J.E. *Applied Catalysis* 77(1991)9.
5. S. E. Voltz, C. R. Morgan, D. Liederman and S. M. Jacob, *Industrial & Engineering Chemistry Product Research and Development,* 12(1973)294.

377116 Characteristics of Pt-BaO/Mgal Mixed Oxides for NOx Storage-Reduction Catalysis

Abstract:

Efficient NOx abatement catalyst operated under lean condition must be developed to promote the use of more fuel-efficient diesel and/or lean-burn engines since conventional three-way catalyst does not work properly. For this purpose, selective catalytic reduction (SCR) catalyst or NOx storage-reduction (NSR) catalyst were commercially being used. NSR technology, which is more appropriate for the small-sized engine, primarily consisted of noble metals such as Pt, alkali earth metal (e.g. Ba) and support (e.g. Al_2O_3 or $MgAl_2O_4$). It was demonstrated that $MgAl_2O_4$ support provided the improved NOx storage activity at higher temperature compared with Al_2O_3 support, implying that support material can play a crucial role in affecting the stability of barium nitrate species.

Hydrotalcite (HT) material, which can be precursors to $MgAl_2O_4$ spinel structure, has received more attention due to its various advantages, such as large surface area, controllable acid-base properties, and the improved thermal stability. In this work, HT materials with different Mg/Al ratios were prepared, resulting in the different amount of $MgAl_2O_4$ and Al_2O_3 in the support after calcination. Then, Pt-BaO active components were loaded on these HT-derived supports, and the catalysts were extensively investigated by using NOx storage measurement and various analytical techniques. In addition, the effect of the promoter on the NOx storage activity in Pt-BaO/MgAl mixed oxides was also investigated.

Combination of activity measurement and characterization results indicates that MgAl mixed oxides supports give rise to the higher thermal stability of barium nitrates and Pt species as well as superior activity than one supported on Al_2O_3, moreover, can be optimized further for the high temperature NSR application by changing Mg/Al ratios in HT support. This work provides fundamental understanding for optimizing HT-derived $MgAl_2O_4$ support, which would be a strong candidate for the high temperature NSR application.

HYDROGEN PURIFICATION FOR FUEL CELLS:

Preferential CO Oxidation in Excess of Hydrogen

Oscar Eduardo Amador [a,*], Luis Fernando Córdoba [b], Julio Cesar Vargas [c]

[a] *Department of Chemistry and Environmental Engineering, Faculty of Engineering, National University of Colombia, Cr 30 Cll 45, Bogotá, Colombia.*
[b] *Department of Chemistry and Environmental Engineering, Faculty of Engineering, National University of Colombia, Cr 30 Cll 45, Bogotá, Colombia.*
[c] *Department of Chemistry and Environmental Engineering, Faculty of Engineering, National University of Colombia, Cr 30 Cll 45, Bogotá, Colombia.*

Abstract

In the present work we have studied the effects of the gold load, as well as the O2/CO ratio, using gold catalyst supported on cerium and zirconium oxides, on the catalytic performances in PROX reaction. The catalysts of Au-CeO2-ZrO2 were prepared by the pseudo sol-gel method from the thermal decomposition of a mixed resin precursor metal propionates. The synthesized catalysts were characterized by various techniques in order to determine the main physic-chemical and structural properties and then related with the results of selective CO oxidation reactions. The specific surface area of the catalysts was determined by N2 adsorption desorption at 77K and interpreted using the Brunauer - Emmet -Teller (BET). The crystalline phases and the determination of the degree of crystallinity of the synthesized materials were identified by X-ray diffraction (XRD). The texture, shape and size of the materials was studied by scanning electron microscopy - SEM and analyzed by elemental microcomposition energy dispersive spectroscopy X-ray - EDXS coupled to SEM.
The catalytic activity and selectivity measurements in PROX over the studied gold catalysts were carried out using a quartz glass U-shaped reactor, equipped with a temperature programmed controller. The reactor effluent was analyzed on-line by an assembly of two, in series, gas chromatographs. The reaction was evaluated in a temperature range of 70-300 ° C, the catalytic tests were performed at ensuring higher GHSV 40000 h-1 hourly space velocity, the activity, hydrogen yield and products distribution was analyzed.

Keywords
ProxCO, Preferential CO Oxidation, Direct Anionic Exchange Method, Gold Catalyst.

1. Introduction

Most of the energy used today comes from fossil fuels that cause serious environmental problems, one of the most attractive options that is being considered is the use of H_2 as fuel, both for issues of power generation (fuel cells) and for direct use in internal combustion engines (heat engines H_2. The production of electrical energy from H_2 feed directly electric motor engines, either by direct combustion or use in fuel cells. The H_2 is a clean way to generate power, as the only product obtained in the process is H_2O [1].

Combustion Engine.
$H_2 + \frac{1}{2} O_2 \rightarrow H_2O$ + Heat generation

Fuel Cell
$H_2 + \frac{1}{2} O_2 \rightarrow H_2O$ + Electric Power

The reforming of methane with water vapor is one of the most important used for the production of H_2 with 95% usage. There is other methods, but in all these processes always involve the simultaneous CO formation [2]. Although the water gas shift reaction is effective to oxidize CO, efficient elimination of trace amounts of CO is required due to small amounts of CO present in the H_2 streams poison the Pt electrocatalysts of PEM fuel cells: an objective which can be achieved by the preferential oxidation reaction of CO (PROX), because its low-cost, effective, available method for reducing CO to a desired level (less than 100 ppm of CO) without excessive hydrogen consumption [3-4].

* Corresponding Author.
oeamadorc@unal.edu.co (O. Amador)

In this paper we present the influence of the active phase integration on the catalytic behavior of impregnated gold at two concentrations over a bimetallic mixed oxide support with formula $Ce_{0.65}Zr_{0.35}O_2$ in processes of selective oxidation of CO. Further, a trimetallic Nickel doped (Ce–Zr) mixed oxide ($Ce_{0.65}Zr_{0.3}Ni_{0.05}O_2$), were prepared and studied as catalysts. In order to understand the influence of the active phase and the role of this concentration on the selective oxidation of CO, the synthesized fresh mixed oxides were characterized by XRD and TPR and carbonaceous deposits produced during the reaction was quantified by TPO.

2. Experimental

2.1.Preparation of Catalysts

Ce–Zr support and Ce–Zr –Ni mixed oxide was synthesized using the pseudosol–gel method based on the thermal decomposition of metallic propionates[5]. The salts used were cerium(III) acetate hydrate, zirconium(IV) acetylacetonate and nickel(II).The support and mixed oxide prepared could be represented with next the formulas:

$Ce_{0.65}Zr_{0.35}O_2$ (CeZr)

$Ce_{0.65}Zr_{0.35}Ni_{0.05}O_2$ (Ni/CeZr)

The starting salts were dissolved separately in boiling propionic acid (0.12 mol/L). The solutions were mixed and the solvent was evaporated until obtaining a resin, which was calcined in air at 600 °C over 3 h (2 °C /min).

The impregnation of gold on the support of CeZr was conducted by direct ion exchange method, since according to reported results [6] through this method of preparation predominant presence of gold metal on the surface.

Because this method does not change the pH, the deposition of gold species is a function of the concentration of absorbable species. Analysis of mass spectrometry and UV-Visible (UV-VIS) spectroscopy reported [7] show that is favored and favorable deposition prevails species [8] in solutions of concentration 10^{-4} M and 10^{-3} M temperatures near 70 ° C, under these conditions we have obtained the best results for this support.

The crystalline structure of mixed oxides was determined by X-ray diffraction with Brucker D8 Advanced apparatus, Cu Kα radiation, in an interval $2\theta = 10°$ up to $2\theta = 100°$.

The reducibility was studied by thermo-programmed reduction (TPR) of a 0.10 g sample, in a hydrogen/argon mixture (50 mL/min, 3.85% (v/v)). The temperature was increased at 15 °C/min from room temperature to 900 °C.

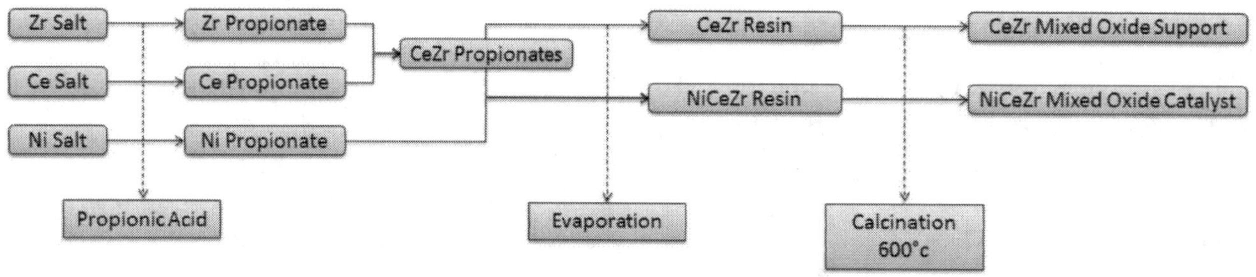

Fig1: Global catalyst preparation procedure following the sol-gel like method.

Fig2: Global catalyst preparation procedure following the DAE like method.

2.2.Catalyst Test

The CO selective oxidation was conducted in a tubular quartz reactor of fixed bed with 6 mm inner diameter, at atmospheric pressure in the temperature range 25 ° C - 250 ° C with 0.100 g of catalyst, the heating system consists of a cylindrical electric furnace. The reactant feed was a mixture of helium (purity 99.9995%), hydrogen, CO and synthetic air (80% N2 + 20% O2). In the experiment a total gas flow of 100 ml / min was handled. In the first test a concentration of 1.5% O2 in flux and for the second test of 3.0% was handled. The catalyst was heated to 300 ° C under a flow of 60 ml / min of H2 for one hour, this in order to simulate a reducing atmosphere and generate the activation of the catalyst, then with this reducing atmosphere the temperature was initial catalytic test, once obtained this temperature, the mixture of gases (He, CO, H2 and air) was introduced and the temperature was increased to that of the catalytic test. The temperature was increased in stages 25, 50, 100, 150, 200 and 250 ° C. At the end of the catalytic test gas flow was stopped and the catalyst was cooled under a stream of He, and stored for further characterization.

The gas reaction products were analyzed on two chromatograph connected in series; the first was a HP 5890 series II GC equipped with a HP-PLOT / Q column (0.53 mm ID and 30 m in length, Agilent Technologies, USA) in which CO2 is analyzed, and the second column GC equipped with HP 5890 A 5 A Molecular grid (Agilent Technologies, USA) for H2, CH4, and CO.

The catalytic tests were performed at ensuring higher GHSV 40000 h-1 hourly space velocity. To analyze the catalytic activity, selectivity and yield to hydrogen production and distribution of products, considering the effect of the charge on the metal (Au) and the effect of the concentration of O2 in the feed mixture.

2.3.Catalytic activity and selectivity

The analysis of the effluent gas was performed by online gas chromatography.

The conversion of CO to CO2 is calculated from the following equation:

$$X_{CO}(\%) = \left(1 - \frac{[CO]_{out}}{[CO]_{in}}\right) * 100$$

Where CO_{in} and CO_{out} are the initial and outlet CO concentration, respectively. The selectivity of oxygen to oxidize CO is defined as follows:

$$S(\%) = 0.5 * \frac{[CO]_{in} - [CO]_{out}}{[O_2]_{in} - [O_2]_{out}} * 100$$

3. Analysis and Discussion

The structural and morphological properties of catalysts Gold supported on Cerium-zirconium oxides were investigated and discussed in detail in a previous work by Laguna et al. Characterization was carried by several techniques (N2 adsorption,

XRF, XRD, Raman Spectroscopy, Techniques SEM and H2-TPR. [11] also found that the DAE method is more suitable than those methods as DP, IWI for the preparation of gold catalysts supported on oxides of cerium-zirconium, allowing a higher fraction of the species of nano-sized gold to be exposed on the catalyst surface [8].

In the case of the Au / CeZr10, there is a high-temperature reduction peak with maximum at 746 ° C. This signal has a very broad shoulder at lower temperature, which extends until 300 ° C. Finally, a low-temperature reduction process, around 145 ° C, Which is not observed for the CeZr support, Appears [11]. The resulting behavior is related to the method of preparation, which leads to the different sizes of the gold particles and improving the capacity reduction of the mixed oxide.

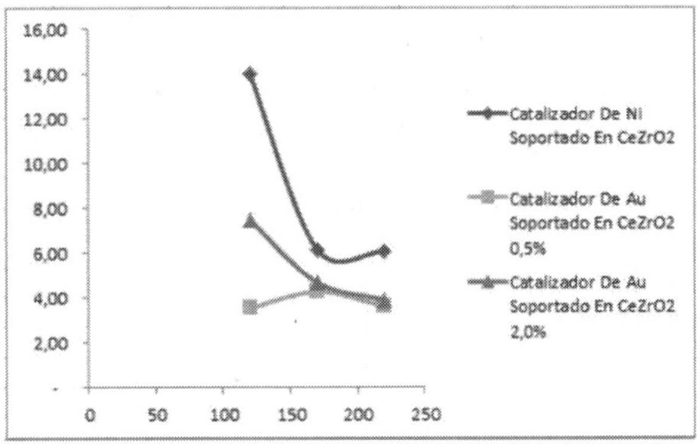

Fig 3. Selectivity (%) vs. Temperature (°C). Feed: 1.0% CO, 1.5% O2, 50.0% H2 and He of Balance..

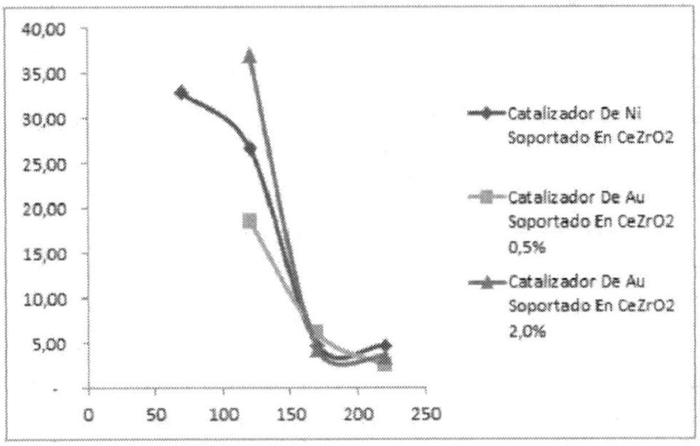

Fig 4. Selectivity (%) vs. Temperature (°C). Feed: 1.0% CO, 3.0% O2, 50.0% H2 and He of Balance.

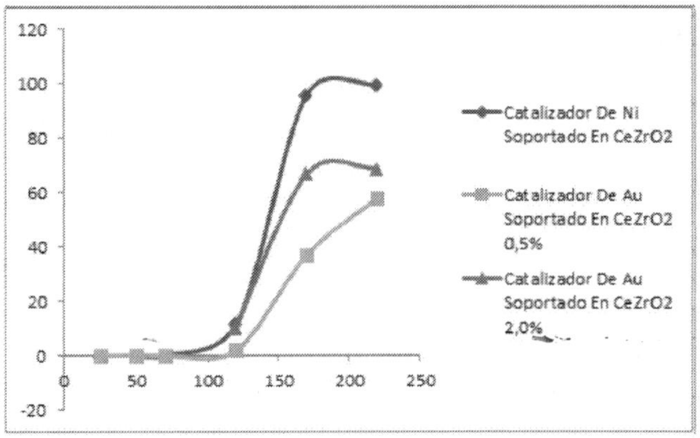

Fig 5. Conversion (%) vs. Temperature (°C). Feed: 1.0% CO, 1.5% O2, 50.0% H2 and He of Balance..

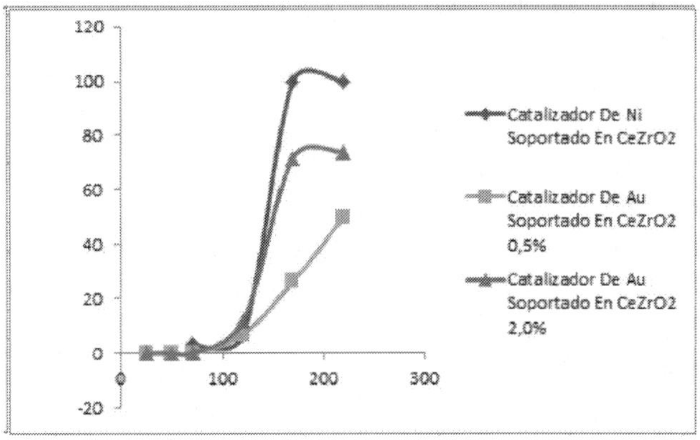

Fig 6. Conversion (%) vs. Temperature (°C). Feed: 1.0% CO, 3.0% O2, 50.0% H2 and He of Balance.

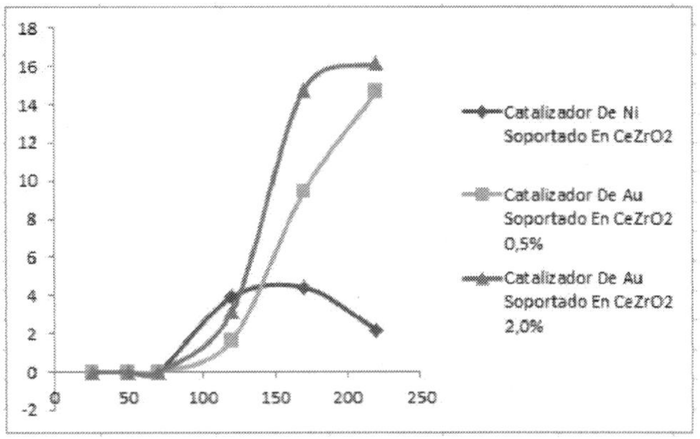

Fig 7. Conversion (ppm) vs. Temperature (°C) of CO_2 Catalyzed. Feed: 1.0% CO, 1.5% O2, 50.0% H2 and He of Balance.

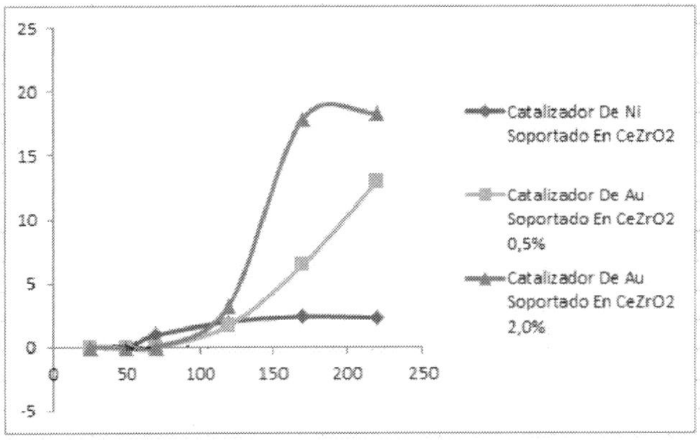

Fig 8. Conversion (ppm) vs. Temperature (°C) of CO_2 Catalyzed. Feed: 1.0% CO, 3.0% O2, 50.0% H2 and He of Balance.

Figures 3 and 4 show selective activity toward the production of CO_2 0.5% Au/CeZr, 2.0% Au/CeZr and NiCeZr catalyst prepared by the method DAE for the selective oxidation of CO. Figures 5 and 6 show the obtained conversion to CO_2 production of 0.5% Au/CeZr, 2.0% Au/CeZr and NiCeZr catalyst prepared by the DAE method for the selective oxidation of CO. Likewise by Figures 7 and 8 we can observe the behavior of the oxidation of CO to CO_2, and this varies as a function of increasing temperature and the change in oxygen concentration in the feed streams.

It can be observed that the conversion and selectivity depend strongly on the nature of the catalyst. In fact, none of the catalysts produces measurable catalytic activity at temperatures below 120 °C. It was observed that the selectivity decreased progressively from 14% for the first cup of food stabilizing close to 4% and for the second cup had to be decreased from 40% to re-stabilize around 4%, this phenomenon presented with increasing progressive temperature.

Among the three catalysts tested, the NiCeZr is the most active, while the Au/CeZr is the most selective for the oxidation of CO and that the largest concentration of tested gold. The method of preparation strongly influences the catalytic performance and greater efficiency to the production of CO_2 was obtained with the DAE method.

The maximum conversion of CO achieved on the sample 2.0 Au / CeZr was 74% with 4% selectivity, 220 ◦C, the highest selectivity of 37% was achieved for a CO conversion of 12% and the test with the highest concentration of O2 (3%). For the sample of 0.5 Au / CeZr the highest conversion obtained was 58% with a selectivity of 4%, this for a feed of 1.5% in oxygen flow and greater selectivity was 19% at a conversion of 7% for run with higher oxygen concentration (3.0%). As the catalyst NiCeZr obtained the highest conversion was 100% at a selectivity of 4.75% and the maximum selectivity obtained was 33% to 3% conversion results for both oxygen supply flow of 3%.

4. Conclusions

Considering that high concentrations of hydrogen show no significant effects on the selective oxidation of CO [12], we can assign the significant increase in selectivity (passing from 7% to 37%) to increase the oxygen concentration in the feed stream.

The temperature increase during the tests developed to catalysts where the active phase is gold causes a significant increase in the selective oxidation of CO to CO2 increase, likewise the increase is also benefited by the increased concentration of O2 in the feed stream. (Figure 7 and 8).

As the catalyst of NiCeZr which has 5% of active phase (Ni) very low selectivity was found in the oxidation of CO to CO2, because due to the high concentrations of H2 this catalyst produced the oxidation of hydrogen to water and the methanation of carbon monoxide to methane. Although this catalyst showed the best results in terms of conversion is concerned, the results of selectivity to CO2 is very low.

Acknowledgements

Thanks to the Department of Chemical and Environmental Engineering of the National University of Colombia by the flanged financial support for the project, also thanks to Mr. Ricardo Cortez, for their cooperation in conducting the investigation.

References

[1] Linares, J. Moratilla, B. Ins. de Ing. de España (2007) 25-86.

[2] A. Iulianellia, G. Manzolinib, M. De Falcoc, S. Campanarib, T. Longoa, S. Liguoria, d, A. Basilea. International Journal of Hydrogen Energy. 20 (2009) 11514-11524

[3] Jenny E. Hu, Zhufang Liu, Bryan W. Eichhorn, Gregory S. Jackson, International Journal of Hydrogen Energy. 15, (2012), 11268-11275

[4] Seyyed M. Mousavi, Siew H. Chan, Electrochimica Acta. 93, (2013) 334-345

[5] F. Romero-Sarria, Julio C. Vargas, Anne C. Roger, Alain Kiennemann , Catalysis Today. (2008) 149-153.

[6] Ivanova, S., Petit, C., Pitchon, V., Applied Catalysis A: General, 2004. 267: p. 191-201.

[7] Ivanova, S., Formation de nanoparticles d'or supportèes : De la préparation à la réactivité catalytique.Tesis Doctoral, 2004, Université Louis Pasteur de Strasbourg. Strasbourg. p. 221.

[8] Aguirre, L. Estudio de la influencia del método de síntesis en las propiedades catalíticas del oro soportado en la activación del glicerol. Tesis de Maestria, 2010, Universidad Nacional de Colombia. P 35-36.

[9] T.R. Reina, E. Papadopoulou, S. Palma, S. Ivanova, M.A. Centeno, T. Ioannides, J.A. Odriozola. Applied Catalysis B: Environmental, 150–151 (2014) 554-563.

[10] Xuemei Liao, Wei Chu, Xiaoyan Dai, Véronique Pitchon. Applied Catalysis A: General 449 (2012) 131-138.

[11] O.H. Laguna, F. Romero Sarria, M.A. Centeno, J.A. Odriozola. Journal of Catalysis. 2 (2010) 360-370

[12] Cécile Rossignol, Sandrine Arrii, Franck Morfin, Laurent Piccolo, Valérie Caps, Jean-Luc Rousset. Journal of Catalysis. 2 (2005) 476-483.

[13] G. Avgouropoulos, T. Applied Catalyst. A:Gen. 244 (2003) 155-167.

[14] D.L. Trimm, Z.I. Onsan, Catalyst Rev. Sci. Eng 43 (2001) 31-84.

Chemical Engineering for Renewable Energy

Rakesh Agrawal, Emre Gencer and Gautham Ramapriya

School of Chemical Engineering

Purdue University, West Lafayette, IN 47907

In the long run, it is likely that all the basic human needs will be met by renewable sources like solar energy. However, there are several challenges associated with harness, storage and use of solar energy to meet our daily needs for food, chemicals, heat, electricity and transportation. In a sustainable future, all these usage must coexist.

We will first present some results from our energy systems modeling to produce hydrogen from solar energy and our modeling results for energy storage at giga Watt-hour levels.

We will touch upon the continuing advancements made in the traditional fields such as separations that will still play a vital role in the processes slated for any renewable energy future.

375531 Some Early Meanderings in Unsteady-State Catalysis

Abstract:

Research work on unsteady-state reaction systems began at Union Carbide in the early 1970s. During this period we discovered the catalytic oxidative coupling of methane to form ethylene and other species. A number of other reactions were also studied. This paper will cover how we did the experimentation, what some of the findings were, and how I began to formulate the principle of optimal dumbness as a valuable research tool.

367576 Bridging Heterogeneous Catalysis and Electrocatalysis:Catalytic Reactions Involving Oxygen

Abstract:

Catalytic reactions that involve oxygen can be found in a large number of processes, including those in energy-related applications, in emission control and in processes important for the chemical industry. Whether the catalytic reaction is an oxygen insertion step as in a selective oxidation reaction, or an oxygen removal step as in a hydrodeoxygenation reaction, oxygen has proven to be a very challenging component, often determining the selectivity of the reaction. Some examples from our laboratories that bridge catalysis and electro-catalysis will be discussed, ranging from oxidative dehydrogenation of alkanes to oxygen reduction reaction in fuel cells.

369549 Solid Acid Catalyzed Alkylation of Isobutane with Light Olefins

Abstract:

Alkylate is an ideal gasoline blending component because of its high octane number, low vapor pressure and absence of sulfur and aromatic compounds. Over the last six decades, production of high-octane alkylate via alkylation of isobutane with light olefins has relied on hazardous liquid acid (HF or H2SO4) catalysts, which pose safety risks and have caused unfortunate refinery accidents. Although solid acid catalysts (SACs) have been explored for nearly four decades as safer alternatives, they deactivate rapidly and have largely been economically unattractive when compared to liquid-acid catalyzed alkylation. Over the past decade, Exelus has designed a new solid-acid catalyst -ExSact - with significantly higher stability, which makes solid acid alkylation finally competitive with conventional liquid acid technologies. The unique design of the ExSact catalyst allows a wide variety of feedstocks to be processed while maintaining high overall product quality and yield. A simple regeneration scheme using hydrogen allows the design of a dual fixed bed process. The viability of this approach has been ascertained through 1000+ hour testing using refinery feedstocks. In this talk, the development of this process will be described, placing it in the context of other recent studies.

378983 Challenges in Multiphase Reaction Engineering

Abstract:

The key challenge for reaction engineering of multiphase systems is in the implementation of science - based scale-up methodology for transfer of molecular discoveries to sustainable, non-polluting technologies that can meet the future energy, environmental, food and materials needs of the world. Available scientific foundations for the needed multiscale approach to process development are briefly reviewed and progress made in replacing empiricism with science discussed. It is recognized that progress in implementing the desired environmentally benign processes depends, in addition to overcoming the remaining technical challenges, on achieving changes in the socio-economic and political arena. Global regulations and peer pressures may be needed to provide the economic incentives for cleaner and sustainable processes.

Aerosol Reaction Engineering: Functional Materials & Devices

Sotiris E. Pratsinis
Particle Technology Laboratory, Department of Mechanical and Process Engineering, ETH Zurich (Swiss Federal Institute of Technology), CH-8092 Zurich, Switzerland
http://www.ptl.ethz.ch/

Aerosol manufacturing routes are advantageous over wet chemistry ones as they offer fewer process steps, easier particle collection and no liquid by-products requiring costly cleaning. In addition, aerosol-made particles and films have unique morphology and high purity (e.g. optical fibers) and can even form metastable phases (e.g. low temperature $BaCO_3$ for NOx storage-reduction catalysts). So they easily form mixed oxides, salts and even pure metals and highly porous but stable films resulting in novel catalysts, micropatterned sensors, phosphors, battery electrodes, dental prosthetics and even nutritional supplements[1]. This has led to synthesis and sales of new materials (e.g. iron nanoparticles for nanofluids) and construction of pilot scale facilities by catalyst manufactures (e.g. Johnson Matthey).

So renewed emphasis is placed on the fundamentals of flame aerosol synthesis of aggregates and agglomerates of particles[2]. For example, primary particles in an aggregate can have a narrower size distribution as it is predicted by self-preserving theory[3] and verified by experiments at Cabot[4].

So recent advances in aerosol reaction engineering allow now synthesis of portable, highly selective sensors[5] consisting of flame-made, metastable ε-WO_3 for on- and off-line monitoring of breath acetone, a tracer for diabetes type-1. These sensors accurately detected breath acetone with fast response-recovery times and a high signal-to-noise ratio of human subjects in collaboration with the Medical University of Innsbruck, Austria. Best correlations were found after overnight fasting (morning) between sensor response and blood glucose and breath acetone.

Furthermore complex material architectures[6] were prepared that were highly active under visible light and comprising of titanium suboxide (e.g. Ti_4O_7, Ti_3O_5) layers onto nano-Ag on nano TiO_2. Abundant combustion intermediates present during flame synthesis of these materials partially reduce TiO_2 and induce strong metal-support interactions (SMSI) resulting in *crystalline* Ti-suboxides that are *stable* upon annealing in air, at least, up to 350 °C for two hours. Under visible light ($\lambda > 400$ nm), these particles exhibit strong photo-reduction of cationic species (Cr^{6+}) & photo-oxidation of methylene blue (15-min half-life).

1. Aerosol-based Technologies in Nanoscale Manufacturing: from Functional Materials to Devices through Core Chemical Engineering, *AIChE J.*, **56**, 3028-3035 (2010).
2. M.L. Eggersdorfer, S.E. Pratsinis, Agglomerates and Aggregates of Nanoparticles made in the Gas Phase, *Adv. Powder Technol.* **25**, 71-90 (2014).
3. M.C. Heine, S.E. Pratsinis, "Polydispersity of Primary Particles in Agglomerates made by Coagulation and Sintering" *J. Aerosol Sci.*, **38**, 17-38 (2007).
4. D. Boldridge, Morphological Characterization of Fumed Silica Aggregates. *Aerosol Sci. Technol.* **44,** 182-186 (2010).
5. M. Righettoni, A. Schmidt, A. Amman, S.E. Pratsinis, "Correlations between blood glucose and breath components from portable gas sensors and PTR-TOF-MS", *J. Breath Res.*, **7**, 037110 (2013).
6. K. Fujiwara, Y. Deligiannakis, C.G. Skoutelis, S.E. Pratsinis, Visible-light active black TiO_2-Ag/TiOx particles, *Appl. Catal. B: Environ.* **154-155**, 9-15 (2014).

Micro-kinetics of NO_x Storage and Reduction with $H_2/CO/C_3H_6$ on $Pt/BaO/Al_2O_3$ Monolith Catalysts

Arun Kota, Vemuri Balakotaiah and Dan Luss

Department of Chemical and Biomolecular Engineering,

University of Houston, 4800 Calhoun Rd, Houston, TX - 77204

Abstract:

Steady state NO_x storage and reduction on $Pt/BaO/Al_2O_3$ by $H_2/CO/C_3H_6$ reactant mixtures is studied using an isothermal, 1+1 dimensional two-phase model that accounts for washcoat diffusional limitations. It predicts the storage and reduction performance over a wide range of inlet feed concentrations and temperatures. Global kinetic models have a major limitation of not being able to predict accurately the performance of the catalyst with changing feed composition. To account for such variations, micro-kinetic models elucidating the role of individual reductants have been developed. The temperature dependence of the rate constants has been quantified to reflect the surface chemistry that is consistent with experimental observations. Figure 1 shows the comparison between the experimental and model predicted data for NO/H_2, NO/CO and NO/C_3H_6 systems. The model predicts the experimentally observed trends in individual systems reasonably well and is further used to predict the conversions and selectivities for a combined $NO/H_2/CO/C_3H_6$ system.

Inclusion of washcoat diffusion enables the model to accurately predict the NO_x conversion and scale-up performance of the catalyst over a wide temperature range. Figure 2 shows that the model accurately represents different controlling regimes during NO/H_2 reaction. It can be seen that the system goes from being kinetic controlled to washcoat diffusion controlled as temperature increases. At higher temperatures most of the reaction is restricted to a thin layer of the washcoat close to

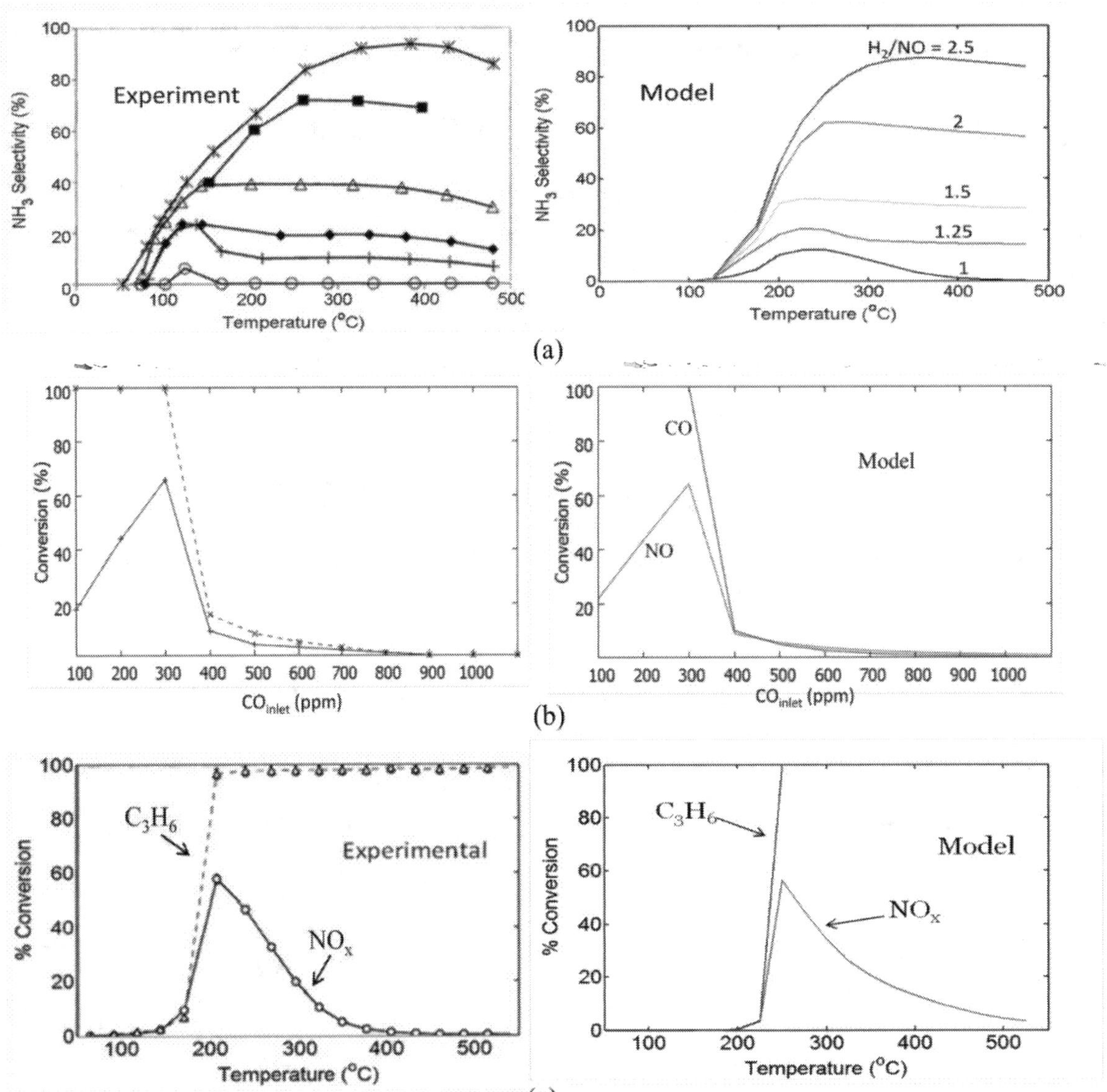

(c)

Figure 1: Comparison of the experimental and model predicted conversions and selectivities (a) NH_3 selectivity in NO/H_2 system (b) NO and CO conversions in NO/CO system and (c) NO and C_3H_6 conversions in NO/C_3H_6 system.

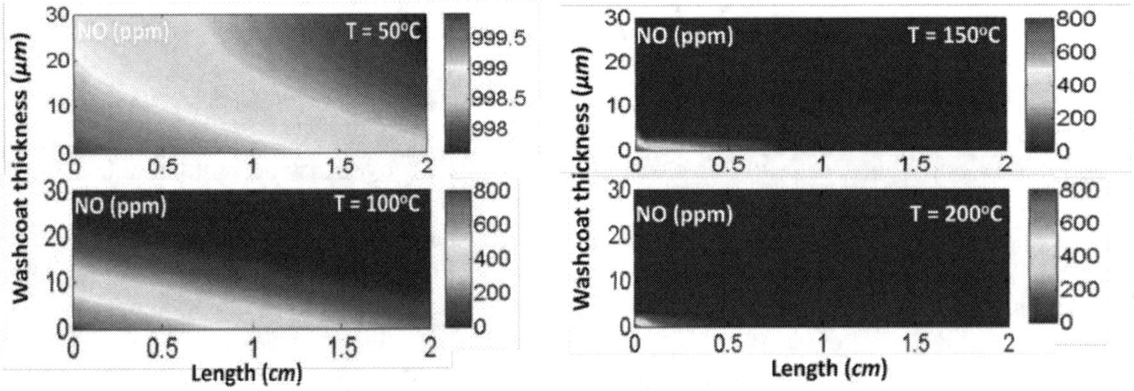

Figure 2: NO concentration inside the washcoat at different temperatures for NO/H_2 system [NO_{inlet}=1000ppm, H_{2inlet}=2500ppm; Pt loading = 2.2 wt%].

the fluid-washcoat interface. The sensitivity of the estimated kinetic parameters to the effective diffusivities in the washcoat is also analyzed by pore scale comparison between the time constants of adsorption ($C_0/R_{ads}(C_0)$) for $H_2/CO/C_3H_6$ and pore scale diffusion time constants ($\frac{r_{eff}^2}{D_{eff}}$). Results indicate that the estimated micro-kinetic parameters are influenced by the washcoat pore structure and the properties of the precious metal and storage components.

385847 Deep SO2 Adsorption at Parts per Billion Level By Alumina-Based Mn/Ce Mixed Oxides for SOFC Cathode Protection

Abstract:

Gas phase contaminant removal is a critical process for current and future power supply systems including fuel cells, gas turbines (GT), or a combination on ships. Because of high energy efficiency, low emissions, and fuel flexibility, Solid Oxide Fuel Cells (SOFCs) are promising energy converting devices. Recently, however, researchers have shown a significant degradation caused by SO_2 reacts with cathode side materials and thus hinders the active sites available for oxygen reduction. The major SO_2 source is the emission from marine fuels such as JP-5 and JP-8 whose sulfur content can be up to 1200 ppmw. The cell exposed to as low as 100 ppbv of SO_2 can result in a current loss during the test. Therefore, it is necessary to remove SO_2 from cathode side. In this work, alumina-supported/alumina-ceria supported manganese oxides were used as adsorbents operated at high space velocity. Breakthrough capacity and saturation capacity were evaluated by varying Mn loadings, namely 2.5, 5.0, 7.5, 10, and 15 wt %. The structural and surface properties of the adsorbents were further characterized by XRD, N_2 physisorption, H_2-TPR, CO_2-TPD, and TEM. The SO_2 adsorption capacity of the mixed oxides correlated closely with their metallic dispersion, surface basicity, and average oxidation state. Mn (5 wt %)/CeO_2-Al_2O_3 was found to be an optimum adsorbent in a wide temperature range of 20-500°C. H_2-TPR technique revealed that the average oxidation state for 5 wt % Mn was around +3.8; while Mn loading (<2.5 wt % or >15 wt %) upon impregnation of nitrate precursor solution was likely to lower the oxidation state to around +3.3, which was related to catalytic activity for S^{4+} oxidation. Furthermore, the reducibility of mixed oxides can be enhanced by introducing Ce into Al_2O_3, facilitating the regeneration of spent Mn/CeO_2-Al_2O_3.

A first-principles study of the catalytic hydrodechlorination of 1,2-dichloroethane

L. Xu, E. Stangland, M. Mavrikakis

[1]Department of Chemical & Biological Engineering, University of Wisconsin-Madison

[2]Core Research and Development, the Dow Chemical Company, Midland MI

Chlorinated hydrocarbons are common intermediates and by-products from various chemical processes. Industrial chlorocarbon wastes are serious environmental pollutants, and an efficient treatment method is highly desired [1]. The catalytic hydrodechlorination, which utilizes hydrogen to convert chlorinated species into hydrogen chloride and valuable hydrocarbon products, is an attractive approach from both the economic and environmental standpoints. Catalytic hydrodechlorination of 1,2-dichlorethane produces both ethylene and ethane, and thus poses an additional challenge of selectivity, where ethylene, due to its higher economic value, is the desired product over ethane.

Here we aim to investigate the hydrodechlorination of 1,2-dichloroethane over transition metal surfaces. Past studies have shown that bimetallic catalysts yield good catalytic performance in terms of both the dechlorination activity and the ethylene selectivity [2]. The lack of fundamental understanding of the reaction mechanism and the catalytic behavior, however, hinders catalyst design and optimization. Therefore, we adopted a first-principles approach in our study. Through density functional theory calculations, we evaluate the energetics of adsorbates and reactions over metal surfaces. The results offer valuable insights into the surface reaction mechanism and will potentially facilitate the rational design of optimal catalysts. This presentation will demonstrate the trend of hydrodechlorination reactivity over a series of monometallic surfaces, followed by a detailed mechanistic study on platinum catalysts.

[1] E. D. Goldberg, Halogenated hydrocarbons: past, present and near-future problems, *Science of the Total Environment* **100**, 17-28 (1991).
[2] V. I. Kovalchuk and J. L. d'Itri, Catalytic chemistry of chloro- and chlorofluorocarbon dehalogenation: from macroscopic observations to molecular level understanding, *Applied Catalysis A: General* **271**, 13-25 (2004).

Ultra-low loading, highly active and stable Ru/γ-Al_2O_3 catalyst for low temperature solar steam reforming of methane

David S. A. Simakov and Yuriy Román-Leshkov

Department of Chemical Engineering, Massachusetts Institute of Technology, Cambridge MA 02139, USA

Converting solar irradiation into chemical energy via solar thermal steam reforming of natural gas is an attractive method to increase the energy content of methane, consequently reducing both natural gas consumption and greenhouse gases emission. Of particular interest is collecting solar radiation through the parabolic trough technology, which is mature and relatively inexpensive, but limited to temperatures ranging from 400-600 °C. One of the major challenges is developing catalysts that are highly active and stable in this low temperature range. Commercial nickel-based catalysts do not have sufficient activity, are prone to fast oxidation deactivation and coke excessively. Catalysts based on platinum group metals do not suffer from these drawbacks, but their cost is a prohibiting factor. In the present work, we demonstrate that Ru/γ-Al_2O_3 catalyst prepared by sonication-assisted wet impregnation with the ultra-low loading of Ru of 0.15 wt% has superior activity (Fig. 1). The catalyst outperforms commercial Ni catalyst (12 wt% of Ni) by about two orders of magnitude in terms of methane conversion rate per amount of active metal (Fig. 2), while also providing excellent resistance to sintering, oxidation and coking. The systematic evaluation of catalytic activity over a range of Ru nanoparticles loading and under various operating conditions (pressure, temperature and space velocity) will be presented, along with a comprehensive characterization study using a variety of analytical techniques to investigate particle size, dispersion, and metal-support interaction effects.

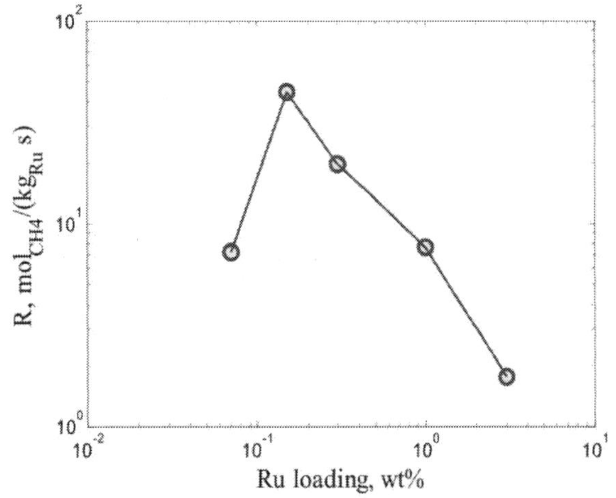

Figure 1: Steady state rate of methane conversion per amount of active metal as a function of Ru loading. P=1 bar, T=600 °C, H_2O/CH_4=3 in the feed, CH_4 GHSV=7500 mL/(g h)

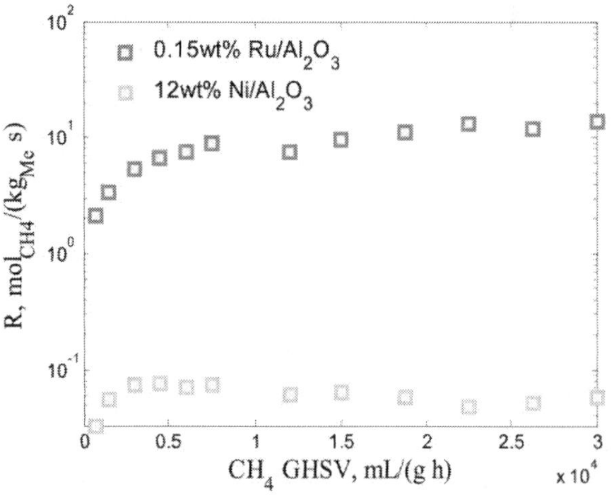

Figure 2: Steady state rate of methane conversion per amount of active metal as a function of GHSV. P=1 bar, T=500 °C, H_2O/CH_4=3 in the feed

Spatio-temporal Features of NOx reduction on LNT+SCR System:

Effect of Hydrocarbon Reductant and Space Velocity

Mengmeng Li, Vencon G. Easterling, Michael P.Harold

Previous studies have shown, in order of effectiveness, that H_2, CO, and hydrocarbons (HCs) are capable of reducing NOx present in diesel exhaust across a lean NOx trap (LNT) catalyst. Additionally, HCs and NH_3 have been utilized to reduce NOx to N_2 using selective catalytic reduction (SCR) catalysts. Recently, the synergy of combining LNT with SCR has attained NOx reduction levels that equal to or greater than stand-alone LNT or SCR while avoiding the need for the urea feed and reducing the loading of the expensive precious metal over LNT catalyst. Although simpler reductants such as H_2 and CO are capable of attaining higher NOx reduction levels, the study of more complex reductants in a combined LNT+SCR system is warranted for real world application on a vehicle.

In this study, NOx reduction using hydrocarbons (HCs) during the rich was investigated over a lab made LNT-SCR dual brick catalyst system. The effects of three reductants were studied in separate and blended combinations of CO, C_3H_6, C_3H_8, and CO+C_3H_6+C_3H_8. Figure 1 shows the overall performance of using the belended gas as reductant. At temperatures below 300°C, NO_2 was found re-reduce back to NO during the rich phase across the SCR catalyst. NOx reduction with HCs becomes active above 300°C, especially for C_3H_6, which could achieve 90% NOx conversion across the dual brick catalyst system. While C_3H_8 is not as active as C_3H_6 when fed in a mixture with C_3H_6 it starts to react at lower temperature than when C_3H_8 is the sole reductant. Figure 2 shows the comparison of three different reductants. The yield of NH_3 was ~40% greater when a C_3H_6 was used as compared to CO at higher temperature.

The gas hourly space velocity (GHSV) effect on NOx reduction was also investigated in this study. Figure 3 shows the GHSV was increased in order to achieve a lower NOx conversion to N_2 over the LNT which could provide a more desirable NH_3 to NOx ratio (ANR) across the SCR. It was found out at high temperature, GHSV in the range of 105,000 h^{-1} to 120,000 h^{-1} could achieve nearly 50% NOx conversion and an ANR of 1.0 from LNT catalyst which is the ideal gas mixture to the downstream SCR catalyst. The recent research shows that the LNT-SCR dual brick catalyst system could also benefit from the HC-SCR mechanism which not only reduced the HC slip from the system but also improved NOx reduction over the combined catalysts. The HC steam reforming and partial oxidation were carried out by spatially resolved mass spectrometry (SpaciMS), which was utilized to decouple the regeneration reaction from NOx reduction. This method enabled a detailed study of NH_3 formation over LNT catalyst and utilization over SCR catalyst while using a reductant feed that approach real world conditions.

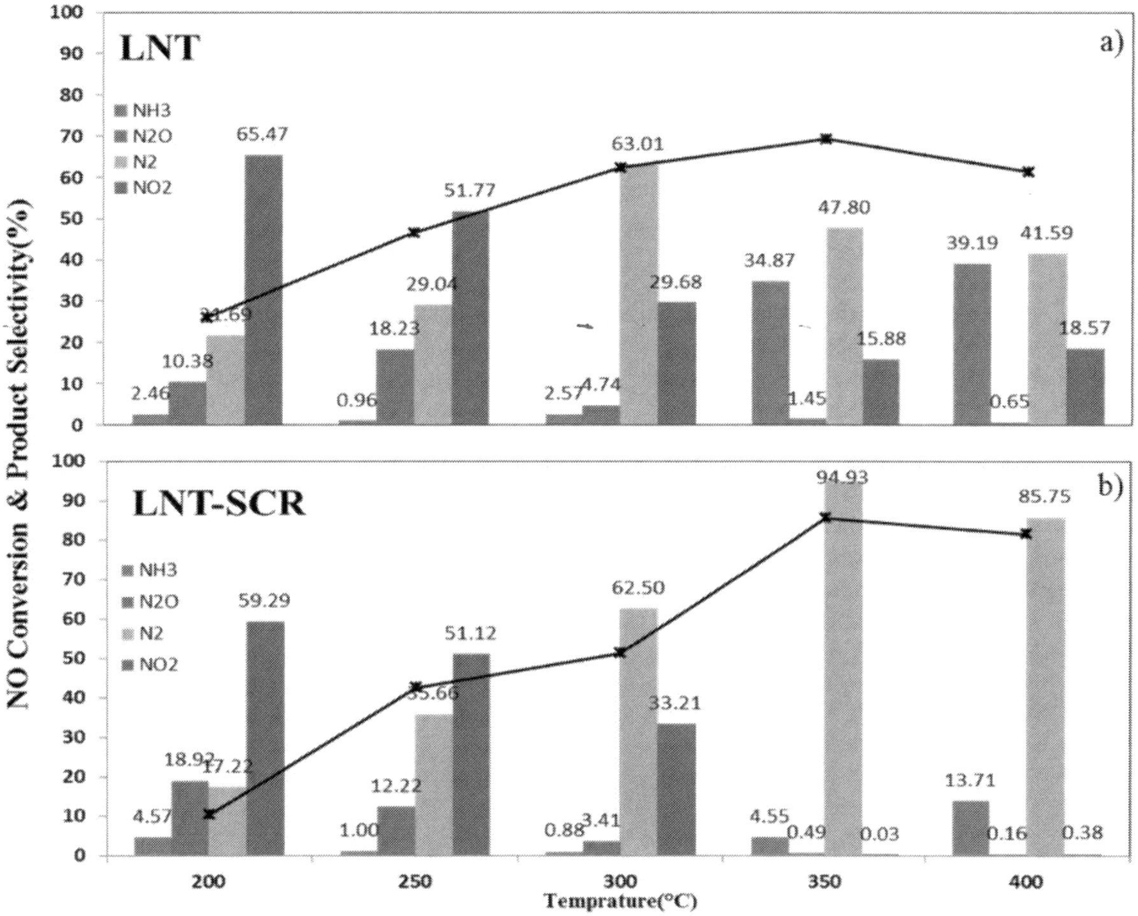

Figure 1. NO conversion and product selectivity using 1.5% CO, 500ppm C_3H_6, 3500ppm C_3H_8 and 1% O_2 as reductant over a) LNT A and b) LNT A-SCR A catalyst (NO conversion in black line).

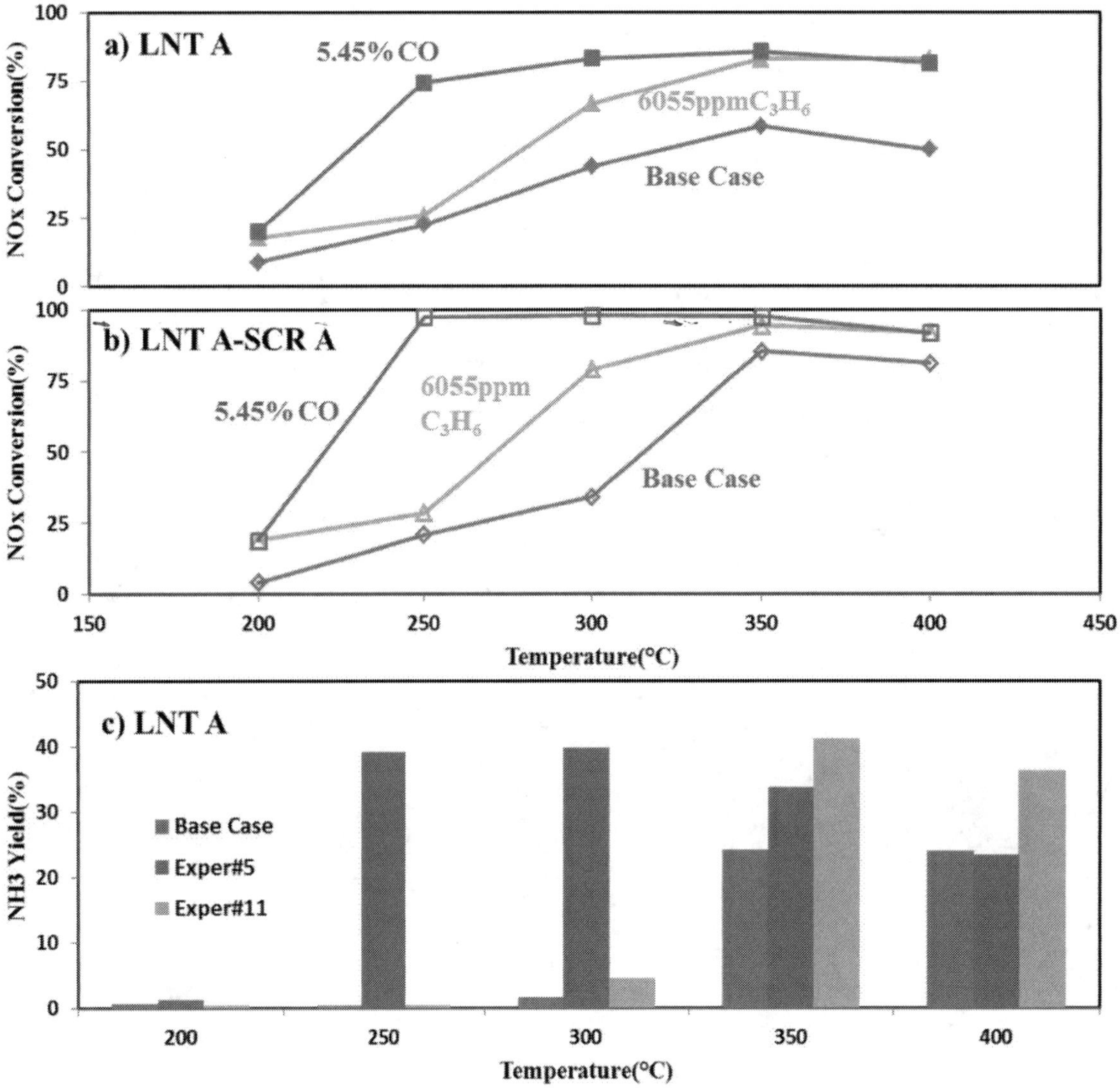

Figure 2: Cycle averaged NOx conversion and NH_3 Yield for Base case (500ppm NO, 1.5% CO, 500ppm C_3H_6, 3500ppm C_3H_8 and 1% O_2), Exper #5 (500ppm NO, 5.45%CO and 1% O_2), and Exper #11 (500ppm NO, 6055ppm C_3H_6 and 1% O_2). a) NOx conversion for LNT A b) NOx conversion for LNT A-SCR A c) NH_3 Yield for LNT A. GHSV=90,000 h^{-1}.

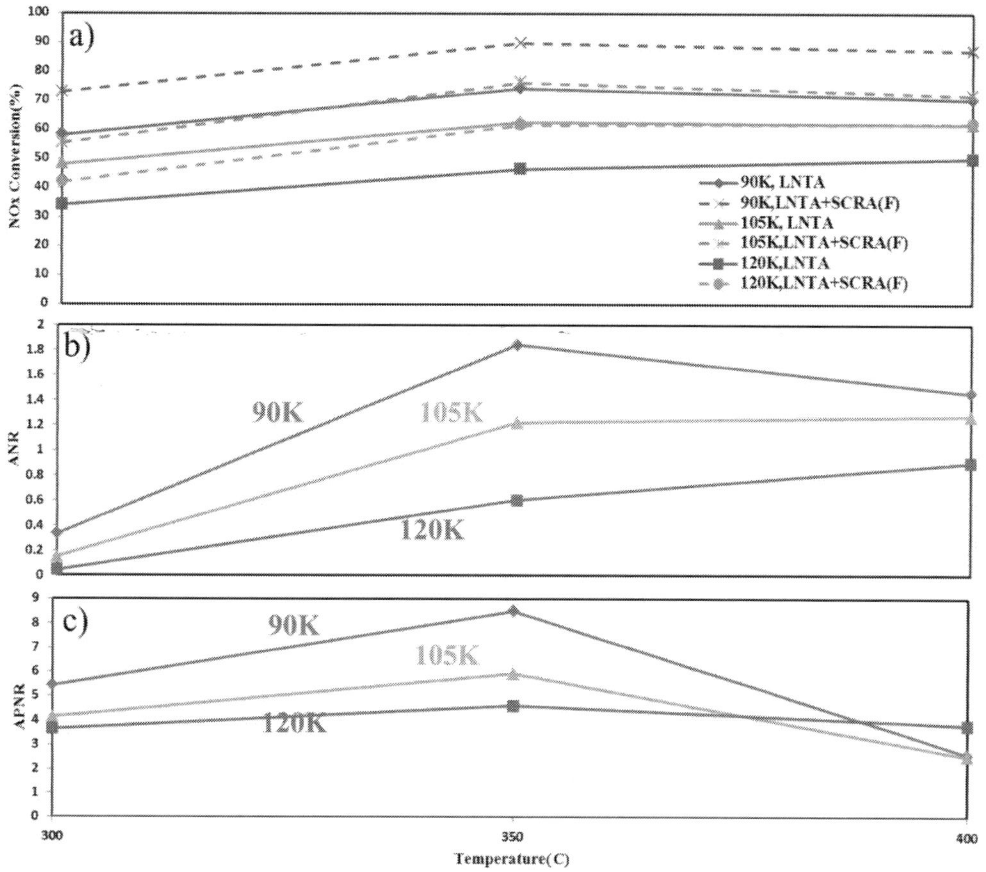

Figure 3: Cycle-averaged a) NO conversion and b) ANR c) APNR using 9300ppm C_3H_6 and 1% O_2 as reductant over LNT A by varying different GHSV (90,000 - 120,000 h^{-1}).

H_2SO_4 Catalysis: Perspective and Opportunities for Reducing SO_2 Emissions Using Particulate and Monolith Catalysts

Anuradha Nagaraj[1] and Patrick L. Mills[2*]
[1]Department of Environmental Engineering
[2]Department of Chemical & Natural Gas Engineering
Texas A&M University-Kingsville TX-78363-8202
*Corresponding author: patrick.mills@tamuk.edu

Abstract: Development of next-generation chemical processes that have zero emissions is a key environmental objective for sustainable development. The manufacture of H_2SO_4 by the air oxidation of SO_2 to SO_3 is an important technology where an opportunity exists for new catalyst development and process innovation by reducing emissions of unconverted SO_2 in process reactor tail gases owing to the sheer number (>1500) and scale (*ca.* 500 to 4500 metric tons/day) of typical plants. Development of new catalysts having higher activity, lower pressure drop, low dusting, and sufficient mechanical crush strength to meet the anticipated reduction in SO_2 emissions from H_2SO_4 manufacturing plants will potentially benefit by using this more realistic approach for particle-scale shape modeling. Transport-kinetic interactions in commercial porous catalyst shapes used for SO_2 oxidation are analyzed using the Wilke, Wilke-Bosanquet, Maxwell-Stefan, and Dusty-Gas flux models.

Keywords: catalyst particle shape, SO_2 oxidation, monolith, flux models.

1. Introduction

The manufacture of H_2SO_4 by the air oxidation of SO_2 to SO_3 is an important technology where an opportunity exists to develop new innovations in environmental catalysis and reaction engineering for an important technology that has a rich and long history with positive economic growth. Best Available Control Technologies (BACT) for controlling SO_2 emissions from stack gases of H_2SO_4 plant involves additional processing resulting are cost intensive and well-studied. However, emissions control technologies using front-of-pipe technology, such as new catalyst technology, improved reactor designs, and process operational strategies have greater potential to be more economical versus end-of-pipe scrubbing technologies.

The primary objectives of this paper are to develop a rigorous modeling framework that accounts for diffusion and non-isothermal reaction in various realistic particulate catalyst shapes and monolith catalysts, and to employ this framework to compare the performance of these catalysts for typical multi-pass SO_2 convertor operation.

2. Catalyst Shapes

From a catalyst design perspective, specification of catalyst composition, catalyst size, catalyst shape and internal pore structure are interlinked parameters that have received notable attention by catalyst scientists and engineers since these properties affect the catalyst effectiveness factor, catalyst bed pressure drop, rate of catalyst attrition, catalyst mechanical integrity, catalyst life, amongst other key measures of performance. It is evident that specification of preferred catalyst design parameters requires careful optimization since tradeoffs exist in how the observed reaction rate and other measures of performance may be affected by altering one or both of these parameters.

2.1 Particulate Catalysts

Commercial particulate catalysts for the air oxidation of SO_2 to SO_3 are available in various shapes, such as solid cylindrical pellets, hollow rings, and multi-lobed rings. A typical composition of these catalysts is based on alkali metals, such as potassium, sodium or cesium, vanadium pentoxide salts, and non-reactive silica support.

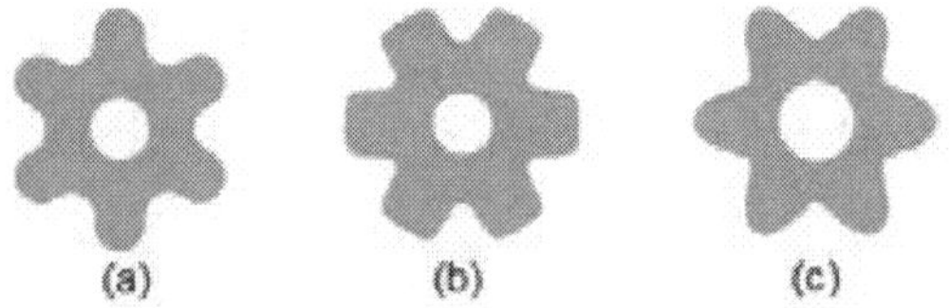

Figure 1. Catalyst particle shapes. (a) Circle lobe, (b) Rounded step, and (c) Trigonometric lobe.

Various complex catalyst shapes based on multi-lobed shape technology used in this study are illustrated in Figure 1.

2.2 Monolith Catalysts

Limited open literature is available on monolith synthesis and modeling for SO_2 oxidation in H_2SO_4 manufacturing process.

In early 1990s, two patents were issued to Felthouse and coworkers at Monsanto Enviro-Chem Systems (MECS) (Felthouse, 1992; Felthouse and Vavere, 1993) for synthesizing monolith catalysts for SO_2 oxidation. In 1991, Bespalov and coworkers at Moscow Chemical Engineering Institute developed a model for SO_2 oxidation in a monolith catalyst (Bespalov *et al.*, 1991). No experimental data was provided, however.

The monolith catalyst channel used in this study is shown in Figure 2. A single channel with 75 mm length, 1.5 mm channel width and washcoat of 0.625 mm thickness is used.

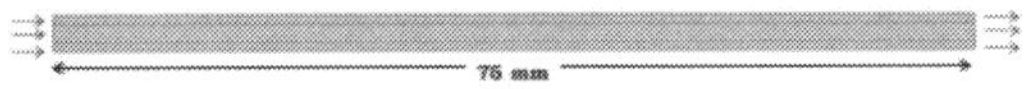

Figure 2. Washcoated single monolith channel

3. Transport-Kinetics Model

The modeling framework used to describe transport-kinetic interactions in both particulate and monolith catalyst is presented in this section.

The reaction kinetics for SO_2 oxidation are described by the following equation which is based on a comprehensive set of experimental measurements (Collina *et al.*, 1970).

$$-r_{SO2} = \frac{k_1 p_{O2} p_{SO2}\left(1 - \dfrac{p_{SO3}}{p_{SO2}\sqrt{p_{O2}}\,K_P}\right)}{22.414\,(1 + K_2 p_{SO2} + K_3 p_{SO3})^2}$$

3.1 Particulate transport model

The specie mass and energy balance equations for particulate shape are:

Mass balance

$$\nabla \bullet N_i = \upsilon_i \, r \, \rho_p$$

where i = SO_2, O_2, SO_3 and N_2 and υ_i = -1, -0.5, 1 and 0 for SO_2, O_2, SO_3 and N_2 respectively.

Energy balance

$$\nabla \bullet \bar{q} = -(\Delta H_{rxn})\, r \, \rho_P$$

3.2 Monolith transport model

The flow through monolith channel is modeled as free and porous media flow. The transport equations used for this monolith channel include balances for species, energy and momentum.

Mass balance

$$\nabla \bullet \bar{N} - u \bullet \nabla \bar{C} = \upsilon \, r \, \rho_p$$

Energy balance

$$\nabla \bullet \bar{q} - \rho C_p u \bullet \nabla T = -(\Delta H_{rxn})\, r \, \rho_P$$

Momentum balance

For channel flow:

$$\rho(u \bullet \nabla)u = \nabla \bullet [-\rho I + \eta(\nabla u + (\nabla u)^T) + F$$

$$\nabla \bullet u = 0$$

For flow through the catalyst layer:

$$(\eta / \kappa + Q)u = \nabla \bullet [-\rho I + (1/\varepsilon_p)\{\mu(\nabla u + (\nabla u)^T) - (2\eta/3 - \kappa_{dv})(\nabla \bullet u)I\}] + F$$

$$\nabla \bullet u = Q/\rho$$

4. Diffusion flux models

Transport-kinetic interactions are analyzed using the Wilke, Wilke-Bosanquet, Maxwell-Stefan, and Dusty-Gas flux models. Particle effectiveness factors derived from the various flux models can differ for otherwise identical values for kinetic and transport parameters.

Wilke model

$N_i = -D_{ei,m} \nabla C_i$ where

$$D_{ei,m} = \frac{1}{\sum_{j=1, j\neq i}^{n} x_j / D_{ij}^e}$$

Wilke-Bosanquet model

$N_i = -D_{i,eff} \nabla C_i$ where

$$\frac{1}{D_{i,eff}} = \frac{1}{D_{ei,m}} + \frac{1}{D_{ei,k}}$$

Maxwell-Stefan model

$$N_i = \frac{-\nabla C_i + \sum_{j=1, j\neq i}^{n} \dfrac{x_i N_j}{D_{ij}^e}}{\sum_{j=1, j\neq i}^{n} \dfrac{x_j}{D_{ij}^e}}$$

Dusty-Gas model

$$N_i = \frac{\sum_{j=1, j\neq i}^{n} \frac{x_i N_j}{D_{ij}^e} - \frac{C_i v^*}{D_{ei,k}} - \nabla C_i}{\sum_{j=1, j\neq i}^{n} \frac{x_j}{D_{ij}^e} + \frac{1}{D_{ei,k}}} \quad \text{where}$$

$$v^* = -\frac{\varepsilon\, d_{pore}^2}{32\,\tau\,\mu}\nabla P$$

5. Results

5.1 Particulate catalyst

The 1-D diffusion-reaction system was first solved for a solid cylindrical pellet with a diameter 6 mm. The specie and energy equations are solved using two of the COMSOL Multiphysics™ modules: (1) diffusion, and (2) heat transfer by conduction. Typical molar composition profiles for all four flux models are compared in Figure 3. The mean pore radius used in this study is 319 nm with a pellet porosity of 0.44 (Davis *et al.*, 1982).The profiles suggest that the Knudsen diffusion is not significant.

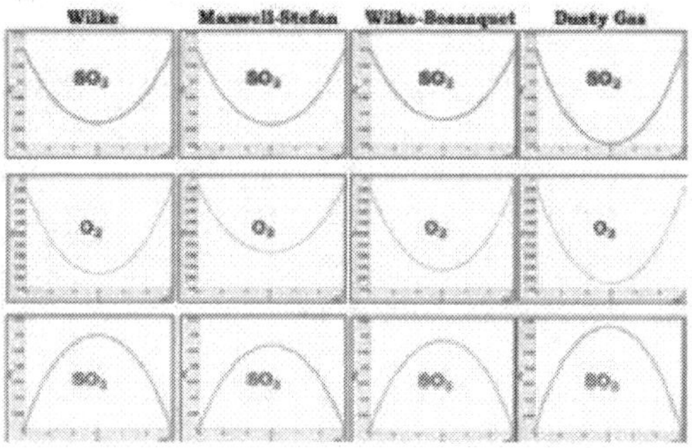

Figure 3. Comparison of the steady-state molar composition profiles for different flux models (T_B = 420°C, P = 1 atm, % SO_2 = 11%; % O_2 = 9%).

Figure 4 shows the 3-D concentration and temperature profiles for the rounded-step catalyst illustrated in Figure 1b for the Wilke and Dusty gas flux models. The boundary conditions correspond to specified bulk concentrations and temperatures so no external transport resistances are present. The effectiveness factor was obtained by integration of the reaction rate at the local values for the particle temperature and specie concentrations over the particle volume and then dividing this result by the reaction rate evaluated at the external bulk conditions.

The profiles for both flux models show that the highest particle temperature and the lowest concentrations for both SO_2 and O_2 are reached in the center of the lobes. This is expected owing to the longer diffusion path length and hence longer characteristic time for transport of mass and energy in the lobe regions. The effectiveness factor for the Wilke model is lower (η = 0.89) than the one obtained when the Dusty gas model is used (η = 0.92). This difference is attributed the presence of a viscous flux term in the latter model that likely occurs owing to the non-equimolar stoichiometry that is not accounted for in the Wilke model. The contribution of Knudsen diffusion is not important since the mean pore radius is 319 nm.

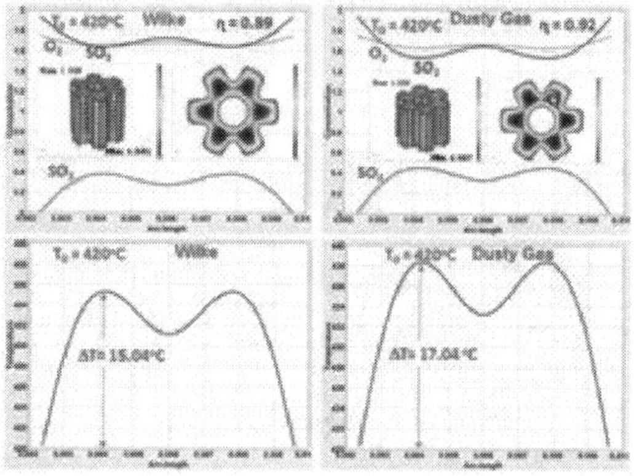

Figure 4. Comparison of the 3-D rounded step shape concentration profiles for Wilke and Dusty Gas flux models.

5.2 Monolith catalyst

The 2-D convective diffusion-reaction system was solved for the single monolith channel illustrated in Figure 2. The momentum, specie and energy equations are solved using the following COMSOL Multiphysics™ modules: (1) Navier-Stokes (channel flow) and Brinkman (catalyst flow); (2) convection and diffusion; and (3) convection and conduction.

Figure 5 shows the 2-D temperature profiles for the monolith channel using different flux models.

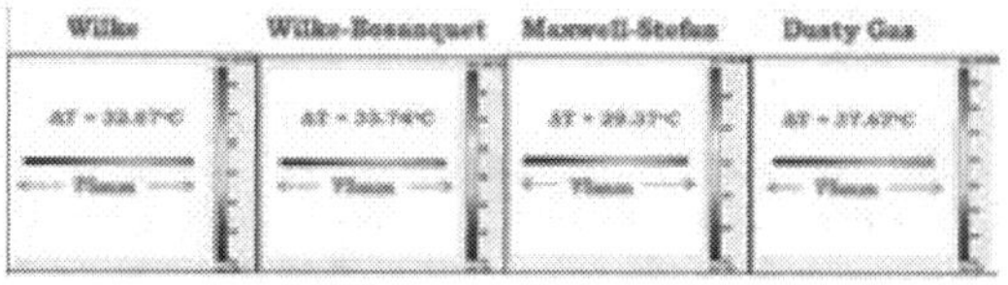

Figure 5. Comparison of the temperature profiles for different flux models (T_B = 420°C, p = 1 atm, % SO_2 = 11%; % O_2 = 9%).

The overall temperature gradients are largest for the Dusty gas model while the smallest occurs for the Maxwell-Stefan model. Since the latter model does not account for convective transport or Knudsen diffusion, these must account for the difference in the predicted gradient. Additional analysis that provides more insight into the differences between the various models will be provided in a future manuscript (Nagaraj and Mills, 2014).

6. Conclusions

COMSOL Multiphysics™ provides an efficient platform for modeling transport-kinetic interactions in particulate and monolith structures. Wilke specie flux model is sufficient when molecular diffusion is controlling, otherwise more complex models are needed. Synthesis of tailored porous supports could lead to more active catalysts for SO_2 oxidation to SO_3. The initial modeling results suggest monoliths are a potential catalyst platform SO_2 oxidation if the structure can be tailored to provide high activity with attractive economics.

7. Nomenclature

C_i	Concentration of species 'i' (mol/m^3)
C_p	Specific heat capacity (J/kg-K)
$D_{ei,m}$	Wilke molecular diffusivity (m^2/s)
$D_{ei,k}$	Knudsen diffusion coefficient for species i (m^2/s)
$D_{i,eff}$	Bosanquet diffusivity (m^2/s)
D^e_{ij}	Effective binary diffusivity of species 'i' (m^2/s)
d_{pore}	Mean pore diameter (m)
k	Rate constant (kmol/kg cat-hr)
K_p	Equilibrium constant (atm$^{1/2}$)
K	Adsorption Constant (atm)
N_i	Molecular diffusion flux (mol/m^2-s)
q	Heat flux (W/m^2)
Q	Heat source (W/m^3)
P	Pressure (atm)
p_i	Partial pressure of component i (atm)
r	Reaction rate (mol/m^3-s)
r_{so2}	Reaction rate of SO_2 (kmol/kg cat-hr)
T	Temperature (^{o}C)
u	Velocity (m/s)
v^*	Dimensionless velocity
x_i	mole fraction of component i

Greek Letters

η	Effectiveness factor
κ	Permeability (m^2)
μ	Viscosity (Pa-s)
ρ_p	Density of catalyst pellet (kg/m^3)
ε	Catalyst pellet porosity
ν_i	Stoichiometric coefficient of component i
τ	Catalyst tortuosity

8. References

Bespalov, A. V., E. M. Chechetkina, et al. (1991). "Numerical modeling of sulfur dioxide reaction in through-channel of a block catalyst with a honeycomb structure." Zhurnal Prikladnoi Kimii 64(10): 2048-2053.

Collina, A., D. Corbetta, et al. (1970). Europeam Symposium on "Use of computers in the design of chemical plants", Firenze.

Davis, M. E., G. Fairweather, et al. (1982). "Analysis of SO2 oxidation in non-isothermal catalyst pellets using the dusty-gas model." Chemical Engineering Science **37(3): 447-452.**

Felthouse, T. R. (1992). Monolithic catalysts for conversion of sulfur dioxide to sulfur trioxide. US, Monsanto Company.

Felthouse, T. R. and A. Vavere (1993). Monolithic catalysts for conversion of sulfur dioxide to sulfur trioxide. US, Monsanto Company.

Nagaraj, A. and Mills, P. L. (2014). Comparison between particulate and monolith catalyst performance for the oxidation of SO_2 using various specie transport flux models," Chemical Engineering Science (in preparation).

Synthesis, Characterization of Vanadia/titania-based Catalysts and their Performance in Ammonia Remediation

Hao Chen[1,3], Raquel Portela[4], Wei Han[1], Pedro Avila[4], Miguel A. Bañares[4], King Lun Yeung[1,2*]

[1]Department of Chemical and Biomolecular Engineering, [2]Division of Environment, [3]Nano Science and Technology Program, The Hong Kong University of Science and Technology, Clear Water Bay, Kowloon, Hong Kong, P.R. China

[4] Instituto de Catalisis y Petroleoquimica, CSIC, Marie Curie 2, 28049, Madrid Spain

*Corresponding author: kekyeung@ust.hk

Division: 20 Catalysis and Reaction Engineering Division

Subdivision: 20037 Applied Environmental Catalysis

Format: Oral Presentation

Abstract:

Malodor is an important class of air pollution that poses a serious threat to personal health and quality of life. Ammonia gas is a major cause of malodor complaints and is produced by microbial decompositions of organic wastes. A catalyst capable of attaining better than 99 % conversion of 50 ppm NH_3 in air at room temperature was successfully prepared. The E_A of the reaction is dramatically decreased by using the catalysts when the ratio of VO_x/TiO_x increases. The NH_3 was oxidized to nitrogen and water during the reaction without producing NO_x. The catalyst was characterized by BET, micro-Raman and X-ray photoelectron spectroscopy to identify the active sites to be monomeric VO_x on TiO_x.

Introduction:

Indoor air quality has captured much attention of many researchers recently because of its effect on human health. Ammonia is one of the most odor origin gases in the world that can be easily emitted in wash room, fish markets, meat treatment industries, piggery, landfill and agricultural activities. They have a sharp pungent smell which can be detected by human nose at even several ppm levels. The damages caused by Ammonia mainly role in eye, respiratory and skin.[1] Selective catalytic oxidation gives a new way to remove ammonia in oxygen or air in the recent years, ammonia reacts with air can produce nitrogen, nitrous oxide and nitrogen monoxide. This technology has reported in literature that different kinds of catalysts have been used for selective catalytic oxidation of ammonia into nitrogen, such as noble metals (Pt, Rh and Pd) iron exchanged into zeolite ZEM-5[2] or supported on gamma alumina[3], Ni oxides[4], copper oxides and Ag oxides supported on γ-Al_2O_3[5], V_2O_5/TiO_2 and CuO/TiO_2[6], Fe exchanged ZSM-5[7], Fe_2O_3/TiO_2.[8]

Experimental:

2.5g Ammonia Vanadate Meta (NH_4VO_3, 98%, Nacalai Tesque INC) is dissolved in 500 ml DDI water, then 38.87g UV100 (TiO_2, Sachtleben) is added into 100 ml the above solution to prepare 1wt% of V_2O_5/TiO_2 catalyst. The mixed suspension is dried via rotary evaporator under 65°C with 120 r/min. After drying, the powder is put inside oven and further dried at 115°C for 24 hours. The obtained powder is then calcined at 200°C for 4 hours. A serial of 0.1wt%, 5wt% and 15wt% catalysts are also prepared with this method.

Results and Discussion:

The catalysts used in the experiments in this report are 1wt%, 5wt% and 15wt% V_2O_5/TiO_2 which are calcined in air at 200°C for 4 hours. To obtain the morphology of catalyst, SEM is measured. Figure 1 shows SEM images of 15wt% V_2O_5-TiO_2 catalyst. Figure 2 shows the Micro-Raman Spectra of 1wt%, 5wt% and 15wt% V_2O_5/TiO_2. The four main peaks of 160cm^{-1}, 390cm^{-1}, 520cm^{-1} and 640cm^{-1} are assigned to anatase phase of TiO_2. The intensity of peaks at 810cm^{-1} and 940cm^{-1} increase as the V_2O_5/TiO_2 loading increases from 1wt% to 5wt% which are associated to polymeric form of V_2O_5, when the loading increases to 15wt%, the peak of 940cm^{-1} shifts to 1020cm^{-1}, which is associated to monomeric form of surface of V_2O_5. In the four catalysts, the peak of crystal V_2O_5 isn't seen in the Micro-Raman spectra at 995cm^{-1}. We can run the experiments at different temperature, to observe the relation between T and conversion shown in Figure 3. To design the experiment condition, the reaction flow rate is 400 SCCM containing 10 SCCM 8.73% NH_3 in N_2 and 390 SCCM N_2, so the concentration of the inlet NH_3 is 2175ppm. The reaction temperature is ranging from 25°C to 250°C. Then the Arrhenius equations of three V_2O_5-TiO_2 catalysts are obtained, shown in Table 1. It is found that 15wt% one has the lowest Ea, that means the reaction barrier decreases a lot as the V_2O_5/TiO_2 ratio increases.

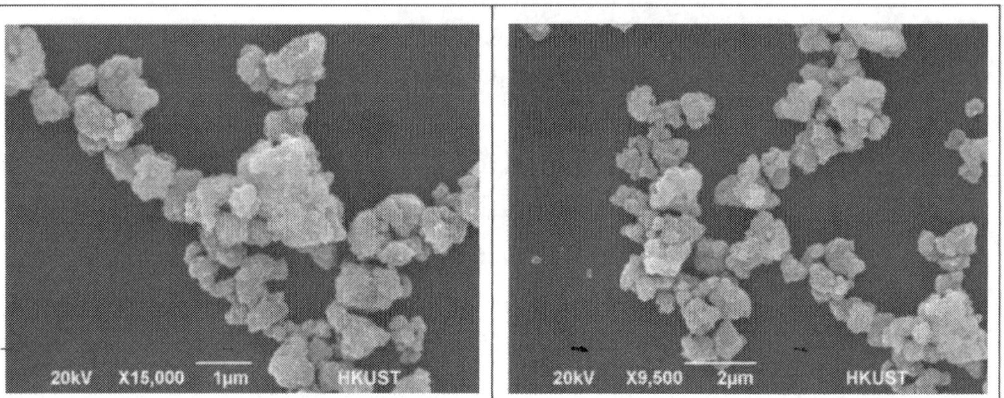

Figure 1 SEM images of 15wt% V_2O_5/TiO_2

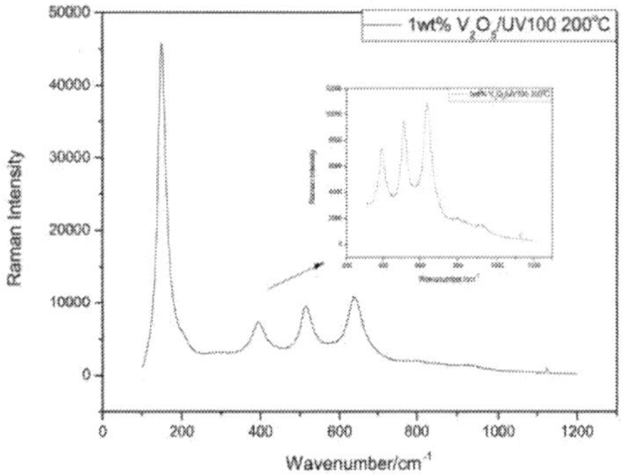

(b)

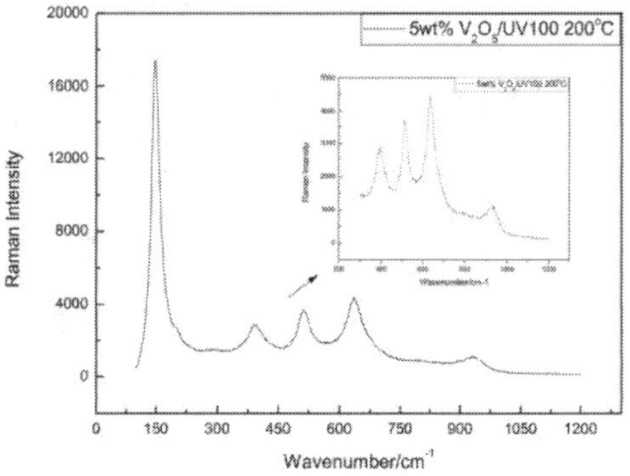

(c)

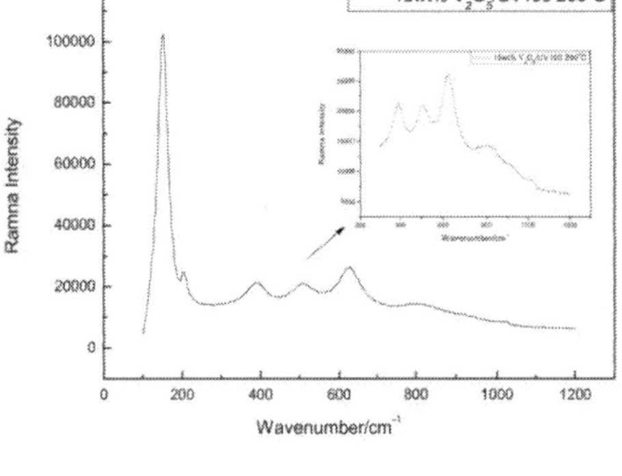

(d)

Figure 2 Micro-Raman Spectra of 1wt%, 5wt% and 15wt% V_2O_5/TiO_2.

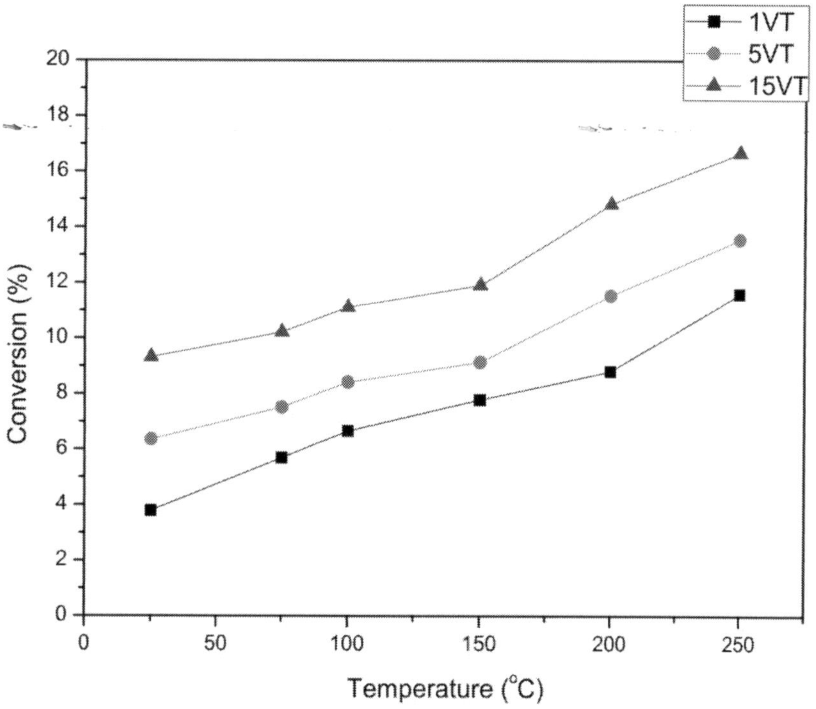

Figure 3. Figure 5 NH_3 conversion using 1wt%, 5wt% and 15wt% V2O5-TiO2 catalyst

Table 1. Arrhenius equations and Ea of 1wt%, 5wt% and 15wt% V2O5-TiO2

Catalyst	Arrhenius equation	$Ea/kJ*mol^{-1}$
V_2O_5-TiO_2-1wt%	lnk=-1.081-722/T	6.00
V_2O_5-TiO_2-5wt%	lnk=-1.087-513/T	4.27
V_2O_5-TiO_2-15wt%	lnk=-0.827-406/T	3.37

Conclusion:

The obtained V_2O_5-TiO_2 catalysts dramatically decrease the reaction barrier of NH_3 oxidation, where the catalysts achieve a 99% conversion for 50ppm NH_3 gas in air at room temperature.

Acknowledgement:

The authors gratefully acknowledge financial supports from the Hong Kong Research Grant Council and Innovation & Technology Fund.

Reference:

1. ***Agency for Toxic Substances and Disease Registry***, toxicological profile for NH_3

2. Li, Y., Armor, J. N. ***Appl. Cata. B*** 1997(13) 131-139

3. Long, R. Q., Yang, R. T. ***Catal. Lett.*** 2002(78) 353-357

4. Amblard, M. Burch, R. Southward, B. W. L. ***Appl. Catal. B*** 1999(22) L159-L166

5. Gang, L., van Grondelle, J., Anderson, B. G., van Santen, R. A. ***J. Catal.*** 1999(186) 100-109

6. Sazonova, N. N., Simakov, A. V., Nikoro, T. A., Barannik, G. B., Lyakhova, V. F., Zheivot, V. I., Ismagilov, Z. R., Veringa, H. ***React. Kinet. Catal. Lett.*** 1996(57) 71-79

7. Qi, G. S., Gatt, J. E., Yang, R. T. ***J. Catal.*** 2004(226) 120-128

8. Long, R. Q., Yang, R. T. ***J. Catal.*** 2002(207) 158-165

389448 Renewable Aromatic Chemicals By Diels-Alder Cycloaddition of Biomass-Derived Furans

Abstract:

Diels-Alder cycloaddition of dimethylfuran (DMF) and ethylene to produce *p*-xylene is a promising method of sustainably producing valuble plastic precursors [1,2]. This step completes the entire process of (a) glucose dehydration to hydroxymethylfurfural (HMF), (b) reduction of HMF to dimethylfuran, and (c) cycloaddition/dehydration of DMF to *p*-xylene. The DMF/ethylene reaction proceeds via two steps: a symmetry-allowed [4+2] Diels-Alder cycloaddition of DMF and ethylene to produce an oxa-norborene cycloadduct intermediate, and subsequent dehydration of the intermediate to form *p*-xylene. In this work, the kinetics of the combined cycloaddition/dehydration chemistry was experimentally evaluated, and the rate-limiting phenomena were identified. The apparent activation energy for the formation of Diels-Alder products from DMF and ethylene was experimentally measured as 17±3 kcal/mole using DMF in heptane with an H-Y catalyst in the temperature range of 200-250°C. Research also focused on characterizing the effects of catalyst loading and determining reaction orders with respect to DMF, ethylene and *p*-xylene.

References

[1] Williams, C. L.; Chang, C.-chih; Do, P.; Nikbin, N.; Caratzoulas, S.; Vlachos, D. G.; Lobo, R. F.; Fan, W.; Dauenhauer, P. J.*ACS Catalysis* **2012** , *2*, 935-939

[2] N. Nikbin, P.T. Do, S. Caratzoulas, R.F. Lobo, P.J. Dauenhauer, D.G. Vlachos, "A DFT study of the acid-catalyzed conversion of 2,5-dimethylfuran and ethylene to p-xylene," *Journal of Catalysis*, **2013** , 297, 35-43

389501 Mechanistic Understanding of Surface Barriers in Transport through MFI Structured Catalysts

Abstract:

Mechanistic understanding of surface barriers in transport through MFI structured catalysts

Zeolites are widely used in the conversion and upgrading during traditional petrochemical processes as well as biomass to renewable fuels and chemical pathways.[1] . The combined high surface areas, shape selectivity, and catalytic functionality of the zeolites have made them ideal for the high temperature thermal pathways. To maximize turnover frequencies, microporous and mesoporous materials are now being synthesized with length scales approaching that of a single lattice (pillared, nanosheets, membranes, etc.) in an attempt to reduce transport time. This driving force has led to the development of hierarchical materials that have been the focus of recent synthesis advances. Traditionally, bulk crystals have been synthesized on the order of one to one hundred microns, however mesoporous materials have now been developed with the specific aim of decreasing microporous transport lengthscales.

Diffusion in MFI has presented significant challenges. Zeolite particles with the MFI structure are known to exhibit significant structural complications including intergrowths, subcrystals, and surface defects. A combination of the diffusion anisotropy and structural effects has led to a complex transport scenario, as exhibited in the apparent diffusivities which have been experimentally observed to vary over three orders of magnitude. In small particles, mass transfer becomes dominant and the apparent diffusivity appears to be orders of magnitude smaller than through large particles.[2, 3] This has led to the general phenomenon referred to as 'surface barriers,' which describes the hindrance to mass transfer at the surface of zeolite particles. While this timescale is often negligible when compared to the diffusion or reaction times, we show that it does contribute significantly to overall turnover in small particles.[2] Experimental evidence is presented with two independent systems that measure both the desorption profiles (Zero Length Chromatography), and sinusoidal steady state of the uptake/desorption system (Frequency Response). Chemical species exhibiting differing kinetic diameters and adsorption affinities were used to probe diffusion within MFI-structured zeolites to elucidate the role of guest molecule size and binding energy. Kinetic Monte Carlo simulations are used to mechanistically probe the effect of physical structures on overall transport timescales, including regular surface restrictions and total pore blockages. The relative trends with particle size are compared to experimental measurements and potential mechanisms are proposed.

1. Williams, C. L.; Chang, C.-C.; Do, P.; Nikbin, N.; Caratzoulas, S.; Vlachos, D. G.; Lobo, R. F.; Fan, W.; Dauenhauer, P. J., Cycloaddition of Biomass-Derived Furans for Catalytic Production of Renewable p-Xylene. *ACS Catalysis* **2012,** 2, (6), 935-939.

2. Teixeira, A. R.; Chang, C.-C.; Coogan, T.; Kendall, R.; Fan, W.; Dauenhauer, P. J., Dominance of Surface Barriers in Molecular Transport through Silicalite-1. *The Journal of Physical Chemistry C* **2013,** 117, (48), 25545-25555.

3. Chang, C.-C.; Teixeira, A. R.; Li, C.; Dauenhauer, P. J.; Fan, W., Enhanced Molecular Transport in Hierarchical Silicalite-1. *Langmuir* **2013,** 29, (45), 13943-13950.

Low Temperature Oligomerization of 1-Butene over H-Ferrierite

Zhuoran Xu[1], Yong Tae Kim[1], Joseph P. Chada[1], Yomaira J. Pagán-Torres[2], Devon C. Rosenfeld[2], Bill L. Winniford[2], Eric G. Schmidt[2] and George W. Huber[1].

[1]Department of Chemical & Biological Engineering, University of Wisconsin-Madison

[2] The Dow Chemical Company, Freeport, TX

The advent of fracturing technology has resulted in abundant and inexpensive natural gas resources. Several companies are building steam crackers which produce olefins from these natural gas resources. These olefins can then be used as a cost effective feedstock in the chemical industry. Zeolites are ideal catalysts for olefin conversion because of their low cost, ease of regenerability, and shape selectivity. Zeolites could ideally produce a range of industrial chemicals from olefins. However, the detailed product distribution, reaction pathway, and catalyst deactivation for olefin conversion over zeolites is not fully understood.

Here, we will present results for butene oligomerization over H-Ferrierite at low temperatures (100-250°C). 1-butene first undergoes hydride shift to produce an equilibrium mixture of 1 and 2 butenes. These butenes then undergo oligomerization reactions. The products from the oligomerization reaction follow the Anderson-Schulz Flory (ASF) distribution. The chain growth probability increases with increasing butene conversion. The oligomerization products also undergo more skeletal isomerization with increasing butene conversion. At reaction temperatures above 200°C cracking and hydride transfer results in the formation of aromatics, naphthalenes and coke. Low partial pressure of 1-butene leads to fast deactivation due to deposition of heavier olefins on the surface of ferrierite. Increasing the partial pressure of 1-butene decreases the rate of deactivation. We will demonstrate the effects of various reaction conditions on the activity, selectivity and stability for 1-butene conversion over H-Ferrierite.

Work with stannosilicates and Sn-BEA zeolite over the last decade has shown remarkable catalytic activity for a range of promising Lewis-acid-catalyzed chemistries including the Baeyer-Villiger oxidation,[1] the Meerwein–Ponndorf–Verley reaction,[2] isomerization of glucose[3] and xylose,[4] and conversion of sugars to furfural,[4] hydroxymethylfurfural,[5] and lactic acid.[6] FTIR characterization of these materials has followed precedents set for the study of titanium-containing zeolites, particularly associating an absorption band at 960 cm^{-1} with the antisymmetric stretch of Si-O-Sn.[7] In the present work, we aimed to analyze and confirm the existence of this peak with as much rigor as was applied to titanium-containing zeolites. Sn-BEA zeolites were analyzed using FTIR, thermogravimetric analysis (TGA) and probe reaction kinetics. It was determined that the peak at 960 cm^{-1} is the result of interaction between water and defect sites in the zeolite. Further, immersion of the zeolite in liquid water was found to generate a tenfold increase in the number of internal silanol groups within the zeolite. These hydrated moieties are stable to temperatures as high as 400 °C and have a significant impact on the catalytic activity of the zeolite. At temperatures between 400 and 600 °C, the silanol groups condense and the zeolite is regenerated to its earlier state. Because this transformation occurs at room temperature upon immersion in water, it has great importance for understanding catalytic activity in aqueous phase reactions, for which the hydrated form of the zeolite will be in use. The difference in activity will also be examined more broadly in an effort to understand the various structures observed for the active site in Sn-BEA and their influence on catalytic properties from activity to hydrophobicity.

1. Corma, A.; Nemeth, L. T.; Renz, M.; Valencia, S., *Nature* **2001,** *412*, 423-425.
2. Corma, A.; Domine, M. E.; Nemeth, L.; Valencia, S., *J. Am. Chem. Soc.* **2002,** *124*, 3194-3195.
3. Moliner, M.; Román-Leshkov, Y.; Davis, M. E., *Proceedings of the National Academy of Sciences* **2010,** *107*, 6164-6168.
4. Choudhary, V.; Pinar, A. B.; Sandler, S. I.; Vlachos, D. G.; Lobo, R. F., *ACS Catal.* **2011,** *1*, 1724-1728.
5. Nikolla, E.; Román-Leshkov, Y.; Moliner, M.; Davis, M. E., *ACS Catal.* **2011,** *1*, 408-410.
6. Holm, M. S.; Pagan-Torres, Y. J.; Saravanamurugan, S.; Riisager, A.; Dumesic, J. A.; Taarning, E., *Green Chemistry* **2012,** *14*, 702-706.
7. Ricchiardi, G.; Damin, A.; Bordiga, S.; Lamberti, C.; Spanò, G.; Rivetti, F.; Zecchina, A., *J. Am. Chem. Soc.* **2001,** *123*, 11409-11419.

386358 Facile Synthesis of Sn-BEA for Isomerization of Cellulosic Sugars

Abstract:

Isomerizations of triose sugars (dihydroxyacetone to glyceraldehyde), pentose sugars (xylose to xylulose), and hexose sugars (glucose to fructose) are key intermediate steps for converting biomass into fuels and platform chemicals. Sn-BEA, a stannosilicate microporous material with zeolite BEA topology, exhibites unique Lewis acidity and recently reseives considerable attention due to its high catalytic activity and selectivity in the isomerization of sugars[1]. However, the conventional synthesis of Sn-BEA requires a long crystallization time (11 to 40 days) and utilization of toxic and environmentally unfriendly hydrofluoric acid. In the present work, we developed a series of methods for the facile synthesis of Sn-BEA catalyst.

In this study, we highlight that the Sn-BEA catalysts with a tunable particle size can be synthesized in 3 hours using a seeded growth method[2]. Furthermore, Sn-BEA was successfully synthesized in a fluorine-free medium for the first time. The crystals obtained from both methods were characterized by XRD, SEM, nitrogen sorption experiments, and probe molecules adsorbed FT-IR. It is shown that the catalytic activity of the synthesized Sn-BEA is closely related to coordination environment of Sn and the hydrophobic nature of the samples. The study provides a set of Sn-BEA samples with different structural properties allowing for fundamentally understanding the catalytic activity of Sn sites within the molecular sieves and developing seletive Lewis acid catalyst for biomass conversion.

References:

1. Moliner, M.; Roman-Leshkov, Y.; Davis, M. E., "Tin-containing zeolites are highly active catalysts for the isomerization of glucose in water". *Proc. Natl. Acad. Sci. U. S. A.* **2010,** *107* (14), 6164-6168.

2. Chang, C. C.; Wang, Z. P.; Dornath, P.; Cho, H. J.; Fan, W., "Rapid synthesis of Sn-Beta for the isomerization of cellulosic sugars". *RSC Adv.* **2012,** *2* (28), 10475-10477.

380643 Active Sites in Sn-Beta for Glucose Isomerization to Fructose and Epimerization to Mannose

Abstract:

Framework tin sites located in a hydrophobic, pure-silica molecular sieve with the zeolite beta framework topology (Sn-Beta) are known to predominantly catalyze the isomerization of glucose to fructose via a Lewis acid-mediated 1, 2 intramolecular hydride shift in water [1]. Additionally, a Lewis acid-mediated 1, 2 intramolecular carbon shift that leads to the direct epimerization of glucose to mannose has been reported with Sn-Beta in methanol [2].

In this work, we show that, in both water and methanol solvents, alkali-free Sn-Beta predominantly isomerizes glucose to fructose via the 1, 2 intramolecular hydride shift. However, increasing degree of post-synthetic Na^+ exchange of Sn-Beta is shown to progressively alter the reaction pathway towards the 1, 2 intramolecular carbon shift that forms mannose from glucose. Na^+remains exchanged onto these silanol groups during reaction in methanol solvent, leading to a nearly complete shift in selectivity towards glucose epimerization to mannose. In contrast, decationation occurs during reaction in aqueous solutions and gradually increases the reaction selectivity to isomerization at the expense of epimerization. The decationation and the concomitant changes in selectivity can be eliminated by addition of NaCl to the aqueous reaction solution.

Pre-adsorption of ammonia onto Sn-Beta leads to nearly complete suppression of infrared (after deuterated acetonitrile adsorption) and ^{119}Sn NMR spectroscopic signatures attributed to open Sn sites, and of fructose yields from the reaction of glucose in water and methanol. These data provide evidence that the open Sn site with a proximal silanol group is the active site for the 1, 2 intramolecular hydride shift in glucose to fructose isomerization, while the open site with Na-exchanged silanol groups is the active site for the 1, 2 intramolecular carbon shift in glucose to mannose epimerization.

[1] Bermejo-Deval, R.; Assary, R. S.; Nikolla, E.; Moliner, M.; Román-Leshkov, Y.; Hwang, S.-J.; Palsdottir, A.; Silverman, D.; Lobo, R. F.; Curtiss, L. A.; Davis, M. E. *Proc. Natl. Acad. Sci. U. S. A.* **2012**, *109*, 9727–9732.

[2] Bermejo-Deval, R.; Gounder, R.; Davis, M. E. *ACS Catal.* **2012**, *2*, 2705–2713.

Acknowledgements

This work was financially supported as part of the Catalysis Center for Energy Innovation, an Energy Frontier Research Center funded by the U.S. Department of Energy, Office of Science, Office of Basic Energy Sciences under Award Number DE-SC0001004. M.O. wishes to acknowledge funding from the National Science Foundation Graduate Research Fellowship Program under Grant No. DGE-1144469. Any opinions, findings, and conclusions or recommendations expressed in this material are those of the author(s) and do not necessarily reflect the views of the National Science Foundation.

R.G. current address: School of Chemical Engineering, Purdue University, 480 Stadium Mall Drive, West Lafayette, IN 47907, USA.

M.O. and R.B.D. contributed equally to this work.

377367 Benzene Methylation Catalysed By Hierarchically Porous Zeolite: An Effective Way to Promote Xylene Selectivity and Catalyst Lifetime for Large Scale Commercial Use

Abstract:

Benzene Methylation Catalysed by Hierarchically Porous Zeolite: An Effective Way to Promote Xylene Selectivity and Catalyst Lifetime for Large Scale Commercial Use

Xuan He[1], Kake Zhu[1], Xuedong Zhu*[2]

1 College of Chemical Engineering, East China University of Science and Technology, Shanghai, China, 2 Engineering Research Centre of Large Scale Reactor Engineering and Technology, East China University of Science and Technology, Shanghai, China

Abstract

Methylation of benzene is an alternative low-cost route to produce xylenes, but selectivity to xylene remains low over conventional zeolitic catalysts. In this work, a combined dry-gel-conversion and steam-assisted-crystallization method is used to synthesize hierarchically porous zeolite ZSM-5 with varied Si/Al ratios. N_2 physisorption, NH_3- TPD, SEM and FT-IR are employed to characterize the texture and acidity of both hierarchically porous zeolites and their conventional counterparts. The method is found to be applicable to ZSM-5 with Si/Al ratios from 20 to 180. Catalytic properties are test on benzene methylation at 460 °C to investigate the effect of additional porosity and Si/Al ratios. At low Si/Al ratios, the benzene conversions over conventional and hierarchical ZSM-5 are close, and selectivity to toluene is high over hierarchical ZSM-5. It is found that hierarchical porosity markedly enhances the utility of zeolite and selectivity towards xylenes via improved mass transport at higher Si/Al ratios. Under optimized hierarchical ZSM-5 catalyst, xylene selectivity reaches 34.9 % at Si/Al ratio of 180, and a remarkable lifetime of 3000 h is accomplished in industrial sideline evaluation.

Key words: benzene methylation, hierarchical porous zeolite, process intensification

References

[1]Van der Mynsbrugge, J., Visur, M., Olsbye, et al. Methylation of benzene by methanol: Single-site kinetics over H-ZSM-5 and H-beta zeolite catalysts[J]J. Catal., 2008, 292: 201-212

[2]Mirth, G., Lercher, J.A.. In Situ IR spectroscopic study of the surface species during methylation of toluene over HZSM-5[J]J. Catal., 1991, 132: 244-252

[3]Odedairo, T., Al-Khattaf, S.. Comparative study of zeolite catalyzed alkylation of benzene with alcohols of different chain length: H-ZSM-5 versus mordenite[J]Catal. Today, 2013, 204: 73-84

THE GRAPHIC ON THIS PAGE WAS NOT AVAILABLE ON THE E-MEDIA VERSION OF THIS PAPER.

Fig.1 Crystallization mechanism on hierarchical porous ZSM-5 in this work

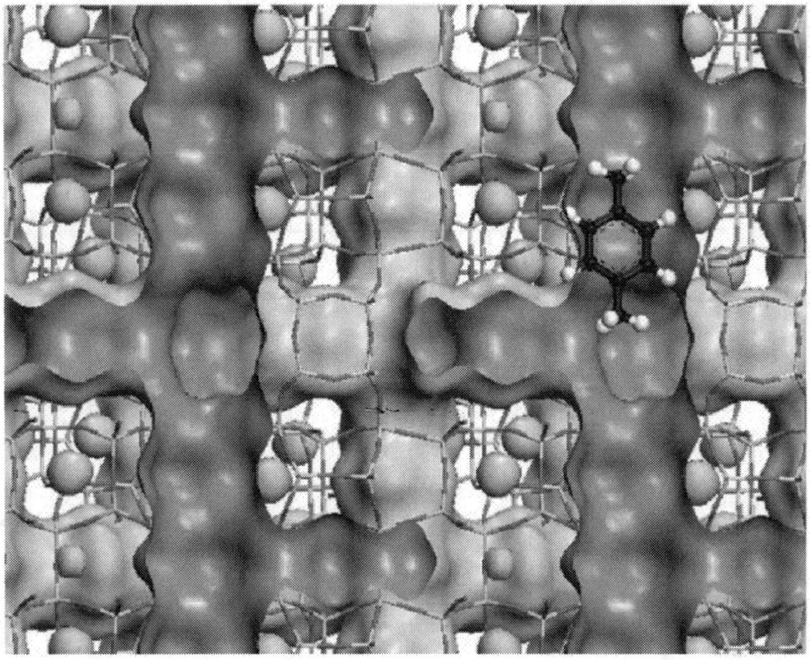

Fig.2 MFI topology and para-xylene

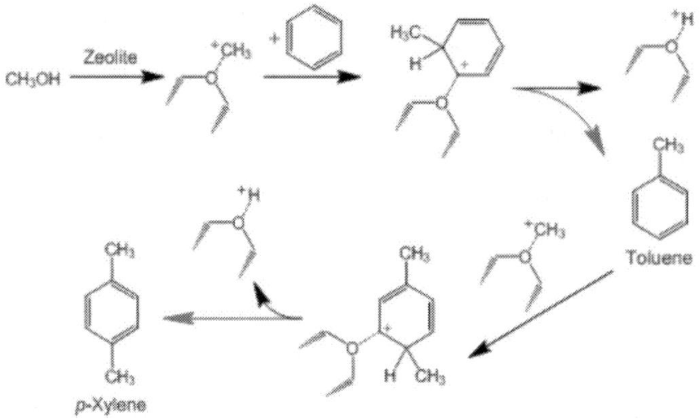

Fig.3 Mechanism on benzene alkylation with methanol

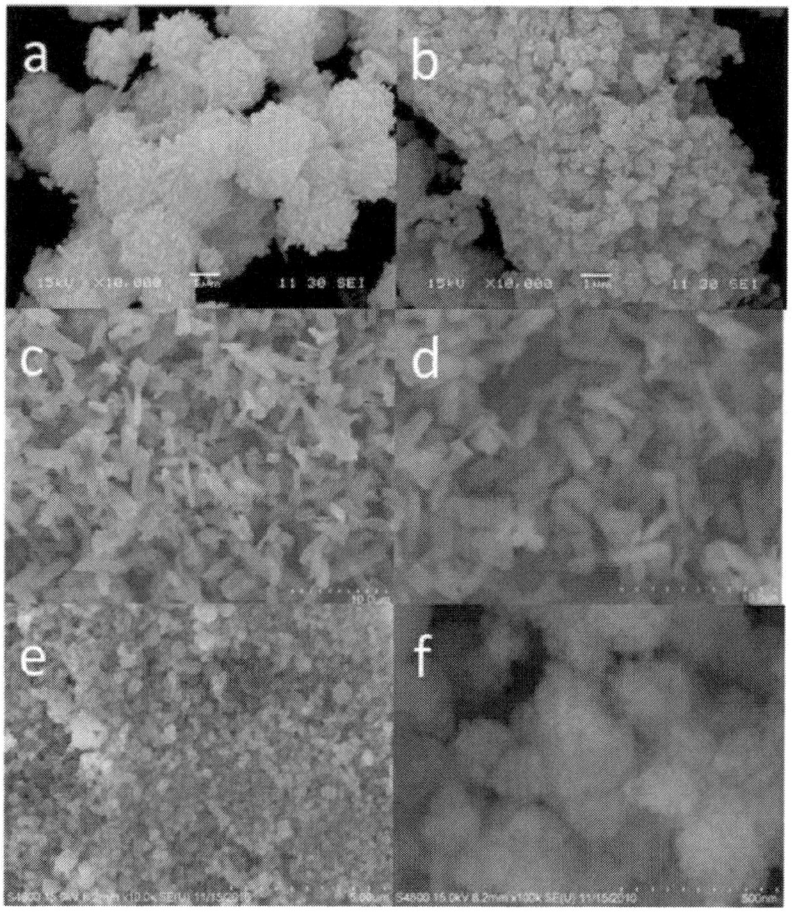

Fig.4 SEM image of conventional ZSM-5 (a, c, d) and hierarchically porous (b, e, f) ZSM-5-HTS with Si/Al ratios of 20 (a, b) and 180 (c-f)

370097 Layered Double-Hydroxide Catalysts for Lignin Depolymerization

Abstract:

A critical step in the conversion of lignin to renewable value-added aromatic chemicals is selective depolymerization of the biopolymer to its constituent monomers. For optimal process economics, this depolymerization should occur over heterogeneous catalysts in the absence of high-pressure hydrogen, but few catalysts exist that perform well and meet these criteria. One promising class of catalyst is layered double-hydroxide materials, such as hydrotalcite. Our group has recently shown that nickel-loaded hydrotalcite selectively cleaves the β-O-4 bond in a lignin model dimer, and reduces the molecular weight of two types of whole lignin. Here we expand on this finding by varying the many tunable parameters of hydrotalcite catalysts, including the type and location of metal and the identity of the charge-compensating anion(s), as well as the reaction solvent. Application of these catalysts to a variety of lignin model dimers and real lignin streams shows that these catalysts do not rely only on basic or metal sites for activity, and that the solvent has an impact on catalyst activity and product spectrum.

Catalytic Depolymerization of Lignin Extracted from Kraft Black Liquor and Lignin Model Compound 4-phenoxyphenol

Bethany D. Carter and David A. Bruce*

Department of Chemical and Biomolecular Engineering, Center for Advanced Engineering Fibers and Films, Clemson University, Clemson, SC, 29634-0909, USA.

* *Corresponding author. E-mail:* dbruce@clemson.edu

Lignin is the only renewable source of aromatic species, consisting of a network of guaiacyl, syringyl, and p-hydroxyphenyl units connected through various types of ether linkages forming an alkyl aryl ether backbone functionalized with methoxyl, benzyl alcohol, and phenolic hydroxyl species. Lignin conversion into lower molecular weight, monomeric type species with valuable functionality, such as phenols and aromatic acids, would shift dependence away from fossil fuels to a renewable source of aromatics.

Lignin was extracted from Kraft black liquor via the continuous Sequential Liquid Lignin Recovery and Purification (SLRP) process whereby the pH of the incoming black liquor (pH ~ 13.5) was lowered to a pH ~ 9.5 by acidification with CO_2 resulting in lignin precipitation. The precipitated, liquid lignin phase is then acidified with sulfuric acid until a final pH ~ 2.5 is achieved, whereby the lignin is fully acidified.

Catalytic conversion of SLRP lignin and lignin model compound 4-phenoxyphenol was investigated using a Pd/SBA-15 heterogeneous solid catalyst. The high surface area, large pore diameter, and hydrogenation sites enable the cleavage of ether linkages in lignin and lignin model compounds as well as stabilization of the resulting product thus preventing repolymerization which would result in low conversion. The nanometer length scale of the catalyst particles also contributes to these advantages. The heterogeneous nature of the catalyst allows for separation and reuse of the catalyst.

Lignin model compound depolymerization experiments were carried out in small batch reactor in the presence of the catalyst and under hydrogen pressure. Characterization of depolymerized lignin fractions was done using gas chromatography-mass spectrometry to analyze lighter molecular weight fractions. Catalyst reusability experiments were also performed to assess potential industrial application.

377869 Conversion of Guaiacol on Noble Metal Catalysts

Abstract:

Catalytic hydrodeoxygenation (HDO) is a rapidly developing technology for oxygen removal from bio-oils obtained from pyrolysis of biomass. In this study, Guaiacol, a phenol derived compound produced by the thermal degradation of lignin, was selected as a model compound to study the HDO process. Guaiacol is among the major components of pyrolysis bio-oils; however it is thermally unstable which leads to catalyst deactivation. Four noble metal catalysts (Pt, Pd, Rh and Ru) supported on activated carbon were tested in a fixed-bed reactor at atmospheric pressure and their performance for the guaiacol HDO process was compared. The screening criteria were as follows: (1) High degree of deoxygenation, (2) Low hydrogen consumption, (3) High carbon recovery in liquid phase, and (4) Long catalyst lifetime. Our results showed that, among the tested catalysts, Pt has higher deoxygenation activity and shows relatively little deactivation. Additionally, the operating temperature for Pt catalyst was optimized and found to be 300 °C. Finally, results from catalyst characterization demonstrated that polyaromatic deposits, especially the condensed ring compounds, are the most likely cause for catalyst deactivation. [1]

[1]D. Gao, C. Schweitzer. H. Hwang and A. Varma, Conversion of Guaiacol on Noble Metal Catalysts: Reaction Performance and Deactivation Studies, *Ind. Eng. Chem. Res*., DOI: 10.1021/ie500495z

384936 2,5-Dimethylfuran As a Model Compound to Investigate Hydrodeoxygenation of Complex Furanyl Compounds

Abstract:

Ying Lin Louie, Konstantinos Goulas, and Alexis T. Bell

Department of Chemical and Biomolecular Engineering,

University of California-Berkeley, Berkeley, California 94720

Several innovative pathways have been explored to produce diesel-range fuel additives via C-C bond coupling reactions of furanyl platform molecules, 5-hydroxymethylfurfural (HMF) and furfural (FUR), obtained from the dehydration of sugars [1, 2]. In each case, however, a final hydrodeoxygenation (HDO) step is required in order to produce a product that is oxygen-free, has a high cetane number, is compatible with conventional diesel fuel, and has minimal impact on the performance current diesel engines. While a number of studies of the hydrodeoxygenation of biomass-derived fuels have been reported [3], little is understood about the reaction pathways and how catalyst composition affects these pathways. The aim of this work is to develop such knowledge based on studies conducted with model compounds containing structural motifs representative of those found in second-generation biomass-derived diesel.

To understand the HDO chemistry of furanyl compounds and develop suitable HDO catalysts for this task, 2,5-dimethylfuran (DMF) was chosen as a model compound. The sequence of reactions by which DMF undergoes HDO was then investigated in both gas and liquid phase over various bi-functional metal/acid catalysts. Screening studies with carbon-supported Pt, Pd, and Rh revealed that Pt/C exhibits the highest C-O hydrogenolysis activity. Hydrogenation of DMF over Pt/C was found to proceed via parallel pathways of direct ring opening and ring saturation existed. It was observed that under mild reaction conditions, C-O bond hydrogenolysis of DMF to 2-hexanone is preferred over saturation of the furan ring to form 2,5-dimethyltetrahydrofuran (DMTHF). The resulting tetrahydrofuran ring is resistant to ring-opening to form ketones and alcohols. The distribution of products was found to be sensitive to the reaction conditions used. High temperatures and low H2 pressures were found to favor the furan ring-opening pathway, and under optimal conditions, we were able to obtain greater than 95% selectivity toward C-O hydrogenolysis products. Removal of oxygen from 2-hexanone and DMTHF occurred very effectively when Pt was supported on an oxide, such as TiO2, Nb2O5, and NbOPO4, which contain Lewis acid centers. Of the acid catalysts screened, niobium based oxides had stronger acidity and the highest ratio of Lewis acid to Brønsted acid sites [4], measured by infrared spectroscopy of adsorbed pyridine absorption. Pt/NbOPO4, in particular, was found to be exceptionally active for promoting the HDO of DMF to hexane. The results obtained from our studies of DMF have enabled us to optimize the HDO of more complex substrates, such as bis(5-methylfuran-2-yl)methane (BMFM, C11) and tetra(5-methylfuran-2-yl)methane (TMFM, C16). Greater than 81% yield to C11 was achieved during the HDO of BMFM at 250oC and 500psi H2, conditions that are much milder than those previously reported in the literature.

1. West, R.M., et al., Liquid alkanes with targeted molecular weights from biomass-derived carbohydrates. ChemSusChem, 2008. 1(5): p. 417-24.

2. Corma, A., O. de la Torre, and M. Renz, Production of high quality diesel from cellulose and hemicellulose by the Sylvan process: catalysts and process variables. Energy & Environmental Science, 2012. 5(4): p. 6328.

3. Zhong He, X.W., Hydrodeoxygenation of model compounds and catalytic systems for pyrolysis bio-oils upgrading. Catalysis for sustainable energy, 2012: p. 28-52.

4. Ziolek, M., Niobium-containing catalysts—the state of the art. Catalysis Today, 2003. 78(1-4): p. 47-64.

Catalytic Oligomer/Polymerization of Furfuryl Alcohol over Homogeneous and Heterogeneous Catalysts

Xiaojun Chan[1], Taejin Kim[1,2]

[1]*Department of Material Sciences and Engineering, Stony Brook University , Stony Brook, NY 11794*
[2]*Department of Chemical and Molecular Engineering, Stony Brook University , Stony Brook, NY 11794*

2014 AIChE Annual Meeting, Nov 17, 2014, Atlanta, GA

Introduction

- ***Lignocellulosic biomass (non-edible biomass resources)***
  - Renewable & Recyclable
  - CO_2 neutral
- ***Furfuryl alcohol ($C_5H_6O_2$, FA)*** is an attractive intermediate chemical for the production of levulinic acid and various useful green polymer products.

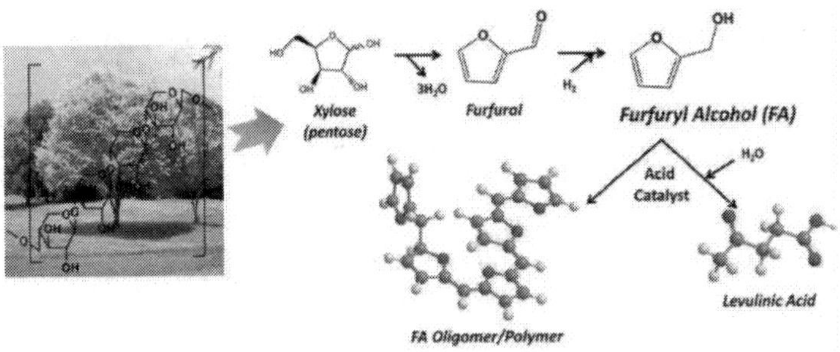

Objects

FA monomer

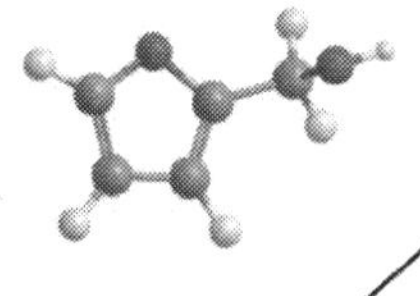

$C_6H_5O_2$

FA Oligomer
(Dimer, Trimer, Tetramer…)

Investigate the FA Oligomerization over Homogeneous and Heterogeneous Catalyst

- ***Intermediate species (Dimer, Trimer)***
- ***Feasibility to use Heterogeneous catalysts***

Experiment

- ***Vibrational Spectroscopy Analysis:***
  FT-IR, Raman(Complimentary methods)
- ***Product Composition Analysis:***
  GC, GC/MS(Quantitive calculation based on Response Factor theory)

FT-IR Spectra of Homogeneous/Heterogeneous Catalystic Furfuryl Alcohol Oligomerization under Room Temperature

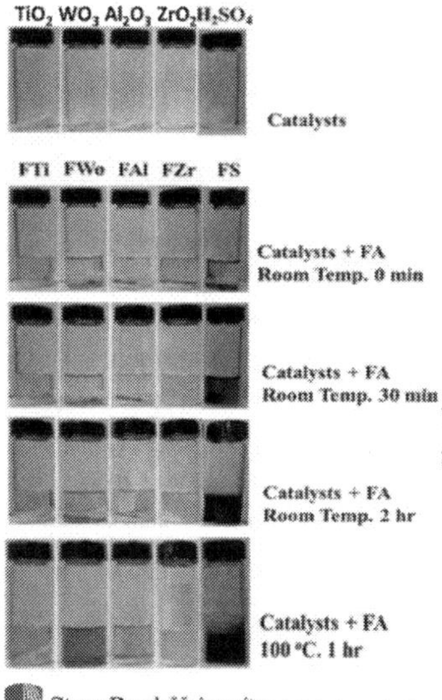

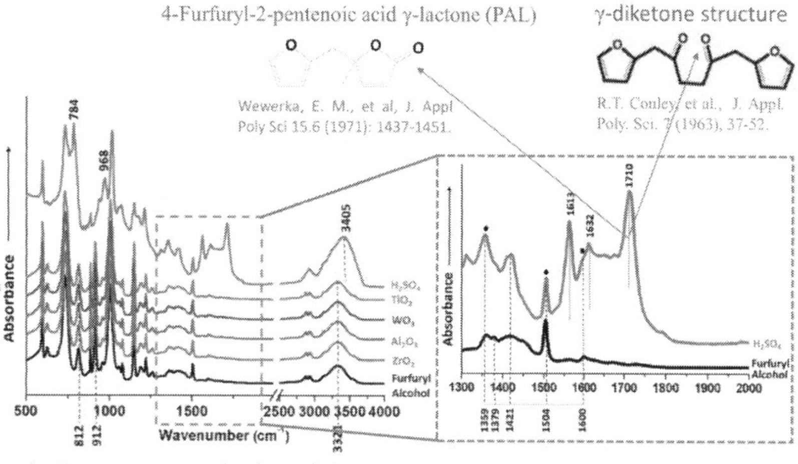

- Homogeneous catalyst (H_2SO_4) shows tremendous performance under room temp. and at 100°C.
- Among Heterogeneous catalysts, WO_3 shows sign of reaction at 100°C

WO_3 Catalystic Furfuryl Alcohol Oligomerization under 100°C

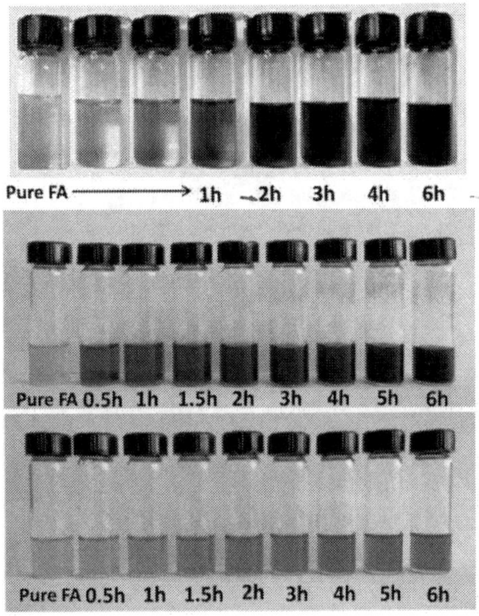

Sulfuric acid + FA
1:100
R.T.

WO_3+ FA
1:10
100°C

FA
100°C

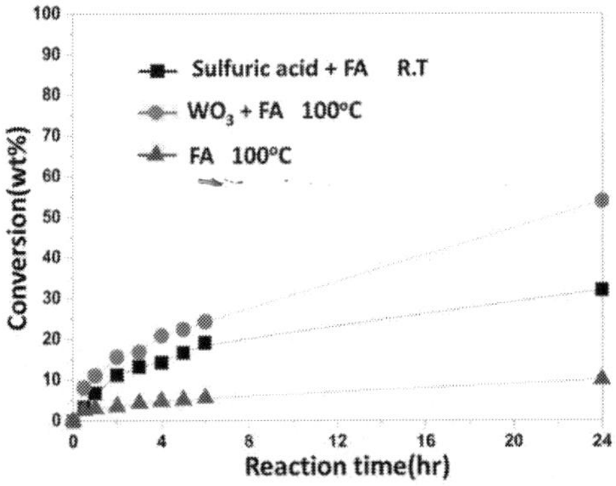

- WO_3 reaches similar conversion at 100°C compared with conversion obtained by Sulfuric acid at room temperature

Kim, T.J., et al, Cata Com, 46 (2014): 66-70.

5

Compare of fresh and used WO_3 for Furfuryl Alcohol Oligomerization

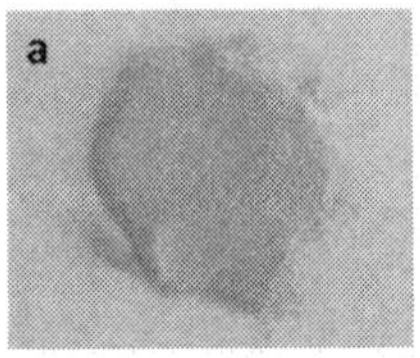

Fresh WO_3

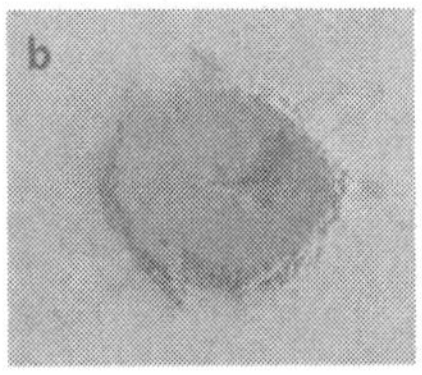

Used WO_3

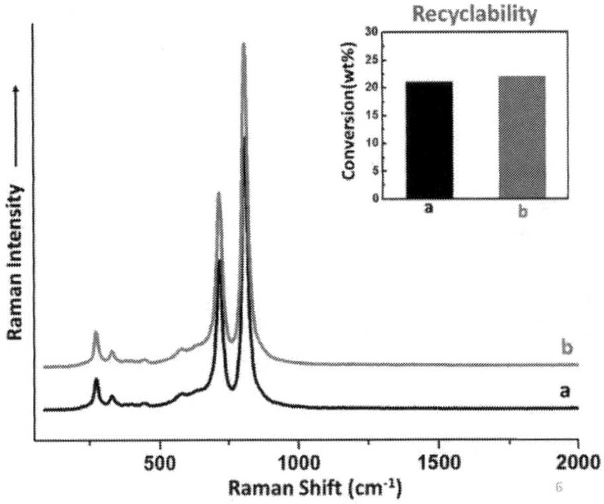

6

Stony Brook University | The State University of New York

Oligomers obtained from Homogeneous(Ho)/Heterogeneous(He) Catalystic Furfuryl Alcohol Oligomerization

	Oligomer	Name	Structure	
6 Dimers	D1	2,2'-Difurylmethane		(Ho, He)
	D2	2-(2-furylmethyl)-5-methylfuran		(He)
	D3	2-Furanethanol, α-2-furanyl-		(Ho)
	D4	Difurfuryl ether		(Ho, He)
	D5	4-Furfuryl-2-pentenoic acid γ-lactone		(He)
	D6	5-Furfuryl-furfuryl alcohol		(Ho, He)
2 Trimers	T1	2,5-Difurfurylfuran		(Ho,He)
	T2	2,2'-(Furylmethylene)bis(5-methylfuran)		(He)

Concentration and Selectivity of oligomers obtained from WO_3 catalyzed Furfuryl Alcohol Oligomerization under 100°C

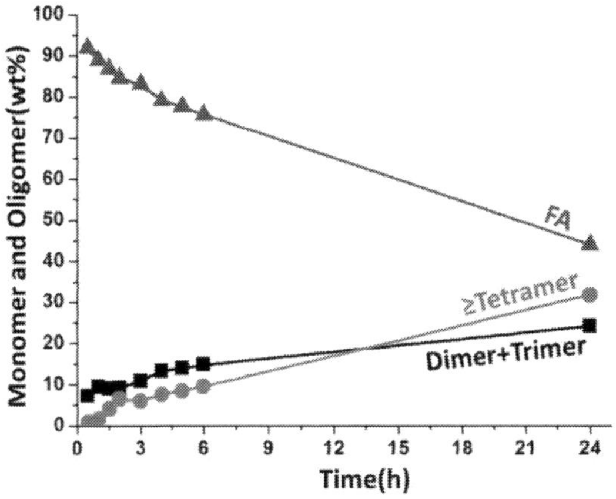

At the beginning, dimer and trimer have higher concentration. At the last several hours, concentration of heavy oligomer exceed dimer and trimer.

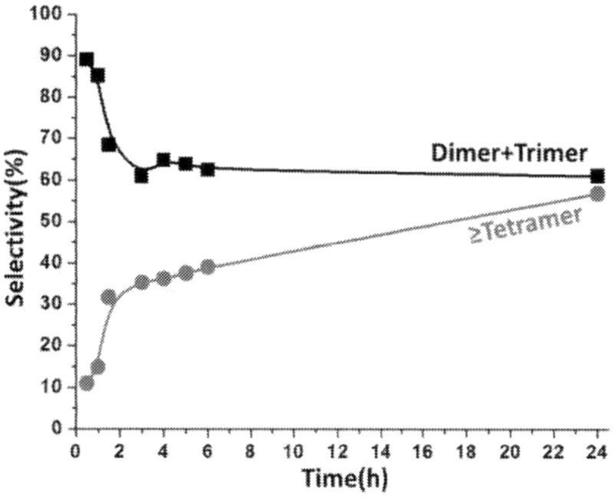

Selectivity to dimer+trimer decrease, while to heavy oligomer increase as a function of time.

FT-IR Spectra of WO_3 catalyzed FA Oligomerization under 100°C

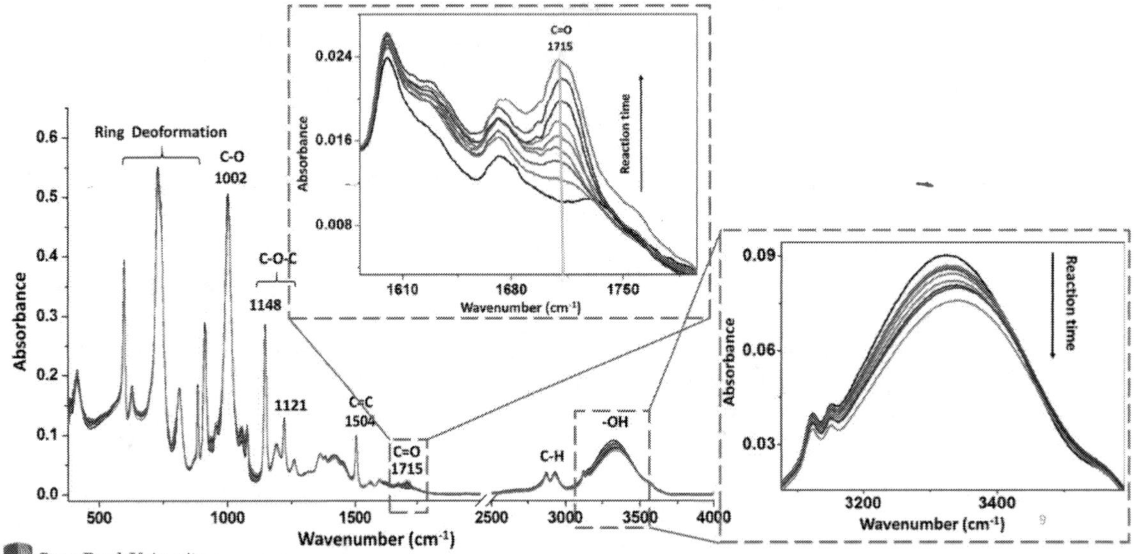

Raman Spectra of Homogeneous/Heterogeneous Catalystic FA Oligomerization

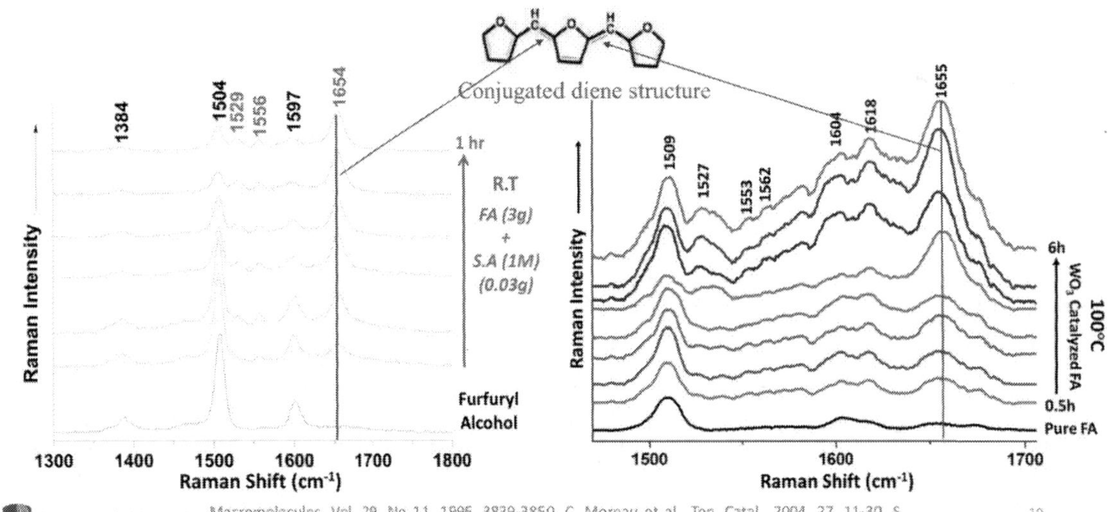

Macromolecules, Vol. 29, No 11, 1996, 3839-3850, C. Moreau et al., Top. Catal., 2004, 27, 11-30, S. Spange et al., Makromol. Chem., 194, 1537-1544 (1993), C.C. Burket et al., Carbon 44 (2006) 2957-2963

Proposed Initiation Mechanism of Homogeneous/Heterogeneous Catalystic FA Oligomerization

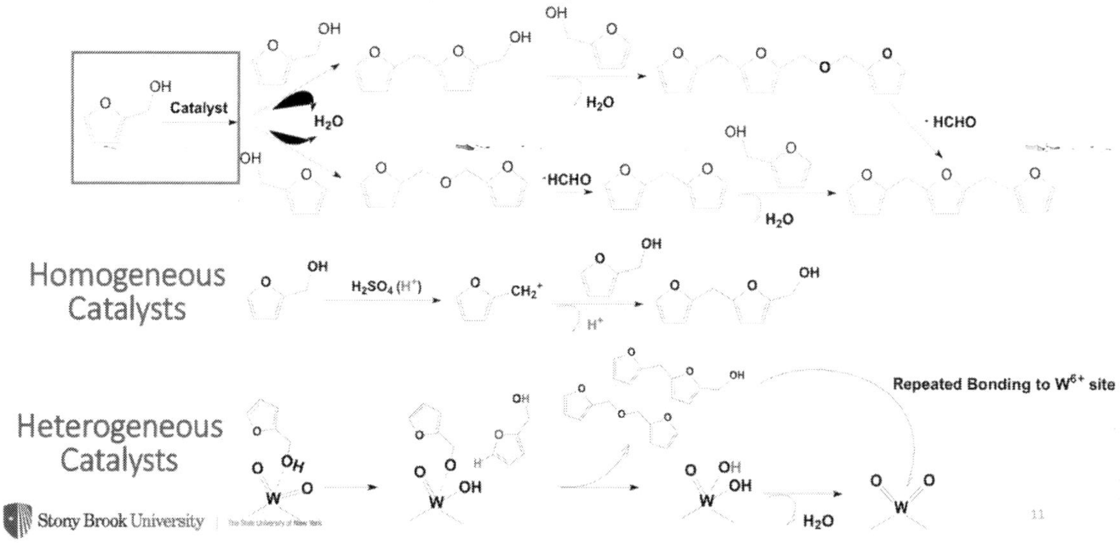

Conclusion

- In this work, we investigated the products and intermediate species for the liquis-phase oligomer/ polymerization of furfuryl alcohol monomer using homogeneous/heterogeneous catalysts
- Six different kinds of dimers, Two different kinds of trimers were observed; most of the maintained functional group, such as methylene, ether and CH_2OH.
- A strong C=O (γ-diketone) and aliphatic C=C band (conjugated diene) observed in the IR and Raman spectra, respectively, provided evidence to understand the oligomer polymerization reaction mechanism.

- Two different oligomerization initiation mechanism for homonegeous/heterogeneous were proposed respectively. Further conformation can be fulfilled through In-situ/Operando Spectroscopy.

Thank you!

Acknowledgements

Prof. Israel E. Wachs

Prof. Devinder Mahajan

Dr. Christopher Keturakis

Dr. Wei Nan

Vitoria Barbosa

STONY BROOK UNIVERSITY

ADVANCED ENERGY RESEARCH AND TECHNOLOGY CENTER

Stony Brook University | The State University of New York

13

389211 Selective Oxidation of Sugars and Polyols to Disaccharic Acids on Pt-Based Catalysts in One Pot Process

Abstract:

Catalytic conversion of biomass-based sources to high value chemicals has attracted increasing interest because it provides renewable alternatives to those derived from petrochemical feedstocks. Sugars and polyols, such as glucose, xylose and glycerol are abundantly available from biomass resources. Oxidation of these oxygen functional feedstocks to disaccharic acids such as glucaric acid and tartronic acid shows promising industrial applications, because these acids are valuable pharmaceutical intermediates and food additives. More importantly, disaccharic acids are potential feedstocks for the production of various megaton everyday chemicals such as adipic acid, succinic acid and malonic acid.

Disaccharic acids were conventionally produced from biomass or petroleum crudes via nitroxidation, cyanide and biological routes, which generate large amounts of toxic side products with poor overall productivity and chemoselectivity toward desired products. Catalytic oxidation of biomass in the presence of molecular oxygen on metal catalysts (*e.g.* Pt, Pd) could potentially yield disaccharic acids in one step without producing any toxics. Selective oxidation of sugars and polyols to disaccharic acids is however challenging because of significant side reactions of feedstocks to form smaller mono-acids, CO_2 as well as humic substances under reaction conditions (T = 40~80 °C in the alkaline medium). In addition, the reaction network of sugars to $C_{1\sim6}$ mono- and di-acids under oxidative conditions is not clearly understood at the current stage. Therefore, design of active and selective oxidation catalysts and fundamental understanding of reaction pathways/mechanism is the key to improve the overall yield of disaccharic acids from biomass. In particular, heterogeneous Pt-based catalysts have been evaluated and characterized for aqueous phase conversion of glucose, xylose and glycerol in the presence of molecular oxygen. Detailed kinetic studies taking into account the effects of Pt particle size, pH, temperature and oxygen pressure on the rate of oxidation, C-C cleavage and dehydration will be presented based on Pt-based catalysts.

368973 Catalytic Oxidation of Cellobiose over Gold-Based Bimetallic Nanoparticles Supported on TiO2

Abstract:

A series of bimetallic catalysts Au-M (M = Cu, Co, Ru, Pd and Ni) were supported on TiO_2 via deposition-precipitation (DP) method using urea as precipitating agent. The resulting catalysts were used in the catalytic oxidation of cellobiose into gluconic acid. The interactions between the metals in the bimetallic catalysts were carefully investigated in order to unravel the effect of the second metal on the chemical and electronic properties of the active sites and their impact on the catalytic performance. Cu and Ru were found to be promising promoters for the oxidation of cellobiose into gluconic acid, promoting oxygen activation in gold-catalyzed oxidation reactions, while Co and Pd were found to enhance the retro-aldol condensation of fructose into glycolic acid. In this case, the catalytic activity is dominated by the second metal added to Au. Hence, Au behaves more like a promoter of the second metal, preventing over oxidation and poisoning by the reaction intermediates and products. Ni specifically promoted the C-C scission, yielding a mixture of lower carbon-containing compounds and glycolic acid as the main products.

The major reaction products obtained were gluconic acid, glycolic acid, ethylene glycol and erythritol. The catalytic activity of the catalysts followed the order: Au-Cu/TiO_2 > Au-Pd/TiO_2 > Au-Ru/TiO_2 > Au-Co/TiO_2 > Au/TiO_2 > Au-Ni/TiO_2. Upon comparison, Au-Ru/TiO_2 was found to be an active and selective catalyst towards the formation of gluconic acid. The nature of the metal interactions in the bimetallic systems greatly influenced the product distribution.

On the nature of the active sites for HCOOH decomposition on supported Au catalysts

Sha Li, Suyash Singh, Ronald Carrasquillo-Flores, Ana C. Alba-Rubio, James A. Dumesic, Manos Mavrikakis*

Department of Chemical & Biological Engineering

University of Wisconsin-Madison

Formic acid (HCOOH) is a major byproduct in biomass processing[1] and a promising liquid hydrogen carrier that can be used for *in situ* hydrogen production. Well-dispersed Au nanoparticles are found to be highly reactive for HCOOH decomposition with 100% selectivity toward CO_2 and H_2.[2-3] In this work, an integrated approach that combines state-of-the-art density functional theory (DFT) with reactivity experiments and microkinetic modeling is presented to unravel the detailed reaction mechanism and the nature of active site on Au catalysts. DFT calculations were performed on Au(111), Au(100) and Au(211) model surfaces to obtain initial guesses for the surface reaction energetics needed for the microkinetic model. Then by adjusting the DFT derived binding energies and activation energies in the microkinetic model, good agreement is achieved between the model predicted reaction rates and reaction orders with the experimental results collected on Au/SiC catalysts. The reaction is predicted to proceed entirely through the formate (HCOO) mediated route which is comprised of three elementary steps: $HCOOH^* \rightarrow HCOO^{**} + H^*$, $HCOO^{**} \rightarrow CO_2^* + H^*$, $H^* + H^* \rightarrow H_2$. The large adjustments made to the DFT derived parameters in order to obtain good fits with the experimental data suggest that none of Au(111), Au(100) and Au(211) provide an accurate representation of the active site for this reaction. In fact, the active site might be an even more under-coordinated Au site with coordination number ≤ 6. Then, *ab initio* molecular dynamics simulations were carried out to determine the structures of small gold clusters which generally contain a high concentration of under-coordinated sites. Theoretical investigations of HCOOH decomposition on several Au cluster models further confirm our conclusion regarding the importance of highly under-coordinated Au sites in this chemistry.

References:

(1) Bond JQ, Alonso DM, Wang D, West RM, Dumesic JA. *Science* **2010,** *327*, 1110.

(2) Ojeda M, Iglesia E. *Angew. Chem. Int. Ed.* **2009,** *48*, 4800.

(3) Yi N, Saltsburg H, Flytzani-Stephanopoulos M. *Chemsuschem* **2013**, 6, 816.

First-Principles Based Kinetic Monte Carlo Simulation of Catalytic Systems over Metal Particles

Eric Dybeck and Matthew Neurock
University of Virginia
AICHE Annual Meeting November 17, 2014

1

Background

Macro Scale:

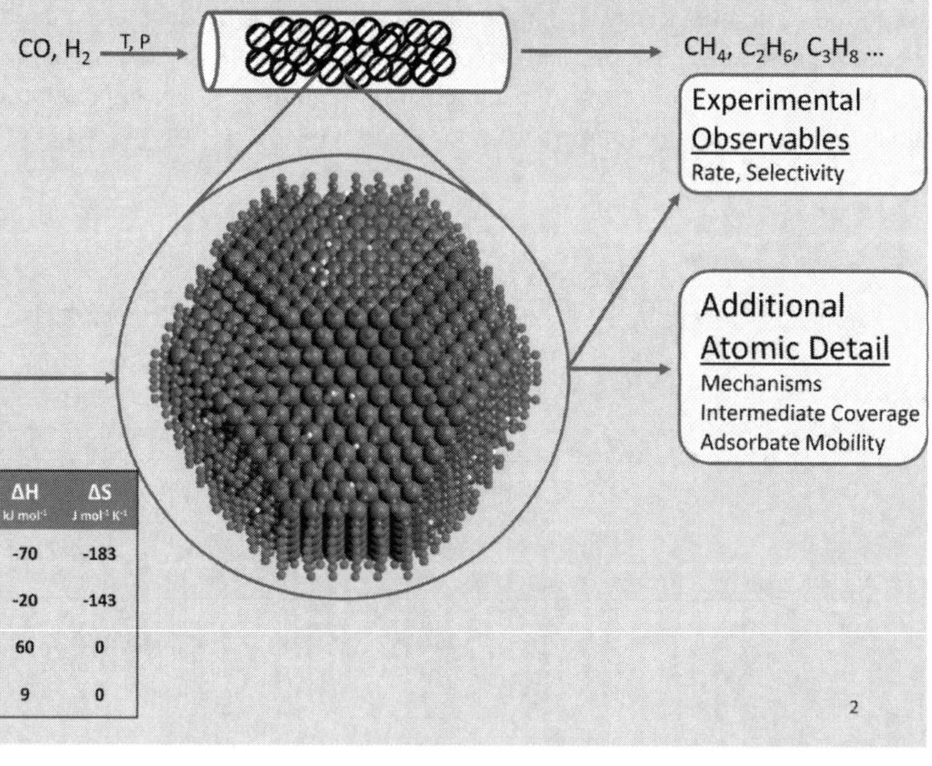

Micro Scale:

Event	ΔH kJ mol^{-1}	ΔS J mol^{-1} K^{-1}
CO -> CO*	-70	-183
H2 -> 2H*	-20	-143
CO*+H* -> HCO*	60	0
HCO*+H*->CH*+OH*	9	0

2

2D-Surface KMC 3D-Particle KMC

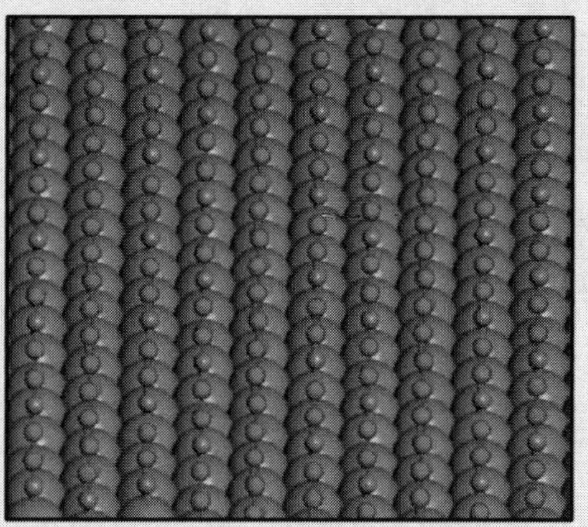

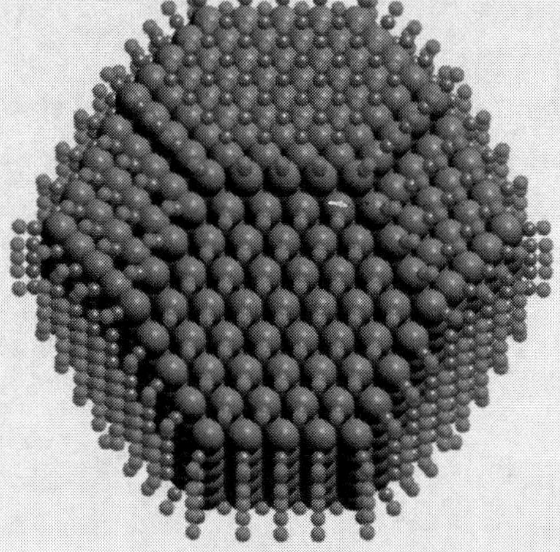

3

2D-Surface KMC 3D-Particle KMC

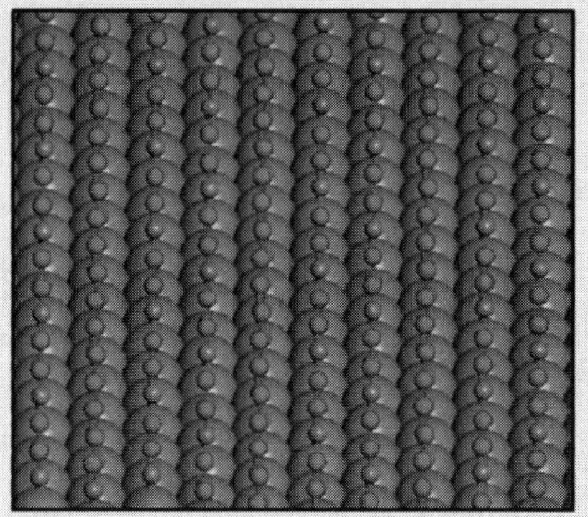

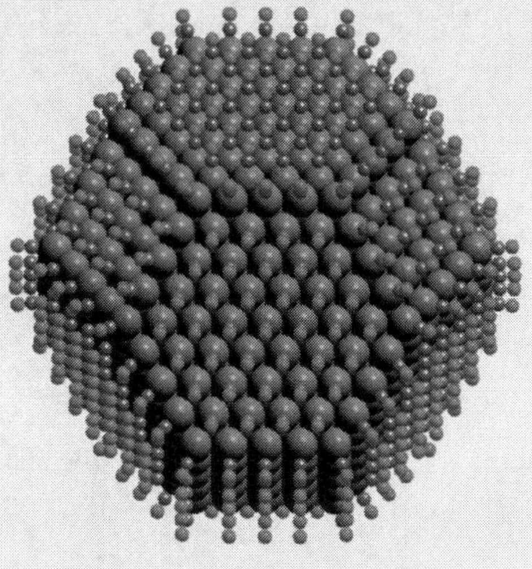

Adlayers on surfaces are constricted to perpendicular orientation

Adlayers on particles can realign to reduce repulsion

4

2D-Surface KMC 3D-Particle KMC

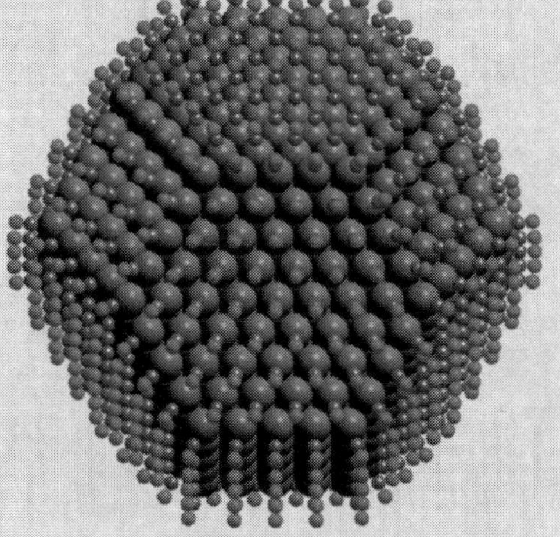

Adlayers on surfaces are constricted to perpendicular orientation

θ_{CO} < 0.75 ML

Adlayers on particles can realign to reduce repulsion

θ_{CO} > 1.0 ML

5

Coordinatively Unsaturated Atoms

Example Reactions

Fischer-Tropsch Synthesis
Ammonia Synthesis
Methane Reformation
CO Oxidation

Particles contain coordinately unsaturated sites which exhibit markedly different activity and binding properties.

6

Coordinatively Unsaturated Atoms

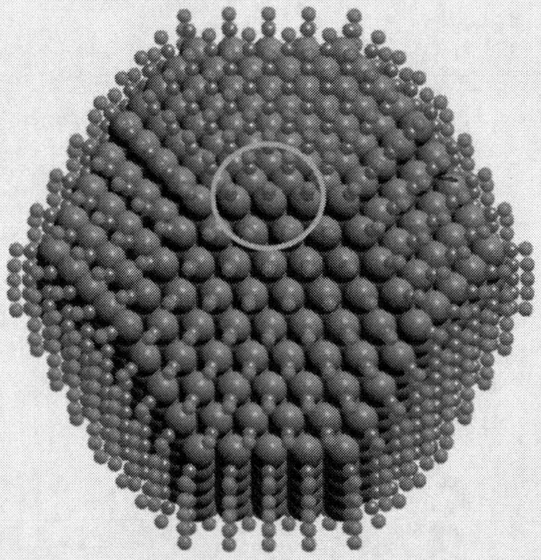

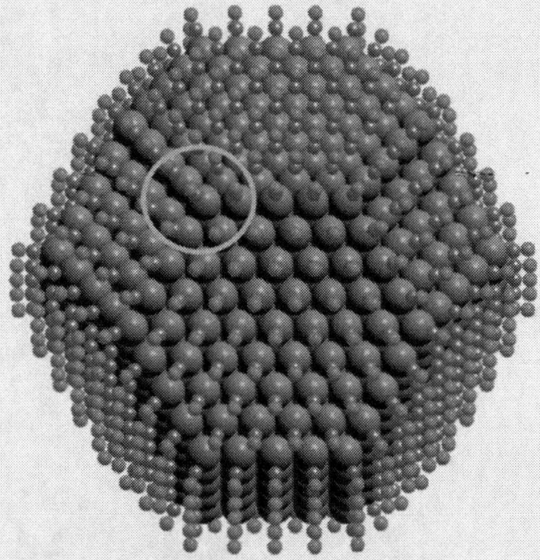

Unsaturated sites on corners and edges can bind to more than one adsorbate

7

Coordinatively Unsaturated Atoms

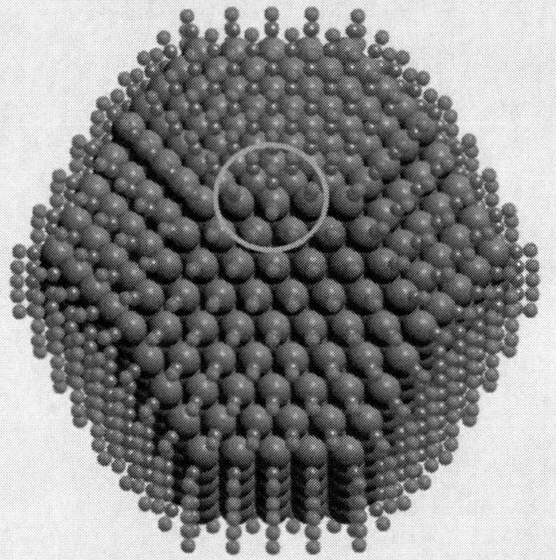

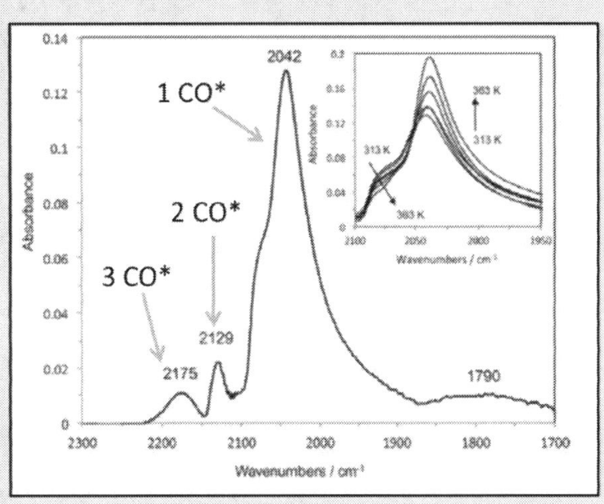

Unsaturated sites on corners and edges can bind to more than one adsorbate

Loveless, B, et al *Journal of the American Chemical Society* (2013) **135** (16) 6107-6121

8

Building the Particle

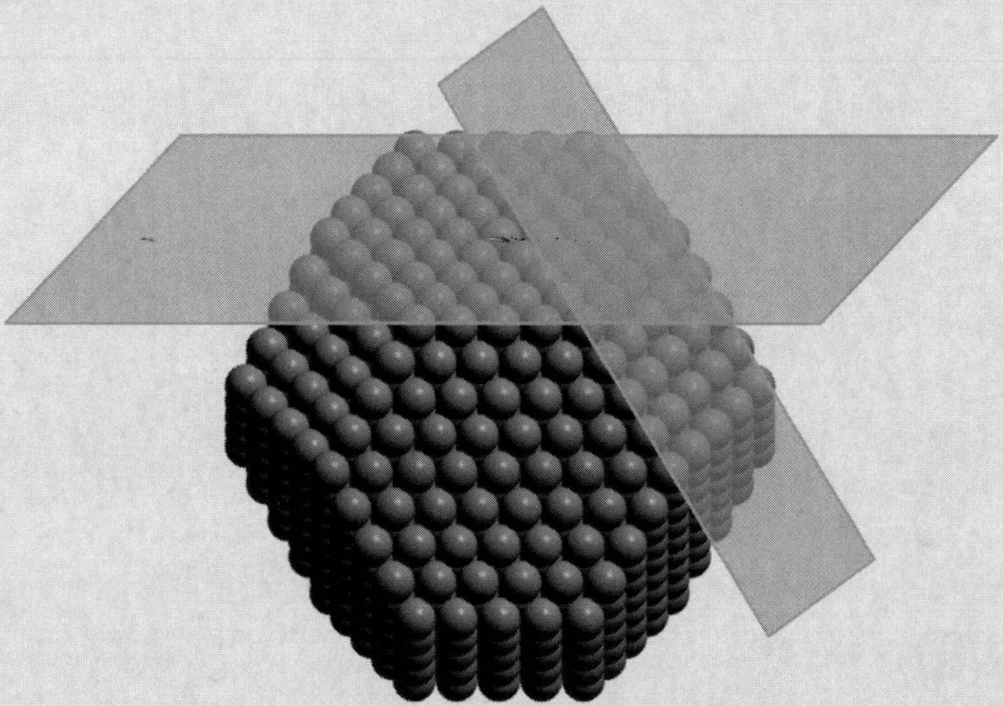

9

Bond Order Conservation (Surface)

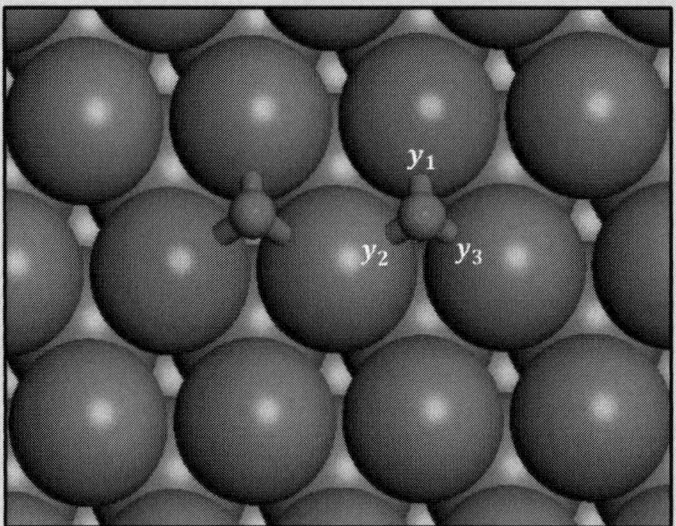

Adsorbate bonds are conserved to unity

$$Q_{A,M} = Q_o(A, M)(2y - y^2)$$

$$1 = \sum y_i$$

Shustorovich E., Sellers H., *Surface Science Reports* (1998) **31** (1) 0167-5729

10

Bond Order Conservation (Particle)

Functionality extended to coordinatively unsaturated atoms.

$$y_i = 1 - \frac{\prod_{j=1}^{n} Q_o(A_j, M_j, C_j)^{1-\delta_{ij}}}{\sum_{k=1}^{n} \prod_{p=1}^{n} Q_o(A_p, M_p, C_p)^{1-\delta_{kp}}}$$

Shustorovich E., Sellers H., *Surface Science Reports* (1998) **31** (1) 0167-5729

11

Intermolecular Interactions

Merck Molecular Force Field 94

$$E_{int\,ij} = \varepsilon_{ij}\left(\frac{1.07R_{IJ}^*}{R_{ij} + 0.07R_{ij}^*}\right)^7 \left(\frac{1.12R_{IJ}^{*\,7}}{R_{ij}^{\;7} + 0.12R_{IJ}^{*\,7}} - 2\right) + \frac{332.716\mathrm{q_i q_j}}{\mathrm{D}(\mathrm{R_{ij}} + \delta)^n}$$

Van der Waals **Electrostatic**

Halgren, T. A. *Journal of Computational Chemistry* (1996), **17**, 490-519

12

Kinetic Monte Carlo Algorithm

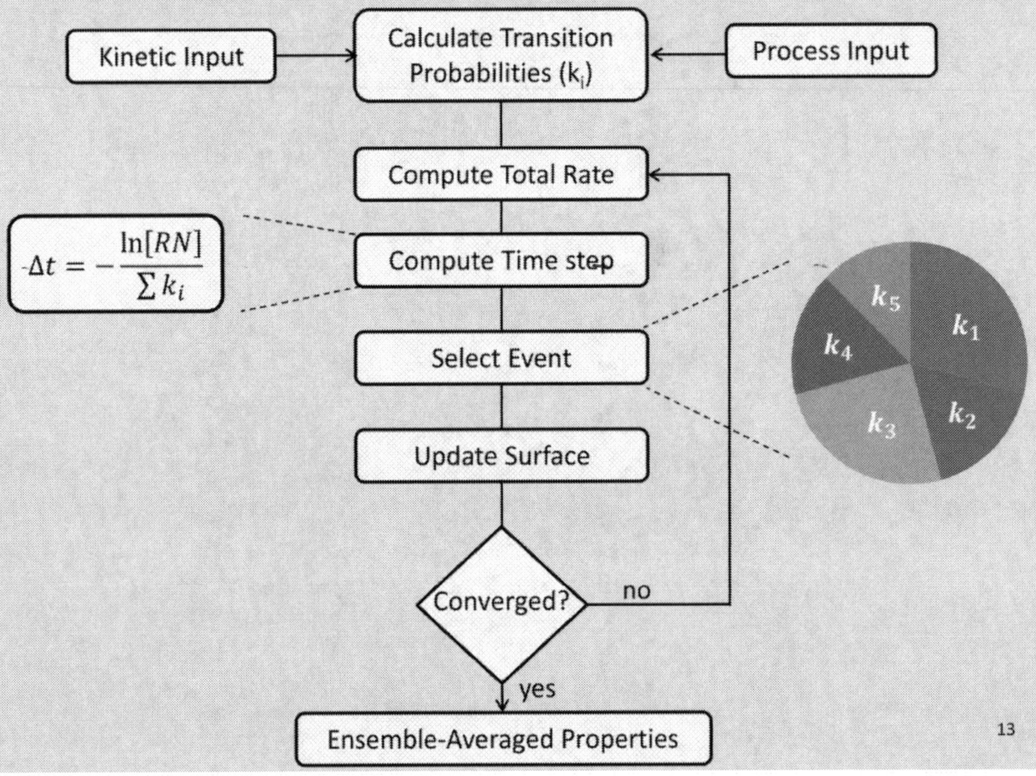

13

Kinetic Monte Carlo Simulation

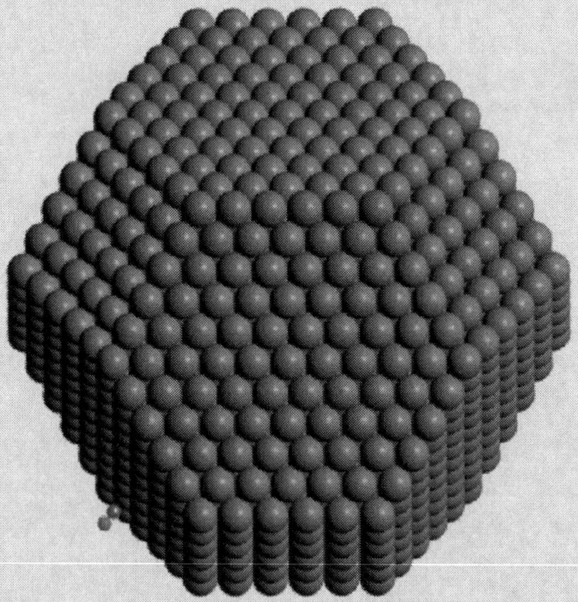

14

CO Oxidation on Platinum Nanoclusters

$$CO + * \leftrightarrows CO^*$$

$$O_2 + CO^* + * \leftrightarrows CO_2 + O^*$$

$$O^* + CO^* \leftrightarrows CO_2 + 2*$$

Allian, A., et al. *Journal of the American Chemical Society* (2011), **133** (12) 4498-4517

15

Model Binding Affinity Fitted to DFT

CO*

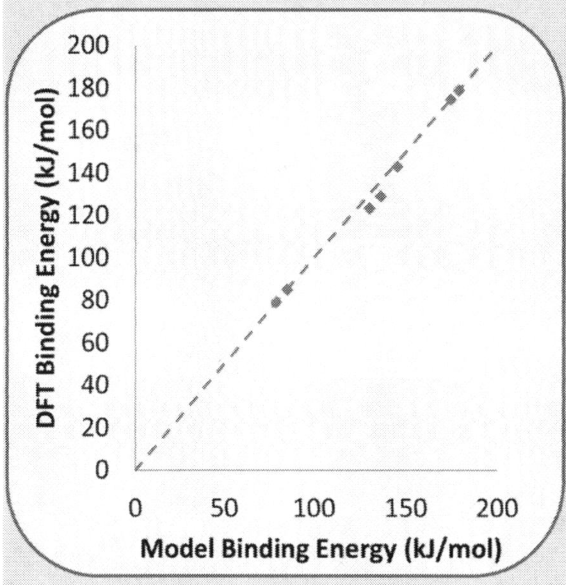

O*

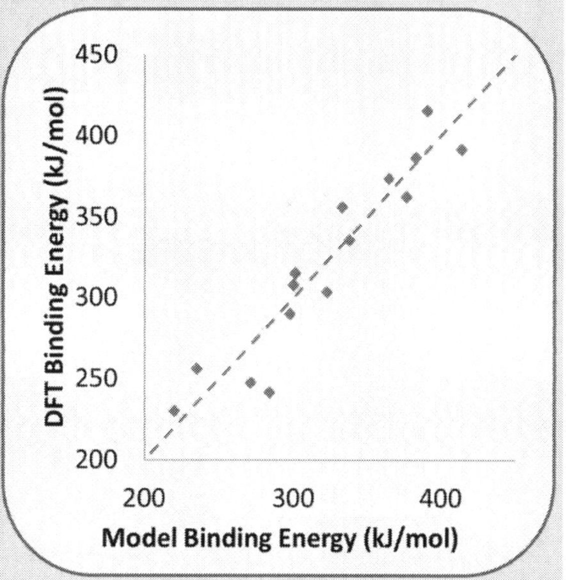

16

Simulated Pressure Effects

P_{CO} **P_{O2}**

$$r_{CO2} = k_{eff}\frac{P_{O2}}{P_{CO}}$$

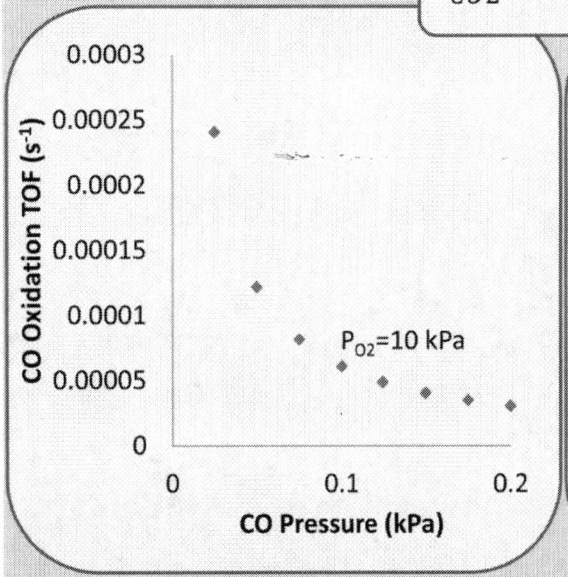

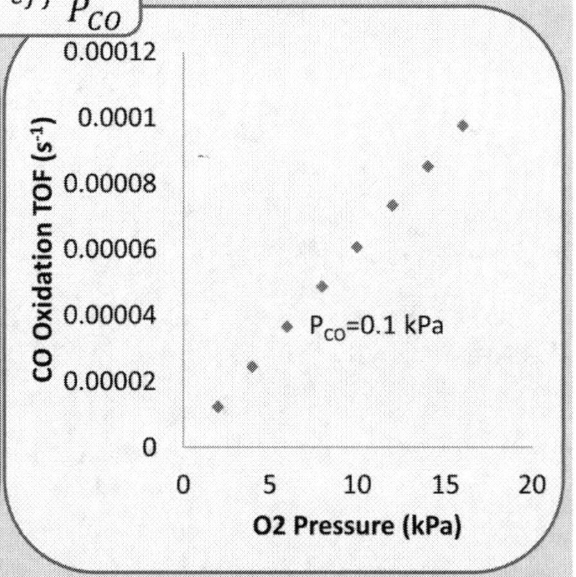

Reaction decreases with CO pressure and increases with O_2 pressure due to high coverage

17

CO Relaxation Effect (1.8 nm)

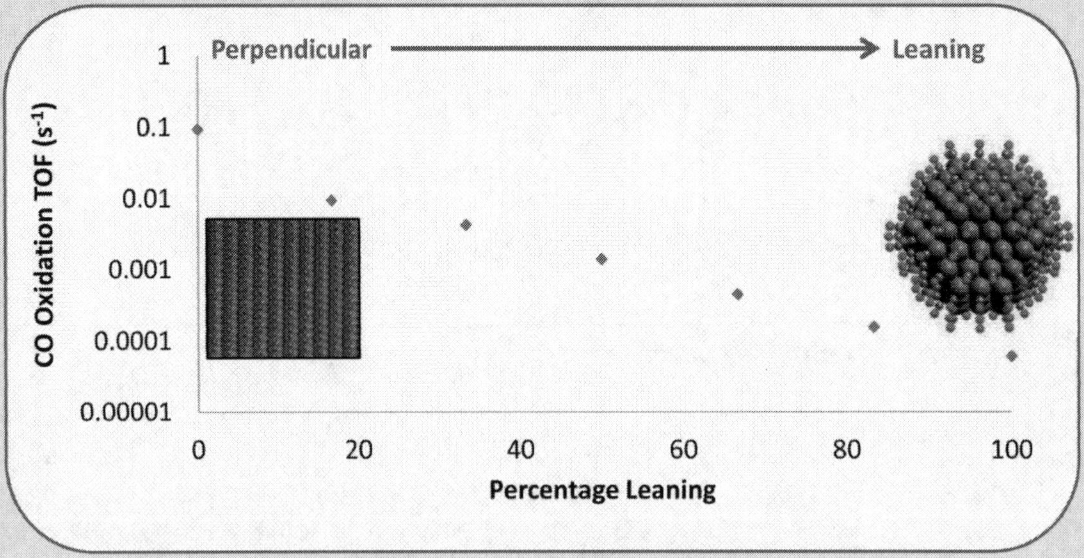

The effect of the CO adlayer relaxing on the particle alters the reaction rate by as much as 3 orders of magnitude on small particles.

18

Particle Size Effects (Simulation)

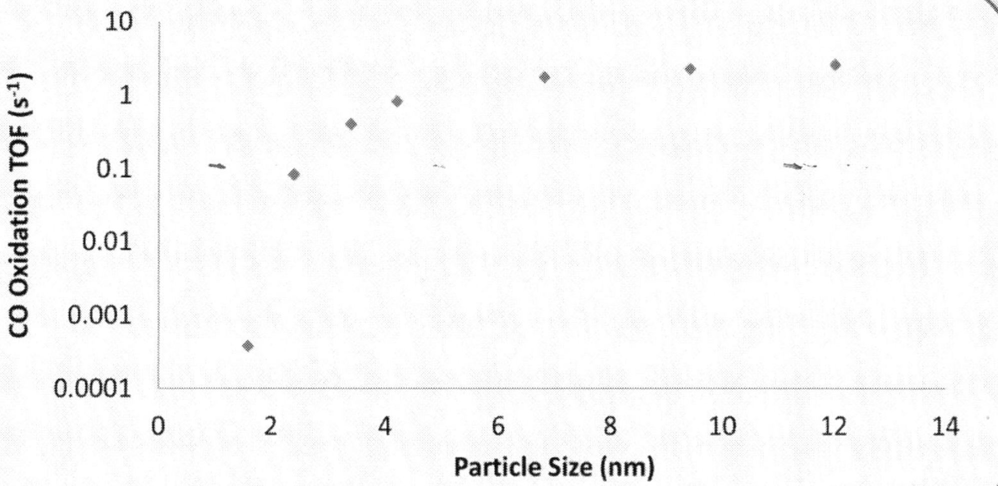

Without di-carbonyl sites, the reaction rate increases sharply with particle size.

19

Poisoning Leads to Structure Sensitivity

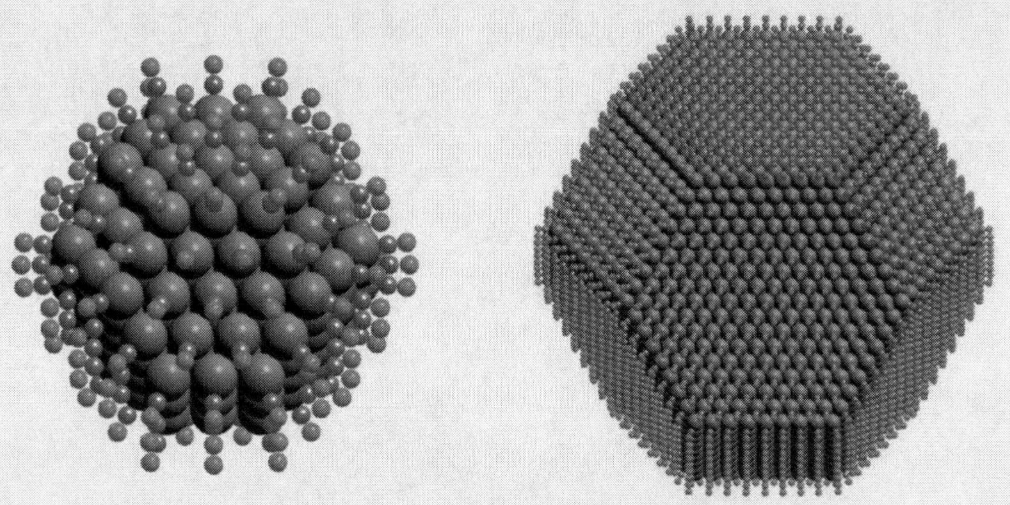

This structure sensitivity arises because the corners and edges are poisoned during the reaction, and because the CO adlayer has more room to fan out on smaller particles.

20

Incorporating Di-carbonyl Groups

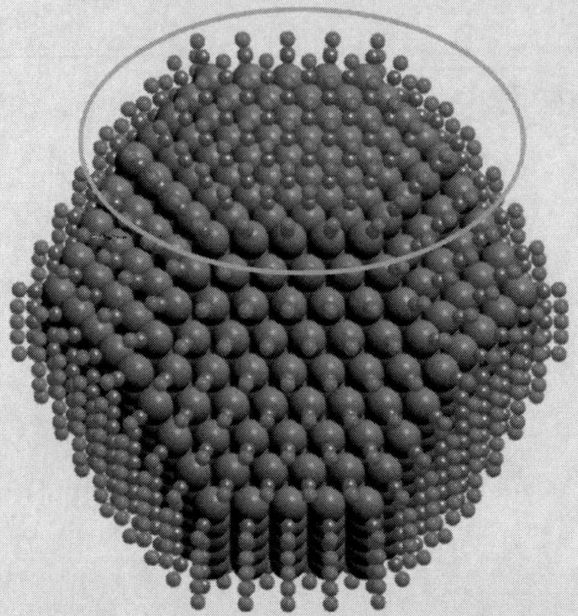

The formation of di-carbonyl groups destabilizes the CO* on the terrace

21

Incorporating Di-carbonyl Groups

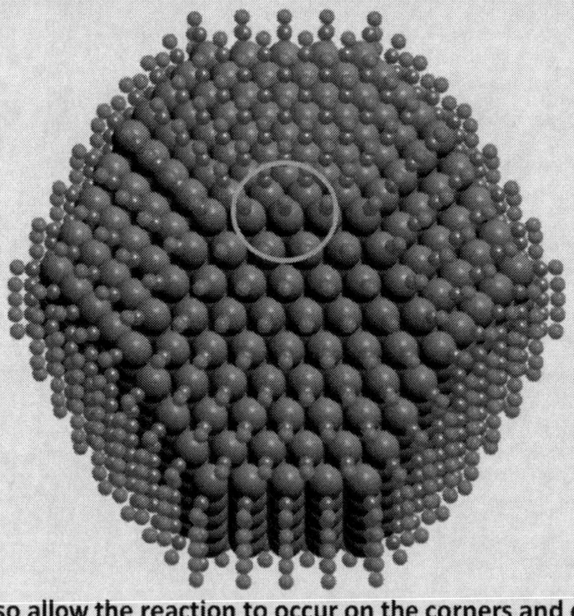

These extra sites also allow the reaction to occur on the corners and edges

22

Particle Size Effects (Simulation)

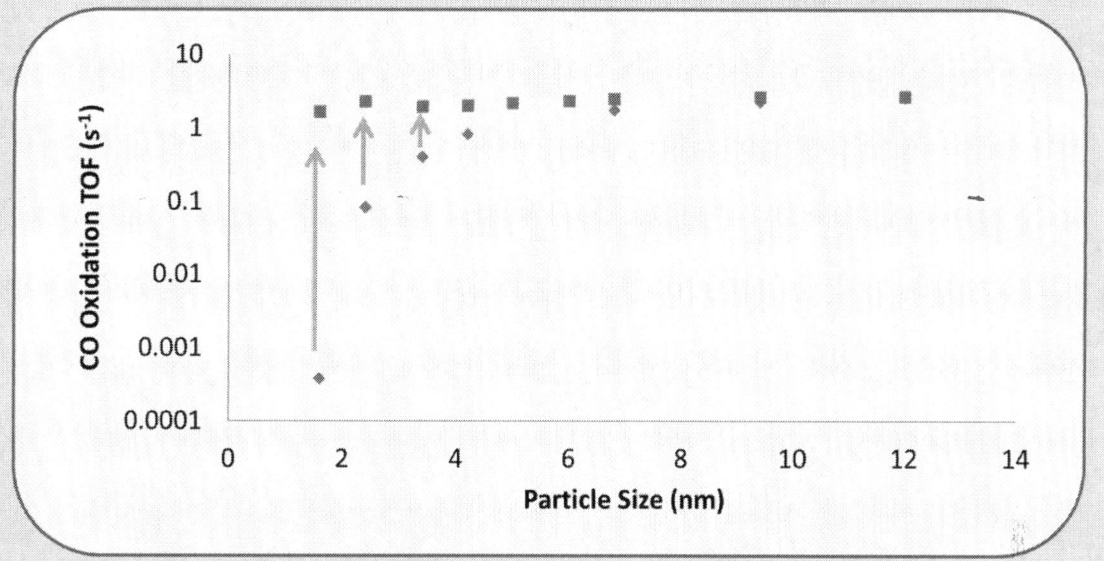

Without di-carbonyl sites, the reaction rate increases sharply with particle size.
However, this structure sensitivity largely disappears when dicarbonyl sites are included

23

Particle Size Effects (Experiments)

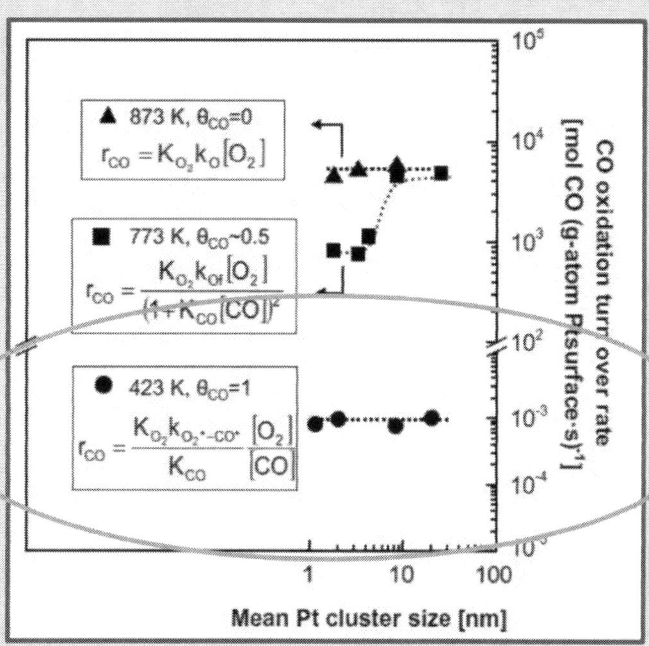

Garcia-Dieguez, M., Iglesia, E. *Journal of Catalysis* (2013) **301** 198-209

24

Summary

Particle Curvature Effects

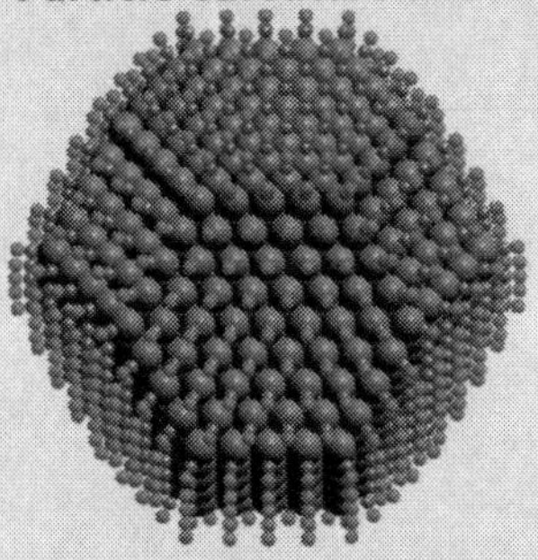

Coordinatively Unsaturated Sites

Generic Catalyst Surface

Extended Bond Order Conservation

25

Acknowledgements

Neurock Research Group

BP XC^2 Program

26

379663 Ab Initio Kinetic Monte Carlo Simulations of H2 Spillover Control on a Bimetallic Catalyst

Abstract:

The heterogeneous catalysis of hydrogenation reactions plays a significant role within numerous areas such as the petrochemical, pharmaceutical and food industries. Precious metals including Pt, Pd and Ru are often employed as hydrogenation catalysts owing to their high reactivity towards hydrogen. Using such metals on a commercial scale can prove costly and so cheaper alternatives are being sought after.

Experiment has shown that doping a less reactive, cheaper host metal such as Cu, with a low concentration of a more reactive metal, such as Pd, increases the surface reactivity towards hydrogen with respect to the pure host metal [i]. The dopant metal provides a low energy pathway for hydrogen dissociation and facilitates spillover onto the host metal. Furthermore, control over the spillover, and hence the surface coverage, may be established through the selective binding of a spectator CO molecule at the minority metal site. The low energy pathway is selectively blocked, thereby inhibiting the dissociation or re-association of hydrogen, trapping the latter on the surface. This phenomenon has been termed the 'molecular cork effect' and may have implications in mediating the kinetics of hydrogen uptake in bimetallic catalysis. Though observed in experiment, the underlying cause of the molecular cork effect is still under investigation [ii].

First-principles kinetic Monte Carlo (KMC) was employed to elucidate the origin and underlying kinetics of the molecular cork effect. This multiscale theoretical approach entails a comprehensive molecular-level analysis at the nanoscale before extending to the catalyst level at the microscale. Density functional theory (DFT) has been used in order to calculate the ground state energies of a variety of hydrogen and CO binding configurations on the (111) surface of Pd, Cu and a Pd/Cu alloy. The binding energies computed support the observation that hydrogen binding at the minority Pd site within the Pd/Cu alloy is more stable than binding to pure Cu. Furthermore the affinity of CO for Pd during 'corking' is also investigated, with CO binding more strongly to Pd than Cu within the alloy. Additionally, Nudged Elastic Band (NEB) calculations were performed to compute the transition state energy for hydrogen dissociation and diffusion across the Pd/Cu alloy surface. The barriers calculated corroborate the experimental findings; diffusion barriers were found to be low, thereby promoting facile spillover. Moreover, the dissociation barriers specifically at the Pd minority site were found to be lower than on pure Cu or elsewhere on the alloy surface.

The binding and lateral interaction energies along with NEB barrier heights and gaseous formation energies are taken from our DFT calculations to parameterise KMC simulations. Subsequently, first-principles KMC was able to reproduce the observations seen in experiment, with the resulting data extracted from the simulations providing valuable insight into an unknown mechanism and also aiding in the design of improved hydrogenation catalysts.

---

[i] Tierney, H. L., Baber, A. E., Kitchin, J. R., Sykes, E. C. H., *Phys. Rev. Lett.*, 2009, **103**, 246102.

[ii] Marcinkowski, M. D., Jewell, A. D., Stamatakis, M., Boucher, M. B., Lewis, E. A., Murphy, C. J., Kyriakou, G.., Sykes, E. C. H., *Nat. Mater.*, 2013, **12**, 523.

Density Functional Theory (DFT) simulations have been used to successfully describe heterogeneously catalyzed reaction mechanisms in the gas phase, but describing liquid phase heterogeneously catalyzed reactions is still hindered due to a lack of understanding of the relationship between the solvent and the surface-bound species. This relationship has at least three components: 1) liquid-phase molecules interact with the surface species through dispersion interactions and/or chemical binding, 2) liquid molecules can adsorb to the catalyst, altering its chemical composition, and 3) the liquid environment can alter the adsorbate surface interaction[1,2,3,4]. These effects can also influence the reactant surface coverage on the catalyst by stabilizing adsorbates through hydrogen bonding or dispersion forces[5,6], thus increasing adsorbate chemical potentials and coverages. Both solvent molecules and coadsorbed molecules can influence catalytic activity through interactions with the reactants, which can alter binding energies and activation energies[7]. Therefore, solvation and surface coverage are intimately linked, and this link must be defined in order to understand catalysis in the liquid phase and, more importantly, to effectively design catalysts for liquid-phase reactions. Given that many catalysts for liquid-phase reactions are carried out over expensive transition metal catalysts, catalyst design focused on incorporation of cheaper catalyst materials would enable more widespread utilization of systems such as fuel cells.

In this work, we study the link between solvation and surface coverage in terms of Pt-catalyzed CO oxidation with OH, which is thought to be the rate-limiting step at direct methanol fuel cell (DMFC) anodes[8], and which is also important in the aqueous phase reforming of glycerol[9,10]. Simulations are conducted using the Vienna Ab-Initio Simulation Package[11] (VASP) with Projector-Augmented Wave[12] (PAW) pseudopotentials and the Perdue, Burke, and Erzenhof[13] (PBE) exchange/correlation functional. The DFT-D2[14] method is used to compute dispersion forces. Kinetic calculations are performed using the climbing image-nudged elastic band (ci-NEB) method[15]. We compare interaction energies between several explicit solvation models at various adsorbate coverages of CO and OH to determine the optimal solvation model. We consider a 2-D ice-like monolayer, a 3-D Ih ice lattice, a liquid-like 3-D water environment formed by "melting" the ice lattice at various temperatures using molecular dynamics (MD), and a 3-D water environment formed by randomly placing water molecules in the simulation box to form a vapor-like model. We also consider adsorbate coverages up to 1/3rd of a monolayer (ML).

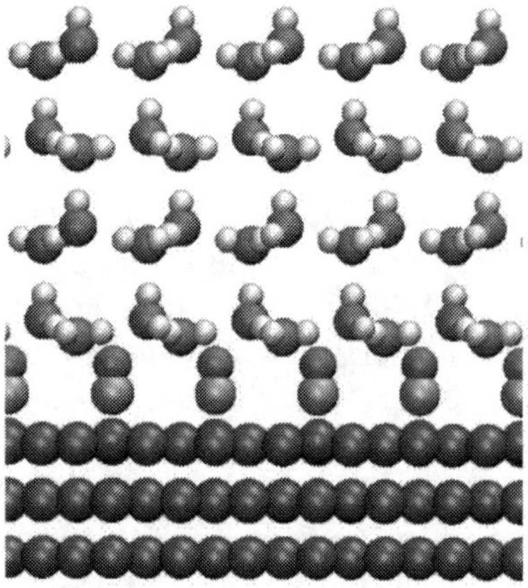
Figure 1: CO adsorbed on the Pt(111) surface under the 3D Ih ice lattice at a coverage of 1/9th ML.
Color code: Pt=grey, C=teal, O=red, and H=white.

Results show that adsorption site, solvation model, water configuration, adsorbate coverage, and dispersion forces all heavily influence the water-adsorbate interaction energies, which we calculate using

$$\text{Pt (aq)} + \text{Pt-CO (vacuum)} \rightarrow \text{Pt (vacuum)} + \text{Pt-CO (aq)}$$

Other important findings are that

1. Interaction energies cannot be correlated solely to the adsorbate dipole. For example, the interaction energies for FCC- and HCP-bound CO are 2.0 eV stronger than for atop-bound CO, despite that the dipole for atop-bound CO is more than three times larger (7.7 Debye versus 2.2 Debye).
2. Coverage has a remarkable effect on the interaction energies. For example, the interaction energies for CO adsorbates decrease by up to 1.0 eV as the coverage increases from 1/9th to 1/3rd ML.

3. Including dispersion corrections has a significant effect on the calculated interaction energies. For example, the interaction energy of atop CO at 1/9th ML increases by 0.8 eV when the D2 method is turned on.

We summarize our findings by providing suggestions about how to explicitly include solvation in computational catalysis at moderate computational expense.

References

1. Zope, B. N.; Hibbitts, D. D.; Neurock, M.; Davis, R. J.; *Science*, 2010, **330**:74-78.
2. Wang, H. F.; Liu, Z. P.; *J. Phys. Chem. C*, 2009, **113**:17502-17508.
3. Okamoto, Y.; Sugino, O.; Mochizuki, Y.; Ikeshoji, T.; Morikawa, Y.; *Chem. Phys. Lett.*, 2003, **377**:236-242.
4. Behtash, S.; Lu, J.; Faheem, M.; Heyden, A.; *Green Chemistry*, 2014, **16**:605–616.
5. Nørskov, J.; Rossmeisl, J.; *J. Phys. Chem. B*, 2004, 17886–17892.
6. Taylor, C. D.; Neurock, M.; *Current Opinion in Solid State and Materials Science*, 2005, **9(1-2)**, 49–65.
7. Getman, R. B.; Schneider, W. F.; Smeltz, A. D.; Delgass, W. N.; Ribeiro, F. H. *Phys. Rev. Lett.*, 2009.
8. H.A. Gasteiger, N. Markovic, J.P.N. Ross, E.J. Cairns, *J. Phys. Chem.*,1994, **98**:617.
9. Huber, *G. W.;* Shabaker, J. W.; Dumesic, J. A.; *Science* 2003, **300**, 2078.
10. Cortright, R. D.; Davda, R. R.; Dumesic, *J. A.; Nature* 2002, **418**, 964.
11. Kresse, G.; Furthmüller, J.; *Comput. Mat. Sci.*, 1996, **6**:15.
12. Blochl, P. E.; *Phys. Rev. B*, 1994, **50**:17953.
13. Perdew, J. P.; Burke K.; Ernzerhof, M.; *Phys. Rev. Lett.*, 1996, **77**:3865.
14. Grimme, S.; *J. Comp. Chem.* 2006, **27**:1787.
15. Henkleman, G.; Uberuaga, B. P.; Jonsson, H.; *J. Chem. Phys.*, 2000, **113**:9901.

376366 Ethylene Glycol Reforming over Pt (111): Modeling of C-C Versus C-O Bond Cleavage Selectivity in Vapor and Aqueous Environments

Abstract:

Catalytic conversion of biomass-derived oxygenates to fuels and value-added chemicals is a promising strategy in the search for renewable and sustainable energy sources. Decomposition chemistry on a metal catalyst is a complex network of multiple series and parallel pathways leading to formation of hydrogen, alkanes, and lighter oxygenates. The final product distribution ultimately depends on the sequence and competition of C-C, C-O, C-H, and O-H bonds scissions. Ethylene glycol (EG) is the simplest such molecule to contain all of C-C, C-O, C-H, and O-H bonds and a C/O stoichiometry of 1:1, and has received significant attention as a model molecule of larger biomass-derived polyols. While the reaction mechanism of EG reforming is to some degree understood at the metal-gas interface, lack of a well-established methodology for describing the influence of a complex liquid phase on a reaction across a solid-liquid interface has hindered similar theoretical studies in an aqueous environment.

In this talk, we present an application of our recently developed iSMS[1] methodology to study the mechanism of EG reforming over Pt (111) in water. Reaction free-energies and free-energy barriers for all elementary reactions in vapor phase are computed within the framework of periodic planewave DFT calculations. The effect of water is then included through a combination of Gaussian-type orbital calculations and COSMO-RS calculations as required by iSMS. Finally, detailed microkinetic models are developed for both vapor and aqueous phases to provide insights into the effect of an aqueous environment on the activity and selectivity of the metal catalyst.

1. Faheem, M.; Suthirakun, S.; Heyden, A. *J. Phys. Chem. C* **2012,** *116*, 22458-22462.

Semi-empirical thermochemical property estimation for biomass derivatives on transition metal surfaces

Density functional theory (DFT) has become an essential tool in estimating thermochemistry and activation barriers in heterogeneous catalysis. For small molecules, we have been able to elucidate reaction mechanisms and parameterize kinetic models that reproduce experimental data with high accuracy[1]. A shift towards modeling larger biomass derivatives interacting with metal surfaces introduces new levels of complexity, including longer computational time, multiple adsorption conformations, and a greater number of intermediates and elementary steps due to multiple pathways encountered in this chemistry. For example, furfural, an important biomass derivative, can undergo hydrogenation, hydrogenolysis, decarbonylation, and even ring opening depending on the metal catalyst[2], making it impractical to compute all possible intermediates, conformations, and reaction barriers with DFT, even on a single metal surface.

In this work, we present a DFT-parameterized semi-empirical scheme capable of predicting thermochemical properties of furan-related intermediates on metal surfaces. This model extends prior group additivity for open-chain molucles[3] to account for furan ring corrections, such as ring deformation. The properties include the heat of formation, entropy, heat capacity, and the dispersion contribution to the heat of adsorption, estimated using dispersion-corrected DFT (D3) and statistical mechanics formulae. For the proposed model, the vibrational and rotational contributions of gas-phase-like groups are estimated using G4-level calculations. We find that the group additivity scheme developed for open-chain molecules is inaccurate for furanics. We report 17 new groups, involving heteroatom-nearest neighbor interactions and 4 new corrections that account for furan ring deformation and the level of ring hydrogenation. Furthermore, we extend the thermochemistry scheme so the thermochemistry estimation can be applied to multiple metal catalysts as a generalization of the linear scaling relations[4] introduced for small molecules. We demonstrate how to construct energy diagrams for various catalysts by combining these methods with linear free energy relations. Comparison of the pathways of furfural chemistry on a select number of metals will be shown.

References

1. Salciccioli *et al.*, *Chemical Engineering Science* **2011,** *66* (19), 4319-4355.
2. Sitthisa, Resasco, *Catalysis Letters* **2011,** *141* (6), 784-791.
3. Salciccioli *et al.*, *The Journal of Physical Chemistry C* **2012,** *116* (2), 1873-1886.
4. Abild-Pedersen *et al.*, *Physical Review Letters* **2007,** *99* (1), 016105.

372257 CO Oxidation on O-Rich Pd(111): How Does Adlayer Structure Affect Reactivity?

Abstract:

The CO oxidation reaction is crucial in the function of catalytic converters found in personal vehicles. Despite its simplicity, several aspects of this reaction are still under investigation. Thus, a recent experimental study by Nakai *et al.* [J. Chem. Phys. 124, 224712 (2006)] proposed that CO oxidation on O-precovered Pd(111) exhibits markedly different rates at different temperatures as a result of different phases in the O overlayer exhibiting disparate reactivities. Thus, the *p*(2x2) phase appears as inert, whereas the ($\sqrt{3}$x$\sqrt{3}$)R30° and *p*(2x1) are reactive and exhibit distinct apparent activation energies and reaction orders with respect to O coverage. It still unclear however, whether there is a causal relation between the changes in the overlayer structure and the different reactivity, and if this is the case what is the mechanism giving rise to this effect.

To shed light into this question we perform first principles-based kinetic Monte Carlo (KMC) simulations using *Zacros*, our software package implementing the Graph-Theoretical KMC approach. These simulations incorporate a detailed cluster expansion Hamiltonian and Brønsted-Evans-Polanyi relations, in order to explicitly account for lateral interactions and their impact on the activation energies of elementary processes. Our simulations reproduce quantitatively the main features of the experimental measurements, and in particular the temperature dependence of CO adsorption and surface reaction rates. The ordering of the adsorbate layer is found to strongly depend on the strength of the lateral interactions but does not seem to have a significant role on the catalytic properties of the system. By analysing the appearance and realisation statistics of the elementary CO oxidation processes, we show that only a small subpopulation of events with rather low activation energies contribute to the observed rate, in agreement with previous observations on a different system, namely the NO oxidation reaction on Pt(111) [Wu et al. J. Catal. 286, 88 (2012)].

380527 Probing the Active Site Requirements and Mechanistic Details of NOx Selective Catalytic Reduction with NH3 on Cu-SSZ-13

Abstract:

Cu-exchanged SSZ-13 (CHA) zeolites have been adopted commercially in automotive exhaust deNO_X applications for selective catalytic reduction (SCR) with NH_3 because they show preferred characteristics regarding performance at low temperatures and stability under hydrothermal conditions. Here, we use an integrated experimental and theoretical modeling approach to probe the active site requirements and mechanistic details of standard NO SCR with NH_3 on Cu-SSZ-13 zeolites. The number of isolated Cu ions (per total Cu) in 6-MR locations correlate linearly with standard NO SCR rates (473 K, per total Cu), while the number of Cu species located in clusters with Cu-O-Cu connectivites (per total Cu) in 8-MR cages correlate linearly with dry NO oxidation rates (573 K, per total Cu). The distribution of isolated and clustered Cu species in SSZ-13 depends on the density of paired Al atoms in the framework and, in turn, the framework Si/Al ratio. *Operando* X–Ray absorption spectroscopy (XAS) of Cu-SSZ-13 samples under working conditions indicates that standard SCR proceeds via a redox mechanism involving isolated Cu(II) and Cu(I) species, with oxidation and reduction half-reactions predominating under different gaseous mixtures that enable isolation and probing of Cu in either the Cu(II) or Cu(I) oxidation state. We discuss a plausible redox mechanism consistent with experiment and density functional theory (DFT), and the different mechanistic roles of Brønsted acid sites also present in Cu-SSZ-13 samples that mediate standard SCR with NH_3.

392233 A New Method for Exhaust Aftertreatment Design - Integrated Engine and Aftertreatment Analysis

Abstract:

Currently, the engine and aftertreatment (catalytical converter) design are separate process. This might lead to several empirical interactions to finalize the engine calibration and aftertreatment layout for a specific project (tailpipe emission target together with engine performance target).

This report describes a 'back of the envelope" methodology to quickly assess the tailpipe emission of a given engine and aftertreatment system. This allows to do combination system (engine and aftertreatment) selection at the early stage of development to reduce time and cost.

There are three main benefits from this methodology:

- It can define the optimal operating range for a given powertrain, with the consideration of engine out emission, fuel economy targets and aftertreatment conversion efficiency.
- It can estimate the catalyst light-off time with provided engine out emission data, or vice versa.
- For a given engine out exhaust gas temperature, species concentration and tailpipe emission target, this methodology can provide the maximum allowable aftertreatment system thermal inertia with defined catalyst light-off time, or vice versa, to meet a specific tail-pipe emission target.

Analysis of ash-loaded DPF performance during drop-to-idle test

Diesel particle filter (DPF) has been widely used in the automobile industry to reduce particle matter emission. The soot reduction process in DPF requires the cycling of two operation modes. During the collection phase, particle matters, including soot and ash, are filtered and accumulated in the wall-flow DPF. The accumulated soot leads to a slow increase in DPF backpressure and triggers the O2-based soot regeneration after reaching the DPF backpressure limit. DPF inlet temperature is then increased to enable the soot oxidation in a short time and the DPF backpressure falls back into the normal range. This process has two innate deficiencies: 1. DPF collects the incombustible ash, which cannot be removed during soot regeneration and increases the DPF pressure drop; 2. High temperature excursion during soot regeneration could damage DPF. Based on the DPF failure observed in the field, these two failure modes could interact and compound with each other, making a more complex situation for analysis.

In this work, we evaluate the exotherm from an ash-loaded DPF using a 1D transient model. Drop-to-idle condition was used as a simulated high temperature event. During this test, flow rate drops during the soot regeneration, while inlet temperature and oxygen concentration increase. A variety of ash-loaded DPFs with different sizes, materials and cell structures were assessed in this test. Simulation results showed that DPF design could change the temperature profile along the channel. In some cases, temperature could be high enough to cause the ash sintering inside DPF channel and thus reduce the effective length, leading to a higher DPF backpressure than prediction.

390165 Transient Kinetic Study of Sulfur Oxide Adsorption/Desorption over Al2O3 and Pt/Al2O3 Catalysts

Abstract:

Background

Sulfur oxides in the engine exhaust interact with diesel aftertreatment catalysts and deactivate them. SO_2 is the main sulfur species exiting the engine, which in turn can be further oxidized to SO_3 over the diesel oxidation catalyst (DOC). SO_2 and SO_3 may have different impacts on the DOC activity, and downstream catalysts, with SO_3 leading to more severe deactivation. In the present work, in order to study the activity recovery of a Pt/Al_2O_3 DOC after sulfur poisoning, sulfur storage/release was studied using temperature-programmed desorption (TPD) experiments and a transient kinetic model was developed based on the experimental data. Diffuse reflectance infrared spectroscopy (DRIFTS) experiments were also utilized to identify the surface species.

Methodology

To separate out the support effect, TPD experiments were performed in a flow reactor system using Al_2O_3 and Pt/Al_2O_3 samples. An Al_2O_3-coated monolith with a loading of 1.60 g/in^3 and a Pt/Al_2O_3-coated monolith with a Pt loading of 50 g/ft^3 were used in the experiments. During the TPD experiments, the catalyst was saturated with a specific concentration of either SO_2 or SO_3, and a temperature ramp from the saturation temperature up to ~900°C was used to desorb the surface species. For the SO_3 TPD experiments, a separate reactor upstream of the main reactor was used where SO_2 was completely oxidized to generate SO_3 required in the feed. SO_2 and SO_3 TPD experiments were carried out under four different sets of operating conditions where saturation temperatures of 120°C and 160°C and approximate concentrations of 50 and 100 ppm for the adsorbing gas species were employed. Temperature ramps of 20 and 10°C/min were used in the SO_2 and SO_3 TPD experiments, respectively. For all flow reactor experiments, the outlet gas concentrations were measured using a MKS MultiGas MG-2030 FT-IR analyzer. In the DRIFTS experiments, pellets of Al_2O_3 and Pt/Al_2O_3 were prepared from the monolith washcoat and IR studies were performed with SO_2 and SO_3 concentrations similar to those employed in the flow reactor experiments.

Results and Discussion

According to the DRIFTS and TPD results, three different sulfur species were identified during SO_2 adsorption on Al_2O_3. A small high temperature peak in the TPD spectra, i.e. above 850°C, was assigned to bulk aluminum sulfate, whereas the other species with less stability were identified as weakly adsorbed SO_2 (with its desorption peak centered at 250°C) and surface sulfites or sulfates (with a peak maximum in the range of 400-500°C). Similar surface species were identified in the case of SO_2 TPD on Pt/Al_2O_3, however, with a higher contribution of the high temperature peaks indicating the promoting effect of Pt on the sulfate formation as well as on the spillover of the surface sulfates to the bulk. 1.45×10^{-4} mol SO_2 adsorbed was calculated based on the SO_2 TPD for Pt/Al_2O

sulfur oxides significantly increased, to 1.50×10^{-3} mol, when exposed to SO_3 instead, suggesting a higher level of sulfur storage upon exposure to SO_3. Also according to these results, SO_3 exposure resulted in the formation of stable sulfate species on the catalyst. The SO_3 TPD results obtained with the Al_2O_3 and Pt/Al_2O_3 samples indicated that SO_2 and SO_3 release patterns were strongly affected by thermodynamic equilibrium. The important role of the Al_2O_3 support in the catalyst sulfation, as previously reported in the literature [1-3], was also confirmed in the experimental study.

A three site adsorption mechanism, with S1, S2 and S3 representing the adsorption sites with an increasing order of strength, was assumed to develop a transient kinetic model for the SO_2 storage/release over Al_2O_3 and Pt/Al_2O_3. First order kinetics were assumed for the SO_2 adsorption on the S1 (weak) and S3 (strong) sites, whereas higher order kinetics, i.e. second or third order depending on the catalyst, was assumed for SO_2 adsorption on the S2 site. The S2 kinetics are explained by the structure of the surface sulfates being formed as a result of SO_2 adsorption on the under-saturated oxygen anions on the surface of alumina. These oxygen species were recently proposed as possible sites for the strong adsorption of SO_2 in the form of surface sulfates [4].

A uniform temperature profile was assumed for the reactor and mass transfer limitations were neglected in the radial direction of the monolith channel. A set of governing equations were solved to obtain the concentration and surface coverages as a function of time and axial position. Some of the kinetic parameters were optimized to describe the experimental data with thermodynamic constraints imposed upon the optimized parameters. A similar methodology was also applied in the development of the SO_3 TPD model over Al_2O_3 and Pt/Al_2O_3. The kinetic model was validated using other sets of experimental data which were not included in the fitting process. The transient models were able to accurately predict the experimental behavior of the catalyst. According to the modeling results, the binding energy of the adsorbed species on the S1 and S2 sites are coverage dependent, indicating surface interactions between the adsorbed species as well as the heterogeneous nature of the alumina surface sites. Coverages of different sulfur species as well as their variation as a function of temperature were also predicted by the model indicating differences in the SO_2 and SO_3 storage mechanism.

REFERENCES

1. O. Saur M. Bensitle, S. Mohammed, J.C. Lavalley, C.P. Tripp and B.A. Morrow, *Journal of Catalysis* 110(1986)104.
2. F. Gracia, S. Guerrero, E. Wolf, J. Miller and A. Kropf, *Journal of Catalysis* 233(2005)372.
3. J.H. Pazmiño, C. Bai, J.T. Miller, F.H. Ribeiro and W.N. Delgass, *Catalysis Letters* 143(2013)1098.
4. M.Y. Smirnov, A.V. Kalinkin, A.V. Pashis, A.M. Sorokin, A.S. Noskov, K.C. Kharas and V.I. Bukhtiyarov *The*

Journal of Physical Chemistry B 109(2005)11712.

Lean NO_x Trap catalysts are a promising potential strategy for reducing NO and NO_2 emissions from lean burn engine exhausts. Operating in the lean burn regime generates higher NO_x and oxygen concentrations than are found in traditional stoicheometric engines, while virtually eliminating reducing components in the exhaust gas. Under these conditions traditional three-way catalysts (TWC) are rendered virtually ineffective. The traditional LNT catalysts formulation consists of a noble metal and alkali earth metal on a high surface area support. The addition of cerium to the traditional support has been shown to increase the reduction of stored NO_x during rich phase operation [1]. For novel catalysts to be effectively implemented it is important to understand the effect of new catalyst formulations on stability, and degradation, specifically the effect of degradation on lean phase oxidation and NO_x storage.

In the current study LNT catalysts have been produced by incipient wetness impregnation of gamma alumina, to yield the desired weight percentage loading. The catalysts investigated are: Pt/Ba/Al, Pt/Ba/Ce/Al with weight ratio loadings of 1/20/100 and 1/20/5/100 respectively. The effect of ceria addition has been characterized by long time NO_x storage experiments, and steady state NO oxidation measurements. Surface characterization has been completed via XRD, EDAX, TEM and FTIR. Further characterization has been conducted through CO chemisorption, BET analysis and H_2-TPR. Ceria addition shows a significant effect on surface morphology, storage component surface species and overall NO_x storage with thermal degradation when compared to the standard alumina supported catalysts. As a result degradation and stability of other novel LNT catalysts will be evaluated in the future.

1. Infantes-Molina, A., et al., *Characterization and reactivity of Ce-promoted PtBa lean NOx trap catalysts.* Catalysis Today, 2012. **197**(1): p. 178-189.

Preliminary Study of Low Temperature Oxidation of Carbon Monoxide Produced By Diesel-Ignited Methane Dual Fuel Low Temperature Combustion in a Single-Cylinder Diesel Engine

Navid Zanganeh[1], Mostafa S. Raihan,[2], Kalyan K Srinivasan[2], Sundar Rajan Krishnan[2], Hossein Toghiani[1] and Jason M. Keith[1]
(1)Dave C. Swalm School of Chemical Engineering, Mississippi State University, Mississippi State, MS
(2)Department of Mechanical Engineering, Mississippi State University, Mississippi State, MS

14AIChE
Annual Meeting, Atlanta, GA

MISSISSIPPI STATE
UNIVERSITY

Presentation Outline

- ❑ Introduction
- ❑ Diesel-Ignited Dual Fuel Combustion
- ❑ Experimental Section- Engine
- ❑ Result and discussion
- ❑ Catalyst preparation and characterization
- ❑ Summary
- ❑ Future work

Introduction

- Diesel engine exhaust is a mixture of thousands of different gases and tiny particles.

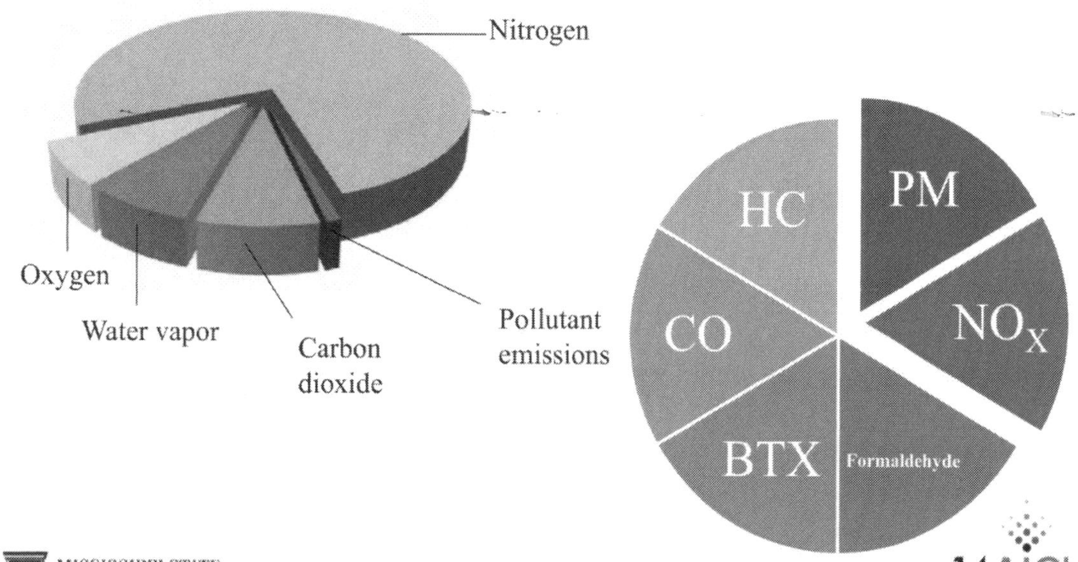

Introduction

- Diesel Exhaust Health Effect
- Diesel Exhaust Environmental Effect
- US Environmental Protection Agency (EPA) Emission Standard

Year	HC (g/bhp-hr)	CO (g/bhp-hr)	NO_x (g/bhp-hr)	PM (g/bhp-hr)
1990	1.30	15.5	6.0	0.60
1991	1.30	15.5	5.0	0.25
1994	1.30	15.5	5.0	0.10
1998	1.30	15.5	4.0	0.10
2002	0.50	15.5	2.4	0.10
2007+	0.14†	15.5	0.2	0.01

† 0.14 is given for non-methane hydrocarbons
Regulation for heavy duty trucks www.epa.gov

Introduction

- ❑ Reducing Diesel Exhaust Emissions
  - ❖Engine treatment
    - ❑ Diesel engines with exhaust gas recirculation
    - ❑ Low temperature combustion strategies
  - ❖Change the fuel
    - ❑Reformulated diesel fuels
    - ❑Natural gas
    - ❑Electricity
  - ❖Install vehicle accessories that reduce diesel exhaust
    - ❑Diesel Oxidation Catalyst
    - ❑Diesel Particulate Filter

Diesel-Ignited Dual Fuel Combustion

Video Presentation in Original e-media

www.youtube.com

Center for advanced vehicular systems (CAVS)

MISSISSIPPI STATE UNIVERSITY

CAVS

Advanced Combustion Engines Laboratory

14AIChE Annual Meeting, Atlanta, GA

Experimental Setup

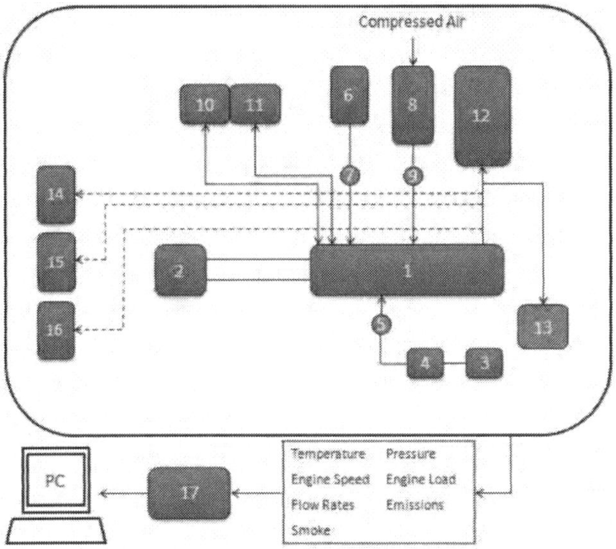

1-Single cylinder engine
2-AC regenerative dynamometer
3-Diesel supply tank
4-Diesel level tank
5-Coriolis flow meter
6-Methane supply tank
7-Coriolis flow meter
8-Intake tank
9-Sonic flow meter
10-Engine water temperature conditioner
11-Engine oil temperature conditioner
12-Exhaust tank
13-Smoke meter
14-Engine exhaust particles sizer (EEPS)
15-AVL FTIR
16-Six-gas emission bench
17-Data acquisition system

MISSISSIPPI STATE UNIVERSITY

CAVS

Advanced Combustion Engines Laboratory

14AIChE Annual Meeting, Atlanta, GA

Experimental Setup

Six-gas emission bench

FTIR

Smoke meter

Test Matrix

Engine Load (IMEP)	Varied Parameter	Constant Parameters
5.1 bar IMEP at 1500 RPM	PES : 0% to 90%	SOI = 355 CAD Boost pressure = 1.5 bar Rail pressure = 500 bar
	SOI : 355 CAD to 280 CAD	PES = 80% Boost pressure = 1.5 bar Rail pressure = 500 bar
	Rail pressure : 200 bar to 1300 bar	PES = 80% Boost pressure = 1.5 bar SOI = 310 CAD
	Boost pressure : 1.1 bar to 1.8 bar	PES = 80% Rail pressure = 500 bar SOI = 310 CAD

PES: Percentage Energy Substitution
SOI: Start of Injection
CAD: Crank Angle Degree
IMEP: Indicated Mean Effective Pressure

Mostafa S. Raihan. A comparative study of diesel ignited methane and propane dual fuel low temperature combustion in a single cylinder research engine. Diss. MISSISSIPPI STATE UNIVERSITY, 2014.

PES Sweep Results

Apparent Heat Release Rate (AHRR)

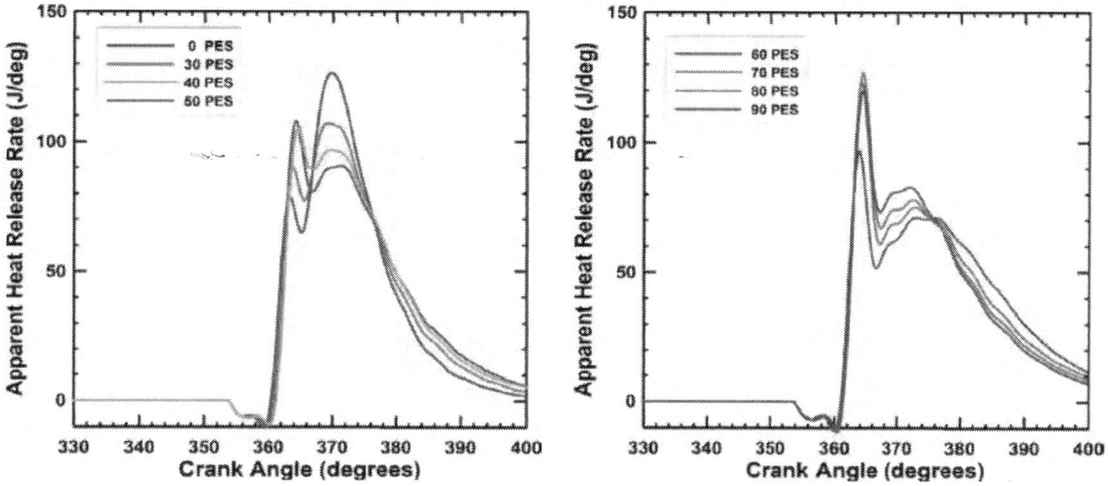

AHRR for 0 to 90 PES at 5.1 bar net IMEP, 1500 RPM, SOI = 355 CAD, Prail = 500 bar.

MISSISSIPPI STATE UNIVERSITY | CAVS | Advanced Combustion Engines Laboratory | 11 | 14AIChE Annual Meeting, Atlanta, GA

PES Sweep Results

In-cylinder Pressure

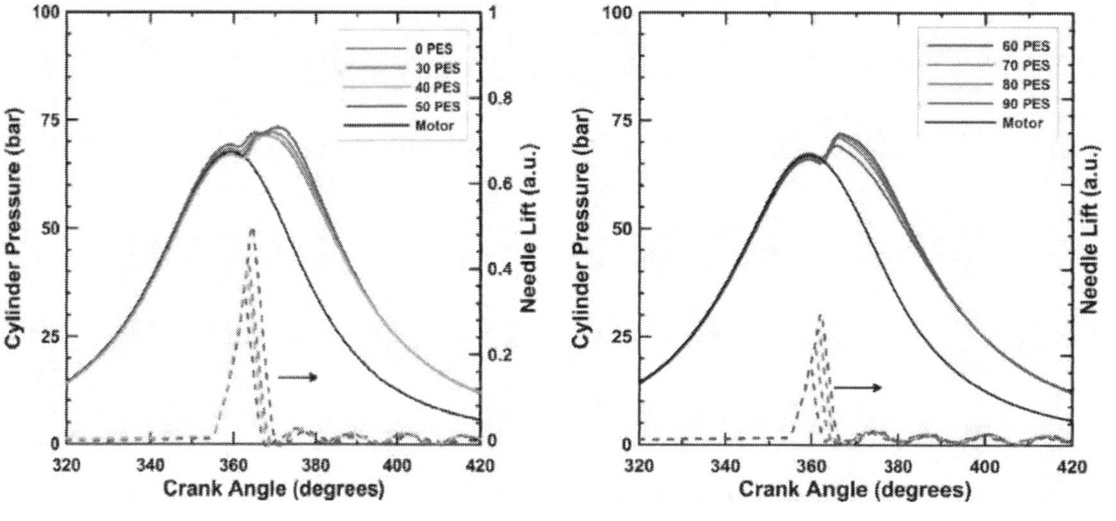

Cylinder pressure for 0 to 90 PES at 5.1 bar net IMEP, 1500 RPM, SOI = 355 CAD, P_{rail} = 500 bar.

MISSISSIPPI STATE UNIVERSITY | CAVS | Advanced Combustion Engines Laboratory | 12 | 14AIChE Annual Meeting, Atlanta, GA

PES Sweep Results

Global temperature

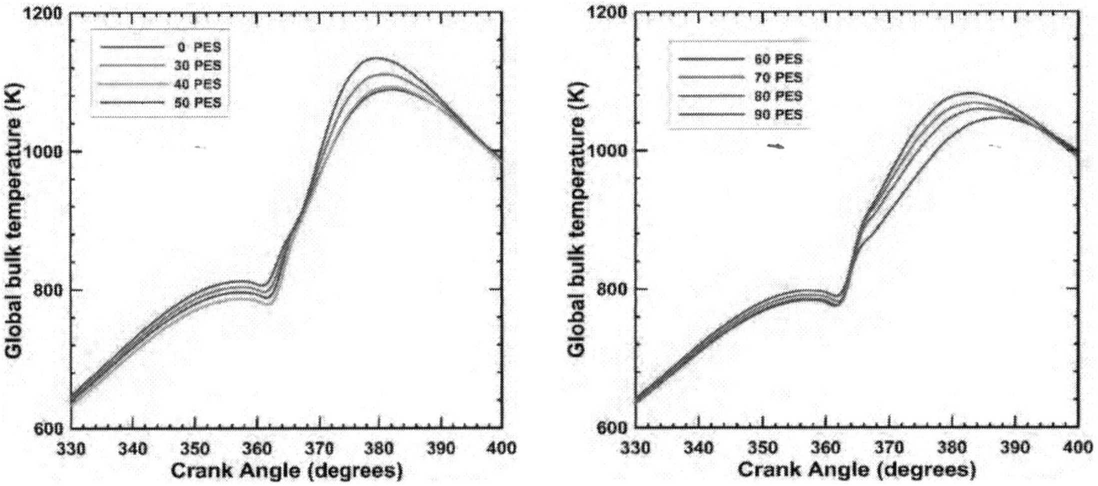

Global temperature profiles for various PES at 5.1 bar net IMEP, 1500 RPM, SOI = 355 CAD, P_{rail} = 500 bar.

MISSISSIPPI STATE UNIVERSITY
CAVS
Advanced Combustion Engines Laboratory

14 AIChE Annual Meeting, Atlanta, GA

PES Sweep Results

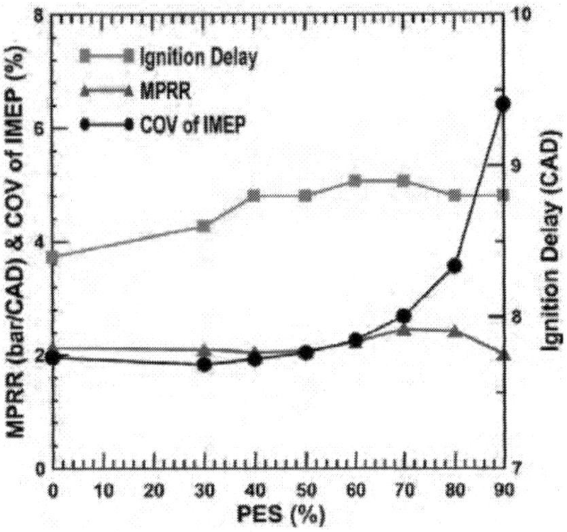

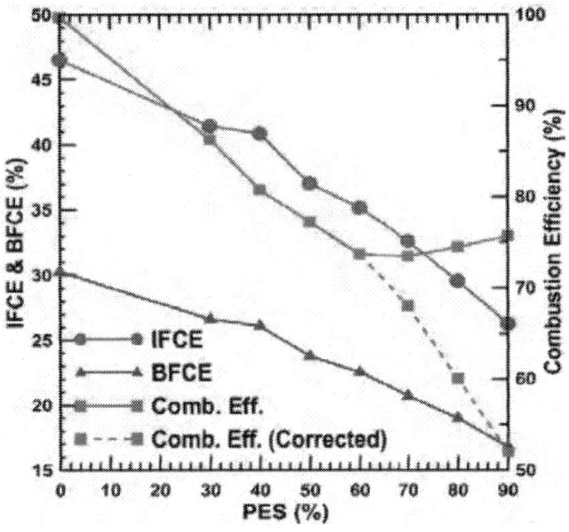

Coefficient of Variance (COV) of IMEP = Combustion instability

MISSISSIPPI STATE UNIVERSITY
CAVS
Advanced Combustion Engines Laboratory

14 AIChE Annual Meeting, Atlanta, GA

PES Sweep Results

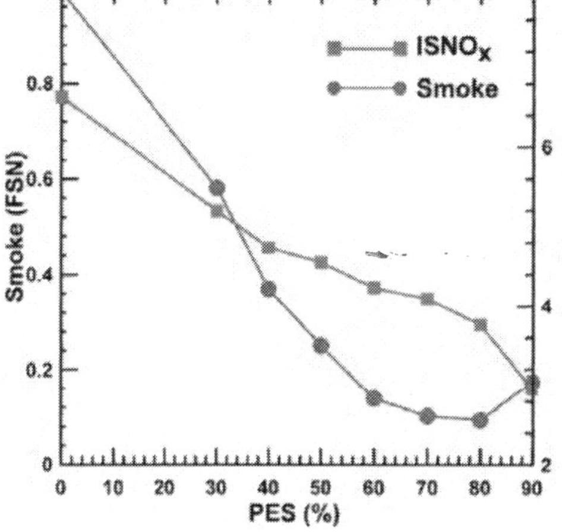

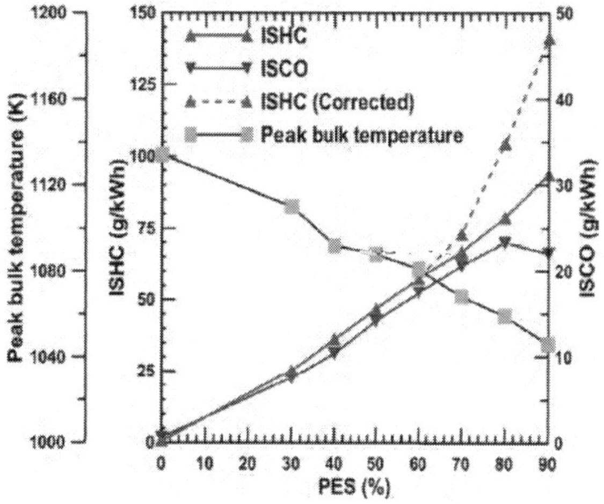

NOx: form by thermal mechanism; extremely dependent on **high local temperature**

HC : Combustion chamber crevices, Partial or bulk quenching, Exhaust valve leakage

CO: **Intermediate combustion species** formed during the oxidation of hydrocarbon fuel

- CO oxidation reaction ($CO + \mathbf{OH} \rightarrow CO_2 + H$) depends on **bulk temperature**

MISSISSIPPI STATE UNIVERSITY

CAVS

Advanced Combustion Engines Laboratory

15

14AIChE Annual Meeting, Atlanta, GA

Summary

- ☐ Diesel ignited dual fuel low temperature combustion is successful to reduce PM and NO_X emissions simultaneously.

- ☐ Diesel ignited dual fuel low temperature combustion results higher CO and HC emissions.

Exhaust Temperature Variation (°C)			
PES Sweep Experiment	SOI Sweep Experiment	Boost Pressure Experiment	Rail Pressure Experiment
324 - 348	231 - 272	186 - 311	234 - 247

MISSISSIPPI STATE UNIVERSITY

16

14AIChE Annual Meeting, Atlanta, GA

Gold Catalysis

- In the late 1980's Haruta et al. demonstrated that Au nanoparticles (less than 5 nm) are highly active for the oxidation of CO at low temperatures (200 K)*

- Gold Precursor

$HAuCl_4$	$Au(en)_2Cl_3$
$[Au(OH)_XCl_{4-X}]^-$	$[Au(en)_2]^{3+}$

MISSISSIPPI STATE UNIVERSITY

*M. Haruta, N. Yamada, T. Kobayashi and S. Iijima, J. Catal.,1989, 115, 301

14AIChE Annual Meeting, Atlanta, GA

Catalysis Laboratory

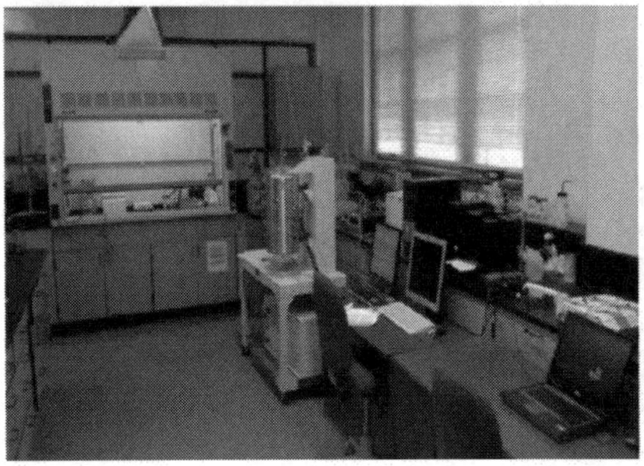

Reactor bench and monitoring

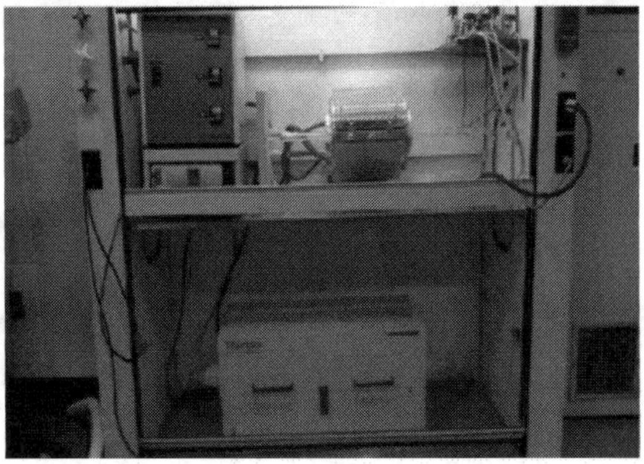

Furnaces which are able to increase the temperature to 1200 °C

MISSISSIPPI STATE UNIVERSITY

14AIChE Annual Meeting, Atlanta, GA

Catalysis Laboratory

Continuous flow tubular reactor

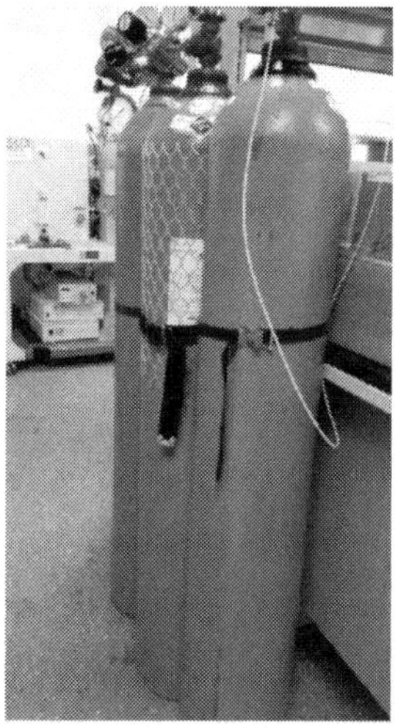

Calibration and feed gas cylinders

MISSISSIPPI STATE UNIVERSITY

19

14AIChE Annual Meeting, Atlanta, GA

Catalysis Laboratory

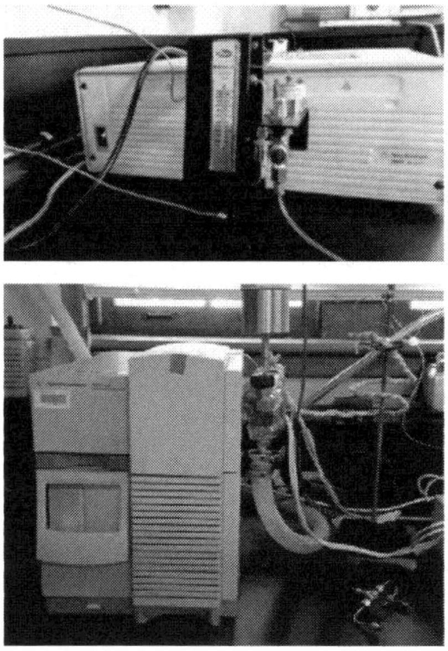

Micro GC and Mass spec

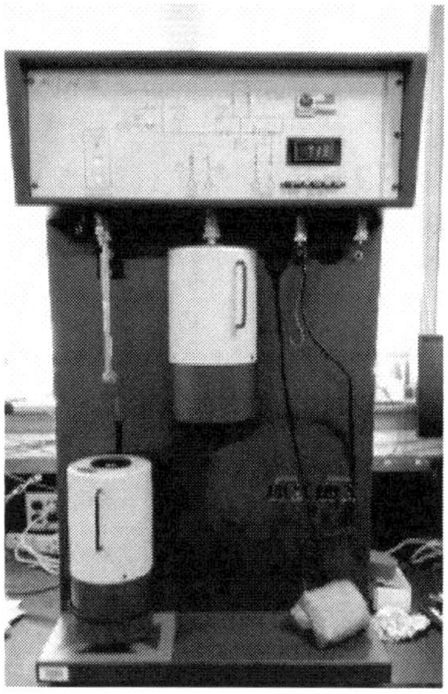

BET

MISSISSIPPI STATE UNIVERSITY

20

14AIChE Annual Meeting, Atlanta, GA

Gold Nanoparticles

Turkevich method*

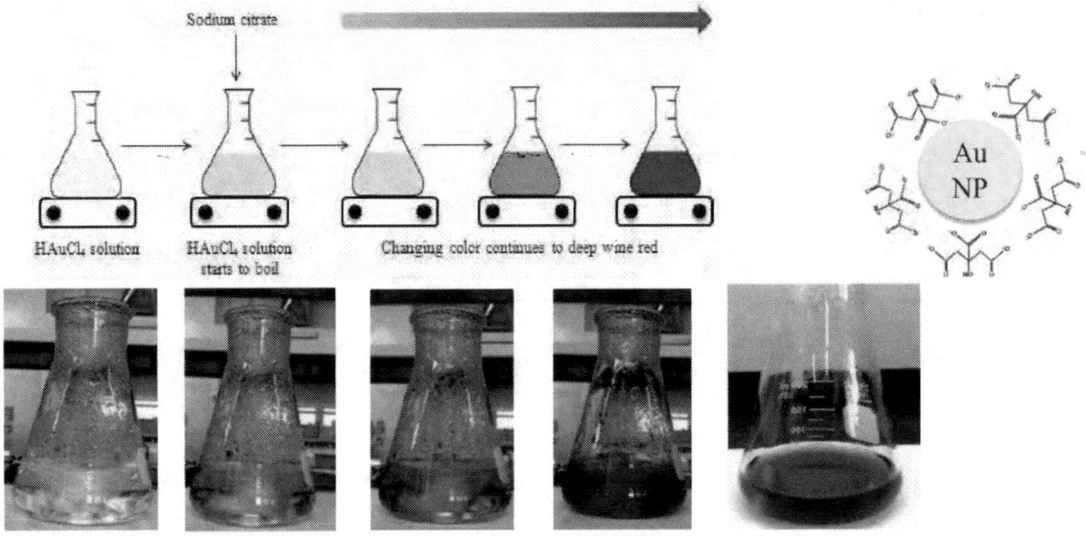

*Turkevich, J., P. C. Stevenson and J. Hillier (1951). "A study of the nucleation and growth processes in the synthesis of colloidal gold." Discussions of the Faraday Society **11**(0): 55-75

MISSISSIPPI STATE UNIVERSITY

14AIChE Annual Meeting, Atlanta, GA

Gold Nanoparticles

Turkevich method

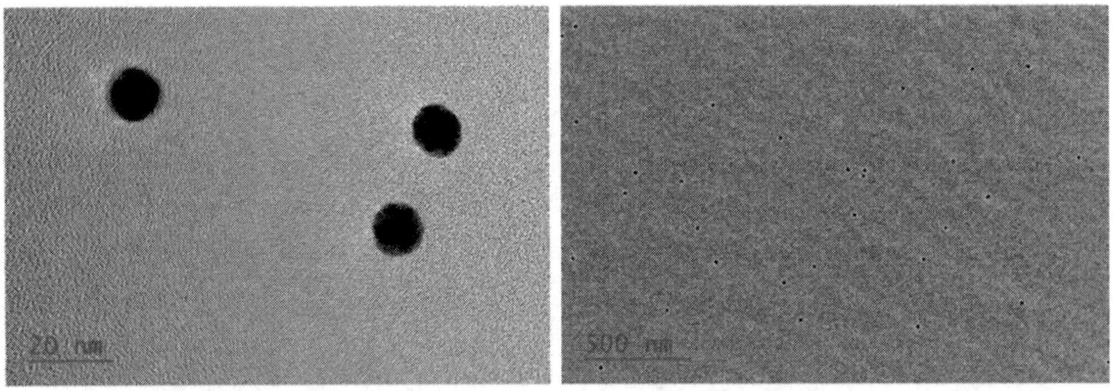

TEM images for sodium citrate stabilized gold nanoparticles

MISSISSIPPI STATE UNIVERSITY

14AIChE Annual Meeting, Atlanta, GA

Au/SiO_2 Catalyst Preparation

Deposition-Precipitation

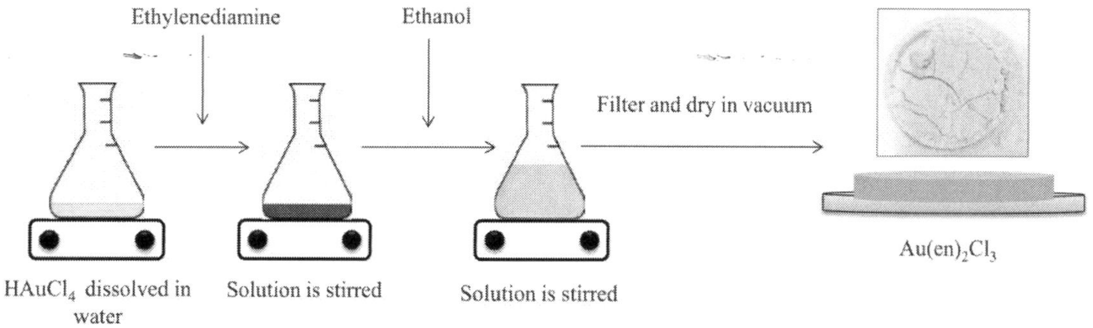

$$HAuCl_4.3H_2O + C_2H_4(NH_2)_2 \rightarrow Au(en)_2Cl_3 + HCl$$

MISSISSIPPI STATE UNIVERSITY

Zhu, H., et al., *Low-temperature CO oxidation on Au/fumed SiO2-based catalysts prepared from Au(en)2Cl3 precursor.* Applied Catalysis A: General, 2007. **326**(1): p. 89-99

23

14AIChE Annual Meeting, Atlanta, GA

Au/SiO_2 Catalyst Preparation

Deposition-Precipitation

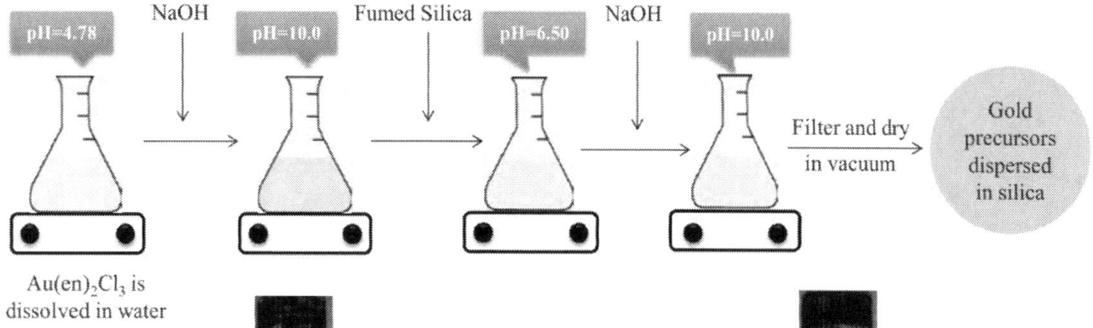

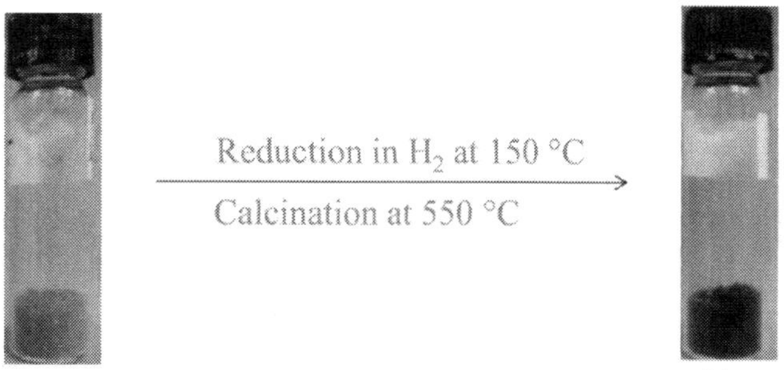

MISSISSIPPI STATE UNIVERSITY

24

14AIChE Annual Meeting, Atlanta, GA

Au/SiO_2 Catalyst Characterization

X-ray Diffraction Spectroscopy (XRD)

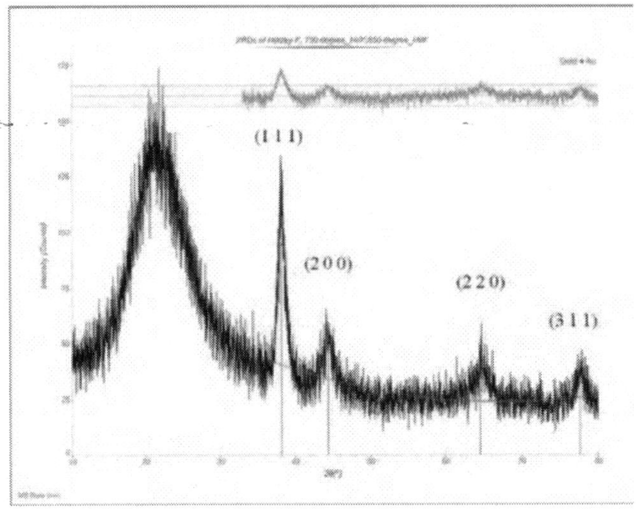

MISSISSIPPI STATE UNIVERSITY

25

14AIChE Annual Meeting, Atlanta, GA

Au/SiO_2 Catalyst Characterization

Energy Dispersive Spectroscopy (EDS)

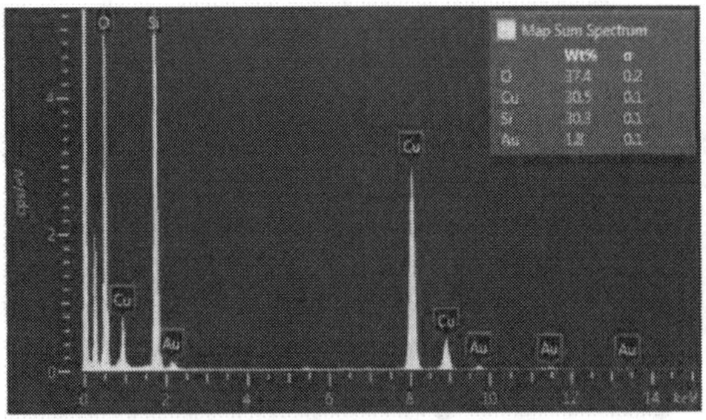

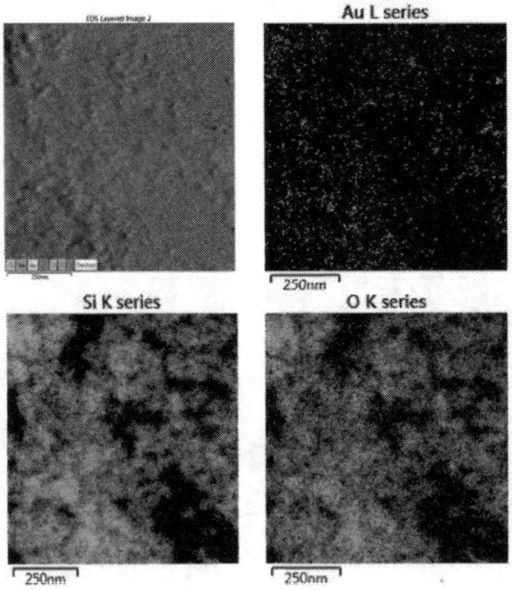

EDS results confirms the formation of Au/SiO_2 Catalyst.

MISSISSIPPI STATE UNIVERSITY

26

14AIChE Annual Meeting, Atlanta, GA

Au/SiO_2 Catalyst Characterization

Transmission Electron Microscopy (TEM)

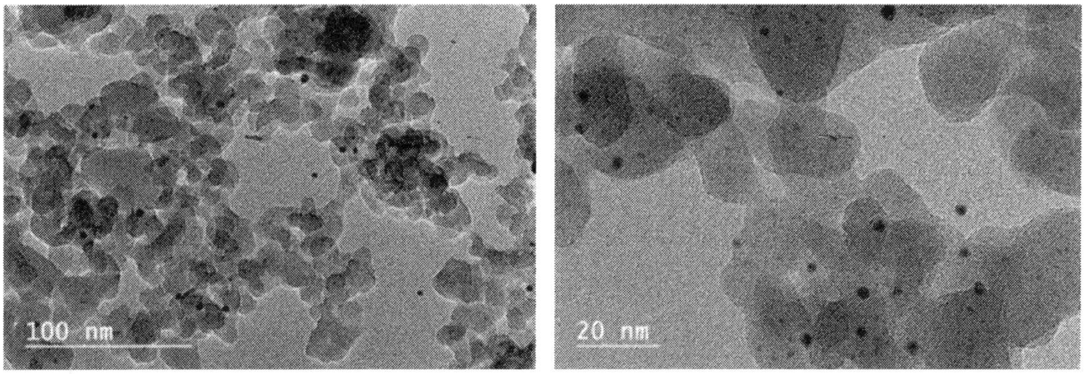

TEM images for Silica supported gold NPs

MISSISSIPPI STATE UNIVERSITY

27

14AIChE Annual Meeting, Atlanta, GA

Thiolate-protected Gold Clusters

- Compounds with a core of gold atoms and organic groups covalently bound to the surface gold atoms.

- Thiolate groups are widely used as stabilization of gold cluster In solution.

- $Au_{25}(SHC2H4)_{18}$

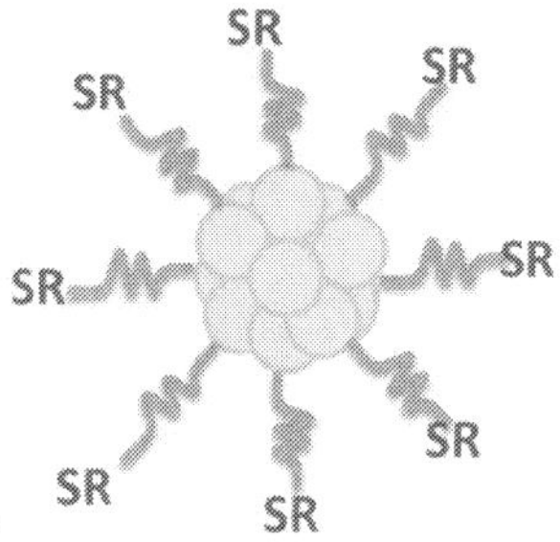

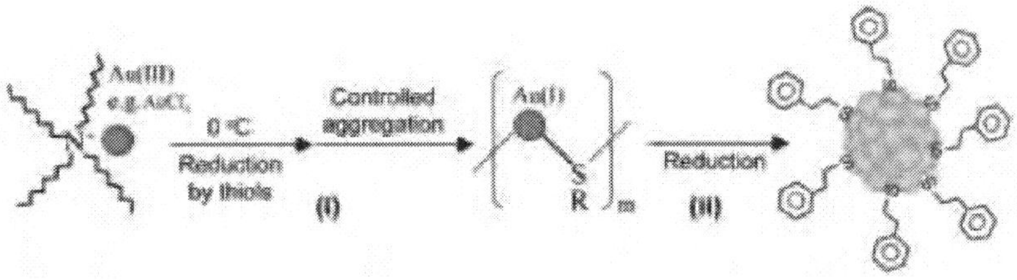

MISSISSIPPI STATE UNIVERSITY

Zhu, Manzhou, et al. "Kinetically controlled, high-yield synthesis of Au25 clusters." *Journal of the American Chemical Society* 130.4 (2008): 1138-1139. 28

14AIChE Annual Meeting, Atlanta, GA

$Au_{25}(SHC_2H_4)_{18}$ Clusters

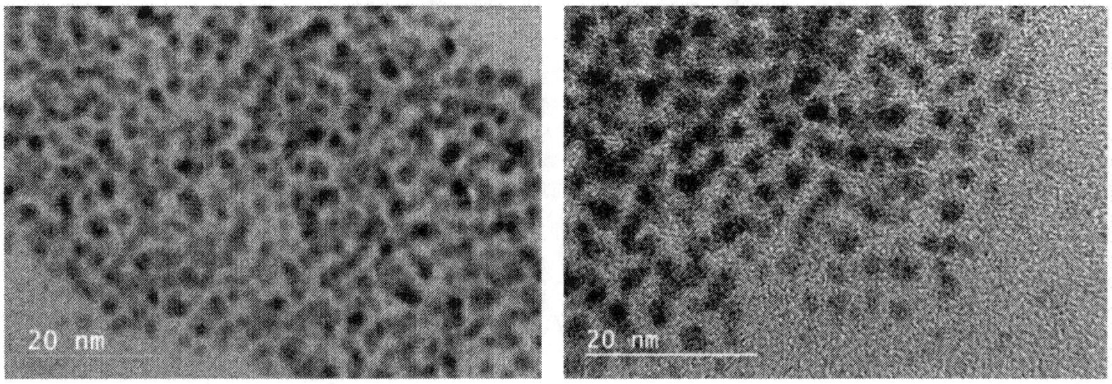

TEM images for gold clusters

$Au_{25-n}Cu_n(SHC_2H_4)_{18}$ Clusters*

$HAuCl_4$ $CuCl_2$ | Tetrabutylammonium bromide | phenylethanethiol | Sodium borohydride

Negishi, Yuichi, et al. "Effect of copper doping on electronic structure, geometric structure, and stability of thiolate-protected Au25 nanoclusters." *The Journal of Physical Chemistry Letters* 3.16 (2012): 2209-2214.

$Au_{25-n}Cu_n(SHC_2H_4)_{18}$ Clusters

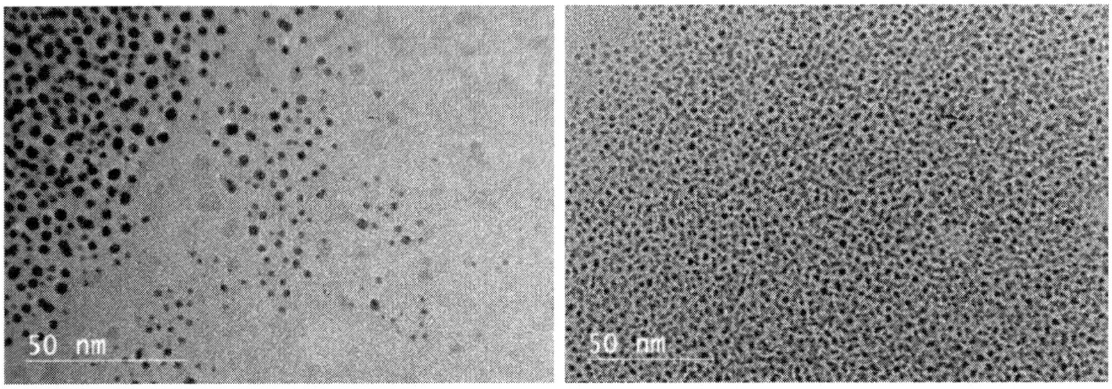

TEM images for gold-copper clusters

$Au_{23}Cu_2(SHC_2H_4)_{18}/SiO_2$

- $Au_{23}Cu_2(SHC_2H_4)_{18}/SiO_2$ was produced by stirring certain amount of Au-Cu clusters with Cab-O-Sil EH5 dispersed into toluene.
- Product was centrifuged and dried.
- Final catalyst obtained after calcination at 300°C for 3h.

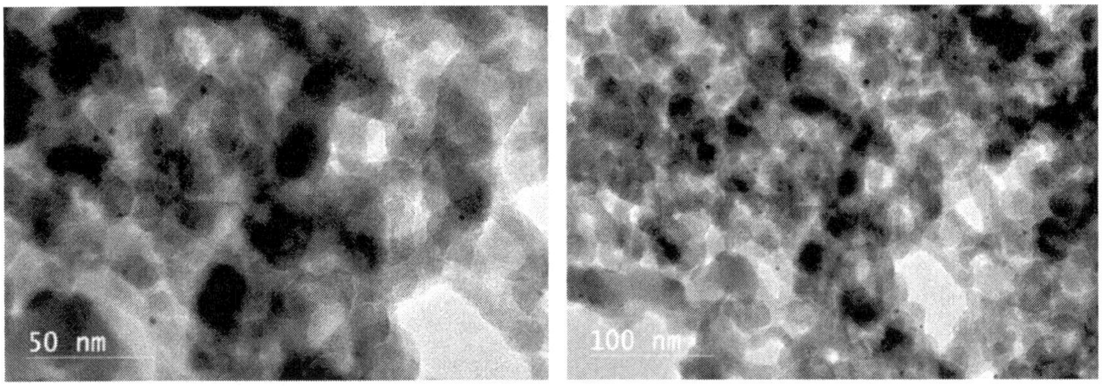

TEM images silica supported $Au_{23}Cu_2(SHC_2H_4)_{18}$ thiolate-protected nano alloy

$Au_{23}Cu_2(SHC_2H_4)_{18}/SiO_2$

EDS results confirms the formation of $Au_{23}Cu_2(SHC_2H_4)_{18}/SiO_2$ Catalyst.

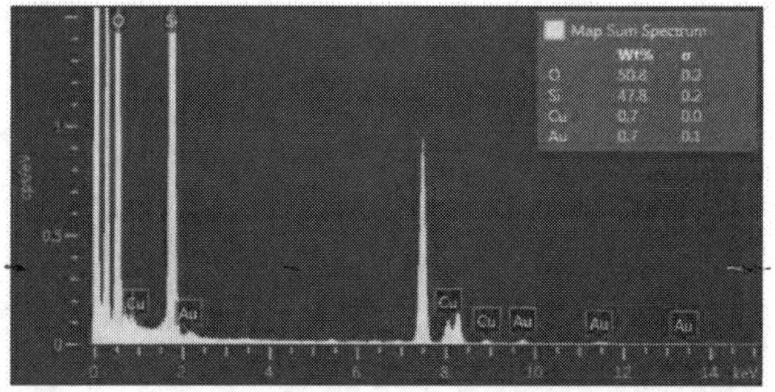

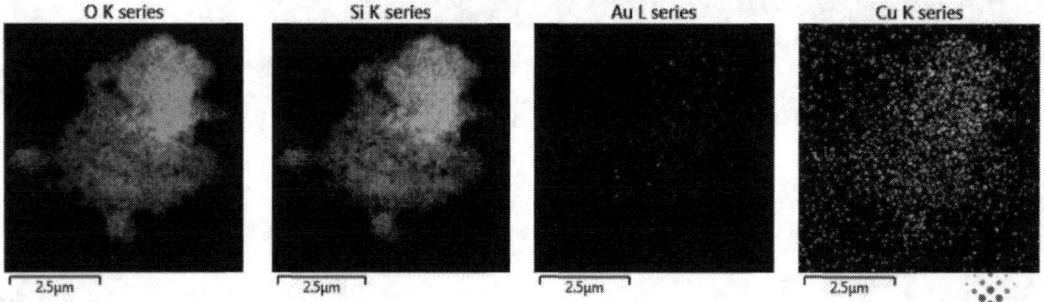

MISSISSIPPI STATE UNIVERSITY

33

14AIChE Annual Meeting, Atlanta, GA

Summary

- Using $Au(en)_2Cl_3$ salt produces stable and active Au/SiO_2 catalyst via DP method which gold NPs diameter are less than 5nm.
- Gold NPs with diameter less than 2 nm can be produced by $HAuCl_4$ salt assisted by thiol groups as stabilizers.
- Silica supported gold-copper nano alloy with diameter less than 2 nm was successfully produced by using thiol ligands.

MISSISSIPPI STATE UNIVERSITY

34

14AIChE Annual Meeting, Atlanta, GA

Future Work

- Low temperature oxidation of CO using $Au_{23}Cu_2(SHC_2H_4)_{18}/SiO_2$
- Investigating the effect of thiol ligands and support on activity of catalyst
- Investigating the CO oxidation in presence of exhaust species (e.g. NO and Hydrocarbons)
- In-situ oxidation of CO produced by diesel-ignited methane (Propane, Gasoline) dual fuel low temperature combustion

MISSISSIPPI STATE UNIVERSITY

35

14AIChE Annual Meeting, Atlanta, GA

Acknowledgement

The authors gratefully acknowledge the financial support from

Bagley College of Engineering
Earnest W. Deavenport, Jr. Chair
Thomas B. Nusz Professorship
Sustainable Energy Research Center (US DOE Award # DE-FG36-06GO86025)
Facilities support from the Center for Advanced Vehicular Systems at Mississippi State University.

JAMES WORTH BAGLEY COLLEGE OF ENGINEERING MISSISSIPPI STATE UNIVERSITY

CAVS Center for Advanced Vehicular Systems

U.S. DEPARTMENT OF ENERGY

MISSISSIPPI STATE UNIVERSITY

36

14AIChE Annual Meeting, Atlanta, GA

Thanks for your attention

MISSISSIPPI STATE UNIVERSITY

37

14AIChE
Annual Meeting, Atlanta, GA

Designing Novel Electrocatalysts for Hydrogen Oxidation in Alkaline Environments

Elizabeth G. Mahoney[a], Wenchao Sheng[b], Yushan Yan[a], Jingguang G. Chen[b]
[a]Department of Chemical Engineering, University of Delaware, Newark, Delaware 19716 USA
[b]Department of Chemical Engineering, Columbia University, New York, New York 10027 USA

Hydroxide exchange membrane fuel cells offer a number of advantages over proton exchange membrane fuel cells. One advantage is increased stability of materials in an alkaline environment, which allows for cheaper, non-platinum (Pt) group metals to catalyze the oxygen reduction reaction and the hydrogen oxidation reaction (HOR). However, even on Pt, the activity of the HOR is approximately two orders of magnitude lower in base than in acid.[1] Non-Pt metals also exhibit a lower HOR activity at higher pH values. The cause of the decrease in catalyst activity at higher pH is currently disputed in literature. Specifically, Strmcnik et. al. hypothesized that adsorbed hydroxyl plays a direct role in the hydrogen oxidation mechanism.[2] Furthermore, Durst et. al. proposed that increased hydrogen binding energy at high pH may be responsible for the decrease in catalyst activity.[3] We hypothesize that the HOR mechanism may be further understood using bimetallic catalysts consisting of Pt and non-Pt metals,[4] which also decrease the amount of Pt needed for the catalysis of the HOR. The focus of this presentation will be the development of bimetallic electrocatalysts for the alkaline hydrogen oxidation reaction, namely a monolayer of Pt deposited on gold substrates (Pt ML/Au). We determined that Pt ML/Au catalysts exhibit similar alkaline HOR activity as bulk Pt and far higher activity than bulk Au. Future work will focus on supporting ML Pt on other substrates to further understand the mechanism of the alkaline HOR.

References:

[1] W. Sheng, H. A. Gasteigert, Y. Shao-Horn, *J. Electrochem. Soc* **2010**, *157*, B1529–B1536.
[2] D. Strmcnik, M. Uchimura, C. Wang, R. Subbaraman, N. Danilovic, D. Van Der Vliet, A. P. Paulikas, V. R. Stamenkovic, N. M. Markovic, *Nat. Chem.* **2013**, 1–7.
[3] J. Durst, A. Siebel, S. Simon, F. Hasche, J. Herranz, H. A. Gasteiger, *Energy Environ. Sci.* **2014**, DOI 10.1039/C4EE00440J.
[4] W. Sheng, A. P. Bivens, M. Myint, Z. Zhuang, R. V. Forest, Q. Fang, J. G. Chen, Y. Yan, *Energy Environ. Sci.* **2014**, *7*, 1719–1724.

377757 Pt-Pd/CeO2-Omc Bimetallic Catalysts for Formic Acid Oxidation

Abstract:

The effects of CeO_2 modification and variation of Pd/Pt ratios on the performance of PtPd/CeO_2-OMC electrocatalysts in oxidation of formic acid is studied. N_2 adsorption analysis shows that the CeO_2 modification increases the surface area of OMC support from 1005 m^2/g to 1119 m^2/g. SEM, XRD and TEM analysis reveal that the presence of CeO_2 enhances the uniform dispersion of active metal(s) on the support surface. Average sizes of the dispersed metal crystals decrease with the increase of Pt mass fraction on CeO_2-OMC. Electrochemical evaluation of CeO_2 modified Pd/CeO_2-OMC provides 12 % higher anodic current activity than that of unmodified Pd/OMC catalyst. The peak potential is also shifted more than 83 mV in the negative direction. In the bimetallic samples, a small amount of Pt with Pd improves the performance of the catalysts significantly. Among the bimetallic catalysts, Pd_3Pt_1/CeO_2-OMC displays superior current density (74.6 mA/cm^2), which is 1.61, 1.67 and 28.3 times higher than that of Pd/CeO_2-OMC, Pd_1Pt_3/CeO_2-OMC and Pt/CeO_2-OMC catalysts, respectively. The bimetallic Pd_3Pt_1/CeO_2-OMC catalyst also shows persistent activity in extended period of runs without any indication of CO poisoning effects.

Alkali cation effects on the hydrogen oxidation reaction in alkaline electrolytes

Ionic species in electrolyte solutions can affect the rate and selectivity of electro-catalytic reactions. For example, (bi)sulfate and organic sulfonate can reduce the rate of the fuel cell relevant oxygen reduction reaction (ORR) on platinum, through their specific adsorption to the platinum surface. Halides, such as bromide, can reduce the rate of hydrogen oxidation (HOR) on platinum through specific adsorption as well. Recently, alkali metal cations were shown to affect the rate of HOR, ORR, and methanol oxidation at high potentials on Pt(111), through their interaction with specifically adsorbed hydroxide[1]. In this talk, we will discuss experimental evidence which suggests alkali metal cations affect the rate of HOR at low potentials in alkaline solutions on a polycrystalline Pt electrode. We find that lithium and sodium cations can promote HOR relative to potassium, giving a HOR rate closer to that measured in acid. A Density Functional Theory (DFT) analysis suggests alkali cation specific adsorption is favorable at low potentials and could be competitive with hydrogen adsorption in alkaline electrolytes on both the Pt(111) and Pt(110) surfaces. The adsorption of a low coverage of cations on the Pt(110) surface, the most active low index surface, can promote the adsorption of hydroxide, matching our experimental trend of rate dependence on cation identity. Specifically adsorbed hydroxide is known to be an important intermediate in hydrogen oxidation in alkaline solutions.

1. D. Strmcnik, K. Kodama, D. van der Vliet, J. Greeley, V.R. Stamenkovic, N.M. Marković. *Nature Chemistry* **1**, 466-472 (2009).

Combined DFT and Microkinetic Modeling of H_2 and CO electrooxidation on the $Sr_2Fe_{1.5}Mo_{0.5}O_6$ (001) Perovskite Anode

Salai Cheettu Ammal, Suwit Suthirakun, Eric Walker, and Andreas Heyden
University of South Carolina
Department of Chemical Engineering
300 S. Main Street, Columbia, SC 29208, USA

$Sr_2Fe_{1.5}Mo_{0.5}O_{6-\delta}$ (SFMO) perovskite has recently been proposed as a promising anode material for solid oxide fuel cells (SOFCs). Despite the advantages of SFMO such as exhibiting good ionic and electrical conductivity, its relatively low electro-oxidation activity is a major drawback which contributes to an overall low cell performance. It has been shown that adding a small amount of Ni to the anode improves the overall cell performance, but the presence of Ni decreases the sulfur tolerance of SFMO. Considering that understanding the fuel oxidation mechanism on perovskite surfaces such as SFMO is crucial for the development of alternative oxide based anode catalysts, we employed DFT+U methods and microkinetic modeling techniques in the present study to investigate the electrochemical oxidation of hydrogen and syngas fuels on the SFMO (001) surface under realistic fuel cell operating conditions. The role of Ni in improving the catalytic activity of SFMO and the sulfur poisoning mechanism of SFMO surface has also been examined.

Constrained *ab initio* thermodynamic calculations have been employed to construct various surface models of SFMO under SOFC operating conditions. Three surface models with varying Mo concentration in the surface layer were identified which were then used to investigate the thermochemistry and reaction barriers of the elementary steps involved in the H_2 and CO electro-oxidation mechanisms. In order to include the effect of anode bias potential, we included additional elementary steps in our microkinetic model considering the possibility of one- or two- electron charge transfer processes at various elementary steps. Simulated polarization curves reflect the kinetic activity of SFMO surfaces and suggest that the SFMO surface with higher Mo concentration on the gas exposed layer is more active towards both H_2 and CO oxidation. When both H_2 and CO are present in the fuel stream (syngas) together with some water, water gas shift (WGS, $CO + H_2O \rightarrow CO_2 + H_2$) is rapid on these surfaces. Increasing the CO_2 concentration promotes the reverse WGS and decreases the electrochemical performance. Further analysis revealed that at relevant operating voltages and reaction conditions, water/CO_2 desorption is rate controlling for both H_2 and CO oxidation reactions and stabilizing the oxygen vacancy structure increases the overall rate. In addition, an increase in electrochemical activity and resistance towards sulfur poisoning is observed with increasing Mo content in the surface layer. However, according to our thermodynamic analysis, the surface Mo concentration tends to be very low under anodic SOFC conditions. Our calculations with small Ni particles on the SFMO surface suggest that the presence of Ni dramatically reduces the surface oxygen vacancy formation energy which could improve the overall electrochemical performance of the SOFC cell. However, the sulfur resistance will likely be reduced in the presence of Ni.

377392 High Performance Molybdenum Dioxide-Based Anode for Liquid-Fueled SOFCs

Abstract:

This paper describes the performance of a solid oxide fuel cell (SOFC) fueled by directly feeding liquid fuels (e.g., premium gasoline) to the anode without using external reformers. The novel component of the fuel cell that enables such operation is the mixed conductivity MoO_2-based anode. Using this anode, a fuel cell demonstrating a power density > 3.0 W m^{-2} at 0.6 V under the premium gasoilne feed condition was successfully fabricated. Power density of ~2.0 W cm^{-2} at 0.6 V was still measured after 24 h on stream with a continuous feed of gasoline. Scanning electron microscopy (SEM) examination of the anode surface pre- and post-testing showed no evidence of coking, which hints at the reason for the observed stability under the harsh cell operating conditions. The effect of sulfur compounds on both chemical and physical stabilities of MoO_2-based anode under the SOFC operating conditions will be investiaged using various analytical tools. The implication of this preliminary study is that an SOFC using a MoO_2-based anode has potential for generating electrical power from gasoline for future hybrid electric vehicles.

Understanding Pt Particle Growth and Distribution on ITO: Impact on ORR Activity and Stability

Though proton exchange membrane fuel cells (PEMFCs) have drawn significant attention as a 21st century alternative to environmentally unfriendly power sources, there are several issues that must be addressed to make PEMFCs economically viable. One of the main problems to be solved is finding electrocatalysts with higher stability, while concomitantly improved activity to reduce loading and cost, at the oxygen reduction cathode.

Our recent work has focused on investigating the intrinsic activity of Sn-doped indium oxide (ITO) [1] as well as its stability in ORR environment [2]. This work has even included the investigation of the influence of Pt loading on ITO and its effect on the oxygen reduction activity and stability [3]. One interesting finding for Pt is that low Pt loadings yield a single Pt morphology with preferential {111} faceting, which yields its characteristic high activity. However, at higher loading secondary Pt particles with a completely different structure emerge. And it is really important to understand how these two particle types influence the activity and stability of the Pt/ITO.

In this work, we present a statistical model that utilizes a modified Poisson probability distribution function, which enable us to understand how the primary and secondary Pt nanoparticles grow during the galvanic displacement. This model also helps to predict the specific and mass activity of the Pt/ITO as a function of Pt loading. The model prediction and experimental results for the specific activity and mass activity are shown in Figure 1, and it is clearly that the excellent agreement is observed. Also, an interesting oscillatory behavior is predicted, which also appears corroborated by the experimental data. The behavior is due to the counterbalancing of the high activity of the preferentially faceted Pt/ITO nanoparticles and the secondary particles that show an activity similar to Pt supported on carbon. The number of primary particles also clearly plays a role [4], which is shown in the model.

This talk will focus on the model development, calculation of model parameters, and linking critical elements of the model to the activity and stability of Pt/ITO electrocatalysts for the oxygen reduction reaction.

References

1. Y. Liu and W.E. Mustain, "High stability, high activity Pt/ITO oxygen reduction electrocatalysts", *J. Am. Chem. Soc.*, **135** (2013) 530
2. Y. Liu and W.E. Mustain, "Stability limitations for Pt/Sn–In_2O_3 and Pt/In–SnO_2 in acidic electrochemical systems", *Electrochimica Acta*, **115** (2014) 116-125
3. Y. Liu and W.E. Mustain, "High Stability, High Activity Pt/ITO Oxygen Reduction Electrocatalysts". Abstract 86, 223rd Meeting of the Electrochemical Society, San Francisco, CA.
4. M. Shao, A. Peles and K. Shoemaker, *Nano Letters*, **11** (2011) 3714.

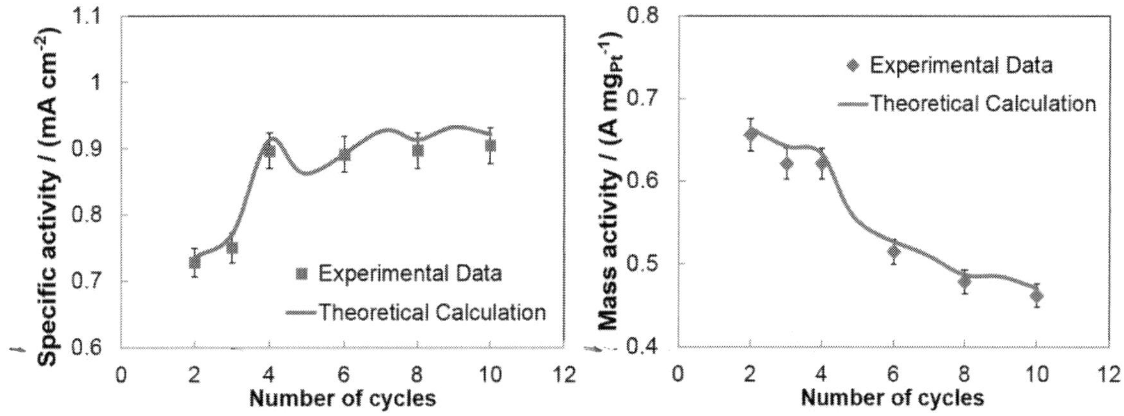

Specific activity / (mA cm-2)
1.1
1
0.9
0.8
0.7
0.6
Experimental Data
Theoretical Calculation
0
2
4
6
8
10
12
Number of cycles
Mass activity / (A mgPt-1)
0.8
0.7
0.6
0.5
0.4
Experimental Data
Theoretical Calculation
0
2
4
6
8
10
12
Number of cycles

Catalysis and Reaction Engineering
Practice Award Session
In Honor of
Kurt VandenBussche, UOP

Professor Marc-Olivier Coppens
Department of Chemical Engineering
Torrington Place
London WC1E7JE
United Kingdom

Diffusion in (Real) Zeolites

Our research revolves around the theme of nature-inspired chemical engineering (NICE). By studying natural systems – trees, lungs, and cell membranes – from the perspective of a chemical engineer, we draw clues to design better performing, more sustainable and efficient chemical processes and materials for technological applications. We seek guidance from nature as a starting point in theoretically and computationally based, experimentally validated designs. In this presentation we apply our approach to simulating the diffusion in zeolites.

379524 Design of Mesoporous Silica-Supported Alkylamine Catalysts for the Gas-Phase Condensation of Aldehydes and Ketones

Abstract:

Highly efficient solid-base organocatalysts for the gas-phase aldol self-condensation of *n*-butanal to 2-ethylhexenal have been developed by grafting site-isolated amines on tailored silica surfaces. The catalytic activity depends largely on the nature of amine species, the surface concentration of amine and silanol groups, and the spatial separation between the silanol and amine groups. In situ FT-IR measurements shows that the formation of nucleophilic enamines leads to the enhanced catalytic activity of secondary amine catalysts, whereas the formation of imines (stable up to 473 K) leads to the low activity observed for silica-supported primary amines. This work demonstrates that the spatial separation of the weakly acidic silanols and amines can be tuned by the controlled dehydration of the supporting silica and by varying the linker length of the amine organosilane precursor used to graft the amine to the support surface. A mechanism for aldol condensation is proposed and then analyzed using density functional theory (DFT). DFT analysis of the reaction pathway suggests that the rate-limiting step in aldol condensation is carbon-carbon bond formation, which is consistent with the observed kinetics. The calculated apparent activation barrier that follows from this analysis agrees reasonably with that measured experimentally. Subsequent studies have demonstrated that the co-operative acid-base properties of secondary amines catalysts can be significantly altered by introducing an optimal amount of Al or Ti into the silica matrix. DFT analysis of the reaction pathway suggests that in contrast to the C-C bond forming rate limiting step observed with silica supported amine catalysts, the rate-limiting step is iminium hydrolysis over silica containing Ti-OH or Al-OH groups, which is consistent with the observed kinetics. We have recently demonstrated that ketones, such as 2-butanone and 2-hexanone can be dimerized with high selectivity over silica-supported acid-base catalysts. The lower activity observed for ketone versus aldehyde condensation is explained by DFT calculations.

389413 Highly Selective H2 Production Via Methanol Steam Reforming Using Cu-MCM-41 Catalysts

Abstract:

Steam reforming of methanol was studied over a series of Cu-MCM-41 catalysts with Cu loading varied from 5% to 20% synthesized using one pot hydrothermal procedure. The catalysts were characterized using N_2 adsorption-desorption isotherms (BET), small and wide angle X-ray diffraction (XRD), thermo-gravimetric and differential calorimetric analysis (TGA-DSC), transmission electron microscopy (TEM), inductively coupled plasma optical emission spectroscopy (ICP-OES), energy dispersive X-ray spectroscopy (EDX), electron paramagnetic resonance (EPR) and temperature programmed reduction (TPR) techniques. Small angle XRD and N_2 adsorption-desorption studies indicated the optimum loading of copper to be 15%, where the characteristic ordered structure of the MCM-41 was retained with high specific surface of about 662 m^2/g. The catalysts exhibited outstanding activity, selectivity and stability for steam reforming of methanol (SRM) reactions. The activity and selectivity of the Cu-MCM-41 catalysts were significantly influenced by the Cu-loading and the reaction temperature. The SRM results revealed that 15%Cu-MCM-41 showed optimum performance in terms of methanol conversion (~ 90%), H_2 selectivity (100%) and CO selectivity (0.79%) at 300 °C. While both the reactant space velocity and methanol/water molar ratio strongly influenced the activity of the catalysts, selectivity of the catalysts remained unaffected. The Cu-MCM-41catalyst showed strong resistance to deactivation and maintained consistent performance over a period of 48 h. The excellent stability of the catalyst in the SRM reaction can be attributed to the uniform distribution of the Cu particles in a very high surface area MCM-41 matrix, inhibiting the sintering of Cu particles. The highlights of this work will be discussed in the presentation.

359676 Hierarchical USY Zeolites for Removing of Trace Olefins from Aromatics : Effect of Acid Leaching in the Dealumination of Zeolites

Abstract:

Hierachical USY zeolite prepared by citric acid treatment was demonstrated to be an effective catalyst for removing trace olefins from aromatics with the suitable ratio of Lewis/Brønsted acid. The samples were investigated by X-ray diffraction, N_2 adsorption, and fourier transform infrared spectroscopy. The results indicated that the mesopores might be generated by skeletal reconstruction and the AlO^+ change the distribution of acid sites in the samples. It can improve the catalytic activity by facilitating diffuse out the polyaromatic components and thus avoided the coke formation. Optimization of pores and acid sites of ultrastabilized USY zeolite modified with citric acid might result in an industrial process to design novel catalysts for removing trace olefins.

Dealuminated zeolite HY was used to support $Ir(CO)_2$ complexes formed from $Ir(CO)_2(C_5H_7O_2)$. Infrared and X-ray absorption spectra and atomic-resolution electron microscopy images identify these complexes, and the images and ^{27}Al NMR spectra identify impurity amorphous regions in the zeolite where the iridium is more susceptible to aggregation than in the crystalline regions. The results indicate the value of high-resolution electron microscopy in characterizing the both the crystalline and amorphous impurity regions of zeolites and a significant stability limitation of metals in the amorphous regions of zeolite catalyst supports.

377062 Durability Study of Low-Al Zeolite Beta through Steaming Process Modification As a Heterogeneous Catalyst for Biodiesel Production

Abstract:

In this study, sodium-modified Zeolite Beta was adopted to catalyze both transesterification of triglycerides and esterification of fatty acids for biodiesel production. Dealumination of Zeolite Beta by steaming process was conducted to give zeolites with different Si-to-Al atomic ratios. Subsequently, ion-exchange with Na^+ was performed on as-obtained Zeolite Beta. Interestingly, Zeolite Beta with a higher atomic ratio of Si-to-Al possesses better catalytic durability. Coincidently, more defect sites where sodium could be stored were found on low-Al Zeolite Beta after the high temperature steaming treatment. Consequently, a biodiesel yield over 80% was resulted in less than 5 hours from transesterification of palm oil and triolein in thirteenth and eighth cycles, respectively, catalyzed by the sodium-modified Zeolite Beta. The results indicated sodium stored in defect sites are able to rapidly supply the surface reactivity of catalyst to effectively catalyze transesterification and esterification.

Kinetic study of gas phase cyclohexene epoxidation over mesoporous TS-1 with H_2O_2 vapor

Stephanie Kwon[1], *Neil M. Schweitzer*[1], *Randall Snurr*[1*] *and Peter C. Stair*[1,2*]
[1]*Northwestern University, Evanston, Illinois 60208*
[2]*Chemistry Division, Argonne National Laboratory, Argonne, IL 60439*

Epoxidation of alkenes is an important chemical reaction, and hydrogen peroxide is one of the most appealing oxidants, as the reaction produces only water as a by-product. Epoxidation reactions reported so far in the literature with hydrogen peroxide were mostly performed in the liquid phase with batch or semi-batch reactors. However, liquid-phase reactions often suffer from a limited range of accessible temperatures, complications from solvent effects, and problems due to active site leaching. Here, we present our work on gas phase cyclohexene epoxidation over mesoporous TS-1 with H_2O_2 vapor. The catalytic activity was very stable, with high selectivity towards cyclohexene oxide (maximum epoxide selectivity over 90%). The stable catalytic behavior allowed us to perform a comprehensive analysis of the kinetics of gas phase cyclohexene epoxidation with H_2O_2 vapor for the first time. An Arrhenius temperature dependence was observed with an apparent activation energy of 34 ± 2 kJ/mol. The reaction orders of reactants were investigated and fitted to a power rate law, yielding $r_{epoxide} = k(C_6H_{10})^{0.37} (H_2O_2)^{0.49}(H_2O)^{-0.35}$. The similarity in the kinetics observed for gas phase epoxidation and liquid phase epoxidation from the literature suggests that the catalyst operates via the same mechanism. Detailed kinetic study will provide more insight to understand the gas phase epoxidation mechanism which will allow us to design a better catalyst. We anticipate that this work will further open doors to a diverse range of gas phase oxidation reactions with H_2O_2 at elevated temperature with high activity and selectivity.

Selective oxidation of ethylbenzene catalyzed by *N*-hydroxyphthalimide (NHPI) and ZSM-5 supported Co(II), Mn(II), Cu(II), Zn(II), Fe(III) with molecular oxygen under mild conditions

Lei Chen, Bindong Li, Dabin Liu

Nanjing University of Science and Technology,

Nanjing, 210094, China.

2014 AIChE Annual Meeting, Nov. 17, 2014

南京理工大学
NANJING UNIVERSITY OF SCIENCE & TECHNOLOGY

Outlines

1. Background of industrial oxidation
2. Sythesis and characteristics of M-ZSM-5
3. Catalytic oxidation of ethylbenzene
4. Catalysis mechanism discussion

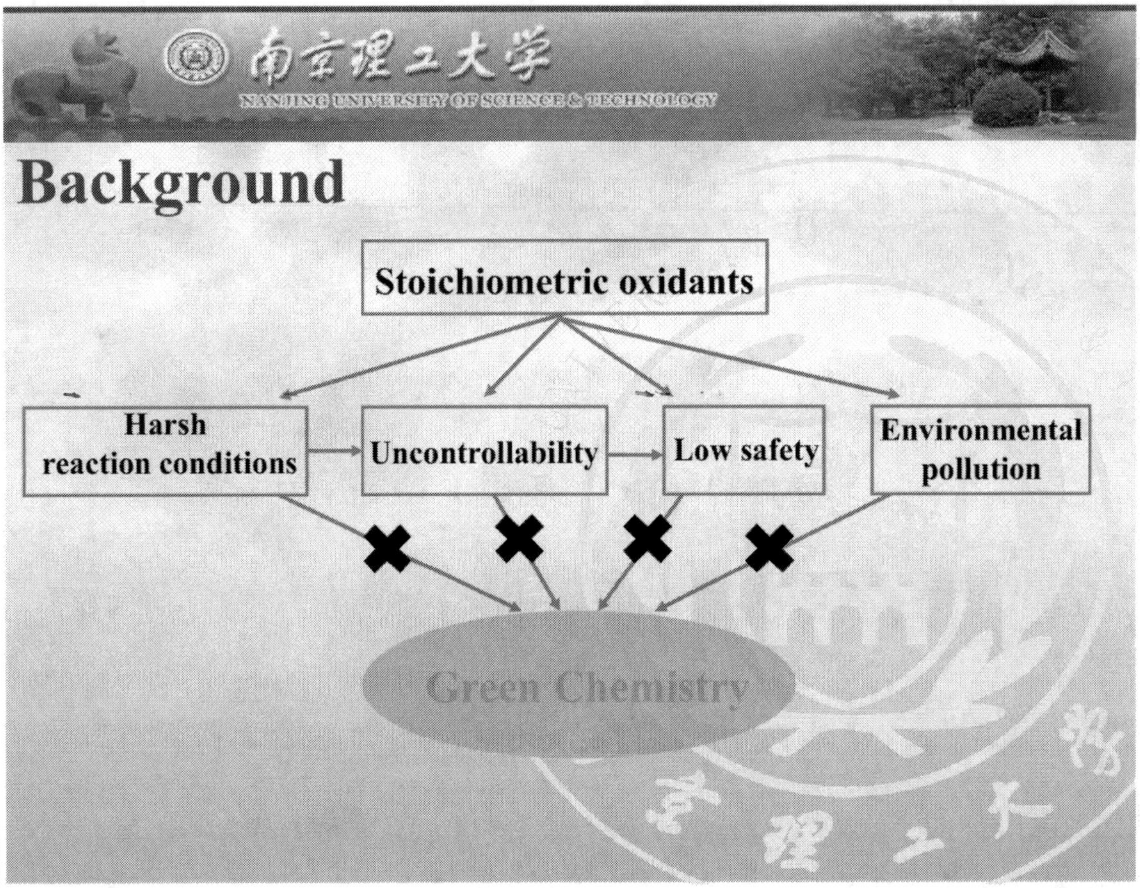
南京理工大学
NANJING UNIVERSITY OF SCIENCE & TECHNOLOGY
Background
Stoichiometric oxidants
Harsh reaction conditions
Uncontrollability
Low safety
Environmental pollution
Green Chemistry

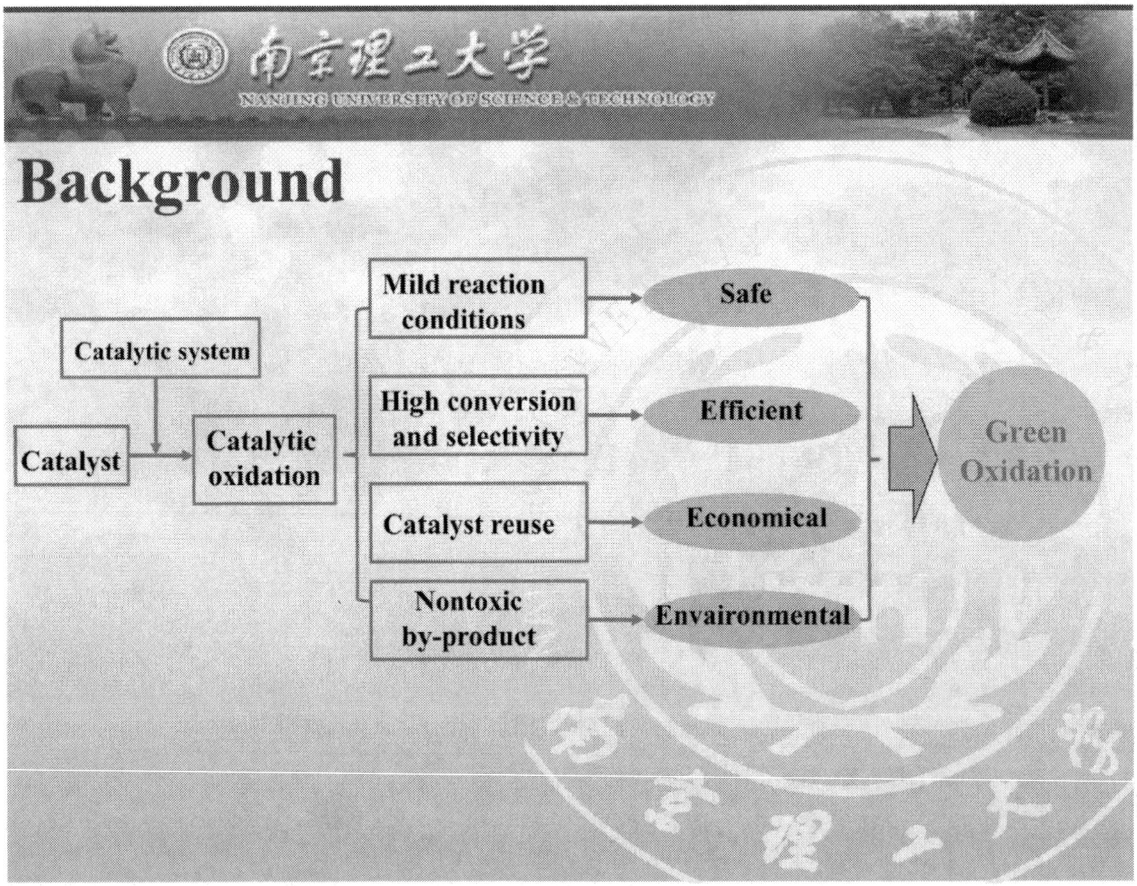
南京理工大学
NANJING UNIVERSITY OF SCIENCE & TECHNOLOGY
Background
Catalytic system
Catalyst
Catalytic oxidation
Mild reaction conditions
High conversion and selectivity
Catalyst reuse
Nontoxic by-product
Safe
Efficient
Economical
Envaironmental
Green Oxidation

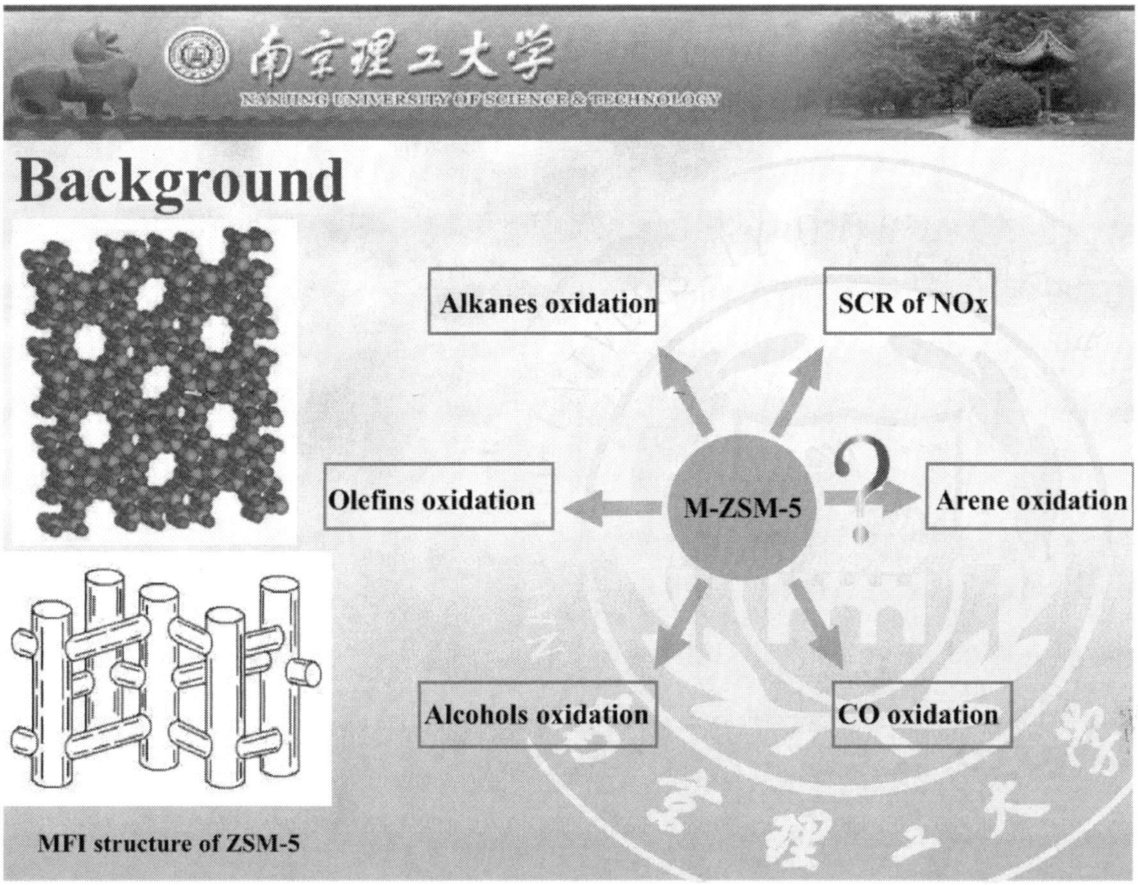
南京理工大学
NANJING UNIVERSITY OF SCIENCE & TECHNOLOGY
Background
Alkanes oxidation
SCR of NOx
Olefins oxidation
M-ZSM-5
?
Arene oxidation
Alcohols oxidation
CO oxidation
MFI structure of ZSM-5

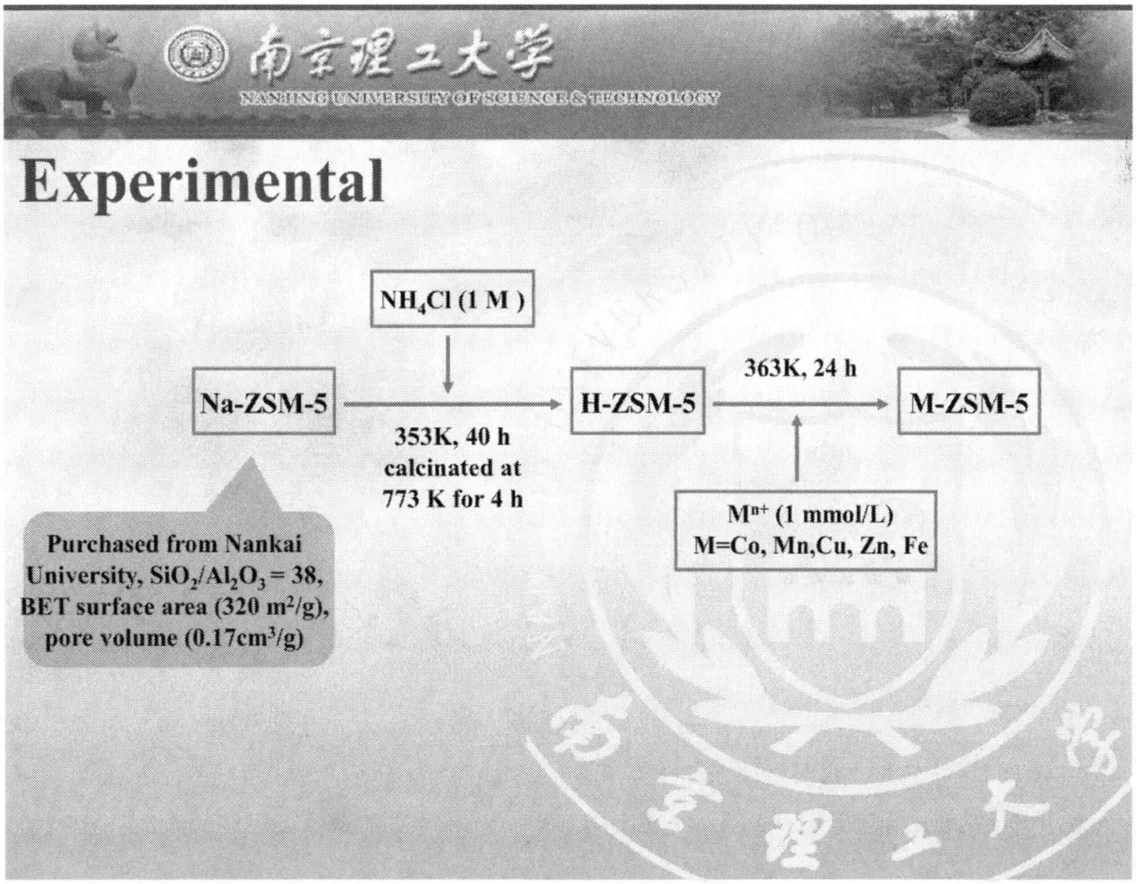
南京理工大学
NANJING UNIVERSITY OF SCIENCE & TECHNOLOGY
Experimental
NH_4Cl (1 M)
Na-ZSM-5
353K, 40 h
calcinated at
773 K for 4 h
H-ZSM-5
363K, 24 h
M-ZSM-5
M^{n+} (1 mmol/L)
M=Co, Mn,Cu, Zn, Fe
Purchased from Nankai
University, SiO_2/Al_2O_3 = 38,
BET surface area (320 m^2/g),
pore volume (0.17cm^3/g)

南京理工大学
NANJING UNIVERSITY OF SCIENCE & TECHNOLOGY

Structure

Fig.1 SEM images of catalysts

南京理工大学
NANJING UNIVERSITY OF SCIENCE & TECHNOLOGY

Structure

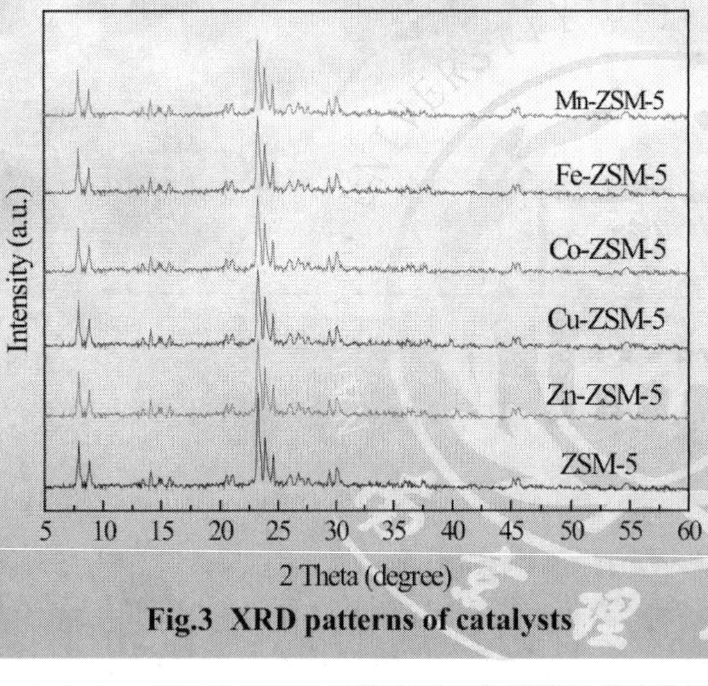

Fig.3 XRD patterns of catalysts

南京理工大学
NANJING UNIVERSITY OF SCIENCE & TECHNOLOGY

Structure

Table 1 Chemical and physical properties of catalysts

catalyst	BET tests		Metal contents (mmol/g)
	Surface area (m^2/g)	Pore volume (cm^3/g)	
ZSM-5	320	0.17	--
Mn(II)-ZSM-5	323	0.16	0.031
Fe(III)-ZSM-5	325	0.16	0.024
Co(II)-ZSM-5	320	0.17	0.030
Cu(II)-ZSM-5	324	0.17	0.027
Zn(II)-ZSM-5	327	0.15	0.022

南京理工大学
NANJING UNIVERSITY OF SCIENCE & TECHNOLOGY

Catalytic activity

Table 2 Comparison of catalytic activity of various catalyst/catalytic system

Entry	Catalyst	Conversion (%)[a]	Selectivity (%)[a]	
			alcohol	acetophenone
1	NHPI	8	69	31
2	Mn-ZSM-5	1.5	28	72
3	Fe-ZSM-5	Trace	-	-
4	Co-ZSM-5	2.0	26	74
5	Cu-ZSM-5	0.8	32	68
6	Zn-ZSM-5	Trace	-	-
7	Mn-ZSM-5 ·NHPI	80	10	90
8	Fe-ZSM-5 · NHPI	66	20	80
9	Co-ZSM-5 · NHPI	84	7	93
10	Cu-ZSM-5 · NHPI	75	15	85
11	Zn-ZSM-5 · NHPI	58	21	79

Reaction conditions: ethylbenzene (10 mmol), NHPI (5 mol%), M-ZSM-5 (0.5 mol%), acetic acid (20 ml), O_2 (1 atm), 100 °C, 12h. [a] The products were identified by GC–MS and quantified by GC.

Catalytic activity

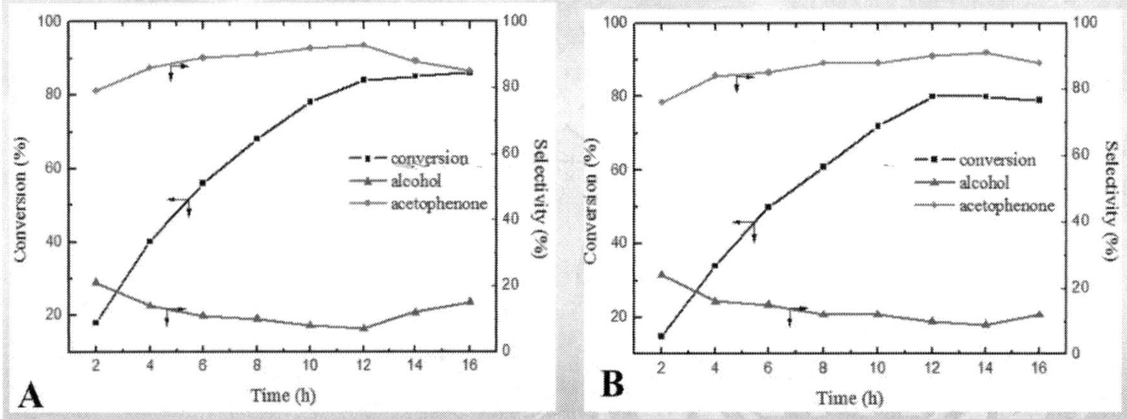

Fig.4 Effect of reaction time on oxidation of ethylbenzene. ethylbenzene (10 mol), NHPI (5 mol%), M-ZSM-5 (0.5 mol%), acetic acid (20 ml), O_2 (1 atm), 100 °C. A: Co-ZSM-5 · NHPI; B: Mn-ZSM-5 · NHPI.

Catalytic activity

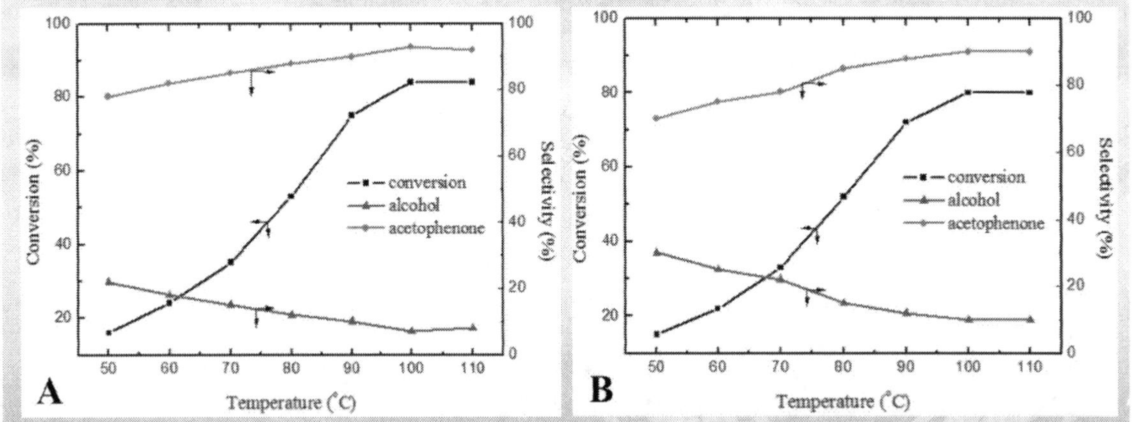

Fig.5 Effect of reaction temperature on oxidation of ethylbenzene. ethylbenzene (10 mol), NHPI (5 mol%), M-ZSM-5 (0.5 mol%), acetic acid (20 ml), O_2 (1 atm), 12h. A: Co-ZSM-5 · NHPI; B: Mn-ZSM-5 · NHPI.

南京理工大学
NANJING UNIVERSITY OF SCIENCE & TECHNOLOGY

Catalyst reuse

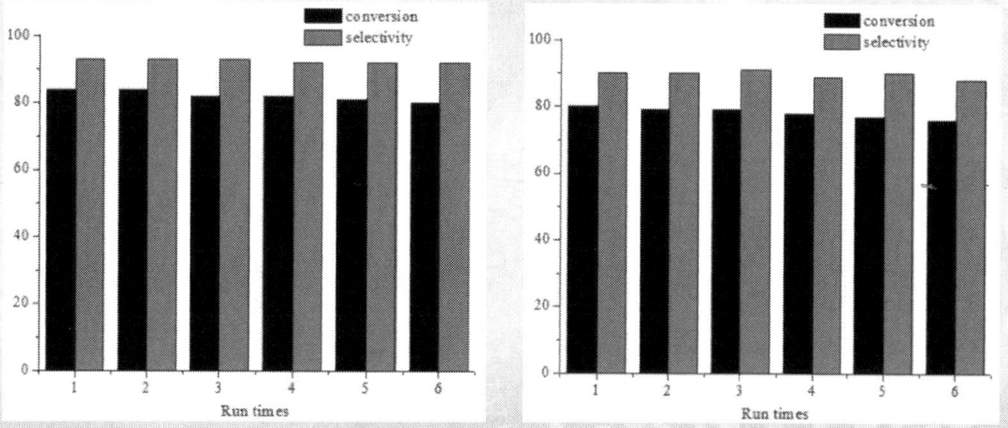

Fig.6 Recycling study of catalysts (A) Co-ZSM-5 and (Mn-ZSM-5). ethylbenzene (10 mol), NHPI (5 mol%), M-ZSM-5 (0.5 mol%), acetic acid (20 ml), O_2 (1 atm), 100 °C, 12h.

南京理工大学
NANJING UNIVERSITY OF SCIENCE & TECHNOLOGY

Mechanism discussion

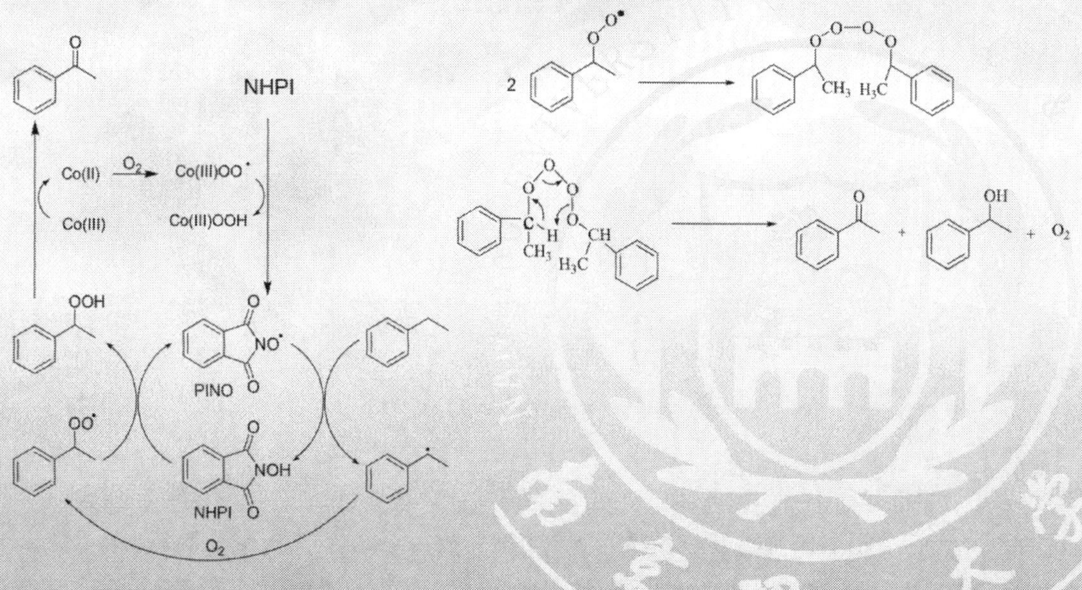

南京理工大学
NANJING UNIVERSITY OF SCIENCE & TECHNOLOGY

Conclusion

1. A series of ZSM-5 catalysts containing transition metal ions was prepared via ion-exchange method and characterized by XRD, SEM, BET, ICP-AES.

2. All catalysts showed good catalytic activity on the selective oxidation of ethylbenzene with molecular oxygen as oxidant in the presence of NHPI.

3. Compared to the other three catalysts, Co-ZSM-5 and Mn-ZSM-5 showed relative high catalytic activity and could be reused at least six times without significant loss of activity.

南京理工大学
NANJING UNIVERSITY OF SCIENCE & TECHNOLOGY

Acknowledgement

This investigation was supported by National Nature Science Foundation of China Academy of Engineering Physics (No. 10976014) and Nature Science Foundation of Jiangsu Province (No. BK2011697).

南京理工大学
NANJING UNIVERSITY OF SCIENCE & TECHNOLOGY
Thank you!

CATALYTIC DIOXYGENATION OF CATECHOL BY NON-HEME IRON(III) COMPLEXES AS A SUSTAINABLE ROUTE TO DIMETHYL ADIPATE

Robin Jastrzebski, Emily J. van den Berg, Bert M. Weckhuysen, Pieter C.A. Bruijnincx*

Inorganic Chemistry and Catalysis, Debye Institute for Nanomaterials Science, Utrecht University, Universiteitsweg 99, 3584 CG Utrecht, The Netherlands

The atom-efficient, direct conversion of functionalized biomass-derived platform molecules to bulk chemicals is an important challenge in the valorization of biomass.[1] Catechols constitute an interesting class of renewable platform molecules, as these functionalized aromatics may be obtained from biomass in multiple ways, including fermentation of glucose, pyrolysis of woody biomass or catalytic depolymerization of lignin. In nature, catechols are metabolized by intradiol catechol dioxygenase enzymes, to afford derivatives of *cis,cis*-muconic acid. Although bio-inspired non-heme iron(III) complexes have been investigated as models for this enzymatic reaction, the *catalytic* dioxygenation of pyrocatechol using such systems has not received much attention.[2] This reaction holds considerable promise, however, as simple hydrogenation of the obtained *cis,cis*-muconic acid would yield adipic acid, an important bulk chemical consumed on a billion kg/year scale for the production of nylons.

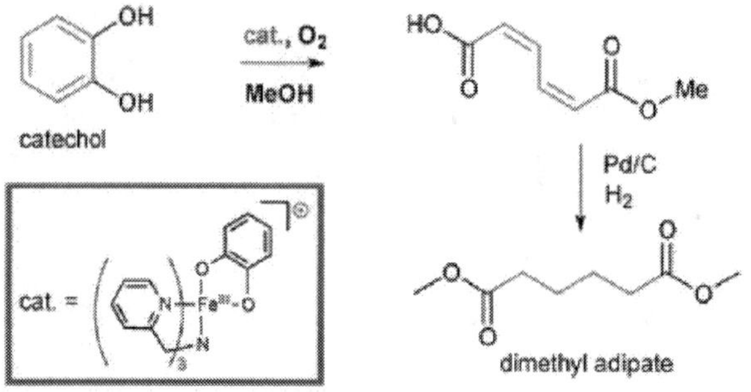

Scheme 1: The dioxygenation of pyrocatechol affords O-*methyl*-cis,cis-*muconic acid as the major product, which can be readily converted to dimethyl adipate.*

Here, we report on the first chemocatalytic, selective intradiol dioxygenation of pyrocatechol using iron(III) complexes of tris(2-pyridylmethyl)amine (tpa).[3] The active catalyst was readily prepared *in-situ* from simple iron(III) salts, tpa and a weak base. The reaction uses molecular oxygen as the oxidant and is run under moderate conditions in methanol (5 mol% Fe, 50 °C, 1 atm air). In the first step, the ring is selectively cleaved to yield a mixture of intradiol muconic acid products with *O*-methyl-*cis,cis*-muconic acid being the major product. Optimization of reaction conditions in a semi-batch autoclave, which enabled reactions at more severe conditions (up to 80 °C and 30 bar 5vol% O_2 in N_2), afforded a more than ten-fold increase in turnover numbers. The muconics acids are subsequently hydrogenated over commercial hydrogenation catalysts such as Pd/C and Raney Ni. A final acid-catalyzed trans-esterification step then fully resolved the complexity of the product mixture and gave dimethyl adipate, the preferred nylon precursor, as the single end product with high selectivity.

References:

1. Zakzeski, J.; Bruijnincx, P. C. A.; Jongerius, A. L.; Weckhuysen, B. M. *Chem. Rev.* **2010**, *110*, 3552–3599.
2. Yamahara, R.; Ogo, S.; Masuda, H.; Watanabe, Y. *J. Inorg. Biochem.* **2002**, *88*, 284–294.
3. Jastrzebski, R.; Weckhuysen, B. M.; Bruijnincx, P. C. A. *Chem. Commun.* **2013**, *49*, 6912–6914.

367347 Greening Paints, Plastics, and Fibers: Renewable Phthalic Anhydride from Biomass-Derived Sugars

Abstract:

Every year, 3 million tonnes of phthalic anhydride are manufactured worldwide for the production of paints, plasticizers, and fabrics. However, phthalic anhydride is currently produced by the catalytic oxidation of o-xylene, a chemical derived from petroleum. Here, we suggest a renewable route to phthalic anhydride from biomass-derived furan and maleic anhydride. Furan and maleic anhydride were converted to phthalic anhydride in two reaction steps: Diels–Alder cycloaddition followed by dehydration. Excellent yields for the Diels–Alder reaction between furan and maleic-anhydride were obtained at room temperature and solvent-free conditions (SFC) yielding 96% exo-4,10-dioxa-tricyclo[5.2.1.0]dec-8-ene-3,5-dione (oxanorbornene dicarboxylic anhydride) after 4 h of reaction. It is shown that this reaction is resistant to thermal runaway because of its reversibility and exothermicity. Subsequently, the dehydration of the oxanorbornene was investigated using mixed-sulfonic carboxylic anhydrides in methanesulfonic acid (MSA). An 80% selectivity to phthalic anhydride (87% selectivity to phthalic anhydride and phthalic acid) was obtained after running the reaction for 2 h at 298 K to form a stable intermediate followed by 4 h at 353 K to drive the reaction to completion. The structure of the intermediate was determined. This result is much better than the 11% selectivity obtained in neat MSA using similar reaction conditions.

Molybdenum based polyoxometalates as efficient catalysts for C-C rearrangement in the epimerization of carbohydrates

FeifeiJu[†], David VanderVelde[‡], and Eranda Nikolla*[†]

[†]Department of Chemical Engineering and Material Science, Wayne State University, Detroit, Michigan 48202, United States
[‡]Division of Chemistry and Chemical Engineering, California Institute of Technology, Pasadena, California 91125, United States

Polyoxometalates have been extensively studied for many years as catalysts for oxidation and photochemical processes[1]. While there exists a wealth of literature on the oxidative and photocatalytic properties of POMs, there are no reports on their ability to catalyze the C-C backbone rearrangement in cyclic molecules, such as in the case of the epimerization of aldose, under aqueous conditions. Herein, we demonstrate for the first time the activity and selectivity of molybdenum-based POMs in catalyzing the C-C rearrangement in the epimerization of aldoses over a broad range of reaction conditions[2]. Near-equilibrium conversions and high selectivities were obtained under aqueous conditions at relatively low temperatures. Our studies demonstrated that the molybdenum oxide octahedra on the surface of the Keggin structure POMs were the active sites for the aldose epimerization. Isotope labeling experiments along with ^{13}C NMR studies were conducted and showed that the epimerization of aldoses involves an intramolecular C1-C2 rearrangement, where the bond between C2 and C3 is cleaved and a new bond between C1 and C3 is formed leading to the formation of the epimer.

References:

1. Pope, M. T.; Muller A. Angew. Chem. Int. Ed. 30(1991), 34-48.

2. Ju, F.; VanderVelde, D.; Nikolla, E. ACS Catal. 4 (2014), 1358-1364.

SURFACE-LEVEL INVESTIGATIONS OF BENZYL ALCOHOL REACTION ON THIOLATE-COVERED SURFACES

Chih-Heng Lien and J. Will Medlin*

Department of Chemical and Biological Engineering, University of Colorado Boulder, Jennie Smoly Caruthers Biotechnology Building 598 UCB, Boulder, CO 80309-0596, USA

* Will.Medlin@colorado.edu

Abstract

Benzyl alcohol is a significant biomass-derived intermediate from lignin and a useful probe molecule for aromatic oxygenates. During reaction with hydrogen, benzyl alcohol can be converted to benzaldehyde, toluene and benzene. Recently, self-assembled monolayers-coated (SAM-coated) catalysts, which can improve the reaction selectivity through modifying the catalyst surface and blocking non-selective catalytic sites, have recently received attention[1]. Therefore, in this study, we explored the use of thiol SAMs to modify Pd/Al_2O_3 and enhance the selectivity of benzyl alcohol hydrodeoxygenation to toluene, and used surface science techniques to characterize the mechanism.

On uncoated Pd catalysts, after an initial deactivation period, the selectivity to toluene under steady state conditions was less than 35%, and the dominant product was benzene. After the SAM modification, the selectivity to toluene can reach 85% and the production of benzene is remarkably reduced, which shows that the SAMs are effective in enhancing the selectivity to toluene. Remarkably, on surfaces coated with adamantanethiol, both selectivity and overall catalytic activity were improved. Surface science techniques were employed to further investigate this chemistry, with Auger electron spectroscopy (AES) used to confirm sulfur coverage and temperature programmed desorption (TPD) to monitor the thermal chemistry of adsorbates. On octadecanethiol-coated surfaces, the benzene signal disappeared along with the hydrogen peak around 495K indicating reduced decarbonylation activity. Both of these observations indicate that the flat-lying conformation, which was the conformation found to generate benzene[2], was restricted and the production of benzene was thus decreased. In contrast, production of toluene was still observed on the thiol-coated surface, consistent with toluene formation occurring through a more upright conformation. Thus, the SAM modification strategy could be useful in making production of biomass-derived products more feasible and competitive by lower purification cost and less environmental impact.

Pore Structure and Bifunctional Catalyst Activity of Overlayers Applied by Atomic Layer Deposition on Copper Nanoparticles

Ana C. Alba-Rubio[1], Brandon J. O'Neill[1], Fengyuan Shi[2], Cem Akatay[3], Christian Canlas[4], Tao Li[5], Randall Winans[5], Jeffrey W. Elam[4], Eric A. Stach[6], Paul M. Voyles[2], and James A. Dumesic[1].

[1]Department of Chemical and Biological Engineering and [2]Materials Science Program, University of Wisconsin, Madison, Wisconsin 53706, United States.
[3]School of Materials Engineering, Purdue University, West Lafayette, Indiana 47907, United States.
[4]Energy Systems Division and [5]X-ray Sciences Division, Argonne National Laboratory, Argonne, Illinois 60439, United States.
[6]Center for Functional Materials, Brookhaven National Laboratory, Upton, New York 11973, United States.

For reasons of environmental and economic sustainability, it would be desirable to replace precious-metal catalyst (e.g. Pt) with base metals that are more earth-abundant (e.g. Cu). However, base metal catalysts are especially susceptible to irreversible deactivation through sintering or leaching, especially under liquid phase reaction conditions [1]. Recently, O'Neill et al [2] reported that atomic layer deposition (ALD) can be employed to stabilize a base-metal catalyst for liquid-phase catalytic processing. Herein, we present a model system that can be used to facilitate investigation of the ALD overcoat structure that leads to catalyst stabilization, and we extend the functionality of the ALD overcoated catalyst architecture to include bifunctional catalytic activity.

Cu nanoparticles were highly dispersed on non-porous silica spheres to facilitate characterization of the materials, followed by ALD overcoating with amorphous AlO_x followed by heat treatment to produce porosity. The very small (<1 nm) pores make the direct imaging of the porosity complicated and in this case not possible. To circumvent this complication, we designed a method to image the pores by scanning transmission electron microscopy and electron energy loss spectroscopy (STEM/EELS), which is facilitated by the spherical support. The porosity of the AlO_x overcoat was probed by deposition of NbO_x by ALD. Porosity was also confirmed by small-angle X-ray scattering (SAXS), and the catalyst performed comparably to previous work on conventional supports [2].

In addition to nanoparticle stabilization, we note that the use of ALD overcoating allows for the synthesis of novel and finely controlled catalytic structures. The ability to add bifunctionality with atomic precision is an important and promising future area in heterogeneous catalysis. In this proof of concept example, the addition of NbO_x into the pores of a silica-supported copper catalyst with an ALD-alumina overcoating leads to the formation of new acid sites capable of catalyzing the etherification of furfuryl alcohol with 1-butanol to form furfuryl butyl ether.

In summary, we have employed a model system based on the use of nonporous supports to facilitate the detailed characterization of supported metal catalysts stabilized by ALD, and we have developed a method to indirectly image the pores in the ALD overcoat. This strategy could be useful for better imaging and understanding of other phenomena in catalysis. In addition, we have demonstrated the use of ALD to synthesize novel bifunctional catalysts.

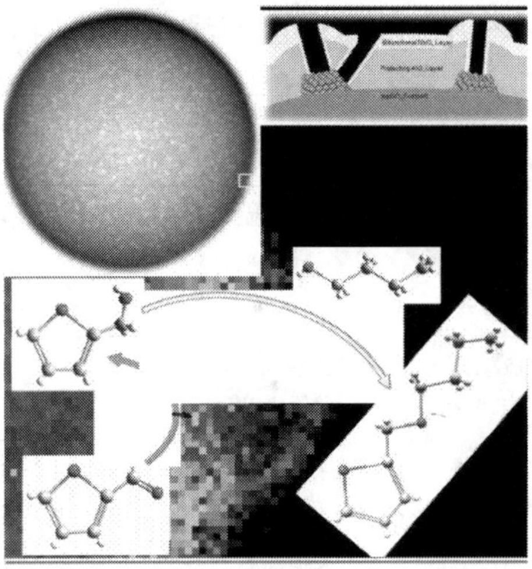

Schematic representation of the ALD layers and reactions that take place

[1] Bullock, R. M. In Catalysis without Precious Metals; Bullock, R. M., Ed.; Wiley-VCH Verlag GmbH & Co. KGaA: Richland, WA, 2010; pp I−XVIII.

[2] O'Neill, B. J.; Jackson, D. H. K.; Crisci, A. J.; Farberow, C. A.; Shi, F.; Alba-Rubio, A. C.; Lu, J.; Dietrich, P. J.; Gu, X.; Marshall, C. L.; Stair, P. C.; Elam, J. W.; Miller, J. T.; Ribeiro, F. H.; Voyles, P. M.; Greeley, J.; Mavrikakis, M.; Scott, S. L.; Kuech, T. F.; Dumesic, J. A. Angew. Chem., Int. Ed. 2013, 52, 13808−13812.

Aqueous-phase hydrogenation of levulinic acid (LA) is the most commonly proposed strategy for the production of γ-valerolactone (GVL) from biomass resources. The reaction is generally carried out over supported metals, most commonly Ru on activated carbon. Ru/C is both active and selective for GVL production; however, it exhibits poor stability in the aqueous phase such that steady state hydrogenation rates decay to a small fraction of their initial values[1].Here, we probe mechanisms leading to the rapid deactivation of Ru/C in this system and propose design criteria for the synthesis of catalytic materials that offer improved performance. Differential rates of LA hydrogenation obtained in packed bed reactors are combined with electron microscopy, X-Ray diffraction, elemental analysis, and physi- and chemisorption to identify two primary modes of catalyst deactivation in this system. First, irreversible loss of activity is attributed to the growth of Ru nanoparticles, which is observed directly by TEM. Although Ru/C appears sinter-resistant in gas phase treatments at significantly higher temperatures (e.g., 673K), metal clusters grow rapidly in aqueous media, even at near-ambient conditions (e.g., 323K). Second, Ru/C displays a reversible loss of activity during LA hydrogenation, and this is attributed to site-blocking via strongly chemisorbed intermediates. We have observed that these two modes of deactivation correlate with support properties, and both can be managed by judicious selection of host materials. For example, Ru/C and Ru/SiO_2 are relatively sinter-prone, while TiO_2 and γ-Al_2O_3 retain high Ru dispersions during extended operation. Active site blocking appears to correlate with the support's point of zero charge (PZC), and supports with lower PZCs, such as amorphous SiO_2, display little reversible loss of activity. The above insights establish a framework for designing stable supported metal catalysts for use in aqueous phase hydrogenation and hydrodeoxygenation, thus addressing a central need in biomass processing.

1 - O. A. Abdelrahman, A. Heyden and J. Q. Bond, *Analysis of Kinetics and Reaction Pathways in the Aqueous-Phase Hydrogenation of Levulinic Acid To Form γ-Valerolactone over Ru/C*, ACS Catal., 2014, 4, 1171–1181.

Chemicals from bioethanol: tailoring copper electronic properties to enhance its performance on the one-pot ethyl acetate synthesis

Isabel C. De Freitas,[1] Jean Marcel R. Gallo,[1] José Maria C. Bueno,[2] Clelia Mara P. Marques[1]

[1]Department of Chemistry, Federal University of São Carlos, São Carlos, Brazil

[2]Department of Chemical Engineering, Federal University of São Carlos, São Carlos, Brazil

The USA and Brazil are responsible for approximately 90 % of the biomass-derived ethanol produced worldwide. Therefore, these countries share the interest in the transformation of bioethanol into chemicals, such as ethyl acetate. This ester is widely used as a replacement for aromatic solvents in paints, inks and dyes; moreover, it can be used in the pharmaceutical, cosmetic, and food industry. Ethyl acetate is commercially obtained from non-renewable sources though multiple reaction processes. Recent studies have shown that ethyl acetate can be obtained from bioethanol though a one-pot reaction catalyzed by copper-based catalysts (for example, $CuCr_2O_4$ and Cu/ZrO_2 catalysts). The reaction mechanism involves the dehydrogenation of ethanol on the metal surface, followed by coupling of the newly formed aldehyde with an ethanol molecule (Step 1: $CH_3CH_2OH \rightleftharpoons CH_3CHO + H_2$; Step 2: $CH_3CHO + CH_3CH_2OH \rightleftharpoons CH_3CO_2CH_2CH_3 + H_2$). Our group has shown that Cu/ZrO_2 is a remarkable catalyst for this reaction and proposed that the reaction takes place in the interface between metal and support. It was also shown that the equilibrium Cu^0/Cu^+ plays a major role in the catalytic performance. However, the low specific surface area of zirconia does not favor the formation of a large extent of interface metal-support.

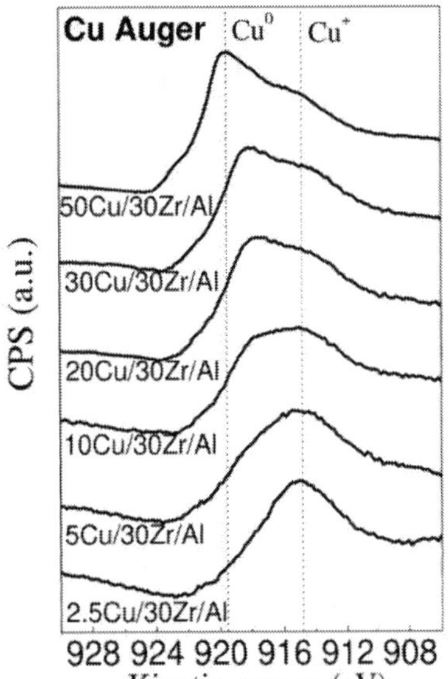

Fig 1. XPS of Cu Auger core level for reduced Cu supported on 30 wt. % Zirconia on Alumina.

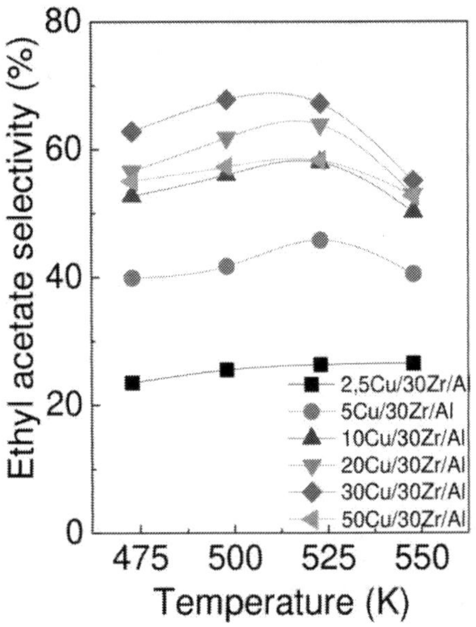

Fig 2. Evolution of the selectivity to ethyl acetate over xCu/30Zr/Al (where x is the wt. % of copper)

For catalysts syntheses, zirconia and copper were dispersed in a high surface area alumina support (370 $m^2 g^{-1}$) in order to increase the interface copper/zirconia. Different ratios alumina/zirconia/copper were studied aiming to find the optimal catalyst composition. The catalysts were studied by several characterization techniques, such as X-ray diffraction, nitrogen physisorption, and temperature programmed reduction. Analyses of infrared of the samples adsorbing carbon monoxide showed that copper is placed primarily on the zirconia surface, however, at high copper loadings a significant amount of copper is also observed in the alumina surface. Another important characterization technique used was X-ray photoelectron spectroscopy (XPS), which showed that increasing copper loading led to an higher Cu^0/Cu^+ ratio on the metal nanoparticle surface (as shown in Fig. 1). Comparing Fig. 1 and 2, it can be observed that the samples containing 20 and 30 % of copper (20Cu/30Zr/Al and 30Cu/30Zr/Al) were the ones with the optimum ratio of Cu^0 over Cu^+, since they displayed the highest selectivity for ethyl acetate (60-70 %). For the samples with lower copper loading (2.5Cu/30Zr/Al, 5Cu/30Zr/Al, and 10Cu/30Zr/Al), the higher Cu^+ concentration on the surface led to an increasing concentration of acetaldehyde on the metal surface and, therefore, facilitated the condensation reactions to heavy byproducts. Increasing the copper loading to 50 % (50Cu/30Zr/Al) led to the formation of bulk copper particles, which are selective to acetaldehyde formation, but lacks in the interface with zirconia, diminishing the formation of ethyl acetate.

It is important to mention, that Cu supported in alumina presents a low selectivity for ethyl acetate, and therefore zirconia is essential on providing active sites for the coupling reaction. If compared with the classical Cu/ZrO_2 catalyst, Cu/30Zr/Al (both at the same copper loading) has similar activity to ethanol conversion, however it shows significant improvement in the selectivity and reaction rate for ethyl acetate. This behavior can be attributed to the higher copper/zirconia interface on the Cu/30Zr/Al catalysts, once both zirconia and copper are highly dispersed in the alumina support.

Authors thank CNPq and FAPESP. JMR Gallo thanks FAPESP for the fellowship (process number 2013/15861-1)

371810 Nonlinear Optimization-Based Approaches for Modeling and Design of Heterogeneous Catalytic Systems

Abstract:

Microkinetic modeling, with input from computational chemistry calculations, is a key step in parameter estimation, experimental design, and rational catalyst design of heterogeneous catalytic systems. All these applications are, however, nonlinear optimization problems with an embedded microkinetic model that captures the dynamics of the reaction system. We, therefore, propose a generic methodology that combines nonlinear programming and differential algebraic systems solution for modeling and first-principles based design of heterogeneous catalytic systems.

For any given system, a comprehensive reaction network and DFT-derived thermochemical parameters of all species and transition states are taken as inputs. Subsequently, three classes of nonlinear optimization problems can be formulated and solved: (i) parameter estimation based on a simultaneous or sequential approach using activation barriers of reactions and binding energies of species as parameters [1, 2] to elucidate the detailed reaction mechanism, (ii) identification of optimal experimental conditions that maximize the coverage of or flux through an experimentally observable and mechanistically relevant surface intermediate, and (iii) determination of "Sabatier-optimal" binding energies of surface species and transition state energies of reactions that maximize turnover frequencies and/or selectivity of desired products.

The formulation of the optimization problems, selection of a relevant set of parameters, solution approaches, and multi-start schemes to identify a near-global minimum will be presented and discussed in the context of methanol synthesis from CO and CO_2 on transition metals.

References:

1. P. Rubert-Nason, M. Mavrikakis, C. T. Maravelias, L. C. Grabow, and L. T. Biegler., AIChE J 2014, 60, 1336 – 1346
2. L. T. Biegler, Nonlinear Programming: Concepts, algorithms, and applications to chemical processes. SIAM 2010

374764 Estimating Reaction Rates from Coverage Dependent Binding Energies

Abstract:

Creating accurate, compact rate models from first principles simulations remains a challenge in computational catalysis. Given a microkinetic description of all the steps in a catalytic mechanism and their sensitivity to their local surface environment, macroscopically observable rates can be rigorously estimated by suitable averaging over the microscopic rates, for instance using kinetic Monte Carlo. This approach can be further simplified by using the variation in apparent surface energetics as an estimator of microscale behavior, allowing for good fidelity using only an excess energy representation for surface species and a Brønsted–Evans–Polanyi (BEP) relationship to correlate overall reaction energies to energies of activation. In this work, density functional theory calculations are used to parameterize excess energy descriptions of adsorbates on a Pt (111) surfaces. These descriptions are used in conjunction with BEP relations to provide a steady-state description of catalytic oxidation networks. Results are compared with equivalent descriptions of the oxidation networks derived from Monte Carlo simulations and weighted distribution sampling.

Predicting Adsorption Energies for Use in Catalyst Design

One of the most successful paradigms for catalyst design is the use of adsorption energies to predict catalytic performance. Hence, understanding and predicting trends in adsorption energies can provide significant insight into catalyst design. We have developed general models for predicting adsorption energies, which can be used in rational design and high-throughput screening.

We have performed density functional theory calculations on adsorbates that bind to transition metal surfaces through various atoms, including C, O, H, N, S, and F. These calculations reveal that the dependence of the adsorption energy on the metal's d band and p band is independent of the identity of the adsorbing atom. However, the dependence on the adsorbate-surface coupling depends on the energy of the highest occupied molecular orbital (HOMO). Adsorbates with higher HOMO energies experience less Pauli repulsion. Combining these behaviors, the adsorption energy of any monovalent radical on any transition metal can be predicted using a single equation. This equation has single adsorbate-dependent fitting parameter that correlates with the HOMO energy, as well as a few universal parameters.

Because the adsorbate's behavior depends on the HOMO energy, adsorbates with similar HOMO energies correlate. Further, for adsorbates with very different HOMO energies, the coupling and d-band filling can be used to predict differences in the adsorption energy. Hence, adsorption energies of all adsorbates can be predicted from the adsorption energy of a single adsorbate.

These findings are useful for both rational design and high-throughput screening. This is particularly true because the surface electronic structure parameters can be related to the alloy structure. Understanding the factors controlling adsorption energies leads to an understanding of the factors controlling catalytic performance. Furthermore, the surface properties can be efficiently predicted, improving prospects for high-throughput screening. We apply our methods to several reactions, and experimentally test a novel catalyst for ethylene epoxidation with higher selectivity than Ag.

Metal nanoparticles find tremendous applications in nanotechnology and especially in modern catalytic technologies for energy generation and conversion. The interactions of the reacting molecules with the nanoparticle surfaces are key in the catalytic performance of the nanoparticles[1]. According to the Sabatier's principle, the metal-adsorbate interactions should be "just right". As an example, strong adsorption of one reactant over another will result to poisoning the catalyst by the preferential adsorption of the strongly bound adsorbate. On the other hand, weak adsorption results to desorption of the adsorbates at elevated temperatures and no reactions occur on the nanoparticle surface. One can fine tune the interactions of adsorbates with the nanoparticles by changing the nanoparticle morphologies[2], such us, their size, shape and metal composition. In this work[3], we use Density Functional Theory calculations to unravel the binding behavior of adsorbates (e.g. CO) as a function of nanoparticle morphologies, such as their size and shape. This work introduces novel adsorption concepts in catalysis and unravels very complex adsorption mechanisms on sub-nanoscale catalysts. At catalyst sizes of less than 1nm diameter, strong electronic effects dominate and govern i) adsorption and ii) catalyst restructuring as a function of surface coverage of the adsorbates.

1. Mpourmpakis, G.; Vlachos, D. G. *MRS Bulletin* **2011,** 36, (3), 211-215.
2. Mpourmpakis, G.; Andriotis, A. N.; Vlachos, D. G. *Nano Letters* **2010,** 10, (3), 1041-1045.
3. Austin, N.; Mpourmpakis, G. *submitted* **2014**.

375019 Uncertainty Quantification of a Computational Study of the Water-Gas Shift Reaction Catalyzed By Platinum Clusters Supported on Titanium Oxide (Pt/TiO2)

Abstract:

The water-gas shift reaction is the primary reaction for generating hydrogen for the chemical and petrochemical industries. In this paper a novel uncertainty quantification study is conducted of a computational model of the water-gas shift reaction occurring at the three-phase boundary of a Pt/TiO_2 catalyst by taking into account the inexact nature of density functional theory and constraining the results to thermodynamic experiments. This uncertainty quantification study provides informative model results in the presence of errors from DFT. The first objective of uncertainty quantification is to make reliable model results given such possible errors. A second objective is constraining density functional theory results to thermodynamic experiments. The thermodynamic constraints are accomplished by causing gas molecule corrections to sum to a constant amount. CO is given a large adjustment compared to other gas molecules due to prior knowledge about the PBE functional being unable to correctly describe the electronic structure of CO. Uncertainty in model results is represented by probability density plots and the dominant catalytic cycle is identified.

360463 Ab Initio Framework for Uncovering Structure-Property Relationships of Amorphous Catalysts

Abstract:

Approaches for modeling catalytic sites on amorphous supports lag far behind approaches for modeling catalytic sites on crystalline materials, e.g., zeolites and metal surfaces. A typical approach to model amorphous supports is to use cluster models with arbitrarily chosen constraints to represent the rigid amorphous support, but these constraints arbitrarily influence catalyst site activity. An alternative strategy is to use no constraints, but this results in catalytic sites with unrealistic flexibility. We present a systematic ab initio method to model isolated active sites on insulating amorphous supports using small cluster models. A sequential quadratic programming framework helps us relate chemical properties, such as the activation energy, to active site structure. The algorithm is first illustrated on an empirical valence bond model energy landscape. We then use the algorithm to model an off-pathway kinetic trap in olefin metathesis by isolated Mo sites on amorphous SiO_2.

390713 Computational Database of Metal-Oxide Surface Reactivities for Catalyst Design

Abstract:

We study surface reactivity of low index facets of rocksalt type Alkaline-Earth and first row transition metal oxide groups using small atoms and molecules (O, OH, CO, NO) .The computed database of adsorption and activation energies will be used to identify possible correlations with other quantities such as surface energies or electronic structure in order to establish scaling relations for future high-throughput screening efforts. A comparison will be made between DFT functionals of various levels of accuracy, e.g., GGA, GGA+U, GGA+vdW and GGA-hybrid, meta-GGA and hybrid meta-GGA, and compared to available experiments. This effort is part of the ``Predictive Theory of Transition Metal Oxide Catalysis" funded through the DOE Materials Genome Project.

381875 High Efficiency Solar Thermal Power and Integrated Chemical Storage Cycles for Continuous Grid Power Supply

Abstract:

Limited fossil fuel reserves and increasing greenhouse gas (GHG) emissions from fossil fuels make it necessary to develop alternative renewable energy sources to meet energy needs. Advancements in renewable power production are especially important since electric power is the largest consumer of primary energy resources with the highest growth rate among alternate energy use sectors, and is currently responsible for >40% of the global CO_2emissions. Among the renewable energy sources, solar energy is prominent due to its abundance. Yet intermittencies and land availability constraints for solar energy collection are the grand challenges for solar thermal power generation and demand a high efficiency solar power generation cycle that synergistically integrates energy storage.

Here, we introduce new alternative processes that produce and store solar thermal power efficiently to enable for round-the-clock power supply. Solar power production processes are evaluated based on the process *sun-to-electricity* (STE) *efficiency* that refers to the fraction of incident solar energy that is directly recovered as the net electricity output. The second metric of interest is the *overall sun-to-electricity* (OSTE) *efficiency*. It is the net STE efficiency averaged over 24 hours, i.e. accounting for energy storage and delivery of the stored energy while solar energy is not available.

The novel solar thermal power cycle has a potential to generate electricity with STE efficiencies greater than 30% at low solar heat collection temperature. The cycle also promises STE efficiencies greater than 40% for the high solar heat collection temperatures.

For the cases with higher solar heat collection temperatures, the solar water with reheat power cycle is modified by coproducing chemicals for storing energy from solar irradiation. When solar energy is not available, we propose the use of the same power cycle. This is achieved by addition of combustors that burn the stored chemicals to provide the high temperature heat. The overall sun-to-electricity efficiency of a twenty-four hour cycle is estimated to be greater than 30%. In summary, we propose a thermodynamic cycle that has the potential to generate uninterrupted electricity for grid distribution around the clock at GWh levels.

380423 Optical Design of Multi-Source High-Flux Solar Simulators

Abstract:

We present a systematic approach to design a class of high-flux solar simulators (HFSS) for solar thermal, thermochemical and materials research. The generic HFSS concept consists of an array of identical radiation modules arranged in concentric rows. Each module consists of a short-arc ion-discharge lamp coupled to a truncated ellipsoidal reflector. The positioning of the radiation modules is based on the overall rim angle of the system, the number of module rows, the number of modules in each row, the diameter of the reflector, and the required minimal spacing between the reflectors. Once the module array configuration has been fixed, the shape of the ellipsoidal reflector, i.e. its eccentricity and focal length, is determined such that it intercepts all radiation emitted by the light source. The radiative flux distribution generated by the array of radiation modules is calculated for selected flat and hemispherical control surfaces by applying the Monte Carlo ray-tracing method. The HFSS design is optimized for maximum radiative flux at the focal plane and to achieve acceptable flux distribution on the concave control surfaces behind the focal plane. For a 45 kW_e HFSS design with a 45° rim angle, consisting of 18 radiation modules arranged in two concentric rows, each consisting of a 50-cm diameter reflector and a 2.5 kW_e Xe short-arc lamp, the following characteristics at the focal plane are predicted: peak radiative flux of 11 MW/m^2; total radiative power and radiation transfer efficiency of 15.4 kW and 81%, respectively. The radiative flux, radiative power and radiation transfer efficiency for a 30-mm diameter circular flat target at the focal plane are 7.5 MW/m^2, 5.3 kW, and 28%, respectively. For a 50-mm diameter circular flat target, the radiative flux, radiative power and radiation transfer efficiency are 5.0 MW/m^2, 9.8 kW, and 52%, respectively.

384735 High Operating Temperature Transfer and Storage (HOTTS) System for Concentrated Solar Power Generation and Metals Production

Abstract:

High Operating Temperature Transfer and Storage (HOTTS) is a solar energy-based process currently under development, designed to provide high-quality heat to industrial processes. This system includes a dry, particle-based heat transfer fluid (pHTF) that can be heated to temperatures of 1,100°C, some 400°C higher than molten salt-based systems. The pHTF can store thermal energy to enable continuous operation through day and night cycles. The HOTTS system also contains Mo-Si-B composites and coatings as the material of construction (MOC), which resists oxidation and particle erosion at these elevated operating conditions.

In this study, we report the heat-transfer performance of different materials as pHTFs in a low temperature (up to 200°C) bench-scale unit. Heat-transfer coefficients are determined for mass fluxes up to 75 kg/m^2-s. We also correlate the flow rate of the particles to the outlet size of the bench-scale unit. The flowability of the pHTFs after high-temperature thermal cycling is also reported. In addition, we discuss the oxidative resistance of hot-pressed, Mo-Si-B composites at temperatures up to 1650°C in air. One Mo-Si-B composition is examined after long-term (100 h) isothermal conditions, as well as thermal cycling to simulate solar-thermal operation. Engineering approaches to implement the pHTF and Mo-Si-B MOC into a pilot-scale HOTTS process are also discussed.

Thermochemical energy storage using strontium carbonate/strontium oxide system for solar energy utilization

Thermochemical energy storage is a potentially effective method to store incident solar energy during the day and utilize the stored solar energy for generating power on-demand. However, it is imperative that the selected thermochemical energy storage scheme to be efficient and economical to have a viable path towards commercialization. Power costs are linked to power block efficiency, which is highly dependent on the temperature of the heat supplied. One way of ensuring high power block efficiencies on a 24/7 basis is to store solar thermal energy through a reversible reaction that releases the stored energy at high temperature. One such method is the decomposition-carbonation scheme involving strontium carbonate ($SrCO_3$) and strontium oxide (SrO):

$$SrCO_3 \leftrightarrow SrO + CO_2(g)$$

With this thermochemical storage scheme, porous $SrCO_3$ is subjected to concentrated solar radiation, which raises its temperature to a suitable point above 1250°C to accommodate the endothermic decomposition to SrO and carbon dioxide (CO_2), the latter of which is stored for the exothermic stage. To release this stored chemical energy, CO_2 is purged into the system to perform the exothermic carbonation of SrO. Heat generated during this exothermic stage is utilized to generate power by coupling the system with a Brayton or Rankine heat engine. Since the exothermic step can occur at high temperatures approaching 1200°C, this chemistry offers a chance to generate high quality thermal energy for high efficiency, dispatchable power generation, ultimately translating to cheaper solar-driven electricity production. The non-explosive, non-toxic nature of CO_2, the catalysis-free reversible reaction, and the lack of side reactions or products make this concept especially attractive.

Upwards of 200 kJ/mol can be stored and retrieved using this method should complete conversion be realized. If full utilization of the material is achieved, this corresponds to a volumetric energy density upwards of 2000 MJ/m^3 for a bed whose porosity is 50%.

However, when simply using $SrCO_3$ powders, this ideal situation is not completely realized due to sintering of the reactants at high temperatures, causing reduction in surface area for reactions to occur. The objective of this work is to find the most effective way to reduce sintering to maximize the reaction extent of the porous material. Investigations using TGA/DSC have been carried out to ascertain the extent to which the material can be repeatedly cycled. The effects of temperature and pressure are described. Using a combination of packed bed studies and solid state characterization (SEM/EDS) the efficacies of the various synthesis methods of porous networks as well as doping strontium carbonate with sintering inhibitors will be presented.

A Model of Transient Thermal Transport Phenomena Applied to the Carbonation and Calcination of a Sorbent Particle for Calcium Oxide Looping CO_2 Capture

Lindsey Yue and Wojciech Lipiński, The Australian National University, Canberra, ACT, Australia

Abstract

A numerical model coupling transient radiative, convective, and conductive heat transfer, mass transfer, and chemical kinetics of heterogeneous solid–gas reactions has been developed for a semi-transparent, non-uniform, and non-isothermal particle undergoing cyclic thermochemical transformations. The carbonation–calcination reaction pair for calcium oxide looping is selected as the model cycle because of its suitability for solar-driven carbon dioxide capture. The analyzed system is a single, porous particle in an idealized, reactor-like environment. The model is used to investigate two cases distinguished by the length of the carbonation and calcination periods. The carbonation–calcination process for a single particle is shown to become periodic after three cycles.

Introduction

Thermochemical cycles are found in important industrial applications, including thermal energy storage, chemical-looping combustion, and production of commodity materials. Calcium oxide looping is a thermochemical cycle of recent interest for carbon dioxide (CO_2) capture [1]. The cycle is well suited for use with concentrated solar irradiation [2]. The high-temperature capture of CO_2 via solar calcium oxide looping promises high efficiencies and capture rates [3]. Concentrated CO_2 is a commodity with applications in the pharmaceutical, medical, environmental, food and beverage, petrochemical, oil recovery, metallurgy, and manufacturing industries [4, 5], as well as in the production of synthetic transportation fuels [6].

In the first step of calcium oxide looping, CO_2 is chemically absorbed from a dilute source by a calcium oxide (CaO) sorbent to form calcium carbonate ($CaCO_3$) in the exothermic, non-solar carbonation reaction,

$$\mathrm{CaO} + \mathrm{CO_2} \rightarrow \mathrm{CaCO_3}, \quad \Delta\bar{h}^0_{298\mathrm{K}} = -178\,\mathrm{kJ\,mol^{-1}} \tag{1}$$

The calcium carbonate is then thermochemically decomposed into concentrated CO_2 and regenerated CaO sorbent in the endothermic, solar-driven calcination reaction,

$$\mathrm{CaCO_3} \rightarrow \mathrm{CaO} + \mathrm{CO_2}, \quad \Delta\bar{h}^0_{298\mathrm{K}} = 178\,\mathrm{kJ\,mol^{-1}} \tag{2}$$

In systems for calcium oxide looping, intra-particle heat transfer, mass transfer, and chemical kinetics are all important considerations. The intra-particle phenomena were studied using a numerical model of coupled heat and mass transfer to calcination kinetics in [7]. In the present work, the chemical kinetics model and consequently, the transient thermal transport model are extended to the full carbonation–calcination cycle. This research aims at understanding the role of transport processes in the complete, two-step calcium oxide looping cycle for CO_2 capture driven by concentrated solar irradiation. A solar reactor for the implementation of the cycle (1)–(2) is currently under development [8].

Numerical Model

The model system is a single, semi-transparent, porous particle with non-uniform composition subjected to high-flux solar irradiation in an idealized, reactor-like environment. The system consists of two solid-phase species, $CaCO_3$ and CaO, and two fluid-phase species, CO_2 and air. The numerical model is based on five volume-averaged, governing conservation equations: four conservation of mass equations for the four species of the system and one conservation of energy equation.

Conservation of mass—Solid phase

The volume-averaged conservation of mass equations for $CaCO_3$ and CaO read, respectively:

$$\frac{\partial\left[(1-\phi)\left\langle\bar{\rho}_{\mathrm{CaCO_3}}\right\rangle^{\mathrm{s}}\right]}{\partial t}=\left\langle\bar{r}'''\right\rangle \tag{3}$$

$$\frac{\partial\left[(1-\phi)\left\langle\bar{\rho}_{\mathrm{CaO}}\right\rangle^{\mathrm{s}}\right]}{\partial t}=-\left\langle\bar{r}'''\right\rangle \tag{4}$$

where the reaction rate term on the right-hand side is defined as:

$$\bar{r}''' \stackrel{\mathrm{def}}{=} -\frac{\mathrm{d}N_{\mathrm{CaCO_3}}}{\mathrm{d}V\mathrm{d}t}=\frac{\mathrm{d}N_{\mathrm{CaO}}}{\mathrm{d}V\mathrm{d}t}=\frac{\mathrm{d}N_{\mathrm{CO_2}}}{\mathrm{d}V\mathrm{d}t} \tag{5}$$

$N_{\mathrm{CaCO_3}}$, N_{CaO}, and $N_{\mathrm{CO_2}}$ are the amounts of substance of $CaCO_3$, CaO, and CO_2, respectively.

The volumetric reaction model is used for both carbonation and calcination. The reaction rate expression is formulated in terms of Arrhenius-like rate constants,

$$\bar{r}'''=k_{\mathrm{carb}}s'''_{\mathrm{carb}}f_{\mathrm{CO_2,carb}}-k_{\mathrm{calc}}s'''_{\mathrm{calc}}f_{\mathrm{CO_2,calc}} \tag{6}$$

$$k_i=k_{0,i}\exp\left(-E_{\mathrm{a,i}}/\bar{R}T\right),\ \text{for } i=\text{carb, calc} \tag{7}$$

where k_i is the Arrhenius-like rate constant, s''' is the local, volume-specific reaction surface area, and $f_{\mathrm{CO_2}}$ is a function describing the CO_2 partial pressure dependence. The kinetics for carbonation of fully calcined Strassburg limestone [9] and calcination of Blanca limestone [10] are selected for reactions (1) and (2), respectively.

Carbonation kinetics

The volume-specific reaction surface area for carbonation is given by:

$$s'''_{\mathrm{carb}}=s'''_{\mathrm{CaO}}X_{\mathrm{local}}^{2/3} \tag{8}$$

where s'''_{CaO} is the volume-specific surface area of 56%-porous CaO and X_{local} is the local extent of the calcination reaction, $X_{\mathrm{local}}=1-N_{\mathrm{CaCO_3}}/N_{0,\mathrm{CaCO_3}}$. The CO_2 partial pressure dependence for carbonation is:

$$f_{\mathrm{CO_2,carb}}=\begin{cases} p_{\mathrm{CO_2}}-p_{\mathrm{eq}} & \text{for } p_{\mathrm{CO_2}}>p_{\mathrm{eq}} \\ 0 & \end{cases} \tag{9}$$

where $p_{\mathrm{CO_2}}$ is the partial pressure of CO_2 and p_{eq} is the equilibrium partial pressure of CO_2,

$$p_{\mathrm{eq}}=A_{\mathrm{eq}}\exp\left(-\frac{B_{\mathrm{eq}}}{T}\right) \tag{10}$$

Values for parameters used for the carbonation rate constant and in Eqs. 8 and 10 are given in Table 1. The carbonation reaction rate normalized by reaction surface area for selected CO_2 concentrations is shown in Fig. 1a.

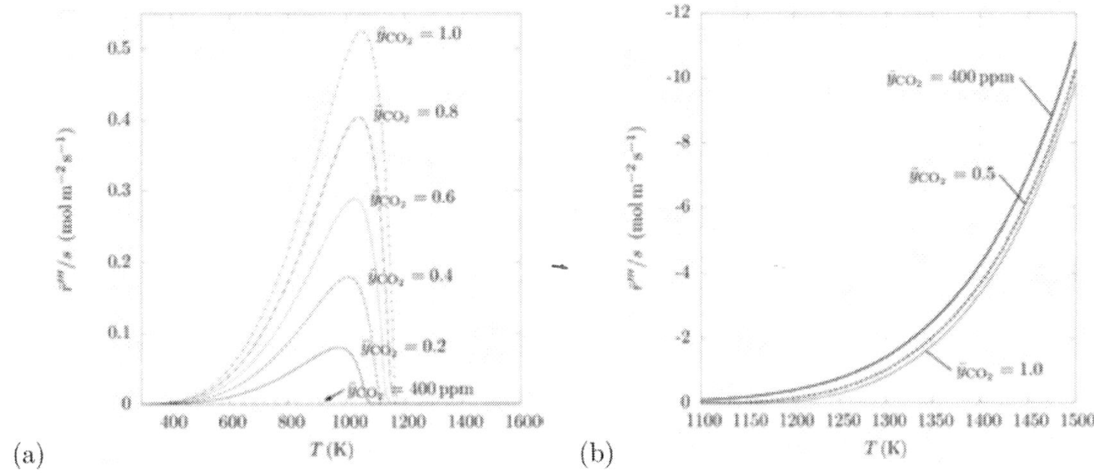

Figure 1: Reaction rate versus temperature for selected CO_2 concentrations for (a) carbonation and (b) calcination

Calcination kinetics

The volume-specific reaction surface area for calcination is given by:

$$s'''_{\text{calc}} = \frac{[s_{\text{CaCO}_3}(1 - X_{\text{local}}) + s_{\text{CaO}} X_{\text{local}}] \times (1 - \phi)}{\dfrac{X_{\text{local}}}{\rho_{\text{CaO}}} + \dfrac{1 - X_{\text{local}}}{\rho_{\text{CaCO}_3}}} \tag{11}$$

where s_{CaCO_3} is the mass-specific surface area of unreacted, 3%-porous $CaCO_3$ and s_{CaO} is the mass-specific surface area of 56%-porous CaO. The Langmuir–Hinshelwood mechanism model with the Freundlich isotherm presented by García-Labiano et al. [10] is used to predict the dependence of calcination on CO_2 partial pressure:

$$f_{\text{CO}_2,\text{calc}} = \begin{cases} (1-\theta)\left(1 - \dfrac{p_{\text{CO}_2}}{p_{\text{eq}}}\right), & \text{for } 0 < \theta < 1 \text{ and } p_{\text{CO}_2} < p_{\text{eq}} \\ 0 \end{cases} \tag{12}$$

Table 1: Reaction rate parameters

Variable	Value	Unit	Variable	Value	Unit
$k_{0,\text{carb}}$ [a]	1.67×10^{-4}	mol m^{-2} s^{-1} kPa	s'''_{CaO} [b]	2.8006×10^{7}	m^2 m^{-3}
$k_{0,\text{calc}}$ [b]	6.7×10^{6}	mol m^{-2} s^{-1}	s_{CaO} [b]	1.9×10^{4}	m^2 kg^{-1}
$E_{\text{a,carb}}$ [a]	2.9×10^{4}	J mol^{-1}	s_{CaCO_3} [b]	300	m^2 kg^{-1}
$E_{\text{a,calc}}$ [b]	1.66×10^{5}	J mol^{-1}	ρ_{CaO}	3350	kg m^{-3}
c_0 [b]	1.8×10^{-7}	Pa$^{-1/2}$	ρ_{CaCO_3}	2730	kg m^{-3}
E_{c} [b]	-9.3×10^{4}	J mol^{-1}	A_{eq} [c]	4.137×10^{12}	Pa
			B_{eq} [c]	2.0474×10^{4}	K

[a] [9]
[b] [10]
[c] [11]

$$\theta = cp_{\mathrm{CO_2}}^{1/2} \tag{13}$$

$$c = c_0 \exp\left(-E_c/\bar{R}T\right) \tag{14}$$

where θ is the fraction of active sites occupied by CO_2 and c is the adsorption constant. Values of parameters used for the calcination rate constant and in Eqs. 11 and 14 are given in Table 1. The calcination reaction rate normalized by reaction surface area for selected CO_2 concentrations is shown in Fig. 1b.

Conservation of mass—Fluid phase

For the fluid phase, the volume-averaged conservation of mass equations for CO_2 and air read, respectively:

$$\frac{\partial\left(\phi\langle\bar{\rho}_{\mathrm{CO_2}}\rangle^{\mathrm{f}}\right)}{\partial t} + \nabla\cdot\left(\langle\bar{\rho}_{\mathrm{CO_2}}\rangle^{\mathrm{f}}\langle\vec{\mathbf{u}}_{\mathrm{f}}\rangle\right) = \nabla\cdot\left(D_{\mathrm{CO_2,eff}}\nabla\langle\bar{\rho}_{\mathrm{CO_2}}\rangle^{\mathrm{f}}\right) + j''_{\mathrm{CO_2}}s'''_{\mathrm{fs}} \tag{15}$$

$$\frac{\partial\left(\phi\langle\bar{\rho}_{\mathrm{air}}\rangle^{\mathrm{f}}\right)}{\partial t} + \nabla\cdot\left(\langle\bar{\rho}_{\mathrm{air}}\rangle^{\mathrm{f}}\langle\vec{\mathbf{u}}_{\mathrm{f}}\rangle\right) = \nabla\cdot\left(D_{\mathrm{air,eff}}\nabla\langle\bar{\rho}_{\mathrm{air}}\rangle^{\mathrm{f}}\right) \tag{16}$$

where the molar concentrations of CO_2 and air change due to bulk advection, species diffusion, and for CO_2, chemical reaction and transport across the phase boundary. Air is modeled as a single species composed of 79% nitrogen and 21% oxygen. Closure equations for bulk advection and diffusion are presented in [7].

Conservation of energy

The solid and fluid phases are assumed to be at local thermal equilibrium, yielding one conservation of energy equation for the system,

$$\begin{aligned}\frac{\partial\left(\langle\bar{\rho}\rangle\langle\bar{h}\rangle\right)}{\partial t} &+ \nabla\cdot\left(\langle\bar{\rho}_{\mathrm{f}}\rangle^{\mathrm{f}}\langle\bar{h}_{\mathrm{f}}\rangle^{\mathrm{f}}\langle\vec{\mathbf{u}}_{\mathrm{f}}\rangle\right) = \nabla\cdot\left(\langle\bar{h}_{\mathrm{CO_2}}\rangle^{\mathrm{f}}D_{\mathrm{CO_2,eff}}\nabla\langle\bar{\rho}_{\mathrm{CO_2}}\rangle^{\mathrm{f}}\right)\\ &+ \nabla\cdot\left(\langle\bar{h}_{\mathrm{air}}\rangle^{\mathrm{f}}D_{\mathrm{air,eff}}\nabla\langle\bar{\rho}_{\mathrm{air}}\rangle^{\mathrm{f}}\right) + \nabla\cdot\left(k_{\mathrm{eff}}\nabla\langle T\rangle\right) - \langle\nabla\cdot\vec{\mathbf{q}}''_{\mathrm{rad}}\rangle\end{aligned} \tag{17}$$

where latent heat changes due to bulk fluid advection, fluid species diffusion, conduction, and radiative heat transfer within the particle. Closure equations and properties used for Eq. 17 are given in [7].

Boundary conditions

Time-periodic boundary conditions are applied to model thermochemical cycling. The particle is initially 100% $CaCO_3$, and the cycle starts with a calcination period followed by a carbonation period. One calcination period t_{calc} and one carbonation period t_{carb} comprise one cycle $t_{\mathrm{cycle}} = t_{\mathrm{calc}} + t_{\mathrm{carb}}$. A cycle j begins at time instant t_j and the carbonation period for cycle j begins at time instant $t_{j,\mathrm{calc}} = t_j + t_{\mathrm{calc}}$.

During the calcination period, high-flux solar irradiation drives particle heating and the endothermic calcination reaction. During the carbonation period, the particle cools by reradiation and convection to the carbonation temperature.

$$q''_{\text{surf}} = \begin{cases} q''_{\text{solar}}, & \text{for } 0 < t - t_j \leq t_{\text{calc}} \\ 0, & \text{for } t_{\text{calc}} < t - t_j \leq t_{\text{cycle}} \end{cases} \tag{18}$$

where q''_{surf} is the flux at the particle surface and q''_{solar} is the magnitude of the concentrated solar irradiation.

CO_2 is absorbed from a dilute source and released in a concentrated stream. Therefore for carbonation, the ambient fluid is modeled as a mixture of air and CO_2 with the CO_2 concentration representative of coal-fired flue gas. For calcination, the ambient fluid is modeled as a stream of commodity-grade concentrated CO_2 with non-CO_2 impurities modeled as air,

$$\bar{y}_{\infty,\text{CO}_2} = \begin{cases} \bar{y}_{\text{CO}_2,\text{calc}}, & \text{for } 0 < t - t_j \leq t_{\text{calc}} \\ \bar{y}_{\text{CO}_2,\text{carb}}, & \text{for } t_{\text{calc}} < t - t_j \leq t_{\text{cycle}} \end{cases} \tag{19}$$

where $\bar{y}_{\infty,\text{CO}_2}$ is the ambient CO_2 concentration, $\bar{y}_{\text{CO}_2,\text{calc}}$ is the calcination CO_2 concentration, and $\bar{y}_{\text{CO}_2,\text{carb}}$ is the carbonation CO_2 concentration. Transition between periods for both conditions occurs as a step change. Values of the parameters used in Eqs. 18 and 19 are given in Table 2.

Table 2: Selected boundary and initial conditions

Variable	Value	Unit	Variable	Value	Unit
q''_{solar}	200	kW m^{-2}	T_0	800	K
$\bar{y}_{\text{CO}_2,\text{calc}}$	0.99		T_{w}	800	K
$\bar{y}_{\text{CO}_2,\text{carb}}$	0.15		T_∞	800	K

The ambient CO_2 concentration interacts with the system through the surface boundary conditions for Eqs. 15 and 16, and the surface radiative flux interacts with the system via the surface boundary condition for Eq. 17. With the exception of the conditions listed in Table 2, all other boundary and initial conditions are taken from [7].

The numerical model does not take into account particle degradation and loss of sorbent activity due to cycling or irreversible morphological changes such as sintering that occur at higher temperatures. These phenomena have been experimentally shown to occur in particles undergoing cyclic calcination and carbonation [12], and should be considered in the realization of solar calcium oxide looping CO_2 capture.

Results

The model is used to investigate two cases: (1) with period lengths resulting in full calcination of the particle referred to as the full-calcination case, and (2) with period lengths resulting in full carbonation of the particle referred to as the full-carbonation case. The period lengths for both cases have been identified in preliminary calculations using the model presented in this study. For the full-calcination case, $t_{\text{calc}} = 300\,\text{s}$ and $t_{\text{carb}} = 800\,\text{s}$. For the full-carbonation case, $t_{\text{calc}} = 30\,\text{s}$ and $t_{\text{carb}} = 190\,\text{s}$. Six cycles were simulated for both cases. After two cycles, periodicity is achieved, and the third and subsequent cycles for both cases are identical. All other physical and numerical simulation parameters are taken from [7] with the exception of Δt and physical parameters discussed in the preceding sections. In the cases presented here, $\Delta t = 2 \times 10^{-6}\,\text{s}$.

Full-calcination case

Overall reaction extent versus time is shown in Fig. 2a. Initially, $X_{\text{overall}} = 0$, and the cycle starts with a calcination period. The particle heats to the calcination temperature and begins decomposing into CaO and CO_2 at $t = 6.6\,\text{s}$ for the first cycle and $t = t_j + 4.4\,\text{s}$ in subsequent cycles. Calcination continues until all $CaCO_3$ is consumed and $X_{\text{overall}} = 1$ at $t = 253\,\text{s}$ for the first cycle and $t = t_j + 92.4\,\text{s}$ in subsequent cycles. When the carbonation period begins, the particle cools until the carbonation temperature is reached and the particle begins to carbonate at $t = 303.6\,\text{s}$ for the first cycle and $t = t_{j,\text{calc}} + 3.6\,\text{s}$ in subsequent cycles. In the full-calcination case, the particle does not completely carbonate during the first or subsequent carbonation periods, reaching a minimum reaction extent $X_{\text{overall}} = 0.407$.

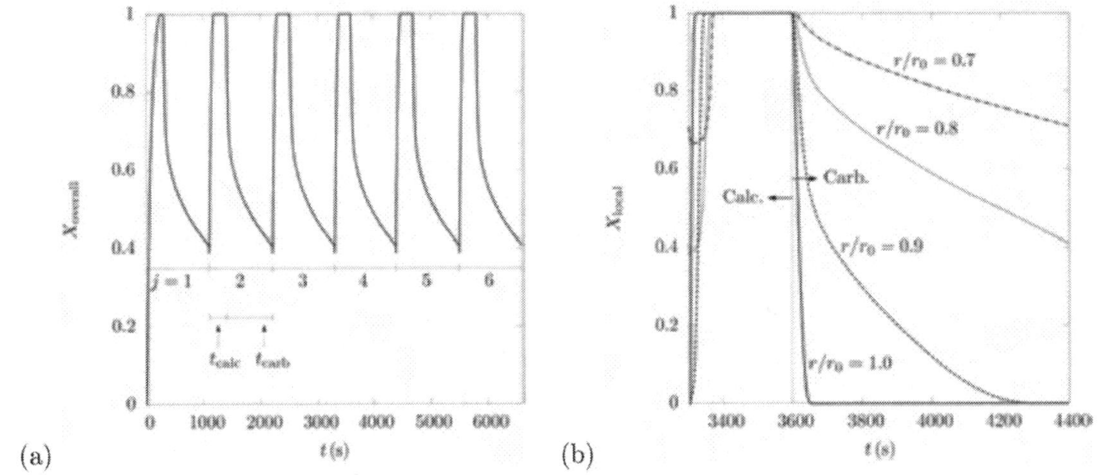

Figure 2: Reaction extent versus time for the full-calcination case for (a) the entire particle undergoing six cycles and (b) selected locations for cycle $j = 4$

During the carbonation period, two regimes are observed. The change in particle composition in the first regime, $t_{j,\text{calc}} < t - t_j < t_{j,\text{calc}} + 65\,\text{s}$, is rapid, while the change in particle composition in the second regime, $t_{j,\text{calc}} + 65\,\text{s} < t - t_j < t_{j+1}$ is slower. The first regime is significantly shorter but accounts for approximately 2/3 of the change in X_{overall}. The second regime comprises the majority of the time of the carbonation period, but only accounts for 1/3 of change in X_{overall}.

When the calcination period starts at the beginning of each cycle, the particle briefly continues to carbonate before calcination begins. This brief amount of carbonation can been seen as a sharp dip in the overall reaction extent at the beginning of each cycle in Fig. 2a. Changing CO_2 concentration in the particle due to the changing ambient CO_2 concentration elevates the transition temperature between the calcination and carbonation reactions. The changing CO_2 concentration and elevated tranition temperature prolong carbonation. For example, at the location $r/r_0 = 0.9$, $T = 810.1\,\text{K}$ and $\bar{y}_{CO_2} = 0.043$ at the beginning on the cycle, with a transition temperature of 990.0 K. The transition temperature for the location $r/r_0 = 0.9$ is reached after 20 seconds, when $T = 1172.4\,\text{K}$ and $\bar{y}_{CO_2} = 0.991$.

Local reaction extent for four selected locations within the particle is shown in Fig. 2b for $j = 4$. At the particle exterior, locations $r/r_0 = 1.0$ and $r/r_0 = 0.9$ begin the cycle with pure $CaCO_3$ ($X_{\text{local}} = 0$) at $t = 3300\,\text{s}$ and then completely calcine to CaO ($X_{\text{local}} = 1$) at $t = 3335.2\,\text{s}$ and

$t = 3361.6$ s, respectively. Locations $r/r_0 = 0.8$ and $r/r_0 = 0.7$ begin the cycle at $t = 3300$ s when $X_{\text{local}} = 0.410$ and $X_{\text{local}} = 0.709$, respectively and completely calcine to CaO at $t = 3377.0$ s and $t = 3383.6$ s, respectively. At the onset of each cycle, these interior locations experience periods of brief carbonation before calcination begins.

When the carbonation period begins at $t = 3600$ s, all locations returns to their initial states, completely carbonated for $r/r_0 = 0.9$ and $r/r_0 = 1.0$ and partially carbonated for $r/r_0 = 0.7$ and $r/r_0 = 0.8$. The $r/r_0 = 1.0$ location only exhibits a rapid carbonation regime, but all other locations shown in Fig. 2b exhibit both rapid and slower regimes, the slower regimes becoming longer with increasing distance away from the particle surface.

Full-carbonation case

Overall reaction extent versus time is shown in Fig. 3a. The calcination periods end at $t = t_j + 30$ s, corresponding to $X_{\text{overall}} = 0.302$. The following carbonation proceeds until the particle is completely carbonated, $X_{\text{overall}} = 0$ at $t = t_{j+1}$. During the carbonation period, rapid and slow regimes are observed. The fast regime, $t_{j,\text{calc}} < t - t_j < t_{j,\text{calc}} + 30$ s accounts for a larger fraction of the carbonation period time by approximately 50% and a larger fraction of the change in X_{overall} by approximately 20% as compared to the full-calcination case.

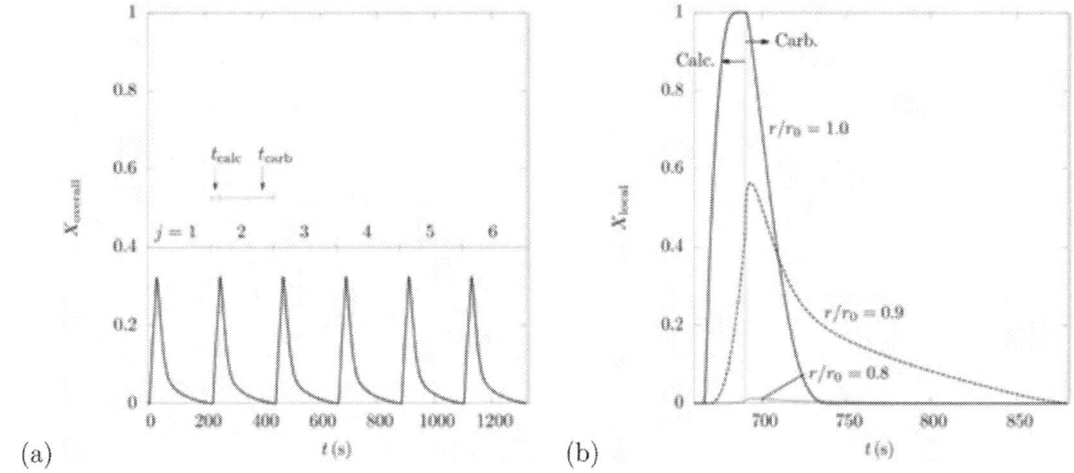

Figure 3: Reaction extent versus time for full-carbonation case for (a) the entire particle undergoing six cycles and (b) selected locations for cycle $j = 4$

Local reaction extent for three selected locations in the particle for cycle $j = 4$ is shown in Fig. 3b. At the start of the calcination period at $t = 660$ s, all locations are nearly $X_{\text{local}} = 0$. When the calcination period begins, the particle surface $r/r_0 = 1.0$ is the first location to reach the transition temperature and start calcining at $t = 666.2$ s. When the carbonation period starts, the $r/r_0 = 1.0$ location immediately begins carbonating and continues to carbonate at a nearly constant rate until full carbonation is reached at $t = 742.3$ s. Locations near the particle surface begin calcining after the particle surface and only reach a partially calcined state. Locations $r/r_0 = 0.9$ and $r/r_0 = 0.8$ begin calcining at $t = 670.1$ s and $t = 672.8$ s, respectively. At the end of the calcination period $t = 690$ s, locations $r/r_0 = 0.9$ and $r/r_0 = 0.8$ calcine to $X_{\text{local}} = 0.421$ and $X_{\text{local}} = 0.004$, respectively.

At the onset of the carbonation period at $t = 690\,\mathrm{s}$, locations where $X_{\mathrm{local}} < 1$ briefly experience calcination before carbonation starts. Residual latent heat and the decrease in the transition temperature due to decreasing CO_2 concentration extend calcination into the carbonation period until the temperature decreases below the transition temperature, and carbonation begins. Interestingly, the location $r/r_0 = 0.9$ experiences the largest increase in local reaction extent, by 34%, when calcination occurs during the carbonation period, resulting in a local reaction extent maximum of $X_{\mathrm{local}} = 0.563$.

Case comparison

The two cases are compared in terms of the relative amount of substance and mass of CO_2 captured, the energy requirement for regeneration of the carbonated sorbent relative to the incident energy, and the amount of substance and mass of CO_2 captured relative to the incident energy for each cycle:

$$\bar{\xi}_{\mathrm{c}} = \frac{N_{\mathrm{CO_2,cycle}}}{N_{0,\mathrm{p}}} \tag{20}$$

$$\xi_{\mathrm{c}} = \frac{m_{\mathrm{CO_2,cycle}}}{m_{0,\mathrm{p}}} \tag{21}$$

$$\eta_{\mathrm{e}} = \frac{N_{\mathrm{CO_2,cycle}} \Delta \bar{h}^0_{298\,\mathrm{K}}}{Q_{\mathrm{surf,cycle}}}, \quad Q_{\mathrm{surf,cycle}} = \int_{t_j}^{t_{j+1}} \int_A q''_{\mathrm{surf}}\, \mathrm{d}A\, \mathrm{d}t \tag{22}$$

$$\bar{\chi} = \frac{N_{\mathrm{CO_2,cycle}}}{Q_{\mathrm{surf,cycle}}} \tag{23}$$

$$\chi = \frac{m_{\mathrm{CO_2,cycle}}}{Q_{\mathrm{surf,cycle}}} \tag{24}$$

where $N_{\mathrm{CO_2,cycle}}$ and $m_{\mathrm{CO_2,cycle}}$ are the amount of substance and mass of CO_2, respectively, absorbed during the carbonation period only.

For the full-calcination case for the representative cycle $j = 4$, $\bar{\xi}_{\mathrm{c}} = 0.5926$, $\xi_{\mathrm{c}} = 0.2606$, $\eta_{\mathrm{e}} = 0.0388$, $\bar{\chi} = 0.2178\,\mu\mathrm{mol\,J^{-1}}$, and $\chi = 0.0096\,\mu\mathrm{g\,J^{-1}}$. For the full-carbonation case for the representative cycle $j = 4$, $\bar{\xi}_{\mathrm{c}} = 0.3089$, $\xi_{\mathrm{c}} = 0.1358$, $\eta_{\mathrm{e}} = 0.2020$, $\bar{\chi} = 1.135\,\mu\mathrm{mol\,J^{-1}}$, and $\chi = 0.0500\,\mu\mathrm{g\,J^{-1}}$. The full-calcination case has higher capture capacities than the full-carbonation case, because particle is used more extensively to capture CO_2 in the full-calcination case. However, the full-carbonation case has higher energetic efficiency than the full-calcination case. A longer calcination cycle is required to fully calcine the particle, resulting in a higher solar energy input. For the energetic efficiency of the full-calcination case, the increase in amount of CO_2 captured does not out weigh the higher required solar energy input, resulting in the observed higher amounts of CO_2 captured per unit solar energy input for the full-carbonation case by a factor of four.

Summary and Conclusions

A numerical model coupling chemical kinetics, mass transfer, and heat transfer with time-periodic boundary conditions has been developed for a heterogeneous reacting particle. The thermochemical calcium oxide looping system was selected as the model reaction system because of its application in solar CO_2 capture. The model predicts the interactions of the coupled transient physical phenomena at the intra-particle level.

The model is used to investigate two cases: one in which full calcination is achieved and one in which full carbonation is achieved. Both systems reach periodicity after three cycles. In the full-calcination case, the maximum extent of calcination achieved at the end of the calcination period is $X_{\text{overall}} = 1$, and the maximum extent of carbonation achieved at the end of the carbonation period $X_{\text{overall}} = 0.407$. In the full-carbonation case, the maximum extent of calcination achieved at the end of the calcination period is $X_{\text{overall}} = 0.302$, and the maximum extent of carbonation achieved at the end of the carbonation period is $X_{\text{overall}} = 0$. In the full-calcination case, the core remains porous CaO during cycling, while in the full-carbonation case, the core of the particle remains dense $CaCO_3$ during cycling.

Carbonation and calcination period times were selected by trial and error to fulfill the target case conditions. Further work will include carbonation and calcination period length optimization pertinent to sorbent particle size selection and solar reactor design optimization. Choice of period lengths is anticipated to be influenced by the reactor size and operation.

Notation

A_{eq} = equilibrium partial pressure pre-exponential factor (Pa)
B_{eq} = equilibrium partial pressure constant (K)
c = adsorption constant ($\text{Pa}^{-1/n}$)
c_0 = adsorption constant pre-exponential factor ($\text{Pa}^{-1/n}$)
D = diffusivity ($\text{m}^2\ \text{s}^{-1}$)
E_{a} = activation energy (J mol^{-1})
E_{c} = adsorption constant energy (J mol^{-1})
f_{CO_2} = CO_2 partial pressure dependence
$\bar{h}$ = molar-specific enthaply (J mol^{-1})
$\Delta\bar{h}^0_{298\,\text{K}}$ = molar enthalpy of reaction (J mol^{-1})
j = cycle index
j''_{CO_2} = interphase molar flux ($\text{mol m}^{-2}\ \text{s}^{-1}$)
k = reaction rate constant ($\text{mol m}^{-2}\ \text{s}^{-1}$)
k_{eff} = effective thermal conductivity ($\text{W m}^{-1}\ \text{K}^{-1}$)
k_0 = pre-exponential factor ($\text{mol m}^{-2}\ \text{s}^{-1}$)
N = amount of substance (mol)
p = pressure (Pa)
q'' = radiative heat flux (W m^{-2})
r = radius or radial direction (m)
$\bar{R}$ = universal gas constant, 8.314 ($\text{J mol}^{-1}\ \text{K}^{-1}$)
$\bar{r}'''$ = volumetric reaction rate ($\text{mol m}^{-3}\ \text{s}^{-1}$)
s = mass-specific surface area ($\text{m}^2\ \text{kg}^{-1}$)
s''' = volume-specific surface area (m^{-1})
t = time (s)
T = temperature (K)
u = velocity (m s^{-1})
V = volume (m^3)
X = reaction extent
$\bar{y}$ = molar concentration

Greek Symbols

α_{eff} = effective absorptivity
ε_{eff} = effective emissivity
η_{e} = energetic efficiency
θ = fraction of occupied active sites
ξ_{c} = mass capture capacity
$\bar{\xi}_{\text{c}}$ = molar capture capacity
ρ = density (kg m^{-3})
$\bar{\rho}$ = molar density (mol m^{-3})
ϕ = porosity
χ = mass per energy input ($\mu\text{g J}^{-1}$)
$\bar{\chi}$ = amount of substance per energy input ($\mu\text{mol J}^{-1}$)

Subscripts

calc = calcination
carb = carbonation
eff = effective
eq = equilibrium
f = fluid
fs = fluid–solid
rad = radiative

w = wall
0 = initial
∞ = ambient

Superscripts
f = fluid phase
s = solid phase

Literature Cited

1. Reich, L., Yue, L., Bader, R., and Lipiński, W. (2014), "Towards solar thermochemical carbon dioxide capture via calcium oxide looping: A review," *Aerosol Air Quality Research*, 14:500–514.
2. Stanmore, B. and Gilot, P. (2005), "Review—calcination and carbonation of limestone during thermal cycling for CO_2 capture," *Fuel Processing Technology*, 86(16):1707–1743.
3. Nikulshina, V., Gálvez, M., and Steinfeld, S. (2007), "Kinetic analysis of the carbonation reactions for the capture of CO_2 from air via the $Ca(OH)_2$–$CaCO_3$–CaO solar thermochemical cycle," *Chemical Engineering Journal*, 129(1–3):75–83.
4. Stalkup, F.I. (1978), "Carbon dioxide miscible flooding: Past, present, and outlook for the future," *Journal of Chemical and Petrochemical Technology*, 30:1102–1112.
5. Pierantozzi, R. (2003), "Carbon Dioxide," in *Kirk-Othmer Encyclopedia of Chemical Technology*, John Wiley and Sons, Inc.
6. Kim, J., Miller, J., Maravelias, C., and Stechel, E. (2013), "Comparative analysis of environmental impact of S2P (Sunshine to Petrol) system for transportation fuel production," *Applied Energy*, 111:1089–1098.
7. Yue, L. and Lipiński, W. (2014), "A numerical model of transient thermal transport phenomena in a high-temperature solid–gas reacting system for CO_2 capture applications," in *Proceedings of the 15th International Heat Transfer Conference*, IHTC15-9014.
8. Reich, L., Bader, R., Simon, T., and Lipiński, W. (2014), "Heat and mass transfer model of a packed-bed reactor for solar thermochemical CO_2 capture," in *Proceedings of the 15th International Heat Transfer Conference*, IHTC15-8893.
9. Sun, P., Grace, J., Lim, C., and Anthony, E. (2008), "Determination of intrinsic rate constants of the CaO-CO_2 reaction," *Chemical Engineering Science*, 63:47–56.
10. García-Labiano, F., Abad, A., de Diego, L., Gayán, P., and Adánez, J. (2002), "Calcination of calcium-based sorbents at pressure in a broad range of CO_2 concentrations," *Chemical Engineering Science*, 57(13):2381–2393.
11. Barin, I. (1989), *Thermochemical Data of Pure Substances*, Weinheim: VCH.
12. Manovic, V. and Anthony, E. (2010), "Sintering and formation of a nonporous carbonate shell at the surface of CaO-based sorbent particles during CO_2-capture cycles," *Energy Fuels*, 24:5790–5796.

374713 Computational Screening of Oxygen Vacancy Formation for Solar-Driven Isothermal CO2 and H2O Splitting with Metal Oxide Redox Materials

Abstract:

The isothermal splitting of CO_2 and H_2O with concentrated sunlight offers the prospect of an energy-efficient synthesis of renewable fuels since this approach circumvents large temperature swings between the reduction and oxidation steps as well as mechanical stress arising from rapid temperature changes. In analogy to the Sabatier principle in heterogeneous catalysis, this talk presents a volcano-like correlation of the thermodynamic redox equilibrium energetics of 29 solid bulk metal oxides that can be employed for the rational design of metal oxides redox materials. Since the underpinning thermochemical data is only available for a limited number of complex metal oxides, electronic structure calculations are employed to compute the formation of oxygen vacancies at the surfaces of TiO_2, Ti_2O_3, Cu_2O, ZnO, ZrO_2, YSZ, MoO_3, Ag_2O, CeO_2 and three perovskites. We find that the free energy of formation of the bulk metal oxides scales with the free energy of the oxygen vacancy formation at the surface which can be employed to estimate unknown free energies of formation for complex metal oxide solutions. The tendency of surface oxygen vacancies to migrate into the bulk is discussed based on density functional theory computations of sub-surface oxygen vacancies and the obtained trend is compared to the experimental oxygen self-diffusion constants.

382884 Multi-Layer Monolith Catalysts for Lean NOx Reduction

Abstract:

The challenge of achieving high NOx conversion and N_2 selectivity and minimal release of unreacted NOx and NH_3 in lean burn vehicle emission aftertreatment calls for structured catalysts with multiple functions. We describe experimental and modeling studies of three different multi-layer applications. The first application combines Fe- and Cu-exchanged zeolitic layers to expand the high NOx conversion temperature range. The second application combines periodic NOx storage and reduction with in situ NH_3 generation and selective catalytic reduction. The third application combines a Pt/Al_2O_3 bottom layer and a zeolitic toplayer to oxidize NH_3 with high selectivity to N_2.

A discussion on the fundamental aspects of Urea SCR control for Transient (Automotive) Applications.

Annual AIChE meeting Atlanta GA, Nov. 17, 2014

Session: Development and Applications of Automotive Catalysts

Devesh Upadhyay

Michiel Van Nieuwstadt

Acknowledgements:
Emil Serban
Giovanni Cavataio

Research

Diesel Controls and Diagnostics
Powertrain Controls R&A

Overview

1. Evolution timeline for SCR hardware and control
2. Typical aftertreatment system configuration/s for automotive applications.
3. Storage dynamics and its implications
4. Control
   - Objectives
   - Constraints
   - Architecture
   - Adaptation
   - Stability
5. Storage Observers
6. NH_3 NO_x Signal Separation, need and method.
7. New Directions in Control
8. Diagnostics
   - Threshold selection
   - SCR Aging
9. Conclusions/Outlook

Research

Diesel Controls and Diagnostics
Powertrain Controls R&A

Urea SCR Aftertreatment System time line with Control implications

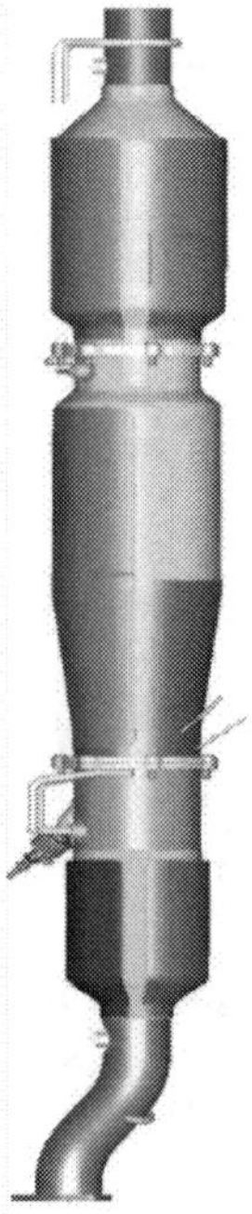

2000-2004: 1st Generation Urea SCR control

- FF control only → FG NO_x following. NH_3 slip not a performance criterion.
- Urea injection with air assist. No spray mixers.
- Cu, F_e Zeolites →US, T_iO_2 /V_2O_5 →EU
- DOE sponsored research at Ford.

2004 -- 2006: 2nd Generation Urea SCR Control

- FF + FBK → FG (model/sensor) and TP NO_x sensor
- Embedded SCR models developed and in use.
- Adaptive control laws considered / in use.
- Direct Urea Injection → Air assist less prevalent.
- Mixers → high γ values >90%
- Urea management for freeze and thaw.
- SCR OBD on the horizon for 2010.
- ZSM-5 in active use, SSZ-13 being assessed.

2006 -- 2010: 3rd Generation Urea SCR Control

- Focus on output signal: NH_3, NO_x signal separation.
- NH_3 slip management is of interest, sensors available for use.
- AMOX catalysts and Hydrolysis catalysts.
- Predictive Control techniques: IMC, GPC explored.
- Complexity of storage mechanisms becoming clear, number and role of various storage sites ? Brøensted, Lewis, strong , weak ....
- SCR OBD is in use → age vs system drift confounding.
- DEF warning standardized
- SSZ-13 zeolites (Cu-CHA) dominant, SAPO studied.

2010 – 2014+ : 4th Generation Urea SCR Control

- Transition to Multi Slice storage models (n x 0D).
- Multi Brick configurations under consideration.
- Investigations into active control via kinetic manipulation.
- Urea Quality sensors available.

Research

Diesel Controls and Diagnostics
Powertrain Controls R&A

Aftertreatment system configuration/s for automotive applications

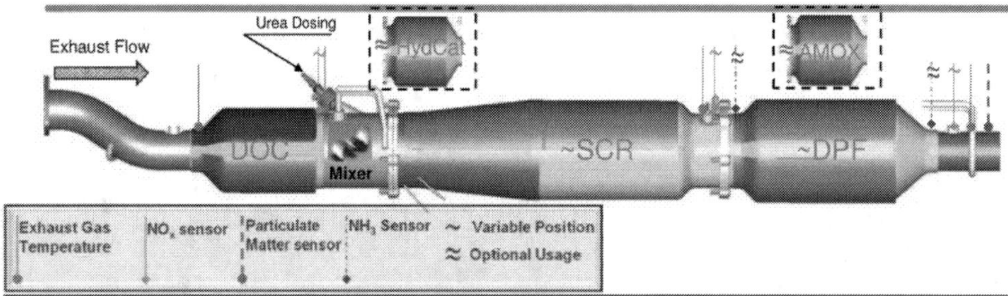

- Most applications use a NO_x sensor in the Feed Gas (FG) location, however, a FG NO_x model has been tried (Ford) and shown to work for control purposes. OBD placer harder constraints on accuracy.
- FG NO_x sensor location can vary between pre or post DOC with varying implications.
- Tail-Pipe (TP) NO_x sensor is necessary for all applications and may be placed either directly after the SCR or after the DPF when the DPF is located after the SCR.
- Due to confounding of the output NO_x signal (TP NO_x sensor) with NH_3 (slip) many applications are resorting to an Ammonia sensor for output signal disambiguation. Others may use a filtering algorithm.
- Ammonia Oxidation or "clean up catalysts" are also becoming an option to reduce TP NH_3 → costly but allows for less constrained control objectives, including relaxed urea cost.
- Some applications may also use Hydrolysis catalysts to enhance the rate and efficiency of NH_3 availability from urea dosage, although this is less prevalent.

The Minimal AT system is the most constrained (cost effective) and uses only the following components: 1. FG NO_x (sensor or model) 2.TP NO_x sensor, 3. Urea Injector 4. mixer 5.DOC, 6. SCR, 7. DPF.

Research

Diesel Controls and Diagnostics
Powertrain Controls R&A

The dynamics of NH_3 storage: Adsorption-Desorption

Critical Storage Identifiers for SCR performance control

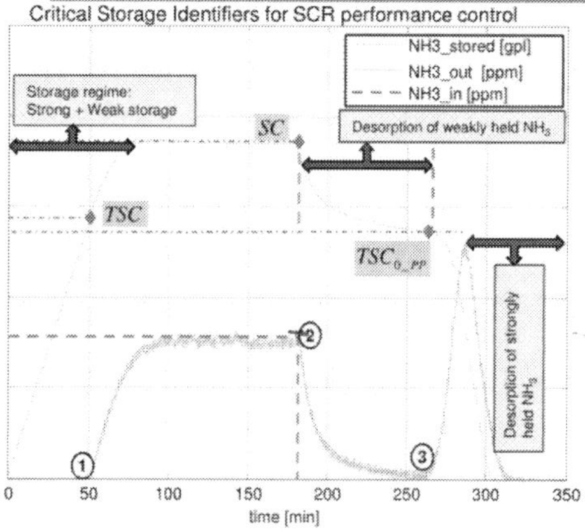

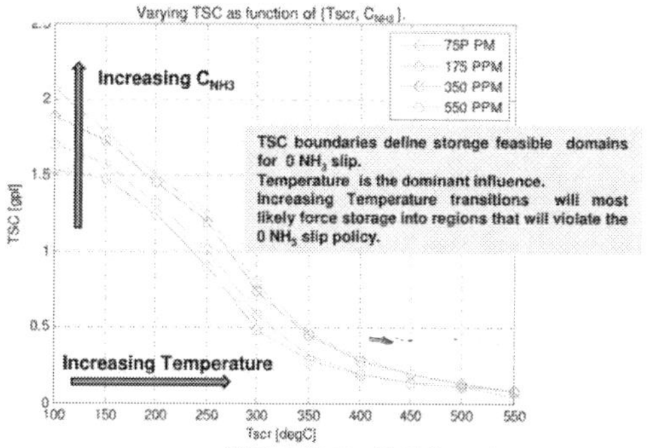

$$TSC = \left(\max\left(m_{NH_3}^{stored}\right)_{C_{NH_3}^{in} > 0} \middle| C_{NH_3}^{out} = 0 \right) \qquad TSC_{0\_PP} = \left(\max\left(m_{NH_3}^{stored}\right)_{C_{NH_3}^{in} = 0} \middle| C_{NH_3}^{out} = 0 \right)$$

- Threshold Storage Capacity (TSC), for the operating pair {Tscr, C_{NH3}}, is the maximum storage possible without any NH_3 slip. For most Conditions the TSC also provides a storage level that allows maximal or near maximal NO_x conversion.
- TSC_{0_pp} is the TSC with 0 NH_3 partial pressure (C_{NH3} = 0) in the gas phase and TSC ≥ TSC_{0_pp} for all operating regimes.
- Markers 1,2 and 3 define the 3 basic storage and NH_3 slip transients that impact all SCR's during continuous operation.

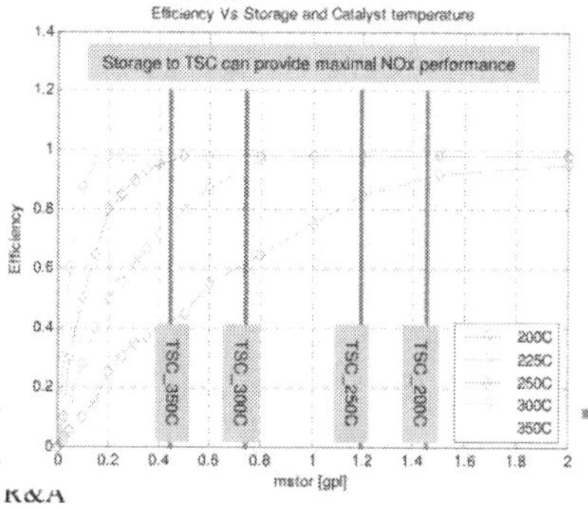

Powertrain Controls R&A

Understanding NH_3 storage on the SCR

- Exact nature of storage remains an active area of research with regard to :
  - The nature and distribution of storage sites → Brøensted, Lewis, I.E. metal-Lewis.
  - Participation of the various stored species in NO_x conversion.
  - Migration of NH_3 within the zeolite pore structure → Strong to weak transition.
  - Role and behavior of Cu.

[010]CHA[4]

- For Control design storage behavior has implications on system response → control bandwidth.

<u>TSC and its correlation with storage: a workable control interpretation</u>

- In traditional terms , TSC may be understood to account for ALL the strongly held NH_3 as well as one or more minimal layer/s of the strongest of the weakly held NH_3. This weakly held layer/s is/are hypothesized to be a result of hydrogen bonded[1] NH_3 as, $NH_4^+.nNH_3$. This layer is a strong function of the gas phase partial pressure of NH_3, and the surface adherence is strong enough to not result in any measurable NH_3 slip.
- TSC_{0_PP} can therefore be logically concluded to account for all the weakly held layer/s that will be released when the partial pressure of NH_3 is reduced to 0 → state transition from 2 to 3 in the previous figure.
- For control purposes it is critical to understand the contributing dynamics of the different types of NH_3 storages with respect to NOx activity and influence on NH_3 slip.
- *<u>Q. Is there a benefit for embedded, control oriented, storage models to account for these various sites ?</u>* Up to 5 storage sites have been explored and reported in literature[2,3]
- We find that the quality of prediction of multi-site storage models does not improve significantly when compared to a single-site model especially when considered in the context of model based control.

1. Rodriguez-Gonzalez, L., Rodriguez-Castellon, E., Jimenez-Lopez, A., and Simon, U., "Correlation of TPD and impedance measurements on the desorption of NH3 from zeolite H-ZSM-5," *Solid State Ionics* 179:1968–1973, 2008, doi:10.1016/j.ssi.2008.06.007.
2. Sjövall, H., Fridell E., Blint, R., and Olsson, L., "Identification of adsorbed species on Cu-ZSM-5 under NH3 SCR conditions," *Top Catal* 42-43(1-4):113–117, 2007, doi:10.1007/s11244-007-0162-6.
3. Skarlis, S., Berthout, D., Nicolle, A., Dujardin, C. et al., "Modeling NH3 storage over Fe- and Cu-zeolite based, Urea-SCR catalysts for mobile diesel engines," *Procedia - Social and Behavioral Sciences* 48:1672–1682, 2012, doi:10.1016/j.sbspro.2012.06.1142.
4. Data base of zeolite structures.

Control Objectives and desired storage

Control Objective

- The ideal control objective for the Minimal AT system may be defined by the following hard constraints:
  - $C_{TP\_NOx} \equiv 0 \forall t$ → achieve 100% NO_x conversion for all time (full control horizon).
  - $C_{TP\_NH3} \equiv 0 \forall t$ → achieve 0 NH_3 slip for all time (full control horizon).

- ***It is obviously impossible to continually maintain NH_3 storage such that both of these objectives are met for all operating conditions.***

- This can be easily verified from the TSC vs η plot shown earlier. As an example consider the following:
  - For a given catalyst temperature, 0 NO_x slip → NH_3 storage $m_{NH3\_k} \approx TSC_k$.
  - Any positive temperature transition will create a storage condition where: $TSC_{k+1} < TSC_k, \rightarrow m_{NH3_k+1} \geq TSC_{k+1}$ such that the NH_3 slip criterion is violated.
  - Or both criteria can be simultaneously violated when the SCR catalyst and/or the AT system has deviated from nominal (biased system) or for unrealistically high conversion demands from a kinetically limited SCR (cold).

- Relaxed control objectives are typically defined requiring maximal NO_x conversion and minimal NH_3 slip.
- Urea injection control objectives translate to storage manipulation as shown below.

NH_3 slip minimizing criterion — NO_x slip minimizing criterion

$$m^*_{NH3} \in \left\{\left\{\max\left(m^{stor}_{NH3}\right) \middle| \left(C^{TP}_{NH_3} < C^{\max}_{NH_3}\right)\right\} \cap \left\{\min\left(m^{stor}_{NH3}\right) \middle| \left(\eta_{NOx} = \eta^*_{NOx}\right)\right\}\right\}, \forall t, T_{SCR}.$$

Note: minimal NH_3 slip → minimal urea consumption.

Pareto point is an intersection set between two objectives.

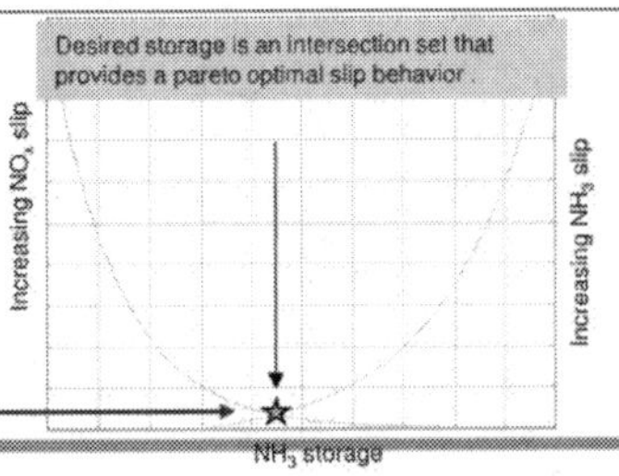

Control Constraints

- The <u>feedback signal</u> from the TP NO_x sensor, is confounded → $Y_{TP\_mes} = C_{NOx} + \alpha C_{NH3}$
  - Unless this confounding is managed ALL closed loop control designs will be prone to instability.
  - Since the system is naturally dissipative the instability will typically emerge as an unstable limit cycle.
- <u>Control authority</u> with respect to storage is single sided → can only add urea, storage depletion is determined by NO_x flow and/or thermal desorptions both of which are un-manipulated events.
- What if we could partition the control authority U = $\begin{bmatrix} U^+ \\ --- \\ U^- \end{bmatrix} \stackrel{\text{def}}{\equiv} \begin{bmatrix} urea \\ ---- \\ NO_x \end{bmatrix}$, such that storage depletion was manipulated via NO_x, ... there are several issues
  - FG NO_x must be increased without a thermal event → EGR intervention → "Defeat Device" management
  - Since $urea \equiv 0$ during this mode it is expected that additional NH_3 slip from weakly adsorbed NH_3 will occur, additionally to achieve even moderate levels of depletion a large NO_x flux may be needed making the Defeat Device management task challenging.
- If temperature (fast) is used as U^- → translates to fuel cost and storage depletion will be from NH_3 slip.
- *It is therefore clear that any active method of fast storage depletion will result in NH_3 slip, defeating the very objective being pursued → deplete storage to avoid potential NH_3 slip.*
- These restrictions also make any preview information (if available) ineffective.
- We have found that maintaining storage via a stable limit cycle, within a narrow boundary of the TSC surface, allows us to nominally maintain storage at desired levels

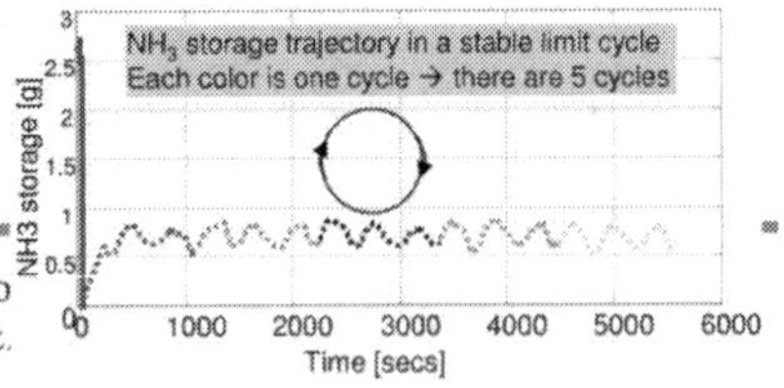

Typical Control Architecture

- Typical control architecture includes Feed Forward (FF), Feed Back(FB) and Adaptive corrections
- Closed loop control adds robustness against modeling error and system and measurement noise.
- Closed loop Control should ideally be some combination of storage control and slip control.
- Storage Control does not have state measurement → hence control to a reference state for ex: TSC as discussed earlier.
- Slip control should be based on disambiguated output signal to influence the correct control direction.
- This may lead to a controller with a structure similar to an IMC structure.

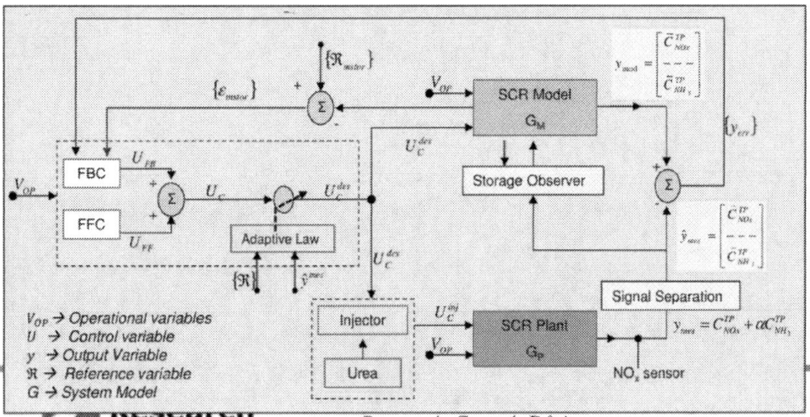

Powertrain Controls R&A

Adaptive correction

- Robustness against diverse set of expected system drifts and uncertainties cannot be addressed by Closed Loop control alone.
- A suitable adaptive algorithm must be used, we use a gradient based update law with a slow update (to avoid instabilities from fast corrections especially for large biases).
- Two options were exercised to isolate the source of efficiency loss:
  1. A game theoretic approach that assess nature of slip from the gradient of performance
  2. A direct approach to distinguish between NH_3 and NO_x as the dominant slip species.

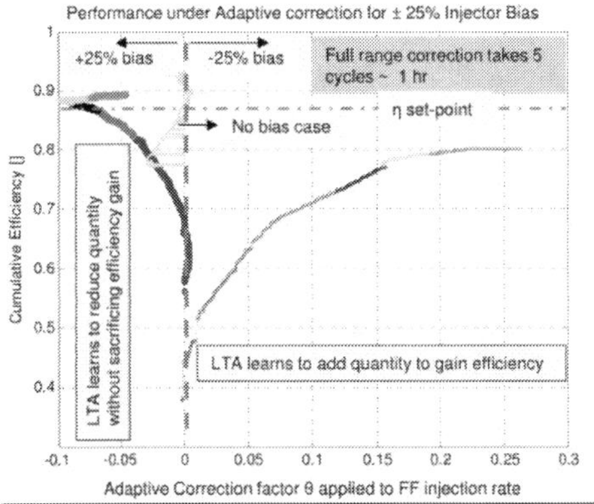

Adaptive law learns out bias errors, although learning is not symmetric to achieve desired target performance. The no Bias case dithers along the 0 correction line to achieve desired performance.

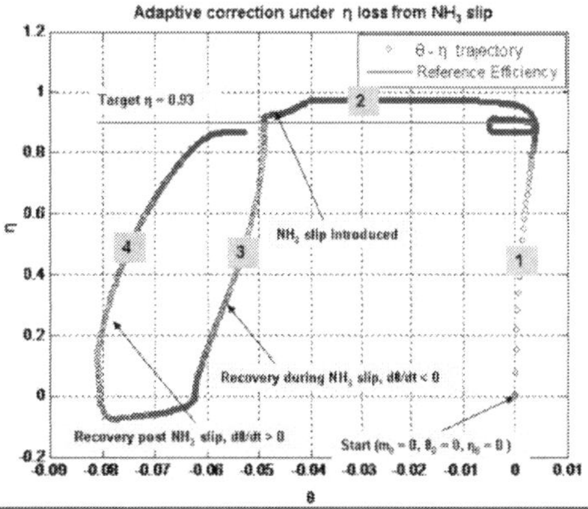

Adaptive law tried in situation of sudden NH_3 release. Update law reduces NH_3 fast after onset of NH_3 slip. Note there was already some –ve correction prior to the onset of slip due to the system over achieving.

What does stability look like for SCR control

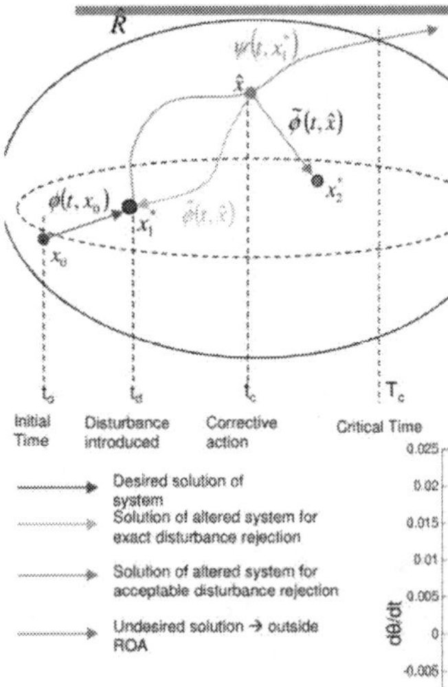

R = Region of attraction
$\hat{R}$ = Relaxed R
ξ = domain of acceptable equilibria
$\phi(t, x_0)$ = solution with initial state, x_0
$\psi(t, x_1^*)$ = solution after disturbance

$$R = \{x \in \Omega | \phi(t, x_o) \to x^* \text{ as } t \to \infty\}$$

$$\hat{R} = \{x \in \Omega | \ \|\phi(t, x_o) - x^*\| < \xi \text{ as } t \to \infty\}$$

- SCR catalysts are dissipative systems with finite ranges for states.
- Instability in SCR's catalysts is observed via sustained (t_∞) oscillation in some control variable.

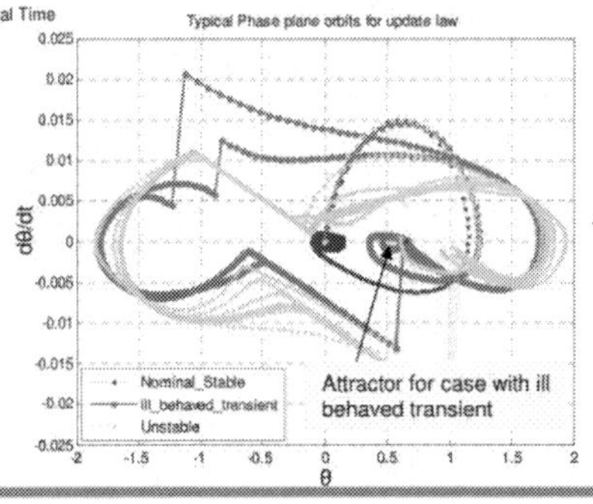

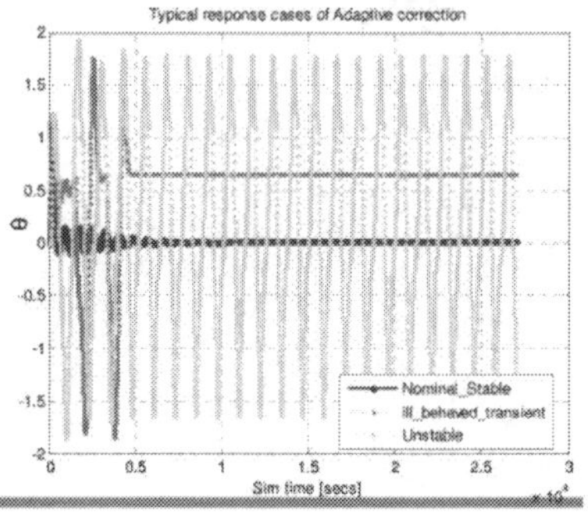

Research

Diesel Controls and Diagnostics
Powertrain Controls R&A

Storage Observer

- All NH_3 storage models are prone to storage error creep over time, hence storage must be periodically adjusted.
- Direct storage measurement is currently not a possible (will probably remain so in the near future).
- A traditional observer may be designed based on measured signal from TP NO_x sensor, provided signal separation into constituent species is available, subject to observability criterion[1].
- A direct reference is available for storage adjustment in the form of TSC. A standard Luenberger observer may be designed with TSC as the reference storage state whenever a deterministic evaluation of NH_3 slip is made. Under this condition it is clear that the current NH_3 storage is ≥ TSC and the observer can make at least a partial correction.

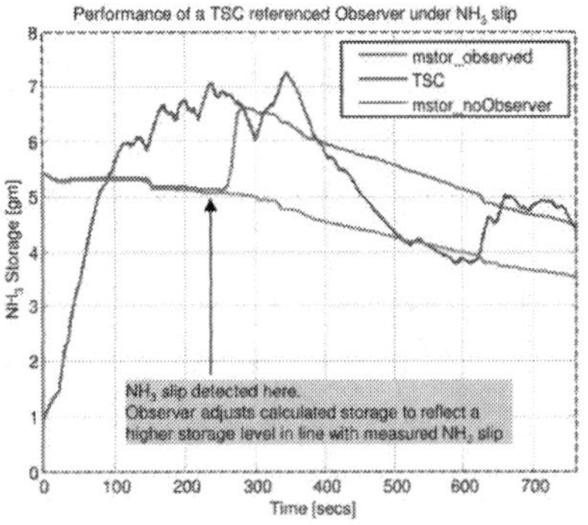

An example of the Observer performance under NH_3 slip is illustrated. The actual calculated storage value was biased low (-2gm) for this test. The observer resets the storage to TSC which recovers most of the storage error.

The same observer is also used to adjust storage downward, when slipping NO_x, by using a reference storage obtained by inversion of a NO_x conversion model, $\eta_{NOx} = f(m_{stor}, T_{SCR})$.

1. Model Based Analysis and Control Design of a Urea-SCR deNOx Aftertreatment System, Devesh Upadhyay and Michiel Van Nieuwstadt, J. Dyn. Sys., Meas., Control 128, 737 (2006)

Research

Diesel Controls and Diagnostics
Powertrain Controls R&A

Signal separation

- Deterministic feedback is critical for robustness of control and OBD performance as well as control stability.
- In the absence of an NH_3 sensor the TP NO_x signal must be separated into its constituent species.
- This has remained an open problem in the Urea SCR community until recently. At Ford we developed a purely Signal Processing based algorithm to accomplish this task → avoids uncertainty form modeling error.
- The fundamental approach relies on convolving the forward difference of the FG and TP NO_x signals over a window defined by the decreasing phase of the FG NO_x signal[1] .

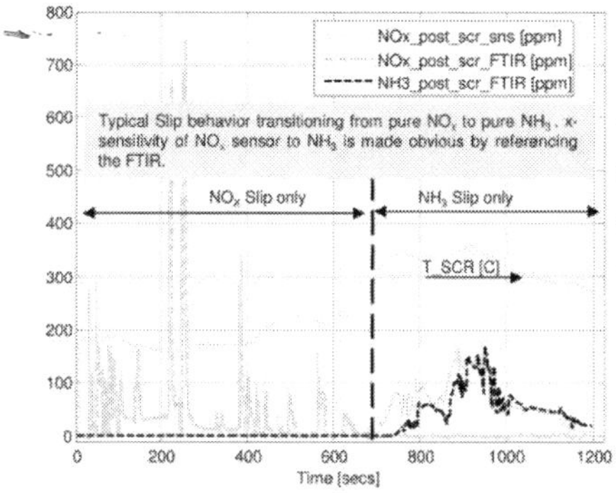

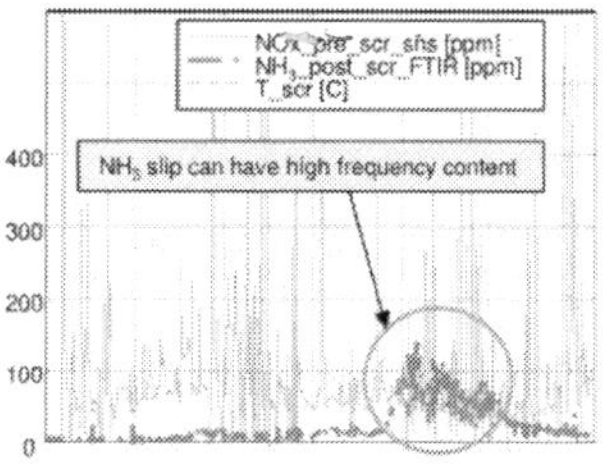

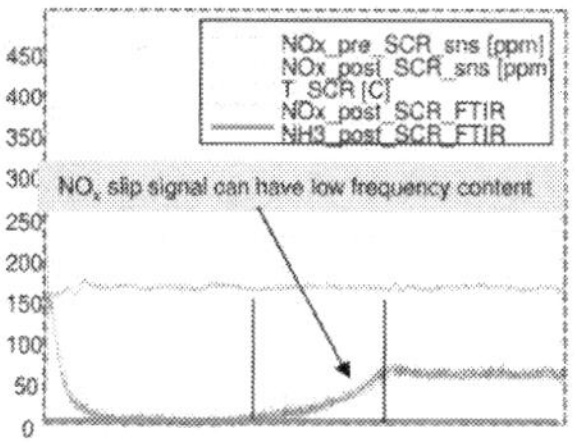

Myths about NH_3 and NO_x slip:

1. NH_3 slip signal will always have a low frequency spectrum
2. NO_x slip will always have a high frequency spectrum

1. Devesh Upadhyay, M van Nieuwstadt, Robust Separation of Signal Domain From Single Channel Mixed Signal Output of Automotive Urea Based Selective Catalytic Reduction Systems, J. Dyn. Sys., Meas., Control 136(1), 011012 (Oct 07, 2013) (9 pages)

Signal slip domain separation Performance

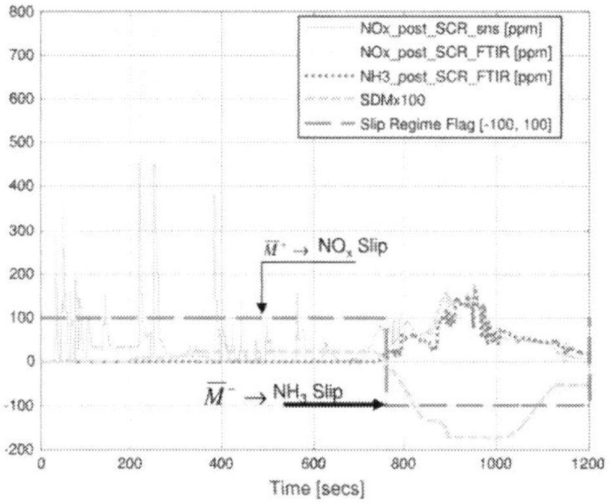

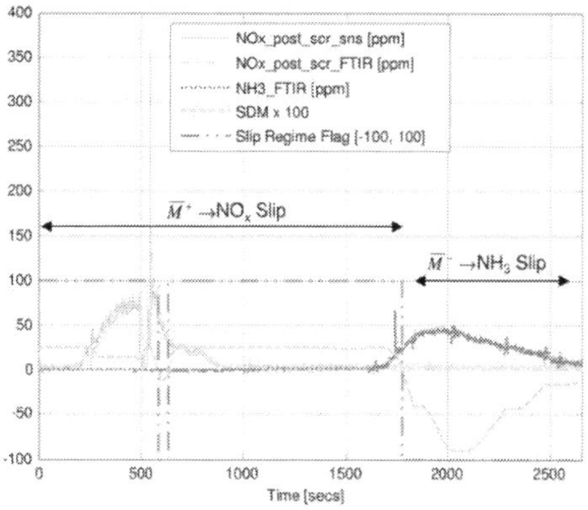

Slip Detection Metric (SDM) identifies NO_x and NH_3 slip domains with correct transition

Slip Detection Metric identifies NO_x and NH_3 slip domains even though both signals have similar low frequency spectra.

New Directions in controls – storage models

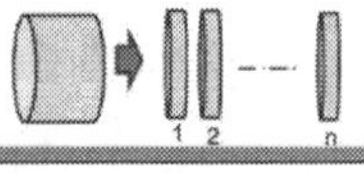

- Trend is to transition from Lumped parameter (0-D) models to n x 0-D models → pseudo 1-D model.
- Thermal Hypothesis: Single node 0-D models will incur inaccuracies from axial temperature gradient. This error maybe mitigated by adopting a n x 0-D model coupled with an axial thermal model of the AT system.
- Scale-Up Hypothesis: Full scale catalyst models, scaled up from SGB reactor core data, may have scale up ratios > 500x. The error from such large scaling factors can also be reduced by using several nodes each with a smaller scale up ratio relative to the SGB core.
- However, It is critical to perform a critical analysis of these assumptions prior to adopting any particular structure for the embedded model. Some points to note are:
  - **Typically only a fraction (~20%) of the catalyst is actively engaged** in catalytic activity, hence the 1x0D → nx0D reconstruction must be done with full understanding of the reactor sizing and usage.
  - Under most conditions axial temperature gradients are small across bricks of nominal dimensions, 6" –12" long → nx0-D benefits will be limited to conditions generating axial temperature gradients → cold starts, very long catalyst bricks, 2 or more separate bricks operating in kinetically distinct thermal regimes.
  - Under certain operating conditions and/or ambient influences the **radial temperature gradients may be more significant** compared to axial gradients. The nx0D approach will **NOT** address this.
  - Observer correction of storage becomes more involved for a n x 0-D model since the **storage adjustment implies a redistribution of the storage correction** between the n-nodes. This is an under-determined problem forcing heuristic schemes for storage adjustment.

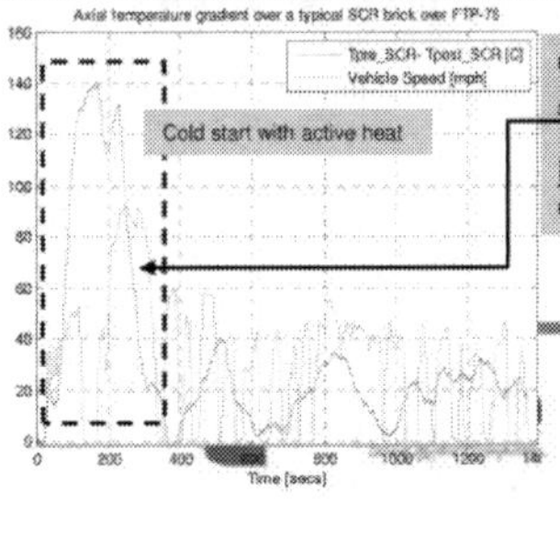

Over FTP cycle (Tamb ~= 25C):
Axially: Front is hotter than back (expected)

Radially: Outer edge can be considerably cooler

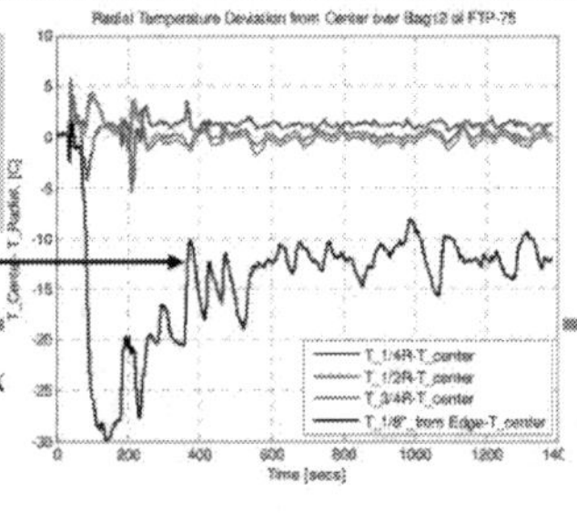

Diagnostics

- At current OBD levels of **0.4 g/mi** TP NOx for CC U10K HD trucks the SCR catalyst must (typically) be flagged as deteriorated at activity levels in the range 70% to 75% over the FTP-75.
- A **70 ~75%** SCR catalyst can return very high efficiencies O(90%+) on a variety of drive cycles/operating conditions.
- Conversely cold ambient (< 5C) can make a healthy SCR return poor performance.
- These behaviors make selection of a robust performance threshold for minimizing Type-1 and Type-2 errors a non-trivial task.
- Robust OBD in the face of **Noise factors** only increases the challenge .
- The primary path is to exploit the characteristic behavior of a deteriorated SCR → propensity to incur dual mode slip → NOx + NH_3. NH_3 slip will lead to a lower measured performance due to the cross sensitivity of the NO_x sensor (discussed before).
- Given the possibility of a healthy SCR to slip NH_3 under poor control or transients, for example, robust detection of a threshold SCR dictates a two step process whenever a failing result is returned. ***This is to ensure that the efficiency loss is persistent and not transient***. And that a healthy catalyst is not flagged as deteriorated (cost implications).

Aging

- OBD testing also entails a catalyst aging procedure. Current generation SCR's Cu SSZ-13 are hydrothermally quite stable → Defining an aging protocol that allows consistent aging to the OBD threshold level involves some degree of trial and error.

Conflict with control

- Since control will attempt to drive the catalyst in the direction of improved performance, a synergistic approach or "decoupling" may become necessary to avoid premature/ delayed detection of an SCR that has deteriorated to the OBD threshold level.

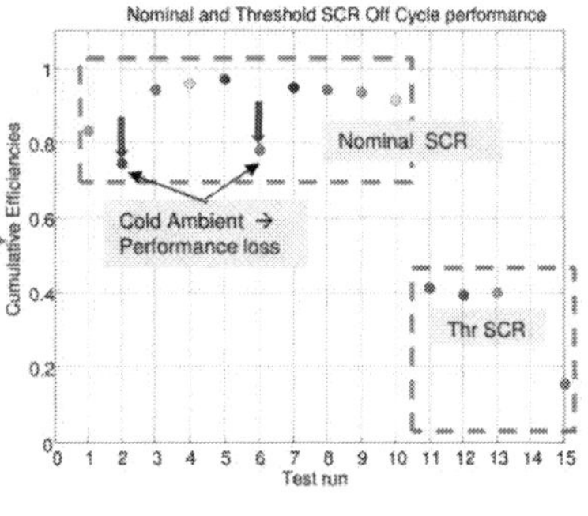

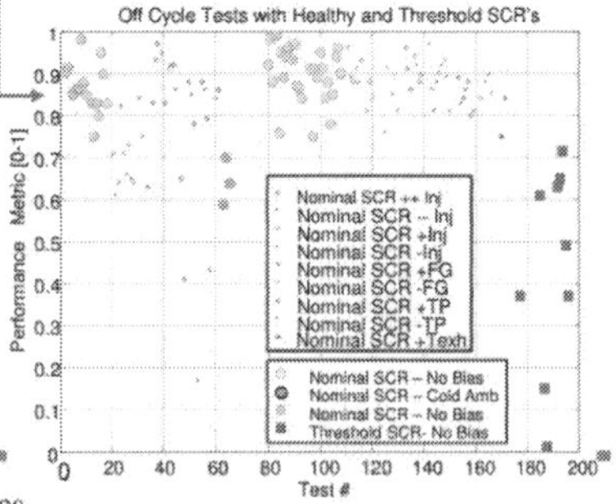

Conclusions/Outlook

- C_u, F_e, Zeolite based urea SCR catalyst will most likely continue to be the dominant diesel aftertreatment system.
- Investigations into chemistry/physics of these catalysts will continue to remain an active are of research until all fundamental behaviors are understood.
- Control of these systems, for on-road transient applications has achieved some degree of maturity with (industry) consensus on the need for:
  - Model based control (storage model)
  - Control architecture → FF+FB+Adaptation
  - Feedback signal separation (or NH_3 sensor)
- Potential areas of continued work in Control/Systems:
  - AT system arrangement variants : 2 brick systems, SCRDPF-SCR.
  - Higher fidelity (distributed), yet reduced order models.
  - Control via active manipulation of kinetics through large gain sensitivities such as substrate temperature.
  - Active management of reactor volume via manipulation of storage distribution.

Research

Diesel Controls and Diagnostics
Powertrain Controls R&A

Thank you for your attention

Research

Diesel Controls and Diagnostics
Powertrain Controls R&A

392229 Investigation of burning mode for diesel particulate oxidation: contrasting O2 and NO2

Abstract:

The physical characteristics of diesel particulates from conventional diesel, biodiesel and blends of the two were investigated at several points along the O2 and NO2 oxidation reactions. BET measurements elucidated the surface area evolution with extent of oxidation for the two oxidizers and revealed differences in surface area evolution in the samples when oxidized with O2. Particulate matter from biodiesel initially has greater surface area than particulate from ULSD, an advantage, which is maintained under O2 oxidation until approximately 40% of the sample has been consumed. HR -TEM and fringe analysis confirm that initial differences in the lamella lengths have disappeared by 40% burnout of the samples. TEM images suggest that surface area evolves mostly internally for O2 oxidation as the primary particle diameters do not follow the shrinking core trajectory. However, for NO2 there are no sample-specific dependencies in either reactivity and surface area. TEM images suggest that NO2 oxidation occurs by a mostly external trajectory.

392232 TWC formulation effects on NH3 generation for passive SCR applications in lean gasoline engine exhaust

Abstract:

Due to the dominance of gasoline vehicles in the U.S. light duty fleet, improvements in gasoline engine fuel efficiency would significantly reduce U.S. petroleum consumption and greenhouse gas emissions. Operating gasoline engines with more air than is required to burn the fuel (known as lean conditions) can improve fuel efficiency by 10% or more. However, lean gasoline vehicles have been unable to meet U.S. emissions regulations for oxides of nitrogen (NOx). The two lean NOx reduction technologies that have been successfully deployed on diesel engines, LNTs (lean NOx traps) and urea SCR (selective catalytic reduction), are both challenged by lean gasoline applications. The high NOx concentrations and exhaust temperatures from gasoline engines result in excessive NOx slip during LNT regeneration and a high fuel penalty due to frequent regeneration events. With SCR systems, high NOx concentrations require increased urea dosing rates, necessitating a large onboard storage tank and/or easily accessible refueling infrastructure. The cost of the urea storage and dosing system may also be prohibitive in gasoline applications. An alternate strategy, termed passive SCR, generates NH_3 from engine-out NOx over a close-coupled catalyst during periodic excursions to slightly fuel-rich conditions. The NH_3 is stored on an underfloor SCR catalyst, where it is used to reduce NOx upon returning to lean conditions. By generating NH_3 over the three-way catalyst (TWC) already present on gasoline vehicles, the passive SCR approach enables use of proven commercial SCR catalysts while avoiding the cost and packaging issues associated with a urea dosing system. Oak Ridge National Laboratory, in collaboration with partners at General Motors, Umicore Autocat USA, and the University of South Carolina, is currently working to identify passive SCR catalyst formulations, system architectures, and operating strategies that would enable lean gasoline vehicles to meet emissions regulations while minimizing the associated fuel penalty and costs. This presentation will describe the results of flow reactor and engine experiments that evaluated the effects of TWC formulation on the NH_3 formation process under both steady state and dynamic operating conditions. The TWC formulations used in the experiments included variations in precious metal composition, oxygen storage capacity, and NOx storage capacity. Correlations between TWC formulation and expected passive SCR fuel penalty will be discussed.

Experimental Investigation and Kinetic Modeling of Diesel Oxidation Catalysts for Emissions Aftertreatment from Diesel Engine Exhaust

Hom N. Sharma

Department of Chemical and Biomolecular Engineering, University of Connecticut, Storrs, CT

Keywords

Diesel engine exhaust, Emission oxidation, Diesel oxidation Catalyst (DOC), microkinetic modeling

Emissions from the diesel engine is one of the major concerns for global warming, human health and environmental pollution, due to many toxic and harmful exhaust gasses like carbon monoxide (CO), carbon dioxide (CO_2), nitrogen oxides (NO_x), sulfur oxides (SO_x) and many other volatile organic compounds (VOCs) [1] and poly aromatic hydrocarbons (PAHs) [2]. Diesel Oxidation Catalyst (DOC) is a predominant component in emissions aftertreatment to meet such regulations [3]. As shown in Figure 1a, In Urea- or aqueous ammonia-based Selective Catalytic Reduction (SCR) of NO_x, the Pt/Pd based DOC is typically placed upstream of the SCR to achieve an optimum NO: NO_2 ratio for improved SCR efficiency. On the other hand, as shown in Figure 1b, in HC-based SCR (e.g., using Diesel/ethanol/mixtures), a downstream DOC would be required to oxidize the toxic byproducts, such as CO, ammonia (NH_3), formaldehyde (HCHO), hydrogen cyanide (HCN), acetonitrile (CH_3CN), and acetaldehyde (CH_3CHO), to relatively 'less' harmful CO_2 and NO_x.

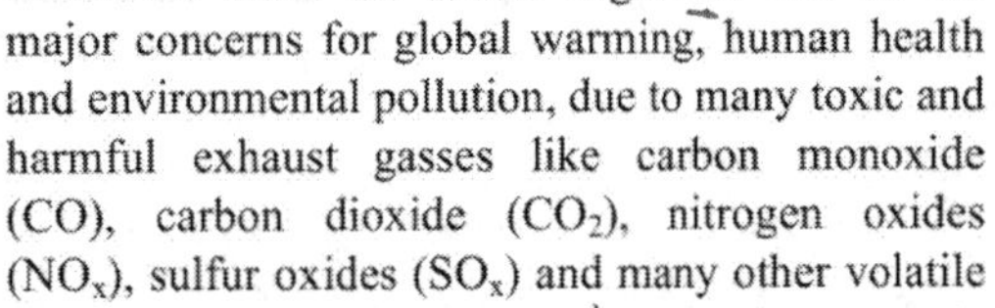

***Figure1** Role of DOC in toxic emissions oxidation.*

The noble metal based DOC is one of the most expensive components of the emissions aftertreatment system, and is typically overdesigned to some extent to ensure complete oxidation of toxic byproducts. Despite the extensive research conducted on DOC performance, comprehensive and predictive kinetic models for simultaneous catalytic oxidation of multiple emissions are limited due to the complex multi-species interactions.

In this work, we have developed a detailed elementary-step microkinetic model for DOC that accounts for (i) CO oxidation, (ii) NO oxidation, (iii) oxidation of other N-containing emissions, such as NH_3 and HCN, (iv) oxidation of toxic aldehydes, such as HCHO, (v) oxidation of hydrocarbons, such as C_2H_4, and (vi) oxidation of toxic sulfur oxides, such as SO_2 [4-6]. Furthermore, A state-of-the-art BenchCAT reactor was be used for the experimental investigation of CO and C_2H_4 oxidation experiments on Pt/Al_2O_3 powder and monolith scale DOC catalysts. Catalyst activity for oxidation of various toxic components was measured as a function of operating conditions, such as temperature, flow rate, and inlet composition. The experimental data was used to rigorously validate the microkinetic model and to probe the ability of the reaction mechanism to capture the interactions among various species during emissions oxidation.

The detailed reaction mechanism development is carried out using various parameter estimation techniques, such as semi-empirical Unity Bond Index-Quadratic Exponential Potential (UBI-QEP) method [7], Transition State Theory (TST), and quantum mechanical Density Functional Theory (DFT). The microkinetic model performs well against various experiments conducted in practically relevant conditions (atmospheric pressure, ppm level concentrations of emission components) and in fixed bed and monoliths reactors. The elementary-step surface reaction mechanism could be utilized for improved design and control of the aftertreatment system as well as appropriate DOC sizing.

References

(1) Kim, Y.; Kim, W. *Ind. Eng. Chem. Res.* **2009**, 48 *(14)*, 6579-6590.

(2) Lim, M. C. H.; Ayoko, G. A.; Morawska, L.; Ristovski, Z. D.; Rohan Jayaratne, E. *Atmos. Environ.* **2005**, 39 *(40)*, 7836-7848.

(3) Environmental Protection Agency *Federal Register* **2006**,Vol. 71, No. 83*(Air Docket No. EPA-HQ-OAR- 2006-0224)*.

(4) Sharma, H. N.; Mhadeshwar, A. B. *Appl. Catal. B-Environ.* **2012**, 127 , 190-204.

(5) Sharma, H. N.; Suib, S. L.; Mhadeshwar, A. B. Interactions of Sulfur Oxides With Diesel Oxidation Catalysts (DOCs). In *Novel Materials for Catalysis and Fuels Processing;* Bravo-Suarez, J. J., Schwartz, V. and Kidder, M. K., Eds.; ACS: 2013; Vol. 1132, pp 117-155.

(6) Sharma, H. N.; Sharma, V.; Hamzehlouyan, T.; Epling, W.; Mhadeshwar, A. B.; Ramprasad, R. *J. Phys. Chem. C* **2014**,.

(7) Shustorovich, E.; Sellers, H. *Surface Science Reports* **1998**, 31 *(1-3)*, 1-119.

A new catalyst for the selective catalytic reduction of NO_x by NH_3 over olivine

Yun Shi, Su-jing Li[1]

Key Laboratory of Biomass Chemical Engineering of Ministry of Education, Institute of Industrial Ecology and Environment; Institute of Environmental Engineering, Zhejiang University (Yuquan Campus), Hangzhou 310027, China.

Abstract:

Selective catalytic reduction (SCR) of NO_x by urea (Urea-SCR), using urea as the ammonia source, is a potential method to remove NO_x from automobile exhausts. Various catalysts, including metal-supported Al_2O_3 and ion-exchanged ZSM-5, have been widely investigated. However, the hydrothermal deactivation, narrow active temperature window, and high cost limited the application of these catalysts. The main challenge for the NO_x removal for automobile exhausts is the low-temperature activity (<300 °C). Researchers reported that transition metal oxides like FeO_x, MnO_x and NiO_x exhibit good activity for Low-temperature (LT) SCR. Olivine, a magnesium iron silicate with the formula $(Mg^{2+}, Fe^{2+})_2SiO_4$, and containing minor amounts of manganese and nickel, is a promising SCR catalysts.

This work reports for the first time the performance of olivine for the selective reduction of NO_x by NH_3. The catalytic activity and parameters that may affect olivine catalytic performance were investigated. It was hoped that this work could provide theoretical foundation for the application of Olivine-SCR.

Catalytic activity tests were carried out at atmospheric pressure in a fixed bed quartz tube reactor (inner diameter 6 mm). A reaction mixture containing 500 ppm NO, 500 ppm NH_3, 5% O_2, and N_2 as the balance gas was fed to the catalyst bed (GHSV = 11,000 h^{-1}). The reaction temperature was increased from 100-450 °C with 10 °C min^{-1}. Each step lasted 20 min to obtain steady state conditions. Without introducing any other elements, olivine itself displays 45-80% for NO_x conversion in the temperature range 100-350 °C, and with a maximum conversion of about 80% at 300 °C. The results also show that very low concentration of N_2O was detected, and N_2 is the main products during olivine-SCR.

The effects of gas hourly space velocity (GHSV) and inlet concentration of O_2 were investigated. The increase of GHSV decreased the NO_x conversion. The inlet concentration of O_2 had an obvious effect on the NO_x conversion. The NO_x conversion was below 30% without O_2 at 300 °C due to lack of active catalytic center that could be regenerated with O_2. The maximum NO_x conversion was obtained when the O_2 concentration was 2%. Further increase of O_2 could decrease NO_x conversion due to competitive adsorption on the active catalytic center between O_2 and NO_x.

As a result, some natural minerals, such as olivine, perovskite and clinoptilolite, are potential catalysts for NO_x reduction due to good activity and stability, low cost and flexible composition.

Key words: NO_x, NH_3-SCR, Olivine, FeO_x

1

[1] Corresponding author: E-mail: sujing-li@zju.edu.cn

375701 Novel Electrocatalysts with Advanced Nanoscale Architectures for Energy Conversion Applications

Abstract:

Highly efficient chemical-electrical energy conversion has become the primary target of the development of renewable energy technologies such as fuel cells and metal air batteries. This to a large extent relies on advanced electrocatalysts with high catalytic activity, high durability, and competitive cost effectiveness. Conventional catalysts comprising platinum (Pt) nanoparticles supported on carbon have rather low catalytic activity, and also expensive and scarce. Moreover, Pt that is generally considered to be chemically inert becomes unstable when exposed to the hostile electrochemical environments. Significant improvements in both catalytic activity and durability, and thus reduced usage of precious metal are thus desired for large-scale applications of these technologies.

We have explored the rational design and synthesis of advanced Pt-based bimetallic (Pt-Ni, Co, Re) and multimetallic (Au/Pt_3Fe) electrocatalysts for the oxygen reduction reaction (ORR) – the key reaction in fuel cells and metal-air batteries. Critical parameters including particle size, shape, alloy composition and element distribution were integrally designed for optimal catalytic performance for the ORR. State-of-the-art electron microscopic and X-ray spectroscopic analysis were applied to characterize the fine core/shell nanostructures, which were further correlated to the catalytic performance revealed by electrochemical studies. The combinational studies of model extended surfaces and practical nanocatalysts, plus additional theoretical calculation and simulation, represent a persuasive approach targeting the fundamental mechanisms of catalytic activity and durability enhancement. Such an approach could also be generalized to the development of other functional materials in a rational way.

MnO_2-based bifunctional oxygen catalyst for rechargeable metal/air batteries: The effect of K^+ intercalation

Pooya Hosseini Benhangi [a], Akram Alfantazi [b], Előd Gyenge [c]

a) Department of Chemical and Biological Engineering & Clean Energy Research Center, The University of British Columbia, 2360 East Mall, Vancouver, BC, Canada V6T 1Z3

b) Department of Materials Engineering, The University of British Columbia, Vancouver, 6350 Stores Road, Vancouver, BC, Canada V6T 1Z4

Extensive research has been undertaken to make clean and renewable energy sources more reliable. Bifunctional oxygen cathodes which can catalyze both ORR (oxygen reduction reaction) and OER (oxygen evolution reaction) are the backbone of rechargeable metal-air batteries as well as regenerative fuel cells [1-3]. Although noble metal family elements and alloys such as Pt , Pt-Au and Pt-Co are known as the best catalysts for ORR in alkaline media, their poor electrocatalytic activity toward OER as well as high price, comparing to non-PGM (non-precious group metal) compounds, limit their usage as a cost effective and active catalyst in bifunctional oxygen cathodes [2, 4]. Manganese oxides have been vastly employed as a robust cost-effective multifunctional and environmental friendly electrode material in battery industry, from primary to rechargeable metal-air batteries, as well as alkaline fuel cells and capacitors [3]. The electrolytic manganese dioxide (γ-MnO_2) is known as the most electrochemically active crystallographic form of MnO_2 for ORR in alkaline media with Tafel slope of 40 mV dec^{-1} and a low overpotential of -375 mV [1, 5]. However, poor OER electrocatalytic activity of manganese oxides (MnO_x) in alkaline media diminish the hope of finding an exclusive bi-functional catalyst for both ORR and OER [1].

Our research is aimed at the development of highly active MnO_2-based catalysts for both ORR and OER with long cycle life by adding non-precious metal compounds as co-catalysts and intercalating potassium ions into the catalyst layer. The mechanisms for OER and ORR of the mixed catalysts and the role of K^+ are investigated by a combination of surface characterization methods and electrochemical techniques.

In the present work, $LaCoO_3$ were synthesized via co-precipitation methods. MnO_2-based gas diffusion electrodes (GDE) consisted of MnO_2, $LaCoO_3$, Vulcan XC-72, Nafion, PTFE and a 40% PTFE treated carbon cloth as the substrate. Each GDE were prepared using the method described elsewhere [1]. In order to intercalate the potassium ions into the catalyst layer, a simple yet effective method has been used. The GDEs were kept in in 6 M KOH solution at open circuit potential while rotating the sample at 400 RPM for 6 days at 313 K.

Cyclic voltammetry tests were performed in O_2 saturated 6 M KOH at 293 K to investigate the electrocatalytic activity of the MnO_2-$LaCoO_3$ catalyst for both OER and ORR. The longer-term durability of the electrodes was also investigated by performing 100 repeated OER-ORR voltammetric cycles. Electron energy loss spectroscopy (EELS) has been also used to further analyze the catalyst layer and study the Mn valance changes during both OER and ORR. X-ray photoelectron spectroscopy (XPS) was also used to confirm the existence of intercalated K^+ ions in the catalyst layer.

Interestingly, huge enhancement is observed in the OER/ORR performance of the activated catalyst comparing to the fresh electrodes without any K^+ activation. Fig 1-a and 1-b show the OER and ORR voltammograms, respectively, of both fresh and activated MnO_2-$LaCoO_3$ GDEs as well as fresh MnO_2 electrodes. The OER overpotential for MnO_2-$LaCoO_3$ decreases to 193 from 367 mV after intercalating potassium ions in the catalyst layer (Fig. 1-a). In the ORR region, activated MnO_2-$LaCoO_3$ electrodes still provides the lowest ORR overpotential of -321 mV (Fig. 1-b), which is far better than the ORR behavior reported in the literature for Core-Corona Structured Bifunctional Catalyst (CCBC), needle-like MnO_x thin film, $CoMn_2O_4$ and even nanostructured Mn oxide thin film [1].

Moreover, the activated catalyst performs almost without any degradation in OER region while suffering up to 61% decrease in the ORR current density at -250 mV (vs MOE) after 100 cycles of severe durability tests. Our results indicate that even 12 hrs of activation in 6 M KOH could enhance the electrocatalytic activity of the degraded catalyst after severe cycling to some point, mentioned before as "healing effect" [1]. The EELS analysis reveals that the Mn valance decreases during severe cycling from about 4 to 2.7 from cycle 1 to 100. This confirms that the loss in the ORR electrocatalytic activity of the activated sample after 100 cycles is associated with the transformation of MnO_2 to Mn_3O_4 during severe cycling.

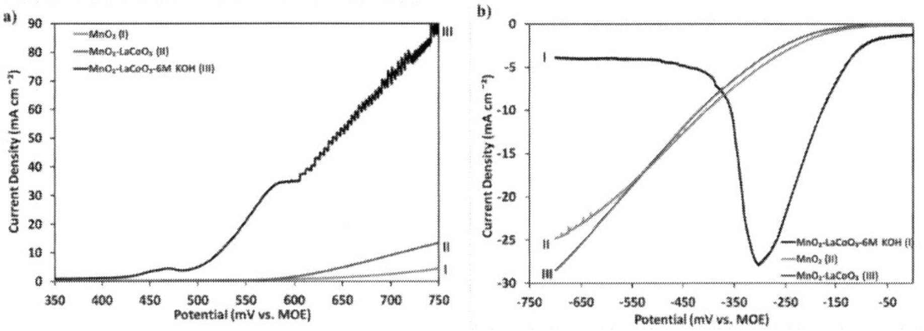

Fig 1. Comparison between the electrocatalytic activity of fresh MnO_2 and fresh/activated MnO_2-$LaCoO_3$ catalysts on 40% PTFE treated carbon cloth in O_2 saturated 6M KOH. The activation consisted of resting each sample in 6 M KOH solutions for 6 days at OCP, 313 k and 400 RPM. The catalyst loading is 0.5 mg cm^{-2}. Cyclic voltammetry tests were performed between -700 and 750 mV vs MOE, starting with anodic polarization. a) OER voltammograms, b) ORR voltammograms.

References:

[1] P.H. Benhangi, A. Alfantazi, E. Gyenge, Electrochimica Acta, 123 (2014) 42-50.

[2] J. Ludwig, J. Power Sources, 155 (2006) 23-32.

[3] A. Serov, A. Aziznia, P.H. Benhangi, K. Artyushkova, P. Atanassov, E. Gyenge, Journal of Materials Chemistry A, (2013).

[4] Y. Chabre, J. Pannetier, Prog. Solid State Ch., 23 (1995) 1-130.

[5] E.L. Gyenge, J.-F. Drillet, J. Electrochem. Soc., 159 (2012) F23-F34.

Deconvoluting the Influences of 3D-Structure on the Performance of Photoelectrodes for Solar-Driven Water Splitting

Daniel V. Esposito[1,2], Youngmin Lee[2], Veronika A. Szalai[2], A. Alec Talin[3], Thomas P. Moffat[2]

[1] Columbia University

[2] National Institute of Standards and Technology

[3] Sandia National Laboratories

Photoelectrodes possessing 3-dimensional (3D) structure offer several advantages for achieving increased efficiency for solar-driven water splitting, including higher surface area for catalysis, enhanced optical absorption, improved carrier collection, and more facile removal of product gases. Unfortunately, the benefits of 3D structuring are often accompanied by detrimental influences such as reduced photovoltage and high surface recombination. Understanding the structure-property relations that give rise to these positive and negative influences of 3D structuring is thus of great importance for optimizing efficiency in photoelectrochemical (PEC) devices. However, complex and interrelated structure-property relationships in 3D photoelectrodes often make this task very difficult. In this work, we demonstrate a systematic approach for deconvoluting the influences of 3D structure using a combination of electroanalytical, spectroscopic, modeling, and solid-state measurement techniques and tools. Measurements were performed on p-Si photocathodes comprised of well-defined micro-pillar arrays and possessing a metal-insulator-semiconductor (MIS) architecture that has shown great promise for achieving high efficiency and good stability.[1-3] By combining these measurements with basic modeling, we compare the optical, catalytic, ohmic, and recombination losses 3D and planar structured MIS photocathodes. The results are summarized in current-voltage loss-analysis diagrams, highlighting the challenges and opportunities for using 3D structuring in PEC devices for water splitting.

References

[1] Munoz, A.G. *et al.*,. *Journal of the Electrochemical Society* 156 (9), D337-D342 (2009).

[2] Y. W. Chen, J. D. Prange, S. Duhnen, Y. Park, M. Gunji, C. E. D. Chidsey and P. C., *Nature Materials*, 2011, **10**, 539-544.

[3] Esposito, D.V., Levin, I., Moffat, T.P., & Talin, A.A.,. *Nature Materials* 12 (6), 562-568 (2013).

385100 Mechanisms of Enhancement By Metal Co-Catalysts and Plasmonic Nanoparticles in the Photo-Catalytic Activity of Semiconductors

Abstract:

In this contribution, we will shed light on the microscopic mechanisms involved in the enhancement of photo-catalytic performance of semiconductors by the joint action of metal co-catalysts and plasmonic metal nanoparticles. In particular, we will focus on the role of Pt co-catalysts and Ag plasmonic nanoparticles in the photo-catalytic activity of Si for the photo-electrochemical evolution of hydrogen.[1–3] In addition to the description of our experimental findings, I will present a mathematical model that can capture all the main features of this fairly complex photo-catalytic system. I will discuss how our findings can be transferred to any co-catalyst/plasmonic-metal/semiconductor photo-catalytic system.

(1) Hou, Y.; Abrams, B.; Vesborg, P. *Nat. Mater.* **2011**, *10*, 434–438.

(2) Lombardi, I.; Marchionna, S.; Zangari, G.; Pizzini, S. *Langmuir* **2007**, *23*, 12413–12420.

(3) Ingram, D. B.; Christopher, P.; Bauer, J. L.; Linic, S. *ACS Catal.* **2011**, *1*, 1441–1447.

Molybdenum Sulfide Materials to Provide Catalytic Activity and Surface Protection for Active and Stable Water Splitting Photocathodes

Thomas F. Jaramillo,[1] Jesse D. Benck,[1] Jakob Kibsgaard,[1] Thomas R. Hellstern,[1] Kara D. Fong,[1] Sang Chul Lee,[2] and Robert Sinclair[2]

1. Stanford University Dept. of Chemical Engineering, Stanford, CA 94305 USA
2. Stanford University Dept. Materials Science & Engineering, Stanford, CA 94305 USA

Photoelectrochemical (PEC) water splitting could provide a sustainable means of hydrogen production.[1] Recent research in PEC water splitting has focused on developing semiconductors and electrocatalysts suitable for application in a dual-absorber device configuration due to the high solar-to-hydrogen efficiencies tandem devices could enable.[2, 3] To develop fully functional water splitting systems, these semiconductors and electrocatalysts must be combined into integrated photoelectrodes.

Silicon is a promising candidate semiconductor for application in a photocathode due to its abundance, relatively low cost, excellent charger carrier transport, and near-ideal band structure.[4-10] However, to make silicon photocathodes efficient and stable, the surface of the semiconductor must be functionalized with a catalyst to reduce the overpotential necessary to drive the hydrogen evolution reaction (HER).[4-10] The silicon surface must also be protected to prevent the formation of an insulating SiO_2 layer, which can degrade device performance.[4, 5]

Molybdenum sulfide materials offer a promising solution to both of these challenges. Nanostructured, crystalline MoS_2 HER catalysts possess excellent activity and stability.[11, 12] We show that MoS_2 can also confer these benefits of high activity and stability to silicon photocathodes. We fabricate conformal protective coatings of MoS_2 on silicon using a simple thermal synthesis. This technique yields silicon photocathodes that show no loss in activity after 100 hours of operation. While the stability of this structure is among the best reported for silicon photocathodes, the low density of HER-active MoS_2 edge sites at the electrode/electrolyte interface limits the performance of the device. To increase the HER activity, we employ a second form of molybdenum sulfide, thiomolybdate $[Mo_3S_{13}]^{2-}$ nanoclusters, which possess high activity due to a structural motif similar to MoS_2 edge sites.[13] As illustrated in the figure below, incorporation of the $[Mo_3S_{13}]^{2-}$ clusters improves the photocurrent onset voltage to within ~150 mV of the best reported Pt/Si photocathodes.[7] Our results demonstrate that molybdenum sulfides can be employed as multifunctional coatings for photocathodes, serving as both catalysts and protecting layers.

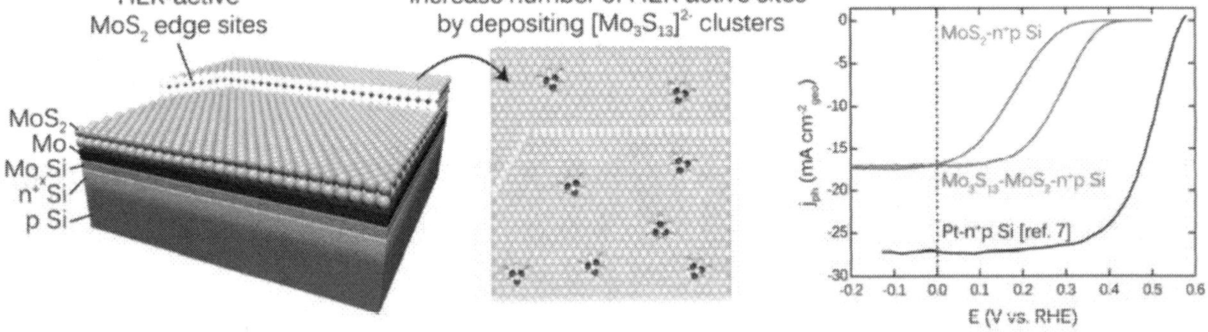

1. M. G. Walter, et al., *Chem Rev*, **110**, 6446 (2010).
2. L. C. Seitz, et al., *ChemSusChem* (2014).
3. M. F. Weber, et al., *Journal of The Electrochemical Society*, **131**, 1258 (1984).
4. A. B. Laursen, et al., *Physical Chemistry Chemical Physics*, **15**, 20000 (2013).
5. B. Seger, et al., *Journal of the American Chemical Society*, **135**, 1057 (2013).
6. J. R. McKone, et al., *Energy & Environmental Science*, **4**, 3573 (2011).
7. S. W. Boettcher, et al., *Journal of the American Chemical Society*, **133**, 1216 (2011).
8. Y. Hou, et al., *Nature Materials*, **10**, 434 (2011).
9. B. Seger, et al., *Angewandte Chemie*, **124**, 9262 (2012).
10. D. V. Esposito, et al., *Nature Materials*, **12**, 562 (2013).
11. Z. Chen, et al., *Nano letters*, **11**, 4168 (2011).
12. J. Kibsgaard, et al., *Nature Materials*, **11**, 963 (2012).
13. J. Kibsgaard, et al., *Nature Chemistry*, **6**, 248 (2014).

Ammonia is among the most important chemicals produced by the chemical industry today, largely because reactive nitrates are key to fertilizer manufacturing and therefore critical to the maintenance of our food supply. However, the traditional Haber-Bosch process that thermochemically converts atmospheric N_2 into NH_3 has significant environmental and energetic drawbacks that make the possibility of alternatives like electrochemical synthesis attractive. Furthermore, natural nitrogen fixation processes in rhizobium bacteria via nitrogenase enzymes suggest that room temperature electrochemical reduction of N_2 may be possible, albeit through a mechanism that differs from Haber-Bosch by delaying breaking the strong N_2 bond until a set of initial proton-electron transfers.

Despite its promise, however, very little is known about whether heterogeneous electrocatalysis of N_2 to NH_3 can be done efficiently or is even possible. In this work, we present theoretical calculations showing that scaling between two key intermediates in the low-temperature mechanism severely limit the theoretical overpotentials of N_2 electroreduction on transition metal surfaces. In particular, we show that the linear correlation of N_2H^* and NH_2^* binding free energies, which represent the reductive adsorption of N_2 and desorption of NH_3 respectively, increase the energy requirement past a threshold which allows for selective conversion of N_2 since the reduction of protons into H_2 is so much more favorable. Ultimately, this trend suggests that materials with different scaling properties or moieties that allow for stabilization of one intermediate relative to another will be necessary to the realization of electrochemical synthesis of NH_3.

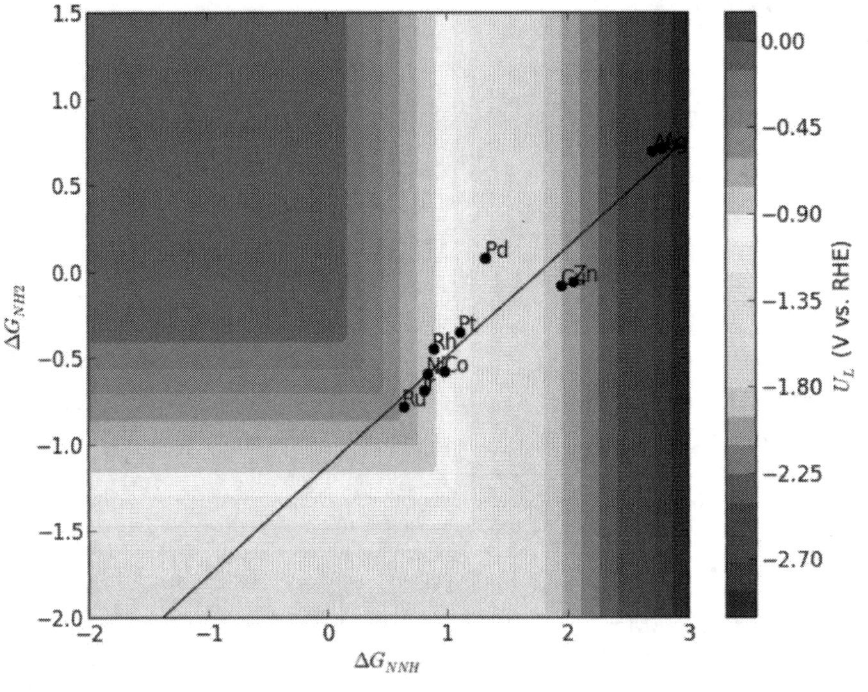

Figure 1 – Heatmap of the theoretical overpotential as a function of adsorption free energies of N_2H^* and NH_2^*, points corresponding to N_2H^* and NH_2^* adsorption free energies for 211 steps on transition metal catalysts with best fit scaling line.

379440 Activity and Stability of Gold and Copper Nanoparticles for Electrochemical Carbon Dioxide Reduction

Abstract:

Although the vast majority of fuels and hydrocarbon products are presently derived from petroleum, there is immense interest in the development of alternate routes for synthesizing these products by hydrogenating carbon oxygenates. Electrochemical methods of reducing carbon dioxide could serve as a method of storing electrical energy derived from intermittent sources like solar and wind if efficient catalysts with high hydrocarbon selectivity are developed. Although metals in the form of foils are increasingly well-characterized as electrocatalysts for carbon dioxide reduction, the activity and stability of their nanoscale counterparts remain poorly understood. We present an understanding of the electrochemical conditions and ligands that afford control over the stability of gold and copper nanoparticles for electrochemical carbon dioxide reduction. Random walk simulations reveal the mechanism by which the nanoparticles lose surface area and assemble into dendrites under polarization conditions. In addition, we have found that the gold and copper nanoparticles exhibit selectivities for electrochemical carbon dioxide reduction that are distinctly different from that of their foil counterparts. The changes in hydrocarbon selectivity for the copper nanoparticles are due to an underlying difference in the mechanism by which electrochemical carbon dioxide reduction proceeds on the nanoparticle surface. Our understanding of the activity and stability of copper and gold nanoparticles has enabled us to design a catalytic network for electrochemical carbon dioxide reduction.

386268 Quantitative Analysis of Light-Off Profiles for Kinetic Parameters

Abstract:

Kinetic parameters for heterogeneous reactions, such as the activation energy, E_a, the Arrhenius pre-exponential factor, A, and the reaction orders for each reactant/product, are typically obtained from a series of experimental rate measurements, each made under isothermal conditions. The accuracy of the resulting activation parameters can be low (although it is often unacknowledged to be so), due to the small number of measurements, the limited temperature range used, and the assumption of differential (i.e., gradient-less) reactor conditions.

Although light-off profiles are quick and easy to generate, and may contain high quality data over conversions ranging from negligible to full, such data are rarely exploited quantitatively. We considered whether curve-fitting non-isothermal data could be used to generate faster and more accurate kinetic information about both the rate law and the activation parameters. Simulation of light-off profiles shows that different rate laws give rise to distinctive profile shapes, influenced primarily by the reaction order of the limiting species.

Light-off profiles for CO oxidation by O_2 over PdO supported on γ-alumina were generated with different residence times and inlet concentration ratios. Our simulated light-off profiles correctly predict the shape and position of each experimental non-isothermal dataset. By fitting the kinetic equations to the data, values for the activation parameters and reaction orders were obtained that are identical to those obtained from traditional isothermal measurements; however, the experimental uncertainty and time required to obtain them were greatly reduced. This method can be used to screen many catalysts rapidly to compare their kinetic behavior and to look for irregularities that may signal transport limitations, a change in rate-limiting step, or a change in surface composition, making it potentially useful as a tool to complement high-throughput catalyst synthesis.

Prevalence of Bimolecular Activation Routes for CO, NO, and O_2 at Catalytically Relevant Coverages on Metal Catalysts

David Hibbitts and Enrique Iglesia*

Department of Chemical and Biomolecular Engineering, University of California, Berkeley, California 94720, United States

*iglesia@berkeley.edu

Metal-catalyzed reactions at steady state often involve high coverages of one or more reactive intermediates which limit the availability of vacant sites and favor bimolecular reactions. Fischer-Tropsch synthesis (FTS), nitric oxide reduction (by H_2), and oxygen reduction reactions (ORR) involve metal-catalyzed activations of diatomic species, which occur either via reactions with vacancies to form strongly-bound atomic adatoms or following initial reactions with chemisorbed H-atoms (H*). Kinetic and isotopic data, and free energy activation barriers obtained from density functional theory, are used here to investigate the mechanism of CO activation during CO-H_2 reactions on Co and Ru catalysts, NO activation during NO-H_2 reactions on Rh and Pt catalysts, and O_2 activation during O_2-H_2 reactions on Pt and Au. Catalyst particle size either increases (during CO-H_2) or weakly affects (during NO-H_2) turnover rates, indicating that low-index planes of catalysts are reactive. Direct CO and NO activations have large enthalpy and free energy barriers on low-index planes, inconsistent with observed rates, leading to the invocation of defect sites being active for CO and NO activation. Hydrogen-assisted routes, in contrast, involve initial reactions of CO* and NO* with chemisorbed H-atoms (H*) to form *HCOH* and *HNOH* species on low-index planes, which then undergo C-O and N-O activations, respectively. These hydrogen-assisted routes are consistent with rates proportional to H_2 pressure (when H* is not a most abundant surface intermediate), and have lower enthalpic and free energy activation barriers than direct dissociations (at any CO* or NO* coverage) on low-index planes of metal clusters. H-addition weakens the multiple bonds in NO and CO, and allows coordination of each atom to metal surfaces, facilitating metal-catalyzed activation. H-addition reactions do not always involve direct reaction with H*; H_2O increases rates of CO activation during FTS on Co and Ru catalysts by mediating O-H bond forming reactions via a H-shuttling mechanism. H* transfers to a nearby H_2O molecule to form a short-lived H_3O^+ intermediate which then protonates CO* (to form COH* which then forms *HCOH*) or HCO* (to form *HCOH*), resulting in lower enthalpic and free energy activation barriers to form *HCOH* intermediates. Such H_2O involvement during O-H bond formation reactions with CO is reminiscent of similar reactions during aqueous-phase oxygen reduction reactions (ORR), which take place at cathodes of fuel cells as well as during aerobic oxidation reactions on Pt and Au catalysts. DFT-derived activation barriers suggest that O_2 reacts with H*, H_2O, or H_3O^+ to form OOH* and *HOOH* intermediates which precede O-O bond activation on Pt or Au catalysts.

DFT-derived activation barriers suggest that in the absence of H_2, NO* activates via bimolecular reactions with NO* to form N_2O and O* in a single elementary step at high NO* coverages, similar to O_2-activation via reactions with CO* to form CO_2 and O* at high coverages during CO oxidation on Pt or NO-activation via reactions with CO* to form CO_2 and N* during NO reduction by CO on Pt. These bimolecular reactions (NO-NO, NO-CO, O_2-CO) prevail over direct dissociation due to the scarcity of vacancies at conditions relevant to steady state catalysis. This scarcity reinforces the preference for bimolecular reactions with H*, described above, during CO-H_2, NO-H_2 and O_2-H_2 reactions where H*-addition also weakens the multiple bonds in CO, NO, and O_2 and coordinates the atoms to the metal surface, critical for bond activations.

378978 Multi-Scale Investigation of Pd/Ceria Catalysts Via Density Functional Theory and Classical Reaxff Simulations

Abstract:

Oxide formation on palladium surfaces impacts the activity and selectivity of Pd/ceria catalysts toward hydrocarbon activation. Results derived from Density Functional Theory (DFT) suggest that mixed $Pd_xCe_{1-x}O_\delta$ surface oxides are stable under oxidizing reaction conditions, and that these oxides form unique Pd-O-Ce linkages that enable Pd to cycle between oxidization states enabling them to facilitate hydrocarbon activation. Although these mixed surface oxides are highly active, there is little consensus as to the exact nature of these sites. Indeed, many conflicting examples can be found indicating that either metallic Pd clusters or oxidized Pd surface species must be present on the catalyst surface to achieve high activity. Since DFT-based methods are limited to highly idealized surface models, they cannot easily assess Pd oxidation state transitions and geometry transformations at the complex cluster/support interface. We have developed a classical Pd/Ce/O interaction potential for the ReaxFF reactive force field to examine the metal-support interface. With this interaction potential, we employ hybrid Monte Carlo/molecular dynamics methods to supported Pd clusters to assess the formation of mixed Pd-Ce interfacial oxides under varying reaction conditions. Initial results indicate that Pd-O-Ce structures similar to those identified in the mixed $Pd_xCe_{1-x}O_\delta$ surface oxide form at the Pd/ceria interface, suggesting that the interface is responsible for hydrocarbon activation in traditionally prepared catalysts featuring ceria-supported Pd clusters.

384622 Catalyst/Support Interactions Between Pt Nanoparticles and Amorphous Silica Using Density Functional Theory

Abstract:

Metal nanoparticles (NPs) play a key role in catalysis due to large surface-to-volume ratios and low-coordination active surface sites. These catalysts are typically immobilized on oxide supports to suppress sintering, which leads to catalyst deactivation and degradation of catalyst selectivity. An atomistic understanding of catalyst systems is essential to the rational design of improved catalyst systems. Catalyst activity and selectivity are highly dependent on shape and size of NPs as well as the nature of the oxide support. Despite significant advancement of experimental characterization techniques, a detailed understanding of the catalyst-support interface for very small NPs (~ 1 nm or smaller) is largely precluded by experimental investigations. Computational methods provide one means to overcoming this challenge. In previous studies catalyst supports have generally been either neglected or treated as highly ideal structures, however, oxide supports are often used in an amorphous state, exhibiting a wide range of surface sites.

Due to its thermal stability and tunable porosity, and hence specific surface area, amorphous silica is widely used as a catalyst support, but is still poorly understood at the atomistic level due to limitations in both experimental and computational methods. We have developed and experimentally validated a method for generating realistic atomistic surface models of amorphous silica for a range of temperatures by simulating the process of surface dehydroxylation.

Using these surface models, we have studied the effects of amorphous silica supports on small platinum NPs (13 atom cuboctahedral structures) with regards to their catalytic properties. Specifically, we have studied the impact of catalyst-support interactions on NP morphology and energetics. We find that the morphology of Pt clusters undergo significant restructuring that depends on the local structure at the Pt-Silica interface. This restructuring typically leads to significantly increased exposure of lower-coordination atoms than are present on the unsupported NP. Additionally, we observe a compressive Pt-Pt bond strain that increases with the number of bonds at the Pt-Silica interface. The energy of Pt-silica interaction depends almost exclusively on the number of oxygen atoms bonding to the Pt cluster, which is in turn a function of surface hydroxylation, and hence, the pretreatment temperature of the silica support. We are currently extending these studies to larger systems (55 and 147 atoms) as well as studying the effects of Pt-Silica interactions on CO binding energies for select systems.

Kinetic study of propane dehydrogenation over chromia-alumina catalyst

Shuai Tan, Seung Won Choi, Sankar Nair, David S. Sholl and Christopher W. Jones*
School of Chemical & Biomolecular Engineering, Georgia Institute of Technology, Atlanta, GA 30332
*christopher.jones@chbe.gatech.edu

Jason S. Moore, Yujun Liu, Ravindra S. Dixit and John G. Pendergast
Engineering & Process Sciences, The Dow Chemical Company, Freeport, TX 77541

With the influx of abundant light gases in North America, technologies for their valorization are again rising to prominence. Propane dehydrogenation (PDH) is a potential pathway for production of valuable olefins from such light gases. Current promising industrial processes are based on Cr (*Catofin*) and Pt (*Oleflex*) catalysts. A key challenge with any catalytic system from PDH is cracking reaction and catalyst deactivation via carbon deposition. Furthermore, PDH is an endothermic reaction that is equilibrium limited. Development of membrane reactors for PDH, combining catalytic reaction and membrane separation in the same device, can significantly improve olefin productivity and decrease the energy cost of PDH by shifting the reaction equilibrium. Among several factors, a detailed understanding of the PDH catalyst is required for the successful PDH intensification by membrane reactors. In this work, we discuss a detailed study of PDH catalysis over Al_2O_3-supported Cr_2O_3, doped with Na_2O. Alkali addition (1 wt%) results in an increase in C_3H_6 selectivity and consequently provides higher C_3H_6 yield compared to the alkali-free catalyst. In addition, the alkali additive suppresses production of CH_4 significantly. Detailed kinetic data are interpreted by several models based on Langmuir-Hinshelwood mechanisms to aid in the design and modeling of the membrane reactor. Details of the catalytic behavior are captured by estimating key parameters (i.e., activation energy, heat of adsorption), and the results are compared to those for related PDH catalysts in the literature.

Experimental and Simulation Study of Selective Ammonia Oxidation on Multi-Functional Washcoated Monolith Catalysts

Sachi Shrestha*, M. P. Harold*[1], K. Kamasamudram**[2] and A. Yezerets**
*Dept. of Chemical and Biomolecular Engineering, University of Houston, Houston, TX 77204-4004, USA
**Cummins Inc., 1900 McKinely Av., MC50197, Columbus, IN 47201, USA
*[1]mharold@uh.edu; *[2]krishna.kamasamudram@cummins.com

Introduction

The emerging technology of choice to mitigate NOx from the exhaust from medium- and heavy-duty diesel engine is NH_3 - selective catalytic reduction, where NH_3,generated on-board by urea hydrolysis, reduces NOx to N_2 on a metal (Cu or Fe)-exchanged zeolite catalyst, which has very high NH_3 storage capacity [1]. One drawback of this technology is the slip or emission of unreacted NH_3 during temperature upswings, dynamical effects, and/or deactivated catalyst. In order to minimize the breakthrough of NH_3 an Ammonia Slip Catalyst (ASC) is used. In the ASC, slipped NH_3 is selectively oxidized to N_2. Pt is a well-known highly active catalyst for NH_3 oxidation, however, the oxidation of NH_3 on Pt also leads to undesired (for this application) byproducts such as N_2O, NO, and NO_2.

Previous studies have demonstrated the benefit of using a bi-functional catalyst to selectively oxidize NH_3 to N_2. The two functions include an oxidation function (Pt/Al_2O_3) and a NOx reduction function, the latter of which comprises a Fe- or Cu- exchanged zeolite catalyst [2, 3]. One ASC configuration that has been proposed is a dual-layer monolith structure, comprising a bottom layer of Pt/Al_2O_3 and a top layer of a metal-exchanged-zeolite [2]. The role of the Pt-containing layer is to oxidize NH_3 to a mixture of NOx and N_2, while the role of the zeolite layer is to selectively reduce the NOx with counter-diffusing NH_3.

In this study we synthesize and evaluate various catalyst architectures that combine Pt/Al_2O_3 with Fe- or Cu- zeolitic catalysts, including a dual layer comprising a Pt/Al_2O_3 bottom layer and a Fe- or Cu-zeolite top layer, as well as mixed and co-impregnated multi-component catalysts. Our objective is to elucidate the effect of the multi-component (Pt/Fe or Pt/Cu) catalyst architecture on NH_3 oxidation performance with the goal of maximizing NH_3 conversion and N_2 yield over a wide range of conditions while minimizing the loading of Pt.

Materials and Methods

A series of catalyst were synthesized in the laboratory in order to study its NH_3 oxidation activity. Pt/Al_2O_3 catalyst was synthesized by incipient wetness impregnation method, and a commercial Fe-zeolite catalyst was provided by Clariant. Slurries containing the catalyst powders were washcoated onto 400 cpsi cordierite monoliths. The washcoat had either a single layer or dual layer architecture with the former comprising mixed layers of Pt/Al_2O_3 and Fe- or Cu- zeolite and the latter comprising a bottom Pt/Al_2O_3 layer and a top Fe- or Cu- zeolite layer. The weight loadings of Pt/Al_2O_3 and Fe-zeolite catalyst were varied systematically to study their effects on NH_3 oxidation activity and product selectivity. The Pt loading in Pt/Al_2O_3 catalyst was varied from 0.7-10.5 g/ft^3, while the zeolite loading was varied from 0.5-1.5 g/in^3. The single layer Pt/Al_2O_3 with the same loading served as the reference.

The bench scale reactor setup consisted of gas feed system, reactor system, and data analyzing system. The analyzing system consisted of FT-IR and QMS to measure the concentration of the effluent gas. The total flow rate was fixed as 1000 sccm, corresponding to the GHSV of 66k hr^{-1} for 2 cm long monolith and 265k hr^{-1} for 0.5 cm long monolith. For all the experiments, the feed concentration of 500 ppm NH_3, 5% O_2, 2.5% H_2O and 2% CO_2 was used with varying levels of NO (0 – 500 ppm).

For simulations, a 1+1 D model was developed which accounts for washcoat diffusion and reaction [4]. The microkinetic reaction network for NH_3 oxidation on Pt/Al_2O_3 catalyst was obtained from Scheuer et al. [2], and global kinetic model for NH_3-SCR was obtained from Metkar et al. [4]. The reaction kinetic for NH_3 oxidation on Pt/Al_2O_3 was fitted to experimental data, and accounts for NO oxidation equilibrium. Second order finite difference method was used to discretize the model equation. The discretized equation was solved in Matlab using odesolver ODE15s.

Results and Discussion

The detailed descriptions on the catalyst used for this study is given in Table 1. Figure 1(a) and (b) shows the NH_3 conversion and product distribution for NH_3 oxidation reaction on Pt/Al_2O_3 and dual layer FeZ(1.5)Pt(10) catalyst respectively at the GHSV 66k hr^{-1}. It was seen that at this space velocity the configuration of the catalyst had no effect on NH_3 conversion capability of the catalyst. Figure 1 (a) shows that NH_3 oxidation on Pt/Al_2O_3 catalyst reached complete conversion by ~ 210 °C and the product yield varied significantly with catalyst temperature. At lower temperature, the reaction showed high selectivity towards N_2 and N_2O, but with increasing temperature the N_2 and N_2O selectivity dropped monotonically giving way to an increase in the NOx selectivity. On the addition of the Fe-ZSM-5 layer resulted in a dramatic increase in the N_2 product yield and decrease in NO yield while having no effect on the NH_3 conversion. These result shows that the Fe-ZSM-5 layer works synergistically with underlying Pt/Al_2O_3 through the lean NH_3 reduction of NOx formed in Pt-layer to N_2 in presence of stored NH_3.

The developed model is effective in predicting the qualitative trends in the data. Further optimization works are in progress to quantitatively fit the experimental results. Highlights of this work include determination of diffusivity in the Fe-ZSM-5 catalyst through the noble experimental technique developed, as well as fitting data at different possible ASC feed condition and GHSVs.

These and other data will be reported and a phenomenological model proposed that explains the main trends.

Significance

This work should benefit in understanding the coupling of oxidation and reduction/storage component of ASCs, thus, helping in further advancement of present commercial slip catalysts.

References

[1] K. Kamasamudram, N. Currier, M. Castagnola, H. Chen, SAE International Journal of Engines 4(1), (2011) 1810-1821.

[2] A. Scheuer, W. Hauptmann, A. Drochner, J. Gieshoff, J., H. Vogel, M. Votsmeier, Appl. Catal. B: Environmental 111–112 (2011) 445-455.

[3] S. Shrestha, M.P. Harold, K. Kamasamudram, and A. Yezerets,Top. in Catal., 56 (2013) 182-186 .

[4] P.Metkar, M.P. Harold, and V. Balakotaiah, Chemical Engineering Science, 87 (2013) 51-66.

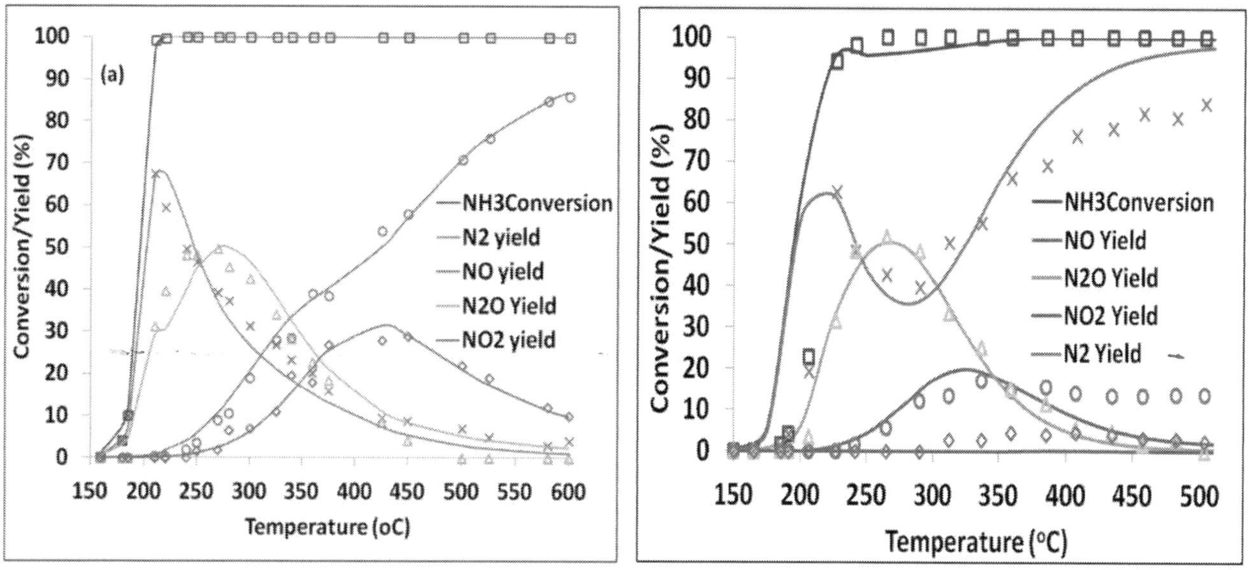

Figure 1: Comparison of activity and product yield of NH_3 oxidation reaction. (a) Pt/Al_2O_3 (b) Dual Layer FeZ(1.5)Pt(10). Reaction Conditions: 500 ppm NH_3, 5% O_2, 2.5% H_2O, 2% CO_2 and 66k hr^{-1} GHSV. Symbols: Experiment, Solid Line: Model Predicted

Table 1. Catalyst Used for evaluation

Catalyst Name	Fe-zeolite Loading (g/in^3)	Pt/Al_2O_3 Loading (g/in^3)	Pt Loading (g/ft^3 Monolith Volume)	Pt Loading (g Pt per 100 g washcoat)
Pt(10)	0	1.4	10.5	0.46
FeZ(1.5)Pt(10)	1.5	1.3	10.5	0.46

363330 A New Alternative Process for Nitric Acid Production Based on Chemical Looping Approach

Abstract:

Abstract

There have been no significant changes in the reaction scheme for the manufacture of nitric acid since the discovery of the process by Ostwald in 1902. This paper reports the theoretical prediction and preliminary experimental investigation for an alternative process of nitric acid production using metal oxide based chemical looping cycles. The first step would be the reduction of metal oxide by ammonia to produce nitric oxide (NO), and the second step would be the regeneration of metal oxide either by air oxidation or/and by water hydrolysis depending on the metal oxide used. The most important aspect of this process is the replacement of highly exothermic ammonia oxidation step in Ostwald's process by the endothermic oxidation of ammonia by metal oxide. The thermodynamic feasibility and phase diagram analysis is carried out for the proposed process using HSC and Factsage packages without considering the catalytic effects of the metal oxides. There is a probability of differences in theoretical and actual process reactions as it is difficult to predict the exact reaction mechanism theoretically. However, at this stage the process seems to be exergetically more efficient with a smaller carbon footprint than conventional nitric acid production. The cases of Cu/CuO, CoO/Co_3O_4, and Fe_3O_4/Fe_2O_3 were experimentally checked for their potential to react with ammonia over a wide range of temperatures at atmospheric pressure. The effect of particle size of metal oxide and the gas residence time in the reactor was also studied for all the three metal oxides. The experiments were carried out in a batch reactor in spouting mode so as to have efficient gas-solid contact. The maximum ammonia to nitric oxide conversions obtained in the batch reactor experiments for CuO, Co_3O_4, and Fe_2O_3 were around 90 %, 70 %, and nearly 50 %, respectively. This paper presents the experimental results and discusses the proposed reaction scheme for the alternative process for nitric oxide production following a chemical looping based approach.

Insights into Solvent Effects in Catalytic Reactions for Biomass Conversion

M. A. Mellmer, C. Sener, J. M. R. Gallo, J. S. Luterbacher, D. M. Alonso, J. A. Dumesic

Department of Chemical and Biological Engineering, University of Wisconsin-Madison, Madison, WI

Biomass conversion into platform molecules and fuels has received increasing interest in our current transition to a sustainable source of chemicals and energy. Recently, the use of organic solvents has been shown to be beneficial in the chemical conversion of lignocellulosic biomass. One such solvent is *gamma*-valerolactone (GVL), which can be produced from both the hemicellulose and cellulose fractions of lignocellulosic biomass and is miscible with water, allowing for monophasic reaction systems. The use of GVL as a solvent for biomass conversion reactions has led to significant improvements in reaction performance compared to conversion in aqueous media, where increased catalytic activity and higher selectivity to desired reaction products have been demonstrated. Furthermore, we have recently reported that the simultaneous conversion of hemicellulose and cellulose can be achieved using GVL as a solvent in a single reactor, eliminating the need for pretreatment and/or separation steps.[1] It was shown that GVL not only increases the selectivity to desired products, but it also solubilizes biomass and humins (*i.e.*, biomass degradation products) at typical biomass processing conditions, thereby avoiding solids handling problems. More recently, we have taken advantage of accelerated rates of cellulose and hemicellulose deconstruction in GVL-H_2O solvent mixtures to develop a processing strategy to produce concentrated streams of C_5 and C_6 sugars from biomass using only dilute concentrations of mineral acids without using enzymes or ionic liquids.[2] More generally, other polar aprotic organic solvents, such as *gamma*-lactones and furans, have shown comparable benefits to GVL in biomass conversion processes.[3]

The aforementioned results indicate that GVL and other polar aprotic organic compounds are promising solvents for biomass processing reactions; however, the fundamental scientific basis for solvent effects in biomass conversion reactions is limited at present. We completed reaction kinetics studies to quantify the effects of polar aprotic organic solvents on the reaction rates and selectivities of acid-catalyzed reactions of relevance for biomass conversion (*e.g.*, xylose dehydration to furfural). Based on kinetic modeling analysis, we have determined that increased selectivities are due to advantageous relative reaction rate increases in the presence of GVL, compared to reactions carried out in water. Furthermore, our results provide a foundation for understanding the effects of aprotic organic solvents on acid-catalyzed biomass conversion reactions. We propose that these increased rates when using polar aprotic solvents stem from changes in the relative stabilization of the acidic proton catalyst versus stabilization of protonated transition states for these acid-catalyzed biomass conversion reactions.

[1] D. M. Alonso, S. G. Wettstein, M. A. Mellmer, E. I. Gurbuz, J. A. Dumesic, *Energy & Environmental Science* **2013**, *6*, 76-80.

[2] J. S. Luterbacher, J. M. Rand, D. M. Alonso, J. Han, J. T. Youngquist, C. T. Maravelias, B. F. Pfleger, J. A. Dumesic, *Science* **2014**, *343*, 277-280.

[3] J. M. R. Gallo, D. M. Alonso, M. A. Mellmer, J. A. Dumesic, *Green Chemistry* **2013**, *15*, 85-90.

385298 Catalytic Gasification: A Green Alternative for Waste Management

Abstract:

Waste management alternatives for ground base applications, such as municipal waste management, are under consideration. One such alternative is the use of Wet Thermal Catalytic Oxidation (WTCO) processes. These processes promote the conversion of long polymeric chains into synthetic or "supply" gas (syngas) at temperatures lower than those required by more traditional gasification technologies. The presence of a catalyst is known to lower the energy required to promote gasification reactions, making WTCO processes an attractive route for waste management and biomass to energy strategies. This technology was examined using a varying degree of fidelity waste simulants (different combinations polymers and other waste species). The conversion of these polymeric chains through catalytic oxidation was studied for two different catalysts (Ru and Pt on Al2O3) combined in a slurry of water and waste polymers. Thermal characterization of simulants and catalytic slurries complemented the WTCO experiments. Data collected in a research-grade laboratory reactor were quantified by gas chromatography, and used for a preliminary kinetic characterization of the gasification process and used for preliminary process model. A phenomenological model of a multi-phase gasification process was formulated and is anticipated to enable the optimization of process configurations for continuous waste management with the additional potential of fuel production.

385930 Reverse Water Gas Shift Chemical Looping for Carbon Dioxide Conversion to Carbon Monoxide on Reduced Perovskite Oxides

Abstract:

Previously, we have shown that from the (La,Sr)CoO_3 perovskite family, $La_{0.75}Sr_{0.25}CoO_3$ was optimum in the reduction of CO_2 to CO because of its structure stability and maximized CO formation at the lowest temperatures. This perovskite was used in the reverse water gas shift chemical looping process (RWGS-CL) for the conversion of carbon dioxide to carbon monoxide that can later be hydrogenated to liquid fuels. In the RWGS-CL process, a stream of H_2 reduces the $La_{0.75}Sr_{0.25}CoO_3$ perovskite. Then, CO_2 is flowed and converted to CO on the reduced oxides. Temperature-programmed (H_2-TPR, CO_2-TPO) and isothermal CO_2 conversion experiments were performed on the perovskite to determine the optimum operating temperatures for the H_2-reduction (500 °C) and CO_2-conversion (850 °C) steps.

The RWGS-CL process was studied for three reaction cycles. After the second cycle, the RWGS-CL exhibited stability of the crystalline structures and constant CO production rates (as high as 172.6 mol of CO/ grams of perovskite /min).

The cyclic nature of the process presents an advantage over the reverse water gas shift (RWGS) reaction, because with separate H_2 and CO_2 streams, methane formation is suppressed. Furthermore, the materials remain stable after the second step, whereas catalysts used for RWGS generally deactivate. The high CO formation rates and the repeatability of the process make RWGS-CL a promising technology for CO_2 conversion that could produce hydrocarbons, if a renewable hydrogen source is available.

386688 Spectroscopic Insights into Cellulose Hydrolysis Utilizing Molten Salt Hydrates As Reaction Media. Understanding Structural Changes at the Molecular Level

Abstract:

The diminishing availability of fossil resources as well as the need of production fuels and chemicals via alternative routes have led to the development of new, sustainable and environmental friendly technologies that utilize biomass as raw material. In this respect, non-edible lignocellulosic biomass is a promising renewable feedstock since it is abundant, does not directly compete with the food chain, and could lead to nearly carbon-free processes with concomitant reduction in CO_2 emissions. Development of methods for cellulose hydrolysis under mild conditions has been a topic of great interest, since they can be considered suitable for large-scale applications for the production of valuable sugar hexoses.

The utilization of molten salt hydrates as cellulose solvents/ modifiers is known to promote cellulose hydrolysis under mild conditions (low acid concentration / temperature). In this respect, the present study focuses on the investigation of the effect of molten salts (LiBr, LiCl, $ZnBr_2$ and $ZnCl_2$) on cellulose structure as well as on the reactivity for the hydrolysis reaction. Our catalytic results revealed that complete cellulose hydrolysis could efficiently take place at temperatures as low as 85°C and low acid concentration (0.05M H_2SO_4) with glucose yields ~85% by using LiBr molten hydrate as reaction media.

We pay special attention to the role of the nature as well as the concentration of each salt on the structure of cellulose by means of vibrational spectroscopies (ex-situ and in-situ ATR/FTIR, and Raman). Careful analysis of pertinent wavenumber regions allowed us to elucidate the effect of salts on the inter/intra-molecular hydrogen bond network, on the strength as well as the conformation of the glycosidic bond. It became clear that the treatment of cellulose, even at room temperature, with molten salt hydrates is effective in breaking the inter- and intra-molecular hydrogen bond network within cellulose chains explaining the enhanced catalytic activity during cellulose hydrolysis. The vibrational spectra also indicate an interaction of the salt hydrates with cellulose around C_6 of each anhydroglucose units. Changes in specific vibrational motions with time under realistic reaction conditions, monitored by in-situ ATR-FTIR spectroscopy, revealed the possibility of two different catalytic steps taking place for cellulose hydrolysis.

389635 Glycerol to Propylene Glycol over Copper Chromite: Byproducts and Pathways

Abstract:

Production of 1,2-propanediol through hydrogenolysis of glycerol over Copper-Chromite is a promising reaction pathway which has large potential markets. However, production is complicated by low catalyst stability and significant difficulty in preventing or removing unwanted reaction byproducts. In collaboration with a commercial partner the authors investigated addition of amine compounds during the reaction which acted as catalyst stabilizers. These extended catalyst lifetimes and reduced production of ethylene glycol, the main reaction by-product. Results of modeling experiments will be presented which characterize the remaining catalysis by-products and attempt to identify potential methods of inhibition during reaction.

394223 Exploiting Polymer-Nanoparticle Interactions to Create a Hydrogel with Biomedical Applications

Abstract:

Injectable hydrogels have recently become a relevant material platform in applications such as in vivo drug delivery or as cell carriers because they can be implanted in a minimally invasive manner and closely mimic biological tissue. Attractive properties for such injectable hydrogels are that they easily flow under applied stress (*shear-thinning*) and rapidly heal upon relaxation of the applied stress (*self-healing*). Previously, we developed a shear-thinning and self-healing injectable hydrogel driven by non-covalent interactions between hydrophobically modified cellulose and PEG-PLA nanoparticles. [1] The nature of this gel enables facile loading with a variety of therapeutics as well as tunable release profiles of multiple therapeutics. Specifically, the PEG-PLA nanoparticles allow for loading with hydrophobic therapeutics while a hydrophilic therapeutic can be entrapped within the aqueous phase of the gel.[2] In this work, we have quantified the factors that affect the mechanical properties and release profiles of the gels, demonstrating the range of their tunability. Additionally, we explored other nanoparticles and biopolymers for the fabrication of similar materials. Lastly, we studied the use of this injectable hydrogel as a cell carrier with osteogenic properties. Human mesenchymal stem cells (hMSCs), which are capable of differentiating into bone, cartilage, and fat cells, were encapsulated in the gel and survived injection through a 28 gauge needle into three-dimensional collagen hydrogels. Additionally, PEG-PLA nanoparticles were loaded with dexamethasone to stimulate osteogenic differentiation of the encapsulated hMSCs. We are currently exploring the use of this drug-loaded cell carrier as an injectable system for bone healing. Overall, we present an injectable hydrogel with tunable properties for the controlled delivery of therapeutics with facile synthesis and minimally invasive application *in vivo* that is capable of serving as a stimulatory cell carrier.

---

[1] Appel, E.A., Tibbitt, M.W., Langer, R., Nature Communications., **2014**, *in revision*.

[2] Greer, J.M., Appel, E.A., Tibbitt, M.W., Langer, R., *manuscript in preparation*.

Validation of High Throughput Electrochemical Gas Sensing Screening System

Zixuan Wang

November 17, 2014

UNIVERSITY OF SOUTH CAROLINA

Background

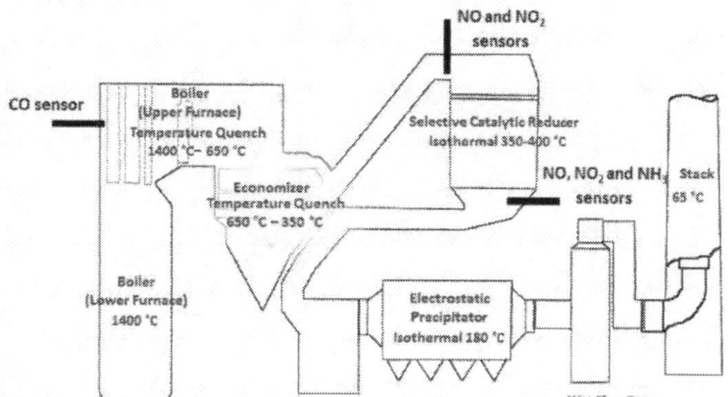

Harsh combustion environment - 800 ° C

Accurate and selective to target gas

UNIVERSITY OF SOUTH CAROLINA

High Throughput Methodology

Subatomic difference affect large scale

Not feasible to individually test material

High throughput synthesis and characterization methods

UNIVERSITY OF SOUTH CAROLINA

High Throughput Methodology

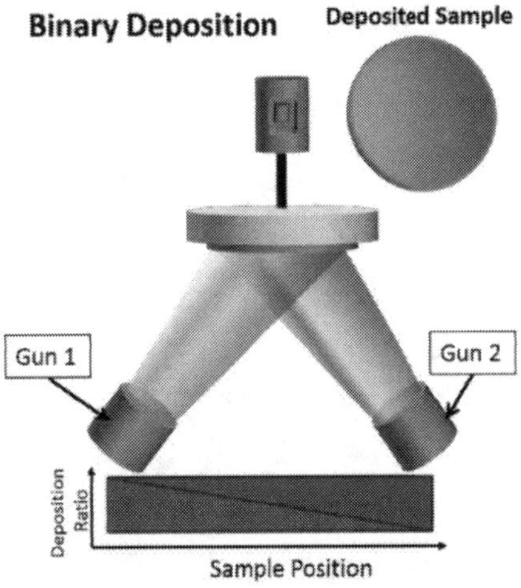

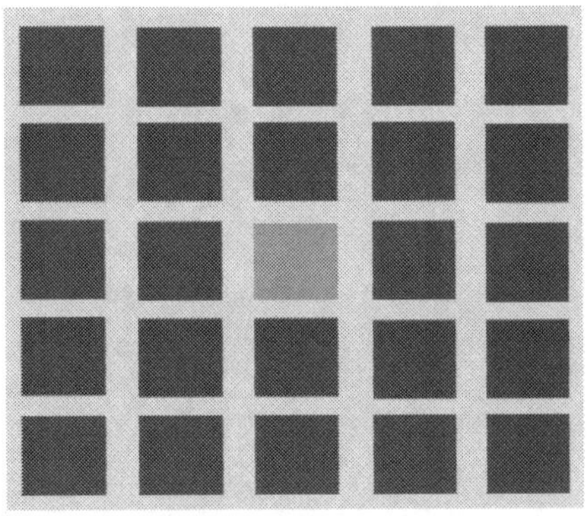

- Normalize affect of distance on the potential

UNIVERSITY OF SOUTH CAROLINA

Development of Uniform Standard Sample

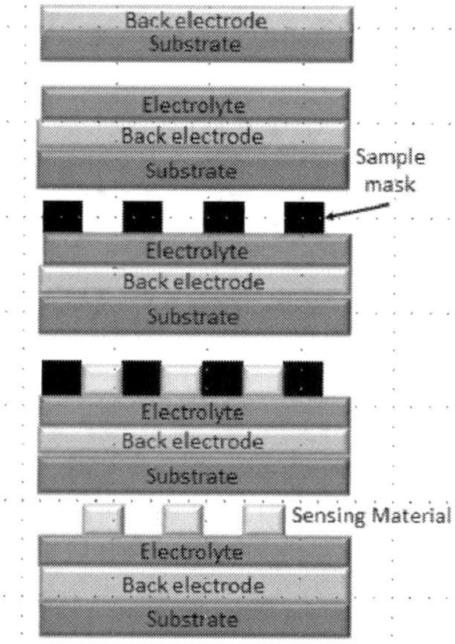

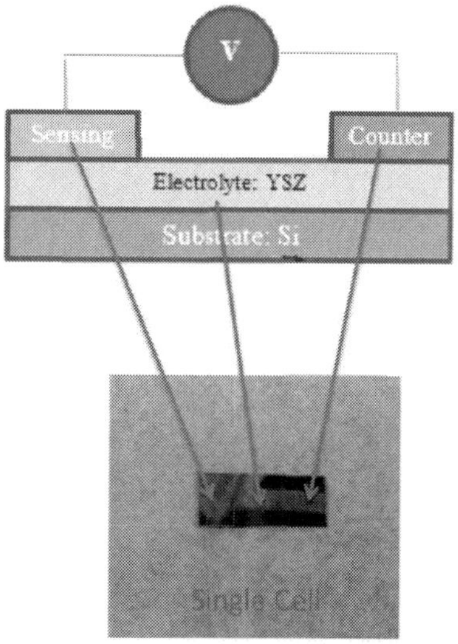

UNIVERSITY OF SOUTH CAROLINA

Surface Mechanisms

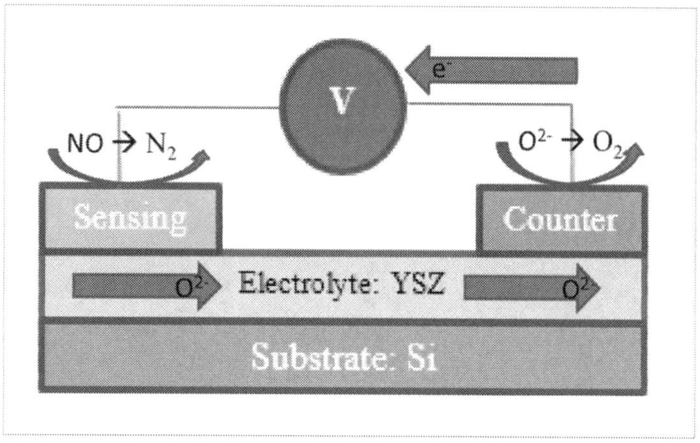

UNIVERSITY OF SOUTH CAROLINA

Sensors and Actuators B 121 (2007) 652-663

Reflectometry Results - YSZ

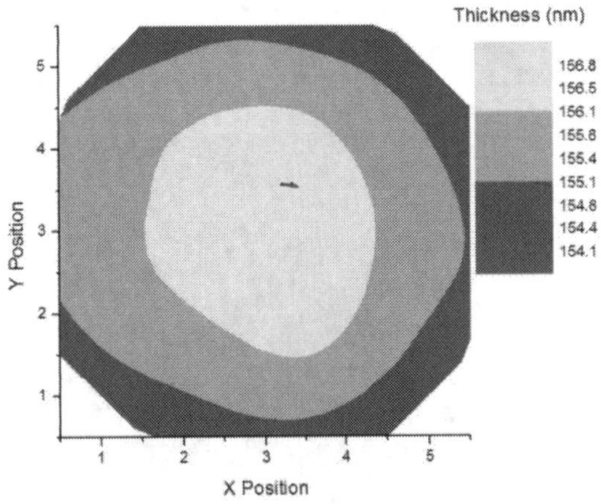

- Results confirm relative uniformity of sample (1.8% difference)

XRD Results - YSZ

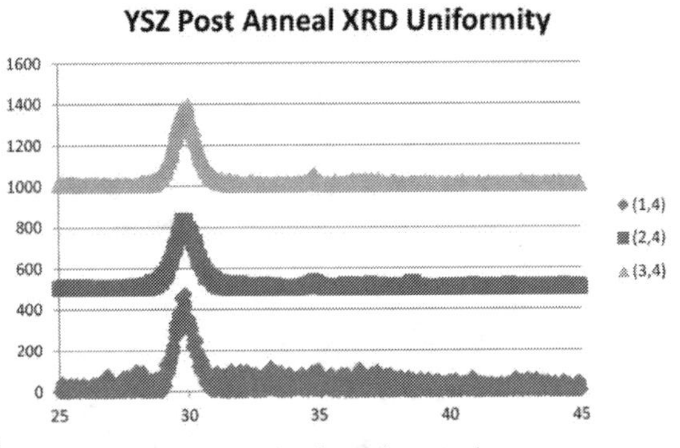

(1,4)	29.745	
(2,4)	29.836	34.848
(3,4)	29.876	34.859

- XRD analysis using Jade
- YSZ Cubic:
  - (111): 29.675°
  - (200): 34.398°
- Confirm identical phases

Reflectometry Results – WO_3

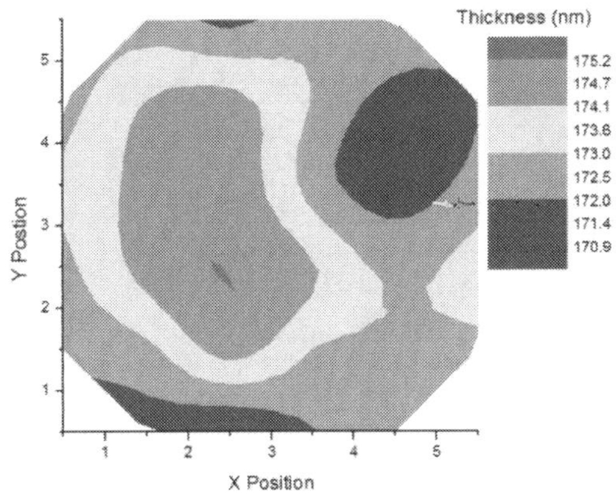

- Confirms relative uniformity (2.5% difference)
- Evidence some parts of the plate that are slightly thicker or thinner

UNIVERSITY OF SOUTH CAROLINA

XRD Results – Post Annealed WO_3

(1,4)	23.143	23.653	24.362
(2,4)	23.140	23.671	24.397
(3,4)	23.127	23.641	24.377

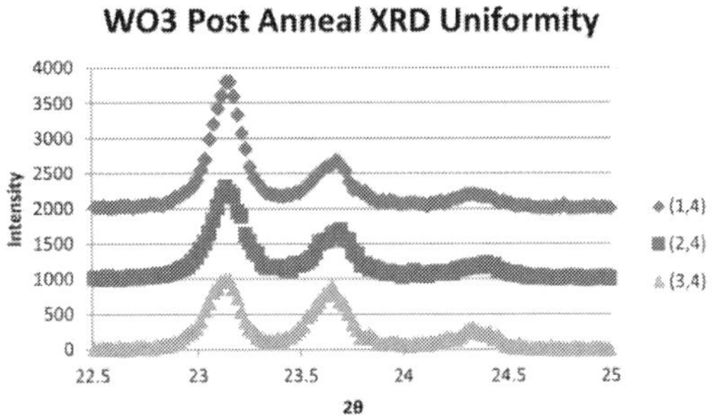

- XRD analysis using Jade
- WO_3 monoclinic:
  - (002): 23.118°
  - (020): 23.582°
  - (200): 24.365°
- Confirm identical phases with phase shifts present

UNIVERSITY OF SOUTH CAROLINA

Reflectometry Results - WO_3-YSZ

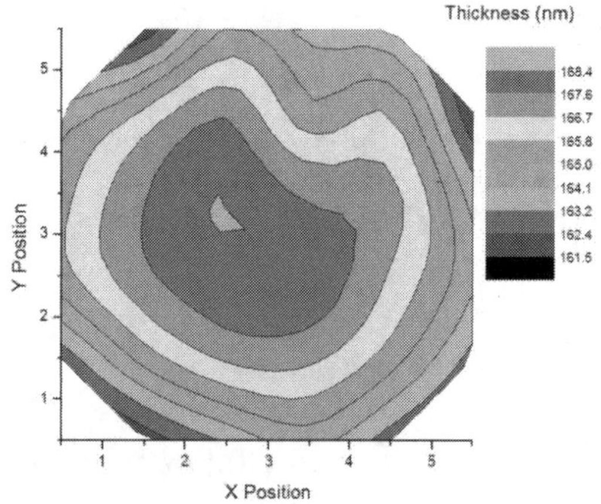

- Sputtering conditions: same
- YSZ held constant
- Confirms relative uniformity (4.3% difference)

UNIVERSITY OF SOUTH CAROLINA

XRD Results – WO_3-YSZ Post-annealed

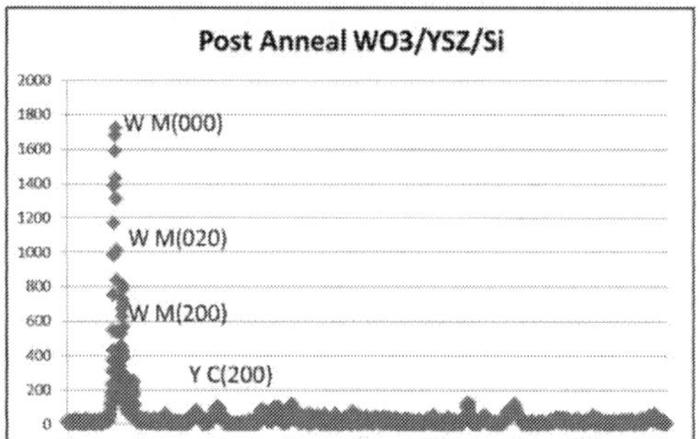

WO_3 monoclinic:
(002): 23.118°
(020): 23.582°
(200): 24.365°
YSZ Cubic:
(111): 29.675°
(200): 34.398°

UNIVERSITY OF SOUTH CAROLINA

Electrochemical Cell Setup

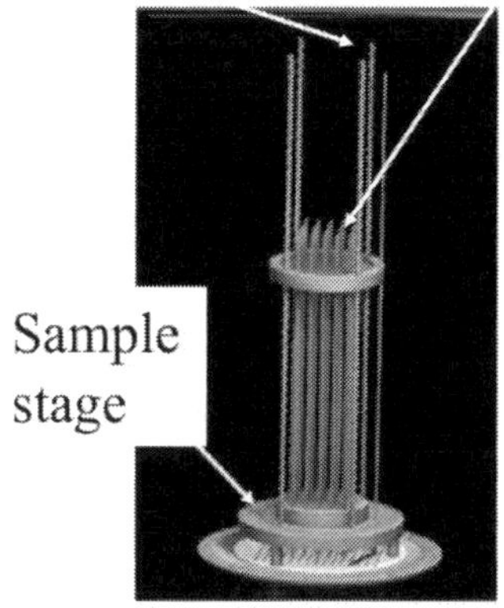

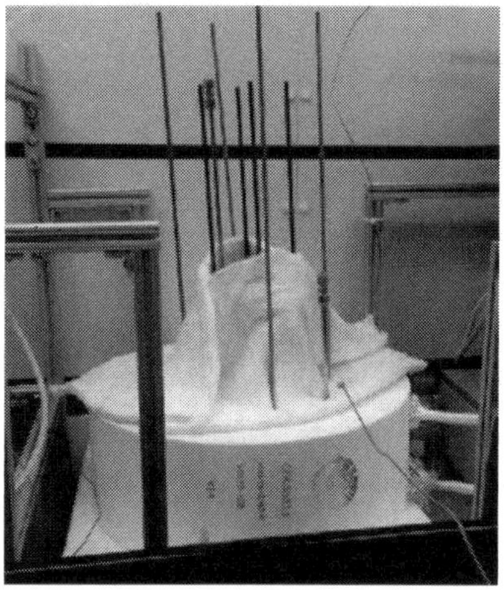

UNIVERSITY OF SOUTH CAROLINA

Single Electrochemical Cell Tests

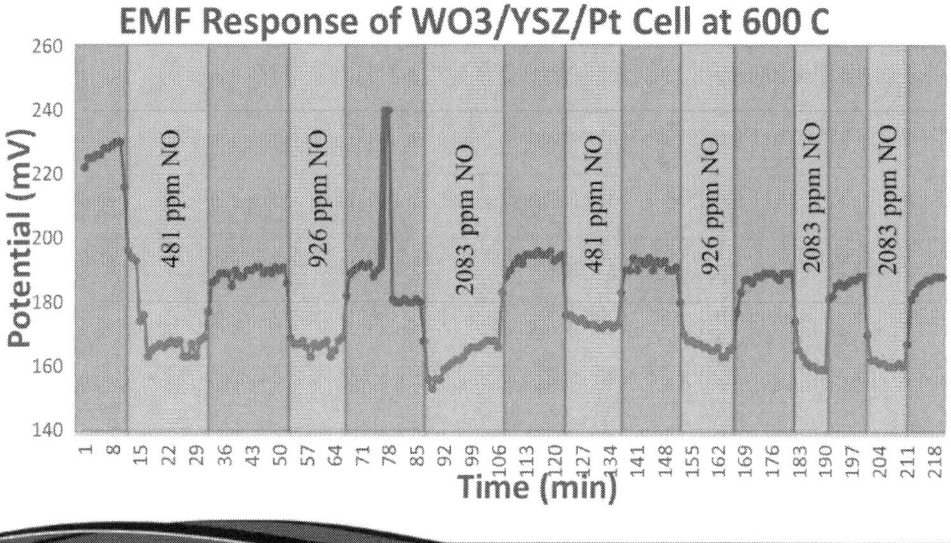

UNIVERSITY OF SOUTH CAROLINA

Electrochemical Cell Tests

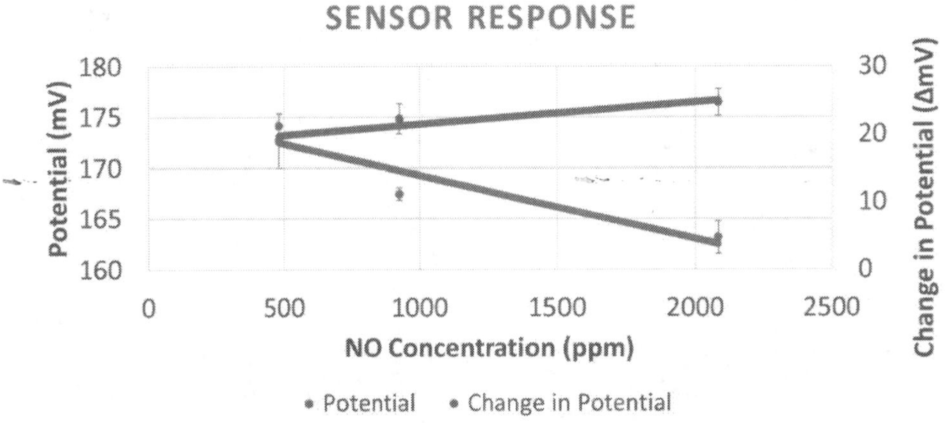

Multielectrode Cell Tests

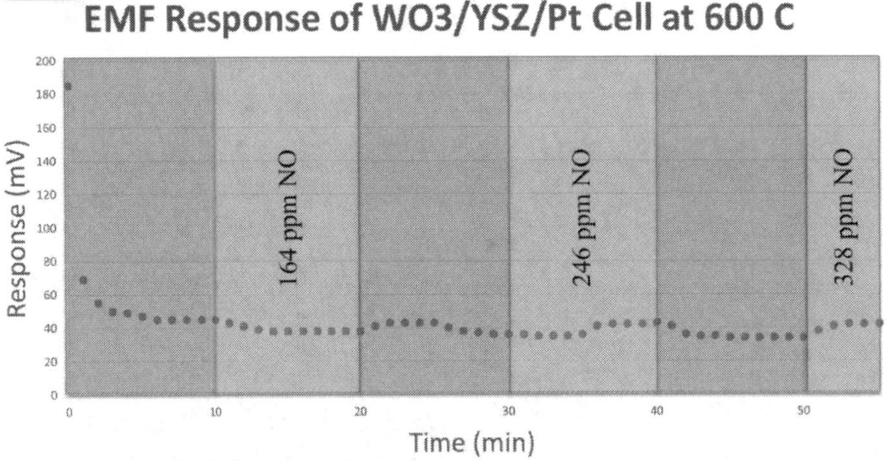

UNIVERSITY OF SOUTH CAROLINA

Conclusions

- Designed planar sensor array
  - Use for combinatorial screening of electrochemical gas sensors
- Confirm deposition uniformity
  - Reflectometry: thickness
  - XRD: phases
- Cell potential
  - Changes valid use of planar sensor design

UNIVERSITY OF SOUTH CAROLINA

Future Work

- Create multielectrode with increased surface area/perimeter of the electrodes
- Normalize potential to distance of electrodes
- Create multielectrode with different composition gradient

UNIVERSITY OF SOUTH CAROLINA

Acknowledgements

- National Science Foundation (NSF)
- University of South Carolina – College of Engineering and Computing
- Strategic Approaches to the Generation of Electricity (SAGE)
- Principal investigators: Dr. Jochen Lauterbach and Dr. Jason Hattrick-Simpers
- Graduate mentor: Ben Ruiz-Yi

UNIVERSITY OF SOUTH CAROLINA

394216 Long-Range Correlations in Liquid Water

Abstract:

The purpose of this research project was to observe water molecules' environments to determine whether or not the system forms identifiable clusters of similar environments. The different clusters aid in proving the hypothesis that different water patches at differing high and low densities values form in a size of 1nm in a system at room and lower temperatures. Molecular dynamics computer simulations were done using the dl_poly classic program with the water model potential TIP4P/2005. The simulations were run at various configurations for density, temperature, and pressure. The results were read by a program, which took the file from the simulations and determined the coordinate positions of the water molecules. Different clusters were found based on the results. These results are indicative of distinctive structures formed in supercooled water that may lead to the water patches of varying densities.

394155 Frictional Study of Polyethylene Glycol Monolayers on Silica Substrate

Abstract:

Lubricating micro- and nano-electromechanical systems (MEMS and NEMS) has become increasingly challenging as devices are continuously miniaturized. As the dimensions of these systems decrease, their surface area to volume ratios increase significantly and frictional forces become dominant where moving parts exist. Solid-state lubrication via self-assembled monolayers presents a promising alternative to traditional macro-scale lubricants which are too bulky to use in such small systems.[1,2] Previous studies have shown that self-assembled monolayers containing alkylsilanes and mixed alkylsilane and perfluoroalkylsilane chains reduce friction coefficients compared to equal length systems. Although, hydrocarbon-based monolayers have been widely studied and shown to improve frictional properties and device life, mixed monolayers impart the beneficial properties of perfluoroalkylsilanes (low critical surface tension and high thermal and mechanical stability) to the film. Furthermore, a mixed length configuration with longer alkylsilane chains relative to the perfluoroalkylsilane chains leads to a liquid-like layer formed by the taller chains that fold over and align with the shearing direction.[3]

In this study, we investigate polyethylene glycol (PEG) monolayers using molecular dynamics simulations as a potential solid-state lubricant for NEMS. The presence of oxygen atoms within the chains render these chains more flexible than their alkyl- and perfluoroalkyl-counterparts. Thus, even sparsely functionalizing a surface leads to a uniform film. We have investigated the tribological properties of such films across a range of surface coverages and chain lengths. Preliminary results show the effects of surface coverage on friction coefficients is diminished as chain length is increased. This suggests that equivalent friction coefficients could be achieved while reducing separation distances by using longer chains on sparser surfaces.

394001 A Machine-Learning Model to Predict Activation Energies of Hydrogenation Reactions

Abstract:

The goal of this research is to develop a machine-learning model to predict activation energies associated to the hydrogenation of organic compounds in the presence of specific catalytic substrates. The current method of predicting activation energies involves a linear relationship with the change in Gibbs free energy of the reaction, which can be very inaccurate. Other methods include experimentation and simulation, both of which have significant cost and/or time drawbacks. This research seeks to improve on these methods to develop a means to predict activation energies quickly, accurately, and with minimal cost.

393764 Microscopic Modeling of the Self Assembly of Poly(ethylene oxide)-Poly(propylene oxide)- Poly(ethylene oxide) (PEO-PPO-PEO) Block Copolymers: Critical Micelle Concentrations

Abstract:

Single chain mean field theory (SCMFT) is applied as a coarse-grain simulation model for Pluronics, a trademarked branch of triblock copolymers. The microscopic model is used to quantify the micellization process of the nonionic class of surfactants at 37 degrees Celsius for drug delivery applications. The simple model with only three interaction parameters is shown to be able to reproduce with high accuracy experimental data for the critical micelle concentrations of different chain lengths of the polymers. These parameters are physically significant and are close to the experimental values from Flory models. In addition, aggregation number of the surfactants and volume fraction profiles are explored. The effects of the chain lengths of hydrophobic tails and hydrophilic heads on CMC values are discussed. SCMFT is transferable in surfactants of different ABA block conformations and other more complex polymer structures where previous simulations have not excelled.

Recommendations for Legislative Actions to Reduce Carbon Emissions in the Electricity Production Sector

Sam White

American Institute of Chemical Engineers
University of Michigan
Washington Internships for Students in Engineering

August 2014

Executive Summary

There is currently a movement in America pushing for environmental effects to be taken into consideration when assessing new and existing business ventures. As emphasized by President Obama's 2014 State of the Union Address, a major area for this movement to make progress is within the energy sector [1]. Electricity production currently produces the most greenhouse gas emissions of any energy sector in the US, making it the energy sector where the largest environmental impact can be made with proper policy action [2].

Luckily, currently-available, low-emission renewable energy technologies are most well suited for electricity production over any other energy sector application. As a result of recent government regulations combined with high coal prices, coal plants, which currently dominate the electricity production landscape, are being retired at an accelerated rate [8]. This, combined with increased electricity demand, has opened up the door for new electricity production technology to be implemented. However, due to the low price of natural gas resulting mainly from the onset of fracking, efforts to implement new renewable-sourced electricity production facilities have been stalled in favor of natural gas power plants. While natural gas plants produce only half the carbon emissions of coal, and do not produce harmful byproducts such as sulfur dioxide or mercury, they are not as environmentally friendly as the other available alternatives, such as nuclear, solar, or wind [11] [12] [13] [14].

The fact that natural gas dominates the shift away from coal is not surprising, however, given the current economic landscape. The figure below shows the levelized cost of electricity (LCOE) for select available electricity production technologies, giving a $/Mwh cost of production by factoring in all applicable costs (construction, fuel, maintenance, operation, etc.) [23].

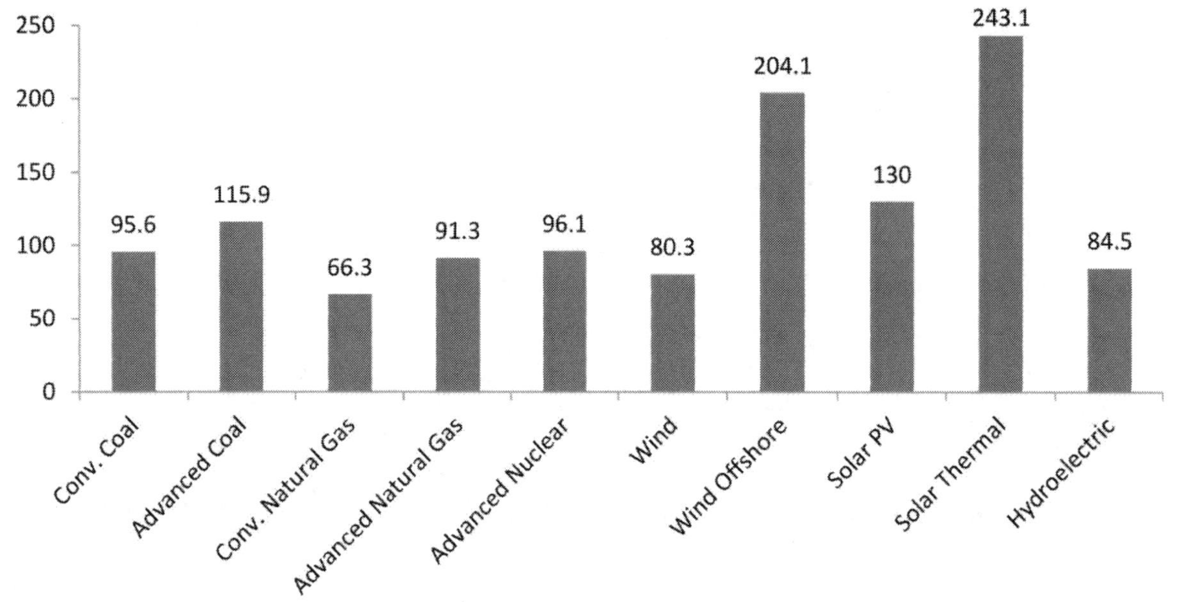

US Average LCOE for Plants Entering Service in 2019, in 2012 $/Mwh [23 EIA LCOE]

As seen in figure above, conventional natural gas currently offers the cheapest form of electricity production. It is important to note, though, that some alternatives like nuclear or wind are already competitive with other conventional electricity production methods like coal. Other alternatives, like solar or offshore wind, are still too expensive to be competitive at the moment, mainly due to the fact that electricity production from renewable sources is still too variable. With the advent of advanced electricity storage technology this issue would be eliminated, but such technology is currently unavailable or too expensive to be implemented economically [19].

Because of their positive environmental impacts, alternative electricity production methods are given some assistance by the federal government in order to make them more economically competitive. These economic aids mainly come in two forms: tax credits and loan programs. On the tax side, an installment tax credit allows solar production facilities a credit equal to 30% of the cost of implementation, while a production tax credit allows wind, hydro, and geothermal plants to receive a .23 cents/kwh credit. However, the installment tax credit is only available until the end of 2016, and the production tax credit already expired at the beginning of 2014 [26]. On the loan program side, the government has extended loans to help both nuclear and renewable power plants, though the renewable plant loan program expired in 2011 [27] [28].

On top of aid being given to alternative electricity production methods, the federal government can also level the playing field by putting a burden on conventional electricity sources. This mainly comes in one of two forms, a carbon tax or a cap-and-trade program, both of which seek to charge electricity producers for the amount of carbon emissions they produce. On top of producing revenue for the federal government, these programs help incentivize electricity producers to innovate new ways to reduce their carbon emissions. Though no such program has ever been implemented, a cap-and-trade program was able to at least pass the House in 2009 [24] [25].

Even small actions by the federal government can have noticeable effects in shifting the production methods used in the electricity production sector. Just a relatively small tax such as $10/metric ton of carbon emissions would cause significant carbon emission reductions compared a scenario in which no government actions are taken [33]. Additional analysis specific to natural gas and renewables shows that any carbon tax would actually increase the amount of natural gas-based electricity production for the next couple of decades, and increase the amount of renewable-based electricity production indefinitely. Additionally, if current favorable tax policies for renewables are renewed long-term, the use of renewables is significantly higher than those projected given current tax policy deadlines, especially after 2030 [34] [35].

As a result, it is recommended that a two-pronged approach be taken to limit the amount of economic disturbance while ensuring the electricity production sector makes a greater shift towards renewables given the reduction of coal use. First, a carbon tax equivalent of $10/metric ton of carbon emissions should be enacted to reflect the social cost of emissions. Each year this tax would increase 5% per year after inflation in order to push companies to innovate new carbon reduction technologies. Second, the production tax credit should be extended for all renewable electricity production methods using the current credit of .23 cents/kwh. This credit would decrease 5% per year after inflation to mirror the carbon tax in order to account for the increased rate of improvement in renewable technologies and the relative advantage given by the increasing carbon tax.

It is important that such a program be long-term in order to allow industry to accurately model the economic landscape their projects will enter. Given the current gridlock in Washington and the long-term nature of implementing new electricity production facilities, a long-term tax plan is the only way to ensure economic stability. Factoring in only revenue from taxes on natural gas use and costs from credits extended to renewables use, such a program would pull in an average surplus of $5.5 billion per year between 2015 and 2040; factoring in coal would increase this figure significantly, particularly in the near future. Feeding at least part of this surplus into loan programs for nuclear plants would ensure that an increased amount of low carbon electricity production facilities are able to come to market.

Table of Contents

Table of Figures

Table of Tables

1.0 Introduction

In his 2014 State of the Union address, President Obama stated that "[O]ne of the biggest factors in bringing more jobs back is our commitment to American energy"[1]. He stressed the importance of "a cleaner energy economy" and acknowledged that "[it] won't happen overnight, and it will require tough choices along the way." Though he touted the success of his "all-of-the-above energy strategy", and highlighted the advantages of natural gas and solar, the President laid out no definitive plans for how the US would approach energy production in the future. However, with the release of recent carbon emission regulations, the executive branch seems to be taking at least some action to direct the country towards a cleaner energy economy. However, if a cleaner energy economy is going to happen in America, the legislative branch needs to get more involved by developing and implementing a long-term energy policy plan.

Of all the areas of energy production in the America, the most promising area for change is the electricity production sector, since a vast majority of alternative energy sources are most viable for electricity production instead of applications like transportation or residential heating. Luckily, this is also the energy sector in which the most impactful changes can be made: in 2012, electricity production was the largest single contributor to greenhouse gas emissions in the US, accounting for 32% of greenhouse gas emissions as seen in Figure 1 below [2].

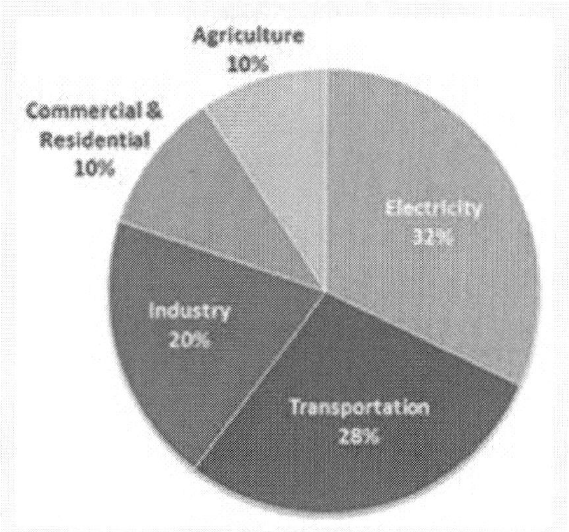

Figure 1. Total Greenhouse Gas Emissions by Economic Sector in 2012 [2]

The implementation of a cleaner energy economy has shown to be a popular policy among voters from both parties. According to a Washington Post – ABC News poll from June 2014, 70 percent of voters say that the federal government should require limits to greenhouse gases from existing power plants. 57 percent of Republicans, 76 percent of Independents, and 79 percent of Democrats support state-level limits on greenhouse gas emissions. Most significantly, voters are willing to pay extra for energy if it means a reduction in greenhouse gas emissions. "Asked whether Washington should still go forward with limits [on energy-sector greenhouse gas emissions] if „they significantly lowered greenhouse gases but raised your monthly energy bill by 20 dollars a month," 63 percent of respondents say yes, including 51 percent of

Republicans, 64 percent of Independents and 71 percent of Democrats" [3]. These results suggest that Americans on both sides of the aisle are in support of implementing a cleaner energy economy.

In addition to the popularity of a cleaner energy economy, it also makes sense from an environmental and health stand point. On top of the fact that the reduction of greenhouse gas emissions reduces the rate of climate change, it brings along health benefits as well. With the release of its Clean Power Plan in June 2014, which seeks to cut carbon pollution from the power sector by 30% by 2030 by imposing stricter emission limits on existing plants, the EPA projects that "annual public health benefits will total somewhere between $55 billion and $93 billion by avoiding up to 100,000 asthma attacks and 2,100 heart attacks each year." By contrast, they predict compliance cost of $7.3 to $8.8 billion per year, significantly lower than the expected public health benefits [4]

However, despite the ongoing executive branch actions regarding energy policy, the legislative branch has taken relatively little action regarding energy policy lately, mainly due to party-politics-driven gridlock. At the beginning of 2014, the Production Tax Credit for renewables was allowed to expire for the first time since 2004, and only the fourth time since its enactment in 1992 [5]. In May of 2014, the Senate allowed a bipartisan energy efficiency bill to die due to disagreement over other energy related issues, such as the fate of the Keystone XL pipeline and possible legislative response to President Obama‟s climate change regulations. In fact, it has been 7 years since Congress passed a major energy bill, which means there has been no major legislative action in response to the changing energy landscape as a result of the shale fracking boom [6].

A plan for implementing a cleaner energy economy becomes even more important as energy demands increase, particularly when it comes to electricity production. As electricity demand increases, more electricity production facilities are built, and how these facilities produce electricity greatly affects their environmental impact. According to the EIA‟s Annual Energy Outlook 2014 report, US electricity demand will increase from anywhere between 14% and 41% between 2012 and 2040, though their best estimates suggest that electricity demand will increase by 29% over this time period. Additionally, the growth of electricity demand is projected to increase sometime around 2025, marking the beginning of an upwards slope in electricity demand growth for the first time in 75 years [7].

In order to meet this demand and fill the gap left behind by plant retirements, new electricity production facilities need to be built. As a result of recent emission regulations like the Clean Power Plan, coal power plants are being retired at an accelerated rate. However, due to the low price of natural gas in recent years thanks to fracking, much of this gap has been filled by natural gas fired plants instead of alternatives, like nuclear or renewables. This is mainly due to economics: even the cheapest alternative energy sources, which are on the verge of being if not already cost competitive to coal, simply cannot compete with the low price of natural gas. While natural gas is a lower-emissions alternative to coal, it has emissions nowhere near as low as alternative electricity sources.

The onset of natural gas is not a bad thing, however: using natural gas as a bridge fuel to reduce emissions in the immediate future while hoping that renewables become more cost competitive in the future is not unreasonable given the rate of development and investment in renewable technologies. Natural gas also offers the unique advantage of "spinning reserve", a source that can quickly be powered up to meet rapid fluctuations in demand. This means natural gas can offer a flexibility that base load electricity sources like coal or nuclear cannot, while acting as a buffer against the variability of renewables like solar or wind.

Embracing natural gas in the here and now is also economically responsible, as it keeps electricity prices low by offering a cheap alternative to coal and helps support the growing American natural gas production market. However, without government action, this shift towards natural gas will dwarf any efforts to make renewables a larger share of our electricity production sector. If the government is serious about using natural gas as a bridge fuel to a cleaner energy economy in the electricity sector, a long-term policy needs to be enacted to slowly guide the market in the right direction.

2.0 Overview of Electricity Production Methods and Their Environmental Impacts

In order to discuss energy policy, it is first important to establish the electricity needs of the country, as well as the current capabilities of various electricity production methods and their effects of the environment. As seen in Figure 2 below, coal and natural gas currently dominate the electricity production sector, with nuclear and renewables making up the bulk of the remaining production.

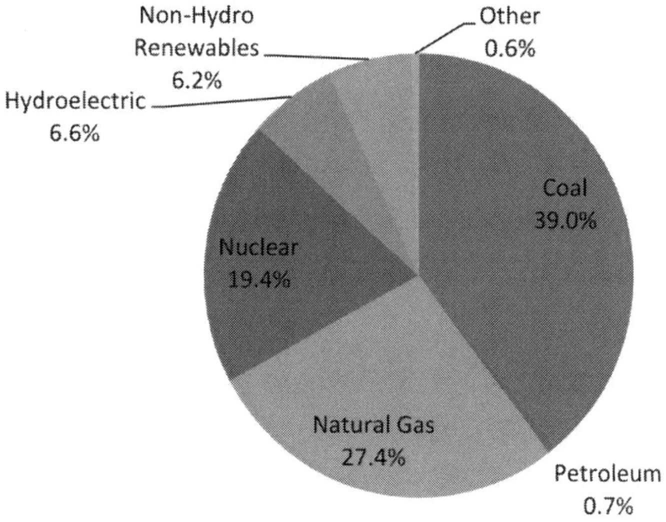

Figure 2. Share of Total US Electricity Production by Source [Adapted from 8]

2.1 Coal and Natural Gas

The two main fossil fuel based / conventional energy producing sources, coal and natural gas power plants are currently the largest source of electricity in America, accounting for nearly two-thirds of the country"s total electricity production [8]. Coal plants are relatively easy to keep in operation over a long time scale, but it is difficult for them to make relatively quick changes to production capacity in response to the changing needs of the grid, particularly during peak-usage hours. As a result, coal plants are very useful in providing a baseline power supply somewhere between peak and off-peak requirements.

Natural gas, on the other hand, allows for a very flexible energy production method. Gas flow rates into a furnace can be controlled with relative precision, unlike solid input streams found in coal plants. As a result, natural gas plants can be very responsive to the changing needs of the energy grid. In fact, natural gas plants are typically used to augment other energy production facilities for this very reason.

From an environmental perspective, the EIA considers conventional electricity sources like coal and natural gas as being the only contributors to CO_2 emissions in the electricity production sector. In 2010, the EIA reported that 1969 million metric tons of CO_2 were released by coal plants nation-wide, while 1285 million metric tons of CO_2 were release by natural gas plants [9].

However, coal tends to have other unique environmental impacts. First, burning coal tends to release nitrogen oxides, which tend to have larger warming potentials than CO_2. For example, N_2O has a SAR 100 year warming potential of 310, meaning that one ton of N_2O has 310 times the impact on global warming compared to one ton of CO_2 over the course of 100 years [10]. From a life cycle assessment standpoint, coal plants produce 974 metric tons of CO_2 equivalent per gigawatt-hour, which accounts for this increased warming potential as well as CO_2 emissions due to non-production factors, such as plant construction or fuel transportation [11]. Coal also tends to produce SO_2, which, along with nitrogen oxides, is the major man-made cause of acid rain. According to the EPA, "In the United States, roughly two-thirds of all SO_2 and one-quarter of all nitrogen oxides come from electric power generation that relies on burning fossil fuels, like coal" [12]. Additionally, coal plants are a major source of mercury, a highly toxic element. According to the EPA, "Coal-burning power plants are the largest human-caused source of mercury emissions to the air in the United States, accounting for over 50 percent of all domestic human-caused mercury emissions" [13].

In the case of natural gas, environmental impacts are lower than that of coal. According to the EPA, "… the burning of natural gas produces nitrogen oxides and carbon dioxides, but in lower quantities than burning coal". In fact, from a life cycle assessment standpoint, natural gas plants produce 469 metric tons of CO_2 equivalent per gigawatt-hour, less than half that of coal [11]. Additionally, the EPA states that "[e]missions of SO_2 and mercury compounds from burning natural gas are negligible". [14]. While the EPA also mentions that unburned natural gas being released either from a power plant or during transportation is a concern, though in 2009 the

EIA stated that "[i]ncreases in energy-related methane emissions [were] largely from underground coal mining", suggesting that methane releases are more prevalent in the procurement of coal than in the use and transport of natural gas [14] [15].

2.2 Nuclear

In the US, nuclear energy makes up nearly one-fifth of all electricity production [8]. According to the World Nuclear Association, "it [nuclear power] is especially suitable for large-scale, continuous electricity demands, which requires reliability (i.e. base load)" [16]. In this way, nuclear power plants are analogous to coal plants in the sense that they are very useful in meeting baseline power needs, but have trouble making quick changes in response to fluctuating peak-hour demand.

From an environmental standpoint, "nuclear power plants do not emit CO_2, SO_2, or nitrogen oxides as part of the power generation process" according to the EPA [17]. However, as with all power production facilities, the construction of nuclear power plants leads to non-negligible greenhouse gas emissions. Based off a compilation of life-cycle emissions studies gathered by the Nuclear Energy Institute (NEI), nuclear energy"s life-cycle emissions are significantly lower than those of coal or natural gas plants, and are in fact comparable to those of renewable like solar, wind, or hydro [18]. From a life cycle assessment standpoint, nuclear produces 15 metric tons of CO_2 equivalent per gigawatt-hour, which is the second-lowest behind wind, and only about 1.6% that of coal plants [11]. Much of these emissions come from the construction of plants, which are significantly larger than other electricity production facilities, and the transportation of fuel.

2.3 Solar and Wind

As shown in Figure 2, only about 6% of electricity production in the US comes from non-hydro renewable sources. Of this 6%, a majority comes from solar (in the form of both photovoltaic cells and solar thermal) and wind [8]. Of all of the available technologies, solar and wind are the most challenging to integrate into the grid due to the fact that they are considered variable sources: if the sun doesn"t shine or the wind doesn"t blow, these technologies are not actively putting energy into the grid, meaning demand has to be met by some other means [19] [20]. In order to offset this, large-scale production facilities are looking into using some form of energy storage system. This would allow excess energy collected during times of sunshine or wind gusts to be used when the sun doesn"t shine or the wind doesn"t blow [19]. However, the storage capacity of current electricity storage technology is limited. As a result, any useful amount of electricity storage is currently too expensive, which limits the viability of renewables.

From an environmental standpoint, neither solar nor wind produce carbon emissions during their operating lifetime, since their energy is derived from the sun and not from burning fossil fuels [19] [20]. However, like nuclear, solar and wind do have some emission impacts due to the processes used to manufacture materials and construct plants: wind is responsible for 14 metrics tons of CO_2-equivalent per gigawatt-hour, while solar photovoltaics are responsible for

39 metric tons of CO_2 equivalent per gigawatt-hour [11]. As with nuclear plants, many of the emissions related to renewables come from construction of production facilities. In the case of solar photovoltaics, higher life-cycle emissions are seen as a result of the emissions that result from the manufacturing process of the solar cells themselves.

2.4 Other Renewables

Solar and wind are not the only renewable sources of energy available to the US. As seen in Figure 2, nearly 6.6% of electricity production comes from hydro plants. The main issue with hydro is that most of the available sites for dams in the US have already been dammed, meaning there is little room for expansion. As a result, the Obama administration has identified a collection of dams that currently do not produce electricity, but outfitting these dams with turbines would only increase the current US hydro capacity by roughly 15% [21]. Another alternative energy source is geothermal, which utilizes naturally-occurring underground sources of heat to power turbines. While this energy is clean, it is only available on a regional basis: Hawaii and the western US have a large geothermal production potential, while areas like New England do not [22]. Since hydropower does not have a lot of potential for growth, and geothermal is regionally specific, they will not be explored as in-depth in this paper.

2.5 Environmental Impact Summary Table

Table 1 below summarizes the life-cycle environmental impact per gigawatt-hour of the main electricity production methods highlighted in this paper.

Technology	Metric Tons of CO_2 released / GWh
Coal	974
Natural Gas	469
Nuclear	15
Solar	39
Wind	14

Table 1. Life-Cycle Environmental Impact Summary of Electricity Production Technologies [Adapted from 11]

3.0 Economics in the Electricity Production Sector

On top of technological limitations and environmental concerns, economics plays a major role in determining which technologies get used for energy production. On the production side, economic factors tend to play a major role in determining which electricity production technologies are utilized by producers. On the consumer side, the electricity production technology used by electricity providers plays a role in the price of electricity sold to consumers.

3.1 Comparing Costs of Electricity Production Technologies

Due the wide range of variables present in energy production, it can be difficult to compare the cost of one electricity production technology to another. However, a useful metric, levelized cost of electricity (LCOE), lays out the cost of given technologies on a per-MWh basis, factoring in such costs as construction, fuel, operating costs, maintenance, and financing costs [23]. Figure 3 below shows the US average LCOE for plants entering service in 2019, in 2012 dollars/MWh, and does not factor in any public financial support (tax credits, subsidies, etc.).

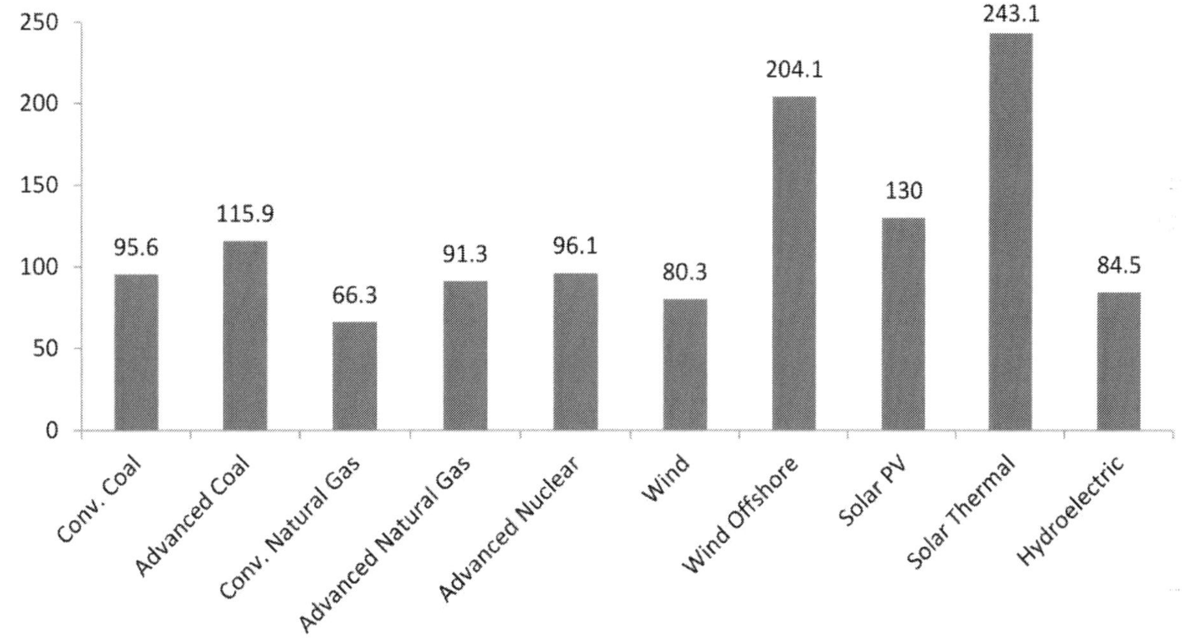

Figure 3. US Average LCOE for Plants Enetering Service in 2019, in 2012$/MWh [23]

As Figure 3 shows, the conventional energy production sources, coal and natural gas, are currently the cheapest available electricity production technologies, which explains why they currently account for such a large portion of US electricity production. Much of this can be contributed to the low price of fuel. Natural gas in particular is benefitting greatly from low fuel prices as a result of the recent American shale fracking boom. However, advanced coal and natural gas facilities, which incorporate carbon capture and sequestration technologies, are significantly more expensive than their conventional counterparts.

While coal and natural gas remain the cheapest electricity production technologies, it is important to note that there are alternatives that are already competitive. For example, the LCOE of nuclear energy is nearly the same as conventional coal and advanced natural gas, and is in fact less expensive than advanced coal. This is particularly noteworthy since nuclear fills the same base load electricity production role as coal, and produces no greenhouse gas emissions during its operational lifetime. It is also important to note that, while many of the renewable energy sources are significantly more expensive than conventional sources or nuclear, wind built on land has the second lowest LCOE, higher than only conventional natural gas.

3.2 Capacity Factor

One factor contributing to the relative expense of renewable sources, such as offshore wind or solar, is their variable nature. Since they are not able to produce electricity at all times of the day, renewable sources suffer from prolonged periods of not operating at peak capacity. The easiest way of quantifying this is through capacity factors, the ratio of actual output to the theoretical output if the plant could operate at full capacity indefinitely. A comparison of capacity factors for each electricity production technology can be seen in Figure 4 below.

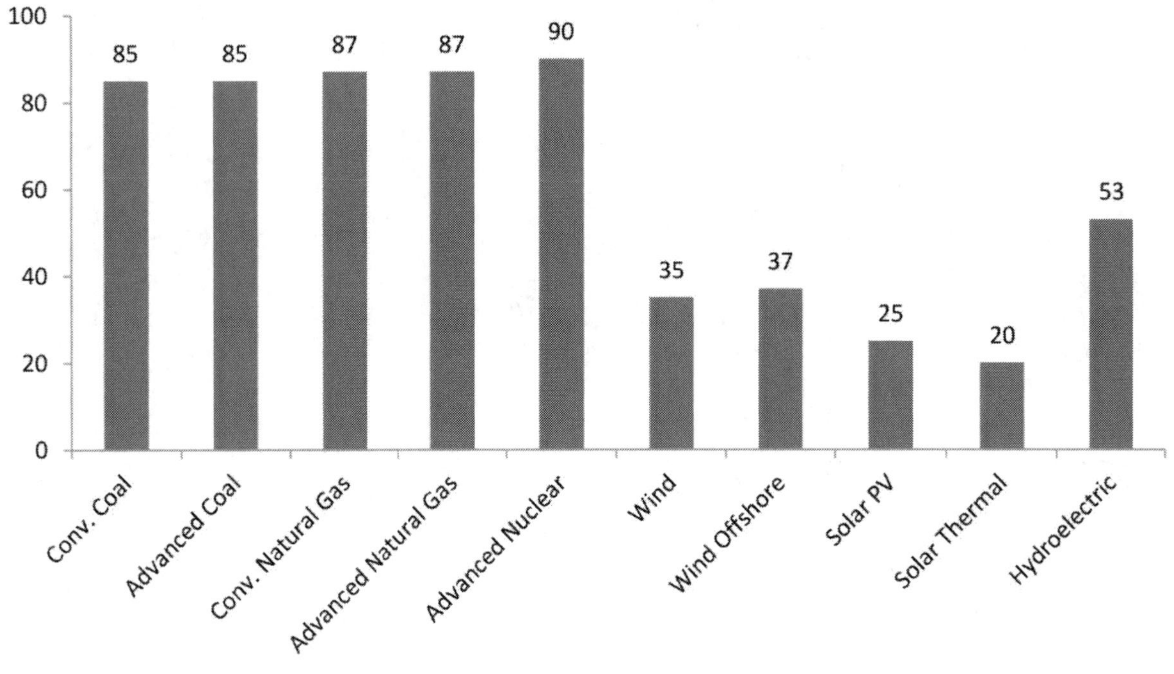

Figure 4. Capacity Factor (%) of Various Electricity Production Technologies [23]

It is important to note that there is no electricity production technology with a capacity factor of 100%: plants do not always run at maximum capacity, and require downtime for things like routine maintenance or system upgrades. However, it is easy to see that renewable sources have significantly lower capacity factors, mainly due to the fact that they cannot produce electricity at all times of the day. Because of this, capital and operating costs of renewable energy plants factor significantly more into their LCOE.

However, with advancements in technology, particularly in the energy storage field, the capacity of renewable sources could significantly increase, and possibly even become comparable with conventional sources. This would cut LCOE by anywhere from 40-70% (depending on current capacity factor). If this electricity storage is cheap enough, even the most currently expensive renewable energy technologies would be competitive with conventional sources.

4.0 Recent Policy Strategies

In recent years, legislators have used a wide variety of strategies to help shape the electricity production sector, most notably by offering some form of economic incentive to desirable technologies. These economic incentives typically come in the form of tax credits, though they may also come in the form of subsidies or guaranteed loan programs. Congress has also made efforts in the past to enact a carbon tax program, but such a program has never been able to make it out of Congress [24] [25].

4.1 Tax Credits for the Electricity Production Sector

The major economic incentive congress has implemented is by extending tax credits to electricity producers, which reduces the amount of taxes they have to pay to the federal government and thus improving their bottom line[1]. Recently, these tax credits have been targeted at renewable electricity resources, most notably solar and wind, though to a lesser extent hydroelectric and geothermal as well [26]. These tax credits have taken on two different forms: an investment tax credit and a production tax credit.

The investment tax credit allows qualifying electricity producers to receive a tax credit equal to a certain percentage of their initial cost of implementation. Solar energy installments are eligible for a 30% credit, while geothermal plants are eligible for a 10% credit. These credits have no limit. As of right now, these tax credits extend to any qualifying facility put in service before January 1, 2017. Wind power facilities put in service between 2009 and 2012 were also eligible for a 30% investment credit, but that has not been renewed [26].

The production tax credit, on the other hand, allows qualifying electricity producers to receive a tax credit based off the amount of electricity produced. As of the end of 2013, this tax credit was 2.3 cents / kWh, and extended to wind, geothermal, and hydropower facilities [5] [26]. If a facility qualifies for both the investment and production tax credit, they are only permitted to claim one of them, at their choosing. The production tax credit extended to any qualifying facility under construction by January 1, 2014, and is available for qualifying facilities for the first 10 years of operation. However, since January 1, 2014, the production tax credit for renewable electricity production facilities has not been extended [26].

4.2 Loan Programs

The federal government also offers loan programs for electricity producers. These loan programs differ from tax cuts mainly in the fact that producers are required to pay back their loan; while there is a cost to the federal government, it is only in the risk of borrowers defaulting,

---

[1] It is important to note that non-profits, such as municipal utilities and utility cooperatives, do not pay taxes, meaning that federal tax cuts do not have an effect on their bottom line. While alternative programs to incentivize non-profits do exist, they are not covered in the scope of this paper.

which has historically been low. Under Title XVII of the Energy Policy Act of 2005, two loan programs were outlined that are still used today: section 1703 and section 1705 [27] [28].

Section 1703 provides loans for "a new or significantly improved technology that is NOT a commercial technology" [27]. Qualifying facilities must also be located in the US. In the current state of the electricity production sector, the only facilities that really meet these requirements are related to nuclear energy; the most recent loans are an $8.3 billion loan for two new nuclear power plants in Georgia and a $2 billion loan for a uranium enrichment plant in Idaho [27]. This program is essential to nuclear power plants, as the most prohibitive cost to entering the market for nuclear plants is upfront capital costs. Due to relatively low operating costs once the plant is built, loan repayment is usually not an issue.

Section 1705, on the other hand, focuses on loan programs for renewable energy production facilities. Unlike section 1703 rules, section 1705 has a time limit, and was only available to projects that had begun construction before September 30, 2011. Since then, this project has not been renewed. Section 1705 loans tend to be lower than their section 1703 counterparts: one of the more expensive section 1705 loans was for $1.3 billion to an Oregon-based wind farm [28].

The major significant difference between section 1703 and 1705 loans is how the Credit Subsidy Cost (CSC) is handled. By law, the CSC, which is "the expected long-term liability to the Federal Government by issuing the loan guarantee", must be paid either by budget appropriations or by the borrower directly. For section 1703, which tend to have more expensive loans and thus larger CSCs, the CSC is paid by the borrower. For section 1705, the CSC was paid for by appropriations [27] [28]. In the American Recovery and Reinvestment Act of 2009, $6 billion was appropriated towards section 1705 CSCs. [29]

4.3 Carbon Tax and Cap-and-Trade Program

On top of offering economic benefits to desirable electricity production methods, Congress can also place economic burdens on undesirable electricity production methods. In the electricity production sector, this can be practically done in one of two ways: either a carbon tax or a cap-trade-program.

A carbon tax works very simply: electricity producers pay an additional tax for each unit of carbon produced during electricity production. This means that carbon-heavy electricity production facilities like coal (and to a lesser extent, natural gas) power plants would pay an additional amount for their electricity production, thus increasing their LCOE. In the same way that tax credits and subsidies seek to level the economic playing field by making renewables cheaper, a carbon tax would seek to level the playing field by increasing the cost of conventional electricity production methods. However, there has never been a carbon tax approved by either chamber of Congress.

An alternative method to a carbon tax is a cap-and-trade program, which seeks to place an artificial limit on the amount of carbon that can be emitted. Once this limit has been determined, electricity producers can buy permits to emit certain amounts of carbon dioxide. The government only sells as many permits as it takes to reach the cap, and then monitors the emissions of each producer to make sure they do not exceed their permitted limit. Producers are able to buy or sell permits to or from other producers as needed.

Cap-and-trade programs create economic incentive to reduce carbon emissions in two ways: first, reducing carbon emissions means that electricity producers need to buy fewer permits, saving them money. Second, if producers reach their permitted cap, their competitors determine the cost of additional permits (since each company now control its own excess permits, not the government). This means exceeding the permitted carbon emission limit can be disproportionately costly, either through the purchase of competitor permits or paying government fines.

In order to ensure long-term carbon emission reduction, cap-and-trade programs also incorporate a permit allowance reduction system. This means that, each year, the amount of permits allowed is reduced, effectively lowering the acceptable amount of carbon dioxide that can be emitted by electricity producers. While a cap-and-trade program did make it out of the House in 2009, the bill was never heard by the Senate, and died when that session of Congress ended [24] [25].

5.0 Projected Electricity Sector Growth: Electricity Sector Capacity Additions for 2013

In order to establish a reference case, it is useful to look at how the electricity production sector has changed recently. Figure 5 below shows the EIA projections for electricity sector capacity additions from 2008.

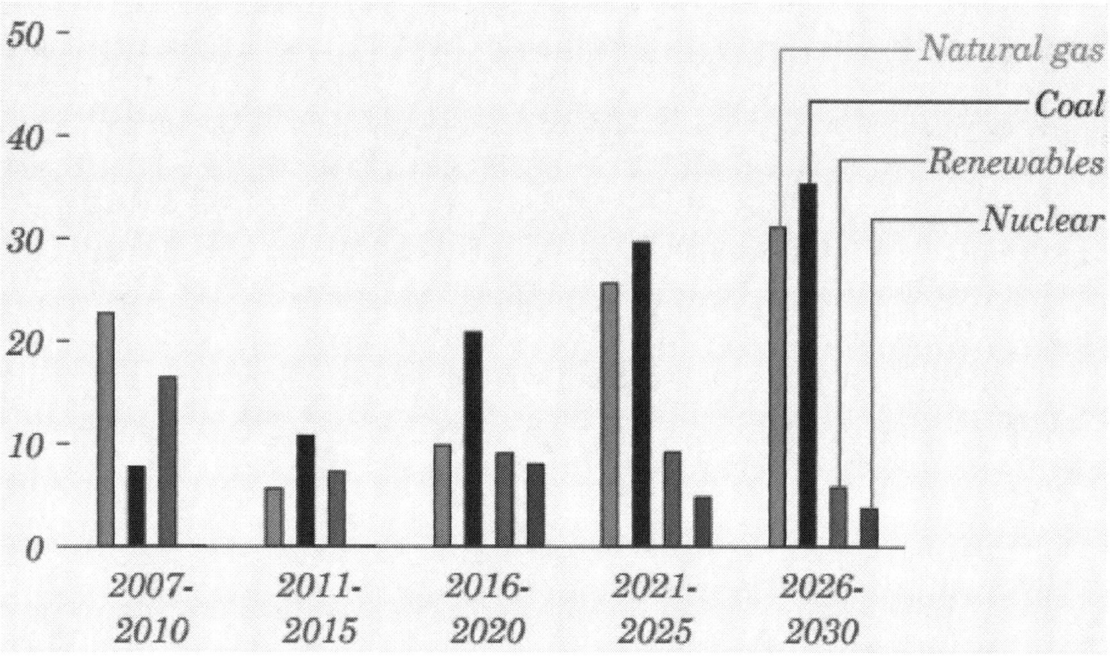

Figure 5. Electricity Generation Capacity Addition Projections by Fuel Type, 2008 (Gw) [30]

As Figure 5 shows, in 2008, it was projected that coal would begin to dominate electricity production additions starting in 2011. Much of this can be attributed to the increasing cost of natural gas, as shale fracking was still in its infancy in 2008. Additionally, renewables, which see a large spike in 2007-2010, were projected to fall by the wayside after 2010 as they begin to be dwarfed by both natural gas and coal.

However, as a result of the shale fracking boom, the reduced cost of renewables, and increased government pressure to reduce carbon emissions, these 2008 projections do not reflect what has actually happened in recent years. Figure 6 below shows the electricity generation capacity additions for 2013.

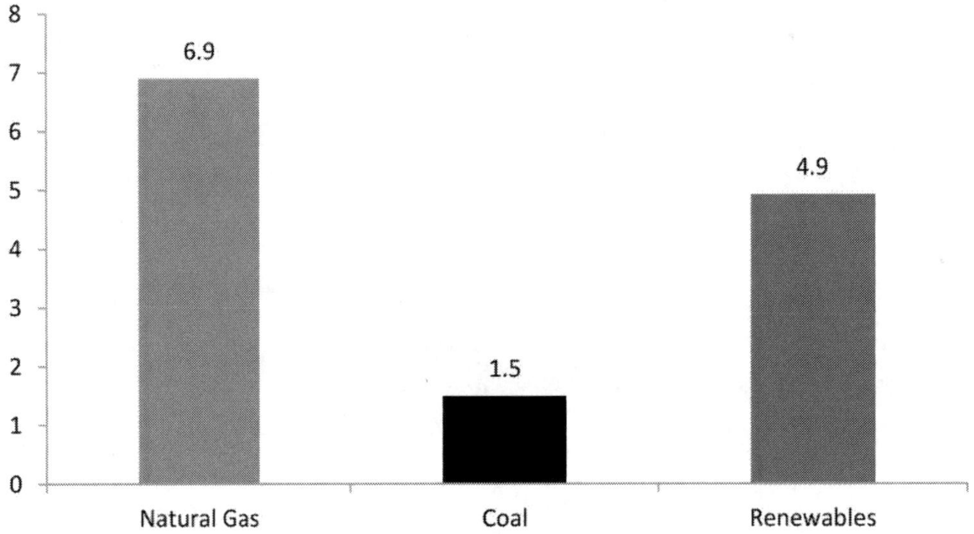

Figure 6. Electricity Generation Capacity Addition by Fuel Type 2013 (GW) [Adapted from 31]

As seen in Figure 6, natural gas dominated new capacity additions, accounting for more than half of the electricity capacity additions in 2013. Renewables made up roughly one-third of new additions, which is attributed to the completion of a number of large scale solar thermal plants in Arizona and California as well as "falling [photovoltaic] costs, aggressive state renewable portfolio standards, and continued federal investment tax credits" [31]. Coal makes up a meager 11% of new capacity additions, in stark contrast to the projections shown in Figure 5.

Long term projections have also changed to reflect the changing economic and political landscape surrounding electricity production. Figure 7 below shows the EIA projections for electricity sector capacity additions from 2014.

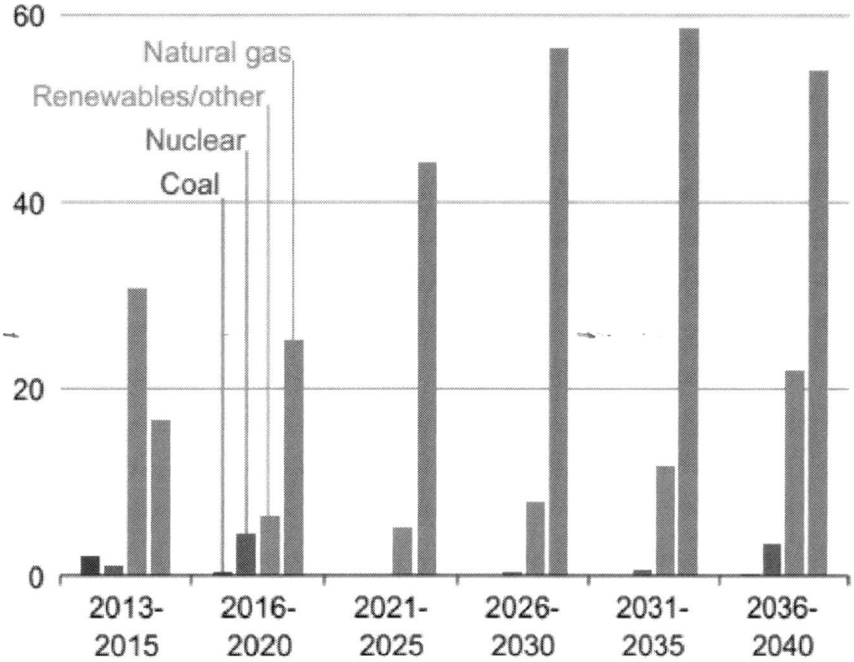

Figure 7. Electricity Generation Addition Projections by Fuel Type 2014 (GW) [32]

As Figure 7 shows, natural gas dominates new additions through 2040, mainly as a result of low gas prices. Coal additions fall virtually to 0 as a result of increased emissions regulations. Most alarmingly, the addition of renewables is actually lower than those projected in 2008 for the period between 2016 and 2035. This is mainly due to the extremely low cost of natural gas relative to any other fuel source, making it take up a larger share of the new additions in the 2014 projections than conventional sources did in the 2008 projections. It is important to note, however, that these projections only show a reference case where no additional government actions are taken.

5.1 Policy Action Effects: Overall Carbon Emissions

Despite the recent transition from coal to natural gas, as well as the increased onset of recently-cheap renewables, carbon emissions from the electricity production sector are projected to increase over the next several decades if current laws and regulations remain in effect. This is mainly attributed to the shift towards natural gas-based electricity production and the shift away from renewables and nuclear as a result of cheap natural gas for the foreseeable future [33]. However, if some form of carbon tax is implemented, carbon emissions could decrease instead increase, as seen in Figure 8 below.

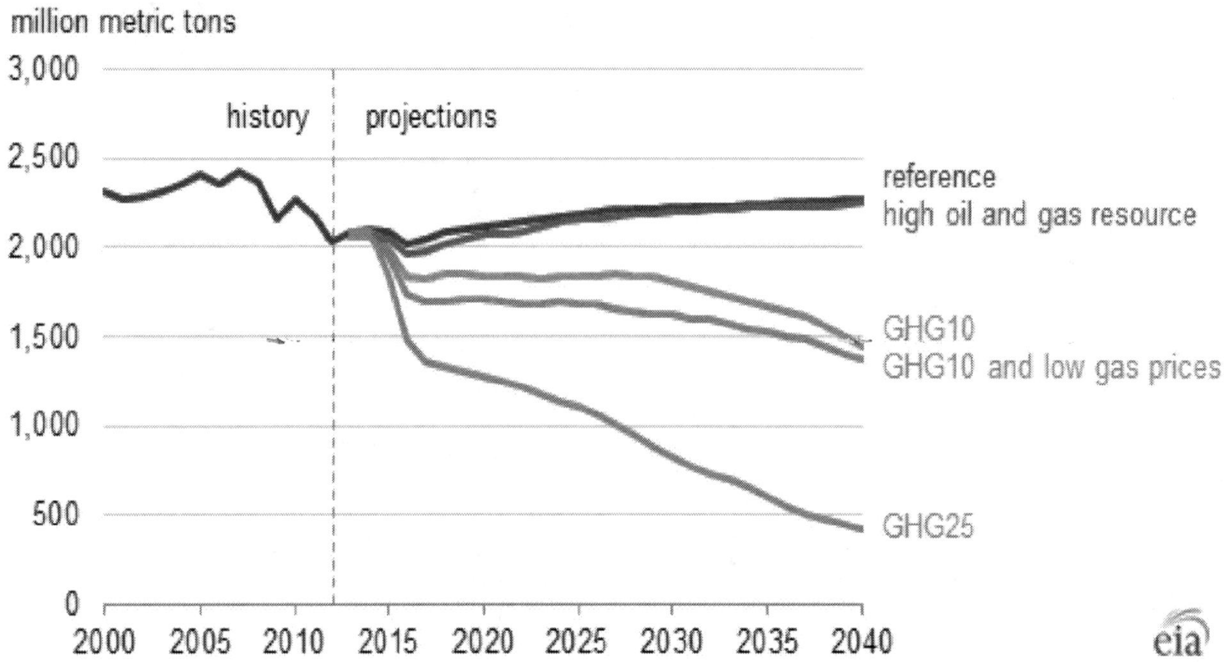

Figure 8. Electricity-Related CO_2 Emissions in Five Cases, 2000-40 [33]

Figure 8 takes into account five different potential cases. The reference case reflects no change in laws or regulations surrounding electricity production and assumes relatively constant economic factors relative to those in 2014. As seen in Figure 8, the reference case shows a 12% increase in emissions up through 2040 [33].

The GHG 10 and GHG 25 cases both refer to the implementation of a carbon tax (or equivalent cap-and-trade permit cost) in 2015 of \$10 / metric ton of CO_2 or \$25 / metric ton of CO_2, respectively. The GHG 10 case shows a 16% lower CO_2 emission rate in 2025 and a 36% lower CO_2 emission rate in 2040 relative to the reference case. The GHG 25 case shows an even greater decrease in CO_2 emission rates, with a 49% reduction in 2025 and an 82% reduction in 2040. In both cases, there is a large dip in emission rates before 2020, partially due to an accelerated shift away from coal and partially due to an increased implementation of renewables [33].

The High Oil and Gas Resource case assumes the same conditions as the reference case, except that oil and gas resources become more abundant than currently projected. In this case, CO_2 emissions decrease ever so slightly more than in the reference case as a result of increased shift towards natural gas, but by 2040 emission rates are roughly the same as the reference case as a result of producers expressing and increased favor towards natural gas relative to renewables [33].

Finally, the GHG 10 and Low Gas Prices case effectively combines the GHG 10 and High Oil and Gas Resources cases. In this case, reduction of CO_2 emissions before 2020 is greater than those seen in the GHG 10 case, mainly due to an even greater shift from coal to natural gas. However, because a greater share of this shift away from coal is given to natural gas and not renewables, CO_2 emissions do not decrease as drastically after 2030 compared to the

GHG 10 case; it can be extrapolated that CO_2 emissions would actually begin to be higher than the GHG 10 case sometime between 2040 and 2050 [33].

Overall, these cases suggest that some form of action needs to be taken in order to produce a marked reduction in carbon emissions from the electricity production sector. If nothing changes, CO_2 emissions will increase slightly over the next 40 years, in contrast to the significant emission reductions we"ve seen since 2005. Emissions get even worse with an increased gas resource and lower gas prices, which may result with the advancement of fracking and other extraction technologies. Based off of this projection, the only way to continue a trend of CO_2 emission reduction is to accelerate the shift away from coal and ensure that an increased share of that shift is made up of renewables or nuclear.

5.2 Policy Action Effects: Conventional Fuels

When assessing the viability of the cases listed in the previous section, it is important to understand the effects of each action on each specific electricity production method. As discussed previously, there is already a shift towards natural gas and away from coal in response to lower gas prices and increased carbon emission regulations. Figure 9 below shows how possible policy actions may augment that shift by showing the projected natural gas electricity generation in the five previous cases.

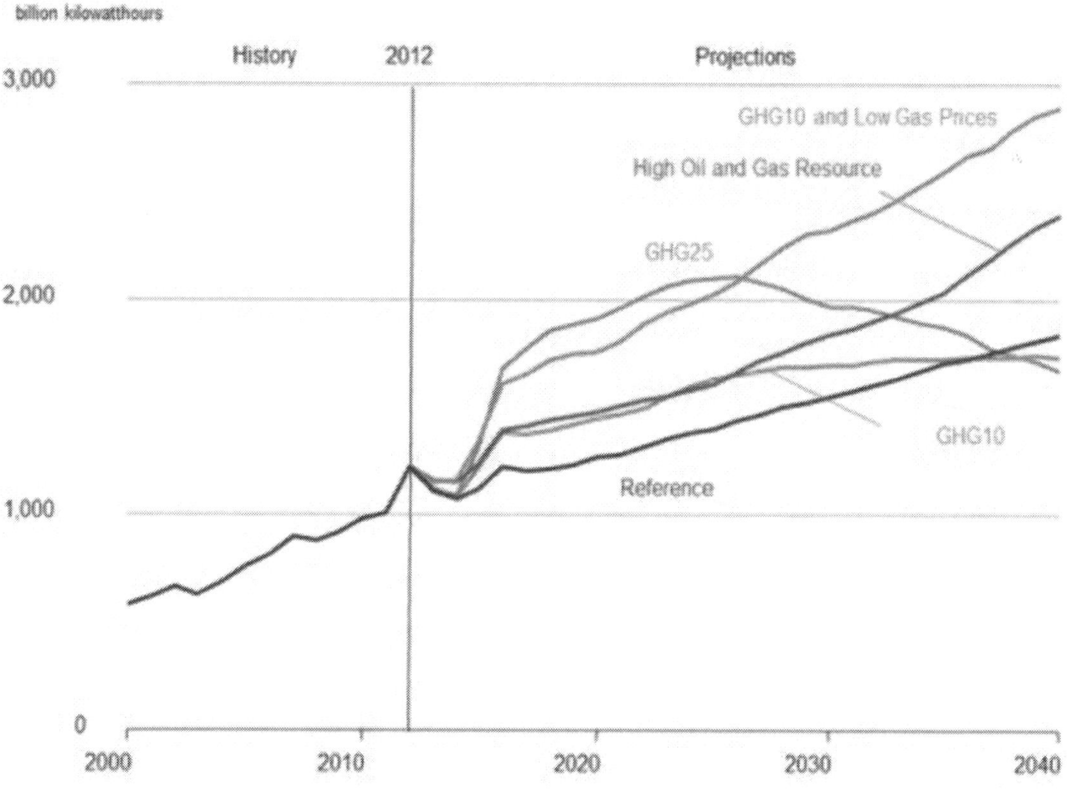

Figure 9. Natural Gas Electricity Generation in Five Cases, 2000-40 [34]

As seen in Figure 9 above, electricity production via natural gas is projected to increase over the next several decades in all cases. In the reference case, we see a relatively stable and constant increase of natural gas fired plants as they replace retiring coal plants; in 2012, 37% of electricity generation comes from coal, and in 2040 that share is 32%, marking a relatively slow shift [34]. With a high gas resource, this shift is more dramatic in the near future, and the long term rate of increase in natural gas plants is also slightly higher than the reference case.

The most interesting part of this projection, however, is the effects of a carbon tax. In all GHG cases, the shift towards natural gas in the near future is increased relative to the reference case. In the GHG 10 case, the share of electricity from coal is 28% in 2025 and 19% in 2040. For the GHG 25 case, the shift is even more drastic, with the coal share being 10% in 2025 and 1% in 2040 [34]. In both cases, the increase in natural gas production is curbed by the increased implementation of nuclear of renewables: in the GHG 10 case, natural gas electricity production levels off around 2030, while in the GHG 25 case, it actually decreases starting in 2025. However, if gas prices remain low, natural gas electricity production continues to increase through 2040, reaching the highest level of any of the five test cases.

5.3 Policy Action Effects: Renewables

Despite the fact that natural gas dominates new electricity production additions, renewables are also seeing an increase in implementation as the electricity production shifts away from coal. However, without some form of government action, the rate of renewable additions will be dwarfed by those of natural gas, leading to the increased carbon emissions seen in Figure 8. Figure 10 below shows the effects of some possible government actions by plotting the projected renewable electricity generation in four cases.

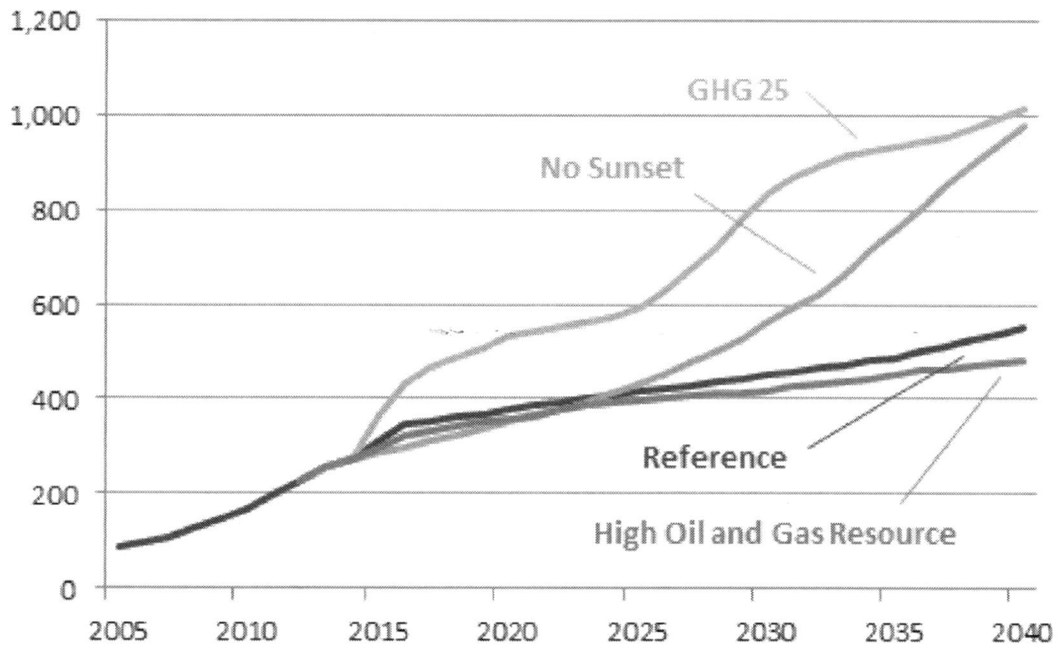

Figure 10. Renewable Electricity Generation in Four Cases, 2000-40 (billion kWh) [Adapted from 35]

In all cases, electricity generation from renewables increases over the next few decades. However, for the reference and high gas resource cases, the rate of increase in renewables is less than that seen between 2000 and 2014, mainly due to the increased implementation of natural gas power plants. The no sunset case shows the effects of extending renewable tax credits (i.e. the installment tax credit and the production tax credit) indefinitely in their current forms. In this case, the growth seen between 2000 and 2014 remains relatively constant through 2025, at which point the rate of implementation of renewables begins to increase steadily. In the GHG 25 case, a large increase is seen just after 2015 in response to the implementation of the carbon tax, after which the rate of implementation of renewables begins to level off. However, around 2025, electricity generation from renewables begins to increase dramatically once again, as more projects come into completion in response to the increased cost of natural gas power plants, whose production begins to decline as seen in Figure 9.

Compared to the reference case, both the no sunset and GHG 25 cases see disproportionate increases in solar and wind relative to other renewables like hydro or geothermal. When it comes to solar, electricity production in both cases is triple that of the reference case in 2040; for wind, it"s double. In both cases, geothermal actually sees slightly lower electricity production in 2040 compared to the reference case, and hydro sees relatively little growth between 2014 and 2040 in all cases [35].

6.0 Policy Recommendations

The electricity production sector in the US is rapidly changing as a result of the shale fracking boom. While the large shift from coal to natural gas will improve carbon emissions on a per kwh basis, the low price of natural gas is also reducing the amount of renewable sources entering the market. If no government action is taken, the growth of electricity demand will still mean an increase in carbon emissions from the electricity production sector over the next 30 years due to the low prices of natural gas. If the government truly wishes to use natural gas as a bridge fuel to a low-carbon electricity sector, a carbon tax needs to be implemented and tax credits for renewables need to be extended.

A carbon tax is necessary to appropriately reflect the societal cost of carbon-emissions-heavy technology. However, a large carbon tax is not necessary, and would cause unnecessary instability as the market acted swiftly to adjust. Instead, a low carbon tax that progressively increases is the most responsible way to implement such a change. Based off the projections in Figures 8 and 9, a carbon tax of $10 / metric ton of CO_2 would lower carbon emissions in the electricity production sector without drastically changing the economic viability of natural gas facilities. This tax would increase at a rate of 5% per year in order to incentivize investment in further carbon emission reduction technology. For a cap-and-trade program equivalent, this means the acceptable level of carbon emissions each year would be reduced, effectively increasing the cost of emissions a proportional amount.

On the other side of things, renewable technology still needs some assistance to be competitive with conventional fuels, as seen in Figure 3. However, the current split between the investment tax credit and the production tax credit makes the development of new and improved renewable technologies difficult, as it is more difficult to predict the economic environment these technologies will emerge into. A single, long-term tax strategy would offer economic stability and predictability that promotes private investment that has been yet unseen. In this way, a single, long-term tax strategy should be used to incentivize the use of all types of renewables in the electricity sector.

Of the two programs currently available, fully embracing the production tax credit makes the most sense, as it incentivizes the implementation of energy efficient and energy dense technologies[2]. It also incentivizes the development of improvements that reduce the upfront cost of these technologies, which help open them up to small scale residential and commercial consumers. The current rate of 2.3 cents per kwh is an appropriate start-point for such a tax program, but with the implementation of a carbon tax, indirect incentive towards renewables increases proportionately. The increasing nature of the carbon tax combined with the increased rate of improvement for renewable technologies actually means that the production tax credit can be slowly reduced. As such, a 5% reduction per year to mirror the carbon tax rate increase would be appropriate.

In both the carbon tax and the production tax credit, both programs need to be comprehensive and long term in order to allow the electricity producers to accurately model the economic environment that they will be entering with new projects. Due to the long-nature of construction of electricity production facilities, as well as the long operating life-time of these

[2] Again, this credit would not incentivize non-profit utility providers. Alternative programs, not covered in the scope of this paper, could be used to incentivize these utility providers.

facilities, stability in the tax structure from the government is paramount. The unexpected expiration of the production tax credit in 2013 mainly to political gridlock is unacceptable, and should be avoided in the future if at all possible.

All in all, if such a two pronged approach of a $10 / metric ton carbon tax combined with an extension of the production tax credit is used, the government will turn a net profit as a result. Based on the projections in Figures 9 and 10, there will be an average profit of $5.5 billion per year between 2015 and 2040 under such a program, with yearly profits ranging between $4.5 billion and $6.1 billion [adapted from 40 and 41]. Keep in mind that these projections only factor in the revenue from natural gas facilities; factoring in the revenue of coal should increase revenue further, despite the increase in cost from the increased use of renewables.

It is further recommended that the money made from this program be reinvested in the loan programs for electricity producers. While the reintroduction of the section 1705 program may be beneficial, it seems more beneficial to use this money to offset or eliminate the CSC cost burden on borrowers of section 1703 loans. With the accelerated retirement of coal, reduced or eliminated CSC costs would incentivize increased use of nuclear power, which acts as a more practical and carbon neutral base load power supply alternative compared to natural gas.

7.0 References

[1] Obama, B. (2014, January 28). *President Barack Obama's State of the Union Address.* Speech presented at the House Chamber, Capital Building, Washington, DC. Retrieved from http://www.whitehouse.gov

[2] EPA. (2014, June 2). *Learn About Carbon Pollution From Power Plants*. Retrieved from http:// www2.epa.gov

[3] Clement, S. (2014, June 2). A huge majority of American support regulating carbon from power plants. And they're even willing to pay for it. *The Washington Post*. Retrieved from http://www.washingtonpost.com

[4] Koch, W. (2014, June 3). EPA carbon emission limits may shift energy mix. *USA Today*. Retrieved from http://www.usatoday.com

[5] Union of Concerned Scientists. (2014, January 31). Production Tax Credit for Renewable Energy. Retrieved from http://www.ucsusa.org

[6] Davenport, Coral. (2014, May 12). Amid Pipeline and Climate Debate, Energy-Efficiency Bill Is Derailed. *New York Times*. Retrieved from http://www.nytimes.com

[7] EIA. (2014, May 7). *Annual Energy Outlook 2014*. "Market Trends: Electricity demand". Retrieved from http://www.eia.gov

[8] EIA. (2014, May 21). *Electric Power Monthly*. "Table 1.1. Net Generation by Energy Source: Total (All Sectors), 2004 – March 2014". Retrieved from http://www.eia.gov

[9] EIA. (2013, May). *State-Level Energy-Related Carbon Dioxide Emissions, 2000-2010.* "Table 2. 2010 state energy-related carbon dioxide emissions by fuel". Retrieved from http://www.eia.gov

[10] EIA. (2011, February). *Emissions of Greenhouse Gases in the United States 2009*. "Table 5. Greenhouse gases and 100-year net global warming potentials". Retrieved from http://www.eia.gov

[11] Meier, P.J. (2002, August). *Life-Cycle Assessment of Electricity Generation Systems and Applications for Climate Change Policy Analysis*. pg 70. Madison, Wisconsin: University of Wisconsin, Fusion Technology Institute. Retrieved from fti.neep.wisc.edu/pdf/fdm1181.pdf

[12] EPA. (2012, December 4). *What is Acid Rain?*. Retrieved from http://www.epa.gov

[13] EPA. (2014, March 10). *Basic Information | Mercury*. Retrieved from http://www.epa.gov

[14] EPA. (2013, September 25). *Natural Gas | Clean Energy*. Retrieved from http://www.epa.gov

[15] EIA. (2011, March 31). *Emissions of Greenhouse Gases in the US*. Retrieved from http://www.eia.gov

[16] World Nuclear Association. (2014, May). *World Energy Needs and Nuclear Power*. Retrieved from http://www.world-nuclear.org

[17] EPA. (2013, October 22). *Nuclear Energy | Clean Energy*. Retrieved from http://www.epa.gov

[18] Nuclear Energy Institute. (2014). *Life Cycle Emissions Analyses*. Retrieved from http://www.nei.org

[19] EPA. (2012, February 28). *Solar Energy | Energy and Global Climate Change in New England*. Retrieved from http://www.epa.gov

[20] EPA. (2014, May 1). *Wind Energy | Energy and Global Climate Change in New England*. Retrieved from http://www.epa.gov

[21] DOE. (2012, April 17). *Energy Department Report Finds Major Potential to Increase Clean Hydroelectric Power*. Retrieved from http://energy.gov

[22] EPA. (2014, May 1). *Geothermal Energy | Energy and Global Climate Change in New England*. Retrieved from http://www.epa.gov

[23] EIA. (2014, April). *Levelized Cost and Levelized Avoided Cost of New Generation Resources in the Annual Energy Outlook 2014*. Retrieved from http://www.eia.gov

[24] American Clean Energy and Security Act of 2009. H.R. 2454 of the 11th Congress. Retrieved from http://www.opencongress.org/bill/111-h2454/actions_votes

[25] Hulse, Carl. (2010, July 22). Democrats Call Off Climate Effort Bill. *New York Times.* Retrieved from http://www.nytimes.com

[26] 26 U.S.C. §§46, 48

[27] DOE. *Section 1703 Loan Program.* Retrieved from http://energy.gov

[28] DOE. *Section 1705 Loan Program.* Retrieved from http://energy.gov

[29] Wilson Sonsini Goodrich & Rosati. *Status Update – Department of Energy Title XVII Loan Guarantee Program.* Retrieved from http://www.wsgr.com

[30] EIA. (2008, June). *Annual Energy Outlook 2008.* "Early Capacity Additions Use Natural Gas, Coal plants Are Added Later". Retrieved from http://www.eia.gov

[31] EIA. (2014, April 8). *Half of power plant capacity additions in 2013 came from natural gas.* Retrieved from http://www.eia.gov

[32] EIA. (2014, May 7). *Annual Energy Outlook 2014.* "Most new capacity uses natural gas and renewables". Retrieved from http://www.eia.gov

[33] EIA. (2014, June 9). *Power sector CO_2 emissions sensitive to policy changes and natural gas supply.* Retrieved from http://www.eia.gov

[34] EIA. (2014, May 7). *Market Trends: Emissions.* "Energy-related carbon dioxide emissions are sensitive to potential policy changes". Retrieved from http://www.eia.gov

[35] EIA. (2014, April 29). *Renewable electricity projections show growth under alternative assumptions in AEO2014.* Retrieved from http:// www.eia.gov

362836 Design of Highly Active and Durable Electrocatalysts By Tuning Nanostructure

Abstract:

Chemical-electrical energy conversion and storage are greatly attractive for the development of sustainable energy. Among the chemical-electrical energy conversion devices, polymer electrolyte membrane fuel cell (PEMFC) is one promising technology for portable applications, from as small as portable electronics, to as large as automobiles. One key for PEMFCs is the electrocatalyst. Development of high-performance electrocatalyst nanomaterials relies on tuning material structures at nanoscale. This is in particular manifested in the design of electrocatalysts demanding both high activity and durability. In this talk, I will present a research system that connects fundamental investigation on well-defined extended surfaces (e.g. single crystal surfaces), extrapolation onto nanocrystals with highly controlled shape and size, exploration of interfacial interaction using novel nanocrystal superlattices as platform, and finally design of high performance catalysts in which all the possible beneficial properties from complex functional structures are implemented.

392059 Aqueous Phase Hydrogenation of Amine Captured CO_2

Abstract:

Carbon dioxide (CO_2) will become increasingly available as a cheap, renewable, and abundant carbon feedstock as commercialization for the carbon capture and storage (CCS) from power plants is coming into play. Development of highly efficient methods for CO_2 reduction would greatly increase the range of possible products derivable from CO_2. The hydrogenation of CO_2 to formic acid appears to be a straightforward approach to utilize CO_2. This reaction in the gas phase is endergonic (DG^o_{298} = +33 kJ mol^{-1}), i.e. thermodynamically unfavourable. However, the same reaction in the aqueous phase is exergonic (DG^o_{298} = -4 kJ mol^{-1}). Moreover, bases in aqueous solutions, e.g. NH_3, enhance the hydrogenation of CO_2 (DG^o_{298} = -10 kJ mol^{-1}). Hydrogenation of amine-CO_2 adducts or ammonium carbonates in aqueous solutions are thus thermodynamically favorable to produce formic acid. However, there are only few studies of hydrogenation of CO_2 to formic acid in the presence of amines with homogeneous catalytic systems. And the homogeneous catalysts were not stable or their performance was suppressed when water was present. Herein, we develop the carbon capture and conversion (CCC) process to convert amine-CO_2 adducts to formic acid with supported metal nano-cluster catalysts in aqueous media. To design the stable heterogeneous catalytic system for this reaction, we aim to directly convert the amine-CO_2 adducts from the CO_2 capture step of CCS process to formic acid, saving the energy and cost for separating, compressing, and transporting CO_2 in the storage step of a CCS process. The effects of the amine structures and the aqueous phase intermediates on the yield of formic acid were also investigated.

$Pd/Sn\text{-}Mn\text{-}CeO_2$ catalyst for low temperature carbon monoxide oxidation

Chao Wang, Erdem Sasmaz, Jochen Lauterbach*
Department of Chemical Engineering
University of South Carolina
Columbia, SC, 29208, U.S.A.
*lauteraj@cec.sc.edu

Supported Pd catalysts are very active for the total oxidation of CO and hydrocarbons and are therefore an essential component of automobile three-way catalysts (TWC). However, during the cold start phase of an automobile, CO and hydrocarbons are emitted untreated into the atmosphere. It is estimated that about 80–90% of all emissions from an automobile are released during the cold start [1]. Hence, promotion of catalytic activity at lower temperatures is strongly desired to achieve complete reduction of pollutants.

CeO_2 has been extensively applied in automobile CO and hydrocarbon emission control due to the high oxidation activity arising from: (i) the excellent oxygen storage capacity on the basis of the redox behavior between $Ce^{3+}$and Ce^{4+}; (ii) the relatively easy formation of labile oxygen vacancies and, (iii) the high mobility of bulk oxygen species [2]. Metallic synergistic effects could increase the amount and activity of oxygen vacancies [3]. When a secondary metal is introduced in the bulk metal, more oxygen vacancy will be produced because of the difference of their oxidation valences and atom radius.

In this study, an oxygen vacancy enriched $Sn\text{-}Mn\text{-}CeO_2$ solid solution was synthesized using a co-precipitation method. CO oxidation reaction test showed that this solid solution catalyst has more than 90% CO conversion at 100°C and above. By impregnating 1wt% Pd onto the $Sn\text{-}Mn\text{-}CeO_2$, over 90% of CO conversion was obtained at room temperature. This high CO oxidation activity at low temperatures could be attributed to the high reducibility of PdO to Pd in combination with the contribution of lattice oxygen of supports having sufficient oxygen release and storage property[4] Detailed oxygen vacancy information and CO oxidation mechanism were characterized by XPS, CO-TPR and in-situ DRIFT to understand the high activity at low temperature.

References

[1] M. Shelef, R. W. McCabe, Catal. Today 62 (2000) 35

[2] B.M. Reddy, et al. J. Phys. Chem. B 107 (2003) 5162

[3] J. Shi, Chem. Rev. 113 (2013) 2139

[4] A. Satsuma et al. Appl. Catal. B Environmental 132 (2013) 511

385488 Novel One Step Synthesis of N Doped TiO2 By Liquid Flame Aerosol Method (LFSP) for Visible Light Photocatalysis of Organic Pollutant

Abstract:

At present, there has been a growing attention in the synthesis of nanomaterials which are active in visible light for the photocatalytic degradation of volatile organic compounds. Nitrogen doped titanium dioxide is attracting a continuously increasing attention because of its potential as material for environmental photocatalysis. Conventional wet synthesis methods need multiple processing steps, high energy input and long processing times, In contrast, flame synthesis, especially liquid flame spray pyrolysis (LFSP), is a versatile one-step process for synthesizing a variety of nanoparticles. In the present work, we have developed a novel, inexpensive, green and highly reproducible method of the making the liquid precursor for the N incorporation in TiO_2 by using the rapid LFSP method. We have studied the effect of the synthesis parameters such as: 1) water to fuel ratio 2) mixing conditions, 3) concentration of the nitric acid, and 4) various N sources. A combination of various physico-chemical characterizations were performed. The photocatalytic activity and stability of the N-doped TiO_2 samples were evaluated through using the decomposition of phenol as model reaction under visible light irradiation. Characterization results show that the nitrogen dopants have a significant effect on the crystallite size and optical absorption of TiO_2. It is found that the N-doped TiO_2 catalysts have enhanced absorption in the visible light region, and exhibit higher activity for photocatalytic degradation of VOCs. We have produced The UV-vis spectroscopy results of N incorporated TiO_2 showed enhancement of light absorption in the visible range by selected composite materials in the UV range (250-400 nm). We have found the Ti/N ratio of 13% for the materials synthesized from our XPS data. The higher surface areas of around (100-120 M^2/g) were observed for the catalysts. These interesting results will be discussed in the presentation.

377535 Solar Photocatalytic Oxidation of NO By Electronspun TiO2/ZnO Composite Nanofiber Mat for Enhancing Indoor Air Quality

Abstract:

Synthetic photocatalyst TiO_2 nanofibers prepared using a sol-gel assisted electrospinning method revealed higher activity than the commercial photocatalyst TiO_2, P25, in the photocatalytic oxidation (PCO) of NO. The inner nanopores structure of nanofibersas observed from the transmission electron microscopy (TEM) image was found to be the vital parameter for the increased activity of the photocatalyst. The TiO_2/ZnO composite nanofiber photocatalysthas lower band gap than TiO_2 nanofiber which favors photo-electron generation. It would have potentially higher photocatalytic efficiency if not for the fact that the introduction of ZnO might cause charge imbalance and formation of the unsaturated chemical bonds on the surface thus changing the surface oxygen vacancies of the catalyst responsible for generating the oxidation radicals. Photocatalytic efficiency was optimized by adding an appropriate amount of Zn (0.10 wt. %) to the TiO_2, and this percentage of Zn added to the composite was in accord with the X-ray photoelectron spectroscopy measurement.

Hetero-atomic Carbon Nanostructures for Carbon Dioxide Reduction

Bijandra Kumar, Mohammad Asadi, Amin Salehi-Khojin

Department of Mechanical and Industrial Engineering, University of Illinois at Chicago, Chicago, IL, 60607, USA.

[*] Corresponding author, salehikh@uic.edu

Abstract:

According to a recent report on the future of carbon dioxide (CO_2) emission[1], it has been predicted that in the coming few decades (2010-2060) approximately 496 gigatonnes of CO_2 will be produced due to fossil fuel combustion by existing infrastructure. High level of CO_2 may drastically effect the global environment due to greenhouse gas effect. However, CO_2 can be electrochemically converted into fuel or commodity chemicals[2,3]. In this study, we found that polyacrylonitrile (PAN) based hetero-atomic carbon nanofibers and 1-ethyl-3-methylimidazolium tetrafluoroborate (EMIM-BF_4) makes an efficient co-catalyst system for CO_2 reduction. Our results indicate negligible overpotential (-0.17 V) for CO_2 reduction and more than an order of magnitude (~13 times) higher current density compared to the highly efficient silver catalyst under similar experimental conditions. The superior performance is credited to the nanofibrillar structure and high binding energy of key intermediates to carbon nanofibers surfaces. The rate of CO_2 reduction also depends on the water mole fraction in the electrolyte due to variation in pH of the solution[4]. In present system, the maximum CO_2 reduction rate was observed in 75 mol% water and 25 mol% EMIM-BF_4 solution and attributed to the high availability of protons. In light of X-ray photoelectron spectroscopy (XPS) analysis, the CO_2 reduction ability of carbon nanofibers is accredited to the reduced carbon atoms rather than electronegative nitrogen atoms[5]. This finding offers a new generation of metal-free and non-precious catalysts with much higher efficiency than the existing noble metal catalysts.

Keywords: Carbon dioxide conversion, catalyst, carbon nanofibers, electrochemical reduction.

References:

1. Davis, S. J., Caldeira, K. & Matthews, H. D. Future CO_2 emissions and climate change from existing energy infrastructure. *Science* **329,** 1330–1333 (2010).
2. Benson, E. E., Kubiak, C. P., Sathrum, A. J. & Smieja, J. M. Electrocatalytic and homogeneous approaches to conversion of CO_2 to liquid fuels. *Chem. Soc. Rev.* **38,** 89 (2009).
3. Rosen, B. A. *et al.* Ionic Liquid-mediated selective conversion of CO_2 to CO at low overpotentials. *Science* **334**, 643–644 (2011).
4. Rosen, B. A., Zhu, W., Kaul, G., Salehi-Khojin, A. & Masel, R. I. Water enhancement of CO_2 conversion on silver in 1-Ethyl-3-Methylimidazolium Tetrafluoroborate. *J. Electrochem. Soc.* **160,** H138–H141 (2012).
5. Kumar, B. *et al.* Renewable and metal-free carbon nanofibre catalysts for carbon dioxide reduction. *Nat. Commun.* **4,** (2013).

Ozone Removal using New Structure of Matter for High Performance Heterogeneous Catalysis with Short Contact Time

Qiang Gu and Bruce J. Tatarchuk
Center for Microfibrous Material Manufacturing (CM3), Department of Chemical Engineering, 212 Ross Hall, Auburn University, AL 36849

Abstract

Pleated Microfibrous Entrapped Catalyst (MFEC) manufactured from 8 μm diameter nickel fibers were tested for ozone decomposition at turbine bleed air conditions to demonstrate the improved catalytic performance over conventional monolith reactors.

In this research, ozone has been used as model compound. MFEC reactors were investigated under aircraft bleed air conditions of high temperature (100-200°C) and high face velocity (10-30 m/s) resulting an interlayer contact time of 67-200μsec. Precious metal (Pd, Ag) and transition metal (Mn) catalysts were impregnated on entrapped particles (e.g. γ-Al2O3) using incipient wetness method. Ozone test concentration was set at a high-demanding 1.5 ppmv. Results showed that a high level of ozone decomposition was achieved with a significant reduction of catalyst consumption. Compared with conventional aircraft filters, this can be a huge advantage in terms of material cost and labor. Because the entrapped small particles (150-250μ m) improve the contact efficiency significantly, this kind of reactor has shown high catalytic activity even though the contact time for the reactant has been shortened to 100μ s range due to the high face velocity. The ozone reaction kinetics has been analyzed using a first order plug flow model, which showed the reaction was external mass transfer limited. Mass transfer rate, effective reaction rate and overall efficiency factor were compared among packed bed, monolith and MFEC reactors. Also catalyst aging tests were carried out according to the frequency of commercial aircraft ozone abater replacement, which evaluated long-term performance of MFEC. XPS, TPR and chemisorption tests have been used to characterize catalyst deactivation mechanisms. Besides, several other aspects including humidity, system pressure been investigated for affection on the reaction.

Keywords: Metal Fiber, Ozone, Reaction

1. Material and Methods

1.1 High Volumetric Test Set Up

Figure 1(a) shows the schematic of the test system. The high velocity test set up includes a 40 horse power blower (Fan Equipment Company Inc.) circulating air inside a closed loop system. A variable drive is used to control the system speed ranging from 10-40 m/s. The system is running at near atmospheric pressure. The MFEC reactor is located near the inlet of the blower.

1.2 Pressurized System Test Set Up

While this smaller test set up (Figure 2) can maintain high temperature and identical ozone level, this system cannot hold any pleat structures. Because of this limitation, it can only test a flat MFEC sheet with a reduced face velocity (up to

15m/s). However, face velocity within each pleat is only 1/n of the system velocity in the pipe (i.e. 1/4 if W structure is used for pleat structure) on a full scale structure. The in-bed face velocities for both pleat structure and flat structure cases are within the same range. Catalytic performance at higher system pressures is compared with lower pressure performance based on the results generated on the same system.

1.3 Set up for Humidity Tests

Humidity plays an important role in catalyst performance, since it can either promote a reaction by providing favorable environment or reduce reaction rate by blocking active catalyst sites. In certain application environments like high altitude ambient air, the humidity level can drop dramatically. Catalytic performance at these environments cannot be predicted based on ground level results. Set up was shown in Figure 3.

1.4 MFEC and Catalyst Preparation

8 m nickel (Ni-200) fibers were acquired from IntraMicron, Inc., AL. -Al_2O_3 was acquired from Alfa Aesear®. In making MFEC sheets, metal fibers was blended with cellulose into a suspension using low speed blending at early stage to avoid chopping of the fibers and high speed at later stage to fully disperse the fibers. The suspension is then transferred into an 8"×8" sheet former. Al_2O_3 is added to the suspension with rapid stirring. Water is drained rapidly, minimizing the different precipitation speed of the fibers and supports. The preform sheets were fully dried (24h). At this stage, cellulose added act us the binding agent of the fibers. However, they can affect the mechanical strength of the sheets if not removed properly before sintering. This was done by leaving the dried sheet at 400 for 1h. The dried sheets were then sintered at 1050 in a reducing environment (H_2) using BTU-1000 furnace for 40 min.

$Pd(NO_3)_2$, $AgNO_3$ and $Mn(NO_3)_2$ were acquired from Alfa Aesear®. These catalysts were added to MFEC by incipient wetness process. MFEC was soaked in catalyst solution for 30 min and drained of excess water after worth. Impregnated sheets were dried overnight and calcined at 673K for 4h.

2. Result and Discussion

2.1 Catalytic Decomposition of Ozone using MFEC

Figure 4 showed the Pd/Al_2O_3 catalyst tested at three temperatures. As observed in the figure, the catalytic performance had a linearly reducing relationship with increasing face velocities. Even though convective heat and mass transfer at these velocities increase with increasing face velocities, the catalytic performance was not affected by these factors, leaving the contact time the crucial factor that determines the performance of the these catalysts. At these velocities, the Reynolds number within the test rig is no doubt in the turbulent flow region ($\sim10^6$). However, considering the small characteristic dimension of the entrapped particles, the Reynolds number within the MFEC is much smaller ($\sim10^2$). Hill et al. has studied the flow type for the beds made of similar size particles and found the flow to be weakly turbulent in this range. Since there is no flow type transition from inlet to the inner part of the media, face velocity is regarded as the only factor affecting the reaction. This is verified by other test results by using different catalyst materials, including Ag, Mn and bi-metal catalysts. The results are shown in Figure 5. For any catalyst tested, they all showed same linear relationship as Pd catalyst. The only difference is their temperature sensitivity. Mn based catalysts are widely used as low temperature ozone decomposition catalysts since they are known to have smaller activation energies. Therefore, Mn based catalysts showed better performance at lower temperatures. Pd catalyst activity has dropped dramatically when running under 150 . In all these tests, thermal decomposition of the ozone by itself is not considered since the temperature is not high enough. An experiment is carried out to verify this assumption. MFEC reactor is replaced with a bare rig. No ozone concentration change is observed when running tests at same conditions.

2.2 Effect of System Pressure Catalytic Performance

In real world, high mass flow applications usually involve using the bleed air of turbo engines. The condition of bleed air usually contains not only high flow rate and high temperature, but also high pressures. Since system pressures affect mass and heat transfer inside the reactor, it has to be evaluated as an important factor that may affect catalyst activity.

In order to compare performance of catalysts at different system pressure, two major quantities, volumetric mass transfer coefficient and effective reaction rate, are used to evaluate performances at different conditions. The volumetric mass transfer coefficient was calculated from Sherwood number,

To estimate Sherwood number used in this calculation, Dwivedi and Upadhyay method for gas-solid system in fixed bed is used, which can be expressed as

This correlation is good for voidage ranging from 0.25 to 0.97 and Re_p from 1 to 10000, which covered the conditions used in this research.

Kalluri et al. has developed the relationship between effective reaction rate and mass transfer coefficient for first order reactions using MFEC material by assuming gas-solid mass transfer rate equals to the reaction rate inside the catalyst. The correlation used for calculating effective reaction rate is

2.3 Aging Performance

Ozone reactor aging performance is one of the most significant aspects when evaluating the overall performance of the reactor, because ozone concentration is of key importance to the health of on-borad passengers on a compact system like a commercial aircraft, especially to airline employees who log far more hours than typical passengers. Heck et al. has reported that noble metal catalysts outperform base metal catalyst in airplane applications because of their better aging performance, even though initial conversion rate of most base metal catalysts for ozone is fairly high. Results were shown in Figure 6.

3. Conclusion

MFEC as a new matter of structure has shown great potential in offering high performance heterogeneous catalysis while maintaining proper weight and volume of the reactor. MFEC is a potential substitute for packed bed and monolith if modified properly.

References

[1] M. Sheng, H. Yang, D. Cahela, B. Tatarchuk, Novel catalyst structures with enhanced heat transfer characteristics, Journal of Catalysis 283 (2011) 254-262

[2] Teknillinen tiedekunta, Prosessi- ja ympäristötekniikan osasto, Oulun yliopisto Deactivation Correlations of Pd/Rh Three-way Catalysts Designed for Euro IV Emission Limits Effect of Ageing Atmosphere, Temperature and Time

[3] Calvin H Bartholomew, Mechanisms of catalyst deactivation, Applied Catalysis A: General

[4] Ryan A. Sothen, Bruce J. Tatarchuk, A semi-empirical pressure drop model: part I- pleated filters, HVAC&R Research, Vol 14, Number 6

[5] Daniel K. Harris, Donald R. Cahela, Bruce J. Tatarchuk, Wet layup and sintering of metal-containing microfibrous composites for chemical processing opportunities, Composites: Part A 32 (2001) 1117-1126

[6] R.M. Heck, R.J. Farratuo, H.C. Lee, Commercial development and experience with catalytic ozone abatement in jet aircraft

[7] R.C. Sullivan, T. Thornberry, J.P.D. Abbatt, Ozone decomposition kinetics on alumina: effect of ozone partial pressure relative humidity and repeated oxidation cycles, Atoms. Chem. Phys., 4, 1301-1310, 2004

[8]R.J. Hill, D.L. Koch, The transition from steady to weakly turbulent flow in a close-packed ordered array of spheres, J. Fluid Mech. (2002), vol 465, pp. 59-97

[9]T. Kameya, K. Urano, Catalytic decomposition of ozone gas by Pd impregnated MnO2 catalyst, J. Environmental Engineering, Vol. 128. No.3, 2002

roll made by paper making machine

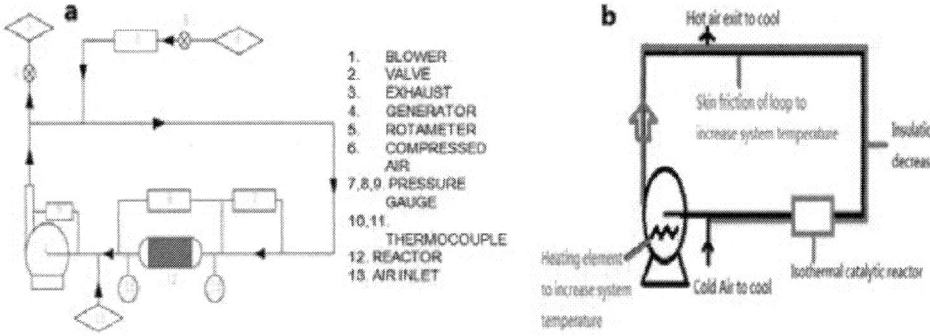

Figure 1 Sketch of high face velocity test setup; (a) Schematic of test set up (b) Heat balance of the system, heat source in red, cooling source in black

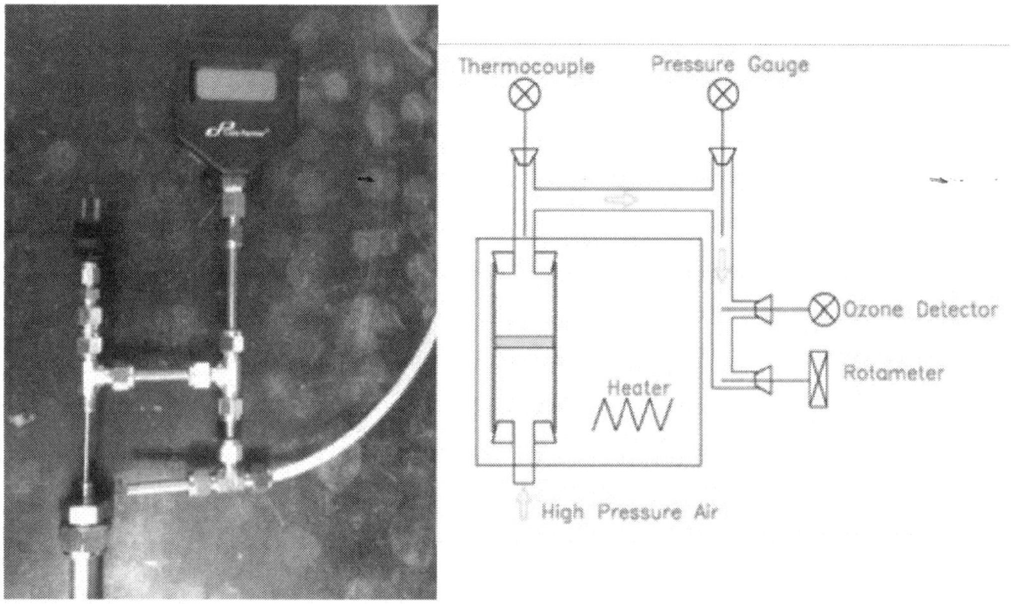

Figure 2 Tube reactor for catalytic performance in pressurized system

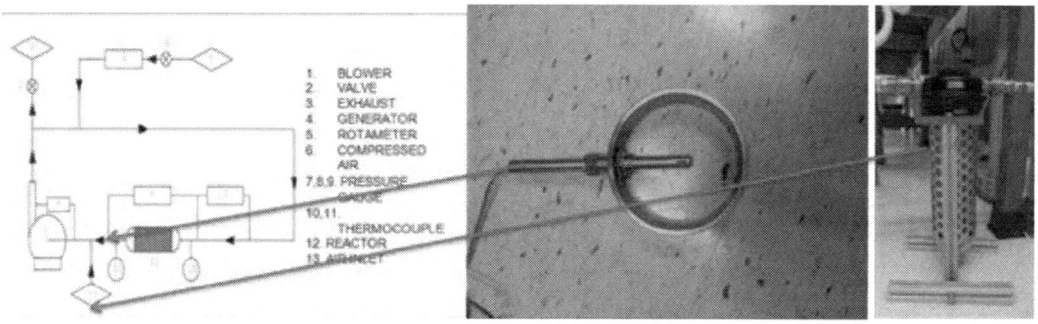

Figure 3 Humidity meter installed on the system

Formatting Issue - Best Available Paper/Graphic

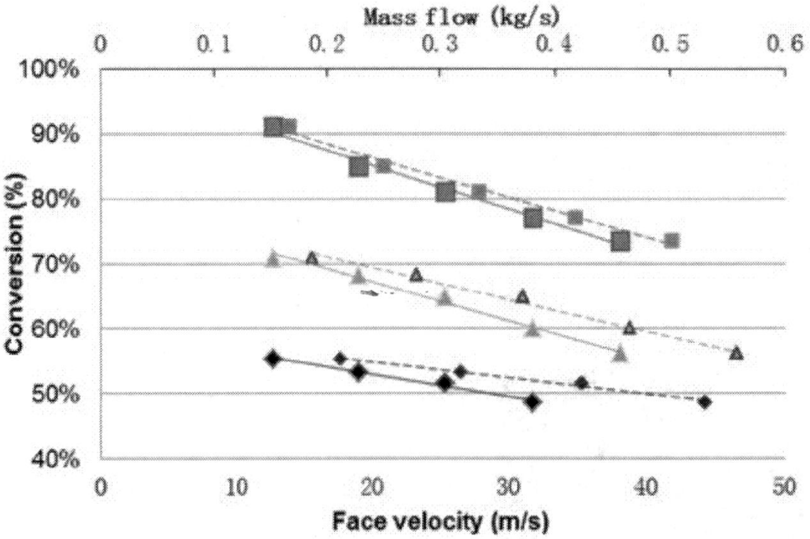

Figure 4 Pd based catalyst for ozone decomposition at high temperatures and high face velocities

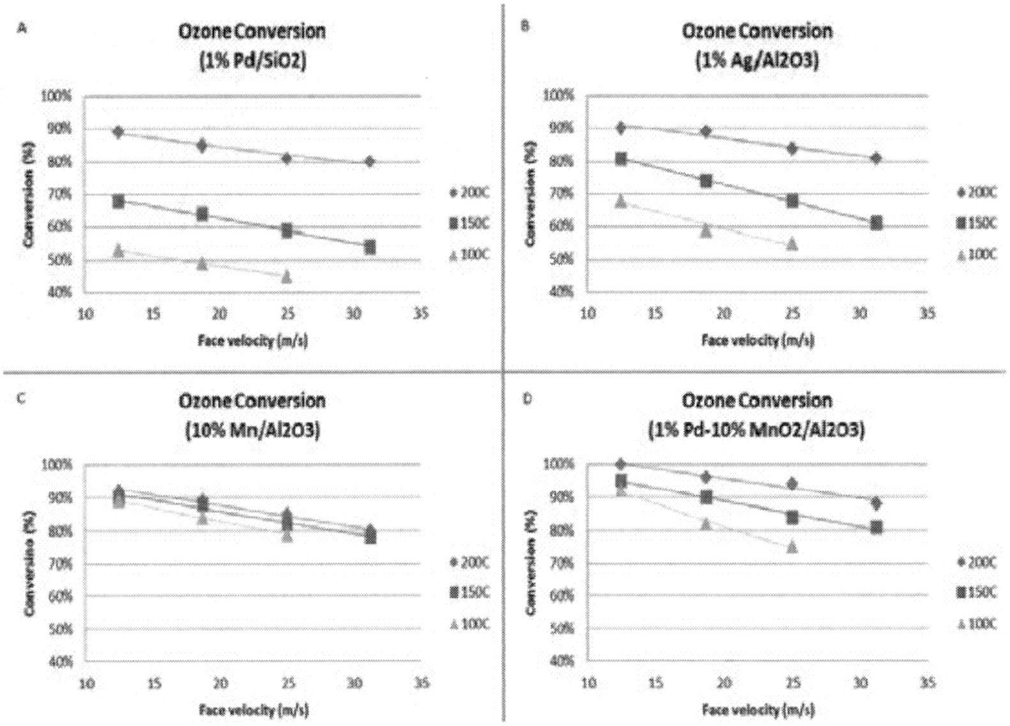

Figure 5 Catalyst screening for ozone decomposition using Pd and Mn based catalysts; A 1% Pd/SiO_2 catalyst at 100, 150 and 200 ; B 1% Ag/Al_2O_3 catalyst at 100, 150

and 200 ; C 10% MnO_2/Al_2O_3 catalyst at 100, 150 and 200 ; D 1%Pd-10%MnO_2 catalyst at 100, 150 and 200

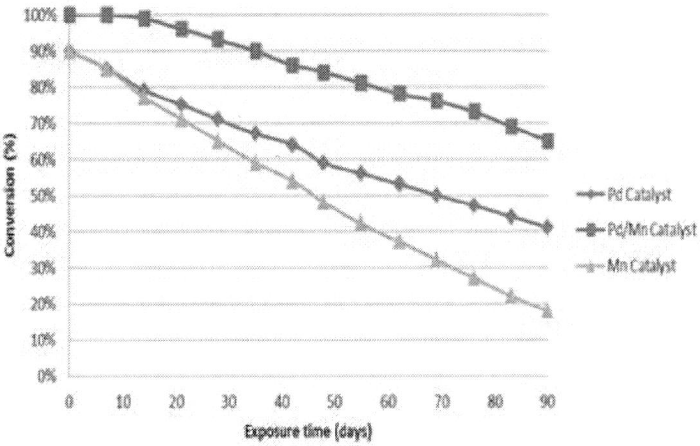

Figure 6 Pd and Mn based catalysts aging performance

Table 1 Ozone decomposition catalyst reaction kinetics analysis using Arrhenius equation

Catalyst	*Rate constant k (1/s)*			*E_a (kJ/mol)*	*TOF (10^{-2} s^{-1})*
	100	150	200		
1% Pd/AlO_3	885	1640	4530	23.6	5.12
1% Pd/SiO_2	574	1463	4403	29.7	5.08
1% Ag/Al_2O_3	1921	3001	4527	12.5	3.37
10% MnO_2/Al_2O_3	3801	4304	4710	3.2	N/D
1% Pd-10% MnO_2	4144	5313	8302	10.0	N/D

Table 2 Two kinds of aluminum supports used for external mass transfer effect test

Formatting Issue - Best Available Paper/Graphic

	pore volume	*surface area*	*particle size*
Aluminum support 1	0.62 cc/g	220 m^2/g	150~250 m
Aluminum support 2	1.14 cc/g	255 m^2/g	150~250 m

Table 3 Humidity level for tests

	without the desiccant	*with the desiccant*
200	4%	15%
150	<1%	2%

A Mechanistic Investigation of Selective Glycerol Hydrogenolysis over Pt-Re Bimetallic Catalysts

Derek D Falcone (derekfalcone@virginia.edu)

John H Hack (jhh5za@virginia.edu)

Robert J Davis (rjd4f@virginia.edu)

Several chemical intermediates can be derived from renewable glycerol, a byproduct of the biodiesel transesterificaton process. In this work the selective hydrogenolysis of the secondary C-O bond in glycerol to produce 1,3-propanediol was studied over Pt-Re bimetallic catalysts in liquid water. In addition to shifting the selectivity of the reaction, Re also promoted the overall rate of glycerol hydrogenolysis by more than an order of magnitude compared to monometallic Pt. The role Re plays in the reaction as well as the oxidation state of Re is still a matter of debate in the literature. One possible role is that a Brønsted acid site is associated with the Re component in liquid water, which catalyzes the dehydration reaction at the secondary carbon of glycerol. In this work, XPS was used to evaluate the oxidation state of Pt and Re after in situ pretreatments in dihydrogen at 393 and 473 K. Infrared spectroscopy was also used to characterize the acidity of Pt-Re nanoparticles supported on silica using pyridine as an acid probe. In addition, batch reaction experiments showed that dilute NaOH substantially decreased the rate of glycerol hydrogenolysis, presumably by the neutralization of acid sites on the catalyst. Propyl acetate hydrolysis was also used as a probe reaction for acid catalysis and the Pt-Re bimetallic catalyst showed rates of hydrolysis comparable to rates in aqueous HCl. These results are consistent with the idea that Brønsted acid sites are present on the bimetallic catalyst in liquid water.

Glycerol hydrodeoxygenation catalyzed by Cu-Pd/Na-TiO_2: Structure of the catalysts

Alba N. Ardila, Erasmo Arriola, Gustavo A. Fuentes*
Dept. of Process Engineering, Universidad A. Metropolitana- Iztapalapa, A.P. 55-534; 09340 Mexico, DF; Mexico.
*- Email: gfuentes@xanum.uam.mx

There is much ongoing research to discover viable catalysts for the conversion of glycerol, a byproduct of biodiesel production, to different high added-value chemicals. On that line, we have developed a novel catalyst, Cu-Pd/Na-TiO_2, that shows very good characteristics during the aqueous phase hydrodeoxygenation of glycerol to 1,2- propanediol (PDO) under basic conditions. This catalyst requires low hydrogen pressure to accomplish that goal and at the same time has very low C-C hydrogenolysis activity, resulting in high yield and selectivity to the diol. It can also be reused a number of times with low decay in activity.
In this work we present results obtained using a combination of techniques, UV-Vis spectroscopy, H_2-TPR, XPS and high resolution STEM, coupled with reaction studies, to analyze the synergistic effect that occurs when Pd and Cu are combined on the same material.
Our results are consistent with the formation of Pd-Cu alloy nanoparticles. A consequence of alloying is the stabilization of Cu in a reduced state, what increases its dispersion and limits Cu leaching to the solution. Pd appears to be also responsible for partial reforming of glycerol to produce H_2 that is used in the hydrogenation of acetol in the absence of feed H_2. The role of the basic sites on the support is also discussed.
This class of materials may be a promising alternative for industrial applications, because the results obtained are better or comparable with those reported in the literature at much higher H_2 pressures.

Co-deposited electroless Cu-Ni-Mo electrocatalyst for glycerol oxidation

Oyidia Elendu, Yaw Yeboah and Egwu E. Kalu

Department of Chemical and Biomedical Engineering

FAMU-FSU College of Engineering

Tallahassee FL 32310

Cu, Ni and Mo have many properties that make them attractive as catalysts. Both the thermochemical catalytic and electrocatalytic activity of many metals are enhanced by the presence of other co-deposited metals. In addition, other synergies arise for co-deposited metals including enhanced deposit thickness and size, luster, wear, recyclability and other mechanical properties. Ni-Cu-Mo electrocatalysts were prepared by electroless deposition technique using a mixed reducing agent for the co-deposition. The electrocatalysts were characterized by XRD, SEM, EDAX, XPS and cyclic voltammetry techniques. The electrocatalysts were evaluated for the electrooxidation of glycerol in alkaline and acid media. The electrocatalytic activity of Ni-Cu-Mo catalysts on glycerol oxidation and product distribution will be presented and discussed.

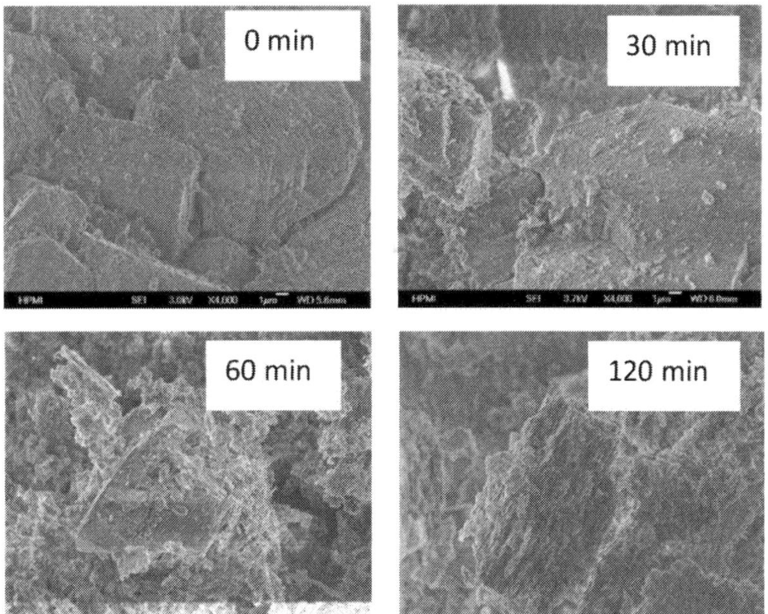

Fig. 1: SEM features of electroless deposited Ni-Cu-No electrocatalyst on Al_2O_3

389637 Base Free, One-Pot Synthesis of Lactic Acid from Glycerol Using a Bifunctional Pt/Sn-MFI Catalyst

Abstract:

Glycerol, an inevitable by-product from biodiesel production, is potentially obtained from non-edible biomass such as microalgae, cellulose and its derivatives. The conversion of glycerol to commodity chemicals not only reduces the chemical waste but also provides a sustainable and green pathway for our society. Lactic acid has been considered as one of the top 14 chemical building blocks produced from biomass conversion by US Department of Energy at 2010. Compared to glycerol, lactic acid (LA) and alkyl lactates have a higher value and offer important applications as green solvents, platform chemicals and precursors for the synthesis of polylactate or polylactic acid (PLA), the second most manufactured bioplastic of the world. LA can be produced from petroleum feedstocks in high yields using mineral acid catalysts. However, the use of toxic chemicals (e.g. hydrogen cyanide) and the high cost for product separation could lead to significant environmental issues. Therefore, currently LA is mainly manufactured by the fermentation of carbohydrates. However, the efficiency and productivity of the fermentation process are still low and need substantial improvements because of the high cost of enzyme catalysts and the need for precise control of operating conditions. There is thus a persistent need to develop efficient thermochemical pathways for the production of LA from new starting materials.

In this study, we report a base free one-step reaction pathway to selectively produce LA with high yields from the oxidation of glycerol over a bifunctional Pt/Sn-MFI catalyst under mild reaction conditions. 80.5% selectivity to LA was achieved at 89.8% conversion of glycerol[1]. In the cascade reaction route, the selective oxidation of glycerol into glyceraldehyde proceeds on Pt catalyst. Sn-MFI exhibits outstanding Lewis acidity for converting glyceraldehyde into dihydroxyacetone and dihydroxyacetone into the final product LA. The superior performance is mainly due to its microporous structure, accessible Sn site and Lewis acidity.

1. Cho, H. J.; Chang, C.-C.; Fan, W., "Base free, one-pot synthesis of lactic acid from glycerol using a bifunctional Pt/Sn-MFI catalyst". *Green Chem.* **2014**.

386098 The Catalytic Effect of H-Beta, Ce- and Ln-Doped Beta Zeolite on the Hydrothermal Transformation of Glycerol

Abstract:

The growth of the biodiesel industry over the last decade has resulted in a price drop of glycerol, making it an attractive feedstock for transformation into value-added chemicals [1]. The three hydroxyl groups of glycerol are highly reactive but their similar activation energies complicate their selective conversions. Among the processes tested for glycerol transformations, "hydrothermal technology" has gained attention due to its potential to overcome the difficulties associated with other methods such as long reaction times of fermentation [2]. High T/P water acts as a potent acid-base medium enabling short reaction times [2, 3] while the addition of a suitable catalyst may increase the selectivity to desired products [2].

Lewis acidic Sn-, Zr-, Ti- doped Beta zeolites exhibit high selectivity to the formation of lactic acid from trioses in both methanol and water at 80 °C and 125 °C, respectively [4]. In this study, the incorporation of even stronger Lewis acidic Ce and Ln into the BEA framework of zeolites was carried out using two different procedures: i) solid state ion exchange (SSIE) between H-Beta zeolite and a metal nitrate at 550 °C and ii) dealumination of H-Beta zeolite by strong mineral acid without collapsing the zeolite structures followed by SSIE [5].

Transformation of glycerol was investigated under hydrothermal conditions (270-360 °C, 55-186 bar, 5-180 min) using H-Beta, Ce- and Ln-doped Beta zeolites in batch tubular reactors. The catalysed glycerol transformation followed mainly two concomitant pathways in all systems: dehydration and/or oxidation. The former took place at faster reaction rates yielding ~17 %C yield of acrolein within the first 10 min. The latter primarily produced lactic acid having yield increased with reaction time; max ~20 %C. The metal-doped Beta zeolites prepared by simple SSIE showed a ~30% increase in selectivity towards lactic acid as compared to H-Beta zeolite. The incorporation of the metals resulted in lower Brønsted acidity of the zeolite surface but higher Lewis acidity which may play an important role in the transformation of pyruvaldehyde (PAL), the final intermediate product which transforms into lactic acid via Cannizzaro reaction as proposed by Bicker *et al.*[6]. The selectivity to lactic acid obtained was slightly higher in the reactions catalysed by Ce- (~30 %C) than those catalysed by Ln-doped Beta zeolite (~25 %C).

Herein, we report for the first time high yield of lactic acid (~ 20 %C, ~ 80 %mol glycerol conversion) achieved directly from glycerol under base-free hydrothermal conditions using high Brønsted-acidic H-Beta and Lewis acidic Ce- and Ln-doped Beta zeolites. The lactic acid yield achieved was competitive to those obtained in the reactions catalysed by Pd/TiO_2, Pt/TiO_2 or Au/TiO_2 at 90 °C in basic environment [7] and greater than in WO_3/TiO_2catalysed reactions under supercritical water conditions [8].

References

1. Yang, F., M. Hanna, and R. Sun, *Value-added uses for crude glycerol - a byproduct of biodiesel production.* Biotechnology for Biofuels, 2012. **5**(13).

2. Long, Y.-D. and Z. Fang, *Hydrothermal conversion of glycerol to chemicals and hydrogen: review and perspective.* Biofuels Bioproducts & Biorefining-Biofpr, 2012. **6**(6): p. 686-702.

3. Akiya, N. and P.E. Savage, *Roles of water for chemical reactions in high-temperature water.* Chemical Reviews, 2002. **102**(8): p. 2725-2750.

4. Taarning, E., et al., *Zeolite-catalyzed isomerization of triose sugars.* Chemsuschem, 2009. **2**(7): p. 625-627.

5. Hammond, C., S. Conrad, and I. Hermans, *Simple and scalable preparation of highly active Lewis acidic Sn-Beta.* Angewandte Chemie-International Edition, 2012. **51**(47): p. 11736-11739.

6. Bicker, M., et al., *Catalytical conversion of carbohydrates in subcritical water: A new chemical process for lactic acid production.* Journal of Molecular Catalysis A: Chemical, 2005. **239**(1-2): p. 151-157.

7. Shen, Y., et al., *Efficient Synthesis of Lactic Acid by Aerobic Oxidation of Glycerol on Au-Pt/TiO2 Catalysts.* Chemistry - A European Journal, 2010. **16**(25): p. 7368-7371.

8. Akizuki, M. and Y. Oshima, *Kinetics of Glycerol Dehydration with WO3/TiO2 in Supercritical Water.* Industrial & Engineering Chemistry Research, 2012. **51**(38): p. 12253-12257.

382616 Methane Conversion to Ethane and Ethylene Using Solid Oxide Membrane Reactors

Abstract:

Over the past few decades, natural gas production in the US has increased due to shale gas, and is projected to continue to increase. This increase in availability and production has driven prices down and prompted research into new ways to use methane, the main component in natural gas. Methane is currently reformed to produce synthesis gas, a mixture of CO, CO_2 and H_2, which is then used to make other chemicals and fuels. Although these processes are economical, they rely on large economies of scale, resulting in very large chemical plants and little use for stranded sources of natural gas. In contrast, direct conversion of methane to chemicals such as ethylene could create an economical use for natural gas that is currently flared.

Methane can be directly converted to C_2 products using oxidative coupling, where CH_3 radicals are formed using oxygen on a catalyst surface then couple to form ethane or ethylene. In this work membrane reactors are fabricated which can control oxygen anion flux through a membrane. These devices consist of three layers of solid oxide and perovskite materials. First a porous layer which dissociates oxygen from air, then a very thin solid layer that conducts oxygen anions, and finally another porous layer which facilitates methane activation. We will show conversion and selectivity results for these devices using several different membranes including: yttria stabilized zirconia (YSZ), scandia stabilized zirconia (ScSZ), and gadolinium doped ceria (GDC). We will also show results for factors that influence methane activation including the addition of transition metals.

357178 High Efficient and Recyclable Heterogeneous Catalysts (Ni-modified zeolites) for Reppe Carbonylation to Acrylic Acid

Abstract:

Reppe Carbonylation is undoubtedly one of the most important reactions in acetylene chemistry for yielding acrylic acid and its esters. Although this route had been replaced by propylene oxidation in the past decade owing to the global availability of inexpensive ethylene from petroleum, the desire for diversified feedstock for bulk chemical commodity is steadily on the increase. Because of the traditional homogeneous catalysts shows disadvantages in the catalyst -product separation and the scaling up of the process, more recent works were focused on the development of heterogeneous catalysts.

The current paper reports, for the first time, a highly efficient heterogeneous catalyst systems for the hydrocarboxylation of acetylene to acrylic acid. The nickel modified Y zeolites with carefully controlled concentrations of metallic and acidic sites were prepared, characterized and tested in acetylene carbonylation performed in a batch reactor. The catalysts were thoroughly characterized by ICP-AES, X-ray diffraction, Temperature-programmed Reduction, Extended X-ray absorption fine structure, Temperature-programmed desorption of NH_3, pyridine-FTIR, SEM, Raman and thermal gravimetric analysis. It was found that all of the exchanged Ni in zeolite were present as charge-compensating cations and remained this state through the overall reaction. The nickel sites play a determinative role in the carbonylation of acetylene, and the acidic sites would interact with acetylene and favor the protonation of acetylene to form active intermediates, but a too high density of acid sites would cause the formation of coke. Base on these findings, it is necessary to design a catalyst with a proper ratio between nickel ion and acid sites. In addition, promoters such as copper salts were required, and the $CuBr_2$/Ni-Y-zeolite has been proved to be a highly effective catalyst system in the hydrocarboxylation of acetylene with 1400 h^{-1} of average turn over frequency after six times of recycle at 235 and 3.6MPa of initial total pressure with 30min of reaction time. The amount of acrylic acids formed with these Ni-exchanged Y catalysts($4\sim5\times10^{\wedge 4}$g/(molNi*h)) was far higher than those reported with Ni supported on silica gel (~3.4 g/(molNi*h)). Furthermore, the coke deposited on the zeolites, locating mainly on the external of external surface of the zeolite, was present as ordered filamentous nature, and much of them was separated from the zeolite crystals and thus keeps the catalysts with a remarkable reusability. The current heterogeneous catalysts system showes the best catalytic activity for synthesizing acrylic acid by carbonylation of acetylene with carbon monoxide and water to date.

Computational tools are beneficial to characterizing the improved catalytic properties of alloys, but computed results are often not reconciled with experiment. In this study, the bulk composition dependent dissociative adsorption energy of hydrogen on CuPd alloys has been measured experimentally and modeled using density functional theory. The hydrogen adsorption energy cannot be simply defined by a single reactive site, nor as a composition weighted average of the pure metal components. Detailed modeling of such systems is difficult due to the numerous possible active sites, the way that adsorbates and sites interact with one another, and accounting for surface phenomena such as segregation. We developed a modeling approach that uses a basis of active sites to estimate the bulk composition dependent adsorption energy, weighted by the site probability distribution. With this method we can explain the composition dependent adsorption energy on Cu-rich alloy surfaces with relatively little computational effort. Deviations from predicted trends in the Pd-rich region can also lend insight into the experimental results at an atomistic level. In Pd-rich alloys a Pd-hydride phase forms which results in deviations from adsorption on the metallic alloy surface.

385729 Computational Study of Dehydrogenation of Light Alkanes over Pt and Pt/Sn Alloys

Abstract:

The effect of platinum tin alloy structure and composition on the kinetics and thermodynamics of dehydrogenation and coke formation pathways during light alkane dehydrogenation have been studied using density functional theory. Light alkane dehydrogenation to olefins can add significant value to hydrocarbon processes that generate ethane and propane by converting low value commodity fuels to high-value chemical and polymer precursors. Supported Pt catalysts are known to be active for ethane and propane dehydrogenation, but the high temperatures required by these endothermic reactions leads to significant coke formation and deactivation. Alloying Pt with Sn and other main group elements has been shown to decrease the amount of coke formed and leads to more stable catalysts. We apply periodic density functional theory to understand how the structure and composition of these alloys affect their ability to suppress coke formation. We investigate the potential energy surfaces from ethane and propane along the desired pathway to their corresponding alkenes, and along the undesired pathways towards surface carbon/coke. The effect of catalyst composition (e.g. Pt/Sn ratio) and surface geometry is investigated. As compared to pure Pt, C-H bond scission is more difficult on the alloys, and dihydrogen desorption is more facile.

390582 Understanding Nobel Metal Sulfation Using First Principles Computations

Abstract:

Understanding Noble Metal Sulfation Using First Principles Computations

Hom N. Sharma[1], V. Sharma[2], and R. Ramprasad[2]

[1]Department of Chemical and Biomolecular Engineering, University of Connecticut, Storrs, CT

[2]Materials Science and Engineering, University of Connecticut, Storrs, CT

Sulfur poisoning augments catastrophic impacts to the noble metal based exhaust aftertreatment catalysts. Sulfur oxides (SO_x) chemisorb onto and react with the active catalyst sites, preventing reactant access and modifying the surface chemistry. Metal-sulfate formation due to the interactions with SO_x with the catalyst metals and supports are associated with changes in various structural, morphological, and electronic properties. In this work, we demonstrate molecular level understanding on sulfur poisoning, which is experimentally observed on Pd-based catalysts but not on Pt-based catalysts in realistic oxidizing/sulfating environment using first principles thermodynamics. In practically relevant temperature and O_2 and SO_3 chemical potentials (or partial pressures), Pt and Pd show significantly different behavior towards oxidation (metal oxide formation) and sulfation (metal sulfate formation). This study revealed a map with various stable phases involving the SO_3 and oxygen on the clean as well as oxidized surfaces and clearly showed the sulfation prone nature of Pd to form $PdSO_4$.

377290 Massive Restructuring of Cobalt Catalysts Under Fischer Tropsch Conditions: Driving Force and Mechanistic Consequences

Abstract:

Fischer-Tropsch (FT) synthesis converts synthesis gas, a mixture of CO and H_2, to long-chain hydrocarbons and water over supported cobalt catalysts. On exposure to synthesis gas under reaction conditions, Co(0001) terraces undergo a massive reconstruction to form nano-islands of about 2 nm diameter [1]. This reconstruction creates defect sites. The strong adsorption of CO and of reaction intermediates is suggested to drive this reconstruction, but a detailed understanding of the structure of the nano-islands and of the thermodynamic driving force for this reconstruction is lacking. Using DFT, we demonstrate that the high CO coverage at the step sites is sufficient to overcome the energy penalty to create additional step sites [2]. The triangular shape of the nano-island is governed by the unusually high thermodynamic stability of square-planar carbon at the B5 step sites and by the increased stability of CO at step sites covered by carbon. Both effects result from the σ-aromaticity of the "Co_4C" subunit [3]. The shape and size of the nano-islands is moreover a strong function of the carbon chemical potential, and depends on the reaction conditions. The electron count in the aromatic system limits the carbon step coverage to 50%, leaving B5 vacancies as the active sites for the hydrogenation of CO to CH.

References.

[1] Wilson J.; de Groot C.; *J. Phys. Chem.*, 99, 7860, **1995**

[2] Banerjee A.; Saeys M.; in preparation

[3] Alexandrova A. N.; Thang Q.T.; Saeys M., submitted.

360459 A First-Principles Study on the Catalytic Roles of Co^0 and Co^{2+} in Ethanol Steam Reforming

Abstract:

The ethanol steam reforming (ESR) reaction is a promising way of producing hydrogen from bio-renewable materials. However the practical application of ESR requires a low-cost catalyst that shows fast kinetics, high selectivity to CO_2 and H_2 as well as long-term stability. As a candidate catalyst, cobalt based materials, such as ceria supported Co, show good performance as judged by the aforementioned criteria. Previous studies have suggested that two oxidation states of Co, namely Co^0 and Co^{2+}, coexist under ESR reaction conditions and they both contribute to the ESR catalytic activity, however their separate roles are still an open question. We used first-principles based methods, such as density functional theory (DFT) and *ab initio* atomistic thermodynamics, to comparatively study the ESR reaction pathways on Co metal and CoO surfaces. Specifically, we examined the initial steps of ethanol decomposition through either dehydrogenation into acetaldehyde or dehydration into ethylene on both the Co metal and the CoO surfaces. We also investigated water gas shift (WGS) reactions on both surfaces since WGS may be important in determining the ESR reaction selectivity to CO_2 or CO. Our results suggest that neither Co^0 nor Co^{2+} is active for all ESR reaction steps, and a collaboration of both oxidation states, for example the existence of a Co^0/Co^{2+} interface, is very likely required for achieving high ESR catalytic activity. We also found that different facets of Co^{2+} surfaces, for example CoO(111) and CoO(100), have distinct activities towards ESR reactions. Controlling the dominant Co facets, for example through tailoring the morphology of the ceria support, may enhance the performance of Co-based catalysts.

366920 Coverage-Dependent Site and Adsorbate Configurational Correlations on (111) Surfaces of Late Transition Metals

Abstract:

Correlations between adsorption energies allow for the prediction of a large set of adsorption energies from a few DFT calculations. One popular correlation that significantly reduces the number DFT calculation one needs to perform to evaluate reaction thermodynamics are correlations between adsorption energies of different adsorbates on a single surface -- oftentimes called scaling relationships. Another equally useful but uncommonly investigated correlation are correlations between adsorption energies on geometrically different sites. These site correlations would reduce the amount of calculations one needs to completely assess adsorption behavior on a single surface.

One aspect of either of these correlations that have not been studied are their coverage dependence. We test the robustness of these two types of correlations by performing over 800 density functional theory calculations of the adsorption of H, C, N, O, F, S, Cl, and Br atomic adsorbates on Cu, Rh, Pd, Ag, Ir, Pt, and Au fcc (111) surfaces at a variety of coverages and configurations. We focus our analysis on the coverage dependence of correlations between adsorption energies on different sites and of different adsorbates. We assess geometric requirements for the existence of correlations by looking at the quality of correlations between adsorption on different sites. We then use linear regression techniques to assess the coverage dependence of correlations between different adsorbates by looking at adsorption energies between elemental adsorbates of the same period or the same group at varying coverages.

We find that correlations between both sites and adsorbates are coverage dependent. Though correlations between geometrically similar sites are independent of coverage, coverage plays a key role in reducing the quality of correlations between adsorption energies at very geometrically different sites. For correlations between adsorbates of the same group and same period, we find that changing the coverage universally changes the correlations and in some cases destroy them. We then use a simple electron counting model to relate our results to previously discovered scaling relationships and discuss the ramifications our conclusions have on materials discovery.

369829 Ethanol from Syngas, an Alternative Route to Produce Alternative Fuel

Abstract:

In the future, the availability of reliable alternative fuels will be crucial for any country to become energy independent. One such alternative is ethanol as it can be used both as a fuel and as a fuel additive. Most of the ethanol produced in the world today is derived from biomass. The biomass feedstocks and fermentation broths used in ethanol production both contain high amounts of water and therefore, the energy efficiency of the process is lessened by product separation processes (azeotropic separation of water and ethanol) that are non-trivial and highly inefficient (due to the evaporation of water). An alternative route to produce ethanol, which negates the need for costly distillation processes, is via the catalytic conversion of syngas (CO and H2) generated from biomass.

Syngas is a mixture of carbon monoxide and hydrogen, which results from the reforming of natural gas, as well as the gasification of coal, biomass, and solid wastes. In theory, syngas can be readily converted to ethanol using chemical catalysts, but to-date no high efficiency, low-cost catalyst has been found.

In early experiments, it was shown that syngas to ethanol reactions are effectively catalyzed by supported rhodium catalysts. However, rhodium is expensive, in relatively low abundance, and its selectivity towards ethanol is limited. To overcome these disadvantages, we propose using bimetallic catalysts that optimize the extents of CO hydrogenation and CO insertion reactions so as to favor the production of low molecular weight alcohols (including ethanol). These bimetallic catalysts consist of a carbon chain growth metal (catalysts to produce long chain hydrocarbons) and an alcohol forming metal (catalysts for methanol production). Of the 16 bimetallic systems initially investigated, the metal combinations that were considered promising for ethanol production were CoPd, CoCu, NiPt, NiIr, and FeCu.

The present study aims at developing a detailed microkinetic model for ethanol production from syngas using catalysts that contain 13-atom bimetallic clusters. In order to understand the reaction mechanism, we are using first principle methods employing Density Functional Theory (DFT) to simulate the catalyst structure, determine favorable adsorption sites, calculate adsorption energies, and find transition states for all of the reactions occurring on the bimetallic catalysts.

One of the first and important steps in the syngas to ethanol conversion process is carbon monoxide (CO) adsorption on the metal catalyst. Therefore, computational models were developed to help understand CO adsorption energetics as well as surface coverage effects on a Co_7Pd_6 catalyst. Key reactions for syngas to ethanol formation were identified from microkinetic model and the influence of CO coverage on activation energies of these key reactions is determined. From these initial studies, we determined that the adsorption energies of CO on both cobalt and palladium as a function of CO surface coverage (where the number of CO species on the catalyst surface was varied from 1 to 6). Further, we calculated the infrared spectra for adsorbed CO species and key bond lengths (metal–carbonyl carbon and adsorbed CO bond lengths) using DFT. Results from the DFT simulations compared favorably with experimental values. These validated simulation results provide valuable insights into the reaction behavior of the studied bimetallic catalysts.

References

[1] L.S.Miranda, Nitin Kumr, J.J. Spivey, The Journal of Physical Chemistry 116 (2012) 7931.

[2] G.Mayank, L.S.Smith, J.J.Spivey, Journal of American Chemical Society 1 (2011) 641.

[3] C.YongMan, L. Ping, Journal of American Chemical Society 131 (2009) 13054.

[4] S.Velu, K.G.Santosh, Energy and fuels 22 (2008) 814.

379363 Engineering Non-Sintered, Metal-Terminated Tungsten Carbide Nanoparticles for Catalysis

Abstract:

Transition metal carbides (TMCs) exhibit catalytic activities similar to platinum group metals (PGMs), yet TMCs are orders of magnitude more abundant and less expensive. However, current TMC synthesis methods lead to sintering, support degradation, and surface impurity deposition, ultimately precluding their wide-scale use as catalysts. We present a method for the production of non-sintered, metal-terminated TMC nanoparticles (NPs) in the 1–4 nm range with tunable size, composition, and crystal phase. The method has been extended to synthesize both monometallic carbides and nitrides as well as various heterometallic combinations from the Group IV-VI block of early transition metals.

We show that carbon-supported tungsten carbide (WC) and molybdenum tungsten carbide ($Mo_xW_{1-x}C$) nanoparticles are highly active and stable electrocatalysts. Specifically, activities ca. 100-fold higher than commercial WC and within an order of magnitude of platinum-based catalysts are achieved for the hydrogen evolution and methanol electrooxidation reactions. Using density functional theory (DFT) calculations, we attribute the higher activity of $Mo_xW_{1-x}C$ NPs to modulation of the hydrogen binding energy (HBE) via bulk heteroatom insertion in the WC lattice. This method opens an attractive avenue to replace PGMs in high energy density applications such as fuel cells and electrolyzers.

Hydrogen Electrocatalysis: Developing Non-precious Metal Catalysts

Wenchao Sheng,[1] Jingguang G. Chen,[2,3*] and Yushan Yan[1*]

[1]Department of Chemical and Biomolecular Engineering, and Center for Catalytic Science and Technology, University of Delaware, Newark, DE 19716
[2]Department of Chemical Engineering, Columbia University, New York, NY 10027
[3]Chemistry Department, Brookhaven National Laboratory, Upton, NY 11973

Hydrogen oxidation and reduction reactions (HOR/HER) are two of the key components in renewable energy conversion and storage devices including fuel cells and electrolyzers. The pH values of the electrolytes have a great impact on the reaction kinetics such that the HOR/HER on the state-of-the-art Pt catalyst in alkaline media decreases by two orders of magnitude than that in acids. The slower HOR kinetics on Pt is projected to impose an additional overpotential (~130 mV at 1.5 A/cm^2_{MEA} with a Pt loading of 0.05 mg_{Pt}/cm^2_{MEA}) in alkaline or alkaline membrane fuel cells (AFCs/AMFCs), and thus the use of Pt catalyst at the anode in AFCs/AMFCs requires a higher catalyst loading.[1, 2] With the development of many non-precious metal catalysts for the oxygen reduction reaction in both acid and alkaline electrolytes, showing comparable activity to Pt, the demand for highly active and cost-effective HOR catalyst in alkaline medium is arising and becomes one of the crucial challenges to make AFCs/AMFCs more practical.

In order to design and develop highly efficient HOR catalysts, it is essential to identify the reaction descriptor for the HOR in alkaline electrolyte. It is shown that the HOR/HER exchange current density in alkaline media on a wide range of monometallic surfaces follows a volcano type of relationship with the calculated hydrogen binding energy (HBE) of the metals with Pt having the optimum HBE value.[3] This discovery indicates that HBE values can serve as the designing principle for developing efficient electrocatalysts for the HOR/HER in base. Motivated by this finding and the fact that Ni exhibits certain HOR/HER activity in base, we calculated the HBE of NiMo bi-metallic and CoNiMo multi-metallic surfaces and found that they have similar HBE values as Pt. We subsequently synthesized NiMo and CoNiMo alloys by electrodeposition and performed HOR/HER tests in 0.1 M KOH at different temperatures. The results show that the HOR/HER on CoNiMo outperforms electroplated Ni by 20 times. At sufficient catalyst loading, this promising material will alleviate the catalyst cost issue raised by higher Pt loading in alkaline electrolytes, and will facilitate the development of the AFCs/AMFCs technology.

References:

1. K. C. Neyerlin, W. B. Gu, J. Jorne and H. A. Gasteiger, *J Electrochem Soc*, 2007, 154, B631-B635.
2. W. C. Sheng, H. A. Gasteiger and Y. Shao-Horn, *J Electrochem Soc*, 2010, 157, B1529-B1536.
3. W. C. Sheng, M. Myint, J. G. G. Chen and Y. S. Yan, *Energy & Environmental Science*, 2013, 6, 1509-1512.

374648 Trends in the Hydrogen Evolution Activity of Metal Carbide Catalysts

Abstract:

The hydrogen evolution reaction (HER) can be employed to generate hydrogen as a potentially carbon-neutral feedstock for the synthesis of ammonia, as a reducing source for the deoxygenation of biofuels, or as a fuel source for hydrogen fuel cells for automotive or stationary applications. In other electrochemical applications, the HER is a major side reaction that affects the product selectivity in electrocatalytic reduction of CO_2 into hydrocarbons or N_2 into ammonia or the local pH and the development of structural properties during metal electrodeposition. Metal carbide catalysts are alternative nonprecious electrode materials for electrochemical energy conversion devices, such as for hydrogen fuel cells or electrolyzers. This talk presents the experimental exchange current densities for the HER on eight mono- and bimetallic metal carbide electrocatalysts (B_4C, TiC, Fe_3C, Co_3C, Ni/C, Mo_2C, WC, and CoWC) and correlates the current densities to the hydrogen binding energies that we have calculated via electronic structure computations. We find these electrode materials to have activities higher than those of their parent metals and intermediate between the catalytic activities of the Pt-group and early transition-metal surfaces. Increased HER activities on metal carbides relative to the activities of their parent metals can be understood with a three-fold higher sensitivity of metal carbides to the coverage-induced weakening of hydrogen adsorption relative to metal surfaces. The discussed trends can be useful for the design of bimetallic carbide electrocatalysts.

379657 In Situ Studies of the Active Sites in Nife Oxyhydroxides for the Electrochemical Oxidation of Water

Abstract:

A key challenge to the development of photoelectrochemical devices that can generate hydrogen from water using solar energy is the discovery design criteria for new, efficient and stable electrocatalysts for the oxygen evolution reaction (OER). Here, we identify the origin of a 500-fold OER activity enhancement that can be achieved with mixed (Ni,Fe)oxyhydroxides ($Ni_{1-x}Fe_xOOH$) over their pure Ni and Fe parent compounds, resulting in one of the most active known OER catalysts in alkaline electrolyte. *Operando* XAS measurements reveal that Fe^{3+} in $Ni_{1-x}Fe_xOOH$ occupies octahedral sites with unusually short Fe-O bond distances, induced by edge-sharing with surrounding $[NiO_6]$ octahedra. Using DFT calculations, we establish that this structural motif results in near optimal adsorption energies of OER intermediates and low overpotentials at Fe sites. By contrast, Ni sites in $Ni_{1-x}Fe_xOOH$ are not active sites for the oxidation of water.

384114 Non-Precious Multimetallic Catalyst with Bimodal Porous Stucture for High-Performance Hydrogen Evolution Activity

Abstract:

Bridging the gap between noble and non-noble metal catalysts for hydrogen-evolution reaction (HER) is a critical step for achieving hydrogen economy for clean and sustainable energy future. Here we found that a proper arrangement of copper-titanium (Cu-Ti) atoms produces highly active surface with a high density of reaction sites exhibiting an optimum hydrogen binding energy similar to that of platinum (Pt). A non-noble element Cu-Ti bimetal catalyst with such surface was designed with monolithic and bimodal pore structure to achieve over doubled improvement in total activity of HER comparing to state-of-the-art carbon supported Pt (Pt/C) catalyst. The morphological control of material pore sets simultaneously enhanced the exposed catalytic surface area and transportation of electrolytic ions and gas bubbles; the intact metallic structure provided high-conductivity and eliminated the necessity of using supporting substrate, both of which enabled the realization of much higher mass loading for industrial application than traditional particle-based electrocatalysts.

387615 Using Nanoscale Confinement to Improve Oxygen Electrochemistry

Abstract:

Chemical engineers have long specialized in the inter-conversion of latent energy and chemical bonds. These skills easily lend themselves to optimizing solar energy storage, which helps facilitate a shift towards sustainable and affordable electricity.

Water splitting is particularly interesting as an energy storage mechanism because water is earth abundant and because the net reduction potential of oxygen gas is near the optimum energy for solar light conversion. From an electrochemical standpoint, water splitting can be divided into two half-reactions: hydrogen evolution and oxygen evolution. Currently the oxygen evolution reaction (OER) is the predominant source of overpotential, as the best catalysts available require applied potentials on the order of 300 mV past equilibrium to run at reasonable current densities.

In this work we use density functional theory calculations to propose a method to increase OER efficiency by creating precisely defined nanostructures. Traditionally catalysis is thought of as a necessarily two-dimensional surface interaction. By expanding our understanding to include three-dimensional confinement effects, we capitalize on the fact that OER intermediates (surface bound O, OH, and OOH) have different sizes and respond differently to an oxygenated bonding environment.

Our study focuses initially on rutile oxide catalysts (RuO_2

materials provided by previous reports and the agreement they demonstrate between experimentally measured and DFT-predicted OER overpotentials. We find that previously potential-limiting scaling between intermediate adsorption energies is broken due to confinement effects and hydrogen bond formation, resulting in a reduction in overpotential. In principle this scheme is capable of improving the efficiency of any OER catalyst on the "left leg" of the previously determined Sabatier-inspired overpotential volcano relation (where ΔG_{OH}-ΔG_{O}<1.6 eV), including more economical nickel and manganese oxides.

388245 Design Principles for Oxygen Evolution Catalysts on Autonomous Water Splitting Particles

Abstract:

The development of approaches that harness solar energy to split water, thereby producing sustainable hydrogen fuel, is appealing as an environmentally sustainable alternative to fossil fuels. Among various approaches, photocatalytic processes that use energetic charge carriers produced from photon absorption by semiconductors to drive hydrogen and oxygen evolution are considered a leading potential technology. A recent technoeconomic analysis by the Department of Energy has identified an autonomous photocatalytic approach, where all necessary functionality exists in a single particle, as being the cheapest process design.

Realization of autonomous water structures requires the design of hetero-structured materials consisting of semiconductors that absorb a significant fraction of the solar spectrum and co-catalysts that efficiently accept charge carriers and have a low inherent overpotential to drive the oxygen (OER) and hydrogen (HER) evolution half reactions. Significant focus has been placed on understanding universal characteristics necessary for driving OER and HER efficiently using electrocatalysis. The binding energy of critical reaction intermediates on catalysts have been related to the required overpotential to drive the reactions, providing a design criterion for optimum HER and OER based on reaction kinetics. However, in autonomous photocatalytic systems, the maximum provided overpotential for each half reaction is defined by the difference between the semiconductor valence or conduction band and the OER and HER evolution potentials, respectively. Furthermore, charge transport across semiconductor-co-catalyst interfaces can cause a loss of overpotential depending on the direction and magnitude of the interfacial band bending.

In this work we examine trends in oxygen evolution rates in autonomous photocatalytic systems for co-catalysts that have previously been identified as optimal in electrocatalytic environments. In addition, we compare how the electronic structure of the semiconductor impacts the co-catalysts performance. Electronic structure measurements based on photoelectron spectroscopy provide mechanistic insight into why the performance of co-catalysts in autonomous water splitting systems cannot be directly correlated to the electrocatalytic performance. Finally we suggest design principles, based on the inherent electrocatalytic activity and co-catalyst electronic structure, for optimal OER co-catalysts in autonomous water splitting systems.

Highly dispersed supported Au catalysts are interesting because of their high activity and selectivity for an increasing number of reactions.[1,2] Despite extensive research, fundamental questions remain regarding the nature of the catalytically active sites and the reaction mechanisms for Au-catalyzed reactions. To this end we adopt a combined theoretical and experimental approach that employs first principles density functional theory calculations, reaction kinetics experiments, and microkinetic modeling to develop an atomic-scale understanding of the surface reactions occurring during formic acid decomposition over Au catalysts. The results of this work suggest that the dominant active sites are highly undercoordinated gold atoms[3]. We have corroborated this conclusion by quantifying the presence of small clusters of gold atoms (i.e., less than 2 nm in size) by scanning transmission electron microscopy, and we have correlated the rate of formic acid decomposition to the gold particle size distribution on a series of supported gold catalysts, generated by employing a diverse set of catalyst preparation protocols.

References

1. Y. Zhang, X. Cui, F. Shi and Y. Deng, Chem. Rev. **2012**, 112, 2467.
2. T. Takei, T. Akita, I. Nakamura, T. Fujitani, M. Okumura, K. Okazaki, J. Huang, T. Ishida and M. Haruta, Adv. Catal. **2012**, 55,1.
3. S. Singh, S. Li, R. Carrasquillo-Flores, A. C. Alba-Rubio, J. A. Dumesic, M. Mavrikakis, AIChE J. **2014**, 60, 1303.

388735 CO Oxidation on Bifunctional Catalysts: The Au/TiO_2 Interface and the Critical Role of Water

Abstract:

Multiple site-specific functionalities can significantly improve the performance of heterogeneous catalysts under steady-state reaction conditions. The most commonly studied systems consist of two distinct sites that catalyze different reaction steps independently and are often referred to as bifunctional catalysts. In previous work we have used density functional theory (DFT) and microkinetic modeling to understand the mechanisms and limitations of bifunctional catalysis over bimetallic alloys. Our results indicate that the existence of universal scaling relations limits the achievable activity improvement and bifunctional catalysts do not necessarily outperform single-site catalysts. More specifically, for CO oxidation on bimetallic systems we found that the overall activity is not significantly altered when bifunctional metal alloy catalysts are considered, but equally active bifunctional catalysts may be tailored from less active and cheaper components.

In contrast, metal/oxide interfaces are not restricted to the same universal scaling relationships inherent to transition metal surfaces, but their complexity requires more detailed studies to gain a fundamental understanding of the roles of each component. The prototype example for such system is room-temperature CO oxidation over Au/TiO_2catalyst, which continues to be one the most intriguing topics ever since its discovery by Haruta and co-workers [1]. Despite the vast amount of research performed on this system, considerable debate remains regarding the nature of the active site and the role of water, which can dramatically increase catalytic activity [2,3].

In order to better understand the roles of water, surface hydroxyls, and the metal-support interface, the detailed reaction mechanism for CO oxidation must be known. Here, we use kinetic measurements, isotopic labeling, *in-situ* IR spectroscopy, and first-principles simulations to study CO oxidation on Au/TiO_2 and our results provide direct evidence for a novel water-mediated reaction mechanism. A large kinetic isotope effect (KIE, k_H/k_D = 1.8) implicates O-H(D) bond breaking in the rate determining step. Kinetics and *in-situ* IR spectroscopy experiments showed that the TiO_2 coverage of weakly adsorbed water largely determines catalyst activity by changing the number of active sites.

Our DFT simulations showed that adsorbed water at the metal/support interface greatly facilitates oxygen adsorption via a barrierless proton transfer. Oxygen activation in the resulting Au-OOH species on the Au cluster is facile; the reaction of Au-OOH with Au-CO to form Au-COOH requires an activation energy of only E_a=0.10 eV. Subsequently, the decomposition of Au-COOH to form CO_2 involves O-H(D) bond scission and a proton transfer step to water. This step was found to be rate determining with a calculated KIE of 2.55. DFT also indicates that direct involvement of surface hydroxyl group is unlikely as the calculated barrier for transferring a Ti-OH group from the support to Au (E_a = 1.63 eV) is too large to be feasible at room temperature. Furthermore, generating Au-COOH from Au-CO and Au-OOH (ΔE = -2.23 eV, E_a = 0.10 eV) is thermodynamically and kinetically far more favorable than the reaction between Au-CO and Ti-OH (ΔE = 0.60 eV, E_a = 0.72 eV). Instead, the more likely role of Ti-OH groups is to anchor and activate water near the metal/support interface. These experimental and theoretical results provide a fresh framework for interpreting the literature, and clearly define the mechanistic roles of water, Ti-OH groups and the metal/support interface.

References

1. M. Haruta, T. Kobayashi, H. Sano, N. Yamada 1987 Novel gold catalysts for the oxidation of carbon monoxide at a temperature far below 0°C. *Chem. Lett.*, 405–408 (1987).
2. B.D. Chandler, S. Kendell, H. Doan, R. Korkosz, L.C. Grabow, C.J. Pursell, NaBr Poisoning of Au/TiO_2 Catalysts: Effects on Kinetics, Poisoning Mechanism, and Estimation of the Number of Catalytic Active Sites. *ACS Catal.* **2**, 684-694 (2012).
3. M. C. Kung, R. J. Davis, H. H. Kung, Understanding Au-Catalyzed Low-Temperature CO Oxidation. *J. Phys. Chem. C* **111**, 11767-75 (2007).

366403 Supported Gold Catalysts Decorated with Ti-Oxo Clusters of Different Nuclearity in the Selective Oxidation of Propane in the Presence of H2 and O2

Abstract:

Effective selective oxidation of light alkanes could provide a promising one-step production of valuable chemicals at low costs, and significantly attenuate the burdens on the environments compared with other oxidation processes. However, the studies in supported gold catalysts in the selective oxidation of propane are only recent and little is known. Ti-containing supports have been widely used in catalysis. Propene is selectively produced over Au/TiO_2,[1] while acetone and isopropanol are the main products over Au/TS-1.[2,3] Although Ti-OOH is generally accepted as the reaction intermediate resulting from the reaction of active Ti sites and H_2O_2 formed on Au, the fundamentals of altering the selectivity over TiO_2 versus TS-1 are not clear. In light of their structures and acid/base properties, we hypothesize that nuclearity of Ti-oxo clusters could play an important role in changing selectivity. We have prepared Ti-oxo complexes of different nuclearity (one to four) and deposited them onto Au particles. The Ti-oxo complexes and the decorated Au catalysts were characterized by NMR, EXAFS, and TEM. Results of characterization and catalysis with these Ti-modified Au catalysts in propane selective oxidation are presented.

1 J. J. Bravo-Suárez, K. K. Bando, J. Lu, T. Fujitani and S. T. Oyama, *J. Catal.*, 2008, **255**, 114.

2 J. J. Bravo-Suarez, K. K. Bando, T. Fujitani and S. T. Oyama, *J. Catal.*, 2008, **257**, 32.

3 J. J. Bravo-Suarez, K. K. Bando, T. Akita, T. Fujitani, T. J. Fuhrer and S. T. Oyama, *Chem. Commun.*, 2008, 3272.

382481 Hydrogen Carriers in Spillover-Mediated Hydrogenation Catalysis

Abstract:

Rate enhancement of hydrogenation reactions on supported metal catalysts has been reported in the literature when γ-alumina instead of silica is used as the catalyst support material [1]. Interpretation of this phenomenon is disputed: some claim that hydrogen atoms spillover from the metal function to the support to provide additional reactive hydrogen atoms [2], while others believe that different concentrations of non-reactive carbonaceous deposits (impurities) on the catalyst supports contribute to the discrepancies in the hydrogenation rates [3].

The possibility of hydrogenating organic molecules using spillover hydrogen produced by encapsulated metal accessible only to H_2 has been used, here, as a strategy to corroborate the existence of atomic hydrogen spillover. Pt clusters encapsulated within FAU were ineffective for spillover-mediated hydrogenation of TIBP, an organic molecule whose kinetic diameter is larger than FAU aperture. Toluene, small organic molecule which can access metal clusters within FAU cages, however, can act as shuttles to transport hydrogen from confined Pt clusters to large TIBP for its hydrogenation, leading to a rate enhancement commensurate to that measured on unconfined Pt clusters supported on mesoporous silica diluted with γ-alumina. The data suggest that partially hydrogenated toluene intermediates, present in the gas-phase only at very low concentrations, can transfer active hydrogen for spillover routes on insulating supports to hydrogenate aromatics that are unable to directly access Pt clusters encapsulated within zeolites.

To address selective removal of non-reactive residues by various supports, we also carried out toluene hydrogenation and CO oxidation reactions, in alternation, on a given catalyst (Pt/SiO_2 or Pt/Al_2O_3) to ascertain whether enhancements were only limited to reactions involving hydrogen, thereby implicating the "spillover" effect and casting doubt on the role of impurities on these observed trends. The turnover rates for toluene hydrogenation and CO oxidation remained same throughout, for up to two hydrogenation-oxidation cycles on both catalysts, with hydrogenation rates on Pt/Al_2O_3 showing an enhancement over Pt/SiO_2 as expected. We conclude from the unattenuated hydrogenation rates, despite intervening CO oxidation reactions, that measured rate differences of hydrogenation reactions on metal catalysts with different insulating supports are unlikely a result of impurities on the support.

References:

[1] R. Prins, Chem. Rev. 2012, 112, 2714.

[2] K. M. Sancier, J. Catal. 1971, 20, 106.

[3] J. C. Schlatter and M. Boudart, J. Catal. 1972, 24, 482.

Abstract - Propene Metathesis on Silica-Supported Tungsten Oxide

Understanding a catalytic system requires knowledge of the structure – activity relationship for the active species. For many heterogeneous catalysts, developing this relationship can be complicated by the heterogeneity of structural centers present on the catalyst surface. These challenges are well exemplified in the case of olefin metathesis, where it has been reported that only 1-2 % of the metal content of WO_x, MoO_x, or ReO_x supported on silica or alumina is active.[1] Despite a long history of industrial use, the nature of the active site is not known definitively, and limited progress has been made in the development of catalysts for large-scale applications. The low percentage of active metal content makes spectroscopic observation of the active site, and reaction intermediates, especially difficult. Another question that remains is the mechanism for the formation of the proposed active site, an isolated metallocarbene surface species, from the metal oxide of the as-synthesized catalyst. The purpose of our work has been to define the structure of active sites and the nature the active species involved in propene metathesis over silica-supported tungstena. Through the application of UV-Vis, Raman, XAS, FT-IR we have been able to characterize the structure of the WO_x species dispersed on silica, and study how the isolated tungstate species are converted into active sites during reaction. This work has also identified pretreatment strategies for enhancing the fraction of dispersed WO_x that becomes active for propene metathesis and reduces the induction time to achieve steady-state activity. Additionally, this research has advanced our ability to distinguish the active species from spectator sites, and led to a clearer understanding of their differences. Initial results on the characterization of carbene intermediates formed on the active sites and their role in catalysis will also be reported.

References:

1. Amakawa, *et. al. JACS.* **134** (2012) 11462.

399182 Effects of the Types and Coverages of Reactive Oxygen on the Catalytic Fate of Propylene during Epoxidation on Silver Clusters

Abstract:

Catalytic epoxidation of propene with di-atomic oxygen is the desired process for propylene oxide production, because this step utilizes oxygen to generate reactive oxygen species that insert into the C=C bond of propylene to form propylene oxide. The yields of this process, in contrast to using ethylene as the feedstock, are low and currently not viable for commercialization. Here, we report the mechanistic pathways that lead to the formation of C3H6O and COx during propylene epoxidation on Ag-clusters. We address the catalytic roles of the reactive oxygen species, as reactive chemisorbed oxygen atoms or hydroperoxyl species, in directing the reaction pathways for C3H6O, CO, and CO2 formation, based on rate measurements in the kinetically controlled regime and active site titrations with O2 and C3H6. The reactive collision frequencies for propylene oxide formation increase proportionally with chemisorbed oxygen atoms. In contrast, the reactive collision frequencies for COx formation are insensitive to C3H6 pressure but increase by five-fold at 443 K as the oxygen coverages vary within a narrow range. This kinetic dependence may indicate that hydroperoxyl species, generated only at higher O2 pressures, from reactions of H2O and O2 on Ag clusters mostly covered with reactive oxygen species, are the reactive intermediates for the undesired combustion reactions. We confirm the catalytic role of H2O in generating the OOH* species by incorporating H2O into the C3H6-O2 feed mixtures. H2O addition leads to larger OOH* concentrations that promote the C3H6 combustion while reduces the number of reactive O* centers available for the desired epoxidation reactions. In summary, we show that the relative amounts of O*, O2* and OOH* species play a defining role on the catalytic fate of propylene during epoxidation reactions.

374434 High Thermal Stability in Anatase Phase of Vanadium Oxide Catalyst Supported on TiO2-B/Anatase Bi-Crystalline TiO2

Abstract:

High thermal stability in anatase phase of vanadium oxide catalyst supported on TiO_2-B/anatase bi-crystalline TiO_2

Licheng Li[1], Tuo Ji[2], Jiahua Zhu[3], Zhuhong Yang[2], Xiaohua Lu[2]

1. College of Chemical Engineering, Nanjing Forest University, Nanjing, 210037, China

2. State Key Laboratory of Materials-oriented Chemical Engineering, Nanjing Tech University, Nanjing, 210009, China

3. Department of Chemical and Biomolecular Engineering, The University of Akron, Akron, OH 44325, USA

The thermal stability of vanadium oxide/titania catalyst is an important factor for application in selective catalytic reduction (SCR) of deNO_x reactions and its further regeneration. However, the catalytic active anatase titania is inevitably transformed to low-active rutile titania under high temperature of above 500 °C. Especially doped with vanadium oxide, the phase transformation of titania will be accelerated which results in an lower anatase-to-rutile transformation temperature. The transformed rutile titania would definitely deactivate the catalytic performance of vanadium oxide/titania catalyst [1]. Different from the general method of introducing structural additive, an isomerism effect is proposed in this work to inhibit the generation of rutile titania. TiO_2-B/Anatase shell-core bicrystal titania (abbr. TiO_2) reported in our previous work was chosen as catalyst support[2]. 1.0 wt% of V_2O_5 was loaded by a facile impregnation process. The composition of crystal phase of V/TiO_2 calcined by various temperatures was studied. It is found that the transforming temperature of rutile titania in V/TiO_2 is 200 °C higher than that of conventional anatase titania loaded with vanadium oxide (abbr. V/Anatase).

XRD and Raman results show that TiO_2 and Anatase has no phase transformation when the temperature is lower 600 °C. At 600~800 °C, the anatase phase of V/Anatase starts to transform to rutile phase. As for V/TiO_2, TiO_2-B turns into anatase and no rutile is generated. The anatase-to-rutile phase transformation of V/TiO_2 occurs when the calcination temperature is higher than 800 °C. The difference in the phase transformation between V/TiO_2 and V/Anatase can be attributed to the present of TiO_2-B: TiO_2-B is located at the outer surface of TiO_2, which prevents the direct contact between anatase titania and vanadium oxide. During the thermal treatment process, TiO_2-B transforms to anatase due to the acceleration effect of vanadium oxide. This process replaces the anatase-to-rutile transformation. As a result, TiO_2 shows the better thermal stability than conventional TiO_2 without the protective TiO_2-B.

Reference:

[1] Wachs I E, Weckhuysen B M, *Appl Catal A Gen*, 1997, 157: 67

[2] Li W, Liu C, Zhou Y, Bai Y, Feng X, Yang Z, Lu L, Lu X, Chan K-Y, *J Phys Chem C*, 2008, 112: 20539

364286 Atomic Layer Deposited Highly Dispersed Platinum Nanoparticles Supported on Multiwalled Carbon Nanotubes for Liquid Phase Reactions

Abstract:

Highly dispersed platinum nanoparticles were deposited on gram quantities of non-functionalized multiwalled carbon nanotubes (MWCNTs) by atomic layer deposition (ALD) in a fluidized bed reactor at 300 °C. (Methylcyclopentadienyl) trimethylplatinum and oxygen were used as precursors. The results of TEM analysis showed that ~1.3 nm Pt nanoparticles were highly dispersed on non-functionalized MWCNTs. The porous structures of MWCNTs did not change with the deposition of Pt nanoparticles. For comparison, the commercial 3 wt.% Pt/C catalyst was also characterized. The ALD-prepared Pt/MWCNT was used for the hydrogenation of xylose to xylitol and hydrogen transfer of ketones. The ALD-prepared Pt/MWCNT showed the best catalytic performance with 100 % conversion of xylose and 99.3 % selectivity to xylitol, compared to commercially available Pt/C, Ru/C, and Raney Ni catalysts. The heterogeneous hydrogen-transfer reactions of acetophenone indicated that an acetophenone conversion of 100% and a 1-phenylethanol selectivity of 99.0% could be obtained with a ketone to Pt mass ratio of 24,690 and a ketone to KOH mass ratio of 22 at 150 °C for 5 h. The stability of ALD produced Pt/MWCNT catalyst was much higher than that of the commercial Pt/C, due to the presence of surface defects on the MWCNTs and the strong metal–support interaction for the ALD-prepared Pt/MWCNT catalyst.

Temporal and Spatial Reactors for Atomic Layer Deposition on Particles

J. Ruud van Ommen, Fabio Grillo, David Valdesueiro,
Dirkjan Kooijman, Mojgan Talebi, Aris Goulas

Delft University of Technology, ChemE, Julianalaan 136, 2628 BL Delft, the Netherlands
j.r.vanommen@tudelft.nl

Whereas most research on atomic layer deposition (ALD) is aimed at depositing ultrathin films on wafers and other flat substrates, ALD can also be used to provide the surface of particles with either an ultrathin continuous coating or a decoration of nanoclusters [1,2,3]. When carried out in a so-called fluidized bed, ALD is an attractive way of producing nanostructured particles, for example for catalytic applications, with excellent scale-up potential. In such a system, the particles are suspended in an upward gas flow. Nanoparticles – contrary to what is typically observed for micron-sized particles – are fluidized as very dilute agglomerates with distinctive fluidization characteristics. We will discuss the challenges related to fluidization and coating of nanoparticles.

Like most conventional ALD reactors, fluidized-bed ALD reactors are typically operated in a temporal mode: the pulses of the different gaseous reactants are delivered subsequently in time. An alternative is to separate the administering of the reactants in space [4]. We recently developed a spatial ALD reactor for particles. We will discuss spatial ALD of Pt on TiO_2 nanoparticles that are pneumatically transported through the reactor with nitrogen at atmospheric pressure. These nanoparticles are blown through a tube as very porous agglomerates, ensuring good accessibility to the precursor/co-reactant ($MeCpPtMe_3$ and O_2), which are injected along the tube. Optimizing the operating conditions, we obtained an even spread of 1-2 nm nanoclusters of Pt over the TiO_2 nanoparticles (see Fig. 1). This material showed to be a very active photocatalyst. Reaction tests demonstrated an increase of the photocatalytic activity with a factor 5, compared to native TiO_2 nanoparticles. With the current setup, we are able to produce nanostructured particles via ALD at a production rate of about 1 g/min. However, the experience with pneumatic transport from other fields opens up promising scale-up prospects for continuous particle ALD.

References

[1] D. M. King, X. Liang, A. W. Weimer, Powder Technol. 221 (2012) 13-25.

[2] D. Longrie, D. Deduytsche, C. Detavernier, J. Vac. Sci. Tech. A 32 (2014) 010802.

[3] A. Goulas, J.R. van Ommen, KONA Powder and Particle J. 31 (2014) 234-246.

[4] W.M.M. Kessels, M. Putkonen, MRS Bull. 36 (2011) 907-913.

Figure 1. TEM image of the Pt nanoclusters, deposited on TiO_2 nanoparticles by ALD in a spatial reactor.

377799 Metallic Co ALD Catalysts for Fuels from Syngas

Abstract:

Direct deposition of a metallic Co by ALD has been demonstrated using low-cost cobaltocene precursor and H_2; this catalyst preparation method eliminates reduction of oxide and improves catalytic productivity. Eliminating high-temperature treatment of catalyst avoids sintering and incorporation of metal into the support (aluminate). Two methods of depositing metallic Co are discussed: on $Al_2O_3$and on W ALD. Both surfaces have exhibited deposition of metallic Co; however, the deposition reactions are vastly different.

Application of the Co catalyst deposited by metallic ALD has multiple advantages. First, cobaltocene is a less-expensive reactant than other proposed chemistries for metallic Co ALD. Second, metallic Co ALD will be less likely to sinter than incipient wetness catalysts because reduction temperatures are avoided. Third, the deposition of metallic Co compared to Co_3O_4 ALD provides higher productivity of Co catalyst because the metal does not form a mixed oxide with the support.

Center for Advanced Materials and Nanotechnology

CardiffCatalysisInstitute

Chemical vapour impregnation- a versatile route to high redox activity nanoparticle catalysts

Michael M Forde PhD.

Graham J. Hutchings, Chris K. Kiely, Qian He, Nikolaos Dimitratos, Ren Su and Lokesh Kesavan

AIChE Meeting (Atlanta) November 2014

Talk Outline

- Overview of our catalyst preparation techniques
- Characterisation of CVI materials- examples
- Catalysis with CVI materials
  - Toluene oxidation
  - Levulinic acid hydrogenation
  - H_2 evolution by photocatalysts

Sol-immobilisation (SI)

IMMOBILISATION OF PREFORMED GOLD SOL

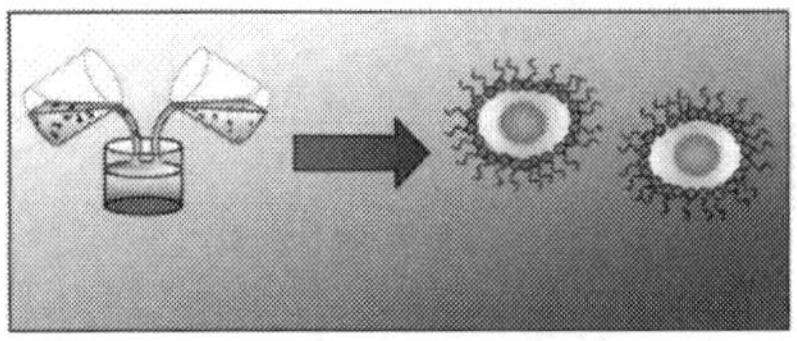

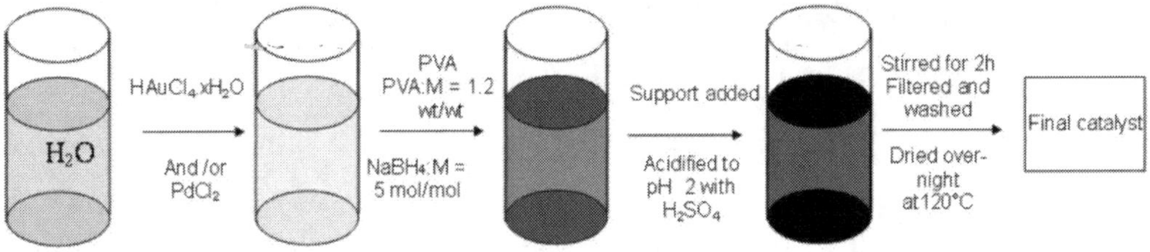

- Final catalyst not heat treated
- Can prepare bi and tri metallics in this way
- Order of metal precursor addition and reduction affects morphology

Core-shell structure by SI

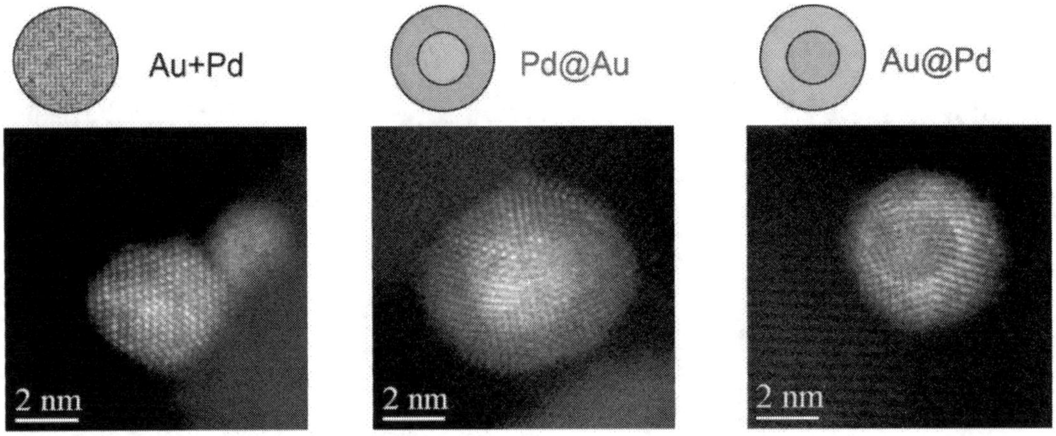

Varying precursor addition order and reduction can be used to change morphology

Go to method for making small nanoparticles

Issues

- ➢ Surface ligands present
- ➢ Spectator species present (chloride etc)
- ➢ Materials not heat stable
- ➢ Lengthy synthesis
- ➢ Reproducibility and general application

Issues common to many techniques!

Chemical Vapour Impregnation (CVI)

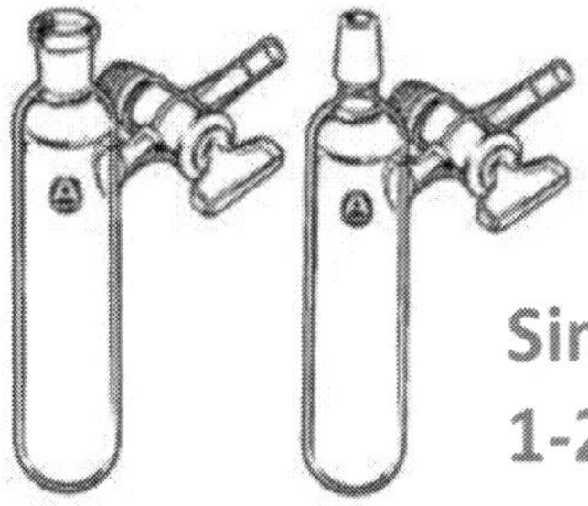

Simple, scalable, reproducible technique
1-2h for deposition

- ✓ Physically mix precursor and organometallic precursor
- ✓ Evacuate tube, heat under constant vacuum
- ✓ Remove vacuum, cool and remove material
- ✓ Heat treat to produce fine dispersion of nanoparticles on
- ✓ support

Features of CVI

- ✓ Can deposit wide range of metals on almost any support
  - ✓ We've done Fe, Cu, Pt, Pd, Rh, Ru, Ag, Zn, Ga
- ✓ Very fine metal dispersion
  - ✓ Size depends on nature of metal, treatment conditions, loading
- ✓ Can be used to prepare bi and trimetallics
- ✓ Can be used to modify one material by selective deposition of a second metal.. Core shell structures
- ✓ Contamination free support
- ✓ Clean surface to NP's

Pd and Pt - examples

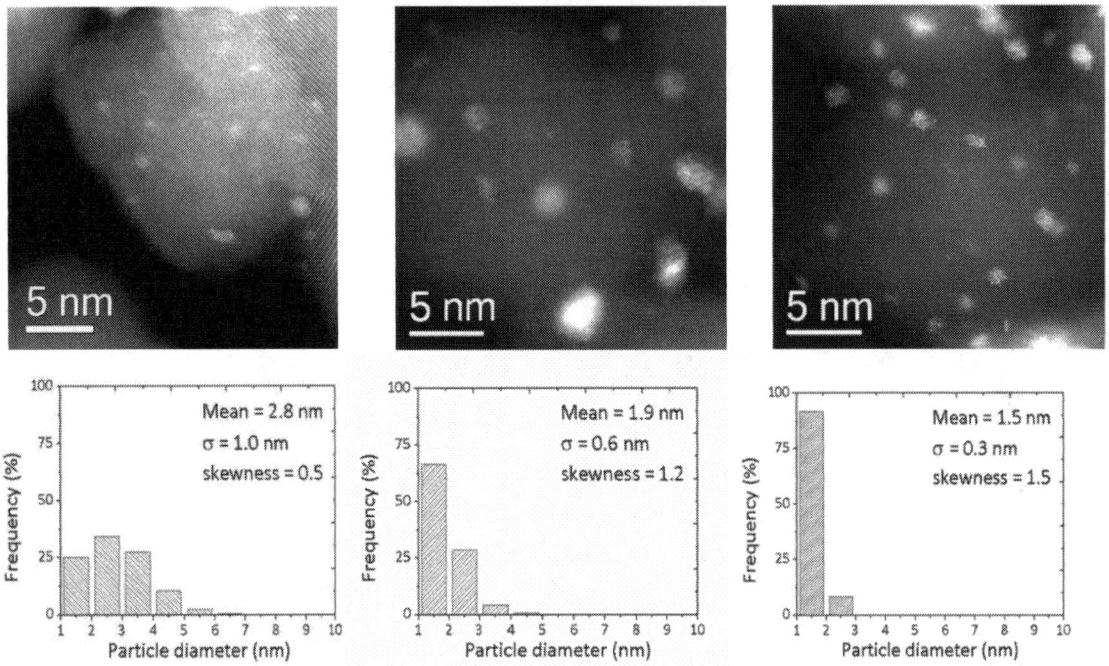

$2.5\%Pd/TiO_2$ $1.25\%Pd$-$1.25\%Pt/TiO_2$ $2.5\%Pt/TiO_2$

Pd and Pt examples

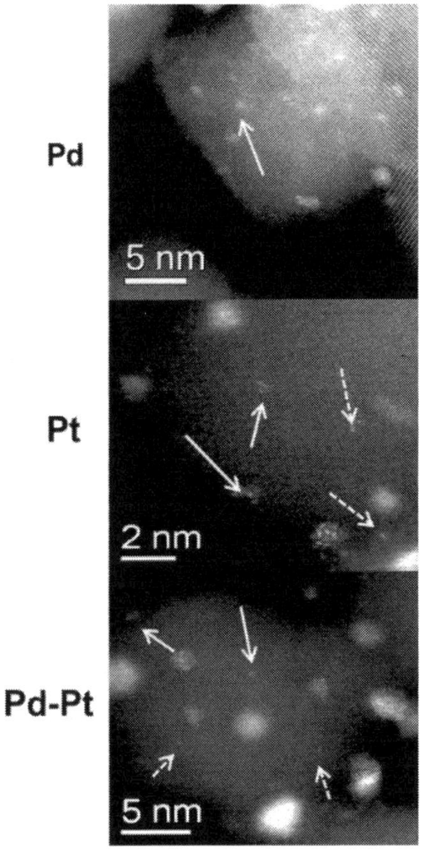

Sub-nm clusters

Single atoms

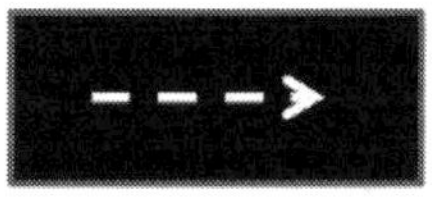

Still observed after 400°C treatment

Characterisation CVI - examples

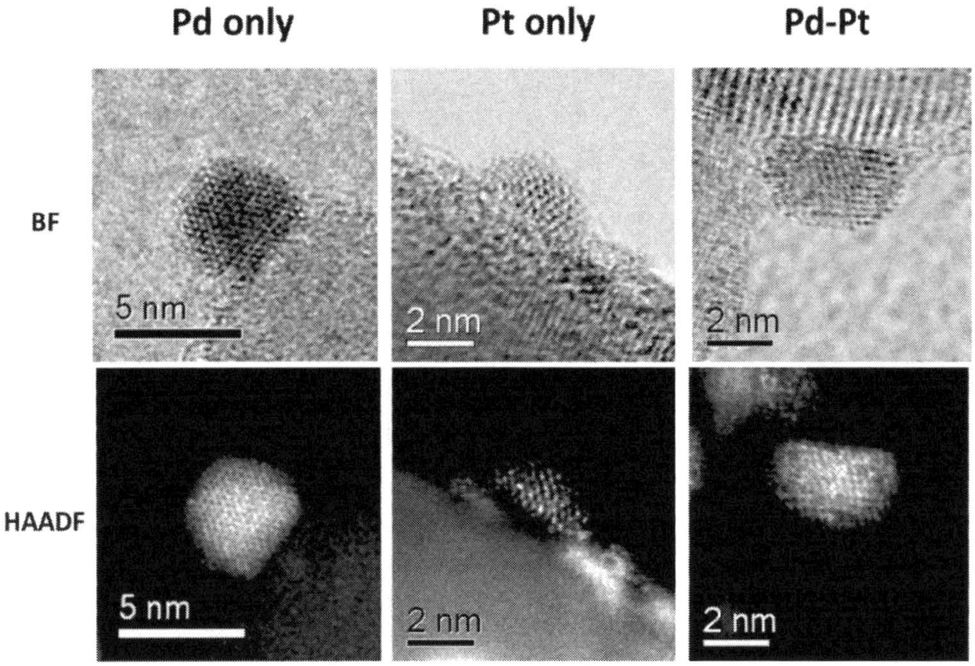

Clean nanoparticles in all cases

Mixed Pd-Pt - example

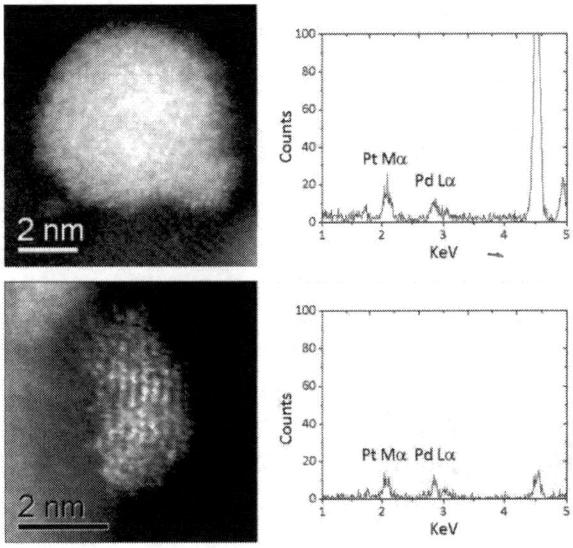

$1.25\%Pd-1.25\%Pt/TiO_2$

One step deposition

Homogeneous alloy expected

Most particles found to be homogeneous alloys

Occasional Pt and Pd only particle observed

Modifying Au catalysts

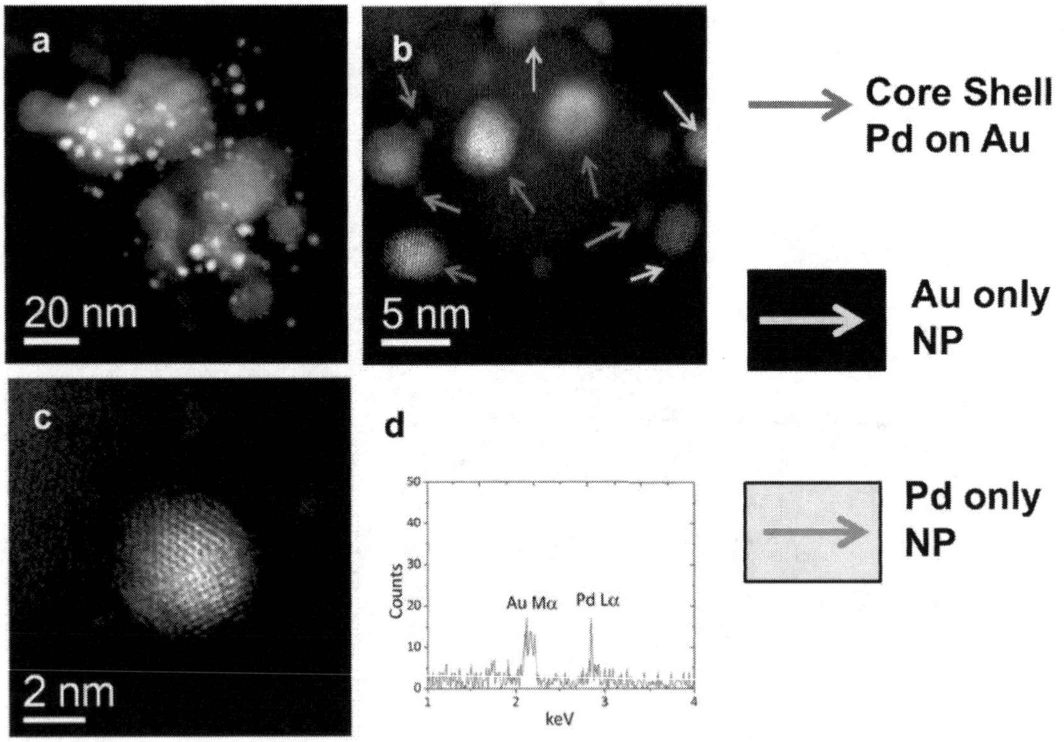

Modifying Au catalysts

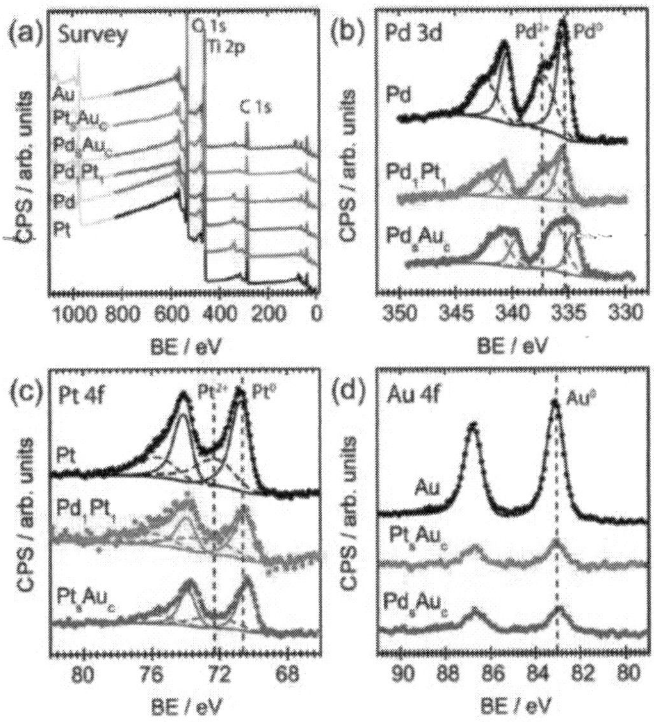

In hybrid Au peak is reduced severely… Evidence of core shell structure

Pd can be found in +2/0 oxidation state (age)

Pt always found in 0 oxidation state

Characterisation- Fe/SiO_2 by CVI

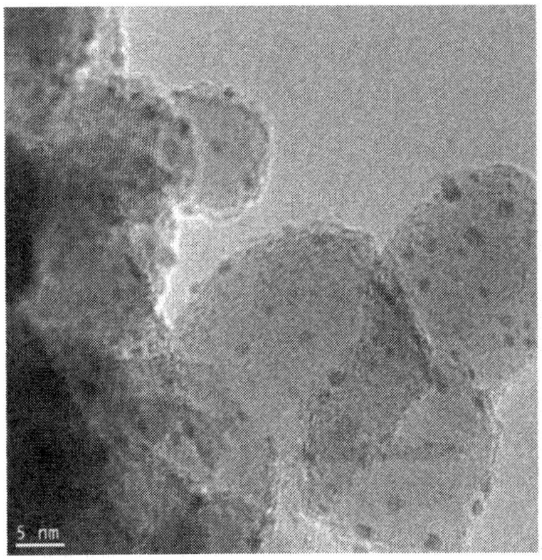

Uniform distribution of FexOy
Particles ~ 3nm

Size affected by metal loading

If heat treatment is done under H_2
All surface Fe disappears

TOLUENE OXIDATION

ACS Nano, 2014, 8, 957-969
Angew. Chem. Int. Ed., 2012, 51, 5981–5985
Phys. Chem. Chem. Phys., 2013,15, 10636-10644
Science, 2011, 331, 195

Catalysis- Toluene Oxidation

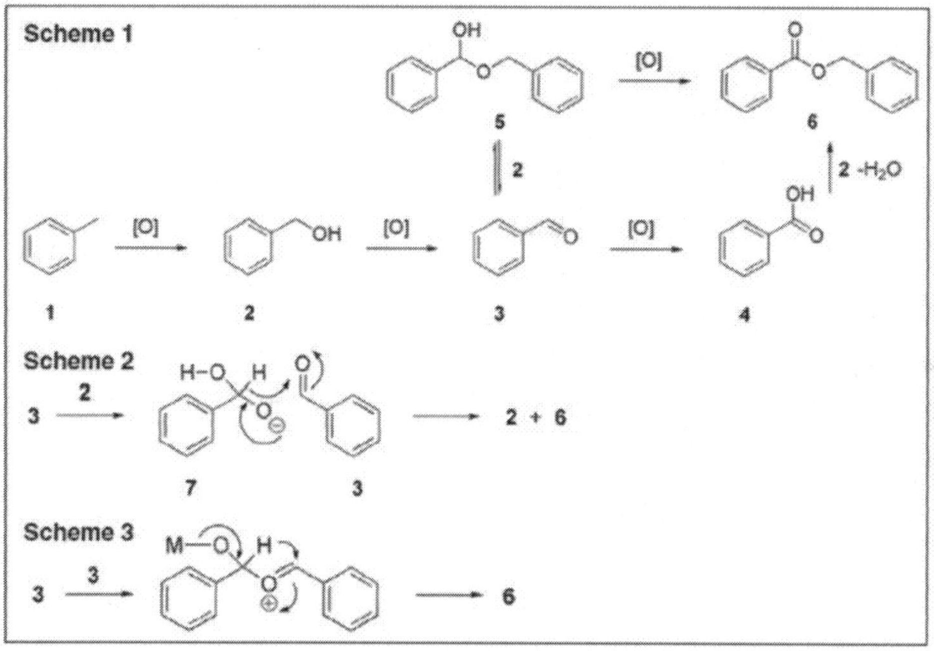

Toluene oxidation- 1wt% (AuPd)/C by SI

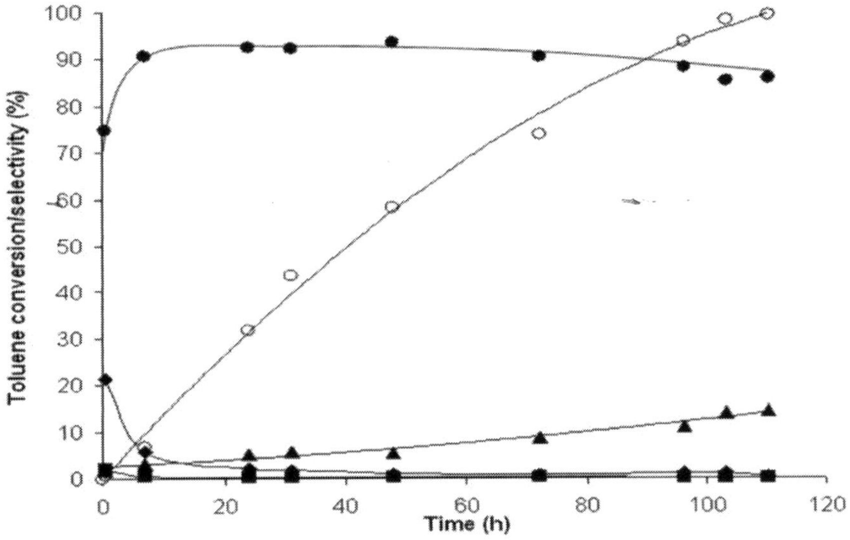

Key result- High conversion to benzylbenzoate
Solvent free, base free, 160°C, 10bar O_2 (continuous)

Toluene oxidation- 1.25%Pd-1.25%Pt/TiO_2 CVI

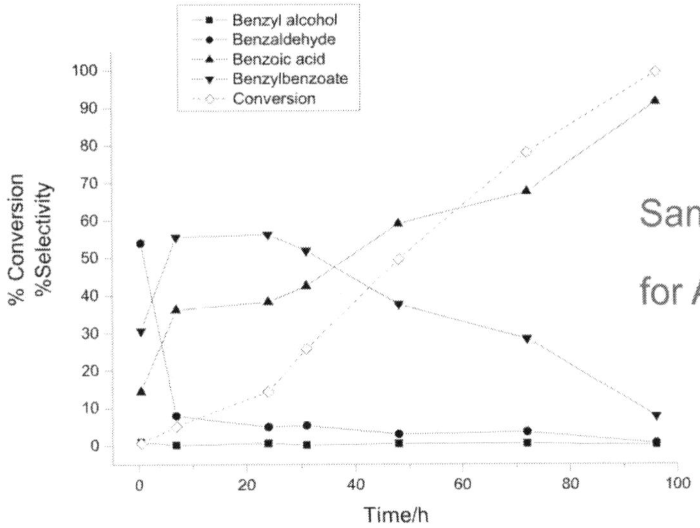

Same conditions as

for AuPd/C (SI) catalyst

Excellent activity
Different product distribution
Pd-Pt/TiO_2 or Pd-Pt/C by SI does not achieve high conversion

Toluene oxidation with TBHP and Pd-Pt/TiO_2

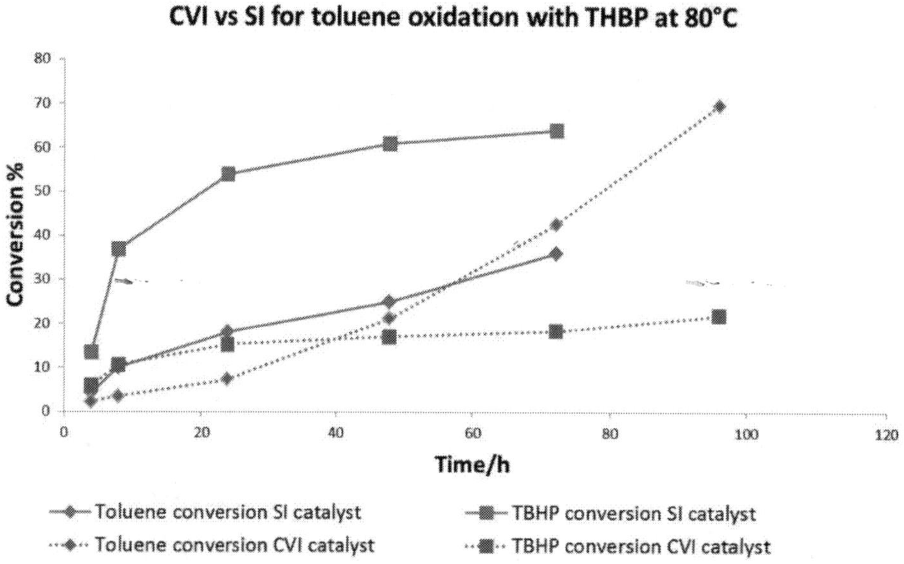

CVI catalyst is activating O_2...under N_2 toluene conversion drops, TBHP conversion increases

Higher conversion achieved with CVI catalyst

Levulinic Acid Hydrogenation

ACS Nano, 2014, 8, 957-969

Hydrogenation Catalysis- Levulinic Acid

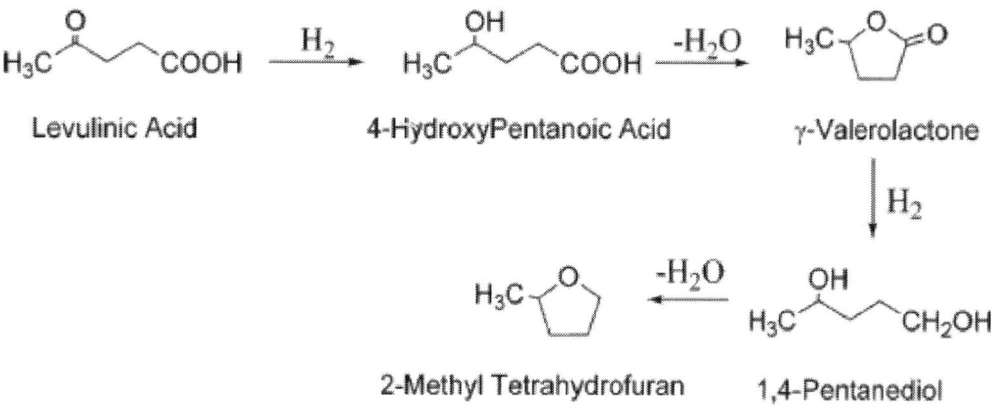

High rate, milder conditions, Lower excess of H_2

Levulinic Acid Hydrogenation

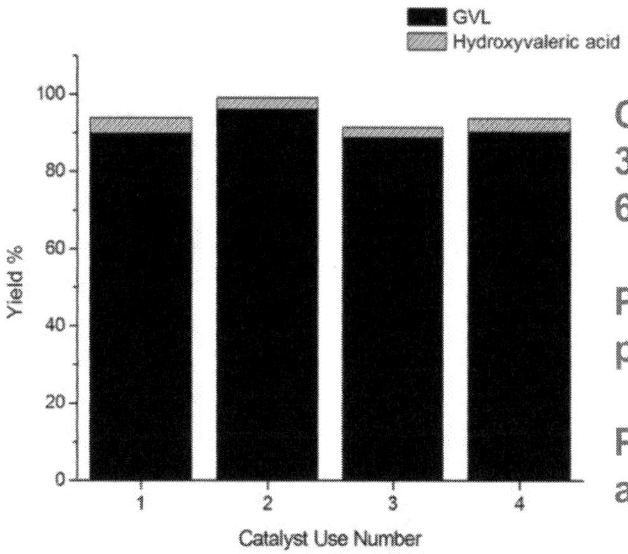

Catalyst even works at 30°C !
60% yield GVL in 20h

Pd-Pt has good performance at 150°C

Pd only performs well above 180°C

Pt/TiO_2 catalyst, 100 °C, H_2:LA= 2:1, 4 h

Photocatalysis

Dalton Trans., 2014,43, 14976-14982

H_2 from ethanol-water

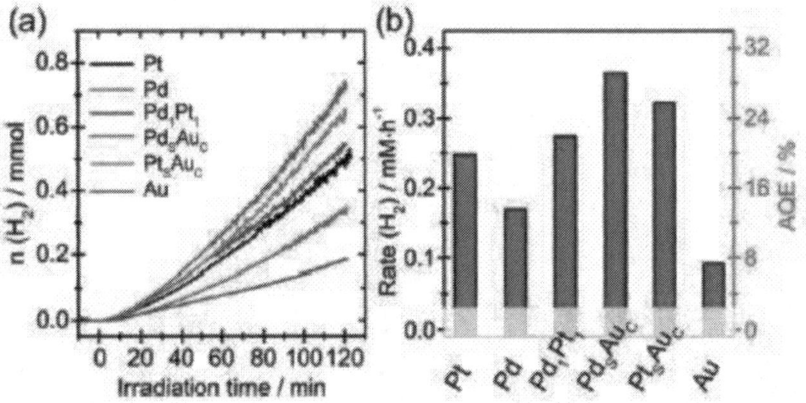

Beneficial effect of core shell structure observed
UV led used as light source
No dark reaction observed

Summary

- Simple, scalable techniques
- Well defined nanoparticles
- Flexibility to make exotic nano-architectures
- Clean, stable nanoparticles
- Wide activity for oxidation, hydrogenation, photo-catalysis

388051 Preparation of Carbon Supported Pt Nanoparticles By Atomic Layer Deposition

Abstract:

Hydrogen fuel cells offer an efficient use of a more energy dense fuel compared to combustion engines using gasoline. However, adaptation of fuel cells to power vehicles depends on several factors, one of which is economic synthesis of each component. Atomic layer deposition (ALD) using a fluidized bed reactor is an economic and scalable method for manufacturing the noble metal nanoparticle catalyst. This work furthers understanding of reported mechanisms of platinum ALD by investigating reactions on a common carbon support, XC72-R. By-products of reactions during ALD using trimethyl(methylcyclopentadienyl)platinum(IV) ($MeCpPtMe_3$) and either oxygen or hydrogen as reactants on un-modified carbon and carbon functionalized using nitric acid are inspected using *in situ* mass spectrometry. This understanding will help fine-tune parameters during a commercial catalyst synthesis process. Platinum weight loading, average particle size and particle dispersion depend on the chemistry of the pre-processing and precursor reactions. Oxygen becomes integrated into the carbon during functionalization and binds platinum tighter than just carbon; therefore particles aggregate less during deposition on functionalized carbon. Combustion of the ligand of $MeCpPtMe_3$ attached to carbon, as well as substrate oxidation during the oxygen dose, contribute to more particle aggregation versus when hydrogen is utilized for ligand removal. Mechanisms for each ALD condition are proposed.

Particle Size and Support Effect of Platinum catalyst in a Proton Exchange Membrane Fuel Cell

Shibely R. Saha[1] and Dongmei Li[1]*
[1]*University of Wyoming, 1000 E. University Ave. Laramie, WY 82071*
*dli1@uwyo.edu

Platinum (Pt) loading in proton exchange membrane fuel cells (PEMFC) is reduced and utilization increased by studying particle size and support effect on Pt catalysis. Generally, smaller Pt particle size in the PEMFC electrode catalyst layer provides increased reaction sites, thus maximizing the utilization and minimizing the loading. Another important factor that contributes to efficient catalyst use is its distribution upon the support, as well as how supported catalysts are dispersed in the gas diffusion layer (GDL). Catalysts that are well-distributed on its support and well-dispersed on the electrode provide desired contact for the reactants and improved mass transport of the reactants to the active catalytic sites [1].

To investigate the impact of support on Pt catalyst, transition metal carbides (TMCs), such as Molybdenum Carbide (Mo_2C), were selected as a support for Pt, due to their similarity in electronic and geometric structure to Pt [2] . Although previous research has used TMCs, including Mo_2C, as Pt supports, low TMC surface area has limited the catalytic performance[3]. To address this challenge, nanosize Pt has been discretely deposited upon the surface of a Mo_2C hollow nanotube support (confirmed by HRTEM) via island growth mechanisms of Atomic Layer Deposition (ALD). Nanosize Pt particles supported onto hollow nanotube Mo_2C (also referred as co-catalyst) with sub-monolayer coverage are presumed to further increase the reactivity of the catalyst, possibly resulting from both increased active Pt sites and enhanced interaction between the Pt particle and support. The reactivity of the co-catalyst towards Hydrogen Evolution Reaction (HER) and Hydrogen Oxidation Reaction (HOR) has been evaluated and compared to that of commercial 20% Pt supported on carbon black powder by electrochemical characterization using a three-electrode cell. This electrochemical characterization showed significantly higher HOR and HER activity from the co-catalyst with 2.4% Pt loading (confirmed by Atomic Absorption Spectroscopy (AAS)), compared to 20% Pt/C, demonstrating its potential applicability as PEMFC anode.

To evaluate co-catalyst performance in a device setting, PEMFC membrane electrode assembly (MEA) was fabricated using the co-catalyst as PEMFC anode. Device performance demonstrated a significant increase in maximum power per mg of Pt, compared with commercial 20% Pt supported on carbon black powder. As such, both electrochemistry cell and device testing results revealed that the sub-monolayer Pt coverage (a.k.a. discrete islands) on nanoscale Mo_2C tubes generated a synergetic effect that requires further systematic theoretical and experimental study to understand.

To investigate how particle size and distribution of the co-catalyst on electrode affects the device performance of PEMFC MEA, ultrasonic AccuMist (Sono-Tek) spray technique was applied, where the catalyst can be atomized, minimizing co-catalyst agglomeration, often observed in conventional brush painting approach. Scanning Electron Microscopy (SEM) confirmed that smaller particle size was generated on the electrode with ultrasonic AccuMist spray approach, with 37% increase in maximum power per mg of Pt in the electrode compared to that by conventional brush painting technique.

References

1. Lister, S; McLeanG., *Journal of Power sources*, p.61-76, 130(2004)
2. Levy, *R.B., Science*, p. 547-549, 181(1973)
3. Esposito, D.V.; Hunt, S.T.; Kimmel, Y.C.; and Chen, J.G.; *J. Am. Chem.Soc.,* p. 3025−3033, 134(2012)

390688 Carboxyl Multi-Walled Carbon Nanotubes Stabilized Palladium Nanocatalysts for Fuel Cell Application

Abstract:

Carboxyl Multi-walled carbon nanotubes (MWNTs-COOH) supported palladium (Pd) nanoparticles (NPs) catalysts (Pd/MWNTs-COOH) were prepared by a one-pot solution-based method, i.e., direct decomposing palladium acetylacetonate ($Pd(acac)_2$) in a refluxing xylene solution in the presence of MWNTs-COOH without any additives. Pd/MWNTs-COOH with different Pd loadings were prepared by altering the amount of $Pd(acac)_2$. Different annealing temperature effect was also investigated on Pd/MWNTs-COOH in a reducing atmosphere. X-ray diffraction (XRD) results indicate a successful deposition of Pd NPs on MWNTs-COOH. Transmission electron microscope (TEM) micrographs revealed that Pd NPs with increasing size were uniformly grown onto MWNTs surface with the increase of Pd loadings, and varying Pd particle sizes were also observed implying varying steps of Pd NPs as diffusing, contacting and coercing on tube wall surface with the increase of annealing temperature. The loading of Pd was calculated from the thermal gravimetric analysis (TGA), the variation of –COOH was investigated by Raman spectroscopy and X-ray photoelectron spectroscopy (XPS). The electrocatalytic activities of the Pd/MWNTs-COOH catalysts were evaluated for methanol oxidation and ethanol oxidation in KOH solutions. For methanol oxidation, bare Pd NPs and Pd/MWNTs (MWNTs without -COOH) was also synthesized and compared, parameters including the concentration of methanol and KOH were investigated, showing that the best catalyst performance was achieved by the balance between the adsorption of hydroxyl and methanol. A pronounced influence of temperature on methanol oxidation has been manifested, which indicates the positive role of increasing temperature. For ethanol oxidation, Cyclic voltammetry (CV), chronoamperometry (CA), electrochemical impedance spectroscopy (EIS) and Tafel were introduced to characterize the electro-performances, Pd/MWNTs-COOH with initial precursor ratio of 2:1 gave the best performance, 100 °C annealed Pd/MWNTs-COOH also gave best performance compared with other temperature annealed ones

376607 Effect of a Supercritical Isooctane Reaction Medium on the Conversion of Methanol to Hydrocarbons over H-ZSM-5

Abstract:

Optimizing the production of fuels and chemicals from carbon resources other than petroleum has been an area of ongoing investigation for more than eighty years. By steam reforming natural gas or by gasifying coal or biomass, a mixture of carbon monoxide and hydrogen known as syngas is produced, which can be subsequently converted to hydrocarbon fuels, heavy waxes, or oxygenated compounds, including methanol. Methanol synthesis from syngas proceeds with high selectivity, and further upgrading of methanol over solid-acid catalysts can selectively produce dimethyl ether, light olefins or fuel-range hydrocarbon species, depending upon the catalyst type and operating conditions. High yields of high-octane gasoline have been achieved by carrying out the methanol-to-hydrocarbons (MTH) reaction over zeolite catalyst H-ZSM-5. However, catalyst deactivation due to coking as well as heat management challenges due to the exothermic nature of the reaction hinder utilization of MTH technology. Our group, among others, have previously demonstrated the benefits of employing supercritical fluids (SCFs) as reaction media in heterogeneous catalysis in order to facilitate heat removal and heavy product extraction from catalyst active sites. This particular study explores the feasibility of using SCF media to enhance MTH reaction performance relative to traditional, gas-phase operation. Conversion of methanol over H-ZSM-5 was carried out in a fixed-bed reactor in the presence of supercritical isooctane. Isooctane was selected as the SCF medium primarily because of its low reactivity over H-ZSM-5. The performance of SCF-assisted MTH was compared with that of gas-phase MTH at both 300°C and 370°C. Relative to gas phase operation at the same conditions, the use of supercritical isooctane resulted in improved catalyst maintenance and increased hydrocarbon productivity with increasing time on stream.

368026 Thermo-Kinetic Modeling of Supercritical Processes for the Conversion of Algal Lipids to Biofuels

Abstract:

Biodiesel, derived from renewable feedstocks like algae, has the potential to replace traditional petroleum-based diesel. However, numerous technical difficulties remain. The extraction of triglyceride oils from algae cells has the greatest economic variability while also being the most poorly understood operation [1]. Traditional hexane-extraction techniques create dirty feedstocks that must be cleaned and purified before proceeding to the transesterification process. Among emerging technologies, supercritical processes appear promising for both the extraction and transesterification steps [2], but the high temperatures and pressures raise energy efficiency and safety concerns.

Supercritical carbon dioxide is a non-toxic, environmentally benign solvent that has the potential to extract and convert algae-oils to biodiesel in a single step [3] at significantly lower temperatures than the supercritical methanol process. Supercritical carbon dioxide mixed with methanol forms a gas-expanded liquid (GEL), which drastically increases the solubility of the algal triglycerides. To optimize conversion and minimize energy use, it is essential that a model be formulated which can accurately predict the reactor conditions in the GEL region. Therefore, an equation-of-state must be used, which can account for the effects of both long and short chain molecules and association, in both VLE and LLE. For this purpose, the statistical associating fluid theory equations-of-state in ASPEN PLUS (PC-SAFT) and gPROMS (SAFT-γ Mie) are being used to calculate fluid-phase equilibria [4]. The necessary parameters are regressed from experimental data, and a thermo-kinetic reactor model is formulated to account for changing phase equilibria as the reaction proceeds.

[1] Silva, C.S.; Soliman, E.; Cameron, G.; Fabiano, L.A.; Seider, W.D.; Dunlop, E.H.; Coaldrake, A.K. Commercial-scale Biodiesel Production from Algae. *Ind. Eng. Chem. Res.* 2014, 53 (13), 5311-5324.

[2] Patil, P.D., Gude, V.D., Mannarswamy, A., Deng, S., Cooke, P., Munson-McGee, S., Rhodes, I., Lammers, P., Nirmalakhandan, N., 2011, Optimization of Direct Conversion of Wet Algae-to-biodiesel under Supercritical Methanol Conditions, *Bioresource Technology*, 102, 118-122.

[3] Soh, L., Zimmerman, J., 2011, Biodiesel production: The potential of algal lipids extracted with supercritical carbon dioxide, *Green Chemistry*, 13, 1422 1428.

[4] Gross, J., Sadowski, G., 2001, Perturbed-Chain SAFT: An Equation-of-State Based on a Perturbation Theory for Chain Molecules, *Ind. Eng. Chem. Res.*, 40, 1244 1260.

CATALYTIC CONVERSION OF ETHANOL TO 1-BUTANOL AND HIGHER ALCOHOLS: INVESTIGATION OF VAPOR LIQUID EQUILIBRIA

Tyler L. Jordison and Dennis J. Miller
Department of Chemical Engineering and Materials Science, Michigan State University,
East Lansing, Michigan 48824, USA
(517) 353-3928

Ethanol makes up the largest fraction of renewable fuels in the U.S., but future growth in ethanol production depends in part on improving fuel properties and on finding uses in addition to direct combustion. One potential route to improved ethanol use is the carbon-carbon coupling commonly regarded as the Guerbet reaction [1]. In the ethanol Guerbet reaction, ethanol is oxidized to acetaldehyde, the acetaldehyde undergoes an aldol condensation to crotonaldehyde, and then the crotonaldehyde is hydrogenated to butanol [2]. Product alcohols can subsequently undergo Guerbet reactions with themselves and ethanol, leading to higher alcohols. The low energy density and affinity for water of ethanol are overcome if ethanol is upgraded to 1-butanol and other higher alcohols. Higher alcohols have energy values closer to gasoline and low affinity for water, making them strong candidates for fuels. Higher alcohols also have broad applications in the commercial chemical industry. The current method for producing higher alcohols is the oxo process, which uses petroleum-derived propylene as the feedstock [1]. The oxo process is complicated, requires high energy input, and is costly [1].

Early experiments of condensed-phase ethanol Guerbet reactions over La/Ni/Al2O3 catalyst gave high selectivity (>80%) towards butanol and C6 and C8 higher alcohols, with gas formation, acetaldehyde, and ethyl acetate as primary co-products. Partial removal of water or co-feeding of aldehydes during reaction led to modest improvement in alcohol selectivity, but further progress requires an accurate characterization of system thermodynamics. Because condensed-phase Guerbet reactions take place near the critical temperature of ethanol, both the expanded liquid phase and the dense vapor phase present in reaction contain significant quantities of reactants and products. At these conditions, the vapor liquid equilibria (VLE) of the reaction system is poorly understood, particularly because ethanol decomposes in part to methane and carbon dioxide. This gas formation leads to high reactor pressures at high conversions, and consequently there is a significant amount of gas dissolved in the liquid. There is also a significant quantity of alcohols present in the vapor.

The distribution of species between liquid and vapor phases must be properly accounted for if reaction kinetics are to be accurately modeled. An equation of state model has thus been implemented to describe ethanol Guerbet VLE at room temperature and at near-critical reaction conditions. The SR Polar equation of state was found to most accurately predict vapor pressures for products of the ethanol Guerbet reaction. Binary parameters were regressed for available binary VLE data. The binary parameters were then applied in flash calculations at room temperature and at reaction temperatures for the representative product mixtures. The calculated vapor-liquid compositions were compared with experimental data.

References

[1] T. Tsuchida, S. Sakuma, T. Takeguchi, W. Ueda, Industrial & Engineering Chemistry Research, 45 (2006) 8634-8642.
[2] A.S. Ndou, N. Plint, N.J. Coville, Applied Catalysis a-General, 251 (2003) 337-345.

Hydroformylation of Olefins with Co-based Complexes in Butane-Expanded Liquid

Dupeng Liu,[1,2] Zhuanzhuan Xie,[1,2] Raghunath V Chaudhari,[1,2] and Bala Subramaniam[1,2]

[1]Department of Chemical & Petroleum Engineering, [2]Center for Environmentally Beneficial Catalysis, University of Kansas

Lawrence, KS 66047

Previously, we had reported on the beneficial effects of carbon dioxide expanded liquids for hydroformylation with Rhodium-based catalyst[1-3]. The H_2/CO ratio in the gas-expanded liquid phase (containing substrate and dissolved catalyst complex) was found to be a key determinant of the turnover frequency and regioselectivity. In this talk, we report *n*-butane (Pc =37.9 bar; Tc =151.9 °C) as the expansion gas for performing olefin hydroformylation with cobalt-based catalyst that typically requires higher temperatures (160-180°C) for catalyst activation and reaction initiation. The volumetric expansion of representative hydroformylation reaction mixtures (1-octene and 1-octene+nonanal mixtures) by n-butane at typical reaction temperatures (160-180°C) are measured experimentally and compared with simulation results. It is shown that n-butane significantly expands 1-octene and 1-octene+nonanal mixtures (by up to two-fold at 22 bar, 160°C) that match well with the results simulated using the Peng-Robinson equation of state. Guided by such volumetric expansion data, we performed cobalt-catalyzed 1-octene hydroformylation in butane-expanded media in a stirred reactor operated at constant pressure by replenishing syngas (molar $H_2/CO=2$) from an external reservoir. We found aldehydes and alcohols to be the major products (total selectivity > 60%). Further, at identical syngas partial pressure (45 bar), the regioselectivity is enhanced in butane-expanded liquid or BXL ($V/V_0=1.1$) compared to neat toluene as solvent with no significant difference in olefin conversion between the two cases. Quantitative results on such enhancements along with process optimization to maximize the yield of linear alcohols will be presented.

1. Jin H, Subramaniam B, Ghosh A, Tunge J. Intensification of catalytic olefin hydroformylation in CO2-expanded media. *AIChE Journal.* 2006;52(7):2575-2581.
2. Xie Z, Fang J, Subramaniam B, Maiti SK, Snavely W, Tunge JA. Enhanced hydroformylation by carbon dioxide-expanded media with soluble Rh complexes in nanofiltration membrane reactors. *AIChE Journal.* 2013;59(11):4287-4296.
3. Fang J, Jana R, Tunge JA, Subramaniam B. Continuous homogeneous hydroformylation with bulky rhodium catalyst complexes retained by nano-filtration membranes. *Applied Catalysis A: General.* 2011;393(1–2):294-301.

388598 Molecular Simulation of Supercritical Fluid Solvent Effects on CO Chemisorption

Abstract:

While supercritical fluids (SCFs) have been employed in heterogeneous catalysis for decades across a range of chemistries (oxidation, pyrolysis, amination, Fischer-Tropsch synthesis), there is little fundamental, molecular-level information regarding the role of the SCF on elementary heterogeneous catalytic steps. In this investigation, we use molecular dynamics (MD) simulations to probe the role of supercritical hexane on the adsorption of carbon monoxide on cobalt, as a first step in catalytic Fischer-Tropsch synthesis. The free energy (potential of mean force) landscape for CO adsorption is calculated using umbrella sampling and the weighted histrogram analysis method (WHAM), for a range of bulk hexane densities, ranging from ideal gas conditions (no SCF hexane) to various near- and super-critical hexane densities and pressures. These results allow for the first commentary on SCF solvent effects on the energetic and entropic aspects of chemisorption, and provide improved kinetic and thermodynamic parameters for subsequent microkinetic modeling.

381679 Effect of Pressure, O/Ostoich Ratio and Temperature on the Partial Oxidation of Heavy Oil Model Compound Phenanthrene

Abstract:

An increasing world energy demand and the proportions of heavy oil reserves relative to lighter and more valuable feedstocks have enabled the development of new emerging research fields. A novel method to upgrade heavy oil through partial oxidation in supercritical water (SCW) is being developed in the research group. This method takes advantage of the properties of SCW (T=374°C, P=220 bar and δ=0.32 g/ml and above), in which water changes from a good solvent for polar to non-polar species. At these conditions a change in the ion product, in the viscosity and a decrease in the dielectric constant are observed [1]. Studies with model compounds representing chemical structures found in heavy oils give relevant information about the reactivity in the medium and the reaction mechanisms occurring in the system. Reactions with polycyclic aromatic hydrocarbons have been studied in the presence and absence of an oxidizer [2, 3].

The aim of this work is to investigate and get an understanding about the effect of process conditions pressure, temperature and O/O_{stoich} ratio in the yield and selectivity to different reaction products of the partial oxidation of heavy oil model compound phenanthrene at near and supercritical water conditions.

Methodology:

Partial oxidation experiments were performed in a stainless steel micro-bomb reactor with a volume of 18 ml. The operation of the reactor has been described elsewhere [4]. The reactor was filled with 0.15g of phenanthrene 98% (Sigma-Aldrich), 1.2 ml of H_2O_2 30% (VWR chemicals) and different volumes of deionized water to reach the desired operating pressure. The reactor was purged with helium in order to remove any air remaining in the system. Once purged the reactor is closed and immersed into a fluidized sand bath used to control the reaction temperature. Gas products formed in each reaction were analyzed in a Perkin Elmer Clarus GC with thermal conductivity detector. Liquid products were separated from the remaining water, extracted with a mixture of chloroform/methanol 4:1 and then filtered. Solids obtained after filtering were dried and analyzed in a Perkin Elmer TGA to determine yield to coke. The remaining liquids were analyzed in a Perkin Elmer Clarus GC with flame ionization detector and in a Varian Star 3400/Saturn 2000 GC/MS to identify the main products obtained.

Results and Discussion:

Experiments to assess the effect of pressure in the system were performed at constant O/O_{stoich} ratio of 0.38, 60 min and reaction temperature of 450°C. Conversions and yields to products were obtained at 210 bar, 230 bar, 250 bar and 275 bar. Results show that conversion of phenanthrene remained almost constant near or above the critical point of water. Interestingly, changes in pressure had a great impact on product distribution and yields. Yields to liquid products increased with an increase in pressure till operating pressures near the critical point of water. Further increase in pressure derived in an important decrease in the selectivity to liquid products. In addition, the increase in the system pressure led to a considerable increase in coke yield and a steady increase in the yields to gas. Moreover, analysis of the liquid fraction showed that the composition of the liquid products obtained was greatly affected by pressure. The widest range of liquid products was obtained near the critical pressure of water. At pressures below the critical point lower molecular weight and oxygenated products were obtained. This may be due to the occurrence of cracking and ring opening reactions, which also led to high gas productions. In contrast, higher molecular weight compounds as pyrene were obtained at pressures over the critical point. This may be due to ring annealing and polymerization reactions enhanced at higher pressures. At 230 bar, the highest product distribution was obtained and selectivity to oxygenated products with boiling points between 300 °C and 400 °C was higher.

Based on previous results, the effect of O/O_{stoich} ratio on the conversion of phenanthrene was studied. Reactions at different O/O_{stoich} ratio (0.1, 0.2, 0.38 and 0. 75) were performed at 450°C, 230 bar and 60 min reaction time. It was observed that conversion of phenanthrene increases with an increase in the concentration of oxygen in the system. The yield to liquid products increased with the increase in O/O_{stoich} ratio. However, at higher concentrations of oxygen in the system the amount of gas produced increased considerably while the amount of coke produced remained almost constant. Therefore, it can be said that an increase in oxygen ratio make the liquid products obtained to continue to react towards undesirable gas. Based on observations made on this work an optimum O/O_{stoich} ratio to favor selectivity towards liquid products was found between 0.2 and 0.38.

Furthermore, experiments were performed to assess the effect of temperature in the system. For these set of experiments different temperatures (360°C, 400°C, 425°C and 450°C) were studied at a pressure of 230 bar and an O/O_{stoich} ratio of 0.2. It was observed that an increase in temperature increased the yield to all of the product fractions. However, it was noted that when temperature was raised from 425°C to 450°C, selectivity to liquid products decreased greatly mainly due to an important amount of gas produced. Results showed that when reactions took place in supercritical water, temperature has no major effect in the product distribution within the liquid fraction. However when temperature was below the critical point, two liquid products were mainly observed (fluorene and fluorenone) which suggest that the reaction occurs at a considerably slower rate than it occurs at supercritical conditions.

Conclusions:

System pressure, O/O_{stoich} ratio and temperature have been proven to have a great influence in the partial oxidation of heavy oil model compound phenanthrene with water at near critical and supercritical conditions. It was found that the overall conversion of phenanthrene increased with an increase in O/O_{stoich} ratio and temperature and that these two variables greatly affect yields and selectivities to different reaction products. Conversion remained constant with an increase in pressure although, selectivity to liquid products reached a maximum at 230 bar near the critical pressure of water. Yields to gas and coke also depended greatly on pressure. GC/MS analysis of the liquid product fractions showed that product distribution in the liquid fraction also depended on the system pressure and the O/O_{stoich} ratio. It has been observed that product distribution remain fairly constant with changes in temperature when operating above the critical point of water. These results are of great importance as they show that different target products can be obtained with slight variations of the main process conditions.

References

1. H. Weingärtner, E.U. Franck, Angew. Chemie Int. Ed. 44 (2005), 2672–2692.
2. J.A. Onwudili, P.T. Williams, Int J Energ Res, 30 (2005), 523–33.
3. J.A. Onwudili, P.T. Williams, J Supercrit Fluid, 39 (2007), 399–408
4. J.L. Pinilla, P. Arcelus-Arrillaga, H. Puron, M. Millan, Appl. Catal. A 459 (2013) 17-25.

380316 Supercritical Water Reformation of Oxygenated Hydrocarbons

Abstract:

Decomposition of lignocellulosic biomass in near-critical and supercritical water is a fast and effective means of converting a less useful solid feedstock into a mixture of pumpable liquid products. This process, known as liquefaction, can yield a wide array of smaller compounds (typically ranging from C_1 to C_7) from the macromolecules present in the biomass. This type of feedstock is highly oxygenated, and as a result, the products from liquefaction are also oxygenated. If the liquefaction products have little value on their own, they can be converted to a gaseous product via supercritical water reformation. This process can produce synthesis gas for transportation fuel synthesis, or a high purity hydrogen/methane mixture that can be used to produce energy. It is important to understand how the individual oxygenated functional groups present in a biomass derived feedstock affect the reformation process. Gasification rate, product selectivity, susceptibility to coke formation, etc. can all vary depending on the oxygenated functional groups present in the feedstock. This study attempts to elucidate these effects by examining the gasification of four oxygenated hydrocarbons in supercritical water. 1-Propanol, propionaldehyde, propionic acid, and acetone were selected as feedstocks, because they contain some of the most common oxygenated functional groups present after biomass liquefaction. Experiments were performed in a custom built, continuous, supercritical water reformer at temperatures between 500°C and 700°C, residence times between 10 and 90 seconds, and water-to-carbon molar ratios of 8 and 15. In addition to process effects, reaction pathways and kinetics were developed.

387855 A Combined Experimental and Theoretical Study for the Conversion of Carbon Dioxide to Carbon Monoxide on Noble Metal Impregnated Perovskite Structures

Abstract:

The last decade has seen lots of interest in achieving efficient conversion of carbon dioxide to carbon monoxide, which is a major energy currency towards generating hydrocarbon energy fuels via Fischer Tropsch synthesis. Amongst the many different types of material being studied for achieving the conversion of carbon dioxide, mixed perovskites have garnered a significant prominence in this field. Novel perovskite materials, $La_{(1-x)}Sr_xCoO_3$(x=0.25, 0.5) in particular, impregnated with noble metals like Pt, Ag have been studied for the same purpose. The conversion process involves heating the materials in inert atmosphere of helium gas, resulting in the generation of oxygen vacancies on the surface of these materials. These heated oxygen vacant materials are proposed as active sites for the dissociation of carbon dioxide to carbon monoxide.

$La_{0.75}Sr_{0.25}CoO_3$ and $La_{0.5}Sr_{0.5}CoO_3$ samples had been prepared by Pechini process. They revealed the presence of mixed crystal phases like cubic, hexagonal and orthorhombic of which cubic phase had the major prominence as was evident from the X-ray diffraction (XRD) patterns. Controlled temperature programmed reactions were conducted to probe the two stage carbon dioxide conversion cycle. The reaction acutely depends on the various structural and electronic states of the material surface. Hence a theoretical analysis of the materials and the process was undertaken using density functional theory (DFT) calculations in order to hypothesize the reaction mechanism involved and to propose the best material candidate for this reaction. Based on the DFT calculations, the most stable bulk phase crystal structure of these materials have been proposed and the various surface states along the major crystal planes (100) and (110) were investigated. Dissociative chemisorption energy is a key representative parameter for studying these reactions. Henceforth a systematic study has been pursued towards obtaining a trend in the dissociative chemisorption energy as a function of the structural and electronic properties of the material.

Modified Sn surfaces for the electrochemical reduction of CO_2

Jeremy Feaster, Etosha Cave, Toru Hatsukade, David Abram, Kendra Kuhl, Chris Hahn, and Thomas F. Jaramillo
Department of Chemical Engineering, Stanford University

Electrochemical reduction of carbon dioxide to high-value fuels and chemicals is an increasingly attractive solution to partner with renewable, carbon-neutral energy sources by mitigating intermittency and energy storage problems inherent in solar and wind power. Sn, an appealing catalyst for this reaction, has been shown to primarily produce formic acid from CO_2; however, there is a lack of understanding as to why Sn exhibits such a high selectivity for formic acid, particularly when compared to carbon monoxide. Our work reports the exploration of CO_2 reduction on Sn electrodes by utilizing a highly sensitive electrochemical cell for identification and quantification of reduced products. Polycrystalline Sn electrodes were studied over a range of potentials, and formic acid, carbon monoxide and hydrogen were confirmed as the major products formed on Sn. By comparing the experimental activities for formic acid of Sn and several transition metals, we suggest that the OCHO bidentate intermediate is the likely intermediate for formic acid production on metal surfaces, whereas the COOH binding intermediate commonly cited in previous works is the likely intermediate for carbon monoxide production.

Using this knowledge, we were able to modify the Sn electrodes to introduce more undercoordinated sites. While no other products were observed on polycrystalline Sn, methane was detected at potentials as early as -1.15V vs RHE on the modified Sn electrodes. Furthermore, characterization of the Sn electrodes via electrochemical impedance spectroscopy (EIS) suggests that the production of methane does not result from an increase of the surface area of the electrodes. These results suggest that the introduction of undercoordinated sites on Sn metals may lead to a stronger binding energy with CO, a key intermediate for CO_2 reduction, and allow for the formation of more reduced products.

Plasmon-Enhanced Reverse Water Gas Shift Reaction over Au/TiO_2 Catalyst

Insoo Ro[1], Aniruddha A. Upadhye[1], Xu Zeng[1], Hyung Ju Kim[1], Isabel Tejedor[2], Marc A. Anderson[2], James A. Dumesic[1] and George W. Huber[1,*]

1. Department of Chemical and Biological Engineering, University of Wisconsin, Madison, WI
2. Department of Civil and Environmental Engineering, University of Wisconsin, Madison, WI

Environmental concerns over anthropogenic CO_2 emissions combined with increased petroleum costs motivate our society to develop new technologies that enable the recycle of CO_2 to useful fuels and chemicals. Due to its abundance, availability and sustainability, solar energy has long been considered as the ideal source of energy. In spite of numerous efforts such as doping[1, 2] and defect creation[3], conventional semiconductor photocatalysts have two major drawbacks leading low efficiency: 1) high rates of electron-hole (e^-/h^+) pair recombination and 2) limited utilization of visible light spectrum in the solar energy. An alternative approach for solar-to-chemical energy involves plasmonic catalysts which are able to absorb visible light through phenomena called surface plasmon resonance.[4] Localized surface plasmon resonance (LSPR) exhibited by certain metal nanoparticles can effectively couple of visible light radiation to drive chemical reactions. Since LSPR depends upon size, shape and type of metal, we can engineer plasmonic catalyst that can utilize most of the solar spectrum for solar fuels applications.

Here, we present our results on plasmon-enhanced CO_2 reduction with hydrogen through Reverse Water Gas Shift (RWGS) reaction. We studied effect of LSPR to enhance catalytic activities of different oxide-supported gold and copper catalysts for RWGS reaction. The plasmonic behavior of these catalysts was characterized by the increased absorption in UV-Vis spectra. Oxide-supported Au and Cu catalysts showed 30 to 1300% higher activity under visible light compared to dark conditions. Out of the tested catalysts, Au/TiO_2 showed the highest activity for RWGS reaction. We obtained a maximum 13 times enhancement in CO_2 conversion rate and an estimated 5% light to chemical energy efficiency for this reaction with Au/TiO_2 catalyst. We performed detailed kinetic experiments over Au/TiO_2 to understand the mechanism of plasmonic enhancement. The apparent activation energy decreased by 25% when light was applied to the catalyst system. Moreover, the apparent rate order with respect to CO_2 changed from 0.5 under dark to 1 under light. These results indicate that the LSPR changes the surface coverage of the adsorbed species, thereby changing the energetics of the reaction mechanism. Here, we will also present our results with a microkinetic modeling, which elucidates the changes in mechanism due to localized surface plasmon resonance. These results will help researchers in the field of plasmonic catalysis understand how surface plasmons affect the kinetics of chemical reactions. This understanding of plasmonic catalysis guides design of improved plasmonic catalysts.

1. A. K. L. Sajjad, S. Shamaila, B. Z. Tian, F. Chen and J. L. Zhang, *J. Hazard. Mater.*, 2010, **177**, 781-791.
2. X. H. Wang, J. G. Li, H. Kamiyama, Y. Moriyoshi and T. Ishigaki, *J. Phys. Chem. B*, 2006, **110**, 6804-6809.
3. M. Takeuchi, H. Yamashita, M. Matsuoka, M. Anpo, T. Hirao, N. Itoh and N. Iwamoto, *Catal. Lett.*, 2000, **67**, 135-137.
4. S. C. Warren and E. Thimsen, *Energy & Environmental Science*, 2012, **5**, 5133-5146.

SURFACE COVERAGE OF REACTION INTERMEDIATES DURING CO_2 ELECTROREDUCTION AND THEIR IMPACT ON CATALYST PERFORMANCE

Sneha A. Akhade[1]*, Wenjia Luo[2], Xiaowa Nie[2], Nicole Bernstein[1], Aravind Asthagiri[2] and Michael J. Janik[1]

[1]Department of Chemical Engineering, Fenske Laboratory, Pennsylvania State University, State College, PA 16802
[2]Department of Chemical and Biomolecular Engineering, Koffolt Laboratories, Ohio State University, Columbus, OH 43210
*email: saa243@psu.edu

The generation of "solar-fuels" or renewable hydrocarbon (HC) fuels, such as CH_4, C_2H_4, CO, HCOOH, CH_3OH via electrochemical reduction (ER) of carbon dioxide may render flexibility in renewable energy storage and transportation applications. The selection of suitable metal electrocatalysts for CO_2 ER to HCs is non-trivial with Cu reported as most selective while other metals such as Pt, Fe, Ni and Co produce low to zero HCs [1]. We employ Density Functional Theory (DFT) to develop a mechanistic approach towards explaining the variation in selectivity and performance of the CO_2 ER process for transition metals. Mechanistically, the reduction of adsorbed CO is the most important elementary step and requires the presence of surface adsorbed H to form C–H bonds [2]. A detailed thermodynamic analysis using reaction free energies ΔG coupled to the electrode potential suggests that a high coverage of adsorbed CO at relevant reduction potentials blocks the metal surface sites for H adsorption and prevents C–H bond formation. The potential-dependent energetics of H adsorption and CO formation are highly sensitive to the surface coverage of the adsorbed species. The results of these studies provide a rationale for improved catalyst design and leave room for studying the kinetic implications on the pathway as well as developing more robust models that realistically predict the electrocatalytic phenomena at length and time scales beyond the framework of DFT.

[1] Y. Hori, in *Handbook of Fuel Cells: Fundamentals, Technology and Application* (VHCWiley, Chichester, 2003), Vol. 2, 720-733.
[2] X. Nie, M. R. Esopi, M. J. Janik and A. Asthagiri, Angewandte Chemie International Edition **52** (9), 2459-2462 (2013).

Synthetic natural gas production by catalytic hydrogenation of CO_2 over a Ru/γ-Al_2O_3 catalyst for renewable energy storage

M.S. Duyar[1], R. Farrauto[2]

Earth and Environmental Engineering Department
Columbia University in the City of New York
500 W 120th Street, New York, NY, 10027
[1] msd2154@columbia.edu
[2] rf2182@columbia.edu

The production of synthetic natural gas (SNG) from CO_2 captured from a combustion source using H_2 generated from wind and solar energy can result in significant reductions in CO_2 while allowing these renewable energy sources to be dispatchable and hence more reliable. The Sabatier reaction (eq. 1) is a chemical route that allows for methane to be synthesized from CO_2 combined with H_2. This reaction can be performed catalytically to convert CO_2 directly into a valuable fuel resource[1].

$$CO_2 + 4H_2 \rightarrow CH_4 + 2H_2O \quad \Delta H^0 = -164 \text{ kJ/mol } (CO_2) \qquad \text{(eq. 1)}$$

In this manner, renewable energy that cannot be used in the electricity grid (due to fluctuations in output) is integrated into the natural gas grid. Hydrogenation of CO_2 to SNG can result in significant decrease in annual CO_2 emissions because it enables the recycling of carbon directly to its source.

The highly exothermic nature of the Sabatier reaction is problematic in large-scale applications because it can cause a temperature rise, which introduces equilibrium constraints on the reaction. A large exotherm can also cause deactivation of the catalyst. Our work addresses the challenge of managing the exotherm related to the methanation reaction in large-scale applications by ultimately developing a catalytically coated heat exchanger. Ru/γ-Al_2O_3 was investigated as a candidate catalyst suitable for such an application due to its high activity and stability. Kinetic parameters for CO_2 methanation over Ru/γ-Al_2O_3 at 200-245°C were evaluated and an empirical rate equation developed. TGA-DSC studies were performed at 260°C in CO_2 and H_2, being introduced consecutively, to investigate adsorption and reaction of H_2 and CO_2 on the catalyst. Based on reactor and TGA-DSC studies, an Eley-Rideal mechanism where adsorbed CO_2 reacts with gas phase H_2 was proposed for CO_2 hydrogenation over Ru/γ-Al_2O_3. Results from adsorption studies also suggested that O atoms adsorbed on the catalyst might be an intermediate in the CO_2 hydrogenation reaction.

[1] C. Janke, M.S. Duyar, M. Hoskins, R. Farrauto, Applied Catalysis B: Environmental 152-153 (2014) 184-191.

DFT Studies on Facet Dependence of CO_2 Electroreduction Path and Selectivity

Xiaowa Nie[1], Wenjia Luo[1], Michael J. Janik[2], and Aravind Asthagiri[1]

[1]William G. Lowrie Department of Chemical & Biomolecular Engineering, The Ohio State University, Columbus, Ohio 43210, USA
[2]Department of Chemical Engineering, Pennsylvania State University, University Park, PA 16802, USA

Electrocatalytic reduction of carbon dioxide (CO_2) is a potential candidate process to utilize the renewable sources (hydro, solar, or wind) for energy storage and synthetic fuel production. However, the complexity of the electrochemical environment and inability to control the product selectivity and lower the overpotential are significant barriers for the application and development of this technology in CO_2 utilization and conversion. Electrokinetic experiments reveal that Cu facet plays an important role in impacting CO_2 reduction selectivity and applied potential. We used density functional theory (DFT) to examine the elementary thermochemistry and kinetics of CO_2 electroreduction on various Cu facets. Computational results uncover an obvious facet dependence of CO_2 reduction path and selectivity.

Cu facet shows selective preference for the key hydrogenation intermediates between hydroxymethylidyne (COH*) and formyl (CHO*). Cu(111) predicts a favorable COH* intermediate through which methane and ethylene are produced at the same electrode potential. COH* reduction to C* step has a highest kinetic barrier which determines the reduction rate and potential requirement. The selective preference of COH* than CHO* results in the favorability of methane and ethylene production over methanol on Cu(111). On Cu(100), formation of CHO* is preferred and ethylene formation goes through C-C coupling of CHO* and further reduction of the C_2 species. C-OH bond dissociation steps in C_2 species conversion have significant barriers. Surface coverage and competitive adsorption of key species such as H* and CO* potentially impact the reaction rate and catalytic selectivity. At relatively high overpotentials, CHO* would go through C-C coupling to form C_2 intermediates rather than reduce to methanol through C_1 species involving CH_2O* and CH_3O* as observed experimentally.

Catalytic Hydrogenation of Carbon Dioxide to Methane: Catalyst Synthesis and Kinetic Measurements

F. Koschany, D. Schlereth, O. Hinrichsen,
Technische Universität München, Catalysis Research Center and Chemistry Department, 85748 Garching b. München, Germany

Introduction

With an increasing world population and limited natural crude oil stocks, a research focus has to be set on new ways to secure both, primary energy supply for the world and carbon feedstock for the chemical industry. Additionally, these fossil energy sources produce CO_2, known as a main contributor to the world exhaust gas emissions and responsible for the greenhouse effect. Waste CO_2 can be converted with hydrogen to methane, following the Sabatier reaction [1]. This will not only bind CO_2 out of the atmosphere, it will also provide a C1 feedstock for the chemical industry and primary energy. Furthermore, if hydrogen is produced by the electrolysis of water, using sustainable electricity sources like wind and solar power, this reaction has a high potential as a chemical energy storage technology. The synthesized methane can easily be stored, transported in the already existing gas distribution systems and converted back to energy during low sustainable energy supply times.

For CO_2 methanation several metal catalysts, e.g. Rh, Ru, Pd and Ni have been tested and potential reaction mechanisms have been proposed [2,3]. As supports Al_2O_3, SiO_2, TiO_2, ZrO_2, and CeO_2 have been studied so far. Kinetic investigations and models were developed by *Weatherbee and Bertholomew* [4] and *Xu and Froment* [5]. In an ongoing study in our research group, different Ni/Al_2O_3 catalysts were synthesized by co-precipitation followed by characterization. A deeper insight into the kinetics of the CO_2 methanation over Ni-based catalysts is gained by activity measurements over a wide range of experimental conditions.

Experimental Work

Nickel-Alumina catalysts, with an Nickel weight fraction of 50%, were synthesized by co-precipitation with $NaOH/Na_2(CO_3)$ solution using the constant pH-method at a defined temperature. After aging for several hours, washing, drying at 80°C and calcination at different temperatures the catalysts were characterized by BET, chemisorption, XRD and TEM. For comparison of the catalytic activity, the specific CO_2 conversion of the catalyst was determined stepwise in the temperature range between 175 and 500 °C. Detailed kinetic measurements were conducted with one selected catalyst. Concerning the experimental setup, a low flow line with a maximum flow of 75 sccm and a high flow line with 1500 sccm are used for gas supply. Water can be added to the feed through a temperature controlled vaporizer. The two reactor lines can be heated up to 600 °C, allowing pressures up to 20 bar. To prevent the condensation of water, all tubing of the product gas line is equipped with heating and the product stream is diluted with inert gas. For quantification a gas chromatograph, a mass spectrometry and an online process gas analyzer are used.

The kinetic data pool is recorded with 25 mg catalyst in the temperature range from 250 to 350 °C. The catalyst is diluted with purified SiC in the catalyst-to-SiC ratio of 1/9 using the same particle sizes of 0.15 – 0.20 mm. The catalyst bed is fixed with silica wool plugs in the isothermal zone of a glass lined steel tubing with an inner diameter of 4 mm. The reaction temperature is measured at the fixed-bed outlet. For reduction, the catalyst is heated to 480 °C by 2 K/min in a flow of H_2/Ar = 5/95 and the temperature is kept for 11 hours.

Results

For the synthesized Ni/Al catalysts the quantitative element analysis showed the expected Ni-to-Al ratio of approximately 0.95. Concerning XRD and BET measurements, only the calcination temperature has a significant impact. The varied synthesis parameters with no obvious influence on the prepared catalyst system include the aging time, the precipitation temperature and the precipitation pH in the range between 7 and 9.5. The XRD patterns for the calcined catalyst show reflexes at 2θ = 37.1°, 43.6°, and 62.9° identifying NiO and 44.7° and 65.3° for

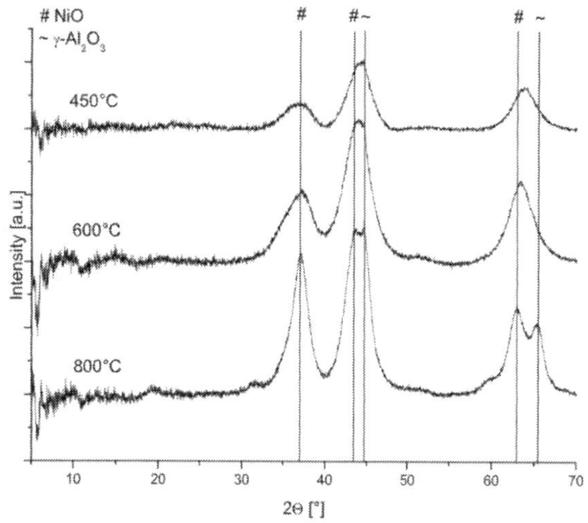

Figure 1: Influence of the calcination temperature on the catalyst`s structure.

γ-Al_2O_3, as shown in Figure 1. These discrete phases are only visible after calcination at 800 °C, for calcination at 450 °C and 600 °C broad reflexes at 37.1°, 44.2° and 63.9° are measured. The BET surface area is between 250-300 m^2/g and it decreases with higher calcination temperature. The BET surface area, measured after reduction of the catalysts, is approximately 20% lower than the one of the calcined, unreduced surface. The specific Nickel surface area of 23±2 m^2/g was obtained by H_2 chemisorption, resulting in a Nickel dispersion of 5.5-7.5 %.

To compare the activity of different Ni catalysts systems, the CO_2 conversion in the methanation reaction was determined at different temperatures in the range between 180 °C and 500 °C. Again, only the calcination temperature led to a significant difference in the activity since the activity decreases with the calcination temperature. However, the catalyst calcined at the higher temperatures is more stable against deactivation caused by temperatures higher than 450 °C. Selectivity considerations showed a selectivity greater than 99.5% to methane for all measured catalyst systems. As the main byproduct, ethane was detected in traces lower than 500 ppm.

Before measuring the kinetic data pool, validation measurements were conducted to ensure the kinetic data pool not being affected by heat or mass transfer limitations. A formation period, during which the catalyst was treated with a gas mixture containing H_2/CO_2 = 4/1, H_2O/CH_4 = 2/1 and argon at 380 °C for 300 h, assured a stable catalyst performance level during the parameter variation. The parameters for generating the kinetic data pool are: p = 2 – 10 bars, $GHSV_{catalyst\ bed}$ = 17000 - 25000 1/h and T = 250 to 350 °C. The feed contains H_2, CO_2, H_2O, CH_4 and Ar in different compositions. For all used data points the carbon, hydrogen and oxygen mass balances were closed within ± 3%. The volume contraction caused by the reaction was included in the calculations. A kinetic model was derived successfully from the first run, which is able to describe the catalyst properties over the mentioned pressure and temperatures between 250 °C and 350 °C. A parity plot comparing these model predictions to the experimental data is shown in Figure 2, with the average absolute residual being about 7.4%.

Conclusion & Outlook

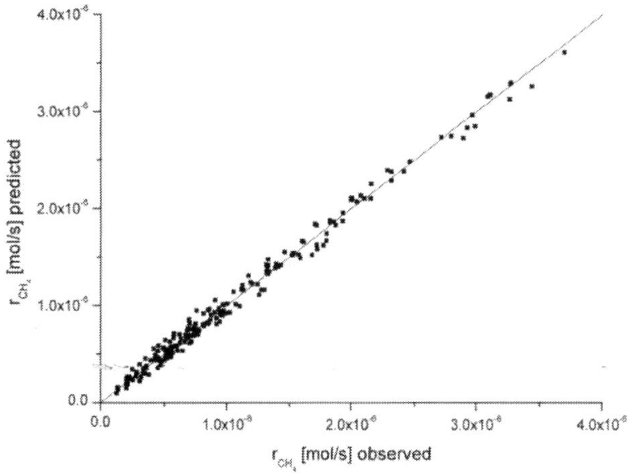

Figure 2: Parity plot for calculated and measured reaction rates.

In this study a Ni/Al catalyst was successfully synthesized by precipitation and characterized by BET, H_2 chemisorption and XRD. Kinetic measurements and modeling approaches resulted in a model describing the catalysts activity over a wide range of industrially relevant reaction conditions. In an additional kinetic study we will analyze the lower temperature range between 180 °C and 250 °C.

In upcoming experiments we will promote our Ni/Al Systems with different other metals and investigate the effects on activity and stability. Furthermore, different Ni/Al ratios will be synthesized and analyzed. An additional focus will be set on identification of structure activity relations to gain a deeper understanding of the catalyst system. Hence, the high selectivity of the synthesized catalyst facilitates a future process design to use CO_2 conversion for energy storage.

Acknowledgement

This project is financed by the BMBF within the iC^4-Integrated Carbon Capture Conversion & Cycling project.

Literature

[1] P. Sabatier, J.B. Senderens, CR Acad. Sci, 1902, 134, 689

[2] W. Wang, *Recent advances in catalytic hydrogenation of carbon dioxide*, Chem. Soc. Rev., 2011, 40

[3] J.-N. Park, E.W. McFarland, *A highly dispersed Pd-Mg/SiO$_2$ catalyst for methanation of CO$_2$*, J. Catal., 2009, 266

[4] G.D. Weatherbee, C.H. Bertholomew, *Hydrogenation of CO$_2$ on Group VIII Metals: II. Kinetics and Mechanism of CO$_2$ Hydrogenation on Nickel*, J. Catal., 1982, 74, 460

[5] J. Xu, G.F. Froment, *Methane steam reforming, methanation and water gas shift. I. Intrinsic kinetics*, AIChE Journal, 1989, 35, 88

Hydrogen-alimented fuel cells (FC) have a strong potential to play a decisive role in the new energy system for the coming years. The production of H_2 pure enough to use it in fuel cells requires the development of very efficient catalysts for the WGS reaction. In our group several gold-ceria based catalysts have been developed presenting very promising results in this process [1,2]. The successful catalytic design makes mandatory an accurate knowledge about the reaction mechanism and the active species involved in the process. In order to address these issues a combination of several *in-situ/operando* characterization techniques is performed in this work using an optimized Au/CeO_2-FeO_x/Al_2O_3 catalyst. Synchrotron-based *in-situ* time-resolved X-ray absorption spectroscopy (TR-XAS) and *operando* DRIFTS during the WGS reaction are employed with the ultimate goal to establish structure-activity relations and to propose the most likely reaction pathways.

The obtained results may be summarized as follows:

1) Oxidation state from *in-situ* XANES: Au remains as metallic gold during the reaction. Ce evolves from Ce^{4+} to Ce^{3+} (actually complete reduction of ceria is not achieved). Fe is present as Fe^{3+} since the profile of the XANES matches to the hematite structure.

2) **Presence of oxygen vacancies on ceria**. Our Raman, XRD and UV-Vis data established the existence of oxygen vacancies in the ceria lattice, especially when iron is present.

3) Gold dynamics. *In- situ* EXAFS analysis indicates the competition between gold particles and water to nucleate in the oxygen vacancies of the support. This process is associated to the observed gold nanoparticles rupture and subsequent agglomeration depending on the reaction temperature.

4) Reaction intermediates. *Operando* DRIFTS data point out that the formation of either bicarbonate or carboxylic species play a key role during the WGS.

Considering these premises, an associative mechanism where the carboxyl intermediates domains the reaction can be proposed:

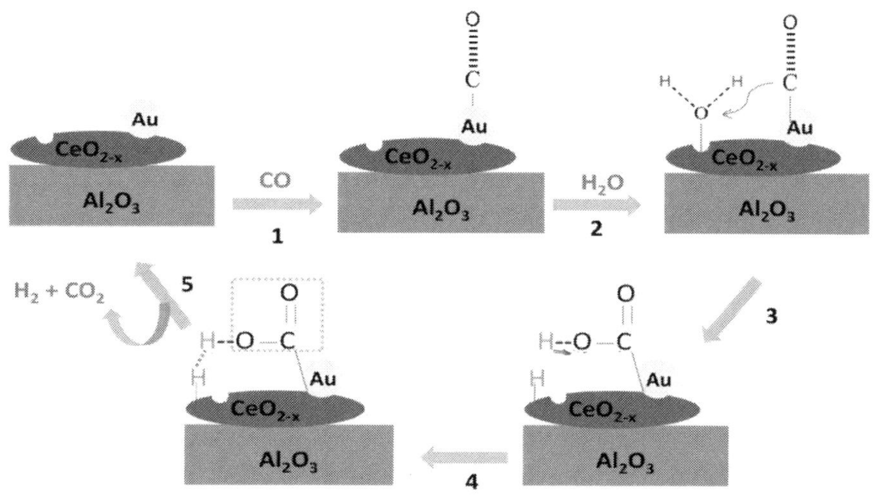

Figure 1*: Simplified WGS mechanism proposal for a Au/CeO_2-FeO_x/Al_2O_3 catalyst*

The main reactions pathways are:

1) Adsorption of CO on gold nanoparticles. 2) Activation of H_2O on the ceria oxygen vacancies. 3) The interaction between CO and activated H_2O on the metal-support interface. More specifically, this step represents the formation of the COOH intermediate due to the combination of CO and OH. This process may happen also in two steps. First the water rupture into OH and H and second the formation of the HO-CO bond. 4) Formation of the CO_2 and H_2. 5) Final release of the products.

Conclusions:

Our *in-situ/operando* study provides some relevant clues regarding the WGS mechanism and towards the catalyst design and future improvements. In particular, an associative route via carboxyl formation seems to control the process. The high catalytic performance is achieved when gold nanoparticles are in close contact with the structural defects of ceria and the oxygen mobility of the later is enhanced by the addition of promoters such as iron oxide.

References

[1] J. A. Odriozola, T. R. Reina, M. A. Centeno, S. Ivanova, V. Idakiev, T. Tabakova, Spanish Patent 201101163.

[2] M. González, T.R.Reina, S. Ivanova, M. A.Centeno, J.A. Odriozola J. Catal. 314 (2014) 1-9

382126 Fe-Promoted Au/Rutile for the Water-Gas Shift Reaction

Abstract:

The water-gas shift reaction (WGS) is an important reaction for hydrogen production and fuel cell applications[1], which are important for renewable energy. Au supported on oxides has been identified as an active catalyst for WGS. Supports play an important role in not only changing the WGS rates, but also modifying the types and chemical nature of the active sites[2,3]. Previous work has shown that Fe doping can promote the CO oxidation rate over Au/TiO_2 catalyst[4]. Here, a series of Fe-doped rutile (TiO_2) catalysts ranging from 0.25wt% to 5wt% Fe were loaded with ~1wt% Au and tested for WGS. WGS rates per mole of Au at 120 °C showed maximum promotion by a factor of 4 for 1wt% Fe content. Significant changes in kinetic parameters were observed with varying Fe content. Apparent orders with respect to CO decreased progressively and apparent activation energy increased progressively with Fe content increasing from 0 wt% to 5 wt%. As CO order decreased, WGS rates first increased, reached maximum at CO order about 0.3 and then decreased. From previous work, WGS rates over supported Au are influenced by the Au particle sizes[2,3], but for the catalysts reported here, similar average particle sizes (2.0±0.5 nm) obtained from Transmission Electron Microscopy (TEM) excluded the particle size effects as the source of the observed changes. Operando Fourier Transform Infrared Spectroscopy (FTIR) was performed on the 1wt% Fe-doped catalyst and the result was compared with Au supported on un-doped rutile. IR peaks at wavenumbers of 2097 cm^{-1}, 2044 cm^{-1}, 1967 cm^{-1} and 1888 cm^{-1} were observed. The peak at 2097 cm^{-1} can be assigned to CO adsorbed on metallic Au while the lower frequencies of 2044 cm^{-1} and 1967 cm^{-1} can be associated with CO adsorbed on partially negatively charged Au. The IR bands at lower frequencies were stronger than the corresponding peaks of Au supported on un-doped rutile and thus suggest a participation of Fe in the neighborhood of the site. Thus, the results showed that Fe-doping can modify the CO adsorption properties of Au/Rutile and that WGS rates can be correspondingly promoted with appropriate Fe content.

Elucidating the Fundamental Structure-Performance Relationships of H_2 Production by Cu/Cr/Fe Water-Gas Shift (WGS) Catalysts

Minghui Zhu,[a] Tulio C. R. Rocha,[b] Robert Schlogl[b] and Israel E. Wachs[a]

[a] *Operando Molecular Spectroscopy and Catalysis Research Laboratory, Chemical Engineering Department, Lehigh University, Bethlehem, PA 18015 USA.*

[b] *Fritz-Haber-Institut der Max-Planck-Gesellschaft, Department of Inorganic Chemistry, Faradayweg 4–6, 14195, Berlin, Germany.*[11]

Industrial production of hydrogen is mostly achieved by the water-gas shift (WGS) reaction ($H_2O + CO \leftrightarrow H_2 + CO_2$). To shift the equilibrium of the exothermic reaction to the right and maximize hydrogen production, the catalytic WGS reaction is performed at two temperature ranges. The high temperature WGS reaction stage was traditionally performed with the unsupported Cr_2O_3-Fe_2O_3 mixed oxide catalyst. In recent years, the addition of Cu to the Cr_2O_3-Fe_2O_3 mixed oxide catalyst was found to increase the performance of the HT WGS catalyst, but the promotion mechanism of Cu is still not well understood because of the lack of characterization studies under reaction conditions.

To advance the fundamental understanding of the promotion of Cu for this important industrial reaction for hydrogen production, a series of CuO-Cr_2O_3-Fe_2O_3 catalysts were synthesized and investigated with modern *in situ* (TPR, TP-DRIFTS and Raman) and *operando* (Raman-MS, XANES/EXAFS-MS and near ambient pressure XPS-MS) spectroscopy studies during the HT WGS reaction. The unsupported Fe_2O_3, Cr_2O_3-Fe_2O_3 and CuO-Cr_2O_3-Fe_2O_3 catalysts were synthesized by the co-precipitation method from the corresponding nitrates. The freshly calcined CuO-Cr_2O_3-Fe_2O_3 catalyst consists of surface Cr^{6+} species and a solid solution of Cr^{+3} and Cu^{+2} in the Fe_2O_3 bulk lattice. During the HT WGS reaction, surface Cr^{6+} reduces to Cr^{3+} and further dissolved in the iron oxide lattice, bulk $Fe_{2-x}Cr_xO_3$ transforms to bulk $Fe_{3-x}Cr_xO_4$ and Cu^{2+} reduces to large metallic Cu^0 particles. Redox experiments confirm that Fe and Cu are active sites during the HT WGS reaction, while the role of Cr is to stabilize the surface area of the $Fe_{3-x}Cr_xO_4$ component by retarding sintering. Addition of Cu promotes both the initial bulk transformation of $Fe_{2-x}Cr_xO_3$ to $Fe_{3-x}Cr_xO_4$ and, more importantly, the redox properties of the catalyst[12].

385402 Water-Gas Shift Reaction on Catalysts Prepared Using Supercritical Fluid Deposition

Abstract:

Catalysts composed of noble metals and reducible metal oxides are of great significance for modern energy applications. These applications include key reactions in hydrogen production such as water-gas shift (WGS) and preferential oxidation (PROX), as well as fuel cell electrodes. A major impediment to implementing hydrogen fuel cell technology is the need for safe, reliable, and efficient catalytic hydrogen generation for logistical small-scale operations. WGS is a crucial step in hydrogen enrichment, and effective catalysts that have been identified are platinum or gold in conjunction with ceria or other partially reducible oxides. For economic feasibility, minimizing the amount of platinum or gold required is vital. Recent research provides evidence that catalysts having extremely low loadings of these metals can still exhibit high activity, as the degree of reaction rate enhancement depends on interfacial area between active metal sites and oxides. Thus, methods of synthesizing supported catalysts with intimate contact between the two are essential.

This paper focuses on the synthesis and characterization of Pt/ceria WGS catalysts prepared using supercritical fluid deposition (SCFD), a technique by which organometallic precursors are adsorbed from a supercritical fluid solvent on to a support. Supercritical fluids (SCFs) offer unique advantages for interfacial deposition processes. Liquid-like density promotes dissolution of metal precursors; viscosity and diffusivities more akin to those of gases facilitate both transfer through the SCF and surface adsorption. These benefits make possible the reproducible, uniform distribution of organometallic precursors on the support. In our work, the support is high-surface area alumina to which ceria has been added by insipient wetness followed by calcination. The platinum is deposited using SCFD of platinum acetylacetonate from carbon dioxide, with subsequent calcination in air and reduction in hydrogen. The nominal loadings are 5% ceria and 1% platinum. Variations in preparation include reductive pretreatment of the ceria before Pt deposition, intended to disrupt the surface with oxygen vacancies that will ultimately enhance the deposition of Pt precursor on ceria-rich areas of the alumina surface.

Results of surface analyses, including temperature programmed reduction and carbon monoxide pulse titrations, indicate that the preparation results in both small nanoparticles (1-3 nm) and highly dispersed platinum. Further evidence for this is obtained using transmission electron microscopy. Catalyst activity testing takes place a flow reactor system using a simulated reformate stream containing CO, H_2, H_2O, and N_2. Turnover frequencies and activation energies for WGS on catalysts will be presented and correlated with variables in synthesis method.

389936 Alkali-Promoted Atomic Pt-Ohx Species on "Active" or "Inert" Supports Catalyze the Low-Temperature Water-Gas Shift Reaction

Abstract:

The water-gas shift (WGS) is an important reaction for hydrogen upgrading during fuel gas processing. Atomically dispersed Pt-O_x-support sites have been found to catalyze the low temperature WGS reaction on "active" supports, such as CeO_x [1]. However, on "inert" supports, such as alumina and silica, precious metal atoms are unstable and easily agglomerate into spectator nanoparticles. For such surfaces, we have recently shown that alkali metal oxides can be used as promoters of the platinum on silica [2], alumina [2], and even on carbon nanotubes [3], such that the high activity of the atomically dispersed Pt-OH_x- species is preserved. Here we have extended this work to other "active" and "inert" supports, including anatase, K type L-zeolites (KL), and mesoporous MCM-41. Through the use of alkali hydroxides, active and stable (up to 673 K) platinum catalysts were developed in this work. While the catalysts show large differences in activity without sodium promotion (Pt/Anatase> Pt/KL> Pt/MCM-41 ≈0 below 573 K), the sodium modified samples show the same, considerably improved activity in both product-free and full gas conditions for the same metal loading (0.5 wt.% Pt) on any of these supports. Modifying only the supports with alkali metals does not improve the catalytic activity if the Pt-O-alkali interaction is not established before calcination. The metal atom-centric Pt-OH_x-O_y-Na_zcluster is the active site, with the same intrinsic activity on any support. XPS analysis shows that platinum in these structures is close to Pt(II), same as for the alkali-promoted platinum on silica and alumina [2]. The addition of alkali metals does not change the apparent activation energy for this reaction. The presentation will include characterization results as well as stability evaluation under various operating conditions.

Acknowledgement

The financial support by the DOE/BES under Grant # DE-FG02-05ER15730 is gratefully acknowledged.

References

1. Fu Q, Saltsburg H, Flytzani-Stephanopoulos M, *Science* **2003**, 301, 935.
2. Zhai Y, Pierre D, Si R, Deng W, Ferrin P, Nilekar AU, Peng G, Herron JA, Bell DC, Saltsburg H, Mavrikakis M, Flytzani-Stephanopoulos M, *Science* **2010**, 329,1633.
3. Zugic B, Zhang S, Bell D, Tao F, Flytzani-Stephanopoulos M, *J. Am. Chem. Soc.* **2014**, 136, 3238.

365037 Density Functional Theory Study of CO2 Electrochemical Reduction on the Fe(100) Surface

Abstract:

Carbon dioxide electroreduction o ers the possibility of producing hydrocarbon fuels using energy from renewable sources. In this presentation, we discuss the use of density functional theory to analyze the feasibility of CO_2 electroreduction on a Fe(100) surface. Experimentally, iron is nonselective for hydrocarbon formation, despite being active for Fischer-Tropsch catalysis and thermal carbon dioxide reduction by H_2. A simplistic analysis of low-coverage reaction intermediate energies for the paths to produce CH_4 and CH_3OH from CO_2 suggests Fe(100) could be more active than Cu(111), currently the only metallic catalyst to show selectivity towards hydrocarbon formation. We consider a series of impediments to CO_2 electroreduction on Fe(100) including O*/OH* (* denotes surface bound species) blockage of active surface sites; competitive adsorption e ects of H*, CO* and C*; and iron carbide formation. Our results indicate that under CO_2 electroreduction conditions, Fe(100) is predicted to be covered in C* or CO* species, blocking any C–H bond formation. Further, bulk Fe is predicted to be unstable relative to FeC_x formation at potentials relevant to CO_2 electroreduction conditions. The analysis of the causes of inactivity of Fe catalysts provides insight into requirements for active and selective CO_2 electroreduction catalysts.

379720 Thermochemical Descriptors for Unraveling Molecular Promoted CO_2 Conversions

Abstract:

There is substantial interest in the development of renewable energy processes that convert CO_2 into more valuable products such as CO, formate, and other hydrocarbons. Independent reports in the recent literature have shown that heterocycles of nitrogen can promote selective CO_2 conversion into these products. While collaborative efforts involving theory and experiment are reaching consensus regarding some aspects these processes, numerous questions remain, including the precise mechanistic pathway that fundamentally explains the role of the heterocycles. Using first principles quantum chemistry calculations, we focus on individual reaction mechanisms involving non-substituted and substituted pyridinium, imidazolium, and their 1- and 2-electron reduced products with CO_2. Our calculated standard redox potentials, pK_as, and kinetic barrier heights are used to elucidate which chemical species may be present under *in situ* electrochemical conditions in solution and on metal electrodes. From this work, we rationalize different mechanistic possibilities warranting further scrutiny and give outlook for molecular-promoted (photo-)electrocatalysis.

A DFT Study of Partial Oxidation of Methanol on MoO_3 (010)

1. Tej Choksi, 2. Jeffrey Greeley[1*]

[1]School of Chemical Engineering, Purdue University, West Lafayette, IN 47907

Reducible metal oxides possess a fascinating range of chemical and catalytic properties and are thought to function as effective catalysts for partial oxidation reactions, among other applications. A particular system that has received attention in the literature is the partial oxidation of methanol on MoO_3 (010), which is believed to proceed via a Mars – van Krevelen mechanism, with lattice oxygen functioning as an oxidant (1), (2). The (010) surface of MoO_3 possesses a bilayered structure of MoO_6 octahedra with the bilayers bound through dispersion interactions (3). The reaction network begins with chemisorption and subsequent dissociation of methanol at a surface oxygen vacancy. Hydrogen released through C-H and O-H scission (4), (5) can either form a H_yMoO_{3-x} bronze (6), (7) or participate in formation of water; desorption of which leads to creation of surface vacancies. Oxygen defects can undergo rearrangement through crystallographic shear planes forming extended surface defects (8).

In this work, methanol oxidation thermochemistry, along with the morphological transformation of (010) MoO_3, have been studied using periodic density functional theory calculations. Methanol dissociation begins with C-H or O-H scission, leading to two distinct pathways, with the O-H route being more favorable. Competition among dehydrogenation of CH_3O^*, CH_2O^*, CHO^* and CH_2O^* desorption and their implications to product selectivity are studied. Rearrangement of species formed through C-H dissociation to those originating from O-H scission remains a distinct possibility due to favorable kinetic barriers. Facile surface diffusion barriers for CH_2O^* and CHO^* further imply that bimolecular reactions may play a role in the overall mechanism. The dependence of choice of functional (optB88, optPBE, PBE, RPBE, BEEF-vdw and HSE06) on the kinetics and thermodynamics of methanol oxidation has also been examined with particular emphasis on understanding the role of van der Waal's corrections. Surface re-oxidation mechanisms involving O_2^* dissociation, diffusion and penetration into the bulk yield relatively high barriers. Finally, low diffusion barriers of hydrogen along the surface and towards the bulk, hints at the plausibility of hydrogen bronze formation.

(1) Farenth, W. E.; Ohuchi, F; Staley, R. H; Chowdhry, U; Sleight, A. W.; *J. Phys. Chem.* **1985**, *89*, 2493.
(2) Brookes, C.; Wells, P. P.; Cibin, G.; Dimitratos, N.; Jones, W.; Morgan, D. J.; Bowker, M.; *ACS Catal.* **2014**, *4,* 243.
(3) Chen, M.; Waghmare, U. V.; Friend, C. M.; Kaxiras E.; *J. Chem. Phys.* **1998**, *109*, 6854.
(4) Greeley, J.; Mavrikakis, M.; *J. Am. Chem. Soc.* **2002**, *124*, 7193.
(5) Kandoi S.; Greeley, J.; Sanchez-Castillo, M. A.; Evans, S. T.; Gokhale, A. A.; Dumesic, J. A.; Mavrikakis, M; *Top. Catal.* **2006**, *37,* 17.
(6) Smith, R. L.; Rohrer, G. S.; *J. Cat.* **1999**, *184*, 49.
(7) Chen, L.; Cooper, A. C.; Pez, G. P.; Cheng, H.; *J. Phys. Chem. C.* **2008**, *112,* 1755.
(8) Smith, R. L.; Rohrer, G. S.; *J. Solid State Chem.* **1996**, *124*, 104.

387164 First Principles Modeling of Surface Plasmon Dynamics and Mechanism for Photo-Catalytic Rate Enhancement

Abstract:

We have recently reported that optically excited plasmonic nano-particles can activate photo-chemical transformations.[1,2] These reactions exhibit a number of unique characteristics fundamentally different than other photo-catalysts and catalytic reactions. In this contribution we have used quantum chemical approaches to model the plasmon-driven photo-chemical reactions on metals.[3–6] I will discuss the time-dependent evolution of the electronic structure of optically excited plasmonic metals. Furthermore, I will show how this electronic structure interacts with adsorbates, ultimately driving chemical transformations of the adsorbates. A number of conflicting mechanisms will be discussed and analyzed in the context of previously reported experimental results.

REFERENCES

[1] P. Christopher, H. Xin, A. Marimuthu, S. Linic, Singular characteristics and unique chemical bond activation mechanisms of photocatalytic reactions on plasmonic nanostructures., Nat. Mater. 11 (2012) 1044–50.

[2] S. Linic, P. Christopher, D.B. Ingram, Plasmonic-metal nanostructures for efficient conversion of solar to chemical energy., Nat. Mater. 10 (2011) 911–21.

[3] L. Genzel, T.P. Martin, U. Kreibig, Dielectric function and plasma resonances of small metal particles, Zeitschrift Für Phys. B Condens. Matter Quanta. 21 (1975) 339–346.

[4] S. Monturet, P. Saalfrank, Role of electronic friction during the scattering of vibrationally excited nitric oxide molecules from Au(111), Phys. Rev. B. 82 (2010) 075404.

[5] V. Krishna, J.C. Tully, Vibrational lifetimes of molecular adsorbates on metal surfaces., J. Chem. Phys. 125 (2006) 054706.

[6] T. Olsen, J. Schiotz, Quantum corrected Langevin dynamics for adsorbates on metal surfaces interacting with hot electrons., J. Chem. Phys. 133 (2010) 034115.

[7] S.I. Anisimov, B. Rethfeld, Theory of ultrashort laser pulse interaction with a metal, in: V.I. Konov, M.N. Libenson (Eds.), Nonreson. Laser-Matter Interact., International Society for Optics and Photonics, 1997: pp. 192–203.

[8] Z. Lin, L. Zhigilei, V. Celli, Electron-phonon coupling and electron heat capacity of metals under conditions of strong electron-phonon nonequilibrium, Phys. Rev. B. 77 (2008) 075133.

388001 Understanding Adsorbate-Specific Photocatalysis on Metal Surfaces

Abstract:

Visible photon excitation of catalytic metal surfaces has been proposed as a potential approach for inducing selective chemical reactions that are difficult to achieve using thermal energy. This approach relies on the selective deposition of photon energy into the activation of targeted adsorbate-metal bonds. Laser photo-excitation has been used for many years to induce chemical reactions on metal surfaces; however the mechanism underlying the photon-induced reactivity is still not well understood. Particularly, there is no understanding of how the efficiency of photon-induced chemical reactivity is impacted by the nature of the adsorbate-metal bond, which is critical to understand approaches to controlling selectivity.

In this work we combine dynamical models of the processes of metal substrate photo-excitation and charge transfer induced activation of adsorbate-metal bonds to understand the impact of the adsorbate-metal bond nature, temperature, illumination intensity and wavelength on the potential for controlling selectivity through photo-excitation of metal catalysts. We have employed an extended two-temperature model (ETTM) to describe the dynamics of non-equilibrium electron distribution in photo-excited metal substrates. The coupling between energetic electrons in the metal and adsorbate electronic states is treated using a first principles inelastic electron scattering model recently developed by Schiøtz[1]. By integrating the two approaches we can calculate the temporal evolution of the system after laser excitation and understand the impact of wavelength on the efficiency of chemical reactions.

We specifically examined the activation of metal-adsorbate bonds formed between CO, NO and O and a Pt (111) surface. Our efforts have been aimed at answering debated mechanistic questions such as the role of non-thermalized electrons (those that do not fall within the Fermi-Dirac distribution) in driving reactions and the time scale of the elementary processes involved in the chemical reactions. We have also addressed the impact of temperature in the photocatalytic processes on metal surfaces. In all of these cases, we focus on how different adsorbate-metal bonds exhibit unique behavior. We are able to explain the previously observed adsorbate specific behavior in experimental studies and provide insights into approaches that could allow for control of selectivity through photo-excitation of heterogeneous metal catalysts.

References:

1- Olsen, T., Gavnholt, J. & Schiøtz, J. Hot-electron-mediated desorption rates calculated from excited-state potential energy surfaces. *Phys. Rev. B* **79,** 035403 (2009)

387617 Dehydrogenation of Water Species on Ni Catalysts from First Principles: Investigations of Electric Field Influence

Abstract:

Solid oxide fuel cells will no doubt find more widespread use in the new millennium in the conversion of chemical to electrical energy offering, as they do, very high efficiencies and a unique scalability in electricity-generation applications [1]. To control the ongoing reactions in such a device one can modify the selectivity of the catalyst itself, which oxidizes the fuel at the anode and dissociates O_2 at the cathode. The selectivity of the catalyst in a solid oxide fuel cell will be altered by the presence of a huge electric field of the order of 1 Volts per Angstrom, which occurs near the electrode-electrolyte interface [2]. Experimentally, Sekine et al. [3-5] studied the electrocatalytic steam reforming of methane in the presence of an electric field. They found that in the presence of an electric field the conversion of methane over Pd/CeO_2, Rh/CeO_2 and Pt/CeO_2 increased. In particular, they found that the conversion of methane occurred at 423K with an external electric field, which is a temperature where a conventional catalytic reaction would hardly proceed. Given this background, it is surprising that the role of the electric field is often ignored in the optimization of such devices.

In order to provide insight into the electrocatalytic mechanisms of the hydrocarbon steam reforming reaction on Ni surfaces, an investigation of how an external electric field influences on the dehydrogenation of water species on Ni surfaces was carried out using density functional theory calculations. We found that an external electric field significantly changed the adsorption structures and charge distribution of water species on the Ni(111) and (211) surfaces. As a result, the electric fields greatly influenced the adsorption energies of water species and reaction barriers of dehydrogenation processes. The electric field effect on the energy barriers is much larger than that of our previous work on the methane dehydrogenation [6, 7]. Particularly for Ni(111), when we applied a positive electric field, the energy barrier of the dehydrogenation of OH increases by 0.3 eV as compared to when an electric field is absent. Our results suggest that a positive electric field significantly suppresses the dehydrogenation of water species on Ni catalysts.

1. Nikolla, E., J. Schwank, and S. Linic, *Direct Electrochemical Oxidation of Hydrocarbon Fuels on SOFCs: Improved Carbon Tolerance of Ni Alloy Anodes.* J. Electrochem. Soc., 2009. **156**(11): p. B1312-B1316.

2. Stuve, E.M., *Ionization of water in interfacial electric fields: An electrochemical view.* Chem. Phys. Lett., 2012. **519-520**: p. 1-17.

3. Sekine, Y., M. Haraguchi, M. Tomioka, M. Matsukata, and E. Kikuchi, *Low-Temperature Hydrogen Production by Highly Efficient Catalytic System Assisted by an Electric Field.* J. Phys. Chem. A, 2009. **114**(11): p. 3824-3833.

4. Sekine, Y., M. Tomioka, M. Matsukata, and E. Kikuchi, *Catalytic degradation of ethanol in an electric field.* Catal. Today, 2009. **146**(1–2): p. 183-187.

5. Sekine, Y., M. Haraguchi, M. Matsukata, and E. Kikuchi, *Low temperature steam reforming of methane over metal catalyst supported on CexZr1−xO2 in an electric field.* Catal. Today, 2011. **171**(1): p. 116-125.

6. Che, F., R. Zhang, A.J. Hensley, S. Ha, and J.-S. McEwen, *Density functional theory studies of methyl dissociation on a Ni(111) surface in the presence of an external electric field.* Phys. Chem. Chem. Phys., 2014. **16**(6): p. 2399-2410.

7. Che, F., A. Hensley, S. Ha, and J.-S. McEwen, *Decomposition of Methyl Species on a Ni(211) surface: Investigations of the Electric Field Influence.* Catal. Sci. and Technol. (submitted), 2014.

Atomistic Simulations of Photocatalytic Water Splitting on Pt/TiO_2

There is a significant amount of experimental research on the production of hydrogen through photocatalytic water splitting. Of existing research, titanium dioxide has been the subject of study as the catalyst due to it being inexpensive and abundant. However, the deposition of more expensive transition metals such as platinum on titanium dioxide, which enhances the photocatalytic activity, made large scale production costly. We develop a nanoscale model of the interface between titanium dioxide and water to simulate the molecular behavior which leads to water decomposition. Rutile and anatase forms of titanium dioxide are simulated for water adsorption using atomistic molecular dynamics in LAMMPS. Platinum atoms supported on the surfaces represent the co-catalyst. Properties such radial distributions functions, diffusion coefficients, and catalyst coordination residence time are used to characterize the transport phenomena of the system. The results are then used to provide parameters and develop a Kinetic Monte Carlo model for the water splitting reaction.

364854 Mechanistic Studies on Hydrodeoxygenation of Furfural to 2-Methyl Furan Via Ring C-H Bond Activations

Abstract:

Removal of excess oxygen-containing functional groups in furanic compounds, e.g. furfural and 5-hydroxymethylfurfural to 2-methyl furan and dimethylfuran, is a key step in the upgrade of cellulosic and hemicellulosic biomass-derived molecules. Mildly oxidized Ru/C catalyst shows remarkable activity and selectivity towards the hydrodeoxygenation reactions; however, molecular level understanding on the catalyst surface-mediated chemical transformation is still lacking. We show that both metal-mediated hydrogenation and Lewis acid-mediated MPV reaction are responsible for hydrogenating C=O to C-OH. In addition, isotopic labeling experiments with detailed mass fragmentation analysis demonstate that ring C-H bond activations are necessary for the C-OH bond scission.

375454 Theoretical Investigation of the Hydrodeoxygenation of Levulinic Acid to g-Valerolactone over Ru Catalysts

Abstract:

Due to diminishing reserves of fossil fuels and a renewed emphasis on green technology for energy production, there is a desire for development of second generation bio-fuels that possess high energy density and offer low net carbon emissions. In the past decades massive efforts were given to conversion of lignocellulosic biomass into usable fuel and levulinic acid (LA) was found to be one of the promising biomass derived platform molecules that can be readily produced and converted into various fuel derivatives. One conversion of significant interest has been the aqueous phase hydrodeoxygenation of LA to g-valerolactone (GVL) since GVL is a similarly versatile platform molecule to LA that is at the same time less water soluble than LA which facilitates its separation from the aqueous solution LA are produced in. Conversion of LA to GVL requires a complex hydrodeoxygenation (HDO) process over transition metal catalysts whose mechanism is experimentally difficult to study due to concurrent catalyst deactivation and mass transfer limitations at practical operating conditions.

In this project, we studied the mechanism of the heterogeneous HDO of LA to GVL over Ru (0001) model surfaces from first principles. First, we studied various elementary processes at the gas-solid interface with PBE-D3 level of theory. Then, we applied our novel implicit solvation method for solid surfaces (iSMS) to determine elementary reaction and activation free energies in aqueous phase environments. Finally, we developed a microkinetic model to predict reaction rates, activation barrier and reaction orders. A challenge during this investigation has been that we significantly underpredicted turnover frequencies in comparison to experimental observations. We identified as the reason for our underprediction the large number of sites occupied by an individual adsorbed LA molecules while only a small number of free sites are available for reaction. A reasonable agreement with experimental observations could be obtained when considering lateral interactions and forcing LA to adsorb on fewer sites.

Controlling Reaction Pathways of C3 Oxygenates using Non-precious Metal Bimetallic Surfaces

MyatNoeZin Myint, Yushan Yan and Jingguang G. Chen

Fundamental studies of reactions of oxygenates on transition metal surfaces are important for better understanding of many areas of heterogeneous catalysis and for numerous industrial processes. The generation of hydrogen from biomass-derived oxygenate compounds is a desirable and viable solution to a sustainable environment.[1,2] Glycerol, commonly referred to as an "alcohol sugar", is a byproduct in the biodiesel production and different types of glycerol reforming processes have been extensively studied.[1,3] However, finding an efficient and stable non-precious metal catalyst for hydrogen production from glycerol remains a great challenge.

Since propanol has been considered as a good surrogate molecule for glycerol, in this work, 1-propanol has been studied as a model compound.[4] Another C3 oxygenate, propanal, has also been studied to determine how the presence of the C=O bond affects the reaction pathway. In this presentation, the adsorption and reaction of 1-propanol and propanal will be compared on different non-precious metal bimetallic surfaces using density functional theory (DFT), temperature programmed desorption (TPD) and high resolution electron energy loss spectroscopy (HREELS) as probe reactions for reforming of oxygenates for hydrogen production. Based on binding energy trends, reactivity, selectivity and reaction intermediates, unique reaction mechanisms for each bimetallic surface are proposed.[5,6]

1. Alonso, D. M.; Wettstein, S. G.; Dumesic, J. A., Bimetallic Catalysts for Upgrading of Biomass to Fuels and Chemicals. Chemical Society Reviews 2012, 41 (24), 8075-8098.
2. Chen, D.; He, L., Towards an Efficient Hydrogen Production from Biomass: A Review of Processes and Materials. ChemCatChem 2011, 3 (3), 490-511.
3. Vaidya, P. D.; Rodrigues, A. E., Glycerol Reforming for Hydrogen Production: A Review. Chemical Engineering & Technology 2009, 32 (10), 1463-1469.
4. Lobo, R.; Marshall, C. L.; Dietrich, P. J.; Ribeiro, F. H.; Akatay, C.; Stach, E. A.; Mane, A.; Lei, Y.; Elam, J.; Miller, J. T., Understanding the Chemistry of H2 Production for 1-Propanol Reforming: Pathway and Support Modification Effects. ACS Catalysis 2012, 2 (11), 2316-2326.
5. **Myint**, M.; Yan, Y.; Chen, J. G., Reaction Pathways of Propanal and 1-Propanol on Fe/Ni(111) and Cu/Ni(111) Bimetallic Surfaces. *The Journal of Physical Chemistry C* **2014**, DOI: 10.1021/jp501208q.
6. **Myint**, M.; Chen, J. G., Tuning C□C and C□O/C=O Bond Cleavage of C3 Oxygenates via Metal-Modified Mo(110). (*In preparation*)

378799 Role of Lewis and Brønsted Acid Sites in the Dehydration of Glycerol over Niobia

Abstract:

Glycerol is a by-product of biodiesel production. It can be dehydrated to form acrolein, which is an important intermediate in the chemical industry. However, there is limited insight on the mechanism to optimize the reaction process. The role of Lewis and Brønsted sites in the dehydration of glycerol on niobium oxide and Na^+-exchanged niobium oxide is investigated using FTIR spectroscopy supported by DFT calculations. Glycerol is impregnated on the catalysts at room temperature using an *ex-situ* method. Under high vacuum conditions, glycerol forms a stable multidentate alkoxy species through its primary hydroxyl groups with the Lewis acid sites. The proton of one of the primary alcohol groups dissociates to form a bridging alkoxy bond with two coordinatively unsaturated metal atoms, while the other primary OH group coordinates non-dissociatively to one of these metal atoms. The remaining secondary OH group forms a hydrogen bond to a surface oxygen atom. This multidentate surface species is irreversible in the presence of water vapor.

When coordinated this way, the primary C-O bond involved in forming the bridging alkoxy bond is activated for dehydration to form 2-propene-1,2-diol. From DFT calculations, the adsorption of 2-propene-1,2-diol interacts with the surface via its primary OH group and alkene group, while its secondary OH group is tilted away from the surface, preventing its interaction with acid sites. In contrast, dehydration of the secondary alcohol group fn glycerol is favored over Brønsted acid sites in the absence of steric constraints. The primary product of this reaction, 1,3-propenediol, tautomerizes to form 3-hydroxypropionaldehyde. The latter is a very reactive and unstable molecule. It can further dehydrate to acrolein with the aid a Brønsted acid site. These pathways are supported by the linear correlation between the selectivity to hydroxyacetone and acrolein, and the acid ratio of Lewis and Brønsted acid sites, respectively. When more than a monolayer of glycerol is impregnated on niobia, monoaromatic compounds are also formed on the surface upon heating.

368719 Oxygenate Chemistry on Platinum-Molybdenum Bimetallic Surfaces

Abstract:

Biomass-derived compounds are highly functionalized, making them difficult to selectively process into a desired product because they can interact with the surface in multiple ways. Addition of oxophilic metals such as molybdenum to supported catalysts has been shown to facilitate C-O bond activation and influence selectivity for a number of deoxygenation reactions. However, the mechanism for the improvements is not known, hampering efforts to design better catalysts. Surface science techniques using single crystals can provide a fundamental understanding of the role of oxophilic metals as promoters to platinum group catalysts for deoxygenation reactions.

Our recent studies have focused on how oxophilic modifiers influence oxygenate chemistry. Temperature programmed desorption (TPD), Auger electron spectroscopy, and X-ray photoelectron spectroscopy have been used to investigate surface chemistry of simple probe molecules (hydrogen, oxygen, CO, water) as well as model compounds for biorefinery streams (2-methoxyethanol, cresol) on unmodified and Mo-coated Pt(111) surfaces. The results indicate that modifier sites influence the adsorption energies of both oxygenates and their decomposition intermediates. These results have been used to help explain observed trends from complementary investigations on supported bimetallic PtMo catalysts that show improved hydrodeoxygenation selectivity.

Conversion of biomass to liquid fuels and valuable chemicals requires a large hydrogen input. Formic acid, produced as a byproduct of many of these upgrading reactions, catalytically decomposes to produce hydrogen and carbon dioxide or water and carbon monoxide. On Pt, reported dehydrogenation selectivity is near 100% (1, 2). In addition to its potential for hydrogen production, formic acid has long been used as a probe reaction on metal and oxide surfaces (3). Most recently, formic acid decomposition has been studied on transition metal surfaces of various crystallographic orientations using first principles calculations and microkinetic modeling (4, 5). In this study we investigated the kinetics of formic acid decomposition during gas phase contact on Pt(111) and (100) single crystals and a polycrystalline Pt foil using an ambient pressure batch reactor connected to an ultra-high vacuum (UHV) chamber. The single crystals and foil were transferred between the reaction cell and UHV chamber without air exposure. Surface composition was monitored before and after reaction by x-ray photoelectron spectroscopy and Auger electron spectroscopy.

Under standard reaction conditions (1% HCOOH, 1.875% H_2, 1.875% CO, total pressure = 800 Torr, 220°C), measured apparent activation energies for Pt(100), Pt(111), and Pt foil were 86 ± 11, 102 ± 14, and 94 ± 12 kJ mol^{-1}, respectively. Measured formic acid orders were fractional: 0.35 ± 0.10, 0.43 ± 0.03, and 0.40 ± 0.07 for Pt(100), Pt(111), and Pt foil, respectively; while CO orders were -0.35 ± 0.03, -0.40 ± 0.03, and -0.43 ± 0.05. H_2 and H_2O orders were approximately 0 on Pt(100) and Pt(111). Average turnover rates at 220°C were 3.7 ± 1.0, 2.6 ± 0.6, and 3 ± 2 s^{-1} on Pt(100), Pt(111), and Pt foil, respectively. The areal rate measured on the Pt foil is within error of the areal rates measured on both Pt(111) and (100). At standard conditions, CO was fed in excess of formic acid. At lower CO initial concentrations (P_{CO} = 2 Torr (0.25%)), the measured apparent activation energy on Pt(100) decreased to 65 ± 7 kJ mol^{-1}, in agreement with measurements between 41-72 kJ mol^{-1} for supported and bulk Pt catalysts reported in literature (1, 2, 6, 7), and the CO order was approximately 0.

Thus, no structure sensitivity between the Pt surfaces tested was detected in this study, within the precision of our measurements. CO was found to be the only product that had a significant impact on the measured kinetic parameters when the initial CO concentration was significant, consistent with the work of Columbia et al., (8) who found that CO blocks sites for formic acid decomposition on Pt(111) in ultra-high vacuum.

References:

(1) Ojeda, M., et al. *Angewandte Chemie*, **2009**, 121, 4894.
(2) Solymosi, F., et al. *Journal of Catalysis*, **2011**, 279, 213-219
(3) Sachtler, W. M. H., et al. *Actes Du Deuxieme Congres International De Catalyse*, **1960**, 831.
(4) Singh, S., et al. *AIChE Journal*, **2014**, 60, 1303.
(5) Yoo, S. Y., et al., *ACS Catalysis*, **2014**, 4, 1226.
(6) Chun, W.-J., et al. *J. Chem. Soc. Faraday Trans.*, **1995**, 91, 4161-4170
(7) Block, J., et al., *Z. Elektrochem.*, **1959**, 63, 3-12
(8) Columbia, M. R., et al., *J. Electroanalyical Chemistry*, **1993,** 345, 93.

390534 A Theoretical Understanding of Amino Acid Interaction with Graphene and Its Derivatives

Abstract:

As the fundamental biomolecular building blocks, amino acids are excellent simple models for understanding the interactions of biomaterials with surfaces and interfaces in biomedical applications. To the best of our knowledge, no theoretical study has been undertaken to study the dissociative interaction and the dynamic properties of amino acids on graphene oxide (GO). In this work, we performed *ab initio* density functional theory (DFT) calculations and reactive molecular dynamics (RxMD) simulations to study the behavior of amino acids on graphene-based surfaces, namely the pristine graphene (G), the graphene functionalized by one amine group (G-NH_2) or one carboxyl group (G-COOH), and the realistic atomistic GO models we previously developed. Pristine graphene is a hydrophobic surface while GO is considered to be hydrophilic due to the presence of planar epoxide and hydroxyl groups. The G-NH_2 and G-COOH surfaces are designed because amino acids are also composed of amine and carboxyl functional groups. The reactive interaction of amino acids with G-COOH and GO has been observed from our calculations, which is also identified in reactive adsorption measurement.

357367 Process for the Production of Tetrachloropropene

Abstract:

Major producers of refrigerant for automobile air conditioning are developing the 4th generation of refrigerants with 2,3,3,3-tetrafluororopene (HFO-1234yf) as a replacement for HFC-134a. This hydrofluoro-olefin has properties that meet European requirements of low global warming potential and no impact on the ozone layer. The most cost efficient feedstock for the production of HFO-1234yf has been identified to be 1,1,2,3-tetrachloropropene (TCPe). One potential economically advantaged raw material for TCPe is PDC (1,2-dichloropropane) which is a byproduct of propylene oxide produced by a chlorohydrin process. PDC can be chlorinated to produce Pentachloropropanes (PCP) intermediates which can be further dehydrochlorinated to TCPe. In this work, the development of a kinetic model of PDC liquid phase chlorination to PCP and a proposed process to produce TCPe will be presented.

365539 Kinetic Analysis for Synthesizing TATB: 18-Crown-6 Catalyzed Amination of 1, 3, 5-Triethoxy-2, 4, 6-Trinitrobezene with Ammonia Water

Abstract:

1, 3, 5-triamino-2, 4, 6-trinitrobenzene, commonly known as TATB, is an attractive insensitive high explosive (IHE). In order to ameliorate the reaction conditions for the synthesis of TATB, this work studied the non-pressurized liquid phase amination of 1, 3, 5-Triethoxy-2, 4, 6-Trinitrobezene (TETNB) with ammonia water using 18-crown-6 (18-C-6) as catalyst in dimethylbenzene. The key component to conduct the substitution reaction is the catalytic intermediate generated from ammonium hydroxide reacted with 18-C-6. The concentration of catalytic intermediate remained nearly constant after the induction period and increased with the increase of the amount of catalyst employed. The reaction between catalytic intermediate and organic reactant is identified as the controlling step. Factors affecting the formation of the catalytic intermediate include speed of agitation, temperature and the relative amounts of catalyst, ammonia water and TETNB. The product yield above 98% was achieved within 2 h at 40°C by using 18-C-6 as the catalyst. A kinetic model was developed for the liquid–liquid reaction. The concluded models accurately predict the amination reaction kinetics and can be used in simulation and design of reactive processes for the production of TATB.

368376 Development of a Langmuir-Hinshelwood Kinetic Model for Bisphenol-a Synthesis Catalyzed By Strong Acid Cation Exchange Resin

Abstract:

Development of a Langmuir-Hinshelwood Kinetic Model for Bisphenol-A Synthesis catalyzed by strong acid cation exchange resin

Dahai Gao, Dan Hickman, Trisha Costeux, Steve Rosenberg

Bisphenol-A (BPA) is an important chemical employed to make polycarbonate plastics and epoxy resins. This compound is synthesized by the condensation of acetone with two equivalents of phenol. A strong acid ion exchange resin catalyzes the reaction. The catalytic resin used in BPA production comprises a polystyrene polymer cross-linked with divinylbenzene and sulfonated to create the acid catalytic sites. To further improve catalytic activities and selectivities, promoters with thiol functional groups are loaded onto ion exchange catalysts. Most BPA catalyst studies have focused on the measurement of activity and selectivity at given conditions. However, limited reports describe the reaction kinetics or propose reaction mechanisms.

This study delivers a kinetic model for BPA synthesis catalyzed by ion exchange resins based on a Langmuir-Hinshelwood mechanism. Bayesian parameter estimation with batch reaction data provides reaction rate constants and adsorption coefficients. The kinetic model 1) provides support for fundamental hypotheses explaining the higher activity and selectivity of the promoted resin; 2) quantifies the extent of water inhibition on BPA synthesis for different catalysts; 3) quantifies the improvement of the promoted resin compared to non-promoted resins; and 4) quantifies changes in the activation energies of different reactions following the addition of promoters.

Isomerization of 1-decene in a liquid multiphase thermomorphic multicomponent solvent system: Model formulation, kinetic experiments and parameter estimation

A. Jörke[1], S. Triemer[1], G. Kiedorf[2], A. Seidel-Morgenstern[1,2], C. Hamel[1,3]

[1] Otto von Guericke University, Institute of Process Engineering, Magdeburg, Germany
[2] Max Planck Institute for Dynamics of Complex Technical Systems, Magdeburg, Germany
[3] Anhalt University of Applied Sciences, Köthen, Germany

Introduction

In this contribution the double bond isomerization of 1-decene was analyzed experimentally and theoretically in terms of kinetic modeling and parameter estimation. This reaction is a very important side reaction in hydroformylation reaction networks, since it is generally believed to be responsible for the production of undesired branched aldehydes [1]. The aim of this work was to reduce and to parameterize a detailed mechanistic kinetic model, which was developed for quantifying the isomerization rate of long chain olefins in complex reaction systems based on selected experiments. The catalyst pre-equilibrium was also included. The used method explained below avoids overparameterization and allows estimating the kinetic parameters of the reduced model with low uncertainties. With the presented method it is possible to design a set of experiments to increase its informational content and the number of sensitive kinetic parameters, respectively.

Experimental

The kinetic experiments were performed with 1-decene as model substance in a thermomorphic multicomponent solvent system (consisting of DMF, dodecane and 1-decene) and a Rh-BIPHEPHOS complex as catalyst. The reactions were realized in a 90 ml high pressure multiple batch reactor system (Parr Instrument Co.) under careful consideration of the catalyst pretreatment at 15-18 bar syngas and a molar Rh-to-ligand ratio of 1:3 to guarantee reproducible initial conditions and a defined catalyst state at the beginning of the reaction. The set of experiments was planned with the method mentioned below so that a maximum number of sensitive parameters can be reached (figure 1). The reactions were carried out at different temperatures, CO partial pressures and initial 1-decene concentrations (figure 2), while the molar Rh-to-olefin ratio was kept constant at 1:10000 as well as the Rh-to-ligand ratio at 1:3.

Methodological

Detailed kinetic models often contain more parameters that can be identified and estimated with low uncertainties because of their linear dependencies. To identify the set of linearly independent parameters and therefore to improve the nonlinear parameter estimation, the method of parameter subset selection (PSS) was developed [2]. The PSS is based on the transient analysis of the Fischer Information Matrix (*FIM*). This matrix is generated with the local parameter sensitivity matrix *S*.

$$S(t) = \begin{bmatrix} \frac{\partial y_1(t)}{\partial \theta_1} & \cdots & \frac{\partial y_1(t)}{\partial \theta_{N\theta}} \\ \vdots & \ddots & \vdots \\ \frac{\partial y_{Ny}(t)}{\partial \theta_1} & \cdots & \frac{\partial y_{Ny}(t)}{\partial \theta_{N\theta}} \end{bmatrix} \qquad FIM(t) = S(t)^T S(t) \tag{1}$$

The matrix $S(t)$ can be easily calculated from the known derivative matrices of the right hand sides $f(y,\theta)$ of the model equations with respect to states F_y and parameters F_θ.

$$\dot{S} = F_y\, S + F_\theta \qquad S_{i,j}(t=0) = 0 \tag{2}$$

General mechanistic kinetic model equations for a similar reaction system in their fully unreduced form derived by [3] are adopted and used in this work. Once the $FIM(t)$ is known, it can be treated with a rank revealing QR-decomposition which includes a singular value decomposition (SVD) as well. This SVD yields a sparse matrix with the singular values σ_i of the decomposed matrix as diagonal elements. Introducing the condition numbers $\kappa_i = \sigma_{max}/\sigma_i$ allows to calculate the transient rank z of the FIM with respect to numerical accuracy. This rank corresponds to the number of identifiable parameters. The QR-decomposition reveals a permutation matrix P which rearranges the columns of the FIM in descending parameter linear independency order. Consequently the first z columns correspond to the most identifiable parameters which can be estimated. The procedure is illustrated in figure 1 (example reaction: isomerization of 1-decene). It can be seen that two not identifiable model parameters in the example reaction of the isomerization of 1-decene exist (K^{III}_{iso}, K_2). The values of this non identifiable parameters can be fixed to an arbitrary physical meaningful value. If possible, this value can be zero and therefore the model can be reduced. Experimental design is also possible with the PSS method via finding experimental conditions that increase the number of identifiable parameters and decrease their uncertainty. Optimization of initial conditions for a different number of experiments was performed in this contribution but optimal control to design dynamic experiments is also possible.

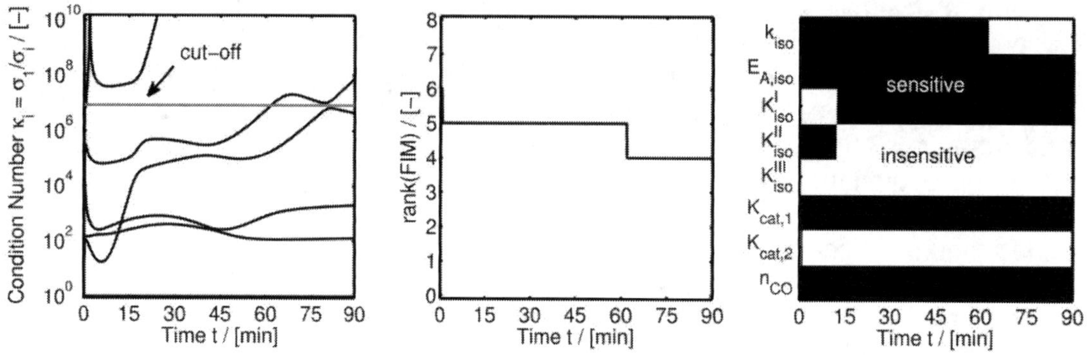

Figure 1: Transient parameter sensitivity analysis over experimental time (t_{exp} = 90 min); left to right: condition number κ_i of the FIM, transient rank of the FIM, model parameter identifiability chart

Results

The PSS method was applied successfully to the Rh-catalyzed isomerization of 1-decene as important side reaction of the hydroformylation. A detailed kinetic model was reduced and parameterized. In total 6 kinetic parameters were estimated. They show low "95% confidence intervals" between 4 - 25%. A good agreement between the model and experiments was achieved (figure 2). Hence, the method is suited to enhance parameter estimation in chemical reaction engineering and will be applied to all occuring reactions in the hydroformylation scheme, mainly isomerization, hydrogenation and the intrinsic hydroformylation reaction.

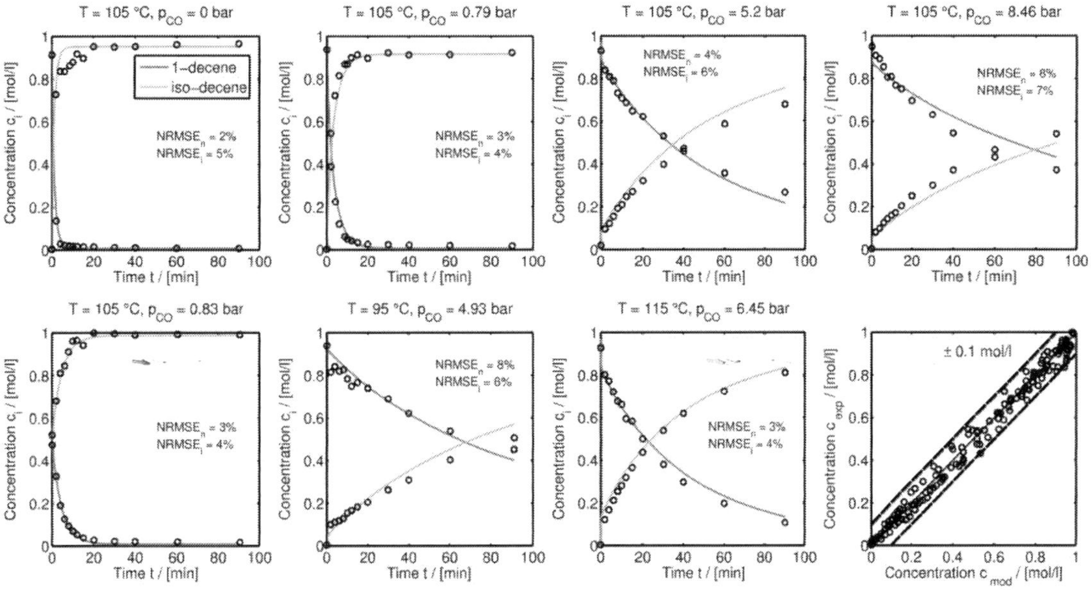

Figure 2: Experimentally observed concentrations of 1-decene and iso-decene at different reaction conditions, fit of the model derived for isomerization in [3] to experimental data, normalized root mean squared error (NRMSE) and parity plot; molar Rh-to-ligand ratio = 1:3; molar Rh-to-olefin ratio = 1:10000; thermomorphic multicomponent solvent system: 49%-mol DMF, 38 %-mol dodecane, 13 %-mol decene

Acknowledgement

This work is embedded in the German Science Foundation collaborative research center „SFB/TR-63 InPROMPT – Integrated Chemical Processes in Liquid Multiphase Systems". The authors gratefully thank the German Science Foundation for financial support.

Literature:

[1] R. Franke et al., Applied hydroformylation, Chem. Rev. 112, 5675–5732, dx.doi.org/10.1021/cr3001803, 2012

[2] M. Fink et al., Subset selection for parameter estimation in an HIV model, Proc. Appl. Math. Mech. 7, 1121501–1121502, doi:10.1002/pamm.200700319, 2007

[3] G. Kiedorf et al., Kinetic description of the hydroformylation of 1-dodecene in a thermomorphic solvent system by using rhodium-biphephos-catalyst, Chem. Eng. Sci., doi:10.1016/j.ces.2013.06.027, 2013

386795 Discovering Kinetic Solvent Effects in Hydrogen Abstraction Reactions for Use in Automatic Mechanism Generation

Abstract:

Liquid phase reaction mechanisms are of great interest within areas of biochemistry and chemical technology. Solvation affects reaction rates, not just due to the solvent's electrostatic properties, but because of the differential solvation of reactants and transition states. Thus, for different reactions of the same class, kinetic solvent effects will depend on the reacting species' functional groups. For radical-radical and radical-molecule reactions, it is difficult to measure all of the necessary rates experimentally, and furthermore, the reaction networks can contain hundreds to thousands of chemical species and reactions. Therefore, we would like to estimate kinetic solvent effects computationally. The Reaction Mechanism Generator (RMG) software [1] automatically generates detailed kinetic models in the gas phase and currently has capabilities for liquid-phase thermodynamics and diffusion in 25 solvents [2]. This work adds intrinsic kinetic corrections to gas phase rates in RMG to account for solvation.

A set of representative hydrogen abstraction reactions was investigated with Gaussian to find reactant and transition state geometries and energies using Density Functional Theory (DFT) and the SMD solvation algorithm. These reactions included OH·, ·OOH, and CH_3· abstracting hydrogen atoms from alkanes and alcohols, which are reactions particularly relevant in both lipid peroxidation and in liquid fuel oxidation. For each reaction, the difference in barrier height between solution phase and gas phase was calculated in eight solvents. Trends were elucidated from this data based upon solvent and molecular structure.

The relationships found were implemented into RMG by adding a solvation module into its kinetics database. Hierarchical trees based on molecular structure, which are already used in RMG for thermodynamic and kinetic data calculations, aid in this effort. The results show promise that kinetic solvent effects can be estimated for other reaction classes in liquid-phase mechanisms. Finding trends for other reaction families, specifically intra-hydrogen migration, is in progress.

References

1. [1] Green WH, Allen JW, Bhoorasingh P, Buesser BA, Ashcraft RW, Beran GJ, Class CA, Gao C, Goldsmith CF, Harper MR, Jalan A, Khanshan FS, Magoon GR, Matheu DM, Merchant SS, Mo JD, Petway S, Raman S, Sharma S, Slakman B, Song J, Geem KMV, Wen J, West RH, Wong A, Wong HW, Yelvington PE, Yee N, Yu J. RMG — Reaction Mechanism Generator-Python. 2013. URL rmg.mit.edu

2. [2] Jalan A, West RH, Green WH. An extensible framework for capturing solvent effects in computer generated kinetic models. J Phys Chem B. 2013;117(10):2955–70. doi:10.1021/jp310824h.

Study of p-Cresol/Tert-Butanol Adduct Formation Using in Situ Raman Spectroscopy and DFT Calculation

Min Huang*, Lin Cao
Department of Chemistry, Tongji University, 1239 Siping Rd., Shanghai 200092, China
minontongji@tongji.edu.cn

In situ Raman spectroscopy and density functional theory (DFT) calculation were used to study the formation of the O-H…O hydrogen bonding complex in p-cresol/tert-butanol system. Raman spectra of different molar ratio were obtained to study the band peak shift, band width broadening and band intensity nonlinear changes caused by the formation of the hydrogen bond. Using CCl_4 as internal reference, quantitative measurements on the intensities of selected bands of p-cresol/tert-butanol system at different temperatures were performed. Gibbs free energies of self-association of p-cresol and tert-butanol were therefore, determined. The results were in good agreement with the literature values. Concentration and temperature dependency of the complex formation rate were also studied, rate constants were identified.

Introduction

Cresol isomers are weak acids, they can form hydrogen bond and be self-associated. At room to moderate temperature, m- and p-cresol were reported to form dimer or trimer self-associates through hydrogen bonding[1,2]. tert-Butanol was also reported to form strong self-associates[3].

It is already known that p-cresol and tert-butyl can form weak compound[4], its detail, however, is not yet been reported. When mixed with alcohol, Krishan[5] et al. found that the line shape and the peak position of the methyl ethyl ketone and acetophenone change with the addition of p-cresol. Biswal[6] et al. also reported that p-cresol is able to form some kind of adduct with methanol and ethanol. Through infrared spectra and theoretical calculations, this adduct was identified to be associated with the formation of the hydrogen bond, p-cresol donated proton and alcohol was the acceptor.

Raman spectroscopy can provide molecular scattering information related to the induced polarization of the molecule. Therefore, change of the molecular structure can be obtained by analyzing the Raman spectra[7,8]. In last contribution, backscattering micro Raman was used to characterize and quantify the hydrogen bond formed between p-cresol and tert-butanol. With the help of density functional theory (DFT) quantum calculations, the hydrogen bond that leads to the formation of the adduct was identified. Raman transition modes of p-cresol, tert-butanol and the adduct were also identified. The stoichiometry of this bonding was determined.

* Correspondence to : Department of Chemistry, Tongji University, 1239 Siping Rd., Shanghai 200092, China
e-mail: minontongji@tongji.edu.cn

In this contribution, In situ Raman spectroscopy and density functional theory (DFT) calculation were used to study the mechanisms of the p-cresol/tert-butanol complex formation. Using CCl_4 as internal reference, quantitative measurements on the intensities of selected bands of neat p-cresol and tert-butanol and their different mix radios were studied at changed temperatures. Gibbs free energies of self-association of p-cresol and tert-butanol were determined. Concentration and temperature dependency of the complex formation rate were also studied, rate constants were identified.

Experimental

CCl_4, tert-butanol AR grade and p-cresol more than 98% pure were purchased from Sinopharm Chemical Reagent Co. Ltd.. The spectra were taken using Renishaw inVia confocal micro Raman spectrometer with 2 cm^{-1} spectral resolution and 5X object lens. For each spectrum, 9 seconds integration time and 6 repeated scans was used to allow the optimal signal to noise ratio. Spectral Physics Ar^+ 514 cm^{-1} single line laser was used for the excitation. For the p-cresol/tert-butanol system, because 459 cm^{-1} line from CCl_4 was strongly interfered by 460 cm^{-1} from p-cresol, 200 cm^{-1} line was used instead as internal reference[9].

Self-association study

Sample prepared according to Table 1 was sealed in a glass capillary tube and placed on a programmable heating unit one at a time. The heater was programed to heat the sample at the rate of 1 K/min from 273K to 333K. On its path, the heater will stop and maintain the temperature for 5 minutes every 5 K for the spectrometer to record the spectra.

Table 1 Compositions used for tert-butanol and p-cresol self-association study

tert-Butanol (g)	p-Cresol (g)	CCl4 (g)
0.419	-	0.133
-	0.541	0.188

Adduct formation study

Samples were prepared according to table 2 were used for concentration dependent study. A fresh new sample was prepared on site then immediately sealed in a glass capillary tube and placed on a programmable heating unit. The heater was programed to heat the sample at the maximum heating rate to its setting temperature. When temperature reached 323 K, the sample was maintained at this temperature for 90 minutes. Spectra were recorded continuously at once every 3 minutes.

Molar ratio of p-Cresol and tert-Butanol at 1:1 was used in the temperature dependent experiments. 303K, 308K, 313K, 318K, 323K and 328K were used for this study. A fresh new sample was prepared on site then immediately sealed in a glass capillary tube and placed on a programmable heating unit. The heater was programed to heat the sample at the maximum heating rate to its setting temperature. After reaching the setting temperature, the sample was maintained at the setting temperature for 90 minutes. Spectra were recorded continuously at once every 3 minutes.

Table 2 Composition used in adduct formation concentration dependent study

Sample number	tert-Butanol (g)	p-Cresol (g)	CCl_4 (g)
1	0.151	0.175	0.118
2	0.116	0.200	0.122
3	0.122	0.160	0.110
4	0.170	0.140	0.127

DFT calculation

Raman spectra of p-cresol, tert-butanol and adducts were calculated with Gaussīon 98 and Nwchem-6.3. Density functional theory (DFT) B3LYP[10,11,12] and M06HF[13] functional, 6-311G++ d p basis set[14] were used. Assignments of the active Raman vibration modes of p-cresol, tert-butanol and adducts were obtained by comparing experimental data with the DFT calculation results.

Results and Discussions

Self-association study

Both p-cresol and tert-butanol are found to form strong self-associates at room to moderate temperature. Their structures were subjected to several studies. The purpose of this study is to verify the method.

p-Cresol

Optimized structure of p-Cresol trimer using B3LYP/6-311 G ++ d p was presented in Figure 1, Figure 2 presents the optimized trimer structure obtained by M06HF/6-311G++ d p, Calculated energy and 650cm^{-1} Raman peak position was summarized in Table 3. It is a symmetric three member ring in this structure, two member from the trimer formed a ring stack

Table 3 Information of monomer and trimmer of p-cresol

method	B3LYP/6-311 G ++ d p		M06HF/6-311 G ++ d p	
species	Energy Ha	Peak position cm^{-1}	Energy Ha	Peak Position cm^{-1}
Monomer	-346.79	656	-346.68	652(5.78)
Trimmer	-1040.35	657.5	-1040.03	648-651(15.6)

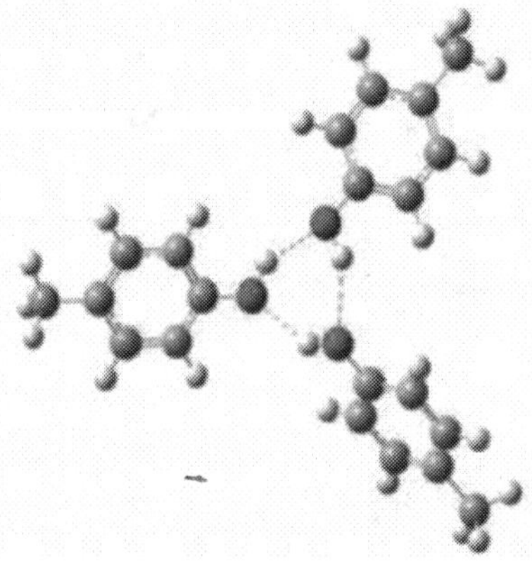

Fig 1 Optimized molecular structure of p-cresol trimmer by B3LYP/6-311G++ d p

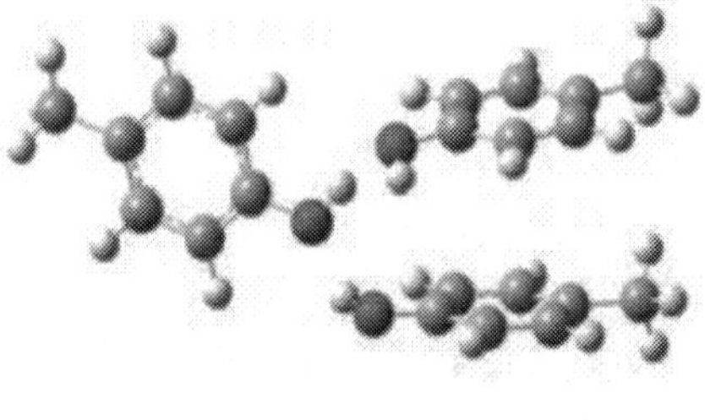

Fig 2 Optimized molecular structure of p-cresol trimmer by M06HF/6-311 G ++ d p

The Gibbs free energy obtained from B3LYP/6-311G++ d p is:

$$\Delta G_{B3LYP-cresol} = -346.79-(-1040.35)/3 = -6.67*10^{-3}Ha = -4.18kcal/mol = -17.5kJ/mol$$

The Gibbs free energy obtained from M06HF/6-311 G ++ d p is:

$$\Delta G_{06HF-cresol} = -346.68-(-1040.03)/3 = -3.33*10^{-3}Ha = -2.1kcal/mol = -8.88kJ/mol$$

By comparing the data in table 3 with the experimental results, it can be concluded that in the p-cresol case, M06HF functional performs better than B3LYP, the Raman peak position of monomer p-cresol is at 652 cm^{-1}, and the Raman peak position the trimer is at 650 cm^{-1}. This information was used for the peak deconvolution. The deconvolution results is presented in Figure 3.

Assume monomer and trimer p-cresol are the stable,

$$M_3 \rightleftarrows 3M \quad \text{(R1)}$$

where M_3 and M represent trimer and monomer respectively,

$$r_{mono} = -\frac{dC_{mono}}{dt} = k_+C_{tri} - k_-C^3_{mono} \quad (1)$$

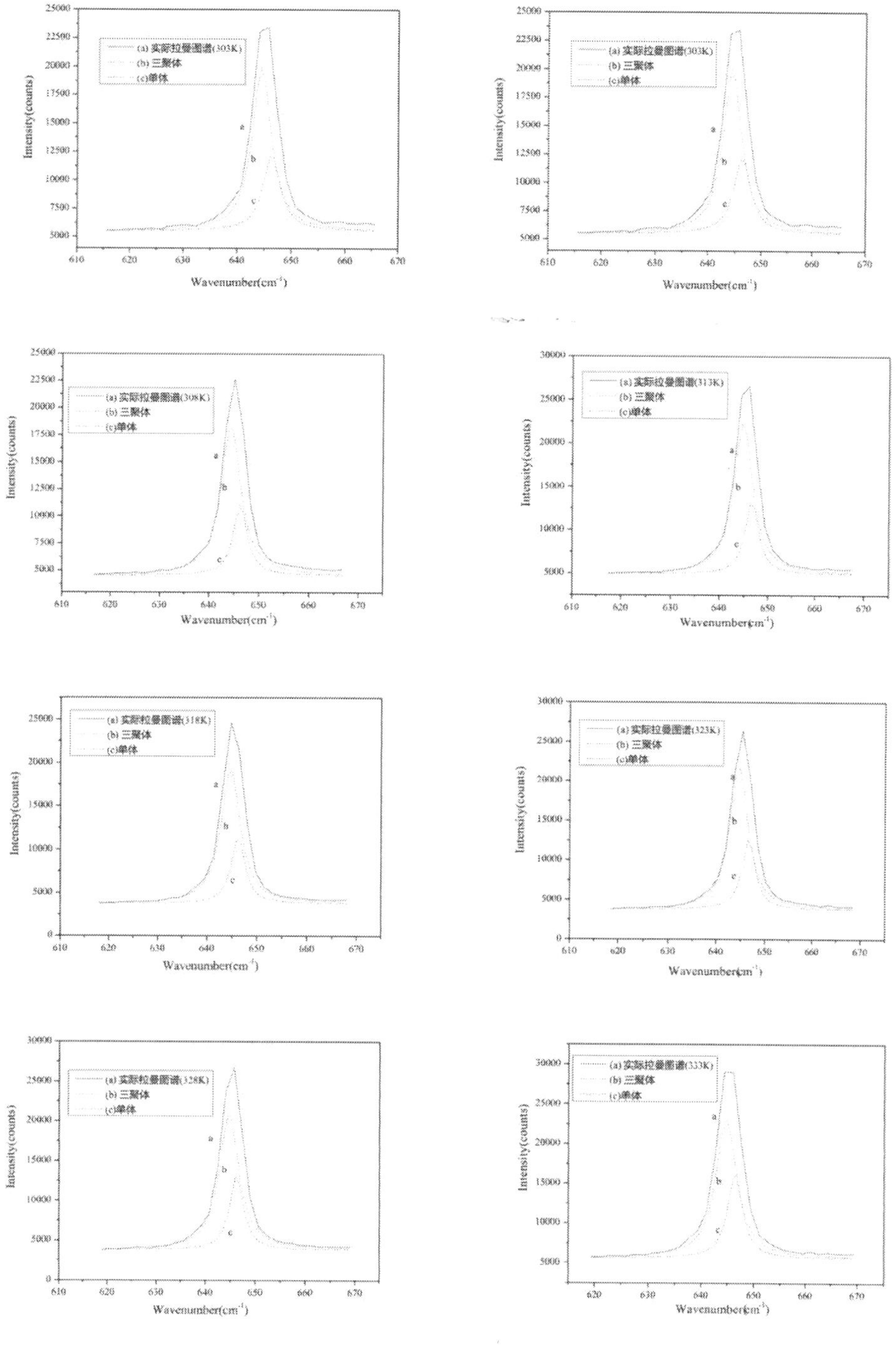

Fig 3 650 cm^{-1} band (p-cresol) recorded at different temperature

The Gibbs—Helmholtz equation becomes,

$$K_e = \frac{k_+}{k_-} = \frac{C_{mono}^3}{C_{tri}} \tag{2}$$

$$\ln K = \ln A + \frac{\Delta G_3}{RT} = \ln(\frac{C_{mono}^3}{C_{tri}}) \tag{3}$$

where ΔG_3 is the Gibbs free energy, C_{tri} and C_{mono} are the concentration of the trimer and the monomer. The Intensity the Raman peak is,

$$I_{RB} = \alpha_{RB} \frac{n_{pC}}{n_{pC} + n_{CCl_4}} \tag{4}$$

$$I_{RB} = \alpha_{RB} \frac{1}{1 + \frac{N_{CCl_4}}{N_{RB}}} \tag{5}$$

$$\frac{N_{CCl_4}}{N_{RB}} = \frac{\alpha_{RB}}{I_{RB}} - 1 \tag{6}$$

$$I_{mono} = \alpha_{RB} N_{mono} \tag{7}$$

$$I_{tri} = 3\alpha_{RB} N_{tri} \tag{8}$$

$$C_{mono} = \frac{N_{mono}}{N_{mono} + N_{tri} + N_{CCl_4}} = \frac{\frac{I_{mono}}{\alpha_{RB}}}{\frac{I_{mono}}{\alpha_{RB}} + \frac{I_{tri}}{3\alpha_{RB}} + \left(\frac{\alpha_{RB}}{I_{RB}} - 1\right) N_{RB}} \tag{9}$$

$$C_{mono} = \frac{\frac{I_{mono}}{\alpha_{RB}}}{\frac{I_{mono}}{\alpha_{RB}} + \frac{I_{tri}}{3\alpha_{RB}} + \left(\frac{\alpha_{RB}}{I_{RB}} - 1\right)\frac{I_{RB}}{\alpha_{RB}}} = \frac{\frac{I_{mono}}{\alpha_{RB}}}{\frac{I_{mono}}{\alpha_{RB}} + \frac{I_{tri}}{3\alpha_{RB}} + 1 - \frac{I_{RB}}{\alpha_{RB}}} \tag{10}$$

$$= \frac{I_{mono}}{I_{mono} + \frac{I_{tri}}{3} + \alpha_{RB} - I_{RB}}$$

$$C_{tri} = \frac{\frac{I_{tri}}{3\alpha_{RB}}}{\frac{I_{mono}}{\alpha_{RB}} + \frac{I_{tri}}{3\alpha_{RB}} + \left(\frac{\alpha_{RB}}{I_{RB}} - 1\right)\frac{I_{RB}}{\alpha_{RB}}} = \frac{I_{tri}/3}{I_{mono} + \frac{I_{tri}}{3} + \alpha_{RB} - I_{RB}} \tag{11}$$

$$\frac{C_{mono}^3}{C_{tri}} = \frac{\frac{I_{mono}^3}{I_{tri}/3}}{\left(I_{mono} + \frac{I_{tri}}{3} + \alpha_{RB} - I_{RB}\right)^2} \tag{12}$$

where I_{RB} is total area of the 650 cm^{-1} peak, I_{mono} is the monomer portion, I_{tri} is the trimer portion, α_{RB} is the relative Raman cross-section, N_{CCl_4} is the number of CCl_4 molecule inside the detection volume, N_{mono} is the number of the monomer molecule inside the detection volume and N_{tri} is the number of the trimer molecule inside the detection volume.

Plot $\ln(\frac{C_{mono}^3}{C_{tri}})$ against 1/T (1/K), a linear relation is expected as in the Figure 4.

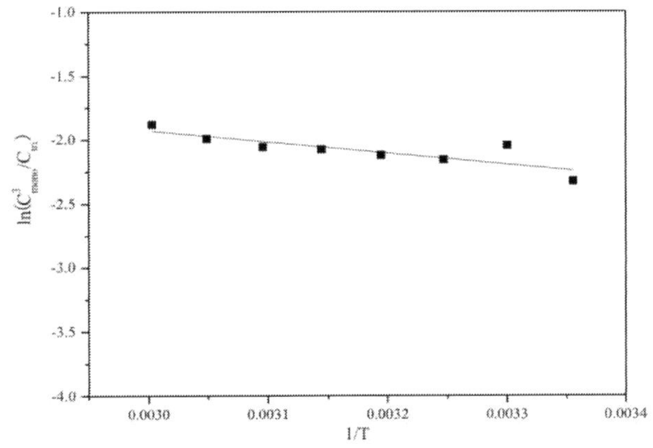

Fig 4 Change of concentration with temperature (p-cresol)

where the slop is found to be -881.3, therefore, ΔG_3= -881.3*8.314= -7.327 kJ/mol= -1.75 kcal/mol.

In the derivation above, Eq.12, the absolute Raman cross-section of trimer was assumed to be 3 times that of the monomer, however, from Table 3, DFT results indicated that it is 2.7 times. Using 2.7 instead, the Gibbs was found to be -1.8kcal/mol, agree well with -2.1kcal/mol by M06HF/6-311G++ d p.

tert-Butanol

DFT optimized structures of trimer and tetramer tert-butanol are presented in Figure 5 and 6. Calculated energy and Raman peak positions are listed in Table 4.

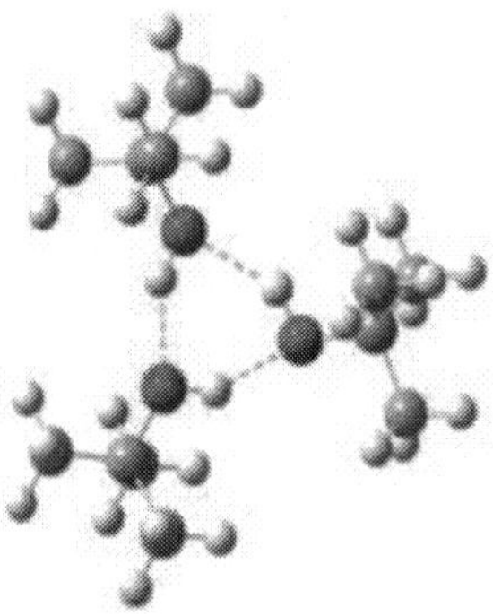

Fig 5 Optimized molecular structure of tert-butanol trimmer by B3LYP/6-311 G ++ d p

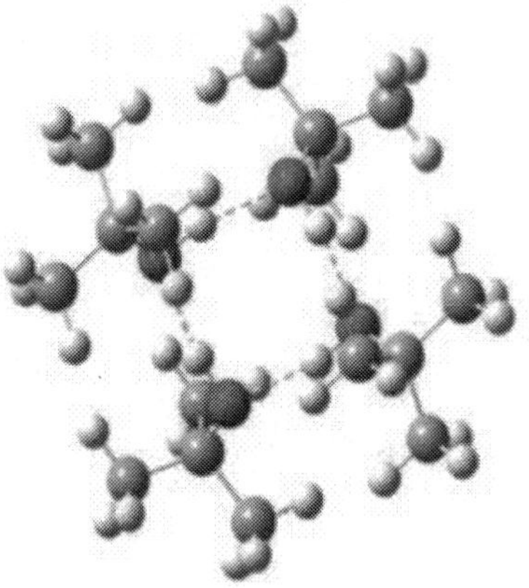

Fig 6 Optimized molecular structure of tert-butanol tetramer by B3LYP/6-311 G ++ d p

Table 4 DFT calculation results of monnmer and trimmer of tert-butanol

method	B3LYP/6-311 G ++ d p		M06HF/6-311 G ++ d p	
Species	Energy Ha	Peak Position cm^{-1}	Energy Ha	Peak Position cm^{-1}
Monomer	-233.65	743.8(16.8)	-233.55	765.9(21.8)
trimmer	-700.65	764-784(72.0)	-700.6	764-770(66)
tetramer	-934.57	746.4(78.3)	-934.205	766-768(91.6)

The Gibbs free energy obtained from B3LYP/6-311G++ d p is:

$$\Delta G_{B3LYP-tert} = -233.65-(-934.57)/4 = -7.5*10^{-3}\,Ha = -4.71kcal/mol = -19.7kJ/mol$$

The Gibbs free energy obtained from M06HF/6-311G++ d p is:

$$\Delta G_{06HF-tert} = -233.55-(-934.205)/4 = -1.25*10^{-3}\,Ha = -0.785kcal/mol = -3.28kJ/mol$$

From the literature, tert-butanol tetramer ring structure is the smallest stable self-associate[3].

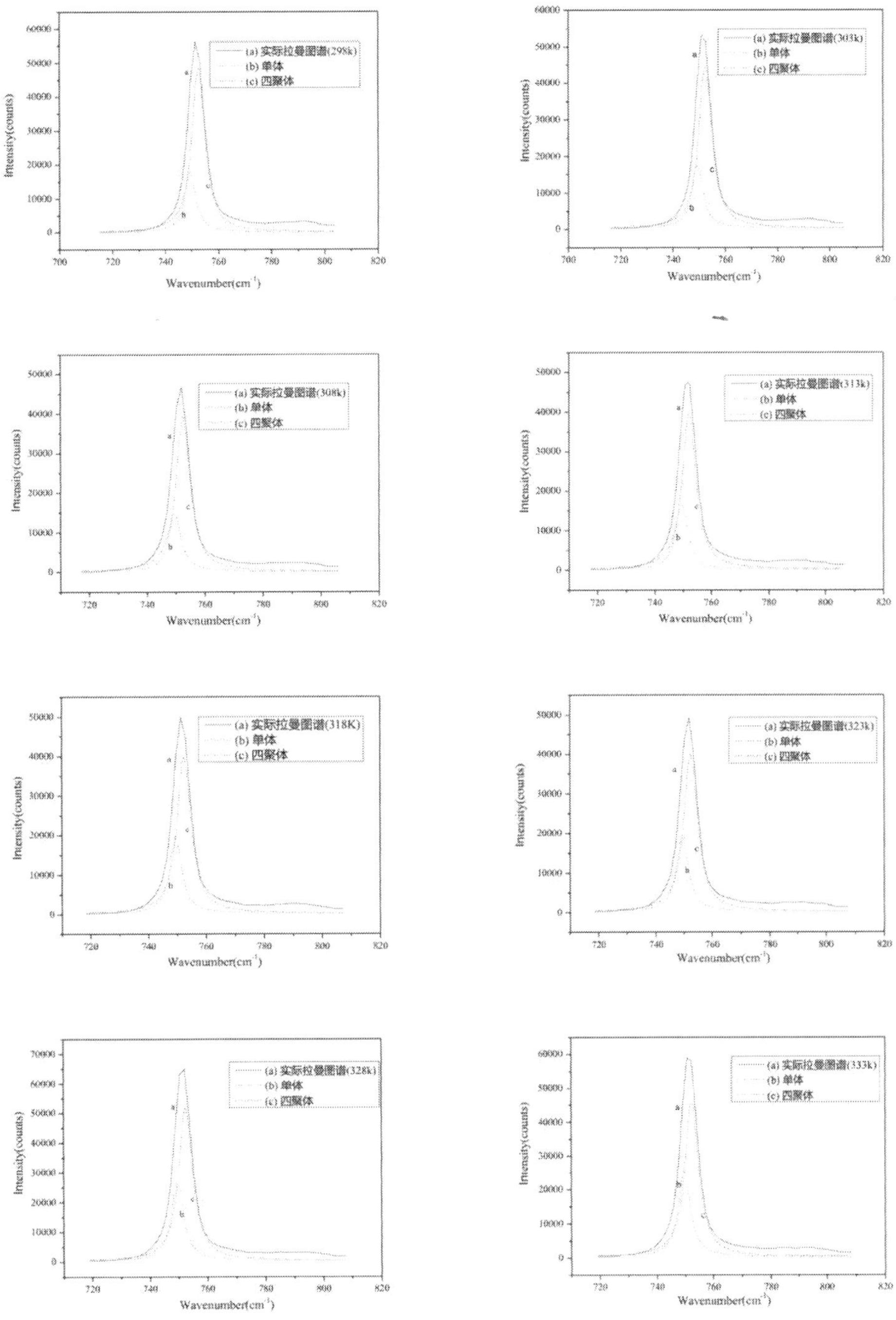

Fig 7 Deconvolution of 750 cm^{-1} band (tert-butanol) from different temperature

From Table 4, the calculated Raman peak of monomer is at 743.77 cm^{-1} and the calculated Raman peak of tetramer is at 746.4 cm^{-1}. The deconvolution results is presented in Figure 7.

Assume the reaction mechanisms is as,

$$M_4 \rightleftarrows 4M \qquad (R2)$$

where M_4 is the tetramer and M represents the monomer tert-butanol, the Gibbs—Helmholtz equation becomes,

$$\ln K = \ln A + \frac{\Delta G_4}{RT} = \ln(\frac{C^4_{mono}}{C_{tet}}) \qquad (13)$$

Using the DFT simulated absolute Raman cross-section,

$$\frac{C^4_{mono}}{C_{tet}} = \frac{\frac{I^4_{mono}}{I_{tet}/4.7}}{(I_{mono} + \frac{I_{tet}}{4.7} + \alpha_{RB} - I_{RB})} \qquad (14)$$

Plot $\ln(\frac{C^4_{mono}}{C_{tet}})$ against 1/T (1/K),

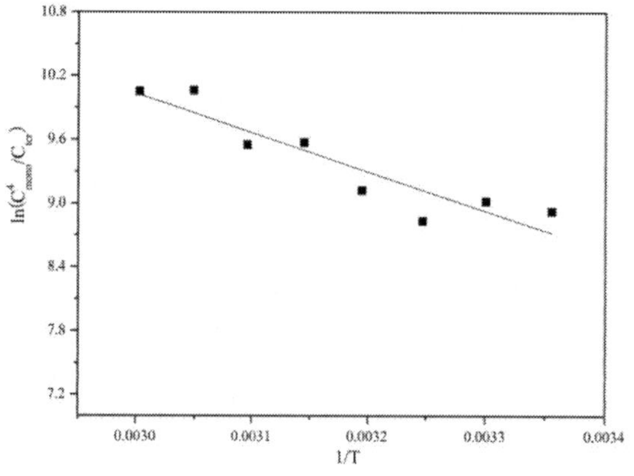

Fig 8 Change of concentration with temperature(tert-butanol)

The slope of the linear fit is -3064.1, therefore, the Gibbs is ΔG=-3064.1*8.314=-25.47kJ/mol=-6.1kcal/mol, compare with the DFT B3LYP/6-311 G ++ d p simulation, -4.71 kcal/mol, M06HF/6-311 G ++ d p simulation, -0.785 kcal/mol. The B3LYP/6-311 G ++ d p is clearly perform better in this case. Srinivasa reported[15] the Gibbs energy is 4.56 kcal/mol for n-butanol.

Adduct formation study

Table 5 presented the DFT B3LYP/6-311 G ++ d p simulation results of tert-butanol monnmer, trimmer and p-cresol/tert-butanol adduct at 750 cm-1.

Table 5 Calculated tert-butanol monomer, trimmer and p-cresol/tert-butanol adduct at 750 cm^{-1}

Species	Peak Position (cm^{-1})	Absolute Raman Cross-section
adduct	739.6	16.5
monomer	743.8	16.8
tetramer	746.4	78.3

Raman intensity is for this study is,

$$I_T = \alpha_{add} N_{add} + \alpha_{tet} N_{tet} + \alpha_{mono} N_{mono} \tag{15}$$

where α_{tert} represents the relative Raman cross-section of monomer, therefore,

$$I_{add} = \alpha_{tert} N_{add} \tag{16}$$

$$I_{tet} = 4.7\alpha_{tert} N_{tet} \tag{17}$$

$$I_T = \alpha_{tert} N_{add} + 4.7\alpha_{tert} N_{tet} + \alpha_{tert} N_{mono} \tag{18}$$

where I_T is the detected total intensity.

The reaction mechanisms of p-cresol and tert-butanol adduct formation may be written as,

$$pc_3 \rightleftarrows 3pc_1 \quad pc_3 \tag{R1}$$

$$tb_4 \underset{k_{1^-}}{\overset{k_{1^+}}{\rightleftarrows}} 4tb_1 \tag{R2}$$

$$tb_1 + pc_1 \underset{k_{2^-}}{\overset{k_{2^+}}{\rightleftarrows}} adduct \tag{R3}$$

where tb_4 and tb_1 represent tetramer and monomer tert-butanol, pc_3 and pc_1 are the trimer and monomer p-cresol, and $adduct$ is the p-cresol/tert-butanol adduct.

In situ Raman spectra of the p-cresol and tert-butanol adduct formation reactions at molar ration=1:1, temperature = 308K, 318K and 323K are presented in Figure 9, 10 and 11. At 308K, Figure 9, or below, p-cresol and tert-butanol deassociate rates and their adduct formation rate are very small, no detectable change in intensity of 750 cm^{-1} band. At around 318K, Figure 10, noticeable change in intensity of 750 cm^{-1} band at early stage were observed. In this temperature range, tert-butanol self-association rate and the adduct formation rates are comparable. During the heating process, tert-butanol gradually transfer from tetramer which has larger Raman cross-section to small Raman cross-section monomer then adduct.

Temperature at 323K, Figure 11, or higher, the intensity of 750 cm^{-1} band increase slowly with time. From Table 5, the Raman cross-section of p-cresol/tert-butanol adduct and the tert-butanol are very similar, therefore, the deassociation process in R2 become dominant adduct formation becomes the control step.

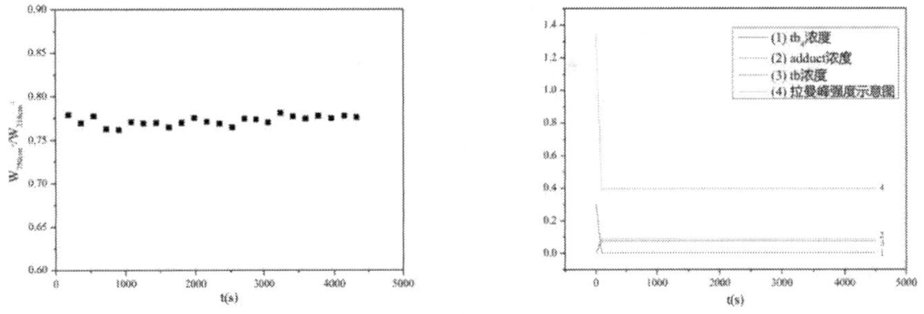

Fig 9 Raman intensity of 750 cm^{-1} band at 308K.
in situ measurement on the left, Polymath simulation on the right

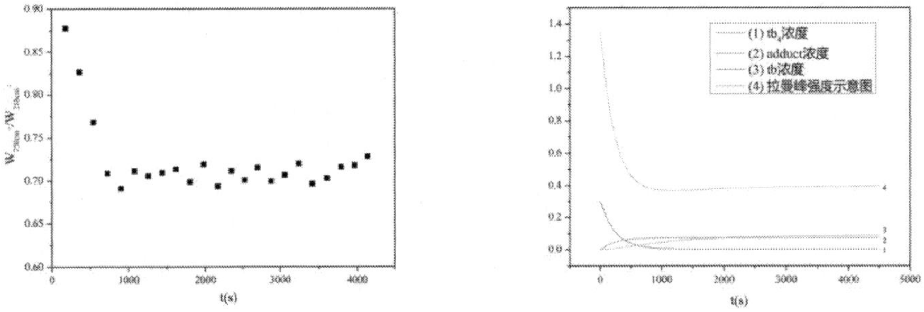

Fig 10 Raman intensity of 750 cm^{-1} band at 318K.
in situ measurement on the left, Polymath simulation on the right

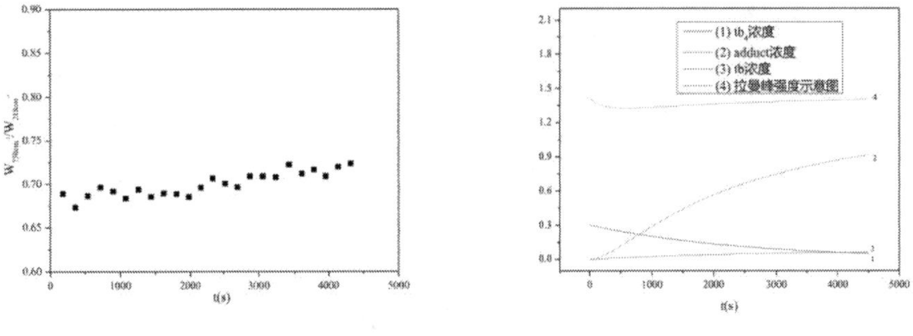

Raman intensity of 750 cm^{-1} band at 323K.
in situ measurement on the left, Polymath simulation on the right

Conclusions

Self-association of p-cresol and tert-butanol were studied using different DFT functionals. Details of the individual transitions that compose a particular Raman band were obtained. The information was used as a bias for Raman band peak deconvolutions. Better fit were obtained by using M06HF functional with 6-311 G ++ d p basis set for p-cresol. Self-associate complex of p-cresol was determined as trimer conformation. Gibbs free energy of self-association from the Raman experiment and DFT calculation were found to be -2.1kcal/mol and -1.80kcal/mol, respectively. On the other hand, B3LYP functional perform better in tert-butanol case, four-membered ring tetramer configuration was determined, the Gibbs energy from Raman experiment and DFT calculation were found, respectively to be -6.1kcal/mol and -4.71kcal/mol.

The adduct formation process was simplified by using the stoichiometric relationship of tert-butyl alcohol and p-cresol association and the Gibbs free energy from which the stability of the self-association can be determined. In situ Raman spectroscopy were used to obtain the spectra at different temperatures and concentrations over time. Relative peak intensity were associated with the reactant and product concentrations using calculated absolute Raman cross-sections from DFT simulations. The experimental data were compared with the concentration profiles from the process simulations.

At low temperature, the rates of tert-butanol tetramer dissociation and the new adduct formation are both low, therefore, no noticeable change can be detected; when the temperature was raised to 318K, tert-butanol tetramer start to dissociate visibly, and the new adduct concentration grows slowly over time; When the temperature continued to be raised to above 328K, tert-butanol tetramer dissociated considerably and the new association complex formation increased significantly over time. These results indicated that the p-cresol and tert-butanol start to form adduct rapidly at above 318K; the molar ratio of tert- butanol and p-cresol should be 1:1 with the tert-butanol be slightly in excess.

References

[1] E. M. Woolley and D. S. Rushforth. "Molecular Association of Hydrogen Bonding Solutes, o-, m-, and p-Cresol, in Carbon Tetrachloride". Can. J. Chem. 1974. 52(4), 653-660

[2] S. P. Casas, L. M. Trejo and M. Costas. "Self-association of phenols in inert solvents. Apparent heat capacities of phenol, substituted phenols and aromatic alcohols in n-heptane". J. Chem. SOC. FARADAY Trans. 1991.87(11):1733-1738

[3] Zimmermann D., Häber T. H., Schall H. et al. Hydrogen bonded rings, chains and lasso: the case of t-butyl alcohol clusters. Molecular physics, 2001, 5(99): 413-425

[4] V. K. Jadhav, M. R. Chlvate and N. S. Tavare. "Separation of p-cresol from its mixture with 2,6-xylenol by adductive crystallization". J. Chem. Eng. Data. 1991. 36(2):249-251

[5] M. G. Krishna Pillai, K. Ramaswamy and R. Pichai. "Influence of solvents on infrared frequencies–The complexing of methyl ethyl ketone and acetophenone with p-cresol". Aust. J. Chem. 1965. 18(10):1681-1685

[6] H. S. Biswal, P. R. Shirhatti and S. Wategaonkar. "O-H...O versus O-H...S hydrogen bonding. 2. Alcohols and thiols as hydrogen bond acceptors". J. Phys .Chem. A. 2010.114(26):6944-6955

[7] R.L. Benevenuto and W. A. Alves. "Acid–base interactions between formamide and tetrahydrofuran molecules: Raman investigation of speciation in a binary system of biological interest". J. Raman Spectrosc. 2008. 39(5):490-494

[8] G. Norbert and W. A. Alves. "Investigating the formation of an adduct of formamide with dioxane by means of Raman spectroscopy". J. Raman Spectrosc. 2008. 39(5):685-688

[9] P.K. Kipkemboi, P. C. Kiprono and J. J. Sanga. "Vibrational spectra of t-butyl alcohol, t-butylamine and t-butyl

alcohol + t-butylamine binary liquid mixture". Bull. Chem. Soc. Ethiop. 2003. 17(2):211-218

[10] Becke A. D. Density-functional exchange-energy approximation with correct asymptotic behavior. Physical Review A, 1988, 38 (6): 3098-3100

[11] Chengteh L., Weitao Y. and Robert G. P. Development of the Colle-Salvetti correlation-energy formula into a functional of the electron density. Physical Review B, 1988, 37 (2): 785-789

[12] Vosko S. H.,Wilk L. and Nusair M. Accurate spin-dependent electron liquid correlation energies for local spin density calculations: a critical analysis. Canadian Journal Physics, 1980, 58 (8): 1200-1211

[13] Zhao Y., Schultz N.E., Truhlar D.G. Exchange-correlation functional with broad accuracy for metallic and nonmetallic compounds, kinetics, and noncovalent interactions. Journal of Chemical Physics, 2005, 123 (16): 161-173

[14] Ditchfield R., Hehre W. J, Pople J. A. Self-consistent molecular-orbital methods. IX. an Extended Gaussian-type basis for molecular-orbital studies of organic molecules. The Journal of Chemical Physics, 1971, 54 (2): 724-728

[15] Srinivasa R. A., Arulmozhi V., Rajalakshmi S. Mutal interactions between hydrogen bonds in alcohols- nuclear magnetic relaxation study. General & Introductory Chemistry, 1993, 102(11): 693-698

Hydrogenation of Hydroxymethylfurfural using a Ruthenium Coated Polymeric Membrane Reactor

John Stanford, Peter Pfromm, and Mary Rezac*

Department of Chemical Engineering, Kansas State University
Manhattan, Kansas, United States

Catalytic membrane reactors afford an alternative and potentially more efficient method for performing three-phase heterogeneous chemical reactions. Traditional three-phase reactors often present mass transfer limitations, namely gas solubility in the liquid phase. Membrane reactors can alleviate the inherent mass transfer limitations by directly and abundantly supplying gas to the catalytic sites on the membrane surface, which acts as a gas/liquid contactor. The focus of this study is the application of a polymeric catalytic membrane reactor to the hydrogenation of hydroxymethylfurfural (HMF), which is a biomass-based platform chemical derived from fructose. The various partially hydrogenated HMF intermediates have a variety of manufacturing uses, and a notable product of hydrogenated HMF, 2,5-dimethylfuran (DMF), has many attractive characteristics as a liquid fuel.

Polyimide polymers are used for membrane synthesis because of their good chemical and high temperature resistance. The polymer solutions are cast as asymmetric integrally-skinned flat sheet membranes and are then coated with a ruthenium catalyst on the non-porous surface. Post-fabrication modifications such as thermal annealing and chemical crosslinking are being explored to enhance the chemical and temperature resistances of the membranes. The completed membrane is positioned in a flow over configuration maintaining liquid contact on the metal coated surface while allowing hydrogen gas to permeate from the porous support side to the catalytic sites on the non-porous surface. This study explores reaction kinetics and the influence of reaction temperature and gas phase pressure on system performance. Reaction products are identified and quantified using HPLC and GC-MS. Scanning electron microscopy is employed to examine the membrane and catalyst morphology before and after the reaction. An important parameter of this study is the reaction solvent choice, which is crucial for maintaining a functional membrane reactor as the necessary elevated temperatures for the chemical reaction also increase the solubility of the solvent in the membrane, leading to swelling and possible membrane failure. Solvent solubility in the polyimides of this study at different solvent vapor activities and temperatures is measured using a quartz crystal microbalance. Solvents of interest include water and short-chain alcohols. Sorption isotherms are described using a diffusion-relaxation model and concentration-dependent diffusion coefficients and relaxation parameters are calculated. This research on the development and characterization of a functional polymeric catalytic membrane reactor for HMF hydrogenation may serve as an example of the methodology needed for developing membrane reactors for other heterogeneous catalysis applications.

Material Properties and Operating Configurations of Membrane Reactors for Propane Dehydrogenation

Seung-Won Choi, Christopher W. Jones, David S. Sholl, and Sankar Nair

School of Chemical & Biomolecular Engineering, Georgia Institute of Technology, Atlanta, GA 30332-0100

Jason S. Moore, Yujun Liu, Ravindra S. Dixit, and John G. Pendergast

Engineering & Process Sciences, The Dow Chemical Company, Freeport, TX 77541

Abstract

We consider the problem of defining physically realistic and technologically interesting ranges for catalytic and transport properties, and operating configurations, of packed-bed membrane reactors (PBMRs) for propane dehydrogenation (PDH). This work employs an isothermal PBMR simulation with simplified kinetics and transport terms to investigate the influence of a broad range of reaction and transport parameters on the PBMR performance, and thereby determine operating windows for such reactors when constructed with industrial PDH catalysts and nanoporous (*e.g.*, zeolitic) membranes. First, dimensionless analysis of the simulation results is carried out using Damköhler (Da) and Péclet (Pe) numbers to represent the material catalytic properties, transport properties, and the reactant throughput. We also consider the effects of operating parameters such as different sweep flow rates and countercurrent versus cocurrent modes. An increased Da and a decreased Pe are useful for enhanced conversion in the PBMR. However, the conversion reaches a maximum and then decreases with further decrease in Pe if the membrane allows permeation of the propane reactant in addition to hydrogen, due to reactant depletion along the length of the PBMR. This is significant because of the consideration of nanoporous zeolite membranes as a leading candidate for PDH PBMRs. A highly hydrogen-selective, small-pore zeolite membrane (*e.g.*, SAPO-34 or DDR) will show better performance over a wide range of weight-hourly space velocity (WHSV) conditions than a higher-flux, lower-selectivity medium-pore zeolite membrane (*e.g.*, MFI) or a Knudsen-selective high-flux mesoporous membrane. H_2-selective membranes show a plateau region of conversion at low Pe. This limit can be overcome by using a larger sweep flow rate (which enhances the driving force for membrane permeation) and countercurrent operation. However, due to a complex trade-off between the reaction kinetics and forward/back-permeation across the membrane, the countercurrent mode can be recommended only in a limited (but nevertheless realistic) window of material properties. The preference for small-pore H_2-selective zeolitic membranes requires the fabrication of thin (~1 μm or thinner) membranes with low defect densities in order to achieve an adequately high transport flux. Low-selectivity, high-flux membranes can only provide moderate enhancement of PBMR performance at low WHSV (< 1 hr^{-1}). The above analysis is useful in quickly identifying desired material properties For example, in order to enhance the PDH conversion from 48 % (PBR) to 52 % (PBMR) at a WHSV of 1 hr^{-1} using Pt-Sn catalyst, the required dimensionless numbers in the operating window are Pe ~ 1.4 and Da ~ 28, which can be realized with a membrane surface area in the range of 4.5×10^{-3}-4.5×10^{-4} m^2/g catalyst and a small-pore zeolite membrane with H_2 permeance in the range of 10^{-8}-10^{-7} $mol.m^{-2}.s^{-1}.\ Pa^{-1}$.

392000 An economic performance evaluation study under uncertainty for Pd-based catalytic membrane reactor modules used for H2 production

Abstract:

A comprehensive economic performance evaluation framework for palladium-based catalytic membrane reactor (CMR) modules potentially integrated into hydrogen production (HP-CMR) through methane steam reforming process systems has been developed. The HP-CMR technology option offers a promising technical pathway towards hydrogen production with enhanced environmental performance in a carbon-constrained world, and since no significant operating experience has been accumulated at the industrial/commercial scale any preliminary attempt to assess its economic viability is certainly justified. Within the above context, the development of comprehensive baseline models for fixed capital investment (FCI), total capital investment (TCI), as well as total product cost (TPC) in order to evaluate the economic performance of the above industrial scale membrane reactor module represents a natural starting point in our study. Various sources of uncertainty (raw material market prices, utilities costs, labor costs, membrane lifetime and maintenance costs, financing costs, etc.) are explicitly recognized and the effect of these uncertainty drivers on FCI, TCI and TPC are explicitly taken into account using Monte-Carlo techniques. As a result, insightful distribution profiles of FCI, TCI and TPC are derived rather than single-point value estimates, and a more realistic distribution of membrane reactor module economic performance outcomes is generated through which "risks and opportunities" can be identified in the presence of uncertainty. Furthermore, preliminary results have shown that while expected values of the FCI and PC obtained from the above distribution profiles remain higher than the ones associated with the traditional/conventional methane steam reforming-based hydrogen production option (within a meaningful range of permeabilities, membrane length as well as input flow rates), FCI and TPC profiles for HP-CMR become appealing by reducing palladium thickness (in terms of their respective expected values when comparatively assessed against the above conventional option in the presence of uncertainty). Finally, within the proposed economic performance evaluation framework Tornado diagrams have been also developed revealing and establishing the predominant role and effect of financing interest cost, palladium unit price and palladium membrane lifetime on the TCI and PC profiles.

378205 Membrane-Based Thermocatalytic Destruction of Chemical Warfare Agents

Abstract:

The possibility of use of chemical weapons has increased in recent years, both as a result of potential terrorist attacks as well as of ongoing international conflicts. The successful application of a flow-through catalytic membrane reactor (FTCMR) as an individual protection (IP) system for the destruction of dimethyl methylphosphonate (DMMP), which is known as a chemical precursor (and used to simulate its characteristics) for Sarin (GB), a toxic chemical warfare agent (CWA) was previously investigated and reported by this group[1]. As part of the study, the effect of the DMMP concentration in the feed on its complete combustion in the FTCMR was investigated. The studies indicate that the protection (complete conversion) time is a function of the DMMP concentration, with longer times observed for the lower concentrations, and shorter protection times associated with the higher concentrations.

In this current study, the focus is on collective protection (CP) applications. As part of this effort multi-tubular membrane bundles are developed and rendered reactive for the proposed application. The current efforts involve the development and testing of a 1 cfm multi-tubular FTCMR unit, with the eventual goal the development and testing of a 200 cfm full-scale CP system.

In addition, a novel hybrid system is being developed that combines the FTCMR with a surface-flow membrane system (SFMS)[2]. The main advantage of this hybrid system, combining the SFMS (capable of the physical removal of a large fraction of the CWA from contaminated air streams) and the FTCMR (which completely oxidizes the remaining amount) is the continuous CWA destruction for extended time periods which are appropriate for both IP and collective protection (CP) applications. As part of this effort, we also investigate the performance of the SFMS. Experimental and modeling investigations indicate that a high removal rate can be achieved with a proper choice of (feed rate/membrane surface area) and the sweep ratio. The performance of the combined system is currently under investigation, and experimental results and the modeling of the system performance will be presented at the meeting.

References:

1. Motamedhashemi, M.M.Y., Egolfopoulos, F.N., and Tsotsis, T.T., "Application of a Flow-Through Catalytic Membrane Reactor (FTCMR) for the Destruction of a Chemical Warfare Simulant," *J. Membrane Sci.*, 376, 119, 2011.
2. Lee, H. C., Monji, M., Parsley, D., Sahimi, M., Liu, P., Egolfopoulos, F., and Tsotsis, T., "Use of Steam Activation as a Post-Treatment Technique in the Preparation of Carbon Molecular-Sieve Membranes," *Ind. Eng. Chem., Res.*, dx.doi.org/10.1021/ie300261r, 2013.

Performance of reduction-tolerant $Sr_{0.7}Ba_{0.3}Fe_{0.9}Mo_{0.1}O_{3-\delta}$ multichannel hollow fiber membrane reactor for partial oxidation of methane

Jiawei Zhu, Zhengkun Liu, Wanqin Jin *
State Key Laboratory of Materials-Oriented Chemical Engineering
Membrane Science & Technology Research Center,
Nanjing Tech University, Nanjing 210009, P. R. China
* wqjin@njtech.edu.cn

Due to the potential impacts on energy and environment, mixed-conducting dense membranes have attracted considerable interests in partial oxidation of methane (POM) reaction process. Such membrane reactors coupled the POM reaction and oxygen separation in one unit, which makes this reaction more commercially feasible. However, the mixed-conducting membranes applied in POM reaction should meet the following requirements: high oxygen permeation flux, good stability under reducing atmosphere and attractive membrane configuration for industrial application.

So far, large research efforts focused on using single-channel hollow fiber membranes for membrane reaction process because they possess a thin wall and can obtain high oxygen permeation flux. Nevertheless, the conventional single-channel hollow fiber membrane may not meet the requirements of industrialization, suffering from low mechanical strength. In our previous work[1], a mixed-conducting multichannel hollow fiber (MCMHF) membrane with high oxygen permeation flux and superb mechanical strength was proposed to overcome some drawbacks of single-channel hollow fiber membranes.

In this work, to accelerate the industrial application of membrane reactor for POM reaction, a MCMHF membrane was used to construct this membrane reactor successfully. By using phase inversion and sintering technique, the MCMHF membrane has being prepared, and $Sr_{0.7}Ba_{0.3}Fe_{0.9}Mo_{0.1}O_{3-\delta}$ (SBFM)[2], the reduction-tolerant oxide, was selected as the membrane material. The morphology, mechanical strength and oxygen permeability of the MCMHF membrane, the effects of temperature, methane concentration and methane flow rate on the performance of MCMHF membrane reactor and the long-term stability of the membrane reactor were investigated in detail. This work demonstrated that the multichannel hollow-fiber can be an attractive membrane configuration for membrane reactor in the viewpoint of industrial application.

Acknowledgments

This work was supported by the Project of Priority Academic Program Development of Jiangsu Higher Education Institutions (PAPD) and the Innovative Research Team Program by the Ministry of Education of China (No. IRT13070).

References

1. Zhu J, Dong Z, Liu Z, Zhang K, Zhang G, Jin W. Multichannel mixed-conducting hollow fiber membranes for oxygen separation. *AIChE J.* 2014;60(6):1969-1976.
2. Dong X, Jin W, Xu N. Reduction-Tolerant Oxygen-Permeable Perovskite-Type Oxide $Sr_{0.7}Ba_{0.3}Fe_{0.9}Mo_{0.1}O_{3-delta}$. *Chemistry of Materials.* 2010;22(12):3610-3618.

The development of non-precious transition metal catalysts is of great interest as next-generation technologies such as fuel cells and electrolyzers continue to grow in the alternative energy commercial market. Many commercial fuel cells, and in particular, acidic polymer electrolyte fuel cells, use platinum or other precious metals as the catalyst material. In addition, research that supports industries that have traditionally used costly, high-energy processes for chemical production is now considering alternative methods that would use less energy. For example, ammonia is currently produced using the Haber-Bosch process, where nitrogen gas is reduced to ammonia at high temperature and high pressure. For both energy conversion and chemical production applications, electrocatalysis under alkaline conditions avoids high energy conditions of elevated temperature and pressure, and alkaline electrocatalysis can be performed with non-precious metal catalysts. However, key challenges in the use of non-precious metal catalysts include material performance, as well as material stability and lifetime.

Here, we will present on a set of iron-nickel core-shell nanoparticles that have shown promise as electrocatalysts under alkaline conditions for both fuel oxidation and water electrolysis. The nanoparticle performance can be tuned for either fuel oxidation or oxygen evolution during water electrolysis through modification of the oxygen content and oxide morphology of the nanoparticles. Furthermore, we are now working on optimizing this metal combination for electrocatalytic production of ammonia from nitrogen. While the performance appears to be tailorable through optimization of the nanoparticle synthesis procedure, nanoparticle lifetime continues to be a challenge. Over a period of several months, the performance of the nanoparticles decreases, and an increase in the oxygen content and presence of oxides is observed. The nanoparticles appear to be polycrystalline with a nanocrystalline domain size of approximately 2-3 nm, and portions of the nanoparticles may be amorphous. The presence of nanocrystallinity and/or amorphous character suggests that the nanoparticles are thermodynamically metastable, and increased stability is likely necessary to retain catalytic performance over time. However, the crystalline/amorphous nature may also be the cause of the observed performance in electrocatalysis applications. An annealing study will be discussed as a possible method to stabilize the nanoparticle structure, as well as the effect of annealing on catalytic performance.

376598 Controlled Deposition of Iron Oxide Nanoparticles Using a Novel Gas-Expanded Liquid (GXL) Process to Generate Supported Fischer-Tropsch Catalysts

Abstract:

Fischer-Tropsch synthesis is a catalytic process used in industry to convert carbon monoxide and hydrogen to heavier hydrocarbons. Active metals like iron, cobalt, nickel and ruthenium are required as catalysts for this process. Supporting these metals on an oxidic support (silica/alumina) has many advantages including an influence in mass selectivity due to the pore structure of the support and stability of the support. Attrition or fracturing inside a fixed-bed or slurry reactor is also reduced. Nano-sized particles also offer several advantages due to their easy dispersability in organic solvents which makes homogeneous loading on supports easier. The usual methods of supported catalyst preparation (ex. incipient wetness impregnation, ion exchange and coprecipitation) rely on various solvent-based transport phenomenon including diffusion and capillary action to disperse the iron into the catalyst pores. Gas expanded liquids (GXLs) have various advantages over conventional solvents due to their pressure tunable transport properties. The diffusion of various molecules is usually enhanced in a GXL due to the lower density/viscosity of solvents at high pressure. We have developed a novel method to controllably deposit iron oxide nanoparticles on a support in the presence of a GXL. This method allows for a more uniform deposition of the metal particles into the pores of the support hence improving the performance of the catalyst in Fischer-Tropsch synthesis. In this study we have carried out preliminary experiments on the synthesis of these supported iron catalysts and tested the effect of various operating parameters on the performance of this catalyst under Fischer-Trosch synthesis conditions.

379001 Controlling Metal Nanoparticles in Fischer Tropsch Synthesis

Abstract:

Controlling Metal Nanoparticles in Fischer Tropsch Synthesis

Jorge Gascon, Freek Kapteijn

The growing concerns about oil depletion have spurred worldwide interest in finding alternative feedstocks for important petrochemical commodities and fuels. Fischer-Tropsch synthesis (FTS) is a surface polymerization reaction where syngas (a mixture of CO and H_2), derived from natural gas, coal, or biomass, is catalytically converted to hydrocarbons over Co, Ru, or Fe catalysts.[1,2] This technology has long been considered as an alternative process to produce long chain paraffins *via* the so-called gas-to-liquids (GTL) process. Owing to their high activity and selectivity to long chain hydrocarbons, Co-based catalysts are the preferred choice. However, earth abundant and cheap Fe displays superior properties over Co-based catalysts, especially for syngas derived from coal (CTL) or biomass (BTL). As compared to Co, Fe is more resistant to poisoning by contaminants that may be present in the syngas feed such as sulfur and ammonia. In addition, Fe-based catalysts are active in the water-gas-shift reaction under typical FTS conditions, enabling the *in-situ* re-adjustment of the H_2/CO molar ratio for the conversion of hydrogen lean syngas originating from coal or biomass.[1] Furthermore, the product slate of Fe-based catalysts is more directed to short chain (unsaturated) hydrocarbons and short chain oxygenates, both among the most important chemical building blocks.

The main challenge in the design of stable FTS catalysts lies in overcoming the high deactivation rate due to sintering, carbon deposition and phase changes. During this lecture, we will highlight different synthesis strategies able to deliver very stable, active and selective FTS supported catalysts.

References

(1) Abelló, S.; Montané, D. ChemSusChem 2011, 4, 1538.

(2) Sun, B.; Xu, K.; Nguyen, L.; Qiao, M.; Tao, F. ChemCatChem 2012, 4, 1498.

Coke-Resistant Core-Shell Catalysts for Dry Reforming of Methane

*Junshe Zhang, Fanxing Li**

Department of Chemical and Biomolecular Engineering
North Carolina State University, Raleigh, NC 27606
*fli5@ncsu.edu

Dry reforming of methane (DRM) represents a promising approach for carbon dioxide utilization and syngas generation. One of the key challenges that hinders the commercialization of DRM resides in catalyst performance. While supported noble metal catalysts are active and less susceptible to coke formation, they are highly expensive. Compared with noble metals, Ni-based catalysts are more commercially attractive. However, industrial application of Ni catalysts in dry reforming reactions is limited by its poor stability, caused by sintering of active metallic particles and carbon deposition on the metal surface.

Previous studies have indicated that coke formation is primarily catalyzed by step and edge sites. In addition, coke formation over small Ni particles is not energetically favored. To date, effective options to inhibit coke formation include poisoning coke-formation sites with sulfur, introducing alkali, alkaline earth or transition metal promoters, chemically or physically dispersing Ni in other metal oxides, confining Ni particles in mesoporous materials, and encapsulating catalyst particles in an inorganic shell. In the current study, we demonstrate that core-shell structured $NiO@SiO_2$ nanoparticles can be fabricated from NiO nanoparticles by the sol-gel method. The core-shell catalysts exhibit excellent catalytic performance toward dry reforming of methane with carbon dioxide at 850 °C; carbon deposition is negligible in 39 h on stream.

390429 Reactions Catalyzed By Metal Particles Supported on Functionalized Carbon Nanotubes in Biphasic Systems

Abstract:

Carbon nanotubes in biphasic systems have recently gained attention as Pickering emulsion stabilizers in the areas of underground oil recovery and heterogeneous catalyst supports due to their ability to undergo functionalization and tune the surface chemistry while maintaining the nanotube's unique electronic and physical properties [1-4]. While a great deal of literature exists for nanoparticle stabilized Pickering emulsions, very little is known about the fundamental behavior of nanotubes as Pickering emulsion stabilizers. In this contribution, Pd nanoparticles are supported on multiwalled carbon nanotubes with varying degrees of defects and hydrophilic functional groups. A combination of detailed emulsion characterization is coupled with phase specific chemical reactions to quantify both the location of functionalized carbon nanotubes at an oil/water interface, as well as the role of diffusion across the interface for enhanced reaction control.

A combination of acoustic spectroscopy, optical microscopy, SEM and TEM images of thin cross sections of the Pickering emulsions is used to quantify the influence of various surface functional groups on the interfacial area stabilized by carbon nanotube emulsions, as well as their orientation at the oil/water interface. This information is coupled with chemical reactions of olefins and small oxygenates with exclusive solubility in either the oil or water phase to measure the number of catalytic particles in each phase, as well as the rate of diffusion across the oil/water interface. Finally, this information is used to prepare novel hybrid emulsions containing blends of carbon nanotubes with various functional groups to create a gradient in catalytic particles along the oil/water interface enabling the enhanced control of chemical reactions in biphasic systems.

References:

[1] Crossley, S., Faria, J., Shen, M., & Resasco, D. E. (2010). Solid nanoparticles that catalyze biofuel upgrade reactions at the water/oil interface. *Science*, *327*(5961), 68-72.

[2] Shen, M., & Resasco, D. E. (2009). Emulsions Stabilized by Carbon Nanotube– Silica Nanohybrids. *Langmuir*, *25*(18), 10843-10851.

[3]Drexler, S., Faria, J., Ruiz, M. P., Harwell, J. H., & Resasco, D. E. (2012). Amphiphilic nanohybrid catalysts for reactions at the water/oil interface in subsurface reservoirs. *Energy & Fuels*, *26*(4), 2231-2241.

[4]Shi, D., Faria Albanese, J. A., Pham, T. N., & Resasco, D. E. (2014). Enhanced Activity and Selectivity of Fischer-Tropsch Synthesis Catalysts in Water/Oil Emulsions. *ACS Catalysis*.

388691 Solid State Chemistry Mass Production of Platinum Group Metal Catalysts with Tailored Particle Morphology

Abstract:

Catalytic property of platinum group metal (PGM) nanoparticles can be altered significantly by the nature of facets exposed. For example, Pt (100) plane excels in activity/selectivity compared to other planes in many reactions, including ring-opening hydrogenation of pyrroles, benzene hydrogenation, methanol oxidation (MOR), and electro-oxidation of ammonia (AOR). Pt-Ni (111) surface can exhibit exceptionally high activity towards oxygen reduction reaction (ORR) for polymer electrolyte membrane fuel cells (PEMFCs). The findings have stimulated the exploration of new methods for preparing PGM catalysts with tailored particle morphology, because the usage of PGMs and thus the cost can be largely decreased. However, to date, there have been no feasible methods for cost-effective and mass production of the shaped PGM catalysts.

We realize mass production of PGM catalysts with tailored particle morphology by developing a robust, green, and low-cost solid state chemistry approach. A variety of shaped PGM catalysts, including cubic Pt/SiO_2, cubic Pt/C, tetrahedral Pd/C, octahedral Pt-Ni/C, and cubic Pt-Cu/C, have been demonstrated for preparation. The experiments suggest that the formation of shaped PGM nanoparticles is resultant of employing both CO and H_2 gases, wherein H_2 aids transportation and reduction of the metal precursors on support and CO is responsible for the particle morphology formation. The prepared octahedral $Pt_{1.5}Ni/C$ catalyst exhibits high ORR activities of 3.99 mA/cm^2 Pt and 1.96 A/mg Pt at 0.90 V vs. RHE, which are about 20 and 10 times the values for commercial Pt/C specifically. The cubic Pt/C catalyst shows 1.44 mA/cm^2 at 0.6 V vs. RHE in AOR, which is five times that of 0.30 mA/cm^2 using commercial Pt/C.

OPTIMIZATION OF GREEN SYNTHESIS OF GOLD NANOPARTICLES USING *DELONIX REGIA* LEAF EXTRACT AND EVALUATION OF THEIR CATALYTIC ACTIVITY

*Preeti Dauthal, Mausumi Mukhopadhyay**

S.V. National Institute of Technology, Surat, Gujarat, India
Tel.: +91 261 2201645; Fax: +91 261 2227334
*Email: mmu@ched.svnit.ac.in; mausumi_mukhopadhyay@yahoo.com

Introduction

Noble metals such as of gold, palladium, platinum and silver with nano-size dimensions have received considerable attention due to their endless promises in the field of material science. Among these noble metal nanoparticles (NPs), particularly gold, nanoparticles (Au-NPs) plays a significant role in various catalytic reactions due to high surface area to volume ratio, energetic surface atoms and great surface energy. Biofabrication methods for the synthesis of these noble nanoparticles are not only a noble approach to synthesize benevolent nano-structures, but also a footstep to diminish the utilization of toxic chemical and generation of perilous by-products, which are hazardous to environment and public health. These methods are considered as benign, sustainable, cost-effective and ecologically sound. Biofabrication of nanoparticles using microorganisms is reported during recent studies (1, 2). However, use of the plant extracts as the potential resource for fabrication of NPs is gaining significant importance as unlike microbial fabrication these methods do not involve isolation and culture preparation techniques.

In a recent study, antioxidant potential of *Delonix regia* (*D. regia*) has been utilized for cost-effective and eco-benign fabrication of palladium nanoparticles (3). However, literature survey revealed that there is no report available for biofabrication of Au-NPs using *D. regia.* Thus, the present study is intended to fabricate Au-NPs by utilizing the metal reducing potential of *D. regia* plant extracts. Catalytic potential of fabricated nanoparticles is evaluated for reduction of anthropogenic pollutant 2-Nitroaniline (2-NA). Use of biofabricated nanoparticles as a catalyst in bioremediation reaction is a cost effective, sustainable and eco-friendly alternative (3-6).

Results and discussion

The effects of the concentration of *D. regia* leaf extracts, reaction temperature, reaction time and pH were evaluated for biofabrication of Au-NPs. Optimization of above reaction conditions was carried out based on surface plasmon resonance (SPR) using UV-visible spectroscopy analysis. Further biofabrication of Au-NPs were carried out using following optimized conditions: 30 °C reaction temperature, 4 mL of 1×10^{-3} M $HAuCl_4.3H_2O$, 0.5 mL of the leaf extract, pH 6 and 10 min reaction time.

Characterization of nanoparticles

Dynamic Light Scattering (DLS), ζ potential, Transmission Electron Microscopy (TEM), X-ray diffraction (XRD), Energy Disperse X-ray Spectroscopy (EDX) and Fourier Transform Infrared Spectroscopy (FTIR) techniques were adopted for evaluation of size, stability, morphology, crystal nature and purity of the biofabricated Au-NPs at optimized condition. UV-visible spectra represented SPR peak at 542 nm which suggested the reduction of gold ions (Au^{3+}) into zero-valent gold (Au^{0}). The results were rapid for the biofabrication of spherical Au-NPs of about 4-24 nm (by TEM) size within 10 min of reaction time. In the TEM micrograph, the edges of the nanoparticles were lighter compared to that of the centre, suggested capping and adherence of bio-organics of leaf extracts on the surface of Au-NPs (7). This observation was further supported by the FTIR, EDX and DLS analysis. Crystalline nature of nanoparticles was evident to a bright circular spot in the Selected Area Electron Diffraction (SAED) pattern and broad diffraction peaks in the XRD pattern (Figure 1).

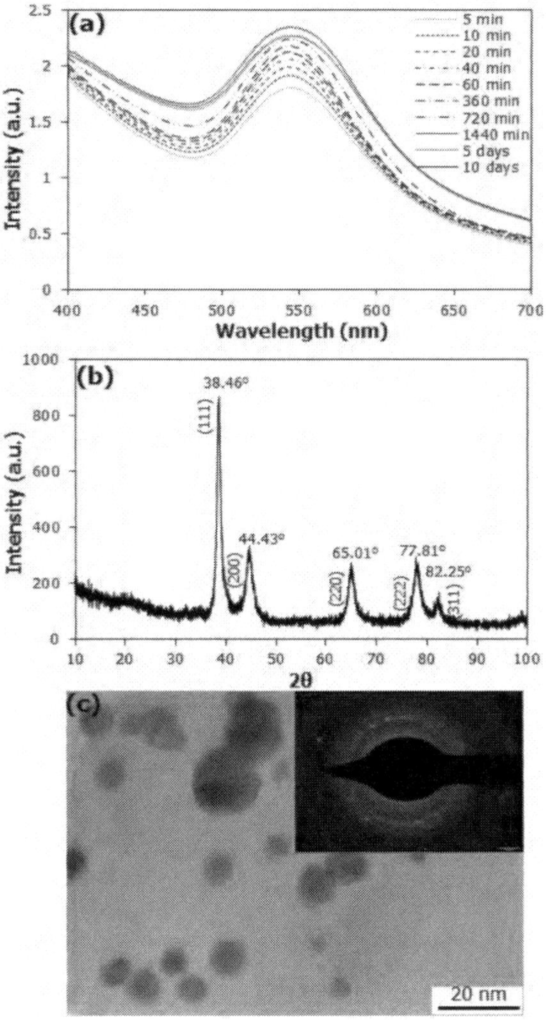

Figure 1. (a) UV-visible spectra (b) TEM micrograph (c) XRD spectra of biofabricated Au-NPs.

XRD pattern represented the five distinctive peaks present at 38.46° (111), 44.43° (200), 65.01° (220), 77.81° (311) and 82.25° (222) positions (JCPDS No. 04–0784), suggested the presence of crystalline gold at zero oxidation state. The lattice constant calculated from these patterns was 4.069 Å, confirmed the face centered cubic (fcc) geometry oriented along the (111) plane. The average particle size was calculated as 25.77 nm with the line width of the maximum intensity (111) diffraction peak by Debye Scherer equation. Thus, the XRD pattern proved strong evidence in favor of the TEM analysis for the presence of crystalline Au-NPs.

The appearance of optical absorption peak at 2.2 keV in EDX spectra further confirmed the presence of Au-NPs. High negative ζ potential (-15 mV) on the surface of biofabricated Au-NPs suggested long-term stability of colloidal Au-NPs. Particle size distribution of colloidal Au-NPs in DLS analysis was found in the range of 12-55 nm with an average particle size of about 39 nm with polydispersity index (PdI) 0.279, which appears to be higher as compare to TEM and XRD measurements. This was due to interference created by electric double layer on charged particles and overlapping NPs. Particles size observed in DLS is usually higher as the various forces of interaction in the solution like van der Waals forces of attraction is prominent (8). However, aggregation of NPs was avoided in the TEM imaging technique lead more precise observation (3).

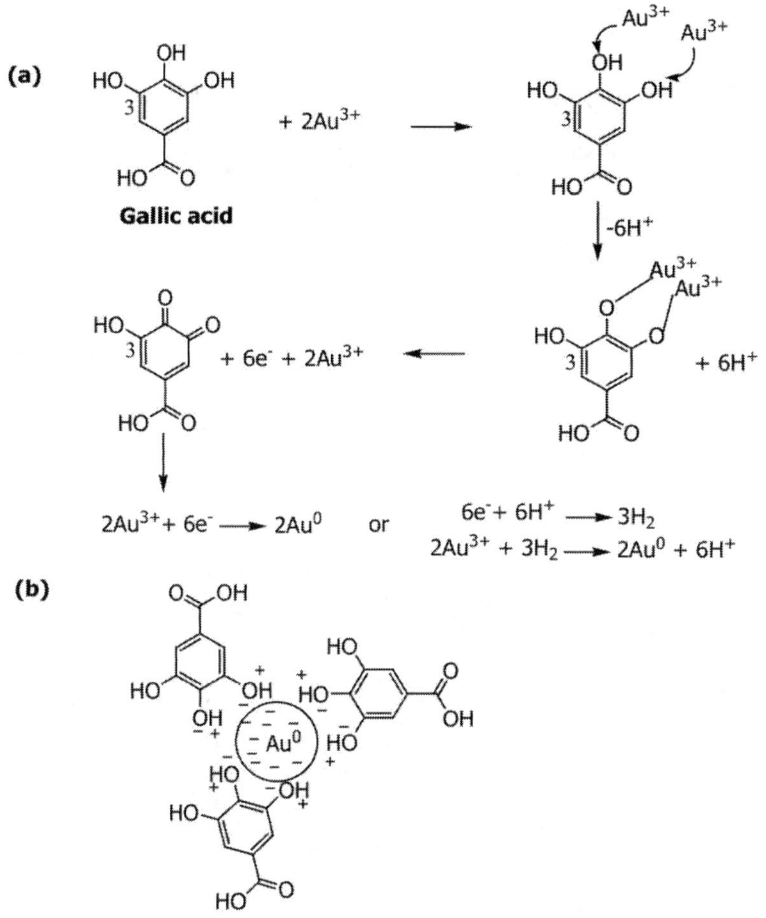

Figure 2. Probable reaction mechanism for (a) bio-reduction and (b) stabilization of Au-NPs using gallic acid.

FTIR analysis confirmed further the role of polar polyphenolic compounds of *D. regia* leaf extracts for biofabrication of Au-NPs and their chemical frameworks were also provided effective wrapping around the nanoparticles and robustness against agglomeration. Presence of water soluble polyphenolic compounds, including protocatechuic acid, gallic acid and hydroxybenzoic acid (phenolic acid) and catechins (flavonoids) are reported earlier in leaf extract of *D. regia* by HPLC analysis (3). Schematic representation of gallic acid mediated reduction of Au^{3+} ion and formation of respective quinone are shown in Figure 2.

Long-term stability (upto 10 days) of the biofabricated Au-NPs was also confirmed by UV-visible and ζ potential analysis. No significant change was observed in the SPR position and intensity, suggested long-term stability of nanoparticles. The ζ potential was found to be -14.3 mV even after 10 days of the incubation period.

Catalytic activity of nanoparticles

Catalytic activity of biofabricated Au-NPs was evaluated for the reduction of anthropogenic pollutant 2-NA. These nanoparticles were demonstrated to have a good catalytic activity for the reduction of 2-NA to 1, 2 Benzenediamine. Figure 3 represented reduction of 2-NA with $NaBH_4$ in aqueous environment at the surface of biofabricated Au-NPs as the catalyst.

The reduction of reaction followed pseudo first-order kinetics with respect to the substrates. It was also found that high concentration of Au-NPs and high temperature reduced the reaction time for catalytic reduction and increased the reduction rate. This observation suggested that catalytic reduction usually took place at the surface of catalyst (9). Catalytic reduction of 2-NA was found to increase by a factor of 1.6-1.9 for every 10 °C rise in temperature. The pseudo first-order rate constant was estimated to be 12.28×10^{-2} min^{-1} at room temperature (28 ± 2 °C) for this catalytic reduction reaction. The activation energy (E_a) for this catalytic reduction was calculated as 44.26 kJ mol^{-1}.

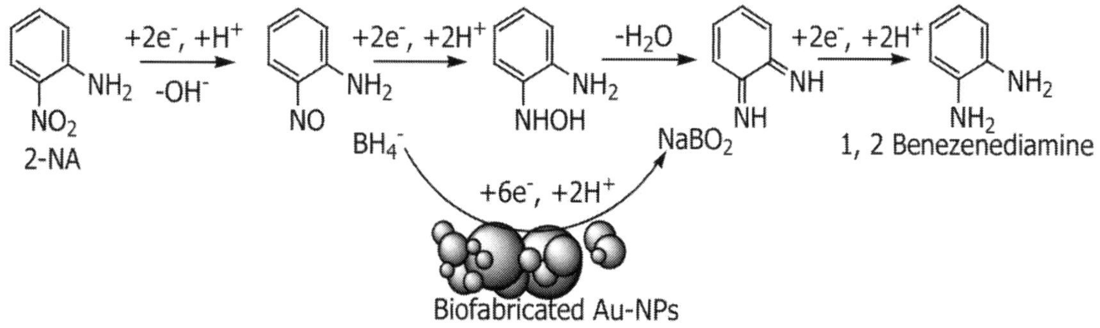

Figure 3. Schematic catalytic reduction of 2-NA to 1, 2 benzenediamine in the presence of $NaBH_4$ and biofabricated Au-NPs as the catalyst.

Au-NPs were recycled up to 10 cycles. It was observed that no significant change observed in the catalytic activity of the recycled catalysts up to 10 cycles. The reaction rate constant obtained at seventh cycle was found to be 10.57×10^{-2} min^{-1}. However, catalytic activity of biofabricated Au-NPs decreased slightly after 7 cycles of reduction reaction. This was due to the agglomeration of NPs (10, 11) or not fully recovered during recycling process. Thus, biofabricated Au-NPs proved as a potent recyclable catalyst for the industrial applications. These results suggested the fabrication of stable Au-NPs with the available biological resource *D. regia*, which offers an inexpensive, sustainable and eco-friendly alternative for fabrication of catalytically active Au-NPs.

References

1. Srivastava, Nishant and Mausumi Mukhopadhyay (2013), "Biosynthesis and Structural Characterization of Selenium Nanoparticles Mediated by *Zooglea ramigera*," Powder Technology, 244, pp. 26-29.
2. Srivastava, Nishant and Mausumi Mukhopadhyay (2014), "Biosynthesis of SnO_2 Nanoparticles Using Bacterium *Erwinia herbicola* and their Photocatalytic Activity for Degradation of Dyes," Industrial & Engineering Chemistry Research, 53, pp. 13971-13979.
3. Dauthal, Preeti and Mausumi Mukhopadhyay (2013), "Biosynthesis of Palladium Nanoparticles using *Delonix regia* Leaf Extract and its Catalytic Activity for Nitro-aromatics Hydrogenation," Industrial & Engineering Chemistry Research, 52, pp. 18131-18139.
4. Dauthal, Preeti and Mausumi Mukhopadhyay (2012), "*Prunus domestica* Fruit Extract-Mediated Synthesis of Gold Nanoparticles and Its Catalytic Activity for 4-Nitrophenol Reduction," Industrial & Engineering Chemistry Research, 51, pp. 13014-13020.
5. Sharma, Nilesh C., Shivendra V. Sahi, Sudip Nath, Jason G. Parsons, Jorge L. Gardea-Torresdey, Tarasankar Pal (2007), "Synthesis of Plant-Mediated Gold Nanoparticles and Catalytic Role of Biomatrix-Embedded Nanomaterials," Environmental Science & Technology, 41, pp. 5137-5142.
6. Dauthal, Preeti and Mausumi Mukhopadhyay (2014), "Biofabrication, Characterization, and Possible Bio-reduction Mechanism of Platinum Nanoparticles Mediated by Agro-industrial Waste and their Catalytic Activity," Journal of Industrial and Engineering Chemistry, DOI: 10.1016/j.jiec.2014.07.009.
7. Suresh, Anil K., Dale A. Pelletier, Wei Wang, Michael L. Broich, Ji-Won Moon, Baohua Gu, David P. Allison, David C. Joy, Tommy J. Phelps and Mitchel J. Doktycz (2011), "Biofabrication of Discrete Spherical Gold Nanoparticles using the Metal-Reducing Bacterium *Shewanella oneidensis*," Acta Biomaterialia, 7, pp. 2148-2152.
8. Dauthal, Preeti and Mausumi Mukhopadhyay (2013), "In-vitro Free Radical Scavenging Activity of Biosynthesized Gold and Silver Nanoparticles using *Prunus armeniaca* (Apricot) Fruit Extract," Journal of Nanoparticle Research, 15, pp. 1366-1376.
9. Murugadoss, Arumugam and Arun Chattopadhyay (2008), "Surface Area Controlled Differential Catalytic Activities of One-Dimensional Chain-like Arrays of Gold Nanoparticles, " The Journal of Physical Chemistry C, 112, pp. 11265-11271.
10. Kuroda, Kyoko, Tamao Ishida and Masatake Haruta (2009), "Reduction of 4-Nitrophenol to 4-Aminophenol over Au Nanoparticles Deposited on PMMA," Journal of Molecular Catalysis A: Chemical, 298, pp. 7-11.
11. Koga, Hirotaka and Takuya Kitaoka (2011), "One-Step Synthesis of Gold Nanocatalysts on a Microstructured Paper Matrix for the Reduction of 4-Nitrophenol," Chemical Engineering Journal, 168, pp. 420-425.

376743 Development of Low-Cost Catalysts for the Destruction of Tars Generated in the Biomass Synthesis Gas (syngas) Production Process

Abstract:

In recent years the process for the production of synthesis gas (syngas) from bio-mass has grown in popularity. Unfortunately, the syngas produced from biomass gasification generally in contaminated with tar which is a complex mixture of polynuclear aromatic hydrocarbons. These tar can cause problems with the downstream processing of the syngas such as clogging filters, deactivating syngas conversion catalysts, and coating and clogging the process piping as the gas cools.

Techniques to treat tars in the crude syngas are a necessary step for the biomass gasification process. In the project, syngas was spiked with naphthalene (used as the model compound for tars) and it destruction was investigated using low cost catalysts. Two clays with different surface area were screened to be catalyst supports. Montmorillonite has a relatively large surface area around 239 m^2/g while kaolinite has a surface area of 10~50 m^2/g were investigated. Catalysts were prepared with/without the presence of sucrose. The catalysts were characterized using TEM, XRD, TGA and TPR. It has been found out the inclusion of sucrose during the impregnation step greatly improved the dispersion of the nickel catalysts for both clays evaluated in this study.

The activity of prepared catalysts was evaluated with 60 g/N^3 naphthalene in the syngas (CO, 20%, CO_2,9%, CH_4, 3%, H_2, 20%, balance, N_2). The steam to carbon ratio was 5. The catalysts were evaluated at 650 and 750 °C. Preliminary results indicate that the addition of sucrose changed the reducibility profile of the catalysts. The catalysts prepared with sucrose showed higher initial activity for 3 hours comparing to around 1 hour for the catalysts without inclusion of sucrose in the impregnation step. In this presentation, details of the catalyst preparation method as well as its effectiveness for tar destruction will be presented.

382727 Progress in Automated Modeling of Biomass Derived Fuels: Updating Reaction Families and Rate Calculation Rules

Abstract:

Thermal conversion of biofuels is very sensitive to the fuel chemistry, and the major challenge in modeling fuels derived from biomass is the presence of a wide range of cyclic oxygenated species. Since manually generating detailed chemical models for biofuels is complex, it is prefer- able to use computers instead. In this study, Reaction Mechanism Generator (RMG)[1], an open- source software, has been used to build detailed kinetic models for gasification of bio-oil, derived from the pyrolysis of biomass. Furthermore, some progress has been made in RMG in order to improve the automated chemical modeling of all fuels derived from biomass.

Biofuels contain many derivatives of lignin, hemicellulose, and cellulose, and it is important to include all reaction classes related to the primary decomposition of these components in order to propose a comprehensive mechanism. In recent work we have investigated that RMG is missing some ring opening reaction families for cylic oxygenated molecules, such as isomerization reac- tions through simultaneous H-migration and C–C or C–O bond breaking. We have now updated RMG's kinetics database with reaction recipes for the new reaction families. We must also specify rules to predict Arrhenius rate parameters for the new reaction classes. However, the number of possible reactions in each reaction family is massive, and applying high-level electronic structure calculations for each would be prohibitively expensive. Instead, rate calculations were performed for a smaller set of reactants belonging to the particular reaction class, then the roles of the different functional groups were deliberated, and group-based rate rules were derived to estimate Arrhenius parameters for any reaction in the new reaction classes.

A detailed kinetic model for bio-oil gasification at high temperature was previously built in RMG, and comparison with literature experiments showed that the RMG-built model couldn't predict CO and CO_2 formation properly. Since products from primary ring opening reactions through C–O bond breaking are major contributors to CO and CO_2 formation, the bio-oil gasification mech- anism has been updated after adding these new reaction classes and associated kinetic parameter estimation rules. There are some significant differences in simulation results between the RMG- built models before and after updating the database, demonstrating the importance of these reaction families and their kinetic features when studying the thermal conversion of biofuels.

1- W. H. Green, J. W. Allen, P. Bhoorasingh, B. A. Buesser, R. W. Ashcraft, G. J. Beran, C. A. Class, C. Gao, C. F. Goldsmith, M. R. Harper, A. Jalan, F. S. Khanshan, G. R. Magoon, D. M. Matheu, S. S. Merchant, J. D. Mo, S. Pet- way, S. Raman, S. Sharma, B. Slakman, J. Song, K. M. V. Geem, J. Wen, R. H. West, A. Wong, H.-W. Wong, P. E. Yelvington, N. Yee, J. Yu, RMG — Reaction Mechanism Generator-Python, rmg.mit.edu (2013) RMG — Reaction Mechanism Generator.

385198 Hydropyrolysis of Lignin Using Pd/HZSM-5

Abstract:

The aim of this work was to study the formation of cycloalkanes from the hydropyrolysis of lignin with HZSM-5 and Pd/HZSM-5 catalysts. Cycloalkanes are important components of jet-fuels, and hydropyrolysis is proposed as a potential route for the conversion of lignin because: 1) the hydrogen medium promotes hydrodeoxygenation reactions, leading to hydrocarbons, and 2) the hydrogen caps reactive radicals resulting from the decomposition of lignin. These radicals would, otherwise, recombine to form char. We observed that palladium supported on HZSM-5 catalyzed hydrogenation and deoxygenation reactions that converted phenolic compounds into aromatic and cycloalkane molecules. This study analyzed the effect of the catalyst-to-lignin ratio, H_2 partial pressure, and temperature on the yields of hydrocarbons with HZSM-5 and Pd/HZSM-5. Pd/HZSM-5 produced 44% more aromatic hydrocarbons than HZSM-5 at a catalyst-to-lignin ratio of 20:1, 650°C, and a constant H_2 partial pressure of 1.7 MPa. The presence of palladium led to significant difference in yields only at 1.7 MPa H_2 partial pressure. In addition, the hydropyrolysis temperature played a substantial role in the equilibrium conversion of hydrogenation reactions that led to cycloalkanes directly from lignin. In addition, a comparison of in-situ and ex-situ hydropyrolysis on hydrocarbon yield will be presented, with further details on the effect of temperature on the mechanistic pathways of these two upgrading pathways.

Continuous Operation of a Hydothermal Biorefinery

An Abstract for the 2014 National Meeting of AIChE

Michael T. Timko and Geoffrey Tompsett

Worcester Polytechnic Institute

Conversion of biomass feeds into fuels and chemicals requires breaking down their recalcitrant components; strategies including high temperatures, acids, alkali, oxidants, liquid ammonia, and ionic liquids have been evaluated. To date, the majority of these techniques have focused on batch-wise processing. Many of these techniques are either extremely water intensive or require expensive reactants that must be removed prior to downstream biochemical or thermocatalytic upgrading. Continuous-flow processing of biomass slurries in a hydrothermal reactor may afford advantages in terms of water and reactant use. Many engineering challenges remain unsolved, including how to deal with reactor plugging, whole biomass utilization, and high-pressure slurry pumping. Here, we have focused our efforts on developing a continuous flow reactor to achieve rapid heating rates in a turbulent mixer to break-up the lignocellulosic matrix, followed by reaction in a secondary zone to maximize simple carbohydrate production. The two-reaction configuration aims to reduce or eliminate degradation of the primary simple carbohydrate products by carefully controlling their residence times in the high-temperature zone. We present time-on-stream data for the prototype reactor with different model feeds. Data include temperature performance of the reactor and chemical analysis of the products using liquid chromatography and for total organic carbon content. We compare continuous-reactor results with similar results obtained in a high-temperature batch reactor where we test the effects of particle size and heating rate on the characteristics of the bio-oil product.

Catalytic epoxidation of propylene glycol and its acetates

Motivation:

Propylene oxide (PO) is a valuable intermediate chemical with a global market of more than 9.2 million tonnes in 2014, expected to steadily increase in the future.[1] PO is used to make many everyday products such as defoamers, anti-freeze, and lubricants in the form of polyglycols, polyurethanes and polyglycol ethers. Conventionally, PO is manufactured by the epichlorohydrin process or by peroxidation; both processes rely heavily on the availability of petroleum derived propylene. The chlorohydrin process involves the formation (and therefore requires the disposal) of stoichiometric amounts of sodium or calcium chlorides. The hydroperoxide route produces undesirable by-products that are difficult to separate, such as methyl formate. In the interest of reducing our reliance on fossil sources and shifting to more environmentally friendly processes, alternate routes to make propylene oxide are being investigated.

Biodiesel production by transesterification of triglycerides produces glycerol as a side product, providing the opportunity to build reaction networks with glycerol as a building block. Instead of the traditional pathway to propylene glycol from propylene derived propylene oxide, the pathway to propylene oxide from bio-glycerol derived propylene glycol becomes an attractive option. Producing propylene glycol from the dehydroxylation of glycerol has been extensively studied and is now a commercial process.[2-5]

Studies focusing on producing propylene oxide from propylene glycol in literature are broadly of two categories. One route being explored is the direct dehydration of propylene glycol to propylene oxide.[6-8] The other is indirect; propylene glycol is converted to its esters, and the vicinal hydroxyl esters formed are then converted to propylene oxide.[9-12] The optimized yield reported in multiple studies from both these routes using a variety of catalysts and processes is at ~32%. Of these, the highest reported conversions were for Cesium on silica at 52%.[7] The reaction of PG and its acetates to PO occurs at high temperature (>300°C) in the gas phase and proceeds with the formation of various side products. Most prior studies screen a number of catalysts, but good process design requires a thorough understanding of the effect of macro process parameters as well as fundamental catalyst properties. In the current work, a few candidate catalysts were chosen and a detailed investigation of the effect of reaction temperature (340°C to 420°C), feed material (PG vs. PG acetates), contact time, and the loading of catalytic salt on the support was conducted.

Methods:

Catalyst preparation: The procedure followed for preparation of catalysts is as described in a previous work.[7] Base salt is impregnated on silica and dried at 100°C for 2 hrs. Catalyst is then calcined for 6 hrs at 550°C and stored in a desiccator.

Feed preparation: Dried PGDA and PG were mixed in different molar ratios with 2% by weight Amberlyst 15 catalyst (dried in a vacuum oven at 100°C for well over 2 days), in a batch reactor set up at 70°C and allowed to equilibrate. Feed was stored with drierite granules to ensure minimal concentration of water, leaving little chance for hydrolysis of PGA esters.

Fixed bed reactor:

The reactor is a stainless steel tube of 0.5 in outer diameter and 43 cc volume. A thermoprobe is fitted axially along the length of the reactor facilitates temperature readings at different lengths during the course of the reaction. Either borosilicate beads (2mm) or silicon carbide were used as packing material, and loaded below and above the catalyst bed. Liquid feed is introduced at the top of the reactor through a narrow tube just above the pre-heating zone. Liquid flow rate is maintained by an HPLC pump (Varian). Nitrogen gas flows concentrically around the feed tube. Nitrogen gas flow rate is maintained by a mass flow controller. The bottom of the reactor is connected to a two chamber condenser system fitted with a three way valve to direct product mixture to either chamber. The collection vessels are fitted with dip tubes and valves to collect samples during reaction. The vessels are maintained at ~-79°C in dry ice-acetone baths.

Analysis:

Samples collected were diluted 10× in a solution of acetonitrile and decanol as the internal standard. All products were quantified by a Varian 450 GC fitted with a thermal conductivity detector. Carbon recoveries were between 98-102 mole %, PG backbone and acetic acid balances were within 94-98 and 90-95 mole% respectively.

Results:

Preliminary experiments showed that PGA feed under purely thermal conditions undergoes little further esterification when catalyst in the reactor was replaced by inert packing material. When justsupport silica gel with no base added was placed in the reactor, only a little propanal was observed, as would be expected, due to the low concentration of acidic silanol groups on the surface. Initially, attempts were made to match the reported highest conversion of PG with cesium nitrate on silica gel[7], but reaction temperatures higher than 380°C resulted in plugging of the reactor. At 360°C, the highest yield of PO was 5%. This is roughly the same as the result obtained from a molten KOAC catalyst at the same temperature.[13] In order to go to higher reaction temperatures, potassium hydroxide catalyst was studied.

Although PG feed conversion increased with temperature, the maximum selectivity to PO of 80% was found at ~400°C, after which the elimination reactions to propanal, acetone and allyl alcohol were more dominant. With increasing contact time, conversion of PG feed increased, but selectivity of PO decreased, indicating that secondary conversion of PO to its isomers is important. The selectivity of PO was found to be at first directly proportional to the loading of basic salt catalyst on the support, but plateaued when the surface of catalyst became saturated. Current efforts focus on understanding the chemical nature of the catalyst on the surface during the reaction. Thermogravimetric analysis shows that potassium acetate catalyst after calcination becomes potassium carbonate. A comparison of KOH and KOAc gave similar results of selectivity and conversion under identrical reaction conditions, indicating that either the anion salt is not a factor for catalyst activity, or that a common acetate intermediate is involved in the formation of actual catalyst species under reaction conditions. In the presence of acetate species in the feed, KOH might be converted to KOAc and then might decompose to potassium carbonate. To confirm this, post reaction FTIR of catalysts are being analysed to evaluate the nature of KOH on silica gel catalysts under reaction conditions.

References:

1. http://energy.gov/articles/lab-breakthrough-adm-leads-petroleum-free-glycol-production-facility
2. Zuwei, X.; Ning, Z.; Yu, S.; Kunlan, L., Reaction-Controlled Phase-Transfer Catalysis for Propylene Epoxidation to Propylene Oxide. *Science* **2001,** *292* (5519), 1139-1141.

3. Robert C. Binning, H. R. N. Production and recovery of propylene oxide by plural stage distillation and caustic addition. 1967.

4. Meiers, R.; Dingerdissen, U.; Halderich, W. F., Synthesis of Propylene Oxide from Propylene, Oxygen, and Hydrogen Catalyzed by Palladium-Platinum Containing Titanium Silicalite. *Journal of Catalysis* **1998,** *176* (2), 376-386.

5. Clerici, M. G.; Bellussi, G.; Romano, U., Synthesis of propylene oxide from propylene and hydrogen peroxide catalyzed by titanium silicalite. *Journal of Catalysis* **1991,** *129* (1), 159-167.

6. M. Constantini, M. J. Production of epoxides from lower alkane 1,2 diols. 1981.

7. Yu, Z.; Xu, L.; Wei, Y.; Wang, Y.; He, Y.; Xia, Q.; Zhang, X.; Liu, Z., A new route for the synthesis of propylene oxide from bio-glycerol derivated propylene glycol. *Chemical Communications* **2009,** *0* (26), 3934-3936.

8. Ryu, J. Y. Process for epoxidizing vicinal diols. 1985.

9. M.B. Sherwin, M. E. F. Process for preparing oxirane compounds. 1979.

10. M.B. Sherwin, J. P. Cracking of mixtures containin hydroxyesters. 1977.

11. H-M Weitz, J. H., R. Platz Manufacture of propylene oxide. 1978.

12. G. Fouquet, F. M., K. Baer Manufacture of propylene oxide. 1978.

13. Gusevskaya, E. V., Karandin, A. V., Likholobov, V. A., Kondratiev, A. V., & Markevitch, V. S. Propylene oxide synthesis via propene acetoxylation over supported palladium and platinum catalysts followed by cracking of glycol acetates in a melt of potassium acetate. *Applied Catalysis A: General*, **1993**, *97*(1), 1-15.

383604 Reactive Distillation Process for the Production of Furfural from Xylose Using Solid Acid Catalysts

Abstract:

There is growing interest in the production of chemicals and fuels from renewably sourced feedstocks due to rising petroleum prices, limited availability, and increasing consciousness of customers about environmentally-friendly processes. Furthermore, fossil fuels used to manufacture chemicals and fuels have an increased environmental footprint relative to renewable feedstocks. The overall concept of converting biomass into chemicals provides several advantages to make processes inherently safer and more sustainable.

Furfural, an intermediate derived from the hemicellulose portion of lignocellulosic biomass, is considered as one of the key renewably sourced feedstocks for products that can compete with petroleum-based chemicals [1]. Furfural currently has numerous commercial applications, including the manufacture of furfuryl alcohol, but the availability of commercial quantities of economically-priced furfural could lead to an expansion to other products and markets. The current commercial furfural manufacturing processes use corn cob and bagasse with ~50% molar yield [2]. Based on this low yield and other drawbacks of the current technology, we undertook a research endeavor to develop a novel process using reactive distillation with solid acid catalysts to convert xylose into furfural.

A comprehensive experimental study was carried out using synthetic xylose feeds containing a mixture of pure xylose, water, and sulfolane as a solvent. A number of different solid acid catalysts were tested in a reactive distillation process for their ability to produce furfural from xylose. The reactive distillation process provides a unique advantage of immediate separation of furfural from the reaction zone as it is formed, thus minimizing the formation of undesired by-products (e.g., humins) from subsequent furfural reactions. During this process, the aqueous feed was introduced at the top of the column, and xylose was dehydrated to furfural which was co-distilled with water out of the top of the column. The high-boiling sulfolane, with dissolved humins, was drained through the bottom of the column. Among various catalysts screened, the zeolite H-mordenite (Si/Al = 10) gave the highest furfural yield of about 80%.

This study was further extended to using authentic aqueous process feeds containing C5-sugars. Salts present in these feeds resulted in significant catalyst deactivation. We present a systematic study to show how we resolved this problem, which resulted in extended catalyst lifetime and furfural yield, similar to that obtained with synthetic xylose feeds.

References

[1] J.J. Bozell, G.R. Petersen, Green Chemistry 12 (2010) 539-554.

[2] R. Karinen, K. Vilonen, M. Niemelä, ChemSusChem 4 (2011) 1002-1016.

Kinetic modeling of cellulose deconstruction via acid hydrolysis

Sri Bala Gorugantu and R. Vinu

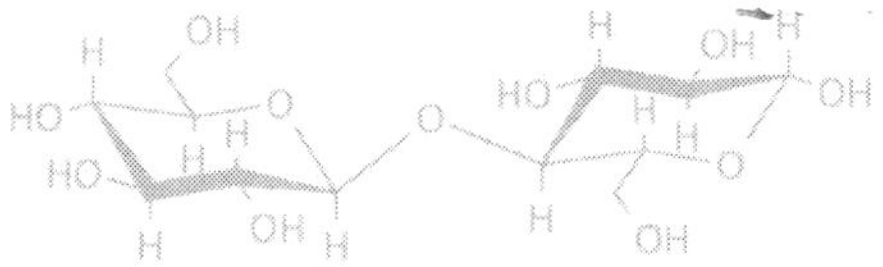

Complex Reaction Kinetics Group
Department of Chemical Engineering, IIT Madras, India

Acid hydrolysis - An Introduction

- One of the oldest pretreatment methods in biomass conversion
- Works for both pretreatment and hydrolysis
- Dilute acid hydrolysis:
  - No cellulose dissolution. Hence *heterogeneous*
  - Carried out at very high temperatures
  - Biomass with low-lignin content
- Concentrated acid hydrolysis:
  - Cellulose gets completely dissolved. Hence *homogeneous*
  - Carried out at relatively low temperatures
  - Holocellulose hydrolysis

Acid hydrolysis - An Introduction

- **Saeman model[1]**: simple series reaction

$$\text{Cellulose} \xrightarrow{k_1} \text{Glucose} \xrightarrow{k_2} \text{Degradation products}$$

$$k_i = k_{i0} A^m \exp\left(\frac{-E_a}{RT}\right)$$

k_{io} = pre-exponential factor (*time*$^{-1}$)
A = acid concentration
E_a = activation energy (*kJ/mole or kcal/mole*)
T = temperature in *K*
R = universal gas constant (*kJ/mol.K*)

- **Abatzoglov et al. Model[2]** : Includes intermediate oligomer formation

$$\text{Cellulose} \xrightarrow{\alpha k_1} \text{Oligo-derivatives} \underset{\gamma k_1}{\overset{\beta k_1}{\rightleftarrows}} \text{Glucose} \xrightarrow{k_2} \text{Degradation products}$$

α= 0.6, β= 0.9, γ= 0.0

1. Saeman, J. F. *Ind. Eng. Chem.* **1945**, 37, 43-51.
2. Abatzoglov, N.; Bouchard, J.; Chornet, E. *Can. J. Chem.* Eng. **1986**, 64, 781-786

Acid hydrolysis - An Introduction

Substrate	Reaction conditions*	k_{10} (min^{-1})	E_{a1} (kJ/mol)	m_1	k_{20} (min^{-1})	E_{a2} (kJ/mol)	m_2	Ref.
Microcrystalline cellulose	Acid: 30-70 wt.% Temp.: 25-40 °C	2.946×10^{10}	127.2	6.0	7.98×10^{26}	166.9	0.77	1
Paper refuse	Acid: 0.2-1.0 wt.% Temp.: 180-240 °C	28×10^{19}	188.7	1.78	4.9×10^{14}	137.2	0.55	2
Municipal solid wastes	Acid: 1.3-4.4 wt.% Temp.: 200-240 °C	1.94×10^{16}	171.61	1.0	1.148×10^{12}	142.3	0.66	3
Douglas fir	Acid: 0.4-1.0 wt.% Temp.: 170-190 °C	1.73×10^{19}	179.5	1.34	2.38×10^{14}	137.5	1.02	4
Hardwood	Acid: 4.41-12.2 wt.% Temp.: 170-190 °C	6.6×10^{16}	165.34	1.64	6.4×10^{12}	128.9	1.10	5

*Acid used in all the experiments was H_2SO_4. The kinetic parameters correspond to equation as defined by Saeman model

1. Camacho, F.; Gonzalez-Tello, P.; Jurado, E.; Robles, A. *J. Chem. Technol. Biotechnol.* **1996**, 67, 350.
2. Fagan, R. D.; Grethlein, H. E.; Converse, A. O.; Porteous, A. *Environ. Sci. Technol.* **1971**, 5, 545.
3. Malester, I. A.; Green, M.; Shelef, G. *Ind. Eng. Chem. Res.* **1992**, 31, 1998.
4. Saeman, J. F. *Ind. Eng. Chem.* **1945**, 37, 43-51.
5. Song, S. K.; Lee, Y. Y. *Biomass.* **1984**, 6, 93.

Acid hydrolysis - An Introduction

Substrate	Reaction conditions*	k_{10} (min^{-1})	E_{a1} (kJ/mol)	m_1	k_{20} (min^{-1})	E_{a2} (kJ/mol)	m_2	Ref.
Microcrystalline cellulose	Acid: 30-70 wt.% Temp.: 25-40 °C	2.946×10^{10}	127.2	6.0	7.98×10^{26}	166.9	0.77	1
Douglas fir	Acid: 0.4-1.0 wt.% Temp.: 170-190 °C	1.73×10^{19}	179.5	1.34	2.38×10^{14}	137.5	1.02	4

*Acid used in all the experiments was H_2SO_4. The kinetic parameters correspond to equation as defined by Saeman model

1. Camacho, F.; Gonzalez-Tello, P.; Jurado, E.; Robles, A. *J. Chem. Technol. Biotechnol.* **1996**, 67, 350.
2. Fagan, R. D.; Grethlein, H. E.; Converse, A. O.; Porteous, A. *Environ. Sci. Technol.* **1971**, 5, 545.
3. Malester, I. A.; Green, M.; Shelef, G. *Ind. Eng. Chem. Res.* **1992**, 31, 1998.
4. Saeman, J. F. *Ind. Eng. Chem.* **1945**, 37, 43-51.
5. Song, S. K.; Lee, Y. Y. *Biomass.* **1984**, 6, 93.

Acid hydrolysis - An Introduction

Substrate	Reaction conditions*	k_{10} (min^{-1})	E_{a1} (kJ/mol)	m_1	k_{20} (min^{-1})	E_{a2} (kJ/mol)	m_2	Ref.
Microcrystalline cellulose	Acid: 30-70 wt.% Temp.: 25-40 °C	2.946×10^{10}	127.2	6.0	7.98×10^{26}	166.9	0.77	1
Douglas fir	Acid: 0.4-1.0 wt.% Temp.: 170-190 °C	1.73×10^{19}	179.5	1.34	2.38×10^{14}	137.5	1.02	4

P 1: Ambiguity in rate parameters

P 2: Lumping of reactions is not favourable to understand the exact kinetics

*Acid used in all the experiments was H_2SO_4. The kinetic parameters correspond to equation as defined by Saeman model

1. Camacho, F.; Gonzalez-Tello, P.; Jurado, E.; Robles, A. *J. Chem. Technol. Biotechnol.* **1996**, 67, 350.
2. Fagan, R. D.; Grethlein, H. E.; Converse, A. O.; Porteous, A. *Environ. Sci. Technol.* **1971**, 5, 545.
3. Malester, I. A.; Green, M.; Shelef, G. *Ind. Eng. Chem. Res.* **1992**, 31, 1998.
4. Saeman, J. F. *Ind. Eng. Chem.* **1945**, 37, 43-51.
5. Song, S. K.; Lee, Y. Y. *Biomass.* **1984**, 6, 93.

Sri Bala Gorugantu | AIChE Annual Meeting 2014 | November 18, 2014

Objective and Methodology

Develop a unified kinetic model for cellulose acid hydrolysis

Objective and Methodology

Develop a unified kinetic model for cellulose acid hydrolysis

Methodology

- Include formation of LMWPs like cello-biose, -triose,.....-hexaose*
- Modified Arrhenius equations:

$$k_i = k_{i0} A^m \exp\left(\frac{-E_a}{RT}\right)$$

All reactions except glucose degradation

$$k_g = k_{g0} A^n T^y \exp\left(\frac{-E_{ag}}{RT}\right)$$

Glucose degradation reaction

- Fit the rate parameters *m, n, y* to match the experimental data
- Test the model applicability to biomasses like walnut green skin, yellow poplar wood

* Xiang. et al. *Appl. Biochem. Biotechnol.* **2004**, 113-116, 1127

Sri Bala Goruganti | AIChE Annual Meeting 2014 | November 18, 2014

Model development : Mechanism

- Random chain scission $C(x) \xrightarrow{k_r} C(x')+C(x-x')$

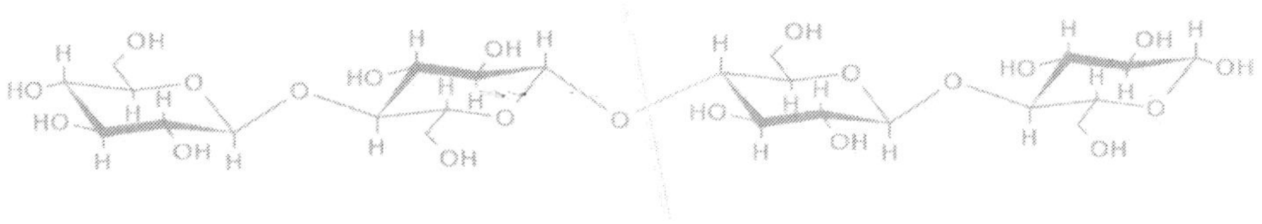

Model development : Mechanism

- Random chain scission $C(x) \xrightarrow{k_r} C(x')+C(x-x')$
- Chain-end scission $C(x) \xrightarrow{k_c} C(x-x_s)+Q_s(x_s)$

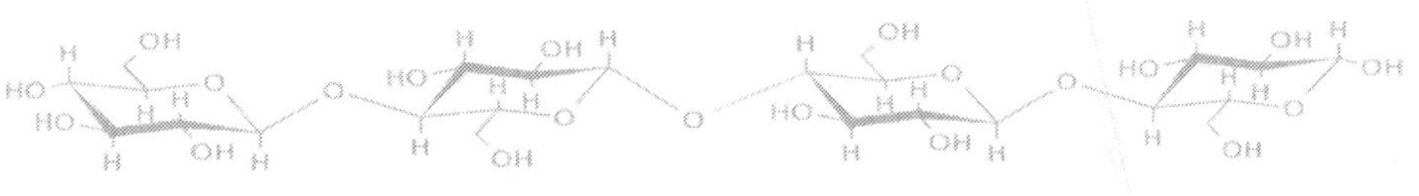

Model development : Mechanism

- Random chain scission $C(x) \xrightarrow{k_r} C(x') + C(x-x')$
- Chain-end scission $C(x) \xrightarrow{k_c} C(x-x_s) + Q_s(x_s)$
- Oligomer decomposition $Q_i(x_i) \xrightarrow{k_q} Q_{i-j}(x_{i-j}) + Q_j(x_j)$
- Glucose degradation $Q_1(x_1) \xrightarrow{k_g}$ Degradation products

Cellohexose $\xrightarrow{k_{13}}$ Cellopentose+Glucose
Cellohexose $\xrightarrow{k_{14}}$ Cellohexose+Cellobiose
Cellohexose $\xrightarrow{k_{15}}$ 2Cellotriose

Cellopentose $\xrightarrow{k_{11}}$ Cellotetraose+Glucose
Cellopentose $\xrightarrow{k_{12}}$ Cellotriose+Cellobiose

Cellotetraose $\xrightarrow{k_9}$ Cellotriose+Glucose
Cellotetraose $\xrightarrow{k_{10}}$ 2Cellobiose

Cellotriose $\xrightarrow{k_8}$ Cellobiose+Glucose

Cellobiose $\xrightarrow{k_7}$ 2Glucose

Model development : Mechanism

- Random chain scission $C(x) \xrightarrow{k_r} C(x') + C(x-x')$
- Chain-end scission $C(x) \xrightarrow{k_c} C(x-x_s) + Q_s(x_s)$
- Oligomer decomposition $Q_i(x_i) \xrightarrow{k_q} Q_{i-j}(x_{i-j}) + Q_j(x_j)$
- Glucose degradation $Q_1(x_1) \xrightarrow{k_g}$ Degradation products

16 Reactions
8 species

Model development : Assumptions

- All reactions are pseudo-first order and irreversible*
- Pre-exponential factors and activation energies for the formation of all oligomers are considered to be same
- Pre-exponential factors for formation of glucose from oligomers (except cellobiose) are considered to be same*
- Degree of polymerization of initial cellulose = 847*
- Number average molecular weight, *Mn* = 135554 g/mol*
- Polydispersity index, PDI = 2.25*

* Pilath et al. *J. Agric. Food Chem.* **2010**, 58, 6131.
* Freundenberg, K.; Blomqvist, G. **Ber. Dtsch. Chem. Ges.** **1935**, 68, 2070.
* Vinu, R.; Broadbelt, L. J. *Energy Environ. Sci.* **2012**, 5, 9808.

Model development : Model equations

PBEs

$$\frac{\partial c(x,t)}{\partial t} = -k_r c(x,t) - \left(k_\alpha + k_\beta + 4k_\gamma\right)c(x,t) + 2k_r\int_x^\infty c(x',t)\Omega_r(x,x')dx' + k_\alpha\int_x^\infty c(x',t)\Omega_1(x-x_1,x')dx' + k_\beta\int_x^\infty c(x',t)\Omega_2(x-x_2,x')dx' + \left(\sum_{s=3}^{6} k_\gamma\int_x^\infty c(x',t)\Omega_s(x-x_s,x')dx'\right)$$

Method of Moments

$$c^{(n)} = \int_0^\infty x^n c(x,t)dx$$

$\Omega_r(x, x') = 1/x'$
$\Omega_s(x'-x_s, x') = \delta(x(x'-x_s))$
$\Omega_s(x_s, x') = \delta(x-x_s)$

ODEs

$$\frac{dc^{(n)}}{dt} = -\left(\frac{n-1}{n+1}\right)k_r c^{(n)} - \left(k_\alpha c^{(n)} + k_\beta c^{(n)} + 4k_\gamma c^{(n)}\right) + \sum_{j=0}^{n}\binom{n}{j}\left(k_\alpha x_1^{\,j} + k_\beta x_2^{\,j} + 4k_\gamma\sum_{s=3}^{6} x_s^{\,j}\right)c^{(n-j)}$$

Methodology

- Reaction conditions:
  - low temperatures (25 °C) to high temperatures (225 °C)
  - Extremely low conc. (0.07%) to high conc. (70%)
- Acid considered: Sulphuric acid
- Experimental data were extracted from various sources
- ODEs obtained for all the species were solved using *ode45* solver in MATLAB®
- Parameters m, n, y are fitted using *lsqcurvefit* toolbox

Results : Dilute acid- High temperature regime

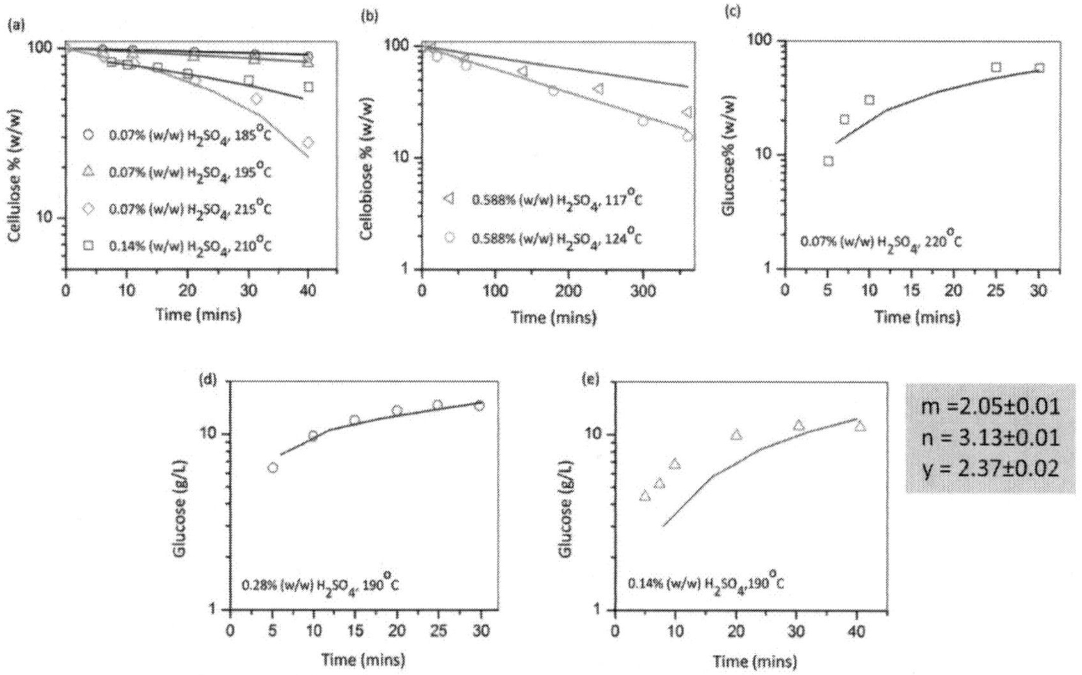

m =2.05±0.01
n = 3.13±0.01
y = 2.37±0.02

Expt. data: *Gurgel et al. *Ind. Eng. Chem. Res.* **2012**, 51, 1173. *Xiang et al. *Appl. Biochem. Biotechnol.* **2003**, 105-108, 505.
*Kim et al. *Appl. Biochem. Biotechnol.* **2001**, 91-93, 331. *Dadach and Kaliaguine. *Can. J. Chem. Eng.* **1993**, 71, 880

Results : Conc. acid- low temperature regime

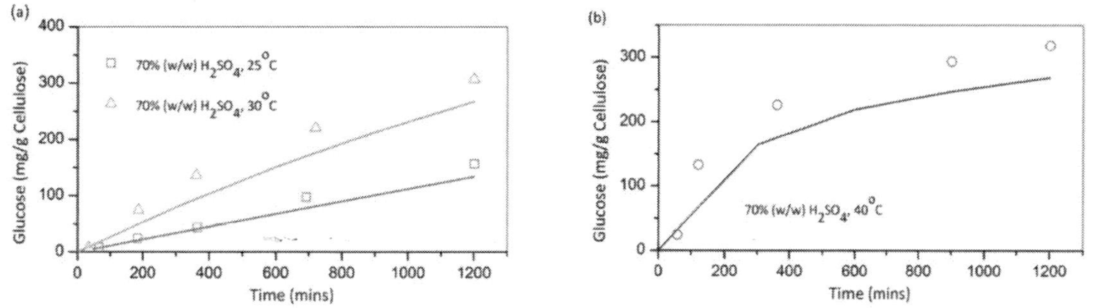

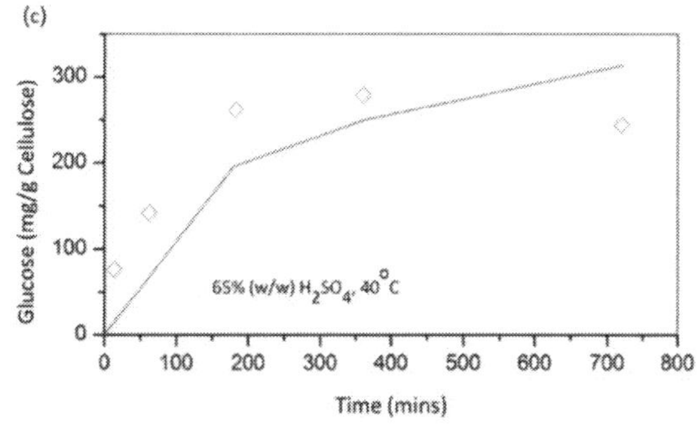

m = 2.59±0.05
n = 2.36±0.02
y = 2.40±0.01

Expt. data: *Camacho et al. *J. Chem. Technol. Biotechnol.* **1996**, 67, 350

Results : Model validations

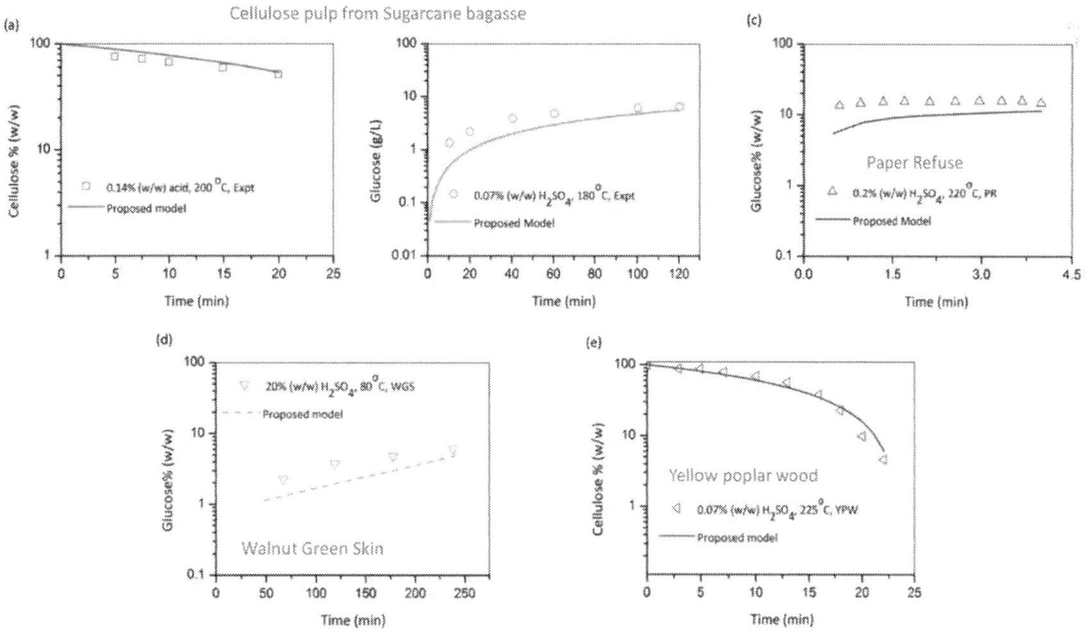

Expt. data: *Gurgel et al. *Ind. Eng. Chem. Res.* **2012,** 51, 1173. *Fagan et al. *Environ. Sci. Technol.* **1971,** 5, 545. *Arastehnodeh, A, et al. *Afr. J. Biotechnol.* **2012**, 11, 878. *Torget, R. W. et al. *Ind. Emg. Chem. Res.* **2000**, 39, 2817.

Results: Effect of DP of Cellulose

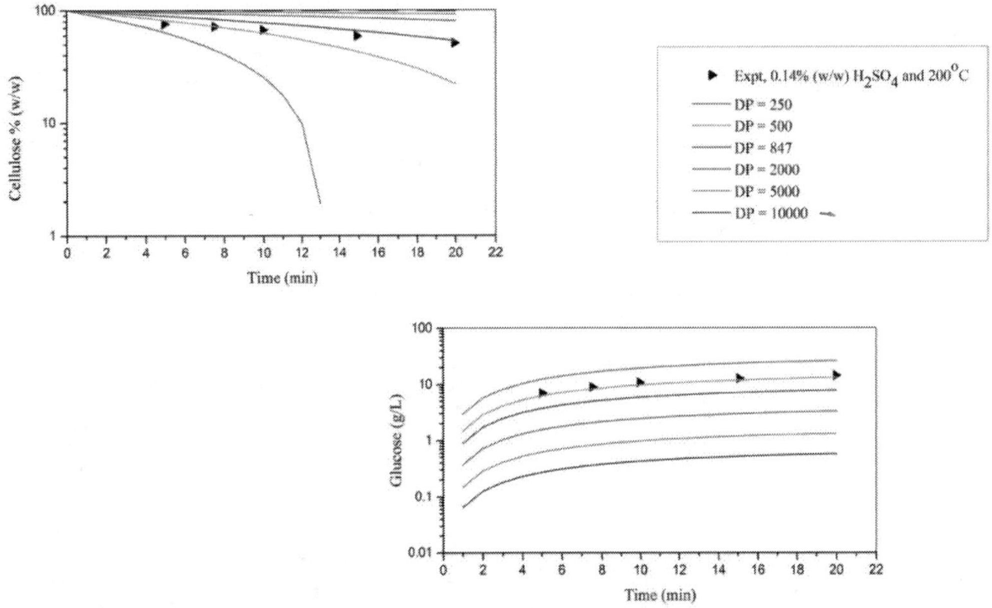

To Summarize!

- Unified kinetic model was developed successfully
- Effects of acid and temperature have been very well reflected by the model
- It is possible to vary just three parameters, *m*, *n*, *y*, to obtain reaction profiles under different reaction regimes.
- Model was applicable for actual biomass hydrolysis like walnut green skin and yellow poplar wood

Acknowledgements

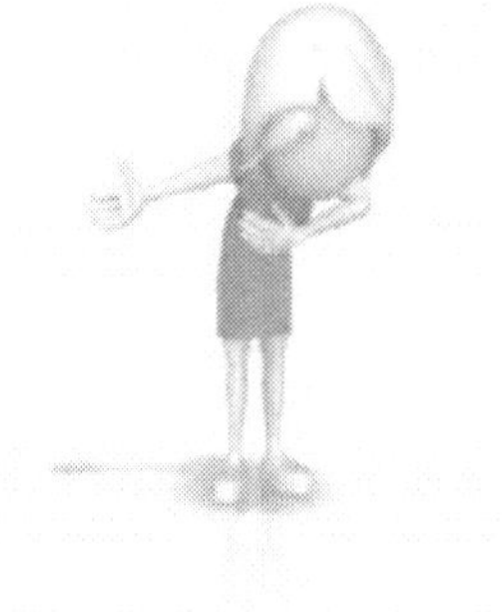

Complex Reaction Kinetics

390325 Catalysts for Electroreduction of CO2 to CO

Abstract:

The study of electrochemical reduction of CO_2 to various value-added chemicals has intensified over the last several years because this approach may help reduce atmospheric CO_2 levels. Multiple approaches need to be implemented to curb the atmospheric CO_2 levels, including switching to renewable energy sources, increasing the energy efficiency of buildings, increasing the fuel efficiency of vehicles, and underground carbon sequestration.[1,2] Electrochemical reduction of CO_2 into useful chemicals such as carbon monoxide (CO), formic acid, methane, and ethylene is another approach to address this challenge, as we recently summarized in a review [3]. Coupled to renewable energy sources such as wind and solar, this process can produce carbon-neutral fuels or commodity chemicals by using CO_2 as the starting material, while this approach may also provide a method for storage of otherwise wasted excess energy from intermittent renewable sources.

For this process to become economically feasible, more active and stable catalysts as well as better electrodes are necessary such that CO_2 electrolyzers can be operated at sufficient conversion (current density >250 mA/cm^2), reasonable energetic efficiency (>60%), and sufficient product selectivity (Faradaic efficiency >90%).

For CO production, a key reactant in the Fischer-Tropsch process, the best performance reported to date is current densities on the order of 90 mA/cm^2 and energy efficiencies up to 45%, when operating at ambient conditions [4]. This presentation will focus on new catalysts systems for efficient conversion of CO_2 to CO: (i) Ag nanoparticles supported on TiO_2 [5]; (ii) Au nanoparticles supported on multiwall nanotubes; and most interestingly (iii) metal-free N-doped carbons. These catalysts have been characterized in a 3-electrode cell and in an electrolyzer. Current densities of between 100 and 250 mA/cm^2 as well as energy efficiencies of up to 70% were obtained. The electrodes in all these cases are prepared using automated airbrushing [4], which reduced catalyst loadings to 0.75 mg/cm^2 for Ag and 0.17 mg/cm^2 for Au. These performance levels, together with the lower cost due to low precious metal loading (due to the use of catalyst supports), or even the elimination of precious metals altogether (N-doped carbons), brings electrochemical reduction of CO_2 to CO closer to economic feasibility.

References

[1] S. Pacala, R. Socolow, Science 305 (2004) 968.

[2] M.I. Hoffert, Science 329 (2010) 1292.

[3] H.R. Jhong, S. Ma, P.J.A. Kenis, Current Opinion in Chemical Engineering 2 (2013) 191.

[4] H.R. Jhong, F.R. Brushett, P.J.A. Kenis, Advanced Energy Materials 3 (2013) 589.

[5] S. Ma, Y. Lan, G.M.J. Perez, S. Moniri, P.J.A. Kenis, ChemSusChem 7 (2014) 866.

382601 Electrocatalysts for CO_2 Electroreduction in Ionic Liquids

Abstract:

Carbon dioxide utilization through electroreduction to CO and/or other chemicals offers opportunities for re-harnessing some of the carbon released through combustion when paired with a carbon-neutral electricity source. CO_2 electroreduction can lead to a wide variety of products ranging from CO to C_2 and C_3 hydrocarbons, alcohols and beyond. Significant strides are still needed before commercialization of CO_2 electroreduction can be realized. Aqueous-based CO_2 electroreduction suffers from parasitic H_2 production where the reduction of water is favored over that of CO_2 at the potentials required for CO_2 activity. The non-H_2 product distribution is commonly broad making selectivity a challenge as well.

The challenges of product distribution and H_2 generation could be addressed by replacing the aqueous system with an ionic liquid (IL) system. ILs, ionic salts with melting points below 100°C, generally have minimal vapor pressures, are conductive, and have moderate-to-high viscosities. Many ILs have electrochemical potential windows substantially wider than water. This allows for CO_2 electroreduction potentials to be reached without the degradation of the solvent like what occurs in water. The IL anion and cation can be tuned to have enhanced CO_2 solubility which increases reactant availability at the electrode surface. Additionally, the lack of available protons in IL systems limits the products to primarily CO and oxalate, making selectivity much more controllable.

While the benefits of using ILs for CO_2 electroreduction are promising, they are not drop-in replacement solvents/electrolytes. The ILs can be passivating towards metals, ligating effects of the ILs can block active sites and the extended double layers can inhibit mass transfer. Electrocatalytic systems have to be designed for IL use to be successful.

We have investigated copper catalysts for CO_2 electroreduction in ILs with high CO_2 solubilities such as 1-butyl, 3-methylimidazolium bis(trifluoromethylsulfonyl)imide ([Bmim][NTf_2]). The impact of electrochemical conditions and morphology of the catalyst on activity, faradaic efficiency and product distribution will be reported.

Electrochemical Reduction of Carbon Dioxide on Molybdenum Disulfide Edges

Mohammad Asadi, Bijandra Kumar, Amirhossein Behranginia, Amin Salehi-Khojin

Department of Mechanical and Industrial Engineering, University of Illinois at Chicago, Chicago, IL, 60607, USA.
Corresponding author, salehikh@uic.edu,

Abstract:

During the last few decades, the amount of carbon dioxide (CO_2) has reached the highest level (396.80 ppm) of the last 20 million years[1], causing radical and largely unpredictable changes in the environment[2,3,4]. Recent efforts have revealed that CO_2 can be converted by electrochemical reduction processes using renewable energy sources into energy-rich modules, offering an efficient path for both CO_2 remediation and an alternative energy source[5,6,7,8,9]. In this study, we found that MoS_2 catalyst with molybdenum (Mo) terminated edges exhibits higher CO_2 reduction performance compared to the noble metals (e.g. gold and silver) with a significantly high current density and low overpotential (54 mV). The catalysts' overall performance by multiplying current density and faradaic efficiency as the two major electrochemical parameters at different overpotentials shows that MoS_2 exhibits the highest performance at each overpotential. The exceptionally high catalytic activity of MoS_2 is attributed to the Mo terminated edges, where the Mo atoms possess approximately one order of magnitude higher electronic (d orbital) density than noble metals. Moreover, the lower work function (3.9 eV) also promotes the advanced performance of the MoS_2 catalyst. The performance of the MoS_2 catalyst is further improved by designing an atomic edge terminated surface via synthesizing vertically aligned MoS_2. This finding exhibits a non-precious and earth abundant MoS_2 catalysts can successfully replace expensive noble metal electrodes, with the promise of higher CO_2 conversion rates and selectivity.

Keywords: Carbon dioxide reduction, Molybdenum Disulfide, electrochemical reduction.

References:

1. Ed Dlugokencky and Pieter Tans, *NOAA/ESRL* (www.esrl.noaa.gov/gmd/ccgg/trends/).
2. Davis, S. J., Caldeira, K. & Matthews, H. D. Future CO_2 Emissions and Climate Change from Existing Energy Infrastructure. *Science* **329**, 1330-1333 (2010).
3. Brook, E. J. Leads and Lags at the End of the Last Ice Age. *Science* **339**, 1042–1043 (2013).
4. Parrenin, F. *et al.* Synchronous Change of Atmospheric CO_2 and Antarctic Temperature During the Last Deglacial Warming. *Science* **339**, 1060–1063 (2013).
5. Whipple, D. T. & Kenis, P. J. A. Prospects of CO_2 Utilization via Direct Heterogeneous Electrochemical Reduction. *J. Phys. Chem. Lett.* **1**, 3451–3458 (2010).
6. Kumar, B. *et al.* Renewable and metal-free carbon nanofibre catalysts for carbon dioxide reduction. *Nat. Commun.* **4,** (2013).
7. Rosen, B. A. *et al.* Ionic Liquid-Mediated Selective Conversion of CO_2 to CO at Low Overpotentials. *Science* **334**, 643–644 (2011).
8. Rosen, B. A. *et al.* In Situ Spectroscopic Examination of a Low Overpotential Pathway for Carbon Dioxide Conversion to Carbon Monoxide. *J. Phys. Chem. C* **116**, 15307–15312 (2012).
9. Salehi-Khojin, A. *et al.* Nanoparticle Silver Catalysts That Show Enhanced Activity for Carbon Dioxide Electrolysis. *J. Phys. Chem. C* **117**, 1627–1632 (2013).

The oxidation state of Cu is reported to have a significant effect on its product selectivity during CO_2 electroreduction. For metallic Cu, it is widely accepted that CO and HCOOH are the main products formed at low overpotentials, while hydrocarbons (notably CH_4 and C_2H_4) become the dominant products at higher potentials. However, a few groups have reported CH_3OH as a major product using pre-oxidized Cu electrodes or supported Cu_2O nanoparticles. It has also been reported that supported Cu_2O layers, although being reduced to Cu metal under CO_2 reduction conditions, result in a decreased overpotential for CO formation, relative to Cu metal.

In this study we examine the stability and product selectivity of CO_2 electroreduction using colloidally prepared Cu_2O nanoparticles. The starting solution of aqueous $CuCl_2$ with polyethylene glycol (PEG) surfactant is reduced using L-ascorbic acid sodium (LAAS), aged for 6 hours, then rinsed with deionized water. The resulting slurry is mixed with a Nafion/ethanol solution, sonicated for 10 minutes, and dried onto Toray carbon paper. The CO_2 reduction experiments are performed using a two compartment cell with a Nafion membrane separator, a Pt wire counterelectrode, and Ag/AgCl reference electrode. The working electrolyte is aqueous 0.5M $KHCO_3$ that is bubbled continuously with CO_2. Product concentrations are determined using gas chromatography. Gas phase products (H_2, CO, CH_4, and C_2H_4) are sampled directly from the CO_2 purge gas leaving the reactor. Liquid products (mainly CH_2H_5OH) are measured by taking syringe samples of the electrolyte at 15 minute intervals.

The SEM images of freshly prepared electrodes show the cubic morphology expected for Cu_2O particles. X-ray diffraction measurements confirm that only peaks assigned to the Cu_2O phase are present (i.e., no peaks assignable to metallic Cu are observed). After being used for CO_2 reduction (i.e., 120 minutes at −1.745 V(SCE), the XRD measurements indicate that nearly all of the Cu_2O is reduced to Cu metal. The SEM images show the reduced electrode material retains the same cubic morphology as the starting Cu_2O nanoparticles, but the surface of the cubes become decorated with much smaller clusters.

The CO_2 reduction experiments are performed at a single potential, −1.745 V(SCE). This potential is selected to be well into the region for CH_4 and C_2H_4 formation on Cu electrodes. For our supported Cu_2O nanoparticle electrodes at this potential, the largest CO_2 reduction product is C_2H_4, with a Faradaic efficiency of 34%. We also observe the formation of C_2H_5OH in parallel to C_2H_4, with a Faradaic efficiency of about 6%. In contrast, the formation rate of CH_4 is much lower, with a Faradaic efficiency of only 1%. (The largest overall product in these experiments is H_2, which we attribute to H_2O reduction on the incompletely covered carbon support. Control experiments using both unmodified and Nafion-treated carbon paper confirm very low activity for CO_2 reduction in the absence of Cu_2O nanoparticles on these supports.)

The most noticeable feature of the CO_2 reduction behavior using the supported Cu_2O nanaparticle electrodes appears to be the large C_2H_4/CH_4 product ratio. Our present results support a model in which the Cu_2O particles are converted to metallic Cu during CO_2 reduction. The morphological evolution of the starting Cu_2O nanoparticles during this reduction leads to the formation of more highly dispersed Cu clusters on the surface of the converted particles. The dispersed Cu clusters are expected to contain a higher concentration of more open crystal faces and lower co-ordination surface atoms, which leads to the observation of a higher C_2H_4/CH_4 ratio, relative to low-index planar Cu surfaces. Additional work is underway to further characterize these dispersed Cu clusters, and to explore methods to improve their stability during prolonged electrochemical operation.

Molybdenum carbide as alternative catalysts to precious metals for highly selective reduction of CO_2 to CO

Marc D. Porosoff and Jingguang G. Chen

To sustain future population and economic growth, the global energy supply is expected to increase by 60% by 2040,[1] but the associated CO_2 emissions are a major concern. CO_2 capture and conversion must become a significant industry to mitigate the threat of climate change and ocean acidification.[2] Efforts have already been put forth to capture and sequester CO_2;[3] however, a process that can efficiently convert CO_2 into a commodity increases utility over sequestration.[4] A substantial reduction of CO_2 levels requires a large scale CO_2 catalytic conversion process, which in turn requires the discovery of low cost catalysts. Results from the current study demonstrate the feasibility of using a non-precious metal material, molybdenum carbide (Mo_2C), as an active and selective catalyst for CO_2 conversion by H_2.

Transition metal carbides (TMCs) are attractive alternatives because they have similar properties to precious metal catalysts in reforming, hydrogenation, water-gas shift (WGS) and CO oxidation reactions. The high activity of TMCs compared with their parent metals originates from a modification of the electronic properties from the addition of carbon,[5] which in turn affects the binding energy and the reactivity of adsorbates. Mo_2C is particularly interesting for CO_2 conversion because of its low cost, dual functionality for H_2 dissociation and C=O bond scission, and potential to behave similarly to reducible oxides.

In CO_2 conversion by H_2, CO is a desired product for its ability to be used in the Fischer-Tropsch process, while CH_4 is undesired because of its low volumetric energy density and abundance.[6] The current study first investigates CO_2 activation over well-defined Mo_2C surfaces with temperature programmed surface reaction (TPSR) and ambient pressure X-ray photoelectron spectroscopy (AP-XPS) experiments to confirm the active nature of the Mo_2C surface toward CO_2 conversion to CO. Then, model surface results are extended to flow reactor experiments over porous Mo_2C catalysts, and further explained by in-situ X-ray absorption near edge spectroscopy (XANES) experiments. The findings show that Mo_2C outperforms precious bimetallic catalysts and is highly active and selective for CO_2 conversion to CO. The activity, selectivity and stability of Mo_2C can be further improved by modifying the carbide with another metal, cobalt (Co). Because Mo_2C is made from much more abundant elements than precious metals, the catalyst can be manufactured at much lower cost, which is critical for achieving a substantial reduction of CO_2 levels by large-scale CO_2 catalytic conversion processes.

References

[1] U. E. I. Administration, (Ed.: DOE), Washington DC, **2013**.
[2] T. R. Knutson, R. E. Tuleya, *J. Clim.* **2004**, *17*, 3477-3495.
[3] S. E. Strand, G. Benford, *Environ. Sci. Technol.* **2009**, *43*, 1000-1007.
[4] G. Centi, S. Perathoner, *Catal. Today* **2009**, *148*, 191-205.
[5] P. Liu, J. A. Rodriguez, *J. Chem. Phys.* **2004**, *120*, 5414-5423.
[6] C. deLeitenburg, A. Trovarelli, J. Kaspar, *J. Catal.* **1997**, *166*, 98-107.

Nanoporous Ag for Electrocatalytic CO_2 Reduction

Feng Jiao
Assistant Professor, Department of Chemical and Biomolecular Engineering, University of Delaware, Newark, DE 19716, E-mail: jiao@udel.edu

Reduction in greenhouse CO_2 emissions from fossil fuel utilization is critical for human society. Ideally, one would like to convert CO_2 produced in power plants, refineries and petrochemical plants to fuels or other chemicals through renewable energy utilization. This desired solution imposes major technological challenges because CO_2 is a fully oxidized and thermodynamically stable molecule. A suitable catalyst for CO_2 reduction is essential to achieve a cost effective process with high efficiency and selectivity. In the past two decades, electrocatalytic CO_2 reduction has attracted much attention because the required electricity may be obtained at a low cost from renewable energy sources, such as wind, solar, and wave. Researchers have identified several potential catalysts that are able to reduce CO_2 electrochemically in aqueous electrolytes. For example, Hori et al. have shown that at a potential of about -0.7 V vs. the reversible hydrogen electrode (RHE) a polycrystalline gold electrocatalyst can deliver a current at 5.0 mA/cm^2 with an efficiency of 87% toward CO production, whereas polycrystalline copper exhibits a poor selectivity and requires a potential of about -1.0 V (vs. RHE) to achieve the same current density (i.e. CO_2 reduction reaction rate). However, gold is not suitable for large scale applications due to its low abundance and high cost. Searching for abundant catalysts with high selectivity is crucial for commercializing CO_2 reduction processes by reducing costs associated with catalyst fabrication and product separation.

The selective conversion of CO_2 to CO is a promising route for clean energy. The CO product can be used as feedstock in the Fischer-Tropsch process, a well-known and well-characterized process that has been used in industry to produce chemicals and synthetic fuels from syngas ($CO + H_2$) for many decades. By coupling the catalytic reduction of CO_2 to CO with the Fischer-Tropsch process to produce synthetic fuels and industrial chemicals, the estimated maximum reduction of atmospheric CO_2 emissions is 40%.

Silver is an attractive CO_2 reduction electrocatalyst, because it is able to reduce CO_2 to CO with a good selectivity (~81%) and it also costs much less than other precious metal catalysts. In this talk, we will present a nanoporous silver (np-Ag) catalyst, which is able to reduce CO_2 electrochemically to CO in a highly efficient and selective way. Not only the porous structure creates an extremely large surface area for catalytic reaction, but also the curved internal surface generates a large number of highly active step sites for CO_2 conversion, resulting in an exceptional activity that is over three orders of magnitude higher than that of the polycrystalline counterpart at a moderate overpotential of < 500 mV. More importantly, such a remarkable activity for CO_2 electroreduction has been achieved with a CO Faradaic efficiency of 92%.

Reference:

Lu, Q.[#], Rosen, J.[#], Zhou, Y., Hutchings, G. S., Kimmel, Y. C., Chen, J. G., & **Jiao, F.*** A selective and efficient electrocatalyst for carbon dioxide reduction. *Nature Communications* 5:3242 (**2014**).

Design of Frustrated Lewis Pair Functionalized Metal Organic Frameworks for CO_2 Hydrogenation

Jingyun Ye[1], Karl Johnson[1,2]
[1]Department of Chemical & Petroleum Engineering, University of Pittsburgh, Pittsburgh, Pennsylvania, 15261, USA
[2]National Energy Technology Laboratory, 626 Cochrans Mill Road, Pittsburgh, PA 15236, USA

Metal organic frameworks (MOFs) are highly versatile nanoporous materials because of their structural and functional tailorability. We use density functional theory (DFT) methods to design functionalized MOFs that are capable of chemically binding and hydrogenating CO_2. We employ UiO-66 [shown in Fig. 1(a)] as our starting base material because it is a MOF that has chemical and thermal stability, is highly selective toward CO_2 and can be modified (functionalized) post-synthetically. We design functional moieties for CO_2 hydrogenation based on frustrated Lewis pairs (FLPs) because these have been shown to be very active for chemisorption of CO_2 and the heterolytic dissociation of H_2. One candidate FLP is 1-[bis(pentafluorophenyl)boryl]-3,5-di-*tert*-butyl-1*H*-pyrazole[1,2] [**1** shown in Fig. 1(b)]. We have developed a family of functional groups based on a modified version of **1**, which we denote **2**, as shown in Fig. 1(c). Structure **2** has both Lewis acid and base sites embedded within the molecule, but has much less steric hindrance than the FLP **1** because internal steric hindrance is not needed once the Lewis pair moiety is chemically bound within a MOF. We have examined the properties of UiO-66 functionalized with derivatives of **2** [Fig. 1(d)]. We have replaced the BF_2 group on **2** with the following moieties: $-BH_2$, $-B(CN)_2$, $-B(CF_3)_2$, $-B(NO_2)_2$ and $-AlCl_2$. We computed the electronic structure, charge distribution, and CO_2/H_2 binding energies and found that chemisorption of CO_2 and dissociation of H_2 are both energetically favorable with many of the functional groups. We have also investigated UiO-66 having two functional groups per cell in order to provide both CO_2 binding sites and H_2 dissociation sites in close proximity. We have identified possible scenarios for CO_2 hydrogenation through an enzyme-like pathway where one of the functional groups in the pore binds and activates CO_2, while the second functional group heterolytically cleaves H_2 and supplies the resulting hydride and proton to the activated CO_2.

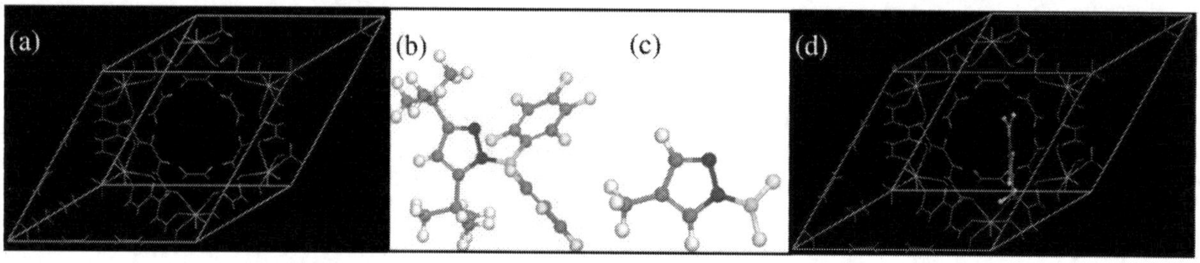

Fig 1. (a) UiO-66 primitive cell. (b) Structure of 1-[bis(pentafluorophenyl)boryl]-3,5-di-*tert*-butyl-1*H*-pyrazole, **1**, a frustrated Lewis pair. Color code: gray=C, white=H, blue=N, cyan=F, pink=B. (c) Structure of 1-(difluoroboranyl)-4-methyl-1*H*-pyrazole, **2**, a modified version of **1** (d) UiO-66 functionalized with single instance of **2**.

References

1. Theuergarten, E., et.al., *Chem. Commun.* **2010,** *46* (45), 8561-8563.
2. Theuergarten, E., et.al., *Dalton Trans.* **2012,** *41* (30), 9101-9110.

STEAM REFORMING OF METHANOL OVER NANOSTRUCTURED Pt/TiO_2 AND Pt/CeO_2 CATALYSTS

Alexandra Paxinou[1,2], Joan Papavasiliou[1], Stylianos Neophytides[1] and George Avgouropoulos[1,2*]

[1] *Foundation for Research and Technology-Hellas (FORTH), Institute of Chemical Engineering and High Temperature Processes (ICE/HT), 26504 Patras, Greece*
[2] *Department of Materials Science, University of Patras, 26504 Patras*

Abstract

Methanol may offer much higher energy densities than either batteries or fuel cells operating on stored H_2, making it an attractive source for advanced portable power. Methanol can be easily stored, transported and dispensed, and, at the same time, it can be efficiently produced from a wide variety of (renewable) sources. However, a methanol fuel cell would be the preferred energy source for a portable power system if a rugged, reliable and lightweight fuel processor was available to efficiently convert the alcohol to H_2 and drive a polymer electrolyte membrane (PEM) fuel cell, in a small volume. Recently, Avgouropoulos et al. [1-3] demonstrated the functionality of an Internal Reforming Methanol Fuel Cell (IRMFC), where the methanol gets reformed by a catalyst incorporated into the anode compartment of PEM fuel cell (internal reforming). Integration of the reformer into the fuel cell eliminates the need for additional heat exchangers and a separate WGS and PrOx fuel processors. A challenge for fuel processing with respect ot the operation temperature of IRMFC (200-220°C) is the development of highly active and selective catalysts, which will be able to operate at temperatures as low as possible with minimal CO formation. Commercially available copper-based catalysts, typically with composition Cu-ZnO-(Al2O3) have been widely used for generating hydrogen from methanol. Even though these catalysts are widely used in H_2 plants, several drawbacks limit their application in small stationary or mobile fuel processors: (a) slow start-up response due to the slow kinetics, (b) irreversible deactivation if exposed to liquid water formed during shutdown. Especially, the pyrophoric behaviour of conventional SRM catalysts has to be controlled when reduced Cu is abruptly exposed to air after turning off the feed of reactants, since major local temperature spikes can occur due to fast copper oxidation, which may lead to sintering and deactivation of Cu particles. Noble metal-containing catalysts, especially Pt or Pd supported on reducible oxide supports, such ZnO and CeO_2, have been also proposed as an alternative to copper for the conversion of methanol to H_2 [4-6].We have reported on the application of CuMnOx and (Pt,Pd,Rh)/CeO_2 catalysts prepared by a combustion method in methanol reforming [6-8].

In view of the proposed application, important improvements of reforming catalysts are required regarding: (i) their activity at temperatures around 200-220°C and (ii) the ability to operate without preactivation steps. In this work hydrothermal method was employed for the preparation of titania nanotubes and ceria nanorods as catalyst supports. Platinum nanoparticles were dispersed on the oxide nanostructures following three different methods: (a) impregnation, (b) deposition-precipitation and (c) combustion. In this work, the catalytic performance of Pt/TiO2 and Pt/CeO2 nanostructures has been investigated for the steam reforming of methanol and discussed on the basis of the physicochemical characterization with a variety of techniques.

References

[1] G. Avgouropoulos, T. Ioannides, J.K. Kallitsis, S. Neophytides, Chem. Eng. J. 176-177 (2011) 95.

[2] G. Avgouropoulos, J. Papavasiliou, M.K. Daletou, J.K. Kallitsis, T. Ioannides, S. Neophytides, Appl. Catal. B: Environ. 90 (2009) 628.

[3] G. Avgouropoulos, J. Papavasiliou, M. Daletou, M. Geormezi, N. Triantafyllopoulos, T. Ioannides, J. Kallitsis, S. Neophytides, US Patent 200861/095779, September 10 (2008).

[4] K. Sun,W. Lu, M.Wang, X. Xu, Appl. Catal. A: Gen. 268 (2004) 107.

[5] Y. Suwa, S. Ito, S. Kameoka, K. Tomishige, K. Kunimori, Appl. Catal. A: Gen. 267

[6] G. Avgouropoulos, J. Papavasiliou, T. Ioannides, Chem. Eng. J. 154 (2009) 274.

[7] G. Avgouropoulos, J. Papavasiliou, T. Ioannides, J. Catal. 251 (2007) 7.

[8] J. Papavasiliou, G. Avgouropoulos, T. Ioannides, Int. J. Hydrogen Energy 37 (2012) 16739.

AUTHOR INDEX

AUTHOR INDEX

AUTHOR INDEX

AUTHOR INDEX

AUTHOR INDEX

AUTHOR INDEX

AUTHOR INDEX

AUTHOR INDEX

AUTHOR INDEX

AUTHOR INDEX

AUTHOR INDEX

AUTHOR INDEX

AUTHOR INDEX

AUTHOR INDEX

AUTHOR INDEX

Catalysis and Reaction Engineering Division 2014

Core Programming Area at the 2014 AIChE Annual Meeting

Atlanta, Georgia, USA
16-21 November 2014

Volume 2 of 2

ISBN: 978-1-5108-1252-9

Printed from e-media with permission by:

Curran Associates, Inc.
57 Morehouse Lane
Red Hook, NY 12571

CURRAN ASSOCIATES INC. proceedings.com

Some format issues inherent in the e-media version may also appear in this print version.

Printed by Curran Associates, Inc. (2015)

For permission requests, please contact AIChE
at the address below.

AIChE
120 Wall Street, FL 23
New York, NY 10005-4020

Phone: (800) 242-4363
Fax: (203) 775-5177

www.aiche.org

Additional copies of this publication are available from:

Curran Associates, Inc.
57 Morehouse Lane
Red Hook, NY 12571 USA
Phone: 845-758-0400
Fax: 845-758-2634
Email: curran@proceedings.com
Web: www.proceedings.com

TABLE OF CONTENTS

VOLUME 1

VOLUME 2

Dry reforming of methane comes as a great option to handle the current environmental and sustainability energetic concerns, because, as it converts two greenhouse gases, produces H_2, which is a widely known energy carrier.

Between the most used catalysts, the ones based in Ni are the most known. They have an excellent activity/price ratio but coke deposition or/and sintering of the Ni particles usually deactivates them.

One of the techniques to improve its resistance to deactivation, and even to improve its catalytic activity is their modification with a second metal like Ru.

The present work explores two different synthetic routes based on impregnation methods that lead to modified Ni catalyst that even with the same metal composition display a very different catalytic behavior, revealing that the interaction metal-metal (Ni-Ru) plays a very interesting role in the performance of this kind of catalyst.

The Ni-Ru catalysts were prepared by impregnation following two different routes. The ***Simultaneous*** route (S samples) (C samples): Alcoholic solutions of $Ni(NO_3)_2.6H_2O$ and $Ru(NO)(NO_3)_3$ are mixed and then impregnated on the modified alumina support. After impregnation, the catalyst was dried at 120°C overnight and finally calcined in air at 500°C/5 h. The ***Consecutive*** route: The modified alumina support is first impregnated with an alcoholic solution of $Ni(NO_3)_2.6H_2O$, dried ovenigth at 120°C and calcined in air at 500°C for 5h. Afterwards, the alcoholic solution of $Ru(NO)(NO_3)_3$ was impregnated. Finally, the catalyst is dried and calcined as previously described. The nominal values of Ru are 0,5 and 2%, the Ni content is always 15%. The catalytic tests were carried out in a fixed bed quartz reactor using a feed mixture of $CH_4/CO_2/N_2$ 35/35/30 working at 110 $L \cdot h^{-1} \cdot g^{-1}$. The reforming tests were performed at 750 °C for10h.

The extent of incorporation of Ni and Ru was similar in both routes and the desired values of metal phase were achieved.

The addition of Ni barely changes the textural properties of the modified alumina used as support, and the addition of Ru doesn't have any influence in the textural properties. All solids present a Type IV isotherm, which is typical for mesoporous solids, with a H1 hysteresis, which indicates a cylindrical-interconnected type of pores.

The XRD patterns of all solids are dominated by those peaks of the support, characteristics of a Mg and Al spinel ($MgAl_2O_4$). In all samples Ni is detected as NiO, and Ru as RuO_2, except in the S_0,5Ru sample, where no RuO_2 peaks are detected. This could be owed to a better dispersion of RuO_2, or to the formation of a mixed-oxide Ni-Ru, or even to a possible inclusion of Ru^{+4} in the support network. This result suggests a stronger Ni-Ru interaction in the samples prepared by the simultaneous route, preferably at lower ruthenium contents.

The existence of such Ni-Ru interaction is highlighted by Raman spectroscopy. The Raman spectra of the samples synthesized by the consecutive route show typical RuO_2 bulk vibrations, while a Raman spectrum characteristics of low-dimensional RuO_x species is presented for the ones synthesized by the simultaneous route.

An *in situ* XRD study in presence of H_2 was performed over both series of catalysts (Figure 1). As the temperature increases, all Ru samples, except S_0,5Ru, present a clear diffraction peak at 44° 2θ, which correspond to the [101] plane of Ru^0. Again, the no appearance of the metallic ruthenium diffraction peaks could be indicative of a singular interaction between Ru and Ni or to a smaller particle Ru size owed to a better dispersion.

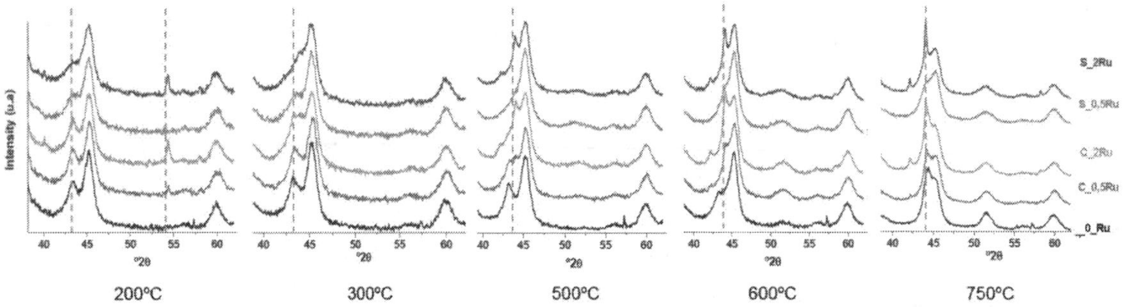

Figure1. XRD in reductive conditions.

Finally, the stronger Ni-Ru interaction in the samples synthesized by the Simultaneous route is confirmed by the TPR analysis. Although all Ru modified samples present a better and easier reducibility of the Ni species, shifting the nickel reduction processes to lower temperatures, a higher percentage of reducibility is obtained over the samples prepared by the simultaneous route. Besides this, a new reduction process at low temperatures (<220°C) due to the ruthenium reduction is detected in all samples, such process (temperature, H_2 consumption and profile) differing for both series indicating the different ruthenium-nickel contact.

Concerning the catalytic activity, those samples prepared by the simultaneous route present higher CH_4 and CO_2 conversion values than those of the consecutive one. Besides this, the S-0,5Ru sample has the best catalytic results, despite the lower amount of Ru. The stability test carried out over these two samples during 50 hours also evidences the lower deactivation rate showed by the S-0,5Ru catalyst, which agrees with its lower amount of coke determined over the reacted sample by TPO and Raman studies (Figure 2).

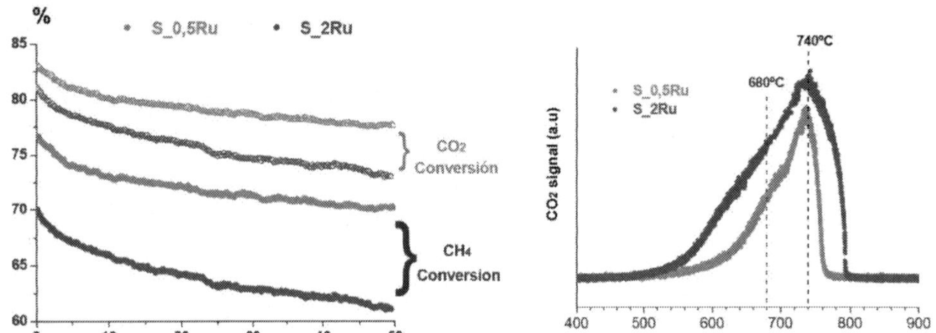

A. B.

Figure 2. A. CH_4 and CO_2 conversions during 50h. B. TPO of the post reaction samples.

Compared to the consecutive impregnation procedure, the simultaneous impregnation of Ru and Ni over a modified alumina support leads to catalysts with a higher Ru-Ni interaction, specially at low ruthenium content, resulting in more active and resistant to coke deactivation catalyst for the dry reforming of methane.

Solution Combustion Synthesis of Active Catalysts for Hydrogen Production from Ethanol

Anand Kumar*, Anchu Ashok, Yussuf Kuty, Rahul R Bhosale, LJP van den Broke

Department of Chemical Engineering, Qatar University, Doha, Qatar

**Corresponding Author: akumar@qu.edu.qa*

Abstract

Solution combustion synthesis method was used to synthesize Cu and Ni based catalysts active for hydrogen production from ethanol decomposition reaction. The catalyst synthesis parameters (such as fuel to oxidizer ratio, combustion temperature) were monitored and correlated with the synthesized materials properties. A thermodynamic calculation was also conducted to get product distribution during ethanol decomposition reaction. The amount of fuel content in the combustion solution was found to greatly affect the phase of the synthesized catalyst. Catalysts were characterized by other methods such as XRD and their activity for ethanol decomposition reaction and hydrogen selectivity.

Introduction

Solution combustion synthesis (SCS) method along with other variations of combustion synthesis (CS); being energy efficient methods for nanomaterials synthesis, economical and suitable for scale up for synthesizing large amount of nanomaterials, are being used in various applications [1, 2, 3]. A typical CS system contains a redox mixture of precursor materials (in the form of a solid pellet or homogeneous solution) that can be heated to initiate an exothermic reaction capable of self-sustaining. In general, two modes of combustion propagation are observed on the basis of physical appearance of reaction completion and product formation. In the first way, the reaction is started locally at one point that propagates gradually throughout the reaction mixture and completes the formation of synthesis products. In the second way, reactive mixture is heated uniformly until the entire volume appears to react simultaneously. Both the modes of synthesis are experimentally verified, and numerous works have been reported on the mechanism of combustion front propagation [1, 2, 3, 5]. The former mode of combustion is referred as self-propagating high temperature synthesis (SHS) while the latter is termed as volume combustion synthesis (VCS) mode of CS. The SCS method uses water soluble precursors which are dissolved in water to achieve molecular level mixing. This mixing is expected to generate products with uniform properties. A beaker containing precursor's solution and a hot plate heater are the only required equipment for synthesizing nanopowders using SCS method. Due to the ability of SCS for synthesizing nanopowders in a short period of time (approx. 30 min.) it is being applied in various areas such as pigments, catalysis, electronic and magnetic materials, drug delivery etc. [4]. SCS mixture is composed of metal nitrates and fuels (e.g. glycine, urea etc.) which are dissolved in water. This solution is heated over a hot plate heater, and after the evaporation of water, the CS reaction takes place according to the scheme suggested in equation 1:

$$\begin{aligned} &M^{v}(NO_3)_v + \left(\frac{5}{9}v\varphi\right)CH_2NH_2COOH + v\frac{5}{4}(\varphi-1)O_2 \\ &\rightarrow M_vO_{\frac{v}{2}}(s) + \left(\frac{10}{9}v\varphi\right)CO_2(g) + \frac{25}{18}v\varphi H_2O(g) + v\left(\frac{5\varphi+9}{18}\right)N_2(g) \end{aligned} \quad (1)$$

where M^v is a v-valent metal. The parameter φ, the fuel to oxidizer ratio, is defined such that $\varphi = 1$ corresponds to a stoichiometric oxygen concentration, meaning that the initial mixture does not require atmospheric oxygen for complete oxidation of the fuel, while $\varphi > 1$ (<1) implies fuel-rich (or lean) conditions.

The SCS could adopt any of the reaction completion modes, SHS or VCS, depending on the reaction conditions. The amount of gases released during the SCS (as indicated in Eq-1) contributes towards the porosity of the synthesized nanopowders, which could be controlled just by controlling the parameter φ. Recent publications [1, 2, 5] indicate that controlling the fuel content (i.e. the parameter φ) during SCS process also affects the phase of the solid component synthesized. The parameter φ can be optimized to synthesize pure metals and bimetal nano-powders as well. Experimental conditions and a reaction pathway for the synthesis of pure Ni, Cu metals and CuNi bimetal using SCS can be found in the reported literature [6-9]. There are numerous other modifications possible using SCS method, such as, synthesizing nanomaterials by impregnating the solution on a thin media (e.g. cellulose paper, carbon nano-tubes or graphite sheet etc.) or on a porous support to decrease the energy density for CS, and increase the cooling rate of the products formed [8-10]. The CS reaction taking place in a thin layer with relatively low temperature and fast cooling rate generates a unique microstructure of the products with fine particles and high surface area suitable for catalytic applications [11-14]. An experimental work to synthesize Cu and Ni based nanoparticles for catalytic applications are presented in this report. A short review on ethanol reforming is discussed to explain the choice of material for ethanol decomposition reaction.

Ethanol (C_2H_5OH) containing a high ratio of H/C (3/1) is looked as a prospective source of hydrogen for fuel cell applications for clean energy production. Another important aspect of considering ethanol as a hydrogen source is the possibility of a carbon neutral energy pathway, as there are substantial active research efforts on producing ethanol from bio-resources [12-14,15]. The carbon dioxide produced during hydrogen generation from ethanol is recycled back into the production of carbohydrates and biomass products by natural photosynthesis process making the entire cycle a carbon zero energy cycle. Hydrogen can be abstracted from ethanol in various ways, e.g. ethanol dehydrogenation, decomposition, steam reform or partial oxidation as described by the reaction scheme below with corresponding heat of reaction (ΔH):

$$C_2H_5OH \rightarrow CH_3CHO + H_2 \qquad -\Delta H^0_{298} = -68 KJmol^{-1} \qquad (2)$$

$$C_2H_5OH \rightarrow CH_4 + CO + H_2 \qquad -\Delta H^0_{298} = -49 KJmol^{-1} \qquad (3)$$

$$C_2H_5OH + 3H_2O \rightarrow 2CO_2 + 6H_2 \qquad -\Delta H^0_{298} = -174 KJmol^{-1} \qquad (4)$$

$$C_2H_5OH + \frac{3}{2}O_2 \rightarrow 2CO_2 + 3H_2 \qquad -\Delta H^0_{298} = 509 KJmol^{-1} \qquad (5)$$

The ΔH values clearly indicate that except ethanol partial oxidation (eq. 5), all other reactions are endothermic and require an external energy source. Ethanol decomposition (eq. 2) and ethanol dehydrogenation (eq. 3) reactions require less energy compared to ethanol steam reforming (eq. 4), which is a highly endothermic reaction involving high temperature and external steam supply to sustain the reaction. A comprehensive review on the catalyst used for ethanol reforming reaction can be found elsewhere [16], which form the basis for choosing Ni and Cu as catalyst materials. Out of various pathways (scheme 2 – 5), ethanol decomposition is relative less explored, although it has a potential for low temperature hydrogen production.

Experimental

Catalyst synthesis

A mixture of metal nitrates (Alfa Aesar) $M^{v}(NO_3)_v \cdot xH_2O$ (where Me = Ni, Cu) and glycine ($C_2H_5NO_2$, as fuel) were dissolved in water to prepare a 75ml of active solution for volume combustion synthesis (VCS). The molar ratios of the fuel to metal nitrates were adjusted according to the scheme 1 to synthesize lab scale amounts (1.5 - 3 gm.) of catalysts sufficient for catalytic activity and characterization measurements. The homogeneous aqueous solution of metal nitrates and fuel is heated over a hot plate until the solution reaches a temperature known as ignition temperature after the evaporation of water. Once ignited, the reaction proceeds very fast, which can lead to either an explosive combustion mode or to the SHS mode inside the beaker itself. The fuel/oxidizer ratio φ was varied from 0.5 to 2 to study the effect of φ on combustion synthesis temperature. A φ value of 1.75 was selected for further studies to produce a reduced bimetallic NiCu phase. The φ=1.75 is the optimized value reported in literature for the production of pure metallic Ni nanoparticles using combustion synthesis [10, 17, 18]. After synthesis, the resulting powders were milled using a planetary ball mill (Retsch PM 100) at 650 rpm for two minutes to get a mixture with uniform particle size. The ratio of catalyst sample and milling balls was kept at 1:2 in all the cases. The time temperature profile during combustion synthesis was recorded using a high speed data recorder (Omega, OMB-DAQ-3000).

Catalyst characterization

Multiple techniques were used to characterize the Ni, Cu and CuNi catalysts. XRD was used to identify the bulk phases present in the catalyst. XRD measurements were carried out in air at room temperature using a X-ray diffractometer with Cu-Kα radiation of 1.54056 Å wavelength available at Gas Processing Center (GPC) in Qatar University. Details of ethanol activity measurements are described below.

Activity Measurements

The activity and selectivity of catalysts containing Cu and Ni were measured in a tubular flow reactor. The reactor is made of quartz having an internal diameter of 10.5 mm and total length 30 cm. In each experiment, the freshly prepared and milled powders were pressed to 35MPa for 4 min to form an agglomerate. The resulting agglomerate was broken up and sieved to obtain a particle size between 0.6 and 1mm. 400mg from the sieved agglomerates were used in activity measurement experiments. Each catalyst was reduced *in situ* by heating it from room temperature to 300^0C at a rate of 5^0C min^{-1} and held at this temperature for one hour under a pure hydrogen flow rate of 50 cc·min^{-1}. After the reduction, the flow was switched from hydrogen to nitrogen to purge out any remaining hydrogen, while the reactor was cooled to room temperature in about 30 min. Then a feed flow rate was introduced consisting of 140 cc·min^{-1} of nitrogen saturated with ethanol, resulting in a stream containing 1.8 cc/min of ethanol.

Two gas chromatographs were connected in parallel to analyze on line the reactor effluents. The first (Varian 3300) contained a 3.7m HAYESEP Q column for separation and detection of carbon dioxide, ethanol, water, and acetaldehyde. The second (Varian 920) contained a 1.8m molecular sieve 5Å and a 5.0m HAYESEP D connected in series to separate nitrogen, oxygen, hydrogen, and carbon monoxide. Product gases were measured at 12 minute intervals while the catalyst bed was heated from room temperature to 400°C over the course of 3 h and 30 minutes. In our calculations, the ethanol conversion and hydrogen selectivity are defined as follows:

$$X_{EtOH} = \frac{F_{EtOH-in} - F_{EtOH-out}}{F_{EtOH-in}}$$

$$S_{H_2} = \frac{F_{H_2-produced}}{3 \times F_{EtOH-converted}}$$

Results and Discussion

Solution combustion synthesis (SCS) method was used to synthesize bimetallic CuNi nanopowders using various fuel to oxidizer (φ) ratio. Detailed mechanistic studies for synthesizing Ni, Cu and CuNi catalysts using SCS method is described elsewhere [8, 9], indicating parameter φ can be used to control the amount of metallic and metal oxide phases in the synthesized catalyst.

Fig. 1 shows the maximum combustion temperature for Ni-Cu system for three different values of φ (φ = 0.5, 1 and 2). A typical pattern is observed, where an optimum value of φ =1 gives highest combustion temperature. These observations are expected in an exothermic reaction system where relative amount of reactants will affect the energy released during reaction. XRD pattern of the synthesized materials at various φ values, as shown in Fig. 2, indicates a graduate shift in the oxide and pure metallic systems wrt φ. A lower value of φ=0.5 indicates mostly oxides of nickel and copper metals which gradually reduces to give mostly bimetallic NiCu at φ=1.75 value. Further increase φ results in synthesis of oxide materials. These patterns indicate that an optimum value of φ is required to create a reducing environment for synthesizing pure metallic/bimetallic phases. A lower value of φ will not release enough amount of reducing agent resulting in oxide phase synthesis, and an excessive amount of φ would indicate incomplete combustion due to presence of large amount of glycine.

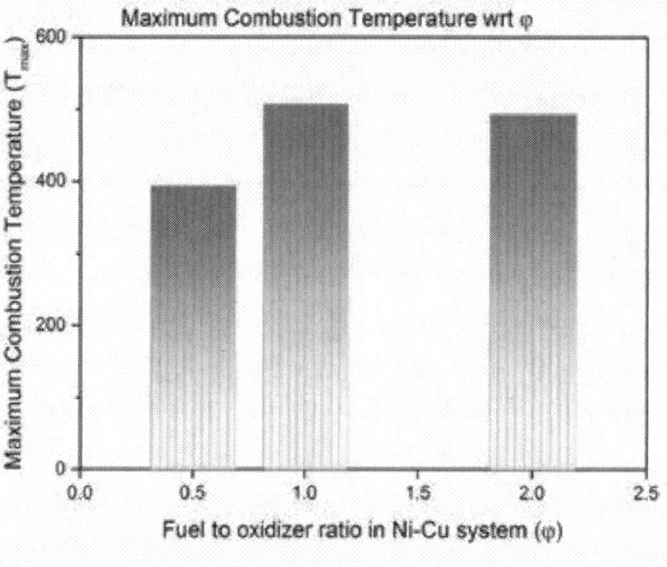

Figure 1: Maximum combustion temperature for Ni-Cu system in VCS for various ϕ values

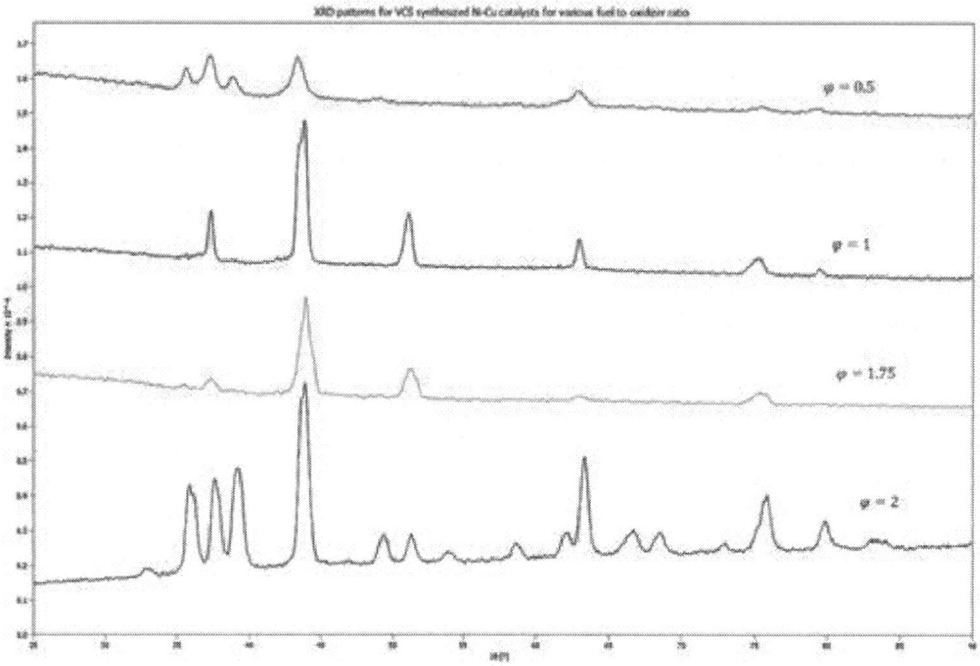

Figure 2: XRD pattern observed during SCS of Ni-Cus system wrt φ

The optimized catalysts containing Ni-Cu bimetallic phase, along with individual Cu and Ni catalysts were used for ethanol decomposition/dehydrogenation reaction. A thermodynamic calculation using software "THERMO" was conducted to get equilibrium product distribution during ethanol decomposition reaction. The equilibrium product distribution, as presented in Fig. 3, shows hydrogen to be more selective at higher temperature values along with CO formation. Low temperature values are mainly suitable for methane and H_2O generation. CO_2 is also found at low temperatures. The concentration of CO_2 increases with temperature, giving a maximum at ~500°C followed by a steady decrease. Throughout the temperature range investigated, carbon formation is observed as a stable product, which could cause catalyst deactivation if proper regeneration techniques are not used. Higher temperature values are more sensitive to carbon formation, with a maximum around 500°C giving approx. 60% selectivity for carbon.

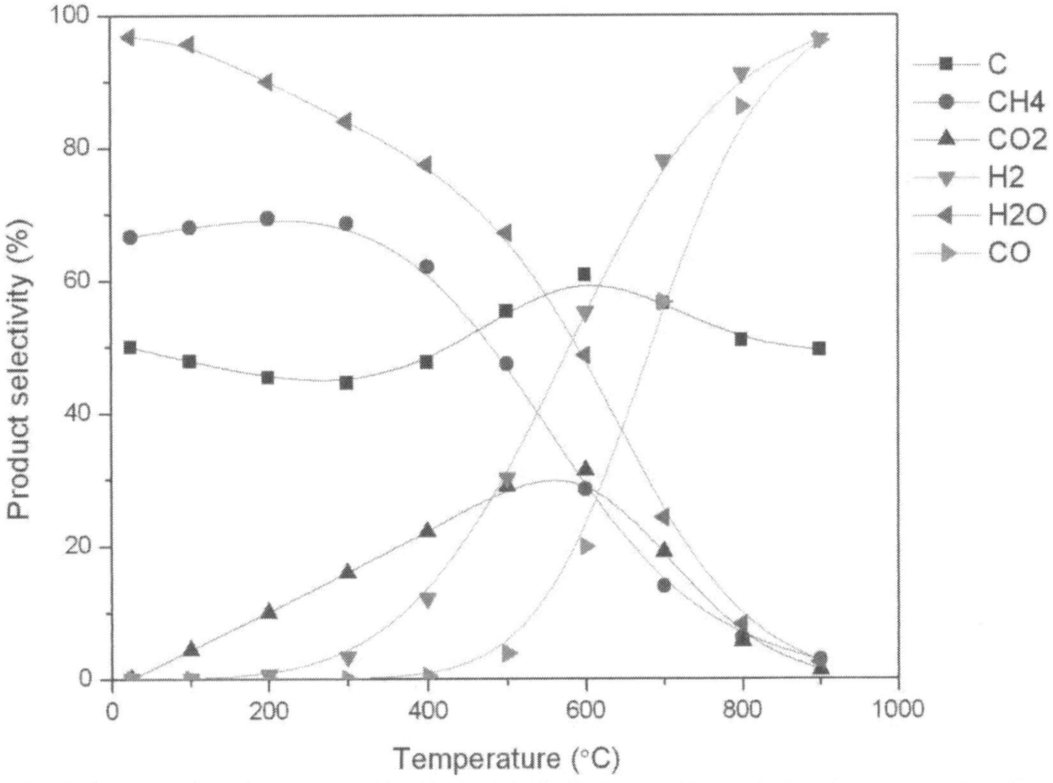

Figure 3: Thermodynamic calculation revealing equilibrium product distribution during ethanol decomposition reaction

Three catalysts Ni, Cu and Ni-Cu were selected for ethanol decomposition reaction to investigate their activity and selectivity for hydrogen generation. A summary of the catalytic performance is presented through Fig. 4 by comparing the hydrogen selectivity at various activity values for these three catalysts. Throughout the selected range of activity, Ni-Cu based bimetallic catalyst shows higher hydrogen selectivity than the individual Cu and Ni metal catalysts synthesized by same method. The hydrogen selectivity for Cu and Ni catalysts are almost similar at higher conversion values, however, Cu seem to be more selective at conversion less than 40%. These values indicate that a bimetallic catalyst could show better catalytic performance that individual constituent metals in ethanol dehydrogenation reaction.

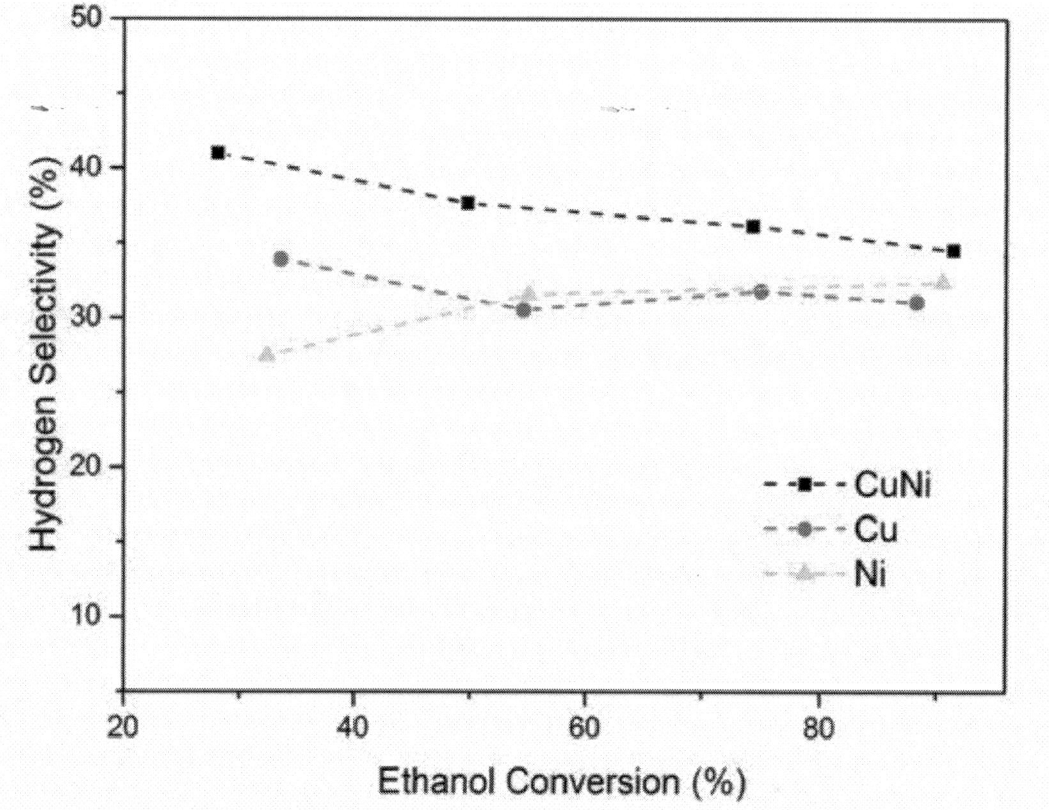

Figure 4: Hydrogen selectivity at various values of ethanol conversion on three catalysts

Summary

Solution combustion synthesis was used to synthesize catalytically active nanoparticles containing copper and nickel elements. These catalysts were characterized by the maximum temperature observed during SCS process, XRD and their activity towards hydrogen production from ethanol. Based on the results discussed, it can be concluded that, the amount of fuel and metal precursors during SCS process affect the combustion temperature, and phases of the synthesized nanoparticles. It was observed that an optimum value of φ is required to synthesize bimetallic compounds of Ni-Cu. The synthesized bimetallic catalyst shows better hydrogen selectivity than individual constituent metals during ethanol decomposition reaction. Thermodynamic calculations show that higher temperatures are more favorable for hydrogen production; however coke generation could be a significant problem for catalysts leading to catalyst deactivation by coking.

Acknowledgments

The authors gratefully acknowledge the funding from Qatar National Research fund (QNRF) under the JSREP-05-004-2-002 grant to support this work. The statements made herein are solely the responsibility of the authors. The authors would also like to acknowledge the support from Gas Processing Center (GPC) at Qatar University for conducting the XRD experiments.

References

1. K. C. Patil, M. S. Hegde, R. Tanu and S. T. Aruna, Chemistry of Nanocrystalline Oxide Materials : Combustion Synthesis, Properties and Applications, World Scientific Publishing Co. Pte. Ltd, Singapore, 2008.

2. S. T. Aruna, A. S. Mukasyan, Combustion synthesis and nanomaterials, Current opinion in solid state materials science 12 (2008) 44.

3. P. Roth, Particle synthesis in flames, Proceedings of the Combustion Institute 31 (2007) 1773.

4. J. J. Kingsley, K. C. Patil, A novel combustion process for the synthesis of fine particle α-alumina and related oxide materials, Materials letters 6 (1988) 427.

5. A. G. Merzhanov, I. P. Borovinskaya, A. E. Sytchev, SHS of Nano-powders, in: Baumard J F (Ed.), Lessons in Nanotechnology from Traditional Materials to Advanced Ceramics, Techna Group Srl, Dijon, France, 2005, pp. 1-27.

6. C. H. Jung, H. G. Lee, C. J. Kim, S. B. Bhaduri, Synthesis of Cu–Ni Alloy Powder Directly from Metal Salts Solution, Journal of nanoparticle research 5 (2003) 383.

7. C. H. Jung, S. Jalota, S. B. Bhaduri, Quantitative effects of fuel on the synthesis of Ni/NiO particles using a microwave-induced solution combustion synthesis in air atmosphere, Mater Lett 59 (2005) 2426.

8. A. Kumar, E. E. Wolf, A. S. Mukasyan,, Solution Combustion Synthesis of Metal Nanopowders: Nickel- Reaction Pathways , AIChE Journal 57 (2010) 2207.

9. A. Kumar, E. Wolf, A. Mukasyan, Solution combustion synthesis of metal nanopowders: Copper and copper/nickel alloys , AIChE Journal 57 (2011) 3473.

10. A. Kumar, J. T. Miller, A. S. Mukasyan, E. E. Wolf, In situ XAS and FTIR studies of a multi-component Ni/Fe/Cu catalyst for hydrogen production from ethanol, Applied Catalysis A: General 467 (2013) 593-603.

11. A. S. Mukasyan, P. Dinka, Novel Method for Synthesis of Nano-Materials: Combustion of Active Impregnated Layers, Advanced engineering materials 9 (2007) 653.

12. A. Kumar, E. E. Wolf, A. S. Mukasyan,, Impregnated layer combustion synthesis method for preparation of multicomponent catalysts for the production of hydrogen from oxidative reforming of methanol, Applied catalysis.A, General 372 (2010) 175-183.

13. A. Kumar A. S. Mukasyan, E. E. Wolf, Modeling Impregnated Layer Combustion Synthesis of Catalyts for Hydrogen Generation from Oxidative Reforming of Methanol, Industrial engineering chemistry research 49 (2010) 11001-11008.

14. A. Cross, A. Kumar, E. E. Wolf, A. S. Mukasyan, Combustion Synthesis of a Nickel Supported Catalyst: Effect of Metal Distribution on the Activity during Ethanol Decomposition 2012, 51 (37), pp 12004–12008, Ind Eng Chem Res 51 (2012) 12004-12008.

15. K. *Öhgren, A. Rudolf, M. Galbe, G. Zacchi*, Fuel ethanol production from steam-pretreated corn stover using SSF at higher dry matter content, Biomass Bioenergy 30 (2006) 863.

16. Y. Sun, Hydrolysis of lignocellulosic materials for ethanol production: a review, Bioresour. Technol. 83 (2002) 1.

17. A. Kumar, A. Mukasyan, E. E. Wolf, Combustion synthesis of Ni, Fe and Cu multi-component catalysts for hydrogen production from ethanol reforming, Applied Catalysis A: General 401 (2011) 20.

18. G. Carotenuto, A. Kumar, J. Miller, A. Mukasyan, E. Santacesaria, E. Wolf, Hydrogen production by ethanol decomposition and partial oxidation over copper/copper-chromite based catalysts prepared by combustion synthesis, Catalysis Today 203 (2013) 163.

387834 High Heat Transfer, Low Pressure Drop Structured Catalytic Reactors for Steam Methane Reforming

Abstract:

Steam methane reforming is conventionally carried out in packed bed reactors. The reactor performance is determined by heat transfer and intra-particle diffusion limitations. In structured catalytic reactors, a thin catalyst layer is coated on the reactor internals, which allows reducing or eliminating intra-particle diffusion limitations. Different structured catalytic reactor designs have been studied. The design considered in the present work is based on the ZoneFlow reactor casing consisting of alternating and connected sectors with blades guiding the flow towards and away from the wall. Previous work [1,2] showed that compared to packed bed reactors, ZoneFlow reactors offer an improved performance due to a combination of reduced intra-particle diffusion limitations and intensified heat transfer with the wall. Furthermore, the pressure drop can be significantly reduced.

In the present work, detailed Computational Fluid Dynamics simulations are carried out to study the influence of the most important ZoneFlow casing design parameters on the heat transfer and the pressure drop in a wide gas flow rate range. The flow pattern explains how a combination of low pressure drop and high heat transfer can be achieved and how the reactor design can be optimized. A comparison with packed bed reactors is also made showing significant process intensification is possible.

[1] De Wilde J., Froment G.F., Fuel, 100, p. 48-56, 2012.

[2] De Wilde J., Froment G.F., Ind. Eng. Chem. Res., 52, p. 14055-14065, 2013.

A Bayesian Approach to Reduced Order Modeling in Catalytic Steam Reformation

Evan D. Ford
David S. Mebane, Ph.D.

With an ever-increasing demand for power, research and development into clean, renewable energy sources is vital given the uncertain longevity of traditional means. Hydrogen energy has promising potential given its natural abundance and use in fuel cells and other mechanisms. However, these devices require gaseous hydrogen while the majority of this element is contained naturally within water, various hydrocarbons, and other chemical compounds. In order for these devices to be practical, some refining or other hydrogen extraction process is required. Steam reformation is just one of the ways hydrogen can be obtained by reacting steam with methane in a high temperature environment. This creates hydrogen gas at the cost of additional energy input, and the use of a catalyst can limit this to only a certain degree. Maximizing hydrogen production in the reformation process is way to mitigate the impact of the losses.

However, the performance of a catalytic reaction is impacted by a bevy of reactor, environmental, and initial conditions. Time and resources used in process optimization can be shortened through simulation.

It is the goal of this project to develop a reduced order model for catalytic steam reformation. This tool will allow for accurate process-scale predictions across a wide spectrum of possible conditions, both steady state and transient. Not only will this have the capability of performing the aforementioned optimization, but can also prove beneficial for reactor design based on space or performance requirements.

A Bayesian statistical calibration method that has been proven effective by Mebane et al.[1] for modeling CO_2 uptake in amine sorbents will be applied. The process uses a detailed reaction network to create two separate models: one that is full scale and another that is of reduced order. This high fidelity model evaluates the network in its entirety, tracking the creation and depletion of each unique chemical species. Then the reaction network is simplified by examining functional relationships to remove all intermediate species; the remaining network serves as the basis for the reduced order model. Using the reaction chain below as an example, a full model would be made from all four reactions tracking the concentrations of 8 species. Then linking reactants to products from each equation as shown in Figure 1 creates a network graph.

$$A = B + C$$
$$C + D = E + F$$
$$B + F = G$$
$$E + G = H$$

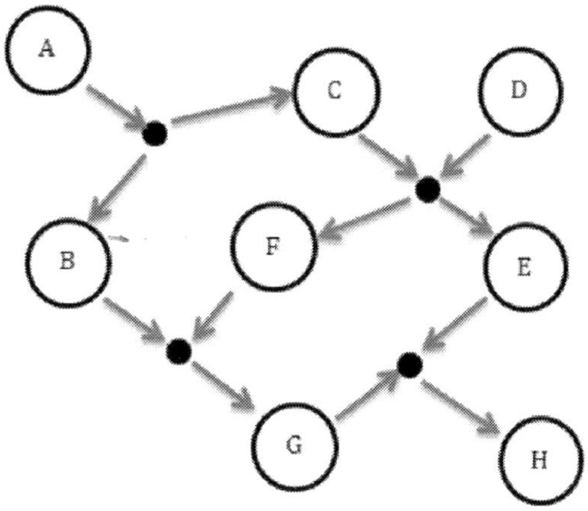

Figure 1: Full Reaction Network Graph Example

Simplifications are made by examining relationships contained in the network. Examining Figure 1 shows that a functional relationship exists between species A and G allowing for species B to be removed. These simplifications continue until the bare minimum of species remain to describe the reaction process, shown in Figure 2.

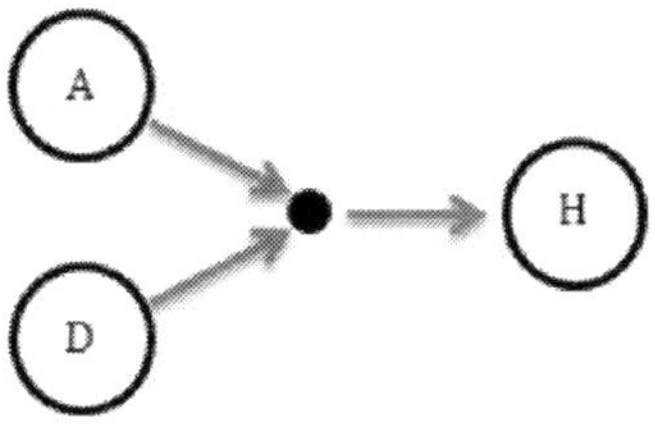

Figure 2: Reduced Reaction Network Graph Example

This new network and corresponding chemical equations are used to develop the reduced model. Removing these intermediate species reduces the dimensionality of the problem, allowing for this model to be less computationally intensive, saving on both time and resources. Model variability that is lost in removing these degrees of freedom is recovered through the use of a Gaussian process-based stochastic function that is defined on an input space consisting of the remaining chemical state variables. The reduced model is thus a stochastic model, meaning that a Bayesian approach to calibration is required. The result of a Bayesian approach to calibration is a stochastic model, all of the realizations of which plausibly reproduce the high-fidelity training data. This approach to reduced order modeling is called dynamic discrepancy, and has been demonstrated in the context of solid sorbent-based carbon capture.[3]

The models and associated reaction network are created based of Xu and Froment's[2] 13-equation reaction chain describing the reformation process. It demonstrates how steam and methane are adsorbed onto open catalyst sites to form various adsorbed radicals which, through several surface reactions, produce hydrogen as well as carbon monoxide and carbon dioxide.

Xu-Froment Steam Reforming Reaction Network:

$$H_2O + L = O^* + H_2 \tag{1}$$
$$CH_4 + L = CH_4^* \tag{2}$$
$$CH_4^* + L = CH_3^* + H^* \tag{3}$$
$$CH_3^* + L = CH_2^* + H^* \tag{4}$$
$$CH_2^* + O^* = CH_2O^* + L \tag{5}$$
$$CH_2O^* + L = CHO^* + H^* \tag{6}$$
$$CHO^* + L = CO^* + H^* \tag{7}$$
$$CO^* + O^* = CO_2^* + L \tag{8}$$
$$CHO^* + O^* = CO_2^* + H^* \tag{9}$$
$$CO^* = CO + L \tag{10}$$
$$CO_2^* = CO_2 + L \tag{11}$$
$$2H^* = H_2^* + L \tag{12}$$
$$H_2^* = H_2 + L \tag{13}$$

This network equates to 16 chemical species coexisting during the steam reformation process needing balanced. A system of ODEs is necessary to track the concentration of each unique species, and because a high number of adsorbed radicals exist in the network, a site limited model is necessary to restrict reaction rates based on the availability of the catalyst surface.

Simulating the environment of a plug flow reactor is achieved through a finite volume method of differential spatial increments governed by the kinetics of a constant stirred tank reactor (CSTR). CSTRs allow for the assumption of a well mixed solution varying only in time. By placing multiple of these in succession such that the exit of one serves as the inlet to the next, a system is created with time and spatial dependence similar to the PFR.

Reducing the Xu/Fromennt network will follow the same process as demonstrated in the example. Optimally, the simplified network will only contain the system's five reactants and products. The reduced model representing the simplified network can then be developed similarly to the prior model. Lastly, Bayes' Theorem combined with the MCMC calibrates the two models finalizing the reduced order model.

References

[1] Mebane, David S., et al. "Bayesian calibration of thermodynamic models for the uptake of CO_2 in supported amine sorbents using ab initio priors." Physical Chemistry Chemical Physics 15.12 (2013): 4355-4366.

[2] Xu, Jianguo, and Gilbert F. Froment. "Methane steam reforming, methanation and water-gas shift: I. Intrinsic kinetics." AIChE Journal 35.1 (1989): 88-96.

[3] Miller, David C., et al. "Carbon Capture Simulation Initiative: A Case Study in Multiscale Modeling and New Challenges." Annual Review of Chemical and Biomolecular Engineering 5.1 (2014).

389268 Characterization and H2 Production Performance of Cu and Zn Encapsulated MCM-41, TiO2 and CeO2 Catalysts

Abstract:

One-pot hydrothermal procedure was used for the synthesis of mesoporous high surface area SiO_2 (MCM-41), TiO_2 and CeO_2 containing Cu and Zn nanoparticles. The effect of metal (Cu and Zn)-support (SiO_2, TiO_2 and CeO_2) interactions on steam reforming of methanol (SRM) reactions to produce high quality H_2 has been investigated. Catalysts are thoroughly characterized using N_2 adsorption-desorption isotherms (BET), X-ray diffraction (XRD), thermo-gravimetric and differential calorimetric analysis (TGA-DSC), transmission electron microscopy (TEM), inductively coupled plasma optical emission spectroscopy (ICP-OES), and temperature programmed reduction (TPR) techniques. Preliminary characterization results indicate that the synthesized MCM-41 and TiO_2 catalysts yield high surface area in excess of 800 and 250 m^2/g, respectively. While wide angle XRD studies confirmed the stability of hexagonal ordered mesoporous structure of the MCM-41, N_2 physisorption studies revealed the Type IV isotherms for TiO_2 indicating its mesoporous nature. ICP and EDX studies demonstrated that the expected amounts of active metals are successfully retained in the support matrices showing the effectiveness of our one pot synthesis procedure. Initial SRM studies showed that both the type of active metal and the support have strong influence on the SRM activity and selectivity. For example, 10%Zn-TiO_2 and 10%Cu-TiO_2 showed similar methanol conversion of 90 % at 300 °C and 1:3 methanol mole ratio. However, 10%Zn-TiO_2 showed considerably lower CO selectivity of 2.35% compared to 69.92% exhibited by 10%Cu-TiO_2. The results from our ongoing experiments will be presented.

389941 Molybdenum Based-Catalysts for the Reforming of Liquid Hydrocarbons for Fuel Cell Applications

Abstract:

In the present study, we have studied the performance of molybdenum-based catalysts for the production of H2 to power solid oxide fuel cells (SOFCs). In previous work, molybdenum dioxide (MoO2) was found to exhibit high activity as well as coking resistance and sulfur tolerance during the reforming of liquid fuels such as jet fuel and gasoline. On the other hand, molybdenum carbide (Mo2C) has been previously reported as highly active for the reforming of methane. However, the use of Mo2C for the partial oxidation of higher hydrocarbons has not been reported in the literature. To overcome this lack of information, we have investigated the performance of molybdenum carbide (Mo2C) as potential catalyst for the reforming of liquid hydrocarbons via partial oxidation. Our results indicate that at 850°C, 1 atm and a stoichiometric oxygen to carbon molar ratio, the use of MoO2 leads to a higher production of syngas as compared to that obtained when using commercial Mo2C. However, stability issues prevent the use of MoO2 for long periods of time. This problem seems to be related to the formation of hot spots in the catalyst bed that may deteriorate the crystalline structure and, hence, the oxygen mobility that is crucial to maintain the catalytic activity of this transition metal oxide. Different approaches are discussed to increase the stability of the oxide phase such as doping and the use of supporting materials. Our recent findings indicates that the doping of MoO2 with titanium enhances the stability of this oxide whereas supporting MoO2 on silicon dioxide (SiO2) leads to the formation of MoO2 nanoclusters with high resilience to sintering.

389511 Computational Mechanistic Investigation of the Phillips Polymerization Catalyst

Abstract:

The Phillips catalyst, chromium oxide supported on amorphous silica, produces half of the world's annual supply of high-density polyethylene. [1] However, the mechanism by which the catalyst polymerizes ethylene is unknown, because only a small fraction of chromium sites are active and are therefore difficult to detect experimentally. In this study, we used density functional theory to evaluate the feasibility of potential mechanisms. As in previous studies [2], our calculations rule out propagation by either metallacycle ring expansion or alternating carbene/chromacyclobutane intermediates: both mechanisms have activation barriers much greater than what is measured experimentally. However, we discovered two new pathways with low propagation barriers: oxachromacycle ring expansion, and chain growth from a vinylchromium(II) site. Unfortunately, the oxachromacycle sites initiate very slowly, while the vinylchromium(II) sites terminate rapidly; both are therefore unlikely as propagating intermediates in polymer production. We also found that a vinylchromium(III) site recently proposed to arise via direct C-H activation of ethylene [3] also terminates quickly. The only propagation mechanism whose computed kinetics are comparable to the experimental kinetics and which does not terminate to produce oligomers instead of polymer chains is Cossee-Arlman polymerization by a monoalkylchromium(III) site without a neighboring proton source. While the precise nature of the initiation step remains unknown, experiments and calculations based on homogeneous organochromium compounds suggest the active sites may arise via a radical mechanism.

[1] McDaniel, M. P., Adv. Catal., 2010, 53, 123-606.

[2] Espelid, O.; Borve, K. J., J. Catal. 2000, 195, 125-139.

[3] M.P. Conley, M.F. Delley, G. Siddiqi, G. Lapadula, S. Norsic, V. Monteil, O.V. Safonova, C. Coperet, Angew. Chem. Int. Ed. 2014, 53, 1872 –1876.

371079 Computational Design of Zr and Hf Organometallic Catalysts for Direct Propene Epoxidation Using Molecular Oxygen Oxidant

Abstract:

Propylene oxide (PO) is a key intermediate in the chemical industry and one of the most produced chemicals worldwide by mass. All of the existing commercial methods for PO production consume H_2 as a co-reactant (e.g., H_2O_2 and cumene processes) or produce co/by-products. Direct propene epoxidation utilizing O_2 as oxidant without co-reactants or co-products could potentially give economic and environmental benefits. We performed computational screening using density functional theory (DFT) calculations to identify suitable catalysts for this process.

Seven Zr/Hf-based catalysts have been studied. Complete reaction cycles for each catalyst were computed, including ground state and transition state energies. Chemical potential diagrams were calculated that describe the relative energies of different catalyst forms as a function of the oxygen atom chemical potential. Energetic spans have been computed and possible side reactions have been tested for all seven catalysts. The enthalpy energetic span is an estimate of the effective activation barrier for the entire catalytic cycle. Our best direct propylene epoxidation catalysts have a computed enthalpy energetic span of <30 kcal/mol.To the best of our knowledge, this is lower than any of the previously reported direct propene epoxidation catalysts.

These catalysts operate via a completely new selective oxidation mechanism that passes through η^3-ozone intermediates. This catalytic cycle involves: (a) the η^3-ozone group releases an O atom to form substrate oxide and generate a peroxo or weakly adsorbed O_2 group, (b) the peroxo or weakly adsorbed O_2 group releases an O atom to form substrate oxide and an oxo group, (c) an oxygen molecule adds to the oxo group to form an η^2-ozone group, (d) the η^2-ozone group rearranges to regenerate the η^3-ozone group.

We also used DFT calculations to investigate potential catalyst deactivation processes and other side reactions. While these are important for some of the catalysts, others show very little tendency for side reactions and catalyst deactivation processes.

371672 Adsorption of Organonitrogen Compounds on Promoted Molybdenum Sulfide Catalysts: A DFT Study on Nimos and Comos

Abstract:

Hydrodenitrogenation (HDN) of heteroaromatic nitrogen-containing compounds is concomitant with hydrodesulfurization (HDS) of organosulfur compounds in crude oil fractions such as vacuum gas oil. Organonitrogen compounds, consequently, inhibit HDS by competitively adsorbing on relevant active sites. On promoted molybdenum sulfide, which is the most commonly used class of catalysts for hydrotreating, the active sites have been shown to be located on or close to the two types of exposed edges – metal and sulfur [1]. These edges are further 'decorated' by sulfur atoms and sulfohydryl groups depending on the reaction conditions. The detailed mechanism of HDS and the inhibition by organonitrogen compounds have, however, not been completely resolved to date. Adsorption (and, hence, inhibition) can depend on the nature of the promoter (Ni, Co, etc.), basicity of the adsorbate, steric hindrance, and the specific catalyst edge and its decoration. In this study, we computationally investigate plausible modes and strength of adsorption of basic and non-basic heteroaromatic nitrogen-containing compounds such as derivatives of pyridine and pyrrole vis-à-vis those of thiophene and dibenzothiophene. Specifically, we identify favored adsorption sites of these classes of organic compounds and estimate their adsorption equilibrium constants on Ni- and Co- promoted molybdenum sulfide (NiMoS and CoMoS) to provide insights into the inhibition of HDS.

Periodic DFT and vdW-DF methods [2] have been used to systematically study the adsorption of over twenty compounds relevant to hydrotreating on the most stable catalyst configurations for NiMoS and CoMoS under typical HDS reaction conditions, viz., nickel-promoted metal edge with bare Ni atoms and cobalt-promoted sulfur edge with 50% coverage of S along with the presence of sulfohydryl groups. It is seen that basic compounds adsorb stronger than thiophene on both types of edges, adsorption is stronger on Ni than on Co except for large basic molecules, adsorption on Co-edges involves the Brønsted proton, and the dominant adsorption on Ni and Co is through the edge and brim sites respectively, while dispersion forces can significantly alter both the strength and mode of adsorption.

The catalyst models, chemisorption modes of different compounds, and the effect of van der Waal's dispersion interaction will be presented and discussed.

References:

1. Topsøe, H. Appl. Catalysis A: General (2007), 322, 3 – 8
2. Klimeš, J. et al. Phys. Rev. B (2011), 83, 195131

Acknowledgements: Financial support from BP is gratefully acknowledged.

Effects of porphyrin composition and axial ligand on the kinetics of di-olefin epoxidation catalyzed by Fe-porphyrin complexes: A DFT study

Ying Yu and Linda J. Broadbelt, Department of Chemical and Biological Engineering, Northwestern University, Evanston, IL

Epoxy resins are important industrial materials commonly used in paintings, coatings and electronic packaging. Recently, a family of biomimetic catalysts, Fe-porphyrin complexes, has been discovered to have a wide range of catalytic reactivity and selectivity towards di-olefin epoxidation with a clean oxidant, hydrogen peroxide. However, the underlying mechanism accounting for different performance of Fe-porphyrins is not well understood. In this study, we investigated the effects of substituents on the porphyrin ring and of the axial ligand. We adopted the most general mechanism proposed in the literature to date by Stephenson and Bell[1], and conducted density functional theory (DFT) calculations to elucidate the detailed mechanism of the two bifurcating rate-determining steps which lead to epoxidation and hydrogen peroxide decomposition, respectively. Experimental results were accurately reproduced in the calculations. Our results show that different substituents on the porphyrin ring can result in orders of magnitude difference in the ratio of reaction rates of the two bifurcating steps. It is suggested that the dramatic kinetic difference can direct the core catalysis to or away from epoxidation, which impacts the selectivity of different catalysts, which is in excellent accord with the experimental observations. Our results also suggest that the axial ligand composition has an influence on catalytic performance. The coordination of methanol as an axial ligand depends on the substituents on the porphyrin ring and the spin states. The presence of this extra axial ligand influences the energy barriers of the studied reactions. Ultimately, these results provide a basis for the design of new biomimetic catalysts for di-olefin epoxidation.

[1] Stephenson, N. A.; Bell, A. T., Mechanistic insights into iron porphyrin-catalyzed olefin epoxidation by hydrogen peroxide: Factors controlling activity and selectivity. J Mol Catal a-Chem 2007, 275 (1-2), 54-62.

383138 First Principles Simulation of NO Oxidation on Cu-SSZ-13 Zeolite Catalyst

Abstract:

Nitrogen Oxides (NO_x) are pollutants generated in oxygen rich combustion. In an increasingly more stringent emission regulations, ammonia Selective Catalytic Reduction (NH_3-SCR) of NO_x into N_2 and H_2O is the leading technology in diesel and stationary sources. However, the active sites and reaction mechanism of the commercially used Cu-exchanged zeolite SCR catalyst, i.e. Cu-SSZ-13, are still being investigated.

Past research has shown that the presence of NO_2 affects NH_3-SCR reaction mechanism and activity. Hence, we began our study by modeling the NO oxidation reaction on Cu-SSZ-13 using first-principles calculations to model two different active sites: single and dimeric Cu.

We examine the preferred locations of single and dimeric Cu species within the confines of SSZ-13 microporous structure in a periodic supercell. On these two sites, we compute GGA and hybrid adsorption energy of relevant intermediates (O, O_2, NO, NO_2, NO_3) involved in NO oxidation. We explore co-adsorption of these intermediates in the two active sites to probe Cu-SSZ-13 capability of accommodating more than one adsorbate. We then employ the use of thermodynamic phase diagram under NO oxidation condition to examine the most stable species. Comparing the computed NO oxidation pathway under reaction conditions on both sites, we conclude that Cu dimer provides a more accessible pathways, consistent with experimental observations.

360461 Elucidating the Role of Water in Olefin Epoxidation Catalyzed By Methyltrioxorhenium

Abstract:

The activation of MTO by H_2O_2 forms two discrete peroxorhenium complexes, $CH_3ReO_2(\eta^2\text{-}O_2)$ and $CH_3ReO(\eta^2\text{-}O_2)_2(H_2O)$, via a sequential and reversible reaction. Past reports on the thermodynamics and kinetics of these reactions are inconsistent with each other. New experiments and calculations using density functional theory were conducted to better understand these reactions and to provide a robust experimental foundation for benchmarking computational studies involving MTO and its derivatives. Including solvation and tunneling contributions to the free energies, we compute negative reaction enthalpies for each reaction and correctly predict the hydration state of all complexes in aqueous acetonitrile. New rate constants for each of the forward and reverse reactions were both measured and computed as a function of temperature, providing a complete set of consistent activation parameters. Computed rate constants for a direct ligand-exchange mechanism, and for a mechanism in which water(s) assists ligand-exchange via proton transfer in the transition state, differ by at least seven orders of magnitude. The water-assisted mechanism is predicted to be faster, and is consequently in closer agreement with the experimentally-measured kinetics. Experiments confirm the catalytic role of water: the kinetics of both steps are strongly dependent on the water concentration.

367168 Elucidating the Mechanism of Phenol Deoxygenation to Benzene on an Fe Surface: A Detailed Mechanistic Study from First Principles

Abstract:

The production of directly applicable liquid fuels from renewable resources is an area of intense research as we rush to find more sustainable energy sources than traditional fossil fuels. One promising technique is the fast pyrolysis of biomass to bio-oils as this technique is one of the simplest and most cost-effective methods for producing crude bio-oil and it can use the entirety of the biomass feedstock [1]. A major obstacle to the widespread use of these bio-oils is the high level of oxygen, up to 50 wt%, contained within the bio-oils in the form of acetic acid, furans, ketones and phenolics [2]. These oxygenates cause the bio-oils to be corrosive, thermally unstable, and highly viscous [3]. In order to produce usable bio-oils using fast pyrolysis, it is necessary to further upgrade the bio-oils by decreasing the oxygen content.

Bimetallic PdFe catalysts have been shown to be highly active towards the hydrodeoxygenation of the phenolics found in crude bio-oil and our previous work has shown that the active catalytic surface is likely the exposed Fe surface with the surface Pd playing a support role during reaction [4, 5]. We have further investigated Pd's support role and the nature of the Pd-Fe synergy by studying the reducibility of Fe_2O_3 with and without surface Pd in a combined experimental and theoretical study [6]. This work showed that the Pd stabilizes the metallic Fe surface by a partial donation of electrons to the surface Fe, making the surface electrons more delocalized. With additional experimental studies, we were able to propose a possible mechanism to account for the high activity and HDO selectivity in Pd-Fe catalysts, which included Pd assisted H_2 dissociation, Pd facilitated stabilization of metallic Fe surface, and Pd enhanced product desorption [7].

In order to better understand and control the catalytic deoxygenation reaction of phenolics, the detailed surface reaction mechanisms must be known. Using density functional theory, we have examined five distinct reaction pathways for the deoxygenation of phenol on the Fe (110) and Pd (111) surfaces so as to identify the most likely deoxygenation mechanism. For each elementary step in each mechanism, we have calculated the transition state structures, stable reaction intermediate structures, activation energies, reaction energies, and rate constants using transition state theory. A comparison of the resulting energies of reaction on the Pd (111) and Fe (110) surfaces show that the deoxygenation of phenol is highly endothermic on Pd (111) while the same reactions are exothermic on Fe (110). Work performed by Wang et al.[8] has shown that the activation energy barrier can be linearly related to the reaction energy. This suggests that the deoxygenation reaction is energetically unfavorable on Pd (111) for all mechanisms. The mechanistic studies on Fe (110) have shown that the most favorable reaction pathway occurs via the direct cleavage of the C-O bond, as the partial hydrogenation of the aromatic ring is significantly unfavorable. These results provide significant insight into the deoxygenation of phenolic compounds on promoted Fe bimetallic catalysts and allow for the further tailoring of the catalyst surface for the promotion of the deoxygenation reaction.

1. Wang, H., J. Male, and Y. Wang, ACS Catal., 2013. **3**: p. 1047-1070.

2. Mohan, D., C.U. Pittman, and P.H. Steele, Energy & Fuels, 2006. **20**: p. 848-889.

3. Maggi, R. and B. Delmon, Biomass & Bioenergy, 1994. **7**(1-6): p. 245-249.

4. Hensley, A.J., R. Zhang, Y. Wang, and J.-S. McEwen, J. Phys. Chem. C, 2013. **117**(46): p. 24317-24328.

5. Sun, J., A.M. Karim, H. Zhang, L. Kovarik, X. Li, A.J. Hensley, J.-S. McEwen, and Y. Wang, J. Catal., 2013. **306**: p. 47-57.

6. Hensley, A.J., Y. Hong, R. Zhang, H. Zhang, J. Sun, Y. Wang, and J.-S. McEwen, ACS Catal. (submitted), 2014.

7. Hong, Y., H. Zhang, J. Sun, A.M. Karim, A.J. Hensley, M. Gu, M.H. Engelhard, J.-S. McEwen, and Y. Wang, ACS Catal. (submitted), 2014.

8. Wang, S., V. Petzold, V. Tripkovic, J. Kleis, J.G. Howalt, E. Skúlason, E.M. Fernández, B. Hvolbæk, G. Jones, A. Toftelund, H. Falsig, M. Björketun, F. Studt, F. Abild-Pedersen, J. Rossmeisl, J.K. Nørskov, and T. Bligaard, Phys. Chem. Chem. Phys., 2011. **13**(46): p. 20760.

Fundamental mechanistic studies of H_2O_2 direct catalytic synthesis and decomposition

A. Plauck[1], A. C. Alba-Rubio[1], E. Stangland[2], J. A. Dumesic[1], M. Mavrikakis[1]

[1]Department of Chemical & Biological Engineering, University of Wisconsin – Madison

[2]Core Research and Development, The Dow Chemical Company, Midland MI

Hydrogen peroxide (H_2O_2) is a versatile and environmentally benign oxidizing agent that has the potential to decrease the environmental impact of numerous manufacturing processes – both chemical and non-chemical. Currently, the anthraquinone process accounts for over 95% of worldwide H_2O_2 production but is complex and only implemented on the large scale[1]. This has limited the introduction and expansion of new processes that would benefit from on-site H_2O_2 production.

Here we focus on an alternative process, the direct synthesis of H_2O_2, wherein H_2 and O_2 are combined in a single reactor over a transition metal catalyst. Although Pd-based catalysts for the direct synthesis of H_2O_2 have been a subject of intense study, especially those modified with Au[2,3], there is still an inadequate understanding of the fundamental details of the active site(s) and elementary transformations. To that end, we combine experimental reaction kinetics studies on a promoter-free Pd catalyst with periodic, self-consistent Density Functional Theory (DFT) calculations on multiple Pd facets to construct a mean-field microkinetic model. This presentation will focus on fundamental mechanistic aspects determining synthesis selectivity to H_2O_2, and the H_2O_2 decomposition reaction on Pd-based catalysts.

[1] Centi, G., Perathoner, S., & Abate, S. Direct Synthesis of Hydrogen Peroxide: Recent Advances in *Modern Heterogeneous Oxidation Catalysis: Design, Reactions and Characterization*. Wiley-VCH, Weinheim, 2009.
[2] Edwards, J. K. *et al.* Switching Off Hydrogen Peroxide Hydrogenation in the Direct Synthesis Process. *Science* **323**, 1037-1041 (2009).
[3] Campos-Martin, J. M., Blanco-Brieva, G. & Fierro, J. L. G. Hydrogen Peroxide Synthesis: An Outlook beyond the Anthraquinone Process. *Angewandte Chemie-International Edition* **45**, 6962-6984 (2006).

379737 Role of Adsorbate-Adsorbate Interactions in Dynamics of Surface Bond Breaking

Abstract:

Dynamics of surface reactions, though closely related to kinetics, is primarily dealing with the motion of individual atoms and molecules. Unraveling dynamics of elementary surface steps has been a long-term goal in surface science because of its significance in many technological applications, such as semiconductor processing, molecular sensing, and heterogeneous catalysis. Developing a fundamental understanding about important physical interactions that govern dynamics of surface bond breaking and forming is crucial for design of chemical processes with unprecedented energy efficiency and selectivity.

Recently, a transiently populated, weakly adsorbed CO precursor state on Ru(0001) has been identified with real-time observation of surface dynamics induced by a optical laser. The selective pump-probe experiments with intense femtosecond x-ray pulses and electronic structure calculations that include van der Waals interactions allow us to unambiguously probe the electronic signature of this species and distinguish it from other modes of excited molecules. In this talk, we will focus on discussing effects of various adsorbates (H, O) on CO desorption dynamics from Ru(0001) as a first step toward understanding reaction mechanisms of CO oxidation and hydrogenation reactions. We show that the long-ranged adsorbate-adsorbate interactions on metal surfaces have dramatic effects on the CO desorption mechanism, i.e., precursor-mediated vs. direct pathways, predicted by ab-initio molecular dynamics simulations. The observation is supported with ultrafast pump-probe experiments and elucidated with free energy surfaces of CO desorption obtained from the potential of mean force and metadynamics simulations. We emphasize that probing dynamics of surface bond breaking contributes to understanding the nature of molecule-surface interactions and developing detailed kinetics.

1. Beye, M. et al. Selective Ultrafast Probing of Transient Hot Chemisorbed and Precursor States of CO on Ru(0001). Phys. Rev. Lett. 110, 186101 (2013).

2. Dell'Angela, M. et al. Real-Time Observation of Surface Bond Breaking with an X-ray Laser. Science 339, 1302–1305 (2013).

390137 The Impact of Photon Excitation on Catalytic Processes at Metal Surfaces

Abstract:

Catalytic activity and selectivity of metallic nanoparticles are intrinsic material properties controlled by catalyst composition and geometry. The intrinsic nature of catalytic properties dictates that the outcomes of chemical reactions are relatively static in chemical reactors and cannot be manipulated on the fly. In addition, while scaling and Bronsted-Evans-Polanyi relations have facilitated the optimization of metal catalysts, they also suggest inherent limitations in metal catalysts for achieving 100% selectivity in chemical transformations of emerging importance.

Photon excitation of operating metal catalysts has been considered as a potential approach to overcome inherent limitations of metal surfaces for driving selective reactions and to allow this functionality on demand. Over the past 30 years it has been found in a few cases that photon excitation of adsorbate covered metal surfaces induces unique reactivity channels that cannot be accessed through thermal excitation. However, the mechanism of photon energy conversion into chemical reactivity is poorly understood and there exists no generalized understanding of how the nature of the adsorbate or adsorbate-metal bonds impact their potential for being activated selectively through photon excitation.

Photon induced activation of bonds formed between strongly chemisorbed adsorbates and metal surfaces can occur through indirect and direct photoexcitation processes. In the indirect mechanism photons are absorbed by metal states followed by the transient transfer of hot electrons into normally unpopulated antibonding orbitals, thereby depositing vibrational energy into the system through inelastic Frank Condon transitions. In the direct photoexcitation process photons induce electronic transitions localized at hybridized metal-adsorbate states, which can drive chemical reactions with resonant wavelength dependent cross-sections. In this talk we combine experimental and theoretical studies to show how selectivity can be controlled through both the indirect and direct photoexcitation of metal catalysts. We focus specifically on reactions that involve the activation of bonds formed between CO, NO and O on Pt surfaces, but will draw general conclusions on the potential of this approach for achieving high selectivity in processes where heterogeneous catalysis driven by thermal excitation has failed.

TOWARD THE DESIGN OF EFFICIENT CATALYSTS FOR OXYGEN REDUCTION: STRUCTURE/ACTIVITY RELATIONSHIPS FOR LAYERED METAL OXIDES

Xianfeng Ma[1], Hao Qin[1], Hongliang Xin[2,3], Ayad Nacy[1] and Eranda Nikolla[1]

[1]Department of Chemical Engineering and Materials Science, Wayne State University, Detroit, MI

[2]SUNCAT Center for Interface Science and Catalysis, SLAC National Accelerator Laboratory, Menlo Park, CA

[3]Department of Chemical Engineering, Stanford University, Stanford, CA

Oxygen reduction processes play an important role in many energy conversion and storage technologies.[1,2] Layered metal oxides with K_2NiF_4-type structure (generally expressed as A_2MO_4) have attracted considerable interest for these applications due to their versatile oxygen activation and transport properties.[3,4] A_2MO_4 oxides belong to Ruddlesden-Popper series structures and consist of alternating rocksalt-like and perovskite-like layers. While these oxides have shown to exhibit high oxygen exchange rates and oxygen reduction activity, there exists limited understanding on the physical factors that govern their performance. In this work, we use various experimental tools and first-principles electronic structure calculations to identify the underlying electronic descriptor that describes reactivity trends of A_2MO_4 oxides for oxygen reduction. The catalytic performance of these materials is determined using well-controlled $^{18}O/^{16}O$ kinetic studies.[5] We find that their catalytic activity toward oxygen exchange exhibits a volcano-shaped relationship with respect to the electronic fingerprint of the surface transition metal site (*A*-site), i.e., the upper d-band edge. We also find that interstitial oxygen located in the layered structure of these oxides plays a significant role in the surface oxygen exchange activity. The structure/performance relationship identified in this study furthers our understanding of surface reactivity of transition metal oxides and provides physical insights for optimizing the structure and composition of these materials toward surface oxygen exchange.

References

1. Whitesides, G. M.; Crabtree, G. W., *Science.* **2007,** 315, (5813), 796-798.
2. Read, M. S. D.; Islam, M. S.; King, F.; Hancock, F. E., *Journal of Physical Chemistry B.* **1999,** 103, (9), 1558-1562.
3. Amow, G.; Davidson, I. J.; Skinner, S. J., *Solid State Ionics.* **2006,** 177, (13-14), 1205-1210.
4. Boehm, E.; Bassat, J. M.; Dordor, P.; Mauvy, F.; Grenier, J. C.; Stevens, P., *Solid State Ionics.* **2005,** 176, (37-38), 2717-2725.
5. Shannon, S. L.; Goodwin, J. G., *Chemical Reviews.* **1995,** 95, (3), 677-695.

Electrochemical denitrification is a potentially interesting technology for the removal of toxic nitrate and nitrite species from groundwater due to this process's environmental compatibility, energy efficiency, safety, and product selectivity[1]-[3]. In spite of these applications, however, little is known about the molecular mechanisms by which nitrate-derived species are electrocatalytically converted to ammonia, hydroxylamine, and N_2. Experimental studies have demonstrated that adsorbed NO is generally considered to be a selectivity-determining species during the reaction[4], [5], but other information about kinetically significant steps is lacking.

Molecular-level studies using periodic density functional theory (DFT) calculations have recently emerged as a powerful tool for the elucidation of elementary reaction information in electrocatalytic systems. Straightforward thermodynamic corrections may be applied to standard DFT calculations to predict potential-dependent kinetics of these electrocatalytic processes, and in this work, we apply such techniques to examine the electrochemical reduction of NO on platinum surfaces in acidic solutions. We first determine the adsorption structures and energetics of NO and its related reaction intermediate species on different single crystal platinum surfaces ((111), (100), and (211)), and we then determine kinetics by calculating the activation barriers for N-O dissociation, protonation, and dimer formation. The initial studies are performed at low NO coverages (ca. 0.1ML) and include analysis of combined proton/electron transfers to adsorbed NO, cleavage of N-O bonds in the resulting reduction intermediates, and desorption of the resulting products. Water-assisted hydrogen transfer[6] is shown to strongly affect the kinetics of electrochemical hydrogenation and, in some cases, N-O bond cleavage, and stepped (211) surfaces are shown to exhibit higher overpotentials for this chemistry than are the low-index surfaces.

At higher NO coverages (ca. 0.5ML), the relative energy states of the intermediates generally show a similar trend for the energetics when compared with the resulting values at lower NO coverages, albeit with overall weaker binding. In addition, we demonstrate that N_2O formation through NO dimers is thermodynamically and kinetically more favorable at higher NO coverages, and dimer formation is most favorable on the Pt(211) surface, providing a possible explanation for observed N_2O formation in electrochemical systems.

[1] M. Duca and M. T. M. Koper, Energ. Environ. Sci. 5 (2012) 9726.
[2] V. Rosca, M. Duca, M. T. de Groot, M. T. M. Koper, Chem. Rev. 109 (2009) 2209.
[3] S. E. Bae, K. L. Stewart, A. A. Gewirth, J. Am. Chem. Soc. 129 (2007) 10171.
[4] A. Rodes, R. Gomez, J. M. Perez, J. M. Feliu, A. Aldaz, Electrochim. Acta 41 (1996) 17
[5] A. Cuesta, M. Escudero, Phys. Chem. Chem. Phys. 10 (2008) 3628.
[6] X. Nie, M. R. Esopi, M. J. Janik, A. Asthagiri, Angew. Chem. Int. Edit. 52 (2013) 2459.

Study of surface chemistry and properties of multicomponent materials is hampered by the need to perform materials preparation, characterization and property measurement across composition space. Composition Spread Alloy Films (CSAFs) enable such studies to span composition space comprehensively. CSAFs are thin multicomponent films that have composition gradients parallel to their surfaces, $A_xB_yC_{1-x-y}$ with $x = 0 \rightarrow 1$ and $y = 0 \rightarrow 1-x$. There are many otherwise intractable fundamental scientific problems in thin alloy films that can be addressed using CSAFs. Examples include the composition dependence of alloy oxidation and catalysis.

In order to correlate catalytic activity with alloy composition, a multichannel microreactor array has been developed to allow reactions to be catalyzed on 100 different compositions of the alloy in parallel. The relationship between alloy catalyst activity and electronic structure has been investigated experimentally across a broad, continuous span of Cu_xPd_{1-x} composition space, x = 0.30 – 0.97. Cu_xPd_{1-x} CSAFs were used as catalyst libraries with a multichannel microreactor to measure the H_2-D_2 exchange kinetics at 100 discrete Cu_xPd_{1-x} compositions and over a temperature range of 333 – 597 K. The H_2-D_2 exchange activity exhibits a monotonic decrease over the composition range x = 0.30 – 0.97. A steady state, microkinetic model that has been validated using a number of single component Cu-Pd catalysts in a fixed bed reactor was used to estimate the energy barriers to dissociative adsorption, $\Delta E^{\ddagger}_{ads}$, and associative desorption, $\Delta E^{\ddagger}_{des}$, of H_2 as functions of alloy composition, x. Their values fall in the ranges $\Delta E^{\ddagger}_{ads}$ = 0.4 – 0.15 eV and $\Delta E^{\ddagger}_{des}$ = 0.7 – 0.5 eV. Spatially resolved UV photoemission spectra were obtained from the Cu_xPd_{1-x} CSAF and used to estimate the energy of the d-band center as a function of alloy composition, $\bar{E}_{d-band}(x)$. The d-band center shifted monotonically from $\bar{E}_{d-band}$ = -2.0 to -3.5 eV across the composition range x = 0.30 – 0.97. The correlation of $\Delta E^{\ddagger}_{ads}(x)$ with $\bar{E}_{d-band}(x)$ across alloy composition space yields $\Delta E^{\ddagger}_{ads}(\bar{E}_{d-band})$ and shows that the barrier to dissociative adsorption of H_2 decreases as the energy of the d-band increases. This represents the most comprehensive experimental correlation of electronic structure with catalytic reactivity.

INSTITUT FÜR ELEKTROCHEMIE

ulm university universität uulm

Bimetallic Alloys in Action:

Adsorbate-induced surface segregation and ensemble formation in Pt_3Ni and Au_4Pd nanoparticles

2014 AIChE Annual Conference
Atlanta, November 18, 2014

Jonathan E. Mueller
Institut of Electrochemistry, University of Ulm

Bimetallic Catalysts

Promising for numerous applications.

Catalytic activity depends on surface structure.

Identifying the *in-operando* surface structure is critical for understanding catalytic behavior.

2

Outline

What is the structure of the catalyst surface *in operando*?

1. Pt_3Ni nanoparticles under electrochemical conditions in the hydrogen UPD region.

2. Au_4Pd nanoparticles under conditions for hydrogen adsorption and evolution.

3

PtNi Catalysts

Pt_3Ni (111) Surface

- Pt_3Ni alloys are one of the most promising catalyst for the ORR in acidic media.[1]
- Experiment and theory show that the $Pt_3Ni(111)$ surface is surface segregated with a Pt-skin and a Ni-rich second layer.[2]
- The high catalytic activity of the $Pt_3Ni(111)$ surface has been atributed to this surface segregation. [2]

Pt_3Ni Nanoparticles

- Nanostructured Pt_3Ni alloys including nanoparticles and thin films also show great catalytic promise.[3]
- Understanding the structure and stability of Pt_3Ni is vital for understanding and further improving their performance.

1) *Science*, **2007**, *315*, 493–497. 2) *Electrochimica Acta*, **2008**, *53*, 6076–6080. 3) *Electrochimica Acta*, **2011**, *56*, 8695– 8699; . *Langmuir*, **2013**, *29*, 9046-9050..

4

Pt_3Ni Nanocatalysts

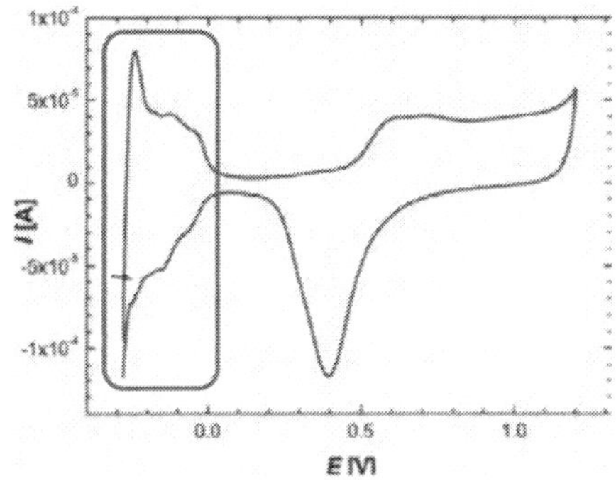

Cyclic voltammogram of a Nafion-stabilized carbon-supported nanocrystalline Pt_3Ni electrode in Ar-saturated 0.05 M H_2SO_4, recorded at a polarization rate of 20 mV/s.

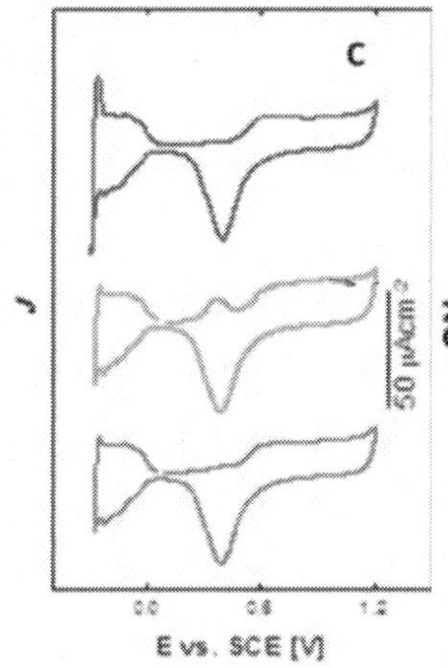

Voltammetric curves recorded before (blue) and after (green and red are 1st and 2nd cycles respectively) 2 hours of cycling in H_{UPD} region.

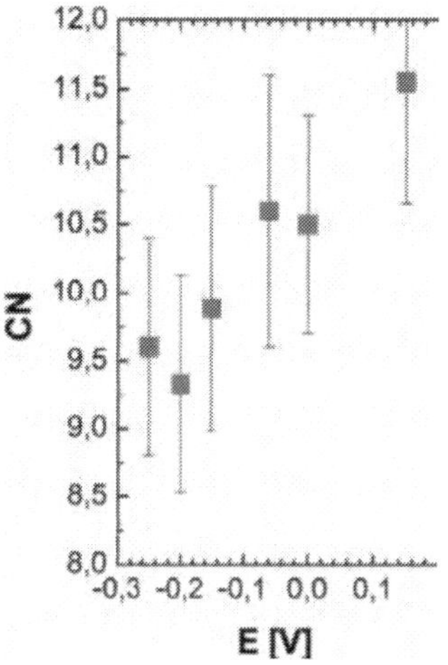

Average coordination number (CN) for Ni in Nafion-stabilized nanoparticulate Pt_3Ni electrode, polarized at various potentials in the H_{UPD} region.

Nickel segregates to the surface in the H_{UPD} region (-0.4 V < E[V] < 0.0 V)

Hoffmannová, H; Okube, M.; Petrykin, V.; Krtil, P.; Mueller, J.E.; Jacob, T. *Langmuir*, **2013**, *29*, 9046-9050.

5

The Atomistic Picture (DFT)

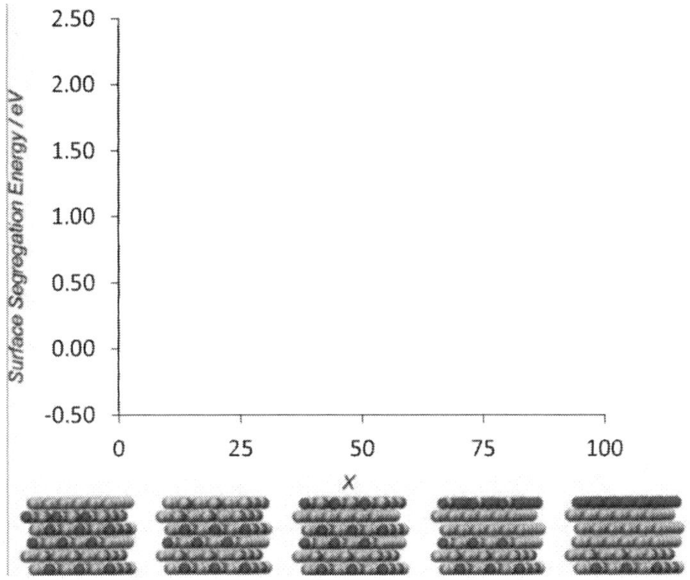

Surface segregation energies in eV per 2 × 2 simulation supercell for Pt_3Ni slabs with no adsorbates, 1/2 ML H, 1 ML H, and 3/2 ML H.

Hoffmannová, H; Okube, M.; Petrykin, V.; Krtil, P.; Mueller, J.E.; Jacob, T. *Langmuir*, **2013**, *29*, 9046-9050.

6

The Atomistic Picture (DFT)

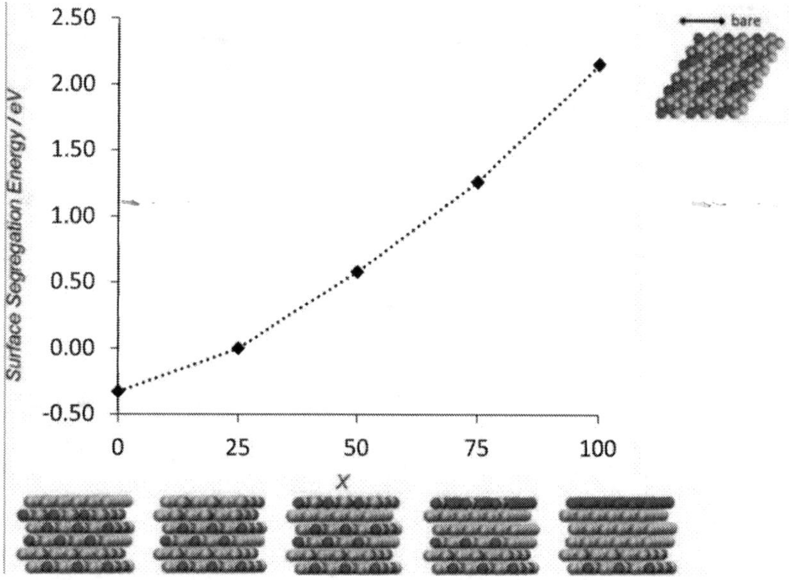

Surface segregation energies in eV per 2 × 2 simulation supercell for Pt_3Ni slabs with no adsorbates, 1/2 ML H, 1 ML H, and 3/2 ML H.

Hoffmannová, H; Okube, M.; Petrykin, V.; Krtil, P.; Mueller, J.E.; Jacob, T. *Langmuir*, **2013**, *29*, 9046-9050.

7

The Atomistic Picture (DFT)

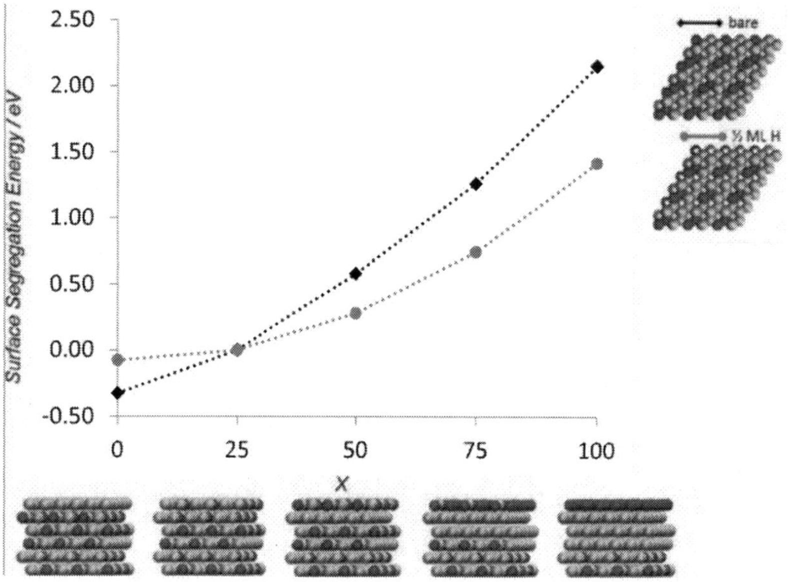

Surface segregation energies in eV per 2 × 2 simulation supercell for Pt_3Ni slabs with no adsorbates, 1/2 ML H, 1 ML H, and 3/2 ML H.

Hoffmannová, H; Okube, M.; Petrykin, V.; Krtil, P.; Mueller, J.E.; Jacob, T. *Langmuir*, **2013**, *29*, 9046-9050.

8

The Atomistic Picture (DFT)

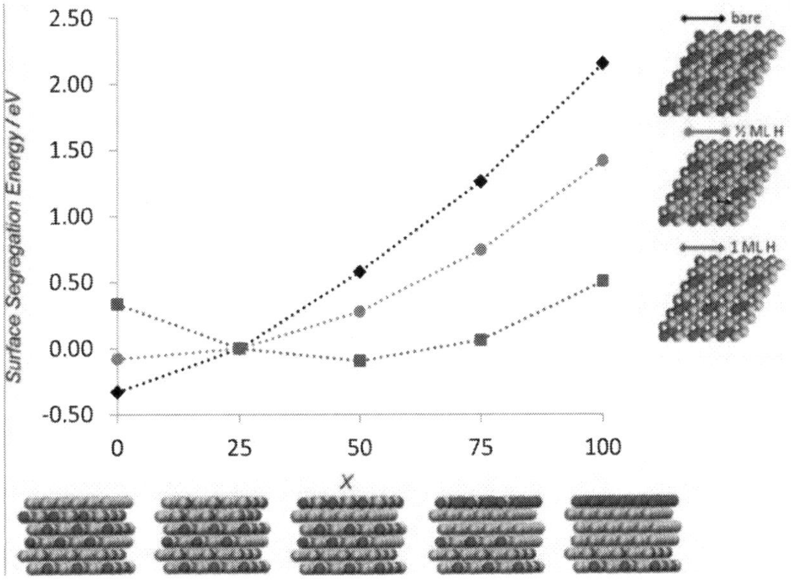

Surface segregation energies in eV per 2 × 2 simulation supercell for Pt_3Ni slabs with no adsorbates, 1/2 ML H, 1 ML H, and 3/2 ML H.

Hoffmannová, H; Okube, M.; Petrykin, V.; Krtil, P.; Mueller, J.E.; Jacob, T. *Langmuir*, **2013**, *29*, 9046-9050.

9

The Atomistic Picture (DFT)

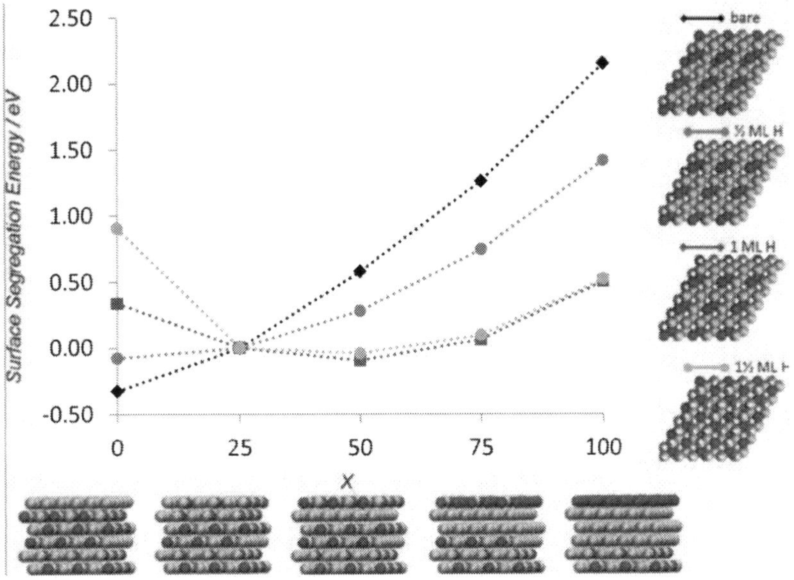

Surface segregation energies in eV per 2 × 2 simulation supercell for Pt_3Ni slabs with no adsorbates, 1/2 ML H, 1 ML H, and 3/2 ML H.

No net surface segregation for H coverage between 0.5 and 1.0 ML.

Hoffmannová, H; Okube, M.; Petrykin, V.; Krtil, P.; Mueller, J.E.; Jacob, T. *Langmuir*, **2013**, *29*, 9046-9050.

10

Adsorbate-Induced Surface Segregation

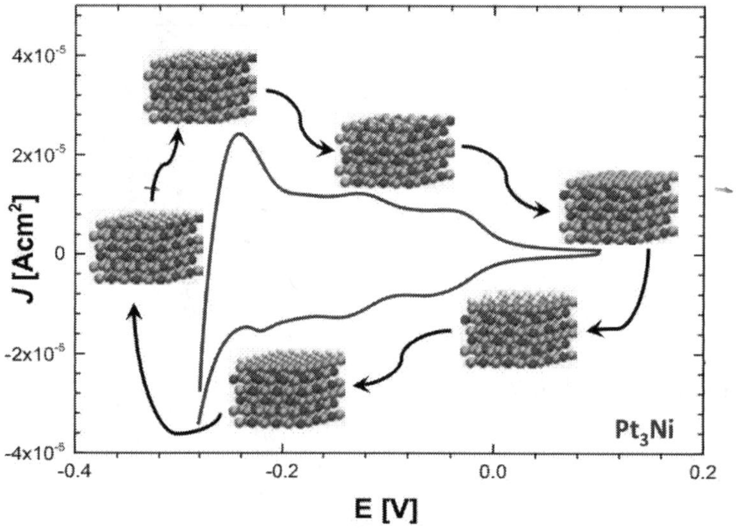

H-adsorbtion induces Ni segregation to the surface resulting in Pt_3Ni nanocatalysts, with dynamic surface structures.

Hoffmannová, H; Okube, M.; Petrykin, V.; Krtil, P.; Mueller, J.E.; Jacob, T. *Langmuir*, **2013**, *29*, 9046-9050.

11

AuPd Catalysts: Ensemble Mechanisms

Ensemble Effect

Experiments suggest:

- CO adsorption: strength depends on Pd surface ensemble (TPD)[1].
- H adsorption: Pd dimers (STM)[2].
- H_2 evolution: Pd monomers (CV)[3].
- Vinyl acetate catalysis: pairs of Pd monomers (IRAS, LEISS)[4].

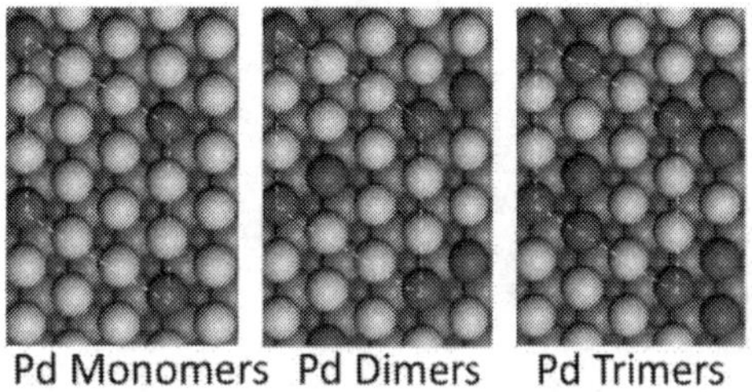

Pd Monomers Pd Dimers Pd Trimers

1) *Phys. Rev. B*, **1994**, *49*, 8367. 2) *Science*, **2001**, *293*, 1811. 3) *PCCP*, **2008**, *10*, 3684. 4) *Science*, **2005**, *310*, 291.

Ensemble Formation

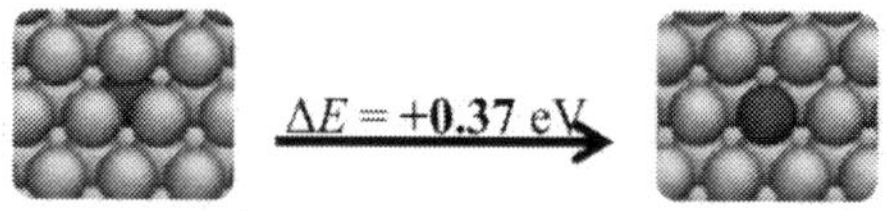

- Pd prefers the subsurface

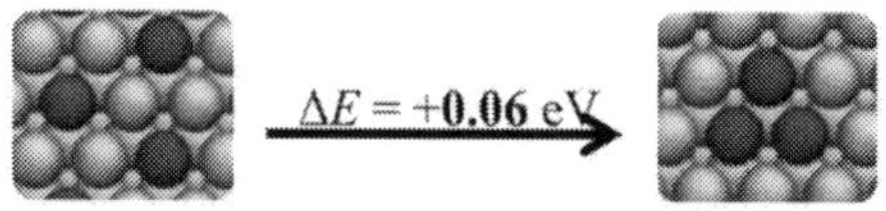

- Pd atoms avoid each other

For alloys with small amounts of Pd we expect Pd subsurface monomers.

ChemElectroChem, **2014**, *1*, 207–212.

12

Au_4Pd Electrocatalysts

CV & EXAFS characterization of Au_4Pd nanoparticles in 0.05 M H_2SO_4.

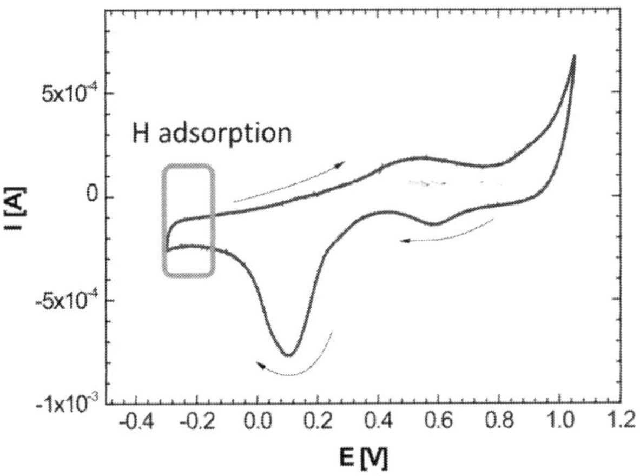

Cyclic voltammogram of nanoparticulate Au_4Pd alloy catalyst (5 nm particle size; polarization rate is 10 mV/s).

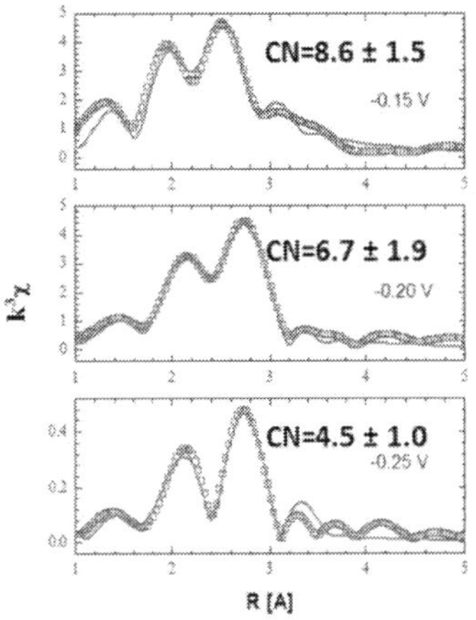

k3-normalized EXAFS functions extracted from X-ray absorption spectra collected at Pd K-edge at controlled potentials.

Pd coordination number decreases dramatically as potential decreases.

13

Okube, M.; Petrykin, V.; Mueller, J.E.; Fantauzzi, D.; Krtil, P.; Jacob, T., *ChemElectroChem*, **2014**, *1*, 207–212.

H-adsorption and Pd ensembles

<u>H adsorption</u>

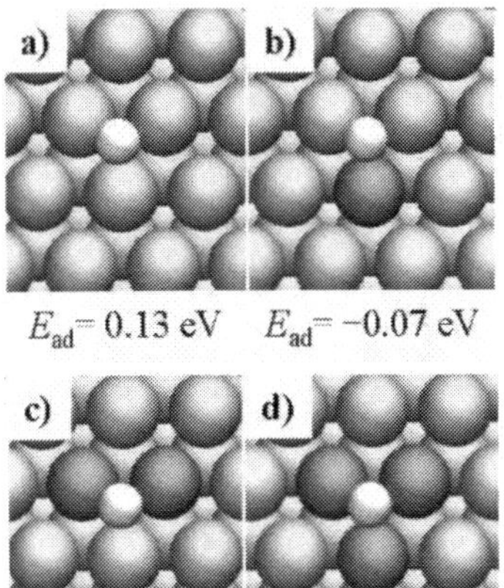

H adsorbs more strongly on Pd than on Au

<u>Influence of H on subsurface Pd</u>

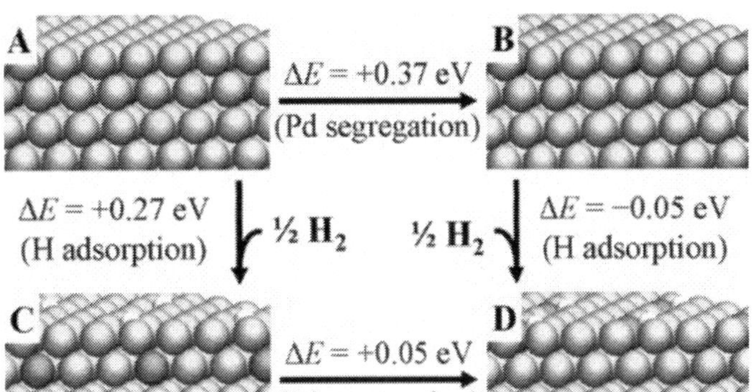

A single adsorbed H is unable to draw a single subsurface Pd to the surface.

14

Mueller, J.E.; Krtil, P.; Kibler, L.A.; Jacob, T. *PCCP*, **2014**, *16*, 15029–15042 .

Au-Pd Catalysts: Breathing Cycles

High concentrations of H_{ad} — H-induced Pd-trimer formation

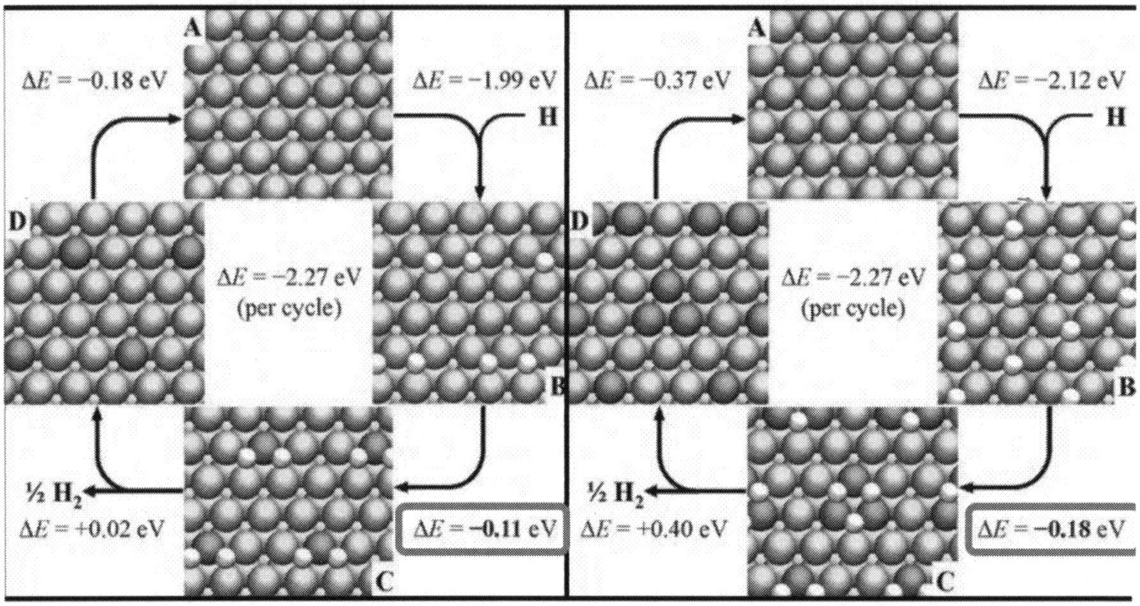

Adsorbed hydrogen is able to draw Pd to the surface!

Mueller, J.E.; Krtil, P.; Kibler, L.A.; Jacob, T. *PCCP*, **2014**, *16*, 15029–15042 .

15

Pd-induced H-absorbtion?

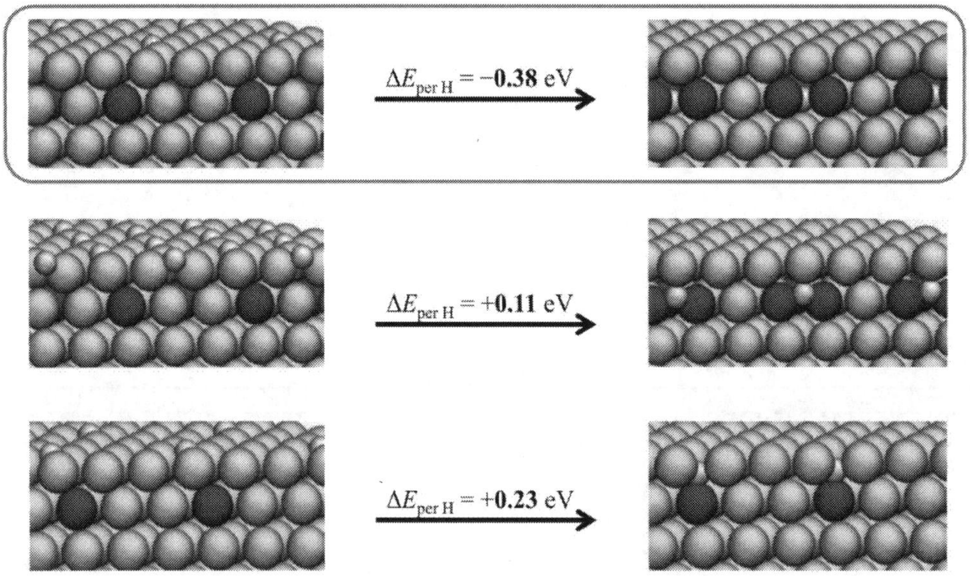

Pd induces H-absorption only for low H:Pd ratios.

Okube, M.; Petrykin, V.; Mueller, J.E.; Fantauzzi, D.; Krtil, P.; Jacob, T., *ChemElectroChem*, **2014**, *1*, 207–212.

16

H-Induced Pd Breathing and Ensemble Formation

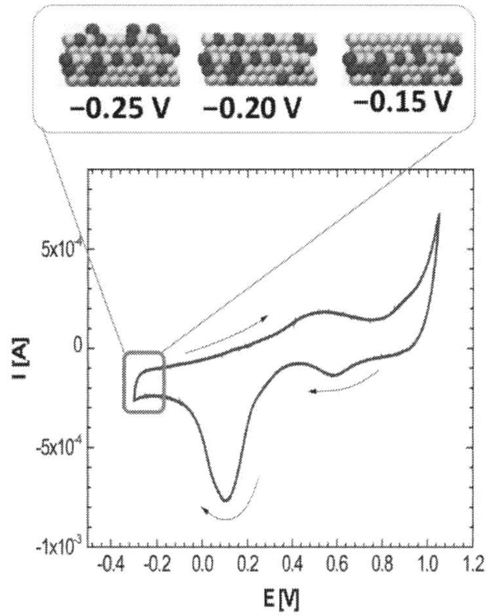

- No hydrogen → Pd subsurface monomers
- Low H coverage → Pd subsurface trimers
- Intermediate H coverage → Pd surface trimers
- High H coverage → Pd surface monomers
- Higher H coverage → Pd at step/kink sites/adatoms

Okube, M.; Petrykin, V.; Mueller, J.E.; Fantauzzi, D.; Krtil, P.; Jacob, T., *ChemElectroChem*, **2014**, *1*, 207–212.

17

Summary

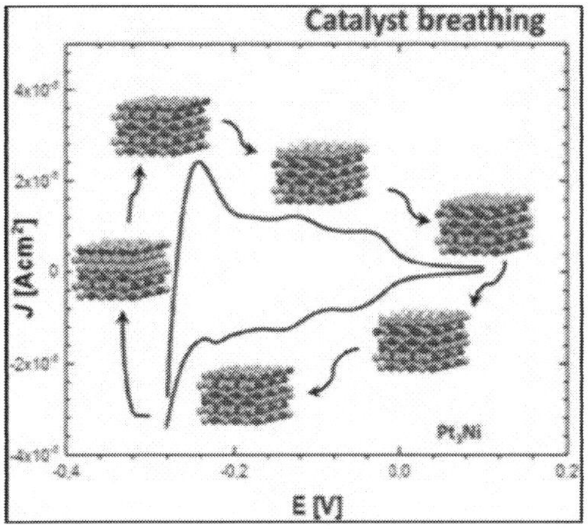

PtNi: Influence of adsorbates on surface segregation.

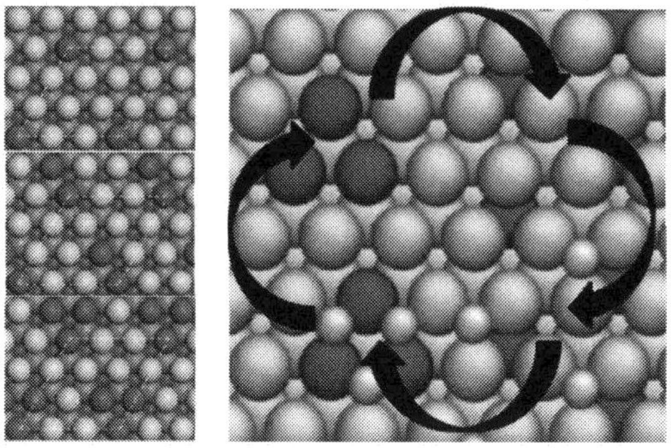

AuPd: Influence of ensembles on adsorption & catalytic activity. Adsorbate-induced catalyst breathing & ensemble formation.

18

Acknowledgements

Prof. Timo Jacob,
Universität Ulm

Dr. Petr Krtil,
Heyrovsky Institute,
Prague

Institute of Electrochemistry, University of Ulm

Unterstützt von / Supported by
Alexander von Humboldt
Stiftung / Foundation

19

Thank you for your attention!

Questions?

20

386395 Flow Chemistry with Microchemical Systems for Chemicals, Energy, Healthcare, and Sustainability

Abstract:

Flow chemistry, important to the global chemicals, energy, and healthcare segments, depends on the symbiosis between chemical reaction engineering and organic synthesis. The design of microchemical systems for flow chemistry impacts a broad cross-section of societal issues. As an example, studying chemical reactions on-chip has advantages in fine chemicals and pharmaceuticals research and manufacturing. Using microchemical systems to understand the science reduces discovery times, solvent waste consumption, and the energy requirements for their preparation at laboratory and production scales, as compared to conventional batch-wise processing. A key area is aqueous-phase catalyzed reactions that require understanding of reaction and transport time scales using classical chemical reaction engineering principles. Aqueous-phase catalyzed Sonogashira couplings yield aryl heteroatoms for many different synthetic and natural applications. On-chip investigations of natural macromolecular aromatics, as examples asphaltenes, directly impact conventional and unconventional energy productions. High-throughput packed-bed microreactors with online analytics generate knowledge on the accumulation mechanisms in unconsolidated, siliceous materials. Flow and reaction in porous media is an important micro-scale, laminar flow area that remains central to green chemistry and environmental sciences. Methane-water interactions with naturally derived macromolecules, as an example methane-cyclodextrin thermodynamics, are key to methane as a resource for chemicals, its natural sequestration, and polysaccharide aqueous chemistry that catalyzes the formation and dissociation of inclusion compounds. Transport limitations render kinetics of inclusion materials, such as hybrid cyclodextrin-gas hydrates, challenging to study. Their on-chip formation and dissociation directly impact the energy sector, yet the potential exists for use in chemo-selective reactive separations. Partnerships between scientists and engineers are needed to drive innovations in flow chemistry using microchemical systems and chemical reaction engineering is key to sustainability.

384395 Enhancing Catalyst Effectiveness and Loading in Microreactors

Abstract:

Microreactors allow for increased viability of multiple energy & fuels applications through the ability to achieve breakthroughs in thermal efficiencies through process integration. By reducing the characteristic transport lengths from millimeters to microns, order-of-magnitude improvements in heat transfer rates per unit reactor volume are achieved. This enables breakthroughs in reactor capacity and performance for highly endothermic processes including methane steam reforming. The reduced hydraulic diameter in turn necessitates the use of catalyst washcoatings in lieu of conventional packed-beds. This catalyst configuration inherently alters the conventional design rubric for catalyst thickness selection, owing to the fact that catalyst films experience internal heat addition, in contrast to conventional catalyst particles which must rely upon external solid-fluid contacting for heat addition.

In this presentation, a 1-D model was used to determine the possibility of gaining higher effectiveness factors for conventionally-determined film thicknesses, or conversely achieving comparable effectiveness at higher than conventional film thicknesses, through exploiting internal heat addition. Mass and energy balances were used with new boundary conditions to allow for this internal heat addition. The conditions for methane steam reforming were tested using this system, showing high thermal efficiencies and higher effectiveness factors at higher catalyst thickness than conventionally used. This hypothesis was then tested using an experimentally validated computational fluid dynamics model of an Annular Microreactor (AMR) device, currently under development by Power+Energy, Inc., to test the sensitivity of the system to catalyst thickness. The effectiveness factor for each catalyst thickness was compared to the isothermal system for the cases of a flow rate resulting in 95% of equilibrium conversion of methane, and a flow rate much greater than equilibrium conversion. The thermal efficiency was calculated for the non-isotheral system for each case. In all cases the non-isothermal case with external heating had a higher relative average effectiveness with a thermal efficiency greater than 100%. This investigation proves that a thicker catalyst can be used to increase hydrogen production with little loss in effectiveness, thus improving the reforming capacity in existing microreactors.

392608 Effect of Process Conditions on Oleic Acid Hydrotreating in a Novel Micropost Plate Reactor

Abstract:

In hydroprocessing of vegetable oils to diesel range products, conversion of triglycerides to free fatty acids is fast compared to free fatty acids to diesel range products. Oleic acid was used as model compound, to study the effect of operating conditions on reaction products in a previously developed novel micro post plate reactor. Temperature, Pressure and residence time was studied. Temperature range was 275-325°C, pressure range 200-500psig and residence time 120-180 seconds. Increase in temperature, pressure and residence time favored production of n-alkane target products and oleic acid conversion. Reduction in temperature, pressure and residence time favored production of intermediate oxygenates products and low oleic acid conversion.

392606 Electric Discharge Activation of Oxidation Reactions in Aqueous Media

Abstract:

Amongst the many ways to activate non spontaneous chemical reactions we have recently focused on the use of (non-thermal) electric glow discharges implemented in microscale reactors. Implementing such reactions in aqueous media have their own issues due to the high electric conductivity of aqueous solutions, potentially leading to electrochemical processes. By applying nanosecond scale duration discharges in liquid in a continuous microdrip configuration we are able to achieve significant oxidation of dissolved organic molecules as measured by colrimetry. The electrical and observed chemical kinetics of these microplasmas will be presented, as will the conversion efficiency of the process.

Electrical Discharge Enabled Oxidative Desulfurization of Diesel Fuel

Peter Kreider, Justin Pommerenck, Kevin Caple, Yousef Alanazi, Alex Yokochi

The primary objective of this work was to examine how non-thermal plasmas generated through electrical discharge (i.e. corona discharge) may be used to drive chemical reactions in microscale reactors. Our work has investigated reactions in both gaseous and liquid media. We will limit the discussion here to liquid phase reactions only, particularly to the oxidative desulfurization of refractory sulfur compounds in fuel-like media. The oxidation of dibenzothiophene to dibenzothiophene sulfone and sulfoxide in fuel-like media has been chosen as a model reaction that is representative of this system's performance. Dibenzothiophene is a refractory sulfur compound in traditional hydrodesulfurization processes related to the oil and gas industry. Efficient oxidation to the sulfone and sulfoxide form allows for advanced separation techniques to be employed in sulfur removal from products like ultra-low sulfur diesel fuel. It has been shown that oxidative desulfurization processes offer improved fuel product quality and energy savings.

Investigating the performance of the non-thermal plasma microreactor in the model chemical reaction will allow for general conclusions to be drawn about the systems performance and will allow for preliminary mathematical modeling. Preliminary work in our labs has demonstrated that the corona discharge activated oxidation of sulfur compounds in fuel-like media takes place. The development of an efficient corona discharge activated microchannel reactor could lead to an efficient approach to diesel desulfurization through advanced oxidation. Additionally, this novel approach to the general activation of chemical reactions through electrically generated plasmas in microreactors could open new processing options for various applications.

377916 Supercritical Water Hydrocarbon Mixing in Complex 3-D Mixer Geometry Under Partially-Miscible Conditions

Abstract:

In this study, the mixing of supercritical water (SCW) and model hydrocarbons in a 3-D mixer geometry under partially-miscible conditions is investigated using 3-D multiphase CFD simulations. This mixing process is of considerable significance in the process of supercritical water upgrading and desulfurization of crude oil. A Volume of Fluid (VoF) method is applied to track the water-hydrocarbon interface followed by phase equilibrium computations to determine the concentrations of the water and oil-rich phases on either side of the interface. The species and energy fluxes across the interface are also computed using consistent modeling procedures. The near-critical thermodynamics of the water-hydrocarbon mixture is modeled using a cubic Peng-Robinson equation of state. The numerical method is validated against previously reported 1-D simulation results of a spherically symmetric case of a hydrocarbon (n-decane and toluene) droplet in SCW. Thereafter, the mixing of SCW and two model hydrocarbons (n-decane and toluene, with a significant difference in upper critical sloution temperature (UCST) in water) in a 3-D cylindrical tee mixer is studied. The impact of the persistence of the water-hydrocarbon phase interface on the retardation of the rate of mixing is quantified. The impact of the water inlet temepraturе (far below, close to and far above the UCST) on the rate of mixing is also investigated

382423 Towards More Accurate Transport Models for Non-Ideal Mixtures of Hydrocarbons and Supercritical or Near-Critical Water Using Generalized Maxwell-Stefan Expression

Abstract:

The mixing of a hydrocarbon droplet, containing a mixture of three pseudocomponents of the crude oil constructed based on Lee-Kesler correlations and predictive Peng-Robinson 78 equation of state model, in a reservoir of supercritical or near-critical water (SCW/NCW) is modeled. This study is an extension of our previous works on the mixing of a single-component or two-component hydrocarbon droplet in SCW. As in our previous works, transport, thermodynamics, and phase equilibrium sub-models are used to estimate the relevant physical properties. The current study uses a generalized Maxwell-Stefan (MS) expression to model the multi-component mass transfer process. The diffusion driving forces in Maxwell-Stefan equations, e.g. terms of fugacity gradients, account for effects of the mixture's non-ideality on mass fluxes, and provide microscale information, such as multi-component partitioning, and insight into large scale mixing in applications that include supercritical water oxidation, supercritical water desulfurization, etc. Results show that fractionation (i.e. the preferred dissolution of a specific hydrocarbon component into the water-rich phase), which characterizes the transport and mixing process at various water temperatures, is predicted to be noticeably enhanced using the MS model with complete driving force expressions including non-ideality terms (referred as model 2) compared to that using the MS model with one single driving force term computed from the species' concentration gradients (model 1). We find that when the water temperature is greater than the lowest water-HC binary upper critical solution temperature (UCST) but lower than the overall UCST, model 2 predicts longer mixing time and higher molar fractions of heavy HC species on HC-rich phase than those predicted using model 1 during the entire mixing process. This study shows that, because of high non-ideality, a generalized MS model with complete driving force expressions is important to accurately predict the mixing process of SCW/NCW and HCs.

Experimental study on catalytic hydroprocessing of oxi-solubilized asphaltenes in water

Parsa Haghighat and Pedro Pereira-Almao, University of Calgary, Calgary, AB, Canada

Abstract

Catalytic hydroprocessing of solubilized asphaltenes in water was studied in a batch reactor. To prepare the feed for hydroprocessing (HP), asphaltenes were solubilized in water by wet air oxidation (WAO). The effect of solubilisation condition was investigated on subsequent HP reaction. At higher WAO temperatures, the quality of liquid distillates improves after HP and the boiling point distribution of liquid products shifts to lighter materials. Pre-sulfide NiMo/γ-Alumina was chosen as a catalyst for conducting the experiments. Experimental results of reaction at different residence time from 1 h to 6 h indicate that first order reaction can be applied to the hydroprocessing reaction of solubilized asphaltene in water.

Keywords: Asphaltene, Hydroprocessing, Aqueous Phase, Catalyst, Decarboxylation

Introduction

In today's world, heavy oil and residua are playing increasing roles by decline of conventional oil and enhance of energy demand. Canada has the world's third proven oil reserves which mainly consist of heavy oil called bitumen [1]. In 2012 the total Canadian oil production was 3.2 MBPD where 2.3 MBPD came from oil sand sources. In 2020 Canada's oil production is expected to reach 4.9 MBPD out of which 3.2 MBPD will be heavy oil [2]. C_5 insoluble asphaltenes constitute the 14-17% of Athabasca bitumen [3]. Therefore, upon expansion of the oil sand extraction activity, Asphaltenes's pitch production will reach 0.5 MBPD and 1 MBPD in 2020 and 2030 respectively.

Asphaltene is the heaviest portion of crude oil with high molecular weight. It contains high amount of heteroatoms such as Sulfur and Nitrogen and metals such as V and Ni[4]. Asphaltenes are insoluble in paraffinic compounds while soluble in aromatics. Precipitation of asphaltenes creates serious problems in pipelines during crude oil's transportation, demanding further expenses to solve that problem[5]. Viscosity of oil has a direct relationship to the amount of asphaltenes in the crude oil, which is a transportation disadvantage of the presence of these components[4]. Another drawback caused by asphaltene is found in catalytic processes of upgrading and refining. Asphaltene particles deposit on the hydrotreating catalyst, block their pores and decrease the available surfaces of catalyst for reactant molecules and finally deactivate them [6].

Up to date, different methods such as thermal cracking, thermal and catalytic hydroprocessing and gasification have been studied for converting asphaltenes to more valuable forms. Among those, delayed coking process (thermal cracking) is commercialized and used vastly in industry[7]. Asphaltenes thermal or catalytic hydrocracking improves the yield and quality of products and diminishes the coke formation compared to thermal cracking. However delayed coking is still the industry's first choice due to economic reasons [8-11].

In this research a potential alternative path to deal with asphaltenes targeting to convert them into higher quality product is presented. In a first step, asphaltenes separated via DAO from vacuum residue fractions are oxy-cracked and solubilized in water with Wet Air Oxidation (WAO) reaction. Asphaltene's wet air oxidation seems to proceed via two parallel reactions. One reaction is the direct conversion to CO_2 and the other one is consecutive reaction in which first asphaltenes are converted to water soluble intermediate products and then to CO_2 (final product)[12]. Investigation on catalytic hydroprocessing of this solubilized asphaltenes in water (WSA) is the subject of this paper.

Water at high temperature, close to its critical condition (374°C, 22.1 MPa) is a favorable solvent for many reactions. At critical condition, properties of water such as density, viscosity and solubility changes dramatically and make it an interesting medium for upgrading hydrocarbons such as biomasses and residua from refineries[13]. In several studies, catalytic and non-catalytic hydrogenation of lignin in water under different reaction conditions has been investigated. Lignin is a natural polymer component with high molecular weight which is an essential component in the wood-based biomasses[14]. Lignin is an appropriate representative component for WSA due to its large molecular weight and high Oxygen/Carbon ratio. Hot compressed water and particularly water under supercritical condition presents promotional effects on increasing the conversion and liquid yield of lignin. Bases such as NaOH and KOH in absence of catalyst increase the solubility and conversion of lignin during hydrogenation. Oxygenated functions of the lignin may react with bases during the hydrogenation process and finally creates soluble salts in water. Amongst many catalysts, Ru/ɣ-Al2O3 represent the best performance in terms of conversion and products quality distribution [15, 16]. Products distribution of lignin hydroprocessing have complex nature with numerous types of components mainly consisting of phenols and minor amount of long alkanes[15].It has been reported that by adding alcohols to water medium, the hydrogenating reaction could be promoted. Presence of alcohols will avoid the further repolymerization of lignin during the hydrogenation. In EtOH medium, the conversion of lignin can reach up to 90% even in more moderate reaction conditions compared to pure water medium[15, 16]. In this work HP of WSA in the absence of ethanol is presented.

Experimental Setup

Wet Air Oxidation of asphaltenes was conducted in a 550 ml batch Parr reactor. The C_5 asphaltenes used for this purpose were precipitated from Vacuum Residue of Athabasca Bitumen provided by SUNCOR. 4 grams of these asphaltenes with 156 gram of water were placed in the reactor vessel and the system was pressurized with oxygen to 1000psi at room temperature and then heated up to reaction temperature.

For hydroprocessing of water solubilized asphaltene a 100 ml Micro 4598 Parr batch reactor with the possibility of controlling temperature, pressure and stirring rate was used. 25 g of WSA with 0.5 gram of catalyst was placed in the reactor. Then, leak test with nitrogen at 200 psi above the final pressure of the reaction was performed and in the next step the reactor was pressurized with hydrogen at room temperature. Finally the reactor was heated up by pre-defined temperature ramp to reach the set point temperature. For the kinetic study, it is assumed the reaction starts when temperature reaches to the set point though some conversion occurs during the heating up period. After the reaction is finished, the reactor is cooled down to initial room temperature and the products are collected for further analyses. The TOC (Total Organic Carbon) analyzer was used to find the amount of soluble carbon in water before and after reaction. The produced gases are measured by GC (HID and TCD detectors). The liquid products are extracted by adding the known amount of CS_2 to reacted WSA and later analyzed by High-Temperature Simulated Distillation System (SimDis) following ASTM D7169 method. The mass balance, selectivity, yield and conversion are all calculated based on the amount of carbon in the system. Equation (1) demonstrates how the conversion is determined.

$$Conversion(\%) = \frac{Amount\ of\ carbon\ in\ water\ before\ reaction - amount\ of\ carbon\ in\ water\ after\ reaction}{amount\ of\ carbon\ in\ water\ before\ reaction} * 100 \qquad (1)$$

The liquid selectivity is determined by Equation (2).

$$Liquid\ Yield(\%) = \frac{Amount\ of\ carbon\ in\ liquid\ phase\ after\ reaction}{Amount\ of\ carbon\ in\ water\ before\ reaction} * 100 \quad (2)$$

Finally, The Ideal Gas Low was used for gas phase calculations.

Catalyst Preparation

3% NiO, 14% MoO_3/Al_2O_3 catalyst was used in this set of experiments. Aqueous solution of Nickel (II) Acetate (Ni $(CH_3COO)_2$) and Ammonium Heptamolybdate ($(NH_4)_6Mo_7O_{24}.4H_2O$) provided the source of active agents and added to the gamma alumina support (Al_2O_3) via dry impregnation method. The catalyst was then calcined for 6 hours at 500°C. Physical characterization of the prepared catalyst is shown in Table1.

Table 1-*Characterization of in-house synthesized NiMo/Alumina Catalyst*

Catalyst	BET surface area	Pore Volume	Pore Size
	m^2/g	cm^3/g	A
(3% NiO, 14% MoO_3)/Al_2O_3	138	0.36	104.12

The catalyst was pre-sulfided before the reaction. 30 ml solution of 10% CS_2 in Decalin and 2 gram of catalyst were poured in the 100 ml batch reactor and pressurized with hydrogen to 200 psi at room temperature. The sulfiding reaction was performed at 350°C for 6 hours with stirring rate of 500 rpm.

Results and Discussion

Effect of asphaltenes solubilisation condition (WAO reaction) on subsequent hydroprocessing

Wet Air Oxidation reaction at different residence times and temperatures was conducted to prepare solubilized asphaltenes in water. Table 2 provides the results of solubilizing reaction under different conditions. The level of conversion and liquid yield are in the range of 80% and 50% for all studied conditions.

Table 2-*Results of solubilizing asphaltenes in water by oxidation reaction.* PH of the liquid phase after solubilisation reaction*

Product denomination	Temperature	Residence Time	Initial amount of Carbon	Solubilized Carbon in water	Conversion	Liquid Yield	PH*
	°C	h	g	g	%	%	
WSA I	190	6	3.32	1.59	80	48	7.4
WSA II	215	4	3.33	1.66	69	50	9.0
WSA III	230	3	3.32	1.43	82	43	9.2
WSA VI	240	2	3.33	1.69	83	51	7.7

WSA prepared under these different conditions were used for hydroprocessing reaction. The hydroprocessing conditions were the same in all runs to ensure the only reaction variable is the WSA feed. The results of the conducted runs are presented in Table 3 and Figure 1.

Table 3-Results of hydroprocessing different feeds under the same condition

Feed	Hydroprocessing Conversion	Hydroprocessing Liquid Yield
	%	%
WSA I	46	15
WSA II	55	17
WSA III	43	15
WSA VI	43	19
WSA V	39	19

The reaction conversion for different WSA feed are in the range of (40 to 50%). The conversion is not distributed widely and it doesn't follow the increasing or decreasing trend. The same scenario exists for the liquid products. Liquid yields are all located in the range of 15% to 20 %. Figure 1 shows the distribution of liquid products which can help to understand the effect of WSA preparation conditions on the hydroprocessing. The distribution of liquid products from hydroprocessing evolved toward lower molecular weight products when the temperature of oxidation reaction was increased. For instance, the amount of naphtha fraction achieved for WSA produced at 240-250°C is about 50% of the distillates, while for the WSA produced at 215°C, just 10% of distillates is within the naphtha fraction. For WSA prepared under higher temperature (240°C or 250°C) and less residence time (2h) the selectivity to lighter products increases after hydroprocessing reaction.

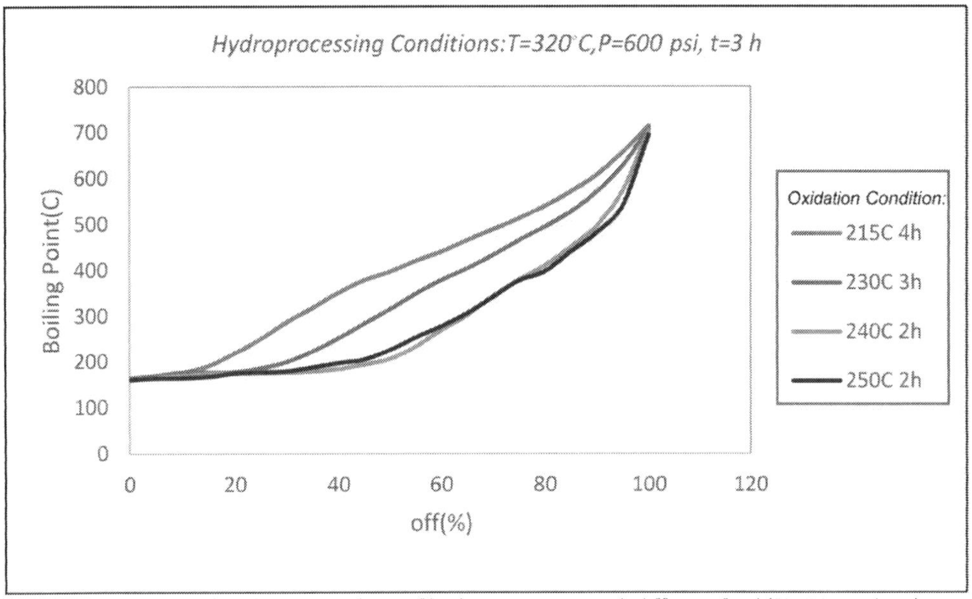

Figure 1-Liquid distribution products of hydroprocessing with different feed (SimDis analysis)

Effect of residence time on WSA hydroprocessing

Figure 2 presents the results of conversion and liquid yield for the residence time between zero to 6 hours where slow increase in conversion and liquid yield is observed. The distribution of liquid products is presented in Figure 3. Changing the residence time doesn't affect the quality of liquid products and keeps them within the same range.

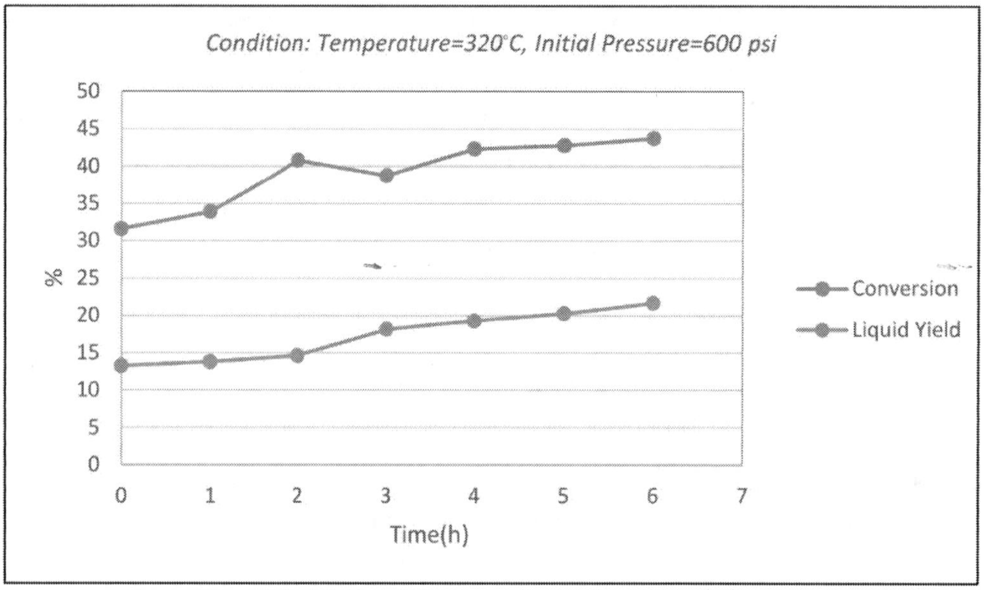

***Figure 2**-Conversion and liquid yield vs residence time in WSA hydroprocesssing*

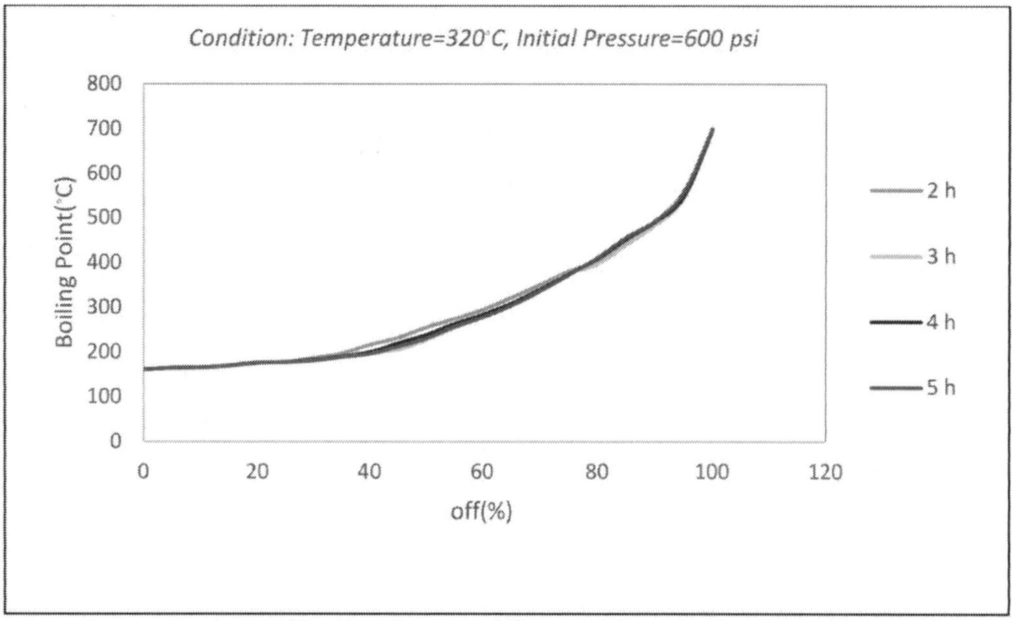

***Figure 3**-Distribution of liquid products with different residence time (Determined by SimDis)*

In general, hydroprocessing is described as a first order reaction. For the case of heavy oil hydroprocessing that the feed is a complex component with wide range of boiling points, different lumped kinetics models were proposed and reported all consisting of series of first order reactions [18-20]. In Figure 4 experimental results of different residence time were fitted to equation of first order reaction in batch reactor, suggesting that kinetics of asphaltenes hydroprocessing in water medium is similar to other HP reactions.

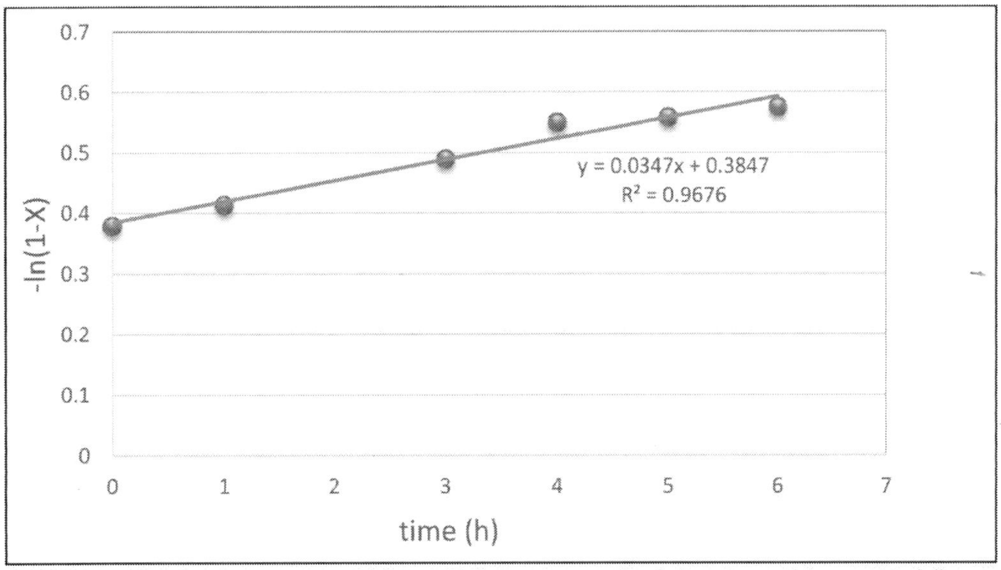

***Figure 4**-Fitting the experimental data with first order reaction (•Experimental Data, —Model)*

Conclusions

This work showed that WSA hydroprocessing is the first order reaction and the conversion and liquid yield increase by extending the residence time. In WSA hydroprocessing, conversion and liquid yield aren't affected by the condition of WSA solubilisation within the explored range. However the liquid selectivity shifted to lower boiling point products after hydrotreating, for WSA prepared at higher oxidation temperature. During oxidation, the large molecules of asphaltenes crack into smaller molecules, which can be referred as Oxy-Cracking of Asphaltenes. The concept of this process is depicted in Figure 5 from a molecular point of view.

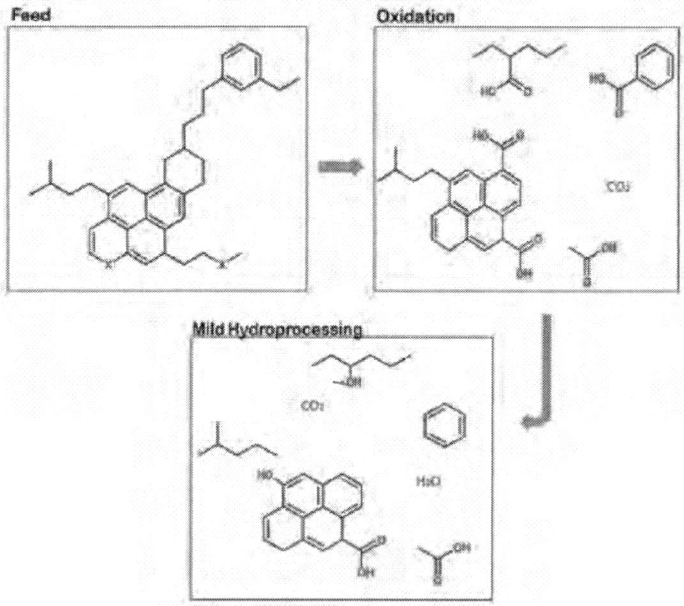

Figure 5-The idea of molecules conversion in the WSA research project

References

1. *Organization of Petroleum Exporting Countries, OPEC Anuual Statistical Bulletin.* 2013.
2. *Canadian Association of Petroleum Producers,2013 CAPP Crude Oil Forecast, Markets & Transportation.* June 2013.
3. Zerpa, N., *Creating More Value from Asphaltenes-The Innovation Challenge.* Nexen Inc., 2012.
4. Kozhevnikov, I.V., A.L. Nuzhdin, and O.N. Martyanov, *Transformation of petroleum asphaltenes in supercritical water.* The Journal of Supercritical Fluids, 2010. 55(1): p. 217-222.
5. M. Branco, V.A. and G.A. Mansoori, *Asphaltene flocculation and collapse from petroleum fluids.* Petroleum Science and Engineering, 2001. 32: p. 217– 230.
6. Martinez, M., A. M. Benito, and M. A. Callejas, *Thermal cracking of coal residues Kinetics of asphaltene decomposition.* Fuel, 1997. 79: p. 871-877.
7. Bonilla, J.A. and J.D. Elliot, *Asphalt Coking Method.* European Patent, 24 July 1991. Patent #: 0209225 B1.
8. Soodhoo, K. and C. R. Phillips, *Non-catalytic hydrocracking of asphaltenes 1. Product distributions..* Fuel, 1988. 67: p. 361-374.
9. Soodhoo, K. and C. R. Phillips, *Non-catalytic hydrocracking of asphaltenes 2. Reaction kinetics.* Fuel, 1988. 67: p. 521-529.
10. Zhao, Y., et al., *Kinetics and selectivity of asphaltene hydrocracking.* Fuel, 2011. 90(5): p. 1900-1906.
11. Mitake, M. and T. Kagayjo, *Hydrocracking of asphaltene from coal using ZnCl2 as catalyst.* Fuel Processing Technology, 1984. 9: p. 293-306.
12. Sanchezoneto, J., et al., *Wet air oxidation of long-chain carboxylic acids.* Chemical Engineering Journal, 2004. 100(1-3): p. 43-50.
13. Akizuki, M., et al., *Effects of water on reactions for waste treatment, organic synthesis, and bio-refinery in sub- and supercritical water.* Journal of Bioscience and Bioengineering, 2014. 117(1): p. 10-18.

14. Yan, N., et al., *Hydrodeoxygenation of Lignin-Derived Phenols into Alkanes by Using Nanoparticle Catalysts Combined with Brønsted Acidic Ionic Liquids.* Angewandte Chemie International Edition, 2010. 49(32): p. 5549-5553.
15. Patil, P.T., et al., *Heterogeneously Catalyzed Hydroprocessing of Organosolv Lignin in Sub- and Supercritical Solvents.* Energy & Fuels, 2011. 25(10): p. 4713-4722.
16. Carbognlani, L., *Literature review on oxidative processes, water solubilization of petroleum heavy fractions/asphaltenes, characterization of solubilized hydrocarbons and production of light ends from solubilized asphaltenes.* NSERC / NEXEN / ALBERTA Innovates Energy and Environment Solutions Industrial Research Chair in Catalysis for Bitumen Upgrading., April 2012.
17. *Kleinert, M. and T. Barth, Phenols from Lignin. Chemical Engineering & Technology, 2008. 31(5): p. 736-745.*
18. *Ancheyta, J., S. Sánchez, and M.A. Rodríguez, Kinetic modeling of hydrocracking of heavy oil fractions: A review. Catalysis Today, 2005. 109(1-4): p. 76-92.*

387460 Hydrocarbon Products from Supercritical Water Partial Oxidation of Benzene Between 375 to 440 °C

Abstract:

Supercritical water is a highly energetic partial oxidation medium for a wide variety of chemical processes ranging from jet fuel reformation to gasification of lignocellousic materials. Supercritical water can also be used to partially oxidize benzene with several studies evaluating the reaction kinetics and mechanisms of benzene partial oxidation ultimately leading to gasification. Notable intermediates of the partial oxidation of benzene include commodity chemicals such as phenol. This study focuses on optimizing the conversion of the supercritical water partial oxidation of benzene to phenol and other non-gaseous species near water's critical temperature. The effects of reaction time, temperature, and benzene proportion were evaluated with an agitated 500 mL Hastelloy C276 batch reactor with 30 wt-% hydrogen peroxide as the oxidant. In a factorial experimental design varying temperature between 390 and 420 °C, reaction time from 15 to 23 minutes, and benzene from 5 to 10 wt-%, it was found that conversion of phenol is favored at shorter reaction times, lower weight fractions of benzene, and higher temperature. In a separate study, the reaction time was varied from 0 to 30 minutes and temperature from 375 to 440 °C. In addition to unreacted benzene, the principal non-gaseous hydrocarbon products were phenol, benzoic acid, and biphenyl with lesser amounts of acetophenone, xanthone, and dibenzofuran. The optimal process conditions and potential reactions for the principal products will be discussed.

385742 Catalytic Hydrothermal Liquefaction of a Microalgae in a Two-Chamber Reactor

Abstract:

We carried out catalytic hydrothermal liquefaction of Nannochloropsis sp. in a novel two-chamber reactor, which physically separates algae and catalyst by a porous metal frit. The experiments were carried out at 350 °C for 1h. Algae was loaded in one chamber and catalyst was loaded in another chamber. Water was loaded in both chambers. We explored the effect of different headspace gases and pressure, catalyst loading, and reaction time on the reaction. GC-FID, FC-TCD, GC-MS, ^{13}C NMR, and elemental analysis were applied to analyze the products. We determined the biocrude yields and elemental composition, heavy vs. light biocrude fractions, and gas distribution.

The two-chamber reactor produced higher oil yields, a higher light biocrude fraction, a more hydrocarbon-rich gas and more light compounds than did a single-chamber reactor. The initial addition of catalyst promoted formation of hydrocarbon gases and light liquid-phase compounds compared to runs with no catalyst. The further increase of catalyst loading in the two-chamber reactor decreased the biocrude yield, which may be caused by cracking of hydrocarbons to gas products. More catalyst had the desirable effect of giving a higher light biocrude fraction at a low H_2 pressure. Using a longer reaction time of 4 hr in the two-chamber reactor decreased oil yield and light compounds but doubled the hydrocarbon percentage in the gas phase. After catalytic hydrothermal liquefaction for 1 h with 10 bar H_2, the carbon and hydrogen content in the light biocrude increased to 75.8 and 9.9%, respectively, compared to 52.6 and 7.4% in the dry algae. Longer reaction time and higher H_2 loading led to further increases in the carbon and hydrogen content.

382439 Advances in Fast Hydrothermal Liquefaction of Microalgae

Abstract:

A majority of the liquid petroleum consumed globally each year is directed to the transportation sector. The combustion of petroleum fuels releases carbon dioxide into the atmosphere, contributing to global climate change and other environmental problems. To reduce the negative impact of transportation on the environment, the development and implementation of liquid fuels with reduced life cycle carbon dioxide emissions is important. The production of biofuels from algae may yield liquid fuels with fewer carbon dioxide emissions than their petroleum alternatives, especially when processed such that energy-intensive drying steps are avoided. Fast hydrothermal liquefaction (FHTL) is a process that relies on the properties of liquid water at high temperature and pressure to convert biomass into energy-dense biocrude oil in 60 seconds. Since FHTL was recently discovered, a great deal about this biomass conversion process is not yet known.

This presentation describes two approaches for understanding FHTL. First, we used Fourier Transform Ion Cyclotron Resonance Mass Spectrometry (FT-ICR MS) to characterize the aqueous phase and biocrude products of FHTL and isothermal hydrothermal liquefaction (HTL), a better studied biomass conversion process. The ultra-high resolution of FT-ICR MS allowed for the identification of individual compounds by their molecular formulas. This detailed level of characterization provides a great deal of information about the FHTL and HTL processes. Second, we performed empirical kinetic modeling to develop a single model that describes both FHTL and HTL processes. Taken together, these approaches provide information of paramount importance to the scale-up and implementation of the FHTL biomass conversion process.

390510 Facilitation of High Value Products during Liquefaction of Wastewater Cultivated Algae

Abstract:

Hydrothermal liquefaction (HTL) of algae has shown to be a promising conversion technology to produce a renewable crude oil. Subcritical water chemistries allow geomimetic conditions to breakdown biomolecules within algae cells to produce a black tar similar to what we drill out of the ground. Key issues to commercializing HTL of algae include removing heteroatoms, particularly oxygen and nitrogen, increasing lower boiling point distillate fractions, and appropriating co-products into optimized value streams. We have shown that wastewater cultivation provides a sustainable strategy for biomass production while achieving these goals. This paper will describe the high value products we can exploit based off the growth conditions used for algae cultivation which provide the appropriate mechanisms to produce higher quality biocrude.

377321 Multiscale CFD-DEM Modeling of the Mixing of Viscous Liquids and Large Concentrations of Spherical Particles in Stirred Tanks

Abstract:

Despite the importance of laminar solid-liquid mixing operations in industry for the production of various goods such as pastes, greases, paints, pharmaceuticals and food products, a large proportion of the reported results in this field has been obtained in the turbulent regime. The intrinsic multiscale nature of solid-liquid flows in stirred tanks and the complexity arising from both the geometry of the impeller and the rheology of such suspensions, particularly in the case of high solids contents, create complex flow patterns within the tank, which may drastically affect both the yield and the quality of the desired product. Furthermore, the estimation of key mixing parameters such as the impeller torque, the just-suspension speed (N_{js}) and the cloud height remains unclear in all flow regimes. In particular, the Zwietering correlation has several limitations that may lead to poor predictions of N_{js}. Therefore, more modeling and experimental work is needed in order to shed light on these issues.

A variety of models have been developed to simulate solid-liquid flows. These include the classical Eulerian-Eulerian (or two-fluid) model and the combination of the Discrete Element Method (DEM) for the particles and CFD methods for the liquid (CFD-DEM). Despite their relative advantages and drawbacks, which will be briefly discussed during this talk, we believe that the CFD-DEM approach can be both efficient and accurate for the simulation of such flows. The objective of this work is then to introduce a multiscale CFD-DEM model, applicable to concentrated suspensions, wherein the particles are not resolved explicitly on the CFD grid but taken into account through the so-called volume-averaged Navier-Stokes equations. This entails projecting the particles onto the grid so that the solid-liquid momentum transfer term appearing in these equations, such as the drag force exerted by the particles on the liquid, can be locally evaluated. After describing the proposed model in detail, simulation results obtained with it for the laminar mixing of concentrated solid-liquid suspensions in stirred tanks will be presented.

383196 Applications of Mean Age Theory to Mixing in Wastewater Treatment Facilities

Abstract:

Mixing operation is essential to wastewater treatment facilities. In most such facilities efficient mixing is required to homogenize mixtures, maximize contacts between constituents, prevent solid settling, etc. Different types of mixing devices are used to achieve these objectives, such as agitated impellers, jet mixers, floating mixers, static mixers, and other custom designed mixing systems. The design and selection of a mixing system has been mainly based on past experience and a quantitative method has been lacking. More recently, CFD modeling techniques have been used to facilitate the design process but most of such applications have been limited to fluid flow patterns.

In this talk, a quantitative method using CFD tools to design a mixing system will be discussed. The method is based on the 3D solution of mean age distribution. This method can be used to quantitatively measure either a batch system or a continuous system. The spatial distribution of mean age reveals the non-uniform distribution of local blend time thus any bypassing or dead zones can be easily located and quantified. The statistical analysis of mean age distribution provides quantitative comparisons of mixing efficiency of different mixing designs and guides the selection of an optimal design. Several examples of applying mean age method to mixing problems in wastewater treatment will be discussed in the talk.

CFD Simulation of Mixing in Wastewater Treatment

Binxin Wu

Philadelphia Mixing Solutions Ltd., Palmyra, PA 17078

Abstract

Aerated lagoons and anaerobic digesters have been widely used to treat wastewater. In this study, a computational fluid dynamics technique is used to simulate mixing in aerated lagoons and anaerobic digesters, in which the rotation of a mechanical impeller is modeled using multiple reference frame method and wastewater is assumed to be a Newtonian or non-Newtonian fluid depending on total solids (TS) concentration. Numerical simulations of mixing in a small lagoon (173.74 m × 129.54 m × 3.66 m) with 4 aerators and in a large lagoon (471.77 m × 129.54 m × 3.66 m) with 12 aerators are conducted, and quantitative comparisons of the average velocity and biological oxygen demand versus placement of aerators in each lagoon are made, assuming wastewater to be water. In a digestion tank with a diameter of 20 m and a liquid height of 10 m, mixing by two submersible mixers is simulated, in which the substrate is assumed to be a pseudoplastic or Herschel-Bulkley non-Newtonian fluid with TS ranging from 2.5% to 12.1%. The results show that the mixing energy level increases nonlinearly with the increase of TS to achieve the same average velocity of the flow field.

Keywords: Bioreactors, Mixing, Computational fluid dynamics, non-Newtonian fluid, biological oxygen demand.

382148 CFD for the Study of the Multiphase Flow and Mass Transfer in a Stirred Tank Reactor for the Screening of Shaped Catalysts

Abstract:

Many reactions in the petrochemical, refining and biofuel industries occur in the presence of shaped solid catalysts, in the form of beads or extrudates. These particles have typical dimensions of the order of 1 to 3 mm, catalysing reactions occurring in liquids or gas-liquid mixtures. Before the selection of a catalyst for an industrial process, primary screening tests need to be performed previously in laboratory-scale reactors to assess the catalyst efficiency in its commercial shape, preserving the spatial distribution of active sites. These catalyst testing reactors need to fulfil some essential requirements to allow for a correct assessment of the kinetics of the reactions [1]: isothermal operation; appropriate flow distribution and mixing rates; and the absence of interphase mass transfer limitations. If these requirements are not met, the measured rate will not be equal to the reaction rate but instead to the rate of the limiting transport phenomenon, i.e., mixing or mass transfer. With the continuous development of new and more efficient catalysts, and the growing interest in the use of complex fluids like heavy crude feedstock or biomass-derived, catalysts testing and selection in lab-scale reactors become often hindered by interphase mass transport limitations or by hydrodynamics.

One of the existing types of reactors used for catalyst testing are stirred tank reactors with annular stationary catalytic baskets, also known as Robinson-Mahoney (RM) catalyst testing reactors [2]. In RM reactors, the shaped catalyst is introduced in an annular basket made from a metallic net or a perforated metal sheet, and immersed in the liquid. An axial flow impeller is then placed in the center of the annular basket to promote simultaneously the flow through the catalyst walls and particles, and the mixing of the bulk. Although RM reactors have been reported to offer less mass transfer limitations comparatively to laboratory fixed bed reactors [3], its way of operation is not yet completely understood and mass transfer limitations can still occur. Small changes in catalytic basket reactor operations (rotation speed, liquid viscosity, foam formation,...) may greatly reduce the reactor performance, and recommend measuring mass transfer coefficients for the reaction conditions before performing the kinetic measurements [3]. In addition, geometrical changes to the basket or impeller designs, or the selection of a different catalyst filling, of different shape or size, can also lead to different reactor performances, even when similar operating conditions are chosen. Measurements of mass transfer rates on RM reactors are not abundant in the literature, and are always specific for reactor geometry, working fluid and catalyst shape and size [4-6]. This happens because the extensive experimental study of RM reactors at the laboratory becomes neither practical nor economical when a large number of geometrical or operational parameters needs to be studied.

In this work, Computational Fluid Dynamics (CFD) is used to simulate the gas-liquid flow in a commercial lab-scale RM reactor. A multiphase 3D CFD model of gas-liquid flow in the reactor with the stationary catalytic basket has been developed with ANSYS Fluent. The studied reactor is a cylindrical, flat-bottomed tank with a diameter of 58 mm and a total volume of 300 cm^3, from *Top Industrie*, designed for the kinetic testing of shaped catalysts. Mixing is promoted by a stainless steel radial flow self-inducing impeller, which simultaneously promotes de induction and dispersion of the gas phase in the liquid. A static annular basket is placed inside the tank around the impeller to support the shaped catalyst particles. The basket walls are made from a perforated steel sheet with a thickness of 0.5 mm and perforations with a diameter of 0.8 mm, corresponding to an open area of 45%. The radial flow turbine promotes the flow across the basket annular section, around the catalyst particles, from the impeller site until the tank wall. Four baffles with a width of 10 mm and a thickness of 1 mm are placed above the basket. The Eulerian-Eulerian multiphase flow model was chosen for the simulation of the gas-liquid flow. Turbulence in the flow, exteriorly to the basket, was simulated with the Standard Multiphase k-εModel, and the Multiple Reference Frame (MRF) approach was chosen to deal the impeller motion. The flow in the basket filling was simulated as a homogeneous and unconsolidated porous medium with the Brinkman-Forchheimer equations, and the perforated basket walls were simulated as semi-permeable zero-thickness surfaces with a pressure drop. The numerical results were compared with experimental data from Particle Image Velocimetry, a bubble tracking imaging technique and gas absorption measurements.

The CFD simulations allow knowing the velocity field and gas volume fraction in each point of the domain. This information is used to estimate the mass transfer coefficients between the solid-liquid and gas-liquid phases, k_S and k_La, respectively, and compare it to the characteristic time of the catalysed reaction. Simulations for different operational conditions are used to identify the operational regimes and the range of conditions that allow a correct assessment of reaction rates with the existing setup. The developed CFD flow model will be used as well in the future as an optimisation tool for the design of enhanced reactor configurations.

References

[1] C. Perego, S. Peratello, Experimental methods in catalytic kinetics, Catalysis Today, 52 (1999) 133-145.

[2] J.A. Mahoney, K.K. Robinson, E.C. Myers, Catalyst evaluation with the gradientless reactor, Chemtech, 8 (N12) (1978) 758-763.

[3] I. Pitault, P. Fongarland, M. Mitrovic, D. Ronze, M. Forissier, Choice of laboratory scale reactors for HDT kinetic studies or catalyst tests, Catalysis Today, 98 (2004) 31-42.

[4] M. Mitrovic, I. Pitault, M. Forissier, S. Simoens, D. Ronze, Liquid-solid mass transfer in a three-phase stationary catalytic basket reactor, AIChE Journal, 51 (2005) 1747-1757.

[5] I. Pitault, P. Fongarland, D. Koepke, M. Mitrovic, D. Ronze, M. Forissier, Gas–liquid and liquid–solid mass transfers in two types of stationary catalytic basket laboratory reactor, Chemical Engineering Science, 60 (2005) 6240-6253.

[6] M. Braga, Étude des phénomènes de transfert et de l'hydrodynamique dans des réacteurs agités à panier catalytique, in: Ph.D. Dissertation, Claude Bernard University, Lyon, 2013.

386338 Steady State CFD for Determining Just Suspended Speed: Theory & Results

Abstract:

A new method for determining the just suspended speed (N_{jss}) in stirred tank reactors was previously developed using steady state CFD simulations. While preliminary results were somewhat empirical based, the method has been improved with more rigorous theoretical backing that is analogously aligned with mean age theory.

Scalar transport equations were solved to track the solid phase independently from the bulk fluid. In this way, N_{jss} was determined based on the mean age of particles in a settling zone along the bottom of the tank. Plotting average age for the solid phase in this region versus impeller speed reveals two distinct magnitudes, with the transition corresponding to N_{jss}. Above N_{jss}, the average age is low due to large convective velocities. Below N_{jss}, the convective scalar transport is negligible, so the average age is several orders of magnitude higher due to the very low rate of diffusive scalar transport.

Simulations have been performed for a range of solids concentrations and various tank geometries. Computational results compare very favorably with published experimental data. This method can be used for determining N_{jss} for systems in which it would be difficult with traditional experimental techniques. It also removes the subjectivity associated with traditional experimental visualization techniques. Most importantly, since the technique employs steady state simulations, it is advantageous over time-dependent simulations, which require much longer computational times due to the multiple phases and transient nature.

393903 Minimum Speed for Off-Bottom Solids Suspension: Zwietering or P/V

Abstract:

Experiments were made to study the effects of solids suspension parameters on correlations such as Zwietering and a constant P/mass approach. Tests were made in 3 geometrically similar tanks with tank diameters of 0.3m, 0.5m, and 1.0m with various axial flow impellers. CFDs of each condition were also made to determine what the conditions were near the locations of the last suspended particles.

There are many articles published about solids suspension in mixed tanks. Many are confusing, and many contradict each other. We decided to launch an experimental and computational study to better understand how disagreements between researched results can happen. The most commonly used correlation for scale-up is based on the work of Zwietering. But many people have criticized it. More recently, there have been claims that the scale-up should be based on constant P/V, and others that it should be a hybrid between the two models.

Our study looks at 12", 20" and 40" tanks and a variety of hydrofoil impeller designs and pitched bladed turbines of various impeller to tank diameter ratios and various off bottom clearances. We varied the solids concentration from 1 to 65%. We varied the particle sizes from 80 to 1000 microns. Every experiment was duplicated with CFD to better understand the hydro-dynamics along the tank base and the walls. Based on our results we plan on expanding our work to a 96" tank and investigate the effect of solids as small as 20 microns on solids suspension.

The impeller design does have an impact on solids suspension, but we also found that for geometrically similar setups, solid suspension was fundamentally different between the 12", 20" and 40" tanks. We also discovered a "sweet" spot for each impeller design, where the power of suspension was at a minimum for that tank geometry and slurry characteristics. An amazingly simple little "trick" will be demonstrated to improve solids suspension in many full-scale mining applications. Finally, we will also show how constant P/V and Zwietering can both be correct scale-up methods.

381029 Mixing Efficiency of Longitudinal Muscle and Segmentive Contractions in the Small Intestine

Abstract:

It is largely recognize in literature that muscular contractions of the small intestine improve nutrient absorption by facilitating the mixing of intestinal contents during digestion and the continuous renewal of nutrients at the intestinal wall.

Two specific motilities which have been observed in the rabbit during ex-vivo experiments are longitudinal motions and segmentive contractions. Longitudinal motions act to move the organ axially with no change in the diameter of the intestine. Segmentive contractions act locally to partially occlude the intestinal core.

The goal of this work was to numerically investigate how these two different types of contraction affect the dynamics and mixing of intestinal fluids of different viscosities. The rabbit intestine was modelled as an elastic solid (Young's modulus = 3000 Pa, Poisson's ratio = 0.35) and was simplified to be of cylindrical geometry. The motion of the intestinal wall was imposed based upon video data from ex-vivo observations of the rabbit ileum.

The governing system of mathematical equations were solved in the CFD software package Polyflow (ANSYS). The behaviour and stretching of two Newtonian fluids (10-3 Pa.s and 1 Pa.s) and one non-Newtonian fluid (of Bird-Carreau type with exponent 0.41, zero shear viscosity 0.65 Pa.s, and infinite shear viscosity 0.501 x10-8 Pa.s) were numerically analysed. The mixing efficiency was quantified using an in-built post-processor by measuring the stretching and folding of differential fluid elements.

It was found that the shear rates that develop as a result of these motions are too small to cause significant pseudoplastic behaviour of the fluid under study. Stretching rates were found to be dependent on both the properties of the fluid and the contraction type. High rates of stretching were observed along the entire wall during a longitudinal motion, but the dynamics suggest that flow is mostly axial with little mixing between the core and wall. The stretching of fluid elements was much more localised during a segmentive contraction, but more movement between the core and the wall was observed. Viscosity was shown to be an important factor with much higher stretching rates in the low viscous fluid. The non-Newtonian fluid behaved similarly to a viscous-Newtonian fluid, due to the low shear experienced.

Simulations also showed that blocking one end of the intestine increased the rate of element stretching during a segmentive contraction, with the rate of stretching increasing the closer to the blockage the contraction occurred.

Results from this study illustrated the effect of different intestinal motility patterns of the wall on the dynamics and mixing efficiency of intestinal contents during digestion

STRUCTURING Pd/ZnO CATALYST FOR METHANOL STEAM REFORMING

Methanol is often considered the best choice of a liquid fuel to generate hydrogen due to its high energy density, low cost and easy transportation. In addition, the reaction of steam reforming of methanol does not require breaking carbon-carbon bonds needing therefore moderate temperature. In recent years the Pd/ZnO catalyst has emerged as an interesting alternative to the classic Cu/Zn/Al_2O_3 catalyst for this reaction. Iwasa et al [1-3] were the first to discover the excellent catalytic activity and selectivity of Pd/ZnO in MSR and they observed that this catalyst exhibit similar reactivity and better stability than commercial Cu-ZnO catalysts [4].

The use of micro-reactors for hydrogen production gains an increasing importance from the point of view of creating a portable and low cost hydrogen source for numerous applications, in particular as feeding for fuel cell. Micro-reactors have the advantages of high surface-to-volume ratio, fast response time and ease integration, which are ideal for portable power systems. In addition, enhanced mass and heat transport properties are also recognized as advantages of micro-reactors.

The aim of this work is to develop a Pd/ZnO catalyst and coating with this catalyst monoliths and michochannel reactors for hydrogen production by methanol reforming.

In the preparation of the Pd/ZnO catalyst it is very important to use a ZnO support with high surface area to produce a high dispersion of the metal. ZnO was prepared using the method proposed by Cubeiro et al [5] based on the thermal decomposition of Zn oxalate. The ZnO synthesized with this procedure presented high surface area (S_{BET} 36 m^2/g).

The powder of Pd/ZnO was prepared by drying and calcination of the same slurry used for washcoating the structured systems. The slurry contained 22.5 wt. % of synthesized ZnO, 2.5 wt. % of colloidal ZnO dispersion (NYACOL® DP5370) and the required amount of $Pd(NO_3)_2$ solution (Johnson Matthey) to prepare 2.5 wt. % Pd/ZnO [5].

Catalytic tests were carried out at atmospheric pressure in a Computerized Microactivity Reference Catalytic Reactor from PID Eng&Tech. The catalyst was reduced at 500 °C and the reaction was carried out at 350 °C wiht steam-to-carbon molar ratio (S/C) of 1.5. The methanol reforming conversion of the powder Pd/ZnO, with a space velocity of 25 mmoles MeOH/min·g_{cat}, was 85 % and stable selectivity was obtained, producing a very low CO content (<1 %) in the reformate stream.

Once the catalyst was optimized and before deposited it on the micro-channel walls, the Pd/ZnO was deposited on corrugated monoliths (1330 cpsi), of the same alloy and channel size of the microchannel block. In this way, the coating procedure was optimized and the structured catalyst was tested. The coating was additive and lineal and the adhesion obtained was excellent (> 95%). The performance of the monoliths was good with a high methanol conversion, 85 % and a very low selectivity to CO (< 5 %).

A study of the diffusion (internal and external) on the monoliths was carried out. For the study of the internal diffusion, monoliths of different length (1, 2, 3 and 4 cm) were coated with the same catalyst loading (320 mg) and consequently different thickness, thereby maintaining the space velocity of 7.5 mmoles MeOH/min·g_{cat}. This study showed the same conversion indicating the absence of internal diffusional limitations. To test the external diffusion, monoliths with the same dimensions (L= 3 cm) were coated with different loadings of catalyst (220-500 mg) and tested at the same space velocity per gram of catalyst (7.5 mmoles MeOH/min·g_{cat}) and consequently at different linear velocity. The constant conversion obtained evidenced the absence of external difusional limitations.

Finally, Pd/ZnO catalyst was deposited on the microreactor walls using the coating procedure developed for micro-monoliths. The coating was additive and lineal and the adhesion obtained was excellent (> 90%). The methanol reforming conversion of the microreactor, was high (> 80 %) and stable selectivity was obtained, producing a very low CO content (<1 %) in the reformate stream with different types of space velocity (15-45 mmoles MeOH/min·g_{cat}). Moreover, the microreactor achieved excellent isothermicity.

[1] N. Iwasa et al, Applied Catalysis A: General 125, 145 (1995).

[2] N. Iwasa et al, Bulletin of the Chemical Society of Japan 71 ,1451 (1998).

[3] N. Takezawa, N. Lwasa, Catalysis Today 36, 45-56 (1997).

[4] T. Conant et al, Journal of Catalysis 257, 64 (2008).

[5] F. J. Echave, O. Sanz, M. Montes, Applied Catalysis A: General 474 (2014) 159.

Nowadays the scientific interest in fuel cell (FC) technology is concentrated on the use of portable electronic devices to replace or supplement batteries for portable applications. FCs provides instant, environmentally clean and safe way to produce power with high energy densities (four to six times larger) as compared to Li-ion batteries. However, they have limited power capacity and cannot respond to the sudden changes in the load that may occur in some applications. These problems could be overwhelmed by the use of polymer electrolyte membrane fuel cells (PEMFCs) considered to enhance the energy density of the batteries, to possess a rapid start up time and low temperature operation regime. The PEMFCs are generally fueled by H_2 produced from hydrocarbons, ammonia or chemical hydrides, depending on the power requirements of the FC and between them the hydrocarbon reforming is the most viable H_2 production process. Usually, the steam reforming is preferred since it provides the highest hydrogen carbon monoxide (H_2/CO) molar ratio of 3. Although the fuel processing reactors are designed to maximize H_2 production through steam reforming, partial oxidation or autothermal reforming reactions also occur giving rise to other compounds such as H_2O, CO, CO_2 and some other molecules coming from hydrocarbons incomplete transformation [1].

The use of the FC for portable applications requires at first place diminution of the volume and weight of the reactors and when the PEMFCs are selected an additional and very important requisite takes the advantage, the lowering of the CO concentration to trace levels. Although, the hydrogen production is well known process and universally used in fixed bed catalytic reactors, the process intensification and the resolution of the volume and weight reactors limitation for a on board H_2 production is possible by applying the microreactor technology. The microreactors have the advantage of fast response time, easy integration, and small footprint, all of them ideal for portable power systems. The second and the most important problem of CO concentration could be solved by the application of the H_2 purification reaction, as water gas shift (WGS) followed by the preferential CO oxidation (CO-PROX) and methanation reaction if needed.

Nevertheless, when speaking about portable devices, generally the methanol is selected as a fuel for on-board reforming/purification H_2 production, due to the possibility for low CO production and the low temperature operating range of the reforming reaction. The use of methanol includes also and additional value the possibility to omit the WGS CO purification unit, known as the highest volume requiring unit. Then the CO-PROX reaction unit becomes the principal purification unit.

In this context, this work presents the description of the microreactor technology application to the CO-PROX reaction. The reactor using conventional $CuOx/CeO_2$ is applied and the results are discussed in comparison to the use of fixed bed reactor and as a function of the methanol concentration in the reaction feed.

The $CuOx/CeO_2$ coated microreactor was successfully tested on the CO-PROX reaction under a simulated reformate feed. The maximum CO conversion was obtained at around 140°C for both fixed and microreactor, but the oxygen selectivity to CO_2 was higher when microreactor is used (Fig 1).

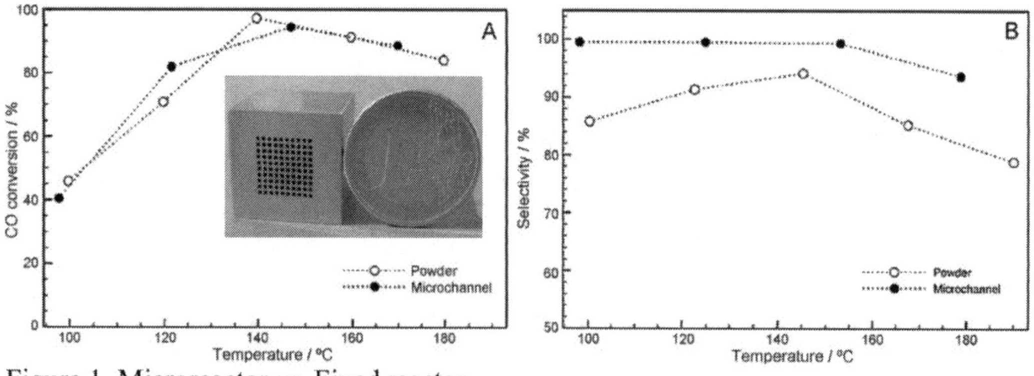

Figure 1. Microreactor *vs.* Fixed reactor

The modeling and the simulation of the microreactor using a computational fluid dynamics (CFD) was successfully validated for the first time with the experimental catalytic results and the influence of the catalysts loading was also considered. The kinetic parameters were included in the model and corroborated with the experimental results.

The integration of methanol reforming and PROX reaction involves the presence of non-converted methanol and water in the inlet stream of the PROX feed. Therefore, the effect of these products on the performance of the catalyst is essential for the integration of both processes. The influence of the methanol (MeOH) to the CO-PROX activity of the microreactor is presented in Figure 2. A negative effect on the CO-PROX activity was observed with the increase of the methanol concentration. A model CuOx-CeO_2 catalysts was also studied by in situ DRIFTS in order to explain its catalytic behavior in the presence of methanol and water. This correlation helped the identification of the reaction intermediates and allows the modification of the catalyst formulation, in order to achieve more robust catalysts in this process.

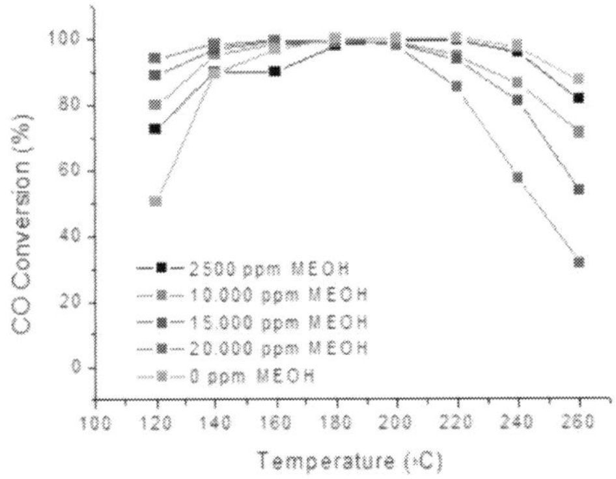

Figure 2. CO-PROX activity vs. methanol concentration

In summary this study evolves all the steps of the microreactor technology: from the microreactor arming through the catalyst washcoating to the evaluation of its activity in the harsh real conditions of potential on board application.

REFERENCES

[1] S. Ivanova, O.H. Laguna, M.A. Centeno, A. Eleta, M. Montes, J.A. Odriozola, in RENEWABLE HYDROGEN TECHNOLOGIES, (Eds. L. Gandia, G. Arzamendi, P. Dieguez), Elsevier 2013.

Evaluation of iron oxide supported reactive materials for thermochemical water splitting

Samantha Roberts, Justin Dodson, and Helena E. Hagelin-Weaver, Chemical Engineering, University of Florida, Gainesville, FL 32611

Two-step solar thermochemical water splitting offers a viable solution for employing renewable energy pathways for production of high-purity hydrogen, which can be used in fuel cells or for the production of liquid fuels. One promising thermochemical cycle uses iron oxide as the reactive material and splits water in two relatively simple steps. In the first step, an oxygen-deficient iron oxide is generated by heating materials to temperatures on the order of 1,500 °C until oxygen desorbs. Hydrogen is produced in the second step when steam is passed over the oxygen-deficient iron oxide. The oxygen-deficient iron oxide is then regenerated by heating to 1,500 °C again. Due to the high operating temperatures of this thermochemical cycle, stable materials which do not undergo severe detrimental sintering are needed.

Our reactive materials development focuses on employing high temperature ceramic oxides, such as zirconia (ZrO_2), as supports to increase stability and thermal shock resistance. The reactive materials were prepared by depositing iron oxide onto nanoparticle zirconia (n-ZrO_2) or yttria-stabilized zirconia (YSZ). After heat treatment, these FeO_x/ZrO_2 materials are surprisingly stable, and yield stable hydrogen production over several cycles. In this study, we have investigated in detail the differences in materials properties and hydrogen production between reactive materials supported on monoclinic ZrO_2 versus YSZ, and the effect of iron loading (between 5 and 20 weight %) has been examined.

The hydrogen yield per cycle for the FeO_x/YSZ material is consistently higher for all iron loadings than for FeO_x/ZrO_2 materials. In addition, the YSZ supported materials exhibit a smaller deviation between cycles, indicating that the YSZ support has a higher stability than n-ZrO_2. Generally as iron loadings increase, the hydrogen production in terms of per gram of Fe in the material decreases. The results indicate that there is an optimal iron loading around 10 wt%, as higher iron loadings lead to less stable materials and less effective iron utilization. It also appears that the active iron utilization for FeO_x/YSZ materials is higher and more stable than the FeO_x/ZrO_2 materials. This may be due to iron incorporation into the YSZ lattice, as indicated by X-ray diffraction measurements, allowing for the iron to be more available than bulk iron oxide.

381700 Graphene Oxide Quantum Dots As Media for Hydrogen Generation from Photocatalytic Water Splitting

Abstract:

Graphene oxide (GO) derived from graphite oxidation is a semiconductor that can absorb light to generate electron-hole pairs for photocatalytic reactions. GO is suitable media for photocatalytic water splitting because it can be well dispersed in water. The electronic properties of GO can be tuned by varying its size and the embedded functionalities. As-received GO is a p-type semiconductor. Doping nitrogen on GO by ammonia treatment can convert GO to an n-type semiconductor. The size effect resulting from quantum confinement also influences the electronic structures of GO. Size modulation along with chemical modification represent a means to tune the electronic properties of GO for effective photocatalytic water splitting. We synthesized nitrogen-doped graphene oxide quantum dots (NGO-QDs) that exhibited promising water-reducing activities. Simultaneous presence of oxygen and nitrogen functionalities in single QDs might have eliminated defect states of the QDs to suppress charge recombination. The NGO-QDs had a band gap of approximately 2.3 eV. The Pt-deposited NGO-QDs effectively catalyzed H_2 evolution from an aqueous solution containing triethanolamine as a sacrificial electron donor. The quantum yield for H_2 evolution reached a value as high as 12 % under irradiation at 420 nm.

Investigation of NiO based Oxygen Transfer Materials: A promising material for Chemical Looping Steam Methane Reforming

Ni-based Oxygen Transfer Materials (OTMs) have been promising catalysts for methane reforming reactions due to their superior reactivity, stability and negligible volatility. This study focuses on investigation of redox activity and stability of NiO based OTMs in a thermal gravimetric analyzer (TGA) instrument.

40%NiO-OTMs were synthesized by wet impregnation of four different supports: alumina, zirconia, titania and silica. Physicochemical characterizations like XRD, BET and isothermal H_2-reduction in TGA were performed to study structural, textural and reduction properties of the OTMs. The redox activity and stability of NiO-OTMs was studied during 20 successive CH_4 reduction and air oxidation cycles in the TGA connected to a mass spectrometer (MS).

BET surface area of the 40%NiO-OTMs after NiO loading decreased compared to the pure supports. XRD confirmed the presence of NiO phase in all OTMs, and metal-support compounds were identified in Al_2O_3- and TiO_2- supported OTMs. Total degree of reduction (RD) of 40%NiO-OTMs obtained from isothermal H_2-reduction at 650 °C in TGA is 100% for all OTMs but Al_2O_3-OTM. Lower RD of Al_2O_3 supported OTM is due to presence of metal-support compounds ($NiAl_2O_4$) as observed by XRD.

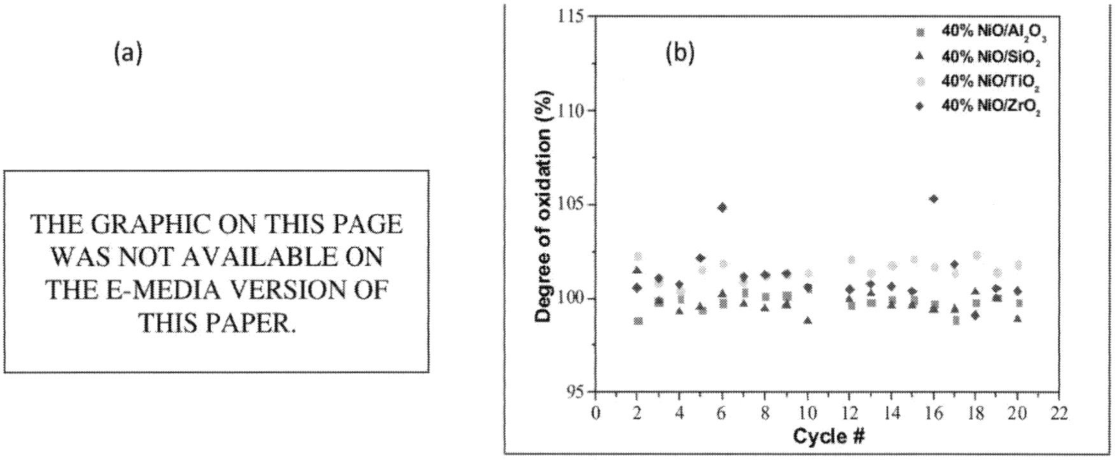

Figure 1 (a) Degree of reduction and (b) degree of oxidation of 40%NiO-OTMs during 20 redox cycles in TGA

Fig. 1a shows the reduction degree during 20-cycle experiments with all four OTMs. Reduction degree (RD) of the alumina-based OTM differs from others in terms of its profile change to reach the steady state. RD of 40%NiO/Al_2O_3 slowly increased until cycle 11 reaching a maximum steady state value of 86%, which is lower than that of the other OTMs. RD profile in the case of 40%NiO/TiO_2 increased slightly from cycle 1 (~85%) until cycle 7 (~100%) and remained rather steady for further cycles which is qualitatively similar to that of Al_2O_3-based OTMs. All four OTMs underwent complete re-oxidation in oxidation steps of cycling experiments, reaching the same oxidation degree level of 100% in all cycles (Fig. 1b). This also means complete removal of deposited carbon resulting in clean catalyst surface for the next reduction cycle. Fig. 1 shows that all 40%NiO-OTMs are active and stable towards redox behaviour during entire 20 redox cycles. TGA data along with mass signals of H_2, CO, CO_2 and H_2O during 20 redox cycles confirm that 40%NiO-OTMs are promising CH_4 reforming catalysts in terms of redox activity and regenerability.

Carbon amounts deposited during reduction steps of 20 cycles, derived from the observed weight gain, are shown in Fig. 2a. 40% NiO/Al_2O_3 OTM exhibited a volcano curve which is caused by changes in catalyst structure with time. For all OTMs trends in hydrogen formation followed the same trend as C-deposition (Fig. 2b) indicating that there is a relationship between deposited carbon and H_2 formation. The amount of H_2 released in all cases is significantly higher than theoretically possible (0.107 mmol H_2 per 10 mg of OTM containing 40 wt% of NiO) from NiO reduction to Ni. This indicates that H_2 is not only produced as a result of the reduction of NiO by methane but also due to decomposition of methane on metallic nickel leading to hydrogen and carbon (coke) formation.

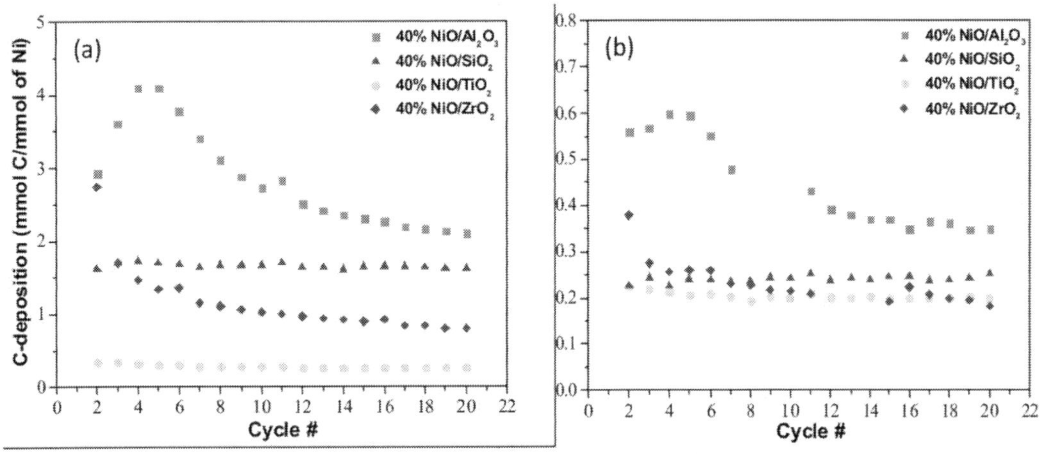

Figure 2 (a) C-deposited and (b) H_2 formed during reduction of 40%NiO-OTMs as a function of number of cycles

An attempt was made to correlate hydrogen production and carbon deposition for all OTMs as shown in Fig. 3. The result confirmed previous hypothesis of prevalent hydrogen production via CH_4 split, whereas the intercept indicates the amount of hydrogen produced independently of the deposited carbon, i.e. through NiO reduction. Numerical value of the intercept (0.14) is reasonably close to the theoretical value of 0.107 mmol, whereas the observed slope (1.77) is somewhat smaller than the corresponding theoretical value (slope equal to 2) based on stoichiometry of methane splitting reaction. This deficit of evolved hydrogen relative to the deposited carbon could be due to errors in MS quantification of H_2 and possible existence of side reactions (e.g. Boudourd and reverse water-gas-shift).

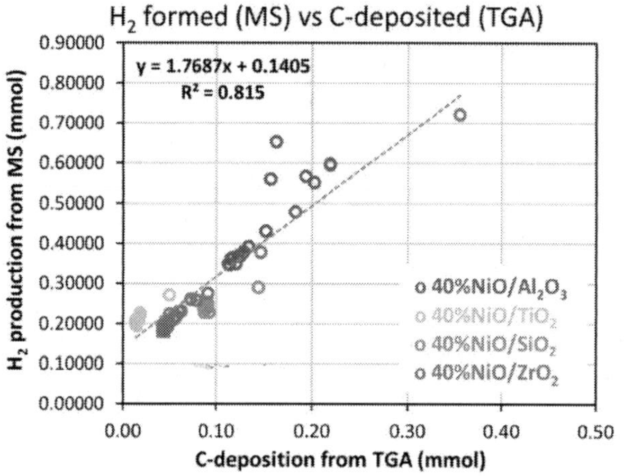

Figure 3 Correlation between H_2 and C formed during reduction of 40%NiO-OTMs

383626 Transition Metal Carbide-Graphene Nanocomposites for Hydrogen Evolution Reaction

Abstract:

Transition metal carbide-graphene nanocomposites for hydrogen evolution reaction

Seok Ki Kim, Yang Qiu, Yin-Jia Zhang, Ronald Michalsky, Robert H. Hurt, Andrew A. Peterson

Hydrogen has received considerable attention as a renewable energy carrier replacing fuels derived from petroleum resources. The direct electrochemical water splitting using a wind or solar energy is one of the most promising technologies for the hydrogen production due to its straightforward and sustainable characteristics. Up to the present, platinum has been used as the most efficient electrocatalyst. But its high cost and rarity have retarded a large-scale application, thus the development of efficient electrocatalyst with earth-abundant material is the key technological challenge. Among various candidate materials suggested for replacing platinum, transition metal carbides(TMC) were recognized as a potential alternatives because of their similar electronic structures to those of Pt-group metals, resulting in Pt-like behaviors in hydrogen-involved catalytic reactions. The focus of this talk will be on the development of TMC/graphene nanocomposites for the hydrogen evolution reaction in acidic electrolyte. We selected four TMC materials including Mo_2C, TiC, Fe_3C, and WC, based on density functional theory calculations and synthesized graphene-supported TMC nanoparticles. Supercritical-solvothermal method allowed the uniform and incorporative deposition of Mo_2C, TiC, Fe_3C and WC nanoparticles on the graphene surface. High electrochemically active surface area, low overpotential and enhanced catalytic stability were achieved by synergetic effect between TMC nanoparticles and graphene.

A Hybrid Solar-Redox Process for Hydrogen and Liquid Fuel Co-Production – Redox Material Development and Process Analyses

Feng He and Fanxing Li*
Department of Chemical and Bimolecular Engineering
North Carolina State University
Raleigh, NC 27695
* Email: fli5@ncsu.edu

Efficient and cost-effective production of clean energy carriers such as hydrogen is of critical importance for sustained growth of the modern economy. Among the various approaches for sustainable hydrogen generation, solar-thermal water-splitting represents a potentially attractive and environmentally friendly option. Typical solar-thermal water-splitting schemes involve cyclic redox reactions of transition metal oxides to indirectly convert solar energy and water into separate streams of hydrogen and oxygen. In its simplest configuration, a solar-thermal water-splitting cycle involves two redox steps. In the first step, solar energy is used to decompose a metal oxide at high temperature. In the subsequent step, the decomposed metal/metal oxide is reoxidized with water, producing hydrogen. The key for the aforementioned process is the metal oxide decomposition step. To date, the endothermic decomposition reaction for most, if not all, solar-thermal water-splitting processes are conducted at temperatures above 1300 °C. Therefore, novel solar-thermal schemes that can effectively promote metal oxide reduction at lower temperatures are highly desired in order to achieve improved efficiency and process economics. The current study investigates a cyclic redox process for cogeneration of liquid fuels and hydrogen using methane and solar energy. Mixed-conductor enhanced Fe_3O_4/Fe redox couple is used for methane oxidation and water-splitting. In this process, syngas from methane oxidation is used for liquid fuel synthesis whereas H_2 is co-produced from water-splitting. Nearly 100% conversion of methane is achieved in the reduction step at 900 °C. Steam to hydrogen conversion in excess of 60% is also observed. Results from experimental studies are then used to evaluate the energy conversion efficiency of the proposed scheme.

390058 The Conversion of Various Phenolic Compounds over Ru/TiO2: Active Site Requirements

Abstract:

One of the greatest challenges associated with the conversion of lignocellulosic biomass to fuels and chemical lies in the stable and selective conversion of lignin derived phenolics. Metals supported on reducible oxides such as Ru/TiO_2 have demonstrated the potential to preserve the carbon contained in methoxy groups and avoiding excessive hydrogen consumption of the ring upon deoxygenation of phenolic OH groups. Ru supported on TiO_2 also exhibits activity and stability enhancements for the conversion of phenolics when compared with Ru supported on other non-reducible supports. In this contribution, we have systematically varied the Ru/TiO_2 interfacial area, as well as the TiO_2 phase, and measured the initial rates of conversion of several model compounds containing hydroxyl and methoxy groups including, guaiacol, anisole, catechol and cresol. The implications of these results will be discussed in the context of kinetic models to describe the conversion phenolics such as guaiacol, where one must take into account several factors. These include the rate of conversion of each produced product (for example catechol and cresol) over the reduced sites, the interface, and the various metal sites. In-situ titration is also used to discriminate deoxygenation reactions that occur through the continuous production and healing of defects on the support vs. those facilitated by active sites located at the Ru/TiO_2 interface.

383593 First-Principles Investigation of the Mechanism and the Active Site for Hydrodeoxygenation over Ru/TiO_2

Abstract:

Using periodic density functional theory (DFT) we have explored the hydrodeoxygenation (HDO) pathways of acetaldehyde and phenol, two surrogate molecules for some of the over 400 different oxygenates from biomass fast pyrolysis [1], over supported Ru/TiO_2 catalysts. The main functional groups represented by acetaldehyde and phenol are aldehydes, ketones, and aromatic alcohols. These groups are abundant output products of fast pyrolysis, and catalytic conversion of these products to oxygen free compounds is of significant importance for bio-oil upgrade. There are three major HDO pathways: (1) hydrogenation (HYD), (2) direct deoxygenation (DDO), and (3) decarbonylation (DCN). Among these, DDO is the most desired pathway because it requires less hydrogen than HYD, and in contrast to DCN, the number of carbon atoms during HDO remains constant. In a systematic study of HDO pathways for acetaldehyde and phenol we have calculated the thermodynamic and kinetic parameters for a large number of possible elementary surface reactions with specific focus on product selectivity, i.e. the preference for C-H, C-O, and C-C bond scission reactions at each reaction intermediate.

In experimental studies Ru/TiO_2 has shown promising results in terms of selectivity and catalytic activity for HDO of different molecules (e.g. furfuran, acetone, phenol) [2-4]. To represent the supported Ru/TiO_2 catalysts we tested various active site models including metallic Ru(0001), pure rutile metal-oxide surfaces [RuO_2(110), TiO_2(110)], modified metal-oxide catalysts [1.0 ML RuO_2/TiO_2(110), Ru_1/TiO_2(110)] and a 10-atom cluster of Ru supported on TiO_2(110) [Ru_{10}/TiO_2(110)]. Our results show that the metallic Ru(0001) surface is not suitable for DDO. It preferentially catalyzes HYD for phenol, resulting in the fully hydrogenated products cyclohexane or cyclohexanone, or DCN for acetaldehyde. The dominant DCN pathway for acetaldehyde can be described as: adsorption

(CH_3CHO) → αC-H bond breaking (CH_3CO*) → ßC-H bond breaking (CH_2CO*) → C-C bond breaking (CH_2*). At high hydrogen pressure CH_2* may leave the catalyst surface as CH_4, while at low hydrogen pressure stepwise dehydrogenation leads to CH* and eventually surface carbon (C*), causing catalyst deactivation due to coke formation. HYD and DDO may occur on Ru(0001) and the formation of ethane or ethylene is feasible, but the removal of surface oxygen requires significant energy barriers ($E_a > 1.7$ eV) and limits the overall reaction. In turn, the accumulation of surface oxygen can lead to high oxygen coverages, partial oxidation or even the formation of a surface oxide.

Oxygen vacancy sites are expected to be important for HDO catalysis on metal-oxide surfaces. Hence, we initially derived a constrained thermodynamic phase diagram for the rutile TiO_2(110), Ru_1TiO_2(110), 1.0 ML RuO_2/TiO_2(110), and RuO_2(110) surfaces. Under typical HDO conditions (T > 500 K, P(H_2) > 200 bar) these surfaces are partially reduced and exhibit bridging hydroxyl groups (HO^{br}). Using this reduced state as reference surface we investigated vacancy formation pathways on each metal-oxide. Upon further hydrogen adsorption the surface hydroxyl groups can be removed as water along three different reaction pathways: (1) the common disproportionation pathway $HO^{br} + HO^{br} \rightarrow H_2O^{br} + O^{br}$, and two pathways involving molecular hydrogen adsorption at the coordinatively unsaturated Ti site (Ti^{cus}) (2) $Ti^{cus}\text{-}H_2 + O^{br} \rightarrow Ti^{cus}\text{-}H + HO^{br} \rightarrow Ti^{cus} + H_2O^{br}$, and (3) $Ti^{cus}\text{-}H_2 + HO^{br} \rightarrow Ti^{cus}\text{-}H + H_2O^{br}$. Among these pathways, reaction (3) occurs with the lowest energy barrier on Ru_1TiO_2, 1.0 ML RuO_2/TiO_2, and RuO_2. Due to weak interactions between molecular hydrogen and the Ti^{cus} site of TiO_2, reactions (2) and (3) are not availableand reaction (1) is favored instead [5]. In all cases, the desorption of the resulting H_2O^{br} forms the surface oxygen vacancy. This active vacancy site provides the preferred adsorption site for acetaldehyde and phenol, and plays an important role in the preferential activation of C-O scission over C-C scission reactions. Hence, the DDO pathway occurs most likely on the considered metal-oxides surfaces and follows the sequence of elementary steps given by: adsorption on surface vacancy (CH_3CHO*-s) → isomerization (CH_3CHO*-d) → αC-O bond breaking (CH_3CH*) → ßC-H bond breaking (CH_2CH*) → αC-H bond formation ($CH_2CH_{2(g)}$), leading to the desired product ethylene. In addition to the DDO pathway, metal-oxide surfaces can also catalyze the HYD pathway resulting in the intermediate formation of ethanol, which can subsequently be dehydrated to ethane.

The highest selectivity for ethylene is predicted on the TiO_2(110) surface; however, the dissociation of hydrogen molecules and the formation of HO^{br} groups, the precursor to the formation of a required oxygen vacancy, are kinetically limited. The presence of Ru clusters on TiO_2 may facilitate hydrogen delivery from the gas-phase to TiO_2(110) through hydrogen spillover, and we investigated this possibility using a Ru_{10} cluster model supported on TiO_2(110). DFT results suggest that spillover of H atoms originating from the Ru cluster is kinetically hindered ($E_a \approx 1$ eV), but the Ru/TiO_2 interface is very active for hydrogen dissociation ($E_a \approx 0.1$ eV) and enables the formation of oxygen vacancy sites as required for C-O bond breaking reactions during HDO on TiO_2. The interface also provides sites with unique bifunctional properties: for example, the C-O bond in phenol adsorbed in a vacancy site at the interface can be selectively broken in a concerted step where hydrogen is provided from the Ru_{10} cluster and the eliminated OH group heals the bridging oxygen vacancy site on TiO_2(110): $C_6H_5O^{br}H + H\text{-}Ru_{10} \rightarrow C_6H_{6(g)} + HO^{br} + Ru_{10}$.

In conclusion, detailed DFT calculations suggest that DCN is preferred over DDO and HYD on the metallic Ru(0001) surface, leading to surface carbon or methane, while on the investigated metal-oxide surfaces DDO and HYD are more likely than DCN. Oxygen vacancy formation on TiO_2 is limited by hydrogen delivery but can be facilitated at the Ru/TiO_2 interface. These vacancy sites selectively catalyze the DDO pathway of acetaldehyde and phenol, leading to the desired products, ethylene and benzene, respectively.

[1] Huber, G. W.; Iborra, S.; Corma, A. Synthesis of transportation fuels from biomass: chemistry, catalysts, and engineering. *Chem. Rev.* **2006**, *106*, 4044–98.

[2] Omotoso, T.; Boonyasuwat, S.; Crossley, S. P. Understanding the role of TiO_2 crystal structure on the enhanced activity and stability of Ru/TiO_2 catalysts for the conversion of lignin-derived oxygenates. *Green Chem.* **2014**, *16*, 645.
[3] Pham, T. N.; Shi, D.; Sooknoi, T.; Resasco, D. E. Aqueous-phase ketonization of acetic acid over Ru/TiO_2/carbon catalysts. *J. Catal.* **2012**, *295*, 169–178.
[4] Newman, C.; Zhou, X.; Goundie, B.; Ghampson, I. T.; Pollock, R. A.; Ross, Z.; Wheeler, M. C.; Meulenberg, R. W.; Austin, R. N.; Frederick, B. G. Effects of support identity and metal dispersion in supported ruthenium hydrodeoxygenation catalysts. *Appl. Catal. A Gen.* **2014**, *477*, 64–74.
[5] Rekoske, J.; Barteau, M. Isothermal reduction kinetics of titanium dioxide-based materials. *J. Phys. Chem. B* **1997**, *101*, 1113–1124.

389047 Acid/Metal Balance in Bifunctional Conversion of Cresol to Hydrodeoxygenated Compounds Via Alkylation, Dehydration, and Hydrodeoxygenation

Abstract:

The hydrodeoxygenation of phenolics has been a matter of great interest in recent studies as part of the biomass conversion efforts. We have recently proposed a mechanism that includes a fast enol-keto tautomerization step that precedes the ring hydrogenation. An unstable ketone intermediate forms over the catalyst surface and it can be hydrogenated via two possible paths. a) when the CO of this intermediate gets hydrogenated instead of the ring, a very reactive unsaturated alcohol (3-methyl-3,5-cyclohexadienol) is formed and it can be readily dehydrated to toluene, b) when the C=C bonds in the ring are hydrogenated, methylcyclohexanone is formed. The latter can be further hydrogenated to methylcyclohexanol. Over acidic supports, this alcohol dehydrates to methylcyclohexene; sub-sequent dehydrogenation produces toluene. A detailed kinetics study of the conversion of each of the possible intermediates yields strong support to the proposed mechanism.

Interestingly, when the reaction is conducted in the liquid phase on zeolite-supported metals (Pt or Pd), alkylation of the aromatic ring by the saturated alcohol becomes an important reaction pathway that produces C11 hydrocarbons of potential value as fuel components. In this contribution, we will discuss the importance of maintaining a suitable acid/metal balance and a favorable zeolite porous structure to maximize the yield of alkylated compounds.

384294 Catalytic Hydrodeoxygenation on Metal Carbides

Abstract:

Stoichiometric removal of oxygen is required for the chemical transformation of renewable biomass feedstock to energy carriers and other substitutes for petroleum derived products. Oxygen removal from biomass can be performed under catalytic hydrotreatment conditions (25-70 bar H_2, 300 – 400°C) using noble metal catalysts, but oxygen removal is often concurrent with excessive H_2 usage through hydrogenolysis and/or alkene/aromatic hydrogenation, producing low-value short-chain alkanes as opposed to light alkenes and aromatics.

We report vapor phase hydrodeoxygenation (HDO) of acetone, anisole, guaiacol, and m-cresol, representative of cellulose and lignin pyrolysis products over Mo_2C-based non-precious metal catalyst formulations that enable selective HDO at ambient pressure and low temperatures (98 – 200°C). The dominant HDO reaction products were unsaturated, deoxygenated hydrocarbons including propylene, benzene, phenol, and toluene with carbon selectivity ranging from 60-90%, and <1% selectivity to C-C bond scission products.

The HDO kinetics of these probe molecules were investigated as functions of contact time, reactant partial pressures, and temperature. This work proposes HDO mechanistic models to fit experimentally measured rates, and these kinetic models were further probed using transient isotopic experiments and the kinetic isotope effect. The HDO site requirements were probed via *in situ* titration experiments using pyridine, CO and CO_2 to assess the number and identity of active sites. The active catalytic phase for HDO was investigated as a function of catalyst oxygen content via *in situ* O_2 dosing experiments and systematic oxygen inclusion during catalyst synthesis to probe the effects of near-surface and bulk oxidation on catalytic HDO activity. The kinetics, mechanism, and site requirements of low pressure HDO on Mo_2C formulations will be presented.

388957 Surface Modification and Catalytic Activity of MoO_3 during Hydrodeoxygenation (HDO) of Lignin-Derived Model Compounds

Abstract:

Hydrodeoxygenation (HDO) is a major route of upgrading bio-oil. MoO_3 has been shown to be active for HDO of bio-oil model compounds. Specifically, the catalyst is selective towards cleavage of C-O bonds at temperatures of 320-350°C and H_2 pressures <= 1 bar. Importantly, this catalyst selectively cleaves stronger phenolic Ph-OMe over weaker PhO-Me bonds. Detailed reactivity studies were performed to gain insight into the nature of the active sites. An initial induction period was observed for the catalyst to achieve peak performance,which suggested a surface modification was required to render the catalyst active. The catalyst surface undergoes partial carburization as evidenced by the presence of oxycarbide- and oxycarbohydride-containing phases (i.e., $MoO_xC_yH_z$) at lower temperatures. Mo^{5+} species are hypothesized to be responsible for the HDO chemistry on the catalyst. The transformation of bulk phases and the surface modification of MoO_3 by carbon/H_2 are also investigated to understand the role of surface carbon in the stabilization and enhanced activity of the partially reduced MoO_3 surface. The partial surface carburization during reaction was found to play a role in stabilizing Mo^{5+} state on the catalyst surface and slowing down over-reduction of MoO_3 into inactive MoO_2 phase.

387495 Online Monitoring of Lignin Liquid Phase Conversion Processes By in-Situ ATR-IR Spectroscopy

Abstract:

Lignocellulosic biomass is becoming an increasingly important alternative resource for biofuels production. Lignocellulosic biorefinery operations must produce chemicals as well as energy vectors in order to be competitive, making use of *all*the major components of lignocellulose including lignin. Indeed, this aromatic polymer holds great potential for the sustainable production of renewable aromatics and phenolics. [1] Typical lignin valorization efforts entail the study of depolymerization and defunctionalization steps and almost inevitably involve complex reaction mixtures. In order to develop more efficient catalytic lignin valorization processes, more insight is clearly needed in the chemistry of lignin linkage cleavage and defunctionalization.

Our group has, for instance, reported on the solubilization and catalytic conversion of a wide variety of lignins in aqueous or ethanolic solvents. [2] A two-step route involving lignin depolymerization by liquid phase reforming step followed by hydrodeoxygenation of the resulting lignin-oil to yield, amongst others, lignin-derived BTX. [3] Improvements in the efficiency of this two-step conversion process requires further insight into the changes in molecular weight upon dissolution and depolymerization as well as insight in the changes in the chemical structure of lignin upon conversion. The application of in situ spectroscopic techniques for online monitoring of the liquid phase lignin conversion processes can prove very useful in this respect. Here, we report on our efforts to further develop the use of in situ spectroscopic techniques such as ATR-IR for on line monitoring of catalytic lignin conversion processes. Liquid phase reforming reactions are run in the presence of a heterogeneous Pt catalyst in ethanol/water at 225°C and 58 bar in a semi-batch autoclave equipped with an ATR-IR crystal. Changes in the fingerprint region provided insight into the rate, nature and extent of lignin depolymerization. The approach taken is expected to be of interest to the various liquid phase lignin depolymerization strategies that are currently being developed.

[1] Zakzeski, J., Jongerius, A. L., Bruijnincx, P. C. A. and Weckhuysen, B. M., Chem. Rev. 110 (2010) 3552.

[2] Zakzeski, J., Jongerius, A. L., Bruijnincx, P. C. A. and Weckhuysen, B. M., ChemSusChem, 5 (2012) 1602.

[3] Jongerius, A. L., Bruijnincx, P. C. A. and Weckhuysen, B. M., Green Chem., 15 (2013) 3049.

389285 Liquid Phase Hydrodeoxygenation of Phenolic Compounds on Platinum and Palladium Catalysts

Abstract:

The hydrodeoxygenation of phenolic compounds, which constitute a substantial portion of bio-oils, was investigated. Noble metals have been reported to be suitable catalysts for vapor phase or aqueous phase processing of phenolic compounds, but have not been tested in more complex reaction environments. The goal was to reveal the influence of other bio-oil components on catalyst stability and reaction kinetics. In situ attenuated total reflection IR spectroscopy was used to monitor the course of the reaction with high time resolution. Platinum and palladium supported on carbon served as the catalysts; typical reactions temperatures were 80 to 115 °C at a H_2 pressure of 250 psig.

The addition of small amounts (as little as 2 wt%) of alcohols to the aqueous reaction mixture suppressed conversion of phenol and of its hydrogenation product cyclohexanone. This effect can be ascribed to decomposition of the alcohol and poisoning of the catalyst and was more pronounced for primary than secondary alcohols. Conversion of phenol was more susceptible to this poisoning effect than conversion of cyclohexanone, which can be explained by different demands of these molecules on the reaction site. Another observation was that the performance of the noble metal catalysts varied strongly with the conditions of the reductive pretreatment, indicating that significant catalyst restructuring in water can occur.

The data demonstrate that competitive adsorption and structure-sensitivity are the dominating factors in the noble metal-catalyzed conversion of phenolic compounds and their hydrogenation products.

386381 Stability, Activity and Selectivity Study of a Precipitated Iron Catalyst for Biomass to Liquid Fuels Via Fischer-Tropsch Synthesis

Abstract:

The production of clean fuels from biomass via Fischer-Tropsch synthesis (BTL-FT) has become more and more attractive. FT, as a proven technology to transform under-utilized hydrocarbon resources into valuable alternative fuels and chemicals, is at the heart of the BTL process. Our research group of Material and Process Synthesis (MaPS) has developed a simplified small scale BTL-FT process which sets its own requirements of the catalyst to make the FT process more feasible and operable, such as the catalyst should have a high water-gas shift (WGS) activity to adjust the feed gas H_2/CO ratio to near to two. Because the Iron-based catalysts are low cost and demonstrate high activity for both FT and WGS reactions, they match the requirements of our new BTL-FT process.

The viability of the FT process depends on three key factors, the life, the activity and the product selectivity of the catalyst. The present work focuses on the investigation of the catalytic performances of an iron based FT catalyst for the conversion of renewable fuels from biomass over both short term and long term operations. A precipitated iron based catalyst was prepared. And a large number of FTS experiments have been conducted using the syngas - a mixture of $CO/H_2/CO_2/N_2$ - which is simulated as a syngas derived from biomass gasification. Three fixed bed reactors were used in this study: (1) Reactor One was used to expose the effect of operating conditions (different temperature, pressure and flow rate) on the catalyst activity and product selectivity during the short term FTS; (2) Reactor Two and Reactor Three were used to investigate the stability of catalyst. Reactor two was reduced by syngas and the other was activated by H_2. After catalyst reduction, the same FT reaction procedures were carried out and the reactions have been running for a long period of more than 7 month for both of Reactor Two and Reactor Three.

The results of this investigation are intended to provide valuable information, such as, if this catalyst can remain active at a relatively high CO_2 and N_2 contents syngas; whether this catalyst have a high WGS activity; How does the catalyst activity and selectivity changing during the life time running? Furthermore, if the catalyst can be activated by the same syngas used for FTS, we can further simplify our new BTL-FT process. The results would make a valuable contribution to the design of BTL-FT process and make it more economical and industrially applicable.

Effects of CO and H_2 Partial Pressures and Temperature on the Deactivation by Carbon of Cobalt Fischer-Tropsch Catalysts

Kamyar Keyvanloo, McCallin J. Fisher, William C. Hecker, Calvin H. Bartholomew*

Department of Chemical Engineering, Brigham Young University, Provo, UT 84602

**hecker@byu.edu*

Fischer-Tropsch (FT) synthesis is a promising technology to produce clean fuels. However, catalyst decays with time-on-stream affecting CO conversion and productivity which can cause economic obstacles for future FT investment. Cobalt catalyst deactivation is a combination of several simultaneous mechanisms including poisoning, sintering, oxidation, and carbon deposition. Long-term deactivation is due to carbon deposition, which can be affected with different process conditions. The goal of this paper is to systematically investigate the effect of the partial pressures of hydrogen and carbon monoxide and temperature on the deactivation by carbon for a 25%Co-0.25%Pt/ 5%SiO_2-Al_2O_3 catalyst. FT rate data were obtained while controlling the CO conversion at less than 24% to favor carbon deposition and lower partial pressure of water to prevent oxidation. Starting cobalt crystallite sizes of 12-14 nm were used to presumably eliminate the possibility of deactivation by sintering; however, XRD on spent catalysts will be used to test this hypothesis. The results from rate measurements taken over 700 h show that higher partial pressures of both hydrogen and carbon monoxide result in faster deactivation. The latter leads to higher carbon deposition, which may block the catalyst pores and active sites. Temperature-programmed hydrogenation (TPH) data on wax-extracted catalyst after 48 h of FT reaction show three peaks centered at 250, 320 and 410 °C (See Fig. 1). The spent catalysts will be characterized by different techniques such as BET, XRD, TPR, EOR, CO-uptake, and TPH to give more insight into the observed deactivation behavior and mechanism.

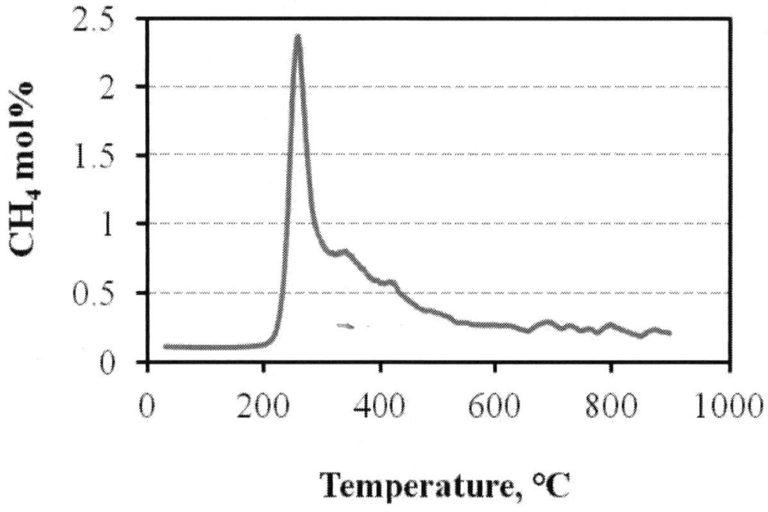

Fig. 1. Temperature-programmed hydrogenation of wax-extracted cobalt catalyst after 48 h of FT reaction.

384399 Use of a Nanoscale Catalyst and Supercritical Reaction Medium in Fuel Production Via Fischer-Tropsch Synthesis

Abstract:

Fischer-Tropsch Synthesis (FTS) is surface-catalyzed polymerization reaction which produces hydrocarbons, oxygenates, heat, water, and CO_2 from syngas, a mixture of hydrogen and carbon monoxide. While FTS produces many of the compounds in liquid transportation fuels, the additional production of light gas, heavy wax, and a large amount of waste heat make the economics of large scale fuel production challenging. This research aimed to improve the reaction in two ways. First, a nanoscale iron-based carbon-supported catalyst was utilized. Nanoscale catalysts have high surface area, which translates to higher FT activity. The carbon nanotube support has high thermal stability and chemical resistance. Additionally, the use of a carbon support eliminates some of the difficulties in reduction often seen with iron nanoparticles on oxidic supports and is typically thought to increase selectivity to olefins. Secondly, the reaction was conducted in a supercritical hexane reaction medium. Such a medium has been shown to improve product selectivity and activity maintenance by alleviating some of the heat and mass transfer challenges of the reaction. The simultaneous use of both the nanoscale catalyst and supercritical fluid reaction media yielded an active, stable catalyst with relatively high selectivity towards valuable hydrocarbon and oxygenate products.

In-situ DRIFTS studies for the effect of H_2 pre-adsorption on CO adsorption over a reduced catalyst

Nitin Kumar, J.J. Spivey

Department of Chemical Engineering, Louisiana State University, Baton Rouge, LA 70803, USA

Syngas (CO + H_2) can be derived from many sources, including those from both renewable and non-renewable materials. The conversion of syngas to ethanol is of key interest because ethanol can be used as energy carriers or as fuel additives. However, catalytic conversion of syngas to ethanol is a major challenge due to the low activity/selectivity of the catalyst, particularly to C_1 byproducts. Most researchers have found that rhodium-based catalysts gives the highest selectivity for C_2^+-oxygenates. However, the high cost of rhodium makes it uneconomical to scale-up the system. The work presented here focuses on alternatives, such as cobalt-based catalysts, which has low water-gas shift activity and high activity for CO hydrogenation.

In order to synthesize a better catalyst, it is important to understand the role of surface species in CO hydrogenation and C-C bond formation and, how they affect catalytic activity/selectivity. Diffuse Reflectance Infrared Fourier Transform Spectroscopy (DRIFTS) is a useful tool to probe the active sites, adsorbed species, and reaction intermediates. The work presented here reports the in-situ DRIFTS results for cobalt-based catalyst and discusses the observed active sites modifications as a result of H_2 pre-adsorption.

Cobalt-based catalyst supported on silica and modified by rhenium and zirconium was prepared by the conventional incipient wetness impregnation method. The catalyst was characterized by ICP, XRD, and H_2-TPR. In-situ DRIFTS experiments were carried out on a reduced catalyst under two different surface conditions: clean surface under helium flow and, surface pre-adsorbed with H_2. The experiments were carried out at room temperatures as well as at higher temperatures (230 °C, 270 °C). The results showed that at room temperature CO adsorption on a clean surface was higher, and over a wide range of wavenumbers compared to that of a H_2 pre-adsorbed surface under similar conditions. It was concluded that at room temperature H_2 occupies some of the CO adsorption sites, thereby reducing the amount of adsorbed CO. The results at higher temperatures, however, were found to be the opposite, i.e. much larger amount of CO was found to be adsorbed over a H_2 pre-adsorbed surface compared to the clean surface. This suggests that the active sites are modified by H_2 such as to facilitate the CO adsorption at higher temperatures.

Mesoporous Carbon Supported Iron Catalyst for Fischer-Trospch Synthesis

Cun Wen, University of South Carolina, Columbia, SC
Jason Hattrick-Simpers, University of South Carolina, Columbia, SC
Jochen Lauterbach, University of South Carolina, Columbia, SC

Catalytic selectivity is central to industrial chemical plants, and can be tuned via adjusting the electronic and geometric properties of not only the metal nanoparticles, for example, but also through the support properties.[1] Because of the unique electronic properties of carbon nanotubes (CNTs), they recently have drawn great a deal of interest in catalysis, including Fischer-Tropsch synthesis.[2] CNTs are composed of graphene sheets rolled up with a variety of layers into nano-scale cylinders with high surface areas. In the rolled up graphene sheets, the sp^2 C-C bonds are bended and the electron density is shifted to the outer surface of the CNTs. However, the high cost and low production-rate of CNTs currently prevents large scale applications. In addition to the high cost, the potentially hazardous nature of the CNTs adds another barrier to large scale utilization.

Recent reports showing that mesoporous carbon has graphitic sp^2 C-C bonds similar to those of CNTs, while MC having lower cost and a higher production rate.[3] Herein, we demonstrate MC as an alternative to CNTs as catalyst support with comparable catalytic performance for FTS. MC with ordered pore structures, as shown in Fig. 1a, are synthesized using silica spheres as templates. After etching the silica spheres with sodium hydroxide, void spaces with concave surfaces are formed as the pores of MC, as shown in the insert of Fig. 1a. Those void spaces are interconnected, forming a 3-dimentional (3D) mesoporous system with the pore size of about 17 nm. The interconnected is beneficial for the Fischer-Tropsch synthesis by improving the mass transfer inside the pore. This is reflected higher selectivity (68 %) to C_{5+} hydrocarbon comparing with CNTs supported iron catalyst (42%). In summary, mesoporous carbon is a competitive candidate as support for Fischer-Tropsch catalyst when compared to carbon nanotube catalysts.

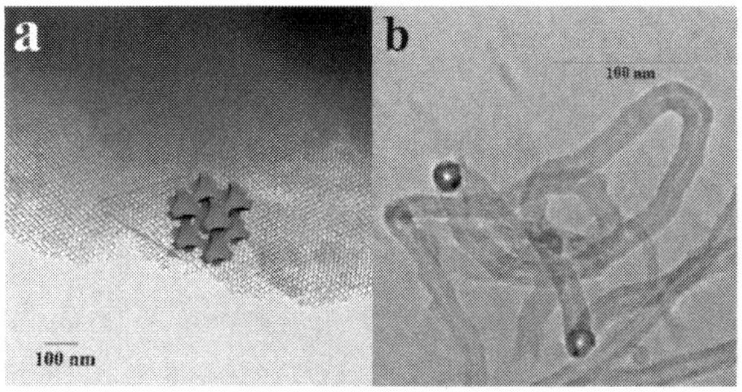

Figure 1. TEM image of a) mesoporous carbon (inset is an illustration of the unit cell of mesoporous carbon) and b) carbon nanotube.

1. Y. Yamada, C.-K. Tsung, W. Huang, Z. Huo, S. E. Habas, T. Soejima, C. E. Aliaga, G. A. Somorjai and P. Yang, Nat. Chem., 2011, 3, 372-376.

2. W. Chen, Z. Fan, X. Pan and X. Bao, J Am Chem Soc, 2008, 130, 9414-9419.

3. Y. Xia and R. Mokaya, Advanced Materials, 2004, 16, 1553-1558.

Propane hydrogenolysis as a probe for the catalytically active phase of the Fischer-Tropsch reaction

Julien Schweicher[1], Adam Bundhoo[1], Alfred Frennet[1], Thierry Visart de Bocarmé[1] and Norbert Kruse[1,2]

[1] Chemical Physics of Materials - Catalysis and Tribology, Université Libre de Bruxelles, CP 243, 1050 Brussels, Belgium

[2] Center of Catalysis for Energy & Environment, Voiland School of Chemical Engineering and Bioengineering, Washington State University, Pullman, WA 99164, USA

After 100 years of research on the Fischer-Tropsch (FT) synthesis, the question for the reaction mechanism is still under debate. Catalysts applied under industrial conditions are based on either Co or Fe. While some authors claim that Co is the catalytically active surface to essentially retain its metallic nature, others favor a chemically reconstructed surface eventually leading to near-surface oxidation or carbide formation during reaction conditions. Previously reported results from our laboratory[1] demonstrated 'crowded surfaces' for Co/MgO, i.e. amounts of atomic oxygen, carbon and hydrogen largely exceeding the monolayer limit for the atmospheric steady-state reaction. The purpose of the present work was to investigate the consequences of such 'non-metallic' surface phase composition using a probe reaction requiring metallic ensemble effects to occur. We selected propane hydrogenolysis which is particularly sensitive to the presence of metallic sites and performed transient experiments by dynamic switching from CO hydrogenation to propane hydrogenolysis (C_3H_8+H_2) under atmospheric conditions using a Co/MgO catalyst (10/1, molar ratio).

More specifically, transient experiments were carried out while keeping constant the temperature, the H_2 partial pressure and the total volumetric flow rate. The gases exiting the reactor were continuously analyzed by mass spectrometry. Such relaxation-type measurements demonstrate in a time-dependent manner how products of the CO hydrogenation reaction are replaced by those of the hydrogenolysis reaction.

The results show that despite the C_3H_8+H_2 reaction is not associated with water formation, large amounts of H_2O continue to be produced after switching from CO hydrogenation to hydrogenolysis. Furthermore, we find continued water production to be associated with hydrogen consumption; however, the hydrogenolysis product (methane) is not occurring in the initial stages after switching the reactant mixtures. We conclude that the FT-active Co/MgO surface phase is largely covered with oxygen. As surface oxygen (or hydroxyl) is progressively removed from the FT surface, metallic Co patches are being made available. Only under these conditions does methane as a product of propane hydrogenolysis form. More specifically, we observe methane - a product of CO hydrogenation and hydrogenolysis - to first decrease after switching and then to increase again. Thus, while methane is produced on an oxidic phase during FT it originates from a metallic surface during hydrogenolysis.

The findings will be discussed in context with present mechanistic models of hydrocarbon formation in the FT reaction.

(1) Schweicher, J.; Bundhoo, A.; Kruse, N. *Journal of the American Chemical Society* **2012**, *134*, 16135.

371223 Effects of Structural Promoters on Cobalt Copper Catalyst for the Conversion of Syngas to Ethanol

Abstract:

Catalytic conversion of syngas derived from natural gas is an efficient way to convert natural gas to high value products. Ethanol is one of the promising alternative fuels that can be synthesized from syngas. Among the studies on monometallic and bimetallic catalysts for CO hydrogenation, Cu-Co bimetallic catalysts have been studied for ethanol and higher alcohols synthesis. In these catalysts, cobalt and copper in close proximity are designed to dissociate CO at Co sites and activate associative CO at Cu site. This forms the critical C-C bond, creating intermediates that can convert to ethanol and higher oxygenates. In addition, structural promoters can increase ethanol selectivity with improved stability, higher dispersion, and enhanced interaction between metals and promoters.

Here we present an experimental study of the Co-Cu catalysts that are synthesized with different structural promoters (ZrO2, Al2O3, MoOx and ZnO) using co-precipitation method. We characterize the catalysts using BET, TPR-H_2, TPD-NH_3, XPS and XRD to compare the effect of the structural promoters on surface area, catalyst dispersion, surface acidity, and reduction behavior. We examine the selectivity of the catalyst in CO hydrogenation to quantify product selectivity and to study thermal stability at temperatures at 270 C, 2/1 H2/CO, 40 bar in a fixed bed reactor to simulate the syngas derived from natural gas.

Site-dependent Lewis acidity of γ-Al_2O_3 and its impact on ethanol dehydration and etherification

Glen R. Jenness, Matthew A. Christiansen, Stavros Caratzoulas, Dionisios G. Vlachos, Raymond J. Gorte

Abstract:

Due to petroleum depletion and an increasing concentration of greenhouse gases, there is a considerable interest in the use of renewable chemical feedstocks generated from renewable biomass. Biomass derivatives are often oxygen rich, which results in a need for catalysts that can remove the extraneous catalysts and create value-added chemical feed stocks. One promising catalytic material for such reactions is γ-Al_2O_3, which, due it's solid Lewis acidity, has been shown to be an effective catalyst for the dehydration of alcohols. Owing to the underlying crystal structure, γ-Al_2O_3 surfaces expose a set of heterogeneous Lewis acid Al^{+3} sites, which creates a challenge in the characterization and optimization of the Lewis acidity for this material.

In the current study, we examine the heterogeneity of the Lewis acidity on the (100) and (110) facets of γ-Al_2O_3 by computing the binding energies of various oxygenates, in addition to the reaction barriers of dehydration and etherification reactions of ethanol. We show that the ethanol dehydration barrier is moderately affected by site heterogeneity (barriers between 1.2-1.6 eV); in contrast, a nearly threefold change in the ethanol etherification barrier is found among the various Al^{+3} sites. In order to rationalize these results, the s-conduction band mean of the Al^{+3} sites is introduced as a descriptor to characterize the ability to transfer electron charge from the adsorbate to the Lewis acid site. It is shown for the first time that this descriptor quantitatively correlates the oxygenate binding energies and the ethanol dehydration reaction barriers. However, for the ethanol etherification reactions the s-conduction band mean of the Al^{+3} sites describes barriers only qualitatively due to the bimolecular nature of this reaction, which results in a change in the nucleophilicity of the ethoxy species by a nearby adsorbed ethanol. As a result, the strength of the Lewis acid sites is not the only descriptor for etherification chemistry.

Overall, our results clearly indicate that while the binding energy is an excellent descriptor of Lewis acidity strength and dehydration chemistry on the clean alumina surfaces, cooperative phenomena (i.e. modulation of the nucleophilicity of the ethoxy by the nearby oxygen) are key issues that lead to a break down in the correlation between Lewis acid strength in terms of the binding energy or the s-conduction band mean and the reaction barriers.

Ethanol Dehydrogenation on γ-Al_2O_3: Site Requirements and Mechanism of Acetaldehyde Synthesis

Joseph F. DeWilde, Christopher J. Czopinski, and Aditya Bhan
University of Minnesota – Twin Cities Minneapolis, MN 55455

Steady state kinetic measurements of ethanol dehydrogenation on γ-alumina confirm acetaldehyde synthesis to be independent of water partial pressure and inhibited only by reactive precursors at temperatures between 623 K and 673 K. Measured kinetic isotope effects for ethanol dehydrogenation demonstrate that either C_α-H bond cleavage or the removal of surface hydrogen species is rate limiting in acetaldehyde synthesis. Ethane is produced at a rate equivalent to that of acetaldehyde synthesis at all measured temperatures (623 – 673 K) and ethanol pressures (0.9 – 8.0 kPa) and independent of co-fed ethylene pressure (0- to 1.5 kPa) at ethylene pressures up to 1.5 kPa, indicating that acetaldehyde and ethane are co-products of the reaction. Pyridine was observed to inhibit the syntheses of ethylene, acetaldehyde, and ethane to the same extent, demonstrating that the synthesis routes to form these molecules require acidic surface sites and share at least one common reactive intermediate. An indirect hydrogen transfer mechanism for the synthesis of acetaldehyde from the dehydrogenation of ethanol on γ-alumina consistent with the steady state kinetic measurements, formation of ethane, observed pyridine inhibition, and determined kinetic isotope effects is presented.

---

[*] Equal contribution to this work has been provided by these authors

[†] To whom correspondence should be addressed

367944 Nature of Surface Metal Oxide Sites Present for Supported MOx/ZSM-5 Catalysts

Abstract:

The well-dispersed group V-VII transition metal oxides supported on zeolites have received significant attention in recent years due to their industrial potential as catalysts for numerous chemical reactions. In spite of the extensive characterization studies on these catalyst systems, a general consensus has not been reached about many fundamental details such as the anchoring sites and the molecular and electronic structures of the surface MOx species on ZSM-5 supports. This study aims to resolve the debates on the anchoring sites and electronic and molecular structures of surface MOx species (M=V,Cr, Mo, W and Re) on ZSM-5 supports with the aid of modern *in situ* spectroscopy (FT-IR, UV-vis, Raman and XANES/EXAFS). *In situ* Fourier Transform Infrared (FT-IR) Spectroscopy showed that Al-$(OH)^+$-Al Brønsted acid sites are the preferred anchoring site of the ZSM-5 support for surface MO_x species, but external silanol and extra framework aluminum hydroxyls also become anchoring sites at high metal oxide loadings. *In situ* UV-vis diffuse reflectance spectroscopy (DRS) demonstrated that all the supported MOx species are fully oxidized and present as isolated sites. *In situ* Raman and *in situ* extended X-ray absorption spectroscopy (XAS) revealed that only one surface MOx structures are present for supported V_2O_5/ZSM-5 and Re_2O_7/ZSM-5 catalysts (mono-oxo O=VO_3 and trioxo $(O=)_3ReO$, respectively), two surface CrOx structures are present for supported CrO_3/ZSM-5 catalysts (mono-oxo O=CrO_3 and dioxo $(O=)_2CrO_2$), and four distinct surface MoOx and WOx structures are present on supported MoO_3/ZSM-5 and WO_3/ZSM-5 catalysts dependent of the Si/Al ratio of the ZSM-5 support(dioxo $(O=)_2MO_2$ species anchored to two Brønsted acid sites at low MOx loading and high Al content, dioxo $(O=)_2MO_2$ species anchored to one Brønsted acid site at increasing MOx loading, dioxo $(O=)_2MO_2$ species anchored to two Si-OH sites at low Al content and high MOx loading and mono-oxo O=MO_4 anchored at extra framework Al-OH sites. These new studies demonstrate how systematic, modern *in situ* spectroscopic studies are able to resolve the fundamental issues of complex metal oxide catalytic systems.

390172 Speciation and Reactivity of Small Fe, Cu, or Co Oxide Clusters Supported on Ceria and Silica

Abstract:

The redox activity of supported metal oxides determines their utility in emissions abatement and their role in activation of H_2 or O_2 for chemicals synthesis. This talk describes two routes to controlling the redox activity of dispersed metal oxides. In the first part, we support Fe(III), Co(II), and Cu(II) oxides by incipient wetness impregnation of aqueous EDTA complexes of Fe, Cu, and Co on neat and alkali-modified CeO_2, followed by calcination to give small clusters, following prior work on SiO_2. [1] The catalysts are characterized principally by TPR, *in situ* XANES, and diffuse reflectance UV-visible spectroscopy (DRS), and it is seen that a larger fraction of the oxide is redox active in reversible 1-electron cycles than compared to impregnation with more typical precursors. This improvement in activity is hypothesized to be due to the presence of more uniform interactions between CeO_2 and the oxide afforded by the bulky, anionic precursor. These materials are active in crotonaldehyde hydrogenation and H_2-SCR of NO, with rates of the latter quantitatively correlated to the fraction of redox active sites.

In a second example, we deposit pre-formed clusters of Cu(I) oxides onto SiO_2. The L_xCu_y clusters were previously synthesized as potential models for active sites in Cu-exchanged zeolites and in oxidase enzymes, where L are bridging siloxane ligands. [2] The Cu(I) oxidation state and some degree of the cluster structure is stabilized by incipient wetness impregnation onto SiO_2, such that they are resistant to oxidative decomposition into bulk CuO up to ~400°C. These Cu(I) materials activate O_2 at lower temperatures, as we demonstrate through their ability to carry out oxidative dehydrogenation of cyclohexane.

[1] Prieto-Centurion, D.; Boston, A. M.; Notestein, J. M., Structural and electronic promotion with alkali cations of silica-supported Fe(III) sites for alkane oxidation. *J. Catal.* **2012,** *296*, 77-85.

[2] Schax, F.; Limberg, C.; Mugge, C., Copper(I) Siloxides - Aggregated Solid-State Structures, Cu-Cu Interactions and Dynamic Solution Behavior. *Eur. J. Inorg. Chem.* **2012,** 4661-4668.

383297 Understanding the Active Sites on Mn/Na2WO4/SiO2 for Oxidative Coupling of Methane (OCM)

Abstract:

With the development of shale gas, natural gas becomes an abundant energy and chemical source for the United States, providing solutions to the energy supply and environmental challenges in a sustainable way. As a major component in natural gas, methane conversion to liquid fuels and valuable chemicals is very attractive and critically important. Among various methane conversion routes, oxidative coupling of methane (OCM) provides a direct way to activate methane for ethylene production, which can be further used to produce chemicals and liquid fuels. $Mn/Na_2WO_4/SiO_2$ has been reported to give the highest C_{2+} yield (26%) among OCM catalysts (Lunford, J. H. *Catal. Today* 2000, *63*, 165). To further improve or design more active catalysts, we need to identify the nature of the active sites. In this context, this study aims to understand the active sites on $Mn/Na_2WO_4/SiO_2$ for methane activation by doing kinetic measurements and detailed characterizations of the catalyst properties.

Catalysts (x%Mn/y%W/0.8%Na/SiO_2) with varying Mn (0~2 wt%) and W (0~3.1 wt%) loadings prepared by incipient wetness impregnation were used for the kinetic study. When changing Mn loading from 0 to 2 wt% (W constant at 3.1 wt%), ethylene formation rate, normalized by catalyst surface area, increased linearly with Mn loading until 0.05 wt%, and did not increase when Mn loading was further increased. The rate increased continuously with W loading from 0 to 3.1 wt% (Mn constant at 2 wt%). Ethylene formation rate for the catalyst with only W (3.1%W/0.8%Na/SiO_2) is (0.9×10^{-7} mol C_2H_4 $(m)^{-2}$ $(s)^{-1}$) at 730°C under 40% CH_4, 20% O_2 and balance N_2, 7 times lower compared to the areal rate of 0.05%Mn/3.1%W/0.8%Na/SiO_2. For the catalyst with only Mn (2%Mn/0.8%Na/ SiO_2), the rate is (0.4×10^{-7} mol C_2H_4 $(m)^{-2}$ $(s)^{-1}$) at 730°C under 40% CH_4, 20% O_2 and balance N_2, 11 times lower compared with 2%Mn/3.1%W/0.8%Na/SiO_2. These results have shown that ethylene formation rate increased significantly with the addition of the 2nd metal oxide, indicating more active sites are generated when both Mn and W present on the catalyst surface.

384690 Doped Rare-Earth Oxides As Catalysts for the Oxidative Coupling of Methane

Abstract:

The oxidative coupling of methane (OCM) is a potential pathway to directly convert methane to highly valuable ethylene, as well as other hydrocarbon products. However, it is very difficult to obtain high yields of C_{2+} products (i.e. products such as ethane, ethylene and higher hydrocarbons), due to the formation of higher oxidation products, such as CO and CO_2 (CO_x). Li/MgO is a simple and active OCM catalyst with a relatively high C_{2+} product selectivity, but it suffers from severe deactivation due to the volatility of lithium under OCM reaction conditions. Rare earth oxides (REOs), such as samaria (Sm_2O_3) and lanthana (La_2O_3), are amongst the most efficient single component oxide catalysts in the methane coupling reaction, while reducible REOs (e.g. ceria, terbia, and praseodymia) are typically more selective towards CO_x products than C_{2+} products. In this study the effects of lithium on samaria and terbia catalysts supported on nanoparticle magnesia (n-MgO) were investigated in detail. It was shown that addition of lithium to terbia supported on n-MgO is not only a more active and selective OCM catalyst compared with the reference Li/MgO, Sm_2O_3/MgO and Li-Sm_2O_3/MgO catalysts, it is also more stable with time on stream. The higher stability appears to be due to a stronger interaction between lithium and terbia compared with lithium and samaria, as evidenced by both X-ray diffraction and X-ray photoelectron spectroscopy measurements. Terbia appears to be unique amongst the reducible rare earth oxides, as Li-doped praseodymium or cerium oxides are not as active as the Li-TbO_x/MgO catalyst. Amongst Li, Na, Ca and Mg dopants, Li is the most effective, but due to its volatility Na-doped REO catalysts can outperform the Li-doped REO catalysts after extended time on stream.

Location of active sites on the surface of MoVTeNbOx crystals: morphology matters

M. Sanchez-Sanchez, P. Xu, D. Hartmann, N. D. Browning, J. A. Lercher

1) Technische Universität München, Lichtenbergstr. 4, 85747 Garching, Germany

2) University of California, Davis One Shields Avenue, Davis, CA 95616 (USA)

3) Pacific Northwest National Laboratory, 902 Battelle Boulevard, Richland, WA 99352 (USA)

Abstract

Complex Mo and V mixed oxides have been extensively studied as catalysts for selective oxidation. Among Mo-V oxides, the quaternary Mo-V-Te-Nb complex mixed-oxide has shown remarkable ability to activate alkanes to produce olefins and other oxidation derivatives. The availability of ethane in shale gas, as well as the interest in adding value to previously under-utilized refinery fractions, makes the activation of alkanes via oxidative dehydrogenation (ODH) an attractive alternative to the industrially established processes. As a consequence, renewed interest has arisen in improving synthetic methods for MoVTeNb oxide, as well as better understanding its catalytic performance.

MoVTeNbO crystallizes mostly into two phases, known as M1 and M2 [1]. The performance of Mo-V-Te-Nb mixed-oxide as selective oxidation catalyst is mainly attributable to the crystalline phase M1, which has been found to be the only phase able to perform alkane ODH [2, 3]. However, depending on the preparation conditions, the final catalyst usually crystallizes as a phase mixture of M1, M2 and/or amorphous phase. M1 is a crystalline orthorhombic phase consisting of octahedra with molybdenum and vanadium centres connected via corner sharing oxygen atoms that form a structure with periodic layers and channels perpendicular to the [001] plane. Niobium is located inside the smallest channels, formed from five octahedra, whereas tellurium occupies channels formed from six or seven octahedra. It is generally accepted that the active site of M1 involves V units connected to Mo and Te species which are believed to enhance the reduction and re-oxidation of the active site [4]. The active center is contained in the group of 5 octahedra (red and green in Figure 1) which are interconnecting two hexagonal channels, provided that vanadium is present in some of the positions and that the combination of Mo and V oxidation states is the right one to generate activity [4]. The function of the crystalline structure itself is to provide the spatial configuration and redox interaction between the species involved in the active site. Most of the metal positions in the M1 lattice can be partially occupied by Mo, V or even Nb cations. This fact gives M1 phase certain flexibility in the chemical composition. In spite of the knowledge gained over the last 10 years about M1 phase, it is difficult to find in the literature any clear correlation between structure and reactivity for this catalyst. In particular, it seems that, when different synthetic methods are applied, the M1 crystals obtained can be chemically similar but they show different catalytic performance [5]. There are indeed still many unanswered questions about how bulk size, surface structure and chemical composition of M1 particles affect the activity.

In this work, we have prepared a series of MoVTeNb oxide and tested in the oxidative dehydrogenation of ethane. Catalytic tests showed that all the catalysts were able to selectively oxidize ethane to ethene at moderate temperatures (350-400 °C) with minor production of COx. Rietveld analysis of the X-ray diffraction data allowed us to quantify the amount of MoVTeNb oxide that has crystallized as M1. In this way, it was possible to find a linear correlation of the reaction rate with the abundance of M1 in the solid. Kinetic studies led to similar energy of activation values for all the M1-containing samples, however significant differences in the pre-exponential factor were noticed, even when the reaction rates were normalized per M1 gram. This points to differences in the total number of active sites that are available per mass of M1 for each sample.

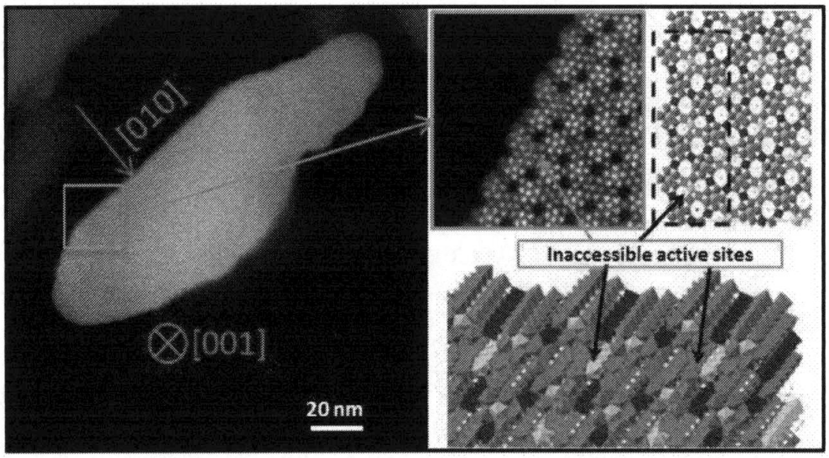

Figure 1. Left, micrograph of a M1 particle. Right, detail of the surface and schematic illustration of the surface termination for plane [010]

Atomic resolution aberration-corrected HAADF-STEM imaging was used to directly image the termination of crystals of M1 phase in MoVTeNb complex oxides. Over 50 random particles were examined and it was possible not only to identify and quantify the most frequently exposed facets but also to obtain an average number of active sites exposed for each facet. It was observed that flattened large particles consist of extensive facets that expose almost no active sites, as it is shown in Figure 1. An immediate conclusion extracted from these observations is that samples with a predominant morphology of large flattened M1 particles must have a lower activity than other morphologies. This prediction was experimentally validated by the results of catalytic tests obtained for two MoVTeNbOx samples showing different morphologies. Average size and shape of the M1 particle were calculated for each sample by statistical analysis of Helium Ion Microscopy images. In this way, it was possible to estimate the density of active sites that are exposed for a given morphology and quantitatively compare different MoVTeNbO samples. Furthermore, the analysis of a MoVTeNb oxide with high content in M1 allowed the calculation, for the first time, of the true turnover frecuency (TOF) for this type of catalysts.

In summary, STEM imaging provided an atomistic picture of the surface termination of the complex MoVTeNb oxide. Statistical analysis of STEM and HIM images allows us to go from the atomic to the macroscopic level and predict the activity of a sample from the morphology of its particles. This work shed some light on the relationship between catalytic activity and crystal morphology for complex mixed oxide catalysts and provides new insights for the design and optimization of synthetic methods.

References:

[1] P. DeSanto, D.J. Buttrey, R.K. Grasselli, C.G. Lugmair, A.F.V. Jr., B.H. Toby, T. Vogt, Z. Kristallogr. 219 (2004) 152–165.

[2] P. Botella, E. García-González, J.M. López Nieto, J.M. González-Calbet, Solid State Sciences 7 (2005) 507-519.

[3] D. Vitry, Y. Morikawa, J.L. Dubois, W. Ueda, Topics in Catalysis 23 (2003) 47-53.

[4] R.K. Grasselli, D.J. Buttrey, J.D. Burrington, A. Andersson, J. Holmberg, W. Ueda, J. Kubo, C.G. Lugmair, A.F. Volpe, Topics in Catalysis 38 (2006) 7-16.

[5] Y. V. Kolen'ko, W. Zhang, R. Naumann d'Alnoncourt, F. Girgsdies, T. W. Hansen, T. Wolfram, R. Schlögl, A. Trunschke, ChemCatChem 3 (2011) 1597 – 1606.

Application of 4-Point Optical Probe to Study Bubble Dynamics and Mass Transfer Coefficients in Slurry Bubble Column

Onkar N. Manjrekar, Milorad P. Dudukovic
Chemical Reaction Engineering Laboratory (CREL),
Department of Energy, Environmental and Chemical Engineering,
Washington University in St. Louis, Missouri, 63130

Abstract:

In the present work, a 4-point optical probe is applied to a slurry bubble column to assess the effect of solids on bubble dynamics and mass transfer coefficient. All experiments are performed in a 8 inch diameter column in the churn turbulent regime, with 10% (by weight) solid concentration of aluminum oxide particles with 60 μm average diameter. The superficial gas velocity ranges from 20 cm/sec to 45 cm/sec. 4-Point optical probe was also used to assess bubble dynamics and mass transfer coefficients in a photo-bioreactor containing micro-algae. Local gas hold-up and bubble frequency are reduced in the presence of solids and micro-algae. Effect of solids and micro-algae on distribution of bubble velocity and bubble chord length is studied. The parameters evaluated in this work can be used in phenomenological models for scale-up and in validation of Computational Fluid Dynamics (CFD) results.

Key Words: Slurry Bubble Column, Solids, Gas hold-up, Bubble dynamics, Photo-bioreactor, Micro-algae.

387646 Flow Regime Transition in Trickle Bed Reactors

Abstract:

The hydrodynamics of a trickle bed impact reactor performance through changes in mass transfer with flow regime [1, 2]. Depending on the reaction chemistry (liquid limiting or gas limiting), operation in one flow regime or another may have a positive or negative influence on reaction rate and yield of the desired product. Accurate characterization of the boundary between low interaction and high interaction regimes is therefore imperative. While a significant number of studies have been reported in the literature, none are available which experimentally discern the independent impacts of void fraction and particle size/shape on the transition. In this work, the flow regime transition is measured by visual observation and analysis of the standard deviation of the pressure drop for a range of packing media. A number of different particle shapes and sizes are investigated and the void fraction is varied by inclusion of fines. In this way, the effects of particle shape, size, and bed void fraction on flow regime are determined.

[1] Wilhite, B.A.; Wu, R.; Huang, X.; McCready, M.J.; Varma, A. *AIChE J.* 2001, 47 (11) 2548.

[2] Al-Dahhan, M.H; Larachi, F.; Dudukovic, M.P.; Laurent, A. *Ind. Eng. Chem. Res.* 1997, 36 (8) 3292.

Bed Height and material Density Effects on Local Time-Averaged Gas Holdup in Fluidized Bed Reactors

Abdelsalam Efhaima, Muthanna H. Al Dahhan

Dept. of Chemical and Biochemical Engineering, Missouri University of Science and Technology, Rolla, MO, USA

ABSTRACT

The hydrodynamic behavior of fluidized bed reactors is complicated due to complex interaction phenomena among the phases, gas and solids. The hydrodynamic must be understood to improve fluidized bed design, scale-up, operations and process efficiencies. Gas hold-up is one of the most important parameters that characterize the fluidization quality with minimum fluidization velocity. Gas hold-up describes the volumetric fraction of gas present within the bed materials, and is also useful in estimating the interfacial area between the dense and dilute phases that would be useful in heat and mass transfer calculations. In this study the effect of material density on the hydrodynamics of gas-soli fluidized beds was investigated and the local time-averaged gas and solid holdups were determined by using a 0.14 m diameter and 1.82 m height Plexiglas cylindrical fluidized bed column, two different Geldart type -B particles were used, glass bead and copper particle with material density of 2500, 5300 Kg/m^3 *respectively with same mean particle size of 210* μm*, Three different superficial gas velocities were investigated* U_g*=25 cm/s, 30 cm/s and 35 cm/s . Also the effect of bed height was determined, Three different bed height to diameter ratios were investigated H/D = 0.286, 0.64 and 1.7 at specific superficial gas velocity Ug = 25 cm/s. Gamma ray computed tomography (CT) has been used as non-invasive technique to measure the time –averaged cross-sectional gas and solid holdups distribution The results show that, local time –average gas hold-up increases by increasing (H/D) ratio, decreasing the bed density increasing the gas hold-up in the bed.*

Keywords : Fluidized beds hydrodynamics, Computed Tomography, optical probe, Gas hold-up.

1 INTRODUCTION

Contacting solid particles with gases is often a necessity in many industrial operations. The gas–solid fluidized bed reactor (FB) is one of the widely employed reactors as gas-solid contactor, due to many advantages such as high rates of heat and mass transfer, low pressure drop, approximately uniform temperature distribution and the ability to fluidize many particle types of different densities and sizes, fluidized bed reactors are extensively used in many industrial applications like drying of granular materials,cooling of fertilizers, pharmaceutical industries, coal combustion and gasification , chemical process, petroleum, mineral, roasting, gas adsorption operations and gas phase polymerization (McCabe et al., 2001). The application of fluidization is well recognized in nuuclear engineering as a unit operation for example, in uranium extraction, nuclear fuel fabrication.

In spite of these advantages, fluidized beds have disadvantages which limit their industrial applications, Particle agglomeration and gas velocity reduction which cause de-fluidization and changing in the hydrodynamics of the fluidized bed, sudden change the hydrodynamics due to unwanted change in the excess gas velocity are some of these problems. In addition, fluidized bed hydrodynamic behaviour is very complex, lack of reliable knowledge about the hydrodynamics behaviour, has limited their application, therefore, any attempt for better understanding the fluidization phenomena, which would result in a more reliable design of fluidized reactors.Until recently, most experimental measurements of gas holdup in fluidized columns were global in nature. For example, overall gas holdup can be obtained via measurements of bed expansion, pressure drop, hot film anemometer, and quick closing valves. Techniques based on absorption and scattering of radiation to reconstruct the local holdup distribution have also been reported in the literature (Schrock VE, 1969).

Grassler and Wirth (2000) used X-CT to determine the solids concentration in a 0.19 m of diameter circulating fluidized bed with (50-70 μm) glass beads as the bed material. Tests were carried out in two different systems. In the first, solid concentrations were calculated with an up flow system. Results for this system showed that radial solid concentration exhibited a parabolic shape with a maximum concentration close to the wall of the reactor and a minimum concentration in the centre of the bed. For the second, the solid concentration was calculated with a down flow system. For this case, the solid concentration distribution was much more complex and depended upon the gas-solids distributor operating conditions. Results showed various solid concentration distributions from a homogeneous distribution with a parabolic profile to concentrated stands in the centre of the bed.

X- ray computed tomography (CT) imaging was used by Escudero et al. 2011, to determine bed height and material density effects on fluidized bed hydrodynamic in a 10.2 cm fluidized bed. They used three different materials Geldart-type-B particle (glass beads, ground walnut shell and ground corncob) with material densities of 2600, 1300 and 1000 Kg/m^3 respectively. Results showed that decreasing the bed density increased the gas holdup in the bed.

Franka and Heindel 2009 studied the effects of side air injection, superficial gas velocity, and bed material on the local time-average gas holdup of a 10.2 cm fluidized bed, using X-ray computed tomography. Using different materials (glass beads, ground corncob, and ground walnut shell), superficial gas velocities (Ug), and side air injection flow rates (Qside) they determined the variations on the fluidization hydrodynamics of the bed. They found that with side air injection, the side air flow rose near the wall but then expanded into the bed as height and Qside increased. As Ug increased the effects caused by the side air injection were less pronounced.

Escudero et al 2011 also studied the profiles of solids holdup at various superficial gas velocities at specific H/D ratio and found that solid holdup decrease by increasing superficial gas velocity.

Mabrouk et al 2005 studied the axial and radial profiles of the solids holdup using an optical fiber needle probe and radioactive particle tracking (RPT) technique. The axial solids hold-up profiles obtained by an optical fiber needle probe and radioactive particle tracking technique show a quasi linear profile. Many intrusive and nonintrusive measuring methods have been used to study the hydrodynamics of fluidized beds. In this paper, the effects of bed height and material density on local time-averaged gas and solid holdups in fluidized bed are investigated by using gamma ray computed tomography technique (CT).

2 EXPERIMENTAL SETUP

Cold flow fluidized bed was used in this study, with outside diameter 0.14 m and a height of 1.82 m. A schematic diagram of the setup used in this study is illustrated in Fig.1. The fluidized bed column was constructed from Plexiglas and consisted of two pieces (column and cone) attached to a plenum base. Connected from the top with an upper section that had a diameter of 0.45m and is 0.91m tall this upper section of the fluidized bed had a larger diameter to reduce the superficial gas velocity of the gas phase and thus enhance the solids separation. The column sat at the top of a stainless steel base. Gas distributor (made of a porous polyethylene sheet with a pore size of 40 μm was used. The plenum was located at the bottom,. Fluidized bed column was electrically grounded to minimize electrostatic effects. Air under ambient conditions was the fluidizing gas. The gas flow rate to the unit was controlled by rotameters.

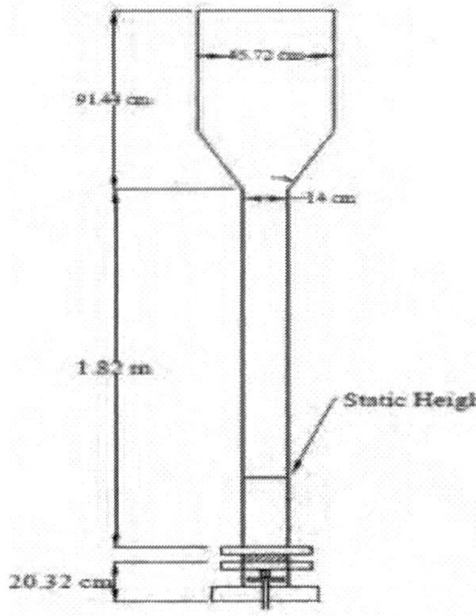

Figure1: Schematic of the cold –flow fluidized bed reactor used in this study.

Table 1. Summarizes the bed materials and their properties

Properties	**Glass Beads**	**Copper Particle**
Particle Diameter (μm)	210	210
Bed Height (H/D)	0.286, 0.64, 1.7	0.286, 0.64, 1.7
Particle Density (Kg/m^3)	2500	5300
Minimum Fluidization velocity (m/s)	0.12	0.20

3 MEASUREMENT TECHNIQUES

3.1 Computed Tomography (CT)

3.1.1 CT Facility and Measurement Procedure

The DSCT scanner at Missouri University of Science and Technology was used in this work to determine the time-averaged cross-sectional variation of gas and solid holdups at the operating conditions outlined earlier.

Computed Tomography (CT) Facility
The CT used in this study is based on a newer generation of double fan-beam scanning configuration. The details of the hardware and software have been described elsewhere (Kumar, 1994; Kumar et al. 1995, Chen et al.1999). In this work, the Alternating-Minimization (AM) method was used for image reconstruction; see Varma, R et al.2007 for Alternating-Minimization (AM) method. A photograph of

the CT facility used in this study is shown in Figure 2. The configuration of the scanner consists of two arrays of NaI detectors of 5 cm in diameter (15 detectors were used for each side) and two encapsulated sources Cs -137 (~200mCi) and Co-60 (~150mCi), located opposite to the center of the each array of detectors, placed 120 cm apart. In this study, only the Cs -137 source is used as we work with two phases in time. The detectors and the source are mounted on a plate which can be rotated 360° around the object. Moreover, the whole assembly can be moved up and down along the column to scan different axial levels of the column.

Figure.2.The (CT) available in (MS&T) with 6 inch Fluidized bed reactor.

Each of these detectors is collimated with a lead collimator that is about 6.35 cm Thick and has an open aperture of dimension 2 mm × 5 mm. This aperture reduces the effective exposed area of the crystal to a rectangular region of dimension 2 mm × 5 mm. The counts received by the detectors are limited to what is incident on this aperture. This detector array is moved with a stepper motor 21 times at an angle of 0.13° from the source, thereby creating 315 detector positions effectively for each source position. The projection is modelled as a fine line between the source and the detector as the open area of the detectors is very small. A total of 197 views or source positions are considered; hence, gamma ray counts data for 315 × 197 (total 62055) projections passing through the domain are collected. The gamma ray counts data recorded are energy threshold; hence, un-attenuated gamma ray photon counts of 660 Kev were recorded. A 80 × 80 pixels resolution is used to reconstruct the image. Hence, each pixel represents an area of 2 mm × 2 mm of the object.

3.1.2 Gas Holdup and CT image

In this study, CT experiments were performed under a two-phase condition (solid and gas) to measure the cross - sectional distribution of each phase, the following CT scans were performed 1- Scanning the column empty as reference CT scan. 2- Scanning the column filled with solids. 3- Scanning the column at normal gas- solid operations at the desired conditions. The beam attenuations were measured along a number of beams paths through the column, originating from different angles. Based on Beer Lambert's Law, the intensity of a beam that is transmitted through an object is expressed as follows:

$$T = \frac{I}{I_0}.\exp(-\mu\rho l) \qquad (1)$$

Where T is the transmission ratio, I_0 is the incident radiation, I is the detected radiation, μ is the mass attenuation coefficient, ρ is the medium density, l is the path length through the medium.

The measured quantity ln (I/I_0) (called A, for simplicity) is equal to the integral sum of the attenuation through the material along the beam path.

$$A = \ln\frac{I}{I_0} = \mu\rho l \qquad (2)$$

For computed tomography, attenuations are measured by the detectors along a number of such beam paths through the object from different directions around it. To reconstruct the holdup, the cross-section of the column is configured inside a square whose sides are equal to the column diameter. This square is divided into a number of square pixels. The attenuation in each pixel can be obtained (reconstructed) by using a reconstruction algorithm. In this work, the Alternating-Minimization (AM) method was used for image reconstruction. To obtain statistically significant results and to reduce the effect of position, the CT scans were obtained by scanning 360 degrees around the column for a total scanning time of about four hours, if the scanned cross-section is divided into pixels or cells and the medium is comprised of two materials (gas and solid), mass attenuation coefficients, μ_s is mass attenuation coefficient for solid, μ_g is mass attenuation coefficient for gas, ρ is the medium density, ρ_s solid density, ρ_g gas density, and thickness l_g , l_s for gas and solid phases respectively , the total attenuation

$$A = \rho_s\mu_s l_s + \rho_g\mu_g l_g \qquad (3)$$

Where L= l_s+ l_g, l_s=ε_sL and l_g=ε_gL, , Where L is the total length between the source and the detector. ε_g , ε_s are the holdups (volume fractions) for the gas and solid phases.

$$A = [\rho_s\mu_s\varepsilon_s + \rho_g\mu_g\varepsilon_g].L \qquad (4)$$

The summation of the holdups is equal to unity (i.e. ε_g=1-ε_s) .

$$A = [\rho_s\mu_s\varepsilon_s + \rho_g\mu_g(1-\varepsilon_s)].L \qquad (5)$$

$$A_{s-g} = [\rho_s\mu_s\varepsilon_s + \rho_g\mu_g(1-\varepsilon_s)].L \qquad (6)$$

Since ρ_g<<ρ_s, the attenuation caused by the gas phase is negligible compared to the solids, and *L* is common for all *A*s. Hence, solids holdup for the line averaged measurement can be written as follows:

$$\varepsilon_s = A_{s-g}/L\rho_s\mu_s \qquad (7)$$

$$A_{s-g} = \rho_s\mu_s\varepsilon_s L \qquad (8)$$

The linear attenuation coefficient of solids (μ_s) was determined by using standard tables (such as NIST Physical Reference Data), since the material composition of solids was known. Finally, the gas holdup was determined using the expression

$$\varepsilon_g = 1 - \left(\frac{A_{s-g}}{L\rho_s\mu_s}\right) \qquad (9)$$

4. RESULTS AND DISCUSSION

4.1. Effect of particle density

4.1.1. Cross-sectional gas and solid holdups distribution

The reconstructed image, processed from data obtained through CT scans, provides the cross-sectional time-averaged gas and solid holdups distribution at a given axial level. Fig. 4(a, b) shows the cross-sectional gas holdup and cross-sectional solid holdup at 30 cm/s superficial gas velocity for glass bead–gas and copper particle–gas systems, respectively, offering an idea of how gas is distributed through the column. The change in the gas holdup magnitude values were indicated by the colour variation, red indicates a higher gas holdup value while blue indicates a lower value of gas holdup.One can see from the figure that the gas holdup distribution for both systems is almost symmetric, with a higher gas holdup in the center and a lower gas holdup near the wall. While Fig.5 (a, b) shows the cross-section gas and solid holdup for glass bead–gas and copper particle–gas system systems, respectively at 35 cm/s superficial gas velocity. Asymmetric gas holdup profiles of gas at the same conditions were also observed by experiments as will be discussed shortly.

Cross-sectional gas holdup | Cross-sectional solid holdup | Cross-sectional gas holdup | Cross-sectional solid holdup

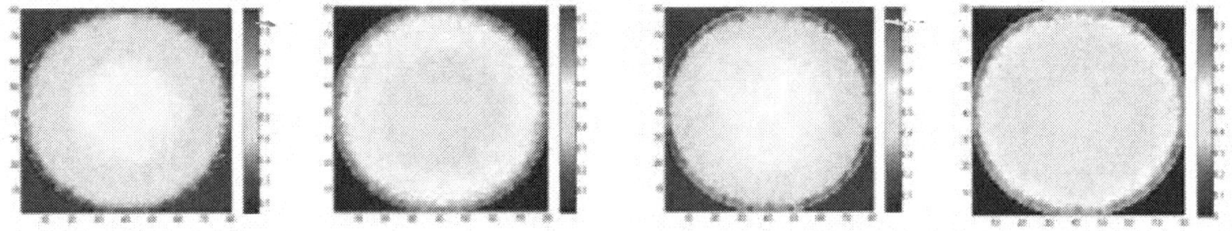

Glass bead- gas at Ug= 30 cm/s | Glass bead- gas at Ug = 30 cm/s | Copper - gas at Ug= 30 cm/s | Copper - gas at Ug= 30 cm/s

Figure 4. Time averaged cross sectional gas and solid holdups distribution for glass bead-gas and copper particle- gas systems respectively at 30 cm/s.

Cross-sectional gas holdup | Cross-sectional solid holdup | Cross-sectional gas holdup | Cross-sectional solid holdup

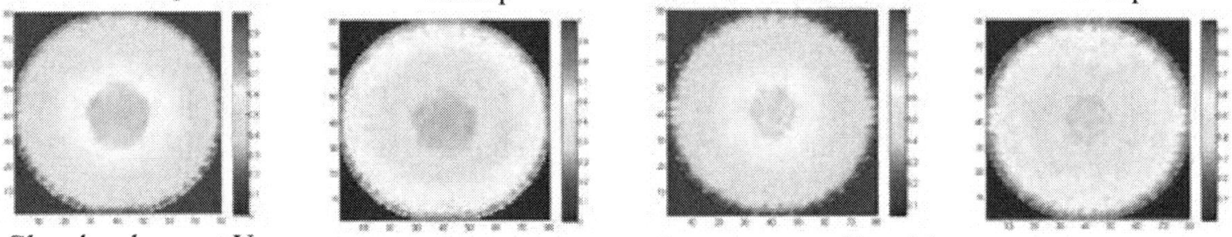

Glass bead- gas at Ug= 35 cm/s | Glass bead- gas at Ug = 35cm/s | Copper - gas at Ug= 35 cm/s | Copper - gas at Ug= 35 cm/s

Figure 5. Time averaged cross sectional gas and solid holdup distribution for glass bead-gas and copper particle- gas systems respectively at 35 cm/s.

4.1.2 Radial gas and solid holdup distributions.

Figure 6(a, b) shows the time-averaged radial gas and solid holdup profiles obtained by averaging of the data at H/D =1.7 (fully development region) for both glass bead–gas and copper particle–gas systems, respectively at 30cm/s superficial gas velocity. Fig. 7(a, b) shows the time-averaged radial gas and solid holdup profiles for both glass bead–gas and copper particle–gas systems, respectively at 35 cm/s superficial gas velocity. One can see that the radial gas holdup profiles are similar in shape. As the density of the material decreases, the gas hold up increases. The copper particle bed is shown to have lower gas holdup than glass bead bed which exhibited a higher gas holdup. The same trends were reported earlier by Nathan P, et al. 2011.

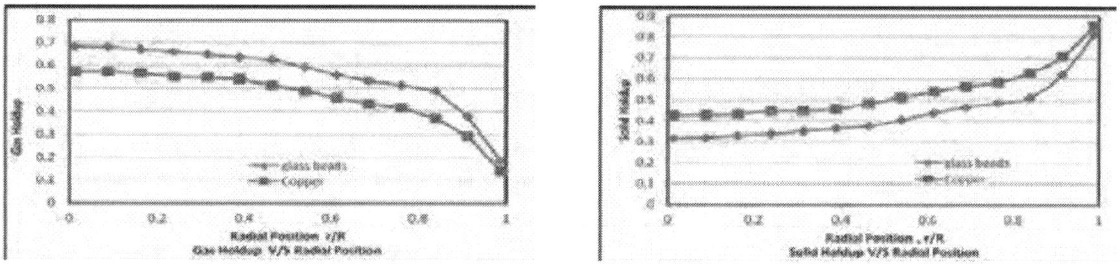

(a) (b)

Figure 6. Radial profile of (a), gas Holdup for glass beads and Copper particle, (b) solid holdup for glass beads and copper particle at Ug = 30 cm/s

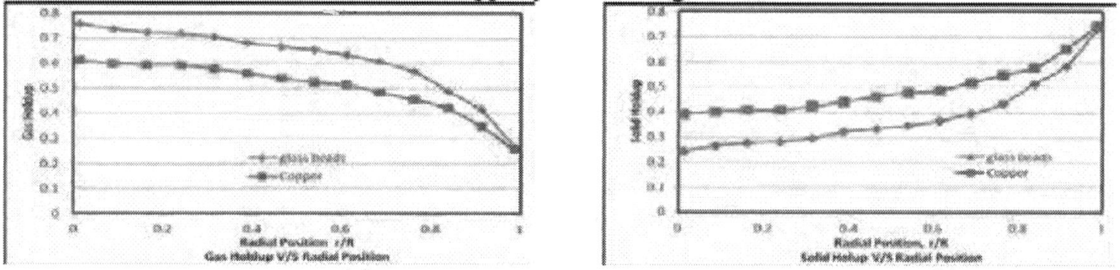

(a) (b)

Figure 7 . Radial profile of (a), gas Holdup for glass beads and Copper particle, (b) solid holdup for glass beads and copper particle at Ug = 35 cm/s.

4.2. Effect of H/D ratios on gas and solid holdups

4.2.1. Cross-sectional gas and solid holdups distribution

Measurements were taken at different heights H/D = 0.286, 0.64 and 1.7 above the distributor for 14cm fluidized bed, using an air- solid system, glass beads, 210μm as bed materials. The variation of the radial gas and solid holdups at various axial positions depended on the gas velocity, when the gas velocity increased, the gas holdup increased. The radial gas and solid holdup profiles obtained at several levels above the distributor at specific superficial gas velocities 25 cm/s are presented in Figure 8. It is clear that the solid holdup profile decreases by increasing H/D ratio and profiles have the same shape as was reported in the literature (Andreux et al., 2005; Mabrouk et al., 2005; Escudero et al. 2011)), with high solids holdup close to the wall and low solids holdup at the centre of the bed, (solid holdup + gas holdup =1).

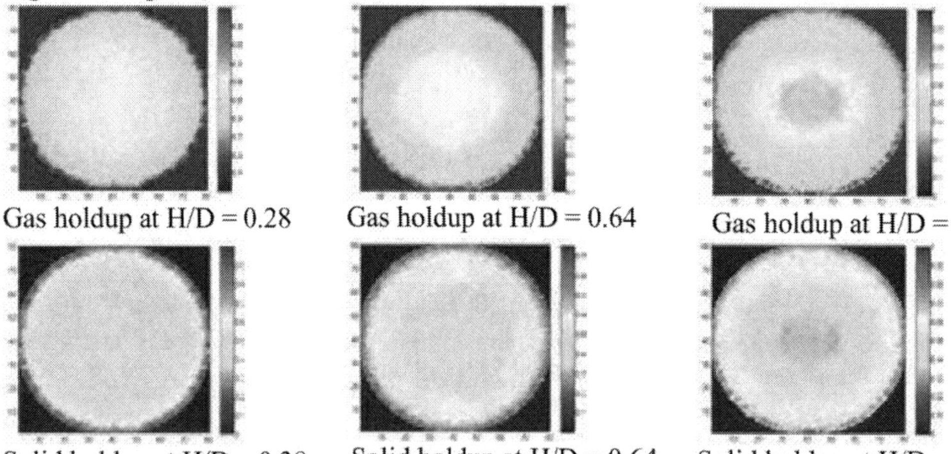

Gas holdup at H/D = 0.28 Gas holdup at H/D = 0.64 Gas holdup at H/D = 1.7

Solid holdup at H/D = 0.28 Solid holdup at H/D = 0.64 Solid holdup at H/D = 1.7

Figure 8. Time averaged cross sectional gas and solid holdup distribution for glass bead-gas at different H/D ratios, at *Ug = 25 cm/s.*

4.1.2 Radial gas and solid holdup distributions.

Figure 9(a, b) shows the time-averaged radial gas and solid holdup profiles obtained by averaging of the data at different H/D ratios for glass bead–gas system, at 25cm/s superficial gas velocity. It is clear that the gas holdup profile increases by increasing H/D ratio and profiles have the same shape as was reported in the literature (Mabrouk et al., 2005; Wang et al., 2008), with high gas holdup at the centre and low gas holdup near the wall, at H/D=0.286 which is the sparger region there was some different in the profile due the entrance effect.

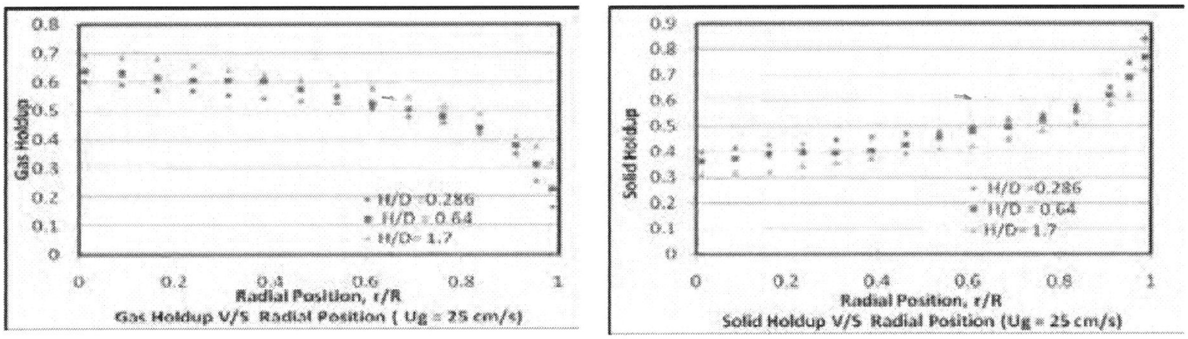

(a) (b)

Figure 9 . Radial profile of (a), gas Holdup for glass beads, (b) solid holdup for glass beads at different H/D ratios at Ug = 25 cm/s.

5. Conclusions

This study is part of much more extensive study and many lab experiments. Time-averaged gas and solid holdup distributions were measured in a 0.14m fluidized bed column using the CT scanner at different H/D ratios. In order to investigate the effect of bed height and particle density on gas and solid holdups distribution, glass bead and copper particle were used as the bed materials. Local time-averaged gas holdup is a function of bed material density, the two materials exhibit a similar fluidization structure, with region of higher gas holdup in the center, while region of low gas holdup (higher solids concentration) near the wall, confirming the results described by Mabrouk et al. 2005. The gas holdup increases with the increase in H/D ratio. It also seems that while H/D ratio significantly affects fluidization hydrodynamics, it appears that changes in H/D ratio do not significantly affect fluidization symmetry as it can be seen in Figures (8 and 9), the copper particle bed is shown to have lower gas holdup than glass bead bed which exhibited a higher gas holdup.

ACKNOWLEDGMENT

The authors wish to acknowledge the Libyan Ministry of Education for sponsoring the primary author's study within the Chemical Engineering Department at Missouri University of Science and Technology. The Gamma ray facility used in this research was funded by Dr. Al-Dahhan.

REFERENCES

1. Martin rüdisüli, tilman j. Schildhauer,serge m.a. Biollaz j. Ruud van ommen, (2012), *Scale-up of bubbling fluidized bed reactors* — A review. Powder Technology Volume 217, pp 21–38.

2. N.P. Franka,(2008), *Visualizing fluidized beds with X-rays*, Master's Thesis, Iowa State University, Ames, IA.

3. Franka, N.P., Heindei, T.J., (2009). *Local time-averaged gas holdup in a fluidized bed with side air injection using X-ray computed tomography*. Powder Technology 193 (1). 69-78.

4. N Hilal, D.J Gunn, (2002), *Solid Hold Up in Gas Fluidized Beds*. Chemical Engineering and Processing: Process Intensification, Volume 41, Issue 4, Pages 373-379.

5. N. Hilal, M.T. Ghannam, M.Z. Anabtawi, (2001), *Effect of bed diameter, distributor and inserts on minimum fluidization velocity*, Chemical Engineering & Technology 24 (2) 161–165.

6. Varma R, Bhusarapu S, O'Sullivan JA, Al-Dahhan MH. A, (2008), *comparison of alternating minimization and expectation maximization algorithms for single source gamma ray tomography*. Measurement Science & Technology, Vol 19(1):015506, 14pp.

7. N. Mosfoufi and J. Chaouki, (2000), *The Axial Movement of Solids in Gas-Solid Fluidized Beds*. I ChemE, Vol 78, Part A.

8. Mabrouk R., Radmanesh R., Chaouki J., Guy C, (2005), *Scale Effects on Fluidized Bed Hydrodynamics*. Inter. J. of Chemical Reactor Eng.

9. King, D.F,(1989), *Estimation of Dense Bed Voidage in Fast and Slow Fluidized Beds of FCC Catalyst.* In Fluidization VI Conference Proceedings", Alberta, Canada, 1-8.

10. Bhusarapu S, Al-Dahhan MH, Dudukovic MP, (2006), *Solids Flow Mapping in a Gas-Solid Riser: Mean Holdup and Velocity Fields*. Powder Technology, Vol 163, (1-2), pp. 98-123.

11. David Escudero, Theodore J. Heindel, (2011), *Bed Height and Material Density Effects on Fluidized Bed Hydrodynamics*.Chemical Engineering Science 66, 3648-3655.

12. Grassler, T., Wirth, K.E., (2000), *X-ray computer tomography- potential and limitation for the measurement of local solids distribution in circulating fluidized beds*. Chemical Engineering Journal 77 (1), 65-72.

13. Kumar, S. B., D. Moslemian, and M. P. Dudukovic, (1995), *γ- ray Tomographic Scanner for Imaging Voidage Distribution in Two-Phase Flow Systems*,' Flow Meas. Instrum, 6(1), 61.

14. Nathan P, Franka, Theodore J. Heindel., (2011), *Local time- averaged gas holdup in a fluidized bed with side air injection using X-ray computed tomography*. Chemical Engineering Science 66, pp. 3648-3655.

15. McCabe W.E., J.C. Smith, and P. Harriott, (2001): Unit Operations of Chemical Engineering, McGraw Hill, New York.

16. Schrock VE.(1969) *Radiation attenuation techniques in two phase flow measurements*. In: Two-Phase Flow Instrumentation, 11th National ASME/AIChE Heat Transfer Conference. Minneapolis, USA. PP. 24–35.

385710 Palladium Theory for Aqueous-Phase Catalyzed Cu-Free Sonogashira Couplings

Abstract:

Flow chemistry, important to the global chemicals and materials sectors, depends on the symbiosis between chemical reaction engineering and organic synthesis. Chemistry and engineering principles are needed to design therapeutics and microscale devices for a broad cross-section of society problems. Studying chemical reactions on-chip has advantages in fine chemicals and pharmaceuticals research and manufacturing. Engineering their continuous manufacture with microchemical systems reduces discovery times, solvent waste consumption, and the energy requirements for their preparation at laboratory and production scales, as compared to conventional batch-wise processing. Significant quantities of chemical waste are generated in fine chemicals and pharmaceuticals manufacture relative to the mass of the reaction products formed that ultimately save lives and improve quality of life. Most reactions involve solids as starting materials or products. Handling solids in microsystems, perhaps the grandest challenge in the field, unveils opportunities for innovations of active particle handling techniques. Alternatively, designing reactions that ensure solubility potentially avoids solids handling challenges. Synthesis by aqueous-phase catalyzed couplings of aryl heteroatoms using palladium and hydrophilic ligand is a broad, versatile, and growing science. As an example, the kinetics of an aqueous phase Pd-catalyzed, Cu-free Sonogashira coupling were investigated. Evaluating the multiphase aqueous-organic reaction requires understanding of liquid-liquid transport, which was accomplished by assessing the relative ratio of the reaction rate to the diffusion rate through thin films: the Hatta number. Our discovery that both deprotonation and carbopalladation mechanisms accurately describe the kinetics undergirds the importance of considering transport phenomena *via* reactor design principles in laboratory-scale batch and flow synthetic chemistry. Discovery of ligand theory for such reactions not only enables accurate kinetic predictions, but it also broadly impacts ligand-transition metal interactions. The palladium to ligand ratio and its magnitude relative to substrates is a global quantity in cross-coupling reactions. Activation barriers similar in magnitude to density functional theory calculations of purely organic synthesis support that water as a reaction solvent could revolutionize the continuous flow manufacture of fine chemicals and pharmaceuticals by accelerating the kinetics and reduce solvent consumptions, while maintaining the solubility of inorganic salt by-products. Partnerships between scientists and engineers are needed to drive innovations in flow chemistry using microchemical systems and chemical reaction engineering is key.

Flow Regime Identification in Moving Bed Upflow Hydrotreating Reactor by Gamma Ray Densitometry

, Vineet Alexander[1], Mohd Fitri Abdul Rahman[1] and Muthanna Al-Dahhan[*]

[1,*]Chemical Engineering and Biochemical Engineering Department
Missouri Science and Technology University, Rolla, MO 65409-1230. USA

Moving bed upflow hydrotreating reactor can be used to process heavier feeds with higher level of contaminants, to increase catalyst cycle and life by better protection from fouling metals of the feedstock and to improve in general the product quality. The problem associated with these reactors is maldistribution, which causes hotspots, sintered carbon deposition and reduces expected conversions. To address such problems, detailed studies to enhance the understanding of the hydrodynamics in the moving bed upflow reactor are still required. In this work, the possibility of flow regime identification and monitoring by use of gamma ray densitometry (GRD) was examined in scale down cold flow moving bed upflow reactor. The experimental work was carried out in 11 inch(ID) Plexiglas column for an air-water system flowing over a packed bed of extrudate catalyst of size 3mm. The Moving bed upflow reactor was scanned by a gamma ray radioactive source of Caesium-137 and detected by a NaI(Tl) detector. The scan was conducted in various axial and radial positions with the superficial liquid flow rate in the range 0 to 0.47 cm/sec and superficial gas flow rate in the range 0 to 8.808cm/sec covering fixed bed, bubbly to turbulent flow regime. The statistical analysis in terms of variance and standard deviation of the measured time series of the received counts (Penetrated gamma ray) has been performed to identify flow regime. This method is very beneficial as it is noninvasive and gives better results than invasive method, such as pressure drop and optical probe. The data analysis is also quite fast and accurate as a matlab program to do the analysis has been developed to perform. The results obtained shows that this technique can be used to monitor and identify flow regime transition in moving bed upflow reactor and other type of packed bed reactors.

Keywords: Upflow moving bed reactor, Gamma Densitometry, Flow regime.

374757 Scalability of Mass Transfer in Gas-Liquid Taylor Flow in Capillaries

Abstract:

Chemical reactions between components present in gas and liquid phase represent an important class of reactions in chemical industry. The rate of such gas-liquid reactions is often limited by mass transfer, which depends on the hydrodynamics of the gas-liquid flow. In order to correctly predict the transport processes in these multiphase systems, a detailed knowledge of the transport mechanisms on different scales is needed.

In the past, batch systems were widely used for this kind of reactions, but the research efforts over the past decade resulted in microstructured devices for flow chemistry, which provide several advantages over these conventional reaction systems. In micro-devices the small dimension of the channels results in a significant higher specific interfacial surface area between the gas bubbles and the liquid phase, which results in an increased mass transfer coefficient compared to conventional systems. However, micro-reactors are characterized by an inadequate throughput for industrial applications, for this reason scale-up of these systems is necessary. In this light, the ACS-GCI-Pharmaceutical Roundtable has specifically identified the demand for novel concepts for continuous reaction systems.

In this work two-phase flow in circular capillaries with three different diameters (0.5 mm, 1.55 mm and 3.2 mm) is studied experimentally, where water is used as the liquid phase and CO_2 as the gas phase. The study of Taylor flow in single straight capillaries is particularly interesting because it allows to obtain accurate data that relate mass transfer coefficients to the fluid-dynamic conditions inside the channel. The range of capillary diameters in this work is chosen in order to investigate system hydrodynamics and transport processes on both micro- and milli-scale (based on the Laplace constant criteria).

The transparent capillaries allow optical access in order to collect images by means of a high-speed camera, which will then provide information about the bubble generation frequency, bubble size distributions, and reaction conversion. The absorption of CO_2 in an alkaline solution is particularly suitable as a model reaction as it occurs in the absence of a catalyst, and the conversion of the reactants can be locally monitored by the decrease in bubble size, and globally through the change in pH.

These measurements allow a quantification of the mass transfer coefficient k_La, and the experimental results are compared with literature results, and in addition we will present correlations to predict the values of k_La for both the milli- and the micro-scale capillaries for a range of Capillary numbers between 10^{-2} and 10^{-4} and Reynolds numbers Re<103. The obtained results will contribute to quantifying the effect of two-phase flow hydrodynamics at each scale on the mass transfer coefficients and their scalability across several orders of length scale.

Invariant Relationships for Heterogeneous Chemical Reaction Systems in Open Reactors

S. Srinivasan, J. Billeter and D. Bonvin

Laboratoire d'Automatique, Ecole Polytechnique Fédérale de Lausanne
EPFL – Station 9, 1015 Lausanne, Switzerland
sriniketh.srinivasan@epfl.ch

Many chemical and biochemical industries utilize chemical reaction processes to get desired products from raw materials. Common examples include reaction systems used to manufacture pharmaceutical drugs, vaccines and other chemicals. A chemical reaction system is a complex combination of various rate processes. Apart from the chemical reactions, these systems may include (i) mass transfers between phases, and (ii) heat transfer due to heating and cooling. Also, the reactor can be operated in a continuous mode, which adds additional mass transport due to the inlet and outlet streams. The measured state variables, namely concentrations, temperature and mass, are functions of the underlying reactions, mass transfer between phases, and mass transport due to the inlet and outlet streams. Since these variables are highly coupled and contain effects of all the phenomena, the analysis of reactor performance based on these variables is highly complex.

Proper understanding of the reaction system is necessary for process design, control and optimization. The analysis of the chemical reaction systems can be simplified, if a transformation can be made from the measured state variables into alternate variables (named *variants*) that each describe the dynamic behavior of the reactions, mass and heat transfers, inlets and outlets. The transformed states also include variables that are *invariant* with respect to time and remain constant during the course of the reaction. A number of applications of the reaction variants and invariants have been studied in the literature [1-7].

Srinivasan et al. [1] have discussed possible applications of reaction and flow variants/invariants for control-related tasks such as model reduction, state accessibility, state reconstruction, and feedback linearizability. Reaction invariants have been used to study the state controllability and observability of continuous stirred-tank reactors [2]. Reaction invariants have also been used to automate the formulation of mole balance equations for the non-reacting part of complex processes (mixing and splitting operations), thereby determining the degrees of freedom for process synthesis [3]. Furthermore, Waller and Makila [4] demonstrated the use of reaction invariants to control pH, assuming that equilibrium reactions are very fast. Gruner et al. [5] showed that, through the use of reaction invariants, the dynamic of reaction-separation processes with fast (equilibrium) reactions resembles the dynamic of corresponding non-reactive systems in a reduced set of transformed variables. Aggarwal et al. [6] considered multi-phase reactors operating at thermodynamic equilibrium and were able to use the concept of reaction invariants, which they called invariant inventories, to reduce the order of dynamic models and to design control strategies accordingly. Recently, Scott and Barton [7] have used reaction invariant relationships to compute bounds for kinetic models of a chemical reaction system operated under batch conditions. While the applications of reaction variants and invariants are clear, the challenge still remains in finding these relationships in the presence of different physical phenomena.

Asbjornsen and co-workers [8] introduced a methodology for computing reaction variants and invariants and used it for reactor modeling and control. However, for open reactors, the reaction variants proposed are not fully representative of reactions since they are also affected by the inlet and outlet flows. Friedly [9] proposed to compute the variants of equivalent batch reactions, associating the remainder to transport processes, and used them to describe the dynamics of flow through porous media accompanied by chemical reactions. For open homogeneous reaction systems, Srinivasan et al. [1] developed a nonlinear transformation of the numbers of moles to reaction variants, flow variants, and reaction and flow invariants, thereby separating the effects of reactions and flows. Later, the same authors refined that transformation to make it linear [10] (at the price of losing the one-to-one property) and therefore more easily applicable. They also showed that, for a reactor with an outlet flow, the concept of vessel extent is most useful, as it represents the amount of material associated with a given process (reaction, exchange) that is still in the vessel. Bhatt et al. [11] extended that concept to heterogeneous G-L reaction systems for the case of no reaction and no accumulation in the film, the result being decoupled vessel extents of reaction, mass transfer, inlet and outlet, as well as true invariants, i.e. identically equal to zero. The use of this linear transformation for the task of kinetic identification has been clearly demonstrated [12] and this transformation has been extended to typical measured signals (calorimetry and spectroscopy) [13, 14].

Srinivasan et al. [15] have recently introduced a new transformation that gives the same variants and invariants as the one proposed in [10]. This new transformation, which is conceptually and computationally much simpler, also gives explicit relationships for reaction invariants in terms of the measured state variables. The invariant relationships from the transformation can now directly be used for process monitoring, control and optimization. This transformation also allows computing the reaction invariants in the presence of additional measurements (e.g. flow rates) and can be extended to heterogeneous systems.

This contribution will first extend the computation of reaction invariants of homogeneous reaction systems when additional measurements (flow rates) are available. The procedure for writing the invariant relationships will then be adapted to F–F reaction systems with (i) reactions in the bulk of both phases, (ii) unsteady–state mass transfer between phases, and (iii) mass transport due to inlet and outlet streams in both phases. The procedure for writing the reaction invariant relationships will be demonstrated with suitable examples.

References

[1] B. Srinivasan, M. Amrhein, D. Bonvin, "Reaction and flow variants/invariants in chemical reaction systems with inlet and outlet streams", AIChE J. 44(8), 1858-1867, **1998**.

[2] M. Fjeld, O. A. Asbjornsen, K. J. Astrom, "Reaction invariants and their importance in the analysis of eigenvectors, state observability and controllability of the continuous stirred tank reactor", Chem. Eng. Sci. 29, 1917-1926, **1974**.

[3] S. B. Gadewar, M. F. Doherty, M. F. Malone, "Reaction invariants and mole balances for plant complexes", Ind. Eng. Chem. Res. 41, 3771-3783, **2002**.

[4] K. V. Waller, P. M. Makila, "Chemical reaction invariants and variants and their use in reactor modeling, simulation, and control", Ind. Eng. Chem. Process Des. Dev. 20, 1-11, **1981**.

[5] S. Gruner, M. Mangold, A. Kienle, "Dynamics of reaction separation processes in the limit of chemical equilibrium", AIChE J. 52, 1010-1026, **2006**.

[6] M. Aggarwal, S. Balaji, B. E. Ydstie, "Invariant-based modeling and control of multi-phase reactor systems", J. Process Contr. 21, 1390-1406, **2011**.

[7] J. Scott, P. I. Barton, "Tight, efficient bounds on the solutions of chemical kinetic models", Comp. Chem. Eng. 34(5), 717-731, **2010**.

[8]O. A. Asbjornsen, "Reaction invariants in the control of continuous chemical reactors", Chem. Eng. Sci. 27, 709-717, **1972**.

[9] J. C. Friedly, "Extent of reaction in open systems with multiple heterogeneous reactions", AIChE J. 37(5), 687-693, **1991**.

[10] M. Amrhein, N. Bhatt, B. Srinivasan, D. Bonvin, "Extents of reaction and flow for homogeneous reaction systems with inlet and outlet streams", AIChE J. 56, 2873, **2010**.

[11] N. Bhatt, M. Amrhein, D. Bonvin, "Extents of reaction, mass transfer and flow for gas-liquid reaction systems", Ind. Eng. Chem. Res. 49, 7704-7717, **2010**.

[12] N. Bhatt, N. Kerimoglu, M. Amrhein, W. Marquardt, D. Bonvin, "Incremental identification of reaction systems - a comparison between rate-based and extent-based approaches", Chem. Eng. Sci. 83, 24-38, **2012**.

[13] S. Srinivasan, J. Billeter, D. Bonvin, "Extent-based incremental identification of reaction systems using concentration and calorimetric measurements", Chem. Eng. J. 207-208, 785-793, **2012**.

[14] J. Billeter, S. Srinivasan, D. Bonvin, "Extent-based kinetic identification using spectroscopic measurements and multivariate calibration", Anal. Chim. Acta 767, 21-34, **2013**.

[15] S. Srinivasan, J. Billeter, D. Bonvin, "Variant and invariant states for reaction systems", 1st IFAC Workshop on Thermodynamic Foundations of Mathematical Systems Theory, Lyon, **2013**.

364284 Nanostructured Size-Selective Catalysts Prepared By Molecular Layer Deposition

Abstract:

Heterogeneous catalysts are widely used. However, heterogeneous catalysts cannot selectively convert specific molecules in the reactant mixture to catalyze only desired reactions. Size-selective catalysts with a metal core and porous oxide shell have a promising structure that can increase the reaction selectivity through reactant molecular discrimination. This paper describes a general method to create porous metal oxide films with well-defined porous structures and precisely controlled thickness down to several angstroms from dense organic/inorganic hybrid metal alkoxide films grown by molecular layer deposition (MLD). A novel nanostructured catalyst with an ultra-thin porous shell obtained from the thermal decomposition of an aluminum alkoxide film deposited by MLD for size-selective reactions was developed. The molecular sieving capability of the porous metal oxide films was verified by examining the liquid-phase hydrogenation of n-hexene versus cis-cyclooctene. The success of making these materials by MLD opens up a new method for preparing size-selective catalysts.

浙江大学
ZHEJIANG UNIVERSITY

Platinum Nanoparticles Supported on Multiwalled Carbon Nanotubes Encapsulated by Cerium Oxide: A Highly Reactive and Stable Catalytic System

Dang-guo Cheng
College of Chemical and Biological Engineering,
Zhejiang University, Hangzhou 310027, China

Content

浙江大学
ZHEJIANG UNIVERSITY

- Background
- Catalyst preparation
- Characterizations and Performance
- Summary
- Acknowledgement

Background

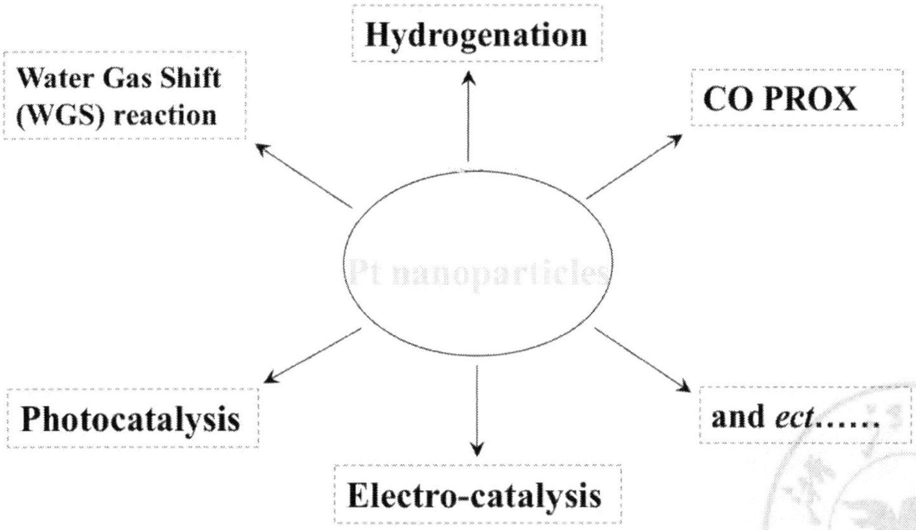

Background

Supported Pt nanoparticles catalyst

- Metal oxide: CeO_2, TiO_2, Al_2O_3

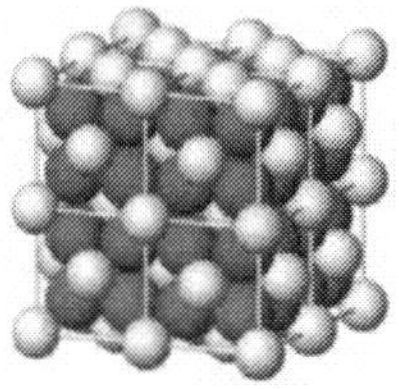

- Zeolites: HZSM-5, HY, SBA-15

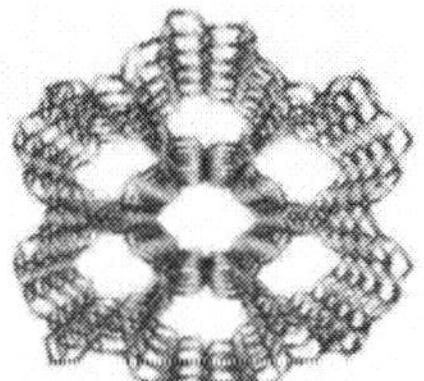

- Carbon material: Nanofiber, Nanotube

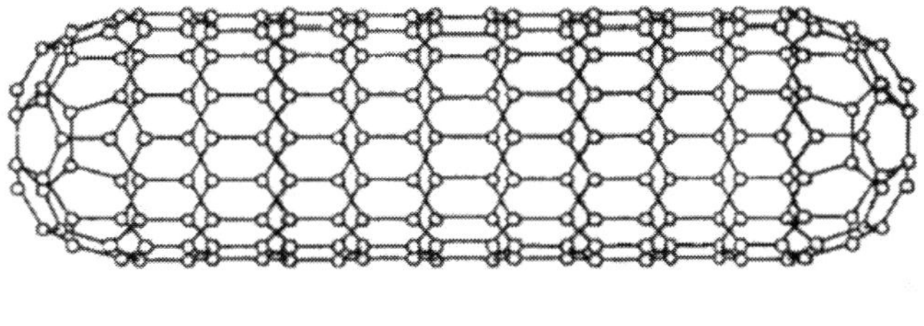

Background

- Aggregation
- Sintering
- Leaching

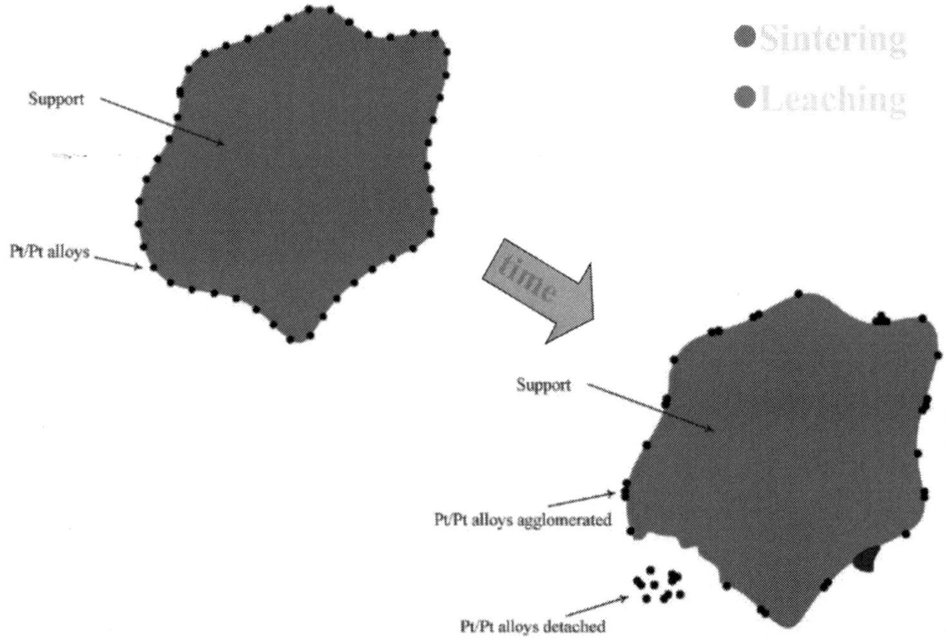

Background

Strategies to prevent the aggregation, sintering, and leaching

- **Exploitation of the strong metal–support interaction**
- **Formation of multi-metallic alloys**
- **Addition of co-catalysts**

Alternative method: Encapsulation

- **To encapsulate NPs in a thermally stable inorganic shell layer to prevents their mobility under reaction condition**

Catalyst preparation

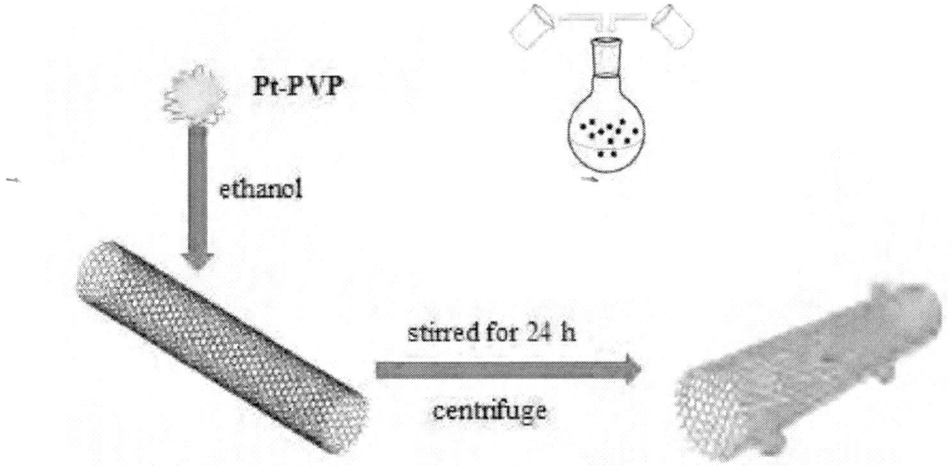

Catalyst preparation

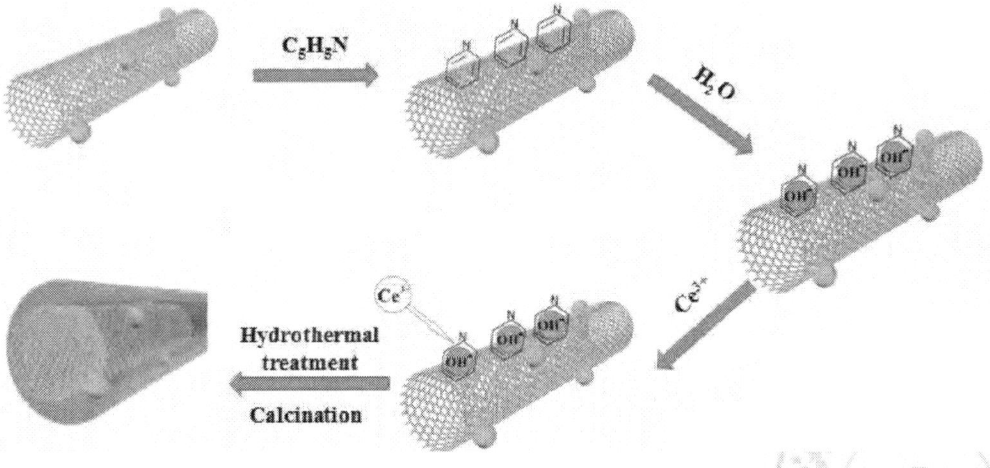

Characterizations and Performance

浙江大學 ZHEJIANG UNIVERSITY

Pt-PVP, Polyol reduction method

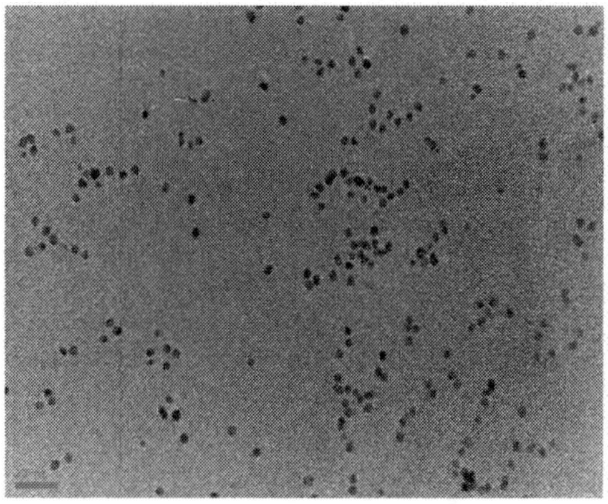

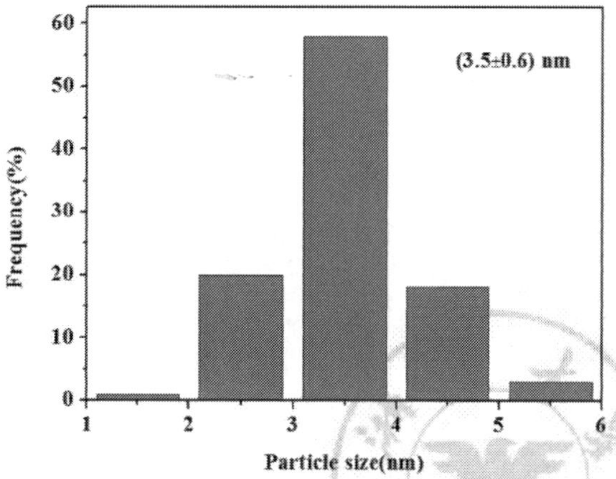

Characterizations and Performance

浙江大學 ZHEJIANG UNIVERSITY

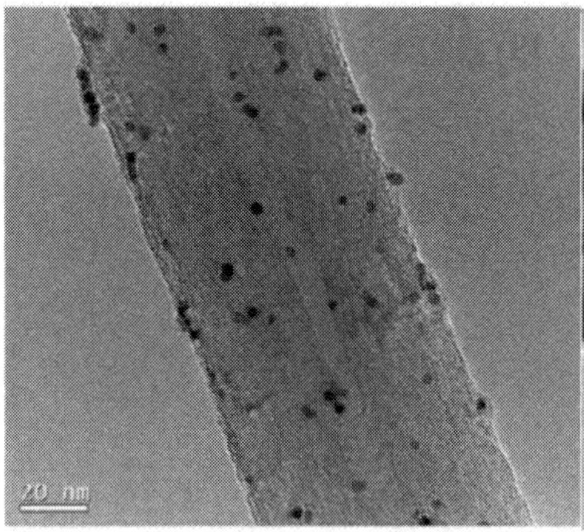

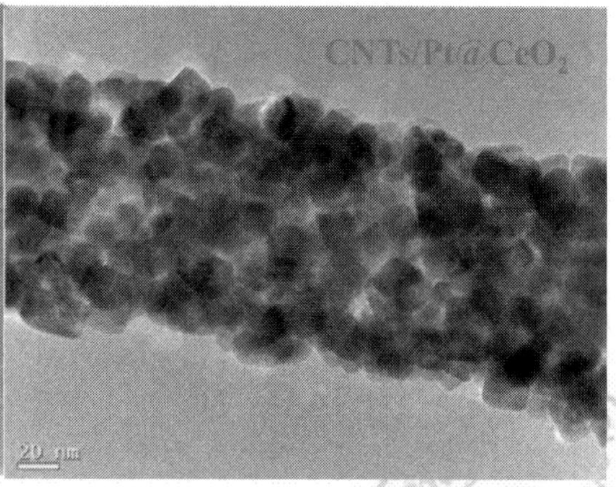

Characterizations and Performance

ZHEJIANG UNIVERSITY

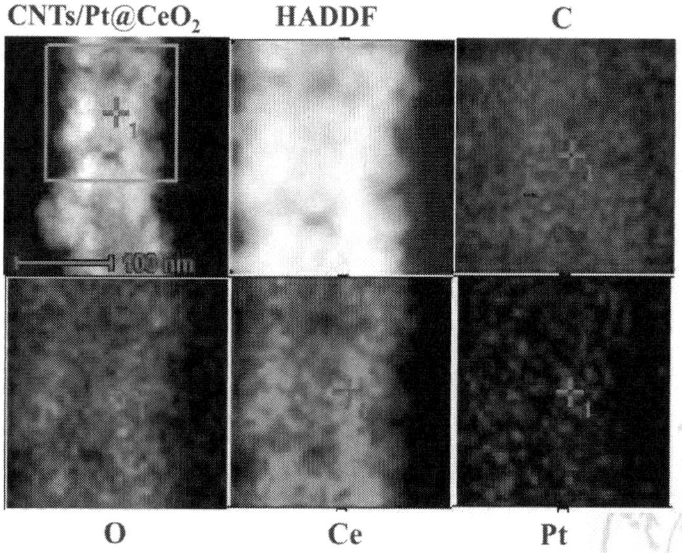

Characterizations and Performance

ZHEJIANG UNIVERSITY

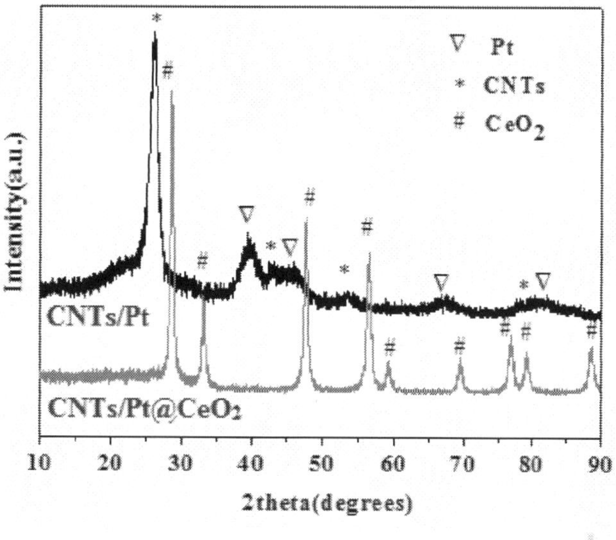

Characterizations and Performance

Sample	$S_{BET}(m^2g^{-1})$	Pore Volume(cm^3g^{-1})	Pore Diameter (nm)
CNTs	110	0.38	3.2
CNTs/Pt	72	0.24	3.4
CNTs/Pt@CeO_2	47	0.17	2.8/3.8

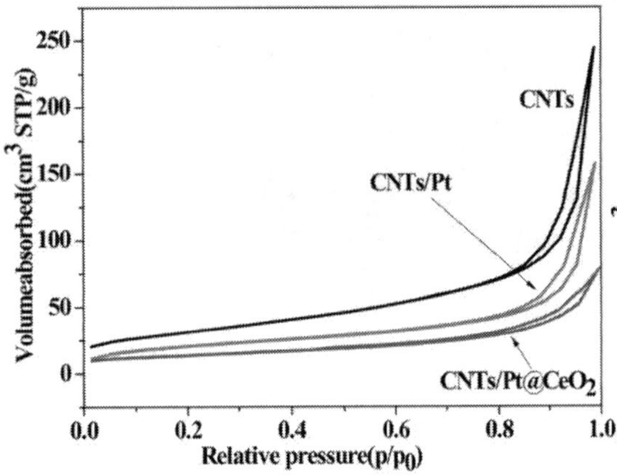

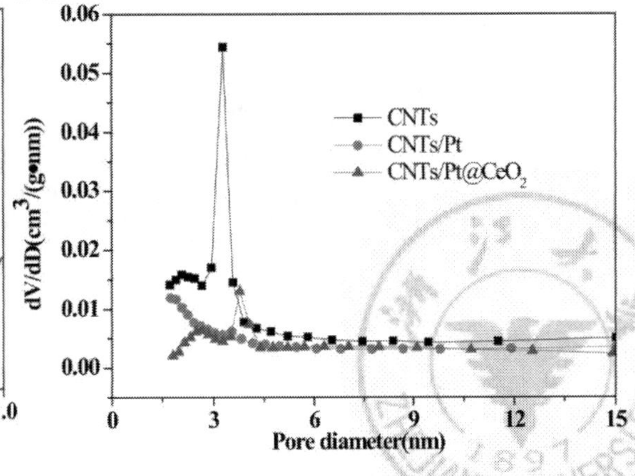

Characterizations and Performance

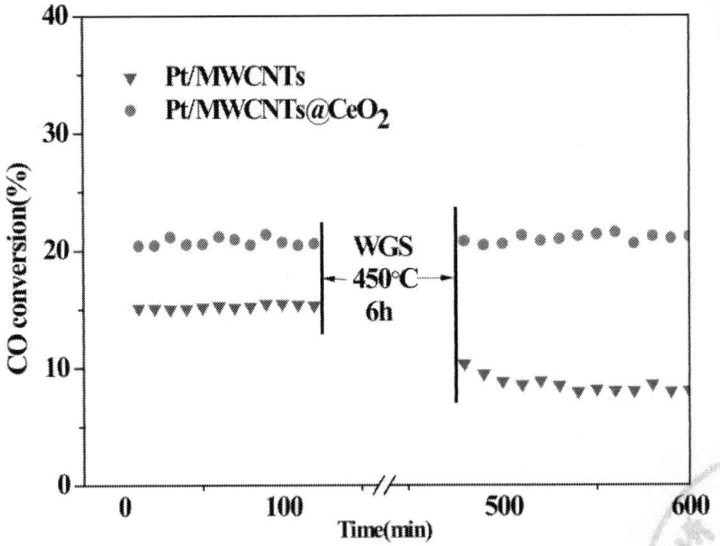

Reaction conditions: 250°C; 0.05 g catalyst; 80 ml/min: 1.0 vol % CO, 3.0 vol % H_2O, diluted in Ar

Summary

浙江大学 ZHEJIANG UNIVERSITY

- CNTs/Pt@CeO_2 catalysts have been synthesized successfully by the Pyridine-thermal synthesis methods.
- Compared with CNTs/Pt, both the activity and stability of CNTs/Pt@CeO_2 catalysts have been enhanced in WGS reaction.
- Further studied need to confirm the existence of the interfacial effect and nanometer effect in CNTs/Pt@CeO_2 catalysts

371281 Using Active Site Confinement and Organic Ligand Approaches to Control Catalysis

Abstract:

The local environment of a catalyst's active site plays a critical role in determining the catalyst's activity and selectivity. This effect is seen most profoundly in enzymatic systems where the environment of the active site is controlled through the chemical entities surrounding the active site and their precise geometric locations.

Inspired by the GroEL/GroES system in *E. coli*, examples with a metal complex/cluster that is immobilized inside of mesoporous materials, particularly SBA-15, will be discussed. These nano-confined active sites demonstrate an increased degree of activity and stability which provide valuable insight into structure-function relationships in catalysis. [1] [2]

Additionally, in different gold cluster systems, we explore approaches to controlling the environment that surround these sites as well as their ultimate effects on catalysis. The use of organic ligands, such as calixarenes, is investigated to confine and to control catalytically active noble metal sites on metal oxide supports and in homogeneous solution, using an approach that is conceptually similar to that used by enzymes to control reactivity through organic side chains acting as ligands. Gold catalysts have been synthesized with calixarene ligands bound to the metal nanoparticle, confining the gold cluster, yet these clusters still demonstrate reactant accessibility to the metal nanoparticle surface as measured by chemisorption probe experiments. [3] [4] Catalytic activity has been demonstrated for these systems using different reaction systems such as reduction reactions on these gold catalysts. Results provide a unique perspective into the structure-activity relationships present in gold cluster catalysis.

References:

[1] F. Marras, J. Wang, M.-O. Coppens, J. N. H. Reek. Chem. Comm. **2010**, *46*, 6587.

[2] L.-C. Sang, A. Vinu, and M.-O. Coppens, *Phys. Chem. Chem. Phys*, **2011**, *13*, 6689.

[3] N. de Silva, J.-M. Ha, A. Solovyov, M. M. Nigra, I. Ogino, S. W. Yeh, K. A. Durkin, A. Katz. *Nature Chem.* **2010**, *2*, 1062.

[4] M. M. Nigra, A. J. Yeh, A. Okrut, A. G. DiPasquale, S. W. Yeh, A. Solovyov, A. Katz. *Dalton Trans.* **2013**, *42*, 12762.

Sinter-resistant metal nanoparticles encapsulated by zeolite nanoshell

Shilpi Saxena[a], Rupesh Singh[a], Raj Ganesh S. Pala[a] and Sri Sivakumar[a,b]

shilpi@iitk.ac.in, rupeshs@iitk.ac.in, rpala@iitk.ac.in, srisiva@iitk.ac.in

[a]Department of Chemical Engineering, Indian Institute of Technology Kanpur, (UP), India.
[b]Material Science Programme, DST Thematic Unit of Excellence on Soft Nanofabrication, Centre for Environmental Science & Engineering, Indian Institute of Technology Kanpur, (UP), India.

KEYWORDS: nanoshell, sinter-resistant, gold nanoparticles, ZSM-5 and MCM-22.

Nanocatalysts have distinct activity and selectivity, which can be achieved by controlling their size, shape[1], surface composition, and electronic structure. In particular, gold nanoparticle possesses unique catalytic properties when they are supported on metal oxides/carbon and are utilized for various reactions such as CO oxidation[2], acetylene hydrochlorination[3], hydrogenation[4], selective oxidation[5] and epoxidation[6]. However, they may suffer from sintering and subsequent loss in activity due to harsh reactions conditions (e.g. high temperature, pressure, solvents etc.). This can be circumvented by encapsulating the gold nanoparticles with different shell materials such as silica, mesoporous carbon, and polymer shell. Although such approaches decrease sintering, the protective shell does not augment the product selectivity of the composite "nanoparticle core"-"protective shell" catalysts. Towards this goal, the present work explores and demonstrates the feasibility of structured encapsulants with well-defined pores to suppress sintering of encapsulated nanoparticles. We propose a general hydrothermal based synthetic strategy to assemble bifunctional catalysts in which a single gold nanoparticle is encapulated by a variety of zeolites-nanoshell (e.g. MCM-22 and ZSM-5). Gold nanoparticles were synthesized by reduction of chloroauric acid by sodium citrate[7]. These nanoparticles were coated with silica (denoted as Au@SiO2) followed by transformation into Au@MCM-22 and Au@ZSM-5 by modified hydrothermal synthesis, by varying amount of Si/Al. The zeolite nanoshells are of uniform shell thickness, morphology, possess high thermal stability and play the dual role of stabilizing the nanoparticles against sintering and their distinct pore structure can facilitate shape selective catalysis.

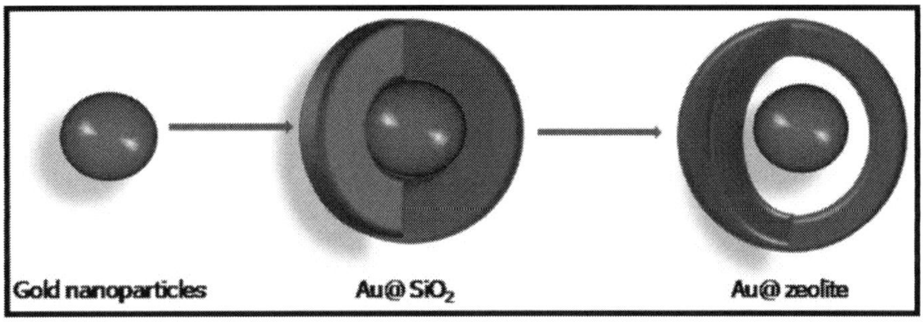

Schematic representation of synthesis of gold nanoparticle encapsulated by zeolite-shell.

The prepared catalyst was characterized using TEM (Transmission Electron Microscopy), SEM (Scanning Electron Microscopy), UV (Ultraviolet–visible spectroscopy), and XRD (X-ray diffraction pattern). TEM images of zeolite-coated gold nanoparticles confirm that gold nanoparticles are present inside the zeolite. XRD spectrum of MCM-22 and ZSM-5 matches with the respective JCPDS XRD datas. While, this general approach can be validated by many reactions, we investigate the selective oxidation of cyclohexane[8] as its product KA-oil (a mixture of ketone and alcohol) is an important intermediate in the production of nylon 6 and nylon-6,6. These gold nanoparticle encapsulated by zeolite nanoshell catalysts showed increased activity/selectivity for cyclohexane oxidation reaction at high temperature and pressure.

Following the same scheme other nanoparticles like Palladium, Platinum, Nickel, and Cobalt will also be coated with different zeolites to cater towards various reactions performed at harsh conditions.

References:

1. *Oliver-Meseguer, J., et al., Small Gold Clusters Formed in Solution Give Reaction Turnover Numbers of 107 at Room Temperature. Science, 2012. 338(6113): p. 1452-1455.*
2. *Haruta, M., et al., Low-Temperature Oxidation of CO over Gold Supported on TiO2, Î±-Fe2O3, and Co3O4. Journal of Catalysis, 1993. 144(1): p. 175-192.*
3. *Nkosi, B., et al., Hydrochlorination of acetylene using carbon-supported gold catalysts: A study of catalyst reactivation. J. Catal., 1991. 128(2): p. 378-386.*
4. *Mitsudome, T. and K. Kaneda, Gold nanoparticle catalysts for selective hydrogenations. Green Chemistry. 15(10): p. 2636-2654.*
5. *Turner, M., et al., Selective oxidation with dioxygen by gold nanoparticle catalysts derived from 55-atom clusters. Nature, 2008. 454(7207): p. 981-983.*
6. *Gajan, D., et al., Gold Nanoparticles Supported on Passivated Silica: Access to an Efficient Aerobic Epoxidation Catalyst and the Intrinsic Oxidation Activity of Gold. Journal of the American Chemical Society, 2009. 131(41): p. 14667-14669.*
7. *Ojea-JimeÌ□nez, I., et al., Small Gold Nanoparticles Synthesized with Sodium Citrate and Heavy Water: Insights into the Reaction Mechanism. The Journal of Physical Chemistry C. 114(4): p. 1800-1804.*
8. *Hereijgers, B.P.C. and B.M. Weckhuysen, Aerobic oxidation of cyclohexane by gold-based catalysts: New mechanistic insight by thorough product analysis. Journal of Catalysis. 270(1): p. 16-25.*

Controlling metal nanoparticle dispersion with 'nanocavity' oxide supports

Controlling growth of metal nanoparticle on oxide support during synthesis is one of the defining challenges of materials synthesis for catalysis. Photodeposition[1] and strong electric absorption (SEA)[2] methods are reported to generate highly dispersed Ag nanoparticles on TiO2 support and Pt nanoparticles on SiO2 support. But none of these methods can intrinsically prevent metal nanoparticles sintering in harsh reaction environment. Previously, researchers have reported using a second oxide overcoat to cover the supported metal nanoparticles and successfully enhancing their thermal stability[3]. However the complete over-coating approach can lead to loss of metal nanoparticle surface area, which eventually decreases catalytic activity of catalyst. Here we present new synthesis strategies in combining these ideas to template and stabilize supported metal nanoparticles inside oxide nanocavity materials. Following method from our previous report[4], oxide nanocavity materials are synthesized by grafting a sacrificial template molecule on a carrier particle (e.g. SiO_2 or TiO_2), then partially overcoating the surface with a second oxide using a self-limiting technique such as atomic layer deposition or sol-gel under reactant-limiting conditions. Finally, the templates are removed to create nanocavities. SiO_2 deposited over templates grafted TiO_2 would be for example, named as SiO_2-number of SiO_2 coating cycles-TiO_2. Recent work in our group shows these nanocavities on the oxide surface have characteristic dimension of ~1nm[4]. The as-synthesized nanocavities are then backfilled with metal nanoparticles by selective photoreduction or SEA. In the photodeposition case, Ag nanoparticles are selectively deposited onto exposed TiO_2 domains on TiO_2 support containing nanocavities by photoreduction with a sacrificial reductant. Results show that SiO_2-8cycles-TiO_2 has better dispersion of Ag nanoparticles and controlled particle size distribution (2.9 ± 0.6 nm by Transmission Electron Microscope (TEM)) compared to Ag nanoparticles on TiO_2 surface without nanocavities (4.7 ± 2.7 nm by TEM). Likewise, Ag nanoparticles on SiO_2-8cycles-TiO_2 show better sintering resistance compared to Ag photodeposited onto unmodified TiO_2. The Extended X-ray Absorption Fine Structure (EXAFS) results confirm Ag particle size increases from 7 to 8.9 nm for unmodified TiO_2 after sintered at 673 K for 3 hours, while Ag particle size decreases from 4.9 to 3.7 nm on SiO_2-8cycles-TiO_2.

In a second route, SEA is used to selective deposit $Pt(NH_3)_4^{2+}$ onto the exposed SiO_2 domains of Al_2O_3-10cycles-SiO_2 and deposit $PtCl_6^-$ onto the exposed TiO_2 domains of SiO2-8cycles-TiO_2. The former takes place at pH~9 and the latter at PH~3. For $Pt(NH_3)_4^{2+}$ deposition, preliminary results show when the nanocavity surface density is changed from 0.08 to 0.24 nanocavity/nm^2, Pt deposition increases from 0.05 to 0.1 complexes/nm^2. With this, we are looking to convert a workhorse technique for catalyst synthesis into tool for precise synthesis of robust, tunable, supported metal nanoparticles.

[1] Chan, S. C., & Barteau, M. A. (2005). Preparation of highly uniform Ag/TiO2 and Au/TiO2 supported nanoparticle catalysts by photodeposition. Langmuir, 21(12), 5588-5595.

[2] Miller, J. T., Schreier, M., Kropf, A. J., & Regalbuto, J. R. (2004). A fundamental study of platinum tetraammine impregnation of silica: 2. The effect of method of preparation, loading, and calcination temperature on (reduced) particle size. Journal of Catalysis, 225(1), 203-212.

[3] Lu, J., Fu, B., Kung, M. C., Xiao, G., Elam, J. W., Kung, H. H., & Stair, P. C. (2012). Coking-and sintering-resistant palladium catalysts achieved through atomic layer deposition. Science, 335(6073), 1205-1208.

[4] Canlas, C. P., Lu, J., Ray, N. A., Grosso-Giordano, N. A., Lee, S., Elam, J. W., ... & Notestein, J. M. (2012). Shape-selective sieving layers on an oxide catalyst surface. Nature chemistry, 4(12), 1030-1036.

Single-Walled Copper Silicate Nanotubes Doped with 3*d* Transition Metals for CO_2 Hydrogenation

Yuan Sheng[†] and Hua Chun Zeng[*, †, ‡]

[†]Department of Chemical and Biomolecular Engineering, National University of Singapore, 10 Kent Ridge Crescent, Singapore 119260. [‡]Institute of Materials Research and Engineering (IMRE), 3 Research Link, Singapore 117602

* Corresponding author, email: chezhc@nus.edu.sg

Transition metal silicate nanotubes have great potential in heterogeneous catalysis due to their large surface area and high metal loading. To achieve a fully accessible surface and versatile catalytic properties, a single-walled structure and tunable composition are highly desirable. However, copper silicate has so far been the only transition metal silicate to form single-walled nanotubes. As shown in Figure 1, in this work we developed a general method to synthesize a series of new single-walled silicate nanotubes (with an average outer diameter and wall thickness of 6 nm and 1 nm respectively) containing other 3*d* transition metal elements (M = Mn, Fe, Co, Ni and Zn) by wet-chemical doping of pre-synthesized copper silicate nanotubes; up to 80% of Cu can be replaced with these metals (M) through complex-assisted ion exchange, while the original tubular structure remains intact. These metal silicate nanotubes can serve as solid precursors to transform into Cu-based nanocatalysts under a reducing environment. Using CO_2 hydrogenation as a model reaction, enhanced activity and methanol selectivity were demonstrated with the doped catalysts at 240°C, 30bar. A specific activity calculated based on total mass of metals which is superior to an industrial $Cu/ZnO/Al_2O_3$ catalyst was also achieved.

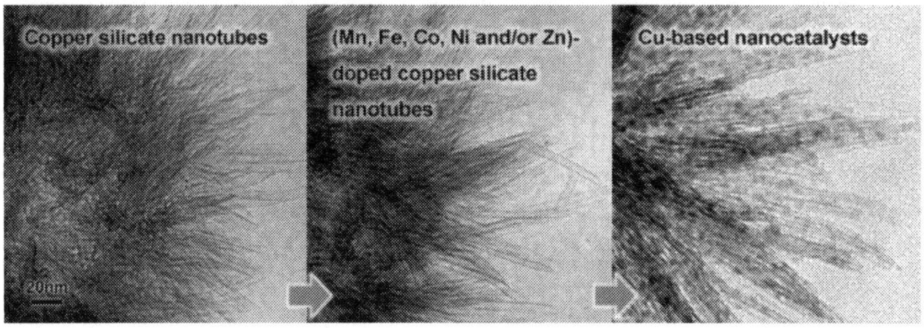

Figure 1. TEM images showing the typical morphologies of pristine and doped copper silicate nanotubes and active catalysts derived from them. The background is colored for visual effects. The scale is the same for the three images.

Driven by increasing costs of waste disposal and raw materials, catalysis research in the 21st century is shifting from a focus on activity to one on selectivity.[1] While breakthroughs in catalyst design research have provided great insights into how the rates of various steps in a reaction mechanism can be tuned by altering the catalyst composition,[2] tuning catalysts to discriminate between competing reactions that are chemically similar remains to be difficult. An example of such a reaction is the oxidation of an alkane to a 1-alcohol with high purity, i.e., with little production of 2- and higher alcohols, polyols, etc., which is important in the pharmaceuticals and specialty chemicals industries. In this contribution, we discuss our work on designing catalysts with unprecedented regioselectivity to produce 1-butanol from butane, accomplished by encapsulating a metal nanoparticle catalyst within a metal-organic framework (MOF).[3] MOFs are porous, crystalline solids built from metal or metal oxide "nodes" connected by organic "linker" compounds. The encapsulation provides a way to discriminate against reagents in a reaction environment and can be engineered to maximize selectivity. For example, the MOF pore diameters can be adjusted to only allow certain sized molecules to access the catalyst; likewise, the pore polarities can be configured to draw in polar molecules and push out non-polar molecules, or vice versa. Additionally, the pore networks can be designed to force reagents to interact with the catalyst in specific geometries (Figure 1). MOF-encapsulated catalysts have much to offer in terms of maximizing catalytic selectivity; however, little is known about the MOF/catalyst interface at the molecular level.

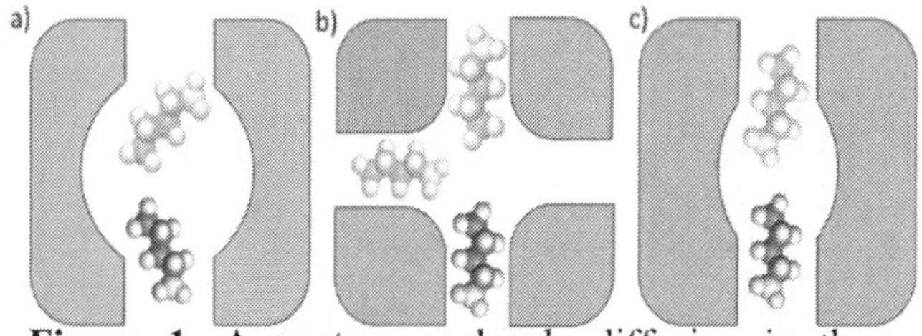

Figure 1. A pentane molecule diffusing in three different MOF pore networks. The desired reaction is pentane oxidation to 1-pentanol, i.e., oxidation of only the yellow carbon. All three scenarios only allow a terminal carbon to access the catalyst, thus preventing production of 2-pentanol, but only scenario c prevents the pentane from "turning around" in the pore, thus preventing over-oxidation to 1,5-pentanediol.

In this work, we use density functional theory calculations to investigate butane oxidation to 1-butanol,

$$2CH_3CH_2CH_2CH_3 + O_2 \rightarrow 2CH_2OHCH_2CH_2CH_3$$

We use "featureless" rings of He atoms to simulate the steric restrictions imposed by the MOF pore (Figure 2). Using climbing image-nudged elastic band calculations, we provide insight on 1) the optimal pore size that is required to activate butane but prevent oxidation of the secondary carbons, and 2) the catalyst composition that is required to activate butane, dissociate oxygen, and promote O-C bond formation, without over-dehydrogenating or over-oxidizing. Our results will feed future synthesis and design experiments to be performed by our collaborators.

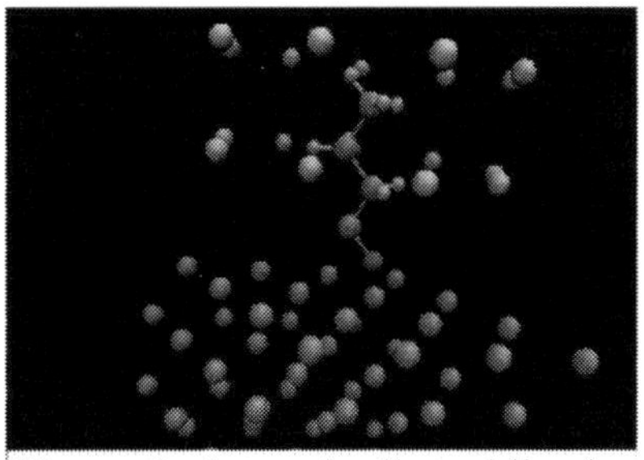
Figure 2. A terminally oxidized butane molecule adsorbed on a transition metal catalyst encapsulated by helium rings (pink=transition metal, red=O, blue=C, white=He, grey=H).

[1] Somorjai, G. A. and Rioux, R. M. "High technology catalysts toward 100% selectivity: Fabrication, characterization and reaction studies." *Catalysis Today*, 100, 2005, p. 201-205.
[2] Grabow, L. C. "Computational Catalysis Screening." In: Computational Catalysis, A. Asthagiri and M. J. Janik, Eds. RSC Publishing: Cambridge, UK, 2005.
and reaction studies. Catalysis Today, 100, 2005, p. 201-205.
[3] G. Lu, S. Li, Z. Guo, O. K. Farha, B. G. Hauser, X. Qi, Y. Wang, X. Wang, S. Han, X. Liu, J. S DuChene, H. Zhang, Q. Zhang, X. Chen, J. Ma, S. Chye, J. Loo, D. Wei, Y. Yang, J. T. Hupp, and F. Huo, "Imparting functionality to a metal–organic framework material by controlled nanoparticle encapsulation." *Nature chemistry*, 4, 2012, p. 310-316.

Predictive modeling for understanding biomass upgrade over metal surfaces

Catalytic conversion of biomass offers a potential route to alleviate the problem of obtaining fuels and value-added chemicals from non-renewable feedstocks. Unlike petroleum-based chemistry, biomass upgrade requires a costly removal of oxygen and is still poorly understood. Of special interest is the hydrogenolysis of furfural and hydroxymethylfurfural (HMF), the products resulting from hydrolysis and isomerization of sugars or biomass pyrolysis[1]. Recent experimental works suggest that transition metals have a profound effect on the hydrogenolysis activity and selectivity[2-3]. Despite these findings, the underlying mechanisms remain hard to pin down with selectivity changes from one metal to another. Density functional theory (DFT) studies helped elucidate the mechanism for furfural hydrogenation, dehydroxylation, and even ring opening on Pd(111) by calculating the associated thermodynamic and kinetic parameters[4-5]. However, DFT quickly becomes infeasible when incorporating furan ring hydrogenation and dehydrogenation into the reaction network, let alone extending these studies to other catalytic surfaces.

In this work, we present a first-principles-based semi-empirical model (FPSEM) capable of predicting the thermochemical properties of furan-related intermediates involved in hydrogenation, hydrogenolysis, decarbonylation, and ring-opening of furfural and HMF on any extended metal surface. The entire energetic profile for a proposed reaction mechanism is estimated by combining this scheme with the transition state scaling relations[6]. Using this scheme, we perform microkinetic modeling simulations, obtain activity and selectivity trends, and provide theoretical insights into furfural and HMF reactivity on transition metal surfaces.

References

1. Lange *et al.*, *ChemSusChem* **2012,** *5* (1), 150-166.
2. Jae, J.; Zheng, W.; Lobo, R. F.; Vlachos, D. G., *ChemSusChem* **2013,** *6* (7), 1158-1162.
3. Sitthisa, S.; Resasco, D., *Catalysis Letters* **2011,** *141* (6), 784-791.
4. Vorotnikov, V.; Mpourmpakis, G.; Vlachos, D. G., *ACS Catal.* **2012**, 2496-2504.
5. Wang, S.; Vorotnikov, V.; Vlachos, D. G., *Green Chemistry* **2014,** *16* (2), 736-747.
6. Wang, S.; Vorotnikov, V.; Sutton, J. E.; Vlachos, D. G., *ACS Catal.* **2014**, 604-612.

387271 Catalytic Hydrothermal Decarboxylation of Unsaturated Fatty Acids with Ptsnx/C

Abstract:

Pt/C is an effective hydrothermal catalyst for decarboxylating saturated fatty acids without added hydrogen, but it has shown low selectivity and activity for decarboxylating unsaturated fatty acids – a common oil precursor in biological sources such as soybeans and algae. We compared three different Sn containing alloys to Pt/C: Pt_3Sn/C, PtSn/C, and $PtSn_3/C$ by performing batch reactions in liquid water at 350 °C for two hours with stearic (C18:0), oleic (C18:1), and linoleic (C18:2) acids. Pt/C gave a 0.70±0.05, 0.16±0.03, and 0.083±0.009 yield of the C_{17} alkanes, as opposed to the unobserved direct decarboxylation products, for the oleic and linoleic acid feeds. Stearic acid was the major byproduct with the unsaturated feeds. In contrast, the $PtSn_x/C$ catalysts yielded two to three times higher C_{17} yield. Experiments in D_2O showed that the source of hydrogen for saturation of the fatty acids was water. These results show that PtSn alloys may give better performance in terms of upgrading biologically derived carboxylic acids to liquid fuels.

James R. Kastner*[,1], Roger Hilten[1], Justin Weber[1], Justin Hargreaves[2]

Continuous Catalytic Ketonization and Hydrodeoxygenation of Water Extracted Fast Pyrolysis Oils Using Iron Oxides

Abstract

Biomass fast pyrolysis is a rapid thermal process (0.5-5 s) that generates energy dense, oxygenated liquid hydrocarbon streams in high yields (60-75%), which in theory can be catalytically converted to drop-in fuels. However, due to the presence of water (30-40%), reactive acids and aldehydes, the resulting oil is acidic (corrosive), unstable, and difficult to catalytically upgrade. In this work, water extracted bio-oil generated an aqueous stream with high concentrations of acetic acid (4-10%), formic acid (4-8%), acetol (1.2-2%), and levoglucosan (2-13%), that could be catalytically converted to acetone, 2-butanone, higher ketones, cyclic ketones, and butyrolactone (GBL), in part via ketonization. Continuous processing of fast pyrolysis oil components was performed in a packed bed reactor using H_2 reduced red mud or RRM (waste product of bauxite processing). Experiments have been performed in which individual model compounds (4% each in water: acetic, formic, 1-hydroxy-2-propanone, and levoglucosan), mixtures of the model compounds, and water extracted fast pyrolysis were catalytically transformed using RRM. Reaction rates for RRM were similar to those reported for CeZrOx (6-10 mmol/g-cat/h), but acetone selectivity was lower (8-16%). Levoglucosan reactivity was lower than the other compounds, but severe plugging was not observed (however, coke was formed) and acetone was formed. Apparently, acetone formation from levoglucosan occurred via the intermediates acetic acid and acetol, since they were identified via GC/MS in the products. Individual conversion of acetol yielded acetone and 2-butanone. Space time conversions mirrored reaction rates and were reasonably high – 330 (acetic or AA), 452 (formic or FA), 355 (acetol or 2HP), and 372 (levoglucosan) g converted/L-cat/h. To better understand the interaction between reactants we have also performed catalytic runs using model compound mixtures (AA-FA, AA-FA-2HP, AA-2MP). Addition of formic acid, 2-HP, and 2-methoxy phenol (2MP) reduced the rate of acetic acid ketonization and conversion, yet the formic acid reaction rate was independent of the other compounds. A subsequent series of experiments, where both the temperature and residence time was increased, raised acid conversions to 100% and increased the total ketone selectivity to 22%. These results prompted us to investigate catalytic transformation of water extracted fast pyrolysis oil in a continuous packed bed reactor (350-425°C). Early results indicate acetone and 2-butanone yields approach those reported with the model compounds at high acid conversions. Interestingly, the presence of H_2 significantly altered product selectivity and methanol was apparently formed from formic acid. The structure-activity relationship between the activated red mud and potential catalytic conversion pathways, the possibility of simultaneous ketonization/hydrodeoxygenation via a reverse Mars and Van Krevelen mechanism, and the possible integration of this step with fast pyrolysis to reduce bio-oil corrosivity will be discussed.

*Corresponding Author

[1] BioChemical Engineering, The University of Georgia

[2] School of Chemistry, The University of Glasgow

James R. Kastner[*,1]: jkastner@engr.uga.edu

Roger Hilten[1]: rog@uga.edu

Justin Weber[1]: jweber90@uga.edu

Justin Hargreaves[2]: Justin.Hargreaves@glasgow.ac.uk

386288 Kinetics and Mechanisms of Lignin Model Compound Reactions over a Hydrotalcite-Derived Copper Catalyst

Abstract:

The reactivity of molecular compounds relevant to the hydrogenolysis of lignin as a renewable source of aromatics was explored in a combined experimental-computational approach. X-ray absorption spectroscopy shows that highly dispersed copper ions are reduced by the solvent, methanol, in the early stages of the reaction. The resulting copper nanoparticles catalyze methanol reforming to generate H2 at ca. 250 °C. The selectivity towards the desired C-O hydrogenolysis reaction was explored for a variety of model compounds, including anisole and dihydrobenzofuran. DFT calculations suggest that the barrier is lowest for C-O cleavage, but the oxide support may play a role in directing substrate adsorption and therefore reactivity. This was confirmed experimentally by observing reaction products in a flow reactor.

386443 An Examination of the Intrinsic Activity and Stability of Various Solid Acids during the Catalytic Decarboxylation of γ-Valerolactone

Abstract:

In this study, we have considered the catalytic decarboxylation of γ-valerolactone (GVL), which occurs over solid acids to yield 1-butene and carbon dioxide. To arrive at a set of governing design parameters for decarboxylation catalysts, we have examined the roles of Brønsted:Lewis site distribution, Brønsted site deprotonation energy, and catalyst morphology in defining the activity and stability of materials employed for GVL decarboxylation. Decarboxylation rates were measured in a gas-phase reactor from 523 – 723K over a series of solid acids including amorphous silica alumina, MFI zeolites, supported phosphotungstic acid, and γ-Al_2O_3. In aluminosilicates, Brønsted sites associated with framework aluminum contribute the majority of decarboxylation activity. Relative to bridging hydroxyls, coordinatively unsaturated aluminum sites are substantially less active, and they do not contribute significantly to butene production rates in materials having both framework and extraframework aluminum. Through comparison of aluminosilicates with supported heteropolyacids, decarboxylation barriers were found to scale with the deprotonation energy of Brønsted acid sites; however, the intrinsic decarboxylation activity of a Brønsted site is not necessarily anticipated by its deprotonation energy. Catalyst deactivation in this system is attributed to coke deposition, and catalyst stability correlates most strongly with pore size such that microporous zeolites deactivate quickly compared to mesoporous analogs.

378881 Mechanistic Pathways and Requirements for Alkanal Deoxygenation on Solid Brønsted Acid Catalysts

Abstract:

Light oxygenate fraction from biomass pyrolysis is a valuable precursor for the production of aliphatics and aromatics via catalytic deoxygenation reactions. Deoxygenation involves condensation, ring-closure, alkylation-dealkylation, and H transfer steps, which may occur in sequence or in parallel on acid [1], base [2], or acid-base bifunctional [3] sites grafted within solid microporous or mesoporous structures. Previous studies have reported a strong dependence of product distributions on the type of solid matrixes and reaction conditions, but mechanistic details, catalytic requirements, and rate dependencies for the individual reaction paths have remained as the subject of active debate.

In this contribution, we apply rate and isotopic analyses, acid site titrations, and temperature programmed techniques to elucidate the primary and secondary reaction pathways for the deoxygenation of alkanals (RCH_2CHO, R=CH_3, C_2H_5, C_3H_7) in the absence of an external H source on solid Brønsted acid sites confined to different extents (H-MFI, H-FAU, $H_4SiW_{12}O_{40}/SiO_2$). Catalytic deoxygenation of alkanals involves the (1) initial inter-molecular C=C bond formation via bi-molecular, aldol type condensation reaction, (2) initial intra-molecular C=C bond formation as a result of sequential H addition and dehydration reaction, (3) direct isomerization and dehydration reaction, (4) secondary ring closure step, and at high temperatures (> 573 K), (5) alkylation-dealkylation reactions [4]. These reactions lead to the formation of larger alkenal, alkene containing the same number of carbon atoms with the alkanal reactant, diene, methyl or ethyl substituted benzenes, and diverse aromatic species (C_6-C_{15}).

We report a distinct reactivity trend across the alkanal homologues and with varying degrees of site confinement. The inter-molecular C=C bond formation is bi-molecular in nature; the rate of this reaction on Brønsted acid sites dispersed on unconfined $H_4SiW_{12}O_{40}$ clusters is more effective for larger than smaller alkanals, because larger alkanals form more stable bi-molecular transition state, according to the prediction from a Born Haber thermochemical cycle constructed using gas phase thermochemical properties. In contrast, the same reactions occur more effective on smaller than larger alkanals on Brønsted acid sites confined within the H-FAU and H-MFI structures, because site confinements destabilize the energy of the bi-molecular transition state. Isomerization-dehydration or inter-molecular H transfer-dehydration steps that evolve the diene and olefin, respectively, are uni-molecular in nature; their relative rates vary with the ability of adsorbed alkanal to accept hydrogen from H donors, which are formed from secondary ring closure of the aldol condensation product and, upon dehydrogenation steps, release H during the inter-molecular H transfer step. As the extent of confinement increases in the order of $H_4SiW_{12}O_{40}/SiO_2$, H-FAU, H-MFI and the rates of bi-molecular C=C bond formation concomitantly decrease, the total fraction of diene and olefin concomitantly increases. Within the diene and olefin fraction, the selectivities towards olefin increase linearly with increasing H donor pressures, because the inter-molecular H transfer from the ring closure product is required for olefin formation. This requirement of inter-molecular H transfer is confirmed from a linear increase in the olefin formation rates with increasing concentration of aromatics, the products of dehydrogenation, and consistent with the increase in the rates by incorporating tetralin, an efficient H donor, into the feed mixtures.

The catalytic effects of reactant size and acid site confinement demonstrated here allow the tuning of the relative rates and reaction specificity during alkanal deoxygenation reactions and thus the average molecular size of hydrocarbon and the relative abundances of aliphatics and aromatics in the reaction products.

Acknowledgement

We acknowledge Natural Sciences Council of Canada (NSERC) and Canadian Foundation for Innovation (CFI) for their financial supports. Fan Lin acknowledge Hatch for their support via a Hatch Graduate Scholarship for Sustainable Energy Research.

References

[1] T.Q. Hoang, X. Zhu, T. Sooknoi, D. E. Resasco, R. G. Mallinson, J of Catal. 271 (2010) 201–208.

[2] K. K. Rao, M. Gravelle, J. S. Valente, F. Figueras, J of Catal. 173 (1998) 115–121.

[3] M. J. Climent, A. Corma, H. Garcia, R. Cuil-Lopez, S. Iborra, V. Fornes, J of Catal. 197 (2001) 385–393.

[4] F. Lin and Y-H. Chin, J of Catal. 311 (2014) 244-256.

Fundamental Insights and Similarities between Homogeneous and Heterogeneous Catalysis in Biomass Conversion

Vinit Choudhary[1], Andrzej Anderko[2], Ana B. Pinar[1], Samir H. Mushrif[3], Vladimiros Nikolakis[1], Nebojsa S. Marinkovic[1], Anatoly I. Frenkel[4], Raul F. Lobo[1], Stanley I. Sandler[1], and Dionisios G. Vlachos[1]

[1]Center for Catalytic Science and Technology and Catalysis Center for Energy Innovation
Department of Chemical & Biomolecular Engineering, University of Delaware, Newark, DE 19716

[2]OLI Systems Inc., 108 The American Rd., Morris Plains, NJ 07950

[3]School of Chemical and Biomedical Engineering, Nanyang Technological University, 62 Nanyang Drive, Singapore 637459

[4]Physics Department, Yeshiva University, 245 Lexington Avenue, New York, NY 10016

Furans derived from lingo-cellulosic biomass can potentially substitute petroleum precursors in producing chemicals, polymers, and fuels. Aldose-to-ketose isomerization is a key reaction in converting hexose and pentose sugars to platform furan derivatives, such as 5-(hydroxymethyl) furfural and furfural, respectively. The heterogeneous catalyst zeolite Sn-beta and homogeneous catalysts, such as various metal salts ($CrCl_3$ and $AlCl_3$), have been reported as very active catalysts for the isomerization reaction in aqueous media. We recently investigated the metal-salt catalyzed aldose-to-ketose isomerization and showed its mechanistic similarities with the Sn-beta catalyzed isomerization. It was found that partially hydrolyzed ions, such as $CrOH^{2+}$ or $AlOH^{2+}$ are the most active for the metal salt catalyzed isomerization that resembles well the active site (-SnOH) in the Sn-beta catalyzed isomerization. Additionally, ^{13}C and ^{1}H NMR studies confirmed that an intra-hydride transfer (C2 to C1) reaction pathway is the dominant reaction channel in the metal salt and Sn-beta catalyzed isomerization. EXAFS and first principles molecular dynamics studies have been employed to understand the complexation of glucose with the Lewis-acid center. Overall, these observations show that the metal salts and the Sn-beta catalyzed aldose-ketose isomerization follow a similar mechanism and a Lewis acid-Brønsted base bifunctional active site is required for the isomerization in aqueous media.

Higher Alcohols Synthesis: Perspectives from experiments and DFT calculations

Burçin Temel, Niels Christian Schjødt, Poul Erik Højlund Nielsen

Haldor Topsøe A/S, Nymøllevej 55, Kgs. Lyngby 2800 DENMARK

bute@topsoe.dk

Higher alcohols synthesis (HAS) has attracted considerable attention in past 30 years— with a level of interest closely tied to oil prices. Higher alcohols can be used as high-value polymer feedstock chemicals, as direct fuels, or as fuel additives.

We have been studying higher alcohols synthesis both experimentally and theoretically. On the experimental side, we are able to produce higher alcohols with an alkali modified methanol synthesis catalyst with space time yields of 0.1 g/(h g.cat).[1,2] In our theoretical studies[3,4] we aimed to find novel alloy catalysts using density functional theory (DFT) and Brønsted-Evans-Polanyi (BEP) scaling relationships. The rate limiting step of higher alcohols synthesis from syngas is the initial C-C coupling and the intermediates in the initial coupling reaction are CH_x and CO. Therefore, we focused on obtaining the reaction barriers going from syngas adsorption to ethanol formation on the pure transition metals such as Ag, Au, Cu, Ni, Ir, Pd, Pt and Rh by using DFT calculations. With the trends found across the calculated metals, we chose carbon and oxygen adsorption as two descriptor parameters for the CH_x and OH_x intermediates and formed BEP scaling relationships. Our collaborators (Nørskov et al.) used these scaling relationships to study the stability and reactivity of new alloys for the higher alcohols synthesis both theoretically and experimentally.[5]

In my talk, I will present results from our experimental work on traditional HAS catalyst as well as results from theoretically predicted novel alloys.

Key Words: Alcohols, DFT, alloys, methanol, ethanol, syngas

References

[1] P. E. H. Nielsen, B. Temel, P. Beato, US 8,637,580 B2 (2014).
[2] B. Temel, P. E. H. Nielsen, P. Beato, US 20130225879 A1 (2013).
[3] S. Wang, B. Temel, J. Shen, G. Jones, L. C. Grabow, F. Studt, T. Bligaard, F. Abild-Pedersen, C. H. Christensen, J. K. Nørskov, Catal Lett 141 (2011) 370-373.
[4] A. J. Medford, A. C. Lausche, F. Abild-Pedersen, B. Temel, N. C. Schjødt, J. K. Nørskov, F. Studt, Top Catal 57 (2014) 135-142.
[5] F. Studt, F. Abild-Pedersen, Q. Wu, A. D. Jensen, B. Temel, J. D. Grunwaldt, J. K. Nørskov, J Catal 293 (2012) 51-60.

C-C Bond Formation during Fischer-Tropsch Synthesis on Ru and Co Catalysts

David Hibbitts[1], Thomas Lawlor[2], Eric Dybeck[2], Matthew Neurock[2*], and Enrique Iglesia[1*]

[1]Department of Chemical and Biomolecular Engineering, University of California, Berkeley, California 94720, United States

[2]Departments of Chemical Engineering and Chemistry, University of Virginia, Charlottesville, Virginia 22904, United States

*iglesia@berkeley.edu, mn4n@virginia.edu

Fischer-Tropsch synthesis (FTS) forms large hydrocarbons from synthesis gas (CO-H_2) derived from biomass, coal, or natural gas. The specific mechanisms for both C-O cleavage and C-C bond formation steps in FTS remain the subject of active debate, even after many experimental and theoretical inquiries. Kinetic and isotopic data on Co and Ru catalysts and enthalpy and free energy activation barriers derived from density functional theory (DFT) calculations on 586-atom Ru and Co particles have shown that at the high CO* coverages prevalent during FTS, C-O activation occurs with assistance from chemisorbed H-atoms (H*) instead of via direct CO dissociation on vicinal vacant sites. Previous studies have also shown that H_2O, a product of FTS, increases FTS turnover rates by a H-shuttling mechanism in which H* reacts with nearby H_2O to form H_3O^+ intermediates which rapidly protonate CO* (to form COH* which then forms *HCOH*) or HCO* (to form *HCOH*), lowering the barriers to form *HCOH*, which then undergoes C-O activation to form CH* and OH*. CH* reacts further to form all observed hydrocarbon products; however, the specific mechanisms for such C-C bond formation steps remain controversial.

CH* reacts with H* to form CH_2* and CH_3* intermediates en route to undesired CH_4. These CH_x* intermediates can couple with each other through a variety of possible elementary steps often referred to as the carbene mechanism because of the possible involvement of CH_2* species (herein referred to more generally as CH_x*-coupling). CH_x* species may also react with CO* in CO*-insertion mechanisms to form CH_xCO** intermediates which then undergo C-O hydrogenolysis. Here, kinetic studies are used to determine the effects of H_2 and CO pressure on chain termination parameters (β, the ratio of chain termination rates to chain propagation) and combined with DFT calculations to mechanistically interpret β and to rule out possible C-C bond formation routes. Isotopic tracer experiments that react ^{13}CO and ^{12}C-containing aldehydes, are also used to investigate the ratio of rates of aldehyde decarbonylation to C-O hydrogenolysis, and the extent of aldehyde incorporation into larger chain products to determine the feasibility and reversibility of CO*-insertion reactions with alkyl species.

389577 DFT Study of the Role of the Promoter in V-Promoted Rh Catalysts for CO Hydrogenation

Abstract:

Unpromoted rhodium catalysts for CO hydrogenation generally produce methane, however promoters like V have shown to dramatically increase selectivity towards oxygenates. Density functional theory (DFT) calculations are performed with the focus on the determination of new pathways at the interface between the promoter (V) and the active metal (Rh). The goal is to understand the physical relationship between the promoter and the catalyst under realistic syngas conversion conditions of high CO surface coverage, incorporation of which has shown to significantly shift the thermodynamics. Promotion is modeled both as an isolated metal atom on Rh(211) step edge, and as an oxide cluster on Rh(111) surface. This work provides a systematic insight into the observed catalytic performance of the promoted catalyst viewed as a combined effect of surface structure, adsorbate coverage and identity of the promoter. Promoter effects on C O cleavage, C H and/or C C bond formation reactions are analyzed via changes to the Rh *d*-band filling. Chemical promotion through an H-assisted CO dissociation mechanism via formyl (HCO) as an intermediate species was found to be a primary pathway for CO activation on V promoted catalyst, due to the high CO dissociation barriers of 1.61 and 1.93 eV for Rh(111)/VO_2 and RhV(211) surface, respectively. The observed decrease in adsorbate binding energy on the undercoordinated Rh step edge could be attributed to both surface coverage and more predominant V promotion (DE = -0.58 eV), the combined effect resulting in a highly reactive surface that facilitates hydrogenation of CO species (E_{act} = 0.76 eV) as well as a particularly low HCO insertion barriers to CH_x of 0.52, 0.32 and 0 eV for x = 1, 2, 3, respectively. H-assisted CO dissociation is thermodynamically favored on Rh(111)/VO_2 with a 0.49 eV energy barrier. The oxide promoted surface provides low dissociation barriers assumed to be due to the undercoordination of vanadium atom of the oxide, achieved upon the oxygen removal via H_2O molecule. HCO insertion to CH_x, assumed to be independent of interface, is exothermic with 0.76, 1.01 and 1.41 eV barriers for x = 1, 2, 3, respectively. V is found to promote H_2O formation due to a repulsive behavior caused by V behaving as a mild Lewis acidic when embedded in Rh surface and OH intermediate (increased O electronegativity).

389338 DFT and Microkinetic Modeling Study of Gas- and Liquid-Phase Fischer-Tropsch Synthesis

Abstract:

Fischer-Tropsch (FT) synthesis has been applied for the production of valuable fuel products from synthesized gas (syngas) for decades. Currently, conventional FT processes in industry are accomplished in both gaseous and liquid phases. In spite of several suggested mechanisms, the actual reaction path and important elementary reaction steps that result in different products under different conditions are still under investigation, especially in the liquid phase. This work involves a density functional theory (DFT) study of an existing microkinetic model for gas-phase FT synthesis on Co(0 0 1) catalyst. In this regard, by utilization of the DFT-GGA method, the kinetic and thermodynamic parameters of each elementary step have been evaluated. Then, using DFT calculations with the COSMO continuum solvent model, FT synthesis has been studied in the presence of n-hexane solvent to find the impact of liquid solvent (i.e. solvation effects) on the kinetics, mechanism, and thermodynamics of this process. In addition, liquid-phase reactor operating conditions (PVT behavior) are evaluated using an equation of state. Finally, by introducing these elementary mechanistic parameters in a simulated reactor, FT product yields and distributions in the different phases are calculated and compared.

Tuning of Higher Alcohol Selectivity and Productivity in CO Hydrogenation Reactions Over K/MoS_2 Catalysts Supported on Mesoporous Activated Carbon and mixed MgAl Oxide

Micaela Taborga Claure[1], Song-Hai Chai[2], Sheng Dai[2], Faisal M. Alamgir[1], Pradeep K. Agrawal[1], and Christopher W. Jones[1]

1. School of Chemical & Biomolecular Engineering, Georgia Institute of Technology, Atlanta, GA, 30332 USA
2. Chemical Science Division, Oak Ridge National Laboratory, Oak Ridge, TN 37831 USA

The use of supports as a means of enhancing both activity and selectivity of MoS_2 has been investigated over the past four decades. Our previous work showed that carbon supports yield high methanol and ethanol productivities.[1] Mesoporous activated carbon, obtained via collaboration with ORNL, when used as a support, is confirmed to show greater alcohol productivity than commercial activated carbon. Hydrotalcite-derived mixed MgAl oxide (MMO) supports yield high $C_{3+}OH$ selectivity influenced by Mo-MMO interaction, but generally yield low productivity. It has been shown that the Mo:MMO ratio greatly affects selectivity while catalyst preparation methods do not.[2] In this work, we seek to evaluate the hypotheses that molybdenum supported on a family of MMO and carbon mixed-supports may allow tuning of higher alcohol productivity (due to carbon support) and selectivity (due to MMO support).

When Mo is first impregnated on the carbon support and the resulting material is physically ground with MMO, the selectivity trends lie in between that of the carbon and MMO single-support catalysts. Although not as selective toward higher alcohols compared to the single-support MMO catalyst, this novel mixed-support catalyst gives an increased selectivity toward higher alcohols compared to the single-support carbon catalyst, and a significant increase in higher alcohol productivity over both single-support catalysts, a result of Mo's ability to migrate from the carbon to the MMO during sulfidation and reaction. In contrast, when Mo is first impregnated on MMO and the resulting material is ground with carbon, the catalyst has selectivity and productivity trends similar to the MMO single-support catalysts, indicating that strong Mo-MMO interactions limit the ability of Mo to migrate to the carbon support.

The interlayer distance between the MoS_2 layers was determined with synchrotron XRD. In addition, STEM was used to evaluate the dependence of Mo-support interaction on interlayer distance and the number of MoS_2 stacked layers. STEM and synchrotron XRD results were used to analyze the effect of the number of stacked MoS_2 layer and K intercalation on reaction performance. Furthermore, STEM and EXAFS results were used to analyze the dispersion of MoS_2 domains. The combined data shed light on the role of MoS2 stacking and domain structure on catalytic properties.

References

1. Morrill, M. R.; Thao, N. T.; Agrawal, P. K.; Jones, C. W.; Davis, R. J.; Shou, H.; Barton, D. G.; Ferrari, D., Mixed MgAl Oxide Supported Potassium Promoted Molybdenum Sulfide as a Selective Catalyst for Higher Alcohol Synthesis from Syngas. *Catalysis Letters* **2012,** *142* (7), 875-881.
2. Morrill, M. R.; Thao, N. T.; Shou, H.; Davis, R. J.; Barton, D. G.; Ferrari, D.; Agrawal, P. K.; Jones, C. W., Origins of Unusual Alcohol Selectivities over Mixed MgAl Oxide-Supported K/MoS2 Catalysts for Higher Alcohol Synthesis from Syngas. *ACS Catalysis* **2013,** *3* (7), 1665-1675.

A systematic study of Mn and Zr oxide promoters for Fischer-Tropsch synthesis using alumina nanosphere-supported model catalysts

Gregory R. Johnson[a], Sebastian Werner[b], Alexis T. Bell
Department of Chemical and Biomolecular Engineering, University of California, Berkeley, CA

Due to the abundance of inexpensive natural gas in the US and various petroleum-producing regions around the world, there are strong motivations for developing technologies for converting this feedstock into liquid fuels and lubricants [1]. Natural gas can be processed into syngas, a mixture of carbon monoxide and hydrogen, from which long-chain hydrocarbons can be formed by Fischer-Tropsch synthesis (FTS) using cobalt-based catalysts. These catalysts form methane as a byproduct, which can be mitigated by including metal oxide promoters in the catalyst formulation [2, 3, 4]. Despite an abundance of literature on the topic of promoters for Co FTS catalysts, it is difficult to form generalizations related to periodic trends. Promoter-induced changes to Co particle size and morphology and inconsistencies in spatial association between the active Co metal and promoter often prevent meaningful comparisons from being made using data from different studies. Here we present a systematic study of the effects of Mn and Zr oxide promoters with the aim of clarifying how metal oxide promoters impact the performance of Co FTS catalysts.

It was found that under identical catalyst synthesis protocols, the maximum C_{5+} selectivity achievable using each promoter occurred at different ratios of promoter to Co. This was attributable to differences in how the readily the promoters associated with the Co versus the support. To assess these differences in promoter spatial distribution, catalysts were prepared using alumina nanospheres (Fig. 1). The uniform background provided by these non-porous spheres allowed for well-resolved imaging of the Co and promoter nanoparticles by STEM-EDS. These materials gave similar activity and selectivity to catalysts prepared using conventional silica supports. Kinetics data obtained at different temperatures and reactant partial pressures did not support a change in the reaction mechanism upon promotion; however, the presence of promoters did lead to a significant increase in the apparent activation energy and heat of adsorption for CO. This was qualitatively consistent with both the heavier product distribution and higher olefin to paraffin ratio of the promoted catalysts. In situ FTIR spectroscopy was used to determine surface coverages of adsorbed CO for promoted and unpromoted catalysts. X-ray absorption spectroscopy was used to probe oxidation state and the local environment of the Co and promoter elements.

In addition to Mn and Zr oxides, other promoters, including La, Ce, and Gd oxides, were tested for impact on catalytic activity and selectivity. It was found that a positive correlation exists between C_{5+} selectivity and the estimated electronegativity of the promoter cations, a proxy for Lewis acidity of the promoter. This finding can be used to guide the identification of candidate promoters to yield further improvements to the FTS product distribution.

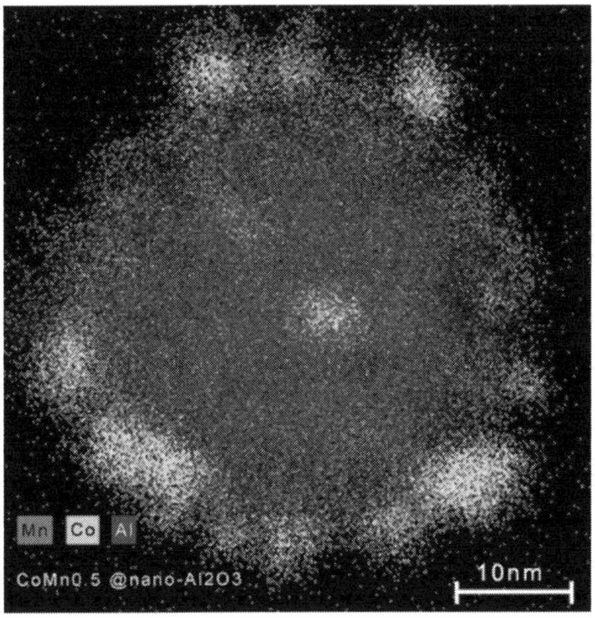

Figure 1. STEM-EDS map of Co-Mn nanoparticles supported on a 40-nm alumina nanosphere. The non-porous alumina provides a uniform background against which the metal nanoparticles are clearly visible. In many cases, nanoparticles sit along the circumference of the projected circle of the spheres, which facilitates high resolution imaging.

References

[1] Khodakov et al. Chem. Rev. 107 (2007) 1692.

[2] den Breen et al. Top. Catal. 54 (2011) 768.

[3] Ledford et al. J. Phys. Chem. 93 (1989), 6770.

[4] Ma et al. Ind. Eng. Chem. Rev. 43 (2004) 2391.

Long-chain Aldehydes and Terminal Alcohols by Catalytic CO Hydrogenation

Yizhi Xiang and Norbert Kruse

Center of Catalysis for Clean Energy & Environment, Voiland School of Chemical Engineering and Bioengineering, Washington State University, Pullman, WA 99164, USA

Long-chain aldehydes and terminal alcohols are important feedstocks for plasticizers, detergents and lubricants. Their large-scale production is based on hydroformylation of terminal C_n alkenes ($n \geq 3$). Long-chain aldehydes provided by hydroformylation are eventually hydrogenated to the respective alcohols, however, the regioselective production of terminal alcohols is limited by the Markovnikov rule. Noble metal recovery continues to be an issue in homogeneous hydroformylation, so the question may be asked if the heterogeneous FT-technology based on a "one-pot" process can be developed to produce strictly long-chain aldehydes or terminal alcohols. We dealt with this question in the recent past and demonstrated highly selective terminal alcohol formation through CO hydrogenation over ternary Co-Cu-Mn catalysts[1]. The Anderson-Schulz-Flory chain lengthening probability could be adjusted in these studies to meet the requirement of optimized C_8-C_{14} alcohols yields. The present paper addresses the question of long-chain aldehydes versus terminal alcohols formation.

Our approach is based on designing CoMn nanosized particles using co-precipitation of metal salts into oxalate precursors and subsequent thermal decomposition, in the absence of a classical support. It turned out that the selectivity of long-chain aldehydes versus terminal alcohols production can be tuned by either adding alkaline metal or Cu to CoMn. Despite the absence of a classical support material, these catalysts showed stable high-pressure catalytic performance in our studies. 3D Atom-Probe-Tomography demonstrated single nanosized "CoCuMn" particles to have a core-shell structure, with Co forming the core, while all three metals are present in a Cu-dominated outer shell of about 2nm thickness. Some of the co-precipitated Mn-oxalate decomposed to MnO_x which was catalytically inactive in CO hydrogenation but played an important role as a 'chemically active' dispersant, thus conferring the catalyst a high BET specific surface area and shifting the product spectrum to the alcohol side.

In particular, when comparing the catalytic performance of 'CoCuMn' with that of 'CoMnK' we found that using the former with a 1:1:1 relative atomic composition produced terminal alcohols with selectivities of more than 50 wt%, eventually reaching more than 90% (including CO_2), at low CO conversion (<10%). More specifically, a unique ASF-chain lengthening probability $\alpha \approx 0.65$ was obtained for $Co_1Cu_1Mn_1$ so as to optimize the C_8-C_{14} terminal alcohol slate with a CO conversion close to 20% (T=240 °C, total pressure of 60 bar ($H_2/CO=2$)). Over CoMnK catalysts with 2:1:0.1 relative composition, oxygenates selectivity were between 40 and 67 wt% (including CO_2, T=200 to 280°C, p=40 bar). Interestingly, different from $Co_1Cu_1Mn_1$, straight chain aldehydes made up to 50~80% of the total oxygenates production. Moreover, the ASF chain lengthening distribution revealed major deviations for short chain oxygenates (C_1-C_3): while almost no C_1 products (methanol+formaldehyde) were formed the C_2 fraction (ethanol+acetaldehyde) largely exceeded expectations from the ASF behavior. Then, again, the C_3 fraction was low while from C_4 onward strictly linear ASF behavior was followed with an α value of 0.4~0.6. The selectivity of aldehyde formation in the C_{4+} total oxygenates production sometimes reached 85 wt%. At high reaction temperatures the CO_2 selectivity increased considerably at the expense of oxygenates selectivity, demonstrating watergas shift reaction to compete with CO hydrogenation under these conditions.

Keywords: CO hydrogenation, long-chain 1-alcohols, aldehydes, CoMn-based catalysts, oxalate co-precipitation.

1. Xiang, Y., Chitry, V. r., Liddicoat, P., Felfer, P., Cairney, J., Ringer, S., and Kruse, N. (2013). *J. Am.Chem. Soc. 135*, 7114-7117.

376715 Ammoxidation of Propene to Acrylonitrile over Bismuth Molybdate

Abstract:

Multicomponent bismuth molybdenum oxide catalysts have been used industrially for over 60 years for the oxidation and ammoxidation of light olefins.[1] Despite extensive research on bismuth molybdates since their discovery in the late 1950s by SOHIO, the mechanistic details of these related reactions have not been fully established. In particular, most of the research in this area has focused on the simpler oxidation reaction of propene to acrolein, despite the fact that these catalysts are generally used for the ammoxidation of propene to acrylonitrile.

The ammoxidation and oxidation reactions are thought to have the same rate-determining step, namely a hydrogen radical abstraction from propene to produce an allyl radical.[2] The ammoxidation reaction is a 6-electron transfer process that requires 3 hydrogen abstractions and a nitrogen insertion, compared to the oxidation reaction that involves a 4 electron transfer, 2 hydrogen abstractions, and an oxygen-insertion from the metal oxide surface. Recent experimental [3] and theoretical [4] investigations in our group have elucidated a likely mechanism for the oxidation reaction over alpha phase bismuth molybdate. The active site for the rate-determining step was found to be an oxygen atom doubly bound to a molybdenum center that is perturbed by the presence of a nearby bismuth atom. Only molybdenum reduces during reaction, however the bismuth alters the band energies around the Fermi level, allowing the reaction to happen.

While the oxidation mechanism is fairly well understood, the ammoxidation reaction remains almost uninvestigated. Researchers have often surmised that findings regarding the oxidation reaction can be extrapolated to the related ammoxidation reaction, however there is little data in the literature to support or refute that claim. Specifically, it is not known how ammonia incorporates into the catalyst, the nature of active site for the rate-determining step, or at what point in the reaction mechanism nitrogen-containing species insert into the hydrocarbon.

In this work, we present a rigorous investigation of the kinetics for the ammoxidation reaction of propene to acrylonitrile over alpha-phase bismuth molybdate. By varying the partial pressures of oxygen, ammonia and propene in the feed and analyzing the ratio of acrolein to acrylonitrile produced, we are able to determine the relative rates of these parallel, competing reactions. From these kinetic studies, we propose a mechanism for the formation of acrylonitrile from propene over bismuth molybdate that builds on that of the oxidation reaction. In addition, we hypothesize a mechanism based on experimental findings and Density Functional Theory calculations for how ammonia inserts into the surface of bismuth molybdate to generate active mono- and di-imido-substituted molybdate species.

[1] Grasselli, R. K. *Topics in Catalysis* **2002**, *21*, 79-88.

[2] Adams, C.R.; Jennings, T.J. *J Catal.* **1963**, *2*, 63-68.

[3] Zhai, Z.; Getsoian, A.; Bell, A.T. *J. Catal.* **2013**, *308*, 25-36.

[4] Getsoian, A.; Shapovalov, V.; Bell, A.T. *J. Phys. Chem. C* **2013**, *117*, 7123-7137.

377694 Importance of Au/Oxide Interface on the Water-Gas Shift Reaction: A Computational Study

Abstract:

Water gas shift (WGS) reaction is a very important industrial reaction, especially in the manufacture of hydrogen, ammonia, methanol and hydrocarbons. Oxide supported Au catalyst has gained great attraction on WGS reactions, especially for its catalytic activity at low temperatures, since the discovery of its unique chemical properties. In this study, we did density functional theory (DFT) calculations at generalized gradient approximation level with exchange-correlation functional in form of PW91 as well as at hybrid level in form of HSE06. A supported gold nano-wire on an extend MgO(100) slab model was used to explore the reaction mechanism of WGS at Au/oxide interface, which was believed to be the active site by several previous experimental studies. Compared to the previous theoretical study on Au(111) itself, the biggest difference of the reaction at Au/MgO interface is that the water dissociation barrier is dramatically decreased from > 2 eV over Au(111) to no barrier at Au/MgO interface. Among the previous proposed redox, carboxyl and formate mechanisms, the carboxyl mechanism is the most plausible one, with the formation of COOH having the highest barrier at ~0.8 eV. Based on these DFT calculated barriers, a two sites micro-kinetic model was built at Au/MgO interface. Kinetic control analysis shows that COOH formation to be rate-determing step. The competitive adsorption between CO and H at the interface also strongly influence the reaction kinetics. In order to improve the performance of Au/MgO, two suggestion can be made based on currently calculations: 1) lower the barrier of rate-determing step, formation of COOH; 2) enhance the binding of water or hydroxyl group to reduce the poison by CO and H.

Base metal oxides are of interest as low-cost replacements for Pt-based NO oxidation catalysts. While the mechanism of Pt-catalyzed NO oxidation is fairly well understood, such is not the case for the oxides. In this work, we report density functional theory (DFT) and kinetic modeling results for one class of potential oxide catalysts, the perovskites. The perovskite $LaCoO_3$ itself has been shown to have good NO oxidation activity, and Sr substitution improves NO oxidation rates and reduces NO_2 inhibition. We used periodic DFT calculations in the Vienna ab initio Simulation Package to compute the energies of vacancies, O, O_2, NO, NO_2, and NO_3 on the (100)-LaO and (100)-CoO_2 terminations of the parent $LaCoO_3$ material. Further, we used first-principles thermodynamic models to determine the most common surface species under NO oxidation conditions. Nitrates and adsorbed NO_2 are most stable on the LaO and CoO_2 terminations respectively.

We replicated the calculations on the (100)-LaO and (100)-MO_2 terminations of the entire first row perovskite series, $LaMO_3$ (M = Sc-Cu). Figure 1 shows that computed vacancy and O energies vary nearly linearly across the series while NO_x adsorption energies are more nearly constant. We used first principles thermodynamics models to contrast coverage behavior across the series. Further, by combining these results with a simple Brønsted-Evans-Polanyi relationship for O_2 dissociation, we compared reaction rates across the series. We find that the preferred pathway depends on surface termination and on oxide, that adsorbed NO_x species inhibit reaction, and further that a number of perovskites should give similar NO oxidation rates.

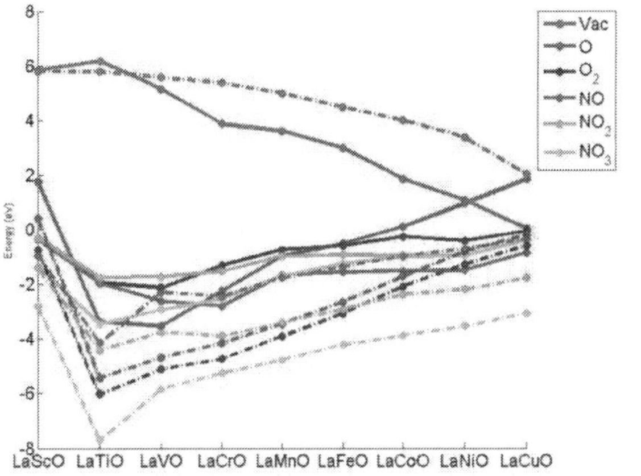

Figure 1. Binding energies of N_xO_y species on $LaMO_3$ (M=Sc-Cu). The CoO_2 and LaO terminations are denoted by solid and dashed lines respectively.

379955 Mechanistic Consequences of Composition and Descriptors of Reactivity for Alkanol Odh Reactions on Polyoxometalate Clusters

Abstract:

Mechanistic consequences of composition and descriptors of reactivity for alkanol ODH reactions on polyoxometalate clusters

Prashant Deshlahra and Enrique Iglesia*

Department of Chemical Engineering, University of California, Berkeley, CA 94720

*iglesia@berkeley.edu

Polyoxometalate (POM) clusters are oxides with Keggin type structure, which consist of central tetrahedral units enclosed in a metal oxide shell and charge balancing protons [H_{8-n} $(X^{n+}O_4)(M_{12}O_{36})$; X= P, Si, Al, Co, M = W, Mo, V etc.]. Protons in these clusters act as catalytic Brønsted acid sites while MO_x addenda species can undergo redox cycles to catalyze oxidative dehydrogenation (ODH) reaction. Such material exhibit well-defined atomic connectivity over diverse compositional range, making them well-suited for combining rigorous kinetic and isotopic and spectroscopic studies with density functional theory (DFT) estimates of their redox and acid properties and the energetics of the relevant elementary steps. We report here mechanistic interpretations of oxidative dehydrogenation (ODH) rates of CH_3OH-O_2 mixtures on Mo- and W-based POM clusters, with different central atoms ($H_3\mathbf{P}Mo_{12}O_{40}$, $H_4\mathbf{Si}Mo_{12}O_{40}$) or addenda atoms ($H_{3+n}P\mathbf{V}_{\mathbf{n=1-3}}\mathbf{Mo}_{\mathbf{12-n}}O_{40}$, $H_4P\mathbf{V}_{\mathbf{1}}\mathbf{W}_{\mathbf{11}}O_{40}$).

ODH reactions involve kinetic coupling between steps that reduce and oxidize the oxide clusters [1]; the dependence of rates on O_2 and CH_3OH pressures show that reduction steps, which hydrogenate the clusters using CH_3OH, typically limit rates for most O_2/CH_3OH reactant ratios. ODH rates showed a Langmuir-type behavior with changing CH_3OH pressures, suggesting saturation of oxide surface with adsorbed CH_3OH with increasing pressure. Initial linear CH_3OH dependence provides a measure of first-order rate constant ($k_{1st,ODH}$), which reflects stability of kinetically relevant transition state (TS) with respect to gaseous CH_3OH reactant. Measured kinetic isotope effects (r_{CH3OH}/r_{CD3OD} = 2.5, r_{CH3OH}/r_{CH3OD} = 1.0) suggest that this TS involves H-abstraction from C-H bond of CH_3OH, but cannot discern between H-abstraction a CH_3OH molecule and that from a dissociated methoxide species; the latter are typically presumed to be prevalent despite lacking spectroscopic evidence of reactive methoxides on several oxide surfaces during ODH conditions [2]. DFT calculations can distinguish between the molecular and the dissociative routes, and we show that H-abstraction from molecularly adsorbed CH_3OH are significantly more favorable than dissociative route and its calculated pathways are consistent with experimental kinetic and isotopic measurements (calculated KIE, r_{CH3OH}/r_{CD3OD} = 2.2). The Measured $k_{1st,ODH}$ values reflect ability of oxide clusters to accept electrons during H-abstraction, in a step mediated by a late transition state (TS). The TS stabilities correlate strongly with H-addition energies (HAE) to catalysts' O-atoms (slope of E_{TS} versus HAE ~ 0.8), which probe energetics of reduction analogously to the H-abstraction step. HAE values change with location on clusters and need exponentially weighted reaction-averaging (HAE_{avg}) over all locations; these HAE_{avg} values provide accurate descriptors of the reduction part of redox cycle. HAE_{avg} values decrease with changing POM clusters' addenda-atoms from W to Mo and their V-substitution suggesting improved ability to accept electrons in reduction step, consistent with higher measured ODH rate constants. ODH rates were independent of O_2 pressure and adding H_2O to CH_3OH-O_2 reactant mixtures decreased rates without changing O_2 pressure dependence, suggesting fast re-oxidation steps resulting in low concentration of reduced centers and inhibition of rates by H_2O blocking the reactive oxidized centers.

Re-oxidation of clusters is kinetically silent but their details are relevant to elementary steps that complete the redox turnovers and are accessible through accurate detection of reduced centers in UV-vis spectra. Spectra from experiments and time dependent DFT calculations indicate small extents of reduction; the concentration of reduced centers decreases with increasing O_2 and H_2O pressures, consistent with increased re-oxidation rates and with blocking of oxidized centers by H_2O, respectively. Mechanistic interpretations of the steady-state concentration of reduced centers and of transient response [3] of change in the concentration after changing reactant pressure are consistent with fast re-oxidation rates; the proportionality of these rates to O_2 pressure increase with the extents of reduction, suggesting non-Langmuirian behavior of the elementary steps involved in re-oxidation. DFT calculation show that re-oxidation steps can proceed via a pathway involving O-vacancy formation in reduced cluster followed by O_2 activation at the vacancy, and another pathway involving formation hydroperoxide radicals by direct reaction of O_2 with the H-atom in the reduced cluster without vacancy formation; prevalence of the two routes is determined by H_2O concentration in the gas-phase. Re-oxidation elementary steps lead to formation of peroxide species on the cluster by addition O_2 to reduced centers, involvement of such species were probed using propene oxidation concurrent with CH_3OH-O_2reactions because propene molecules form epoxide products by reacting with such peroxide species.

These results describe detailed mechanistic interpretations using kinetic, isotopic and spectroscopic methods along with theoretical calculations to establish the rigorous relations between catalyst properties and stability of relevant transition states for reduction and re-oxidation steps ubiquitous in catalysis by oxide materials.

References

[1] J.M. Tatibouet, Appl. Catal. A.: General. 148 (1997) 213; P. Mars, D.W. van Krevelen, Chem. Eng. Sci. 3 (1954) (Special Suppl.) 41.

[2] Seman, M., Kondo, J.N., Domen, K., Radhakrishnan, R., Oyama, S.T., J. Phys. Chem. B, 106, 12965 (2002); Jehng

J.M., Hu H., Gao X., Wachs I.E. Catal. Today 28, 335 (1996).

[3] Argyle, M.D., Chen, K., Iglesia, E., Bell, A.T., *J. Phys. Chem. B*, 109, 2414 (2005).

Effects of CeO_2 Support Facets on VO_x/CeO_2 Catalysts in Selective Oxidation of Methanol

Yan Li[1], Zhehao Wei[1], Feng Gao[2], Libor Kovarik[2], Charles H.F.Peden[2,*] and Yong Wang[1,2*]

1 Gene and Linda Voiland School of Chemical Engineering and Bioengineering, Washington State University, Pullman, Washington, 99164, United states
2 Institute for Integrated Catalysis, Pacific Northwest National Laboratory, Richland, Washington, 99354, United states

Supported vanadia catalysts have received much attention due to their good activity and selectivity in a number of selective oxidation reactions. The underlying supports have shown strong effect on the activity of supported vanadium oxide catalysts. However, the origin of the support effect is still not clear. In this work, well-defined VO_x/CeO_2 model catalysts were prepared by depositing VO_x species on three CeO_2 polymorphs including nanocubes, nanorods and nanopolyhedras, which are dominated by different low index facets, i.e., (100) facet for CeO_2 nanocubes, (110) and (100) for CeO_2 nanorods and (111) for CeO_2 nanopolyhedras. The nature of the active center and VO_x-support interactions were investigated by a variety of microscopic and spectroscopic techniques. The catalytic performance of VO_x catalysts and reaction kinetics were investigated using methanol oxidation as a probe reaction. Our results shows that in the presence of mixed facets, surface vanadia species preferentially deposit on the (100) facet of CeO_2, presumably because of its higher surface energy. At the same surface vanadium densities, VO_x species on (100) facets show better dispersion, followed by (110) and (111) facets. The VO_x species on CeO_2 nanorods display higher ODH activity and lower apparent reaction activation energies compared to VO_x species on CeO_2 polyhedras with dominating (111) facets and CeO_2 nanocubes with dominating (100) facets. The higher activity for $VO_x/CeO_2(110)$ might be related to the more abundant oxygen vacancies present on the (110) facets.

Reference

[1] I.E. Wachs, Catal. Today, 2005, 79,100.
[2] I.E. Wachs, Appl Cata a-Gen, 2011, 36, 391
[3] Y. Li, Z. Wei, J. Sun, F. Gao, C. H. F. Peden, Y. Wang, J Phys Chem C, 2013(117),5722
[4] Y. Li, Z. Wei, F. Gao, L. Kovarik, C. H. F. Peden, Y. Wang, J Catal, 2014, http://dx.doi.org/10.1016/j.jcat.2014.04.013.

Counting accessible and active sites on MO_x-SiO_2 catalysts via selective titration with phenylphosphonic acid

Todd Eaton[1], Andrew Boston[1], Anthony Boone Thompson[1], Nick Thornburg, Kimberly Gray[2], Justin Notestein[1]
[1]Department of Chemical and Biological Engineering
[2]Department of Civil and Environmental Engineering
Northwestern University, Evanston, IL

The rational design of new materials is greatly aided by the quantification of the number and nature of actives sites in supported catalysts. Dispersion in supported metals is typically measured through CO chemisorption, direct particle imaging, or fitting to EXAFS. [1-3] Strong acid sites or the reducible fraction of a supported oxide are also straightforwardly quantified, [4, 5] however quantification of weaker Lewis sites is not routine. Specifically, many supported oxide catalysts undergo significant structural changes as surface densities increase from statistically isolated sites to monolayers and crystallites, so the number of kinetically relevant sites cannot be assumed to be either the total or the geometrically accessible surface sites. Here we utilize the selective titration of SiO_2-supported TiO_x and mixed TiO_2-SiO_2 via phenylphosphonic acid (PPA) to quantify fluid-accessible surface sites as a function of Ti surface density and synthesis method. We also perform *in situ* titration of the catalysts during alkene epoxidation with H_2O_2, known to occur over undercoordinated Ti sites, using discrete amounts of PPA titrant. Our results indicate that the titrant binds most strongly to the catalytically active species, which implies the selective titration of undercoordinated Ti sites. By monitoring the rate per mass of catalyst as a function of titrant loading, we can estimate the number of active sites across catalysts and calculate a site-specific turnover frequency. For example, the intrinsic TOF for cyclooctene epoxidation under these conditions is quite constant at 40 ± 7 hr^{-1} across 22 materials of vastly different characteristic spectra and precursor. While the intrinsic TOF is quite stable, the fraction of Ti atoms participating drops precipitously with increasing surface density. This method gives molecular-level insight into the catalyst structure that would otherwise be hidden using bulk spectroscopies. We further apply this technique to examine subtle differences in Ti coordination environment, specifically in-framework tetrahedral sites (e.g. $Ti(OSi)_4$ as in Ti-Beta) versus ex-framework tetrahedral sites (e.g. $Ti(OSi)_3OH$ as in grafted TiO_x-SiO_2 materials) and present further applications of this technique to other materials, looking specifically at other group IV, V and VI supported metal oxides.

[1] J.W. Niemantsverdriet, Spectroscopy in Catalysis, 2nd ed., Wiley-VCH, 2000.
[2] W.D. Williams, M. Shekhar, W.-S. Lee, V. Kispersky, W.N. Delgass, F.H. Ribeiro, S.M. Kim, E.A. Stach, J.T. Miller, L.F. Allard, Journal of the American Chemical Society 132 (2010) 14018-14020.
[3] J.M. Thomas, W.J. Thomas, Principles of Heterogeneous Catalysis, 1st ed., Wiley-VCH, 1997.
[4] J. Macht, R.T. Carr, E. Iglesia, Journal of Catalysis 264 (2009) 54-66.
[5] D. Prieto-Centurion, J.M. Notestein, Journal of Catalysis 279 (2011) 103-110.

Quantifying the number of active sites in supported catalysts improves our understanding and assists the rational design of new materials. Many supported oxides undergo significant structural changes as surface densities increase from site-isolated cations to monolayers and crystallites, which changes the number of kinetically-relevant sites. Here we perform selective titration of TiOx-SiO_2 via phenylphosphonic acid (PPA) to quantify fluid-accessible Ti sites, and *in situ* titration of the catalysts during *cis*-cyclooctene epoxidaiton with aqueous H_2O_2, know to occur over undercoordinated Ti sites. Results indicate that PPA bonds most strongly to the catalytically-active species and implies the selective titration of undercoordinated Ti sites. By monitoring the rate decrease per mole of PPA we estimate the number of active sites across 22 different catalysts with vastly different electronic properties, and find the intrinsic TOF for *cis*-cyclooctene epoxidation under these conditions is 40±6 h^{-1} across all catalysts examined. This method gives molecular-level insight into catalyst structure that would otherwise be hidden using bulk techniques.

Transition metals
Supported catalysts
Structure-activity relationships
Epoxidation
Titanium

Aldol condensation of n-butyraldehyde in a biphasic stirred tank reactor: Experiments and models

Shinbeom Lee and Arvind Varma

School of Chemical Engineering, Purdue University,
480 Stadium Mall Drive
West Lafayette, IN 47907, USA

ABSTRACT

A biphasic stirred tank reactor contains two immiscible liquids with liquid phase catalysts. To model this type of reactor, intrinsic kinetics and interfacial area are required. In this study, the modeling of a biphasic stirred tank reactor for n-butyraldehyde aldol condensation was conducted, including a study to determine the interfacial area. The interfacial area in the stirred tank reactor was directly measured by borescope system. The experimental range was 0-1.9M NaOH, 25-110 °C and 600-1000 rpm in a 300ml stirred tank. The temperature and high NaOH concentration effects were investigated experimentally and a correlation for interfacial area estimation was established. The reactor modeling was then carried out using a conventional resistance model, combining intrinsic kinetics from our prior work and interfacial area. The modeling was verified with 38 experiments of nBAL aldol condensation in the experimental range. In this verification, as compared with the experiments, the error by the model using interfacial area measurements was ±8%, while it was ±15% when using the correlation.

Keywords: biphasic reactor modeling, stirred tank reactor, interfacial area, temperature effect, aldol condensation

362173 Probabilistic Model for Correcting Directional Sensitivity of Optical Measurements

Abstract:

This work reports findings on the probabilistic model for correcting directional sensitivity of the optical probe technique commonly used in gas-liquid multiphase reactors. Optical signals are collected from the tapered end optical probe[1] in a lab scale gas-liquid stirred tank reactor at various directions under wide range of operating conditions. Based on the fact that the optical probe measurements are reliable when the bubble impaction angles are within 180 degrees,[2] contributions to the overall gas phase dispersions (holdups) from the bubbles traveling at two opposite directions, i.e., from left to right and right to left, are quantified. It is found that significant underestimations of gas holdup can be detected when the probe is not oriented properly. These results demonstrate optical probe technique's ability to accurately measure true gas holdup in regions where the direction of major bubble movement has not been previously determined.

1. Xue J. *Bubble Velocity, Size and Interfacial Area Measurements in Bubble Columns*. Saint Louis: Department of Chemical Engineering, Washington University; 2004.

2. Bombač A, Žun I, Filipič B, Žumer M. Gas-filled cavity structures and local void fraction distribution in aerated stirred vessel. *AIChE Journal.* 1997;43(11):2921-2931.

Electrical Capacitance Volume Tomography for Imaging of Pulsating Flows in a Trickle Bed

Aining Wang, Qussai Marashdeh and Liang-Shih Fan

William G. Lowrie Department of
Chemical and Biomolecular Engineering
The Ohio State University, Columbus, Ohio 43210

Abstract

In this work, experimental results of the air-water pulsating flows in a trickle bed column were obtained using the Electrical Capacitance Volume Tomography (ECVT) system. Detailed 3-D pulse structures in both the fully developed and the not fully developed (transient) conditions were illustrated. Pulse frequency, pulse travelling velocity, average liquid holdup and liquid holdup inside the gas-rich and liquid-rich regions, respectively, were measured. Based on a simplified mass-balanced model, the linear liquid velocities inside the gas-rich and liquid-rich regions were estimated. The results revealed that the gas flow rate was the most important parameter in controlling the pulsating flow properties. Discussion on the physical nature of the pulsating flow was also given.

ECVT is a novel process tomography technology recently developed based on Electrical Capacitance Tomography (ECT) which can provide direct 3D imaging. By designing capacitance plates with inherit 3D features, real-time inter-electrode capacitances can be directly related to the phase distribution inside the column. Its non-invasive nature makes it a suitable tool for multiphase flow research. In the present work, an air-water trickle bed packed with glass beads operating under pulsating regime was investigated using the ECVT. The trickling-to-pulsating transition line was identified first, and then the liquid pulse structure was directly imaged using the ECVT. A body-tail structure was observed for the fully developed pulse, which was composed of a relatively sharp and flat leading edge, a relatively uniform body and a non-uniform and off-centered tail. Not fully developed pulses (transient pulse) were also observed. The quantitative results revealed that the average pulse frequency was relevant to liquid flow only when the superficial liquid velocity was low, whereas the gas flow rate had a constant influence on the pulse frequency. The pulse velocity depended almost entirely on the gas flow rate. The average liquid holdup was influenced by both the gas and liquid flow rates because the inlet flow rates affected the length ratio between the gas-rich region and the liquid-rich region. However, inside a single unit with a gas-rich region and a liquid-rich region, the liquid holdups in these two regions were independent of the inlet liquid flow rate when the gas flow rate was constant. Based on a simplified model, the linear liquid velocities inside the gas-rich and liquid-rich regions were estimated. It was found that the increasing gas flow rate significantly accelerated the linear liquid velocity in the liquid-rich region. Discussion related to the physical nature of the pulsating flow was also given. The results indicated that the inlet gas flow rate was the most important parameter in controlling the pulsating flow properties.

Local Liquid and Gas Velocity Measurement Using Two-tip Optical Probe for Trickle Bed Reactors (TBRs)

Mohd Fitri Abdul Rahman[a], Khairul Anuar Mohd Salleh[b] and Muthanna H. Al Dahhan[a]

[a]Department of Chemical and Biological Engineering, Missouri University of Science and Technology, 143 Schrenk Hall, 400 W. 11th St., Rolla, MO 65409, USA

[b]Department of Mining and Nuclear Engineeri;ng, Missouri University of Science and Technology, Fulton Hall, 310 W. 14th St., Rolla, MO 65409, USA

Abstract

Trickle bed reactors (TBRs) are one of a fixed bed of particles where liquid and gas flow concurrently downward. TBRs are commonly applied in many industries which produce petroleum and chemical base products. Hydrodynamics studies in TBRs are indispensable as it helps process optimization. Travelling liquid velocity between the solid particles is one of the key performance characteristics of TBRs. This work measured the local liquid and gas velocities by two-tip optical probe (OP) measurement techniques in three phase system which are solid-liquid-gas. The experiments was performed in 0.14 m diameter reactor made of acrylic glass which filled with 3 mm glass beads which act as the solid. Water was used as the liquid phase while air will be the gas phase. There are 12 ports for OP with 0.15 m distance from each other. The superficial liquid velocities for both gas and liquid were in the range 0.03 m/s to 0.27 m/s and 0.004 m/s to 0.016 m/s, respectively. Two-tip optical was fabricated to measure the local liquid velocity in the TBR. Two tips were arranged parallel to each other at a distance of 1 mm. Results show that local liquid velocities obtained by two-tip optical fiber probe are in agreement with published observations.

Keywords: Trickle bed reactor; optical fiber probe, local liquid and gas velocity

387594 Kinetic and Mechanistic Study of the MoO_2 Reduction to Mo_2C in Methane Under Pulse-Type Conditions

Abstract:

Molybdenum carbide (Mo_2C), an interstitial transition metal carbide, has been widely used as an advanced material in myriad industrial applications due to its exceptional properties. These include its refractory nature, extreme hardness and strength, and high electrical and thermal conductivity. It also possesses catalytic activity for many chemical processes such as hydrogenation, reforming, water-gas shift and the Fisher-Tropsch reaction with a unique noble metal-like behavior. Solid state reduction between graphitic carbon and molybdenum oxides is commonly employed to produce Mo_2C. However, this carbothermic reaction involves very high temperatures (>1000 °C), leading to a low specific surface area, which is detrimental to catalytic purposes. New synthesis methods have been established to address this concern, such as the carburization reaction of molybdenum oxides with a flowing gas stream. This method yields the metal carbide with a specific surface area up to 200 m^2/g. Therefore, the present work investigates the kinetics of the formation of Mo_2C using a molybdenum dioxide (MoO_2) precursor and methane as the carburizing agent. Moreover, the carburization of transition metal oxides usually undergoes a single step directly to the carbide form, but in the case of MoO_2, transient X-Ray diffraction (XRD) indicates the formation of molybdenum metal as an intermediate species. In order to elucidate this particular reaction mechanism, thermogravimetric analysis, XRD and electron microscopy techniques have been employed. This mechanism was studied under pulse conditions wherein a cyclic flow of CH_4 was reacted with a packed bed of MoO_2.

Hydrogen solubility in the large-scale hydrogenation process of nitrobenzene to aniline

Ch. Meitzner*, F. Mayer*, T. Kwast*, R. Lange*

*Technische Universität Dresden, Germany

Keywords: gas-liquid-solid reactions, hydrogenation, nitrobenzene, hydrogen absorption

Introduction

Annually an amount of about 3 Mio. tons of aniline are produced in industrial well developed processes [1]. The main reaction is performed via a catalytic hydrogenation of nitrobenzene in moving or fixed bed reactors. The operating conditions vary at temperatures between 270 and 475 °C and operating pressure reaches up to 5 bar [2]. One of the most significant process characteristic is the high exothermic behavior caused by an entropy of about 450 kJ/mol [3]. Therefore either the heat production needs to be diminished or the process heat have to be usable for other purposes. For a more energy efficient process the reaction could be performed at lower temperatures in the three-phase regime.

For an optimized mass transfer the absorption or solubility of the reaction gas, hydrogen, needs to be known in various reaction mixtures. This will influence the preferred residence time distribution of the liquid phase at the catalytic surface. But since a lot of studies already exist concerning the kinetics of the reaction or the characterization of new catalysts, almost no data exist for the solubility of hydrogen in these multi-component mixtures, which is essential for a detailed process analysis. Consequently, in this work a method is developed to create usable solubility data. Hence, results will be presented starting from the pure substances until various reaction mixtures by using this experimental method.

Experimental work

Creating valuable data a pressure based technique has been selected. It is based on a stirred tank reactor of a known volume. By measuring the pressure loss of hydrogen the amount of hydrogen is calculated which solutes in the reaction mixture. Following, the absorption in terms of the Henry-Coefficient has been calculated.

It is assumed that the gas behavior is identical to real gas theory and that the temperature remains constant during the solution process. Various validation experiments showed good agreements with published literature data of Turek [4] and Katayama et.al. [5] (Compare Fig. 1).

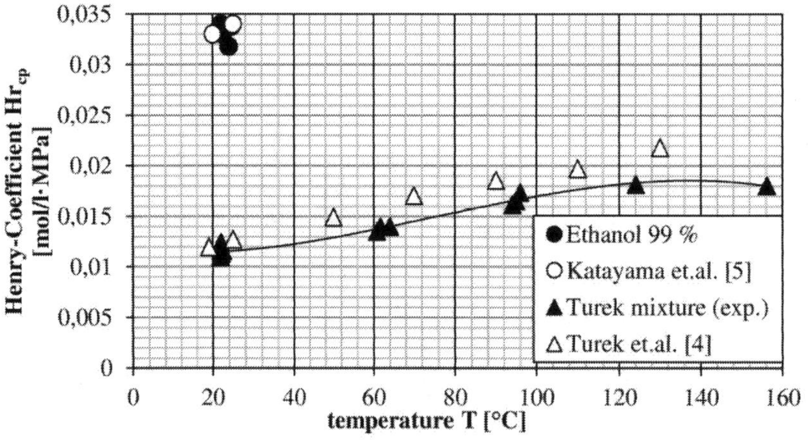

Fig. 1: Validation of measurement technique by literature data from Turek et.al. [4] and Katayama et.al. [5]

Thus a wide-ranging experimental design has been performed starting from pure substances up until real four-component mixtures. Due to high exothermic conditions of the reaction it needs to be considered to use a solvent for the reaction. Therefore, ethanol is chosen as a solvent and mixtures are prepared from high dilutions to pure substances. After the investigation of different starting concentrations for the process several conversion degrees have been observed. Each mixture was investigated at pressures until 30 bar and temperatures up to 160 °C.

Results

Starting from different dilutions it has been found that the percentage of ethanol plays an important role for the hydrogen solubility. In general it can derived that for ambient temperature (about 20 °C) the absorption will be higher with a rising percentage of ethanol.

Furthermore a nonlinear behavior of the Henry coefficient according to the temperature has been found (Fig. 2).

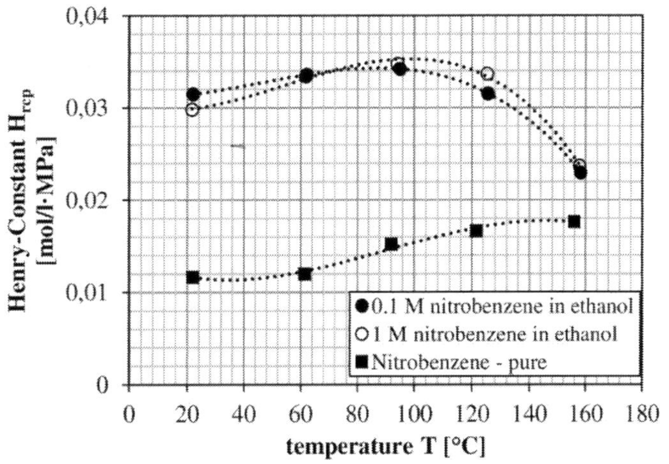

Fig. 2: Solubility of hydrogen in various starting concentrations

Especially for high dilutions a maximum of solubility has been observed at temperatures of about 100 °C. This could be among others an effect of the low steam pressure and the low boiling temperature of ethanol. These two facts were considered in the study and calculations. All components remain in the liquid phase in large part in all experiments and the respective steam pressure of the solution was included in the calculations.

Additionally, the effect of the products has been studied. It has been figured out that especially the presence of water causes interesting effects and influences the solubility during the process. Therefore the effect of aniline and water cannot be neglected since it significantly effects the progress velocity of the chemical reaction.

Acknowledgement

The authors highly acknowledge the Helmholtz-Gemeinschaft e.V. for funding the project "Helmholtz Energy Alliance – Chemical Multiphase Processes". Therein, the nitrobenzene hydrogenation is one of the main focuses of the research.

Literature

[1] Arpe, H.-J.: *Industrial Organic Chemistry*. 2010.

[2] GdCh: *BUA Stoffbericht 171 Anilin.* 1996.

[3] Klemm, E., Amon, B., Redlingshöfer, H., Dieterich, E., Emig, E.: *Kinetics in the hydrogenation of nitrobenzene to aniline – investigations in an isothermal catalytic wall reactor.* Chem Eng Sci 56 (2001). 1347-1353.

[4] Turek, F., Geike, R., Lange, R.: *Liquid Phase hydrogenation of nitobenzene in a slurry reactor.* Chem Eng Process. 20 (1986). 213-219.

[5] Katayama, T., Nitta, T.: *Solubilities of hydrogen and nitrogen in alcohols and n-Hexan.* J Chem Eng Data 21 (1976). 194-196.

370601 Toward Improved Heterogeneous Catalysts for Olefin Metathesis

Abstract:

Propylene ($C_3^=$) is the second largest feedstock, after ethylene, for the petrochemical industry and a shortage is developing that will increase over the next decade. To meet this demand, industry is turning to heterogeneous olefin metathesis catalysis via $C_2^= + C_4^= \rightarrow 2\ C_3^=$. This process still employs inexpensive, simply prepared, and ill-defined, multicomponent heterogeneous catalytic systems. A key challenge is to improve performance of catalysts for this reaction. Unlike the well-defined, single-site homogeneous olefin metathesis catalysts, a lack of fundamental information about the active sites of commercial heterogeneous olefin metathesis catalysts based on supported metal oxides has hindered further development. Heterogeneous catalysts consist of transition metal oxides (WO_3, MoO_3, and Re_2O_7) supported on high surface area oxide supports (typically SiO_2 and Al_2O_3). The catalytic active sites, electronic and molecular structures, and surface reaction intermediates of these catalysts are poorly understood. In this talk, we will elucidate molecular/electronic-catalytic activity/selectivity relationships by forming designed ReO_x surface oxide species on a planar Al_2O_3 single-crystal film support. The surface is characterized by using high resolution X-ray photoelectron spectroscopy (HR-XPS), low energy ion scattering (LEIS), ultraviolet photoelectron spectroscopy (UPS), high resolution electron-energy loss spectroscopy (HREELS), and infrared-absorption spectroscopy (IRAS). ReO_x species on planar Al_2O_3 films were synthesized and characterized with atomic and nanoscale control of the reactive sites present. By controlling the O_2 partial pressure in the background gas and the substrate temperature, we control the oxidation states of the evaporated Re and Re-oxide nanostructures formed. The structures associated with each oxidation state and the initial reactivity of each structure were then determined. The adsorption, desorption, reaction pathways, and reaction activation energies of ethylene at various reactive sites was probed using temperature programmed desorption (TPD). Increased fundamental understanding of the nature and reactivity of active sites will provide a firm basis for the development of improved heterogeneous metathesis catalysts.

375420 Advanced Nanocatalysts from Organic Solution Synthesis

Abstract:

Materials with one or multiple dimensions confined at the nanoscale, i.e., nanomaterials, possess distinctive properties compared to their bulk counterparts, and have been demonstrated to be particularly advantageous in catalysis. While conventional catalyst preparation such as impregnation has derived catalytic materials at nanoscale, synthesis of heterogeneous catalysts with robust control over particle size, shape and composition has been a great challenge for chemical engineers. This talk intends to highlight our progress on organic solution synthesis for the development of advanced nanomaterials for catalytic applications. Examples illustrating the methods and mechanisms of controlling nanocrystal nucleation and growth will be discussed, in terms of control over particle size, shape, composition and surface structure. Ligands removal for surface cleaning and synthesis of heterogeneous composite nanoparticle catalysts will also be discussed. While our focus will be placed on Pt-based catalysts for electrochemical reactions in proton exchange membrane fuel cells (PEMFCs), our approach would have general impact on the development of heterogeneous catalysts based on nanomaterials.

Synthesis of NiMo/γ-Al_2O_3 Catalyst for Hydrodesulfurization using Colloidal Synthesis

Rupesh Singh, Deepak Kunzru, Sri Sivakumar

Department of Chemical Engineering, Indian Institute of Technology Kanpur, Kanpur –208016, India

Abstract

Hydrodesulfurization (HDS) is a common refinery reductive treatment of organic sulfur present in petroleum feedstocks to remove sulfur in the form of H_2S.[1] Nowadays, due to increasing demand of petroleum products, heavier fractions, having high sulfur content, need to be processed. Out of the different types of sulfur compounds present in petroleum fractions, benzothiophene and dibenzothiophene are the most difficult to desulfurize because of their bulky molecular structure and the steric hindrance associated with alkyl substituted groups. [2]

To reduce the sulfur content to within allowable limits, it is necessary to develop hydrodesulfurization (HDS) catalysts with improved activity, selectivity, and stability. Presently, NiMo/γ-Al_2O_3 and CoMo/γ-Al_2O_3 are the most commonly used catalyst combinations for HDS. [3]In these catalysts, Ni-Mo-S/ Co-Mo-S sites are considered to be the active sites with Ni / Co on the edge of MoS_2 phase. HDS activity is related to the number of Co/Ni edge sites and the active sites for HDS are associated with the sulfur vacancies at these edge sites.[4] The activity of HDS catalyst depends upon its particle size, shape, composition as well as on the support. The main drawback with conventional methods (such as catalyst impregnation[5]) is that it is difficult to control the particle size, shape and uniformity of the active metal on the support.

In this study, highly monodisperse NiMo nanoparticles in the size range of 3±2 nm were synthesized using colloidal synthesis. Anionic surfactants were used to arrest the growth of the NiMo nanoparticles during the synthesis. These nanoparticles were subsequently supported on γ-Al_2O_3, followed by drying and calcination to prepare NiMo/γ-Al_2O_3 catalysts suitable for HDS reaction. Four different catalyst batches for NiMo catalyst batches were prepared using same method. The MoO_3 content in the catalyst batches was varied from 4.5 to 13.2 wt% keeping Ni/(Ni+Mo) constant to a constant value of 0.3. Catalyst characterization (TEM, XRD) confirmed the formation of nanoparticles. Dibenzothiophene (a refractory sulfur compound commonly present in diesel) was used as a model compound to study the HDS activity of the prepared catalyst. The HDS activity was tested in a fixed–bed reactor at 30 bar pressure in the temperature range of 275 to 335°C. The feed consisted of 1.68 wt% DBT dissolved in n-decane. Figure 1(a) represents the comparison of DBT conversion for NiMo catalyst prepared using different preparational techniques, i.e. colloidal and conventional impregnation methods with same compositions. For temperature range of 275 to 305°C NiMo catalyst prepared using colloidal synthesis technique shows better catalytic activity. HDS of DBT follows two pathways: (i) direct desulfurization of DBT in which the hydrogenolysis of sulfur bond takes place leading to the formation of biphenyl (BP); and (ii) desulfurization after hydrogenation of one of the aromatic rings followed by hydrogenolysis of the sulfur bond, leading to formation of cyclohexylbenzene (CHB). Figure 2(b) shows the schematic representation of the HDS of dibenzothiophene. For all catalyst samples, the major product was BP and selectivity towards BP decreases with an increase in temperature.

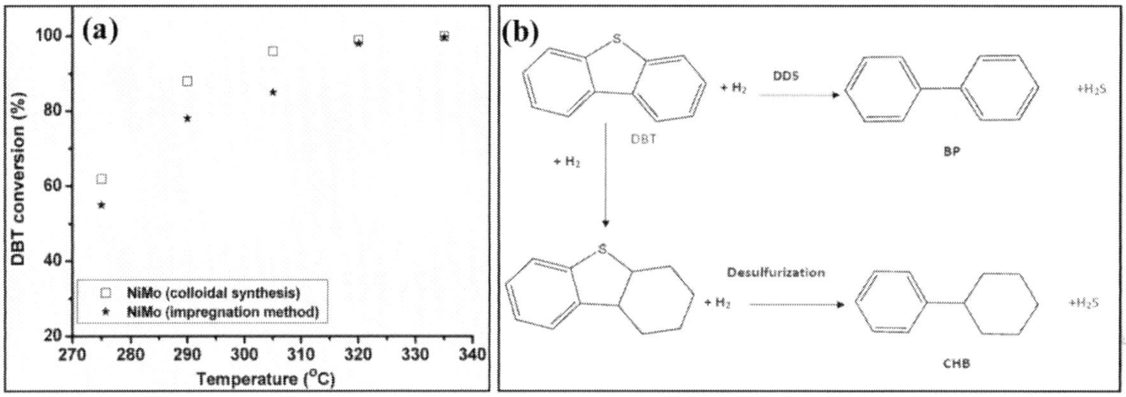

Figure 3: (a) Effect of temperature on DBT conversion for NiMo catalyst. **(b)** Schematic representation of the HDS of DBT

Keywords: Hydrodesulfurization, NiMo/γ-Al_2O_3 catalyst, dibenzothiophene, colloidal synthesis, NiMo nanoparticles

Acknowledgement: The financial support provided by Chevron Corporation, USA and Hindustan Petroleum Corporation Limited, Mumbai for this project is gratefully acknowledged.

References:

1. Vasudevan, P.T. and J.L.G. Fierro, *A Review of Deep Hydrodesulfurization Catalysis.* Catal. Rev.-Sci. Eng.,, 1996. **38**: p. 161.
2. Dengqian Zhang, et al., *Synthesis, characterization and catalytic performance of meso-microporous material Beta-SBA-15-supported NiMo catalysts for hydrodesulfurization of dibenzothiophene.* Catalysis Today 2011. **175**: p. 477– 484.
3. D. Solı´s, et al., *Hydrodesulfurization of gasoils over NiMo/Al2O3–H(or Ni)NaY zeolite hybrid catalysts.* Catalysis Today 2004 **98** p. 201–206.
4. Jeppe V. Lauritsen , et al., *Location and coordination of promoter atoms in Co- and Ni-promoted MoS2-based hydrotreating catalysts.* Journal of Catalysis 2007 **249** p. 220–233.
5. Yinyong Sun, Huamin Wang, and R. Prins, *Hydrodesulfurization with classic Co–MoS2 and Ni–MoS2/g-Al2O3 and new Pt–Pd on mesoporous zeolite catalysts.* Catalysis Today 2010. **150** p. 213–217.

380019 A DFT Study of the Oxygen Reduction Reaction Mechanism over the O-Doped Graphene-Supported Pt4/Pt3Fe and Pt3V Alloy Catalysts

Abstract:

Developing a highly efficient ORR catalyst will have a significant impact on fuel cell applications. The graphene supported Pt_3M alloy nanoparticles as ORR catalysts exhibit 3-4 times higher ORR activity than that of the supported Pt catalyst. Here density functional theory calculations were performed to investigate the pathways of the oxygen reduction reaction over the Pt_4 and Pt_3M (M = Fe, V) clusters supported on the O-doped graphene substrate. The results show that the dangling C atom resulting from oxygen doping becomes an anchor site for metal clusters. Our results also showed that O_2 adsorbs as a di-oxygen species on the supported Pt_4 and Pt_3Fe clusters, but dissociates spontaneously on supported Pt_3V. The di-oxygen species dissociates into co-adsorbed HO* and O* upon reduction on both supported Pt_4 and Pt_3Fe and no stable HOO* intermediates were isolated. On supported Pt_4, further reduction via both HO*+HO* and H_2O+O* routes is possible, with reaction favoring the HO*+HO* route energetically. On supported Pt_3Fe, the HO*+HO* and H_2O+O* routes are competitive. On supported Pt_3V, the reduction reaction is likely to proceed exclusively through the HO* + HO* route as no stable co-adsorbed H_2O and O* state was isolated. The results were discussed in the context of the experimentally observed enhancement of ORR reactivity on the graphene supported Pt_3Cr and Pt_3Co nanocatalysts.

Low-Temperature Catalytic Conversion of Ethanol to Acetone on Au/$ZnZrO_x$ Nanostructures

Chongyang Wang[1], Gabriella Garbarino[2], Faith Wilson[1], Guido Busca[2], Maria Flytzani-Stephanopoulos[1*]

[1]*Tufts University, Medford, MA 02155 USA*

[b]*Università degli Studi di Genova, Genoa 16129 Italy*

**corresponding author: mflytzan@tufts.edu*

Abstract

Nanoscale zirconia (ZrO_2) has been intensively studied as catalyst support. ZrO_2 supported gold catalysts have demonstrated promising metal/metal oxide interaction for a number of reactions[1-4]. In this work, we investigated the Au/ZrO_2 system as catalyst for low-temperature ethanol-to-acetone conversion; moreover, the usage of a novel $ZnZrO_x$ support is used to modify and tune the physical and chemical properties of pure ZrO_2. The nanoscale $ZnZrO_x$ composite metal oxide has superior properties than either bulk ZrO_2 or ZnO in terms of dispersing and stabilizing Au atoms, as prepared and under the ethanol conversion reaction conditions (30-500°C). The $ZnZrO_x$ supports were synthesized by a carbon hard-template method, with gold added to the support by the anion-adsorption method[5]. Selectivity and activity of the catalysts were tested in temperature-programmed surface reaction (TPSR) and under steady-state reaction conditions with various steam-to-ethanol ratios monitored online by mass spectrometry and gas chromatography. Surface acidity titration by isopropanol-temperature programmed desorption (IPA-TPD/MS) and characterization by ICP, XPS, XRD, FTIR and STEM imaging were conducted to follow the changes on the catalyst surface. The work has demonstrated that $ZnZrO_x$ as gold catalyst support can maintain the strong Au-Ox-surface bonds during reaction, which promoted the activity and suppressed undesired ethylene formation compared to pure ZnO or ZrO_2. This is the first time that the Au/ZrO_2 system is exploited for catalytic ethanol conversion, and the full conversion of ethanol was reached below 400°C.

[1] V. Idakiev, T. Tabakova, A. Naydenov, Z.Y. Yuan, B.L. Su, Applied Catalysis B: Environmental 63 (2006) 178-186.
[2] Y. 24-Zhang, Y. Zhan, Y. Cao, C. Chen, X. Lin, Q. Zheng, Chinese Journal of Catalysis 33 (2012) 230-236.
[3] J. Li, N. Ta, W. Song, E. Zhan, W. Shen, gold bulletin 42 (2009) 48-60.
[4] X. Zhang, H. Wang, B.-Q. Xu, The Journal of Physical Chemistry B 109 (2005) 9678-9683.
[5] C. Wang, M. Boucher, M. Yang, H. Saltsburg, M. Flytzani-Stephanopoulos, Applied Catalysis B: Environmental 154-155 (2014) 142-152.

390317 Structural and Environmental Effects on the Activity of Nanoporous Gold Catalysts for Sustainable Oxidation Reactions

Abstract:

Unsupported nanoporous gold (npAu) is currently being investigated as a promising material for more sustainable selective oxidation processes. This material provides a unique catalytic surface that takes advantage of the surface reactivity of pure gold to guide product selectivity – for example, for the selective coupling of alcohols to form esters. The focus of this work is to improve our understanding of the factors that control npAu activity and stability for such reactions. Nanoporous Au, typically prepared by the dealloying of sacrificial Ag from a bulk alloy, is a dilute gold-silver alloy. The residual silver is thought to produce adsorbed atomic oxygen (via O_2 dissociation) which spills over onto the gold and is then consumed in the activation and further reaction of alcohols (e.g. methanol). When used for alcohol oxidation reactions, npAu requires a lengthy activation period prior to use by exposure to reaction gases (e.g. exposure to methanol and oxygen at 70 °C for up to 15 hours). The effect of this treatment on the catalyst has remained unclear. Furthermore, the activated catalysts are prone to deactivation upon exposure to higher alcohols (e.g. ethanol, 1-butanol) or alternative coupling reactants (e.g. acetaldehyde). Here we present recent developments on the activation of nanoporous gold using ozone and its effect on the ligament size, surface composition, and catalyst activity and selectivity. We show that ozone treatment at 150 °C greatly increases the stability of the catalyst, while maintaining high selectivity to methyl formate during methanol oxidation. We provide evidence for the formation of silver oxide features at the catalyst surface, which appear to affect the conditions under which oxygen activation and catalytic methanol coupling occur. By using a combination of electron microscopy, including environmental TEM, and X-ray photoelectron spectroscopy we relate dynamic changes in the material to activity and selectivity. These studies provide detailed insight into the origin of catalytic activity and illustrate a methodology for investigating this class of nanoporous metallic catalysts.

Structure and Catalytic Property of Pt-Mo Nanocatalyst for Glycerol Decomposition

Bin Liu[$] and Jeffrey Greeley[£¢]

[$]Department of Chemical Engineering, Kansas State University, Manhattan, KS 66506,
[£]School of Chemical Engineering, Purdue University, West Lafayette, IN 47907-2100,
[¢]Center for Nanoscale Materials, Argonne National Laboratory, Argonne, IL 60439

ABSTRACT

Biomass derived glycerol is regarded as an important platform compound, and the efficiency for hydrogen and chemical productions strongly depends on the catalytic property that determines the selectivity for C-H, O-H, C-C and C-O bond scissions. In this paper, the catalytic behaviors of these bond scissions associated with glycerol decomposition on Pt-Mo alloy-based catalysts are studied using periodic Density Functional Theory (DFT) and linear scaling relationships. An important consideration is the structure of Pt-Mo bimetallic alloy under aqueous phase environment, where the Mo sites are most likely present oxidized and covered by hydroxyl groups. The surface configurations will be significantly modified as a result of MoO_x formation. The surface reconstruction and segregation of MoO_x from Pt at the nanoscale are investigated using DFT calculations, whereas the catalytic functionalities of oxidized Pt-Mo surfaces during glycerol decompositions are also studied. To gain the mechanistic insight on glycerol decomposition on Pt-Mo alloy surfaces, our bond-order based scaling relationship developed for monometallic surfaces (e.g. Pt, Pd, Ni, Rh, Cu) was validated and extended to this multi-component nanocatalyst system based on the binding energies of atomic C and O on several Pt-Mo alloy surface configurations. This approach can be used to accelerate our catalytic trend analysis. In this paper, we will show that the Pt-skin alloy surfaces exhibit similar glycerol decomposition trends to pure Pt surfaces, with weaker C, O binding, the thermodynamics of glycerol decomposition becomes more endothermic. The weaker CO binding may actually allow more active sites for glycerol dehydrogenation and C-C bond scission. Mo sites may favor O-H, C-O bond scissions; in aqueous phase reforming conditions, water-gas shift reactions could also be affected favorably at such sites. The study reveals a promising bifunctional mechanism exhibited by the Pt-Mo bimetallic alloy as a reforming catalyst.

Experimental and Modeling Study of 2-Butene and Isobutene Pyrolysis

Kun Wang, Stephanie M. Villano, and Anthony M. Dean*

Chemical and Biological Engineering Dept., Colorado School of Mines, Golden, CO 80401

*corresponding author: amdean@mines.edu

Abstract:

Butene is the smallest alkene with isomeric structures (1-butene, 2-butene, and isobutene). The study of 1-C_4H_8 and 2-C_4H_8 pyrolysis kinetics allows analysis of the impact of the double bond location, while isobutene serves as a model compound for larger branched alkenes. Recent investigations on the combustion of butanol isomers[i], and larger alkane fuels[ii] such as heptane and isooctane, have demonstrated the importance of the reactions of butene isomers.

Building on earlier work with 1-butene[iii, iv, v,] the present study focuses on 2-C_4H_8 and i-C_4H_8 pyrolysis. The conversions of 2-C_4H_8 and i-C_4H_8 start at higher temperatures than 1-C_4H_8. This is consistent with the fact that cleavage of the bond C=C-C---C in 1-butene only requires 76 kcal/mol, while breaking the CC=CC---H in 2-C_4H_8 and C=C(C)C---H in iC_4H_8 both require 88 kcal/mol. However, 2-C_4H_8 is more active than the branched i-C_4H_8 due to the different secondary reactions, with the fuel conversion ~89% comparing with ~42% at 725ºC (pressure 0.8 atm and resident time ~5 sec). At similar fuel conversions, 2-C_4H_8 produces more propene, 1-butene, and 1,3-butadiene, while i-C_4H_8 forms more H_2 and C_3H_4 isomers (allene and propyne). Methane and ethene are produced in comparable amounts for both fuels. Pyrolysis of i-C_4H_8 leads to production of more molecular weight products, with the distinction that it forms more C_7 species while 2-C_4H_8 forms more C_5 species.

A detailed kinetic mechanism has been developed that predicts the butane isomer conversions, production of both major and minor species, as well as the formation of molecular weight growth species up to C_{10}. The resonantly-stabilized free radicals (RSFR), 1-methyl-allyl (CC=CC. or C=CC.C) and 2-methyl-allyl (C=C(C)C.) in 2-C_4H_8 and i-C_4H_8 system, respectively, were found to play a critical role in the kinetics. High-level electronic structure calculations were used to generate high-pressure rate constants for these reactions, and the temperature- and pressure-dependent rate constants were obtained using a QRRK/MSC analysis. This approach has allowed significant new insights into the factors contributing to molecular weight growth during pyrolysis of hydrocarbon fuels.

---

[i] S.M. Sarathy, S. Vranckx, K. Yasunaga, M. Mehl, P. Oßwald, W.K. Metcalfe, C.K. Westbrook, W.J. Pitz, K. Kohse-Höinghaus, R.X. Fernandes, H.J. Curran, Combust. Flame 159 (2012) 2028-2055

[ii] Marco Mehl, William J. Pitz, Charles K. Westbrook, Henry J. Curran, Proceedings of the Combustion Institute 33 (2011) 193–200

[iii] Wang K., Al Shoaibi A., Villano M. S., and Dean M. A., "Experimental & Modeling Study of Molecular Weight Growth Species Formation during Olefin Pyrolysis", ACS 2013 National Spring Meeting, New Orleans, Louisiana

[iv] Wang K., Villano M. S. and Dean M. A., "Experimental & Modeling Analysis of 1-Butene Pyrolysis: Characterization of Molecular Weight Growth Kinetics", Western States Section of the Combustion Institute 2013 Fall Technical Meeting, Fort Collins, Colorado

[v] Wang K., Al Shoaibi A., Villano M. S., and Dean M. A., "Experiment and Kinetic Modeling of Molecular Weight Growth Kinetics during Olefins Decomposition: 1-Butene Pyrolysis" (to be submitted)

SESSION: 20003 Reaction Engineering for Combustion and Pyrolysis

Predictive kinetic modeling of formation of aromatics during pyrolysis of alcohol fuels

Shamel Merchant,[1*] Aäron G. Vandeputte,[1] Ruben De Bruycker [2], Marie-Françoise Reyniers[2], Kevin M. Van Geem[2] and William H. Green[1]

[1] Department of Chemical Engineering, Massachusetts Institute of Technology, MA, USA
[2] Laboratory for Chemical Technology, University of Ghent, Belgium
[*]Corresponding author: shamel@mit.edu

Increasing environmental deterioration and energy demands has led to several studies for new alternative fuels. Several hundreds of candidates have been proposed of which C_4 and C_5 alcohols are the leading candidates. They are considered for industrial scale production by a number of companies including BP (Butamax), Gevo and Cobalt Inc. The selection of a good candidate fuel requires the development of detailed kinetic models for its pyrolysis and/or combustion. Many of these networks have been extensively validated with experimental data over a wide range of conditions. Despite the fact that most models succeed to accurately predict speciation, i.e., primary products, and combustion characteristics, a recurring problem is the misprediction of many aromatic products, such as benzene, toluene and naphthalene which are secondary products. As these components are important precursors to PAH, soot, coke and some important carcinogenics, understanding their formation is of major importance for environmental preservation. Some pathways to the aforementioned aromatics have been suggested in literature but are insufficient to describe their formation from alcoholic fuels. The associated chemistry is complex and requires computer-aided modeling to comprehensively study this process and reach definitive conclusions.

In this work we present a detailed kinetic model for butanol (n-, iso- and tert-) and pentanol (n- and iso-) isomers that is able to predict the formation of aromatic compounds. The kinetic mechanism is generated using the open source software package Reaction Mechanism Generator (RMG). The model is validated against extensive speciation data from pyrolysis experiments in a plug flow reactor. Using these kinetic models we are able to provide mechanistic insights as to, a), how fuel structure, specifically, branching of carbon backbone and presence of alcohol group, affects aromatic emissions, and b), which key pathways and precursors (cyclopentadiene) lead to formation of aromatics under engine relevant conditions. This study shows the potential of automated predictive kinetic modeling as a tool for the quick screening of a vast number of fuel candidates which limits expensive and time-consuming engine tests.

The Role of Aerosols in Hypergolic Ignition

Craig D. Needham and Phillip R. Westmoreland

Dept. of Chemical and Biomolecular Engineering, North Carolina State University, Raleigh, NC 27695

In this study, we have used *ab initio* quantum chemistry calculations to investigate the energy benefit from the solvation of gaseous fuel and oxidizer molecules into liquid aerosol droplets. In their studies of the hypergolic bipropellant pair monomethylhydrazine (MMH) and nitric acid (HNO_3), Wang and Thynell [1] hypothesized that aerosol-phase solvation was the dominant energy source for raising the temperature of their systems from the boiling point of the liquids to the ignition temperature. The major role of these relatively poorly studied reactions in the MMH/HNO_3 ignition mechanism suggests that further study could yield insights about hypergolic ignition as a whole.

Given that the interaction of gaseous MMH and HNO_3 leads to aerosol formation, this work simultaneously explores the minimum amount of solvent needed to solvate the gaseous molecules and the kinetics of that solvation. Our calculations start with a single pair of molecules and proceed via the introduction of additional solvent molecules, a method adapted from Tao [2]. Both MMH and HNO_3 were explored as solvents, with the accompanying species acting as the solute. In each case, the geometry of the system was optimized allowing us to study the potential energy and the degree of ionization (i.e., the progression of the proton(s) as they are transferred from anion to cation) as the amount of solvent increases. Analysis of these calculations gives us detailed information about the energetics of solvation in the aerosol phase and the amount of gaseous fuel and oxidizer necessary for aerosol droplets to form in the system.

Acknowledgements: This work was funded by the U.S. Army Research Office under Multi-University Research Initiative grant number W911NF-08-1-0171 with Ralph A. Anthenien, Jr., as program manager, and by the NCSU GAANN Fellowship program.

[1] Wang, S. Q., and Thynell, S. T., "An experimental study on the hypergolic interaction between monomethylhydrazine and nitric acid," *Comb. Flame* **159** (2012) 438-447.

[2] Tao, F.-M., "Gas phase proton transfer reaction of nitric acid-ammonia and the role of water," *J. Chem. Phys.* **108** (1998) 193-202.

Effect of Non-Thermal Product Energy Distributions on Ketohydroperoxide Decomposition Kinetics

C. Franklin Goldsmith[ab], Michael P. Burke[b], Yuri Georgievskii[b], Stephen J. Klippenstein[b]

[a]*School of Engineering, Brown University*

[b]*Chemical Sciences and Engineering Division, Argonne National Laboratory*

Abstract

The decomposition of ketohydroperoxides (OQ′OOH) to two radicals is commonly predicted to be the key chain branching step in low-temperature combustion. The possibility of a direct decomposition of the OQ′OOH from its initially produced energy distribution is studied with a combination of master equation (ME) and direct trajectory simulations. The temperature and pressure dependent rate constants for the thermal decomposition of a ketohydroperoxide, $HOOCH_2CH_2CHO$, to four product channels were computed using RRKM/ME methods. Direct dynamics calculations were initiated from a transition state in the O_2 + QOOH reaction network to understand the fraction of energy in that transition state that is converted into the internal energy of the OQ′OOH. A novel approach to solving the master equation is used to determine the probability that a vibrationally hot OQ′OOH either will be stabilized to a thermal distribution or will react to form new products. Under most conditions, the majority of vibrationally excited OQ′OOH will be quenched into a thermal distribution. At higher internal energies and lower pressures, however, a significant fraction of the hot OQ′OOH will decompose rather than thermalize. Proper interpretation of low-pressure experiments may require inclusion of vibrationally hot intermediates, particularly if a chemical kinetic mechanism is optimized against the low-pressure data.

Figures

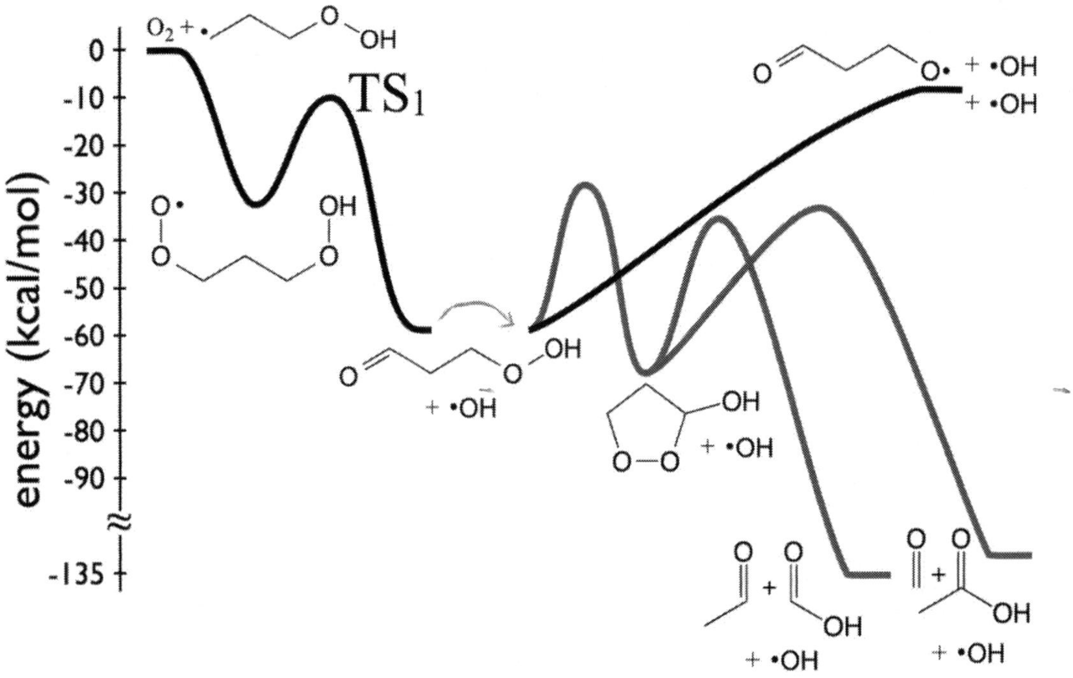

Figure 1: Combined $C_3H_7O_4$ and $C_3H_6O_3$ (+OH) potential energy surfaces. The chain branching pathway is shown in black; the Korcek pathway is shown in blue. Direct dynamics simulations were initiated from TS_1.

378086 Molten Salt Pyrolysis of Waste Polystyrene

Abstract:

In recent decades, the world has been concerned about the environmental impact of waste plastic, and researchers have been focusing on developing new methods to recycle plastic waste. The most common are mechanical recycling techniques, but they fail to treat the majority of waste plastics. Thus, recent studies focus on monomer recovery via thermal pyrolysis. The goal of this project was to develop a novel single pot pyrolysis technique by employing molten salts to attain good yields at varied conditions. The effects of temperature, salt composition and catalyst were studied, and the yield and the composition of liquid products were analyzed. The best temperature and salt composition that result in the highest liquid yield, monomer recovery and monomer selectivity, were identified.

Development of an Improved Kinetic Mechanism for the Pyrolysis of Anisole in H_2 environment

Yogesh Koirala, Stephanie M. Villano, and Anthony M. Dean
Department of Chemical and Biological Engineering, Colorado School of Mines
and
Hans-Heinrich Carstensen, Marie-Francoise Reyniers, and Guy B. Marin
Laboratory for Chemical Technology, Ghent University

Fast pyrolysis of lignocellulosic biomass is considered a promising thermochemical process to produce biofuels or specialty chemicals. However, due to the lack of a fundamental understanding of the complex chemistry, the obtained product distribution is not well understood and cannot be controlled. An improved understanding of the underlying decomposition reactions of the principal components of biomass - cellulose, hemicellulose and lignin – could lead to improved process control and possible mitigation of the formation of undesirable side-products. In this study, the focus is on lignin. Because the structure of lignin is complex and not unique, model compounds are often utilized to gain insight into its reactivity. The simplest compound that can be employed to investigate the weak phenolic-carbon linkage in lignin is anisole ($C_6H_5OCH_3$). Although several studies have already been carried out to understand the thermal decomposition of anisole, there are limited data available at fast pyrolysis process conditions. Furthermore, the existing models are not able to predict the observed product distribution accurately. In this work, we present anisole thermal decomposition data that were collected at conditions that closely resemble fast pyrolysis (T=525-650°C, τ ~0.8s, and P~0.8 atm) using a tubular flow reactor. We compare two sets of data. The first case is 0.75mol % anisole diluted in N_2; in the second case 5% H_2 is added to shift the radical pool concentrations. An improved reaction mechanism, which is based on a hydrocarbon pyrolysis model augmented with reactions of oxygenates, has been assembled. The rate coefficients of latter reactions are obtained from electronic structure calculations. The kinetic model is used to predict the data collected here as well as previously published data.

An important kinetic feature of the reactions of oxygenated aromatics is that resonance-stabilized radicals can be formed. In these, the spin-density is distributed over several sites which leads to multiple products upon recombination with other radicals such as methyl. For example, methyl addition to phenoxy forms anisole and also o/p-methylcyclohexadienone. The o/p-methylcyclohexadienone (o/p-MecyHDOE) products have weaker C-H bonds than anisole, thus favoring hydrogen abstraction reactions from these isomers. There are also multiple unimolecular reactions of the methylcyclohexadienone isomers. Figure 1 shows the potential energy surface for the ortho isomer (1). The stationary points and transition states located on this PES were calculated using the CBS-QB3 level of theory. To improve the accuracy of the $\Delta_f H_{298}$, S_{298}, and C_p values, low frequency vibrational modes that resemble torsions around single bonds are treated as hindered internal rotors; the hindrance potentials were calculated at the B3LYP/6-31G(d) level of theory via relaxed surface scans. The electronic energy of each species was converted to its heat of formation using the atomization method. High-pressure rate coefficients were calculated using canonical transition state theory. These high pressure rate coefficients were used to obtain pressure- and temperature-dependent rate coefficients using Quantum-Rice-Ramsperger-Kassel theory with a modified strong collision approximation.

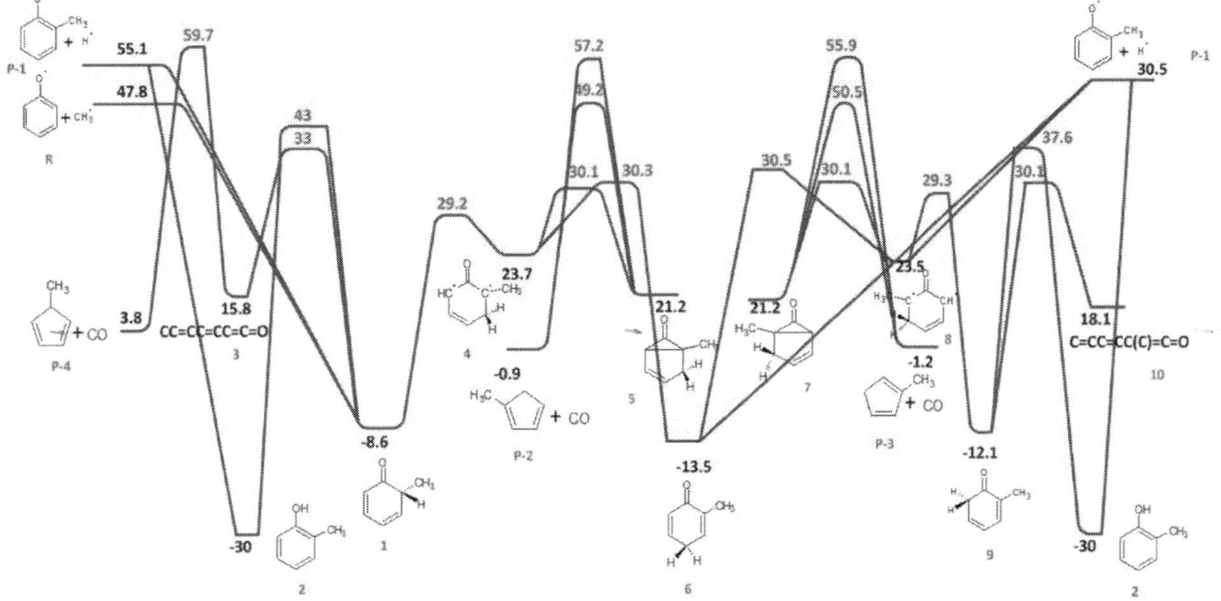

Fig 1. Potential Energy Surface for o-methyl cyclohexadienone (units: kcal mol -1)

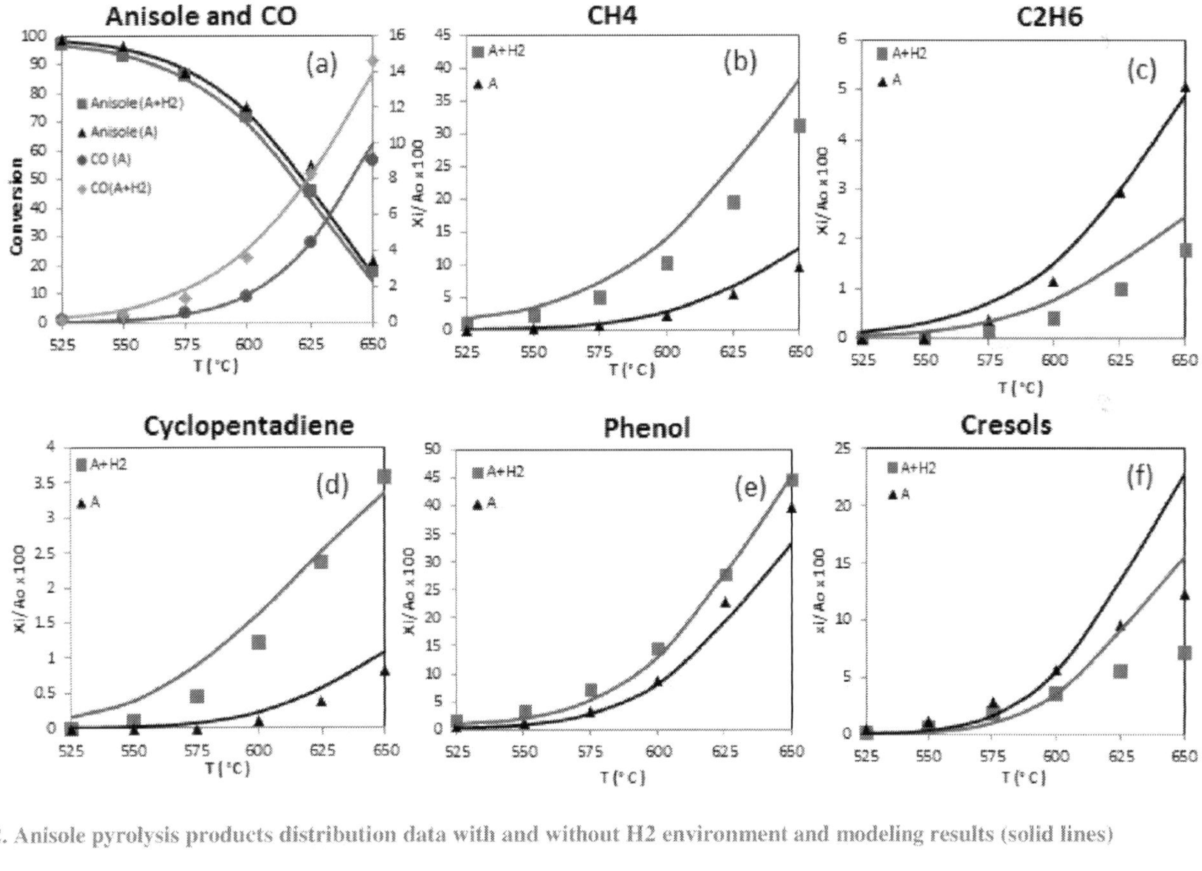

Fig 2. Anisole pyrolysis products distribution data with and without H2 environment and modeling results (solid lines)

Comparisons of the predictions with this model to the experimental data obtained in this study are shown in Fig 2 for some major species. The observed shift in the product distribution due to H_2 addition is well captured by the mechanism. Methyl abstraction from H_2 generates methane plus an H atom. Hence, there is increase in the methane (Fig 2b) and H atom concentrations in the system. The H atoms either abstract from remaining anisole or its isomers or add to the ring, leading to cyclohexadienone via β scission. Cyclohexadienone then isomerizes to phenol. The decrease in methyl radical concentration leads to less ethane production (Fig 2c) as well as a decrease of cresol production (Fig. 2f). This decrease in methyl radical would increase phenoxy concentration, leading to increased mole fractions of CO and cyclopentadienyl radical. The model suggests that abstraction reactions of phenoxy also contribute to some extent to the increased phenol yields. Although this updated mechanism properly accounts for the effect of H_2 addition, it is clear that further improvements are needed, as the cresol yields are still over-predicted at the higher temperature (which correlates to higher conversions).

Recombination and Addition Reactions Kinetics of Resonantly-Stabilized Free Radicals

Kun Wang, Stephanie M. Villano, and Anthony M. Dean*

Chemical and Biological Engineering Dept., Colorado School of Mines, Golden, CO 80401

*corresponding author: amdean@mines.edu

Abstract: Resonantly-stabilized free radicals (RSFR) have the potential to play a critical role in molecular weight growth kinetics during hydrocarbon pyrolysis or oxidation. Due to their stability, RSFR can accumulate to relatively high concentrations and certain recombination reactions have been recognized as important pathways to growth. In this work, we use electronic structure calculations to examine addition and recombination reactions of RSFR. Unlike propargyl recombination, where formation of benzene is important, the calculations for the self- and cross-recombination of allyl and methyl-allyl indicate that the energy barrier for the cyclization of the initially-formed adduct is sufficiently high that this pathway is not favored. A subsequent H-abstraction reaction is required prior to cyclization. Results for resonantly-stabilized radical addition to olefins are very sensitive to the nature of the olefin. In many cases the addition barrier is higher than that for alkyl radical addition and the initially formed adduct is less stable; thus this is not a favorable path to molecular weight growth. However, the situation is different for terminal addition to 1,3-butadiene; the adduct is more stable due to resonance stabilization and the remaining double bond permits a direct cyclization pathway. These types of addition reactions are expected to play an important role in molecular weight growth kinetics during the oxidation and pyrolysis of hydrocarbons. Since addition and recombination reactions produce chemically-activated adducts, we have systematically explored the pressure- and temperature-dependence of the overall rate constants as well as that for the multiple product channels.

Keywords: Resonantly-stabilized free radicals (RSFR), molecular weight growth, kinetic model, recombination reactions, addition reactions

360025 Doped Metal Clusters on Oxides: Rationalization and Design through the Prism of Chemical Bonding

Abstract:

Ultra-small clusters of transition metals deposited on oxides can be highly-tunable and effective catalysts, and they represent an exciting electronic structure problem: every atom counts, and so does every electron. Doping and the choice for the support are two major levers in the design and rational manipulation of such sensitive catalysts. The rationale is rooted in the chemical bonding, i.e. in details of the electronic structure and explication of its precise effect on cluster shape, stability, mobility, and catalytic activity. With this prism in hand, we will present several cases when doping and surface-deposition lead to dramatic, though explainable changes in shapes and properties of clusters. We will show why mixed PtPd clusters exhibit their famous and unusual stability on oxides, as compared to pure Pt and Pd analogues. We will demonstrate how strategic doping of deposited Pd clusters with a small amount of Au leads to greater catalytic efficiency. We will explain why some clusters change shapes from globular to flat upon deposition, and why some do the exact opposite. On the fundamental level, we will deliver the emerging chemical bonding picture for surface-deposited clusters, where effects such as multiple aromaticity, covalency, first and second order Jahn-Teller effects are all in play and define the phenomena of interest. This work is done using Density Functional Theory calculations and our in-house methodology for modeling surface-deposited clusters in realistic conditions, and facilitated by collaboration with spectroscopic and surface-science experiments.

Highly Dispersed Group IV, V Transition Metal Oxide Catalysts from Grafted Metallocalixarenes for Alkene Epoxidation

Nicholas E. Thornburg, Justin M. Notestein

Department of Chemical & Biological Engineering
Northwestern University, Evanston, IL 60208

Many industrially significant selective oxidation reactions are catalyzed by supported and bulk transition metal oxides. Epoxidation catalysts have largely focused on TiO_x supported on or co-condensed with SiO_2 [1], and much of the rest of group IV and V have seen relatively little research. Here, metallocalixarenes of titanium, zirconium, hafnium, vanadium, niobium and tantalum (all of groups IV and V) were grafted onto mesoporous silica to yield a family of highly dispersed supported metal oxide catalysts. This series is used to elucidate intrinsic periodic trends in epoxidation reactivity. The calixarene ligand affords steric bulk and multidentate binding to the metal cation, which ensures site-isolation of metal centers during grafting, removing the confounding factor of different extents of metal oxide aggregation during synthesis.

In this work we examine cyclohexene and styrene epoxidation rates and selectivities, with H_2O_2 oxidant, over supported Ti(IV), Zr(IV), Hf(IV), V(V), Nb(V) and Ta(V), both with and without the calixarene ligand [2,3]. Supported Nb(V) proceeds with greater rate and selectivity than Ta(V) catalysts, followed by Ti(IV). We characterize the bulk structure of these materials with UV-visible and X-ray absorption spectroscopies. Further, we quantify each metal oxide's degree of active site dispersion and geometric accessibility with a novel, *in situ* titration with phenylphosphonic acid. This allows us to calculate intrinsic turnover frequencies for each element and to independently describe periodic trends in reactivity and extent of surface aggregation [4]. This synthesis strategy and novel method for counting accessible, kinetically relevant sites afford an intrinsic comparison of atomically-precise, isolated metal oxide species functioning as active sites for epoxidation, and can be coupled with other synthesis methods for enabling rational design of next-generation heterogeneous metal oxide catalysts.

[1] Gao, X., Wachs, I. E., *Catal. Today,* 51 233-254 (1999).
[2] Notestein, J. M., Iglesia, E., Katz, A., *J. Am. Chem. Soc.*, 126 16478-16486 (2004).
[3] Morlanés, N., Notestein, J. M., *J. Catal.,* 275 191-201 (2010).
[4] Eaton, T. R., Boston, A. M., Thompson, A. B., Gray, K. A., Notestein, J. M., *Ang. Chem.*, submitted.

We report a stable site-isolated mononuclear platinum catalyst with a well-defined structure. Platinum complexes supported in zeolite KLTL were synthesized from $Pt(NH_3)_4(NO_3)_2$, oxidized at 633 K, and used to catalyze CO oxidation. Infrared and X-ray absorption spectra and electron micrographs determine the structures and locations of the platinum complexes in the zeolite pores, demonstrate the platinum–support bonding, and show that the platinum remained site isolated after oxidation and catalysis. The initial TOF characterizing the as-prepared catalyst was 0.0038 s^{-1} at 423 K and 1 bar with a feed of 1% CO and 5% O_2 in helium; that characterizing the oxidized sample was 0.012 s^{-1}; our catalyst is among the most active for CO oxidation under our conditions, as shown by a comparison with literature data.

382685 Dynamic Nuclear Polarization NMR Enables the Analysis of Sn-Beta Zeolite Prepared with Natural Abundance 119Sn Precursors

Abstract:

Pure-silica zeolites containing a small amount of tetravalent heteroatoms with open coordination sites have emerged as highly active, water-tolerant solid Lewis acids for many important reactions. The catalytic activity of tin-containing zeolites, such as Sn-Beta, is critically dependent on the successful incorporation of the tin metal center into the zeolite framework. However, synchrotron-based techniques or solid-state nuclear magnetic resonance (ssNMR) of samples enriched with ^{119}Sn isotopes are the only reliable methods to verify framework incorporation. For ^{119}Sn ($I = 1/2$, natural abundance = 8.6%) magic angle spinning (MAS) NMR, the coupled effects of low natural abundance of the ^{119}Sn isotope, low intrinsic NMR sensitivity, and low Sn loadings in the sample make NMR analysis impractical without ^{119}Sn isotopic enrichment. Unfortunately, the high cost of isotopic enrichment drastically hinders high throughput screening, routine analysis, or analysis of low-yield syntheses with NMR. Dynamic nuclear polarization (DNP) can greatly improve sensitivity by transferring the larger polarization of electron spins, such as those found in stable exogenous radical compounds, to nuclear spins through irradiation with high-frequency microwaves. The target nuclei then become dynamically polarized, and their NMR signals are enhanced by orders of magnitude. This work demonstrates, for the first time, the use of DNP NMR for characterizing zeolites containing 2 wt % of natural abundance Sn without the need for ^{119}Sn isotopic enrichment. To determine and minimize potential undesirable interactions of the radical with the Lewis acid center, four biradical polarizing agents with varying kinetic diameters were impregnated on the Sn-Beta sample. The biradicals TOTAPOL, bTbK, bCTbK, and SPIROPOL functioned effectively as polarizing sources, and the solvent enabled proper transfer of spin polarization from the radical's unpaired electrons to the target nuclei. The larger bCTbK and SPIROPOL radicals did not interact directly with the intrapore Sn sites and bCTbK led to an enhancement (ε) of 75, allowing the characterization of natural-abundance 119Sn-Beta with excellent signal-to-noise ratios in <24 h. Without DNP, no 119Sn resonances were detected after 10 days of continuous analysis.

388123 Molecular Active Sites in Heterogeneous Ir-La/C Catalyzed Carbonylation of Methanol to Acetates

Abstract:

We report that when Ir and La halides are deposited on carbon, exposure to CO spontaneously generates a discrete molecular heterobimetallic structure, containing an Ir-La covalent bond that acts as a highly active, selective, and stable heterogeneous catalyst for the carbonylation of methanol to produce acetic acid. This catalyst exhibits a very high productivity of ~1.5 mole acetyl/mole Ir•s with >99% selectivity to acetyl (acidic acid and methyl acetate) without detectable loss in activity or selectivity for more than one month of continuous operation. The enhanced activity can be mechanistically rationalized by the presence of La within the ligand sphere of the discrete molecular Ir-La heterobimetallic structure which acts as a Lewis acid to accelerate the normally rate-limiting CO insertion in Ir catalyzed carbonylation.

378969 Selective Hydrogenation of Acrolein over Single Atom Alloy Catalysts

Abstract:

The creation of atomically uniform sites in heterogeneous catalysis remains a challenge for the catalysis community. One such possibility would be the creation of an isolated metal atom in a matrix of another element, termed by Sykes et al as a single atom alloy (SAA). [1]

Pd-Ag alloy catalysts with very dilute amounts of Pd have been synthesized. EXAFS results confirmed that at very low concentration of Pd (as low as 0.01 wt %), Pd was completely dispersed as isolated atoms in Ag nanoparticles. Activity was improved by the presence of these isolated Pd atoms due to the creation of sites with lower activation energy for H_2 dissociation. In addition, for the same particle size, the 0.01%Pd/8%Ag alloy nanoparticles exhibited higher selectivity than their monometallic counterparts[2], suggesting that the Pd atom may act as a site for the favorable bonding of the acrolein molecule for facile hydrogenation of the aldehyde functionality. It is also seen the method of preparation is critical to obtaining highly disperse atoms of the active metal in the nanoparticles. For example, co-impregnation of the metal salts is more conducive to obtain a highly dispersive alloying metal compared to sequential impregnation. We are also extending our study using organic precursors instead of inorganic precursors for catalyst synthesis. EXAFS showed synthesis of Pd-Ag catalysts using Silver acetylacetonate, and Palladium acetylacetonate have led to larger particle size which are more selective towards allyl alcohol. [2] In addition, the effect of using different organic solvents (acetone, toluene, benzene, ethanol, chloroform) on particle size, selectivity and activity is studied for the catalysts made using organic precursors.

References:

(1) G. Kyriakou, M.B. Boucher, A.D. Jewell, E.A. Lewis, T.J. Lawton, A.E. Baber, H.L. Tierney, M. Flytzani-Stephanopoulos, E.C.H. Sykes, Science **2012**,335, 1209.

(2) H. Wei, C. Gomez, J. Liu, N. Guo, T. Wu, R. Lobo, C.L. Marshall, J.T. Miller, R. Meyer, J. Catal. **2012**, 298, 18.

388297 Platinum-Copper and Palladium-Gold Single Atom Alloys for Selective Hydrogenation Reactions

Abstract:

Selective hydrogenation of alkynes and dienes to alkenes is important for producing polymer-grade alkenes [1]. As an attractive alternative to conventional bimetallic catalysts, single atoms alloys (SAAs) can act as bifunctional surfaces with individual, isolated, highly active group VIII metal atoms dispersed in inert nanoparticles [2]. The SAAs are capable of dissociating and spilling over the hydrogen to the group IB noble metal areas that weakly adsorb hydrogen atoms and are active for selective hydrogenation reactions [2]. It was recently demonstrated that alloying of a trace amount of palladium in copper increases the reactivity markedly, while maintaining the high selectivity of copper for hydrogenation reactions [3].

Here we have investigated the selective hydrogenation performance of Pt-Cu and Pd-Au SAAs. Pt-Cu and Pd-Au nanoparticles enhance both the activity and selectivity for industrially important hydrogenation reactions for 1, 3 – butadiene and phenylacetylene when compared to their monometallic catalysts. We show both an increase in activity compared to monometallic Cu or Au and an increase in selectivity compared to monometallic Pt or Pd. The Pt atoms were loaded on pre-formed Cu nanoparticles with galvanic replacement reaction, while the Pd-Au bimetallic nanoparticles were formed by sequential deposition of palladium onto the gold surfaces. The preparation, reactivity, and characterization of the alloy nanoparticles with X-ray photoelectron spectroscopy, electron microscopy and X-ray absorption techniques will be presented. Using scanning tunneling microscopy (STM) and temperature programmed desorption/reaction (TPD/R), we show that Pt-Cu(111) SAAs catalyze hydrogenation of butadiene to butene with 100% selectivity. Pt atoms readily alloy into Cu(111) at low coverages and moderate temperatures where the Pt atoms exist as isolated atoms that are well separated from one another. The addition of small amounts of Pt (~1%) into Cu reduces the barrier for H_2 dissociation, allowing for the low temperature dissociation of H_2. Hydrogen atoms spill-over onto the Cu sites increasing the concentration of weakly bound H atoms that readily hydrogenate butadiene to butene. TPR of co-adsorbed H and butadiene shows the exclusive desorption of reactively formed butene, where the reaction extent is limited by the availability of H on the surface.

References

[1] Segura, Y., López, N. & Pérez-Ramírez, J. Origin of the superior hydrogenation selectivity of gold nanoparticles in alkyne- alkene mixtures: Triple-versus double-bond activation. *Journal of Catalysis* **247**, 383-386 (2007).

[2] Kyriakou, G. *et al*. Isolated metal atom geometries as a strategy for selective heterogeneous hydrogenations. *Science* **335**, 1209-1212 (2012).

[3] Boucher, M. B. *et al*. Single Atom Alloy Surface Analogs in $Pd_{0.18}Cu_{15}$ Nanoparticles for Selective Hydrogenation Reactions. *Phys.Chem.Chem.Phys.* **15**, 12187-96 (2013).

مدينة الملك عبدالعزيز
للعلوم والتقنية KACST

COMPARATIVE STUDY OF THE CATALYTIC STABILITY OF BIFUNCTIONAL NANOPOROUS-BASED CATALYSTS IN N-HEPTANE HYDROISOMERISATION

FAISAL ALOTAIBI

KING ABDULAZIZ CITY FOR SCIENCE AND TECHNOLOGY (KACST)
RIYADH, KINGDOM OF SAUDI ARABIA

مدينة الملك عبدالعزيز
للعلوم والتقنية KACST

CONTENT

- Introduction and Objectives
- Zeolite structure
- Deactivation behavior of USY zeolites
- The effect of post-synthesis modification
- Deactivation behavior of Beta zeolite
- Deactivation behavior of Mordenite zeolite
- Conclusions

مدينة الملك عبدالعزيز
KACST للعلوم والتقنية

INTRODUCTION

- Generally, catalysts used in heterogeneous catalytic processes experience a decrease in activity over a period of time.
- The time taken for the activity of a catalyst to fall to an undesirable level depends on several factors. In acidic zeolites the main cause is deposition of coke.
- The cost due to catalyst deactivation is very high in industrial processes; therefore catalysts with superior time stability are very valuable.
- Zeolites with large pore are attractive catalysts for reforming-type reactions, and their time stability would be one of the most significant factors in determining their use.

مدينة الملك عبدالعزيز
KACST للعلوم والتقنية

OBJECTIVES

- To evaluate the (TOS) behavior and deactivation rates of some commercial and in-house catalysts which have been loaded by a metal such as platinum as potential bifunctional catalysts, with a wide range of Si/Al ratios, pore sizes and acidity characteristics in the hydroisomerisation of n-hepane.
- To explore the relationship between zeolite pore structure, acid site strength and coking tolerance in order to rationalise the catalytic performance and stability of the different zeolite catalysts.

OBJECTIVES

- To explore ways to limit deactivation by coking, to establish how zeolite catalysts can survive under alkane hydroisomerisation conditions and to determine how their lifetimes can be extended from minutes to days.
- Therefore, to study further improvement of catalyst lifetimes by post-synthesis modification and to report the results.

THE ZEOLITE Y PORE STRUCTURE

Zeolite Y has a 3-dimensional pore structure with pores running at right angles to each other in the x, y and z planes, composed of SBUs 4, 6, and 6-6. The pore diameter is large at 7.4 Å by a 12-member oxygen ring and leads into a larger cavity of 13 Å diameter, known as a 'supercage'.

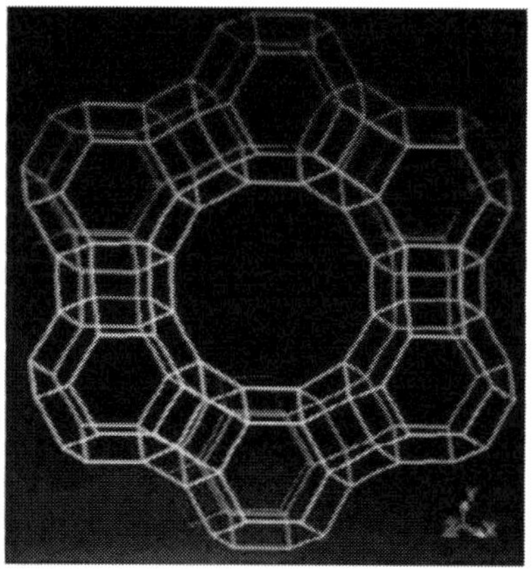

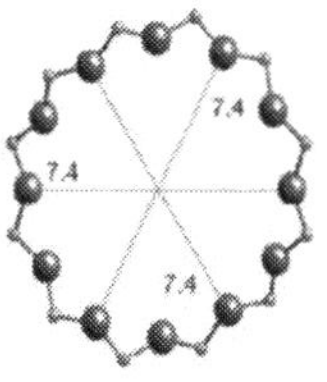

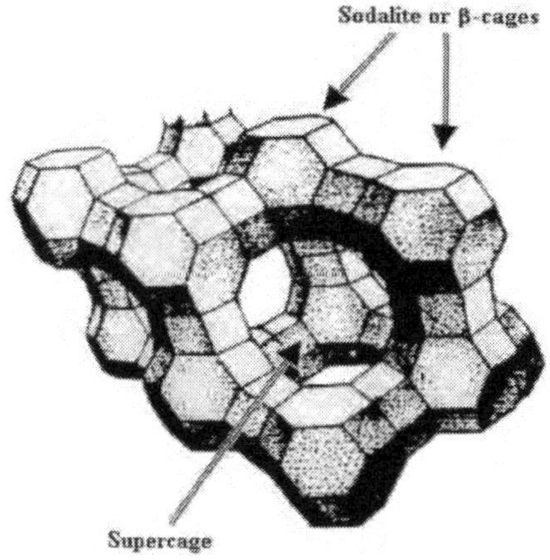

THE ZEOLITE BETA PORE STRUCTURE

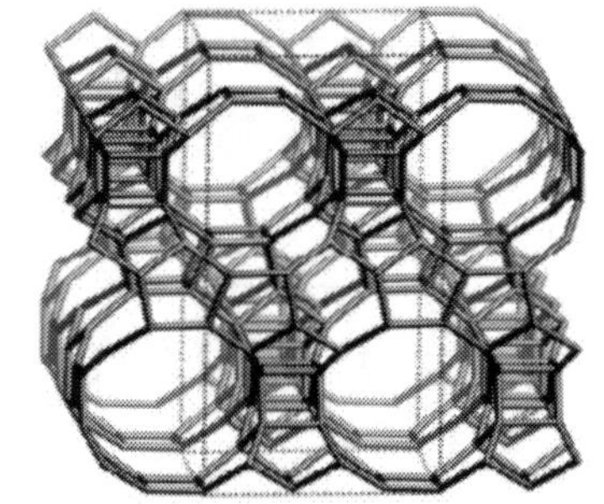

Zeolite beta consists of three interconnecting pore systems: two 12-ring linear channels (5.7 X 7.5 Å) in different crystallographic directions intersect partially, while a third, sinusoidal (tortuous) channel (5.6 X 6.5 Å) is formed by these intersections. The cavities have a diameter of 7.6 Å, indicating that the three-dimensional structure of zeolite beta has no supercages or cavities

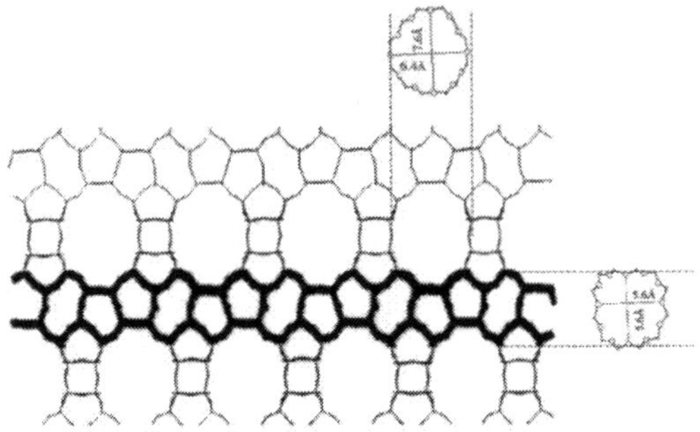

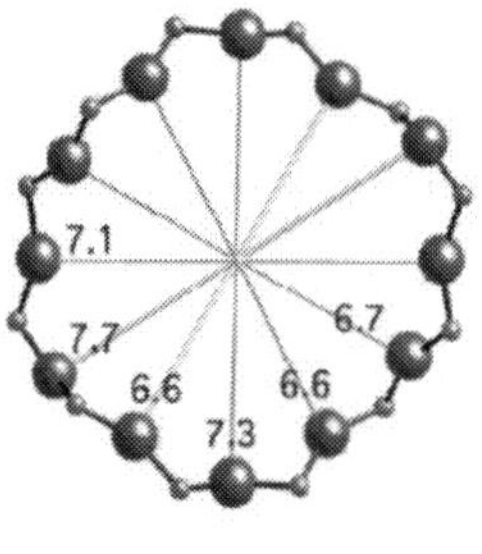

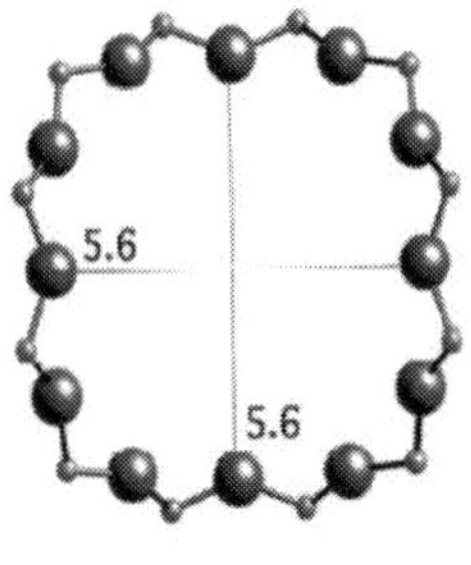

THE ZEOLITE MORDENITE PORE STRUCTURE

The micropore system of mordenite consists of two pore types: 12-ring elliptical channels (6.5 X 7.2 Å) in the c-axis direction and 8-ring (3.4 X 4.8 Å) channels along the direction of the b-axis that are inaccessible to molecular sizes above C3-molecules

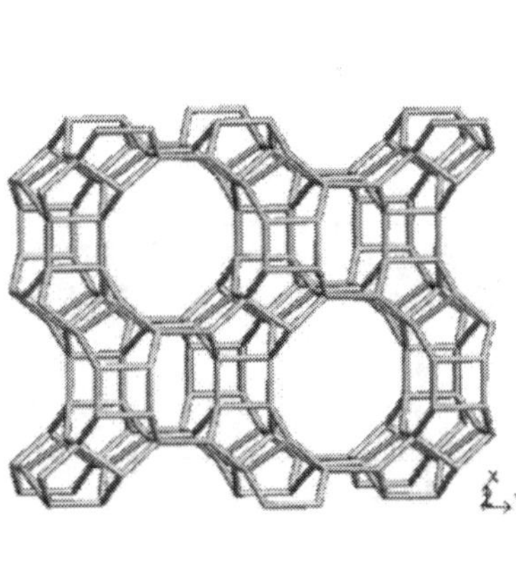

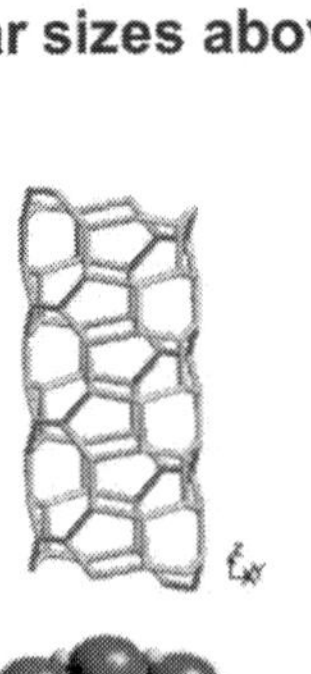

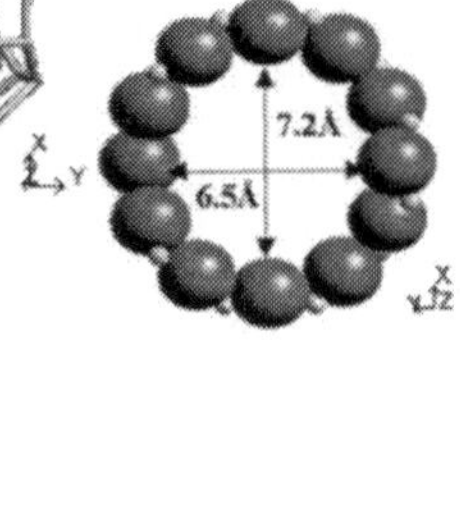

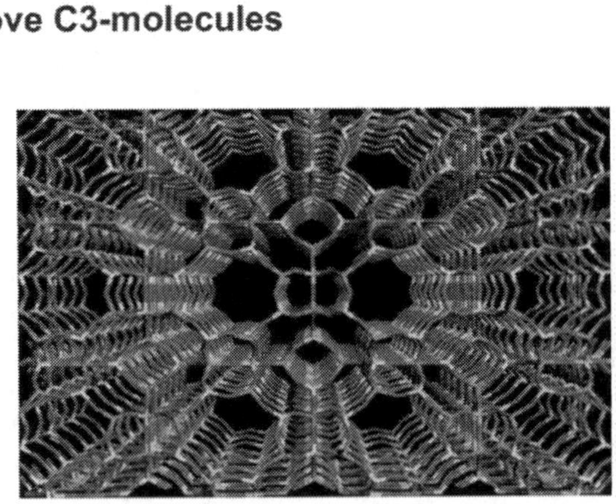

مدينة الملك عبدالعزيز
للعلوم والتقنية KACST

PORE DIAMETERS AND STRUCTURE SCHEMATIC

Zeolite	Pore structure	Pore structure schematic diagram
Y	3-D, 7.4 Å	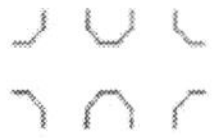
Beta	3-D, 5.7 X 7.5 Å and 5.6 X 6.5 Å	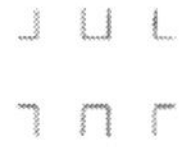
Mordenite	2-D, 6.5 X 7.2 Å and 3.4 X 4.8 Å	

مدينة الملك عبدالعزيز
للعلوم والتقنية KACST

SCHEMATIC OF THE ZEOLITE CHANNEL

a). Coking Favorable Structures

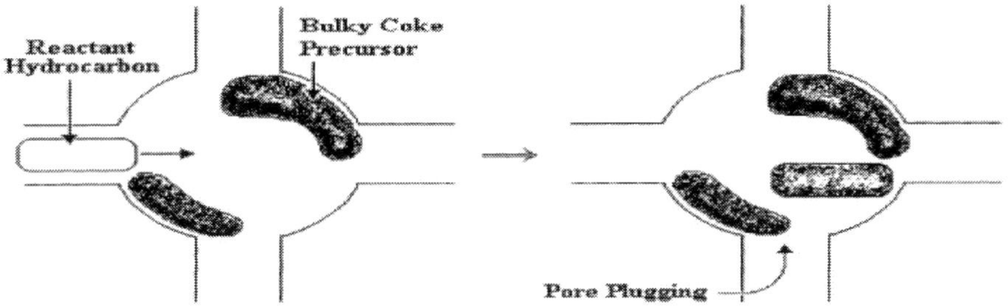

b). Coking Resistant Structures

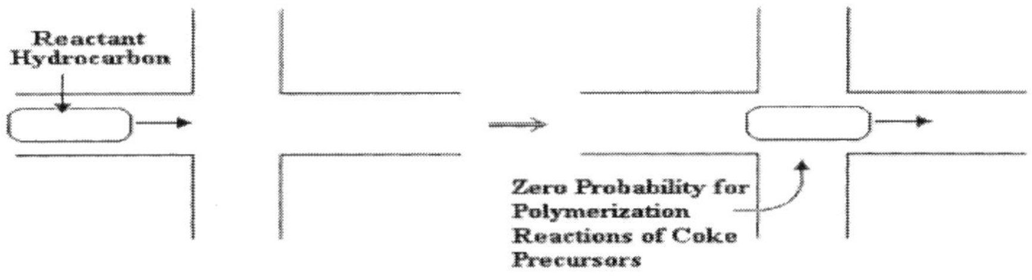

مدينة الملك عبدالعزيز للعلوم والتقنية KACST

ATMOSPHERIC HYDROISOMERISATION UNIT

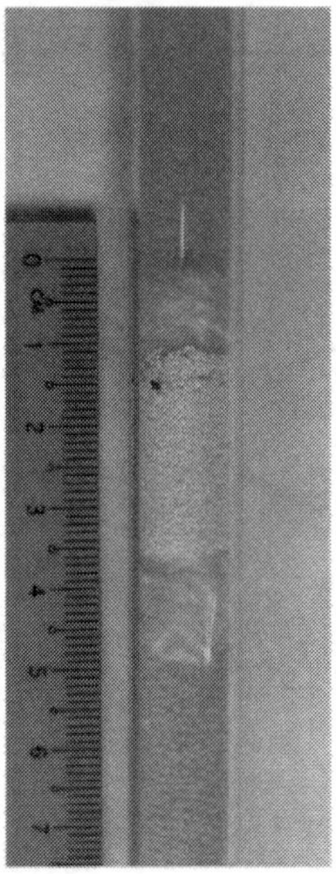

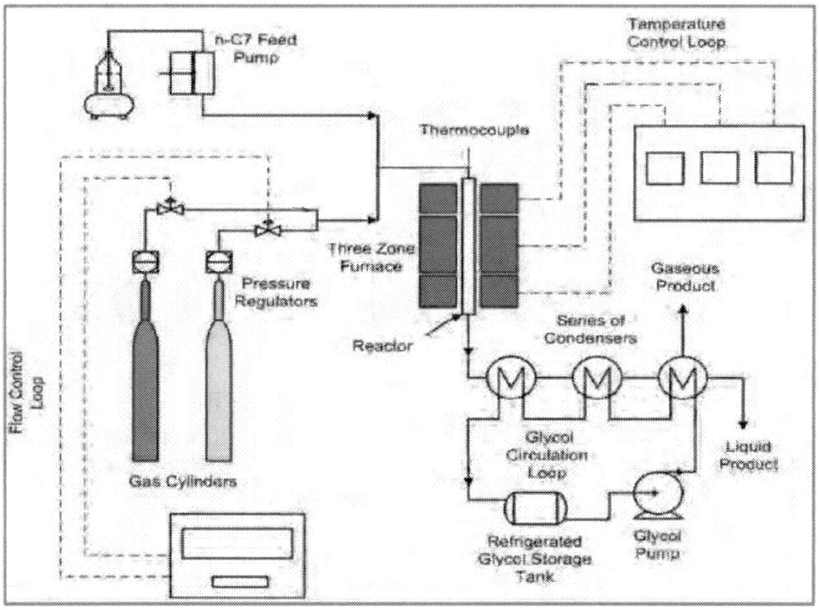

WHSV = 5.13 h^{-1} ; H_2/C7 (mol/mol) = 9; total pressure = 1 atm; temperature = 210, 230, 250 and 270 °C; 1 wt% Pt loading

مدينة الملك عبدالعزيز للعلوم والتقنية KACST

THE CATALYTIC STABILITY OF CBV712 USY

- **This catalyst was supplied by Zeolyst International**
- **It is mildly steamed and acid leached**

Si/Al (mole ratio)	Si/Al framework	Acidity mmol g^{-1}	Surface area $m^2\ g^{-1}$	Pore volume $cm^3\ g^{-1}$	Pore size Å	Pt dispersion %
6.0	8.0	0.70	816	0.45	43	38

Alotaibi et al. Appl Petrochem Res (2014) 4:95-136,

Alotaibi et al. Appl Petrochem Res (2014) 4:189–207

مدينة الملك عبدالعزيز
للعلوم والتقنية KACST

THE CATALYTIC STABILITY OF CBV712 USY

XRD pattern of CBV712 USY

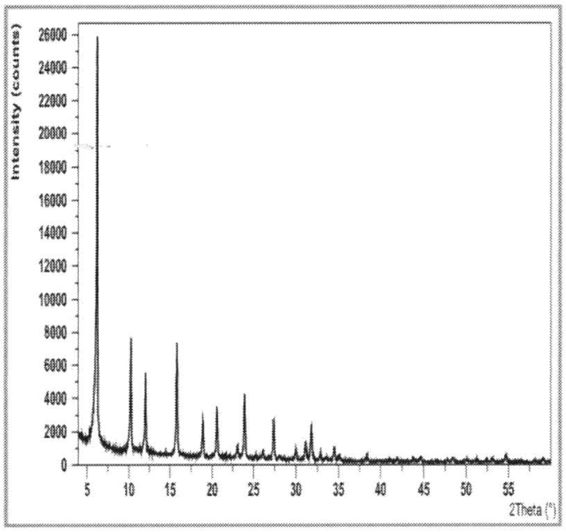

SEM micrograph of CBV712 USY

Alotaibi et al. Appl Petrochem Res (2014) 4:95-136,

Alotaibi et al. Appl Petrochem Res (2014) 4:189–207

مدينة الملك عبدالعزيز
للعلوم والتقنية KACST

THE CATALYTIC STABILITY OF CBV712 USY

^{29}Si MAS NMR spectra of CBV712

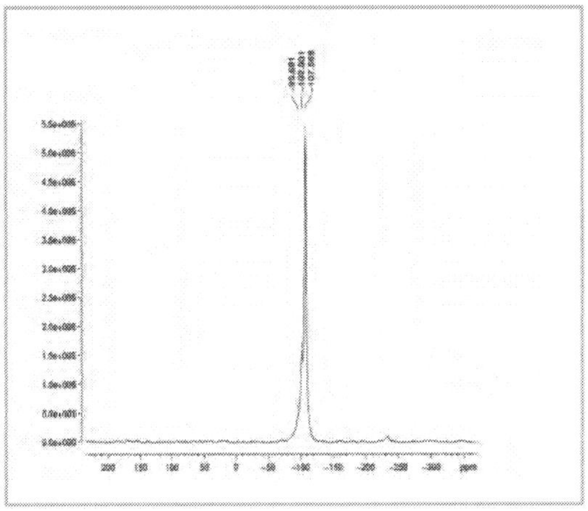

^{27}Al MAS NMR spectra of CBV712

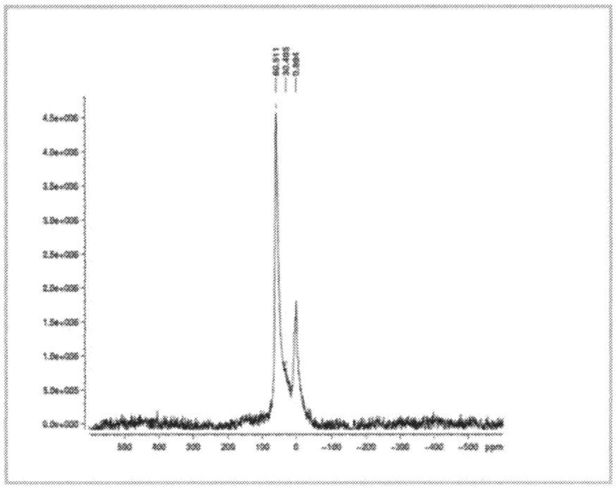

Alotaibi et al. Appl Petrochem Res (2014) 4:95-136,

Alotaibi et al. Appl Petrochem Res (2014) 4:189–207

مدينة الملك عبدالعزيز
للعلوم والتقنية KACST

THE CATALYTIC STABILITY OF CBV712 USY

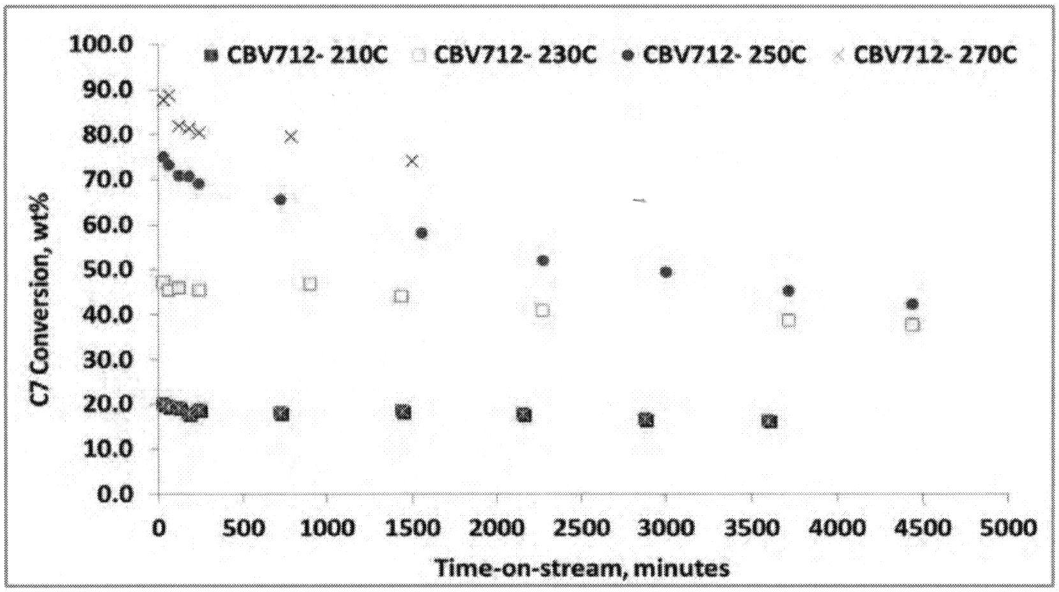

WHSV = 5.13 h^{-1} ; H_2/C7 (mol/mol) = 9; total pressure = 1 atm; temperature = 210, 230, 250 and 270 °C; 1 wt% Pt loading

مدينة الملك عبدالعزيز
للعلوم والتقنية KACST

THE CATALYTIC STABILITY OF CBV712 USY

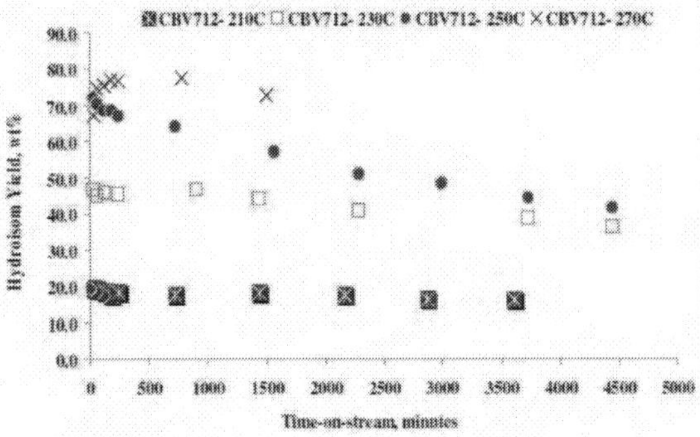

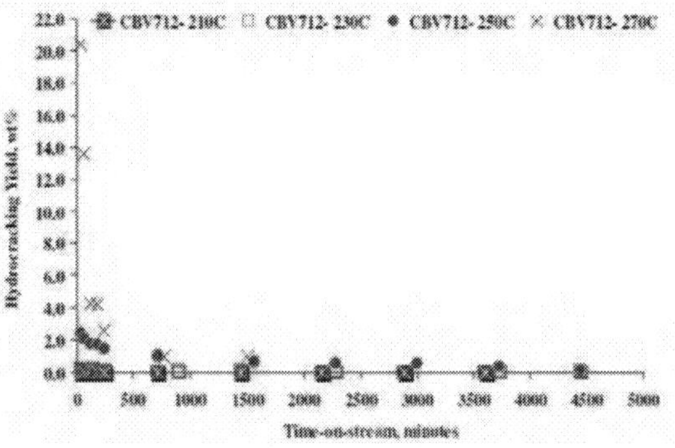

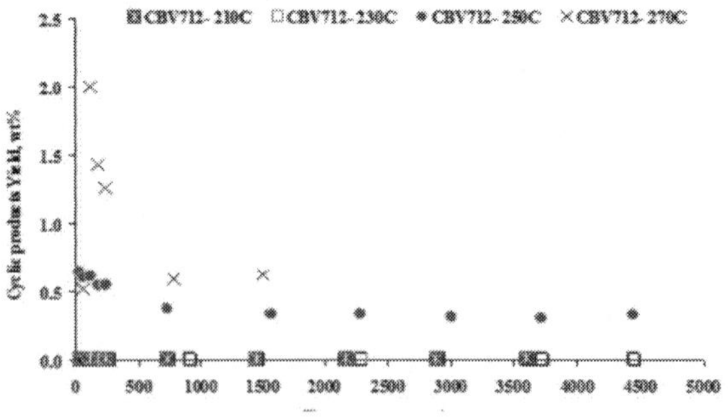

جامعة الملك عبدالعزيز للعلوم والتقنية KACST

THE CATALYTIC STABILITY OF CBV712 USY

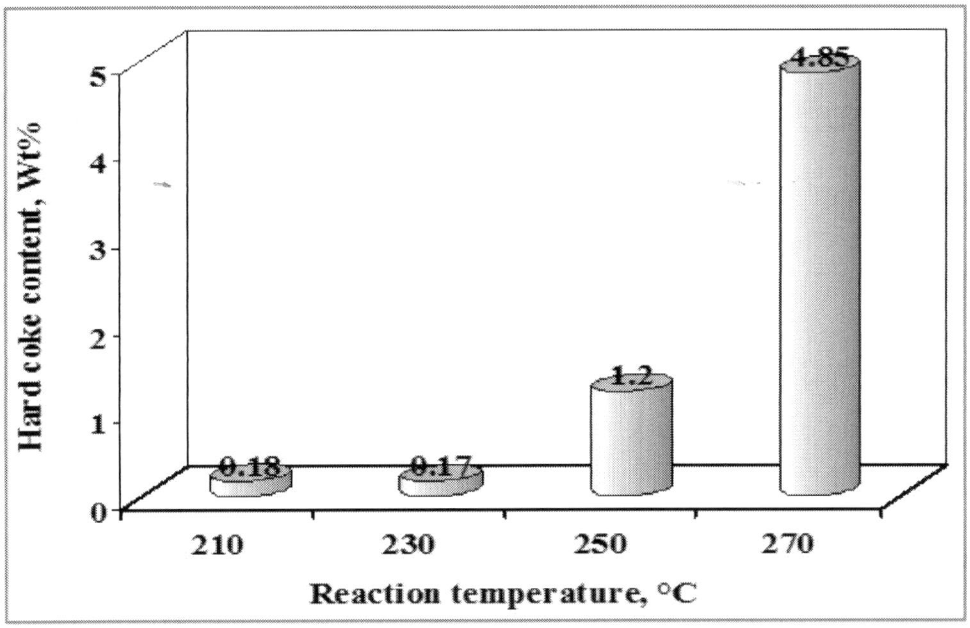

مدينة الملك عبدالعزيز للعلوم والتقنية KACST

THE CATALYTIC STABILITY OF CBV712 USY

- A rapid loss in C7 hydroisomerisation conversion activity that can be explained by considering the Brønsted and Lewis acid sites of the CBV712.
- The Brønsted acidity is predominant resulting in the majority of the aluminium ions incorporated in a tetrahedral coordination, as seen in the ^{27}Al MAS NMR and TPD analysis.
- Additionally, it can be concluded that the catalyst CBV712 has a high acid site density and a mild strength of acidity due to the existence of both types of acid sites, resulting in the catalyst exhibiting high activity, particularly at higher temperatures, resulting in increased activity for the cracked products as coke precursors.

مدينة الملك عبدالعزيز
KACST للعلوم والتقنية

THE CATALYTIC STABILITY OF CBV720 USY

- **This catalyst was supplied by Zeolyst International**
- **It is SEVERLY steamed and acid leached**

Si/Al (mole ratio)	Si/Al framework	Acidity mmol g^{-1}	Surface area $m^2\ g^{-1}$	Pore volume $cm^3\ g^{-1}$	Pore size Å
15	31	0.48	880	0.50	25

Alotaibi et al. Appl Petrochem Res (2014) 4:95-136,
Alotaibi et al. Appl Petrochem Res (2014) 4:189–207

مدينة الملك عبدالعزيز
KACST للعلوم والتقنية

THE CATALYTIC STABILITY OF CBV720 USY

TEM images of 1%Pt/CBV720 USY

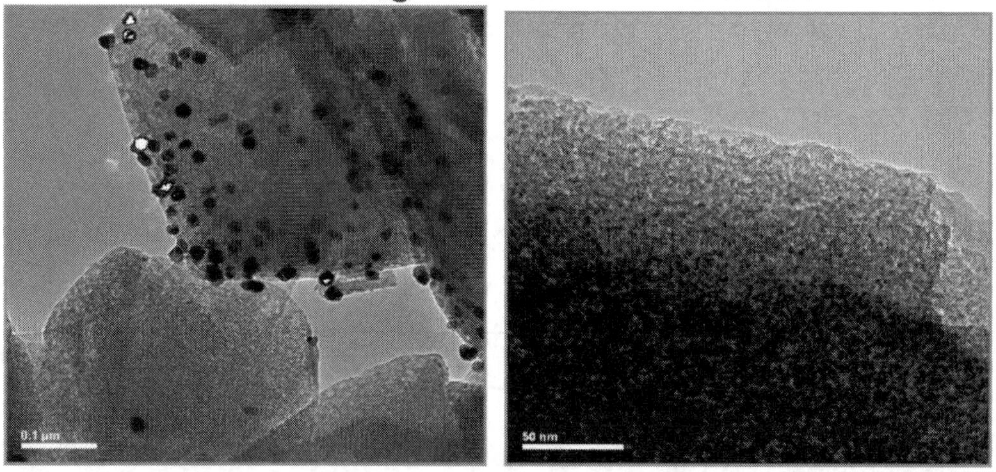

A few very large (~25 nm) Pt particles were seen within the bulk of the zeolite support and at higher magnifications, Pt particles of 2 or 3 nm in diameter were seen in varying quantities throughout the USY support, with slightly smaller particles occurring at or near the surface.

THE CATALYTIC STABILITY OF CBV720 USY

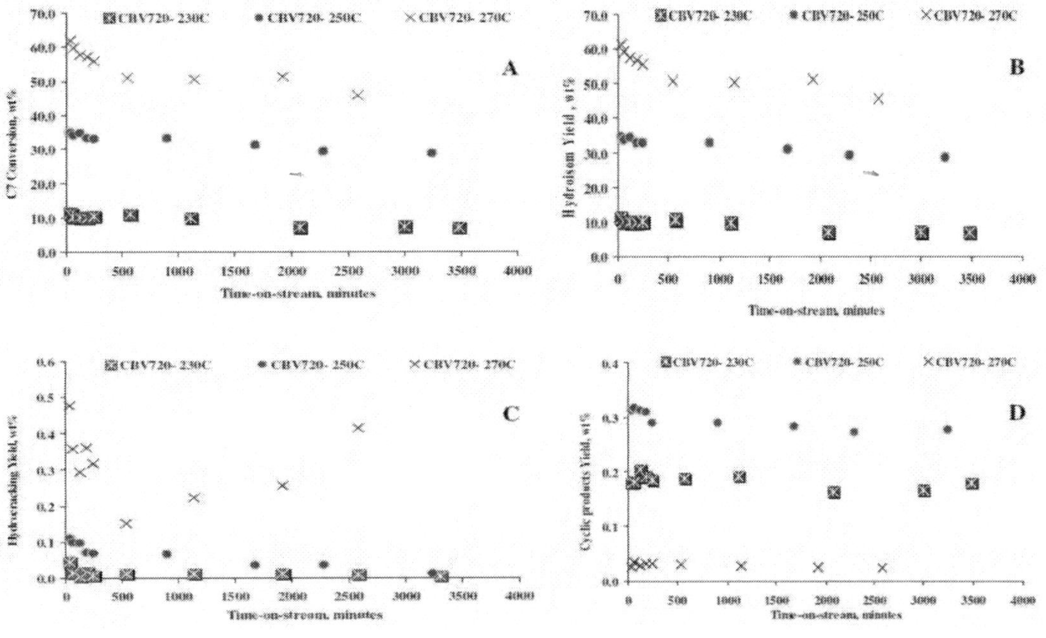

THE CATALYTIC STABILITY OF CBV720 USY

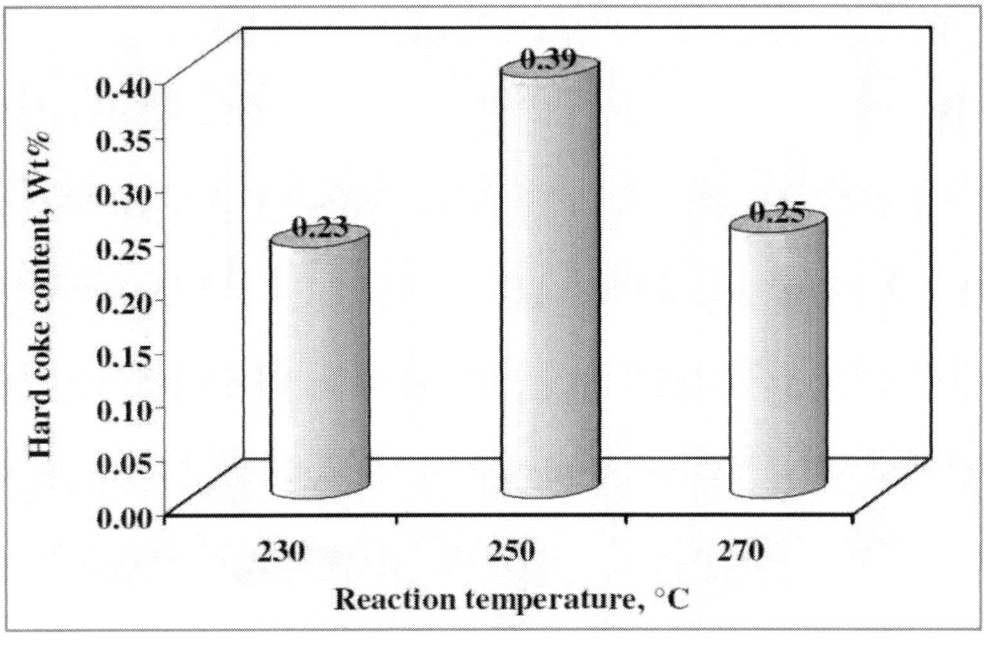

THE CATALYTIC STABILITY OF CBV720 USY

مدينة الملك عبدالعزيز للعلوم والتقنية KACST

- As CBV720 was produced via a severe dealumination process and leached by a mineral acid, the resultant catalyst has a small number of acid sites.
- Due to the low acidity, there was less cracking activity, resulting in the catalyst having a higher stability against deactivation by coking.
- The n_{Pt}/n_A ratio for CBV720 was calculated to be 0.115, indicating a balanced catalyst that performed as an 'ideal' hydroconversion catalyst. The high value of n_{Pt}/n_A meant that the number of chemical steps by which the cracked products could be generated and then oligomerated to form the coke-maker molecules was small, which in turn led to slower coke formation and higher catalyst stability.

THE CATALYTIC STABILITY OF USY-A

مدينة الملك عبدالعزيز للعلوم والتقنية KACST

- This catalyst was supplied by Crosfield, UK
- It is mildly steamed

Si/Al (mole ratio)	Si/Al framework	Acidity mmol g^{-1}	Surface area $m^2\ g^{-1}$	Pore volume $cm^3\ g^{-1}$	Pore size Å
2.87	7.11	0.86	756	0.42	28

Alotaibi et al. Appl Petrochem Res (2014) 4:95-136,

Alotaibi et al. Appl Petrochem Res (2014) 4:189–207

KACST مدينة الملك عبدالعزيز للعلوم والتقنية

THE CATALYTIC STABILITY OF USY-A

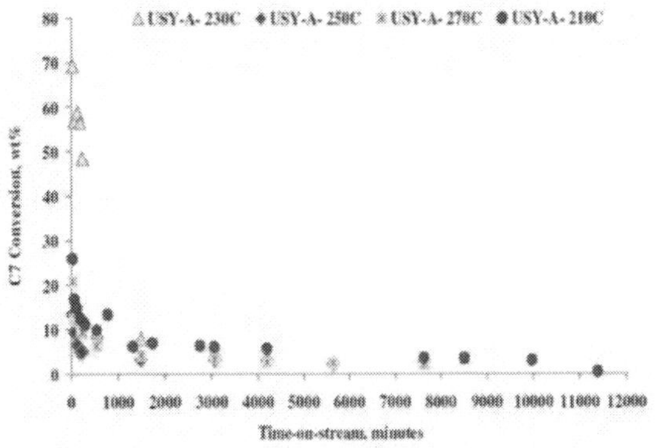

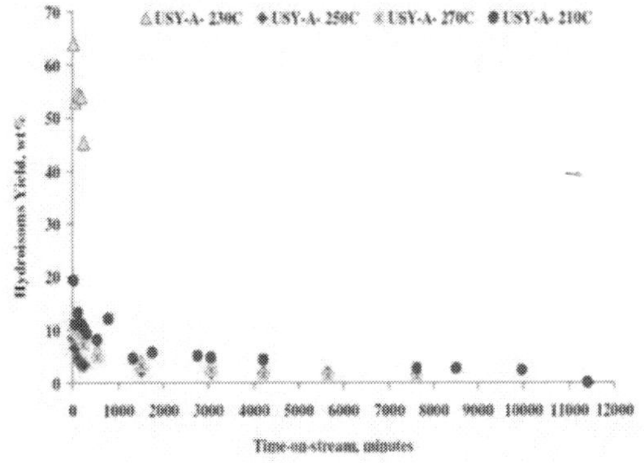

Initially, the overall conversion at 210 °C was 25.8 wt% after 30 minutes, decreasing sharply with TOS to 9.87 wt% after 540 minutes. Conversion decreased steeply with a deactivation rate of around 2 wt% to 0.46 wt% after 190 hours, at which the 1 wt% Pt/USY-A was completely deactivated.

KACST مدينة الملك عبدالعزيز للعلوم والتقنية

THE CATALYTIC STABILITY OF USY-A

TEM images for coked 1%Pt/USY-A

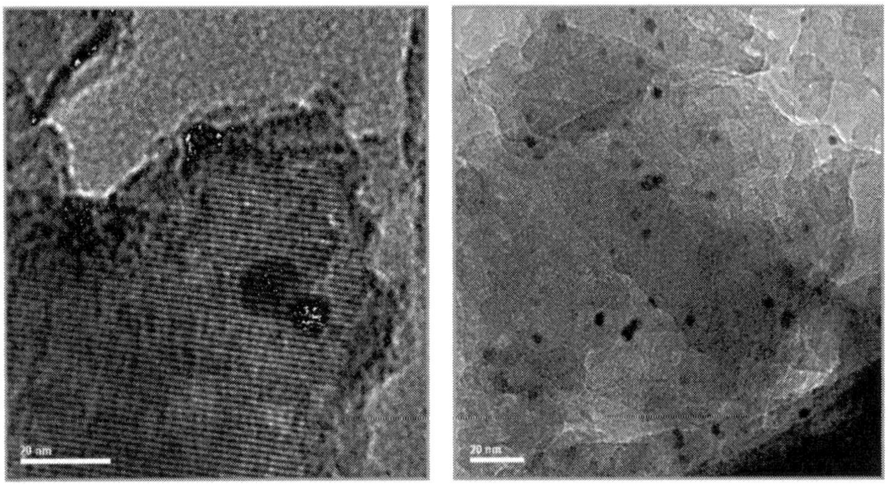

It is likely that the catalytic coke components were essentially formed and trapped within the channels of the catalyst micropores, so the first direct effect on the active sites was either by poisoning or by blocking of access to reactant molecules.

EXTRACTION OF EFAL IN USY ZEOLITES BY AMMONIUM HEXAFLUOROSILICATE POST-SYNTHETIC MODIFICATION

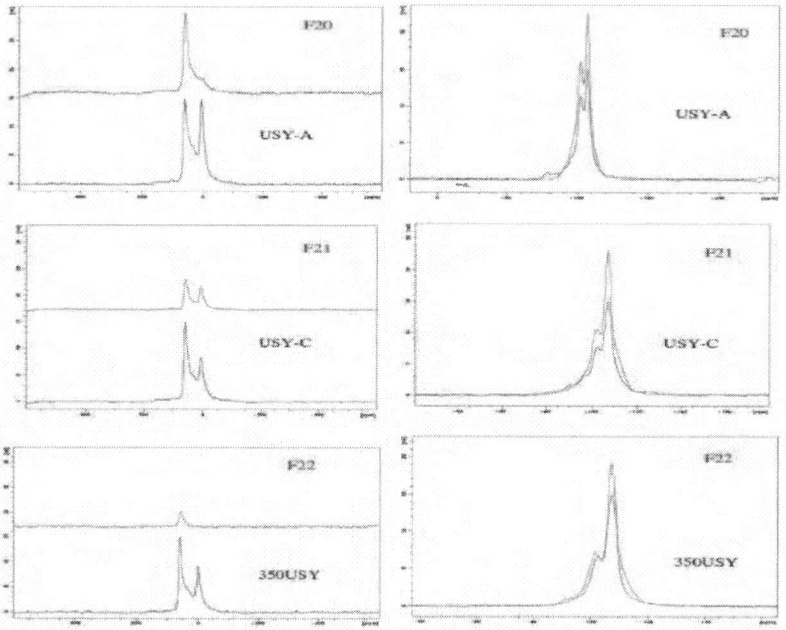

^{27}Al & ^{29}Si NMR for F20, F21 and F22 samples

THE TOS BEHAVIOUR OF AHFS MODIFIED USY

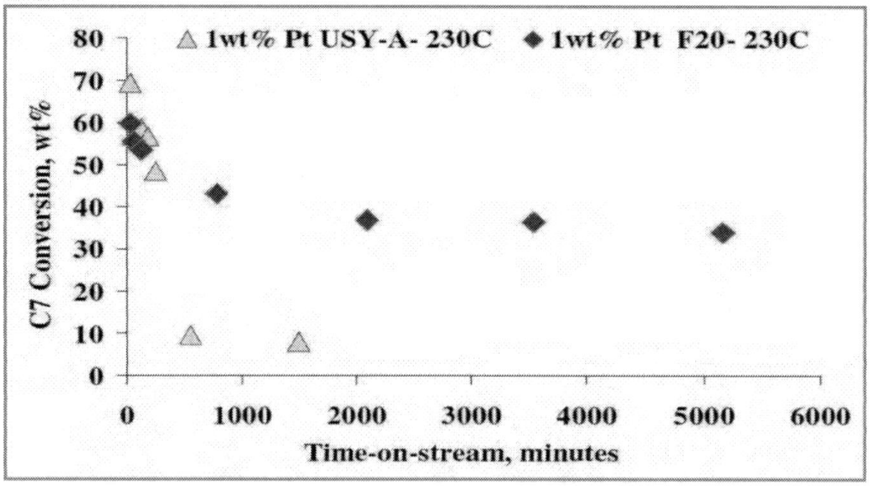

In comparison with USY-A, which was completely deactivated in the initial few minutes, F20 deactivated much more steadily with TOS, which may be explained by the effect of pore blockage by the polymeric EFAL species.

MONOMETALLIC AND BIMETALLIC USY

	Al, wt%	Si, wt%	Bulk (Si/Al), mol/mol	Framework (Si/Al), mol/mol	Crystallinity %	Actual Ni, wt%	Actual Pt, wt%	Theoretical Ni, wt%	Theoretical Pt, wt%	Ni/Pt mol/mol	n_{Pt}/n_A	Hard coke, wt%
350HOA USY	5.89	34.04	5.55	12.38	66	–	–	–	–	–	–	–
Pt/350HOA USY												
0.5% Pt	–	–	–	–	–	–	0.36	–	0.50	–	0.02	0.19
1% Pt	5.89	34.04	5.55	12.38	66	–	0.90	–	1.00	–	0.06	0.66
1.5% Pt	–	–	–	–	–	–	1.17	–	1.50	–	0.08	0.13
2% Pt	–	–	–	–	–	–	1.46	–	2.00	–	0.10	0.14
Pt-Ni/350HOA USY												
1% Pt-0.25% Ni	5.29	27.50	5.00	–	60	0.22	0.90	0.25	1.00	0.81	0.11	0.47
1% Pt -0.5% Ni	5.29	27.60	5.01	–	63	0.44	0.91	0.50	1.00	1.61	0.16	0.62
1% Pt -0.75%	5.31	27.65	5.00	–	61	0.68	0.92	0.75	1.00	2.46	0.21	0.43
1% Pt-1% Ni	5.31	27.97	5.06	11.57	60	0.92	0.91	1.00	1.00	3.36	0.27	1.01
Ni/350HOA USY												
0.5% Ni	–	–	–	–	–	0.44	–	0.50	–	–	0.10	–
1% Ni	–	–	–	–	–	0.88	–	1.00	–	–	0.20	–
1.5% Ni	–	–	–	–	–	1.28	–	1.50	–	–	0.29	–
2% Ni	–	–	–	11.03	–	1.86	–	2.00	–	–	0.42	–
10% Ni	4.74	27.74	5.62	10.86	73	7.48	–	10.00	–	–	1.68	–
20% Ni	4.28	24.83	5.57	10.58	–	11.96	–	20.00	–	–	2.68	–

MONOMETALLIC AND BIMETALLIC USY

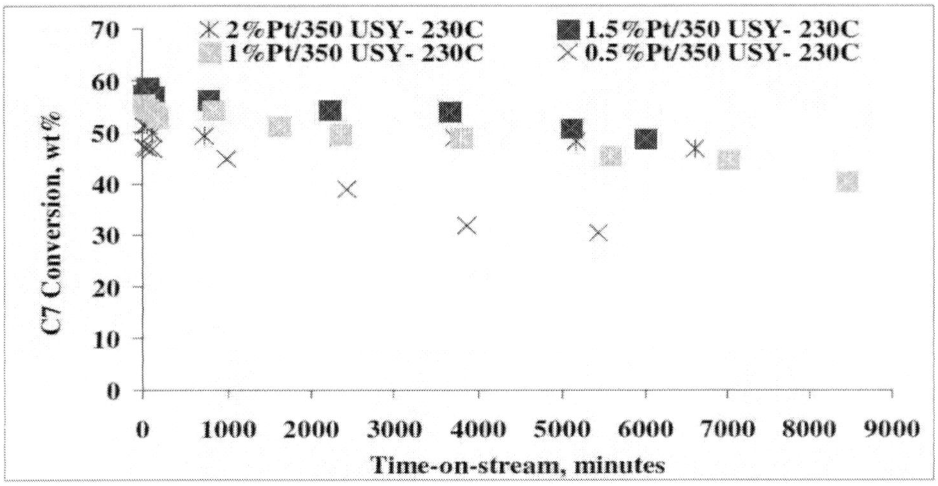

The catalyst containing 2 wt% Pt was found to have the smaller drop in conversion, while that containing 0.5 wt% Pt underwent greatest fall in activity during the TOS. The higher sustainability of the catalyst containing 2 wt% Pt can be explained by the formation of catalytically active metallic particles and better balance between metal and acid sites

MONOMETALLIC AND BIMETALLIC USY

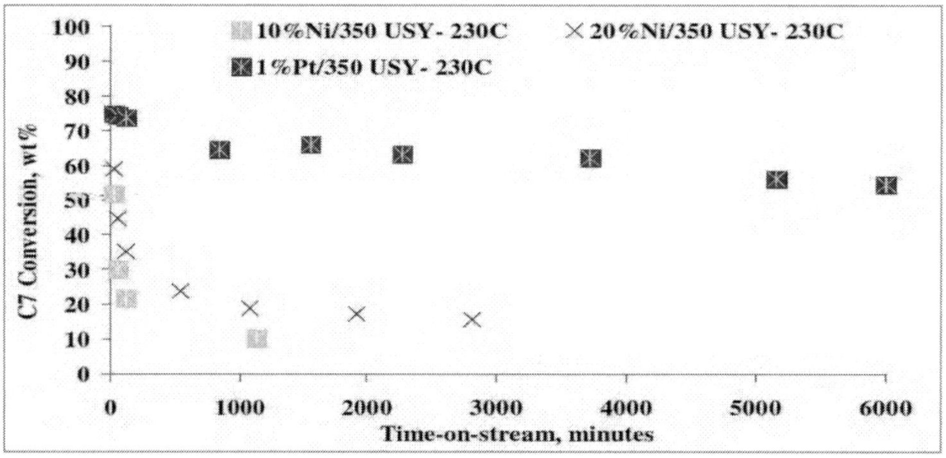

Deactivation was found to be very rapid for the monometallic catalysts containing Ni. This may be explained in terms of the poor hydrogenating capacity of Ni compared to Pt. In addition, the low stability of catalysts with higher nickel content may be explained by the tendency of Ni to polymerise olefins and hence to form coke.

MONOMETALLIC AND BIMETALLIC USY

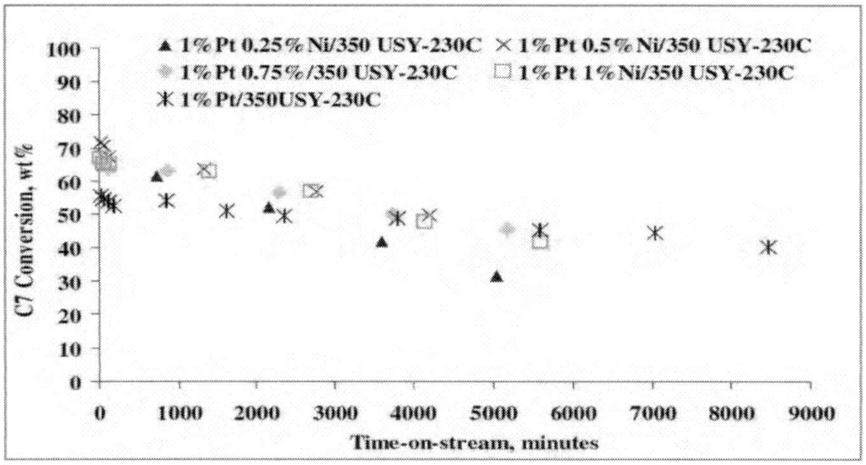

A small addition of Pt motivated a substantial increase in hydrogenating activity when compared with the catalysts containing only Ni and enhanced the reduction of the nickel cations, increasing the dispersion of the two metals as the nickel particles provided support for the Pt atoms, indicating that the Pt atoms were inserted into the nickel crystallite.

THE CATALYTIC STABILITY OF CP814E BETA ZEOLITE

مدينة الملك عبدالعزيز
للعلوم والتقنية KACST

- This catalyst was supplied by Zeolyst International
- It is as-synthesised beta zeolite, and contains nano-sized crystals (50nm).

Si/Al (mole ratio)	Si/Al framework	Acidity mmol g^{-1}	Surface area $m^2\ g^{-1}$	Pore volume $cm^3\ g^{-1}$	Pore size Å
13.21	13.45	0.63	577	0.85	73

Alotaibi et al. Appl Petrochem Res (2014) 4:95-136,

Alotaibi et al. Appl Petrochem Res (2014) 4:189–207

THE CATALYTIC STABILITY OF CP814E BETA ZEOLITE

مدينة الملك عبدالعزيز
للعلوم والتقنية KACST

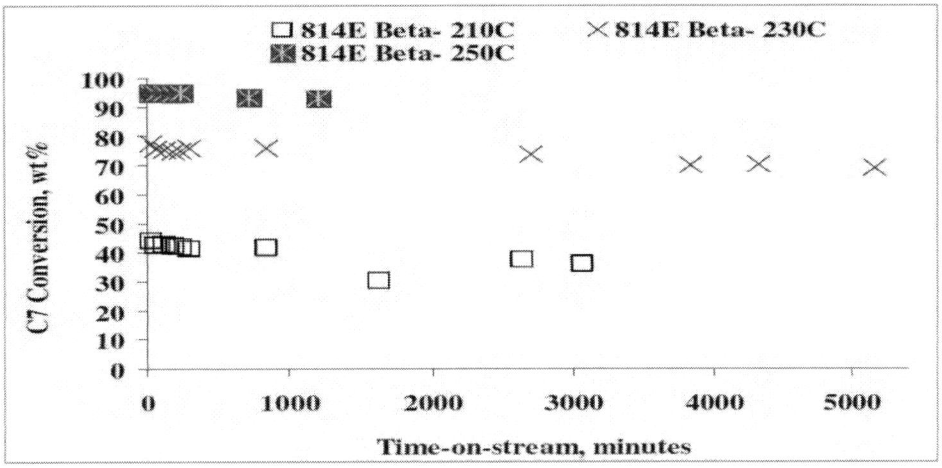

The initial conversion at 230 °C for the newly activated catalyst remained unchanged for about two days on stream, and then slightly decreased over time, with an overall conversion of 77 wt% after 30 minutes decreasing with the TOS to 74 wt% after 2700 minutes, and then decreased slowly within the deactivation rate range of 1 to 2 wt% every 24 hours until the reaction was finally terminated after about 86 hours.

مدينة الملك عبدالعزيز
للعلوم والتقنية KACST

THE CATALYTIC STABILITY OF MORDENITE

SEM micrographs of 640HOA MOR (A), 660HOA MOR (B) and 690HOA MOR(C)

From the SEM images, it can be seen that mordinite zeolites contains larger crystallites up to 500 nm in length, accompanied by some smaller debris and agglomerates. The literature indicates that no synthesis route for nano-crystalline mordenite is so far available and that synthesis parameters such as silicon precursors, water content and $Na_2O/SiO_2/Al_2O_3$ ratios strongly affect mordenite morphology in terms of the size and aspect ratios of the crystals.

مدينة الملك عبدالعزيز
للعلوم والتقنية KACST

THE CATALYTIC STABILITY OF MORDENITE

SEM micrograph of in-house and as synthesized mordenite

SEM microgram shows that the catalyst contained large crystallites of 15-20 μm whose elliptical shape was different from that of commercial catalysts, and it has been previously reported that synthetic mordenite presents several morphologies, as a function of the crystallisation conditions.

مدينة الملك عبدالعزيز
للعلوم والتقنية KACST

POST-SYNTHESIS MODIFICATION OF IN-HOUSE MORDENITE ZEOLITE

SEM micrographs of mordenites steamed at 500 °C (A), 600 °C (B), 700 °C (C) and acid leached (D)

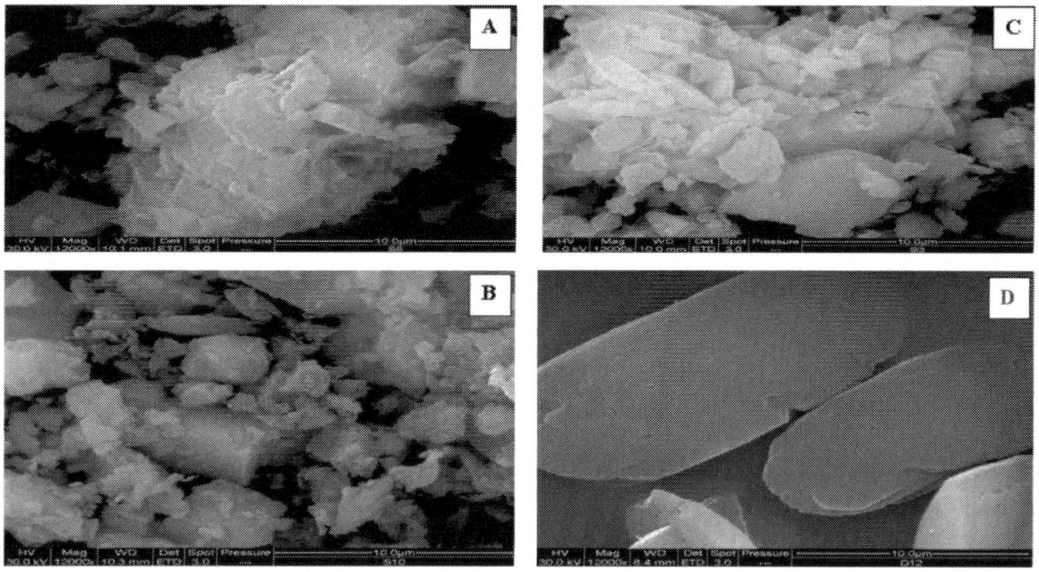

مدينة الملك عبدالعزيز
للعلوم والتقنية KACST

POST-SYNTHESIS MODIFICATION OF IN-HOUSE MORDENITE ZEOLITE

^{27}Al NMR of acid-leached and steamed samples

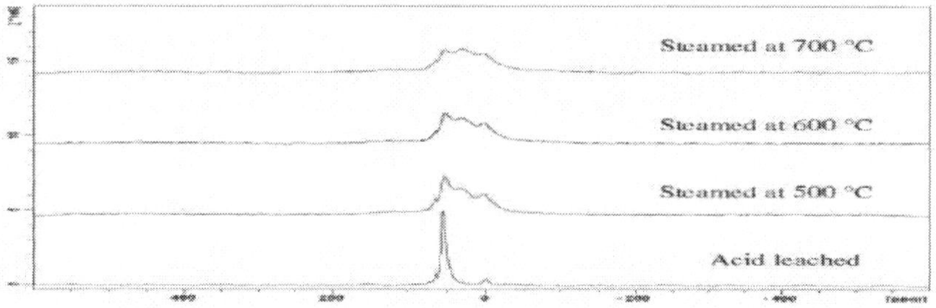

^{29}Si NMR of acid-leached and steamed samples

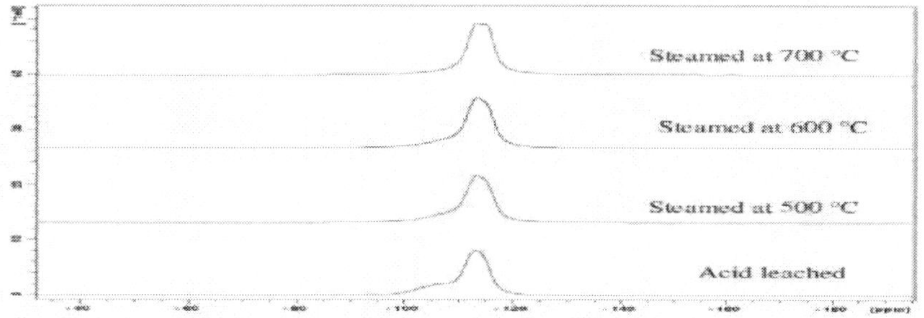

POST-SYNTHESIS MODIFICATION OF IN-HOUSE MORDENITE ZEOLITE

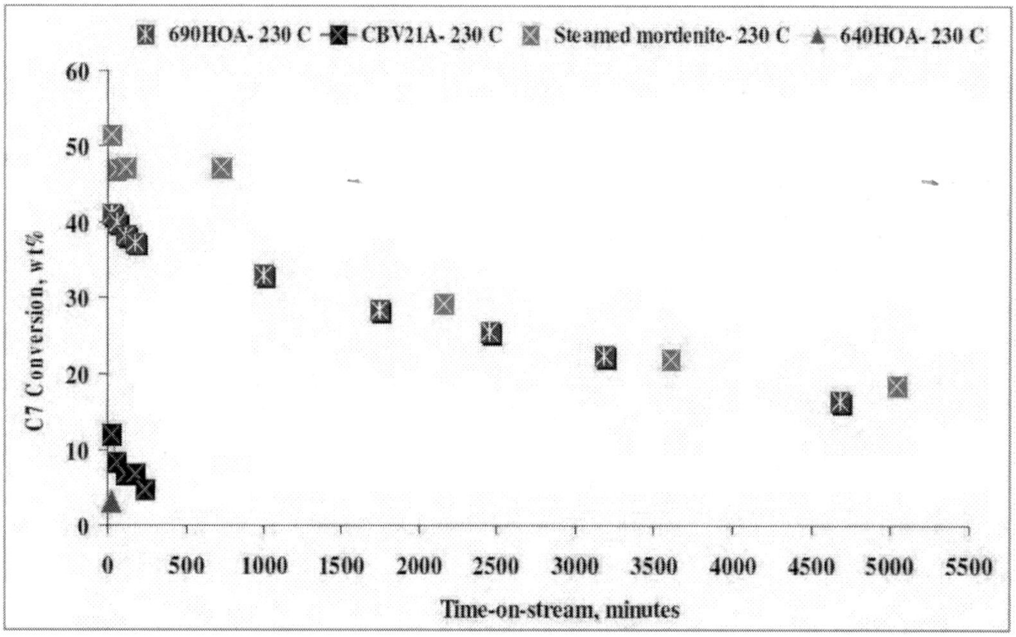

CONCLUSIONS

- Many different parameters control the stability of USY, Not only reaction conditions but also catalyst characteristics play a crucial roles.
- USY zeolite pore structures which can inhibit coke formation in conjunction with careful dealumination to decrease the number of acid sites are necessary for the time-stable reforming zeolite-based catalysts.
- There is a clear correlation between the obtained yield of the cracked products and to the hard coke. Maximum hard coke values were found at 270°C.

مدينة الملك عبدالعزيز
للعلوم والتقنية KACST

CONCLUSIONS

- The greater the value of nPt/nA, the slower the coke formation and the higher the catalyst stability.
- The AHFS acid-leached zeolite had a larger pore volume (smaller pore size and larger surface area), due to the alleviation of the obstruction by EFAL of the mesopore system, thus slowing the formation of coke and providing good catalytic stability.
- The higher Pt loading, up to 2 wt%, improved catalytic stability, due to a better balance and closeness of the catalytic functions.
- Pt-Ni bimetallic catalysts were better than monometallic catalysts containing only Ni because they had higher activity, better selectivity for C7 isomers and more stability.

مدينة الملك عبدالعزيز
للعلوم والتقنية KACST

ACKNOWLEDGEMENTS

- SABIC for funding
- Dr. Arthur Garforth, Manchester University
- Dr. Essam Jamea, SABIC T&I
- Dr. Raed Abudawoud, Saudi Aramco
- Dr. Arthur's research group, Manchester

مدينة الملك عبدالعزيز
للعلوم والتقنية KACST

Thank You

Porous aromatic polymers as heterogeneous catalysts

Anastasiya Bavykina, Maarten Goesten, , Michiel Makkee, Freek Kapteijn and Jorge Gascon

Catalysis Engineering, ChemE, Julianalaan 136, 2628BL, Delft, the Netherlands

The synthesis of porous, tunable materials with predictable properties is highly interesting, as it allows for targeted catalysis by custom-made materials. Metal-Organic Frameworks (MOFs) in particular have provided for a wealth of materials with predictable, and sometimes remarkable properties. The dipolar, and sometimes highly ionic nature of the dative bonds in MOFs does however limit application in catalysis, as lability of the bridging ligand leads to instability or other undesired effects. It is therefore interesting to look at similarly customizable materials that only contain covalent bonds. In this work, we'll present the synthesis of three new, Porous Aromatic Frameworks (PAFs)[1] (α, β and γ), obtained by Suzuki-Miyaura coupling of selected molecular modules.

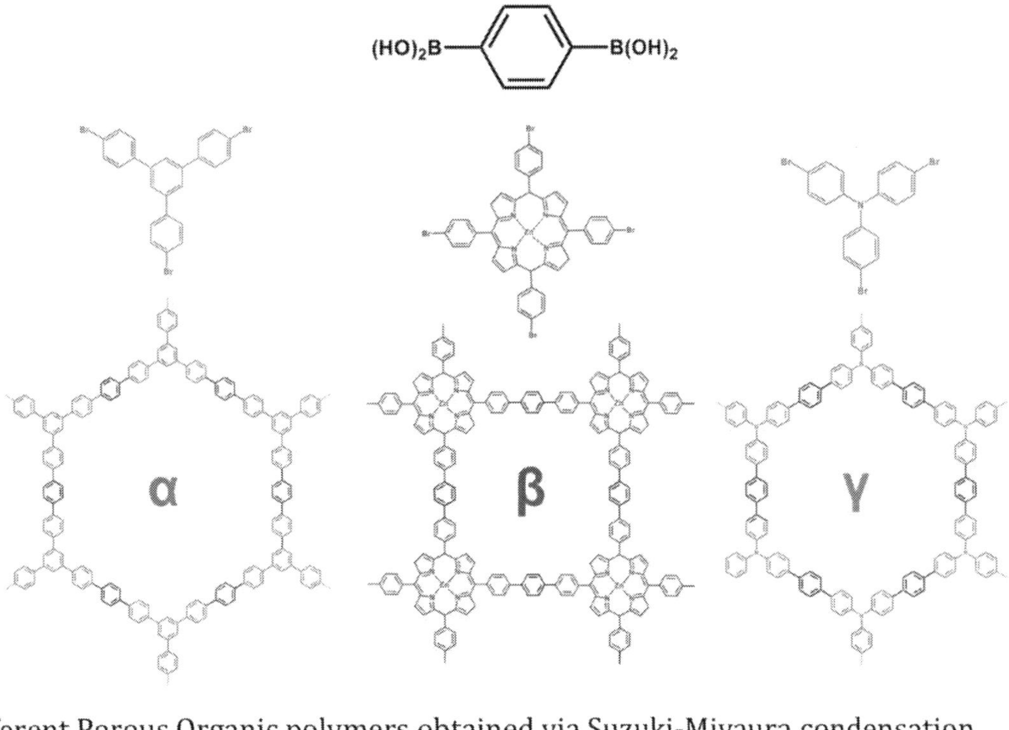

Different Porous Organic polymers obtained via Suzuki-Miyaura condensation.

We demonstrate the pre- and post-synthetic tuneability of these materials through their decoration with catalytically active metal and acid moieties. We further investigate the use of cheap and highly porous Covalent Triazine Frameworks (CTFs)[2] in similar manner, as ligand and stabilizer for highly dispersed atomically dispersed metals. All obtained materials show good to excellent activity in targeted reactions, namely acid catalysis and the production of H_2 from formic acid, along with excellent stability and recyclability.

References

(1) Dawson, R.; Cooper, A. I.; Adams, D. J. *Progress in Polymer Science* **2012**, *37*, 530.

(2) Kuhn, P.; Antonietti, M.; Thomas, A. *Angew Chem Int Edit* **2008**, *47*, 3450.

Different roles of water and liquid phase hydrocarbons to the reaction rate and product selectivity in Fischer-Tropsch Synthesis

Xiaojun Lu, Diane Hildebrandt, and David Glasser

Material and Process Synthesis; College of Science, Engineering, and Technology, University of South Africa

Abstract

Low temperature Fischer-Tropsch Synthesis with catalyst experiences mass transfer limitation in most cases especially in industrial applications as the catalyst is coated with liquid phase products (can be a mixture of hydrocarbons and water in different proportions). Reactants need to transfer through the liquid layer to reach the active sites on the surface of the catalyst to commence the reaction. This affect is often bypassed when a catalyst is synthesised and test in the lab as powder is more often used instead of pelletized catalyst.

Due to nonideality, water formed by reaction is not entirely in gas phase and the proportion of water in liquid phase depends on the operating conditions and the composition of liquid phase hydrocarbons. Our experimental work with a CSTR loaded with cobalt catalyst for FT research suggested that a considerable proportion of water is in liquid phase and temperature has a big effect in the proportion. The effect of water (both water produced or water co-feeding) to the performance of the FT reaction is not always consistent in the open literature and very few explanation is widely accepted [1]. Water intends to increase the solubility of CO while liquid phase hydrocarbons intend to increase the solubility of H_2. This will alter the actual ratio of H_2 to CO next to the active sites of the catalyst, and the actual ratio of the reactants close to the active sites depends on the composition of the liquid mixture.

This work will discuss the different roles of water (including product water and co-fed water) and liquid hydrocarbons to the reaction rate and product selectivity based on the experimental data generated in our lab and reported in the literature. It is found that the deposit of liquid phase hydrocarbons will slow down the reaction rate and increase the selectivity to the paraffins especially CH_4. A flushing experiment [2] has confirmed that the ratio of H_2/CO in the liquid phase product is far different from the ratio in the gas, which might be the reason that the selectivity to the paraffins increases. In the contract, a higher proportion of H_2O dissolved in the liquid phase hydrocarbons will increase the selectivity to olefins and suppress the formation of paraffins.

Keywords: Fischer-Tropsch Synthesis, water, liquid hydrocarbons, H_2 solubility, CO solubility

Reference:

1. Dalai, A. K.; Davis, B. H. Fischer–Tropsch synthesis: A review of water effects on the performances of unsupported and supported Co catalysts. *Applied Catalysis A: General.* **2008**, *348*, 1–15.
2. Lu, X.; Zhu, X.; Hildebrandt, D.; Glasser, D.; Liu, X. A New Way to Look at Fischer-Tropsch Synthesis Using Flushing Experiment. *Ind. Eng. Chem. Res.* **2011**, *50*, 4359–4365.

Reactivity Study of Deasphalted Oil Upgrading via Catalytic Steam Cracking using Ultradispersed Catalysts

Fredy A. Cabrales-Navarro, Luis A. Coy, Pedro Pereira-Almao*

Department of Chemical and Petroleum Engineering, Schulich School of Engineering, University of Calgary, 2500 University Drive N.W., Calgary, AB T2N 1N4 , Canada
* Corresponding author (facabral@ucalgary.ca)

Abstract

The worldwide shortage of light oil reserves has forced the oil sector to move toward enhanced oil recovery methods for light oils, such as reservoir fracturing, and more sustainably to the exploitation of extra-heavy oils, material characterized by having considerable amount of Asphaltene compounds that difficult their processing under conventional refinery schemes. In this way, the implementation of Deasphalting units, where asphaltenes molecules are removed from the hydrocarbon fraction, has become very popular in the industry, demanding processes for upgrading these non-asphaltene containing heavy hydrocarbons, for which an alternative technology involving innovative ultra-dispersed catalysts and steam activation, such as Catalytic Steam Cracking (CSC), may have a very important role.

In this work, the reactivity of deasphalted oil (DAO) obtained from a topped fraction of a South American heavy oil is evaluated via Catalytic Steam Cracking. In the catalytic steam cracking process, an in-house formulated ultra-dispersed nano-catalyst is incorporated in the feedstock through a decomposition reaction performed in a catalyst manufacturing skid designed for this purpose, and the catalyst-loaded hydrocarbon is fed to an up-flow reactor mixed with steam where it is heated to catalytic cracking reaction conditions. Additionally, conventional non-catalytic steam cracking (SC) was also carried out in order to assess the performance of the catalytic process compared to the non-catalytic route. The performance of the process is compared based on the products quality determined through a series of analytical techniques including: Simulated Distillation (SimDist), C5 Asphaltenes by Microdeasphalting, API Gravity, Microcarbon Residue content (MCr), P-value, and Viscosity.

The results of the Steam Cracking reactivity tests showed a typical behavior with slight increase on distillate yields and limited conversion values, increasing trends for the microcarbon content and a decreasing trend on the stability of the asphaltenes when increasing severity of the reaction. On the other hand, it was possible to identify the advantages of the Catalytic Steam Cracking compared to the non-catalytic route in terms of product quality, thus being able to reach higher conversion levels which translates into higher distillate yields and significant viscosity reduction, while keeping a stable product with no significant increase on the production of byproducts such as coke.

Keywords: Deasphalted Oil, Asphaltenes, Catalytic Steam Cracking.

1. Introduction and Objectives

In the last decade, a sharp worldwide depletion of conventional hydrocarbon sources has forced the oil and energy sector to move toward unconventional alternatives to fulfill the increasing demand of petroleum-based fuels currently used as primary energy source. Therefore, the exploitation of heavy and extra-heavy hydrocarbons has become very common in most oil producer countries and appropriate technologies to handle these heavy oils are being developed for their production as well as their upgrading and refining [1, 2]. The development of new more cost efficient upgrading methods has become the center of attention for many research institutions and universities related to the energy sector, especially in countries like Canada where there are enormous reserves of heavy and extra-heavy oils [3]. This work focuses on the study of an alternative upgrading process applicable to particular heavy hydrocarbon fractions using innovative nano-catalytic technology and steam activation processes.

The use of steam as source of hydrogen for hydrogenation of heavy hydrocarbon fractions has been considered as a feasible alternative for upgrading heavy oils for the past two decades. In this steam-based process the catalyst has two main roles, the first one is water splitting to produce hydrogen free radicals (H^{*}, OH^{*}, O^{**}), and the second one is to promote the transfer of these radicals to the hydrocarbon (hydrogenation, primary carbon oxidation). Usually, catalysts for steam conversion processing work at pressures ranging preferably from 100 psi to 300 psi, considerably lower than the required for conventional Hydroprocessing [2, 4], reason for which the development of a suitable catalytic steam cracking catalyst could bring significant economic benefits.

In this research, the reactivity of Deasphalted Oil (DAO) using catalytic steam cracking (CSC) will be studied [2, 5]. In the catalytic steam cracking process, an in-house formulated K-Ni nano-dispersed catalyst is incorporated in the feedstock through a decomposition reaction performed in a catalyst manufacturing skid designed for this purpose [6], and then the catalyst-loaded hydrocarbon is fed to an up-flow reactor mixed with a steam stream where it is heated to catalytic cracking reaction conditions. In addition, conventional non-catalytic steam cracking (SC) is also carried out in order to assess the performance of the catalytic process compared to the non-catalytic route. The dissociation of the steam promoted by the catalytic active sites induces hydrogenation, preventing olefins formation and initial condensation of remaining asphaltene molecules, allowing the process to reach higher temperatures than conventional thermal cracking processes, without compromising the stability of the products and, at the same time, inhibiting the production of considerable amounts of coke as byproduct [7, 8]. As a result, higher amounts of light hydrocarbon fractions are expected, increasing the economic viability of the process. An example of the applicability of Catalytic Steam Cracking using ultradispersed catalysts in a field upgrading technology is proposed in the patent entitled "Systems and Methods for Catalytic Steam Cracking of Non-asphaltene Containing Heavy Hydrocarbons" by Pereira-Almao, et al. [5]

In order to accomplish the experimental evaluations proposed in this investigation, a Reactivity Unit able to process heavy hydrocarbons for a wide variety of process conditions, including Steam Cracking, Catalytic Steam Cracking, and Hydroprocessing was designed, procured, and constructed. For the purpose of this work, the reactivity unit was successfully tested in a series of thermal and catalytic experimental runs.

2. Experimental Methods

2.1. Catalytic Feed Preparation

For the incorporation of the ultra-dispersed catalyst into the hydrocarbon feed, a Skid preparation unit available in the Catalysis for Bitumen Upgrading group at the University of Calgary was used. The first step in the preparation of the catalytic feed is to adjust the viscosity of the oil by using diluent, which could be either gasoline or toluene. Once the feed viscosity is adjusted, it is mixed with the aqueous solutions of the corresponding metals that need to be incorporated, being Nickel Acetate and Potassium Hydroxide the solutions selected for incorporation of Ni and K particles. The target concentration for the K and Ni is 400 ppm and 300 ppm respectively. The emulsion of oil and metal solutions is then passed through a decomposition unit to decompose the metal solutions and finally the stream is sent to a hot separator where the water and diluent are separated from the oil containing the ultradispersed catalyst, and then to a cold separator where the mixture of diluent and water is collected. A general scheme of the preparation of the catalytic feed is shown in Figure 1.

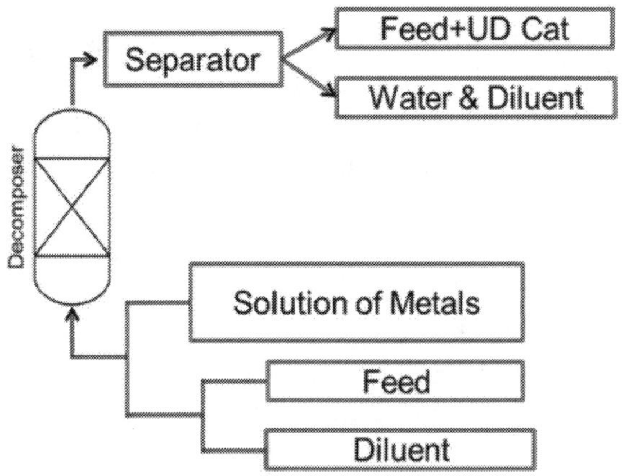

Figure 1. Ultra-dispersed catalyst preparation

2.2. Reactivity Experiments

The reactivity tests were carried out in a reactivity unit (RTU-1) equipped with an up-flow open tubular reactor that resembles a conventional thermal cracking unit, capable of handling high pressures and a wider range of operability conditions. A general scheme of the process carried out in the RTU-1 is shown in Figure 2.

The Reactivity Test Unit - 1 is composed by five sections: the feed section, reaction section, hot separation section, cold separation section, and finally the gas release and depressurization section. All feeds are mixed inline before entering the reaction zone, which is composed by an up-flow open tubular reactor operated at constant temperature and pressure. The products coming out from the reaction zone go to a hot separator in which the water, gases, and a fraction of the lightest hydrocarbons are separated from the product, by keeping the separator at (usually) a constant temperature of 250 °C. The gaseous separated stream is then passed through a condenser in order to liquefy the light hydrocarbons and the water, and the whole stream is sent to the cold separation section where the gases are separated from the liquid product. Finally, the gases coming out from the cold separation zone pass through a back pressure valve, which maintains the operating pressure at the given set point, and the gas stream is sent to the gas release and depressurization zone, equipped with KOH traps for gas sweetening.

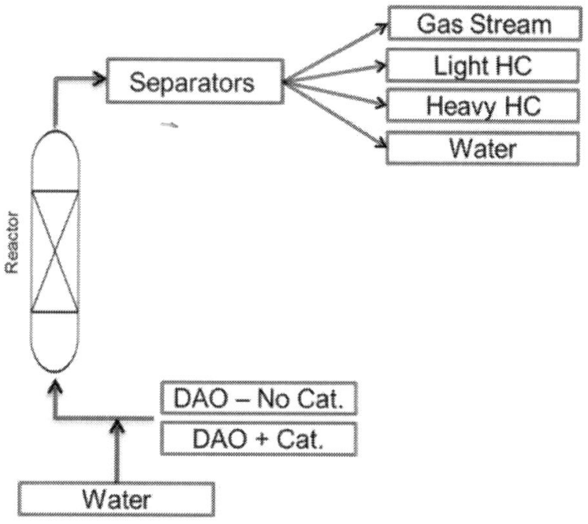

Figure 2. Reactivity Unit

2.3. Characterization of Products

In order to assess and compare the performance of the reactivity tests in terms of quality of the liquid products, the following analytical characterization methods are used: Simulated Distillation, Microcarbon Residue content, P-value, Viscosity, API gravity and C5 Asphaltenes by Microdeasphalting. Gas Chromatography (GC) is used for the characterization of the gas products.

3. Experimental Plan for Reactivity Study

The reactivity evaluation in the new Reactivity Test Unit – 1, in which the operational performance of the unit was assessed, was carried out for a Deasphalted Oil obtained from the 250 °C+ fraction of a South American extra heavy oil, whose characteristics are shown in Table 1.

As baseline to compare with the catalytic process, the non-catalytic upgrading of the DAO via steam cracking using 5 %wt. of water was studied. Four conditions were tested, Condition 1 at 415 °C and LHSV $2h^{-1}$, Condition 2 at 420 °C and LHSV $2h^{-1}$, Condition 3 at 425 °C and LHSV $2h^{-1}$, and Condition 4 at 423 °C and LHSV $2h^{-1}$, taking two mass balances of 60 min for each evaluated condition; the reaction pressure for all tests was 300 psi.

CUTS DISTRIBUTION	
Naphta (28 - 190 °C)	0.00
Kerosene (190 - 260 °C)	2.28
Diesel (260 - 343 °C)	11.66
LVGO (343 - 453 °C)	27.20
HVGO (453 - 560 °C)	23.30
Residue (560°C+)	35.56
PROPERTIES	
Specific Gravity [API]	11.19
MCr [%wt]	9.34
TAN [mgKOH]	0.965
Viscosity @25°C [cP]	19297
Hydrogen (H) [%wt]	11.83
Carbon (C) [%wt]	85.27
Saturates [%wt]	20.83
Aromatics [%wt]	55.09
Resins [%wt]	17.44
Asphaltenes C-5 [%wt]	6.65

Table 1. DAO feedstock characterization

For the catalytic evaluation of the DAO reactivity, the catalyst matrix K-Ni was incorporated into the feed using the skid preparation described previously, targeting a concentration of 300 ppm of Ni and 400 ppm of K. As in the Steam Cracking case, a 5 % wt of water was used, and in this case a total of 5 conditions were tested, using as initial temperature the highest temperature used in steam cracking at the stability limit (423 °C). The other conditions were set along the run based on the conversions obtained. In this way, 4 of the 5 catalytic reaction conditions were established, which are: Condition 5 at 423 °C and LHSV $2h^{-1}$, Condition 6 at 427 °C and LHSV $2h^{-1}$, Condition 7 at 425 °C LHSV $2h^{-1}$, Condition 8 at 430 °C LHSV $2.5h^{-1}$. Also, it was decided to carry out another reaction at 435 °C and LHSV $2h^{-1}$ in a run carried out separately. For each condition two mass balances of 60 min were taken. The pressure of the test (300 psi) was the same for all the evaluated conditions.

Analytical characterization of products was completed for the first mass balance of all the conditions. In order to check the reproducibility of the experimental runs, the second mass balances of Condition 4 and Condition 9 were selected, which are the most severe conditions for the Thermal Cracking and Catalytic Steam Cracking at the stability limit.

4. Results

As can be seen in Figure 3, where the reaction temperature and pressure profiles as a function of time are shown for Steam Cracking, from the very beginning of each set of runs until the last condition (C4) was finished, the operation was very smooth, with no significant variations on neither pressure nor temperature during the experiments, which indicates a satisfactory behavior of the unit in terms of operation and stability. For each condition, a stabilization period equal to 3.5 times the residence time was used, before collecting the two mass balances.

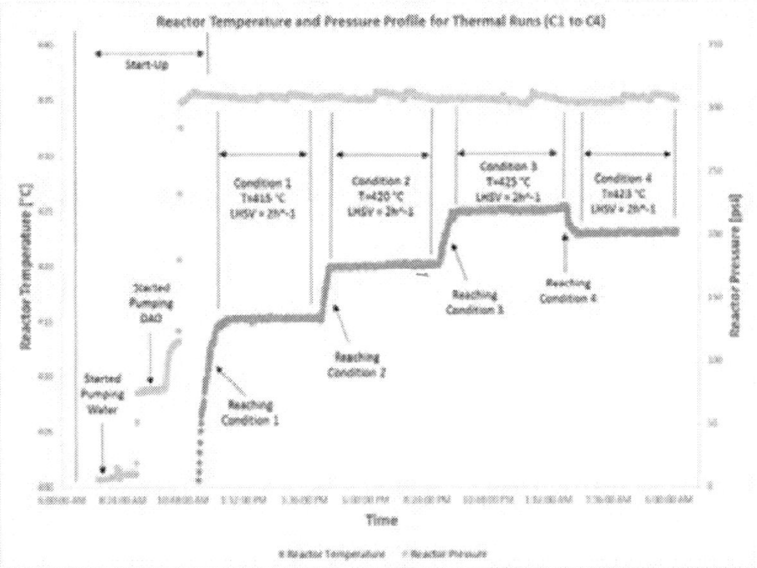

Figure 3. Reactor pressure and temperature profiles for SC runs

Figure 4 shows the reaction temperature and pressure as a function of time for the catalytic runs (C5-C9). In the same way as in the steam cracking runs, there was no significant change on both monitored variables, and the operation of the unit was very smooth.

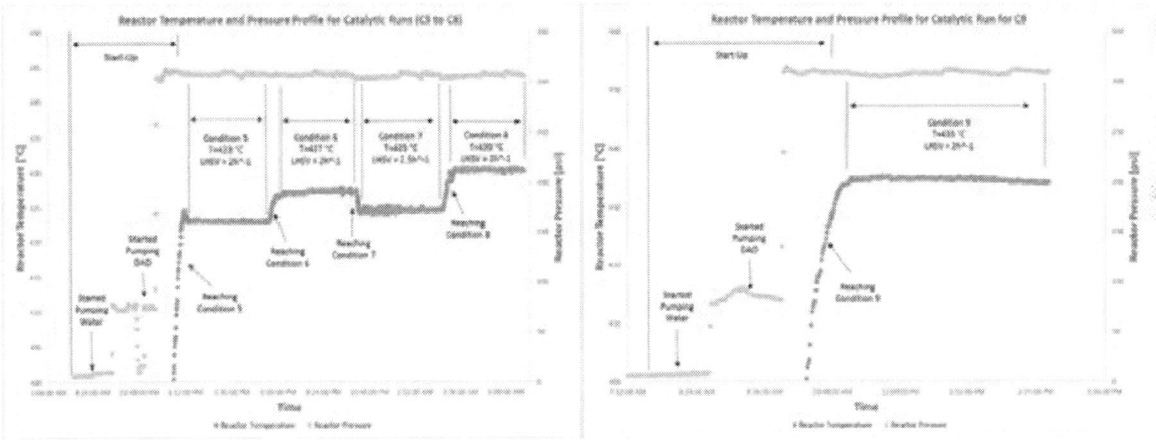

Figure 4. Reactor pressure and temperature profiles for CSC runs (C5-C9)

In Table 2, a summary of the results for both Steam Cracking and Catalytic Steam Cracking is presented, including reaction conditions and some analytical tests determined during the run to monitor the performance of the process. As can be seen in the table below, for the steam cracking reactions it was possible to go from 16.44 % of conversion of the 350 °C+ fraction at the lowest temperature, up to 22.60 % at the highest. Reasonable trends were observed for the microcarbon content, which increased as the temperature increased, reaching a maximum value of 12.62 % at the highest temperature. As expectable, viscosity reduced with the increase of reaction severity. Also, the P-value, which is an indication of the stability of the asphaltenes molecules in the product, showed a decreasing trend with severity, with values lower than the accepted limit of 1.15 for the most severe condition (C3) and, at the limit for C3, indicating the limit of the non-catalytic process at these conditions. For the Catalytic Steam Cracking reactions it was possible to go from 23.93 % of conversion of the 350 °C+ fraction at the lowest temperature (423 °C), up to 34.72 % at the highest (435 °C). Comparing Condition 4 (SC) with Condition 5 (CSC), which are both at the same temperature, an increase of 2 points in conversion was observed, which suggest that the catalyst was not activated at that temperature since the difference is not considerable. However, at 435 °C a significant increase on conversion was observed, probably caused by the activation of the catalyst at that level of severity. Reasonable trends were observed for the Microcarbon content, which increased with temperature, reaching a maximum value of 12.76 % at the highest temperature.

Analysis	Feed	Steam Cracking					Catalytic Steam Cracking					
		C1	C2	C3	C4-1	C4-2	C5	C6	C7	C8	C9-1	C9-2
Reaction Temperature [°C]	N/A	415	420	425	423	423	423	427	425	430	435	435
LHSV [h^-1]	N/A	2	2	2	2	2	2	2	2	2.5	2	2
Hydrocarbon Mass Balance [%]	N/A	97.10	94.81	95.98	95.99	95.36	98.09	97.84	99.66	98.21	97.13	98.41
Conversion (350 °C + HC) [%]	N/A	16.44	18.81	22.60	21.74	20.42	23.93	27.85	25.29	26.35	34.72	33.94
Asphaltenes C-5 [%wt]	6.65	8.34	8.41	11.21	10.60	10.13	10.11	14.75	11.47	11.68	14.5 2	13.66
API Gravity	11.19	10.72	10.99	12.03	11.68	11.21	10.50	11.34	10.63	11.51	12.67	12.36
MCr [%wt]	9.34	10.28	11.69	12.62	12.54	11.71	12.38	12.35	12.22	12.52	12.76	13.51
Viscosity @25°C [cP]	1929 7	3176	1864	936	1184	1060	1181	720	1665	720	238	311

Table 2. Data summary

In Figure 5, the product distribution by cuts as a function of the conversion for the DAO SC and CSC runs is shown, which is directly related to the severity of the reaction. As can be observed for the SC reactions, the yield of the heavier cuts (LVGO, HVGO, and Residue) tend to decrease as the conversion increases, and the yields of the lighter fractions (Naphtha, Kerosene, and Diesel) slightly increase, which is a typical trend for a conventional visbreaking process. The yield of gases did not change significantly with the conversion. Similarly to SC, the yields of the heavier cuts (LVGO, HVGO, and Residue) tend to decrease as the conversion increases for the CSC processing, and the yields of the lighter fractions (Naphtha, Kerosene, and Diesel) tend to increase. However, in this case, the yields of LVGO were higher than the yields of residue, which benefits the formation of a total lighter product, with lower viscosity than thermal cracking products. For the catalytic runs, the gas yield showed an overall increase with the severity of the reaction.

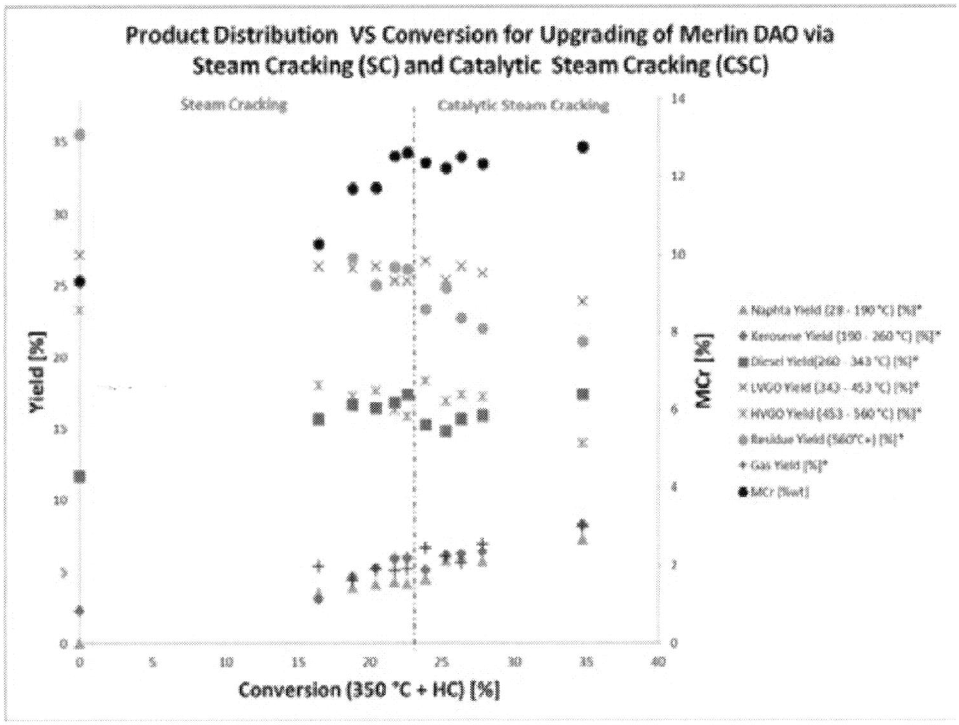

Figure 5. Products distribution for SC and CSC reactions

When determining stability for the liquid products from the catalytic runs, there were some inconvenients due to the presence of solid particles that were hindering the visualization of Asphaltenes by optical microscopy since it was not clear whether the observed particles were precipitated asphaltenes or simply small particles of catalyst or coke. To avoid this, 4 g of product were diluted with $CHCl_3$ (99% Aldrich) in a ratio of 1:50(wt/wt) and the solution was filtered to separate and quantify the solids. The filtered material was rotoevaporated in order to remove the $CHCl_3$ and the cleaned oil was used to determine the presence of precipitated asphaltenes in the product. In this way, the increase in stability when using the K-Ni ultradispersed catalyst was confirmed, allowing to reach higher levels of conversion, thus leading to increased amounts of lighter products compared to thermal processing.

5. Conclusions

Reactivity evaluation evidenced that DAO steam cracking showed typical behavior for thermal processing of heavy feedstocks with limited levels of conversion, but significant viscosity reduction. Catalytic steam processing yields higher levels of light products and an overall higher products quality.

The performance of the Reactivity Test Unit 1 (RTU-1) built for the development of this work was successfully assessed and met the expectations established in the proposed design. It was possible to carry out experimental tests at different reaction conditions for both thermal and catalytic processes with good results in terms of operation of the reactivity unit.

6. References

1. Pereira-Almao, P., et al., *Aquaconversion technology offers added value to E. Venezuela synthetic crude oil production*, in *The Oil and Gas Journal*. 2001. p. 79.
2. Pereira-Almao, P., et al., *Steam conversion process and catalyst*. 1999, United States Patent.
3. *Alberta's Energy Reserves 2012 and Supply/Demand Outlook 2013-2022*. 2013.
4. Ancheyta, J., *Modeling and Simulation of Catalytic Reactors for Petroleum Refining*. 2011: Wiley.
5. Pereira-Almao, P., et al., *Systems and Methods for Catalytic Steam Cracking of Non-Asphaltene Containing Heavy Hydrocarbons*. 2013, United States Patent.
6. Vasquez, A., *Synthesis, Characterization and Model Reactivity of Ultra Dispersed Catalysts for Hydroprocessing*, in *Chemical and Petroleum Engineering Department*. 2007, University of Calgary: Calgary.
7. Trujillo-Ferrer, G., *Thermal and Catalytic Steam Reactivity Evaluation of Athabasca Vacuum Gasoil*, in *Chemical and Petroleum Engineering Department*. 2008, University of Calgary: Calgary.
8. Fathi, M.M., *Comparative Upgrading of Arab Light Vacuum Residuum via Aquaprocessing and Thermal Cracking*, in *Chemical and Petroleum Engineering Department*. 2011, University of Calgary: Calgary. p. 178.

High-yielding Suzuki coupling reactions of basic, amine containing substrates: The role of CO_2, solvent, and pH.

C. Senter, A. Rumple, W. Medina-Ramos, J. Fisk, B. Holden, P. Pollet, C.A. Eckert, and C. L. Liotta

School of Chemical and Biomolecular Engineering
School of Chemistry and Biochemistry
Georgia Institute of Technology
Atlanta, Georgia 30332

Suzuki coupling is a transition-metal catalyzed reaction of an aryl halide and a boronic acid that forms carbon-carbon bonds. In large part due to its versatility, the Suzuki reaction has emerged as an important reaction in industrial synthesis. While the choice of catalyst, ligand, base and solvent system all play key parts in the success of the reaction, the role played by the substrate is just as important. Suzuki reactions involving substrates with basic nitrogen centers often obtain poor yields, a result attributed to the deactivation of the metal catalyst by the coordination of these nitrogen moieties. These poor yields are problematic, for nitrogen containing compounds are important building blocks in many biological and pharmaceutical compounds. Common strategies to achieve high-yielding reactions include adding additional amine protection and deprotection steps or the use of expensive, "designer" ligands. By running reactions under CO_2 pressure and manipulating the amount of water present in the reaction system, we have developed a successful protocol for high-yielding Suzuki coupling reactions of 4-NH_2-2-halopyridines and 2-halopyridines using $Pd(TPP)_2Cl_2/K_3PO_4$ in acetonitrile/water solvent systems. The results of this investigation will be discussed.

Palladium has been extensively used in commercial oxidation processes such as the oxidation of CO in exhaust treatment and the combustion of methane in gas turbines. The metallic Pd is easily oxidized under oxygen-rich conditions, so the PdO phase plays an important role in the surface reactions. We performed density functional theory calculations (GGA and HSE) to investigate the adsorption of CO on the PdO(101) surface. On the pristine surface, GGA predicts the atop and bridging sites are energetically equivalent, but HSE shows that initially CO only adsorbs at the atop sites, in agreement with TPRS experiments where only the atop peaks were found. Further pDOS and charge difference studies indicate that the GGA may overestimate the Pd electron backdonation to the bridging CO that results in its high stability, so we focus on the atop CO in the oxidation study. With increasing temperature, chemisorbed CO is activated and reacts with surface oxygen. Two CO oxidation pathways were found and the barriers are consistent with CO_2 TPD features at different temperatures. The oxidation is facile when CO reacts with its neighboring coordinatively unsaturated O atom. After CO_2 desorbs from the surface, the Pd site next to a vacancy is favored for the CO adsorption and red-shifts the CO stretching band. The charge transfer suggests that the existence of vacancies enhances electron backdonation to the CO and results in this redshift.

Olefins are building blocks of high importance in the petrochemical industry. Dehydration of biomass-derived alcohols using metal-oxide catalysts offers an alternative (to petrochemical feedstocks) and sustainable route for the production of alkenes[1-3]. However, the functionalities of the oxides are not well understood. In this work[4], we use first-principles calculations to investigate the dehydration of various alcohols on different metal oxides. Our results show that the basicity of surface oxygens, the Lewis acidity of the metal centers of the oxides and the carbenium ion stability[3] of the alcohols can be fundamental descriptors to develop structure-activity relationships on oxides. The dehydration predictions of the developed model are in very good agreement with TPD-TGA experiments. This work unravels the role of the surface hydroxyl groups of the oxides in the dehydration mechanism and offers a way to bridge Lewis-acid with Brønsted-acid catalyzed dehydration reactions on metal oxides. Most importantly, the developed models can be used as a screening tool for the dehydration performance of various alcohols over different metal oxides.

1. Christiansen, M. A.; Mpourmpakis, G.; Vlachos, D. G. *Acs Catalysis* **2013,** 3, 1965–1975.
2. DeWilde, J. F.; Chiang, H.; Hickman, D. A.; Ho, C. R.; Bhan, A. *ACS Catalysis* **2013,** 3, (4), 798-807.
3. Roy, S.; Mpourmpakis, G.; Hong, D. Y.; Vlachos, D. G.; Bhan, A.; Gorte, R. J. *ACS Catalysis* **2012,** 2, (9), 1846-1853.
4. Kostestkyy, P.; Yu, J.; Gorte, R. J.; Mpourmpakis, G. *submitted* **2014**.

373646 Engineering Catalyst Microenvironments for Metal-Catalyzed Hydrogenation of Biologically-Derived Platform Chemicals

Abstract:

Mirroring the evolution of traditional petroleum refineries, successful future biorefineries will rely on the production of value-added chemicals to supplement biofuels production. One promising strategy for the production of biorenewable chemicals from biomass uses heterogeneous catalysis to upgrade biologically-derived platform molecules. Such a scheme leverages the efficiency of heterogeneous catalysis and couples it with the ability of biocatalysis to generate molecules not easily accessible by other means. The efficiency advantage is lost, however, if the heterogeneous catalysts used for upgrading are not stable in the presence of biogenic impurities or highly reactive feedstocks. In this work, we use hydrogenation of the biologically-derived intermediate triacetic acid lactone (TAL) to demonstrate that these types of catalyst deactivation can be alleviated by appropriate catalyst design. Specifically, we show that a traditional supported-Pd hydrogenation catalyst is inhibited by biogenic impurities such as amino acids, especially by methionine, which decreases the catalyst activity by 83% after 14 hours of low-concentration exposure (*i.e.*, 0.01 mM). This inhibition is mitigated by formation of a microenvironment surrounding the metal nanoparticles, achieved by overcoating the catalyst with poly(vinyl alcohol) (PVA), and the overcoated catalyst is stable in the presence of amino acids, including methionine. Results from solid-state NMR studies are used to provide molecular-level insight into the interaction between methionine and the Pd surface for both the parent and PVA-overcoated catalysts. Additionally, we show that deactivation of the Pd catalyst by deposition of carbonaceous residues in the presence of TAL is alleviated by formation of bimetallic PdAu nanoparticles. Accordingly, the use of PVA to form microenvironments on a PdAu catalyst allows for an order-of-magnitude improvement in stability for the hydrogenation of biologically-produced TAL recovered from spent cell culture medium.

The ABE condensation has been shown to be a viable pathway to biodiesel precursors [1]. In our recent work, we have shown that magnesium oxide and hydrotalcite-supported metal catalysts are efficient and recyclable catalysts for this reaction [2]. Hydrotalcites and metal oxides are known to change their structure in the presence of water [3]. However, to the best of our knowledge, no analysis of the mechanisms of phase change and deactivation during reaction has been undertaken in the literature. Therefore, in this work, we present an investigation of the dependence of crystalline phase, surface area and metal dispersion and oxidation state as a function of the preparation and reaction conditions.

During gas-phase reactions, we note a rapid decrease of the reactivity with time-on-stream followed by a steady state lasting at least 8 days. Nitrogen physisorption measurements of the catalyst before and after the reaction show a decrease of the surface area from approximately 200 m^2/g to 18 m^2/g, a decrease that is consistent with the loss of activity over the course of the reaction. X-ray diffraction measurements show minimal change in the crystallinity or the domain size of the catalyst, based on the broadening of the diffraction lines. This observation, combined with the fact that nitrogen physisorption shows a loss of small-diameter pores during reaction suggests that the mechanism for deactivation of the PdCuHT catalysts during gas-phase reaction is due to the formation of coke in the pores, resulting in their blockage and loss of surface area and reactivity.

On the other hand, during liquid phase reaction, we observe a change in the crystalline phase of the catalyst. As can be seen in figure 1, the surface area decreases during reaction. The loss of surface area can be attributed to a phase change, from the cubic mixed oxide (periclase) phase to a lamellar structure (hydrotalcite/brucite) and an amorphous phase, as evidenced by the reduction of the area of the peak at $2\theta = 42.7^{\circ}$, attributed to the cubic magnesium-aluminum oxide structure and the increase in a broad feature around $2\theta = 43.4^{\circ}$, tentatively attributed to poorly crystalline Mg-Al hydroxides (figure 2). Features around 22.2, 34.5 and 45.8°, can be attributed to the formation of the lamellar hydrotalcite structure. From the nitrogen physisorption data, it seems that a steady state is reached after about 7h, with no further loss after that point.

The presence of Al in the hydrotalcite materials seems to lend structural stability to the mixed oxide material; even after 22h of batch reaction, there is a clear reflection at $2\theta = 43.4^{\circ}$. On the other hand, as can be seen in figure 3, the reflection from the cubic periclase structure ($2\theta = 42.7^{\circ}$) in PdCu/MgO is lost after about 4h and is replaced by the reflections of the layered brucite hydroxide structure at 38.1, 50.8 and 58.6°. Interestingly, there is no evidence of poorly crystalline (or amorphous) phase formation in the case of MgO-based catalysts, indicating that the presence of Al is required for this. Further evidence towards the stabilizing effect of Al in the mixed metal oxides is given by the nitrogen physisorption measurements. The surface areas of Mg-Al oxides are much higher than the pure MgO structures, at all reaction conditions.

The effects of the reaction on the metal were investigated using X-ray absorption spectroscopy. The results of these experiments indicate that the Pd metal is completely reduced following the reaction. Also, increasing reaction times favor C-H bond activation over C-C bond scission over the Pd catalysts; our current investigations are focusing on the reason for these selectivity shifts via chemisorption and X-ray absorption studies.

1. Anbarasan, et al. *Nature* **491**, 235–239
2. Sreekumar, et al. *ChemCatChem*, submitted
3. Climent, et al. *J. Catal.* **225**, 316-326

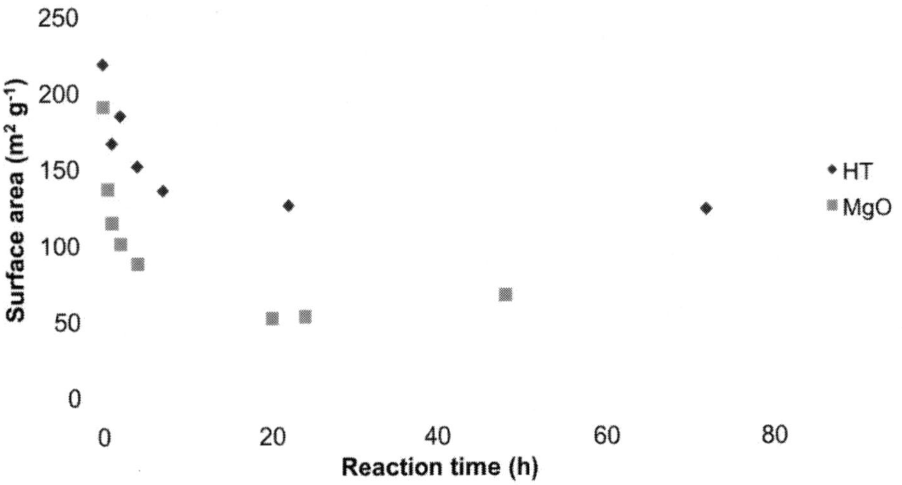

Figure 1: Evolution of surface area of calcined hydrotalcite (HT) and magnesium oxide (MgO) during liquid-phase batch reaction. 160 mg catalyst, 74 mg ethanol, 222 mg acetone, 444 mg butanol, 75 mg dodecane (internal standard), 513 K.

Formatting Issue - Best Available Paper/Graphic

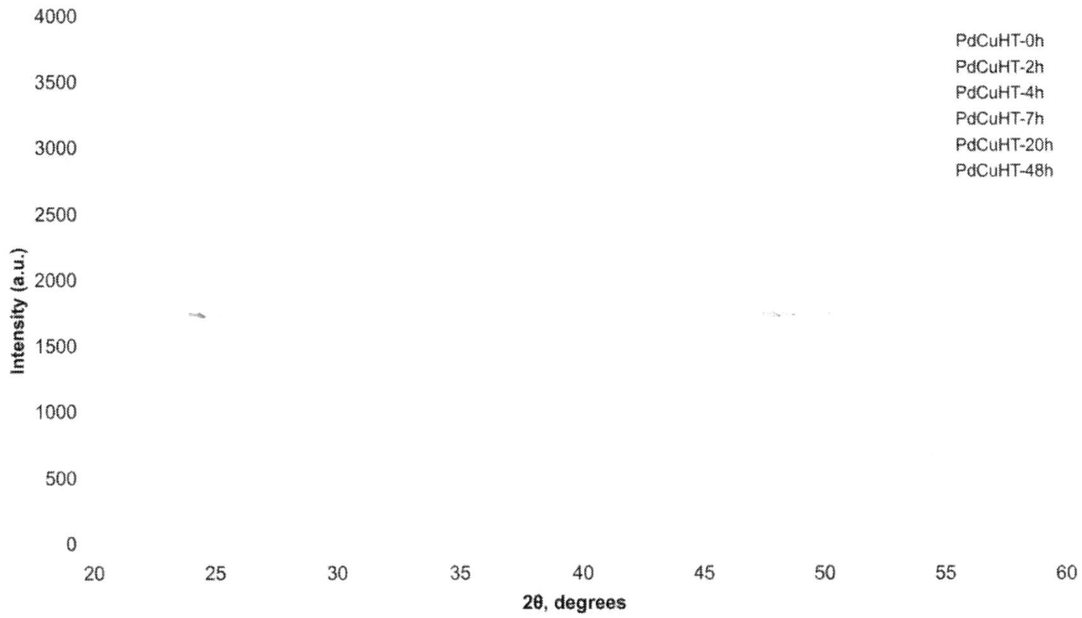

Figure 2: Evolution of PdCuHT catalyst crystal structure during reaction.

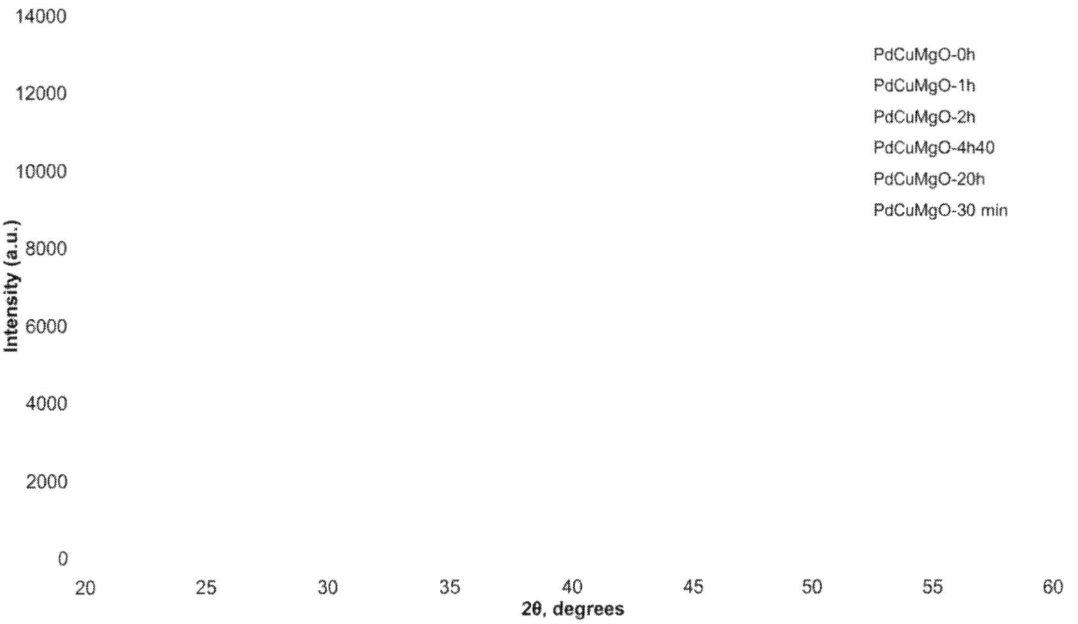

Figure 3: Evolution of PdCuMgO catalyst crystal structure during reaction

385437 Conversion of Bioethanol and Ethylene to Drop-in Fuel Hydrocarbons Under Supercritical Conditions

Abstract:

The conversion of bioethanol to hydrocarbons is a promising technology for the production of drop-in fuels from lignocellulosic biomass. This process is heterogeneously catalyzed by HZSM-5 and it takes place in two steps: 1) dehydration of ethanol to ethylene, 2) oligomerization of ethylene to various lengths of hydrocarbon molecules. To characterize these steps, we will present experimental results which include supercritical conditions. Our results indicate that supercritical conditions decrease selectivity to aromatic hydrocarbons, which is an advantage from an emissions perspective. This, in turn, is expected to reduce coke formation and improve catalyst lifetime. In this talk, we will report results from: 1) the conversion of ethanol to hydrocarbons, 2) the oligomerization of ethylene to drop-in fuel molecules. Our goal is to understand the role of supercritical conditions and develop kinetic models for this process to predict the yields of permanent gases, linear/branched paraffins, olefins, naphthenes, aromatics, and solid char.

Tuning Cellulose Pyrolysis Chemistry: Selective Decarbonylation via Catalyst-Impregnated Pyrolysis

Alex D. Paulsen, Matthew S. Mettler, Dionisios G. Vlachos, and Paul J. Dauenhauer

Abstract: Production of bio-oils and biofuels from lignocellulosic biomass relies on the chemistry of high temperature cellulose pyrolysis.[1] Cellulose is known to react through a short-lived liquid intermediate[2] during pyrolysis to produce a variety of condensable organics.[3-5] However, utilization of biomass as a fossil fuel replacement is hindered by the instability of bio-oil under storage conditions, likely caused by polymerization of aldehydes present in bio-oil. Therefore, we examine cellulose impregnation of supported metal catalysts to tune the chemistry within the liquid intermediate and reduce the production of aldehydes. We show that using a carbon-supported metal catalysts significantly reduce the yield of aldehydic products by converting hydroxymethylfurfural and furfural to carbon monoxide and deoxygenated furans (methyl furan, furan, dimethylfuran) while maintaining the overall bio-oil yield.

1. Paulsen, A. D., Hough, B. R., Williams, C. L., Teixeira, A. R., Schwartz, D. T., Pfaendtner, J., & Dauenhauer, P. J., Fast Pyrolysis of Wood for Biofuels: Spatiotemporally Resolved Diffuse Reflectance In situ Spectroscopy of Particles. *ChemSusChem* **2014,** 7, (3), 765-776.
2. Dauenhauer, P. J.; Colby, J. L.; Balonek, C. M.; Suszynski, W. J.; Schmidt, L. D., Reactive boiling of cellulose for integrated catalysis through an intermediate liquid. *Green Chemistry* **2009,** 11, (10), 1555-1561.
3. Mettler, M. S.; Mushrif, S. H.; Paulsen, A. D.; Javadekar, A. D.; Vlachos, D. G.; Dauenhauer, P. J., Revealing pyrolysis chemistry for biofuels production: Conversion of cellulose to furans and small oxygenates. *Energy & Environmental Science* **2012,** 5, (1), 5414-5424.
4. Mettler, M. S.; Paulsen, A. D.; Vlachos, D. G.; Dauenhauer, P. J., The chain length effect in pyrolysis: bridging the gap between glucose and cellulose. *Green Chemistry* **2012,** 14, 1284-1288.
5. Paulsen, A. D., Mettler, M. S., Vlachos, D. G., & Dauenhauer, P. J., The Role of Sample Dimension and Temperature in Cellulose Pyrolysis. *Energy & Fuels* **2013**.

381028 Syngas Production over Ni/MgAl2O4/Al2O3 Catalyst in Combined Reforming of Methane

Abstract:

Steam carbon dioxide reforming (SCR) of methane is an attractive process for getting a H_2/CO ratio of 2 which is desirable for GTL (gas to liquid) and methanol (MeOH) synthesis processes. It is widely reported that Ni and noble metals are used as reforming catalysts due to good activity and stability in reforming process. Generally Ni is mainly applied for catalysts in reforming process because of relatively low cost and inherent availability. However it has some drawbacks like sintering and deactivation by coke deposition on the catalyst surface. It was also reported that the Ni based catalysts Supported $MgAl_2O_4$ spinel compound layer existed between Ni and Al_2O_3 show good stability for the process. In this work, for improvement of catalyst stability, out-layered $MgAl_2O_4$ spinel for $Ni/MgAl_2O_4/Al_2O_3$ catalyst was investigated in SCR of methane. Supports with $MgAl_2O_4$ contents with different ratio on γ-Al_2O_3 were prepared by impregnation method. The catalyst without out-layered $MgAl_2O_4$ was also prepared for comparison of catalytic activity. Prepared catalysts were characterized by various techniques such as N_2 physisorption, H2-TPR, XRD, CO_2-TPD, SEM, TEM-EDS and TG analysis. Aspen simulation package was used to estimate optimum experimental conditions for SCR of methane. The simulation results were compared with the experimental results. It was found that the catalyst with the out-layer of $MgAl_2O_4$ showed better catalytic activity and stability. It was considered that due to the out-layer magnesium aluminate spinel formation between Ni and Al_2O_3 suppress carbon deposition and inactive $NiAl_2O_4$ or NiO-MgO formation.

382331 A Highly Selective Route to Linear Alpha Olefins from Biologically-Derived Unsaturated Acids and Lactones

Abstract:

We will demonstrate a general catalytic strategy that produces a stream of highly pure (>99%) linear alpha olefin from biomass-derived lactones or unsaturated carboxylic acids. The process employs a single flow reactor and uses inexpensive, robust heterogeneous solid acid catalysts with no precious metal components, thus offering great potential as a practical approach for the production of key bulk chemicals and fuels from renewable resources, which is of great interest for sustainable chemistry. We will present results for linear alpha olefins with carbon chain length ranging from C4 – C10, with both odd and even carbon numbers.

The work to be presented will also provide a clear understanding and control of catalyst properties and product distribution, which will provide guidance for the synthesis of superior catalysts and further improvement in the yields of desired linear alpha olefin products. Finally, our finding that decarboxylation of lactones or unsaturated acids in the presence of water leads to high selectivity to linear alpha olefin represents an intriguing case where the high water content commonly encountered in biomass processing can be utilized to positively influence product selectivity.

383504 High Concentration H2S-Resistant Catalysts for Upgrading of Coal or Biomass Derived Syngas

Abstract:

A major challenge in any coal or biomass gasification application is syngas cleanup which represents 36-41% of the total plant cost. Additionally, syngas derived from coal/biomass is typically lean in H_2 and needs to be upgraded relative to distillate fuels which have hydrogen to carbon molar ratios of approximately 2:1. As part of a program to develop an advanced gasification platform to convert low rank coals to high hydrogen syngas for integrated gasification combined cycle (IGCC) and coal to liquids (CTL) applications, Southern Research Institute is developing, testing, and optimizing catalysts for reforming tars, methane, and C_2+ hydrocarbons, and decomposing NH_3 in the presence of high concentrations of H_2S (up to 1000 ppmv) and other syngas contaminants such as HCl. Catalytic steam reforming prior to low temperature downstream polishing has the potential to increase the overall energy efficiency of the process by utilizing the excess steam from gasification for syngas upgrading. However, based on a comprehensive assessment of pertinent literature, no catalysts are currently available that can operate under H_2S concentration of higher then 100 ppm for sustained time under gasification conditions. In this study, catalysts with high tolerance to sulfur were prepared and tested at lab scale for treating raw simulated syngas containing several hundred ppmv H_2S from two commercially relevant gasifier designs, TRIG (Transport Reactor Integrated Gasifier) and Lurgi FBDB (Fixed-Bed Dry-Bottom). Toluene and methyl naphthalene were selected for testing as the model components of tar. Catalysts were tested at high temperatures > 900 °C and pressures (15 to 200 PSIG), and conversion of methane, ammonia and tar were determined. Catalysts were characterized for BET surface area, pore volume, pore size distribution and XRD phases. Results of the catalyst screening and parametric study for sulfur tolerance and catalyst activity will be presented.

Spatio-temporal dynamics of oxygen storage and release in a three-way catalytic converter with axially non-uniform catalyst activity

Tian Gu[1], Matthew Franchek[2], Karolos Grigoriadis[2] and Vemuri Balakotaiah[1]
[1]Department of Chemical and Biomolecular Engineering, University of Houston, TX-77204
[2]Department of Mechanical Engineering, University of Houston, TX-77204

Abstract

Three-way catalytic converters (TWC) are exhaust after-treatment systems commonly used in stoichiometry operated gasoline vehicles. Within a narrow zone around stoichiometry, the TWC can simultaneously oxidize CO and HC while reducing NOx. In order to buffer slight excursions from stoichiometric operation and abate emission breakthrough, oxygen-storing components, such as ceria, are added in TWC to store oxygen in the fuel lean phase and release it in the fuel rich phase.

Axially non-uniform catalyst activity of a TWC may result from non-uniform platinum-group metal (PGM) and/or ceria loading or non-uniform deactivation. For fixed amounts of total PGG and ceria loading, proper redistribution (with more catalyst near the inlet) favors fast light-off of the TWC and reduces cumulative cold-start emissions substantially [1]. Local temperature peaks due to certain operations (e.g. cold start, sudden acceleration, etc.) may cause non-uniform aging of the catalyst which leads to different TWC cold start dynamics from a uniformly aged one.

In this work, we use a recently developed physics based reduced-order one-dimensional transient model to study the spatio-temporal dynamics of oxygen storage and release in a TWC with axially non-uniform catalyst activity. For a uniform catalyst activity case, the TWC model is validated using emission data collected in a vehicle over multiple drive cycles. The internal mass transfer approximation is confirmed by comparing with a detailed washcoat diffusion-reaction model and is observed to be consistent for all practical purposes. Simulations are conducted for multiple cases such as cold start and fast lean-rich cycling to study oxygen storage dynamics, transient temperature profile and emission breakthrough in the TWC. Cold start and fast cycling simulation results for the uniform catalyst activity case are shown in Fig. 1-2. The results for non-uniform catalyst activity case are compared with the uniform case.

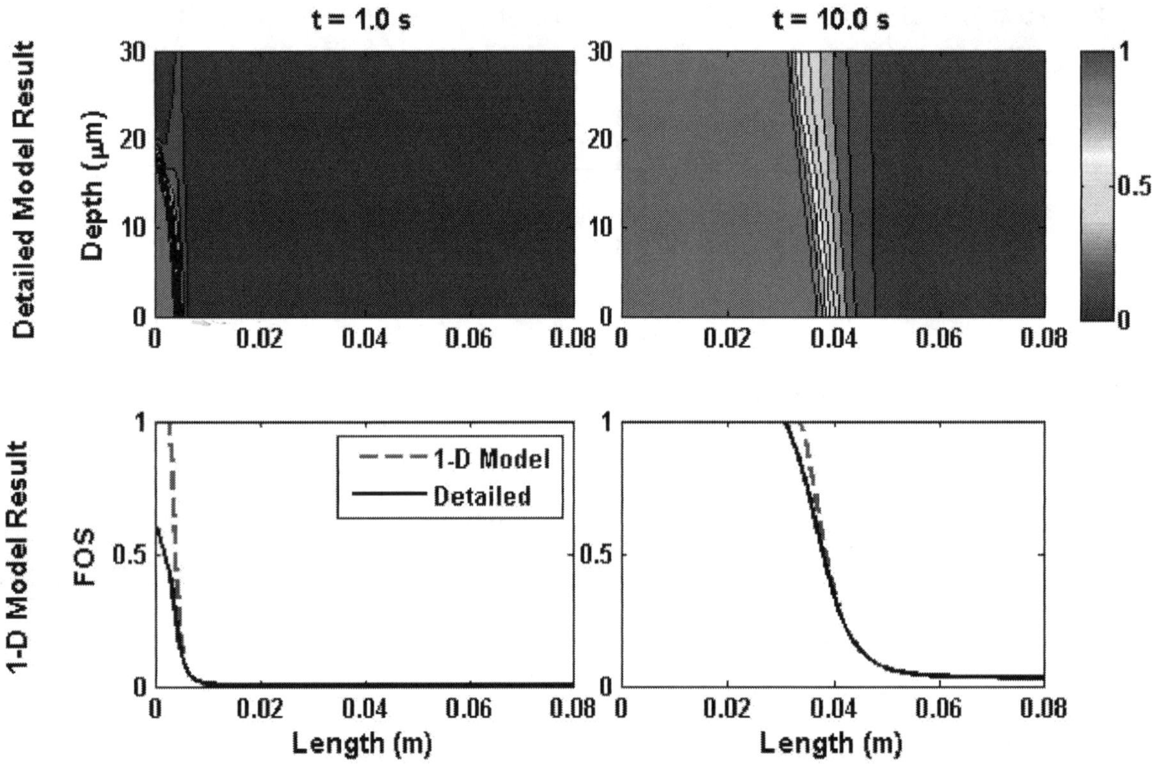

Fig.1 Profiles of fractional oxidation state (FOS) under cold start condition.

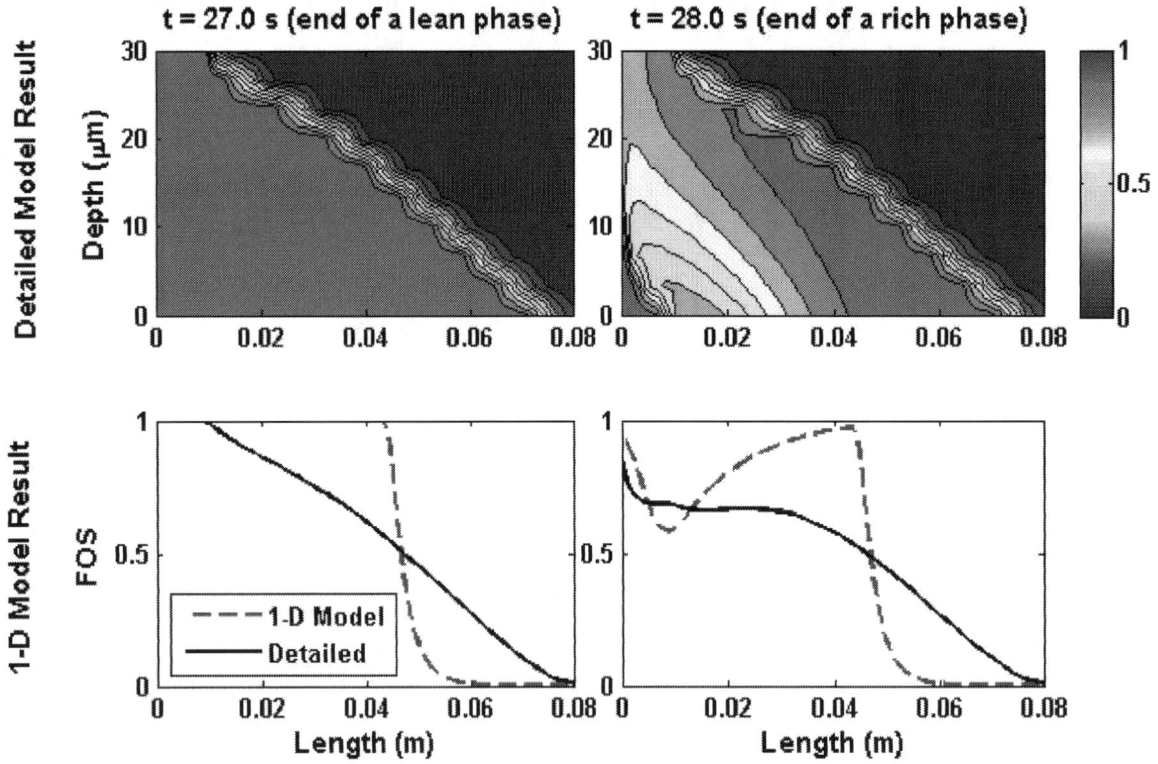

Fig.2 Profiles of FOS during fast lean-rich cycling.

Reference

1. Ramanathan, K., West, D. H., & Balakotaiah, V. (2004). Optimal design of catalytic converters for minimizing cold-start emissions. Catalysis today, 98(3), 357-373.

Pattern Formation and Dynamics of a Model of Simultaneous Oxidation and Reduction in Three-Way Catalytic Converters Involving Cross-Flow

M. Kohout and I. Schreiber

Department of Chemical Engineering,

Institute of Chemical Technology, Prague

INSTITUTE OF CHEMICAL TECHNOLOGY PRAGUE

DEPARTMENT OF CHEMICAL ENGINEERING

2014 AIChE Annual Meeting, Atlanta, Nov 16-21, 2014

Outline:

- relations between kinetic instabilities and patterns due to transport in heterogeneous catalytic reactions
- case study: three-way-catalytic system with cross flow ("cross-flow converter")
- instabilities in complex chemical systems → stoichiometric network analysis
- sources of oscillations in a spatially lumped reduced model
- pattern formation in a reaction-diffusion-convection system with cross-flow

Reaction scheme for oxidation of CO and C_2H_2, and reduction of NO on a three-way catalyst (TWC)

Nibbelke et al. Appl. Catal. B, 19, 1998

Microkinetic reaction scheme used in the model of TWC

No.		Reaction step	Kinetic expression
1	**CO subsystem**	$CO + * \rightleftharpoons CO^*$	$R_1 = k_1^f L_{NM} c^s_{CO} \theta_* - k_1^b L_{NM} \theta_{CO^*}$
2		$O_2 + 2* \rightarrow 2O^*$	$R_2 = k_2 L_{NM} c^s_{O_2} \theta_*$
3		$CO^* + O^* \rightarrow CO_2 + 2*$	$R_3 = k_3 L_{NM} \theta_{CO^*} \theta_{O^*}$
4		$CO + O^* \rightleftharpoons OCO^*$	$R_4 = k_4^f L_{NM} c^s_{CO} \theta_{O^*} - k_4^b L_{NM} \theta_{OCO^*}$
5		$OCO^* \rightarrow CO_2 + *$	$R_5 = k_5 L_{NM} \theta_{OCO^*}$
6	**C_2H_2 subsystem**	$C_2H_2 + * \rightleftharpoons C_2H_2^*$	$R_6 = k_6^f L_{NM} c^s_{C_2H_2} \theta_* - k_6^b L_{NM} \theta_{C_2H_2^*}$
7		$C_2H_2^* + 2* \rightleftharpoons C_2H_2^{***}$	$R_7 = k_7^f L_{NM} \theta_{C_2H_2^*} \theta_*^2 - k_7^b L_{NM} \theta_{C_2H_2^{***}}$
8		$C_2H_2^* + 3O^* \rightarrow 2CO^* + H_2O + 2*$	$R_8 = k_8 L_{NM} \theta_{C_2H_2^*} \theta_{O^*}$
9		$C_2H_2^{***} + 3O^* \rightarrow 2CO^* + H_2O + 4*$	$R_9 = k_9 L_{NM} \theta_{C_2H_2^{***}} \theta_{O^*}$
10		$C_2H_2 + O^* \rightleftharpoons C_2H_2O^*$	$R_{10} = k_{10}^f L_{NM} c^s_{C_2H_2} \theta_{O^*} - k_{10}^b L_{NM} \theta_{C_2H_2O^*}$
11		$C_2H_2O^* + 2O^* \rightarrow 2CO^* + H_2O + *$	$R_{11} = k_{11} L_{NM} \theta_{C_2H_2O^*} \theta_{O^*}$
12	**interactions on oxygen storing sites**	$O_2 + 2s \rightarrow 2O^s$	$R_{12} = k_{12} L_{OSC} c^s_{O_2} \zeta_s$
13		$CO^* + O^s \rightarrow CO_2 + * + s$	$R_{13} = k_{13} L_{NM} \theta_{CO^*} \zeta_{O^s}$
14		$C_2H_2^* + 3O^s + * \rightarrow 2CO^* + H_2O + 3s$	$R_{14} = k_{14} L_{NM} \theta_{C_2H_2^*} \zeta_{O^s}$
15		$CO_2 + \gamma \rightleftharpoons CO_2^\gamma$	$R_{15} = k_{15}^f L_{SUP} c^s_{CO_2} Z_\gamma - k_{15}^b L_{SUP} Z_{CO_2^\gamma}$
16	**NOx subssystem**	$NO + * \rightleftharpoons NO^*$	$R_{16} = k_{16}^f L_{NM} c^s_{NO} \theta_* - k_{16}^b L_{NM} \theta_{NO^*}$
17		$NO^* + * \rightarrow N^* + O^*$	$R_{17} = k_{17} L_{NM} \theta_{NO^*} \theta_*$
18		$NO^* + N^* \rightarrow N_2O^* + *$	$R_{18} = k_{18} L_{NM} \theta_{NO^*} \theta_{N^*}$
19		$N_2O^* \rightarrow N_2O + *$	$R_{19} = k_{19} L_{NM} \theta_{N_2O^*}$
20		$N_2O^* \rightarrow N_2 + O^*$	$R_{20} = k_{20} L_{NM} \theta_{N_2O^*}$
21		$N^* + N^* \rightarrow N_2 + 2*$	$R_{21} = k_{21} L_{NM} \theta^2_{N^*}$
22		$NO + O^* \rightleftharpoons NO_2^*$	$R_{22} = k_{22}^f L_{NM} c^s_{NO} \theta_{O^*} - k_{22}^b L_{NM} \theta_{NO_2^*}$
23		$NO_2^* \rightleftharpoons NO_2 + *$	$R_{23} = k_{23}^f L_{NM} \theta_{NO_2^*} - k_{23}^b L_{NM} c^s_{NO_2} \theta_*$

Stoichiometric network analysis (Clarke, 1980) :

- linear stability analysis can be done with the use of graphs

- identification of positive and negative feedback loops

- decomposition of reaction networks to elementary subsystems

- identification of autocatalytic cycles as sources of instabilities

- combination of autocatalysis and negative feedback leads to bistability or oscillations

Other approaches: Complex balanced networks (Horn, 1976; Feinberg, 1995)

Outline of the SNA

B. L. Clarke, Adv. Chem. Phys. 43 1 (1980)

Assume that there are *m* species taking part in *r* chemical reactions so that first *n* species, $n \leq m$, are entering at least one of the reactions

$$R_j : \sum_{i=1}^{n} \nu_{ij}^{L} X_i \rightarrow \sum_{i=1}^{m} \nu_{ij}^{R} X_i \ , \qquad j = 1, \cdots, r$$

where $\nu_{ij}^{L} \geq 0$ and $\nu_{ij}^{R} \geq 0$ are, respectively, the left hand and right hand stoichiometric coefficients of species X_i in reaction R_j. The first *n* species are assumed to be reactants or intermediates and the remaining $m - n$ are inert products.

$\bar{S} = \{\nu_{ij}^{R} - \nu_{ij}^{L}\}$ is the $(n \times r)$ **stoichiometric matrix**

$X = (X_1, \cdots, X_n)$ is the vector of the chemical species concentrations

$\bar{V}(X) = (\bar{V}_1(X), \cdots, \bar{V}_r(X))$ is the vector of reaction rates

The time evolution of *X* in a flow-through system at constant temperature in a well-stirred reaction cell of constant reaction volume is based on mass balance equations:

$$\frac{dX}{dt} = F(X) = \bar{S}\bar{V}(X) + k_0(X_0 - X) = SV(X)$$

S is the extended stoichiometric matrix including inflows and outflows

Stoichiometric constraints: If the rank of *S* is less than *n*, there is a nonempty null space of S^T of dimension $d_n = n - \text{rank}(S)$ and there are d_n independent linear constraints for *X*.

Closed systems or subsystems, such as enzyme reactions, may satisfy such constraints

Decomposition of the network into subnetworks

A stationary state X_s satisfies the equation:

$$SV(X) = 0$$

Hence $V_s = V(X_s)$ is contained in the null space of *S* that has dimension $d_r = r - \text{rank}(S)$.

All components of *V* must be non-negative numbers which narrows the set of all possible stationary reaction rate vectors V_s (=currents) to an open, convex, d_r-dimensional cone, $d_r = r - \text{rank}(S)$, in the reaction rate space of all *V*'s

The edges of this steady state cone represent sets of steady states that have minimum possible nonzero elements of *V*'s and define a set of major subnetworks (=extreme currents, elementary modes) of the mechanism.

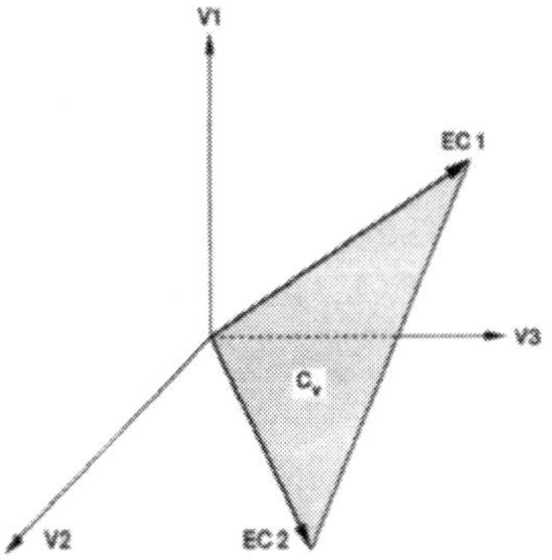

Two-dimensional current cone C_v spanned by two extreme currents EC1 and EC2.
The shaded bounded region of the cone is obtained by applying the normalizing condition $\sum_{k=1}^{K} \alpha_k EC_k \leq 1$

If EC_k, $k = 1, \cdots, K$, are vectors pointing along the edges of the cone then any linear combination $\sum_{k=1}^{K} \alpha_k EC_k$ with nonnegative coefficients is again a current.
Conversely, any current V_s can be expressed as a linear combination of extreme currents

Example:

V1 - inflow: → X
V2 - reaction: 2X + R → 3X
V3 - outflow: X →

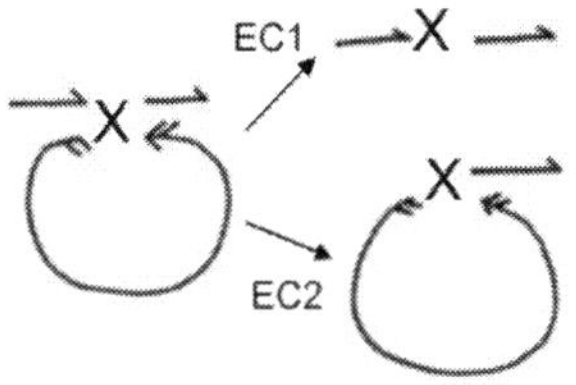

Stability of stationary states associated with subnetworks

The identification of the edges is useful when examining the stability of the subnetwork at a stationary state X_s. The **Jacobian matrix** J of evolution equations at X_s is

$$J = \left.\frac{dF}{dX}\right|_{X=X_s} = S\left.\frac{dV}{dX}\right|_{X=X_s} = S(\text{diag } V_s)\kappa^T(\text{diag } X_s)^{-1}$$

$V_s = \sum_{k=1}^{K} \alpha_k EC_k$ is a linear combination of extreme currents

$\kappa = \{\kappa_{ij}\} = \{\partial \ln V_j(X_s)/\partial \ln X_i\}$ is the **kinetic matrix**

κ_{ij} is the **effective order** of the j-th reaction with respect to the i-th species; if the reaction rates obey power law kinetics then κ_{ij} is independent of Xs.

For power law kinetics, **the stability** of the current Vs **is indicated by principal subdeterminants** β_l of order $l = 1, \cdots, n$ of the matrix

$$B = -S(\text{diag } V_s)\kappa^T$$

Principal subdeterminant of order l contains l diagonal elements of B; there is $\binom{n}{l}$ such determinants for each l

If at least one β_l is negative then at least one eigenvalue of **J is unstable** provided that l species associated with β_l are sufficiently small.

Since any current is a linear combination of the extreme currents, the **stability** of the network's steady states **depends on** the stability of its **extreme subnetworks**.

An unstable EC_k induces instability of the entire network if the corresponding α_k is large enough and X_s satisfies the requirement of small concentration of those species for which the corresponding $\beta_l < 0$.

This instability typically involves a **positive feedback loop** associated with **autocatalysis** → convenient graphical identification in the network diagram

The instability arises via: 1) **saddle-node bifurcation** → multiple steady states
2) **Hopf bifurcation** → periodic oscillations (negative feedback required)

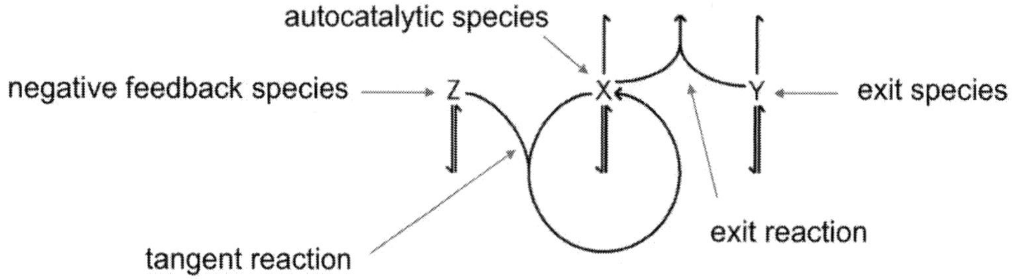

Classification of oscillatory reaction mechanisms into categories

Category 1 – order of autocatalysis = order of exit reaction = 1
Category 2 – order of autocatalysis > 1, no exit reaction

category 1B:

$\square + X \xrightarrow{k_1} 2X + 2Z$
$Z \xrightarrow{k_2} \frac{1}{2}Y$
$Y \xrightarrow{k_3} X$
$X + Y \xrightarrow{k_4}$
$2X \xrightarrow{k_5}$
$X \rightleftharpoons, Y \rightleftharpoons, Z \rightleftharpoons$

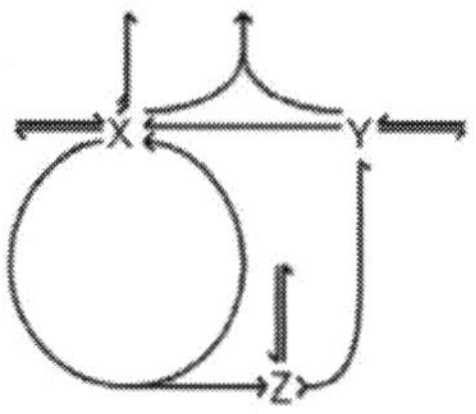

category 1CX:

$Z + X \xrightarrow{k_1} 2X$
$X + Y \xrightarrow{k_2}$
$X \xrightarrow{k_3}$
$Y \xrightarrow{k_4}$
$X \rightleftharpoons, Y \rightleftharpoons, Z \rightleftharpoons$

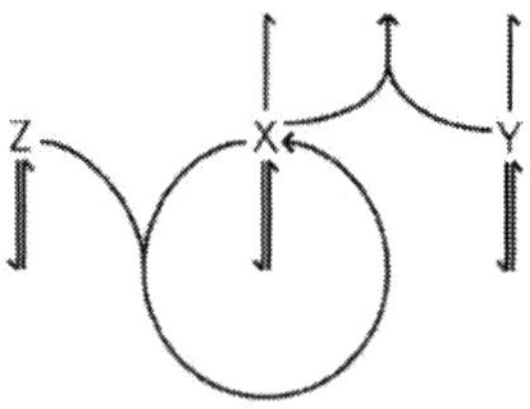

category 1CW:

$Z + X \xrightarrow{k_1} 2X$
$X + Y \xrightarrow{k_2} 2W$
$Y \xrightarrow{k_3} X$
$Y + W \xrightarrow{k_4}$
$X \xrightarrow{k_5}$
$X \rightleftharpoons, Y \rightleftharpoons, Z \rightleftharpoons, W \rightleftharpoons$

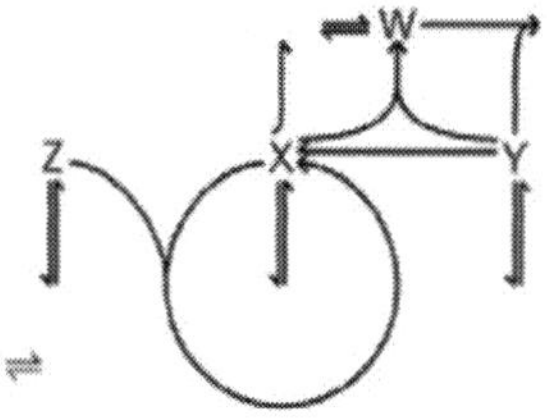

category 2B:

$\square + 2X \xrightarrow{k_1} 3X + Z$
$X + Z \xrightarrow{k_2}$
$Z \xrightarrow{k_3} X$
$2X \xrightarrow{k_4}$
$X \rightleftharpoons$
$Z \rightleftharpoons$

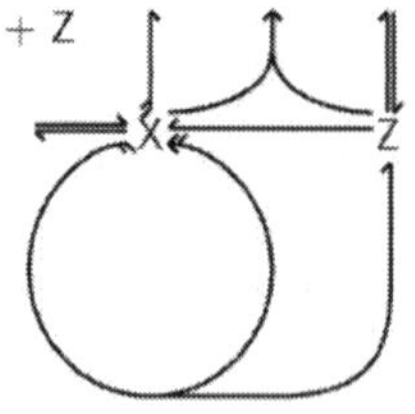

category 2C:

$Z + 2X \xrightarrow{k_1} 3X$
$X \xrightarrow{k_2}$
$X \rightleftharpoons$
$Z \rightleftharpoons$

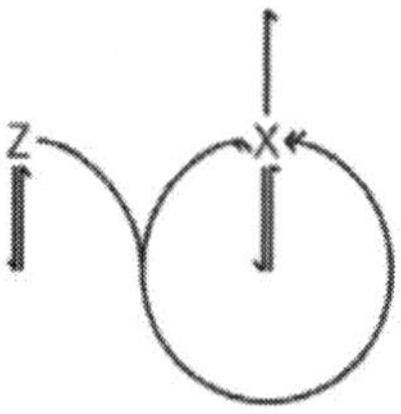

Experimental techniques of category determination

Nonessential species and their type can be determined by:

1) Size of the relative oscillatory amplitude (with respect to mean value)
2) Response to a pulse perturbation at steady state
3) Response to a pulse perturbation at certain phase of oscillations – quenching

Essential species and their type can be determined by:

1) Mutual phase shift in oscillatory regime
2) Concentration shift at steady state due to shift of input concentration
3) Bifurcation diagrams

Lumped isothermal model of the TWC

$$\frac{dc_i(t)}{dt} = u\left(c_i^{in} - c_i^{out}\right) + \frac{k_c a}{\varepsilon^g}\left(c_i^s - c_i\right),$$

$$\frac{dc_i^s(t)}{dt} = \frac{1}{\varepsilon^s}\sum_{j=1}^{J} \nu_{i,j} R_j - \frac{k_c a}{\varepsilon^s\left(1-\varepsilon^g\right)}\left(c_i^s - c_i\right),$$

$$\frac{d\theta_k(t)}{dt} = \frac{1}{L_{NM}}\sum_{j=1}^{J} \nu_{k,j} R_j ,$$

$$\frac{d\xi_m(t)}{dt} = \frac{1}{L_{OSC}}\sum_{j=1}^{J} \nu_{m,j} R_j ,$$

$$\frac{d\chi_q(t)}{dt} = \frac{1}{L_{SUP}}\sum_{q=1}^{J} \nu_{q,j} R_j ,$$

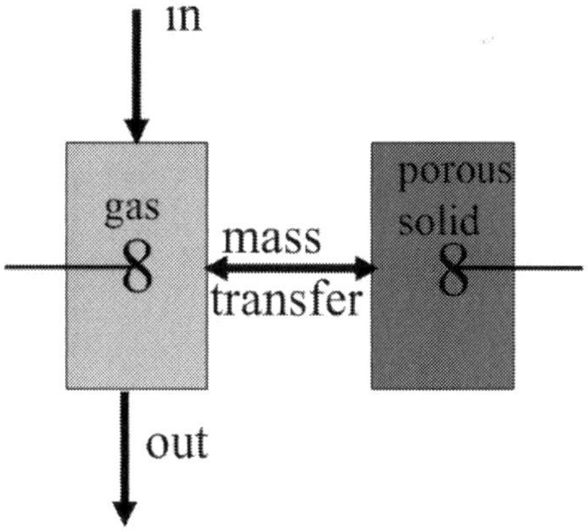

oxygen input y_{O2}^{in} and temperature T are considered as main adjustable (bifurcation) parameters

parameter	value
u	20.8 s^{-1}
ε^w	0.8
ε^g	0.917
$k_c a$	2067 s^{-1}
y_{CO}^{in}	1.22 %(mol)
y_{C2H2}^{in}	0.068 %
y_{NO}^{in}	0.1 %
L_{NM}	20-80 mol m^{-3}
L_{OSC}	0.1-30 mol m^{-3}

CO and C_2H_2 oxidation subsystem

bifurcation diagram in the y_{O2}^{in} – T plane for the lumped system

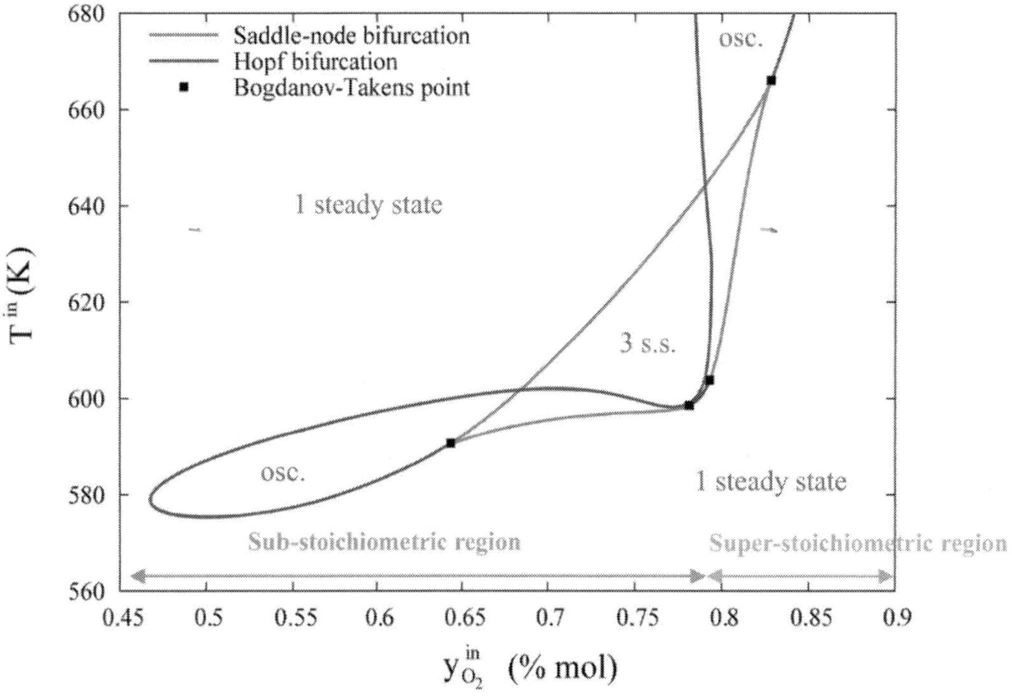

CO and C_2H_2 oxidation subsystem

Network diagram – two major oscillatory subnetworks :

i) turquoise & blue → C_2H_2 adsorbed and oxidized simultaneously

ii) purple → C_2H_2 is first adsorbed and then oxidized

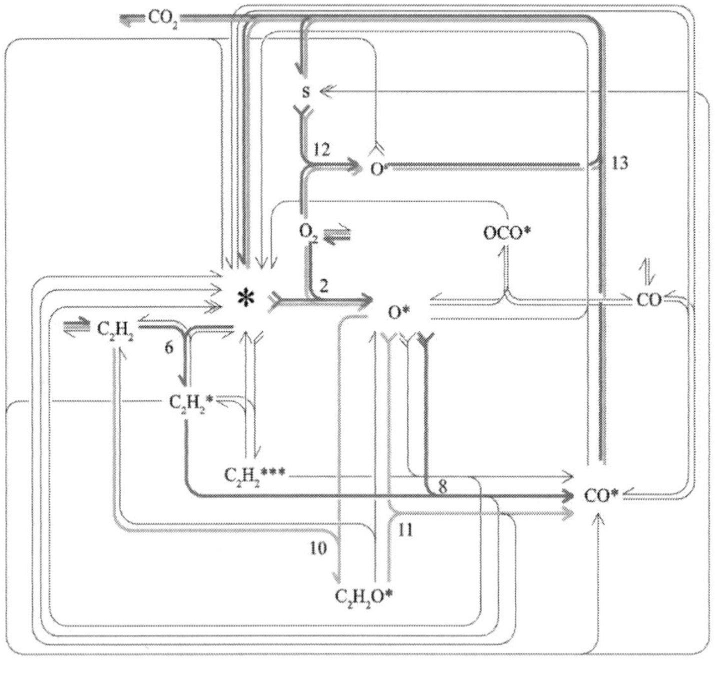

CO and C_2H_2 oxidation subsystem

Phase shifts at the Hopf bifurcation

Subnetwork	Category	CO^*	O_2	O^*	$*$	C_2H_2
Sub-st. region						
E_{20}	1C	3.76 (X)	141.91 (Z)	15.55 (X)	0.00 (X)	-175.37 (Y)
θ_1^{sim}		6.84	178.57	-6.73	0.00	-176.39

Subnetwork	Category	O_2	$*$	s	C_2H_2	$C_2H_2^*$
Super-st. region						
E_{14} (ii)	1C	174.27 (Y)	0.00 (X)	-4.09 (W)	129.88 (Z)	-20.46 (X)
θ_2^{sim}		178.55	0.00	-51.79	-176.51	-3.38

Dominant subnetworks

Sub-stoichiometric (rich) region
[Eley-Rideal]
excess of C_2H_2
O_2 controls the oscillations

Category 1CX

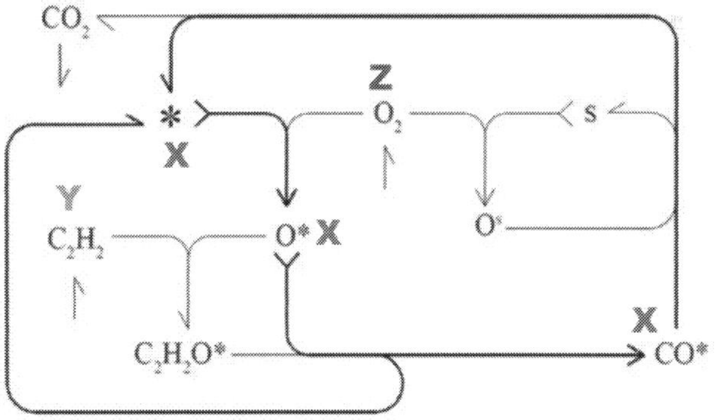

Super-stoichiometric (lean) region
[Lagmuir-Hishelwood]
excess of O_2
C_2H_2 controls the oscillations

Category 1CW

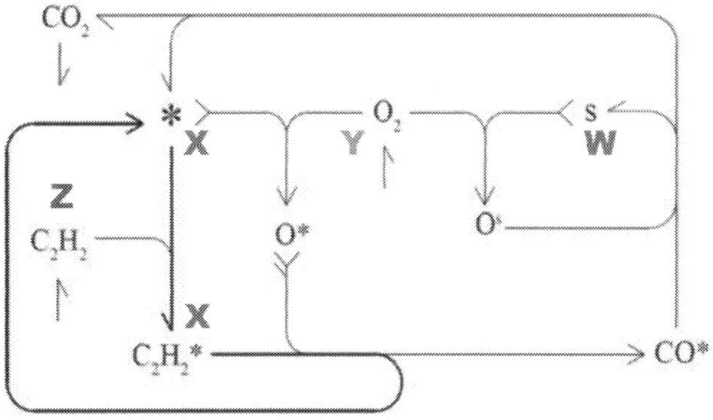

CO oxidation and NO_x reduction subsystem

bifurcation diagram in the y_{O2}^{in} – T plane for the lumped system

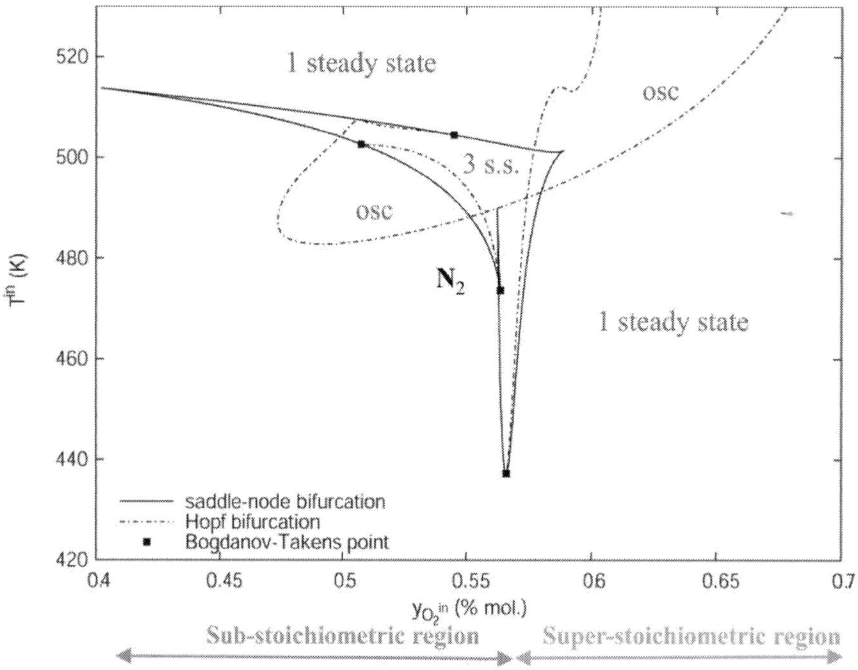

2

CO oxidation and NO_x reduction subsystem

Network diagram turquoise & blue → N_2 producing subnetwork
purple → N_2O producing subnetwork

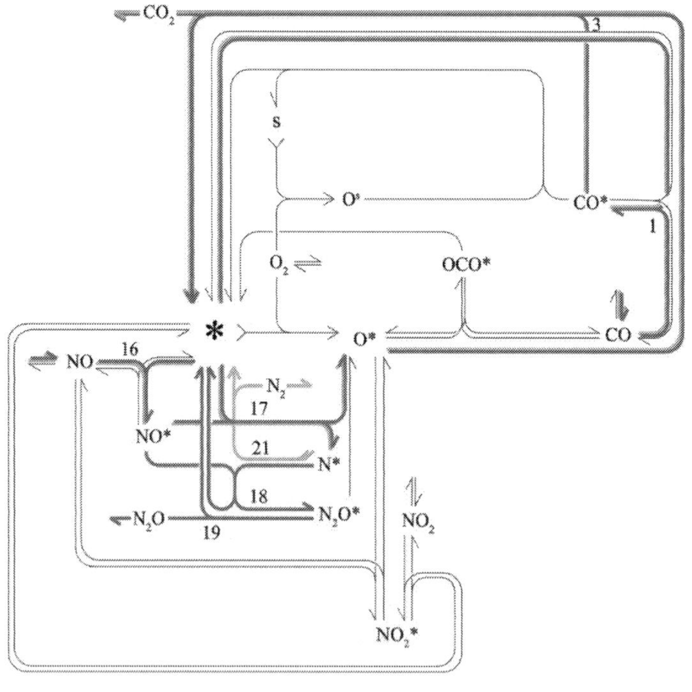

CO oxidation and NO_x reduction subsystem

Phase shifts at the Hopf bifurcation

Subnetwork	Category	CO	CO*	*	NO	NO*
Sub-stoichiometric region						
E_{20} (i)	1C	148.71 (Z)	45.46 (X)	0.00 (X)	174.25 (Y)	170.23 (Y)
θ_1^{sim}		6.84	101.93	0.00	-6.73	0.00

Subnetwork	Category	CO	CO*	*	NO	N*
Super-stoichiometric region						
E_{12} (ii)	1C	151.25 (Z)	81.54 (X)	0.00 (X)	174.89 (Y)	-3.01 (X)
θ_2^{sim}		66.68	128.26	0.00	-52.62	-176.51

Dominant subnetworks

Sub-stoichiometric region (rich phase)
CO controls the oscillations
N_2 is produced

Category 1CX

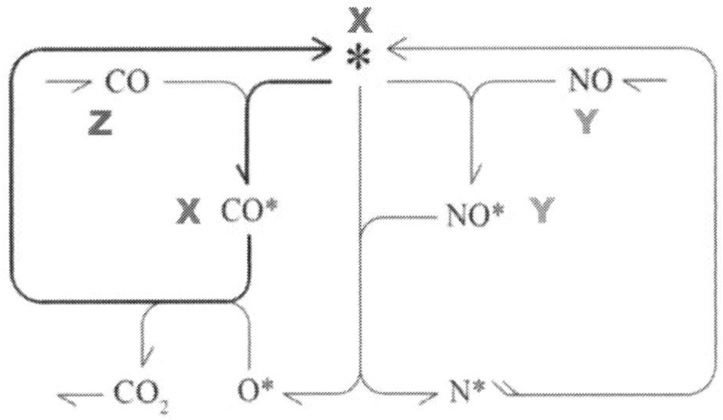

Super-stoichiometric region (lean phase)
CO controls the oscillations
N_2O is produced

Category 1CW

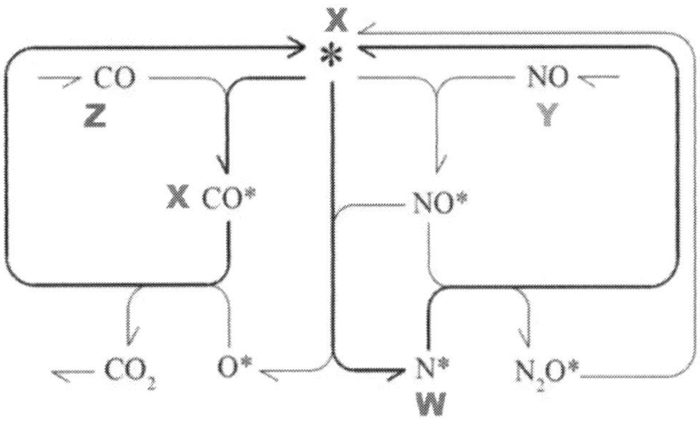

Pattern formation in a 1D cross-flow reaction-diffusion-convection system

$$\frac{\partial c_i(t,z)}{\partial t} + v\frac{\partial c_i(t,z)}{\partial z} = D_i\frac{\partial^2 c_i(t,z)}{\partial z^2} + P(c_i^{\mathrm{res}} - c_i) + \frac{k_c a}{\varepsilon^g}(c_i^s - c_i),$$

$$\frac{\partial c_i^s(t,z)}{\partial t} = \frac{1}{\varepsilon^s}\sum_{j=1}^{J} \nu_{i,j}R_j - \frac{k_c a}{\varepsilon^s(1-\varepsilon^g)}(c_i^s - c_i),$$

$$\frac{\partial \theta_k(t,z)}{\partial t} = \frac{1}{L_{\mathrm{NM}}}\sum_{j=1}^{J} \nu_{k,j}R_j,$$

$$\frac{\partial \xi_m(t,z)}{\partial t} = \frac{1}{L_{\mathrm{OSC}}}\sum_{j=1}^{J} \nu_{m,j}R_j,$$

c_i concentrations in channel

c_i^s concentrations in the pores

Θ_k , ξ_m surface concnetrations

D_i diffusivity in free gas

P cross-flow mass transfer coefficient

v velocity of the gas in channel

isothermal model,

Danckwerts boundary conditions

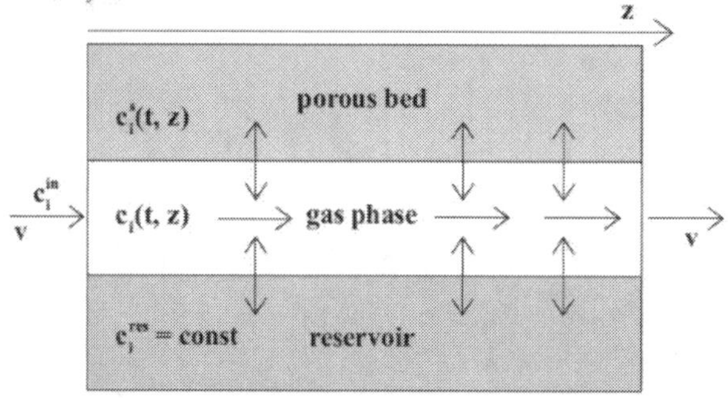

CO and C_2H_2 oxidation subsystem

bifurcation diagram and spatiotemporal dynamics

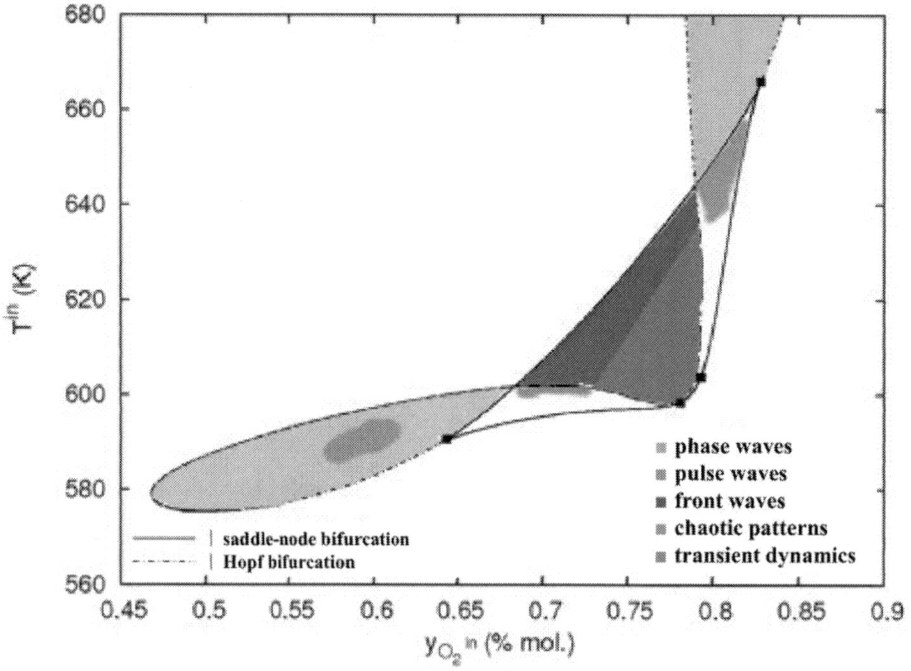

CO and C_2H_2 oxidation subsystem

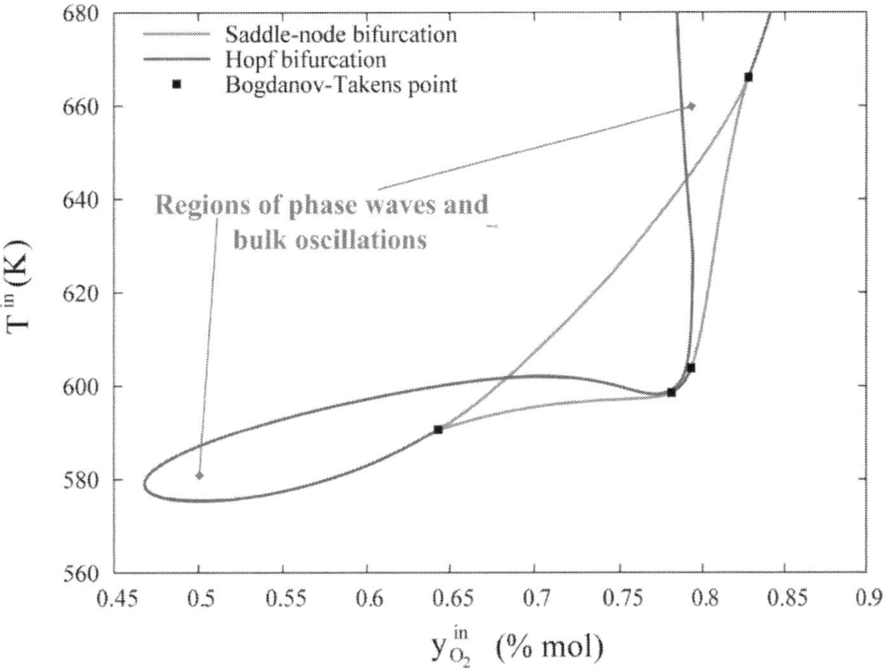

CO and C_2H_2 oxidation subsystem

Phase waves synchronized into bulk oscillations

Two regions of oscillations

substoichiometric:
high frequency of oscillations
low average CO conversion

$y_{O_2}^{CF} = 0.5\, mol\%,\ T = 580K$

superstoichiometric:
low frequency of oscillations
high average CO conversion

$y_{O_2}^{CF} = 0.81\, mol\%,\ T = 658K$

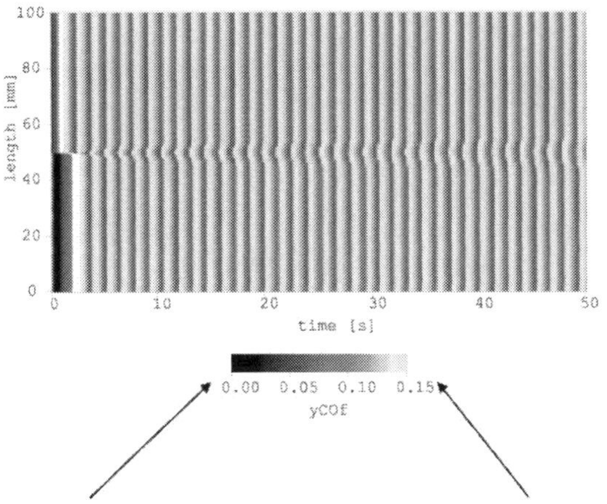

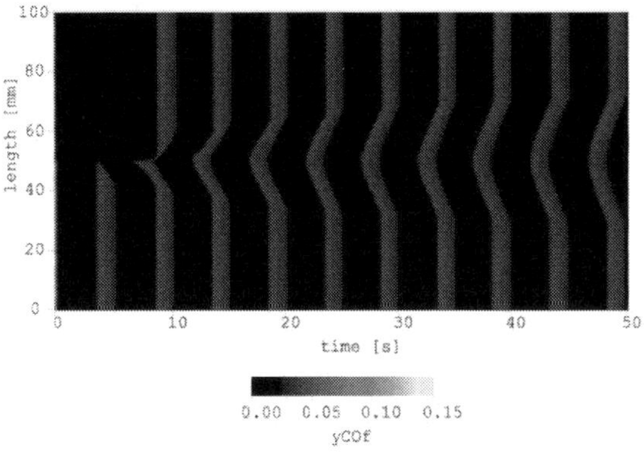

high CO conversion

low CO conversion

convection: weak: no effect
strong: waves wash out

CO and C_2H_2 oxidation subsystem

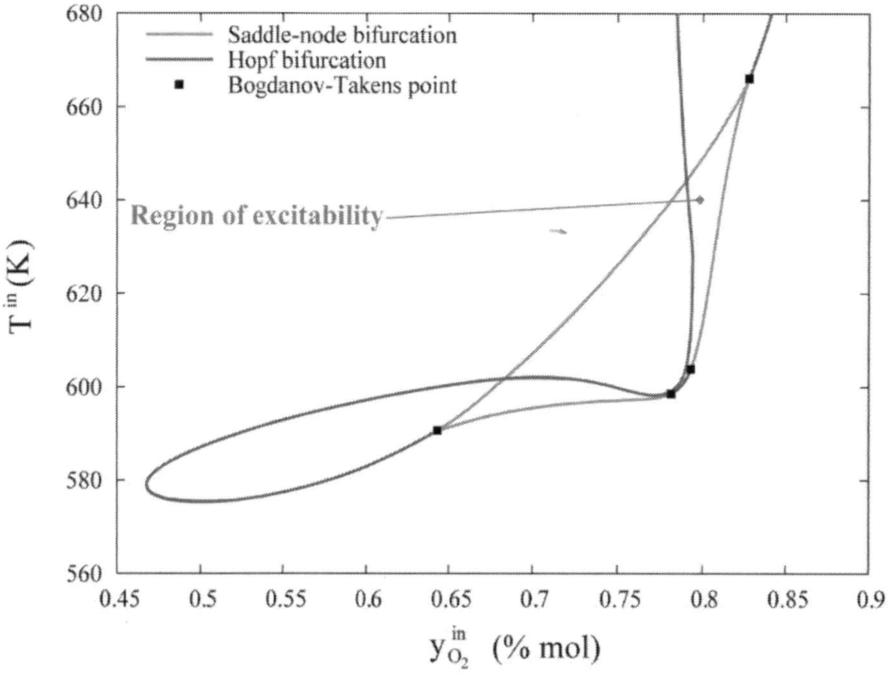

CO and C_2H_2 oxidation subsystem

Traveling pulse waves – effect of convection on wave velocity and on number of pulses

$y_{O_2}^{CF} = 0{,}8\,mol\,\%,\ T = 640K$

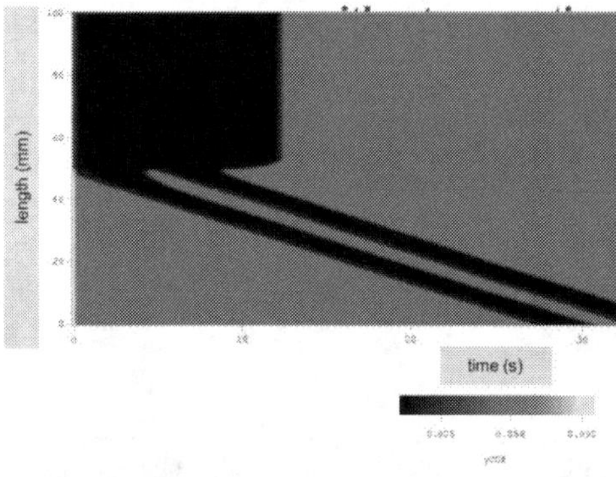

convection, v = -10^{-3} m/s

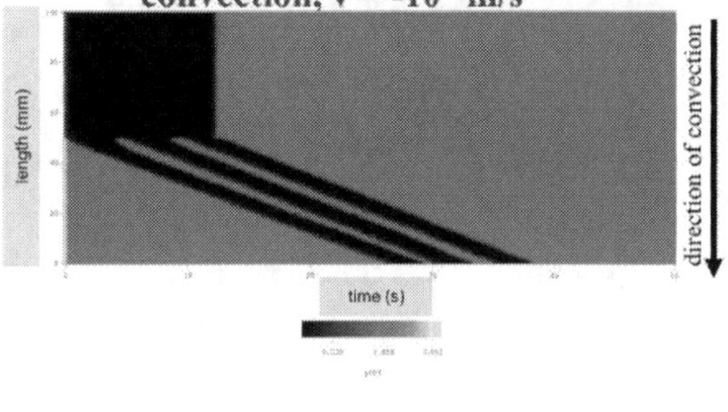

convection, v = 10^{-3} m/s

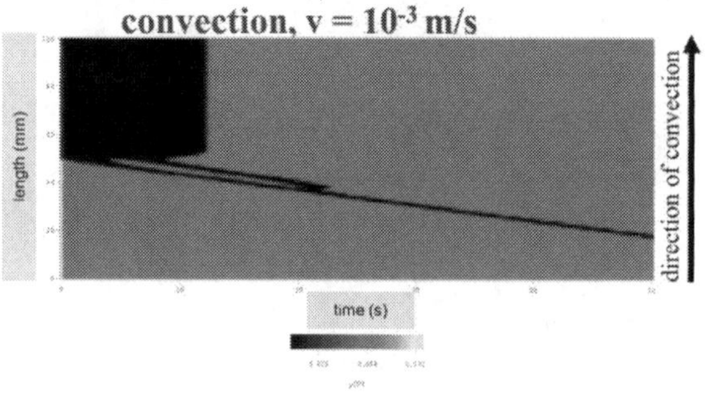

CO and C_2H_2 oxidation subsystem

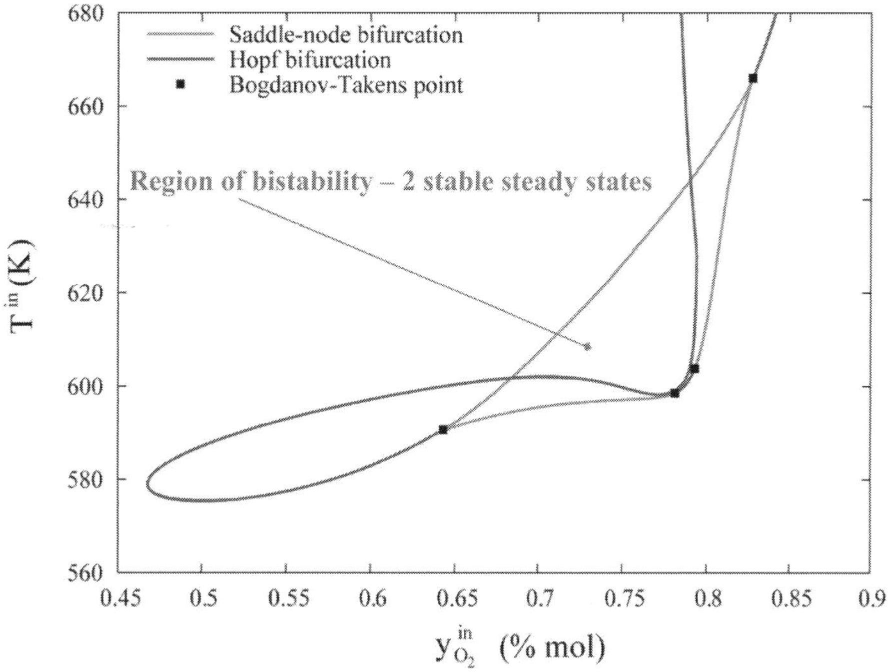

CO and C_2H_2 oxidation subsystem

Travelling front waves

without convection: low/high and high/low CO conversion fronts

direction of wave velocity depends on temperature

$y_{O_2}^{CF} = 0{,}73\,mol\%,\ T = 602K$

"extinction" front

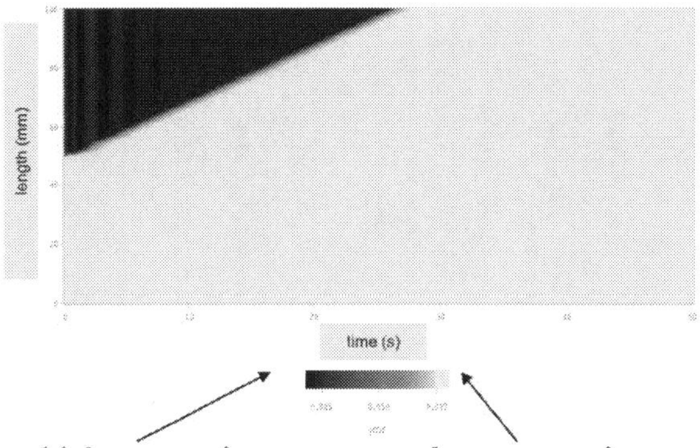

$y_{O_2}^{CF} = 0{,}73\,mol\%,\ T = 608K$

"ignition" front

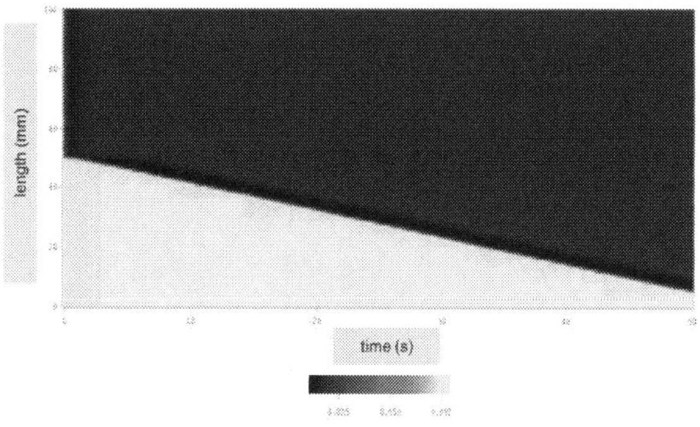

high conversion low conversion

Effects of convection: -positive/negative convective flow - higher/lower wave velocity)

-front wave can be reversed by convective flow

CO and C_2H_2 oxidation subsystem

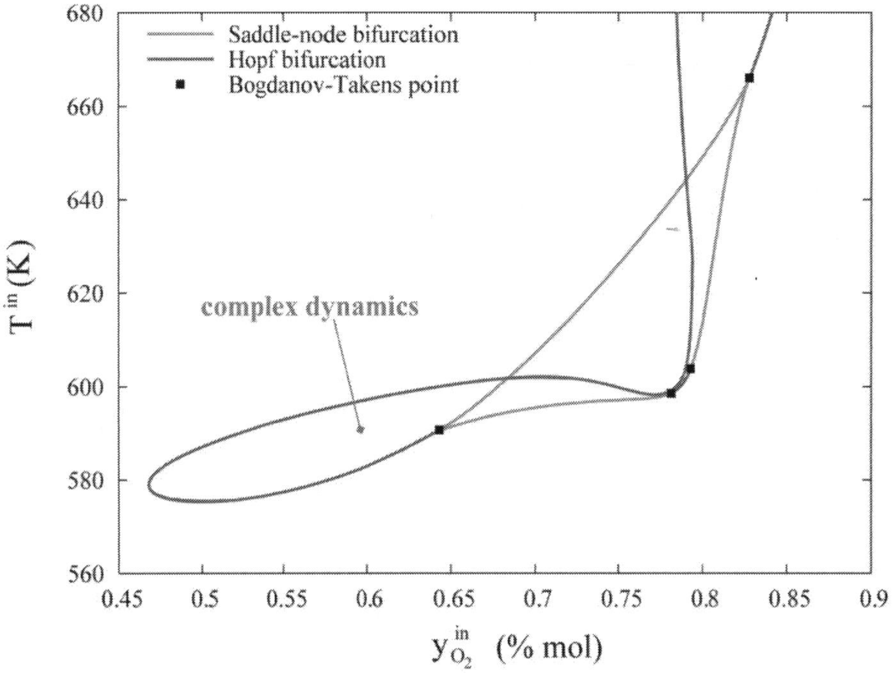

CO and C_2H_2 oxidation subsystem

Chaotic spatiotemporal patterns

$y_{O_2}^{CF} = 0{,}6\,mol\%,\ T = 590K$

without convection **with convection, v = 3.10⁻³ m/s**

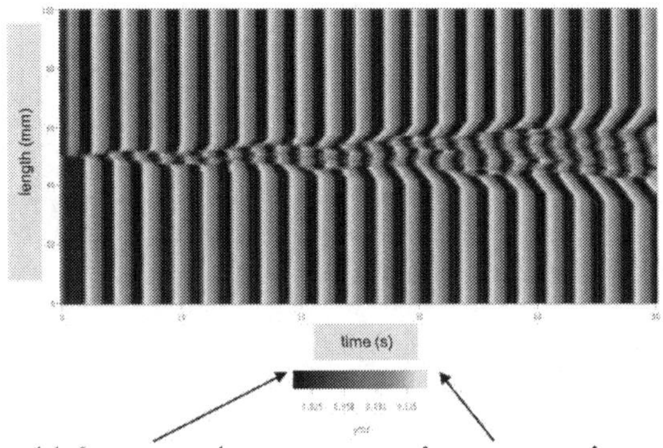

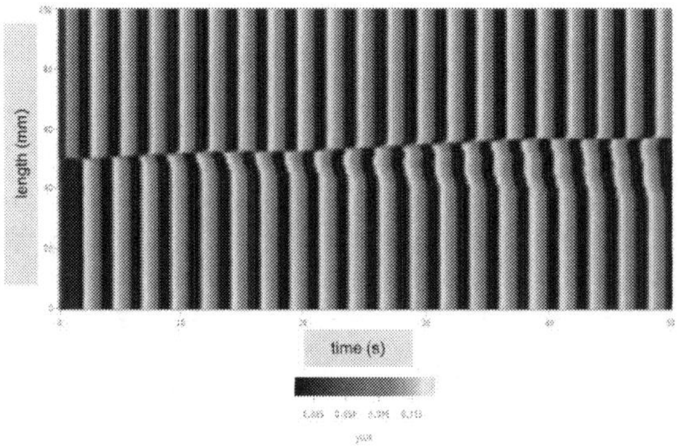

CO oxidation and NO_X reduction

Bifurcation diagram and spatiotemporal dynamics

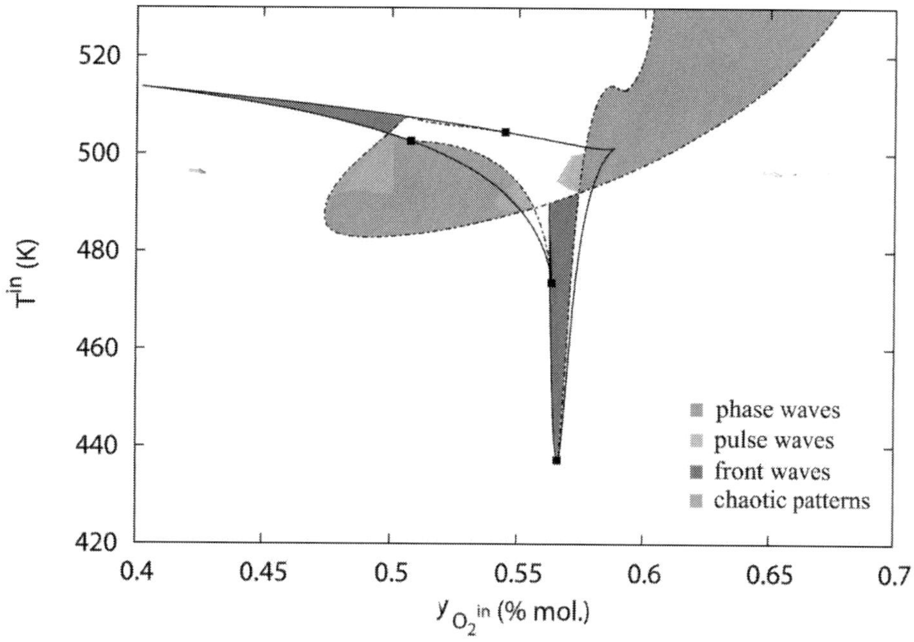

CO oxidation and NO_X reduction

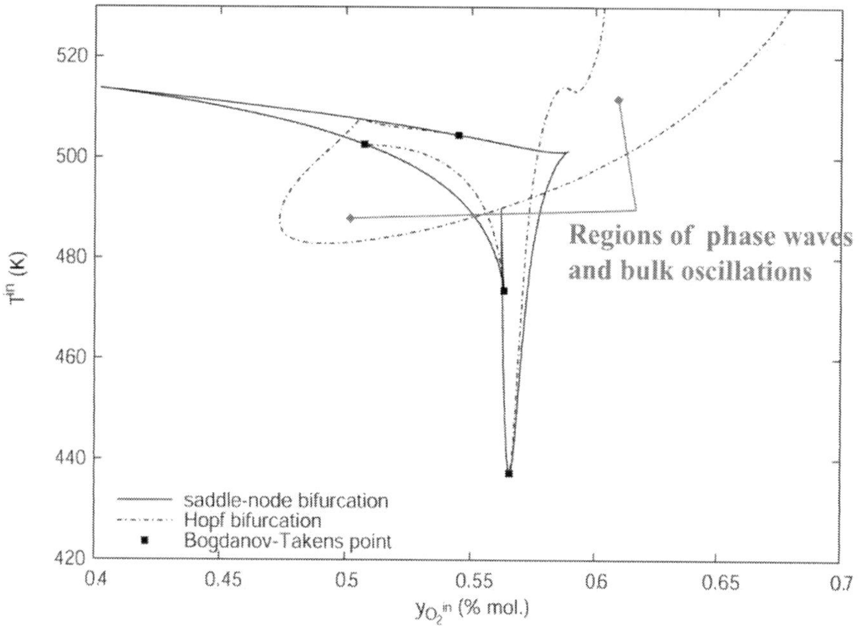

CO oxidation and NO_X reduction

Phase waves synchronized into bulk oscillations

Two regions of oscillations

substoichiometric:
low frequency of oscillations
lower average CO conversion

$y_{O_2}^{CF} = 0.5\,mol\%,\ T = 485K$

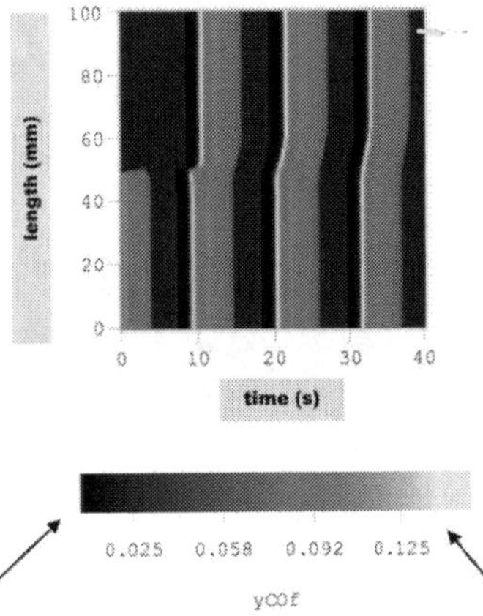

high CO conversion

low CO conversion

superstoichiometric:
high frequency of oscillations
higher average CO conversion

$y_{O_2}^{CF} = 0.625\,mol\%,\ T = 510K$

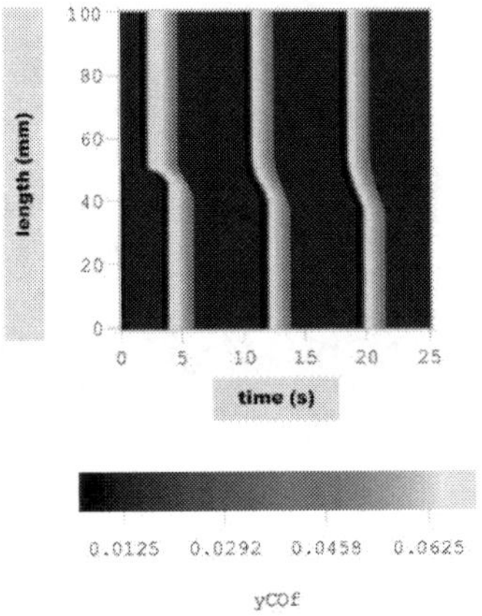

convection: weak: no effect
strong: waves wash out

CO oxidation and NO_X reduction

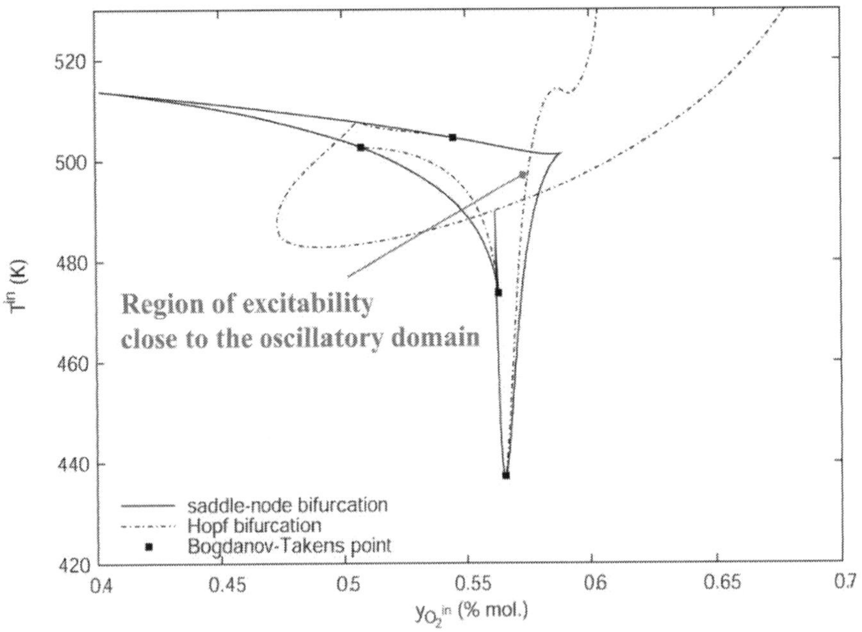

CO oxidation and NO_X reduction

Traveling pulse waves

$$y_{O_2}^{CF} = 0{,}575\,mol\%,\; T = 497K$$

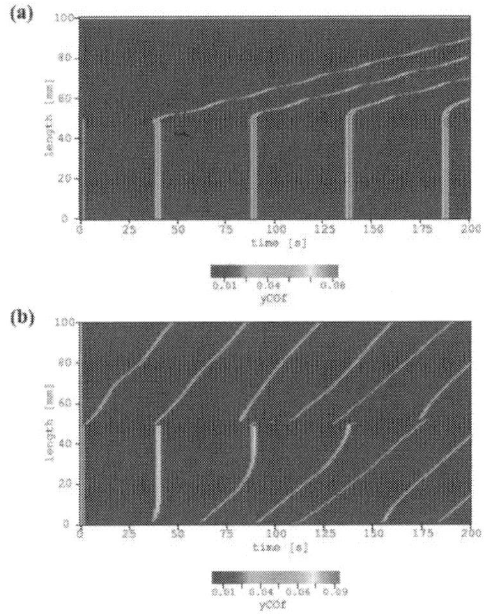

No convection

Positive convection
v = 10mm/s

CO oxidation and NO_X reduction

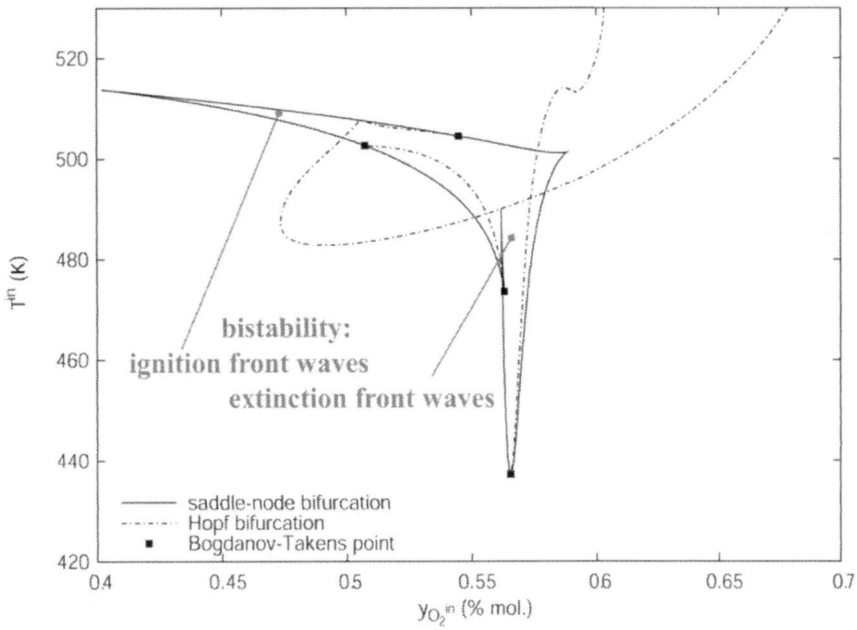

CO oxidation and NO_X reduction

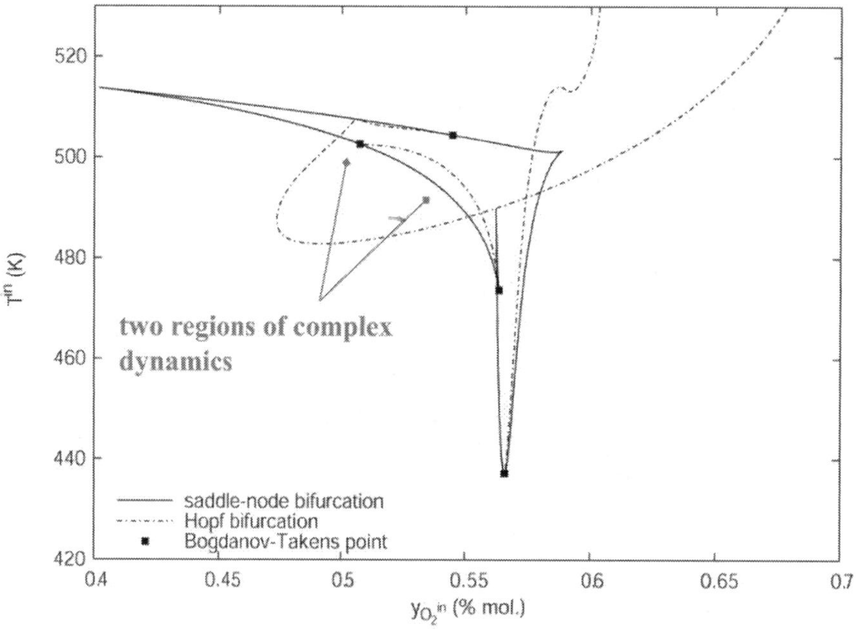

CO oxidation and NO_X reduction

Chaotic spatiotemporal patterns without convective flow

two types of chaos caused by diffusion flow

convection: complex structure is washed out

$y_{O_2}^{CF} = 0{,}537\,mol\%,\ T = 490K$ $\quad$ $y_{O_2}^{CF} = 0{,}5\,mol\%,\ T = 496K$

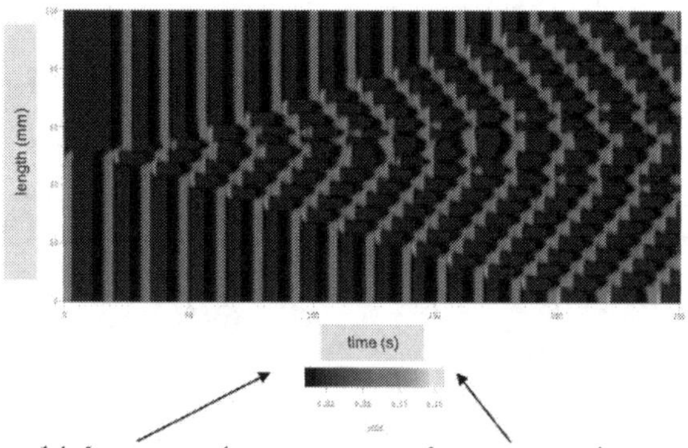

high conversion $\quad$ low conversion

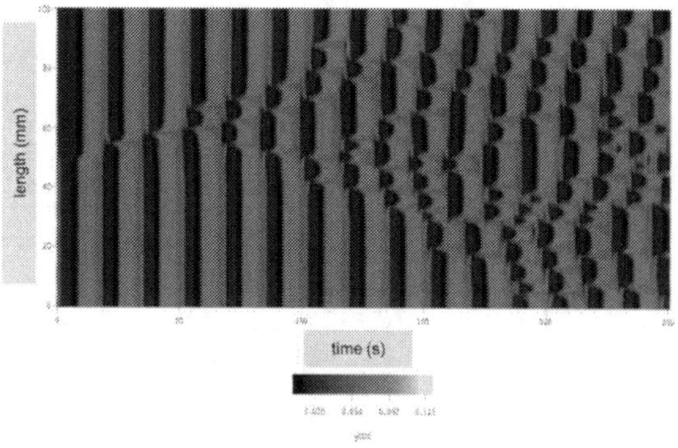

CO oxidation and NO_X reduction

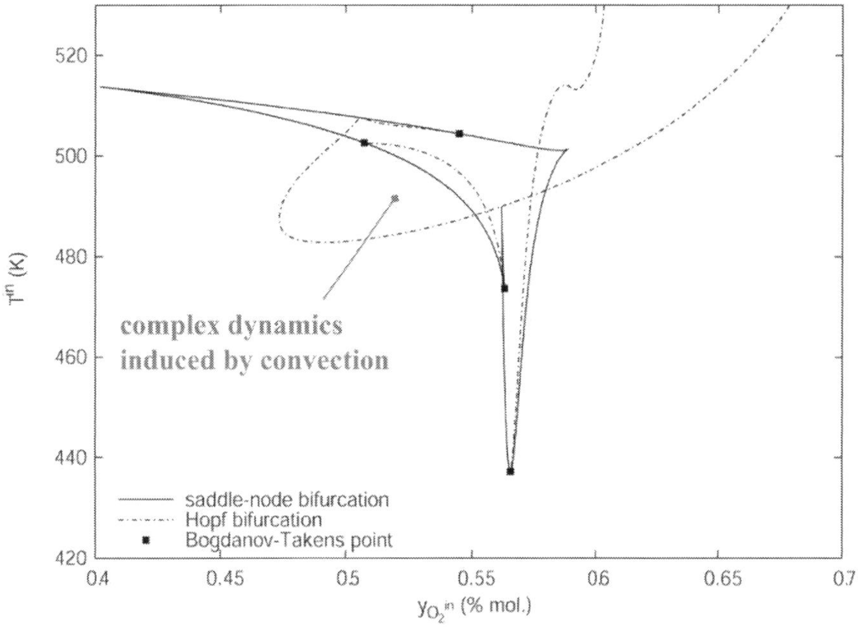

CO oxidation and NO_X reduction

Chaotic spatiotemporal patterns caused by convection

$$y_{O_2}^{CF} = 0{,}531 mol\%,\ T = 490K$$

Pattern without convection | **Pattern with convection, v = 10^{-2} m/s**

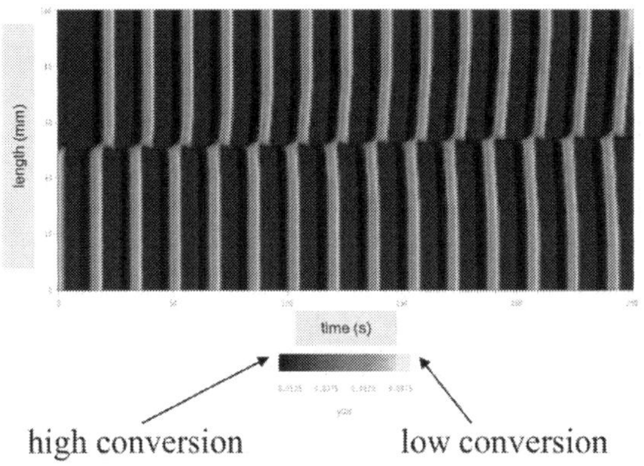

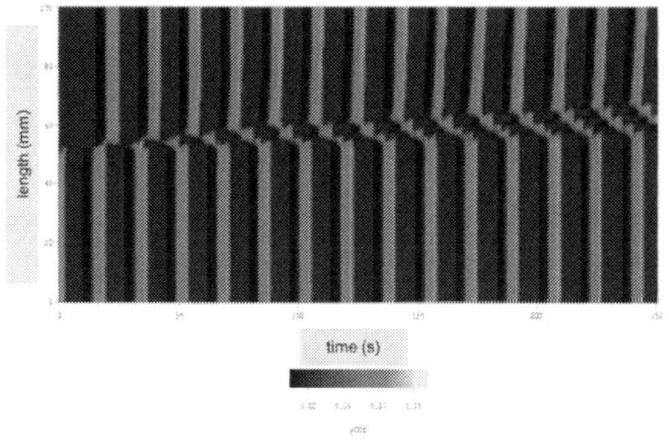

Summary

CO & C_2H_2 oxidation subsystem and CO oxidation & NO_X reduction subsystem were examined separately to find relations between structure of reaction networks and spatiotemporal dynamics

Bifurcation diagrams for the lumped system display two major oscillatory regions separated by a bistable/excitable domain

Subnetworks accounting for oscillatory dynamics in the diagrams are very different in:
- each of the subsystems
- in substoichiometric and superstoichiometric regions

Bifurcation diagrams for the lumped system provide a guideline for various types of spatio-temporal patterns in the reaction-diffusion system with diffusion driven cross-flow.

Spatiotemporal dynamics in the reaction-diffusion-convection-cross-flow system reflect to certain degree differences in instability-causing subnetworks

This work was supported by the grant GACR P105/12/0664 from the Czech Science Foundation

Spatiotemporal Behavior of Pt/Rh/CeO_2/BaO Catalyst during Lean-Rich Cycling

Hoang Nguyen, Michael P. Harold*, and Dan Luss*
Department of Chemical & Biomolecular Engineering
University of Houston, TX 77204-4004
(*Corresponding authors: mharold@uh.edu, dluss@uh.edu)

Abstract:

Due to the strong coupling between the reaction and the transport processes, the temperature and species concentration vary with time and reactor position. Most current models for studying the dynamic oxygen storage capacity (OSC) of ceria-containing catalysts do not take into account these complex spatial temporal patterns, or utilize the kinetic parameters that are obtained from experiments in which these gradients are eliminated. To address this gap, a new experimental method that enables the simultaneous measurement of concentration and temperature profiles within a catalytic reactor has been developed. Species concentration measured by spatially resolved mass spectrometry (SpaciMS) [1] and the temperature profile measured by coherent optical frequency domain reflectometry (c-OFDR) [2] within a monolith channel were conducted to investigate the impact of key operating variables (space velocity, cycle time, and rich pulse intensity) on the oxygen storage and release process of a Pt/Rh/CeO_2/BaO monolithic catalyst during lean-rich cycling operation.

The catalyst was fed periodically for a short time (5s) with a rich feed of 1.2 vol.% propylene and followed by a longer (30s) lean feed of 10 vol. % O_2. The temperature measurements revealed complex spatial temporal temperature features, i.e., the maximum temperature rise after the rich-lean switch was much higher than that following the lean-rich switch. Moreover, in the presence of ceria, hydrogen forms a local maximum due to its downstream reaction with CeO_2.

A 30% increase in the space velocity enables the trapping of more propylene while it inhibits the uptake of oxygen, thus favoring the partial oxidation of propylene. Increasing the lean and rich duration, on the other hand, enable trapping more oxygen but results in the breakthrough of more propylene. Under this operation, the catalyst is poisoned by CO and C_3H_6 derived adspecies, thus the partial oxidation is favor and hydrogen specie is formed continuously along the reactor length. During the rich period, the combustion process is vigorous in the upstream section but due to lower catalyst temperature and surface mobility, the product CO_2 is desorb slower. Whereas increased the rich duration time with less reductant concentration will not affect the break-through time of propylene or the selectivity of the final products. More H_2 is formed inside the channel and less reductant loss at the lean/rich interface.

A mathematical model of the system is under development to simulate the spatio-temporal trends with the goal to provide insight into the complex interplay between the reaction and transport processes during the transient catalytic oxidation of propylene. In summary, the study shows that simultaneous measurements of spatially-resolved temperature and concentration are a powerful tool in elucidating the complex transient features of a periodically-operated catalytic reactor.

References:

1. J.S. Choi, W.P. Partridge, C.S. Daw, Appl Catal A-Gen, 293 (2005) 24-40.

2. H. Nguyen, M.P. Harold, D. Luss, Chem Eng J, 234 (2013) 312-317.

Bifurcation Analysis of Thermally Coupled Homogeneous-Heterogeneous Combustion in Monoliths

Imran Alam[1], David H. West[2] and Vemuri Balakotaiah[1]

[1]Chemical and Biomolecular Engineering Department,
University of Houston, 4800 Calhoun Road, Houston, TX 77004
[2]SABIC Technology Center, Sugarland, TX 77478
Author emails: Imran Alam : imran.alam.iit@gmail.com, David West:dwest@americas.sabic.com ,Vemuri Balakoataiah : bala@uh.edu

ABSTRACT

Introduction

Catalytic partial oxidation is an attractive technology for meeting future energy demands and production of intermediate chemicals. The models describing this process typically involve both catalytic and homogeneous reactions. There have been numerous experimental and computational studies on these systems in the past twenty-five years. Most of these studies have used micro-kinetic models for homogeneous reactions or focused on hydrodynamic aspects. Such analyses lead to models that are not amenable to a bifurcation analysis in the parameter space as they are computationally complex due to existence of exponentially thin boundary layers in space and/or time. Consequently, a clear picture of the essential/qualitative features of the process as the parameters are changed is not entirely clear. The earlier work of Song et al. [1] analyzes bifurcation features for homogeneous-heterogeneous reaction systems using a stagnation flow model. However stagnation flow is not realistic for many industrial processes. We circumvent this situation by basing our analyses on monolith reactors where flow is parallel to the catalytic surface. We use singularity theory techniques to analyze and determine the qualitative features of the thermally coupled homogeneous-heterogeneous combustion process.

Mathematical Models & Analysis

We present a mathematical model for the combustion of propane in monolith reactors. First, we show that the bifurcation features of monolith model exhibit trends quite similar to the stagnation point flow model studied by Song et al. [1]. Next, we present a detailed bifurcation analysis of the monolith model taking into account the interaction between transport phenomena and reactions in both gas and solid phases. For simplicity, we consider a single reaction occurring homogeneously as well as catalytically and the coupling occurs mainly through thermal effects. This reaction, describing propane oxidation proceeds according to the overall stoichiometric relation:
$C_3H_8 + 5O_2 \rightarrow 3CO_2 + 4H_2O.$
For the homogeneous oxidation of propane, we used the rate expression from Westbrook and Dryer [2]. For the catalytic reaction, we use the rate expression from Hiam, et al.[3].

Our calculations use the inlet fluid temperature, T_{in} as the bifurcation variable. We compute several double-S shaped diagrams for the state variables (solid or gas temperature, conversion, etc.) vs. T_{in} . Thus, for higher inlet concentrations, our bifurcation diagrams typically have two ignitions and two extinctions, and we interpret the first ignition as being due to the catalytic reaction and the second due to homogeneous reaction. In practice, the intermediate stable state, between the first extinction and second ignition is important as it represents the lower temperature state and a reasonable conversion of hydrocarbon. However, if the second ignition happens to the left of the first ignition, the system jumps directly to the high temperature state, bypassing the intermediate state. We find that bifurcation diagrams where the second ignition is to the right of the first occur typically at low residence times.

The ignitions of the catalytic and homogeneous reactions depend on a multitude of parameters, for example, activation energies, adiabatic temperature rise (or the inlet concentration), residence time, etc. For instance, for the homogeneous ignition, we have the following relation [4]:

$$\frac{E_h}{RT_{ig}} \frac{\Delta T_{ad}}{T_{ig}} \frac{r_h(c_{in}, T_{ig})}{c_{in}} \tau > 0.368$$

For a stoichiometric feed, the adiabatic temperature rise for propane oxidation is 2044K, and using the above expression we estimate the ignition temperatures for various residence times. The lowest ignition temperature is obtained for a residence time of 0.5s and stoichiometric feed, and this turns out to be around 800K. On the other hand, if the residence time is reduced to 0.5 milliseconds, a much higher ignition temperature of about 1200K is estimated. It becomes clear that for stoichiometric feeds, the high temperature rise makes the search for intermediate branches impractical.

For the catalytic ignition, we use the criterion [5] (neglecting washcoat diffusion)

$$\frac{E_c}{RT_{ig}} \frac{\Delta T_{ad}}{T_{ig}} \frac{r_c(c_{in}, T_{ig})}{c_{in}} [\tau + \frac{\rho_f c_{pf}}{ha_v}] > 0.368$$

We observe that the ignition temperature is not a monotonic function of channel hydraulic radius, R_Ω. For small values of R_Ω, we have a higher catalytic reaction rate and good heat transfer between the solid and the gas phases. However for larger values of R_Ω, the ignition becomes controlled by the heat transfer between solid and gas phases as the temperature difference between the phases becomes significant.

Results and Discussion

With guidance from the above estimates, we computed bifurcation diagrams and transition loci for our model. Below is a typical hysteresis locus of the system for R_Ω of 0.264 mm in the residence time (tau)-inlet mole fraction (x_{in}) plane. For a particular value of residence time, namely 0.16 ms, we then show a bifurcation set (ignition/extinction locus) for the system. Different cross sections of the figure corresponding to different inlet mole fractions of reactant have been indicated. These correspond to different relative locations of ignition and extinction points. The case corresponding to cross-sections 2 and 3 represent situations in which there is an accessible and stable intermediate branch. This means that the partially (or catalytically) oxidized state can be attained without triggering the homogeneous ignition. A comprehensive analysis of the system will be presented.

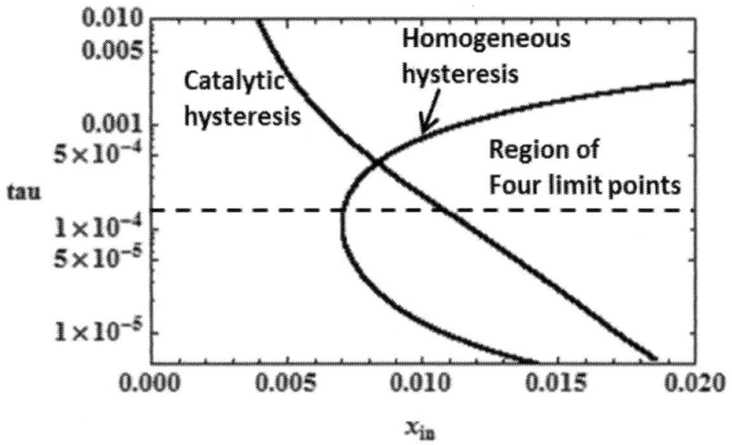

(a)

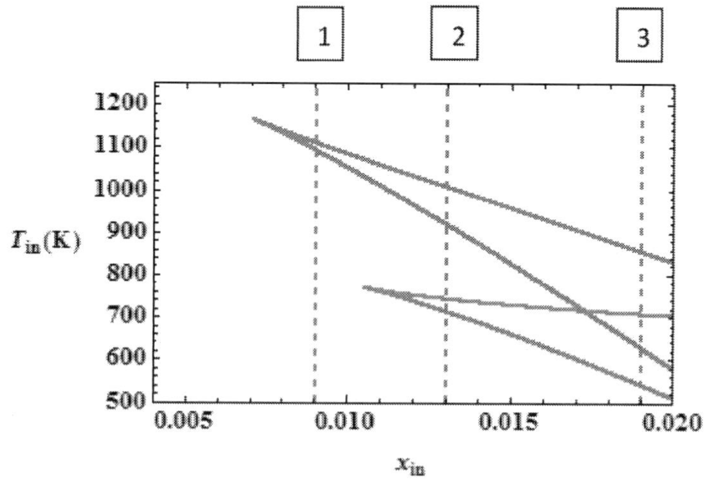

(b)

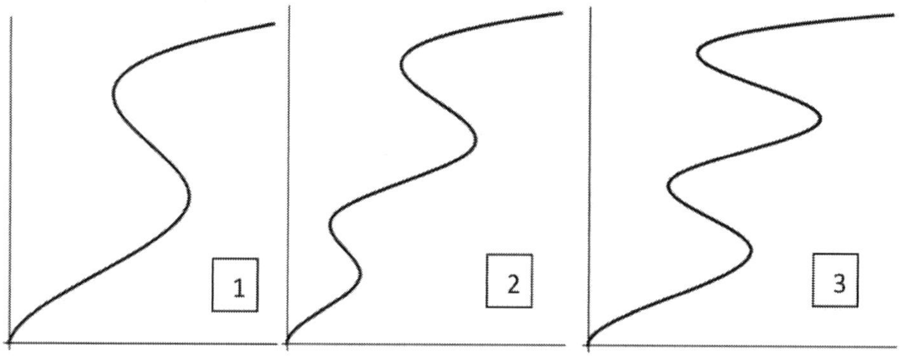

(c)

Figure 1: (a) Computed hysteresis locus in the residence time-inlet mole fraction plane for the case of R_Ω of 0.264 mm, (b) Computed bifurcation set in the feed temperature- mole fraction plane for residence time 0.16 ms and R_Ω of 0.264 mm (c) Schematic bifurcation diagrams of conversion (or solid/ fluid temperature) vs. feed temperature.

References

[1]. X. Song, W.R. Williams, L. D. Schmidt, R. Aris, Combustion and Flame **84** (1991) 292-311.
[2]. C. K. Wesbrook, F.L. Dryer, Combust. Sci. Tech. **27** (1981), 31-43.
[3]. L. Hiam, H. Wise, S. Chaikin, J. Catalysis **9** (1968), 272-276.
[4]. Christoforatou, E.L, V. Balakotaiah, D. H. West, AIChE J. **44** (1998) 394-404.
[5]. K. Ramanathan, V. Balakotaiah, D. H. West, Chem. Engg. Sci., **58** (2003) 1381-1405.

Reaction Kinetics of the Oxidation of Alkene Mixtures over a Chromium Oxide Catalyst and Feasibility Study of a Simulated Moving Bed Reactor

G. Kiedorf [1], *A. Seidel-Morgenstern* [1,2], *C. Hamel* [2,3]

[1] Max Planck Institute for Dynamics of Complex Technical Systems, Magdeburg, Germany
[2] Otto von Guericke University, Institute of Process Engineering, Magdeburg, Germany
[3] Anhalt University of Applied Sciences, Köthen, Germany

Introduction

The use of reactor systems with heat or mass transport integration offers potential to improve the performance of chemical plants. A promising approach is the recovery of heat generated in exothermic reactions. The Reverse-Flow-Reactor (RFR) is a well-established concept [1]. However, non-uniform utilization of the catalyst bed and slips of unconverted reactants after switching the flow direction are disadvantages of the RFR. One alternative is the Simulated Moving Bed Reactor (SMBR), in which exothermic reaction fronts move through a reactor cascade in the gas flow direction (Fig. 1). After passing a segment, the feed position is shifted into flow direction. This periodic operation attempts to trap a self-sustained exothermic front and allows an auto-thermal operation. The concept was realized successfully for processing simple feed gases [2]. However, knowledge regarding processing industrially relevant gas mixtures is insufficient. First experimental results revealed very complex temperature profiles for the oxidation of mixtures of hydrocarbons characterized by different ignition temperatures [2]. To understand the dynamics of such temperature fronts it is first of all essential to quantity the reaction kinetics.

Kinetic Investigation and Feasibility Study

The contribution provides results regarding the oxidation of mixtures of short chain alkenes (C2-C3) using a chromium oxide catalyst (Fig. 2). Starting with measuring and correlating the adsorption isotherms of reactants and products over this catalyst, mixture adsorption isotherms were predict based on the ideal adsorbed solution theory. Based on this information, established reaction rate approaches were formulated and applied. For a broad range of temperatures, oxygen and hydrocarbon concentrations kinetic parameters were estimated. To suppress catalyst coking a

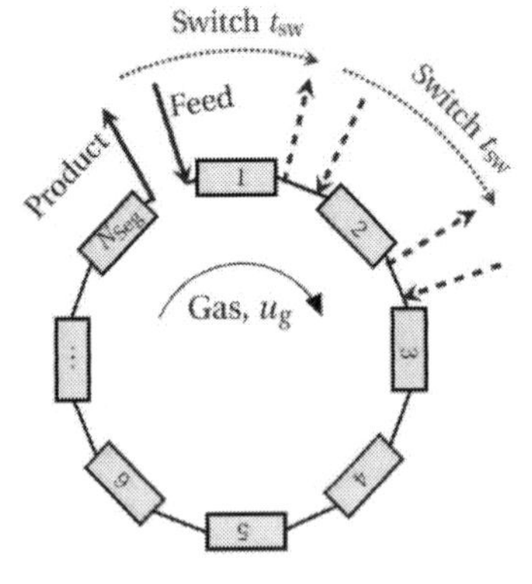

Fig. 1: Schematic illustration of a simulated moving bed reactor, SMBR [2]

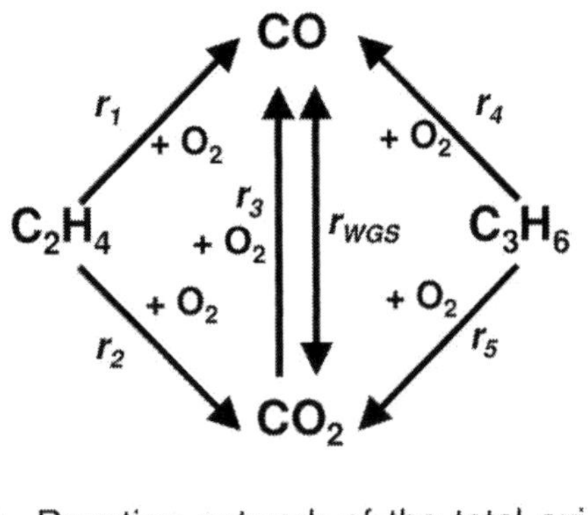

Fig. 2: Reaction network of the total oxidation of ethene and propene

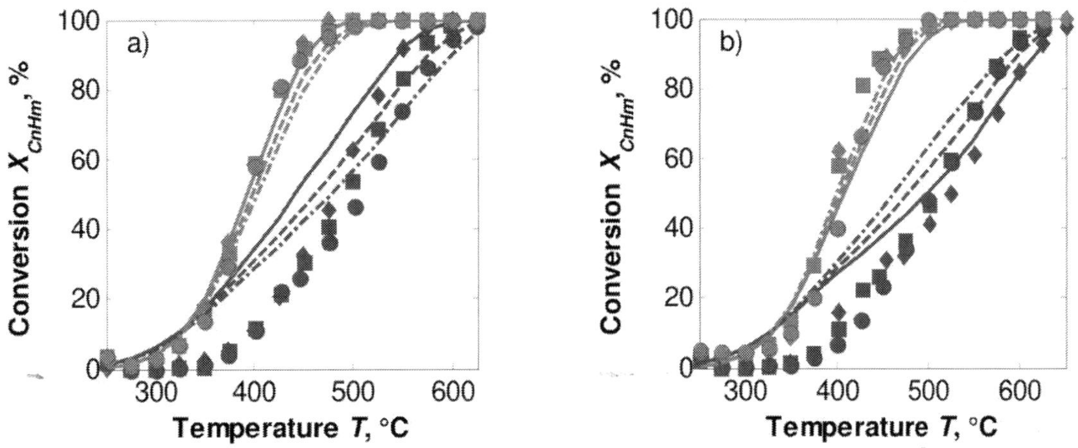

Fig. 3: Ethene and propene conversion over temperature at $W/F = 50$ (kg·s)/m^3
a) The ratio of the ethene (blue) and propene (red) initial volume fractions are: $x_{C2H4} / x_{C3H6} = 1/1$
◆/— $x_{CnHm} = 0.25$ % ■/--· $x_{CnHm} = 0.375$ % ●/-·- $x_{CnHm} = 0.50$ %
b) The sum of the initial volume fractions of ethene (blue) and propene (red) are 1 %
◆/— $x_{C2H4} / x_{C3H6} = 1 / 3$ ■/--· $x_{C2H4} / x_{C3H6} = 1 / 1$ ●/-·- $x_{C2H4} / x_{C3H6} = 3 / 1$

suitable regeneration strategy was applied. This time dependent effect was quantified by a deactivation kinetic as well. For long term experiments a suitable reactivation procedure was established. The estimated free model parameters were evaluated using a subset-selection method to identify significant parameters [3, 4]. With the derived reduced kinetic model a good agreement with a large set of experimental data was achieved (Fig. 3).

Subsequently applying the kinetic model a numerical feasibility study of the potential of the SMBR concept was performed. Different inlet concentrations of the hydrocarbon mixture and oxygen were investigated. Additionally diverse dosing concepts were analyzed to limit the maximum temperatures and to improve the process stability. In these calculations significant potential for the SMBR employing dosing strategies was identified.

Literature

[1] Y.S. Matros, Forced Unsteady-state Processes in Heterogeneous Catalytic Reactors, The Canadian Journal of Chemical Engineering, 74, 566-579, 1996

[2] V.M. Zahn, C.–U. Yi, A. Seidel-Morgenstern, Analysis and demonstration of a control concept for a heat integrated simulated moving bed reactor, Chemical Engineering Science, 66, 4901–4912, 2011

[3] M. Fink et al., Subset selection for parameter estimation in an HIV model, Proc. Appl. Math. Mech. 7, 1121501–1121502, doi:10.1002/pamm.200700319, 2007

[4] G. Kiedorf, D.M. Hoang, A. Müller, A. Jörke, J. Markert, H. Arellano-Garcia, A. Seidel-Morgenstern, C. Hamel, Kinetics of 1-dodecene hydroformylation in a thermomorphic solvent system using a rhodium-biphephos catalyst, Chemical Engineering Science, http://dx.doi.org/10.1016/j.ces.2013.06.027, 2013

Detailed Catalytic Fixed-Bed Reactor Simulations: The Dry Reforming of Methane

GREGOR D. WEHINGER, THOMAS EPPINGER & MATTHIAS KRAUME

Chemical & Process Engineering
Technische Universität Berlin

2014 AIChE Annual Meeting
Atlanta, 19 November, 2014

Agenda

1. Dry reforming of methane (DRM)
2. Motivation for spatially resolved CFD simulations
3. Detailed catalytic fixed-bed reactors simulations
   - Workflow
   - Heat transfer example
   - Full catalytic spherical particle fixed-bed
4. Outlook to complex geometries
   - Non-spherical particles

Dry reforming of methane

- Dramatic increase of atmospheric concentrations of "greenhouse" gases during last decades
- Global concern over current technological practices
- Chemical engineering's field of interest:
  - Process with CH_4 and CO_2 → synthesis gas

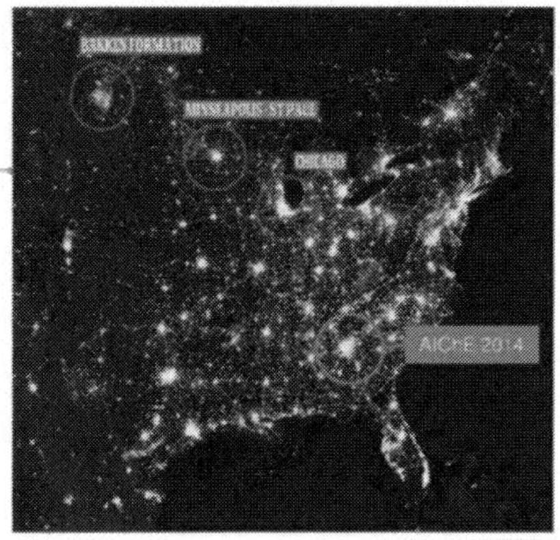

NASA

➢ Dry reforming of methane (DRM):
  - Catalytic endothermic process:

$$CH_4 + CO_2 \leftrightharpoons 2H_2 + 2CO \quad \Delta H \approx 260\ kJ/mol$$

  - Nickel-based or noble metal-supported catalysts
  - Coke formation as drawback
  - Performed at 700-1000 °C in fixed-bed reactors

Fixed-bed reactors for DRM

- DRM: highly endothermic reaction
- Heat transfer into the reactor → small tube diameter
- High gas velocities and small pressure drop → large particles

➢ Small tube-to-particle-diameter-ratios (N < 15)

➢ Conventional descriptions are questionable

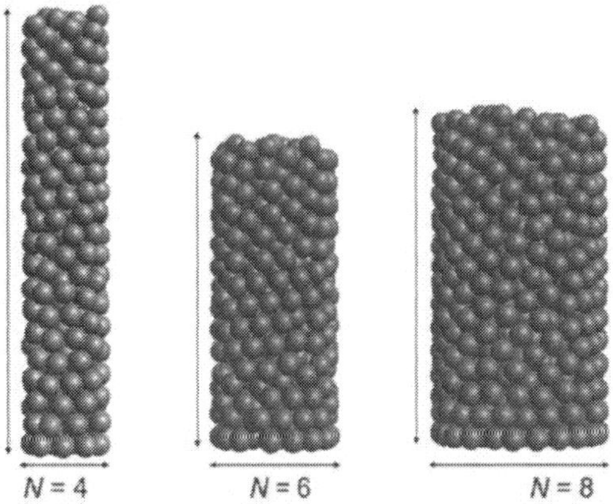

Behnam et al. (2013) *Ind. Eng. Chem. Res., 52*, 15244–15261

Do we describe fixed-bed reactors adequately?

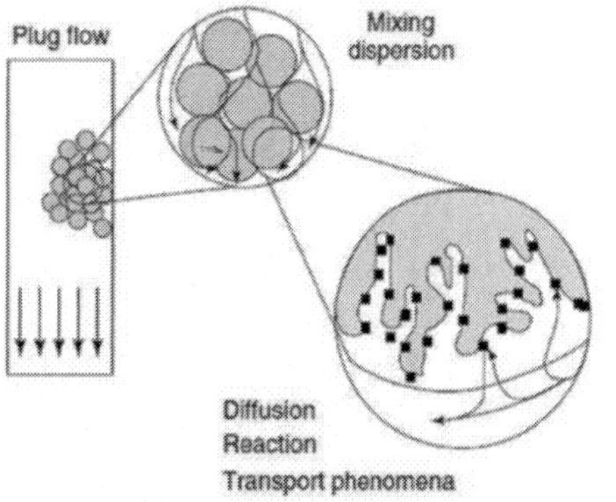

Kapteijn & Moulijn (2008) *Handbook of Catalysis*, Chap. 9.1

- Different time and length scales
- Conventional description based on plug flow and pseudo-homogeneous kinetics
- Inhomogeneous bed structure dominant for small *N*:
  - Significant wall effects
  - Local backflows
  - Large axial and radial gradients

➢ Multiple phenomena

➢ Adequate modeling with full CFD and detailed chemical models

Elements of spatially resolved catalytic fixed-bed reactors

DEM
(discrete element method)
Generation of a random packing

DARS
Calculation of detailed homogeneous and heterogeneous reaction mechanism

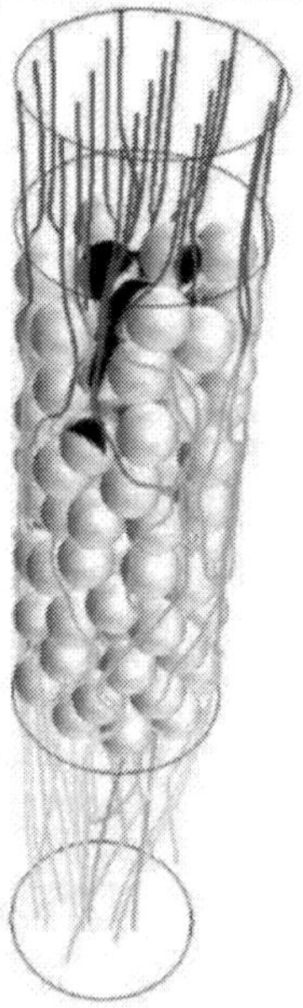

CFD - STAR-CCM+
Calculation of fluid dynamics, temperature and species concentration

Pore model
Approximation of processes within pores

Random bed generation and meshing

Bed generation

- Randomly distributed particles
- With discrete element method (DEM)

Meshing

- Boundary layer resolution
- Flattening of particle-particle contact points (caps method)

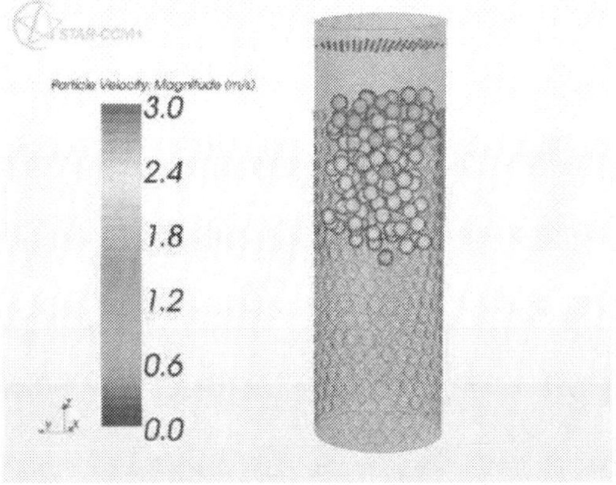

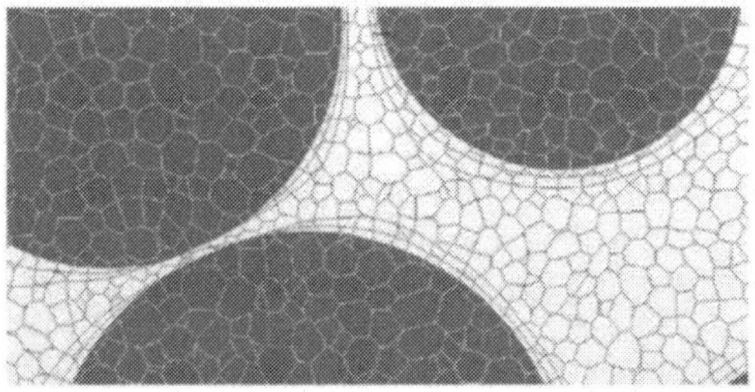

Eppinger et al. (2011) *Chemical Engineering Journal*, *166*(1), 324-331

Heat transfer inside fixed-bed

- Comparison between spatially resolved experimental data with detailed CFD
- Axial temperature measurement with pure argon
- 1 mm α-alumina spheres in quartz tube

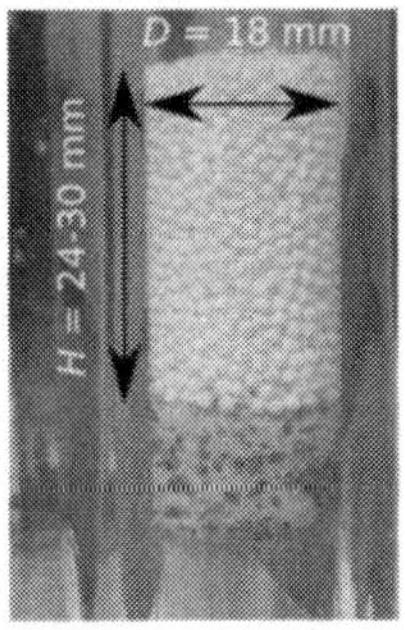

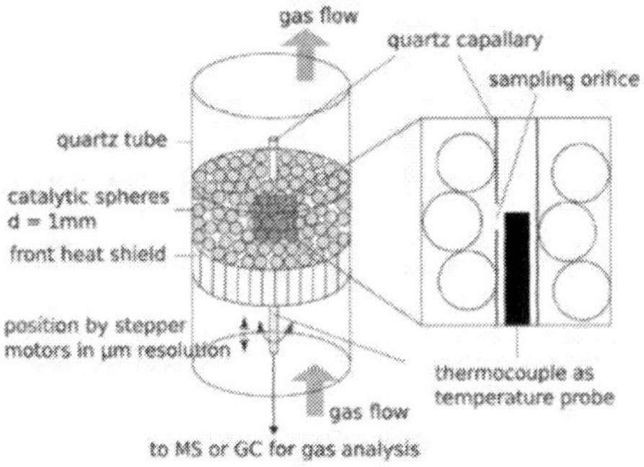

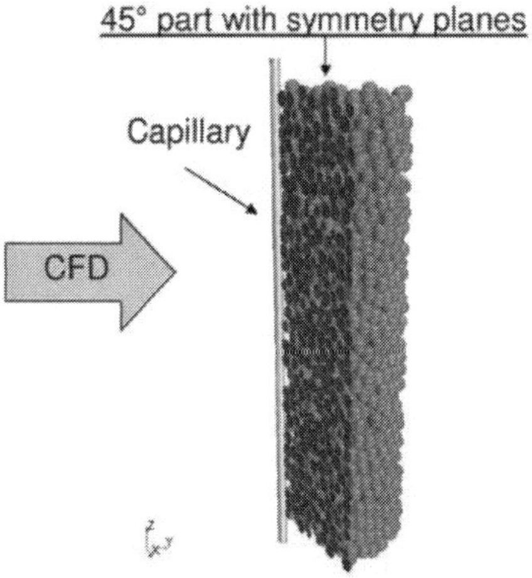

Horn et al. (2006) *Catalysis Letters* 110, 169

Radial porosity, velocity & temperature

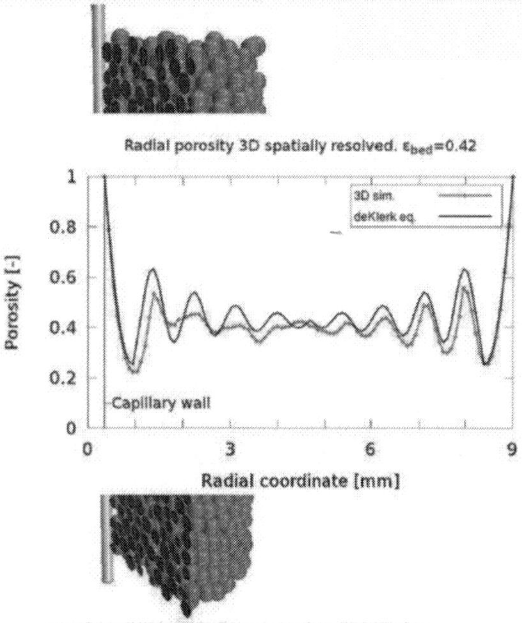

de Klerk (2003) *AIChE journal*, *49*(8), 2022-2029

Capillary wall

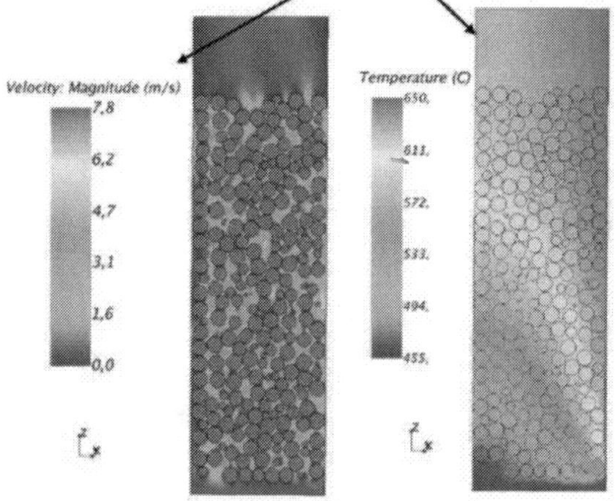

Flow direction

Slide 9

Axial temperature profile

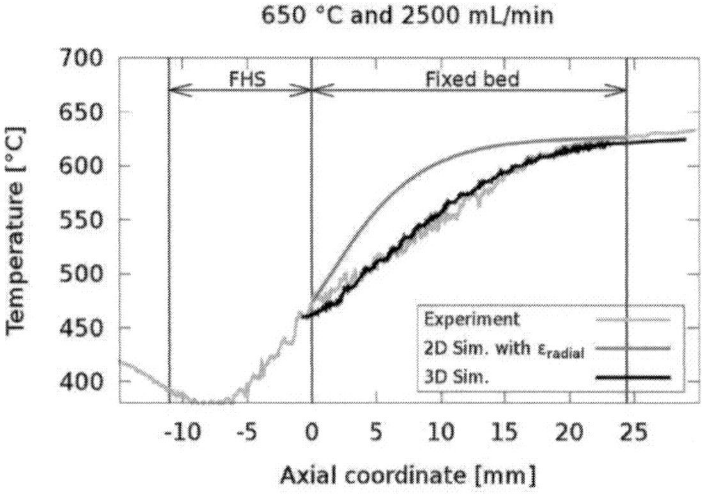

- Great agreement between experiments and 3D
- Superiority of 3D over 2D

Slide 10

Catalytic fixed-bed for DRM

- Detailed DRM kinetics from McGuire et al. (2011)
- $N = 4$, $H/D = 2,5$
- 113 spherical solid particles
- Instantaneous diffusion:

  $F_{cat/geo} = A_{cat}/A_{geo} = 90$
- Approx. 2-3 mio cells
- Realizable k-ε turbulence model
- Inlet:
  - $\mathrm{Re}_P = \frac{v_{in} \cdot d_P}{\nu} = 35$
  - $T_{Wall} = T_{in} = 973\ \mathrm{K}$
  - $x_{CO_2}/x_{CH_4}/x_{N_2} = 0.2/0.1/0.7$

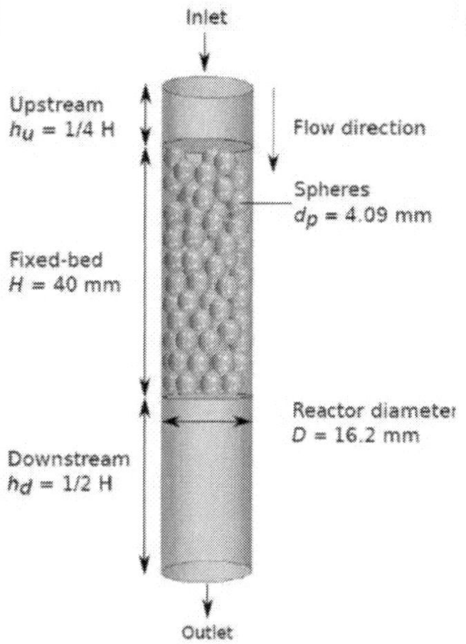

McGuire et al. (2011) *Applied Catalysis A: General, 394*, 257 - 265
Wehinger et al. (2014) *Chemical Engineering Science, 111*, 220-230
Wehinger et al. (2015) *Chemical Engineering Science, 122*, 197-209

Flow field and hydrogen production

$Re_p = \frac{v\, d_p}{\nu} = 35,\ T_{wall} = 973\ \mathrm{K}$

DRM: $CH_4 + CO_2 \leftrightharpoons 2H_2 + 2CO\ \ \Delta H \approx 260 \frac{kJ}{mol}$

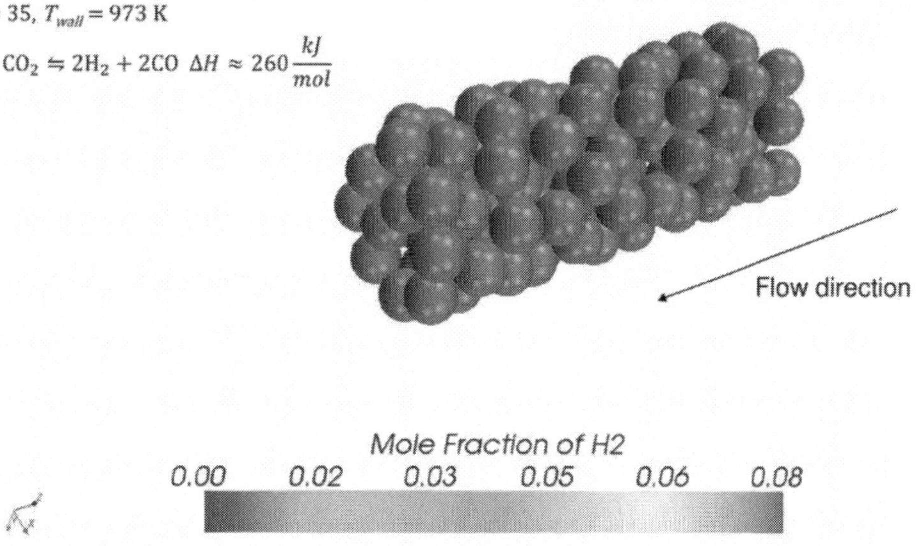

Temperature distribution

$Re_p = \frac{v\,d_p}{\nu} = 35,\ T_{wall} = 973\ \mathrm{K}$

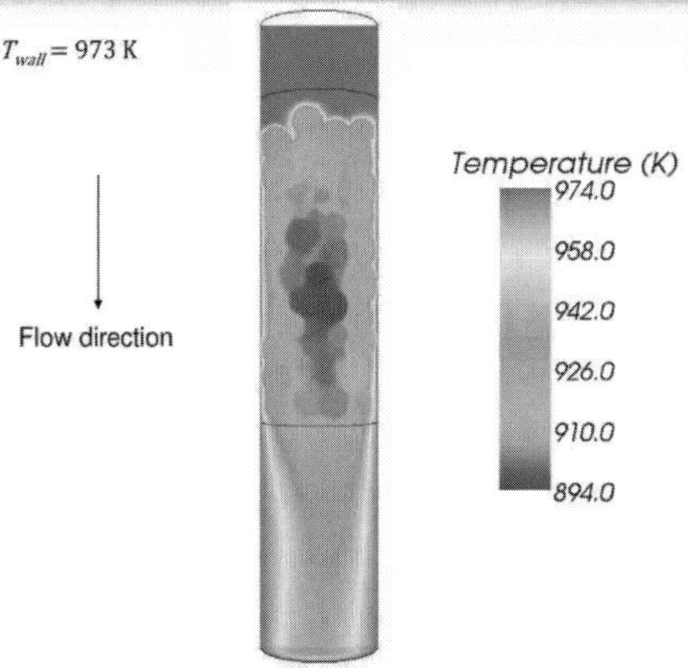

Slide 13

Hydrogen production and carbon deposition

$Re_p = \frac{v\,d_p}{\nu} = 35,\ T_{wall} = 973\ \mathrm{K}$

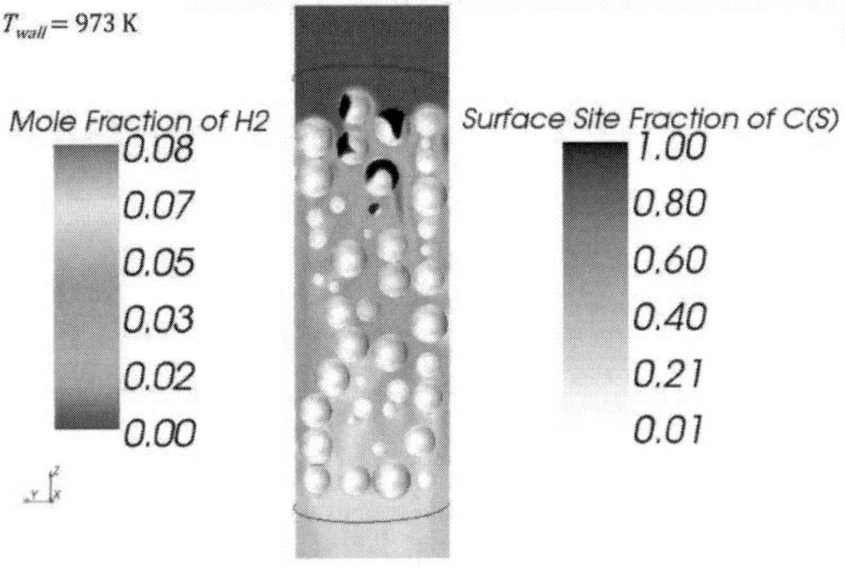

- Interaction between chemical kinetics and transport of momentum, heat and mass

Slide 14

Outlook to complex geometries

cylinders | 1-hole cylinders | more complex geometries

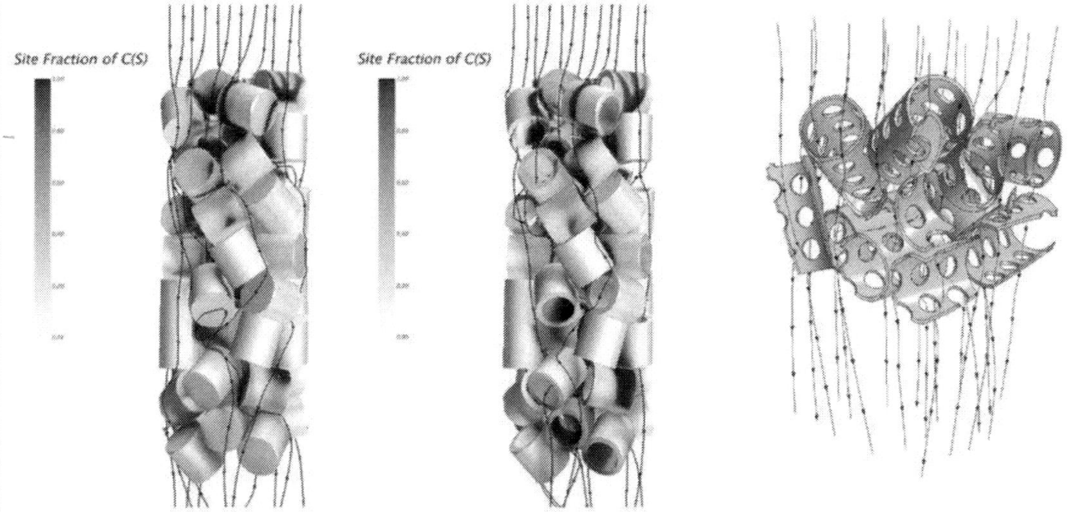

Conclusion

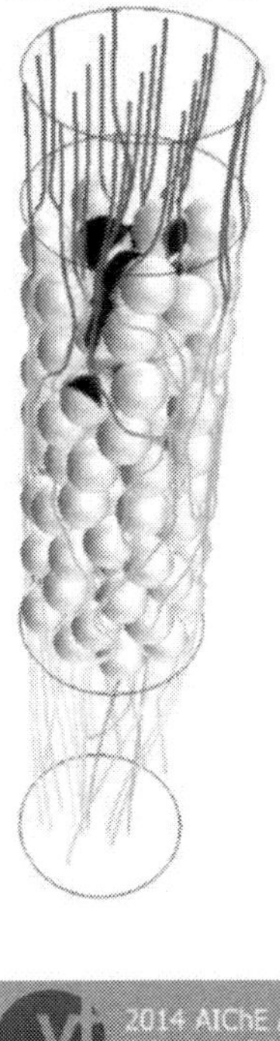

- Conventional plug flow and pseudo-homogeneous models fail in small *N* fixed-bed reactors.

➢ Adequate modeling with detailed CFD in combination with detailed kinetics

➢ Interaction between chemical kinetics and transport of momentum, heat and mass

➢ Better understanding of multi-scale chemical reactors

Thank you for your attention.

Special thanks for financial support go to DFG in the frame work of the Cluster of Excellence "Unifying concepts in catalysis (Unicat)".

Literature:

Behnam et al. (2013) *Ind. Eng. Chem. Res., 52*, 15244−15261
de Klerk (2003) *AIChE journal, 49*(8), 2022-2029
Eppinger et al. (2011) *Chemical Engineering Journal, 166*(1), 324-331.
Horn et al. (2006) *Catalysis Letters 110*, 169
Kapteijn & Moulijn (2008) *Handbook of Catalysis*, Chap. 9.1
Korup et al. (2013) *Journal of Catalysis 297*, 1-16.
McGuire et al. (2011) *Applied Catalysis A: General, 394*, 257 - 265
Mueller (1992) *Powder technology, 72*(3), 269-275.
Wehinger et al. (2014) *Chemical Engineering Science, 111*, 220-230
Wehinger et al. (2015) *Chemical Engineering Science, 122*, 197-209

TU berlin

unicat
Unifying Concepts in Catalysis

Slide 17

2014 AIChE Annual Meeting
Atlanta, 19 November, 2014

TU Berlin – Chemical & Process Engineering
Gregor D. Wehinger

unicat

Heat transfer at the interface of wall and catalyst bed in both lab and bench scales fixed bed reactors for Fischer-Tropsch

Xiaowei Zhu, Diane Hildebrandt, and David Glasser

Material and Process Synthesis; College of Science, Engineering, and Technology, University of South Africa

Abstract

The heat removal in a fixed bed reactor is a combination of effective radial heat transfer within the bed and the heat transfer at the wall. Heat transfer in a fixed bed reactor as a research topic has been studied for many years both within the catalyst bed and at the boundary of the catalyst bed and the reactor wall. When a fixed bed reactor is applied for Fischer-Tropsch Synthesis, it becomes much more complicated as liquid presents when reaction takes place [1]. Compared with the heat transfer study within the catalyst bed, the investigation of the heat transfer at the boundary of the catalyst bed and reactor wall is much less but it is equally important as a considerable temperature chasm takes place at this boundary. The sudden temperature drop at the boundary has been a source of interest to a number of researchers over many years. Several explanations [2] have been offered, although few of them have received general acceptance. One proposal is that the gas at the boundary is reluctant to move because of the wall effect. This slow or "static" flow pattern in the gas decreases the heat transfer coefficient dramatically when compared with the conditions in the catalyst bed and at the reactor wall. In the case of FTS, the heat transfer in this static layer can be more complicated, as the reaction is expected to form liquid, which will eventually coat the internal wall. The liquid holdup and the tempo at which the liquid in this static layer moves can also be affected by the linear speed of the gas in the reactor, which would alter the effective thermal conductivity coefficient at this boundary.

A series of heat transfer experiments have been carried out in both lab and bench scale fixed bed reactors. 8 grams of catalyst is loaded in the lab scale reactor and 100 grams in the bench scale one. The space velocities per volume of catalyst were similar for these two reactors; however the superficial velocities were far away. The effective heat transfer coefficient at the boundary was observed to vary with the superficial velocity in the bench scale reactor but not the lab scale one.

Keywords: FT, Fixed bed reactor, heat transfer, wall to bed

Reference

1. Zhu, X.; Lu, X.; Liu, X.; Hildebrandt, D.; Glasser, D. Heat Transfer Study with and without Fischer-Tropsch Reaction in a Fixed bed Reactor with TiO_2, SiO_2, and SiC Supported Cobalt Catalysts. *Chemical Engineering Journal* **2014**, *247*, 75-84.
2. Dixon, A. G. Fixed Bed Catalytic Reactor Modelling-the Radial Heat Transfer Problem. *The Canadian Journal of Chemical Engineering* **2012**, *60*, 507-527.

386099 Pruning of Catalytic Reaction Networks: A Comparison of Campbell's Degree of Rate Control with Reaction Step Resistance

Abstract:

Many complex catalytic reaction mechanisms start off as being "hypothetical," i.e., they emerge from a researcher's knowledge of the catalytic chemistry on a given catalyst coupled with intuition, or are based on a stoichiometric, or graph-theoretical, generator, but can soon become "real," when modern quantum-chemical methods are used to predict their reaction energies and activation barriers. Whether the imagined molecular and kinetic complexity of a network is justified in reality, however, is an open question; one which we are not yet fully equipped to answer unequivocally.

The tools that are currently available for simplifying complex microkinetic models include: 1) a comparison of the energy landscape of different pathways; 2) a comparison of step reversibility, as proposed by Dumesic, and 3) Campbell's degree of rate control (RDC). The first of these, although the most common, is mainly a qualitative tool, while the second, involves only a thermodynamic criterion, not kinetic. The most useful and rigorous tool so far is Campbell's DRC, which involves a partial differential analysis of the relative change in the rate of overall reaction (OR) for a relative small change in the rate constant of a step, holding constant all other step rate as well as equilibrium constants, i.e., it is a numerical sensitivity analysis. With the help of the water-gas shift reaction network and other examples, we show that our alternate methodology, as a part of our Reaction Route (RR) Graph approach, which involves a comparison of reaction step "resistance," is just as rigorous and, in fact, more revealing than Campbell's DRC, and allows transparent pruning of complex catalytic reaction networks.

367539 Breaking Lignin: First-Principles Steered Dynamics for Studying Lignin Depolymerization

Abstract:

Lignin is a highly heterogeneous biopolymer comprising 20-30% of all biomass by weight. We present here the results of a first-principles steered molecular dynamics approach to studying the depolymerization of representative lignin repeats under force. Our constant force molecular dynamics runs are both representatitive of experimental mechanochemical degradation (e.g. via ball milling) as well as a useful screening tool to identify the dynamics behind depolymerization during pyrolysis or other high temperature, high energy conditions. In our simulations, we pull various points of dimeric, trimeric, and hexameric repeats of coniferyl alcohols (a representative monolignol) and examine how and when select linkages break apart. We then follow the dynamics that occurs after this initial break to see if there are additional bond-breaking events and fragmentation. Finally, we analyze whether this degradation occurs via a homolytic or heterolytic mechanism and hypothesize that heterolytic bond-breaking events could be advantageous to promote in order to ensure that recombination does not occur. Overall, our screening approach, which is accelerated through the use of stream-processors (i.e. graphical processing units), can pave the way for identifying which reaction paths are most critical to catalyze in order to selectively depolymerize lignin into value-added products.

379902 Discerning and Modeling Reaction Mechanisms in the Cellulose Pyrolysis Reaction Network

Abstract:

Biomass fast pyrolysis is a promising technology for the thermal conversion of biomass to liquid fuels. With less-severe operating temperatures than gasification, fast pyrolysis produces a primarily liquid product, which requires less-costly equipment for processing and storage than a vapor product. Currently, improvements to make this technology more economically attractive are impeded by a lack of understanding of the reaction chemistry, which has been difficult to ascertain due to the complex reaction network accessible at the temperatures employed, typically near 500 °C, and the short timescales on the order of seconds. Without this information, we cannot predict how changes in reaction conditions and feeds will affect product yield and composition, knowledge of which are crucial for process design and optimization. To overcome this obstacle, we employ density functional theory to compare competing reaction mechanism proposals and examine fleeting transition states, knowledge of which can aid in rational catalyst design. Specifically, we have focused on reactions central to cellulose fast pyrolysis, including evaluation of several competing theories for the formation of the main product of pure cellulose pyrolysis, levoglucosan, and uncovered a new mechanism that yields results consistent with experiment. We also evaluated several key decomposition reactions of glucose, the cellulose monomer and postulated key intermediate, and how they are perturbed by alkali metal ions naturally present in biomass. We incorporated the results of these studies in the first mechanistic model of cellulose pyrolysis, which offers predictions of detailed product speciation that closely match experimental results.

387062 Automated Kinetics Via Transition State Theory for Improved Mechanism Generation

Abstract:

An automated tool to calculate reaction rate parameters via classical transition state theory has been implemented in Reaction Mechanism Generator (RMG)[1]. The tool uses a group additive approach and distance geometry to derive good estimates of transition state geometries, and automatically optimizes the reactants and saddle point using DFT quantum calculations before calculating the kinetics.

During automatic mechanism generation, group additive methods have typically been used to estimate thermodynamic and kinetic parameters, with varying degrees of accuracy. The accuracy particularly suffers when there is little data from which to derive required group values, or when non-nearest-neighbor effects such as steric hindrance or ring strain become important. As kinetic group data are sparse, many of the used estimates contain several orders of magnitude error. Models in RMG are expanded to include reactions that meet a user-defined flux, so the large kinetic errors lead to unfavorable model expansion, and may lead to important pathways being excluded from the model if their rates are incorrectly underestimated. Our automated TST-based kinetic calculator should reduce these cases, improving models generated in RMG.

Kinetic calculations require reactant and transition state geometries, so we have developed a group additive approach to estimate distances between reacting atoms at the transition state. The distance estimates are used to provide good transition state geometry approximations, which are optimized using semi-empirical or DFT quantum chemistry, then validated via a reaction path analysis. Properties for the reactants and products are determined automatically in a similar manner[2]. Reactant and transition state partition functions are calculated from their molecular properties, while the tunneling correction is calculated via the Eckart model. Point groups are found[3] in order to assign molecular symmetries, allowing the kinetic parameters to be calculated via classical transition state theory.

These steps have now been integrated in RMG to allow automated kinetic calculations for reactions with activated complexes. This has been shown to improve the kinetics used in models generated by RMG, and the overall model generation process.

References

[1] Reaction Mechanism Generator - RMG-Py *http://greengroup.github.io/RMG-Py*

[2] Magoon GR, Green WH Design and implementation of a next-generation software interface for on-the-fly quantum and force field calculations in automated reaction mech- anism generation. *Comput Chem Eng.*2012;52:35–45.

[3] Patchkovskii S, 2003; *http://www.cobalt.chem.ucalgary.ca/ps/ symmetry/*

Severe Pyrolyses of Methylated Decalins, with Rice-Herzfeld and Graph-Theoretic Interpretations.

By

P.S. Virk
Department of Chemical Engineering
Massachusetts Institute of Technology, Cambridge, MA 02139.

High temperature, short residence time pyrolyses of decalin, 1-methyl-decalin, 2,3-dimethyl-decalin and 2,6-dimethyl-decalin were experimentally investigated using a coiled tube flow reactor with outlet temperatures 1050-1150 K, residence times ~0.3 s, and steam/hydrocarbon ratios ~1, to simulate commercial steam cracking of naphthenic feedstocks at severities from 0.8 to 3.3. Decomposition kinetics and product selectivities were interpreted two ways. First by the classical Rice-Herzfeld types of free-radical pathways, initiated by substrate bond fissions and containing the requisite H-abstraction, beta-scission, isomerization, combination and disproportionation types of elementary steps, and second by taking graph-theoretic excursions upon the A(djacency) matrices of each of the bicyclic substrates. The influence of either the absence or presence of methyl substitution, and of the number of methyl substituents, and their peripheral proximities, upon the pyrolysis pathways of each substrate was discerned from the observed yields of specific products of interest in olefin plant operations: namely, ethylene, propylene, butadiene, isoprene and benzene.

The Mechanism and Kinetics of Solid Zn Oxidation by CO_2 in the Presence of ZnO

The two-step thermochemical redox cycle based on zinc oxide is a promising approach for producing renewable H_2 and/or CO (syngas) from solar power. In a first reduction step ZnO is thermally dissociated into a gaseous mixture of Zn and O_2 using concentrated solar radiation. These products are then quenched rapidly to yield a Zn-ZnO mixture as the solid product called "solar Zn". The solar Zn is then re-oxidized in a second, non-solar oxidation step by CO_2 and/or H_2O (steam) to produce syngas. In this work we explore the reaction mechanism and kinetics of the second oxidation step.

Previous work has shown a positive effect of a high Zn to ZnO conversion and a low oxidation temperature on the solar-to-fuel energy conversion efficiency. Achieving a high conversion of pure metallic Zn below its melting point (420°C) is not possible because the produced ZnO rapidly passivates the Zn surface. In contrast, solar Zn was found to convert completely at temperatures as low as 400°C. The latter has been attributed to an additional reaction path according to which Zn sublimates from the surface of Zn particles and reacts on the ZnO surface. Since the ZnO product forms on the ZnO surface, the Zn surface remains exposed thus the supply of Zn to the reaction site is not impaired. The oxidation of solar Zn with CO_2 and H_2O below 420°C is thus considered a viable route for the efficient production of syngas.

In order to select and design a gas-solid reactor for this reaction a kinetic model based on detailed understanding of the reaction mechanism is necessary. The complexity of the mechanism arises from the two parallel reaction paths: 1) the direct Zn surface oxidation by CO_2 producing passivating ZnO scale and 2) the sublimation of Zn from the surface of the Zn particles and its gas phase migration followed by the deposition and/or reaction with CO_2 on the ZnO surface. A general kinetic model thus needs to consider 1) the rate of Zn surface passivation, 2) the rate of Zn sublimation, 3) the decrease in reactive Zn surface area as consequence of both passivation and sublimation, 4) the gas phase migration of Zn and CO_2 to the reaction sites, 5) the rate of Zn oxidation on the ZnO surface, and 6) the change in ZnO surface area with time. Based on different assumptions several simplified scenarios were considered and tested on experimental data.

Modelling the Zn(g) concentration gradients between the Zn surface and the ZnO surface, acting as Zn(g) source and Zn(g) sink, respectively, was found to be important. This gradient creates a diffusional driving force for Zn sublimation and becomes steeper as the ZnO/Zn ratio increases. As Zn sublimation becomes more dominant, the effect of the Zn surface passivation diminishes and higher conversions are attained. This mechanism reflects the experimentally observed effect of ZnO on the Zn conversion well.

Since the major part of Zn reacts on the ZnO surface, the conversion rate is mainly determined by the rate of the ZnO surface reaction. Therefore, this reaction was modelled in detail and a rate law based on the Langmuir-Hinshelwood mechanism was derived. It describes the effect of the Zn(g), CO_2, and CO concentration on the reaction rate.

389718 Insights into the Mechanism of Acetic Acid Ketonization over Zeolites

Abstract:

The corrosive nature of acetic acid leads to storage and transport limitations of biomass pyrolysis oils while its acidity catalyzes the polymerization of other species present in the pyrolysis oil to undesired products. The conversion of acetic acid to the more stable chemical building block, acetone, is of significant interest. The vast majority of mechanistic studies concerning this reaction occur over reducible oxides and basic catalysts. This reaction can readily occur with high selectivity over protonated zeolites as well, but the details of the mechanism are unclear. In this contribution we use a combination of reaction kinetics coupled TPD experiments and in-situ FTIR to separate the role of surface acyl and anhydride species from adsorbed carboxylates on the ketonization reaction over HZSM-5. In addition, a series of reactants with various degrees of substitution and the number of alpha hydrogens are used both pure and co-fed to demonstrate the critical role of the alpha hydrogen for this reaction. We propose a mechanism involving the coupling of an enolate with a dehydrated surface intermediate over protonated zeolites based on these exerimental results. Because most feedstocks that contain acetic acid also contain water, a kinetic study was conducted to identify the rate determining step and the inhibiting role of water on this reaction.

391734 Ni Catalysts with Different Promoters Supported on Zeolite for Dry Reforming of Methane

Abstract:

Ni catalysts with different promoters supported on zeolite for dry reforming of methane

Raja AL Otaibi[1], Feraih Alenazey[1], Faisal Alotaibi[1], Ahmed S. Al-Fatesh[2], Anis H. Fakeeha [2]

[1] P.O. Box 6086 Riyadh 11442, King Abdulaziz City for Science and Technology, Kingdom of Saudi Arabia

[2] King Saud University , Riyadh, KSA

Corresponding author: raletabi@kacst.edu.sa

Abstract

Dry (CO_2) reforming of methane (DRM) is a well-studied reaction that is of both scientific and industrial importance. It combines CH_4, the main component of natural gasused as source of energy, and CO_2, a major challenging greenhouse gas, where by an impact can be made to limit global warming. This reaction produces syngas that can be used to produce a wide range of products, such as higher alkanes and oxygenates by means of Fischer–Tropsch synthesis. The reaction is particularly interesting because both $CH_4$and CO_2 are available feedstocks due to their natural abundance. Therefore it is a good alternative to steam reforming , which is generally used for synthetic gas (CO/H_2) production in the industry because it yields lower H_2/CO ratios. DRM is considered a high endothermic reaction and requires operating temperatures of 700–1000 °C to achieve high equilibrium conversion of CH_4 and CO_2 to H_2 and CO. The most widely used catalysts for DRM are based on Ni. However, many of these catalysts undergo severe deactivation due to carbon deposition. Noble metals have also been investigated and are usually found to be more resistant to carbon deposition than Ni catalysts, but are generally uneconomical. In this investigation, we aim to study the effect of lanthanum and calcium as promoters for Ni catalyst supported on two different zeolite for dry reforming of methane. Several catalysts of promoted Ni/ on Y-Zeolite have been prepared by impregnation method. The physicochemical properties of the prepared catalysts were characterized with (XRD), (SEM), (TEM) (BET) and XRF. These catalyst were tested for dry reforming of methane in a micro tubular reactor using temperature ranges of 700 °C at atmospheric pressure using total flow rate = 40 ml/min (CH_4 =15 ml/min , CO_2 =15 ml/min & N_2 = 10 ml/min). It was found that, Ca -Ni/Zeolite (CBV300) gave the highest conversion (51%) with the average selectivity of hydrogen (47%). Ni/Zeolite showed very low conversion due to rapid deactivation of catalyst compared to La – Ni/Zeolite and Ca –Ni/Zeolite which present the enhancement of the promoters to the Ni/Zeolite .

Nickel Hexaaluminates for Methane Steam and Dry Reforming

Jessica L. Ewbank (jle3@gatech.edu),[1,2] Jeff Miller (millerjt@anl.gov),[3] Eli Stavitski (istavitski@bnl.gov),[4] Carsten Sievers (carsten.sievers@chbe.gatech.edu)[1,2]

[1]School of Chemical & Biomolecular Engineering, Georgia Institute of Technology, Atlanta, Georgia 30332
[2]Institute of Paper Science and Technology, Georgia Institute of Technology, Atlanta, GA 30032
[3]Argonne National Laboratory, Chemical Technology Division, Argonne, IL 60430
[4]Brookhaven National Laboratory, Upton, NY 11973

Syngas is an extremely important feed in the petroleum and chemical industry.[1] _ENREF_3Methane steam reforming is the most common way to commercially produce syngas. Production of syngas by methane dry reforming is also attracting much attention for several reasons: the increased interest in the effective utilization of the greenhouse gas carbon dioxide, a possible way to utilize methane and carbon dioxide resulting from anaerobic digestion of biomass, and syngas production with hydrogen to carbon monoxide ratio near one.[2,3] _ENREF_20 Hexaaluminates of various compositions have been explored in the conversion of methane by steam reforming, dry reforming, and partial oxidation.

We prepared nickel hexaaluminates of three nickel-aluminum-oxygen compositions by the Pechini synthesis: Stoichiometric ($NiAl_2O_4$), Nickel Excess ($NiAlO_3$), Nickel deficient ($NiAl_4O_8$). It was found that all samples are highly active, stable, and robust catalysts for methane steam and dry reforming. $NiAl_4O_8$, $NiAl_2O_4$, and $NiAlO_3$ were characterized by temperature programmed reduction (TPR), temperature programmed oxidation (TPO), x-ray absorption spectroscopy (XAS), H_2 chemisorption, in-situ x-ray diffraction (XRD), temperature programmed desorption of ammonia (NH_3 TPD), and N_2 physisorption. The samples were active for methane steam and dry reforming with and without conventional pre-treatment, e.g. reduction. Hexaaluminates showed stable conversion and limited formation of carbonaceous deposits making them promising candidates for commercial application in methane reforming reactions. Nickel hexaaluminate spinels had moderate surface areas (~$40 m^2/g$) and mesoporous pore structures as determined by N_2 physisorption. It was also found fromN_2 physisorption that the pore structure of hexaaluminates changed when the nickel content was increased. TPR-TPO cycling indicated that all nickel was fully reduced upon reduction and could be mostly re-oxidized. XRD measurements performed after TPR- TPO cycling confirmed the crystal structure can be regenerated in a redox environment. Hydrogen chemisorption performed on the samples after reduction at 900 °C showed low H_2 uptake indicating redox cycles may play a crucial role in reactivity. NH_3 TPD showed low acidity correlating well with the limited coke deposition that was observed during methane steam and dry reforming reactions. TPR combined with XAS measurements showed that for some samples the all nickel contained in the samples can be reduced at elevated temperatures.

Reactivity and characterization results highlight unique properties of nickel hexaaluminates. It is hypothesized that the remarkable stability and activity are due to the ability of the nickel to move back and forth between the B-site in the perovskite like crystal structure when exposed to the reaction gas.[4] Nickel incorporation in the lattice of the spinel structure prevents agglomeration of nickel particles thus decreasing coking and increasing catalyst lifetime.

References

1. E. Ruckenstein and Y. H. Hu, *J. Catal.*, 1996, **162**, 230-238.
2. S. Y. Foo, C. K. Cheng, T. H. Nguyen and A. A. Adesina, *J. Mol. Catal. A: Chem.*, 2011, **344**, 28-36.
3. D. San-Jose-Alonso, J. Juan-Juan, M. J. Illan-Gomez and M. C. Roman-Martinez, *Appl. Catal. A: Gen.*, 2009, **371**, 54-59.
4. Y. Nishihata, J. Mizuki, T. Akao, H. Tanaka, M. Uenishi, M. Kimura, T. Okamoto and N. Hamada, *Nature*, 2002, **418**, 164-167.

377829 Activity and Productivity of Co ALD Catalysts for FTS

Abstract:

Significant advancement of FTS could allow for industrialization of GTL at a capacity having a considerable impact on the energy economy. The gas phase deposition technology of Atomic layer deposition (ALD) is a thin film technology analogous to chemical vapor deposition. Metallic Co ALD was utilized for a new type of highly-productive catalyst and reactor for valuable fuel products. Avoiding reduction of Co_3O_4 to metallic Co eliminates processing temperatures above reaction temperature; this reduces the likelihood of sintering Co catalyst particles, and lowering productivity. Product selectivity, conversion and productivity have been demonstrated for low weight loading Co catalysts having 5-11 wt% Co.

Session: 595 - Syngas Production and Gas-to-Liquids Technology
Wednesday, November 19, 2014: 3:15 PM

14AIChE
Annual Meeting, Atlanta, GA

[AIChE 2014 - 595d]

Challenges and Potential Solutions in GTL FPSO Process Design for Monetization of Offshore Stranded Gas

Hyun-Jin Kim

Daewoo Shipbuilding & Marine Engineering Co., Ltd

DSME

CONTENTS

Slide 2/24

1. **BACKGROUND**
2. **DSME's GTL FPSO DEVELOPMENT APPROACH**
3. **GTL FPSO FEASIBILITY STUDY**
4. **GTL FPSO CONCEPTUAL ENGINEERING**
   1) Overview of GTL FPSO
   2) Topside Conceptual Engineering
   3) Hull Conceptual Engineering
5. **Preliminary Economic Analysis**
6. **SUMMARY**

1. BACKGROUND

- **GTL Production Technology**

  Natural Gas → Syngas (H2/CO) → Clean Liquid Fuels

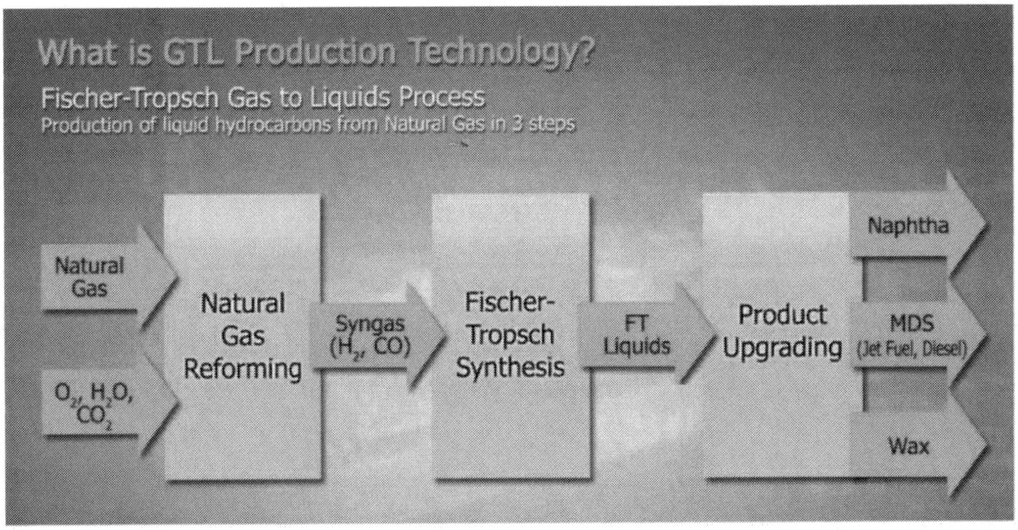

[Source: DSME Offshore Plant R&D Group (2014)]

1. BACKGROUND

- **Main Drivers – Taking GTL Offshore**
  - ✓ Sustained increasing world energy demands (Especially, Transportation fuel demand) & high prices of oil
  - ✓ Monetization of stranded gas fields in offshore (*CO_2 rich gas)
  - ✓ Practicable marine application using proven GTL technology in onshore.
  - ✓ Abundant Stranded Gas & Associated Gas Fields in offshore

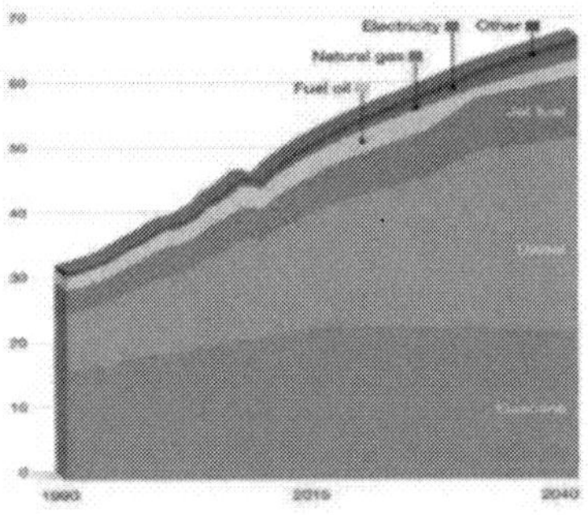

[Source: ExxonMobil Outlook for Energy (2012)]

< Transportation Fuel Demand >

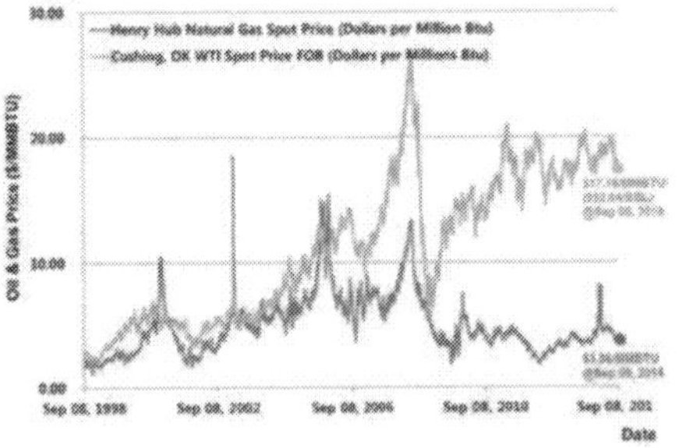

[Source: EIA – WTI & Henry hub price (2014)]

< Historical Oil and NG Prices >

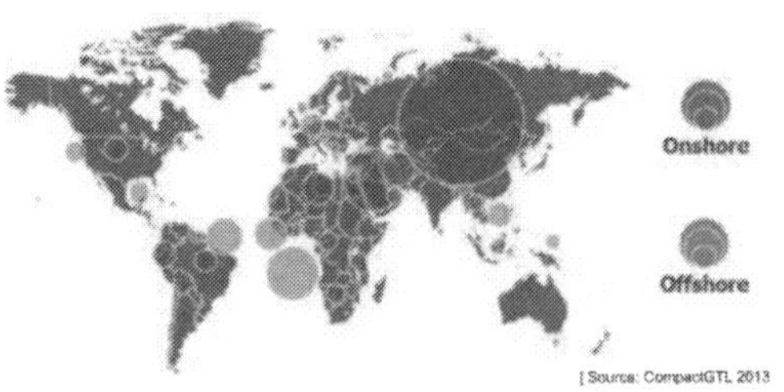

[Source: CompactGTL 2013

< Abundant Stranded Gas & Associated gas Fields >

1. BACKGROUND

Slide 5/24

- **GTL FPSO Production & Supply Chain**

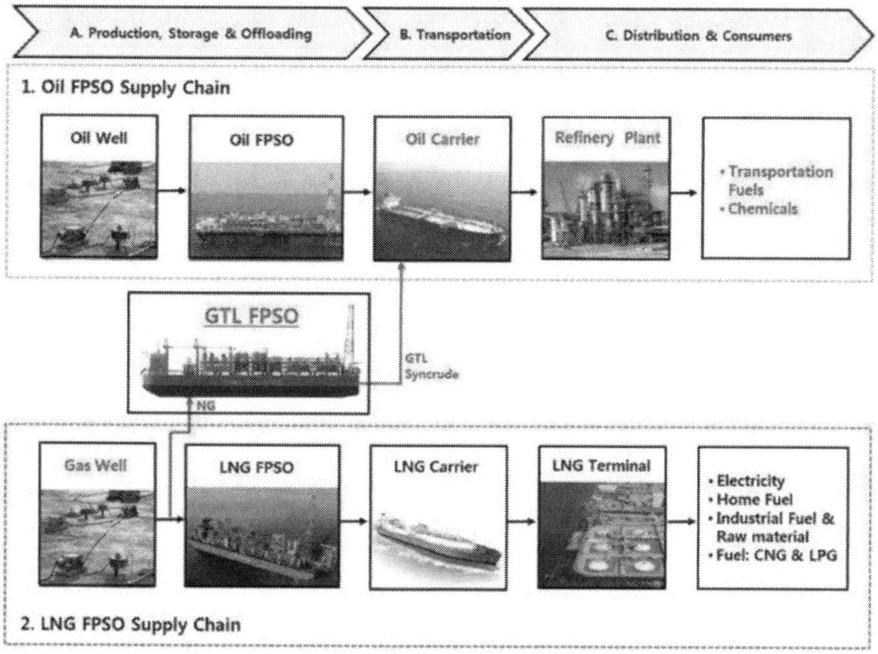

DSME

[AIChE 2014 - 595d] • Challenges and Potential Solutions in GTL FPSO Process Design for Monetization of Offshore Stranded Gas • Hyun-Jin Kim

1. BACKGROUND

Slide 6/24

- **Advances in FPSO at DSME**

  ✓ Agbami FPSO (Chevron, 2008 @Nigeria)
  ✓ Pazflor FPSO (Total, 2011 @Angola)
  ✓ CLOV FPSO (Total, 2014 @Angola)

Agbami FPSO

Storage		2,160,000 BBL
Production	Oil	250,000 BBL
	Gas	450 MMSCFD
Dimensions		320m X 58.4m X 33m
Weight	Hull	69,700 MT
	Topside	33,798 MT
Client		Chevron
Delivery		Jul 2008 (1st Oil)
Scope of Work		EPCIC
Field		Offshore Nigeria

Pazflor FPSO

Storage		2,000,000 BBL
Production	Oil	232,000 BBL
	Gas	155 MMSCFD
Dimensions		325m X 61m X 32m
Weight	Hull	81,019 MT
	Topside	41,102 MT
Client		Total
Delivery		Aug 2011 (1st Oil)
Scope of Work		EPCIC
Field		Offshore Angola

CLOV FPSO

Storage		1,781,000 BBL
Production	Oil	160,000 BBL
	Gas	230 MMSCFD
Dimensions		305m X 61m X 32m
Weight	Hull	76,600 MT
	Topside	31,800 MT
Client		Total
Delivery		June 2014 (1st Oil)
Scope of Work		EPCIC
Field		Offshore Angola

DSME

[AIChE 2014 - 595d] • Challenges and Potential Solutions in GTL FPSO Process Design for Monetization of Offshore Stranded Gas • Hyun-Jin Kim

2. DSME's GTL FPSO DEVELOPMENT APPROACH

Slide 7/24

- **Approach to GTL FPSO development (2011~2016)**

Nov. 2014

	Feasibility Study	Conceptual Design		Application Investigation		
Period	Stage 1			Stage 2		Project Develop. and Biz. Promotion
	Feasibility Study	Conceptual Engineering		Pre-FEED		
	1st year	2nd year (Phase.1)	3rd year (Phase.2)	4th year	5th year	
GTL FPSO Integrated Design (DSME)	• GTL FPSO design basis establishment • GTL Process Technology Screening & Assessment • Technical & Economical Assessment • Hull initial conceptual design	• Preliminary Design Basis • Conceptual Design Report • Preliminary BFD, PFD • Pre. Process Studies • Pre. Hull Design Basis • Preliminary Hull G/A	• Preliminary UFD • Preliminary Process Studies • Preliminary Hull Outline Spec. • Preliminary Weight estimation • Preliminary 3D Animation Model	• Project Design Basis • Conceptual Design Development • Process & Safety Studies • Update Hull Design (G/A, Midship Sec., etc.) • Overall Weight Estimation • Budget Estimation • 3D Animation Hardware Mock-Up • Pre-FEED Package Completion		

Scope of work reported in this presentation

3. GTL FPSO FEASIBILITY STUDY (1/5)

Slide 8/24

- **GTL FPSO Project aims to:**
  - ✓ **GTL FPSO plant: 20,000 BPD**
  - ✓ **Preliminary location: Northwestern Australia or Southeast Asia**
  - ✓ **Target: Offshore Natural gas with medium CO_2 level of 10%**

Feasibility Study

A. Technology Assessment & Selection	B. Preliminary Process Design Concept	C. Market Analysis	D. Economic Analysis

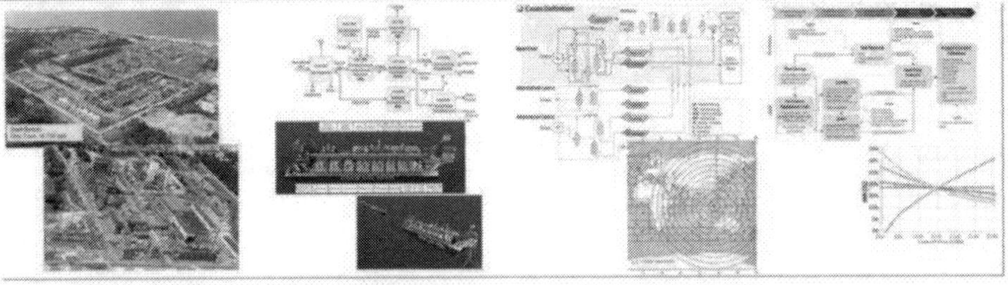

1. Current Status of GTL Technologies(SASOL & Shell Technologies Comparison) 2. Investigating New Technologies	1. Benchmark Technology Selected 2. Process design basic and Process Definition - Block Diagram - Topside Process Concept - Hull Process Concept	1. For Selecting Upgrading Section and Product Distribution - Base Case - Alternative Case 1 - Alternative Case 2 2. Transportation	1. To estimate revenue from Market Analysis - CAPEX, OPEX.. 2. Sensitivity Analysis on Preliminary IRR

3. GTL FPSO FEASIBILITY STUDY (2/5)

- **Special considerations for taking GTL FPSO**

a. Compactness of Design

- Footprint, Weight, Height & Multi-layer Arrangement

b. Robustness to marine motion

c. Safety Issues

d. Cost & Economic Viability

e. Operability and Reliability

f. Fabrication Aspects

g. Certification Requirements - Class Approval

< Ship Motion Effect Test >

< Topside Module>

3. GTL FPSO FEASIBILITY STUDY (3/5)

- **Reforming technology selection for GTL FPSO**

Key Parameters & Issues

1. **CO_2 tolerance**
   - ✓ Numerous stranded gas fields are CO_2 rich
   - ✓ Utilize and handle high CO_2 natural gas (~30%)
2. **Eliminate the need for ASU**
   - ✓ Safety concerns
   - ✓ Required space, weight and height
   - ✓ Electricity requirement
3. **Target H_2/CO ratio: approx. 2.0**
4. **Commercialization/Demo scale readiness**
5. **Minimize ship motion effects**
6. **Required space**

- **SCR**
  (Steam CO_2 Reforming)

  $CH_4 + H_2O \rightarrow 3H_2 + CO$
  $CH_4 + CO_2 \rightarrow 2CO + H_2$

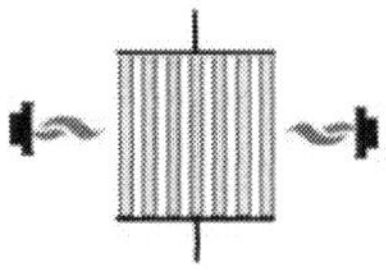

- Conventional SR
- ATR
- POX

3. GTL FPSO FEASIBILITY STUDY (4/5)

- **FT Reactor Technology Selection for GTL FPSO**

Key Parameters & Issues

1. **Space & weight requirement**
   - ✓ SPR is smaller & lighter than MTFB
   - ✓ MTFB is shorter than SPR
2. **Operational Issues**
   - ✓ SPR allows addition/replacement of catalyst without shutdown
   - ✓ SPR has a lower pressure drop
3. **Minimize ship motion effects**
4. **Characteristics of fundamental chemistry and kinetics**

- **SPR**
  (Slurry Phase Reactor)

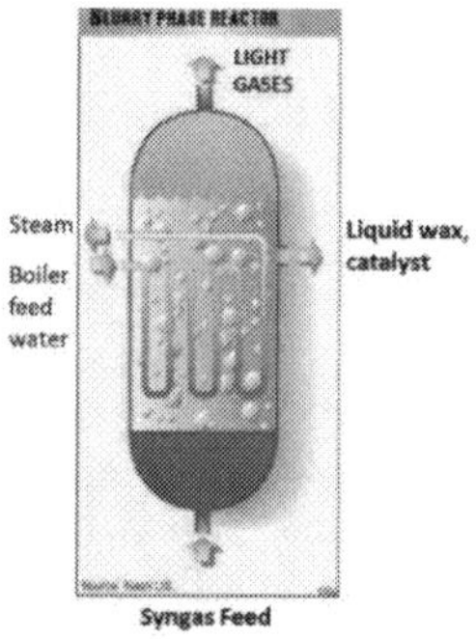

- MTFB
  (Multi-tubular Fixed Bed)

3. GTL FPSO FEASIBILITY STUDY (5/5)

- **Utility Requirements for Self-sufficiency**

1. **Hydrogen Production Unit**
   - ✓ Needed to stabilize the FT products
   - ✓ Amount of hydrogen required will be less than that required for full product upgrading
2. **Power Plant Unit**
   - ✓ Steam from process heat recovery is used in S/T generators to produce power
   - ✓ A GTCC will provide the balance of the power required by the process
3. **Waste Water Treatment Unit**
   - ✓ Remove oxygenated hydrocarbons from the wastewater to meet applicable standards for discharge
4. **Desalination Unit**
5. **Cooling Water Unit**

4. GTL FPSO CONCEPTUAL ENGINEERING

Slide 13/24

1) Overview of GTL FPSO

Conventional Oil & Gas system / Utilities
- Inlet Facility, Power Generation, Utilities
- Flare Tower & Etc.

GTL Process on Topside
- H_2S Removal, Reformer, FT Reactor
- Syngas Conditioning, Upgrading & Etc.

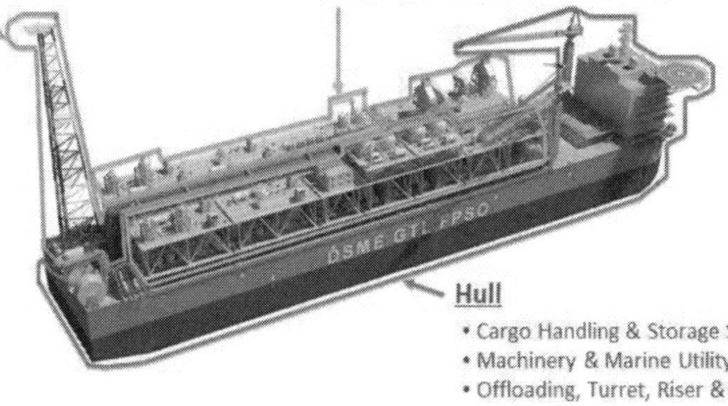

Principal Dimensions

• Length O.A.	320 m
• Length B.P.	320 m
• Breadth MLD	60 m
• Depth MLD	31.5 m
• Draft Design/Scant.	22.5 / 23.5 m
• Cargo Capacity (GTL)	245,000 m^3
• Production (GTL)	20,000 BPD

Hull
- Cargo Handling & Storage System
- Machinery & Marine Utility
- Offloading, Turret, Riser & Etc.

4.1 Topside Conceptual Engineering

Slide 14/24

1) Design Basis

A) General Requirement
a. Life time (20 years) & Availability (approx. 91~92%, 333 days)

B) Target Fields
a. North West Australia / South East Asia

C) Site Specific Information (In-situ Data)
a. Metocean Data (Wind conditions, Wave height & Peak periods, Current spec)
b. Water Depth, Air & Seawater Conditions

D) Feedstock Specifications
a. Feed Gas Compositional data, P & T
b. Contaminants in Feed Gas

E) Product Specifications
a. Target production capacity
b. Product Spec.: GTL Syncrude 20,000 BPD
c. Shuttle Carrier: Conventional Crude Oil Tanker

4.1 Topside Conceptual Engineering

2) Simplified Block Diagram of DSME's GTL FPSO Process

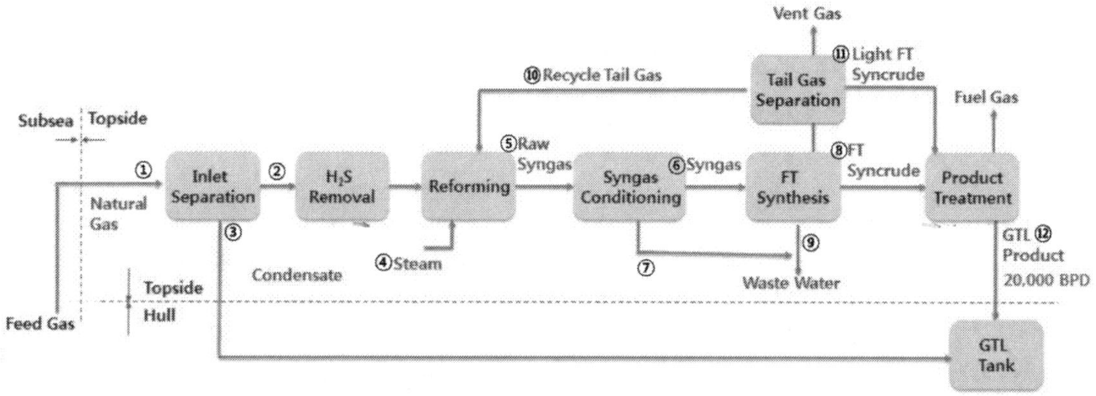

Stream	①	②	③	④	⑤	⑥	⑦	⑧	⑨	⑩	⑪	⑫
Description	Natural Gas Feed	Natural Gas	Condensate	Steam	Raw Syngas	Syngas	Waste Water 1	FT Syncrude	Waste Water 2	Recycle	Light FT Syncrude	GTL Syncrude Product
Standard Volume Flow [MMSCFD]	206	195		928	3,151	970				365		
Total Mass Flow [Kg/hr]	202,155	193,676	1,615	811,815	1,564,350	802,820	645,896	104,803	170,386	494,895	674	103,833
Total Product Flow [BPD]	1,151		345				97,693	20,551	25,770		142	20,000

4.1 Topside Conceptual Engineering

3) Conceptual Topside Plot Plan

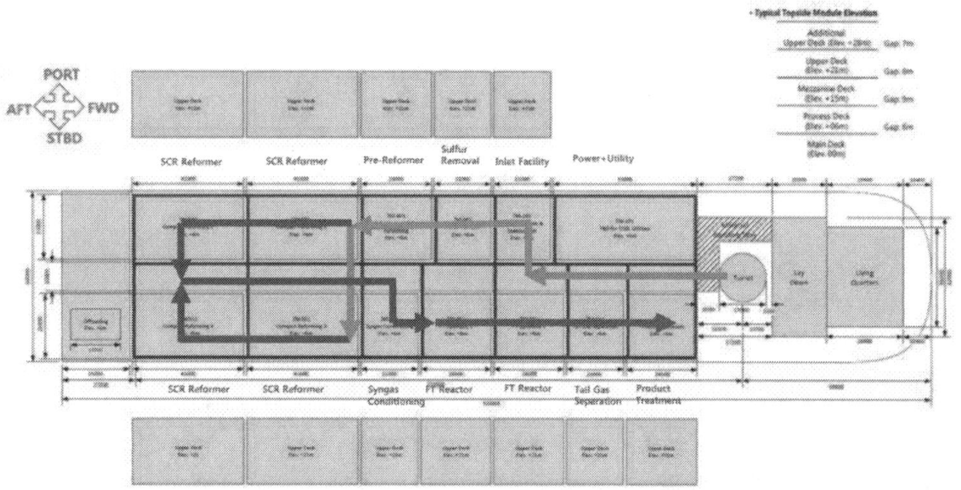

※Process Flow : Natural Gas

Syngas

FT Product

Principal Dimensions

Length	O.A.	320 m
Length	B.P.	320m
Breadth	MLD	60m
Depth	MLD	31.5m
Cargo Capacity	GTL + Crude Oil	1,500,000 BBL
Production	GTL	20,000 BPD

4.1 Topside Conceptual Engineering

4) Preliminary 3D Model

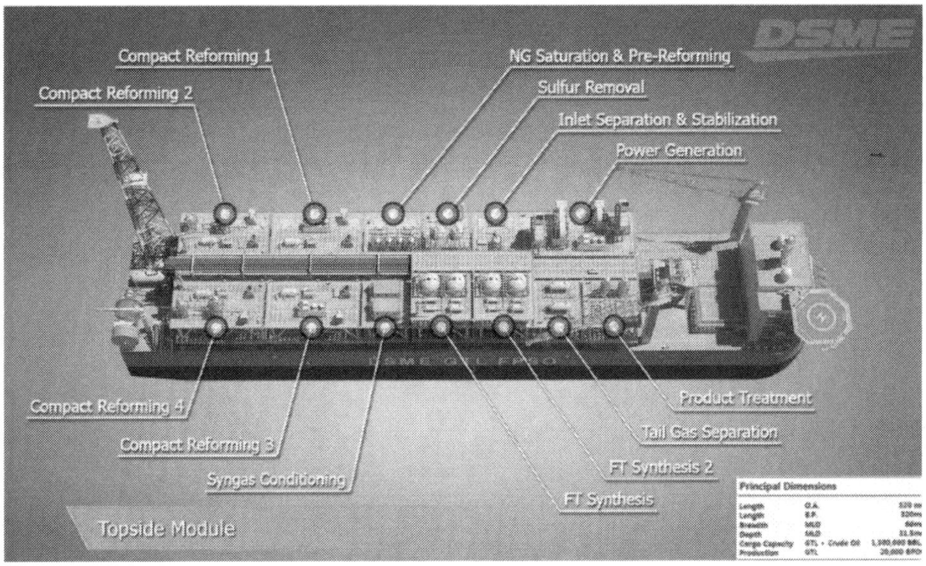

4.2 Hull Conceptual Engineering

1) Preliminary General Arrangement

- Dimension: 320m X 60m X 31.5m
- Storage Capacity: 240,000 m3 (1,500,000 BBL)

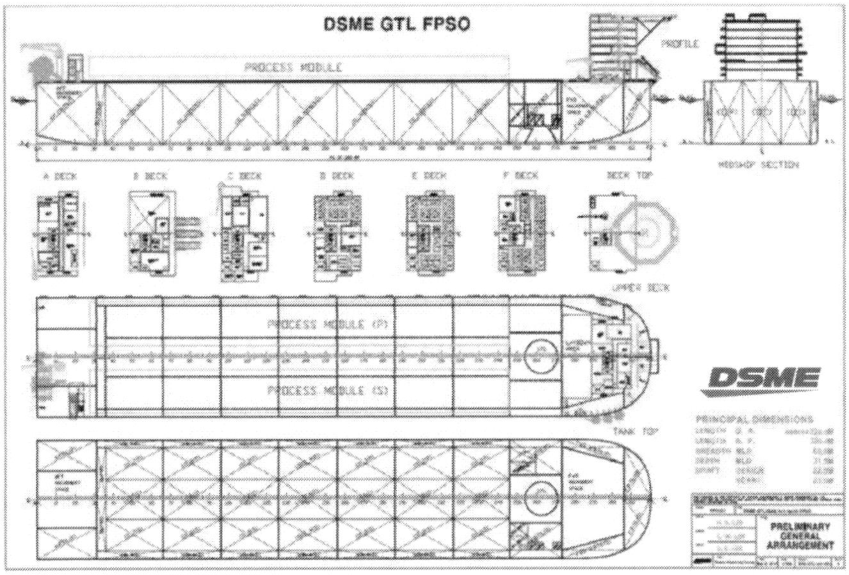

4.2 Hull Conceptual Engineering

2) Mooring system for GTL FPSO

	External Turret Mooring	Internal Turret Mooring
Type	Permanent and Disconnectable Type	
Environments	Mild ~ Moderate	Moderate ~ Harsh
Example	West Africa, South East Asia	North Sea, Northwest Australia
Number of Riser	Small	Large
Water Depth	Shallow ~ Deep Water	
Figure		

- ✓ **Internal Turret Mooring considered as a base case for GTL FPSO**
- ✓ External Turret Mooring considered as an alternative case

4.2 Hull Conceptual Engineering

3) Offloading Concept

- ✓ **Tandem offloading with heating Coil to tankers (70~100℃)**
- ✓ **Offloading Time : ~26 hours (Including ramp up, steady offloading and ramp down)**

[GTL FPSO – Tandem Offloading Concept]

5. Economic Analysis

Slide 21/24

1) Economic Model Workflow

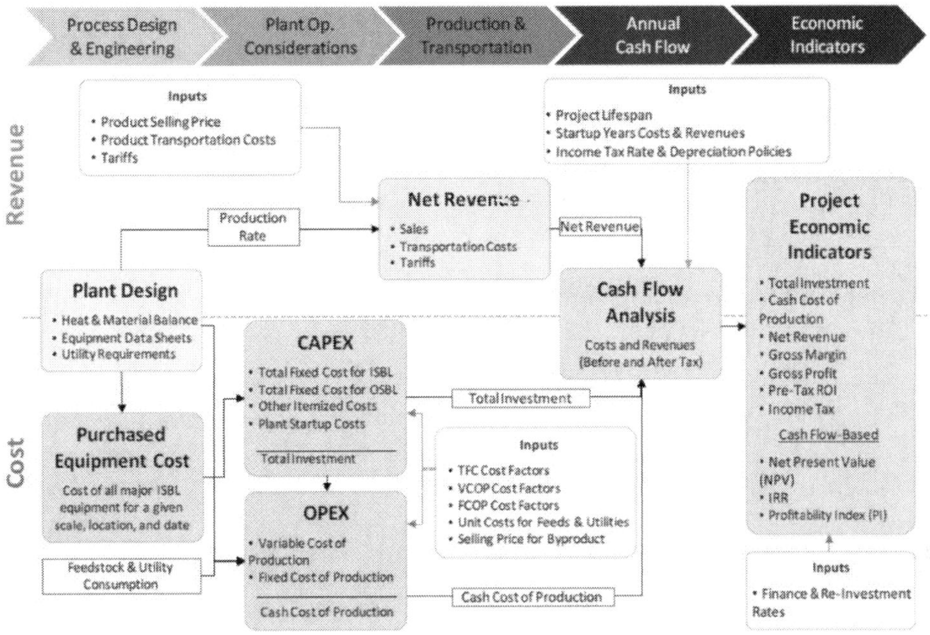

5. Economic Analysis

Slide 22/24

2) Preliminary Economic Analysis

- **Sensitivity analysis of major cost factors**
  - ✓ **Main Variables VS IRR (%)**
    - Crude Oil, Total TIC, Topside TIC, Natural Gas, Shipping Cost
  - ✓ **Variation of factors: +/-50%**
  - ✓ **Most sensitive cost to the Project economics : Crude oil price**
    (If increase of USD 10/bbl in crude price, 2.1% increase in IRR)
  - ✓ **Sensitivity Ranking:**
    Crude Oil > Total TIC > Topside TIC > Natural Gas > Shipping

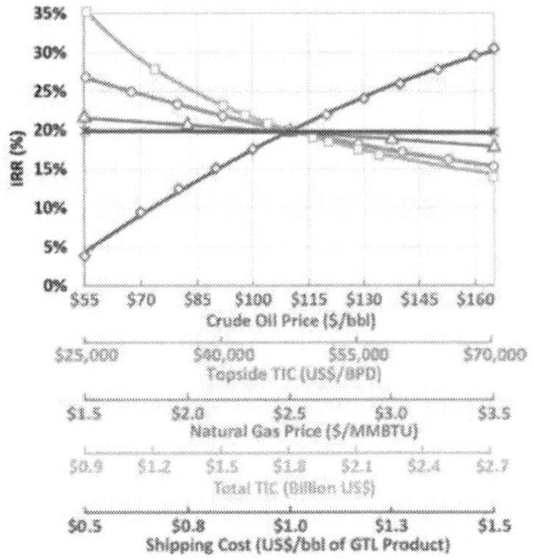

[Sensitivity Analysis on IRR]

Slide 23/24

6. SUMMARY

1. **GTL FPSO offers an attractive solution to monetize remote, offshore and stranded gas fields.**
2. **GTL FPSO was completed Feasibility Study & Conceptual Engineering**
   - 20,000 BPD of GTL syncrude products
   - Principle dimension (320 m(L) X 60 m(B) X 31.5 m(D)) and storage capacity (1,500,000 BBL)
3. **SCR selected for Reforming Technology**
   - Produce required H_2/CO ratio of 2 & Tolerate feed gas with high CO_2 content
   - Elimination of ASU
4. **SPR selected for FT Reactor Technology**
   - Favorable size and weight
   - Easier onboard catalyst handling
5. **DSME performed various activities in conceptual engineering of GTL FPSO**
   - **Topside:** Design basis, Block flow diagram, Process simulation report, Heat & material balance, Topside process simulation, Equipment list, Equipment datasheet, Vendor proposal, Topside layout, and Investment cost
   - **Hull:** General arrangement, Mooring system, Cargo oil system, and Offloading system
6. **Preliminary economic analysis indicates economically competitive IRR**

DSME

DSME

360085 Nitrogen Co-Feeding in Fischer Tropsch Synthesis: The Effect to the Reaction Rate and Product Selectivity at Higher Reaction Rates

Abstract:

Our previous study, "Effects of Nitrogen Co-Feeding in Fischer Tropsch Synthesis At Low CO Conversion", showed that diluting syngas with N_2 has a positive effect on the activity and selectivity of products at low CO conversion. In the current study, we pursue similar kind of experimental work for Fischer Tropsch Synthesis (FTS), with an aim of investigating the effect of nitrogen co-feeding in FTS at high CO conversion.

We started the reaction as per previous experiment at 60 (NTP)ml/min, 20 bar and the temperature of 200 °C (for 48 hours) for syngas FTS. Then, we increased the reaction temperature to 220 °C and started the investigation of the effect of N_2 co-feeding in FTS at this temperature by repeatedly switching between two kinds of feed gases into the reactor: (1) with only syngas at 20 bar, 60 (NTP)ml/min, and (2) with N_2 co-feeding syngas at 25 bar, 75 (NTP)ml/min. In order to keep the "reaction kinetics" the same in all runs, we kept the molar flow rate of syngas as well the partial pressure of the CO and H_2 constant while varying the molar flow rate of N_2 and the reactor pressure. The reactor was run for around 2800 hours at 220 °C while alternating the experimental conditions from syngas to syngas diluted with N_2. The experimental results show that the increase in temperature from 200 °C to 220 °C resulted in the increase of CO conversion from 16% to 46%. However, the catalyst deactivation was observed with time going from a CO conversion of 46% down to a CO conversion of 12%. Thereafter, the CO conversion remained fairly constant approximately 12% for a period of 1800 hours.

Our data demonstrate that N_2 co-feeding in FT reaction decreases the selectivity to the undesired lights hydrocarbons, especially for CH_4. It was also noticed that the rates of light paraffin ($C_2 - C_5$) formation as a function of time on stream (TOS) decreased significantly at high CO conversions and remained fairly constant at CO conversion average 12% for both conditions. However, the rates of formation of light olefins as a function of TOS appear to be quite constant for the high CO conversions and a significant increase in the rates of formation of light olefins ($C_2 - C_5$) as a function of TOS is observed for low CO conversion. Further experimental results, such as product selectivity and paraffin to olefin ratios are calculated and compared.

368342 Optimization of Pyrochlore Catalysts for the Dry Reforming of Methane

Abstract:

The dry reforming of methane (DRM) using CO_2 has long been considered a viable method for converting methane from geological or biological sources into syngas, which can then be readily used in the production of a variety of chemicals and particularly liquid fuels that can more readily be shipped via pipeline.

Though DRM holds great promise, the high temperatures required for the reaction have made it very difficult to find catalysts that exhibit high activity for extended periods. Several factors often lead to the deactivation of these catalysts: the sintering of active metals, the structural rearrangement of the catalyst support causing a reduction in surface area, and the accumulation of carbon on the catalyst surface. To date, many catalyst materials have been investigated for this reaction; for example, unsupported transition metal carbides and sulfides, supported group VIII metals, and more recently perovskites and hydrotalcites have received attention. In this study, however, we have chosen to develop optimized pyrochlore catalyst materials.

Pyrochlores are crystalline oxides having high thermal stability and a general formula of $A_2B_2O_7$, where *A* represents a rare-earth metal and *B* represents a transition metal. Initial experimental efforts by others showed that pyrochlores are active for DRM but the tested catalysts exhibited poor long term stability; however, more recent data suggest that this trend in deactivation may not be applicable to all pyrochlores. $La_2Zr_2O_7$ (LZ) is a pyrochlore structure which has shown good long term stability, so that efforts have been made to tailor its catalytic properties, showing Rh as a promising dopant to enhance catalytic performance for DRM. In order to determine the role of Rh in the reaction performance and understand how the reaction proceeds, we are using first principles methods employing Density Functional Theory (DFT) to analyze structural stability, species adsorption, and calculate transition state energies for the reactions on this catalyst material. To date, DFT simulations are reported for pyrochlore structures but most of them deal with bulk properties of these catalysts. To our knowledge, the work done by Mantz (2011) is the only one that deals with the interactions of a species with pyrochlore surfaces.

DFT simulation results for the 2% Rh doped LZ pyrochlore have shown good agreement with experimental data for CO and CO_2 adsorption. The reactivity at the catalyst surface is studied by calculating the activation energies for all elementary reaction steps. Some activation barriers are calculated using the Climbing Image Nudged Elastic Band (CI-NEB) method (computational approach), while other barriers are found by means of a Brønsted-Evans-Polanyi relationship. These data are employed in the micro-kinetic model for a batch reactor, which outputs the partial pressure profile for the gas phase species and the coverage profile for the adsorbates with respect to residence time. This model provides then a deeper understanding of the DRM reaction mechanism on the catalyst.

References

[1] S. Gaur, D.J. Haynes, J.J. Spivey, Applied Catalysis A: General 403 (2011) 142.

[2] D.J. Haynes, D.A. Berry, D. Shekhawat, J.J. Spivey, Catalysis Today 145 (2009) 121.

[3] D.J. Haynes, D.A. Berry, D. Shekhawat, J.J. Spivey, Catalysis Today 136 (2008) 206.

[4] D.J. Haynes, A. Campos, D.A. Berry, D. Shekhawat, A. Roy, J.J. Spivey, Catalysis Today 155 (2010) 84.

[5] Y.A Mantz, MRS Proceedings 1327 (2011).

381160 Effect of Reducing Agents (Syngas, H2 and CO) on the Catalytic Performance of Iron Catalysts for Fischer-Tropsch Synthesis

Abstract:

Reduction of the iron catalyst precursor is an important stage during Fischer-Tropsch (FT) reaction and reducing gases tend to influence the activity, selectivity and the catalyst life span. Hydrogen is mostly the used reducing gas in FT because of its ability to avoid the buildup of carbon during the FT synthesis (Jia et al. 2007). Reduction with carbon monoxide alone gives very active catalysts but may favor carbon deposition. Economic benefits can potentially be obtained if the same syngas used for the normal FT reaction can also be used to activate the catalyst at the normal FT reaction temperature. The current study focus on the investigation of the effect of reducing agents (syngas, H_2 and CO) on the catalytic performance of a precipitated iron based catalyst for Fischer-Tropsch synthesis (FTS). Three fixed bed reactors (FBR), each loaded with 1 gram of the same iron based catalyst, were set in parallel and simultaneously reduced with syngas (CO/H_2), H_2 and CO, respectively, for 48 hrs at 250 ^{o}C, at atmospheric pressure using a flowrate of 60 (NTP)ml/min. After the reduction, the same FT reaction procedures were carried out among the three reactors for more than 1000 hrs. The experimental data demonstrated that the catalyst activities for FTS followed a decreasing order: syngas > hydrogen > carbon monoxide reduced. More experimental results, such as the FT reaction rates, water-gas-shift (WGS) reaction rates, product selectivity and the paraffin to olefin ratios are calculated and compared. The study provides an indication that the reducing gas used has a long term effect on the catalyst activity and product selectivity.

393526 Direct Synthesis of Dimethyl Ether from CO2 & CO over Bifunctional Methanol Synthesis & Silicotungstic Acid Incorporated Mesoporous Alumina Catalyst Mixtures

Abstract:

Direct synthesis of dimethyl ether was achieved using a bifunctional catalyst pair composed of silicotungstic acid impregnated mesoporous alumina and a copper based methanol synthesis catalyst. Presence of CO_2 in the feed stream significantly enhanced DME selectivity, giving a maximum with the feed composition of $H_2/CO/CO_2$=50/40/10. DME selectivity values approaching to 0.9 were highly promising.

Hydrogen production from stream reforming acetic acid over core-shell-like zeolite capsule NiO/CeO_2-ZrO_2 catalyst

Chang-feng Yan, Qiquan Luo, Rongrong Hu

Guangzhou Institute of Energy Conversion, Chinese Academy of Sciences
No.2 Nengyuanlu Road, Wushan, Tianhe District, Guangzhou 510640
yancf@ms.giec.ac.cn

ABSTRACT

Hydrogen production from catalytic stream reforming of model compound acetic acid of bio-oil was investigated under atmospheric pressure in a bench-scale quartz tube reactor. A series of core-shell-structured catalyst comprising NiO/CeO_2-ZrO_2 as a core and ordered mesoporous silica as a shell were synthesized by means of a surfactant-directed sol-gel process by using cetyltrimethylammonium bromide (CTAB) as a template and tetraethylorthosilicate(TEOS) as silica precursor. The shell thickness of ordered mesoporous silica can be easily controlled by changing the ratio of TEOS to NiO/CeO_2-ZrO_2. X-ray diffraction, SEM-EDS, TEM, BET and TPR were used to analyze the relationship between the core-shell catalyst nature and catalytic performance. The effects of reaction temperature, LHSV, S/C and time-on-stream were studied in detail over core-shell catalyst. The structure of core-shell catalyst was assembled to reduce the concentration of CO and avoid the catalyst deactivation caused by carbon deposition. The results illustrated that the core-shell catalyst for stream reforming of acetic acid, which H_2 selectivity reaches 80% and CO selectivity goes down to 10% at T=650□, LHSV=$4.5h^{-1}$ and S/C=3, was superior to NiO/CeO_2-ZrO_2 catalyst.

Keywords: core-shell catalyst; acetic acid stream reforming; bio-oil; hydrogen production; zeolite.

Understanding the Ostwald ripening dynamics of platinum catalyst with phenomenological modeling and molecular simulation approach

Srinivasan Raman, Jeevan Maddala, and Raghunathan Rengasamy*

Department of Chemical Engineering, Texas Tech University, Lubbock, Texas – 79409
E-mail: raghu.rengasamy@ttu.edu

Abstract:

Fuel cells are expected to serve as a clean and renewable energy source for automotive and stationary power applications. The expected lifetime of fuel cells with consistent performance is about 5000 – 40000 hours for this technology to be commercialized. However, the current scenario is far from satisfactory. Platinum is the most widely used catalyst in fuel cells and the degradation of platinum leads to performance loss in fuel cells, which makes their operation and maintenance expensive. Platinum catalyst in fuel cells degrades over time by various mechanisms such as aggregation, dissolution and Ostwald ripening. The effects of these degradation mechanisms on the loss of electrochemical active surface area (ECSA) needs to be understood at the molecular level in order to create catalysts that can resist degradation and result in fuel cells with longer operational lifetime. Several models and simulations exist in literature which helps in predicting the degradation phenomenon separately, but simulations that help in molecular level understanding and encompassing longer time scales are limited.

In this work, we simulate the degradation of Pt in a 2-dimensional lattice structure constructed based on the catalyst particle information such as particle size and particle density. The lattice was taken to be filled with carbon and the Pt particles studded at each lattice positions. The surface diffusivity of Pt in pristine graphite was taken from molecular simulation literature. The carbon layers were assumed to resemble graphite sheets and the surface diffusivity value was used in the simulations. The Ostwald ripening dynamics was numerically simulated by considering the Pt ion movement in the lattice. The effect of application of voltage cycling patterns is considered and the particle size growth with time is discussed. The possibilities for extending the method towards studying the aggregation phenomena and diffusion in proton exchange membrane fuel cells will be highlighted.

375690 Investigation of Arsenic Poisoning on Novel SCR Catalysts

Abstract:

The reduction of nitrogen oxides (NO_X) using selective catalytic reduction (SCR) through the injection of ammonia is a common practice in coal-fired power plants today. However to meet increasing emission standards, numerous novel SCR catalysts have been developed to make the catalysts more reactive over a wider range of operating conditions, such as temperature. Modern mixed-oxide catalysts used to carry out this reaction are susceptible to poisoning and deactivation over time due to the numerous species present in the flue gas such as arsenic. To increase the catalysts operating lifetime under exposure to such species, sorbents, such as CaO, can be added into the flue gas. Due to these limitations of current SCR catalysts, novel SCR catalysts are currently being investigated. While the deactivation mechanism has been studied on traditional V-Mo-Ti oxide catalysts to an extent, it is unknown what effect the arsenic species might have on novel SCR catalysts.

In this study, small pore zeolites, SSZ-13 and SAPO-34, both with and without metal ion exchange, are evaluated for their resistance to arsenic as they have shown promise as SCR catalysts in other deNO_X applications. They are compared to a commercially available (V-W-Ti oxide) catalyst, as well several large pore zeolites such as ZSM-5 and zeolite-Y. The catalysts are exposed to a simulated flue gas stream that is created by combusting methane, creating an environment representative of the one present in a coal-fired boiler. Prior to combustion, the stream is premixed with arsine (AsH_3), which is converted to arsenic oxide (As_2O_3) as it passes through the oxygen-rich flame. The catalysts are characterized before and after exposure to arsenic using XPS, XRD, SEM/EDX and NH_3-TPD, to determine the specific mechanism of arsenic adsorption on the surface of the catalysts. Finally, the gas stream composition is varied to include SO_2, NO, NO_2, and HCl to study the effect that other flue gas components play in arsenic adsorption.

389699 Interpretation of Electrochemical Impedance Spectroscopy Using a Macrohomogeneous PEMFC Model

Abstract:

We have produced a transient, two-phase, non-isothermal model of the proton exchange membrane fuel cell (PEMFC), which can be used to directly simulate electrochemical impedance spectroscopy (EIS). For interpretation of EIS experiments, physics-based models offer several advantages over equivalent electrical circuit models. Equivalent circuit models require prior assumptions about the dominant processes contributing to impedance, and the relationship between physical and circuit parameters can be difficult to derive. In the physics-based model, all processes contribute to the impedance response, but the dominate process is easily determined. As an example, the PEMFC model reproduces an experimentally observed inductive loop at low frequencies. From the physics-based model, the inductive loop can be assigned to the modulation of membrane conductivity by product water generation. The EIS simulation technique can be applied to any transient electrochemical model without explicitly deriving the small signal response and without the assumption of linearity.

377220 Kinetics of Hydroesterification of Ethylene Using Homogeneous Palladium Complex Catalyst

Abstract:

The hydroesterification of olefins with CO and alcohols is an emerging technology for producing valuable carboxylic acid esters, and it is attracting continuous and increasing academic and industrial interest. Studies on the kinetics of hydroformylation of olefins are not only instructive for improvement of the catalytic system but also provide relevant information for further engineering study.

In this research, the kinetics of hydroesterification of ethylene to methyl propionate has been investigated using a homogeneous $Pd(OAc)_2/PPh_3$/p-toluenesulfonic acid (TsOH) catalyst system. The hydroesterification reaction was carried out in a flat-bottomed, magnetically stirred 150 mL steel autoclave equipped with a 120 mL teflon tube. The effect of concentration of catalyst, the partial pressure of CO and ethylene on the activity and selectivity has been studied in a temperature range of 80-100 °C. Under the conditions of Pd concentration of 4mmol/L, $Pd(OAc)_2$:PPh_3:TsOH (mole ratio)=1:9:10, C_2H_4: CO=1:1, 2 MPa and 100 °C, the selectivity for methyl propionate was greater than 92%. Under these typical hydroesterification conditions, the kinetic order of the mixed gas of ethylene and CO was found to be zero order, and the activation energies was 52.87 kJ/mol in the range of 80-100 °C. The selectivity for methyl propionate was low when the concentration of ethylene was high in the mixed gas, so the kinetic orders below were evaluated under the low ethylene concentration. When the molar ratio of C_2H_4: CO was 1:10, the kinetic order of C_2H_4 was 1.23, so the kinetic order of CO was -1.23, and when the concentration of $C_2H_4$and CO is constant, the kinetic order of methanol was 1.36. On the basis of the initial rate data obtained, a rate equation was proposed. Also, a plausible reaction mechanism for the carboxylation of styrene has been proposed.

372146 Chicken Manure As Heterogeneous CaO Catalysts for Biodiesel Production from Transesterification of Waste Cooking Oil

Abstract:

Due to diminishing of fossil fuel and environmental problems, biodiesel has become one of the environmentally benign and renewable energy fuels due to its raw material being in abundance, its potential for large scale production, and its better physical and chemical properties as compared to the petroleum-based diesel e.g. better lubricating efficiency, lower sulphur contamination, and very few pollutants produced. Calcium oxide (CaO) has been intensively used as a heterogeneous base catalyst for biodiesel production via transesterification due to its high basicity and strong basic strength, and non-toxicity. Recently, the discovery of inexpensive catalysts with high catalytic activity from the abandoned waste materials (e.g. eggshell, animal bone, crab and oyster shells, fly ash and etc.) makes the biodiesel production process highly capable of competing on a cost and quality basis with the diesel fuel produced from petroleum.

In this work, new application of chicken manure, which is usually as a fertilizer, has been explored. Interestingly, chicken manure contains large amount of calcium compounds (over 60 wt. %) which can be used as a new source of CaO catalyst. Calcination process was used to convert the calcium compounds in chicken manure into the active CaO catalyst. Hammett indicator test showed that the basic strength of the CaO catalysts obtained after calcination of chicken manure lie in the range of $15 < H\_ < 18.4$, indicating that the surface basicity of these catalysts could be attributed to the strong basicity of the metal–O groups (Lewis base) which is considered as a strong base for transesterification reaction. Waste cooking oil was then used as a feedstock for biodiesel production by using the CaO catalyst derived from chicken manure. Transesterification was carried out and the oil conversion and FAME (or biodiesel) yield were optimized by varying the molar ratio of alcohol to oil , catalyst concentration, temperature, and time. The chicken manure-derived CaO catalyst exhibits high catalytic performance for the transesterification of waste cooking oil with over 85% of FAME (or biodiesel) yield. The utilization of chicken manure as a source of catalyst not only addresses environmental problem of waste, but also economic aspects of biodiesel production by using waste as catalyst and feedstock.

Effects of Support Properties and SiO_2 stabilizer on Catalyst Performance for Supported-Iron Fischer-Tropsch Catalysts

Kamyar Keyvanloo, William C. Hecker, Calvin H. Bartholomew*

Department of Chemical Engineering, Brigham Young University, Provo, UT 84602

** hecker@byu.edu*

The effects of support properties including pore size (d_{pore} of 9-45 nm), hydroxyl group concentration, and support stabilizer were investigated for six alumina supported Fe/Cu/K Fischer-Tropsch catalysts containing 20 or 40% iron. Catalysts were supported on one of four different aluminas stabilized with La_2O_3 or 5% SiO_2. Large pore supports having a bimodal pore size distribution were found to easily accommodate 40% Fe without pore blockage. Catalyst activity increases with increasing support pore size and pore volume. Also positive correlations were found between lower OH group concentration, higher extent of reduction, and higher activity. SiO_2 was found to be a more effective hydrothermal stabilizer than La_2O_3 as it suppresses the high-temperature transformation of γ-Al_2O_3 to α-Al_2O_3 phase up to 1200 °C allowing SiO_2 stabilized catalysts to be dehydroxylated at higher temperatures. The addition of 5% SiO_2 to Al_2O_3 also inhibits deactivation by sintering and increases activity, possibly due to the presence of surface silicate or aluminosilicate groups, which apparently serve to anchor Fe carbide crystallites. Indeed, the activity of Fe/Si-Al_2O_3 continues to increase after 700 h on stream, while the Fe/La-Al_2O_3 catalysts lose activity over the same time period.

373496 Oleic Acid Decarboxylation and Further Transformation over Pt/SAPO-11 and Pt/Al2O3

Abstract:

Catalytic decarboxylation and further conversion of oleic acid to branched and aromatic hydrocarbons (gasoline/diesel-range hydrocarbons) in a single process step, over Pt–SAPO-11 and Pt/ Al_2O_3 is presented. The degree of decarboxylation and selectivity to the primary intermediate, the linear heptadecane are affected by reaction temperature, time and environment (H_2orCO_2). Higher temperatures, longer reaction times, and the presence of hydrogen increased the degree of decarboxylation (as highas 98%) and selectivity to heptadecane (up to 63%). In the presence of acidic supports, like SAPO-11 or chlorided alumina, the initially formed paraffin intermediate undergoes further conversion to branched, cyclic and alkyl aromatic hydrocarbons. For example,dodecylbenzene, of high interest in the petrochemical industry as a raw material for detergents, could be obtained in high yields.Under appropriate reaction conditions, significant quantities of hydrocarbons boiling in the gasoline range could also beformed by hydrocracking reactions over these bifunctional catalysts.

Conversion of Glycerol in a Zeolitic Microreactor

Siu Ming KWAN[1], Yakub FAM[1], King Lun YEUNG[1,2,*], Vanesa CALVINO-CASILDA[3], Fernando RUBIO-MARCOS[4], José F. FERNANDEZ[4], Miguel A. BANARES[3]

[1]*Department of Chemical and Biomolecular Engineering,* [2]*Division of Environment, The Hong Kong University of Science and Technology, Clear Water Bay, Kowloon, Hong Kong, P.R. China.*

[3]*Catalytic Spectroscopy Laboratory, Instituto de Catálisis y Petroleoquímica, CSIC, Campus Excelencia, CSIC-UAM, Madrid, Spain*

[4]*Electroceramic Department, Instituto de Cerámica y Vidrio, Campus Excelencia, CSIC-UAM, Madrid, Spain*

[*]*Corresponding author: kekyeung@ust.hk*

Session: 20 CATALYSIS AND REACTION ENGINEERING DIVISION
Topic: 20034 Poster Session: Catalysis and Reaction Engineering (CRE) Division
Presentation type: Poster

Abstract

Very fast and highly exothermic reactions including direct fluorination, high temperature combustions and selective oxidations are the typical reactions investigated in microreactors. They take full advantage of the rapid heat and mass transfers in the microreactor to achieve an improved reaction performance (i.e., better conversion, selectivity, yield or safety). It is however agreed that the greatest need and potentially the biggest impact of the microreactor would be in synthesis of high value products including fine chemicals, pharmaceuticals and nanomaterials. Many fine chemical reactions of interest are constrained by unfavorable thermodynamics that could benefit from the membrane reactor operation. Selective product removal could improve the product purity and achieve supra-equilibrium conversions, whereas the selective addition of a reactant could eliminate hot spots and enhance the product selectivity.

Glycerol has been oversupplied in the global market recently because of the huge increase in biodiesel manufacturing. This means a dramatic rise in the availability of glycerol would be happened. Important research is currently being developed in order to find new applications for glycerol from biodiesel plants. By developing heterogeneous catalysts that are particularly efficient for cabonylation of glycerol with urea, one can obtain good yield under more moderate reaction conditions. Banares et al. recently reported a novel hierarchical Co_3O_4/ZnO catalyst prepared by dry nanodispersion for the carbonylation of glycerol. Besides, special interest is also focused on the preparation of acetals and ketals of glycerol by acetalisation with an aldehyde or a ketone, respectively, in the presence of an acid catalyst. In this work, the performance of the membrane-catalyst microreactor that utilizes a zeolite ZSM-5 membrane and catalyst for both the carbonylation of glycerol with urea and the acetalisation of glycerol with acetone is reported.

Preparation, Characterization and Catalytic Performance of Spatially Doped Titania Nanotubes

Shammi Akter FERDOUSI[1], Yakub FAM[1], King Lun YEUNG[1,2,*]

[1]Department of Chemical and Biomolecular Engineering, [2]Division of Environment,
The Hong Kong University of Science and Technology, Clear Water Bay, Kowloon, Hong Kong
Corresponding author: kekyeung@ust.hk

Session: 20 CATALYSIS AND REACTION ENGINEERING DIVISION
Topic: 20029 Fundamentals of Environmental Catalysis

Abstract

One dimensional (1D) nanostructures including titanium nanotubes can exhibit unique physical, chemical, electronic and catalytic properties that fine applications in energy, catalysis, electronics and medicine. The transformation between 2D titanium precursor sheets into 1D nanotubes have been subject of many study. This work postulates that the 1D nanotubes are formed through rolling of the nanosheet driven by hydrogen-bond formation. This mechanism suggests different route for incorporating metals, metal oxides, organic and inorganic compounds at spatially discrete location within the titania nanotubes. This work is first in a series of materials created through spatial doping in the core and interlayer spaces and surfaces of the titania nanotubes.

Here a new synthesis method is developed to prepare the titania nanotubes with doped metal and metal oxide on different spatial locations (i.e., core, surface and interlayer spaces). The precursors were introduced by either *in-situ* or *ex-situ* techniques into the titania precursor materials. Dopants from the Group IB, IIB, IVB, V and VIII elements including Ag, Au, Co, Cu, Pd, Pt and V and their oxides were successfully doped on the surface, in the core and within the interlayer spaces of the titania nanotubes. The location of the dopants was manipulated by controlling the affinity of the dopant precursor with the titania surface during the preparation. Precursor with low affinity will preferentially decorate the nanotube surface, whereas that with strong affinity will lead to metal being embedded between the interlayer spaces of the nanotube. Intermediate affinity allows core-filling to produce metal particles, slugs or nanowires within the core of the nanotubes. Microscopy and spectroscopy studies were carried out to investigate the structural, chemical and electronic properties of these new nanotube materials, and propose the formation mechanism of new materials. These new doped titania nanotubes also exhibit much better catalytic performances on photocatalytic degradation of organic pollutants in water under visible light than commercial P25 powder.

Spatially Ceria-doped Titania Nanotubes for Photocatalytic Conversion of Carbon dioxide to Hydrocarbon Fuel

Yakub FAM[1], Shammi Akter FERDOUSI[1], King Lun YEUNG[1, 2,*]

[1]Department of Chemical and Biomolecular Engineering, [2]Division of Environment,
The Hong Kong University of Science and Technology, Clear Water Bay, Kowloon, Hong Kong
Corresponding author: kekyeung@ust.hk

Session: 20 CATALYSIS AND REACTION ENGINEERING DIVISION
Topic: 20015 Catalysis for CO_2 Conversion

A photocatalytic reduction route of carbon dioxide into hydrocarbon fuels (methane and higher alkanes) and other valuable chemicals (methanol, formic acid) has been taken into consideration in solving the current issues of fossil fuel depletion as well as global warming. Titanium dioxide (TiO_2) is one of the most investigated photocatalysts having been utilized in this so-called artificial photosynthesis process. Despite its inexpensiveness, non-toxicity, abundance, and resistivity to photo-corrosion, TiO_2 is modified into titania nanotube (TiNT) to possess more advantageous characteristics such as large specific surface area, reduced grain boundaries and facile charge transport path. However, TiNT is still in need of other materials as dopants to extend its photo-chemical activity into visible light range. Ceria has been employed in a wide range of applications due to its larger solar light harvesting and redox properties.

Though there are few literatures involving TiNT-based and ceria-based materials in the application of artificial photosynthesis, they have shown to be potential photocatalysts. In this study, ceria nanoparticles were spatially doped into external surface, interlayer and core of TiNT. The type and pH value of cerium precursor solution play an important role in the doping process. As pH increases, the doping of ceria also increases, but the high pH (pH≥9) was proved to bring a negative effect on the doping process due to the instability of cerium precursor. Besides, the chemical state of ceria doped in/onto TiNT was found out to be in the form of CeO_2 on the surface and CeO_{2-x} in the core/interlayers. Furthermore, such doping results in a significant red-shift of light absorption of TiNT from UV to visible range, which is beneficial for its application in artificial photosynthesis.

377841 Non-Oxide Co ALD Catalyst for Fischer-Tropsch Synthesis

Abstract:

Successful Fischer-Tropsch synthesis (FTS) catalytic activity was achieved for metallic Co Atomic layer deposition (ALD) catalyst, producing liquid fuels. Product selectivity, conversion and productivity have been demonstrated for low weight loading Co catalysts having 5-11 wt% Co. The gas phase deposition technology of Atomic layer deposition (ALD) is a thin film technology analogous to chemical vapor deposition. Metallic Co ALD was utilized for a new type of highly-productive catalyst and reactor for valuable fuel products. Avoiding reduction of Co_3O_4 to metallic Co eliminates processing temperatures above reaction temperature; this reduces the likelihood of sintering Co catalyst particles, and lowering productivity.

378157 Rapid Screening of Heterogeneous Catalysts for Biodiesel Production Via Triglyceride Transesterification

Abstract:

Triglyceride transesterification is an important reaction in productions of fatty acid alkyl esters (biodiesel) used as an alternative to petro-diesel fuel. Conventional biodiesel process occurs mainly by homogenous catalysis which adds to the cost of the operation as a result of extra costs of downstream purifications. Therefore, heterogeneously catalysed biodiesel process has the potential to reduce the cost of production. In this study, heterogeneous catalysts were screened both in batch and continuous mode for their activity, reusability and water tolerance in catalysis of triglycerides (rapeseed oil and triacetin) transesterification with methanol. The catalysts investigated were a strontium zirconate-based solid and amberlyst resins functionalised with sulphonic acid (Amberlyst^{TM} 70) and quaternary ammonium hydroxide (Amberlyst^{TM}A26 OH) groups. These catalysts were screened either suspended or in packed bed using a mesoscale oscillatory baffled reactor (integral baffles) as a multiphase mixing reactor. The mesoscale oscillatory baffled reactor also allowed for the investigations of the process parameters in batch and continuous mode using small quantity of reagents which reduces the amount of waste generated. Effects of operating conditions, such as catalyst types and loadings, feed molar ratios (6:1 – 30:1), water content in the range of 0 – 2.5% (v/v), mixing intensity and temperatures (40 – 60°C) were investigated.

The results clearly indicated that methyl esters yield from the rapeseed oil and triacetin transesterification increased with temperature, catalyst loading and methanol molar ratio, e.g. after 2h reaction time with 3wt% strontium-zirconate-based catalyst, the fatty acid methyl esters (FAME) yield increased from 86.3% at 40°C to 91.7% at 60°C at a 12:1 methanol to rapeseed oil molar ratio, and from 85.6% at 6:1 molar ratio to 96.8% at 24:1 molar ratio. However, the FAME yield decreased significantly with water content: it was 92% at 0% (v/v) water but only 12% at 2% (v/v) water after 2h reaction time. Substantial loss in activity of the strontium zirconate-based catalyst was observed as it leached into the methanol phase, for instance the FAME yield after 2h reaction time decreased from 96.8% for fresh catalyst to 0.3% after 2 cycles at 60°C and 24:1 molar ratio. The Amberlyst^{TM} 70 and Amberlyst^{TM} A26 OH resin catalysts were active in catalysing rapeseed oil and triacetin transesterification with methanol. But lower conversions were obtained for the rapeseed oil, e.g. the FAME yields after 60min reaction time at 60°C were < 1% for Amberlyst^{TM} 70 and ~7% for Amberlyst^{TM} A26 OH compared to ~99% for the triacetin reaction. The amberlyst resin catalysts were further screened continuously in dynamic and multi-steady states using methanolysis of triacetin. There was consistency in the triacetin conversions at various residence times for the dynamic and multi-steady state screening, indicating that there was no catalyst leaching or deactivation. This was a demonstration of the catalytic stability and reusability of the Amberlyst^{TM} 70 and Amberlyst^{TM} A26 OH resins for triacetin transesterification. The Amberlyst^{TM} A26 OH resin had substantially higher catalytic activity for triacetin transesterification, even at lower methanol molar ratio (6:1), as >98% triacetin conversion was achieved after 20min reaction time and 60°C compared to 83% for the Amberlyst^{TM} 70 at 30:1 methanol molar ratio. Presence of water in the methanol feed (2.5% (v/v)) resulted in lower triacetin conversion for the Amberlyst^{TM} 70, but the catalyst recovered its activity after the feed line was switched to anhydrous methanol. However, this was not the case for the Amberlyst^{TM} A26 OH which lost its activity when methanol that contained water was used. Finally, this study shows that heterogeneously catalysed biodiesel processes based on the Amberlyst^{TM} 70 and Amberlyst^{TM}A26 OH resins could be developed. The dynamic and multi-steady states screening techniques allows for rapid means of assessing various solid catalysts.

Hydrothermal Production of Aviation Fuels from Microalgae Lipids

Jie Fu[a], Cuiyue Yang[a], Jianghua Wu[a], , Zhaoyin Hou[b]*, Xiuyang Lu[a]*

[a] Key Laboratory of Biomass Chemical Engineering of Ministry of Education, Department of Chemical and Biological Engineering, Zhejiang University, Hangzhou 310027, China

[b] Key Laboratory of Applied Chemistry of Zhejiang Province, Department of Chemistry, Zhejiang University, Hangzhou 310028, China

Abstract: In this contribution, we confirmed that aviation fuels could be synthesized directly from microalgae lipids in water over Pt/C catalyst without additional hydrogen. After decarboxylated at 330 and 370 °C for 120 min, the oxygen content in microalgae lipids was significantly reduced and the heating value of produced aviation fuels was increased obviously. The reaction mechanism of direct decarboxylation of microalgae lipids aviation fuels was further investigated using each of those representative compounds in microalgae lipids, such as methyl laurate, methyl eicosanoate, methyl stearate, ethyl stearate, and tristearin as a single feed under the same conditions. Those reaction conditions, such as solvent, water loading, catalyst loading and reactant loading were optimized. It was concluded that water is more favorable for the selective decarboxylation of methyl stearate among those tested solvents, the catalytic decarboxylation rate of fatty acid ester with larger carbon number in water was faster than those with smaller carbon number, and Pt/C catalyst kept its activity at the third recycle. These results provide a new insight for the direct decarboxylation of the microalgae lipids to aviation fuels.

Keywords: aviation fuels, microalgae lipids, fatty acid esters, decarboxylation, Pt/C

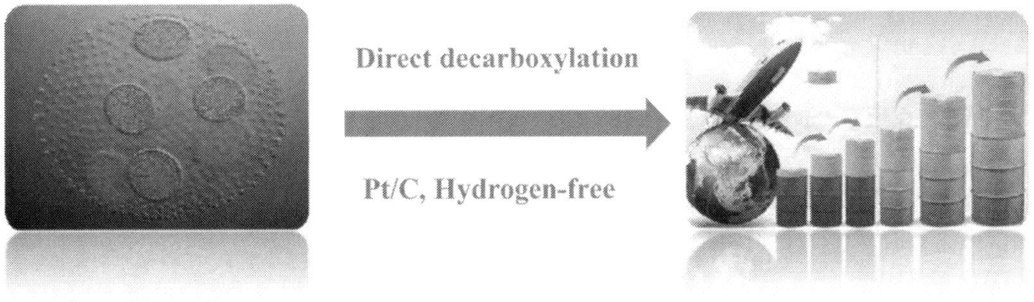

377921 Decarboxylation and Further Isomerization of Oleic Acid over Pt-Supported on Small Pore Zeolites and Hydrotalcite Catalysts

Abstract:

The catalytic decarboxylation and further conversions of oleic acid to paraffins, branched and aromatic hydrocarbons over Pt supported on small pore zeolites and hydrotalcite is demonstrated. The influence of support, platinum source, and reaction temperature on the decarboxylation of oleic acid were investigated. An increase of reaction temperature increased the degree of decarboxylation and selectivity to heptadecane. Pt-SAPO-34 was a very effective catalyst. In addition to a high degree of decarboxylation, Pt-SAPO-34 displayed a high selectivity to heptadecane and dodecylbenzene among the products. Branched isomers, cracked (mostly < C17) paraffins, alkenes such undecene, dodecene and carboxylic acids such nonanoic acid, decanoic acid were observed as side products. The further isomerization of the initially formed linear heptadecane (by decarboxylation of oleic acid) to branched isomers is suppressed in the narrow pores of SAPO-34 due to restricted transition state shape selectivity limitations in the pore system of SAPO-34. Catalyst acidity, dispersion of Pt and pore diameter of the support played crucial roles in determining product selectivity.

CO_2 Hydrogenation Activity and Selectivity Dependency on Crystal Facets of Cobalt Oxides

Cun Wen, University of South Carolina, Columbia, SC
Jason Hattrick-Simpers, University of South Carolina, Columbia, SC
Jochen Lauterbach, University of South Carolina, Columbia, SC

While the previous studies on CO_2 hydrogenation are focused on metallic catalysts, such as Cu, Pt, Ru, and Rh, recent reports showed that metal oxides can be more reactive than their metallic counterparts.[1] For Cobalt supported on TiO_2, the TOF is about five times higher than that of cobalt metal. Besides the higher activity, the cobalt oxides also show higher selectivity (45-60%) to methane than their metallic counterpart (15-35%).[2] We will demonstrate the importance of crystal facets for CO_2 hydrogenation catalyst design. In this study, the catalysts for CO_2 hydrogenation are based on series of cobalt oxide catalysts that we have recently published for CO hydrogenation.[3] The cobalt oxide catalysts have different morphologies and crystal facets on the surface, as shown in figure 1. The Co nanorods predominantly have {110} and {001} facets, while the Co nanoparticles have {111} and {001} facets. The {110} facets have unique atomic structure as the only facets of Co_3O_4 that contains Co^{3+}. The difference in crystal facets have led to different performance during CO hydrogenation, which in many studies share the similar mechanism with CO_2 hydrogenation. The catalytic activity and selectivity to CH_4 will be measured and compared on these two cobalt oxide catalyst. Furthermore, the detailed relationship between faceting and catalytic performance will be revealed with spectroscopic studies, including in situ DRIFTs, Raman, XPS, and EXASFs.

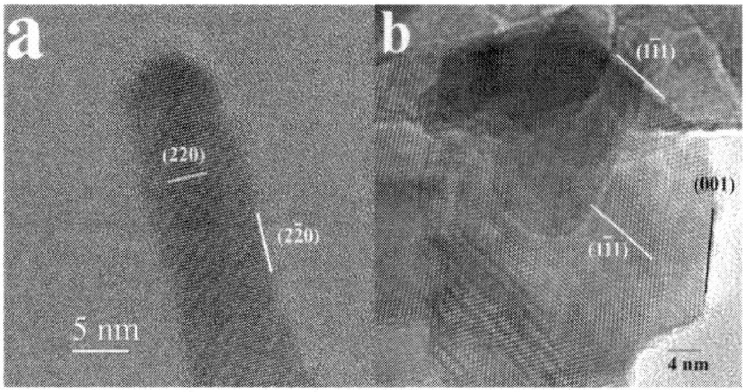

Fig. 1 HRTEM images of cobalt oxide a) nanorods and b) nanoparticles.

1. Wang, W., et al. (2011). Chemical Society Reviews 40(7): 3703-3727.
2. Melaet, G., et al. (2014). Journal of the American Chemical Society 136(6): 2260-2263.
3. Wen, C., et al. (2014). Chemical Communications 50: 4575-4578.

Microwave Thermal Decomposition of Waste Polystyrene in Motor Oil

Tae Won Kang
Department of Chemical Engineering, Kongju National Univ., Kongju, Korea

Abstract

A novel microwave-induced pyrolysis of waste polystyrene in motor oil was performed in quartz tube reactor. Different pyrolysis conditions were investigated, such as time range from 30 minutes to 1 hour and effective microwave power range from 180 to 250 watt. The distillate components were analyzed with GC-MS, and styrene, 1-methyl styrene, toluene, ethyl benzene were the four main products. Among these distillate products, styrene took over 70 percentages. Temperature of pyrolysis using microwave was much lower than that of conventional thermal pyrolysis method.

Key words: pyrolysis, microwave-induced pyrolysis, thermal decomposition, polystyrene

1. Introduction

With the consumption of plastics in the whole world, waste plastics have become an environmental problem because of recycling limitations and their resistance to natural decomposition. A great deal of research has been done over the last two decades into recycling methods to deal with this amount of waste in an economical and ecological way.

The four main routes for plastic waste disposal are: landfill, mechanical recycling, incineration and feedstock recycling processes. The latter are also known as gasification, pyrolysis and thermolysis processes. Pyrolysis, which could recover not just the energetic value but also the chemical value contained within the waste plastics, is a method where heat is applied to waste plastics in an inert environment to break the polymers into smaller molecules. The smaller molecules may then be blended into refinery streams for the production of fuels. In addition, these smaller molecules could be separated and refined into more specialized chemical feedstock [1].

On the pyrolysis of plastics, there were some related literature works. Guffey et al. (1991)[2] used polystyrene and other polymers to model the pyrolysis of tar sand bitumen in the presence of a heavy oil. Their results showed that polystyrene in the presence of a heavy oil would decompose at temperatures near 400℃. Inomata et al. (1974)[3] investigated the pyrolysis of polystyrene and heavy oil in a screw type reactor. The temperature used in their research was 470-600℃. John A. Marsh et al.(1993)[4] investigated the thermal degradation (pyrolysis)of waste polystyrene in heavy oil as the recycling method. the pyrolysis temperature was around 400℃ and distillate products from the reaction were characterized by gas chromatography/mass spectroscopy. Marsh's research provided favorable results compared with the other researchers' for high recovery of polystyrene could be obtained and styrene as the main pyrolysis product could amount to 25 weight percent. Anderson and Berger(1993) investigated the coprocessing of waste plastics and rubber in the presence of heavy oils and bitumens and in their studies, temperatures were between 400 to 450℃.

Microwave, as a kind of electromagnetic wave in the wavelength range of one millimeter to one meter, since the mid-1980s, has been used for chemical applications. The main mechanism for microwave heating is dipolar polarisation where molecules already permanently polarised due to their chemical bonds are realigned in the fluctuating field. This realignment occurs millions of times each second, causing a heating effect within the whole volume of the material. However, Even though microwave heating has many advantages over conventional heating including more even distribution of heat and better control over the heating process, their full potential and growth have not been realized due to a lack of awareness of the resulting benefits and cost savings.

Carlos Ludlow and Howard A. Chase [1] evaluated a novel microwave-induced pyrolysis process by studying the degradation of high-density polyethylene and aluminum/polymer laminates in a semi-batch bench-scale apparatus. Microwave induced pyrolysis is a new process developed initially by Tech-En Ltd. (Hainault, U.K.). The process involves mixing plastics, which are known to have a very high transparency to microwaves, with a highly microwave absorbent material, namely, carbon. When carbon is exposed to microwave field, it can reach temperatures up to 1000℃ in a few minutes. If shredded plastics are mixed with the carbon, prior to or during heating, the energy absorbed from the microwaves is transferred to the plastics by conduction, providing a very efficient energy transfer and a highly reducing chemical environment. The latter avoids formation of undesired hydrocarbon products containing oxygen. Microwave can penetrate deep inside of the molecules and break their bonding structures. Application of this characteristic will lead to high recovery of chemical feedstock from polymeric materials as plastics, and also reduce the thermal energy needed for pyrolysis.

Proposed in this study is a new approach, which involves polystyrene, motor oil, and microwave energy. When the polystyrene dissolved in the oil, the molecules would have greater freedom of motion leading to easier decomposition. Also the motor oil would serve to reduce coke and gas production, increase distillate yield, and decrease process energy requirements.

This study presented a lab-scale microwave-induced pyrolysis apparatus. The distillates were detected by GC-MS.

2. Experiments

2.1. Pyrolysis mechanism

Under around 400℃ condition, dissolved in motor oil, polystyrene could decompose into styrene, toluene, 1-methylstyrene and some other products with cleavage of the polystyrene backbone and the following internal electron rearrangement, hydrogen rearrangement and hydrogen transfer. Figure 1 is the schematic pyrolysis diagram for this process.

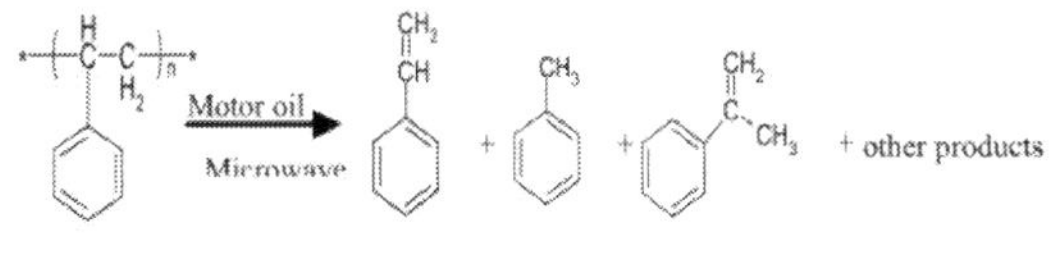

Figure 1. Schematic diagram for decomposition of polystyrene in motor oil

2.2. Experimental Apparatus

The experimental apparatus developed and used for this investigation is shown in Figure 2. The microwave thermal decomposition system consists of microwave generator, remote head, circulator, three stub tuners, waveguide, quartz reactor, and dummy load. Microwave generator (ASTEX applied science and technology inc, USA) used in this study has the 2450MHz frequency and can produce the power ranging from 250 to 2500w. The size of quartz reactor had the dimension of 2cm-diameter and 120cm-length. A nitrogen sweeping gas was used to carry vapor from the reactor to the collection flasks. Infrared thermometer (3M company, USA) was applied to give the pyrolysis temperatures.

Forward power meter (Cober Electronics,

USA) was installed at the end of the waveguide to get the microwave power absorbed by dummy load, with this and the forward power measured by meter and the reflected power on the microwave generator, the effective power for the pyrolysis could be obtained as:

Effective power = Forward power − Reflected power − Absorbed power

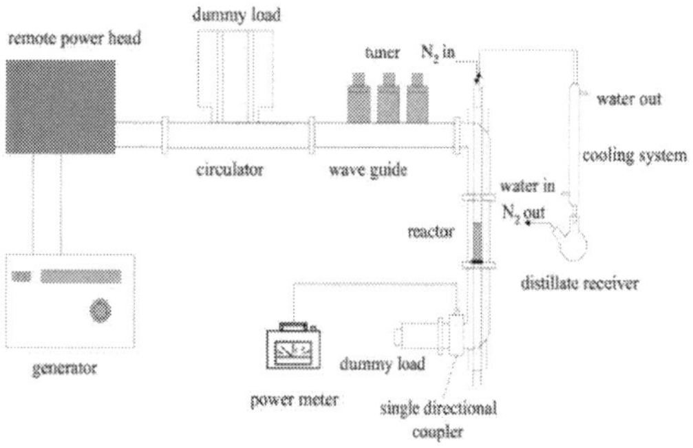

Figure 2. Schematic drawing of the microwave-induced pyrolysis apparatus

2.3. Materials and Procedures

2.3.1. Pretreatments of polystyrene and motor oil

Polystyrene was selected as a model polymer because of its wide use, and because its pyrolysis products are easily distinguishable from those of the oil. Other plastics typically produce a variety of alkanes and alkenes that are difficult to distinguish from the oil products. Polystyrene packaging beads were used as a source for the waste polystyrene. The beads were heated at 150℃ for five minutes to densify them[4]. A 30 weight motor oil (SK Corporation, Korea) was pretreated by heating it with a power around 100 watts higher than the reaction power to insure that any light ends in the oil would not show up in the products and be mistaken for decomposition products.

2.3.2. The importance of motor oil

In most of pyrolysis process, solvents were not chosen for better performance, however, here, in this research, the motor oil was applied for it served two purposes. When the plastic dissolved in the oil, the molecule had greater freedom of motion, which could prevent the broken chains from recombining. Also, the oil could effectively isolate the monomers from polymer chains so that they would not recombine into less desirable coke production.

2.3.3. The importance of silicon carbide

Microwave fields have certain properties. Microwave could be transmitted by good insulator such as quartz which will not heat, reflected by metals and absorbed by materials such as polar rubber compounds and carbon black loaded compounds. In this research, silicon carbide was chosen as the microwave absorbent. When the silicon carbide is exposed to a microwave field, it can reach temperatures up to 1000℃ in a short period of time. When shredded plastic wastes are mixed with the silicon carbide, prior the heating up of the latter, the energy absorbed from the microwaves would be rapidly transferred to the waste by conduction. This results in very efficient energy transfer and a highly reducing environment to avoid formation of undesired products and to enhance production of desired compounds.

2.3.4. Experimental procedures

25 gram of silicon carbide, 2 gram of pretreated polystyrene and 15 gram of pretreated motor oil were put into the reactor accordingly, then, the reactor was installed in the waveguide. As long as the cooling system, dummy load and carrier gas Nitrogen were set ready, the power could be turned on, the microwave generated by the generator could go along the waveguide through the reactor until the dummy load. Silicon carbide heated by microwave energy would transfer its thermal energy to the polystyrene and motor oil by conduction, then, pyrolysis began its first step. With cooling system, the distillate was collected for future GC-MS and data analysis. The pyrolysis system temperature was measured by infrared thermometer.

2.4. Product Analysis

Gas chromatography/Mass spectroscopy (GC/MS) was used to find the chemical composition of the distillate products. The analysis was done on a 6890N GC/ 5973 MSD system (Agilent Technologies). The separation was performed on a 15m fused silica capillary column HP-5MS. The carrier gas was helium. The temperature was ramped from 30 to 200℃ at 3℃ /min.

3. Results and Discussion

3.1. Silicon carbide and motor oil amount

Microwave-induced pyrolysis is different from traditional thermal decomposition as the name indicates. Since the oil and polystyrene could not absorb microwave energy, silicon carbide was taken as the microwave absorbent, and in this research, 25 gram of silicon carbide addition was decided for all the experiments. Table 1 shows the power absorbed by silicon carbide with respect to microwave input power.

Table 1. Silicon carbide performance as microwave absorbent

Group	1	2	3	4	5	6
Forward power (w)	275	300	325	350	375	400
Effective power (w)	158	172	187	201	215	230

In pyrolysis of polystyrene process, motor oil played an important role in preventing the decomposed bonds recombine [4]. In this research, a new 30 weight oil was used to simulate the used motor oil. As for the amount of motor oil, John A. Marsh [4] gave a weight percentage of polystyrene to motor oil of 1:10 as the appropriate one, while in our research, 1:7.5 was chosen as the normal weight percentage of polystyrene to oil since with this ratio it could give comparatively favorable chemical value recorery with the other conditions at the same level.

3.2. Microwave power and time effect on distillate amount

Compared with the importance of temperature in traditional pyrolysis, power played a key role in microwave-induced pyrolysis.

As what a commonsense could be, the distillate amount should increase with the increase of power and reaction time. In order to verify whether this rule would apply to our research, 15 experiments were made according to different effective power as: 180w, 200w, 220w, 240w and 250w, and different pyrolysis time as: 30 minutes, 45 minutes and 60 minutes. Both based on these experiments, Figure 3 and Figure 4 are aimed to give a comprehensive show for the results.

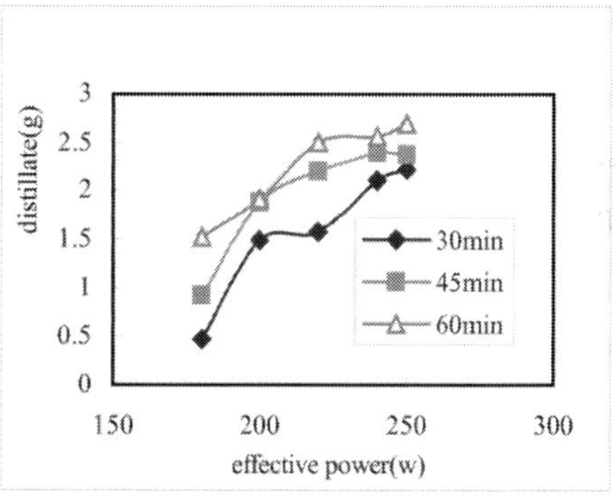

Figure 3. Power effect on distillate for different reaction time

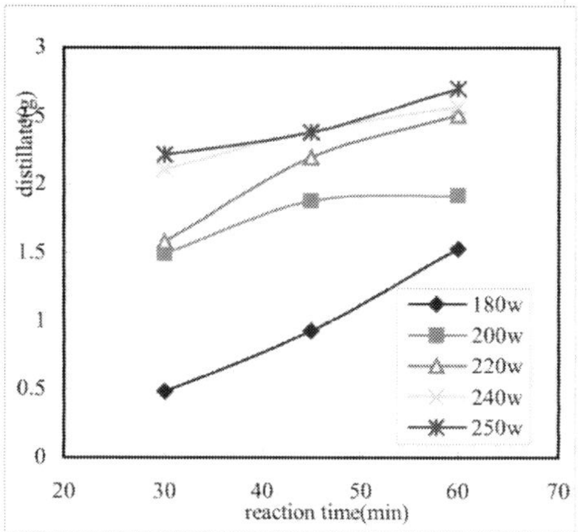

Figure 4. Time effect on distillate for different power

From the two graphs, the preceding commonsense was definitely reasonable and it was possible to completely convert polystyrene to valuable distillate products when a polystyrene/oil mixture was pyrolized at the conditions of microwave power greater than 220W. At those conditions bed temperature of the reactor was around 300℃. Compared with the temperature requirement higher than 400℃[2][3][4] for a conventional pyrolysis of polystyrene, microwave-induced method made the complete pyrolysis at much lower temperature. This may result from the special performance of microwave, which could penetrate deep into the molecules and break their bonding structures.

3.3. GC-MS product analysis

Gas chromatography/mass spectroscopy was used to find the chemical composition of the distillate. In John A. Marsh's conventional polystyrene pyrolysis research [4], styrene, methyl-styrene, toluene, ethyl benzene and cumene were the main components, and styrene took 25 weight percent of the total distillate and 50 weight percent of compounds produced from the pyrolysis of polystyrene. Compared with Marsh's results, ours gave more promising results. The components generated from polystyrene were identified as styrene, 1-methyl-styrenen, toluene, and ethyl-benzene. Styrene was the primary product (over 30 weight percent of the total distillate). When the distillate products were considered by themselves, styrene accounted for approximately over 70 weight percent of the compounds produced from the pyrolysis of polystyrene. Table 2 and Figure 4 show one GC/MS analysis results among the 15 and Figure 5 displays the styrene recovery results under different time and power.

Table 2. Composition of distillate products as determined by GC/MS from the pyrolysis at 220w for 1 hour

Component	Overall weight Percent (%)	Weight percent based on polystyrene products (%)
Styrene	36.80	73.94
1-Methyl-styrene	6.84	13.74
Toluene	5.01	10.07
Ethyl benzene	1.12	2.25
Components from heavy oil	50.23	-

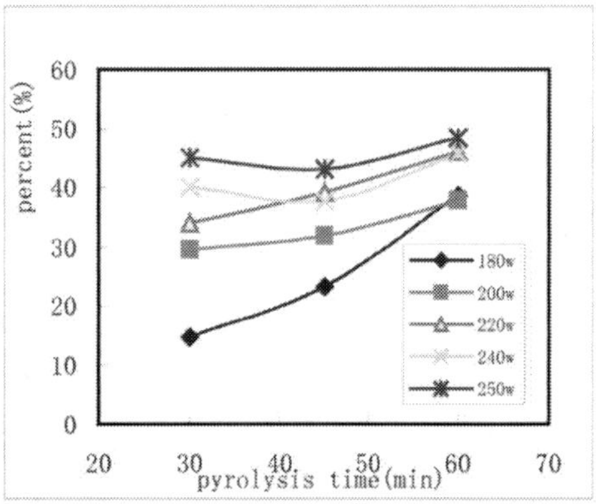

Figure 5. Styrene recovery at various process variables

Styrene, as the main distillate product, gave an objective comparison on pyrolysis under different conditions, such as microwave power and reaction time. The styrene weight percentage was increasing distinctly with increased pyrolysis time and increased microwave power when microwave power was below or equal to 220w. With power higher than 220w, the increasing trend was not apparent, and it seemed complete pyrolysis had been accomplished.

4. Conclusions

This research investigated the microwave-induced pyrolysis of polystyrene in the presence of a motor oil. The following conclusions can be drawn from the results of this study:

(1) Microwave-induced pyrolysis showed favorable performance on the thermal

decomposition of poystyrene in improving the recovery of valuable products from polystyrene and lowering the pyrolysis temperature, compared with previous works.

(2) Waste polystyrene could completely decompose to distillate products at 220w (around 300℃) in the presence of motor oil.

(3) The distillate products from the pyrolysis of waste polystyrene include on the average styrene (71.60 weight %), 1-methyl-styrene (15.42 weight %), toluene (7.84 weight %) and ethyl-benzene (5.14 weight %).

5. References

1. C. Ludlow-Palafox, H. A. Chase, Pyrolysis of Plastic Wastes Using a Microwave Induced Pyrolysis Process, 6th World Congress of Chemical Engineering, 2001.

2. Guffey, F. D., P. A. Holper, and D. E. Hunter. Summary of Laboratory Simulation Studies of the ROPE^{TE} Process. DOE Report. West. Res. Inst. : Laramie, WY. USA, 1991.

3. Inomata, Osamer, Hiroshi Ando, Toshiji Abe, Sadao Matsuzawa, and Yukio Shimizu. "Decomposition of Plastic Wastes. Ⅱ. Pyrolysis of Plastic Wastes by the Solvent method" Nenryo Kyokaishi (Japan). 53(568) .pp. 715-725, 1974.

4. John A. Marsh, Thermal Processing of Waste Polystyrene in Waste Motor Oil, 1993.

378561 Kinetics of Cyclohexanone Ammoximation over Titanium Silicate Molecular Sieves

Abstract:

Cyclohexanone oxime is the key intermediate in the manufacture of e-caprolactam. Conventional production routes involve numerous steps with the use of hazardous chemicals like oleum, halides, and oxides of nitrogen. In addition, large quantities of the low value by-product ammonium sulfate and considerable amount of waste are also produced. An alternative method of cyclohexanone oxime synthesis is the reaction of cyclohexanone with ammonia and hydrogen peroxide in the liquid phase over titanium silicate molecular sieves[1]. Since the main byproduct of the reaction is water and the reaction takes place under mild conditions, it is an environmental friendly, clean and efficient process technology. Since the discovery of the catalytic effect of titanium silicate on the process[2], the literature has mostly dealt with either the identification and characterization of active sites or the experimental determination of conversion and selectivity data under specific conditions[3,4]. However, only a limited number of studies have focused so far on the mechanism and the kinetics of this reaction over titanium silicate sieves with diluted hydrogen peroxide[5]. The objectives of the present study are to examine the reaction framework of cyclohexanone ammoximation and develop more complete kinetic models including hydrogen peroxide decomposition by a more detailed kinetic analysis.

The kinetics of the cyclohexanone ammoximation in the liquid phase over titanium silicate molecular sieves has been studied in a slurry reactor. The reactions are carried out at different temperatures and molar ratios of $NH_3/C_6H_{10}O$ and $H_2O_2/ C_6H_{10}O$ under atmospheric pressure. The concentration of titanium silicate sieves in liquid is kept constant. The reaction products of cyclohexanone ammoximation are analyzed using a gas chromatograph. The H_2O_2 concentration in reaction mixtures is determined by iodometric titration. In this study the initial rates method is used for kinetic analysis. They are obtained by fitting the concentration/time data by Marquardt technique. The Gauss-Newton method is used to estimate model parameters by fitting the kinetic data.

The ammoximation of cyclohexanone is found to proceed with good cyclohexanone conversion and selectivity to form oxime over the titanium silicate sieves. The experimental results show this reaction is actually conducted in a multiphase reaction medium with a complex parallel-consecutive reaction scheme. The major product of the reactions is cyclohexanone oxime, while some by-products are also produced. We have found that more than ten trace amount of by-products appeared in the reaction product. In spite of very small amounts of organic by-products, the catalyst can irreversibly absorb them resulting in a blockage of active sites and a loss of catalytic activity. Most organic side reactions are non-catalytic homogeneous reactions. The possible reaction mechanism obtained from the detailed kinetic analysis suggests H_2O_2 is adsorbed in the active site of catalyst, the synthesis of hydroxylamine by NH_3 and H_2O_2 is irreversible and this reaction rate is quicker than the one of cyclohexanone and hydroxylamine. The analysis of the reaction framework is useful to understanding the deactivation mechanism of the catalyst and increasing the catalyst stability.

On the basis of these experimental data under the reaction conditions investigated, the power law rate equations have been developed and the kinetic parameters are evaluated by simulation the initial rate data. The reactions are between zero and unity order with respect to reactant concentrations. The rate equations show that cyclohexanone ammoximation is no more sensitive than H_2O_2 decomposition with respect to H_2O_2. The activation energy is found to be 95kJ/mol and 82kJ/mol for the ammoximation of cyclohexanone and H_2O_2 decomposition, respectively. These data clearly show a good agreement between the model prediction and experimental data under the present operating conditions. The kinetic models are useful to the design and operation of reactor.

Keywords: kinetics, titanium silicalite, cyclohexanone, ammoximation, cyclohexanone oxime

REFERANCES

[1] Thangaraj, A., Sivasankers, S., Ratnasamy, P., "Catalytic properties of crystalline titanium silicalites III. Ammoximation of Cyclohexanone", J. Catal., 131 (2), 394-400(1991).

[2] Taramosso, M., Perego, G., Notari, B., "Preparation of process crystalline synthetic material comprised of silicon and titanium", U. S. Pat., 4410501(1983).

[3] Roffia, P., Padovan, M., Leofanti, G., Mantegazza, M. A., De Alberti, G., Tauszik, G. R., "Catalytic process for the manufacture of oxime", U. S. Pat., 4794198(1988).

[4] Tonti., S., Roffia, P., Gervasutti, V., "Multistep Process for the Liquid Phase Ammoximation of Carbonyl Compounds", Eur. Pat. 0496385(1992).

[5] Murzin, D. Yu, Kul'kova, N. V., Kotova, V. G., Kvyatkovskaya M. Yu. Kinetics and Catalysis, "Kinetics study of cyclohexanone ammoximation in the liquid phase on titanium silicates", 39(5), 702-705(1998).

Ceria modified Titania supported Pd-Cu Catalysts for Nitrate Reduction in Water

Min-Sung Kim[a] and Kwan-Young Lee[a,b,*]

[a] Department of Chemical and Biological Engineering, Korea University, 5-1, Anam-dong, Sungbuk-gu, Seoul, 136-701, Republic of Korea

[b] Green School, Korea University, 5-1, Anam-dong, Sungbuk-gu, Seoul, 136-701, Republic of Korea

1. Introduction

The increasing nitrate levels in water have been an intensive attention because nitrate caused serious health risk [1]. To remove hazardous nitrate, Vorlop and Tacke suggested a catalytic reduction method which used metal-supported oxide as catalyst and hydrogen as reductant [2]. Many studies presented that bimetallic type of the catalyst including a noble metal (Pd or Pt) and a promoter such as Cu was suitable for the nitrate reduction because the nitrate could be converted nitrite on the promoter by hydrogen spilled over from noble metal and the nitrite was reduced to N_2 or ammonium on noble metal site [3]. Among the catalysts, Pd-Cu-based catalysts showed high nitrate conversion and N_2 selectivity [4].

In our previous work using Pd-Cu/TiO_2 catalysts, we observed that metallic Cu state was the active phase for the nitrate reduction and TiO_2 support could enhance the catalytic activity because the nitrate could be bound to oxygen vacancies formed on the reduced support surface [5]. The metallic states of active species and the oxygen vacancies were generated in the catalyst with the most reduced support (i.e., $Ti^{4+} \rightarrow Ti^{3+}$). Thus, it could be speculated that the reduced state of the support was important factor for the nitrate reduction activity.

Besides titania, ceria has widely known as the reducible support. Using these properties, Li et al. demonstrated that the titania containing ceria had more Ti^{3+} than single titania [6]. From this result, the TiO_2-CeO_2 support might be a promising material. In this study, we prepared Pd-Cu/TiO_2-CeO_2 catalysts with different ceria amount, and investigated the effect of the ceria addition on the catalytic performances and the catalytic properties.

2. Experimental

The TiO_2-CeO_2 supports were synthesized by co-precipitation method, then palladium and copper were impregnated on the supports by deposition-precipitation method. The abbreviations, BET properties and chemical compositions of the catalysts are shown in Table 1. The nitrate reduction reactions were carried out in a semi-batch reactor and performed under 25°C and atmospheric pressure. 0.15 g of catalyst was located in the reactor and hydrogen was continuously fed to the reactor (90 ml/min). 2 mM of nitrate solution prepared from KNO_3 was subsequently added into the reactor.

Table 1. Chemical compositions and BET properties of Pd-Cu/TiO_2-CeO_2 catalysts.

Abbreviation	Chemical composition		BET properties	
	Ti:Ce content (mol ratio)	Pd [Cu] content (wt. %)	Surface area (m^2/g)	Pore volume (cm^3/g)
PCTC1	18:1	3.0 [1.0]	86.1	0.305
PCTC2	5:1	3.1 [1.1]	72.7	0.226
PCTC3	2:1	3.1 [1.1]	62.7	0.316

3. Results and Discussion

Figure 1 displays the concentration profiles of nitrate over all catalysts. For comparison, the activity of Pd-Cu/TiO_2 catalyst (PCTC0) in our previous work [5] is also presented. During the first 20 min of reaction, the nitrate was more rapidly reduced in TiO_2-CeO_2 supported catalysts than in PCTC0. Among the catalysts, the fastest initial reaction rate (calculated as consumed nitrate concentration per amount of catalyst per the first 3 min) was shown in PCTC1 (3.20 mmol/min/g_{cat}) and the value was about 8 times greater than that of PCTC0 (0.41 mmol/min/g_{cat}).

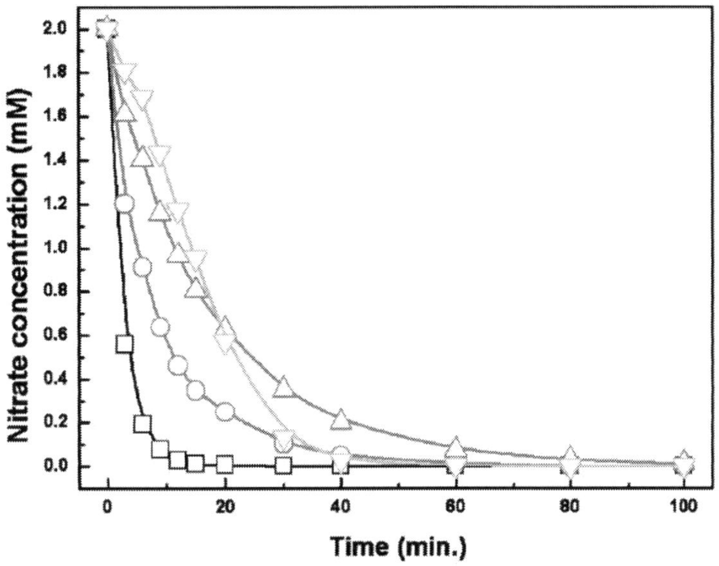

Fig. 1. Nitrate concentrations as a function of reaction time for catalytic nitrate reduction (□: PCTC0,

□: PCTC1, ○: PCTC2, △:PCTC3).

In Ti $2p_{3/2}$ spectrum, the value of lower binding energy of PCTC1 (456.9 eV) was very close to that of Ti^{3+} (456.8 eV). On the other hand, the peak shifted to the higher binding energy with increasing the amount of ceria. For Ce 3d, PCTC1 had the largest portion of Ce^{3+}. Similar to Ti 2p spectrum, the Ce^{3+} contents of PCTC2 and PCTC3 lowered as ceria content increased. The formation of Ti^{3+} and Ce^{3+} indicated that the surface of PCTC1 was rich in oxygen vacancies. In the cases of Pd and Cu, the chemical state of PCTC1 was closer to metallic Pd and Cu than other catalysts. These results suggested that the synergetic effect of oxygen vacancies and metallic Cu states in PCTC1 significantly enhanced the nitrate reduction activity.

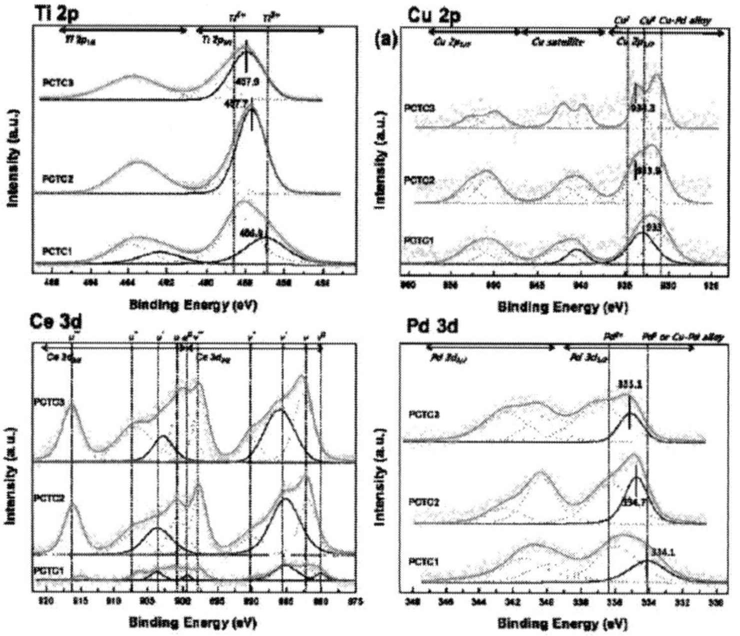

Fig. 2. XPS spectra of the prepared catalysts.

4. References

1. Huang, C.-P.; Wang, H.-W.; Chiu, P.-C. Water Research 1998, 32, 2257.

2. M.-M., Y.I.; Sheintuch, M. Industrial & Engineering Chemistry Research 1998, 37, 309.

3. Soares, O.S.G.P.; Órfão, J.J.M. Applied Catalysis B: Environmental 2011, 102, 424.

4. Soares, O.S.G.P.; Órfão, J.J.M. Applied Catalysis B: Environmental 2009, 91, 441.
5. Kim, M.-S.; Chung, S.-H.; Lee, D.-W.; Lee, K.-Y. Applied Catalysis B: Environmental 2013, 142–143, 354.
6. Li, F.B.; Li, X.Z.; Hou, M.F; Choy, W.C.H. Applied Catalysis A: General 2005, 285, 181.

379908 Sulfur Trioxide Formation during Oxy-Coal Combustion

Abstract:

Increasing demand of energy, due to the ever-growing world population, has ensured coal to be the primary source of electricity generation for many years to come. With a recorded level of 398.03 ppm of CO_2 in the atmosphere, 38% of which can be attributed to power generation sector, US holds the second place in world for CO_2 emissions. Oxy-coal combustion technology is a promising solution for implementing carbon capture and storage (CCS) successfully. Removal of N_2 from the oxidizer lessens the volume of flue gas to be processed; which consists mostly of CO_2. This CO_2 laden flue gas can be easily sequestered, with a CO_2 removal efficiency of 90%~95%. Several power generation plants, based on oxy-coal combustion, are being successfully operated in Australia and Europe. FutureGen 2.0, a joint venture of DOE, Air Liquide and B&W, is expected to be running commercially by 2023.

Reduced flue gas volume causes increase in the concentrations of corrosive species. Increased SO_3 concentration has been reported by different case studies resulting from changed combustion environment. SO_3 is a major concern for power plants as it can cause both high temperature corrosion in boilers and severe low temperature corrosion in flue gas ducts and economizers. Before retrofitting any existing plant for this technology, it is crucial to perform an elaborate study reporting SO_3 concentration under the conditions analogous to an actual power plant.

In this study, SO_3 formation trends in oxy-coal combustion under different operating conditions will be explored. To serve this purpose, an experimental setup was built consisting of a burner followed; a reactor; subjected to temperature profile reflecting actual operating conditions. Samples can be obtained from the reactor at every 5cm interval simultaneously without interrupting the continuity of the experiment. FTIR, coupled with a variable path length cell, is applied for measurement of SO_3 in gaseous phase. From the experiments performed, a clear picture of SO_3 concentration profile along the varying process conditions will be obtained and this data can be used to develop and validate a kinetic model that can predict the concentrations of SO_x species in the flue gas.

Effects of CO, H_2 and C_3H_6 on the NH_3-SCR reactions over a Cu-chabazite SCR catalyst

Yang Zheng, Michael P. Harold*, Dan Luss*

Department of Chemical & Biomolecular Engineering
University of Houston
Houston, TX 77204

Abstract

The effects of reducing agents of CO, H_2 and C_3H_6 on the performance of a Cu-chabazite catalyst were investigated, including NO_2 decomposition, NH_3-SCR reactions, NH_3 storage, as well as N_2O production. Different reductants affect NH_3-SCR reactions to different extents by means of different pathways. Propylene is most effective in promoting NO_2 decomposition to NO via organic intermediates. CO effectively reduces nitrates to nitrites that can then react with NO_2 to release NO, while H_2 has a slightly promoting effect when NO_2 becomes thermodynamically unstable at high temperatures. As for SCR reactions, propylene competes with NH_3 for adsorbed NO_2 or nitrates to release NO and thus increase NO/NOx ratio. This leads to the dominance of either fast (NH_3 + NO + NO_2) or standard SCR (NH_3 + NO + O_2) reaction under slow SCR (NH_3 + NO_2)condition if C_3H_6 present. The competition of CO with NO for nitrates reduction to nitrites, has a negligible effect on both standard and fast SCR reactions but a promoting effect on slow SCR reaction. In contrast to steady-state operation, lean/rich cycling enhances cycle-averaged NOx conversion of all NH_3-SCR reactions when adding either C_3H_6 or CO + H_2 in rich phase. Both high NO_2/NOx ratio and long lean-phase purge can mitigate propene poisoning, enabling the combined contribution of both HC- and NH_3-SCR pathways. In addition, decreased N_2O emission from slow SCR reaction is observed when CO or C_3H_6 present partly due to each reaction with ammonium nitrates. All three reductants do not compete with NH_3 for storage sites and only C_3H_6 inhibits NH_3 oxidation. This study underscores the importance of accounting for the impact of different reducing agents on conventional NH_3-SCR reaction mechanism during the rich regeneration for integrated systems (LNT + SCR, SCR on filter) or cold-start period.

Determining the Oxygen Storage Capacity (OSC) of Ceria Materials by Oxygen Chemisorption

Andrew D. D'Amico[1*], Onaje LaMont[1,2], Sarah McNew Schimming[2], and Carsten Sievers[2]
[1]*Micromeritics Instrument Corp, Norcross, GA 30093 USA*
[2]*Georgia Institute of Technology, Atlanta, GA 30332 USA*
**corresponding author:andrew.damico@micromeritics.com*

Introduction

Ceria and zirconia metal oxides have considerable redox activity, making them valuable in several catalytic applications such as conversion of vehicle exhaust [1], production of CO and H_2 for fuel by solar energy [2], and deoxygenation of bio-oil pyrolysis products [3]. Surface reaction results in the formation of oxygen vacancies. Typical techniques for quantifying oxygen storage capacity (OSC) of ceria materials include dynamic oxygen pulses [4], CO pulses[4], CO temperature-programmed reduction (TPR) [5], or H_2 reduction [4]. The focus of this work is to look at the advantages and utility of oxygen adsorption isotherms to determine OSC—corresponding with oxygen vacancies in the ceria materials.

Materials and Methods

All catalysts were prepared by coprecipitation using cerium nitrate hexahydrate and zirconyl nitrate hydrate, dissolved in deionized water. The coprecipitation was modified by adding the precursor solution to an ammonium hydroxide solution while stirring continuously. The precipitate was filtered, rinsed with deionized water, and dried in an oven overnight at 100 °C. The catalysts were calcined at 500 °C for 4 hours in air with a ramp rate of 5°C/min.

A commercially available volumetric adsorption instrument (Micromeritics 3FLEX, chemisorption option) and dynamic adsorption instrument (AutoChem II 2920) were used to determine the oxygen uptake.

Results and Discussion

Oxygen storage capacity at three reduction temperatures is given in **Table 1** for each catalyst. Oxygen pulse experiments were used along with oxygen isotherms to validate the results. The repeatability and reproducibility were within 3%. The results indicate the addition of zirconia can at least double the OSC at temperatures between 200°C - 450 °C, while no measurable reducibility was measured for pure zirconia in the same temperature range.

Table 1. Oxygen uptake (mmol O_2 per gram) for different ceria-zirconia mixed oxides after reduction at different temperatures.

Ce composition (%)	450°C	300°C	200°C
100	0.138	0.040	--
60	0.390	0.072	--
46	0.328	0.046	--
0	--	--	--

The oxygen storage capacity (OSC) can be determined directly from the oxygen isotherm, as shown in **Figure 1**. The shape of these isotherms allows for confidence in the experimental results. Isotherms should monotonically increase with increasing pressure, and deviations can indicate gas contamination. Repeat experiments at different analysis temperatures (eg 50°C or 100°C) can also be used to validate results, as the irreversible oxygen uptake seems to be independent of analysis temperatures above 50°C (only a function of the reduction temperature). Also, the quality of the system and measurement can be determined based on pressure stability. By using pressure transducers that read in the range of 10^{-6} torr, the sensitivity of the measurement is approximately 0.1 μmole O_2 (0.002 cm^3 STP) below 1.5 torr.

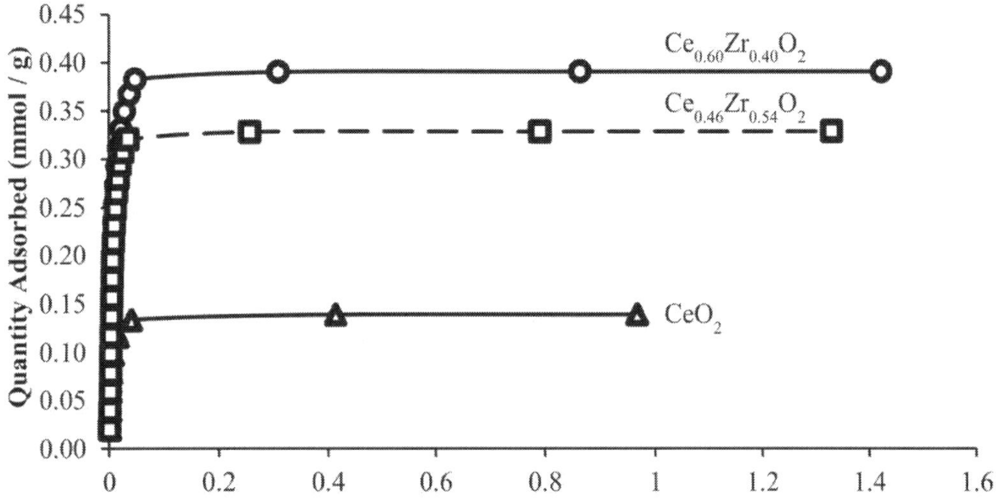

Figure 1. O_2 isotherm at 450°C, after isothermal reduction and purging at the same temperature, for different ceria-zirconia mixed oxides. 60% Ce (●), 46% Ce (■), and CeO_2 (▲).

Significance

Oxygen adsorption isotherms are an effective solution to determining the oxygen storage capacity (OSC) of ceria materials using commercially available instruments and agree well with typical dynamic chemisorption techniques.

References

1. B. Zhao, Q. Wang, G. Li, and R. Zhou. *Journal of Alloys and Compounds* **2010**, 508, 500–506.
2. W. Chueh, C. Falter, M. Abbott, D. Scipio, P. Furler, S. Haile, and A. Steinfeld. *Science* **2010**, 330, 1797.
3. D. Bulushev and J. Ross. *Catalysis Today* **2011**, 171, 1-13.
4. H.C. Yao and Y.F.Y. Yao. *Journal of Catalysis* **1984**, 86, 254-265.
5. E. Mamontov, T. Egami, R. Brezny, M. Koranne, and S. Tyagi. *J. Phys. Chem. B* **2000**, 104, 11110-11116.

381087 Effect of Reaction Parameters on Hydrocracking of Petroleum Vacuum Residue with Activated Carbons in Supercritical Hydrocarbon Solvents

Abstract:

Because of the depletion of conventional light crude oils and rising demand for valuable petroleum products, the petroleum industries have been forced to refine the heavy feedstocks such as oil sands, bitumen and heavy residues. Among the heavy feedstocks, vacuum residue (VR) is the heaviest fraction obtained from the vacuum distillation. It includes mainly hydrocarbons with high boiling points over 520°C, and also about 5% sulfur, 0.5% nitrogen and minor amounts of organometallic compounds such as nickel and vanadium. The objective of this study is to achieve high conversion of vacuum residue (VR) to lower boiling hydrocarbon products with reducing coke formation in supercritical hydrocracking process (SHP). Since the performance of the SHP for VR could be affected by various operating conditions, the parametric experimental study was carried out in a batch reactor using hydrocarbon solvents. Five kinds of activated carbon catalysts (bituminous coal-derived activated carbon (AC), coconuts shell-derived activated carbon (MAC), their acid treated AC and MAC and Fe impregnated acid treated AC) were applied to SHP. The reaction results at various operating conditions, such as temperature (350, 380 and 400°C), hydrogen partial pressure (0.50, 1.38 and 3.45 MPa), reaction time (5, 10 and 30min) were compared with each other. In addition, the effects of additives (H_2O, methanol, ethanol and H_2/CO_2 mixture) on SHP were evaluated. The SHP parametric results were evaluated under conversion, coke formation, and distribution of oil products (naphtha, middle distillate, vacuum gas oil, and residue).

381357 Synthesis of Butenes through Butanol Dehydration over Highly Ordered Crystalline Nanoporous Tungsten Oxide

Abstract:

Metal oxides are important catalysts in the petrochemical industry and fine chemical industry. Recently, mesoporous transition metal oxides as catalysts have received enormous attentions due to their excellent catalytic properties attributed to high surface area and highly ordered crystalline framework. Various methods for the preparation of mesoporous transition metal oxides have been developed. Among them, the nano-replication method has brought forward incredible possibilities in preparing new mesoporous materials. This study focuses on the process to produce butene via dehydration of butanol. There is an imbalance between demand and supply in the butene market due to rising naphtha prices according to rising oil prices. In this study, 2-butanol dehydration is used to demonstrate the potential advantages of the highly ordered crystalline nanoporous tungsten oxide, which was obtained from a mesoporous silica template. BET, XRD, NH_3-TPD, and pyridine-IR were used to study the structural properties of the highly ordered mesoporous WO_3 catalyst. The highly ordered mesoporous WO_3 catalyst prepared from a mesoporous silica template can be an effective catalyst for producing butenes through 2-butanol dehydration.

381793 An Investigation into the Bandgap and Bandedge of Semi-Conducting La-Transition Metal Perovskites for Photocatalytic Applications

Abstract:

The emission of greenhouse gases like CO_2 has posed dire environmental complications which has challenged researchers to find sustainable methods to eliminate CO_2 presence in the atmosphere. The ideal solution is to convert CO_2 to useful and non-toxic compounds like hydrocarbons that can be used as fuel. The photocatalytic reduction of CO_2 facilitated by solar energy has stirred up great interest over the past few decades. The desire of this study is to synthesize highly efficient and stable photocatalyst with narrow band gap energy in order to ease the emission of photoexcited electrons that will reduce CO_2 in the presence of water into energy bearing products.

The design of semi-conducting perovskites heterogeneous photocatalysis was based on the strategy of synergistic effects of B-site transition elements which also determine the catalytic activity. $LaFe_xCr_{1-x}O_3$, $LaFe_xMn_{1-x}O_3$ and $LaMn_xCr_{1-x}O_3$ ($x= \{1, 0.25, 0.5, 0.75\}$) perovskites were synthesized through Pechini process. The produced perovskites were subsequently characterized by X-ray powder diffraction technique. The samples showed high crystallinity and an assortment of cubic, hexagonal and orthorhombic crystal structures with cubic found to be a major component. The band gap energy are to be determined by ultraviolet-visible diffuse reflectance spectrophotometry.

393527 Coke Elimination During Conversion of Biogas to Syngas by Tungsten Incorporation into Ni Based Mesoporous Alumina Synthesized by the One-Pot Route

Abstract:

Dry reforming of methane with CO_2 was investigated over W and Ni incorporated bi-metallic mesoporous alumina catalysts prepared by the one-pot route. High surface W-Ni incorporated mesoporous alumina catalysts synthesized in this work showed excellent performance for the conversion of synthetic biogas to synthesis gas. Tungsten incorporation significantly enhanced stability of the catalytic performance and eliminated coke formation.

382854 Dehydration of Fructose into Furans over Zeolite Catalyst Using Carbon Black As Adsorbent

Abstract:

Due to increasing demands for transportation fuels and commodity chemicals as well as environmental and political concerns associated with diminishing fossil fuel resources, research efforts have been driven towards the efficient utilization of renewable feedstock, such as naturally abundant lignocellulosic biomass[1]. Development of economically viable processes for the syntheses of chemical intermediates from biomass-derived carbohydrates has been considered as an important challenge in this area. One major pathway for converting lignocellulosic biomass into fuels and chemicals involves the production of furans, such as 5-hydroxymethylfurfural (HMF) and furfural, from the dehydration of hexose and pentose sugars[2-3].

The key challenge for the selective production of HMF from the dehydration of fructose is the development of selective heterogeneous catalyst and protection of further reaction of the formed HMF. In this study, formation of HMF from the dehydration of fructose over zeolite catalysts in aqueous phase was studied in a reactive adsorption system using carbon black (BP2000) as an adsorbent[4-5]. The dehydration of fructose over zeolite BEA catalyst in water revealed that selectivity to HMF is largely influenced by the formation of levulinic acid and formic acid from the rehydration of produced HMF as well as subsequent condensation reactions. In order to increase selectivity to HMF, reactive adsorption with carbon adsorbents was used to prevent the further reactions of the produced HMF. It was found that BP2000 carbon black exhibited high selectivity and capacity for the adsorption of HMF and furfural from the aqueous phase. The unique adsorption performance of BP2000 is likely due to the large surface area, hydrophobic nature and micropore structure. With using the carbon adsorbent, furan selectivity (selectivity of HMF and furfural) over zeolite BEA was improved from 27% to 44% with a furan yield of 41%. The furan selectivity and yield obtained in the presence of the carbon adsorbent were similar to those obtained from the aqueous/organic reaction system using methyl isobutyl ketone (MIBK) as an extraction phase.

References:

1. Huber, G. W.; Iborra, S.; Corma, A., "Synthesis of transportation fuels from biomass: Chemistry, catalysts, and engineering". *Chem. Rev.* **2006,** *106* (9), 4044-4098.

2. Chang, C.-C.; Green, S. K.; Williams, C. L.; Dauenhauer, P. J.; Fan, W., "Ultra-selective cycloaddition of dimethylfuran for renewable p-xylene with H-BEA". *Green Chem.* **2014,** *16* (2), 585-588.

3. Williams, C. L.; Chang, C.-C.; Do, P.; Nikbin, N.; Caratzoulas, S.; Vlachos, D. G.; Lobo, R. F.; Fan, W.; Dauenhauer, P. J., "Cycloaddition of Biomass-Derived Furans for Catalytic Production of Renewable p-Xylene". *ACS Catal.* **2012,** *2* (6), 935-939.

4. Dornath, P.; Fan, W., "Dehydration of fructose into furans over zeolite catalyst using carbon black as adsorbent". *Microporous Mesoporous Mater.* **2014,** *191* (0), 10-17.

5. Swift, T. D.; Bagia, C.; Nikolakis, V.; Vlachos, D. G.; Peklaris, G.; Dornath, P.; Fan, W., "Reactive adsorption for the selective dehydration of sugars to furans: Modeling and experiments". *AIChE J.* **2013,** *59* (9), 3378-3390.

Simultaneous CO_2 valorization and DME production using a packed-bed membrane reactor with *in situ* water removal

G. Zarca[1], N. Diban[1*], A.Urtiaga[1], I. Ortiz[1], J. Ereña[2], A.T.Aguayo[2]

[1] *Department of Chemical Engineering, University of Cantabria, Avda. Los Castros s/n, 39005, Santander, Spain*
[2] *Department of Chemical Engineering, University of Basque Country, Apdo. 644, E-48080, Bilbao, Spain*

(*) corresp. author: dibann@unican.es

1 Introduction

The interest for capture and chemical valorization of CO_2 by producing different valuable products (e.g. formaldehyde, acetic acid, propylene, methanol (MeOH), olefins, etc.) is increasing in recent years [1]. Dimehtyl ether (DME) is a key intermediate for the production of automobile fuels and raw materials (olefins and BTX), and a 'clean' fuel for Diesel engines. The synthesis of DME from syngas in the same packed-bed reactor (PBR) unit is a process wit high potential to incorporate CO_2 as co-feed.

A bifunctional catalyst ($CuO-ZnO-Al_2O_3/\gamma-Al_2O_3$) able to catalyze the one-step syngas-to-DME synthesis has been successfully developed previously and the kinetics and thermodynamics of the reactions in a PBR have been well described [2]. Therefore, this DME process becomes a way to promote CO_2 capture, an interesting environmental challenge nowadays. However, CO_2-rich feedings hinder the DME production because the hydrogenation reaction of the CO_2 generates water accumulation that inhibits the methanol dehydration step. Therefore, the motivation for in-situ H_2O removal during DME synthesis is to displace the water-gas shift equilibrium to enhance conversion of CO_2 into methanol to improve DME productivity.

Membrane removal of water in syngas mediating reactions using integrated hydrophilic membranes has been previously tested both experimentally and through simulations. Rhode et al. [3] showed that the incorporation of an inert hydrophilic membrane for *in situ* water removal in a packed-bed membrane reactor (PBMR) promoted CO_2 capture in Fischer-Tropsch (FT) reactions. Simulations studies have demonstrated the potential and limits of in-situ removal of water under DME synthesis conditions to favor the CO_2 catpure in this process [4]. The model predicts that water removal from the reactive media during DME production from a fixed-bed reactor integrating a hydrophilic membrane favours the thermodynamic equilibrium of DME formation, thus an improvement of the reaction yield is envisaged. Among the different hydrophilic membrane materials tested in the literature [5], only zeolite membranes withstood the high demanding operating conditions (225-325 °C, 10-40 bar) required during DME synthesis. The zeolite membranes applied by Espinoza et al. [6] under FT conditions for *in situ* water removal were ZSM5, Mordenite (MOR) and Silicalite (SIL).

In this work, the suitability of these zeolite membranes in a PBMR for CO_2 capture with adequate DME production has been evaluated theoretically. The influence of the flow rate of the sweep gas stream on the CO_2 capture and DME yield has been evaluated.

2 Mathematical model

A mathematical model able to describe the mass transport through the membrane of all the components present in the system has been developed [7]. Using the same experimental conditions as in Diban et al. [7] and selecting the zeolite membrane with low H_2O permeabilities and high H_2O/component i permselectivities reported there, the enhancement of the DME yield was evaluated through the recirculation of the sweep stream. The sweep gas was composed of a mixture of $H_2+CO+CO_2$, in a similar molar ratio as the feed stream (3/0.5/0.5). The influence of the sweep gas flow rate, $F_{CO_x,FSG}$, on the process performance was evaluated in the range between 0.06 to 1.80 $mol_{COx}\,h^{-1}$ and compared to the performance of a PBR where water was not removed from the feed stream. The composition of the feed stream was also $H_2+CO+CO_2$ at the molar ratio 3/0.5/0.5 and the flow rate at the entrance of the reactor was fixed at 0.06 $mol_{COx}\,h^{-1}$.

3 Results and discussion

In this work, the influence of the fresh sweep flow rate, $F_{CO_x,FSG}$, on the process performance was studied. The results showed that CO_2 conversion (X_{CO_2}) to DME was only possible, that is, CO_2 was consumed instead of being produced in the reactor, using a PBMR configuration with *in situ* water removal. In contrast, the PBR configuration, X_{CO_2} was negative, indicating the CO_2 production under those working conditions. The increase in $F_{CO_x,FSG}$ led to higher X_{CO_2} . On the contrary, the tendency observed for DME yield Y_{DME} at higher $F_{CO_x,FSG}$ was the opposite going below the Y_{DME} values in a PBR (16.3 %) when the sweep flow rate was 1.80 mol_{COx} h^{-1}. At a $F_{CO_x,FSG}$=0.06 mol_{COx} h^{-1}, Y_{DME} was above 20%. This was attributed to the high methanol permeance through the membrane that presented high methanol selectivity due to its hydrophilic character. The loss of methanol from the reaction site reduced its possible conversion towards DME.

4 Conclusions

In this work, it was demonstrated theoretically that the use of a zeolite membrane for *in situ* water removal in a PBMR for DME synthesis allowed CO_2 capture. Increasing the sweep gas flow rates enhances outstandingly the CO_2 capture with conversions above 80%, while the DME yield is reduced below that attained in a PBR (16.3%). A compromise sweep gas flow rate is recommended in the range between 0.18 and 0.6 mol_{COx} h^{-1} that improves CO_2 conversion remaining DME yield similar to PBR yields and does not implies high reactants consumption. Further evaluation of the effect of recycling of the sweep gas stream to the feed side on the process performance must be done.

5 Acknowledgements

Postdoctoral grant JCI-2011-10994 and finantial support by projects CTQ2008-00690, CTQ2010-19188 and ENE2010-15585 are kindly acknowledged.

6 References

[1] F. Pontzen, W. Liebner, V. Gronemann, M. Rothaemel, B. Ahlers, CO_2-based methanol and DME-Efficient technologies for industrial scale production, Catalysis Today, 171 (2011) 252-250.

[2] J. Ereña, I. Sierra, A.T. Aguayo, A. Ateka, M. Olazar, J. Bilbao, Kinetic modeling of dimethyl ether synthesis from (H_2+CO_2) by considering catalyst deactivation, Chemical Engineering Journal, 174, (2011) 660-667.

[3] M.P. Rhode, D. Unruh, G. Schaub, Membrane application in Fischer-Tropsch synthesis to enhance CO_2 hydrogenation, Industrial and Engineering Chemistry Research, 44, (2005) 9653-9658.

[4] I. Iliuta, F. Larachi, P. Fongarland, dimethyl ether synthesis with in situ H_2O removal in fixed-bed membrane reactor: model and simulations, Industrial and Engineering Chemistry Research, 49, (2010) 6870-6877.

[5] N. Diban, A.T. Aguayo, J. Bilbao, A. Urtiaga, I. Ortiz, Membrane reactors for in situ water removal: a review of applications, Industrial and Engineering Chemistry Research 52, (2013) 10342-10354.

[6] R.L. Espinoza, E. du Toit, J. Santamaria, M. Menedez, J. Coronas, S. Irusta, Use of membranes in Fischer-Tropsch reactors, Studies in surface science and catalysis 130 A, (2000) 389-394.

[7] N. Diban, A.M. Urtiaga, I. Ortiz, J. Ereña, J. Bilbao, A.T. Aguayo. Influence of the membrane properties on the catalytic production of dimethyl ether with in situ water removal for the successful capture of CO_2, Chemical Engineering Journal 234, (2013) 140-148.

Controlled synthesis of MgAl layered double hydroxide grafted TiO_2 cuboids and their photocatalytic CO_2 conversion activity

Cunyu Zhao [a], Lianjun Liu [a], Huilei Zhao [a], Luhui Wang [b], Jinye Xu [a], Ying Li [a*]

a University of Wisconsin-Milwaukee, Mechanical Engineering Department, Milwaukee, WI 53211, USA

b Zhejiang Ocean University, Chemical Engineering Department, Zhejiang, P.R. China

* Tel: 1-414-229-3716
Fax: 1-414-229-6958
E-mail: liying@uwm.edu

Abstract: A series of MgAl layered double hydroxide (LDH) grafted TiO_2 cuboids materials with various molar ratios of (Mg+Al) to Ti were synthesized by a combination of hydrothermal and coprecipitation method, in which the nucleation and growth of MgAl LDH platelets were controlled effectively. In our experiment, MgAl LDH platelets grew on the assembled TiO_2 cuboids. It is noted that this is the first time that MgAl LDH grafted TiO_2 cuboids novel structured composites has ever been reported. The calcined MgAl LDH grafted on TiO_2 composites successfully kept the same structure as MgAl LDH grafted TiO_2 cuboids and the calcined composites were applied to photocatalytic CO_2 reduction with H_2O. The compositions of MgAl LDH grafted TiO_2 composites were optimized according to the photocatalytic activity of CO production from CO_2 reduction with H_2O. The TiO_2-MgAl LDH sample with 72.2wt.% TiO_2 composition exhibited the highest photocatalytic activity. The possible mechanism was proposed in this work to illustrate that the adsorbent-photocatalyst hybrid materials are beneficial to CO_2 photoreduction.

Keywords: MgAl layered double hydroxide (LDH), TiO_2 rods, CO_2 conversion, photocatalytic

385714 Comparison of in-Situ Hydrogen Oxidation Reaction Kinetics for Various Polymer Hydroxide Exchange Electrolytes

Abstract:

Previous studies using rotational disk electrode (RDE) and in-situ H_2-pump set-ups have shown that the kinetics for the hydrogen oxidation and evolution reactions (HOR/HER) on carbon supported platinum (Pt/C) are on the order of 100x larger in acid than in base. However, no study has been carried out on the difference in i_0 between liquid and solid polymer alkaline systems. In this study, the exchange current density (i_0), a parameter used to quantify the kinetics of the HOR/HER, was calculated for Pt/C interfaced with a number of hydroxide exchange membranes (HEMs) and ionomers using the Butler-Volmer equation and in-situ kinetic studies. Comparison of the exchange current densities in H_2-pump and RDE set-ups allows insight into how the HOR/HER kinetics differ in each system. Initial data with a commercial HEM and ionomer material show the i_0 for Pt/C in a HEM fuel cell to be two times larger than the i_0 measured with RDE, meaning the HEM materials that transport the hydroxide ions to the catalyst surface create a different type of alkaline environment than the liquid electrolyte transporting free hydroxide ions in RDE. Various HEM and respective ionomer materials are substituted into the in-situ H_2-pump to observe how different cationic side chains affect the alkaline environment in a HEM fuel cell.

385852 Insights into Hydrogen Oxidation Reaction Mechanisms in Alkaline Electrolytes through the Use of Catalyst with Tunable Oxophilicity

Abstract:

Hydroxide exchange membrane fuel cells (HEMFC) are an attractive alternative to conventional proton exchange membrane fuel cells (PEMFC) due to the potential of incorporating cost effective materials. The hydrogen oxidation reaction on platinum's (Pt) surface, which requires a near zero over potential in acidic conditions, turns out to be two orders of magnitude slower in an alkaline environment. Two theories that have been proposed to explain this decreased activity. One focuses on a change in mechanism that involves hydroxide ions in the hydrogen oxidation pathway, while the other suggests it is purely a result of the shift in Pt-hydrogen binding energy in a different chemical environment. The first theory emphasizes the need for proper oxophilicity to allow for increased kinetics. Therefore, by tuning a catalyst's affinity for oxygen, an enhanced HOR performance should be observed. In this study we present HOR performance on the PtCu catalyst supported on CuNW. By altering the degree of blending of Pt and Cu, different Pt-oxygen affinities can be achieved. However, investigating the effect of the different PtCu catalysts on HOR yielded no significant difference in performance suggesting that the oxophilic nature of the catalyst is of secondary significance.

386052 Analysis of Perovskites and Layered Perovskite Oxides As Materials for Conversion of Carbon Dioxide to Carbon Monoxide in Thermochemical Cycles

Abstract:

More than 30,000 Mt are emitted each year to the atmosphere, yet less than 1% of the CO_2 produced is re-utilized in industrial processes such as urea and salicylic acid synthesis. Here, an alternative to CO_2 utilization is proposed, where CO_2 can be transformed to high-value chemicals that could be used as a source for C1 chemistry by thermochemical cycles (TC).

Perovskites (ABO_3) are ideal for TCs because they have the ability to form oxygen vacancies while maintaining its crystalline structure, at lower temperatures than similar oxides (~820 °C) such as ceria and spinels. Layered perovskite (A_2BO_4) have similar properties to perovskites and therefore could be potential candidates for adsorption and conversion of carbon dioxide. The goal of the study is to demonstrate an oxide that has equilibrium between adsorption strength and CO_2-conversion capabilities. The stability of the CO_2 molecule posses a challenge because of the energy needed to break the C=O bond.

In this work, different perovskites, layered perovskites and spinels ($La_{0.75}Sr_{0.25}CoO_3$, $La_{0.75}Sr_{0.25}FeO_3$, $La_{0.75}Sr_{0.25}Co_{0.5}Fe_{0.5}O_3$, $LaSrCoO_4$ and $CeCo_2O_4$) were studied for their oxygen formation capabilities under inert environment and their carbon dioxide adsorption strength properties. The material that can convert carbon dioxide to carbon monoxide on its oxygen vacancies with the minimal energy input will be used for further studies on light-induced thermochemical conversion of CO_2 to CO.

Modeling Transport-Kinetic Interactions for SO_2 Oxidation to SO_3 in Particulate and Monolith Catalysts Using COMSOL Multiphysics

Anuradha Nagaraj

Department of Environmental Engineering

Texas A&M University-Kingsville

Patrick L. Mills

Department of Chemical and Natural Gas Engineering

Texas A&M University-Kingsville

Introduction. Development of next-generation chemical processes that have zero emissions is a key environmental objective for sustainable development. The manufacture of H_2SO_4 by the air oxidation of SO_2 to SO_3 is an important technology where an opportunity exists for new catalyst development and process innovation by reducing emissions of unconverted SO_2 in process reactor tail gases owing to the sheer number (> 1500) and scale (ca. 500 to 4500 metric tons/day) of typical plants. The global supply of H_2SO4 is projected to grow from 200 MM tonnes in 2006 to more than 258 MM tonnes in 2015 with a value of > \$10 MMM [1]. An opportunity exists to develop new innovations in environmental catalysis and reaction engineering for an important technology that has a rich and long history with positive economic growth. Best Available Control Technologies (BACT) for controlling SO_2 emissions from stack gases involves additional processing resulting are cost intensive and well-studied. However, emissions control technologies using front-of-pipe technology, such as new catalyst technology, improved reactor designs, and process operational strategies have greater potential to be more economical versus end-of-pipe scrubbing technologies.

From a catalyst design perspective, specification of catalyst composition, catalyst size, catalyst shape and internal pore structure are interlinked parameters that have received notable attention by catalyst scientists and engineers since these properties affect the catalyst effectiveness factor, bed pressure drop, rate of catalyst attrition, catalyst mechanical integrity, catalyst life, amongst other key measures of performance. It is evident that specification of preferred catalyst design parameters requires careful optimization since tradeoffs exist in how the observed reaction rate and other measures of performance may be affected by altering one or both of these parameters.

The primary objective of this study is three-fold: (1) to review the current H_2SO_4 catalysts utilized in the oxidation of SO_2 to SO_3; (2) to evaluate monolith catalysts as potential candidates as SO_2 oxidation catalysts; and (3) to develop modeling framework to compare reactor designs for both particulate and monolith catalysts under typical multi-pass convertor operation with potential for process optimization.

Methods. Typical commercial H_2SO_4 catalyst compositions consist of silica support, vanadium and possibly metal promoters. Typical monolith catalyst compositions and preparation techniques for SO_2 oxidation are described in patents of Felthouse *et al.* [2,3]. The current state-of-the-art of both particulate and monolithic catalysts based upon recent open and patent literature will be reviewed.

COMSOL Multiphysics is used to model non-isothermal diffusion, reaction and transport effects in a monolithic catalyst with compositional and temperature dependence of all model parameters. This builds upon our previous work [4] on modeling of transport-kinetic interactions in particulate catalysts having various complex shapes. There is a knowledge gap in monolith modeling for SO_2 oxidation reaction since the only known open literature on modeling SO_2 oxidation reaction on monoliths was the numerical model developed by Bespalov *et al.* in 1991 [5] and therefore, it is important to develop advanced models for better understanding of the transport-kinetic interactions of this reaction in a monolith channel. A new model that describes transport-kinetic interactions in monolith catalysts is described that is valid for process conditions encountered in multi-pass convertor operation. The reaction kinetic model is based upon the work of Collina *et al.* [6] since it accounts for the dependence of the SO_2 oxidation rate on the partial pressures of SO_2, O_2, and SO_3 with inhibition by both SO_2 and SO_3. The reaction rate and adsorption equilibrium parameters in this model are valid from 420 to 550°C, which is within the commercial operating range of most convertors. COMSOL Reaction Engineering Lab is used to study various adiabatic reactor designs for SO_2 converter performance under ideal plug flow conditions, while a more advanced COMSOL Multiphysics model is described that accounts for non-ideal gas and thermal energy interphase and intraphase transport effects.

Results and Discussion. The concentration profiles shown in Figure 1 are from a case study of catalytic abatement of a volatile organic compound (VOC) to describe monolith reactor performance for contaminant removal from waste gas. A similar approach will be followed to develop monolith catalyst model for SO_2 oxidation reaction.

Additional results that compare the performance of particulate versus monolithic catalyst shapes using typical process conditions will be presented and discussed, thereby showing the incentive for development of monolithic catalyst with higher activity and reduced emissions.

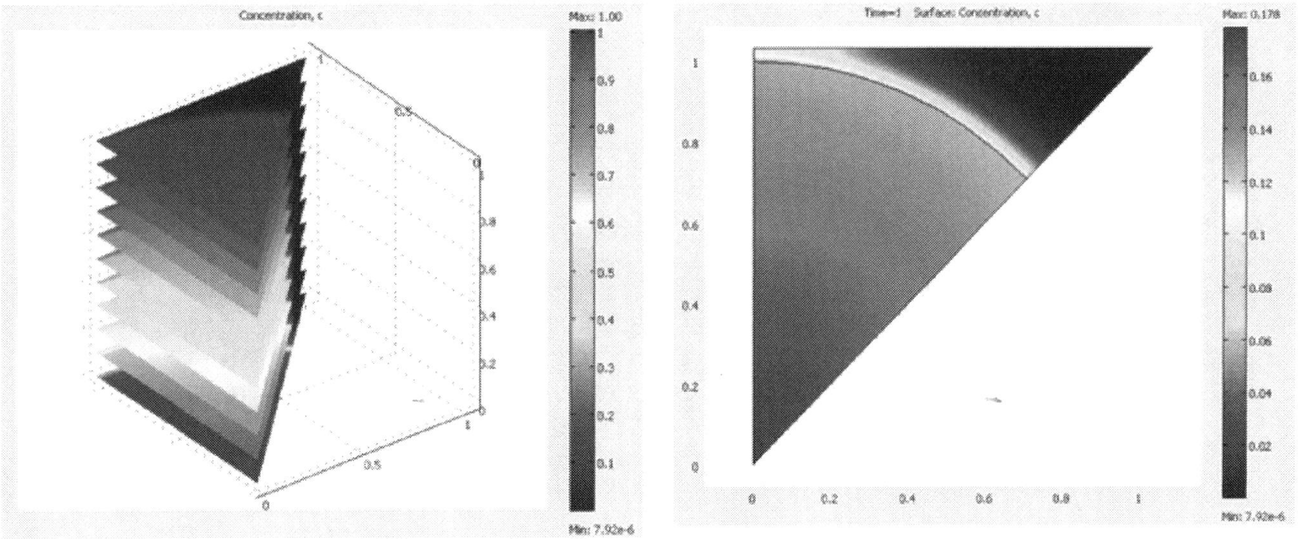

Figure 1. (a) Concentration profile along the length of the reactor. (b) Concentration profile at the outlet.

References

1. British Sulphur Consultants, Sulphuric Acid : Global Supply and Demand in the Next Decade, Topsøe Catalysis Forum – Denmark, August 23rd to 24th 2007.

2. Felthouse, T. R. (1992). Monolithic catalysts for conversion of sulfur dioxide to sulfur trioxide. US, Monsanto Company

3. Felthouse, T. R. and A. Vavere (1993). Monolithic catalysts for conversion of sulfur dioxide to sulfur trioxide. US, Monsanto Company

4. Nagaraj, A., & Mills, P. L. "Analysis of Heat, Mass Transport, and Momentum Transport Effects in Complex Catalyst Shapes for Gas-phase Heterogeneous Reactions Using COMSOL Multiphysics". Paper Presented at the COMSOL Conference 2008 Boston, MA

5. Bespalov, A.V., et al., "Numerical modeling of sulfur dioxide oxidation reaction in through-channel of a block catalyst with a honeycomb structure". Zhurnal Prikladnoi Khimii, 1991. 64(10): 2048-2053.

6. Collina, A., Corbetta. D. and Cappelli, A. "Use of Computers in the Design of Chemical Plants," 97th Event of the European Federation of Chemical Engineering, Firenze (1970).

386434 Catalyst/Support Interactions Between Pt Nanoparticles and Amorphous Silica

Abstract:

Metal nanoparticles (NPs) play a key role in catalysis due to large surface-to-volume ratios and low-coordination active surface sites. These catalysts are typically immobilized on oxide supports to suppress sintering, which leads to catalyst deactivation and degradation of catalyst selectivity. An atomistic understanding of catalyst systems is essential to the rational design of improved catalyst systems. Catalyst activity and selectivity are highly dependent on shape and size of NPs as well as the nature of the oxide support. Despite significant advancement of experimental characterization techniques, a detailed understanding of the catalyst-support interface for very small NPs (~ 1 nm or smaller) is largely precluded by experimental investigations. Computational methods provide one means to overcoming this challenge. In previous studies catalyst supports have generally been either neglected or treated as highly ideal structures, however, oxide supports are often used in an amorphous state, exhibiting a wide range of surface sites.

Due to its thermal stability and tunable porosity, and hence specific surface area, amorphous silica is widely used as a catalyst support, but is still poorly understood at the atomistic level due to limitations in both experimental and computational methods. We have developed and experimentally validated a method for generating realistic atomistic surface models of amorphous silica for a range of temperatures by simulating the process of surface dehydroxylation.

Using these surface models, we have studied the effects of amorphous silica supports on small platinum NPs (13 and 55 atoms) with regards to their catalytic properties. We find that the morphology of Pt clusters undergo significant restructuring that depends on the local structure at the Pt-Silica interface. This restructuring typically leads to significantly increased exposure of lower-coordination atoms than are present on the unsupported NP. In contrast, Pt55 typically undergoes only modest changes in lattice spacing and generally maintains its unsupported structure. The degree of restructuring as well as metal-support interaction strength is strongly correlated with the number of hydroxyl groups at the Pt-silica interface, and hence, the pretreatment temperature of the silica support. Preliminary investigations indicate that metal-support interactions can strongly affect the binding of CO and O for Pt13.

Effect of flow regime on trickle-bed reactor performance

Shinbeom Lee and Arvind Varma

School of Chemical Engineering, Purdue University,
West Lafayette, IN 47907, USA

ABSTRACT

Trickle bed reactors, composed of gas and liquid feeds on solid catalysts in a fixed bed, have four different flow regimes. Among these, the operation in pulsing flow is promising to enhance reaction conversion and yield due to its unique characteristics. This study was carried out to confirm the enhancement of reactor performance in pulsing flow for complex reaction operated at high pressure and temperature. For this purpose, hydrogenation of acetophenone on 1% Rh on alumina catalyst was selected as the target reaction through slurry reactor tests. A 3/8" stainless steel tube filled with 0.6mm catalyst spheres was utilized along with a liquid circulation system. The hydrodynamic tests were conducted to determine the trickle/pulsing flow transition, to establish the reactor operating conditions. In the tests, the effects of various parameters on the regime boundary were investigated using water/N_2 and reactants/H_2 with 0.03-0.3 m/s and 2-20 mm/s for gas and liquid linear velocities respectively at 1-25 bar, 25-110 ^{o}C. Through reaction tests in trickle and pulsing flow, the effect of pulsing flow on conversion and selectivity of acetophenone hydrogenation was determined.

Keywords: pulsing flow, trickle flow, fixed-bed, acetophenone hydrogenation, hydrodynamics

Nobel metal-based bifunctional catalysts for bioethanol conversion into ethyl acetate

Jean Marcel R. Gallo,[1] Murillo C. Ribeiro,[1] Clelia Mara P. Marques,[1] José Maria C. Bueno[2]

[1]Department of Chemistry, Federal University of São Carlos, São Carlos, Brazil

[2]Department of Chemical Engineering, Federal University of São Carlos, São Carlos, Brazil

Bioethanol is a lignocellulosic biomass-derived chemical that can be directly used as a fuel or converted into chemicals. Since 1975, bioethanol is used as fuel in Brazil and in recent years there has been a growing interest in converting ethanol into important chemicals, such as ethylene, butadiene and ethyl acetate. Ethyl acetate, for example, is widely used as a replacement for aromatic solvents in paints, inks and dyes; moreover, it can be used in the pharmaceutical, cosmetic, and food industry. While this solvent is commercially produced from non-renewable sources, in countries with high bioethanol production, such as the USA and Brazil, there is an increasing commercial and environmental interest in its synthesis from bioethanol.

The most studied reaction route for bioethanol conversion into ethyl acetate is in inert atmosphere using copper based catalysts (for example, $CuCr_2O_4$ and Cu/ZrO_2 catalysts) and involves the dehydrogenation of ethanol into acetaldehyde (catalyzed on the copper surface) followed by a coupling reaction between the acetaldehyde and an ethanol molecule (catalyzed by both copper and support). The main advantages of this process are the use of inexpensive metal catalysts, and the coproduction of hydrogen, a valuable chemical for industry. Contrarily, there are several drawbacks: the reaction is endothermic, it requires high temperatures and high metal loadings, catalysts deactivate relatively fast, and it is observed the formation of byproducts that form azeotrope with the ethyl acetate. It is also important to mention that copper catalysts are not active in a more straightforward (and exothermic) one-pot route for ethyl acetate synthesis, i.e., the oxidation of bioethanol to acetic acid followed by the esterification with ethanol. The reaction schemes for both routes of ethanol conversion into ethyl acetate are shown below:

Dehydrogenation/Coupling (inert atmosphere)
Step 1: $CH_3CH_2OH \rightleftharpoons CH_3CHO + H_2$
Step 2: $CH_3CHO + CH_3CH_2OH \rightleftharpoons CH_3CO_2CH_2CH_3 + H_2$
Overall reaction: $2\ CH_3CH_2OH \rightleftharpoons CH_3CO_2CH_2CH_3 + 2\ H_2$

Oxidation/esterification
Step 1: $CH_3CH_2OH + ½\ O_2 \rightleftharpoons CH_3CHO + H_2O$
Step 2: $CH_3CHO + ½\ O_2 \rightleftharpoons CH_3CO_2H$
Step 3: $CH_3CO_2H + CH_3CH_2OH \rightleftharpoons CH_3CO_2CH_2CH_3 + H_2O$
Overall reaction: $2\ CH_3CH_2OH + O_2 \rightleftharpoons CH_3CO_2CH_2CH_3 + 2H_2O$

Herein, gold and silver-based bifunctional catalysts were studied for bioethanol conversion into ethyl acetate in both inert (dehydrogenation/coupling) and oxidant (oxidation/esterification) atmosphere. Catalysts preparation involved the modification of supports such as metal oxides, silica and aluminosilicates with gold or silver.

From the catalysts tested, the best results were obtained using gold supported on m-ZrO_2 (monoclinic phase). In inert atmosphere at 523 K, the selectivities for ethyl acetate and acetaldehyde were, respectively, 63 and 28 % (at 71 % ethanol conversion). The main byproducts formed were crotonaldehyde (4 %), acetone, diethyl ether and methyl ethyl ketone (< 2 % each). Compared to classical copper catalysts (for example, Cu/ZrO_2), the Au/ZrO_2 catalyst led to similar ethyl acetate yield, but lower formation of byproducts (mainly methyl ethyl ketone, which forms an azeotrope with both ethyl acetate and ethanol). In the copper system, several condensation reactions are observed yielding heavy compounds, however it is not fully understood if they happen in the Cu or zirconia surface. In the gold catalysts, these side condensation reactions are minimized and depending on the conditions, ethyl acetate and acetaldehyde accounts to 91-98 % of the products. Replacing monoclinic zirconia by other supports such as silica, alumina, silicalumina, or tetragonal zirconia, led to a significant decrease in the ethyl acetate selectivities. The support nature affects significantly the catalyst performance because the reaction takes place in the interface between metal and support. While the dehydrogenation of ethanol to acetaldehyde happens in the gold surface, the aldehyde has to be coordinated in a metal site close to the interface with the support in order to undergo coupling reaction to form ethyl acetate.

Au/ZrO_2 was also tested in the one-pot conversion of ethanol on ethyl acetate in oxidant atmosphere. In this mechanism, gold catalyze the oxidation of ethanol to acetic acid, while the support is responsible for the esterification of acetic acid with ethanol to yield ethyl acetate. This reaction can be carried out in temperatures as low as 448 K and selectivities for acetaldehyde and ethyl acetate of 49 and 44 % were, respectively, obtained (70 % conversion of ethanol). The main byproducts observed were diethyl ether (5 %) and crotonaldehyde (1 %). The results indicate that zirconia was effective catalyzing the esterification reaction; however, in the conditions used, a significant part of the ethanol is oxidized to acetaldehyde instead of acetic acid.

JMR Gallo thanks FAPESP for the fellowship (process number 2013/15861-1)

387319 Establishing the Connection Between the Geometric and Electronic Structure of Oxygen Species on Ag Surfaces: First-Principles DFT and Monte Carlo Studies

Abstract:

Surface adsorbed species differentiated by local geometric and electronic structures often have distinct reactivity toward chemical transformations. It has been shown that during partial oxidation of ethylene electrophilic oxygen species on silver are responsible for the formation of ethylene oxide while the nucleophilic oxygen primarily leads to total combustion.[1] Although much work has been done on identifying the electronic fingerprint of surface adsorbed species, the link between geometric and electronic structure of these species is still lacking.[2]

Our objective is to establish the connection between the geometric and electronic structure of oxygen species on Ag surfaces (flat or stepped) under relevant catalytic conditions. We have employed an atomistic thermodynamics approach and Monte Carlo simulations to extend the results of DFT calculations to relevant catalytic conditions (T, P). To ensure complete sampling of the large space of possible surface configurations we have used the cluster expansion method to systematically coarse-grain the DFT-calculated energies.[3] The electronic structure (e.g., core-level shift) of oxygen species has been probed using density functional theory (DFT) calculations.[4]

[1] R. B. Grant and R. M. Lambert, "A single crystal study of the silver-catalysed selective oxidation and total oxidation of ethylene," *J. Catal.*, vol. 92, no. 2, pp. 364–375, Apr. 1985.

[2] T. C. R. Rocha, A. Oestereich, D. V Demidov, M. Hävecker, S. Zafeiratos, G. Weinberg, V. I. Bukhtiyarov, A. Knop-Gericke, and R. Schlögl, "The silver-oxygen system in catalysis: new insights by near ambient pressure X-ray photoelectron spectroscopy.," *Phys. Chem. Chem. Phys.*, vol. 14, no. 13, pp. 4554–64, Apr. 2012.

[3] A. Walle and G. Ceder, "Automating first-principles phase diagram calculations," *J. Phase Equilibria*, vol. 23, no. 4, pp. 348–359, Aug. 2002.

[4] A. Nilsson and L. G. M. Pettersson, "Chemical bonding on surfaces probed by X-ray emission spectroscopy and density functional theory," *Surf. Sci. Rep.*, vol. 55, no. 2–5, pp. 49–167, Oct. 2004.

Water-treated $Rh/\gamma\text{-}Al_2O_3$ Catalyst for Methane Partial Oxidation

Xia Xu[1,2], *Alan M. Lane*[1,2*]

[1]*Center for Materials for Information Technology, The University of Alabama, Tuscaloosa, AL 35487 USA.*

[2]*Department of Chemical and Biological Engineering, The University of Alabama, Tuscaloosa, AL 35487 USA.*

Abstract

A water-treatment technique was applied to a conventional $Rh/\gamma\text{-}Al_2O_3$ catalyst. Both standard and water-treated $Rh/\gamma\text{-}Al_2O_3$ catalysts were prepared and their catalytic performance was tested in the methane partial oxidation reaction. The results showed that the water-treated $Rh/\gamma\text{-}Al_2O_3$ catalyst shows higher CO selectivity and lower CO_2 selectivity between 300 and 600 °C, compared with standard-treated catalyst. The enhancement is attributed to the formation of well-dispersed smaller Rh nanoparticles.

Keywords: $Rh/\gamma\text{-}Al_2O_3$ catalyst; methane partial oxidation; water treatment.

1. Introduction

Production of synthesis gas (H_2 and CO gas mixture) by methane (CH_4) partial oxidation (MPO) is an important intermediate step in many existing energy conversion technologies, such as fuel cells and Fischer-Tropsch reaction [1-5]. SDecreasing the reaction temperature and improving reaction efficiency has been a great challenge to researchers. One of the solutions is to develop high-efficiency catalysts. Our group in previous research [6-8] showed that treating a $Pt/\gamma\text{-}Al_2O_3$ catalyst at 500 °C in a humid reducing environment (H_2O/H_2) can significantly decrease the reaction temperature of catalytic CO preferential oxidation in H_2. This is due to the decreased particle size and increased stability of Pt and decreased CO chemisorption strength

* Corresponding author contacted by telephone: +1-205-348-6367, fax: +1-205-348-7558 and email: alane@eng.ua.edu.

after this treatment. In subsequent research by others, a similar water-treatment technique for reducing reaction temperature was also investigated and confirmed [9,10].

In this work, we applied the same water-treatment to Rh/γ-Al_2O_3 catalysts and investigated its catalytic effects on the MPO reaction. The reaction products were analyzed by Mass Spectrometry (MS), and the catalyst morphology was characterized by Transmission Electron Microscopy (TEM). The results show that water-treated Rh/γ-Al_2O_3 catalyst can effectively decrease the size of Rh nanoparticles, and water-treated Rh/γ-Al_2O_3 catalyst shows higher CO selectivity and lower CO_2 selectivity in the temperature range of 300-600 °C.

2. Experiment

2.1 Catalyst preparation

Catalyst standard-treatment: 1 wt% Rh/γ-Al_2O_3 catalyst was prepared by the incipient wetness method. Alumina (γ-Al_2O_3, 150 m^2/g, acid type, ~0.25 mm, Alfa) was first calcined in air at 500 °C for 2 hours to remove volatiles, and then impregnated using an aqueous solution of $Rh(NO_3)_2$ (Alfa). The catalyst was dried in a fume hood over night, followed by calcining in air at 500 °C for 2 hours. The catalyst was finally reduced in H_2 at 500 °C for 2 hours, and cooled down with N_2 protection. This process is considered to be the standard-treatment of the catalyst, as shown in Fig. 1.

Catalyst water-treatment: Water treatment begins after the standard treatment of catalyst. The standard-treated catalyst (0.1 g) was wetted with 5 mL deionized (DI) water, and then reduced in H_2 for 1 hour at 500 °C with heating rate of 10 °C/min, and finally cooled down to room temperature with N_2 protection. This process is considered to be water treatment of the catalyst, as shown in Fig. 1

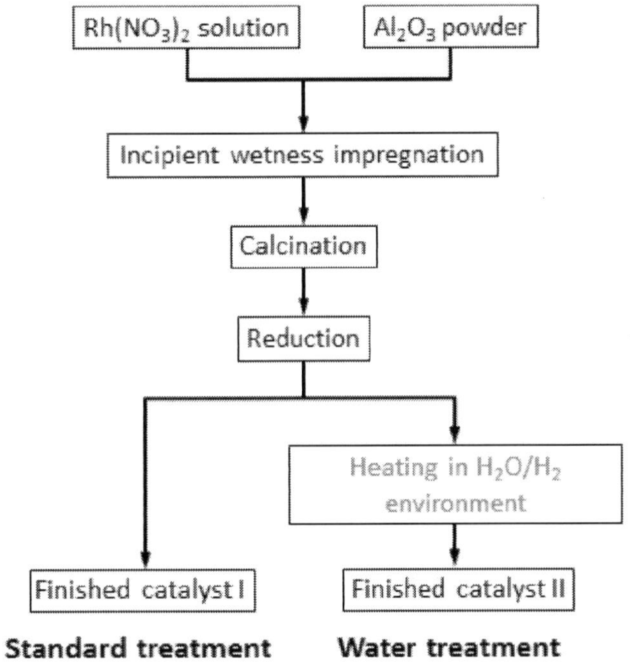

Fig.1. Rh/γ-Al_2O_3 catalyst preparation process: standard treatment and water treatment.

2.2 Catalyst testing and characterization

Figure 2 shows a schematic of MPO system for testing the above catalysts. A Hiden Catlab Microreactor with a vertical quartz tube (6 mm I.D.) was used for the reaction. The gas flow rates were controlled by mass flow controllers (MFC), and the reaction temperature was controlled by a programmable controller. The catalyst (0.1 g) was fixed inside the vertical reactor with quartz fibers. The gas flow rates of CH_4, O_2 and Ar were controlled to 40 ml/min, 20 ml/min and 20 ml/min, respectively. A mass spectrometer (Hiden QIC-20, MS) was used to analyze the reaction products, and Ar was used as an internal standard. A drier was located between reactor and MS to remove the water from the products, to prevent plugging of the MS capillary tube by water. The test conditions were between room temperature and 800 °C at 1 atm, and data was collected every 50 °C from 300 °C to 800 °C, as the initiation temperature of MPO with a Rh/γ-Al_2O_3 catalyst is around 275 °C [11].

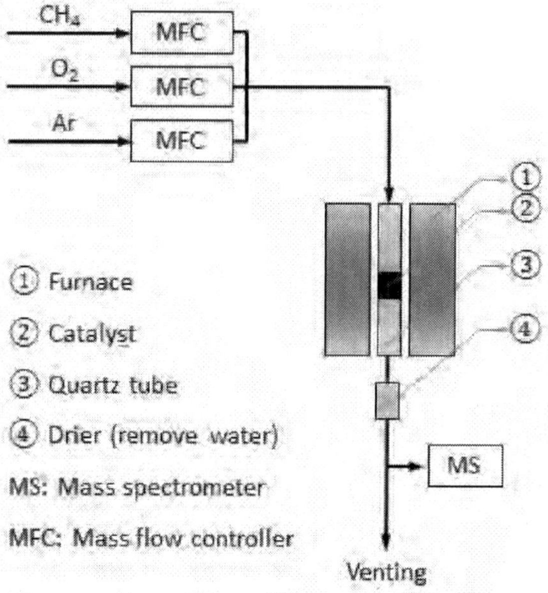

Fig. 2. A schematic of methane partial oxidation system.

The conversion of CH_4 and the selectivity of CO and CO_2 were calculated using the following formulas:

$$CH_4\ \text{Converstion:}\ \frac{CO_{out}+CO_{2\ out}}{CO_{out}+CO_{2\ out}+CH_{4\ out}} \tag{1}$$

$$CO\ \text{Selectivity:}\ \frac{CO_{out}}{CO_{out}+CO_{2\ out}+CH_{4\ out}} \tag{2}$$

$$CO_2\ \text{Selectivity:}\ \frac{CO_{2out}}{CO_{out}+CO_{2\ out}+CH_{4\ out}} \tag{3}$$

Transmission Electron Microscopy (TECHNAI FEI 20, TEM) was used to characterize the size and morphology of the catalysts. Each catalyst was ground in a mortar for 5 min, and then dispersed in methane with ultrasonication. The dispersion was dropped onto a copper TEM grid with a carbon substrate, and then dried for TEM characterization.

3. Results and discussion

Figure 3 shows the CH_4 conversion, CO and CO_2 selectivity of standard-treated ("S") and water-treated ("W") Rh/γ-Al_2O_3 catalysts in the MPO reaction. CH_4 conversion under both catalysts is similar in the temperature range of 300 - 600 °C. However, water-treated catalyst shows higher CO selectivity and lower CO_2 selectivity than standard-treated catalyst between 300 and 600 °C. The selectivity of CO and CO_2 under both catalysts becomes similar when the temperature is higher than 600 °C. This can be attributed to the MPO reaction reaching the thermodynamic equilibrium, and catalysts cannot change the reaction behavior.

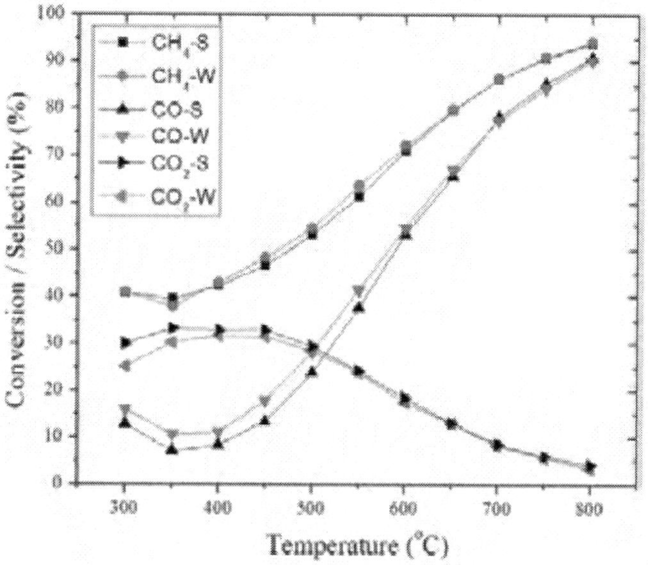

Fig. 3. The CH_4 conversion, CO and CO_2 selectivity of standard-treate d and water-treated 1 wt% Rh/γ-Al_2O_3 catalyts in methane partial oxidation reaction (S: standard-treated catalyst, W: water-treated catalyst).

The TEM images of standard-treated and water-treated catalysts are shown in Fig. 4a and Fig. 5a. The dark parts are Rh nanoparticles, and the brighter parts are γ-Al_2O_3 particles, as Rh has higher electron density than Al because of larger atomic number. The average size of the Rh nanoparticles in the water-treated catalyst is 4.7 nm (Fig. 5b), much smaller than that in standard-treated catalyst of 8.1 nm (Fig. 4b).

The increased selectivity of CO and decreased selectivity of CO_2 with the water-treated catalyst could be attributed to smaller Rh nanoparticles on the alumina support, as they could provide larger surface areas and increase the contact area with reactants, thus prodiving higher reaction efficiency. Unpublished results [12] for Pt/γ-Al_2O_3 suggest that the chemisorption strength of CO onto water-treated catalysts is significantly reduced, allowing more competitive adsorption of other reactants and products. These water-treatment mechanisms require more investigation.

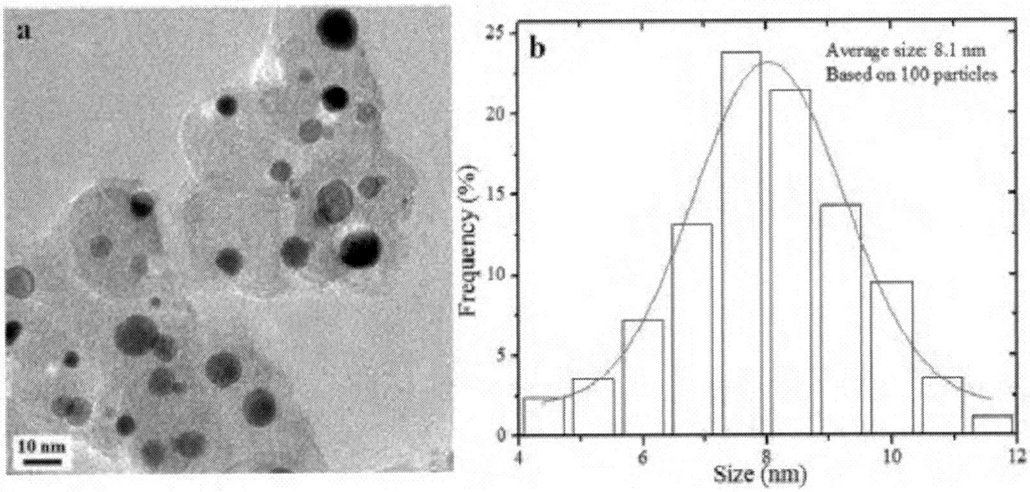

Fig. 4. Standard-treated 1 wt% Rh/γ-Al_2O_3 catalyst: (a) TEM image; (b) Rh nanoparticle size distribution .

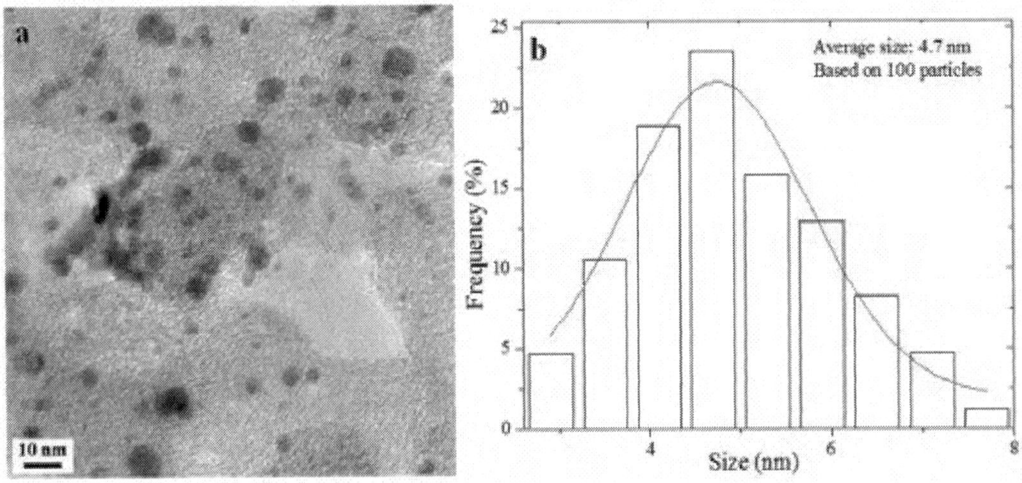

Fig. 5. Water-treated 1 wt% Rh/γ-Al_2O_3 catalyst: (a) TEM image; (b) Rh nanoparticle size distribution .

4. Conclusions

Both standard-treated and water-treated catalysts were tested for themethane partial oxidation reaction. Water-treated catalyst shows higher CO and CO_2 selectivity in the temperature range of 300-600 ^{o}C. The enhancement is attributed to the formation of well-dispersed smaller Rh nanoparticles after water treatment. Further research is needed to explore the mechanism of Rh nanoparticle size decreasing during the water treatment process.

Acknowlegment

This work was supported by Department of Energy (DOE) under contract GR23134, Institute for Sustainable Energy. This work utilized TEM owned and maintained by the Central Analytical Facility, which is supported by The University of Alabama.

References

1. Khine, M. S. S., Chen, L. W., Zhang, S., Lin, J. Y. & Jiang, S. P. Syngas Production by Catalytic Partial Oxidation of Methane over $(La_{0.7}A_{0.3})BO_3$ (A = Ba, Ca, Mg, Sr, and B = Cr or Fe) Perovskite Oxides for Portable Fuel Cell Applications. *International Journal of Hydrogen Energy*, **38**, 13300-13308 (2013).

2. Liso, V., Olesen, A. C., Nielsen, M. P. & Kaer, S. K. Performance Comparison between Partial Oxidation and Methane Steam Reforming Processes for Solid Oxide Fuel Cell (Sofc) Micro Combined Heat and Power (CHP) System. *Energy*, **36**, 4216-4226 (2011).

3. Pinol, S., Morales, M. & Espiell, F. Hydrogen Production by a New Selective Partial Oxidation of Methane in Air Using Reducted $La_{0.5}Sr_{0.5}COO_3$-Delta as Catalyst for Fuel Cell Applications. *Journal of New Materials for Electrochemical Systems*, **11**, 119-124 (2008).

4. Li, B. *et al.* Synthesis Gas Production from Partial Oxidation of Methane over Highly Dispersed Pd/SiO_2 Catalyst. *Fuel*, **103**, 1032-1038 (2013).

5. Dai, X. P. & Yu, C. C. Direct Partial Oxidation of Methane to Synthesis Gas Using Oxygen Carriers in the Absence of Gaseous Oxygen. *Progress in Chemistry*, **21**, 1626-1635 (2009).

6. Son, I. H., Shamsuzzoha, M. & Lane, A. M. Promotion of Pt/Gamma-Al_2O_3 by New Pretreatment for Low-Temperature Preferential Oxidation of Co in H_2 for Pem Fuel Cells. *Journal of Catalysis*, **210**, 460-465 (2002).

7. Jo, M. C., Kwon, G. H., Li, W. & Lane, A. M. Preparation and Characteristics of Pretreated Pt/Alumina Catalysts for the Preferential Oxidation of Carbon Monoxide. *Journal of Industrial and Engineering Chemistry*, **15**, 336-341 (2009).

8. Son, I. H., Lane, A. M. & Johnson, D. T. The Study of the Deactivation of Water-Pretreated Pt/Gamma-Al_2O_3 for Low-Temperature Selective CO Oxidation in Hydrogen. *Journal of Power Sources*, **124**, 415-419 (2003).

9. Tomita, A., Shimizu, K., Kato, K., Akita, T. & Tai, Y. Mechanism of Low-Temperature CO Oxidation on Pt/Fe-Containing Alumina Catalysts Pretreated with Water. *Journal of Physical Chemistry C*, **117**, 1268-1277 (2013).

10. Tomita, A., Shimizu, K., Kato, K. & Tai, Y. Pt/Fe-Containing Alumina Catalysts Prepared and Treated with Water under Moderate Conditions Exhibit Low-Temperature Co Oxidation Activity. *Catalysis Communications*, **17**, 194-199 (2012).

11. Lopez-Fonseca, R., Jimenez-Gonzalez, C., de Rivas, B. & Gutierrez-Ortiz, J. I. Partial Oxidation of Methane to Syngas on Bulk Nial2o4 Catalyst. Comparison with Alumina Supported Nickel, Platinum and Rhodium Catalysts. *Applied Catalysis A-General*, **437**, 53-62 (2012).

12. Sun, I. H. Promotion of Platinum/gamma-aluminum Oxide for Preferential Oxidation of Carbon Monoxide in Hydrogen for PEM Fuel Cells by Cerium Promoter and by Preparation of Stable Platinum Nanoparticles Using a Water-pretreatment. Ph.D. Dissertation, The University of Alabam, 2003.

387563 Optimization of Local Distributor for the Application in Monolithic Structures

Abstract:

In the following work a multi phase distributor was developed which is suitable for the application in monolithic structures. It is based on the injection principle which is often performed in single channel experiments to underline the concept of microchannels for gas-liquid-solid reactions. Therein the gas is injected into a channel by a needle. Below the gas entrance of the needle liquid flows around the needle and enters the channel. This is multiplied to get a distributor for multichannel geometries. Starting from this "injector scale-up" a new concept is created based on a modular structure with a needle matrix suitable for each cell density of the monolithic structure. First of all gas is injected through a module with two tangential inlets. Subsequently, a needle plate guides the gas into each channel. Below these two parts a liquid part divides the liquid between the needles and is guided in the channels directly. Due to the flexible design of the needle plate this concept can be adapted for the respective cell density of the monolith.

In the presented work the fundamental concept of this distributor was refined and optimized by using CFD. Several CFD studies have been processed to optimize each module with the aim of a homogeneous distribution of gas and liquid phase in the beginning of the monolith. Additionally, experimental work has been performed to validate the CFD simulation. By using a gravimetric method as well as a high speed tomography the simulation program was proved. Overall experiments were performed for liquid velocities starting from 0,018 m/s until 0,07 m/s for a 39 cpsi monolith. The gas velocity was varied from 0,1 m/s up to 3 m/s.

It was found that depending on the inlet geometry a more homogeneous phase distribution can be reached the more vortices you create within the distributor. But compared to a common full cone nozzle distributor a specific operating area for gas and liquid velocity exists to get an optimized distribution at the beginning of the catalytic structure.

387687 Identifying Mechanisms of Zeolite Growth: A Pathway to Catalyst Optimization

Abstract:

Crystallization commonly occurs through classical pathways that involve the addition of atoms or molecules to a crystal surface that grows by layer nucleation and spreading. Zeolites can be categorized in a growing list of materials (mostly natural minerals) that form through nonclassical routes [1] involving amorphous precursors that undergo structural disorder-to-order transformation(s). Here we present a case study of zeolite L (LTL framework type) crystallization wherein we track the formation and evolution of amorphous precursors. We show that the amorphous material undergoes a series of structural transformations, forming worm-like particles (WLPs) that progressively decrease in population over the course of zeolite nucleation and growth. The evolution of pre-nucleation amorphous precursors was evaluated on the basis of their temporal change(s) in morphology, particle size, and degree of aggregation (i.e. fractal dimension) to explore the pathway(s) of amorphous-to-crystalline phase transformation. The putative mechanism of growth involves the initial aggregation of colloidal silica particles, which subsequently form WLPs. The latter appear to provide nutrients (i.e. silica and alumina) to growing crystals, potentially serving as interfaces for heterogeneous nucleation and the growth of LTL crystals via solution-mediated processes. We have coupled these studies with a systematic investigation of zeolite growth modifiers (ZGMs), which are site-specific molecules that bind to growing crystal surfaces and alter their size and morphology [2]. Here we will discuss how ZGMs affect the formation of WLPs and their structural transformation to LTL crystals. In addition, we will discuss the broader relevance of nonclassical growth pathways for other zeolite framework types. Collectively, these mechanistic studies provide an improved fundamental understanding of zeolite crystallization, which is paramount to the rational design of microporous materials with tailored physicochemical properties for improved performance in a variety of commercial applications.

References:

[1] Lupulescu, A.I. and Rimer, J.D., In Situ Imaging of Silicalite-1 Surface Growth Reveals the Mechanism of Crystallization, *Science* 2014 (In Press)

[2] Lupulescu, A.I., Kumar, M., Rimer, J.D., A Facile Strategy to Design Zeolite L Crystals with Tunable Morphology and Surface Architecture, *J. Amer. Chem. Soc.* 135 (2013) 6608-6617

Pyrolytic Conversion of Cellulose to Fuels: Tuning the Mechanisms of Cellulose Pyrolysis to Improve Bio-Oil Properties

Alex Paulsen, Matthew S. Mettler, Dionisios G. Vlachos, Paul J. Dauenhauer

Abstract: High temperature cellulose pyrolysis is an important part of converting lignocellulosic biomass to bio-oils and biofuels. Cellulose is known to react through a short-lived (10-100 ms) liquid intermediate during pyrolysis[1], which determines the product distribution in cellulose pyrolysis. Predicting the composition of the bio-oil product is necessary to improve pyrolysis economics. Therefore, our work focuses on identifying the crucial mechanisms of cellulose pyrolysis.[2-4] We find that levoglucosan undergoes secondary reactions within the liquid intermediate to form pyrans, anhydrosugars, and light (C_2-C_3) oxygenates. The mechanisms of these reactions are unknown; however, we have shown the relationship between levoglucosan decomposition and hydrogen transfer within the liquid intermediate by using deuterated carbohydrates.[5] It is likely that hydrogen serves as a Brønsted catalyst for the decomposition of levoglucosan within molten cellulose.

Additionally, we utilize a novel thin-film pyrolysis technique to investigate the effect of temperature on kinetically-limited, isothermal cellulose pyrolysis yields over a range of reaction temperatures (350 - 550 °C).[6] We observed that the total yield of furanic compounds remains constant over this range of temperatures. While the five-membered furan ring remains intact, oxygen functional groups (i.e. the aldehyde in hydroxymethylfurfural) are cleaved at higher temperatures to form deoxygenated furans (such as furan and methyl furan). The presence of these aldehyde groups in bio-oil leads to polymerization, which makes bio-oil unstable at storage conditions. We promoted this chemistry by impregnating cellulose with a carbon-supported metal catalyst to reduce the aldehyde content of bio-oil while maintaining the overall bio-oil yield.

1. Dauenhauer, P. J.; Colby, J. L.; Balonek, C. M.; Suszynski, W. J.; Schmidt, L. D., Reactive boiling of cellulose for integrated catalysis through an intermediate liquid. *Green Chemistry* **2009,** 11, (10), 1555-1561.

2. Mettler, M. S.; Paulsen, A. D.; Vlachos, D. G.; Dauenhauer, P. J., The chain length effect in pyrolysis: bridging the gap between glucose and cellulose. *Green Chemistry* **2012,** 14, 1284-1288.

3. Mettler, M. S.; Mushrif, S. H.; Paulsen, A. D.; Javadekar, A. D.; Vlachos, D. G.; Dauenhauer, P. J., Revealing pyrolysis chemistry for biofuels production: Conversion of cellulose to furans and small oxygenates. *Energy & Environmental Science* **2012,** 5, (1), 5414-5424.
4. Paulsen, A. D., Hough, B. R., Williams, C. L., Teixeira, A. R., Schwartz, D. T., Pfaendtner, J., & Dauenhauer, P. J., Fast Pyrolysis of Wood for Biofuels: Spatiotemporally Resolved Diffuse Reflectance In situ Spectroscopy of Particles. *ChemSusChem* **2014,** 7, (3), 765-776.
5. Mettler, M. S.; Paulsen, A. D.; Vlachos, D. G.; Dauenhauer, P. J., Pyrolytic conversion of cellulose to fuels: levoglucosan deoxygenation via elimination and cyclization within molten biomass. *Energy & Environmental Science* **2012**.
6. Paulsen, A. D., Mettler, M. S., Vlachos, D. G., & Dauenhauer, P. J., The Role of Sample Dimension and Temperature in Cellulose Pyrolysis. *Energy & Fuels* **2013**.

Synthesis of Zeolite Catalysts in the Absence of Organic Structure-Directing Agents

Matthew D. Oleksiak, Marlon Conato, and Jeffrey D. Rimer
University of Houston, Department of Chemical and Biomolecular Engineering
4800 Calhoun Rd, Houston, TX 77204

The unique acidity, thermal stability, and shape-selectivity of nanoporous zeolites are utilized in industrial applications spanning catalysis to ion exchange. The use of organics in zeolite synthesis is cost prohibitive, and avoided when possible; however, difficulties in the formation of zeolites via organic-free synthesis routes are prevalent [1]. The most common challenge is the formation of undesirable crystal phase(s), i.e. impurities. Oftentimes, organic structure directing agents (OSDAs) are used to form phase-pure crystals, though these additives are expensive and must be removed by post-synthesis calcination. In order to both understand and control organic-free polymorphism, we constructed kinetic phase diagrams based on varying synthesis conditions using only inorganic ions (e.g., alkali metals) as the structure directing agents [2]. These kinetic phase diagrams revealed regions of pure and mixed phases of eight zeolite framework types synthesized from colloidal silica and sodium aluminate.

Our systematic study has elucidated the phase behavior of many crystal structures, including for the first time the GIS framework type, a small-pore zeolite that exhibits two polymorphs: zeolites P1 and P2. This small-pore crystal structure, along with others (e.g., ANA type), have recently been shown to be catalytically active when noble metals are incorporated as extra-framework cations during crystallization. [3] By generating phase diagrams over a range of synthesis parameters, we have determined the conditions required to produce pure P1 and P2, as well as their structural analogues MER and PHI. These polymorphs differ by slight changes in the connectivity of their secondary building units – an outcome that can be controlled by judiciously selecting the silicon-to-aluminum and/or silicon-to-hydroxide ratios of the growth solution. We have shown that crystal phase transformations are governed by both thermodynamic and kinetic factors; and we have used a combination of experimental techniques to explore whether the transformation from one polymorph to another occurs by a ubiquitous pathway.

Additionally, we have investigated the growth mechanisms of two common zeolites, FAU and LTA. From time-elapsed studies of crystallization, we have shown that precursor particles are comprised of core-shell structures with a Si-rich core and an Al-rich shell wherein the nucleation of crystals on the exterior surfaces of core-shell particles has been observed. Collectively, these studies seek to identify the relative roles of classical and nonclassical pathways governing zeolite nucleation and crystal growth as a means of more effectively controlling phase purity, and ultimately their physicochemical properties for a wide range of commercial applications.

References:

[1] Oleksiak, M.D. and Rimer, J.D., Synthesis of Zeolites in the Absence of Organic Structure-Directing Agents: Factors Governing Crystal Selection and Polymorphism, *Rev. Chem. Eng.* 30 (2014) 1-49

[2] Maldonado, M., Oleksiak, M.D., Chinta, S., Rimer, J.D., Controlling Crystal Polymorphism in Organic-Free Synthesis of Na-Zeolites, *J. Am. Chem. Soc.* 135 (2013) 2641-2652

[3] Goel, S., Wu, Z., Zones, S.I., and Iglesia, E., Synthesis and Catalytic Properties of Metal Clusters Encapsulated within Small-Pore (SOD, GIS, ANA) Zeolites, *J. Am. Chem. Soc.* 134 (2012) 17688-17695

The activity and stability of dry reforming of methane over Ni-based Catalysts

Zhenghong Bao, Yongwu Lu, Fei Yu*.

Agricultural and Biological Engineering, Mississippi State University, Mississippi State, MS 39762

The dry reforming of methane (DRM) with CO_2 has been widely studied due to its important applications in producing of syngas ($CO + H_2$), removal of two greenhouse gases and upgrading of biogas (mainly composed of CH_4 and CO_2). Compared to the noble metal catalysts, the nickel-based catalyst is suitable for industrial scaled-up DRM due to its high activity and low cost. However, nickel-based catalysts have been encountering a significant challenge that is the active metal sintering and coke deposition, leading to the catalyst deactivation and then the poor stability. The coke formation is mainly caused by methane decomposition, Boudouard reaction and/or reverse carbon gasification reaction. Nevertheless, it has been reported that the deactivation of nickel catalysts can be suppressed by adding promoters like strong Lewis bases with the enhancement of chemisorb CO_2, such as MgO and CaO, and like lanthanide elements with the capacity of oxygen storage and release, such as CeO and LaO. Aluminum oxide is a commonly applied catalyst carrier own to its good pore size dispersion, high specific surface area and high mechanic strength. Herein, we synthesized a series of NiMgCeAl composite oxide catalysts using conventional co-precipitation method and systematically investigated their catalytic properties under wide reaction conditions. We aim not only to obtain and optimize efficient Ni catalysts suitable for industrial applications but also to get scientific understanding on the DRM reaction and process.

Fig. 1 shows the evolution of CH_4 conversion for nickel-based catalysts with different promoters and different nickel content at different reaction temperatures. It shows that CH_4 conversion increases with the reaction temperature, and the 15 wt%Ni-MgCeAl sample has the highest CH_4 conversion during the reaction temperature range of 650 to 800 °C. This sample will be selected to perform the stability test. Fig. 2 shows CH_4 conversion decreases slightly (4%) after 100 h time-on-steam test, indicating a lower deactivation, caused by sintering or carbon deposition, occurred on 15 wt%Ni-MgCeAl and so that this kind of catalyst is in great potential to be commercialized for dry reforming of methane.

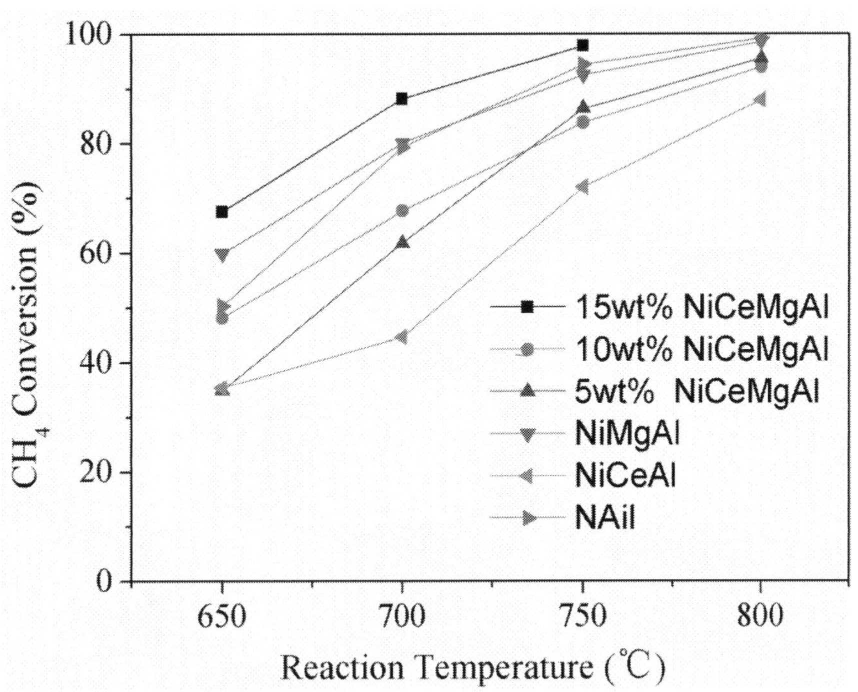

Fig. 1. CH_4 conversion of DRM over Ni-based catalysts at 650~800 °C. Reaction conditions: catalyst loading 0.1mL, P=1 atm, CO_2/CH_4 = 1.04, GHSV=48000 h^{-1}.

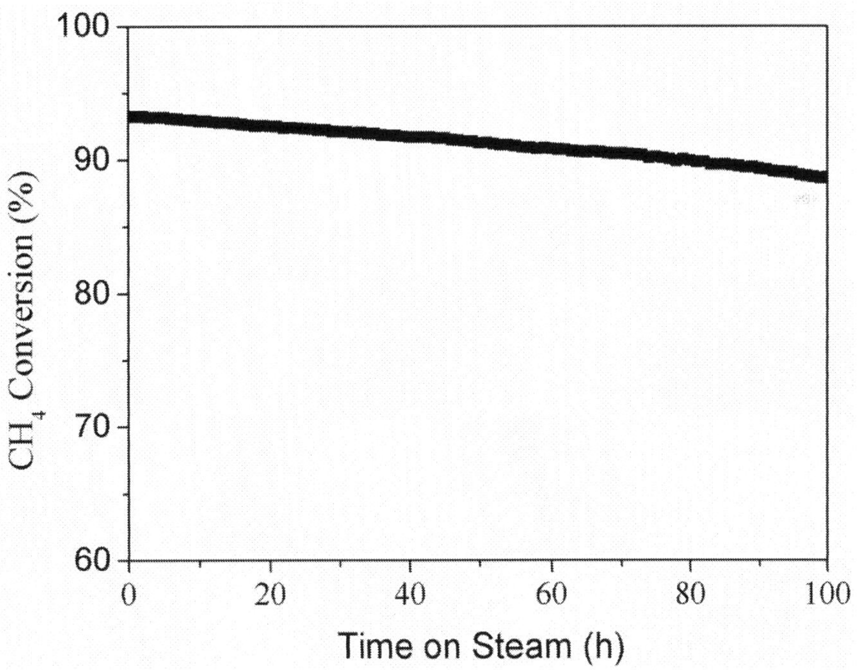

Fig. 2. CH_4 conversion of DRM as a function of time over 15wt%Ni-MgCeAl catalyst. Reaction conditions: catalyst loading 0.1mL, P=1 atm, T=750 °C, CO_2/CH_4 = 1.04, GHSV=48000 h^{-1}.

388344 Fundamental Insights on the Electrochemical Reduction of Water and Carbon Dioxide Using Solid Oxide Electrolysis Cells

Abstract:

Extensive use of fossil fuels and consequential high levels of CO_2 emissions are major contemporary challenges. Many proposed strategies for managing CO_2 from chemical processes involve the conversion of CO_2 back to hydrocarbons. The co-reduction of CO_2 and H_2O to syngas (CO and H_2) is one such process for which no efficient approach currently exists. Solid oxide electrolysis cells (SOECs) are solid-state electrochemical systems that, in principle, can facilitate the simultaneous co-reduction of CO_2 with H_2O to syngas with potentially very high rates due to the favorable kinetics at their high operating temperatures. While electrochemical co-reduction of CO_2 and H_2O using SOECs offers a great deal of promise, these systems are limited by the inability to operate near the thermodynamic reversible potentials due to the activation overpotential losses. In the present work, we combine experimental and theoretical techniques to determine the factors that lead to the overpotential losses in SOEC during electrolysis of H_2O and CO_2. We find that the conventional Ni-based electrocatalysts are efficient at catalyzing electrolysis of H_2O but exhibit significant overpotential losses toward electrochemical reduction of CO_2. Based on these insights, we have devised ways and screened through a number of electrocatalysts in order to identify the ones that lead to the lowest overpotential losses during the electrochemical reduction of CO_2 and H_2O.

388463 Development of Heterogeneous Electrocatalysts for Reversible Li-Air Battery Cathodes

Abstract:

Nonaqueous lithium-air batteries are promising electrochemical storage devices with very high theoretical specific energy and energy density. Li-air batteries are comprised of an anode (pure metallic lithium), an O_2 electrode (cathode, commonly composed of porous carbon), and an aprotic electrolyte. The chemistry at the cathode of Li-air batteries is fairly complex and plays an important role in their performance. The cathode of Li-air batteries is responsible for a number of processes, including (i) the transport of Li^+ throughout the electrode in order to facilitate the formation of Li_2O_2 (ii) the electrochemical reactions (oxygen reduction reaction (ORR) and oxygen evolution reaction (OER)) with low overpotential losses, (iii) storing of Li_2O_2 in the porous media to achieve high capacity, and (iv) conduction of electrons with high efficiency. The commonly used Li-air cathodes are composed of carbon with porous architecture. While carbon cathodes are very good electron conductors with tunable porosity, they suffer from (i) the ability to transporting Li^+ and (ii) poor activity toward the oxygen evolution reaction.

In this contribution, we show our efforts in designing robust heterogeneous electrocatalysts for Li-air cathodes. We find that the incorporation of layered metal oxide electrocatalysts in Li-air cathodes significantly lowers the charging potential of the cells. Our controlled electrochemical and electron microscopy studies show that the decrease in the charging potential is due to an impact in the formation of Li_2O_2 during discharge, and lowering of the activation barrier for the oxygen evolution reaction. We have been able to link the electrocatalytic properties of these materials to their chemical and structural composition. These structure/activity relationships can pave the way for the design of optimal layered metal oxide electrocatalysts for Li-air cathodes.

388828 Controlling Zeolite Synthesis with Molecular Modifiers: A Versatile Approach to Design Catalysts

Abstract:

There are growing opportunities for the use of zeolite catalysts in a wide range of applications. Despite their extensive use in commercial processes, an understanding of their growth mechanism(s) remains elusive. The rational design of zeolite catalysts calls for more versatile synthetic approaches capable of *a priori* tailoring crystal properties, such as crystal size, morphology, and surface architecture. Here we will discuss the design of industrially relevant zeolites, notably LTL, MOR and CHA. Focusing on this class of porous materials presents an opportunity to explore the impact of crystal engineering on targeted catalyst performance (e.g., activity and lifetime). To this end, we will discuss our studies of zeolite crystallization in the presence of zeolite growth modifiers (ZGMs), which are molecules that selectively bind to specific surfaces of zeolite crystals and mediate anisotropic growth rates [1]. We systematically examined a library of modifiers ranging in their structure and functional moieties to assess the physicochemical properties that regulate their efficacy and specificity as zeolite crystal growth modifiers. As we will demonstrate, the judicious selection of ZGMs can markedly alter zeolite crystal morphology from high aspect ratio (needle-like) crystals and micron-sized spheroidal crystals to thin discs and small spheres, respectively, with concomitant reduction in the internal diffusion pathlength and external surface area. Our studies reveal that ZGMs permit the control of zeolite crystal size spanning more than three orders of magnitude. This work also revealed that ZGM hydrophobicity and the spatial sequencing of binding moieties are effective molecular descriptors of its efficacy [2].

Collectively, these studies seek to establish a commercially viable approach to optimize catalyst performance and to provide general heuristics for designing zeolites. We will present these fundamental studies and also discuss how techniques, such as atomic force microscopy [3], can be used to probe the mechanisms of ZGM action. Moreover, we will present time elapsed studies of crystallization in order to elucidate the impact of modifiers on zeolite nucleation and growth.

References:

[1] Lupulescu, A.I. and Rimer, J.D., Tailoring Silicalite-1 Crystal Morphology with Molecular Modifiers, *Angew. Chem. Int. Ed.* 51 (2012) 3345-3349.

[2] Lupulescu, A.I., Kumar, M., Rimer, J.D., A Facile Strategy to Design Zeolite L Crystals with Tunable Morphology and Surface Architecture, *J. Amer. Chem. Soc.* 135 (2013) 6608-6617

[3] Lupulescu, A.I. and Rimer, J.D., In Situ Imaging of Silicalite-1 Surface Growth Reveals the Mechanism of Crystallization, *Science* 2014 (In Press)

388979 The Relationship Between Enhanced Enzyme Activity and Structural Dynamics in Ionic Liquids: A Combined Computational and Experimental Study

Abstract:

Candida antarctica lipase B (CALB) is an efficient biocatalyst for esterification, and polymerization reactions. In order to understand how to control an enzyme activity and stability we performed a combined experimental and molecular dynamics simulation study of CALB in organic solvents and ionic liquids. Our results demonstrate that the conformational changes of the active site cavity are directly related to enzyme activity. We indicates that CALB activity in synthetic solvents decreases with an observed solvent-induced reduction in the size of the substrate binding cavity in the following order: [Bmim][TfO] > *tert*-butanol > [Bmim][Cl]. The size of the active site cavity is modulated by an α-10 helix, which can substantially change its conformation due to specific interactions with solvent molecules. This change is evident in [Bmim][Cl] where interactions of LYS-290 with chlorine anions induced a conformational switch from an α-helix to a turn, which resulted in reduced activity. Moreover, the relation between the enzyme-solvent interfacial energies and initial reaction rates show a linear trend; these results predict that the initial reaction rates increased with a reduction in interfacial energy. Overall, our observations can be used to explain the structure-activity relation of CALB in various solvents, as may be applicable for the selection of appropriate solvents in order to optimize enzymatic reactions.

389044 Mechanism and Kinetics of Diels–Alder Reaction Between Dicyclopentadiene and 1,3-Butadiene in a High Temperature and High Pressure Continuous Tube Reactor

Abstract:

The synthesis of 5-vinyl-2-norbornene(VNB) ,the very critical intermediate of the best third monomer (ethylidenenorbornene, ENB) for EPDM. The most economical way to manufacture VNB is through the Diels–Alder reaction between cyclopentadiene and 1,3-butadiene because of the availability and low cost of the reactant. However, these involves high temperature and high pressure, exothermic reaction, which lead to serious safety concerns and the instability of product's quality. Besides, the existence of various side reactions greatly reduces the selectivity of the reaction and results in tremendous energy and raw meterials waste. Therefore, new means of process enhancing, such as new type of reactors and side-reaction controlling strategies are expected to reinforce stability, safety and economics of process.

In this paper, the combination of previous studies and our investigation of the products of thermal reaction between 1,3-butadiene and cyclopentadiene(CPD) make it possible to describe the process as a set of the parallel and consecutive reactions. The effects of residence time, reaction temperature and ratio between CPD and 1,3-butadiene in feeding on synthesis processes of VNB are investigated. Both the selectivity and conversation of 1,3-butadiene and CPD have been measured. All reactions were conducted in a continuous tube flow reactor. The axial dispersion model has been established to simulate the reaction process, which coincides well with the experimental results. Detailed kinetic parameters are determined, which lay a solid foundation for the optimization of VNB synthesis process. Based on this, the optimum reaction conditions have been evaluated, including residence time, reaction temperature and ratio between CPD and 1,3-butadiene in feeding.

389569 Dispersed Pt/Ceria Catalysts for Water-Gas Shift Prepared Using Supercritical Fluid Deposition

Abstract:

Hydrocarbons are currently the most viable source of hydrogen. Synthesis gas, or syngas, is a fuel gas generated through steam reforming, gasification, or partial oxidation of fossil fuels. Syngas is composed largely of carbon monoxide and hydrogen, along with several other compounds depending on the source of the gas. To use synthesis gas as a feed to a fuel cell, the carbon monoxide concentration must be reduced significantly, approximately 10 – 20 ppm, to prevent poisoning of the cell and maintain performance. The water-gas shift (WGS) reaction reduces the CO content of the stream while simultaneously increasing H_2 content.

The water gas shift reaction (WGSR), $CO + H_2O \leftrightarrow H_2 + CO_2$, is a slightly exothermic reaction that is very important in the processing of feed streams for fuel cells. Industrial application of the WGS reaction is normally done in two steps, both a high temperature and low temperature shift. At low temperatures, high conversions are achievable but are limited due to slow kinetics. At high temperatures, reaction rates are faster; however, conversion is limited due to equilibrium constraints. Current industrial catalysts are not suitable for fuel cell feed purification due to both technical and safety constraints. The operating temperature of a fuel cell is far lower than the temperature required for efficent use of these catalysts. To account for low activity, large amounts of catalyst must be used leading to non-viable reactor volumes for fuel cell applications. The pyrophoric nature of these catalysts as well as their thermal instability also makes them ill-suited for fuel cell use. There is a need to develop a catalyst that can operate nearer normal fuel cell operating conditions, normally in the range of 80°C – 120°C, while maintaining its surface integrity and activity for long term performance.

Noble metals have been shown to catalyze the WGS, and platinum, specifically, shows a higher activity. The addition of ceria to platinum catalysts has been shown to have synergistic effects on the activity for WGS, including both increased activity and longer stability. Research has shown that these enhancements are dependent on the degree of interface between platinum and ceria. In this study, we investigate the use of supercritical fluid deposition (SCFD) to prepare catalysts of highly dispersed platinum on supported ceria. SCFD takes advantage of the excellent transport properties of supercritical fluids to facilitiate the uniform adsorption of organometallic precursors on a support surface. The precursors then undergo calcination and reduction treatments to form metal nanoparticles.

In our work, we investigate the effects of pretreating alumina-supported ceria prior to depositing platinum precursor, with the goal of maximizing the contact area between platinum and ceria. Time required for deposition to occur is related using atomic emission spectroscopy for elemental analysis. Temperature-programmed reduction is used to characterize metal/support interactions, and CO pulse titration is employed to evaluate the dispersion of platinum on the surface. Catalyst activity for WGS is measured in a tubular flow reactor. Both reaction rates and surface analyses are correlated with catalyst preparation variables.

389691 Mechanistic Insights into Rate-Dominant Transport Barriers in MFI Structured Zeolites

Abstract:

Zeolites are used to catalyze chemical reactions and improve selectivity to desired products through well ordered, shape selective microporous channels. MFI structured ZSM-5 has widely been used for catalytic fast pyrolysis of biomass and subsequent upgrading.[1] To maximize turnover frequencies, microporous and mesoporous materials are now being synthesized with length scales approaching that of a single lattice (pillared, nanosheets, membranes, etc.). However, in small particles, mass transfer becomes dominant and the apparent diffusivity appears to be orders of magnitude smaller than through large particles.[2, 3] This has led to the general phenomenon referred to as 'surface barriers,' which describes the hindrance to mass transfer at the surface of zeolite particles. While this timescale is often negligible when compared to the diffusion or reaction times, we show that it does contribute significantly to overall turnover in small particles.[2] Chemical species exhibiting differing kinetic diameters and adsorption affinities were used to probe diffusion within MFI-structured zeolites to elucidate the role of guest molecule size and binding energy. Kinetic Monte Carlo simulations are used to mechanistically probe the effect of physical structures on overall transport timescales, including internal grain boundaries, regular surface restrictions and total pore blockages. The relative trends with particle size are compared to experimental measurements and potential mechanisms are proposed.

1. Williams, C. L.; Chang, C.-C.; Do, P.; Nikbin, N.; Caratzoulas, S.; Vlachos, D. G.; Lobo, R. F.; Fan, W.; Dauenhauer, P. J., Cycloaddition of Biomass-Derived Furans for Catalytic Production of Renewable p-Xylene. *ACS Catalysis* **2012,** 2, (6), 935-939.

2. Teixeira, A. R.; Chang, C.-C.; Coogan, T.; Kendall, R.; Fan, W.; Dauenhauer, P. J., Dominance of Surface Barriers in Molecular Transport through Silicalite-1. *The Journal of Physical Chemistry C* **2013,** 117, (48), 25545-25555.

3. Chang, C.-C.; Teixeira, A. R.; Li, C.; Dauenhauer, P. J.; Fan, W., Enhanced Molecular Transport in Hierarchical Silicalite-1. *Langmuir* **2013,** 29, (45), 13943-13950.

Numerical Simulation for Coating Process of TiN Thin Film By a Thermal CVD Method

Yuya Hatori[1], Omayu Takahoshi[2], Ken-ichiro Tanoue[2], Tatsuo Nishimura[2],
1 UBE Industries, Ltd.
2 Yamaguchi University, Ube, Japan.

Introduction

Thermal Chemical Vapor Deposition (CVD) is one of the methods to make the thin film on a heated substrate during chemical reactions from vapor raw material and has been widely used due to many merits. For example, it is possible to deposit immediately thin film on a large amount of the substrates with complex surface shape. Industrially, a film is hardly deposited on one substrate. It is necessary to deposit a film simultaneously on two or more substrates.

Therefore, it is necessary to examine the growth rate distribution in consideration of temperature distribution and materials concentration distribution. However, within a reactor, the mass transfer accompanied by heat and some chemical reactions happens simultaneously. It is very difficult to perform reaction control. Therefore, sufficient deposition model is not constructed. If the model which can reproduce a growth rate by numerical analysis is constructed after clarifying a reaction mechanism, the ingredient of a thin film and control of film thickness are attained.

Although there are many reports for the reaction mechanism of CVD for titanium nitride (TiN), not only the equations of deposition rate but also activation energy differed from researchers. Furthermore, most of researchers presented that the deposition rate could be controlled by only surface reaction. (Zimerman *et al.* (2000), Tago *et al.* (1998, 2000), Akiyama (2005), Akiyama *et al.* (1995, 2008))

In this study, the growth rate of titanium nitride (TiN) film by thermal chemical vapor deposition (CVD) in horizontal tubular reactor was investigated experimentally. Titanium tetrachloride ($TiCl_4$) was selected as a raw material, and $TiCl_4$ vapor was mixed with nitrogen and hydrogen in the reactor. Furthermore, two types numerical simulations were conducted to predict the growth rate with taking account into heat and mass transfer during CVD process. One was the simple model which was assumed that the gas flow distribution obeyed the Poiseuille's law and not only the convective term along radial direction but also the diffusion term along the flow-direction in heat and mass transfer were ignored. The other was the strict model which didn't have any above mentioned assumptions.

Experimental set up and procedure

Total experimental apparatus consisted of gas supplies, a tank of titanium tetrachloride ($TiCl_4$), a horizontal tubular reactor and an aspirator. In this study, titanium tetrachloride ($TiCl_4$), hydrogen (H_2) and nitrogen (N_2) were used as raw materials. If there is moisture content in gas cylinders, the $TiCl_4$ decomposes to hydrochloric acid (HCl) and titanium oxides (TiO_2). In order to decrease the moisture in the cylinders, packed beds of zeolite particles were set at the outlet of them. The gas flow rates were controlled by the mass flow controllers. A part of N_2 gas was used as the carrier one to transport $TiCl_4$ vapor from the tank into the reactor. The temperature in a tank was set at 313 K. In order to prevent condensation of $TiCl_4$, the pipeline was heated at 333 K with the taped heater. A horizontal reactor was made of quartz with 21 mm in inner diameter, 25.4 mm in outer diameter and 710 mm in length. It was surrounded by the electric furnace. In order to deposit titanium nitride (TiN) film, 22 quartz substrates with 17 mm in inner diameter, 20 mm in outer diameter and about 20 mm in length were set into the reactor. Then the reactor was heated by the electric furnace.

Before the temperature at the center of the reactor wall, T_W, approached to the set value T_S, the moisture at the inner wall was removed as much as possible by feeding N_2 gas. After $T_W = T_S$, The materials ($TiCl_4$, N_2 and H_2) were carried by separate pipeline and then mixed before entering the tubular reactor. The following chemical reaction occurs in the reactor.

$$TiCl_4 + 2H_2 + \frac{1}{2}N_2 \rightarrow TiN + 4HCl \tag{1}$$

The mixed gas in the outlet of a reactor consists of N_2, H_2, and $TiCl_4$ and HCl. In this study, the aspirator was connected into the outlet of the reactor to not only decrease the pressure of the reactor but also cause neutralized reaction with sodium hydroxide (NaOH) aqueous solution in it. The reaction time was 5 hours. The weight of the substrates before and after experiment was measured by the electric balance. Then, the growth rate of a TiN film per unit and the surface area of the quartz substrates were calculated. Furthermore, the temperature of the wall in alignment with a tubular axis was measured every 20 mm, and was recorded on memory of the data logger.

Numerical simulation method for Thermal CVD process

In order to understand heat and mass transfer during thermal chemical vapor deposition of $TiCl_4$-H_2-N_2 system in horizontal tubular reactor, two types of numerical simulations have been conducted as follows.

1) Simple model[4)]

Two - dimensional numerical simulation in the horizontal tubular reactor in has been conducted. The number numbers of the mesh along the radial component and along the axial component are 100 and 2000, respectively.

The gas flow in the reactor was assumed to consist of Poiseuille's law with temperature dependence along the flow axis. The velocity field can be calculated by,

$$U = 2\overline{U_0}\frac{\bar{T}}{T_0}\left(1 - \left(\frac{r}{R}\right)^2\right). \tag{2}$$

Where the average temperature along radial direction is given by,

$$\bar{T} = \frac{\int_0^R 2\pi r T\left(1-\left(\frac{r}{R}\right)^2\right)}{\int_0^R 2\pi r\left(1-\left(\frac{r}{R}\right)^2\right)} . \tag{3}$$

If not only convective term along radial direction and diffusion term along the flow-direction can be neglected, mass balance for $TiCl_4$ is given by,

$$\frac{\partial(UC_{TiCl_4})}{\partial z} = \frac{1}{r}\frac{\partial}{\partial r}\left(rD\frac{\partial C_{TiCl_4}}{\partial r}\right). \tag{4}$$

If not only convective term along radial direction and diffusion term along the flow-direction can be neglected, energy balance is given by,

$$\frac{\partial(\rho C_p UT)}{\partial z} = \frac{1}{r}\frac{\partial}{\partial r}\left(r\lambda\frac{\partial T}{\partial r}\right). \tag{5}$$

These governing equations and boundary conditions were discretized by the control volume method and solved by THOMAS method. All physical properties were taken account in the dependence of the temperature as shown in ref. (4).

The growth rate of TiN film was evaluated by,

$$G_{\mathrm{Cal}} = D M_{\mathrm{TiN}} \frac{\partial C_{\mathrm{TiCl_4}}}{\partial r} \quad \text{at } r = R \, . \tag{6}$$

Using experimental kinetic parameters for the rate constant of surface reaction, heat and mass transfer in the tubular reactor using equations 4 and 5 were solved simultaneously. Then the growth rate was obtained by equation 6. The kinetic parameters for the rate constant were optimized by the nonlinear programming using the Levenberg-Marquardt method and obtained as follows[4].

$$k_{\mathrm{S}} = \begin{cases} 2.0 \times 10^{8} \exp\left(-\frac{201900}{R_g T}\right) & (T < 1163\ \mathrm{K}) \\ 0.1348 & (T > 1163\ \mathrm{K}) \end{cases} \tag{7}$$

2) Strict model

Three - dimensional numerical simulation in the horizontal tubular reactor in has been conducted. Optimized – polyhedral grids of about 860,000 were utilized to solve not only fluid flow, heat transfer and mass transfer.

The equation of continuity at steady state is given by,

$$\nabla \cdot (\rho \boldsymbol{v}) = 0 \, . \tag{8}$$

The equation of motion at steady state is given by,

$$\nabla \cdot (\rho \boldsymbol{v}\boldsymbol{v}) = -\nabla p + \mu \Delta \boldsymbol{v} \, . \tag{9}$$

The equation of continuity of $TiCl_4$ at steady state is given by,

$$\nabla \cdot \left(\rho \boldsymbol{v} Y_{\mathrm{TiCl_4}}\right) = \nabla \cdot \left(\rho D \, \nabla Y_{\mathrm{TiCl_4}}\right). \tag{10}$$

The equation of continuity of H_2 at steady state is given by,

$$\nabla \cdot \left(\rho \boldsymbol{v} Y_{\mathrm{H_2}}\right) = \nabla \cdot \left(\rho D \, \nabla Y_{\mathrm{H_2}}\right). \tag{11}$$

The mass fraction of N_2 is calculated by,

$$Y_{\mathrm{N_2}} = 1 - \left(Y_{\mathrm{H_2}} + Y_{\mathrm{TiCl_4}}\right). \tag{12}$$

The energy balance at steady state is given by,

$$\nabla \cdot \left(\boldsymbol{v}(\rho \textstyle\sum_i Y_i h_i)\right) = \nabla \cdot (\lambda \nabla T) \, . \tag{13}$$

Where the enthalpy of component i (= H_2, N_2, $TiCl_4$) is calculated by,

$$h_i = \int_{T_{\text{ref}}}^{T} C_{\text{p},i} dT \ . \tag{14}$$

Ideal gas law is adapted to solve the gas density as follows.

$$\rho = \frac{P_0}{R_g T} \sum_i \left(\frac{Y_i}{M_i} \right) \tag{15}$$

These governing equations using suitable boundary conditions were calculated by ANSYS Fluent ver.14.5.7. All physical properties were taken account in the dependence of the temperature as shown in ref. (4).

The growth rate of TiN film was evaluated by,

$$G_{\text{Cal}} = \rho D \frac{M_{\text{TiN}}}{M_{\text{TiCl}_4}} \frac{\partial Y_{\text{TiCl}_4}}{\partial n} \ \text{at wall} \ . \tag{16}$$

Results and discussion

Fig. 1 shows the growth rate distributions of TiN film along axial position from heater inlet z. Plots and lines show the experimental results and calculated ones, respectively. At $Q_t = 2.7$ SLM and T_S = 1173 K, the experimental growth rate was almost 0 at $z < 0.10$ m as shown in Fig. 5a. The wall temperature was about 973 K[4]. There were two regions for the growth rate distribution. At 0.10 m $< z <$ 0.15 m, the growth rate increased dramatically with the distance and had a maximum value at $z = 0.15$ m. The wall temperature increased also with the distance[4]. Then, the growth rate increased with the wall temperature in the first region. After that the growth rate decreased exponentially with the distance when the wall temperature was almost constant at 0.16 m $< z <$ 0.320 m in the second region. At $Q_t = 2.7$ SLM, higher setting temperature T_S led to increase in the growth rate as shown in Figs. 5b and 5c because the reaction rates increased with the temperature. Furthermore, at T_S =1273 K, higher total gas flow rate Q_t made the higher growth rate as shown in Figs. 5d, 5e and 5f because the apparent thickness of boundary layer for concentration decreased with the total gas flow rate. The calculated results for not only the simple model[4] but also the strict model agreed almost wholly with the experimental ones. Therefore, the simple model which was proposed by our previous report[4] was very suitable method to predict the kinetics of thermal chemical vapor deposition in horizontal tubular reactor. Detailed observation near the maximum growth rate shows that there are some differences between the experimental results and the calculated ones due to the effect of gas phase reactions.

In the future work, the total simulation with the effect of gas phase reactions will be conducted to predict the more detailed heat and mass transfer during the CVD process.

Conclusions

The growth rate of TiN film by thermal CVD in horizontal tubular reactor has been investigated experimentally and numerically. Some conclusions were obtained as follows.

1) There were two regions for the growth rate distribution along the axis in the tubular reactor. In the first region, the growth rate could be controlled by the surface reaction of $TiCl_4$ because it increased with temperature at the inner wall. On the other hand, in the second region, the growth rate was found to decrease exponentially with axial position in the reactor and could be controlled by the diffusion rate of remaining $TiCl_4$.
2) Two types of simulations were conducted. One was the simple model which was assumed that the gas flow distribution obeyed the Poiseuille's law and not only the convective term along radial

direction but also the diffusion term along the flow-direction in heat and mass transfer were ignored. The other was the strict model which didn't have any above mentioned assumptions. The calculated results for not only the simple model but also the strict model agreed almost wholly with the experimental ones. Therefore, the simple model which was proposed by our previous report was very suitable method to predict the kinetics of thermal chemical vapor deposition in horizontal tubular reactor.

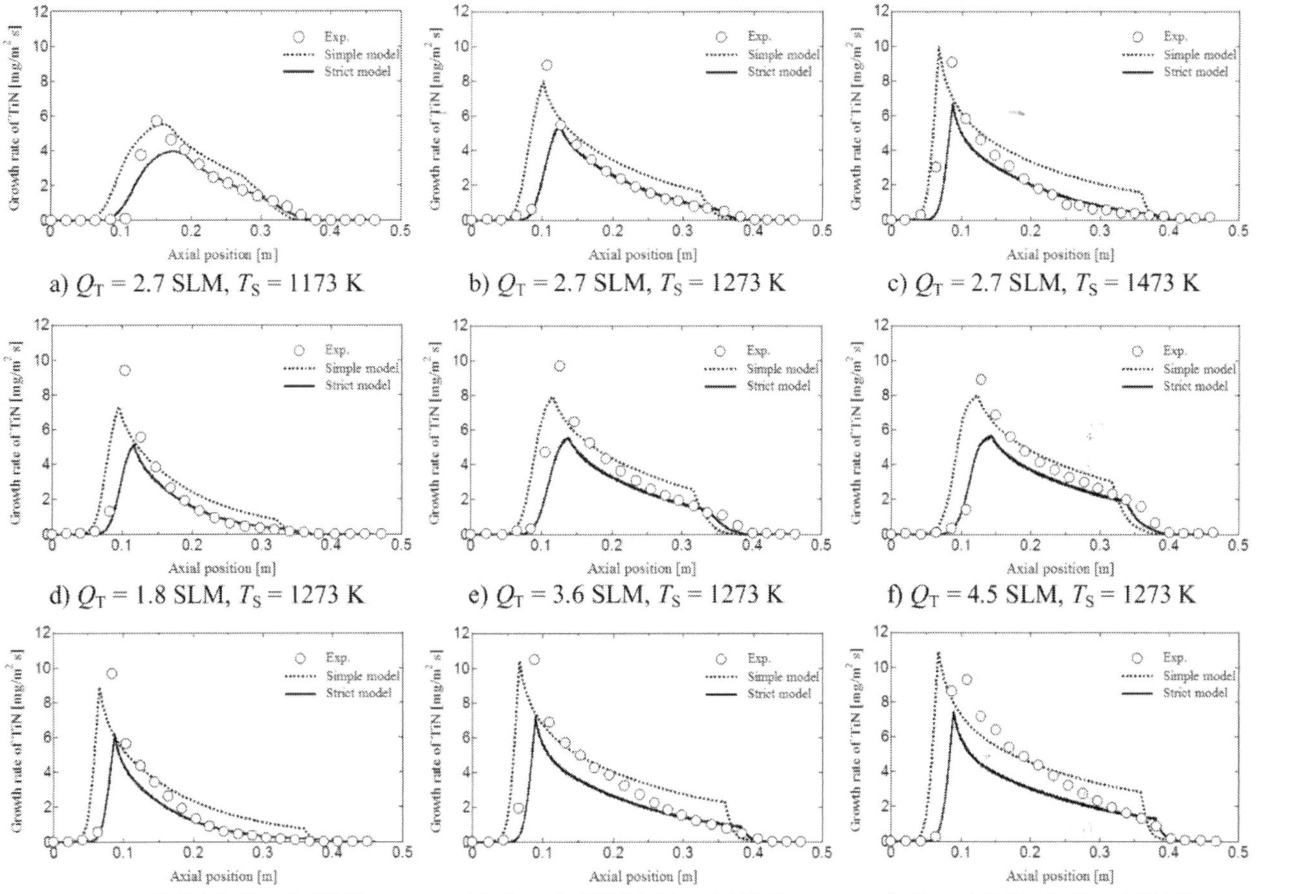

a) Q_T = 2.7 SLM, T_S = 1173 K
b) Q_T = 2.7 SLM, T_S = 1273 K
c) Q_T = 2.7 SLM, T_S = 1473 K
d) Q_T = 1.8 SLM, T_S = 1273 K
e) Q_T = 3.6 SLM, T_S = 1273 K
f) Q_T = 4.5 SLM, T_S = 1273 K
g) Q_T = 1.8 SLM, T_S = 1473 K
h) Q_T = 3.6 SLM, T_S = 1473 K
i) Q_T = 4.5 SLM, T_S = 1473 K

Fig.1 Comparison of calculated growth rate distributions of TiN film along the axial direction with experimental ones.

Nomenclature

C_{TiCl4}	: Concentration of $TiCl_4$	[mol m^{-3}]
C_p	: Specific heat of gas mixtures	[J kg^{-1} K^{-1}]
$C_{p,i}$	: Specific heat of component i	[J kg^{-1} K^{-1}]
D	: Diffusion coefficient	[m^2 s^{-1}]
G_{Cal}	: Calculated growth rate of TiN	[g m^{-2} s^{-1}]
$\underline{h}_i$	: Enthalpy of component i	[J/kg]
k_s	: rate constant for the surface reaction	[m s^{-1}]
M_{TiCl4}	: Molecular weight of $TiCl_4$ (=189.68)	[g mol^{-1}]
M_{TiN}	: Molecular weight of TiN (= 61.89)	[g mol^{-1}]

n	: Normal component at the reactor wall	[m]
P	: Kinetic pressure in the strict model	[Pa]
P_0	: Static pressure in the experimental condition (= 50700)	[Pa]
r	: Radial coordinate in the tubular reactor	[m]
R	: Inner radius of the reactor	[m]
R_g	: General gas constant (= 8.314)	[J mol^{-1} K^{-1}]
T	: Temperature	[K]
$\bar{T}$	: Mixed mean temperature in the radius coordinate	[K]
T_0	: Temperature at the reactor inlet	[K]
T_{ref}	: Reference temperature to calculate the enthalpy (298.15)	[K]
T_S	: Setting temperature of the furnace	[K]
U	: Superficial velocity along the axial coordinate in a tubular reactor	[m s^{-1}]
$\bar{U}$	: Average superficial velocity in a tubular reactor	[m s^{-1}]
$\bar{U}_0$	: Average superficial velocity at the reactor inlet	[m s^{-1}]
$\boldsymbol{v}$	: Velocity vector in the strict model	[m s^{-1}]
Y_i	: Mass fraction of component i	[-]
z	: Axial coordinate in the tubular reactor	[m]
λ	: Thermal conductivity of gas	[W m^{-1} K^{-1}]
μ	: Viscosity of gas	[Pa s]
ρ	: Density of gas	[kg m^{-3}]

References

1. Akiyama, Y.; "Analysis of Film Growth Process via Chemical Vapor Deposition based on Reaction Engineering," *Proc. Schl. Eng. Tokai Univ* **45(2)**, 115-120 (2005)
2. Akiyama, Y., K. Shitanaka and H. Murakami; "Modeling Titanium Oxide Growth by Chemical Vapor Deposition Using Titanium Tetra Isopropoxide," *J. of Chem. Eng. of Japan*, **41(8)**, 779-784 (2008)
3. Akiyama, Y., T. Sato and N. Imaishi; "Reaction analysis for ZrO_2 and Y_2O_3 thin film growth by low-pressure metalorganic chemical vapor deposition using β–diketonate complexes," *J. Crystal Growth*, **147**, 130-146 (1995)
4. Hatori, Y., H. Yamamoto, K. Tanoue and T. Nishimura; "Heat and Mass Transfer Analysis for Coating Process of TiN Thin Film in a Tubular Reactor by Thermal CVD," *Kagaku Kogaku Ronbunshu*, **40**, 425-431 (2014)
5. Tago, T., M. Kawase and K. Hashimoto; "Numerical Analysis Method for Growth Kinetics of Chemical Vapor Deposition of Alumuna Using a Non-Isothermal CVD Reactor," *Kagaku Kogaku Ronbunshu*, **26**, 763-769 (2000)
6. Tago, T., M. Kawase, Y. Masaki and K. Hashimoto; "Growth Kinetics of Chemical Vapor Deposition of Al_2O_3," *Kagaku Kogaku Ronbunshu*, **24**, 81-85 (1998)
7. Zimmermann, E., T. Wang, Ch. Ho Yu and D. Neuschiiz; "Retarding Effect of HCl on Rate of Chemical Vapor deposition TiN from $TiCl_4$-H_2-N_2," *J. Elec. Soc.* **147**, 2206-2209 (2000)

389866 A Fundamental Study of the Reaction and Diffusion of Poly-Aromatic Hydrocarbons in Hierarchical Pore Structure Zeolites

Abstract:

A Fundamental Study of the Reaction and Diffusion of Poly-Aromatic Hydrocarbons in Hierarchical Pore Structure Zeolites

David P. Gamliel, Shoucheng Du, George M. Bollas, and Julia A. Valla

Department of Chemical & Biomolecular Engineering, University of Connecticut, Storrs, CT

Thermochemical conversion of biomass via gasification and pyrolysis has attracted significant scientific attention as a means for converting biomass and waste products to useful energy and chemical feedstocks. A major factor deterring large-scale commercialization of this technology is the presence of high molecular-weight poly-aromatic hydrocarbons (PAHs), also known as tars, in producer streams. Tars deactivate catalysts, cause blockages in transfer lines, and damage downstream units such as compressors, turbines, and fuel cells [1]. Additionally, PAHs are typically carcinogenic in nature. A process to convert PAHs into potentially useful products would be very desirable.

Zeolites are a promising catalyst for PAH transformation. They exhibit unique properties such as acidity, well-defined microporosity, and shape selectivity, which make them ideal for catalytic applications. Zeolites are used as the workhorse of the fluid catalytic cracking (FCC) unit, in which they are used to crack heavy hydrocarbons into lighter, more valuable products. However, low diffusion rates of large molecules through the highly microporous structure results in low conversion and coke formation. Bulky molecules may even be totally excluded from the pore system. The introduction of mesoporosity may be one way to reduce these diffusion limitations and increase access of bulky molecules to catalyst active sites.

Dou et al. [2] have studied the catalytic cracking of 1-methyl naphthalene in a fixed bed reactor co-fed with hydrogen. They found that Y zeolite and Ni/Mo catalyst were the best for cracking the tar, achieving greater than 95% conversion. Buchireddy et al. [3] studied the conversion of naphthalene over Ni impregnated Y zeolite in the presence of syngas. They found that the presence of the Ni increases the reforming capability of the catalyst and that activity increases with an increase in zeolite acidity.

The main focus of our work is the catalytic conversion of PAHs using microporous and mesoporous zeolites with and without impregnated metals. Mesoporous zeolites have been prepared by using two different methods: desilication and the surfactant assisted method [4]. Desilication is accomplished by introduction of random mesoporosity via alkaline treatment. Surfactant assisted desilication is performed by alkaline treatment in the presence of a surfactant, in this case CTAB (cetyltrimethylammonium bromide). Complete characterization of the prepared zeolite includes TEM (Figure 1), XRD, and N_2 adsorption. We found that increasing base concentration increases the mesoporosity (measured by N_2 adsorption), but drastically reduces crystal integrity. These findings are consistent with previously reported results [5-6].

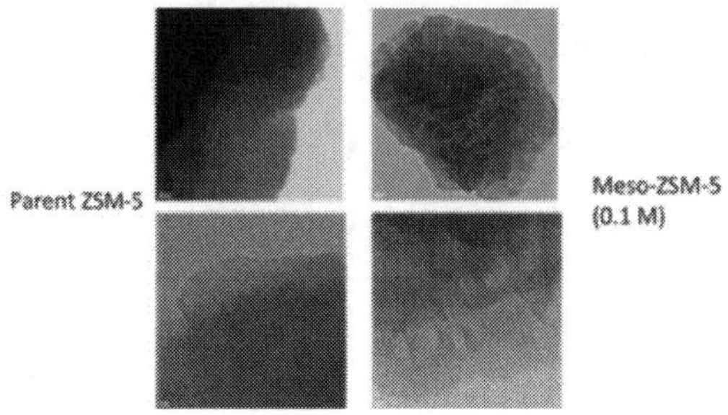

Figure 1 TEM images of parent ZSM-5 (left) and mesoporous ZSM-5, prepared with 0.1 M base treatment

Catalytic experiments have been performed in a pyroprobe microreactor (CDS Analytical). Naphthalene is chosen as a representative tar compound. Experiments are performed at a variety of temperatures, residence times, and catalyst to reactant ratios. For both the Y zeolite and ZSM-5 it was found that conversion of naphthalene is decreased when mesoporosity was introduced. However, selectivity to desirable compounds, those lighter than naphthalene, is increased when performing mesoporous zeolite was used. Figure 2 shows the selectivity of both Y zeolite and ZSM-5 to lighter compounds for both microporous and mesoporous catalyst. This may indicate a decrease in diffusion limitations for the mesoporous catalyst.

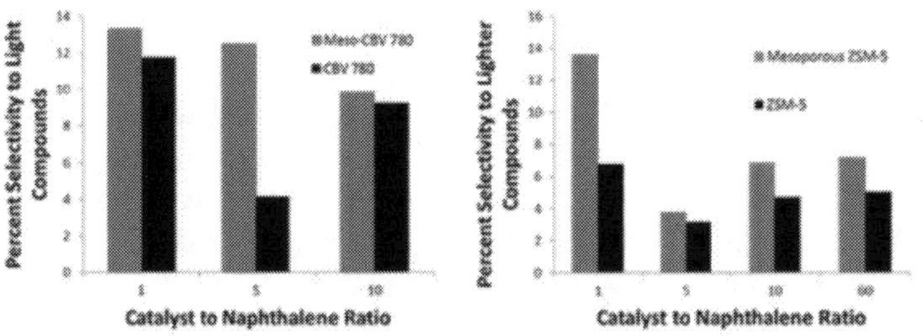

Selectivity to compounds lighter than naphthalene

When the mesoporous and microporous catalysts are compared on an activity basis, as in Figure 3, the two catalysts are essentially equivalent.

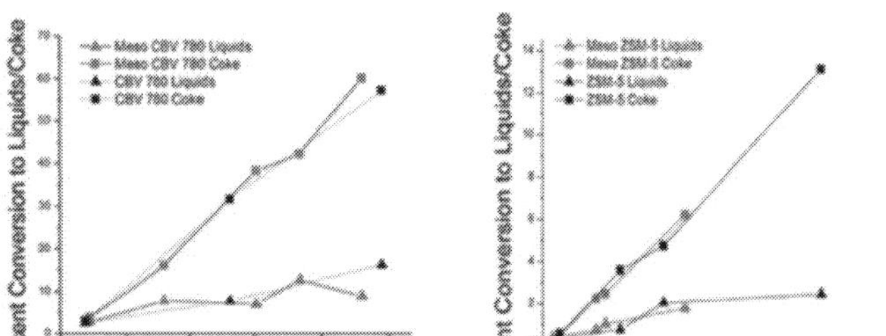

Figure 3 Conversion of naphthalene to coke and liquids prepared on an activity basis

The pyridine adsorption FTIR technique is used in this study to determine the effect of introduced mesoporosity on the Brønsted and Lewis acidity of each zeolite, and their relation to the Si/Al ratio. Additionally, in situ diffuse reflectance infrared Fourier transform spectroscopy (DRIFTs) diffusion and adsorption studies with benzene and naphthalene have been performed to determine the mass transport and kinetic properties of each catalyst. Catalytic reactions have also been performed in reactive syngas, as opposed to inert nitrogen, to promote reforming reactions, and simulate a more realistic producer gas environment. Microporous and mesoporus zeolites impregnated with metals to further increase catalytic activity have also been studied and these results will be presented.

References

[1] S.D. Phillips, *Ind. Eng. Chem. Res.* 46(2007) 8887.

[2] B. Dou, J. Gao, X. Sha, S.W. Baek. *Applied Thermal Engineeirng*. 23 (2003) 2229-2239

[3] P.R. Buchireddy, R.M. Bricka, J. Rodriguez, W. Holmes. *Energy Fuels.* 24 (2010) 2707-2715.

[4] J. Garcia-Martinez, M. Johnson, J.A. Valla, K. Li, J.Y. Ying. *Catal. Sci. Technol.* 2 (2012) 987-994

[5] D. Verboekend, G. Vile, J.Perez-Ramirez. *Advanced Functional Materials* 22 (2012) 916-928

[6] K. Li, J.A. Valla, J. Garcia-Martinez. *ChemCatChem.* 5 (2013) 2-23

Acknowledgement

This work was generously funded by the National Science Foundation (Award CBET-1236738)

389938 Kinetics of Platinum Dissolution in Acidic Electrolyte: Investigation By Rotating Ring-Disk Electrode

Abstract:

Despite the high nobility of platinum metal, the trace dissolution of platinum is a limiting factor for supported nanoparticle catalysts in proton exchange membrane fuel cells. The kinetics of platinum dissolution in acid solutions are not well known due to the very low solubility of dissolved platinum and the simultaneous formation of a surface oxide layer, which prevents measurements at steady state. The rotating ring-disk electrode (RRDE) technique is ideally suited for the transient detection of dissolution under arbitrary potential cycling conditions. In this work, the RRDE technique was applied to the study of the anodic and cathodic dissolution of platinum. Experimental challenges were addressed, including measuring nanoamp ring currents while drawing milliamp disk currents and eliminating impurity effects. The RRDE results were fit to a combined kinetic model of platinum dissolution and surface oxidation, building on the authors' previous work.

390091 Assessment of Catalytic Materials for Hydrothermal Upgrading of Algae Biocrude

Abstract:

Liquefaction of wet biomass such as algae, crop residues, and biowaste in hot compressed water produces a crude bio-oil while obviating the need to first dry these feedstocks. This crude bio-oil is viscous, has a high total acid number, and is rich in heteroatoms such as O, N, and S. Hydrothermal catalytic upgrading of the crude bio-oil may be a pathway to its transformation to a hydrocarbon stream that could be blended with petroleum and refined using existing fuel production infrastructure. Various catalysts have shown promising activity for upgrading these crude oils, but no catalyst has yet proven to be clearly suitable for long-term operation in these harsh conditions. This poster will present results from an international collaborative effort that systematically studied the efficacy of a set of different metals supported on different materials. We will compare various metal catalysts and supports that have shown some resistance to degradation and deactivation in hydrothermal environments. The catalyst supports include anatase TiO_2, barium hexaaluminates, lanthanum hexaaluminates, and zeolites. This careful systematic study provides new information about how best to perform hydrothermal catalytic upgrading of crude bio-oils.

High pressure catalytic conversion of n-pentane on H-ZSM-5

Edward P Schreiner, Shewangizaw Teketel, Raul F Lobo

The flight of aircraft and missiles at hypersonic speeds (> Mach 5) is hindered by overheating of the engine and electronic components caused by fuel combustion and air friction [1]. Currently, aircraft are cooled by transferring thermal energy from the engine and sensitive electronic systems to the fuel; in such cases, the maximum amount of heat that can be removed is determined by the heat capacity, thermal stability, and initial temperature of the fuel. The cooling capacity can be improved by utilizing *endothermic fuels*, fuels that undergo endothermic chemical reactions driven by the thermal energy removed from the aircraft [2,3]. A potential endothermic reaction is the cracking of hydrocarbons on solid catalysts to produce lighter molecules, with olefins being preferred to maximize endothermicity.

In this study, the effects of temperature (633-723K), pressure (10-60 bar) and weight hourly space velocity (WHSV) (400-1500 h^{-1}) on the conversion of pentane on H-ZSM-5 were investigated to determine the potential of this reaction/catalyst system for endothermic cooling. H-ZSM-5 was chosen because of its use in industrial fluidized catalytic cracking units to increase light olefin yield [4]. Pentane was chosen as it is the simplest liquid hydrocarbon at STP and its high cooling potential (pentane cracking to ethane and propylene has a heat of reaction of 1.15 MJ/kg). 5-25% pentane conversion was observed at a pressure of 40 bar and temperatures in the range of 633-723K. At a temperature of 673K, saturation kinetics (i.e. the rate of pentane consumption does not increase with increasing pressure) were observed at pressures above 30 bar, resulting in approximately 14% conversion. At 40 bar and 673K, increasing WHSV (400-1500) resulted in decreased pentane conversion (26-10%). Propane and butane were the major products, followed by heavier C6+ compounds and other light products (C1-C4 paraffins and olefins). An inverse relationship between the production of C6+ and other light compounds was observed with changes in reaction conditions: C6+ carbon selectivity increased from 20% at 10 bar to 27% at 60 bar, while it declined from 28% at 633K to 18% at 723K.

The high selectivity (>70%) to propane and butane, along with the observed changes in the selectivity to light olefins and heavy C6+ compounds, indicate that bimolecular reactions, such as hydride transfer and alkylation, control product distribution. While the product distribution from the high pressure conversion of n-pentane resulted in an endothermic process, hydride transfer reactions led to the transformation of high energy olefins into low energy paraffins and aromatics, resulting in a decrease of the process endothermicity. Hydride transfer occurs readily on H-ZSM-5 and other zeolites, and an alternative catalyst that suppresses these reactions should improve selectivity to light olefins, which will increase cooling capacity of the endothermic fuel.

References

[1] Lander H, Nixon, AC (1971) Journal of Aircraft 8: 200-207.
[2] Gascoin N, Abraham G, Gillard P (2010) J Analytical and Applied Pyrolysis 89: 294-306.
[3] Kim J, Hyeon DH, Park SH, Chun BH, Jeong BH, Han JS, Kim SH (2013) Catalysis Today
[4] Farshi A, Abri HR (2012) Petroleum Science and Technology 30: 1285-1295.

$Au/CuO-CeO_2$ Catalyst for Preferential Oxidation of CO

in Hydrogen-rich Stream.

Ying-Chieh Chou, Yu-Wen Chen*

Department of Chemical and Materials Engineering, National Central University, Chung-Li 320, Taiwan

Nanoscaled gold particle supported on $CuO-CeO_2$ was used for preferential oxidation of carbon monoxide in hydrogen-rich stream (PROX). $CuO-CeO_2$ support was prepared by co-precipiataion method and gold wad added by deposition-precipitation method. Different amounts of CuO and different calcination temperatures of the support were used to develop the best catalyst for PROX reaction. These catalysts were characterized by XRD, TEM, HR-TEM, and XPS. TEM images showed the particle size of gold were around 2-4 nm and Au were dispersed well on the support. The PROX reaction was carried out in a fixed bed flow reactor with a feed of CO: O_2: H_2: He = 1.33: 1.33: 65.33: 32.01 in volume ratios. The results showed that the catalyst with specific Cu/Ce ratio and calcination temperature could reach 100% of CO conversion at the PEM fuel cells operating temperature (65□-100□) even the gold content reduced from 1 wt. % to 0.5 wt. %. The $CuO-CeO_2$ prepared by co-precipitation had higher concentration of hydroxyl groups compared to those by incipient-wetness impregnation method. The higher calcinations temperature for the support resulted in high crystallinity of CeO_2, which had higher oxygen vacancy. Compared to $CuO-CeO_2$ catalysts, which are more active at higher temperature, the $Au/CuO-CeO_2$ catalysts are preferable at lower temperature for PROX reaction. The presence of CuO could suppress H_2 dissociation and retard the oxidation of H_2, resulting the higher CO oxidation activity in the limited amount of O_2.

391385 Machine Learning Approaches to Design Catalysts for C1 Chemistry

Abstract:

Simultaneous carbon dioxide and methane conversion to chemicals and fuels shows great promise in mitigating the environmental impact of greenhouse gas emissions, but requires the activation of the molecules from their energetically stable ground states and the selective cleavage of thermodynamically stable bonds. Density functional theory calculations and micro kinetic modeling are employed to elucidate structure-property-activity relationships for carbon dioxide and methane adsorption and activation on complex oxide catalysts with controlled structure, composition, and electronic properties. The products investigated include methanol, formic acid, formaldehyde, and CO. Efforts to use machine learning approaches to designing superior catalysts and control process conditions will also be presented.

Effect of Transition Metals (M) in M@Pd core@shell nanoparticles for Direct Formic Acid Fuel Cells

Shuozhen Hu[a], Louis Scudiero[b], Su Ha[a]

[a] *The Gene and Linda Voiland School of Chemical Engineering and Bioengineering, Washington State University, Pullman, WA 99164*

[b] *Chemistry Department and Materials Science and Engineering Program, Washington State University, Pullman, WA 99164*

The study of Pd-based core@shell nanoparticles that consist of one transition metal as the core and Pd as the shell has generated great interest in the field of electrochemical catalysis, especially in direct formic acid fuel cells (DFAFCs), because their electronic and chemical properties are completely different from those of pure Pd. The current DFAFC technology uses Pd as the anode catalyst, on which formic acid is oxidized to CO_2 and releases electrons. An enhanced activity has been obtained for these Pd-based core@shell nanoparticles compare to pure Pd for formic acid oxidation. However, there is a lack of understanding of how these added transition metals interact with Pd, namely by influencing their electronic structure and electrochemical property toward electrochemical oxidation of formic acid. It is important to fill this knowledge gap and find out a trend between the effect of adding the transition metals and the electrochemical activity of Pd-based core@shell nanoparticles. This will help developing DFAFCs with higher efficiency. In addition, the cost of DFAFCs can be reduced by using the Pd-based core@shell nanoparticles since less-expensive transition metals can be used as the core and less Pd is required compare to the DFAFCs with pure Pd anodes.

In this study, M@Pd (M = Au, Cu, Ru, Pt, etc.) core@shell nanoparticles with same thickness of Pd are synthesized via a two-step process. Non-supported nanoparticles are used for TEM analysis to characterize the structure and morphology of the nanoparticles. XPS is used to differentiate the chemical shifts due to charge transfer between Pd and M. XPS analysis shows a different binding energy shift for Pd peak for M@Pd core@shell nanoparticles compare to that of pure Pd nanoparticles. In addition, a change of the XPS valence band shape and down-shift of the d-band center are also observed. These different binding energy shifts indicate that the core metals (M) interact with Pd shell and significantly influence the electronic structure of Pd. This electronic perturbation changes the bond strength between Pd and the adsorbate. Thus, the electrochemical performance of M@Pd core@shell nanoparticles is altered, which can be detected by cyclic voltammetry (CV) and chronoamperometry (CA) experiments. Based on the CV and CA results, M@Pd core@shell nanoparticles show different Pd-H bond strength and activity toward formic acid oxidation with different transition metals as the core. The relationship between the electronic perturbation and electrochemical activity can be elucidated.

Particle size effect of Iridium nanoparticles on hydrogen oxidation and evolution reaction activity in alkaline electrolytes

Jie Zheng and Yushan Yan*

Department of Chemical and Biomolecular Engineering
University of Delaware
150 Academy Street, Newark, DE 19716, USA

*yanys@udel.edu

Hydroxide exchange membranes fuel cells (HEMFCs) enable the use of non-noble metal or metal-free catalysts for oxygen reduction reaction (ORR) in alkaline electrolytes. However, the activity of hydrogen oxidation/evolution reaction (HOR/HER) in alkaline electrolytes is about two orders of magnitude lower than that in acid on Pt, Pd and Ir[1,2], which makes HEMFCs less cost-effective since a higher loading of precious metal will be required for HOR in HEMFCs or HER in HEM electrolyzers. The reason for the sluggish kinetics of HOR/HER in alkaline electrolytes remains unclear, and therefore, studies of HOR/HER in alkaline electrolytes would be of great importance in order to gain fundamental understandings as well as to develop catalysts with higher activity.

Strmcnik et al. studied the HOR/HER on polycrystalline Ir in 0.1 M KOH, and showed a HOR polarization curve overlapping with the concentration overpotential curve, indicating a facile HOR kinetics on polycrystalline Ir that was attributed by the oxophilic nature of Ir[3]. Their results suggest that Ir is a promising alternative to Pt as HOR/HER catalyst. However, Durst et al. obtained a smaller exchange current density on Ir/C (0.37 ± 0.12 mA/cm^2 at 313 K) than Pt/C in alkaline electrolytes. The discrepancy between these two works might come from the difference of electrode types (polycrystalline Ir with extended surface for Strmcnik et al. and carbon supported Ir nanoparticles for Durst et al.), which is generally known as particle size effect. Particle size effect for HOR/HER has been studied for Pt and Ru[4,5], but it is yet to be investigated for Ir.

In this work, we obtained Ir/C samples with different Ir particle size by heat treating commercial 20 *wt.%* Ir/C at different temperatures. Their HOR/HER kinetics in alkaline electrolyte was evaluated using RDE measurements with correction of solution resistance. The kinetic currents were fitted into the Butler-Volmer equation to get exchange current densities. Our results showed that the specific exchange current density increases as particle size increases. We suspect that the HOR/HER on Ir is facet sensitive, and the different ratios of facets exposed on Ir nanoparticles with different sizes might be responsible for the particle size effect.

References

(1) Sheng, W.; Gasteiger, H. A.; Shao-Horn, Y. *Journal of the Electrochemical Society* **2010**, *157*, B1529.
(2) Durst, J.; Siebel, A.; Simon, C.; Hasche, F.; Herranz, J.; Gasteiger, H. A. *Energy & Environmental Science* **2014**.
(3) Strmcnik, D.; Uchimura, M.; Wang, C.; Subbaraman, R.; Danilovic, N.; van der Vliet, D.; Paulikas, A. P.; Stamenkovic, V. R.; Markovic, N. M. *Nature Chemistry* **2013**, *5*, 300.
(4) Sun, Y.; Dai, Y.; Liu, Y.; Chen, S. *Physical Chemistry Chemical Physics* **2012**, *14*, 2278.
(5) Ohyama, J.; Sato, T.; Yamamoto, Y.; Arai, S.; Satsuma, A. *Journal of the American Chemical Society* **2013**, *135*, 8016.

The Effect of Graphene As a Support for Non-PGM Bifunctional Oxygen Catalyst in Rechargeable Metal/Air Batteries

Miguel A. Garcia-Contreras [a,b], Pooya Hosseini-Benhangi [a], Amin Taheri Najafabadi [a], Elöd Gyenge [a]

a) Department of Chemical and Biological Engineering & Clean Energy Research Center, The University of British Columbia, 2360 East Mall, Vancouver, BC, Canada V6T 1Z3

b) Department of Chemistry, National Institute of Nuclear Research, Carretera México-Toluca s/n, La Marquesa Ocoyoacac, Estado de México, México 52750

Introduction

Graphene, as the rising star in the world of advanced materials, offers game-changing prospects for many key areas of research including energy storage and conversion, biotechnology, novel electronic devices and so forth [1]. Bifunctional oxygen cathodes which can catalyze both ORR (oxygen reduction reaction) and OER (oxygen evolution reaction) are the backbone of rechargeable metal-air batteries as well as regenerative fuel cells [2-4]. Manganese oxides have been vastly employed as a robust cost-effective multifunctional and environmental friendly electrode material in battery industry, from primary to rechargeable metal-air batteries, as well as alkaline fuel cells and capacitors [4]. The electrolytic manganese dioxide (γ-MnO_2) is known as the most electrochemically active crystallographic form of MnO_2 for ORR in alkaline media with Tafel slope of 40 mV dec^{-1} and a low overpotential of -375 mV [2, 5]. However, poor OER electrocatalytic activity of MnO_x in alkaline media diminish the hope of finding an exclusive bi-functional catalyst for both ORR and OER [2]. This could be avoided by adding another class of catalyst active for OER to the MnO_2. $LaCoO_3$, also known as a perovskite-type oxide, has been reported to demonstrate distinguished electrocatalytic activity for OER in alkaline electrolytes in comparison to Pt or MnO_x, i.e. up to 1 mA cm^{-2} at 1600 mV (vs. RHE) as well as a Tafel slope of 60 mV decade^{-1} with excellent durability over 100 cycles in 0.1 M KOH [2, 6]. A thorough comparison with literature regarding other bifunctional electrodes shows that the MnO_2-perovskite catalytic system is very competitive with other catalysts presented in the literature, while having the advantage of cost efficiency and easy preparation methodology [2].
Our research is aimed at investigating the effect of support on the performance of MnO_2-based catalysts for both ORR and OER with long cycle life. The mechanisms for OER and ORR of the mixed catalysts and the role of support are investigated by a combination of surface characterization methods and electrochemical techniques.

Experimental

- In the present work, $LaCoO_3$ was synthesized via co-precipitation methods. While different methods for graphene synthesis has been proposed in the literature, a novel approach toward electrochemical exfoliation of graphite assisted by ionic liquids (ILs) was used to synthesize both graphene and N-doped graphene. This approach provides three main features: cost efficiency due to low IL content, extended electrochemical stability in a non-aqueous electrolyte, and high exfoliation yield by effective anionic intercalation within the graphitic layers by IL/acetonitrile electrolyte with dramatically lower loads of ionic liquids (~1:50 IL/acetonitrile vol. ratio) [1]. The catalyst layer consisted of MnO_2:$LaCoO_3$:Support:Nafion with a weight ratio of 1:1:1:0.67. The support varies from the electrochemically synthesized graphene, N-doped graphene, multi-walled carbon nanotubes, and vulcan. Cyclic voltammetry tests were performed in O_2 and N_2 saturated 6 M KOH at 293 K to investigate the electrocatalytic activity of the MnO_2-$LaCoO_3$ catalyst for both OER and ORR on different supports. The longer-term durability of the electrodes was also investigated by performing 100 repeated OER-ORR voltammetric cycles. Combination of Raman and X-ray photoelectron spectroscopy (XPS) as well as SEM analysis were employed to further characterize the catalyst layer.

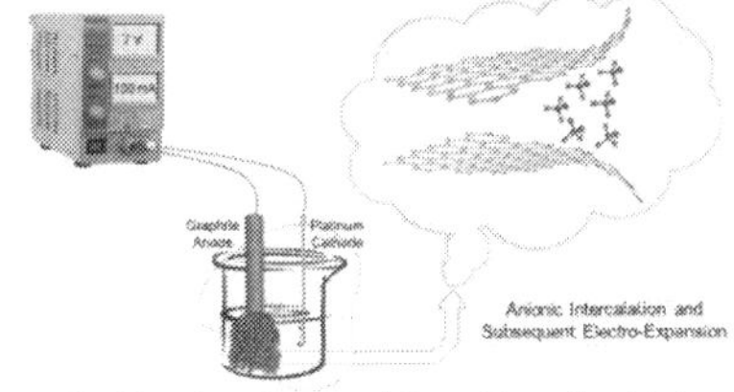

Fig. 1 Experimental setup used for graphite anodic exfoliation.

Characterization

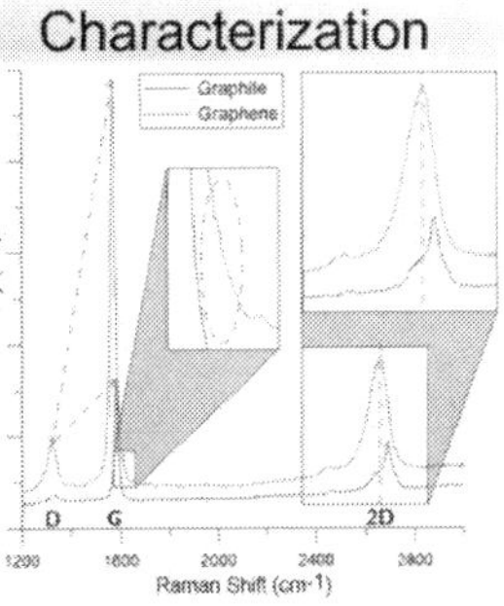

Fig. 2 Raman spectra of graphene.

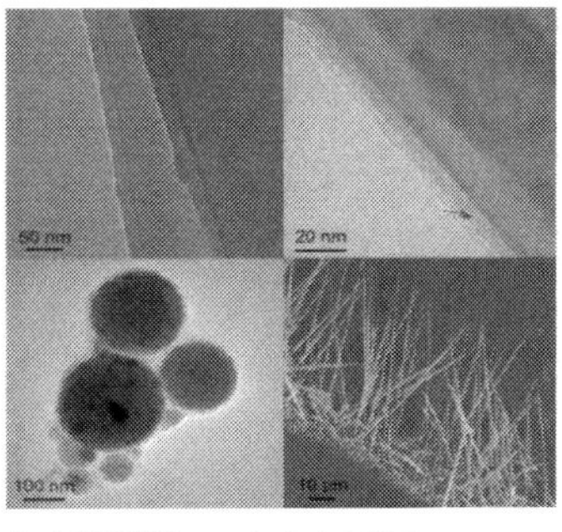

Fig. 3. SEM/TEM images for electro-exfoliation.

- Majority of products
  - ✓ Ultrathin layers of folded/crumpled GN (AFM: < 5 layers)
  - ✓ ~500 nm average length based on 250 sample images
- Minority
  - ✓ Nanotubes: sheet folding due to intercalation
  - ✓ Nanoparticles: breakage of sheets' vertices

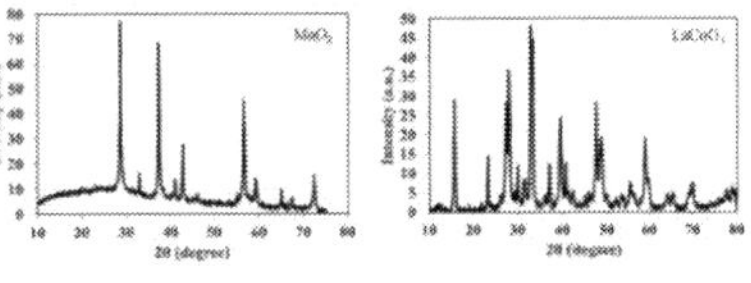

Fig 4. XRD spectra of: a) commercial Sigma-Aldrich MnO_2 powder, b) synthesized $LaCoO_3$ powder.

Fig. 5 SEM images of: a) commercial Sigma-Aldrich MnO_2 powder, b) synthesized $LaCoO_3$ powder.

Catalytic Performance

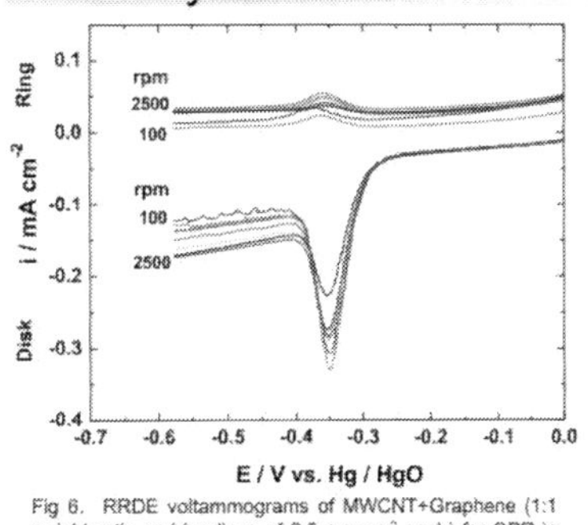

Fig 6. RRDE voltammograms of MWCNT+Graphene (1:1 weight ratio and loadings of 0.5 mg cm^{-2} each) for ORR in O_2-saturated 6M KOH at 293 K.

Support	Number of electrones	Potential range / V vs Hg/HgO	% H_2O_2	Onset Potential for ORR @ 1600 rpm / V vs Hg/HgO
Graphene	2.3-2.0	-0.35-0.4	83-97	-0.031
N-Doped Graphene	2.8-2.2	-0.35-0.4	60-88	-0.043
MWCNT + Graphene	2.3-2.0	-0.35-0.4	80-99	-0.048
Vulcan	2.4-2.0	-0.35-0.4	78-100	-0.030

Table 1. Number of electrons ,% of H_2O_2 and onset potential values during ORR on different supports in O_2-saturated 6M KOH at 1600 RPM and 293 K.

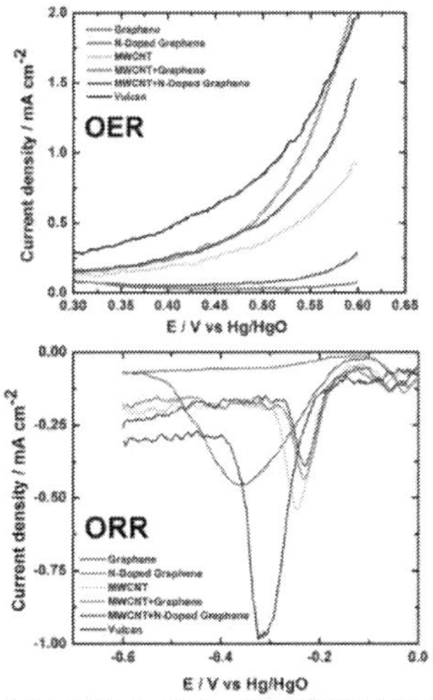

Fig. 7. OER and ORR on supports in 6M KOH at 400 RPM and 293 K.

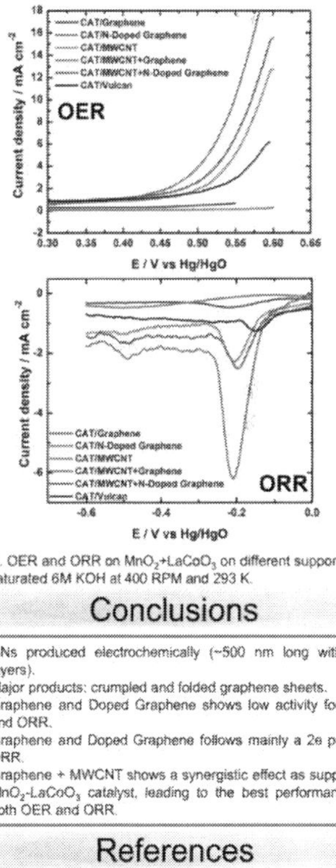

Fig 8. OER and ORR on MnO_2+$LaCoO_3$ on different supports in O_2-saturated 6M KOH at 400 RPM and 293 K.

Conclusions

- GNs produced electrochemically (~500 nm long with < 5 layers).
- Major products: crumpled and folded graphene sheets.
- Graphene and Doped Graphene shows low activity for OER and ORR.
- Graphene and Doped Graphene follows mainly a 2e path for ORR.
- Graphene + MWCNT shows a synergistic effect as support for MnO_2-$LaCoO_3$ catalyst, leading to the best performance for both OER and ORR.

References

[1] A.T. Najafabadi, E. Gyenge, Carbon, 71 (2014) 58-69.
[2] P.H. Benhangi, A. Alfantazi, E. Gyenge, Electrochimica Acta, 123 (2014) 42-50.
[3] J. Ludwig, J. Power Sources, 155 (2006) 23-32.
[4] A. Serov, A. Aziznia, P.H. Benhangi, K. Artyushkova, P. Atanassov, E. Gyenge, Journal of Materials Chemistry A, (2013).
[5] E.L. Gyenge, J.-F. Drillet, J. Electrochem. Soc., 159 (2012) F23-F34.
[6] M. Risch, A. Grimaud, K.J. May, K.A. Stoerzinger, T.J. Chen, A.N. Mansour, Y. Shao-Horn, The Journal of Physical Chemistry C, 117 (2013) 8628-8635.

Acknowledgments

- Generous financial support by:
  - ✓ Natural Sciences and Engineering Research Council of Canada
  - ✓ M. A. Garcia-Contreras is grateful to CONACYT and ININ, MEXICO for financial support to conduct a Research Stay at UBC

364910 Hydrothermal Synthesis of Graphene/Fe^{3+}-Doped TiO_2 Nanowire Composites with Enhanced Photocatalytic Activity Under Visible Light Irradiation

Abstract:

TiO_2 is one of the most widely investigated semiconducting metal oxides that exhibits photocatalytic activity and is used in a wide range of applications. However, with its 3.2 eV electronic band gap, TiO_2 is only sensitive to the light wavelengths below 380 nm which belong to the UV range. Besides, the photogenerated electron−hole pairs, which are mainly responsible for the photocatalytic activity, have faster recombination rates than the rate of chemical interaction between TiO_2 and adsorbed pollutants, reducing the photocatalytic efficiency. To enhance photocatalytic activity of TiO_2, many methods have been employed, such as ion doping, graphene incorporating and so on.

In this study, the hydrothermal method was employed to synthesize Fe^{3+}−doped TiO_2 nanowires (FeNW) and then fabricate graphene/Fe^{3+}−Doped TiO_2 nanowire nanocomposite (GFeNW). Graphene oxide (GO) reduction to graphene and hybridization between FeNWs and graphene by forming chemical bonding was achieved in a one-step hydrothermal process. Graphene/TiO_2 nanoparticle nanocomposite (GNP) and graphene/ TiO_2 nanowire nanocomposite (GNW) were also synthesized as controls. Photocatalytic performance and related properties of TiO_2 nanoparticle (NP), TiO_2 nanowires (NW), FeNW, GNP, GNW and GFeNW were comparatively studied. It was found that by incorporation of graphene, GNP, GNW and GFeNW have higher performance than their counterparts. It was also found that FeNWs and NWs, in comparison with NPs, have more uniform dispersion on graphene with less agglomeration, resulting in more direct contact between TiO_2 and graphene, and hence further improve electron−hole pair separation and transportation. The diffuse reflectance spectra and photocurrent properties of these composites indicated that Fe^{3+} doping could improve the response of TiO_2 nanowire under visible-light irradiation. The result reveals that the relative photocatalytic activity of GFeNW is highest of all.

393953 The Effects of Catalysts Pore Structure on Intraparticle Diffusion in Cobalt-based Fischer-Tropsch Synthesis

Abstract:

Intraparticle diffusion in Fischer-Tropsch synthesis not only decreases the overall efficiency, but also changes the reaction behaviors via affecting syngas diffusion, olefin's second reactions and so on. Due to the capillary effect, the Fischer-Tropsch catalysts pores are always filled with liquid products under steady-state conditions. Always the process is supposed to depend on the characteristics of molecular diffusion in liquid[1-2]. According to Fick law, intraparticle diffusion rate and . In that case, pore structure is closely related to the diffusion, embodying by particle porosity ε and tortuosity factor τ. We prepared two kinds of ZrO_2-supported catalysts, whose reaction behaviors were similar, but greatly different in pore network, especially in structure regularity. Based the relationship between pore structure and surface fractal dimensions, we compared quantitatively the differences in the catalysts pore network via considering coordination number Z, which characterizes particle pore network connectivity. Then, the effects of catalysts pore structure on intraparticle diffusion and following Fischer-Tropsch synthesis reaction behaviors were revealed from a new perspective.

For Fischer-Tropsch synthesis, the results showed the more porous, the worse connectivity, the gap between D_e and D_A should be narrower. The smaller Z, pore network connectivity would be worse. However, its variation effects would be limited when the Z absolute value was high. The calculation also indicated that Z scarcely varied with the pore structure regularity. Considering high Z is common for porous materials, it might not be a good idea to improve the mass-transfer by adjusting the pore structure in the case of certain porosity, especially increasing the pore regularity.

Keywords: Fischer-Tropsch synthesis, Intraparticle diffusion, Pore structure

Reference

[1] M. F. M. Post, A. C. Hoog.VAN'T, J. K. Minderhoud et al, AIChE Journal, 1989, 35(7):1107-1114.

[2] E. Iglesia, Applied Catalysis A: General, 1997, 161(1-2):59-78.

Supported by:

(1) Strategic Priority Research Program, Chinese Academy of Sciences, Grant No. XDA07070700

(2) National Natural Science Foundation of China, Grant No. 21273265

393954 A Novel Catalyst Molding Method to Remove Intraparticle Diffusion Limitation in Fischer-Tropsch Synthesis

Abstract:

The technology of the Fischer-Tropsch synthesis (FTS) in fixed-bed reactor over cobalt-based catalyst, indirect liquefaction of coal to obtain fuel and chemicals, has a promising application due to its easy industrialization. Noticeably, in order to reduce the pressure drop and make sure the reaction could be carried out smoothly, the catalyst particles size should be larger than 1 mm in industrial fixed-bed reactor. However, the intraparticle diffusion limitation of larger catalyst particles becomes serious, which could increase CH_4 selectivity and decrease C_{5+} selectivity, thus significantly reduce the value of products. Small particles could completely eliminate the effect of intraparticle diffusion limitation for short diffusion length. In that case, we intend to enhance mass-transfer in larger catalyst particles (> 1mm in diameter) via secondary molding of smaller catalyst. Specifically, smaller particles (< 0.1mm in diameter) and pore-forming agent would be combined to form larger particles (>1 mm in diameter) by using binders, such as PCS, clay and PMHS. After removing the pore-forming agent, smaller catalyst particles were bonded together. And larger spaces among them, where the pore-forming agent once occupied, were formed. Then, a honeycomb structure catalyst pellets were brought. In this structure, both reactants and products could pass through the interspace quickly during reaction. The new molding method realized negligible diffusion limitations in the catalyst pellets, which meet the need of the industrialization of FTS in fixed-bed reactor. According to the FTS results, the secondary molded catalyst exhibit excellent performance with lower CH_4 selectivity and higher C_{5+} selectivity comparing with those prepared by traditional molding method, such as extrusion.

Keywords: Fischer-Tropsch synthesis, Intraparticle diffusion, molding method

Supported by

(1) Strategic Priority Research Program, Chinese Academy of Sciences, Grant No. XDA07070700

(2) National Natural Science Foundation of China, Grant No. 21273265

Direct synthesis of allyl compound from polyols over Iron-oxide Based Catalyst

Teruoki Tago*, Hirofumi Shitara, Kazuhiro Terai, Yuta Nakasaka, Takao Masuda
Hokkaido University, Sapporo, Hokkaido 060-8628 Japan
*corresponding author:tago@eng.hokudai.ac.jp

The catalytic conversion of glycerol was performed with iron oxide-based catalysts for the production of useful chemicals. The catalytic reaction was carried out in a fixed-bed flow reactor at 623 K under atmospheric pressure. Useful chemicals such as allyl-alcohol, propylene and ketones were obtained. These useful chemicals are expected to be produced from glycerol through two main pathways: one pathway involves the production of allyl alcohol and propylene (Pathway I), and another involves the production of hydroxyacetone and acrolein (Pathway II). Allyl alcohol and propylene are attractive and valuable chemicals obtained from glycerol, which is a characteristic reaction using iron-oxide catalyst. It was found that the addition of potassium to the catalyst was effective in increasing the yield of allyl alcohol. Moreover, the catalyst was applied to produce butadiene from erythritol (C4 polyol). While the self-cyclization occurred, 1,3-butadiene (yield of approximately 16 mol%-C) could be obtained. This paper reports the production of allyl compounds from glycerol and the effects of catalyst composition and experimental conditions on these yields are discussed, based on investigations of the reaction pathway and mechanism.

PREFERENTIAL OXIDATION OF CO OVER Cu-Ce OXIDE NANOSTRUCTURES

Joan Papavasiliou[1] and George Avgouropoulos[1,2*]

[1] *Foundation for Research and Technology-Hellas (FORTH), Institute of Chemical Engineering and High Temperature Processes (ICE/HT), 26504 Patras, Greece*
[2] *Department of Materials Science, University of Patras, 26504 Patras*

Abstract

The development of effective catalyst for the preferential oxidation of CO in H_2-rich streams (PROX process) has attracted much attention during the last decade, due to its application in fuel cell energy systems [1-7]. Polymer Electrolyte Membrane Fuel Cells (PEMFCs) which operate with pure H_2 appear to be an ideal energy solution for portable and mobile energy applications. In order to avoid several technical and safety limitations concerning the use of pure H_2, the utilization a H_2-rich gas mixture is favored. This gas mixture can be produced from hydrocarbons or alcohols (natural gas, gasoline, methanol) as a H_2 carrier, through a catalytic fuel processor. The starting fuel is converted in a H_2-rich gas through a catalytic steam reforming reactor and a water-gas shift reactor. The produced reformate gas still contains a significant concentration of carbon monoxide (~1% CO), thus needing further purification, since this amount of CO poisons the anode electrocatalysts of low temperature PEM fuel cell. Preferential catalytic CO oxidation (PROX reaction) is a simple and cost-effective method for removing CO to less than 20 ppm from reformed fuels.

Cu-Ce mixed oxides have been prepared with different methods (impregnation, co-precipitation, sol-gel, combustion) and have been studied and proposed for the title process [3-7]. Recent years, the research community employs new chemical synthesis methods, aiming to the preparation of nanostructured catalysts with high selectivity, extremely high activity, low energy demands and long life time [8]. These can be achieved only by controlling the size, the shape, the particle size distribution, the composition and the electronic structure of the surface, the thermal and chemical stability of the specific nanocomponents.

In this work hydrothermal method was employed, using citric acid as chelating agent, sodium hydroxide as precipitating agent and copper nitrate and cerium nitrate as copper and cerium precursors, respectively. Synthesis parameters of $CuCeO_x$ catalysts were temperature and duration of hydrothermal process, and sodium hydroxide concentration. Choosing proper combination of these parameters, different nanostructures (rods, cubes, polyhedra, spheres) can be obtained, with attractive physicochemical characteristics. Activity and selectivity of nanostructured catalysts was tested for CO preferential oxidation in the presence of excess H_2.

References

[1] M. Kipnis, Appl. Catal. B 152-153 (2014) 38.
[2] K. Liu, A. Wang, T. Zhang, ACS Catalysis 2 (2012) 1165.
[3] L. Yang, S. Zhou, T. Ding, M. Meng, Fuel Process. Technol. 125 (2014) 155.
[4] S. Zeng, Y. Wang, S. Ding, J.J.H.B. Sattler, E. Borodina, L. Zhang, B.M. Weckhuysen, H. Su, J. Power Sources 256 (2014) 301.
[5] G. Avgouropoulos, T. Ioannides, Appl. Catal. A 244 (2003) 155.
[6] G. Avgouropoulos, T. Ioannides, H. Matralis, Appl. Catal. B 67 (2006) 1–11.
[7] G. Avgouropoulos, T. Ioannides, Chem. Eng. J. 176-177 (2011) 14.
[8] R. Si, M. Flytzani-Stephanopoulos, Angew. Chem. Int. Ed. 47 (2008) 2884.

367768 Mixed-Light Paraffin Dehydrogenation in a Catalytic Multi-Tubular Reactor: Modeling and Simulation

Abstract:

Petrochemicals such as polypropylene and polybutadiene are priced much higher than the petroleum refinery produts. Petroleum refinery units try to maximize their profits by bottom of the barrel upgradation using fluidized catalytic cracking (FCC), which converts the heavier hydrocarbons into lighter ones namely liquified petroleum gas (LPG), which predominantly contains propane and butane. The light gases (propane and butane) are feedstock for petrochemical production. These light gases are converted (dehydrogenated) to highly reactive propylene and butylene (the chief raw materials for the above mentioned petrochemicals). Propane and butane are always available as mixture along with trace quantities of methane and ethane. In any commercial process, the propane and butane are *separated first* and then dehydrogenated in *separate reactors*. This may probably lead to *high fixed and operating costs*. Current study proposes a *process intensification* method via mixed-feed dehydrogenation in a multi-tubular fixed bed catalytic reactor with Pt-Sn suppoted on γ-Al_2O_3. The dehydrogenation takes place at 500- 800°C and is highly endothermic. The fixed bed dehydrogenation reactor was modeled for the mixed feed of propane and butane. The reactor is simualted for various parameters under isothermal as well as adiabatic operating conditions, and the results between the two processes were compared.

362956 Highly Crystalline Multimetallic Nanoframes with Three-Dimensional Electrocatalytic Surfaces

Abstract:

In this presentation, a research system that connects fundamental investigation on well-defined extended surfaces (e.g. single crystal surfaces), theoretical studies using computer-powered simulation, and finally design of high performance catalysts in which all the possible beneficial properties from complex functional structures are implemented, will be demonstrated.

Control of structure at the atomic level can precisely and effectively tune catalytic properties of materials, enabling enhancement in both activity and durability. To mimic the optimal surface structure of $Pt_3Ni(111)$-Pt-Skin for oxygen reduction reaction (ORR), a highly active and durable class of electrocatalysts are prepared by exploiting the structural evolution of Pt-Ni bimetallic nanocrystals. The starting material, crystalline $PtNi_3$ polyhedra, transformed in solution by interior erosion into Pt_3Ni nanoframes with surfaces that have three-dimensional molecular accessibility. The edges of these $PtNi_3$ polyhedra, which were Pt rich, are maintained in the final Pt_3Ni nanoframes. Both the interior and exterior catalytic surfaces of this open framework structure are composed of the nano-segregated Pt-Skin structure that exhibits enhanced oxygen reduction reaction (ORR) activity. The Pt_3Ni nanoframe catalysts achieved over 36 and 22-fold enhancement in mass and specific activities, respectively, for this reaction versus ORR in comparison to state-of-the-art Pt/C catalysts during prolonged exposure to reaction conditions.

Notes and references:

1. Stamenkovic *et al.* ***Science 315, 493-497 (2007).***
2. Kang *et al.* **Science** *DOI: 10.1126/science.1249061*

Superficial Gas Velocity Effects on Local Time-Averaged Phase Holdup in Fluidized Bed Reactors Using Gamma Ray Computed Tomography Technique

Abdelsalam Efhaima, Muthanna H. Al Dahhan

Dept. of Chemical and Biochemical Engineering, Missouri University of Science and Technology, Rolla, MO, USA

ABSTRACT

The gas–solid fluidized bed reactor has been used extensively in many industrial applications in various industrial operations, due to its capability to provide effective mixing and highly efficient transport processes. The hydrodynamic behavior of fluidized bed reactors is complicated due to complex interaction phenomena among the phases, gas and solids. The hydrodynamic must be understood to improve fluidized bed design; scale-up, operations and process efficiencies. Gas or solids holdup is important in understanding fluidization hydrodynamics and process efficiency and describes the volumetric fraction of gas or solids in a system. Gas holdup is also useful in finding the interfacial area between the dense and dilute phases that are used in heat and mass transfer calculation. In this study, the effect of superficial gas velocity on the hydodynamics of gas-solid fluidized beds was investigated by using a 14 cm diameter and 1.82 m height Plexiglas cylindrical fluidized bed column, air at room temperature was used as the fluidizing gas, Three different superficial gas velocities were investigated U_g=20 cm/s, 25 cm/s and 35 cm/s., one Geldart type -B particles was used, glass bead with material density of 2.5 g/cm^3 and mean particle size 210 μm. Gamma ray computed tomography (CT) has been used as non-invasive technique to measure the time –averaged cross-sectional gas and solid holdup distribution, The results show that gas hold-up increases by increasing the superficial gas velocity. These results will be used in future studies to validate computational fluid dynamics (CFD) models of fluidized beds.

Keywords : Fluidized beds hydrodynamics, Computed Tomography, optical probe, Gas hold-up.

1 INTRODUCTION

Fluidization is a process in which solid particles become suspended and fluidized at a high enough gas or liquid velocity (higher than minimum fluidization velocity), due to many advantages such as high rates of heat and mass transfer, low pressure drop, approximately uniform temperature distribution and the ability to fluidize many particle types of different densities and sizes, fluidized bed reactors are extensively used in various industries such as mineral, chemical industries, petrochemical , pharmaceutical industries, drying, coal combustion, gasification, catalytic cracking, calcinations (Kunii &Leven spiel 1991). The performance of these multiphase fluidized bed reactors greatly depends on their hydrodynamic properties.

In spite of these advantages, fluidized beds have disadvantages which limit their industrial applications, Particle agglomeration and gas velocity reduction which cause de-fluidization and changing in the hydrodynamics of the fluidized bed, sudden change the hydrodynamics due to unwanted change in the excess gas velocity are some of these problems. In addition, fluidized bed hydrodynamic behaviour is very complex, lack of reliable knowledge about the hydrodynamics behaviour, has limited their application, therefore, any attempt for better understanding the fluidization phenomena, which would result in a more reliable design of fluidized reactors.Until recently, most experimental measurements of gas holdup in fluidized columns were global in nature. For example, overall gas holdup can be obtained via measurements of bed expansion, pressure drop, hot film anemometer, and quick closing valves. Techniques based on absorption and scattering of radiation to reconstruct the local holdup distribution have also been reported in the literature (Schrock VE, 1969).

Grassler and Wirth (2000) used X-CT to determine the solids concentration in a 0.19 m of diameter circulating fluidized bed with (50-70 μm) glass beads as the bed material. Tests were carried out in two different systems. In the first, solid concentrations were calculated with an up flow system. Results for this system showed that radial solid concentration exhibited a parabolic shape with a maximum concentration close to the wall of the reactor and a minimum concentration in the centre of the bed. For the second, the solid concentration was calculated with a down flow system. For this case, the solid concentration distribution was much more complex and depended upon the gas-solids distributor operating conditions. Results showed various solid concentration distributions from a homogeneous distribution with a parabolic profile to concentrated stands in the centre of the bed.

X- ray computed tomography (CT) imaging was used by Escudero et al. 2011, to determine bed height and material density effects on fluidized bed hydrodynamic in a 10.2 cm fluidized bed. They used three different materials Geldart-type-B particle (glass beads, ground walnut shell and ground corncob) with material densities of 2600, 1300 and 1000 Kg/m^3 respectively. Results showed that decreasing the bed density increased the gas holdup in the bed.

Franka and Heindel 2009 studied the effects of side air injection, superficial gas velocity, and bed material on the local time-average gas holdup of a 10.2 cm fluidized bed, using X-ray computed tomography. Using different materials (glass beads, ground corncob, and ground walnut shell), superficial gas velocities (Ug), and side air injection flow rates (Qside) they determined the variations on the fluidization hydrodynamics of the bed. They found that with side air injection, the side air flow rose near the wall but then expanded into the bed as height and Qside increased. As Ug increased the effects caused by the side air injection were less pronounced.

Escudero et al 2011, also studied the profiles of solids holdup at various superficial gas velocities at specific H/D ratio and found that solid holdup decrease by increasing superficial gas velocity.

Mabrouk et al 2005, studied the axial and radial profiles of the solids holdup using an optical fiber needle probe and radioactive particle tracking (RPT) technique. The axial solids hold-up profiles obtained by an optical fiber needle probe and radioactive particle tracking technique show a quasi linear profile.

Many intrusive and nonintrusive measuring methods have been used to study the hydrodynamics of fluidized beds. In this paper, the effects of gas velocity on local time-averaged gas holdup in fluidized bed are investigated by using gamma ray computed tomography technique (CT).

2 EXPERIMENTAL SETUP

Cold flow fluidized bed was used in this study, with outside diameter 0.14 m and a height of 1.82 m. A schematic diagram of the setup used in this study is illustrated in Fig.1. The fluidized bed column was constructed from Plexiglas and consisted of two pieces (column and cone) attached to a plenum base.

Connected from the top with an upper section that had a diameter of 0.45m and is 0.91m tall this upper section of the fluidized bed had a larger diameter to reduce the superficial gas velocity of the gas phase and thus enhance the solids separation. The column sat at the top of a stainless steel base. Gas distributor (made of a porous polyethylene sheet with a pore size of 40 μm was used. The plenum was located at the bottom,. Fluidized bed column was electrically grounded to minimize electrostatic effects. Air under ambient conditions was the fluidizing gas. The gas flow rate to the unit was controlled by rotameters.

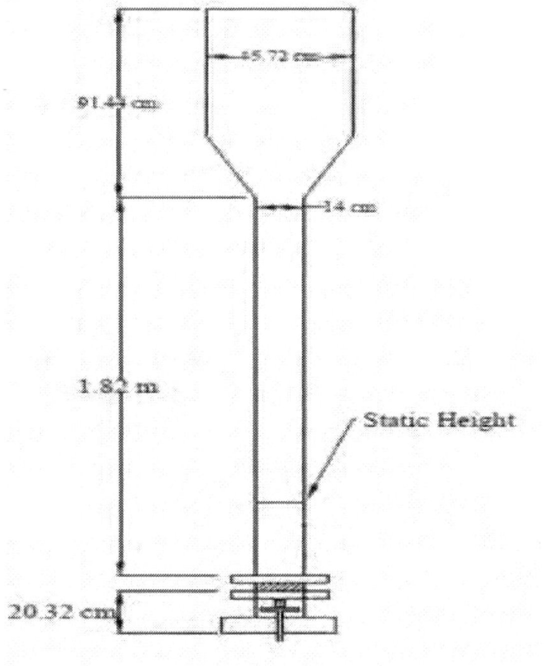

Figure1: Schematic of the cold –flow fluidized bed reactor used in this study.

Table 1. Summarizes the bed materials and their properties

Properties	Glass Beads
Particle Diameter (μm)	210
Superficial Gas Velocity (cm/s)	20, 25, 35
Particle Density (g/cm^3)	2,5
Minimum Fluidization velocity (cm/s)	12

3 MEASUREMENT TECHNIQUE

3.1 Computed Tomography (CT)

3.1.1 CT Facility and Measurement Procedure

The DSCT scanner at Missouri University of Science and Technology was used in this work to determine the time-averaged cross-sectional variation of gas and solid holdups at the operating conditions outlined earlier.

Computed Tomography (CT) Facility

The CT used in this study is based on a newer generation of double fan-beam scanning configuration. The details of the hardware and software have been described elsewhere (Kumar, 1994; Kumar et al. 1995, Chen et al.1999). In this work, the Alternating-Minimization (AM) method was used for image reconstruction; see Varma, R et al.2007 for Alternating-Minimization (AM) method. A photograph of the CT facility used in this study is shown in Figure 2. The configuration of the scanner consists of two arrays of NaI detectors of 5 cm in diameter (15 detectors were used for each side) and two encapsulated sources Cs -137 (~200mCi) and Co-60 (~150mCi), located opposite to the centre of the each array of detectors, placed 120 cm apart. In this study, only the Cs -137 source is used as we work with two phases in time. The detectors and the source are mounted on a plate which can be rotated 360° around the object. Moreover, the whole assembly can be moved up and down along the column to scan different axial levels of the column.

Figure.2.The (CT) available in (MS&T) with 6 inch Fluidized bed reactor.

Each of these detectors is collimated with a lead collimator that is about 6.35 cm Thick and has an open aperture of dimension 2 mm × 5 mm. This aperture reduces the effective exposed area of the crystal to a rectangular region of dimension 2 mm × 5 mm. The counts received by the detectors are limited to what is incident on this aperture. This detector array is moved with a stepper motor 21 times at an angle of 0.13° from the source, thereby creating 315 detector positions effectively for each source position. The projection is modelled as a fine line between the source and the detector as the open area of the detectors is very small. A total of 197 views or source positions are considered; hence, gamma ray counts data for 315 × 197 (total 62055) projections passing through the domain are collected. The gamma ray counts data recorded are energy threshold; hence, un-attenuated gamma ray photon counts of 660 Kev were recorded. A 80 × 80 pixels resolution is used to reconstruct the image. Hence, each pixel represents an area of 2 mm × 2 mm of the object.

3.1.2 Gas Holdup and CT image

In this study, CT experiments were performed under a two-phase condition (solid and gas) to measure the cross - sectional distribution of each phase, the following CT scans were performed 1- Scanning the column empty as reference CT scan. 2- Scanning the column filled with solids. 3- Scanning the column at normal gas- solid operations at the desired conditions. The beam attenuations were measured along a number of beams paths through the column, originating from different angles. Based on Beer Lambert's Law, the intensity of a beam that is transmitted through an object is expressed as follows:

$$T = \frac{I}{I_0} . \exp(-\mu\rho l) \qquad (1)$$

Where T is the transmission ratio, I_0 is the incident radiation, I is the detected radiation, μ is the mass attenuation coefficient, ρ is the medium density, l is the path length through the medium.

The measured quantity ln (I/I_0) (called A, for simplicity) is equal to the integral sum of the attenuation through the material along the beam path.

$$A = \ ln\frac{I}{I_0} \ = \mu\rho l \qquad (2)$$

For computed tomography, attenuations are measured by the detectors along a number of such beam paths through the object from different directions around it. To reconstruct the holdup, the cross-section of the column is configured inside a square whose sides are equal to the column diameter. This square is divided into a number of square pixels. The attenuation in each pixel can be obtained (reconstructed) by using a reconstruction algorithm. In this work, the Alternating-Minimization (AM) method was used for image reconstruction. To obtain statistically significant results and to reduce the effect of position, the CT scans were obtained by scanning 360 degrees around the column for a total scanning time of about four hours, if the scanned cross-section is divided into pixels or cells and the medium is comprised of two materials (gas and solid), mass attenuation coefficients, μ_s is mass attenuation coefficient for solid, μ_g is mass attenuation coefficient for gas, ρ is the medium density, ρ_s solid density, ρ_g gas density, and thickness l_g , l_s for gas and solid phases respectively , the total attenuation

$$A = \rho_s \mu_s l_s + \rho_g \mu_g l_g \qquad (3)$$

Where L= l_s+ l_g, l_s=ε_sL and l_g=ε_gL,

Where, L is the total length between the source and the detector. ε_g , ε_s are the holdups (volume fractions) for the gas and solid phases.

$$A = [\rho_s \mu_s \varepsilon_s + \rho_g \mu_g \varepsilon_g].L \qquad (4)$$

The summation of the holdups is equal to unity (i.e. ε_g=1-ε_s) .

$$A = [\rho_s \mu_s \varepsilon_s + \rho_g \mu_g (1 - \varepsilon_s)].L \qquad (5)$$

$$A_{s-g} = [\rho_s \mu_s \varepsilon_s + \rho_g \mu_g (1 - \varepsilon_s)].L \qquad (6)$$

Since ρ_g<<ρ_s, the attenuation caused by the gas phase is negligible compared to the solids, and L is common for all As. Hence, solids holdup for the line averaged measurement can be written as follows:

$$\varepsilon_s = A_{s-g}/L\rho_s\mu_s \tag{7}$$

$$A_{s-g} = \rho_s\mu_s\varepsilon_s L \tag{8}$$

The linear attenuation coefficient of solids (μ_s) was determined by using standard tables (such as NIST Physical Reference Data), since the material composition of solids was known. Finally, the gas holdup was determined using the expression

$$\varepsilon_g = 1 - \left(\frac{A_{s-g}}{L\rho_s\mu_s}\right) \tag{9}$$

4. RESULTS AND DISCUSSION

4.1. Cross-sectional gas and solid holdups distribution

The reconstructed image, processed from data obtained through CT scans, provides the cross-sectional time-averaged gas and solid holdups distribution at a given axial level (H/D = 1.7, which is fully development region) . Fig. 4 shows the cross-sectional gas holdup and cross-sectional solid holdup at various superficial gas velocities 20, 25 and 35 cm/s. offering an idea of how gas and solid are distributed through the column. The change in the gas holdup magnitude values were indicated by the colour variation. Red indicates a higher gas holdup value, while blue indicates a lower value of gas holdup. One can see from the figure that the gas holdup distribution is almost symmetric It is observed that gas holdup increases as the superficial gas velocity increases; this is due to a higher volume of air passing through the bed, and confirms a trend identified by Mabrouk et al. (2005). Image of cross-sectional are presented in Fig.4, for Ug = 20 cm/s, a high local gas holdup is concentrated in the center of the bed and extended to the region that near the bed wall, this behaviour indicates that the air is flowing throughout the bed.

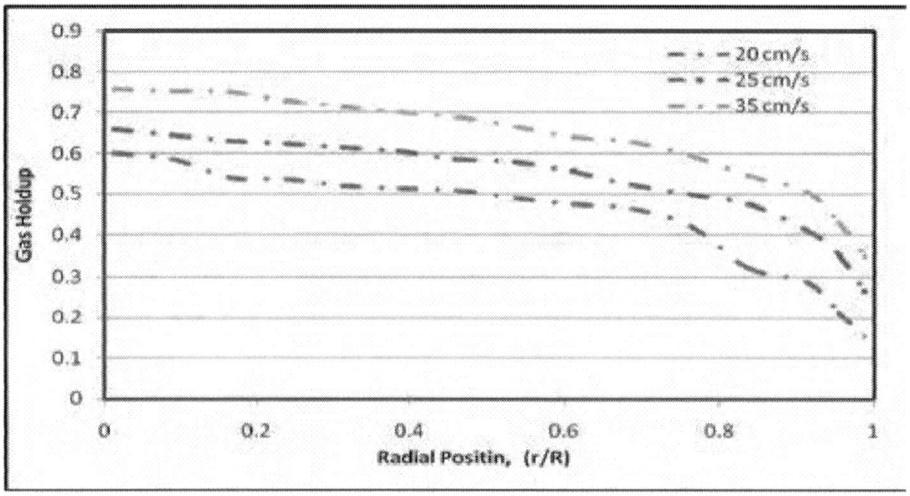

Figure.3. Effect of superficial gas velocity on radial profiles of gas holdup.

As superficial gas velocity increases to 25 cm/s higher gas holdup is located at the centre of the bed, as it can be seen in Figure.3. This behaviour indicates that, the air is flowing mostly through the centre of the bed; the local gas holdup is more symmetrically distributed through the bed.

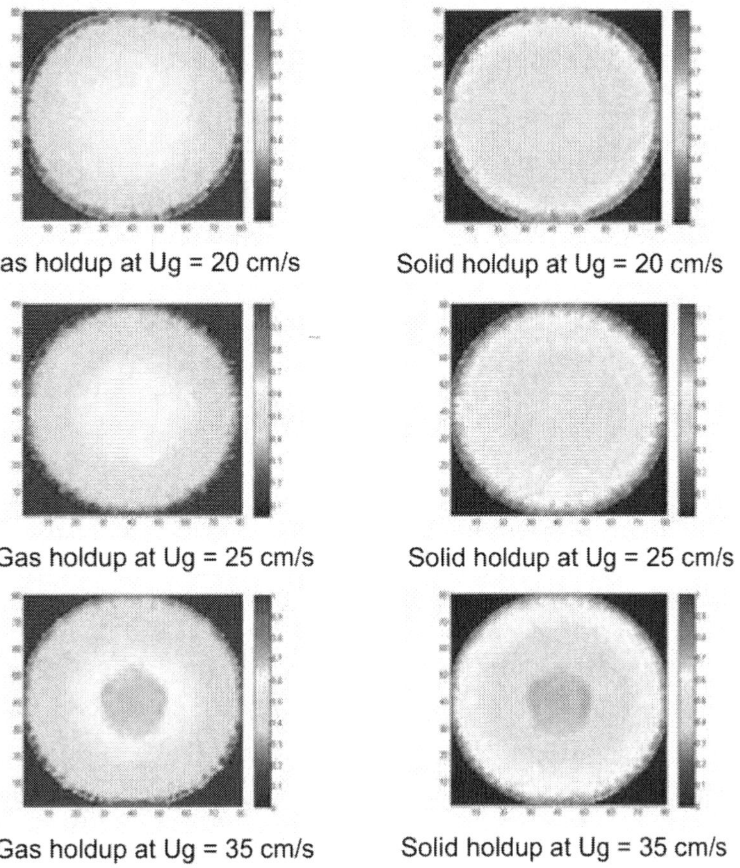

Figure 4.Cross-section gas and Solid holdup for glass beads at different superficial gas velocity.

When the superficial gas velocity further increased, Ug = 35cm/s, the large bubbles erupt from the bed near the centre throwing glass beads against the wall, which fall back into the bed, theses hydrodynamics create high gas holdup region in the centre of the bed , while lower gas holdup regions (higher solids concentration) are found along the bed walls.

CONCLUSION

This study is part of much more extensive study of fluidized bed hydrodynamics conducted with computed tomography technique. Local time-averaged gas holdup values obtained with gamma ray computed tomography revealed that, increases in superficial gas velocity increased the bed expansion and the overall gas holdup in the system, it also seems that while superficial gas velocity significantly affects fluidization hydrodynamics, it appears that changes in superficial gas velocity does not significantly affect fluidization symmetry as it can be seen in Fig.3 and 4.

ACKNOWLEDGMENT

The authors wish to acknowledge the Libyan Ministry of Education for sponsoring the primary author's study within the Chemical Engineering Department at Missouri University of Science and Technology. The Gamma ray facility used in this research was funded by Dr. Al-Dahhan.

REFERENCES

1. Martin rüdisüli, tilman j. Schildhauer,serge m.a. Biollaz j. Ruud van ommen, (2012), *Scale-up of bubbling fluidized bed reactors* — A review. Powder Technology Volume 217, pp 21–38.

2. N.P. Franka,(2008), *Visualizing fluidized beds with X-rays*, Master's Thesis, Iowa State University, Ames, IA.

3. Franka, N.P., Heindei, T.J., (2009). *Local time-averaged gas holdup in a fluidized bed with side air injection using X-ray computed tomography*. Powder Technology 193 (1). 69-78.

4. N Hilal, D.J Gunn, (2002), *Solid Hold Up in Gas Fluidized Beds*. Chemical Engineering and Processing: Process Intensification, Volume 41, Issue 4, Pages 373-379.

5. N. Hilal, M.T. Ghannam, M.Z. Anabtawi, (2001), *Effect of bed diameter, distributor and inserts on minimum fluidization velocity*, Chemical Engineering & Technology 24 (2) 161–165.

6. Varma R, Bhusarapu S, O'Sullivan JA, Al-Dahhan MH. A, (2008), *comparison of alternating minimization and expectation maximization algorithms for single source gamma ray tomography.* Measurement Science & Technology, Vol 19(1):015506, 14pp.

7. N. Mosfoufi and J. Chaouki, (2000), *The Axial Movement of Solids in Gas-Solid Fluidized Beds.* I ChemE, Vol 78, Part A.

8. Mabrouk R., Radmanesh R., Chaouki J., Guy C, (2005), *Scale Effects on Fluidized Bed Hydrodynamics*. Inter. J. of Chemical Reactor Eng.

9. King, D.F,(1989), *Estimation of Dense Bed Voidage in Fast and Slow Fluidized Beds of FCC Catalyst.* In Fluidization VI Conference Proceedings", Alberta, Canada, 1-8.

10. Bhusarapu S, Al-Dahhan MH, Dudukovic MP, (2006), *Solids Flow Mapping in a Gas-Solid Riser: Mean Holdup and Velocity Fields*. Powder Technology, Vol 163, (1-2), pp. 98-123.

11. David Escudero, Theodore J. Heindel, (2011), *Bed Height and Material Density Effects on Fluidized Bed Hydrodynamics*.Chemical Engineering Science 66, 3648-3655.

12. Grassler, T., Wirth, K.E., (2000), *X-ray computer tomography- potential and limitation for the measurement of local solids distribution in circulating fluidized beds*. Chemical Engineering Journal 77 (1), 65-72.

13. Kumar, S. B., D. Moslemian, and M. P. Dudukovic, (1995), *γ- ray Tomographic Scanner for Imaging Voidage Distribution in Two-Phase Flow Systems,'* Flow Meas. Instrum, 6(1), 61.

14. Nathan P, Franka, Theodore J. Heindel., (2011), *Local time- averaged gas holdup in a fluidized bed with side air injection using X-ray computed tomography*. Chemical Engineering Science 66, pp. 3648-3655.

364886 Asymmetric Hydrosilylation of Aromatic Ketones Catalyzed By an Economical and Effective Copper-Diphosphine Complex

Abstract:

Asymmetric hydrosilylation of prochiral ketones catalyzed by transition metal complexes is an easy way for the access to chiral secondary alcohols, which are very important intermediates in organic synthesis. However, most of the catalytic systems for the hydrosilylation reaction require anhydrous anaerobic atmosphere as well as low temperature, which limit the widespread application. We report here the combination of air and moisture stable $Cu(OAc)_2 \cdot H_2O$ and the commercial diphosphine BINAP using the inexpensive polymer siloxane PMHS as reducing agent, which is economical and effective for the asymmetric hydrosilylation of aromatic ketones. The active species of the catalytic system was formed in situ involving $Cu(OAc)_2 \cdot H_2O$, BINAP and PMHS. Basically, 3 mol% Cu^{2+} loading, Cu^{2+}/ligand ratio of 3:1 and 3 equivalents of PMHS were used in these experiments. Although the reaction rate was slower under air atmosphere than nitrogen atmosphere, the enantiometic excesses of (R)-1-arylethanol remained unchanged. The results indicated that the enantiomeric excesses of (R)-1-arylethanol achieved the highest (up to 84% *ee*) and all of conversion of the aromatic ketones exceeded 98% at 15 °C in air atmosphere. Furthermore, the addition of *t*-BuOK and *t*-BuOH to the reaction system at room temperature in air atmosphere was also examined. It came out that the reaction fully completed in shorter time and the enantiometic excesses could go up a little bit (up to 89% ee).

380661 The Characterization of Micro-Structure of Cobalt on Gamma-Al2O3 for FTS:Effects of Pretreatment on Ru-Co/ Al2O3

Abstract:

The production of synthetic fuel by FTS has been received considerable interest due to the lower sulfur and aromatics. The activity and selectivity of supported cobalt catalysts depend on several factors such as particle size, nature of the support, the presence of the promoters, preparation methods including pretreatment conditions. In order to enhance the reducibility, noble metal promoters are introduced. Many researchers focus on the role and the actual structure of the termed alloy catalysts and bimetallic, although the role and the surface composition is still debating with uncertain. The noble metals have been suggested to influence catalyst performance in a number of different ways. They may act to promote hydrogen spillover and the activity by preventing the buildup of carbonaceous deposited.The objective of this work is to study the effect of calcination atmosphere with usage of the direct reduction on the cobalt over the structural support such as granule type γ-Al_2O_3.This work also investigates the impact on the microstructure by introducing ruthenium with different slurry impregnation method. To neglect the consideration of the term as the alloy and bimetallic catalyst, the relative higher weight ratio of Ru to total Co was fixed to 2/10. The effect of promoter was studied with the help of TPR, TEM, XPS and XRD analysis.Through the characterization and activity test of catalysts, The Air calcined catalysts has strong interaction with the support, which induced metallic phases related to fcc phases. In contrast, the catalysts calcined in hydrogen atmosphere having weak interaction with hcp structure showed uniform distribution of active species on the support. These phenomena are more clearly in the Ru promoted catalyst, which was affected by spillover and retained the cobalt cluster sizes. It was also observed that the hydrogen calcined catalyst shows the marginal effect on CO conversion and definitely makes the shift to middle distillation ranges by the presence of hcp structure of metal.

REACTION PATHWAY FOR HYDROTREATMENT OF PALM OIL FOR GREENDIESEL PRODUCTION

J.F. Vélez[1*] K. Sakashita[1, 2], S. Asaoka [1, 3], A. Ishihara [1, 4], D. López[5], G. Hincapié[5], J.D. Tapia[5], A. Molina [1*]

1 Universidad Nacional de Colombia-Sede Medellín, Facultad de Minas, Bioprocesos y Flujos reactivos
2 Office of Ks Processtech, 9-41-2101, Shinonome 1-chome, Kouto-ku, Tokyo, 135-0062, Japan
3 Office Asaoka, 1-23-1, Konandai, Konan-ku, Yokohama, 234-0054, Japan
4 Division of Chemistry for Materials, Graduate School of Engineering, Mie University, Mie Pref., Japan
5 Química de Recursos Energéticos y Medio Ambiente, Instituto de Química, Universidad de Antioquia, Medellín, Colombia
*Contact Information: jfvelezm@unal.edu.co, amolinao@unal.edu.co

INTRODUCTION

Hydrotreatment (HDT) converts vegetable oils in biofuel and light-alkane gases by high-pressure addition of H_2, that reduces the size of the fatty acids molecules, mainly through hydrodeoxygenation and decarboxylation reactions.

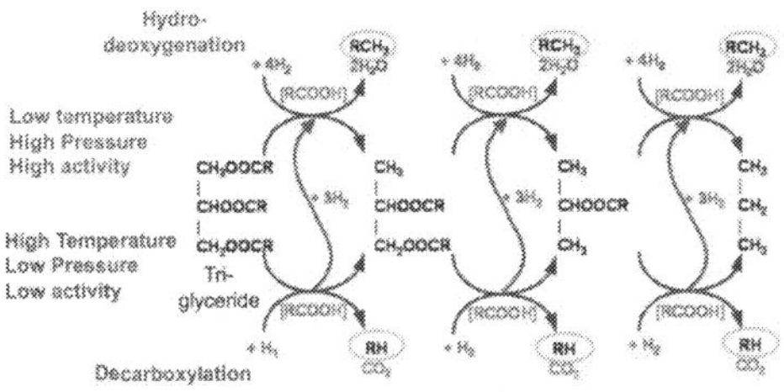

General pathway for the hydrogenation of triglycerides

Most studies on the HDT of vegetable oil (HVO) describe general reaction pathways. However, they lack information about the intermediate species present in reactants and products.

METHODOLOGY

Experimental Work

✓Experiments conducted in a 250 mL semi-batch reactor.

✓The feedstock was refined-bleached-deodorized Colombian palm oil (RBDPO). The hydrogen was grade 5– 99.999%.

✓The experiments were carried out with the commercial solid 1.8 mm trilobite NiMo catalyst (NiMo/γAl_2O_3, Haldor Topsøe TK-561 BRIM®) at 5% w/w concentration.

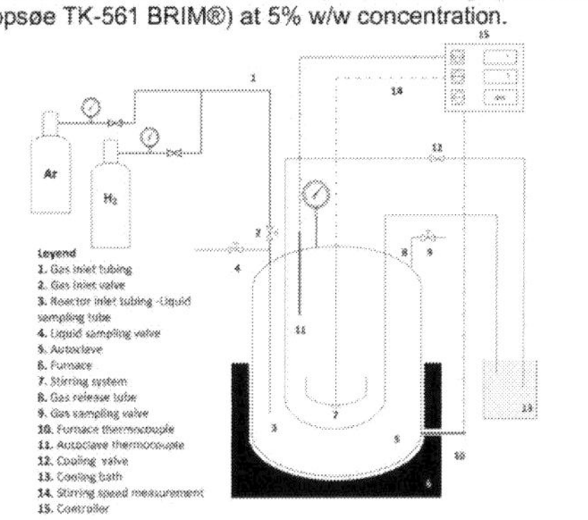

✓The operating pressure was varied between 30-60 atm, the operating temperatures were 335, 350 and 365 °C and the reaction time 15, 30 and 45 minutes.

✓The liquid products, mainly fatty acids and hydrocarbons, were analyzed by GC-MS and the gaseous products with a micro GC.

RBDPO Composition

A RBDPO sample was obtained from a local palm oil company. Table 1 shows the principal components of the sample.

Table 1. Composition of RBDPO

Fatty Acid	Fatty Acid Chain	Triglyceride	% w/w
Oleic	18:1	OOO	39.41
Palmitic	16:0	PPP	35.04
Linoleic	18:2	LiLiLi	13.25
Stearic	18:0	SSS	8.22
Myristic	14:0	MMM	2.21
Arachidic	20:0	AAA	0.91
Linolenic	18:3	LinLinLin	0.32
Eicosenoic	20:1	EEE	0.32
Palmitoleic	16:1	PaPaPa	0.31

RESULTS

Characterization of liquids products

Figure 1 shows the distribution of the identified liquid species, the first column correspond to the concentration fresh oil, it is mainly composed of triglycerides R18 (triestarin, triolein and trilinolein) and R16. (tripalmitin and tripalmitolein). The second column correspond to the concentration of the identified liquid species after the hydrotreatment reactions. The odd-carbon chains originate from triglycerides hydrodecarboxylation and the even-carbon chains from hydrodeoxygenation.

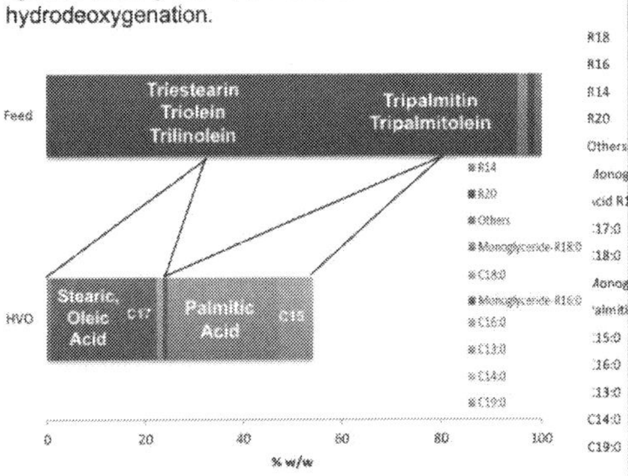

Figure 1. Distribution of species detected in the liquid phase

Proposed Mechanism

The reaction pathway in Table 2,considers the hydrogenation of triglycerides as the initial reaction, the hydrogenolysis of triglycerides to fatty acids and the hydrodeoxygenation and decarboxylation reactions that yield alkanes.

Table 2. Proposed pathway for the hydrotreatment of RBDPO. The numbers in parenthesis represent the concentration of each species. Species with names in red are present in the oil. Species in blue were detected in the reactions products. Species in black are supposed from stoichiometry.

Reaction Type	Reaction
Hydrogenation of Unsaturated triglycerides	$C_{57}H_{92}O_6 + 9H_2 \rightarrow C_{57}H_{110}O_6$ (Trilinolenin (0.32%w) → Triestearin (8%w))
	$C_{57}H_{98}O_6 + 6H_2 \rightarrow C_{57}H_{110}O_6$ (Trilinolein (13%w) → Triestearin (8%w))
	$C_{51}H_{92}O_6 + 3H_2 \rightarrow C_{51}H_{98}O_6$ (Tripalmitolein (0.31%w) → Tripalmitin (35%w))
	$C_{63}H_{116}O_6 + 3H_2 \rightarrow C_{63}H_{122}O_6$ (Trieicosenoin (0.32%w) → Triarachidin (0.9%w))
Hydrodecarboxylation	$C_{45}H_{86}O_6 + 3H_2 \rightarrow 3C_{13}H_{28} + C_3H_8 + 3CO_2$ (Trimyristin (2%w); C13:0 (0.1%w))
	$C_{51}H_{98}O_6 + 2H_2 \rightarrow 3C_{15}H_{32} + C_3H_8 + 3CO_2$ (Tripalmitin (35%w); C15:0 (5.2%w))
	$C_{57}H_{110}O_6 + 3H_2 \rightarrow 3C_{17}H_{36} + C_3H_8 + 3CO_2$ (Triestearin (8%w); C17:0 (6.4%w))
	$C_{63}H_{122}O_6 + 3H_2 \rightarrow 3C_{19}H_{40} + C_3H_8 + 3CO_2$ (Triarachidin (0.9%w); C19:0 (0.2%w))
Hydrogenolysis	$C_{57}H_{104}O_6 + 3H_2 \rightarrow C_{57}H_{110}O_6$ (Triolein (39%w) → Triestearin (8%w))
	$C_{51}H_{98}O_6 + 3H_2 \rightarrow 3C_{15}H_{31}COOH + C_3H_8$ (Tripalmitin; Palmitic Acid (23%w))
	$C_{57}H_{104}O_6 + 5H_2 \rightarrow 3C_{17}H_{33}COOH + C_3H_8$ (Triolein; Oleic Acid (2.8%w))
	$C_{57}H_{110}O_6 + 5H_2 \rightarrow 3C_{17}H_{35}COOH + C_3H_8$ (Triestearin (8%w); Stearic Acid (12.2%w))
	$C_{51}H_{98}O_6 + 2H_2 \rightarrow C_{19}H_{38}O_2 + 2C_{15}H_{31}COOH$ (Tripalmitin; Monogly-R16 (0.8%w); Palmitic Acid)
	$C_{57}H_{110}O_6 + 3H_2 \rightarrow C_{21}H_{42}O_2 + 2C_{17}H_{35}COOH$ (Triestearin (8%w); Monogly-R18(1.0%w); Stearic Acid)
Hydrodeoxygenation	$C_{45}H_{86}O_6 + 3H_2 \rightarrow 3C_{14}H_{30} + C_3H_8 + 2H_2O$ (Trimyristin (2%w); C14:0 (0.1%w))
	$C_{51}H_{98}O_6 + 3H_2 \rightarrow 3C_{16}H_{34} + C_3H_8 + 2H_2O$ (Tripalmitin; C16:0 (1.1%w))
	$C_{57}H_{110}O_6 + 3H_2 \rightarrow 3C_{18}H_{38} + C_3H_8 + 2H_2O$ (Triestearin; C18:0 (1.3%w))

Highlights

Hydrotreating is an alternative technology for the production of greendiesel in Colombia. Experiments with Refined, Bleached, Deodorized Palm Oil (RBDPO) with a commercial solid trilobite NiMo catalyst (NiMo/-Al_2O_3, Haldor Topse TK-561 BRIM®) were carried out in a batch reactor. The hydrogenation reaction produced 14% w/w of C13-C19 alkanes but also intermediates species such as fatty acids and monoglycerides 40% w/w . The results indicate the saturation of triglycerides, the production of fatty acids, and the hydrodeoxygenation (C18, C16 and C14 production), and decarboxylation (C19, C17, C15 and C13 production) as the most important reactions.

Acknowledgement

The authors are grateful to the Japan International Cooperation Agency (JICA) for their coordination of the visit of the Japanese experts. Haldor Topsøe and Saceites S.A. supplied the catalyst and oil samples, respectively. Juan Vélez is grateful to Universidad Nacional de Colombia for the partial financial support to this research under the scholarship "Programa nacional de apoyo a estudiantes de posgrado para el fortalecimiento de la investigación, creación e innovación de la Universidad Nacional de Colombia 2014".

DFT Analysis of the Hydrothermal Stability and Lewis Acidity of Metal-Substituted Zeolites

Brian Montejo-Valencia and Maria Curet-Arana
Department of Chemical Engineering, University of Puerto Rico-Mayagüez

Metal-substituted zeolites have been proven to be effective catalysts for various important reactions involving molecules derived from biomass. These catalysts are unique because they exhibit excellent selectivities at mild conditions. This study aims at obtaining a fundamental understanding on the relationship between structure and catalytic properties of metal-substituted zeolites. We are using density functional theory calculations to analyze the substitution of different metals in the Zeolite-Beta, such as Sn-BEA and Ge-BEC. We have characterized the systems with the LUMO energies, the hardness and the NBO charges. We have also studied the interaction of Sn-BEA, Ge-BEC and the unsubstituted BEC with water in order to determine their hydrothermal stability.

The catalytic surfaces were simulated using clusters of 53 to 80 atoms. We have performed systematic calculations with B3LYP, B3PW91 and ωB97XD exchange-correlation functionals and compared calculated properties, such as NMR with experimental results available in the literature. The basis sets used in all calculations were 6-31+g** for O, Si and H atoms and LANL2DZ for Sn and Ge atoms.

Our preliminary results demonstrate that: 1) All the functionals used in this study accurately described the geometries, but B3LYP and B3PW91 failed in describe the LUMO energies. 2) Based on the LUMO energies, Sn-BEA is more acidic than Ge-BEC. 3) The zeolite functionalized with Sn is less hydrothermally stable than the zeolite functionalized with Ge.

A Novel Membrane Reactor For Methanol Synthesis

Sahar Soltani, Seyed Hamed Barghi, Zhongtang Li, Muhammad Sahimi, and Theodore Tsotsis[†]

Mork Family Department of Chemical Engineering & Materials Science, University of Southern California, Los Angeles, California 90089-1211, United States

Methanol (MeOH) is an important chemical because it could be used as an energy carrier and also a chemical feedstock. MeOH was initially produced at harsh operating conditions such as pressures up to 250-350 bar and temperatures as high as 320-450 °C. The recent safety and economic demands lead to the development of the novel MeOH synthesis processes that operate at milder conditions. This progress is mostly due to the production of more efficient catalysts which can synthesize MeOH at pressures between 35-55 bar and temperatures less than 300 °C.

MeOH is mostly being produced from syngas by an equilibrium-limited reaction. Therefore, in order to make the process more profitable, the unreacted syngas need to be recycled. However, this approach is not very practical because the syngas produced from coal and in particular biomass gasification contain considerable amounts of nitrogen and some other inert components. Moreover, recycling gives rise to the concentration of the inert components in the reactor that has a negative effect on the synthesis reaction rate. The aforementioned drawbacks would encourage the development of more viable methods. Among these novel methods is the membrane reactor (MR) process, which is the concept used in our present work.

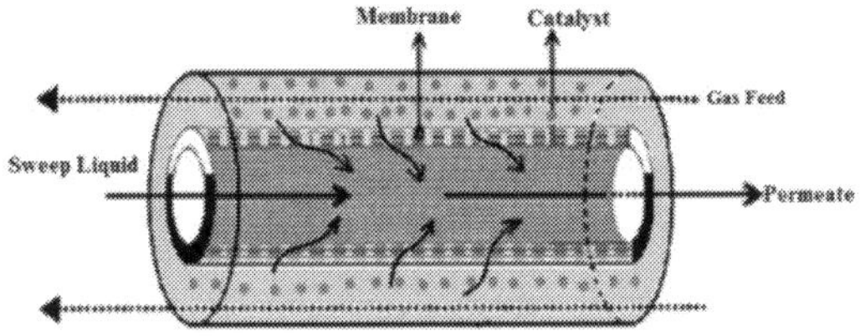

Figure1. Schematic of the tubular membrane reactor

Fig. 1 depicts a schematic of the MR used in this study. In this MeOH synthesis system, a liquid sweep stream flowing inside a highly-permeable ceramic membrane is being used for in-situ MeOH separation. Tetraethylene glycol dimethyl ether (TGDE) is the liquid solvent used for the sweep flow because MeOH has a higher solubility in it compared to the other components involved in the MeOH synthesis process. The catalyst used in this study is a commercial $Cu/Al_2O_3/ZnO$ (MK-121-Haldor-Topsøe) catalyst. Rigorous experiments were carried out for a MeOH system with and without membrane in order to assess

[†] *Corresponding author: ttsotsis@usc.edu*

the liquid sweep impact on the reaction conversion. To determine the reaction rates, kinetics experiments were also performed. In addition, a mathematical model of the system has been developed to investigate the effect of the different parameters such as reaction temperature, reaction pressure, feed composition, catalyst weight to feed flow rate ratio (W/F), and the liquid thickness inside the membrane pores on the reaction conversion. Both the experimental and modeling results indicate the effectiveness of the liquid sweep flow on increasing the carbon conversion. The thickness of the liquid inside the membrane pores also show a substantial influence on the reaction conversion, which encourages further modification efforts to decrease the layer thickness.

380760 Pd-Substituted Zinc Stannate As a New Oxygen Storage Material for CO Oxidation

Abstract:

Extensive work from both scientific and engineering fields have been devoted to CO oxidation reaction, in response both pollution control and renewable energy development. This is extensively used reaction that has various industrial applications in automotive catalytic converter and fuel cells. CeO_2, SnO_2, Fe_2O_3, TiO_2, are commonly utilized for their application in CO oxidation. The metal oxides which are reducible have advantages in releasing the lattice oxygen and promoting the CO oxidation. A key factor is that the material should have high oxygen storage capacity at low temperatures and thus a high activity of CO oxidation. In these ionic-substituted oxides, a weakening of "metal-oxygen" bond results in lattice oxygen release at very low temperatures during reaction and that are better as compared to metal dispersed oxides.[1]One of the limitations of CO oxidation catalyst is that it shows poor performance in hydrogen rich atmosphere. Many efforts have been focused on preferential CO oxidation catalysts that are inactive towards undesired H_2 oxidation.[2] Few studies revealed that compared to other oxides, zinc stannate sensors showed high selectivity towards CO in H_2 atmosphere at low temperatures.[3] Inspired by these studies, we hypothesize that zinc stannate may be an important material for CO oxidation, which has not been reported so far. In this report, we synthesized new and novel oxygen storage Pd-substituted Zn_2SnO_4 through solution combustion method that exhibits good oxygen storage capacity resulting in a low temperature CO oxidation. Encouragingly, this system showed higher activity towards CO oxidation at low temperatures. Our results indeed indicate that, when Pd is substituted in zinc stannatethe metal-oxygen bond weakens, which benefits the lattice oxygen release.We proposed the mechanism based on reaction rates, both CO and O_2 are competitive for the same adsorption sites, according to the data collected in kinetic regime.

References

1. Hegde, M.; Madras, G.; Patil, K., Noble metal ionic catalysts. *Accounts of chemical research* **2009,**42(6), 704-712.

2. Pozdnyakova, O.; Teschner, D.; Wootsch, A.; Kröhnert, J.; Steinhauer, B.; Sauer, H.; Toth, L.; Jentoft, F.; Knop-Gericke, A.; Paál, Z., Preferential CO oxidation in hydrogen (PROX) on ceria-supported catalysts, part I: Oxidation state and surface species on Pt/CeO_2 under reaction conditions. *Journal of Catalysis* **2006,**237(1), 1-16.

3. Yu, J. H.; Choi, G. M., Selective CO gas detection of Zn_2SnO_4 gas sensor. *Journal of electroceramics* **2002,**8 (3), 249-255.

Model Catalysts Studies on Thin Oxide Supports with Polarization Modulation Infrared Reflection Absorption Spectroscopy (PM-IRAS)

Zhiyong Wang, Erdem Sasmaz, Jason Hattrick-Simpers, Jochen Lauterbach*
Department of Chemical Engineering
University of South Carolina
Columbia, SC, 29208, U.S.A.
*lauteraj@cec.sc.edu

Adsorption structure of various adsorbates and their reaction pathways on single crystal catalysts under UHV conditions may not necessarily be the same as those on industrial catalysts at ambient or elevated pressures [1; 2]. In addition, metal single crystal surfaces cannot explain the particle size effect or metal-support interactions of industrial catalysts [3; 4]. To bridge so called "pressure gap" and "material gap" [5], a UHV-compatible high pressure cell equipped with a mass spectrometer and PM-IRAS is used to carry out *in situ* investigations of CO adsorption and NH_3 decomposition reactions on model catalysts, ranging from single crystals, e.g. Pt(100), to model supported catalysts, e.g. Ru/Al_2O_3.

Adsorption of CO is investigated on Pt(100) applying heating and cooling cycles at a wide pressure range from 10^{-6} Torr to 100 Torr. Our results indicate that CO molecules can dissociate at elevated pressures and temperatures leading to carbon contamination on the Pt surface. As the annealing time increases, deposited carbon can form islands on the surface. In addition, CO molecules can cluster together as islands on the annealed sample. Dipole coupling simulations reveal that different CO island geometries can form on the surface without affecting the local CO interactions.

To bridge the "material gap", nanoparticles deposited on metal oxide thin films are synthesized as model supported catalysts for different reaction systems. In particular, Ru nanoparticles supported on alumina thin film are investigated for NH_3 decomposition, which is not only an important reaction for CO_x-free hydrogen production used in fuel cells [6], but also for better understanding the mechanism of selective catalytic reduction of NO_x [7]. Temperature program desorption of NH_3 on Ru/Al_2O_3 model catalyst having different metal particle sizes are performed, and the surface species formed during NH_3 decomposition at different pressures and temperatures are monitored using PM-IRAS. The effect of Ru particle size and pressure on NH_3 decomposition is identified and will be discussed in detail.

References

[1] G.A. Somorjai, Introduction to Surface Chemistry and Catalysis, John Wiley & Sons, Inc., 1994.
[2] K. Christmann, Introduction to surface physical chemistry, Steinkopff Verlag Darmstadt Springer-Verlag New York, 1991.
[3] T. Dellwig, G. Rupprechter, H. Unterhalt, H.J. Freund, Physical Review Letters 85 (2000) 776-779.
[4] G. Rupprechter, T. Dellwig, H. Unterhalt, H.J. Freund, Topics in Catalysis 15 (2001) 19-26.
[5] R. Günther, W. Christian, Journal of Physics: Condensed Matter 20 (2008) 184019.
[6] F. Schuth, R. Palkovits, R. Schlogl, D.S. Su, Energy & Environmental Science 5 (2012) 6278-6289.
[7] P.R. Ettireddy, N. Ettireddy, T. Boningari, R. Pardemann, P.G. Smirniotis, Journal of Catalysis 292 (2012) 53-63.

Conversion of glucose to levulinic by dual solid-acid catalysts

Tapas C. Acharjee and Y. Y. Lee

Department of Chemical Engineering, Auburn University, AL 36849

Levulinic acid (LA) is a versatile specialty chemical usable as a building-block for synthesis of various chemicals. Traditionally, homogeneous acid catalysts such as HCl or H_2SO_4 have been used to produce levulinic acid from glucose. Difficulties encountered in the recovery of the catalyst, environmental concerns, and high equipment costs incurred by strong acids open up options for other routes. Heterogeneous catalyst is a viable alternative because of high selectivity, reusability, and non-corrosive reaction environment. In this study, we have explored the use of solid acids for this purpose. Catalytic conversion of glucose to LA involves multi-step complex reaction sequence. The known reaction pathway includes glucose→fructose→hydroxymethyfurfural (HMF)→LA and various side-reactions. Each of these reactions behaves differently depending on the type of acid catalyst. Specifically, Lewis-acid has much higher selectivity for the first reaction (glucose→fructose) whereas Bronstead-acid favors the latter reactions. The side reactions from glucose, fructose and HMF involve decomposition of these sugars to gaseous and carbonaceous compounds (humins), which limits the yield of the final product. From these considerations, we have applied two different solid acids as a means to facilitate the overall process and to increase the product yield. For this purpose, we have prepared a tin (Sn) catalyst embedded onto a large-pore beta-Zeolite (Sn-beta-zeolite) by solid-state ion exchange (SSIE). Sn catalyst acts as a Lewis-acid in aqueous medium; it is known to catalyze the first reaction efficiently. As a Bornstead-acid, we used Amberlyst 15, which catalyzes the latter reactions, especially, HMF to LA. Use of the dual-catalyst system has improved yield of LA from 33% (in the case of Amberlyst 15 alone) to 42% of the theoretical maximum. This paper presents the details of the reaction experiments, rate data as affected by the reaction conditions, and practical aspects of the process including decay of catalysts upon repeated use. Also presented are the structural and surface properties of the Sn-beta-zeolite prepared in our laboratory by way of XRD, BET, SEM and IR.

381408 WO3 Modified Omc Supported Pt-Pd/WO3-Omc Electrocatalysts

Abstract:

This study presents the experimental investigation of the promotional effects of WO_3 on the bimetallic PtPd/WO_3-OMC catalysts for formic acid oxidation. CV measurement of Pd/WO_3-OMC exhibits 15 % higher anodic current activity than that of Pd/OMC catalyst while the peak potential is also shifted negatively. In the Pt-Pd catalyst samples, the small addition of Pt improves the performances of PdPt/WO_3-OMC catalyst significantly. The atomic ratio2 for Pd/Pt catalysts, displays best catalytic efficiency, which shows current activity 5.2 times higher than that of Pd_1Pt_2/WO_3-OMC catalyst. It also shows persistent activities in extended period of runs with the increase of carbon monoxide (CO) tolerance effects. BET surface area analysis shows that WO_3 modification enhances the surface characteristics of OMC with pore size slightly decreased. Many techniques have used to characterize the synthesized catalysts.

371552 Ab Initio Study of Solvent-Induced Frequency Shifts of 5-Hydroxymethylfurfural

Abstract:

5-Hydroxymethylfurfural (HMF) is a promising platform-chemical for sustainable fuels and plastics. A major issue in the production of HMF is the side reactions that lead to significant loss of sugars and high cost. Use of polar aprotic solvents improves HMF yields from sugar dehydration by preventing HMF side reactions, but the influence of solvent on the structure of HMF is not fundamentally understood. In an effort to understand solvent effects on HMF stability and yield, solvent-induced frequency shifts (SIFS) of the carbonyl stretching vibration ν(C=O) of 5-hydroxymethylfurfural were measured in seven protic, polar aprotic, and non-polar solvents. The Gutmann Acceptor Number (AN), an empirical measure of Lewis acidity, was found to correlate with the measured frequency shifts. The SIFS were then investigated using *ab initio* electronic structure calculations, treating the solvent implicitly and with an explicit solvent ligand interacting with the carbonyl. The conductor polarizable continuum model (CPCM) solvation model predicted that ν(C=O) shifted as a function of the dielectric constant, in agreement with the analytical predictions of the Kirkwood-Bauer-Magat (KBM) theory for a dipole in a dielectric continuum, but in disagreement with experimental ν(C=O). Experimental ν(C=O) were best predicted using gas-phase complexes of HMF and explicit solvent ligand. NBO analysis and Bader's Atoms in Molecules theory were used to investigate the electronic structure and hydrogen bonding in these complexes. SIFS arose from H-bonding interactions, as observed in delocalization of carbonyl lone-pair electrons to H-bonding solvent σ*(X-H) orbitals, and in increased charge density and decrease in local potential energy at the H-bond (3,-1) critical point in the charge density. Consequently, by predicting the experimental SIFS and examining the electronic structure, we find the first theoretical evidence for treating Gutmann's solvent AN as a measure of solvent Lewis acidity.

371826 Impact of Support Size, Shape and Pore Structure on the Metal Distribution of Supported Catalysts

Abstract:

Supported catalysts are essential components in a variety of industrial processes, ranging from catalytic converters to production of new drugs. They are generally required because of their high surface area and high mechanical and thermal stabilities. The performance of a catalytic process is intimately related to the catalyst design - uniform, egg-yolk, egg-shell and egg-white metal profiles. Although catalyst preparation and catalytic processing have been investigated for many years, many aspects of catalyst manufacturing are still not fully understood and in industry the design of catalysts is usually based on trial and error, which is expensive and time-consuming, and does not offer assurances on the final results.

In this work we have examined the effects of support shape, size, and pore sturcture on the metal distribution of Ni/Alumina catalysts. The cylindrical pellets with diameter 1mm, 3mm and 4mm are tested to examine the impact of support size. In general, egg-shell profiles can be enhanced with an increase in the support size. This is because drying is much faster for small supports so the metal precursor doesn't have enough time to migrate during the process. The impact of support shape is examined by testing four high surface area γ-alumina carriers, including rings with 7.9mm diameter, cylinders with 3.2mm diameter, spheres with 3mm diameter and trilobes with 2.5mm diameter. Asymmetric metal profiles are found in triblobes. In order to examine the impact of support pore size, we have tested five alumina supports, where SA1 (sphere, 75μm, $0.05m^2/g$) and SA2 (cylinder, 3μm, $0.18m^2/g$) are low surface area supports with large pore structures, SA3 (cylinder, 1.2μm, $12m^2/g$) and SA4 (cylinder, 20/230 nm, $75m^2/g$) are moderate surface area supports with moderate pore sizes, SA5 (cylinder, 7/500 nm, $257m^2/g$) are high surface area supports with small pore sizes. In general, for low concentrations, such as 0.1M and 0.2M, egg-shell distribution is enhanced by increasing the support pore size. The most significant egg-shell distribution was observed in SA1 and SA2 samples. This trend becomes opposite when we extended our studies to high metal loading conditions, such as 1M and 2M. Under these situations, a W shape distribution was observed in SA2 samples. Detailed explanation will be given in the poster.

We have also developed theoretical models to simulate the impregnation and drying processes of Ni/Alumina systems. The initial input of the drying model comes from the results from the impregnation simulations. Therefore, we can combine impregnation and drying together to analyze their effects on the preparation of supported catalysts. We have compared simulation predictions and experimental results based on cylindrical catalysts. Our simulations and experiments have allowed us to better understand the fundamental mechanisms that occur during impregnation and drying, and to develop a strategy that can generate desired metal profiles. The models used in the present work can capture the essential physics of impregnation and drying while still maintaining a level of generality. Although the results presented are based on a particular metal/support system, they serve to provide physical insight into the fundamentals of the impregnation and drying processes.

Introduction:

Formaldehyde is one of the most important intermediate chemicals and has been produced industrially for many years. In particular, formaldehyde is converted to thermosetting resins used as binders and adhesives in wood-based panel. Formaldehyde condenses with urea, melamine, phenols, urethanes, cynamide, aromatic sulfonamides and amines to produce a number of resins. Formaldehyde is produced through a sequence of reactions – (1) steam reforming of natural gas produces synthesis gas (Eq. 1), which is converted to methanol via methanol synthesis reaction (Eq. 2) and eventually partial oxidation of methanol at 270-400°C leads to formaldehyde production (Eq. 3).

$$CH_4 + H_2O \rightarrow CO + 3H_2 \qquad \Delta \hat{H}_r^o = 206.16\frac{kJ}{mol} \qquad \text{Eq. 1}$$

$$CO + 2H_2 \rightarrow CH_3OH \qquad \Delta \hat{H}_r^o = -90.84\frac{kJ}{mol} \qquad \text{Eq. 2}$$

$$CH_3OH + \frac{1}{2}O_2 \rightarrow CH_2O + H_2O \qquad \Delta \hat{H}_r^o = -159\frac{kJ}{mol} \qquad \text{Eq. 3}$$

In this study, formaldehyde is produced directly from synthesis gas at room temperature using the following reaction –

$$CO + H_2 \rightarrow CH_2O \qquad \Delta \hat{H}_r^o = -5.3\frac{kJ}{mol} \qquad \text{Eq. 4}$$

Initial evidence of formaldehyde production was observed in Temperature Programmed Surface Reaction (TPSR) test using Pd-Ni/Al_2O_3 catalyst at room temperature and atmospheric pressure. Formaldehyde production was confirmed in a pressurised fixed bed reactor operating at room temperature and 20 – 120 bar pressure. Higher concentration of formaldehyde was achieved in aqueous phase reaction in a batch reactor operating at room temperature and 100 bar pressure.

Method:

The catalysts used in this process were produced using wet impregnation method. Appropriate amount of $Ni(NO_3)_2.6H_2O$ and $Pd(NO_3)_2$ 10 wt% in 10wt% nitric acid solution precursors were added to commercial γ-Al_2O_3 suspended in 20 ml water. The mixture was stirred for 5 h at 50°C, followed by drying at 90°C overnight. The dried catalyst was calcined at 500°C for 6 h. The catalyst was reduced (*in-situ for TPSR and fixed bed reactor test and ex-situ for batch reactor test)* by flowing 50ml/min H_2 through the catalyst bed for 5 h at 400°C followed by purging with Ar for 1 h at 400°C and finally cooled to room temperature under Ar flow. In fixed bed reactor tests, H_2 and CO were fed to the reactor in 1:1 mole ratio. 60 ml/min of gas mixture was passed through 1 g catalyst bed for 4 h at room temperature and 20 -120 bar pressure. The gas stream leaving the reactor was passed through a series of two scrubbers filled with 40 ml of 5 v/v% methanol to recover formaldehyde. At the end of the run, formaldehyde concentration was measured using the photometric cell test kit (Merck Millipore) in a spectrophotometer. In the batch reactor tests, 1 g reduced catalyst was added to 40 ml of 5 v/v% methanol in a batch reactor and pressurized to 100 bar with 1:1 mole ratio H_2 and CO and stirred at 200 rpm for 2 – 24 h. At the end of the run, solution was filtered and formaldehyde concentration was measured by using the above method.

Results:

Based on a study done in HSC Chemistry™ software, it was concluded that high temperatures are not favourable for reaction Eq. 4 because the Gibbs free energy of formation increases with temperature. This theoretical result was proven by running two experiments in a fixed bed reactor at room temperature (20°C) and at 40°C at 85bar. The formaldehyde concentrations measured in these tests were 0.17 ppm and 0.08 ppm, respectively. Unlike temperature, Eq. 4 is directly proportional to pressure. Increasing the pressure from 40 bar to 117 bar resulted in increase of formaldehyde concentration from 0.09 ppm to 0.25 ppm. Figure 1 shows that the concentration increases linearly with respect to temperature. All the concentrations reported here are from the first scrubber in the fixed bed reactor setup.

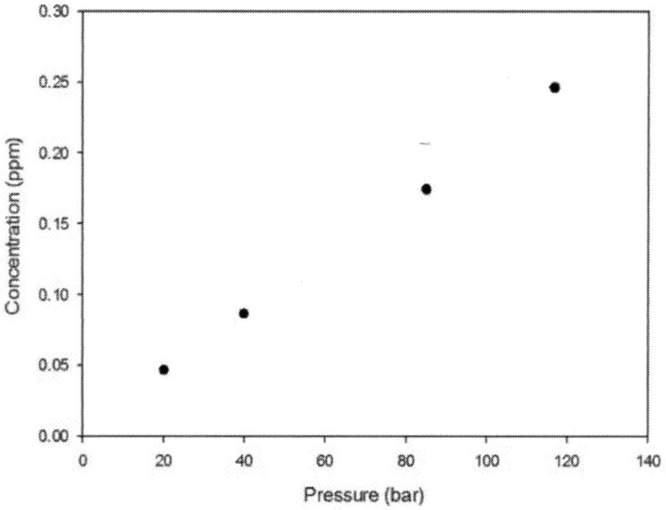

Figure 1: Formaldehyde concentration with respect to changes in pressure in the fixed reactor operating at room temperature

Since the Eq. 4 is thermodynamically limited, an aqueous phase process was tested in the batch reactor. This would shift the equilibrium of the reaction by removing formaldehyde from the catalyst surface and absorbing it in methanol/water mixture. At room temperature (25-26°C) and 100bar, up to 10.9 ppm formaldehyde was achieved using a batch reactor after 24 hours of operation. Figure 2 shows that the formaldehyde concentration increased with time. Interestingly, the formaldehyde concentration achieved at 4 h in the batch reactor is more than the gas phase reaction carried out in the fixed bed reactor.

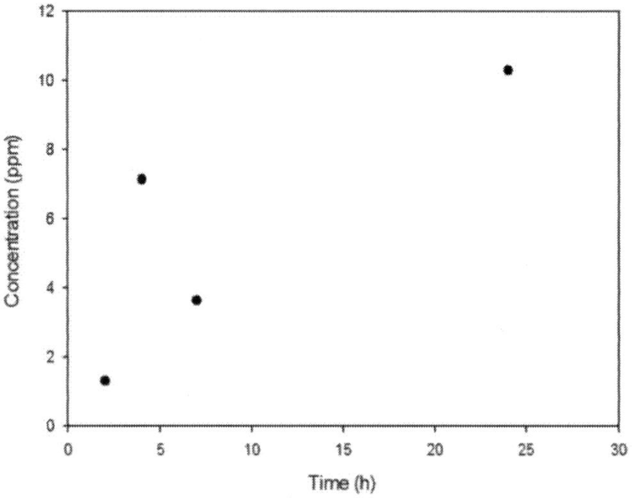

Figure 2: Formaldehyde concentration with respect to changes in time in the batch reactor operating at room temperature and 100 bar

381868 HPA Loaded/ Modified Alumina Catalyst for Transesterification of Jetropha Oil Forming Biodiesel

Abstract:

HPA Loaded/ Modified Alumina Catalyst for Transesterification of Jetropha Oil forming Biodiesel

Sunder Lal[1], Deeptiraj Pant[2] and Richa Tomer[3]

Abstract:

We have studied transesterification of Jetropha oil with heterogeneous catalyst and found that the heterogeneous catalyst HPA with alumina support worked very well for transesterification of oils. The maximum conversion was above 70% and the GCMS analysis showed that the biodiesel consisted of ten major products viz. Methyl Palmitate, Methyl Stearate, Methyl Oleate, Methyl Linolenate, Methyl Linoelate, Methyl Eicosanonate, Methyl Archidate, Methyl Tetra deconate, Methyl Pentyl hexonate and Methyl Behenate. They were found to be present in all oils however their concentration were different for different oils. The first order plots were found to fit the conversion versus time data as well as the yield of different components of the biodiesel versus time data. The catalyst activity remained unchanged at least for 1000 h.

385509 High-Performance Dealloyed PtCu/Cunw Oxygen Reduction Reaction Catalyst for Proton Exchange Membrane Fuel Cells

Abstract:

One major barrier to fuel cell commercialization is the high cost of oxygen reduction reaction (ORR) catalysts. Current catalysts are predominantly supported Pt nanoparticles. These nanoparticles show high surface area but low specific activity and durability. A transition to unsupported catalysts possessing an extended surface should improve both specific activity and durability and in turn, cost-effectiveness when high surface area can be achieved. Platinum-coated copper nanowires (Pt/CuNW) exemplify these advantages. In this study, post-synthetic processing is used to further improve the performance of Pt/CuNW catalyst. Specifically, annealing followed by electrochemical dealloying enhances activity through geometric lattice tuning. The resultant bimetallic PtCu/CuNW catalyst yields specific and mass activities (SA and MA) of 2.65 mA cm^{-2} and 1.24 A mg_{Pt}^{-1}, surpassing the respective DOE benchmarks of 0.72 mA cm^{-2} and 0.44 A mg_{Pt}^{-1}. PtCu/CuNWs demonstrate enhanced durability over Pt nanoparticle catalysts by maintaining 64.1 % of its active surface area after 30,000 cycles between 0.6 - 1.1 vs. RHE at a scan rate of 50 mV s^{-1} in Ar saturated 0.1 M $HClO_4$. Post durability PtCu/CuNWs outperformed the DOE benchmarks with a SA and MA of 1.50 mA cm^{-2} and 0.477 A mg_{Pt}^{-1}.

389066 Kinetic Modeling of Propylene Polymerization with Me2Si[Ind]2ZrCl2/MAO Catalyst System

Abstract:

AIChE-2014

Abstract

Polyolefins are the largest group of thermoplastics, often referred to as commodity thermoplastics, these are polymers of simple olefins such as ethylene, propylene, butenes, isoprene, and pentenes, and their copolymers. Two most important and common polyolefins are polyethylene and polypropylene and they are very popular due to their low cost and wide range of applications. Polypropylene is produced by polymerizing propylene with suitable catalysts and has demonstrated certain advantages in improved strength, stiffness and higher temperature capability over polyethylene. World commodity polymers consumption is estimated to reach 214 million tonnes by 2015, with polypropylene accounting for the largest share. [1]

Metallocene catalyzed propylene polymerization has recently attracted research interest since these catalysts allow the production of tailored macromolecules with properties those can be accurately designed. A broad spectrum of properties and applications of the polypropylene can be attained with metallocenes due to their single types of sites. Kinetic studies of catalytic polymerization provide substantial insight into the mechanism of the reactions and scale-up or commercialization of a polymerization process tremendously depends on the understanding of the kinetic behavior of the system under various operating conditions.

Metallocene catalyst system refers to the combination of bis(cyclopentadienyl)metal complexes of Group 4 (IVB) [especially zirconium, also titanium and to a lesser extent hafnium], or cyclopentadienyl-substituted derivatives thereof, and a cocatalyst, typically methylalumoxane (MAO). Titanocene and zirconocene dichlorides were the first metallocenes studied. [2, 3]

In this work, the mechanistic aspects of metallocene catalyst systems have been canvassed in detail and used in building up mathematical model for solution phase polymerization of propylene with $Me_2Si[Ind]_2ZrCl_2$/MAO catalyst system and kinetic parameters are obtained. Data for the model validation are taken from Marques *et al.* [4]

A novel natural logarithmic differential evolution (NLDE) approach of optimization, a remediated version of differential evolution algorithm [5] is proposed and used to solve parameter estimation problem in this work.

Model equations developed in this study include a set of coupled, nonlinear and stiff ordinary differential equations for the dynamic polymerization. These ordinary differential equations (ODEs) are solved with MATLAB™ 7.0.1 software. [6]

Developed model is able to capture essential polymer properties. Number- and weight average molecular weights, polydispersity index (PDI); fraction of vinyl terminated chains, butenyl-terminated chains, isobutyl-terminated chains and vinylidene-terminated chains in propylene polymerization are determined with the applied kinetic model.

Study on the effects of various parameters like monomer concentration, polymerization temperature, catalyst concentrations, and cocatalyst to catalyst molar ratio etc. upon rate of polymerization, molecular weights and poly dispersity index and stereoregularity is carried out.

References:

1. ICIS, Homepage of ICIS. the URL of which is http://www.icis.com/Articles/2012/03/26/9544287/global-polymers-to-face-heightened-competition.html as on 27th Dec, 2012.
2. Natta, G., P. Pino, G. Mazzanti, U. Giannini, E. Mantica, and E. Peraldo, "The nature of some soluble catalysts for low pressure ethylene polymerization", *J. Polym. Sci.* **26**, 120-123 (1957a).
3. Natta, G., P. Pino, G. Mazzanti and U. Giannini, "Crystallizable organometallic complex containing titanium and aluminum", *J. Am. Chem. Soc.* **79**, 2975-2976 (1957b).
4. Marques, M. F. V., M. Poloponsky and É. G. Chaves, "Comparative study of propylene polymerization with different metallocene catalysts using a statistic experimental planning model", *Polímeros: Ciência e Tecnologia* **12 (1)**, 48-59, (2002).
5. Price K.V., Storn R. (1997). Differential evolution: A simple evolution strategy for fast optimization. *Dr. Dobb's Journal*, 22, 18–24.
6. MATLAB version 7.0.1, computer software, The MathWorks, Inc., Natick, Massachusetts 2004.

Effect of Methoxylated Groups in Sn-Beta Zeolite on Glucose Isomerization and Epimerization

Caterina Tran and Neeraj Rai

Dave C. Swalm School of Chemical Engineering and Center for Advanced Vehicular Systems, Mississippi State University, Miss. State, MS, USA

Abstract

The isomerization of glucose to fructose is a key intermediate step to the formation of fuels and chemicals from cellulosic biomass. Sn-Beta zeolite catalyst has been shown to efficiently catalyze this reaction in aqueous medium. In the presence of organic solvent such as methanol, however, the epimerization of glucose to mannose occurs [Bermejo-Deval et al. *ACS Catal.* **2012**, *2*, 2705-2713]. Previously, the role of the silanol group in Sn-Beta for these reactions was examined via density functional calculations [Rai et al. *ACS Catal.* **2013**, *3*, 2294-2298]. The study indicated that the hydride transfer for the isomerization reaction benefited from the participation of the silanol group. Conversely, epimerization occurred when the silanol group did not actively participate in the transition state. It was speculated that silanol and/or stannanol groups were methoxylated when methanol was used as the medium. In the present work, density functional calculations were used to elaborate on the effect of the methoxylation of the stannanol and silanol groups in Sn-Beta and their roles in the isomerization and epimerization reactions.

399235 Reaction Engineering Alternatives to Waste Gasification for Sustainability

Abstract:

Waste management alternatives for ground base applications, such as municipal waste management, are under consideration. One such alternative is the use of Wet Thermal Catalytic Oxidation (WTCO) processes. These processes promote the conversion of long polymeric chains into synthetic or "supply" gas (syngas) at temperatures lower than those required by more traditional gasification technologies. The main goal of this project is the characterization of reaction pathways as waste management and sustainable energy alternatives for waste management applications. The project is motivated by the need to develop low-cost technologies as environmentally-friendly alternatives for converting waste materials into high-value products. The completion of the study is anticipated to lay out the foundation for scaling-up the technology to a continuous process.

Furfural has been shown to be an attractive platform chemical in biomass upgrading routes [1]. Its combination with aliphatic and aromatic alcohols and diols has the potential to yield useful precursors for gasoline, jet and diesel blendstocks, as well as chemicals. In this work, we show the utility of transfer hydrogenation and aldol-type coupling for this combination and we investigate the mechanism of the transfer hydrogenation and that of the coupling over homogeneous and heterogeneous catalysts. Initial optimization studies show that mixtures of Ni(dppe)Cl_2 and K_2CO_3 are efficient catalysts for the dehydrogenative cross-coupling, giving aldol adduct yields higher than 90% for a variety of alcohols. Furthermore, the presence of both Ni(dppe)Cl_2 and K_2CO_3 is required for alcohol and furfural cross-condensation, the role of the Ni^{2+} complex being the dehydrogenation of the alcohol to the reactive aldehyde and that of K_2CO_3 being the aldol condensation between furfural and the aldehyde. This is shown by the results of reactions in the presence and the absence of either of the two components of the catalytic system and also by a reaction progress study that shows the aldol adduct is a secondary product of the reaction. Also, further reaction studies over this system suggest that dehydrogenation proceeds via a Ni(II)-mediated transfer of a hydride from a butoxide to a furfural.

In order to make this process attractive and relevant from an industrial standpoint, we propose carrying it out in heterogeneous conditions. Heterogenizing this reaction requires the replacement of the base and the metal with heterogeneous versions thereof. For this reason, we chose a nickel-substituted hydrotalcite material and evaluated its reactivity in the gas phase and the liquid phase. Surprisingly, increasing the Ni content results in a decrease of both dehydrogenation and aldol condensation reaction rates. This suggests that the presence of Ni^{2+} sites is not required for the chemistry to take place. Instead, we show that the dehydrogenation of butanol takes place via a transfer of hydrogen from the butanol to the furfural, in a Meerwein-Ponndorf-Verley-type reaction [2]. Both this reaction, as well as the aldol condensation reaction are known to be base-catalyzed [2-4]; our results suggest that the presence of Ni in the catalyst decreases its basicity and hence the reaction rates.

For the dehydrogenation step, kinetic isotopic effect (KIE) studies with 1,1-d_2-butanol and butanol-OD excluded the possibility of O-H bond cleavage as the kinetically relevant step as opposed to C-H bond scission in the dehydrogenation step. These observations also precluded the possibility of a concerted transfer of the surface proton to the carbonyl of the furfural and the hydride transfer from butoxide to the carbonyl functionality of furfural. The first-order dependence of the dehydrogenation rate on butanol and furfural at low pressures and the zero-order dependence at higher pressures of butanol indicate that both furfural and butanol participate in the rate-determining step of the dehydrogenation and that butoxide is the most abundant surface intermediate on the active sites of hydrotalcite.

On the other hand, for the aldol condensation step, enolization of the *in situ* formed butyraldehyde or a C-C bond forming step between the butyraldehyde enolate and furfural have been reported to be kinetically relevant steps in the aldol condensation reactions over basic oxides [3,4]. During gas-phase aldol condensation reactions using butyraldehyde and furfural as reactants, we observed a first-order dependence on furfural and a zero-order dependence on butyraldehyde. While the observed order of the reaction with respect to butyraldehyde is zero, the established aldol condensation mechanism requires a first-order intrinsic dependence on butyraldehyde, since it participates both during the enolization and the C-C bond forming events of aldol condensation. As such, the independence of the rate of aldol condensation from the butyraldehyde pressure indicates the saturation of butyraldehyde enolate on the hydrotalcite surface. On the other hand, the first-order dependence on the furfural pressure indicates that the C-C bond formation is the kinetically relevant step for the aldol condensation under the experimental conditions. Since β-hydroxy aldehyde is not detected in the product stream during entire course of the reaction, dehydration step is believed to proceed rapidly under the reaction conditions.

1. Binder, et al. *ChemSusChem* 3, 1268 – 1272
2. Ivanov, et al. *J. Mol. Catal.* 91, 45 – 59
3. Rekoske and Barteau *Ind. Eng. Chem. Res.* 50, 41 – 50
4. Zhang, et al. *Appl. Catal.* 40, 183-190

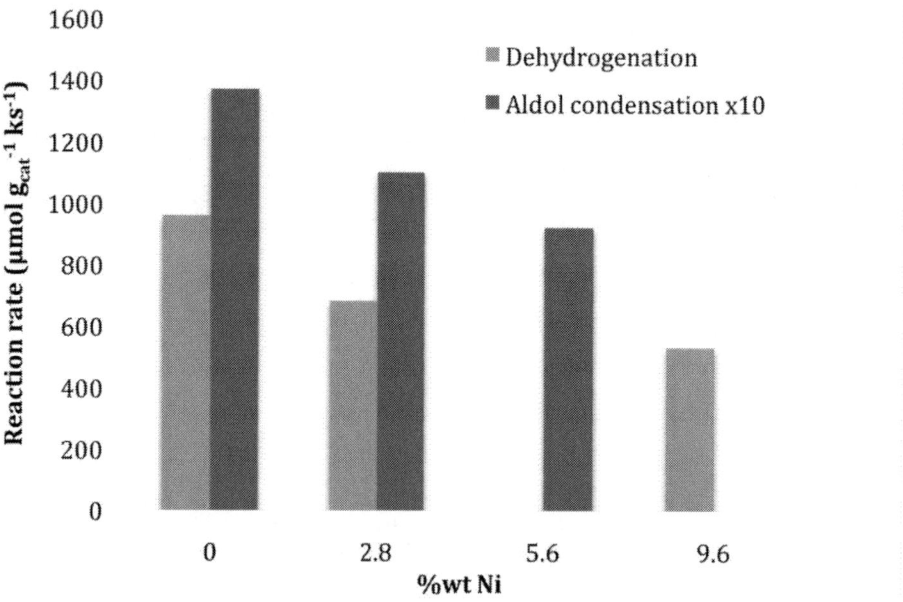

Figure 1: Effect of Ni content on dehydrogenation and aldol condensation rates. 413 K, 150 mg catalyst, 0.95 kPa furfural, 2 kPa butanol, balance He

Hydrodeoxygenation (HDO) of Guaiacol over Ceria-Zirconia Catalysts

Chukwuemeka Okolie, Sarah M. Schimming, Onaje D. LaMont, Michael König, Andrew D. D'Amico, Carsten Sievers

AIChE Annual Meeting
November 20, 2014.

Georgia Institute of Technology
School of Chemical & Biomolecular Engineering

Background - Pyrolysis

Thermochemical decomposition of organic material at elevated temperatures in the absence of oxygen

1000 kg wood

Flash pyrolysis →

- Combustible gas (150 kg)
- Pyrolysis oil (700 kg)
- Char (150 kg)

BTG – http://www.btgworld.com/2005/pdf/technologies/folder-pyrolysis.pdf

Background – Pyrolysis oils

- Higher energy density than biomass
- Complex mixture (Volatile and non-volatile compounds)
- High oxygen content (~ 40 wt.%)
- Water content (~ 20 wt.%)
- Typically less than 40 ppm sulfur
- Immiscible in mineral oil
- Not as stable as mineral oil

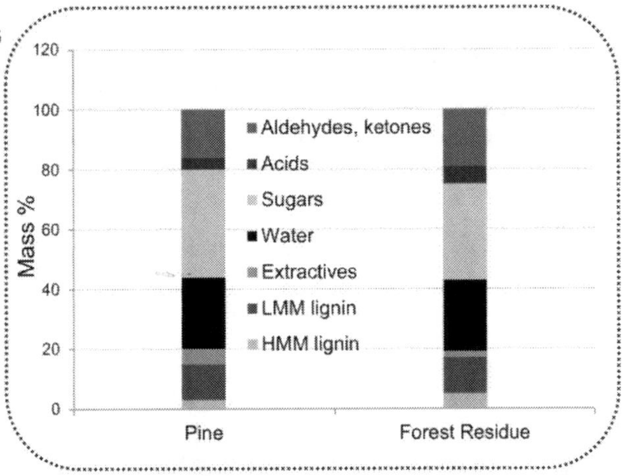

Poor fuel properties

A. Oasmaa, *Biomass & Bioenergy*, 14, 103-113 **1998**.

Georgia Institute of Technology
School of Chemical & Biomolecular Engineering

3

Background – Upgrading Pyrolysis-oils

Pyrolysis-oils are considered as a source for traffic fuels and chemicals

The upgrading of pyrolysis-oils is required to eliminate oxygenated groups, this improves
→ Thermal stability
→ Chemical stability
→ Heating value
→ Volatility

Hydrodeoxygenation (HDO) is an evolving technology needing
→ Moderate temperatures
→ Moderate pressures
→ catalysts

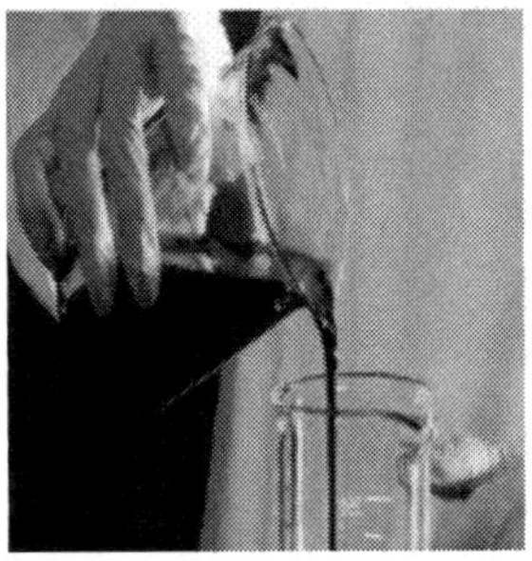

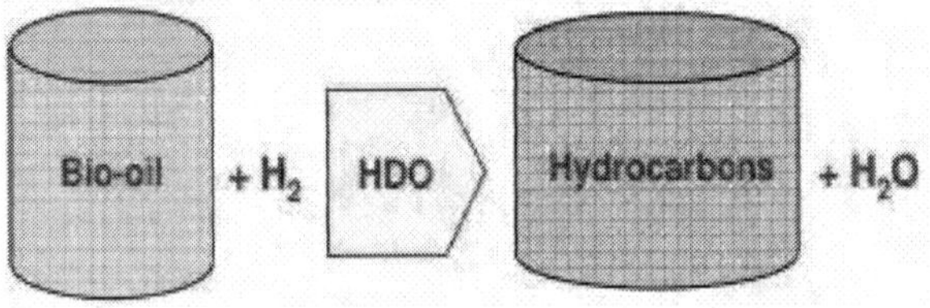

HDO is a reaction in the excess of hydrogen to produce hydrocarbons and water

Georgia Institute of Technology
School of Chemical & Biomolecular Engineering

4

Background – Guaiacol as a Model Bio-oil

Wood based pyrolysis oils are complex mixtures

(Carboxylic acids, alcohols, ketones, carbohydrates, degraded lignin, water etc.)

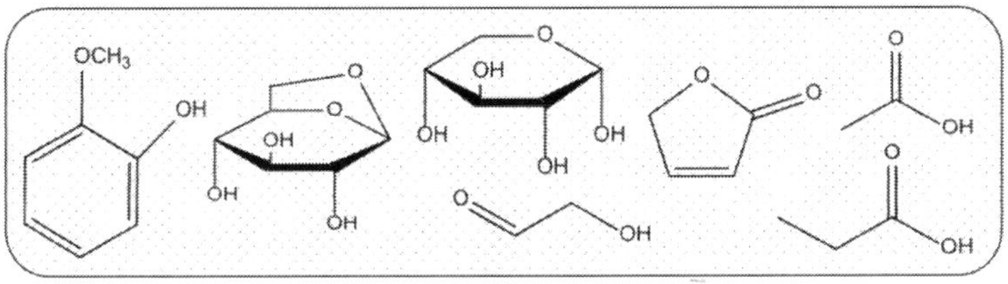

Difficult system to study

- Viscosity problems, large amount of side reactions, solubility problems

Guaiacol is a good model compound that simulates the behavior of oxygenated pyrolysis oil

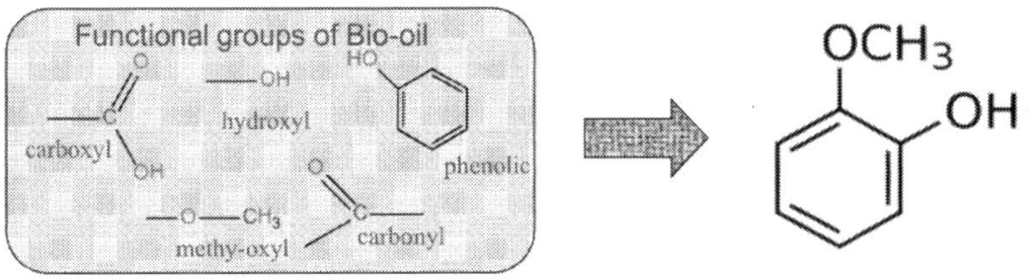

Georgia Institute of Technology
School of Chemical & Biomolecular Engineering

HDO Catalysts

Most studies on HDO of pyrolysis oils use supported noble metals or sulfided NiMo or CoMo catalysts.

- Sulfide catalysts can bind heteroatoms (S, N, O) of organic compounds in defect sites and subsequently remove them from the parent molecule.
- During HDO, sulfide are converted into inactive oxides. Toxic H_2S needs to be supplied to keep them in their active form.

Goal: Development of sulfur-free HDO catalysts with comparable properties.

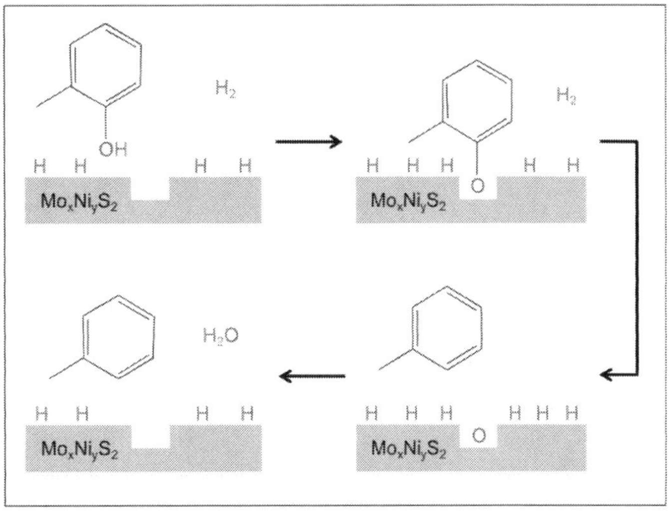

Mechanism: Oxygen atoms are bound to defect sites and subsequently replaced by hydrogen.

E. Furimsky, Appl. Catal. A-Gen. 199 (2000) 147-190.

Georgia Institute of Technology
School of Chemical & Biomolecular Engineering

Ceria-Based HDO Catalysts

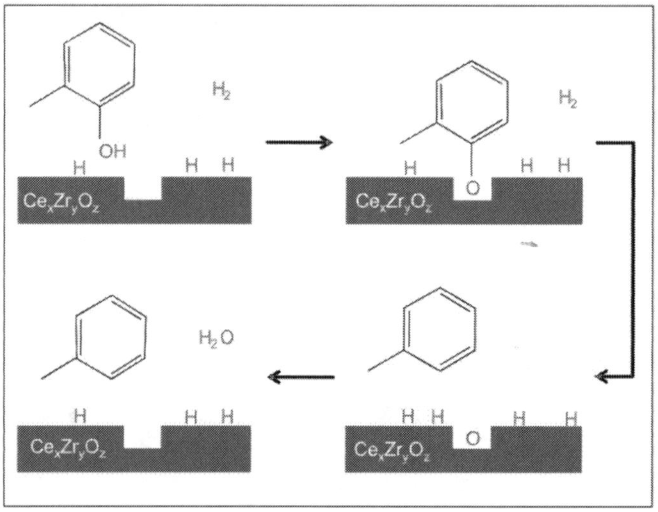

Requirements for oxide-based HDO catalysts:

- Formation of oxygen vacancies at realistic reaction temperatures.
- Ability to activate hydrogen.

Georgia Institute of Technology
School of Chemical & Biomolecular Engineering

7

Catalysts Synthesis

Synthesis by precipitation / coprecipitation:

- CeO_2 – Ce100
- ZrO_2 – Zr100
- $Ce_{0.82}Zr_{0.18}O_2$ – Ce82
- $Ce_{0.60}Zr_{0.40}O_2$ – Ce60
- $Ce_{0.46}Zr_{0.54}O_2$ – Ce46

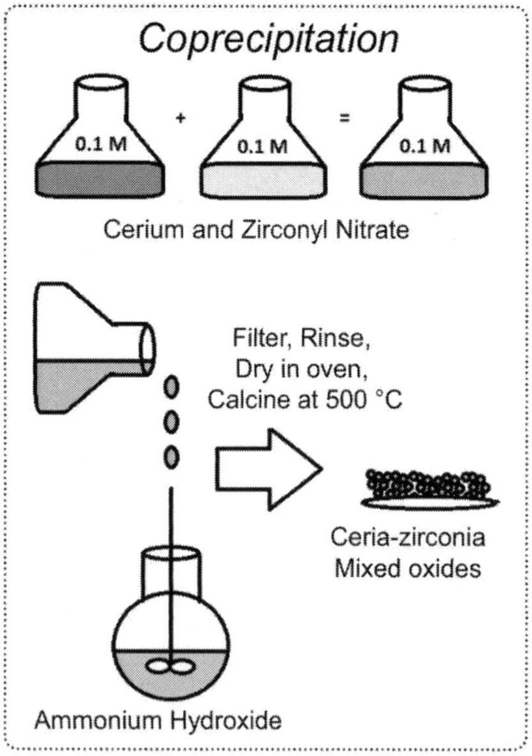

Georgia Institute of Technology
School of Chemical & Biomolecular Engineering

8

X-Ray Diffraction

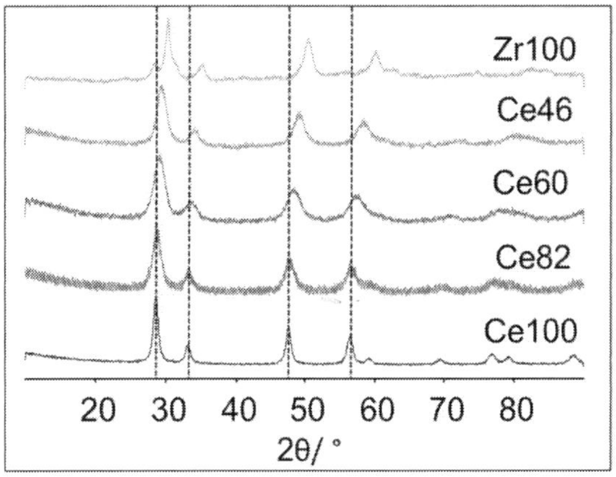

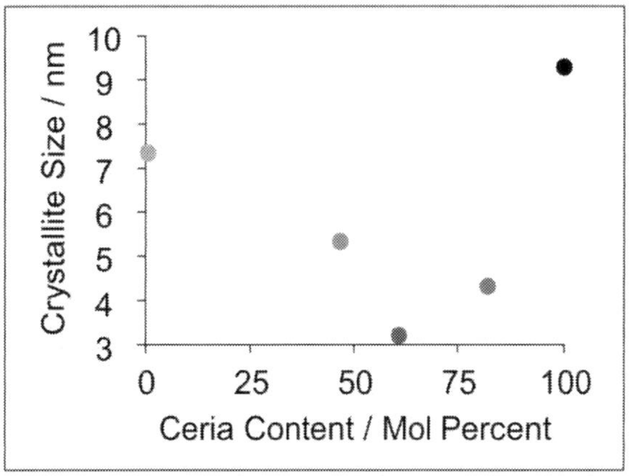

- ➢ Ceria-zirconia retains the CaF_2 structure of ceria.
- ➢ Distance between crystal planes decreases, with increasing zirconium content.
- ➢ All mixed oxides have smaller crystallite sizes than ceria.
- ➢ Crystallite size reaches a minimum at Ce60.

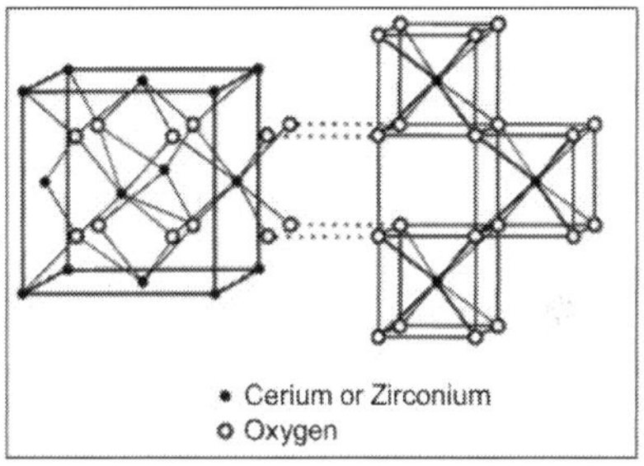

Georgia Institute of Technology
School of Chemical & Biomolecular Engineering

Mamontov, E.; Egami, T.; Brezny, R.; Koranne, M.; Tyagi, S., *J. Phys. Chem. B* **2000,** *104* (47), 11110-11116.

9

N_2 Physisorption and OSC

BET Surface Area:

- ➢ All mixed oxides have a higher surface area than ceria
- ➢ Ce60 has the highest surface area

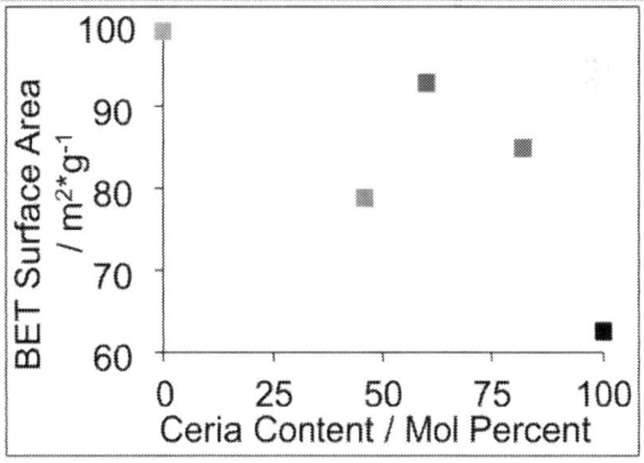

Oxygen Storage Capacity:

- ➢ No formation of oxygen vacancies at 200 °C
- ➢ At 300 °C, oxygen defects form
- ➢ Largest concentrations of vacancies on Ce60 and Ce82

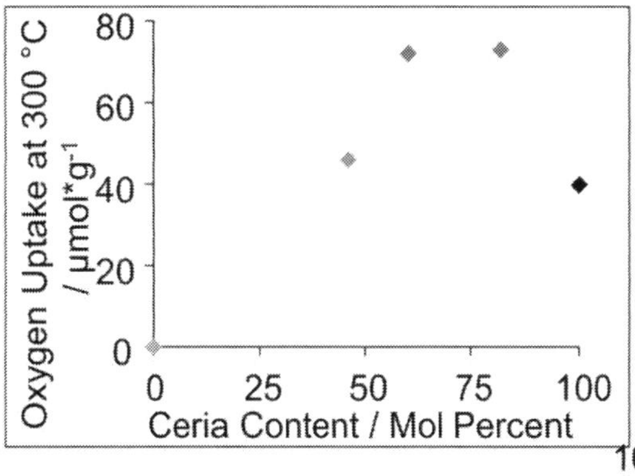

Georgia Institute of Technology
School of Chemical & Biomolecular Engineering

10

Temperature Programmed Reduction

- Surface reduction starts around 200 °C.
- Addition of zirconia reduces the temperature for bulk reduction.
- Ceria-zirconia samples also show increased surface reduction below 300 °C.
- Ce60 has highest H_2 consumption below 300 °C.
- Reducibility of surface sites is essential for HDO activity.

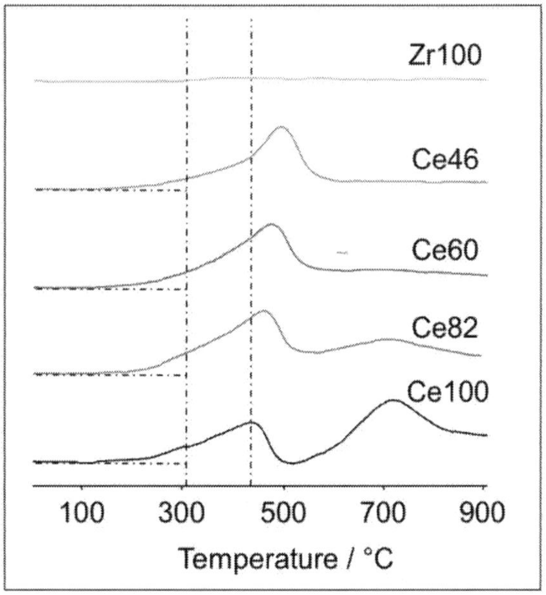

Georgia Institute of Technology
School of Chemical & Biomolecular Engineering

11

H_2-D_2 Exchange on Ceria-Zirconia

$$H_2 + D_2 \leftrightarrow 2HD$$

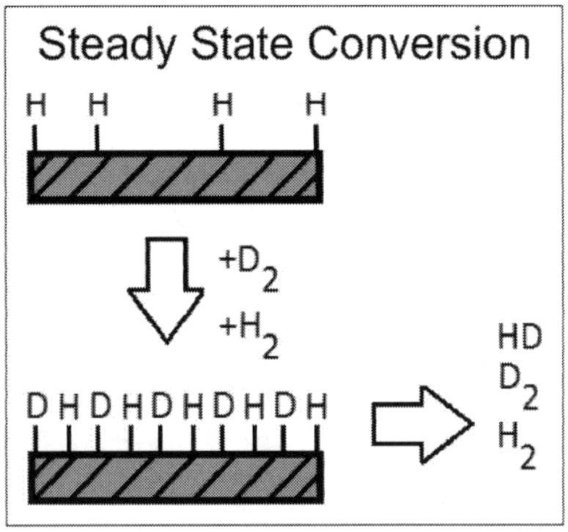

H_2-D_2 exchange:

- Test reaction for hydrogen dissociation
- Hydrogen is readily activated over ceria-zirconia catalysts
- Up to 10.5 hydrogen atoms per nm^2 participated in isotopic scrambling at 300 ^oC
- HDO reactions over catalysts are not limited by the amount of reactive hydrogen on the surface

Georgia Institute of Technology
School of Chemical & Biomolecular Engineering

12

Experimental

Simulation of wood based bio-oils
➔ 4.5 wt.% of guaiacol (GUA) in decane and octane

Trickle bed reactor
➔ 1 bar
➔ H_2 flow 40 mL/min
➔ 300 - 400°C
➔ 72 hours

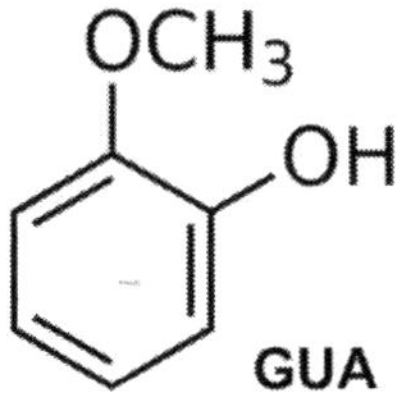

GUA

Catalysts
➔ 500 mg in reactor on quartz wool

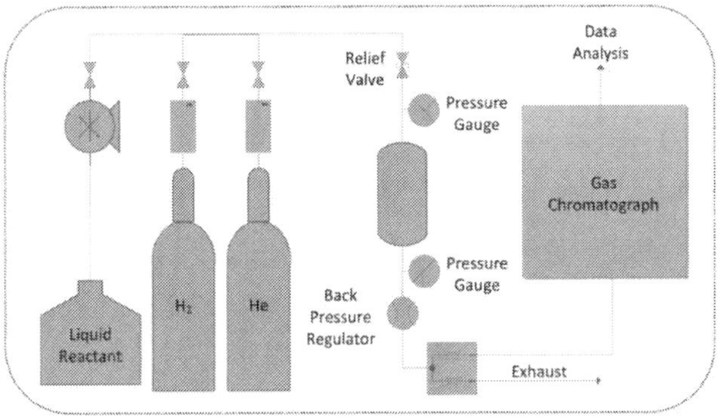

Georgia Institute of Technology
School of Chemical & Biomolecular Engineering

13

Reaction Products - Experimental

❖ Catalysts

Pure ceria catalyst: Ce100
Mixed ceria & zirconia catalysts:
Ce82
Ce60
Ce46

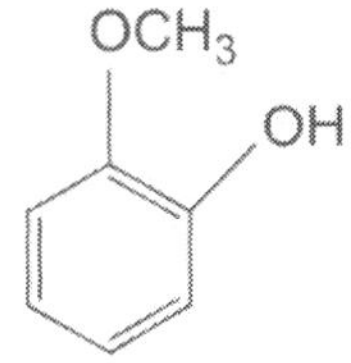

Guaiacol

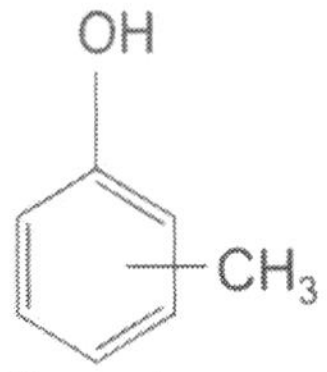

Cresols

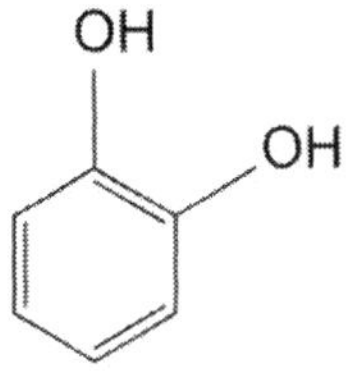

Catechol

Conversion: $X[\%] = \frac{F_{g,in} - F_{g,out}}{F_{g,in}}$ x 100

Product yield: $Y_p\,[\%] = \frac{F_{p,out}}{F_{g,in} - F_{g,out}}$ x 100

Product Selectivity: $S_p[\%] = \frac{Y_p}{X}$ x 100

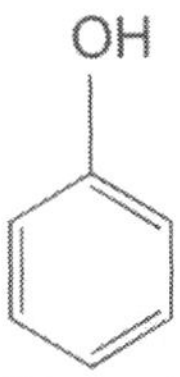

Phenol

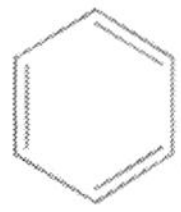
Benzene

Georgia Institute of Technology
School of Chemical & Biomolecular Engineering

14

HDO: Variation of Contact Time

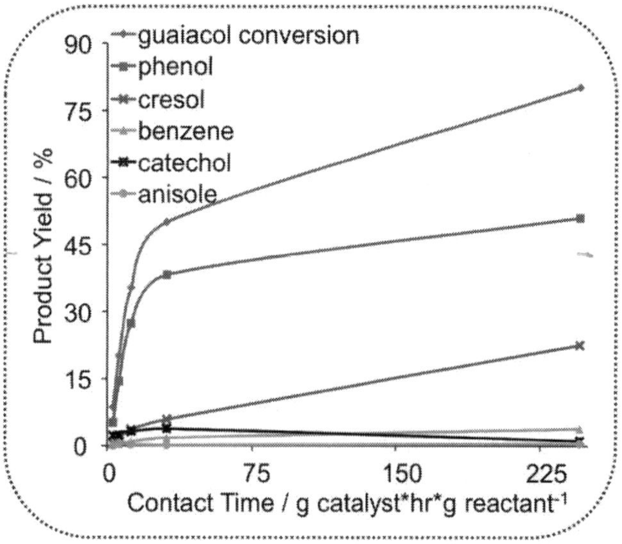

Catalyst: 0.5 g of Ce82
Temperature: 400 °C
Pressure: 1 bar
Hydrogen flow: 40 mL/min
Liquid flow: 0.001 to 0.08 mL/min
0.001 mL/min repeated
Total run time: ~ 72 hours

With increasing contact time:
- Total HDO product yield increases
- Phenol, benzene, and cresol yields increases
- Catechol yield passes through a maximum

Ceria-zirconia has limited activity for conversion of phenol.

Suggested HDO reaction scheme

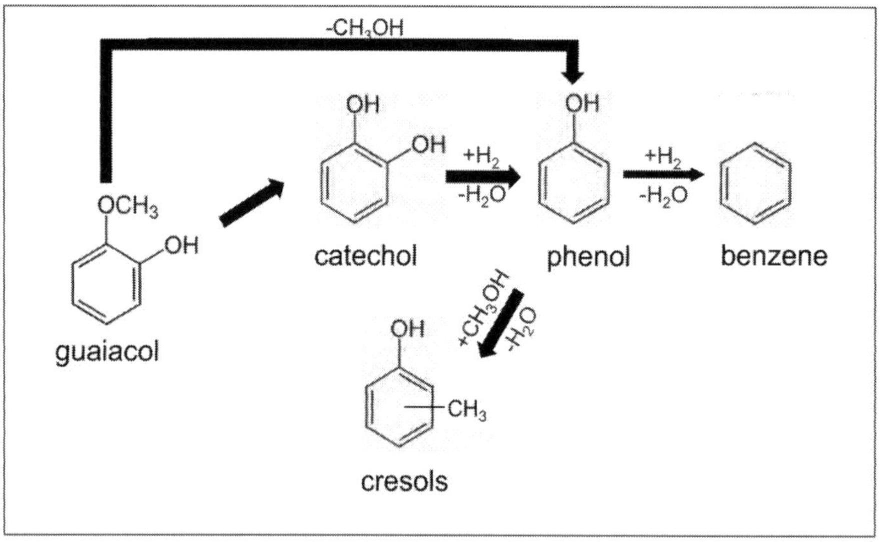

- ➢ Catechol is an intermediate.
- ➢ Cresol is formed by transalkylation reaction between guaiacol and phenol.
- ➢ Ceria-zirconia has limited activity for conversion of phenol.

Variation of temperature

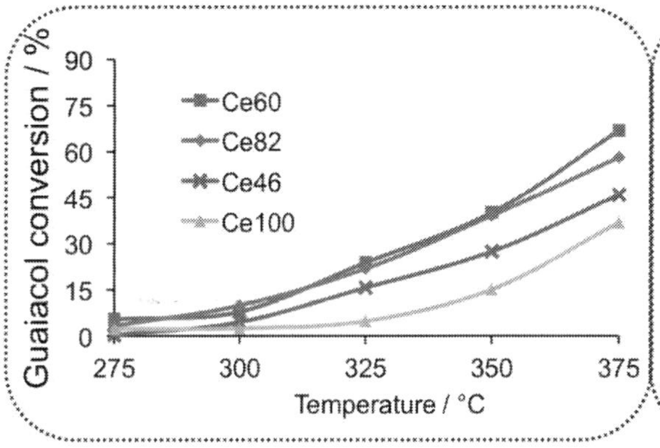

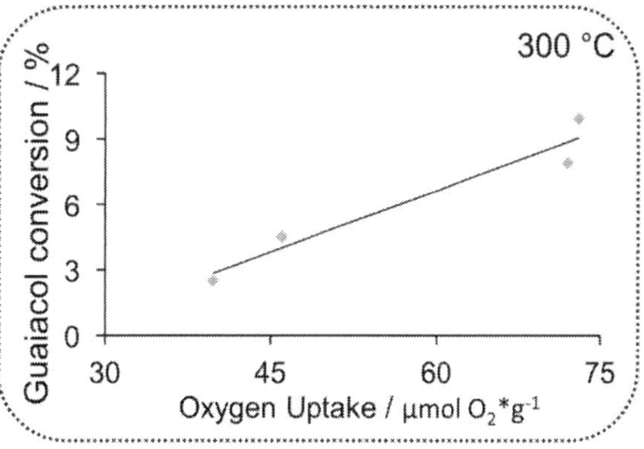

- All mixed oxide catalysts have a higher activity than pure ceria.
- Ce60 and Ce82 have the highest activity.
- Guaiacol conversion at 300 °C scales linearly with the oxygen storage capacity at 300 °C.

Temperature: 275 °C – 375 °C
Pressure: 1 bar
Hydrogen flow: 40 mL/min
Liquid flow: 0.008 mL/min
Total run time: ~ 72 hours

Georgia Institute of Technology
School of Chemical & Biomolecular Engineering

17

Activation Energies for Guaiacol Conversion

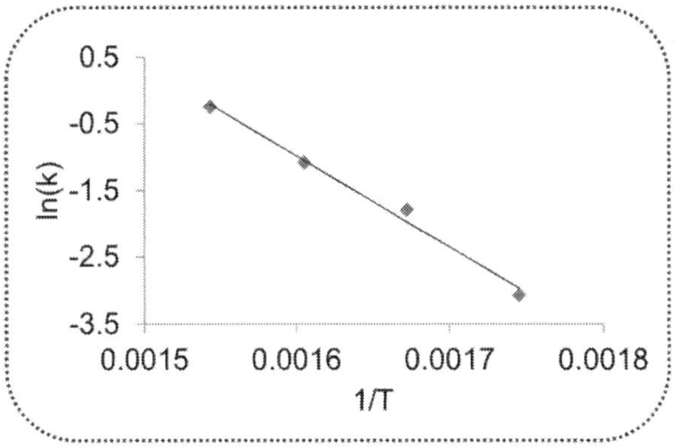

Catalyst	E_A (kJ/mol)
Ce100	131
Ce82	97
Ce60	111
Ce46	114

- Approximation: Guaiacol conversion is a single, irreversible, pseudo-first order reaction.
- All mixed oxides have lower activation energies than pure ceria.
- Ce82 has the lowest activation energy.
- The activation energies compare to values of 50-124 kJ/mol in studies over CoMo or NiMo on Al_2O_3.[1]

Georgia Institute of Technology
School of Chemical & Biomolecular Engineering

[1]E. Furimsky, Appl. Catal., A 199 (2000) 147.

18

Conclusions

- Addition of zirconia to ceria improves reducibility at below 300 °C.
- The crystallite size is the most important descriptor for the performance of ceria-zirconia catalysts.
- Ceria-zirconia catalysts are active for HDO of guaiacol with phenol as the major product.
- Ceria-zirconia catalysts did not deactivate over the course of the reaction due to coking or water in the feed as other common HDO catalysts do

Georgia Institute of Technology
School of Chemical & Biomolecular Engineering

19

Acknowledgements

People

- Allyson Rogers
- Tiorra Ross
- Dr. Sievers' group members at Georgia Tech

Funding

- Renewable Bioproducts Institute
- International Paper
- NewPage Corporation
- Georgia Tech start-up funds

Georgia Institute of Technology
School of Chemical & Biomolecular Engineering

20

THANK YOU!!!

Georgia Institute of Technology
School of Chemical & Biomolecular Engineering

385639 Novel Polymeric Solid Acid Catalysts for Cellulose Hydrolysis

Abstract:

Biofuel derived from lignocellulosic biomass is one of the leading renewable energy to replace fossil-base transportation fuels. The main drawback of the current biochemical processing of biomass is the cost of pretreatment, cellulase enzymes as well as the slow enzymatic hydrolysis process. A synthetic enzyme mimic polymeric solid acid catalyst overcomes all these limitations. Dual functional polymeric solid acid catalysts are synthesized in order not only to catalyze cellulose hydrolysis but also to facilitate its dissolution. The acidic polymeric chain, poly (styrene sulfonic acid) (PSSA) immobilized on a glass/membrane substrate via atom-transfer radical polymerization (ATRP) catalyzes biomass hydrolysis. Its neighboring poly (vinyl imidazolium chloride) (PIL) chain grafted via UV-initiated radical polymerization helps solubilize lignocellulosic biomass and enhance the catalytic activity. UV-initiator 4-ethoxy-5-oxo-4,5-diphenylpentanoic acid (BEE-COOH) was successfully synthesized and immobilized prior to UV polymerization. Cellulose hydrolysis was conducted in 1-methyl-3-ethylimidazolium chloride [EMIM]Cl as well as in aqueous solution. An optimal ratio of the two polymer chain densities as well as optimal polymer chain lengths were found for achieving highest catalytic conversion of cellulose substrates. These polymeric catalysts are stable in ionic liquid solvents and maintain high catalytic activity after more than 12 repeated runs. Our novel polymeric solid acid catalysts demonstrate total reducing sugar (TRS) yield over 94% after 12 runs in [EMIM]Cl. Methods for ionic liquid regeneration and reuse were also developed. More importantly the catalytic activity of these polymeric acids can be further tuned by ring substitution and copolymerization. These polymeric solid acids are reusable, environmental friendly and potentially superior catalysts for many other organic reactions requiring acid catalysts. Theoretical investigations are conducted to obtain fundamental insight into their high catalytic activity.

383348 Continuous Transesterification at Near-Critical and Supercritical Ethanol Conditions

Abstract:

Biodiesel is a renewable petroleum diesel replacement produced predominately through the transesterification of triglycerides using an alcohol, most often methanol, with an alkaline catalyst. Since methanol is typically produced from natural gas, the majority of methanol comes from non-renewable sources. One way to make biodiesel truly renewable is to instead use ethanol in place of methanol; however, reaction times with ethanol are greatly increased over those needed for methanol. It is possible to reduce the reaction time to 30 minutes or less by carrying out transesterification at near-critical and supercritical conditions of ethanol. Since no catalyst is used, saponification is suppressed thus increasing tolerance of water and free fatty acids. For this presentation, a constant molar ratio of 41:1 ethanol to canola oil was used. Temperatures ranging from 250°C to 350°C and residence times of between 2 and 20 minutes were examined to optimize the production of ethyl esters and determine yield and selectivity. Reactions were carried out in a continuous, custom made 101 mL Haynes® 282 Alloy tubular flow reactor.

Hydrocracking of DDGS Corn Oil over Transition Metal Carbides Supported on Al-SBA-15: Effect of Metal Ratio, Preparation Method, and Fractional Sum of Metal Electronegativities

Basem Al Alwan[a], Steven O. Salley[a,b], and K. Y. Simon Ng[a,b,c]

[a] *Department of Chemical Engineering and Materials Science, Wayne State University, Detroit, MI 48202, USA*

[b] *National Biofuels Energy Laboratory, NextEnergy, Detroit, MI 48202, USA*

[c] *Alternative Energy Technology, Wayne State University, Detroit, MI 48202, USA*

Abstract

Economic, ecological, and environmental issues associated with fuels derived from fossil sources have spurred research to develop alternative liquid fuels that are sustainable and renewable. Triglycerides from vegetable oils may be a suitable source for liquid biofuel production (green diesel) via catalytic hydrocracking. Low quality vegetable oils such as distillers dried grains with solubles (DDGS) corn oil has been considered as an economical alternative to using fresh vegetable oil as a biofuel feedstock in order to avoid competition with food markets. The hydrocracking of DDGS corn oil over bimetallic carbide catalysts was explored for green diesel production. A catalyst composed of nickel-tungsten (Ni-W) carbide supported on Al-SBA-15 was designed based on the ability of nickel to adsorb and activate hydrogen and the potential of tungsten for hydrogenation reactions. Four different Ni-W ratios (1:9, 1:1, 2:1, and 9:1) were prepared by impregnation method in order to study the effect of metal ratio on the catalyst structure, activity and selectivity. The catalysts showed significant differences in activity and selectivity, with the catalyst having a Ni-W ratio of 9:1 achieving 100% conversion of corn oil and 100% selectivity to diesel for two days. Results indicate that minimizing metal alloy formation and enhancement of the metal dispersion leads to higher activity, selectivity and durability of the catalysts.

A Dendrimer-Encapsulated-Nanoparticles (DENP) method was employed in order to minimize alloy formation and increase the metal dispersion on the support. The catalysts prepared by the DENP method showed greater activity than the catalyst prepared by impregnation method for the hydrocracking of DDGS corn oil.

The effects of the fractional sums of the electronegativities of the transition metals on the catalyst activities and selectivities were investigated. The closer the fractional sum of the transition metal electronegativities was to the electronegativity range of noble metal catalysts (2.0–2.2), the better was the catalyst performance. The effects of doping a promoter (Ce) on the catalyst electronegativity and activity were also studied.

377104 Ni-Co/r-GO Catalysts for Hydrogen Generation from Chemical Hydride System Containing Ammonia Borane

Abstract:

Renewable energy attracts more and more attention in recent years. Among various green energies, hydrogen is one of the most promising energy due to its higher efficiency and power density, as well as its environmental benignity. Commonly, the hydrogen energy could be effectively extracted through devices such as fuel cells. Nowadays, the researches related to proton exchange membrane fuel cells (PEMFCs) become one of popular topic because PEMFCs have high energy efficiency and low operation temperature. Hence, the transportation of hydrogen to PEMFCs is the main topic in the future. In general, there are four methods of hydrogen storage: (1) high-pressure tanks; (2) liquefied hydrogen; (3) metal hydrides; and (4) chemical hydrides. Chemical hydrides not only have higher hydrogen storage capacity but also generate hydrogen quickly in the presence of adequate catalysts. Ammonia borane (NH_3BH_3) is a prospective compound in chemical hydrides. The hydrogen energy density of NH_3BH_3 is 19.6 wt%. Furthermore, NH_3BH_3 and its spent products are rarely toxic and flammable.

In this work, NH_3BH_3 and Ni-Co/r-GO catalysts were pulverized by using high energy ball milling machine (denoted as solid-state AB composite). Reduced graphene oxide (r-GO) was adopted as a light-weighted catalyst support. The Ni-Co/r-GO catalysts were characterized with SEM, EDS, TEM, VSM and ICP-AES. Hydrogen was generated by injecting adequate water from a syringe to initiate the hydrolysis reaction of solid-state AB composite. The amount of water dosage and loading of as-prepared catalysts will be discussed in this study. Hydrogen profiles were recorded by a mass flow meter (MFM) installed on a computer and temperature profiles were observed by a thermocouple at the same time.

The Nature of Active Copper Species in Cu/SiO_2 Catalysts for Hydrogenation of Methyl Acetate to Ethanol

Yue Wang, Yongli Shen, Yujun Zhao, Xinbin Ma*

School of Chemical Engineering and Technology
Tianjin University

AIChE Annual Meeting
Nov 20 2014, Atalanta

Background

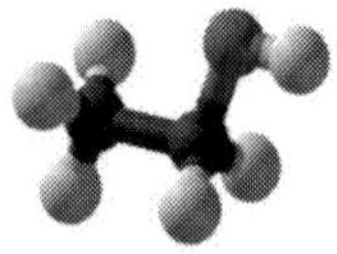

- **High-octane rating; great antiknock performance**
- **High oxygen content; reduce the CO, NO_x emissions**

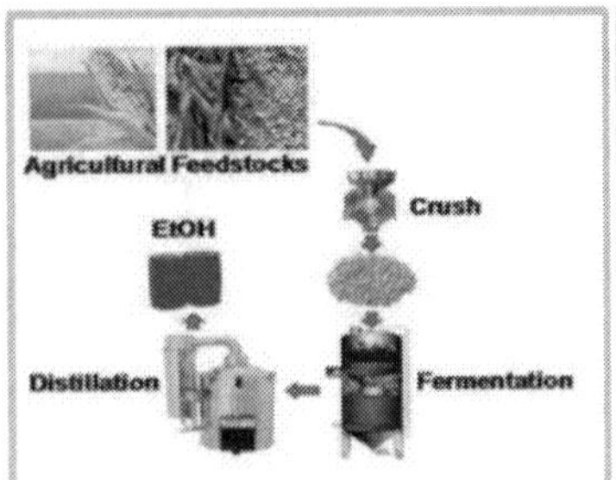

Fermentation
Restriction of cereal grains as raw materials

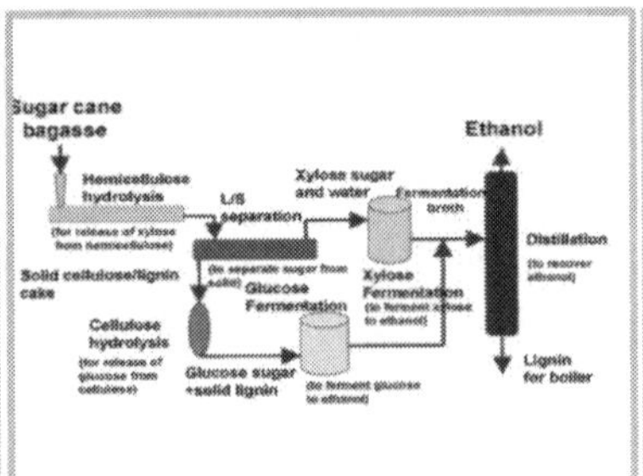

Cellulose hydrolyzation
Low energy - density and poor conversions

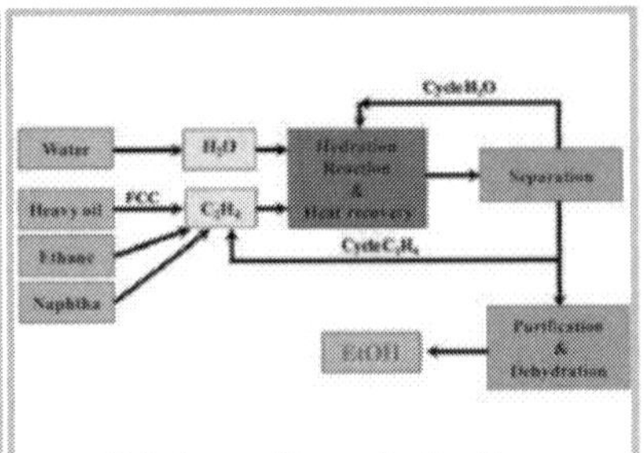

Ethylene direct hydration
Low selectivity, high energy and water consumption

Background

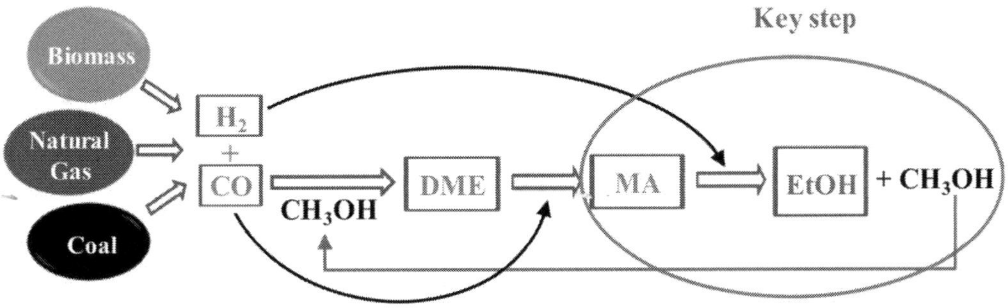

The new route for ethanol synthesis from syngas via dimethyl ether

Process is short, low-cost, low discharge, low energy and water consumption, suitable for characteristics of Chinese energy framework (rich in coal, lack of petroleum and natural gas)

Reviews of literatures related

Cu^0-Cu^I:

- **Cu^0 sites could dissociate H_2;**
- **Cu^+ sites could function as electrophilic or Lewis acidic sites to polarize the C=O bond via the electron lonepair in oxygen**

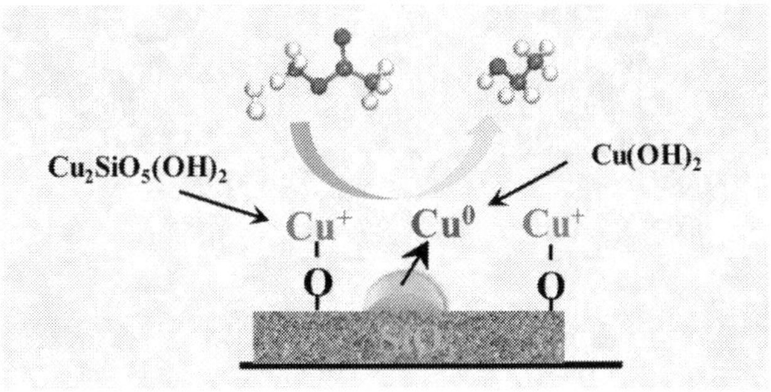

J. Catal., 2000, 193, 16-28 *J. Catal.*, 2008, 257, 172-180 *J. Phys. Chem. C*, 2009, 113, 11003-11013 *Catal. Commun.*, 2011, 12, 412-416

Cu^0:

- **Metallic Cu was the active site for hydrogenation** *Energy & Fuels*, 2009, 23, 2843–2844
- **Highly dispersed metallic Cu is in favor of improvement the catalytic performance** *ACS Catal.*, 2012, 2, 1958–1966

The strategy——Manipulation of Cu^0/Cu^I

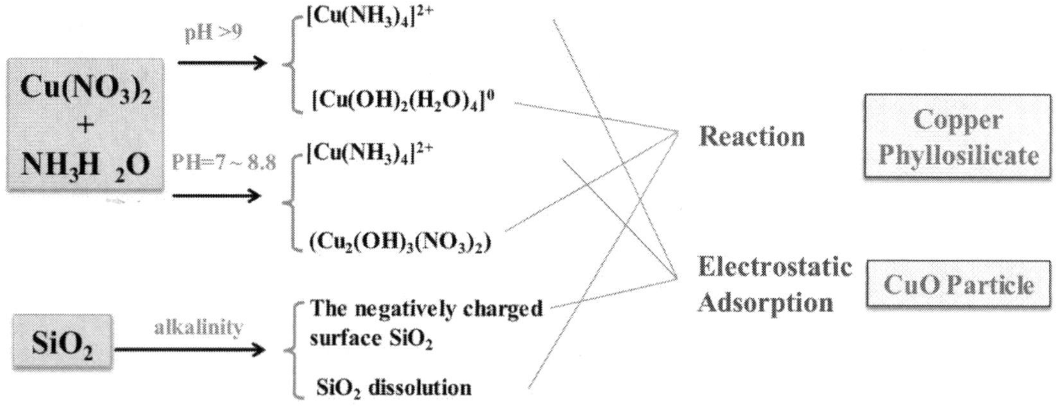

Target: relationship between active species and catalytic activity

Challenges: to exclude influences of other factors (geometry, channel structure, surface area……)

- the amount of Cu loading
- interaction time between cuprammonia and silica sol
- the SiO_2 surface groups

Toupance et al., J Phys Chem B, 2002, 106, 2277-2286; *Chen et al., J Catal, 2008, 257, 172-180.*

The structure of catalysts

Catalyst	Metal loading (wt%)[a]	S_{BET} (m^2/g)	D_{pore} (nm)	Crystallite size (nm)	
				by XRD[b]	by TEM
$5Cu/SiO_2$-303-4h	4.9	254	10.4	ND[c]	2.1
$5Cu/SiO_2$-353-4h	5.0	250	11.7	ND	1.8
$5Cu/SiO_2$-4h	5.0	263	13.6	ND	2.2
$10Cu/SiO_2$-303-4h	9.8	309	7.7	ND	2.9
$10Cu/SiO_2$-353-4h	9.9	302	12.4	ND	2.0
$10Cu/SiO_2$-4h	9.9	294	15.7	ND	2.4
$20Cu/SiO_2$-303-4h	19.7	381	6.8	4.5	4.0
$20Cu/SiO_2$-353-4h	19.6	382	10.7	4.8	3.9
$20Cu/SiO_2$-4h	19.5	409	10.8	ND	2.9
$20Cu/SiO_2$-16h	19.6	414	11.0	4.5	3.5
$20Cu/SiO_2$-24h	19.5	406	9.9	4.7	3.6
$20Cu/SiO_2$-40h	19.6	421	10.5	ND	3.1
$30Cu/SiO_2$-4h	29.0	480	8.6	3.6	4.1
$40Cu/SiO_2$-4h	37.2	558	6.2	3.8	4.7

a Metal loading was determined by ICP-OES.
b Diameter of Cu particles calculated from the XRD data by Scherrer equation.
c ND: not detected.

XRD

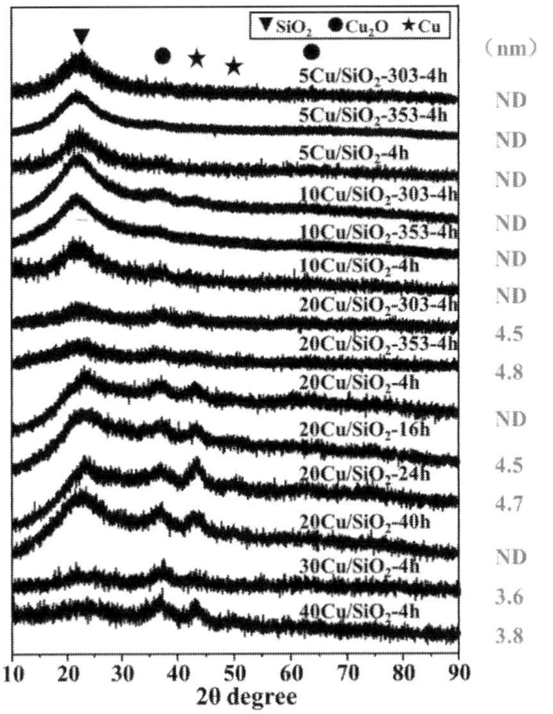

- A diffraction peak at 2θ of 43.3° and 50.4° characteristic of fcc Cu.
- The peaks at 2θ of 36.4° attributed to the Cu_2O (111) plane, indicating that a portion of copper exists as Cu (I) after reduction.
- The broad peaks at around 22° are assigned to amorphous SiO_2.
- The catalysts were successfully prepared with highly dispersed sub-5 nm metallic copper particles

TEM

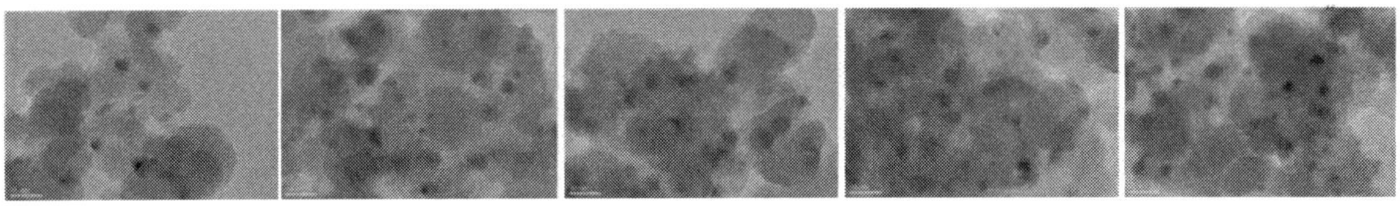

5Cu/SiO_2-303-4h 5Cu/SiO_2-353-4h 5Cu/SiO_2-4h 10Cu/SiO_2-303-4h 10Cu/SiO_2-353-4h

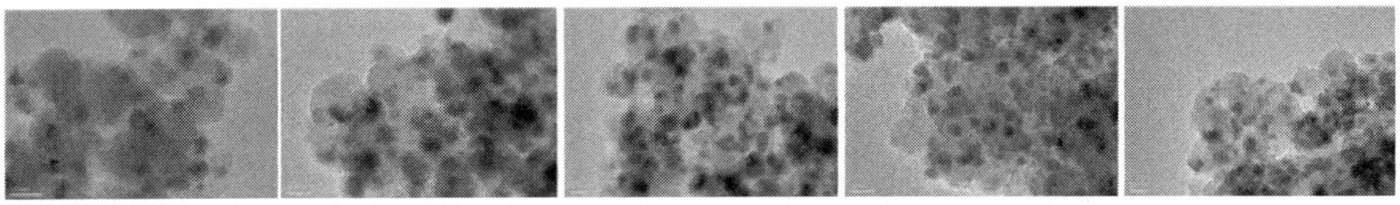

10Cu/SiO_2-4h 20Cu/SiO_2-303-4h 20Cu/SiO_2-353-4h 20Cu/SiO_2-4h 20Cu/SiO_2-16h

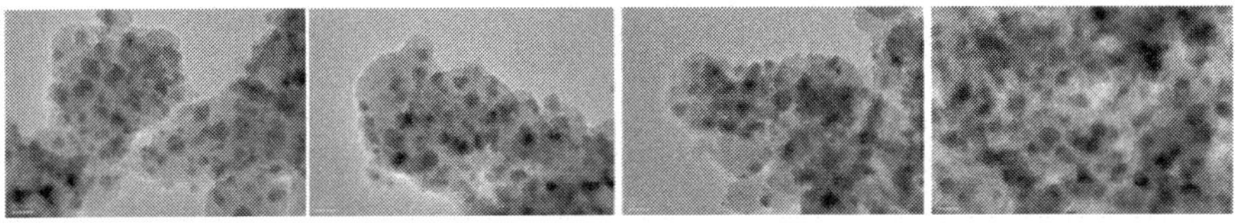

20Cu/SiO_2-24h 20Cu/SiO_2-40h 30Cu/SiO_2-4h 40Cu/SiO_2-4h

Catalytic performance

Catalyst	MA conversion (%)	Selectivity (%)[b]			$S_{Cu(0)}$ (m^2/g)	$S_{Cu(I)}$ (m^2/g)
		EtOH	EtOAc	Others[c]		
$5Cu/SiO_2$-303-4h	12.3	44.3				
$5Cu/SiO_2$-353-4h	22.1	43.0				
$5Cu/SiO_2$-4h	19.8	44.4				
$10Cu/SiO_2$-303-4h	26.4	43.3				
$10Cu/SiO_2$-353-4h	51.2	52.4				
$10Cu/SiO_2$-4h	42.3	50.0				
$20Cu/SiO_2$-303-4h	79.7	76.4	23.5	0.1		
$20Cu/SiO_2$-353-4h	59.9	59.4	40.6	0.0	39.6	9.4
$20Cu/SiO_2$-4h	61.2	62.2	37.7	0.1	24.4	10.3
$20Cu/SiO_2$-16h	73.4	70.9	29.0	0.1	23.9	13.8
$20Cu/SiO_2$-24h	81.7	77.8	22.2	0.0	22.1	20.5
$20Cu/SiO_2$-40h	73.8	71.5	28.4	0.1	23.5	14.8
$30Cu/SiO_2$-4h	62.7	65.3	34.6	0.1	37.6	12.2
$40Cu/SiO_2$-4h	66.3	67.3	32.6	0.1	44.2	13.9

face area, and concentration of surface Cu^+ sites, which essentially enhanced catalytic performance. CO-chemisorption FTIR characterization disclosed that the surface concentration of a special Cu^+ site that can stably adsorb CO molecule reached the maximum on a Cu/SiO_2–3HZ-38 hybrid catalyst with the best catalytic activity. Therefore, Cu^+ sites might play important roles in the catalytic process of DMO hydrogenation to EG. Nonetheless, the presence of both Cu^+ and Cu^0 species on the functioning catalyst was confirmed because the synergetic effects of Cu^0 and Cu^+ surface species were involved in the catalytic cycle.

Appl. Catal. A, 2012, 445, 287-296

a Reaction conditions : catalyst weight = 0.5g, P = 2.5 M Pa, T = 488 K, H_2/MA = 80, and $WLHSV_{(MA)} = 2.0\ gg_{cat}^{-1}h^{-1}$.
b The selectivies were calculated excluded MeOH as a product since it was cycled back to the previous process to be a reactant.
c Others: CH_3CHO, $HCOOC_2H_5$, CH_3COOH, $HCOOCH_3$, C_3H_7OH and C_4H_9OH.

Catalytic performance

Catalyst	MA conversion (%)	Selectivity (%)[b]			$S_{Cu(0)}$ (m^2/g)	$S_{Cu(I)}$ (m^2/g)
		EtOH	EtOAc	Others[c]		
$5Cu/SiO_2$-303-4h	12.3	44.3	55.2	0.5	5.0	8.9
$5Cu/SiO_2$-353-4h	22.1	43.0	56.8	0.2	11.2	6.9
$5Cu/SiO_2$-4h	19.8	44.4	55.3	0.3	7.6	6.7
$10Cu/SiO_2$-303-4h	26.4	43.3	56.5	0.2	12.0	8.6
$10Cu/SiO_2$-353-4h	51.2	52.4	47.5	0.1	19.9	8.0
$10Cu/SiO_2$-4h	42.3	50.0	49.9	0.1	18.6	8.5
$20Cu/SiO_2$-303-4h	79.7	76.4	23.5	0.1	23.1	17.0
$20Cu/SiO_2$-353-4h	59.9	59.4	40.6			
$20Cu/SiO_2$-4h	61.2	62.2	37.7			
$20Cu/SiO_2$-16h	73.4	70.9	29.0			
$20Cu/SiO_2$-24h	81.7	77.8	22.2			
$20Cu/SiO_2$-40h	73.8	71.5	28.4			
$30Cu/SiO_2$-4h	62.7	65.3	34.6			
$40Cu/SiO_2$-4h	66.3	67.3	32.6			

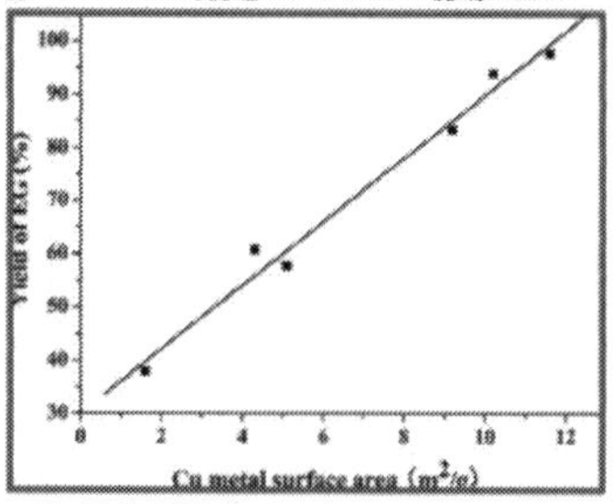

Figure 10. Yield of EG plotted against copper metal surface area for catalysts with different copper loadings.

a Reaction conditions : catalyst weight = 0.5g, P = 2.5 M Pa, T = 488 K, H_2/MA = 80, an
b The selectivies were calculated excluded MeOH as a product since it was cycled back
c Others: CH_3CHO, $HCOOC_2H_5$, CH_3COOH, $HCOOCH_3$, C_3H_7OH and C_4H_9OH.

J. Phys. Chem. C, 2009, 113, 11003-11013

Synergy of Cu^0 and Cu^+ on catalytic performance

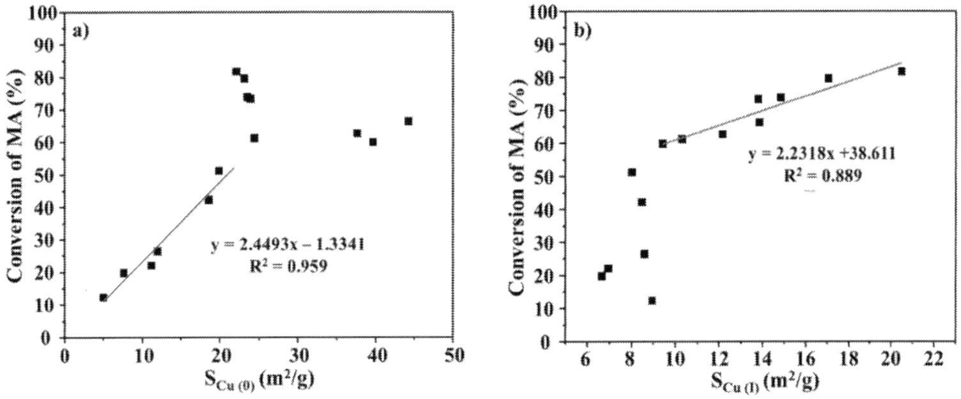

A synergy of active copper species on the hydrogenation of MA :

When there is insufficient of Cu (0) species, the catalytic activity of hydrogenation determined by the activity of H_2, whereas, it would be decided by the adsorption of methoxy and acyl species as the amount of surface Cu (0) achieved to a certain degree.

Conclusion

- **We successfully prepared a series of Cu/SiO_2 catalysts with remarkable differences on state of Cu species and exclued the other factors' influence.**

- **The conversion of MA grew linearly associated with the Cu (0) surface area increasement under a certain value. However, with more Cu (0) active sites, the samples catalytic activity increaed with enrichment of Cu (I) surface area.**

- **These relationships strongly proved there was a synergy of active copper species on the hydrogenation of MA and indicated that when there is insufficient of Cu (0) species, the catalytic activity of hydrogenation determined by the activity of H_2, whereas, it would be decided by the adsorption of methoxy and acyl species as the amount of surface Cu (0) achieved to a certain degree.**

- **The calculated results showed the MA could be adsorbed by $(Cu)_x^{\delta+}/SiO_2$ but not activated by Cu (111). And the H_2 is much easilier dissociated on Cu (111) than on $(Cu)_x^{\delta+}/SiO_2$. These calculatedresults are agreed with the experiments data.**

Progress on scale-up

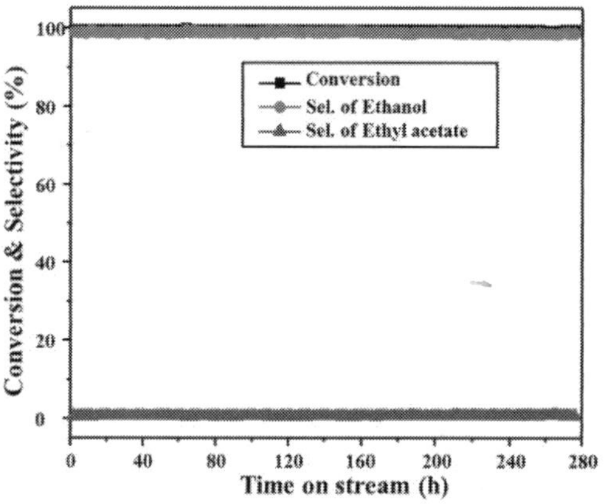

Reaction Conditions: P = 2.5 M Pa, T = 488 K, H_2/MA = 80, and $WLHSV_{(MA)}$ = 1.0 $g g_{cat}^{-1} h^{-1}$

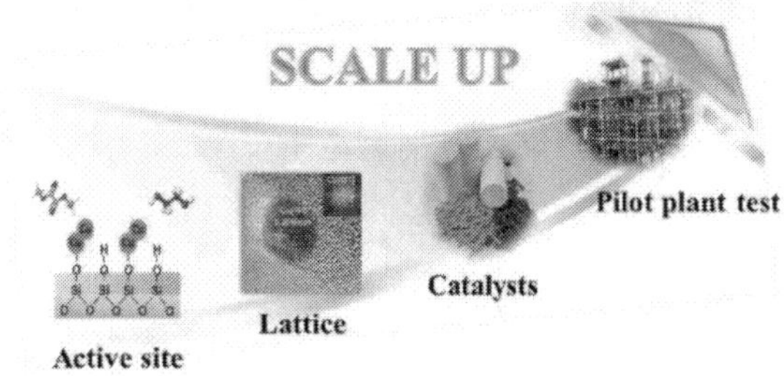

Cooperating with Wison Company for pilot plant test.

Acknowledgement

国家自然科学基金委员会
National Natural Science Foundation of China

WISON惠生

387439 Coupled Simulation of the Fluid Dynamics, Reactions and Electric Field in an Atmospheric Pressure Plasma Deposition Reactor

Abstract:

Atmospheric pressure plasma assisted chemical vapor deposition offers perspectives for the in-line treatment of surfaces, e.g. for the production of high-efficiency solar panels. The technology has been demonstrated on the lab-scale [1], but scale-up turns out to be challenging. Computational Fluid Dynamics can be used to get an improved understanding of the reactor behavior [2]. The present paper addresses the detailed and coupled simulation of the fluid dynamics, reactions and electric field in an atmospheric-pressure low-temperature pure nitrogen plasma reactor. The model equations and solution strategy are discussed and validated. Next, simulations with different reactor geometries are presented and the influence of the electric field on the fluid dynamics and on the species concentration fields in the reactors studied.

References:

[1] Descamps P., Asad S.S., De Wilde J., J. Phys. D: Appl. Phys., 46, 365201 (13pp), 2013.

[2] Descamps P., Kaiser V., Asad S.S., Leempoel P., Plasma Process. Polym., 8, 1116, 2011.

377562 Propylene Selectivity Improvement for Methanol to Olefins Process in a Fixed-Bed Reactor

Abstract:

As a promising technology to convert coal and nature gas into petrochemicals, methanol to olefins (MTO) process has received dramatic attentions from industry and academia. An on-purpose propylene process (MTP) based on fixed-bed reactor was developed by Lurgi Company and has been industrialized in China in 2011 with an annual capacity of about 500kTY propylene. However, the propylene selectivity turned out to be less than 60%, failed to meet the anticipated requirement of 65%[1], and the total selectivity including C_2, C_3 and C_4 olefins is much lower than the fluidized-bed MTO technology (>80%). Our theoretical study showed that the reason for low propylene selectivity is the low methanol concentration and the prolonged residence time and the key to improve propylene selectivity is to increases the methanol concentration in reactor and to reduce the residence time.

The autothermal tubular reactor is a multi-functional reactor combining reaction and heat transfer together in one apparatus, where the reaction heat is continuously removed through tubular heat exchange with the cold reactants[2]. It is advantageous for MTP process since the temperature rise of catalyst bed can be controlled with a higher feed methanol concentration. A two-dimensional autothermal reactor model was developed in the present work. It shows that by increasing the feed methanol to 30%, the autothermal tubular reactor can increase the propylene selectivity significantly from around 60% in a conventional adiabatical reactor to 88.6% and the space velocity from 0.61 in the conventional reactor to 4.68 gMeOH/gcat/hr. By diluting the catalyst bed, the reaction temperature can be restricted between 445-480°C, an reasonable temperature window for MTP reactions, and most of the reaction takes place in 470-480°C, which indicates the temperature was successfully controlled.

References:

1. Agudamu, Sun Y, Zhang F. Product distribution of methanol to propylene reaction found in Shenhua Ningxia Coal Group. *Coal Chemistry Industry (in Chinese).* 2013(01):58-60.

2. Kolios G, Frauhammer J, Eigenberger G. Autothermal fixed-bed reactor concepts. *Chemical Engineering Science.* 2000;55(24):5945-5967.

Removal of organosulfur compounds from real coke oven gas using a mobile and modular test plant

E. Speelmanns[1], M. Rieger[2], H. Thielert[2], G. Wozny[1]
[1] Chair: Process Dynamics and Operation (dbta), TU Berlin,
Straße des 17. Juni 135, D-10623, Berlin, Germany
[2] ThyssenKrupp Industrial Solutions, Uhde-Str. 15, D-44141, Dortmund, Germany

A primal and inevitable step in the production process of steel in blast furnaces is the production of coke. In this carbonization process coal is heated up at high temperatures out of contact with oxygen while volatile matter and coal tar are driven off the coal. Hence, with each ton of coke about 200 kg of coke oven gas are produced. In modern coke plants, this mixture of different gases is then treated and washed to reduce emissions and gain valuable products. Hence, compounds such as tar, hydrogen sulfide, benzene and ammonia are removed. The remaining gas consists of mainly hydrogen (60%) and methane (20%) and several other substances (20%) in varying concentrations. Part of the gas is used in the coke plants to heat the coke ovens or in blast furnaces of integrated steel plants. However, the range of applications for the gas is limited although it offers a great potential as a valuable raw material. Up to now, it is not possible to process the gas further e.g. for the production of methanol or urea. Obstructive are organosulfur compounds such as COS and CS_2 that are found in coke oven gas in small concentrations (100–200 ppm). These substances are known to be toxic for the catalysts that need to be used in downstream processes. Therefore, it is necessary to remove the organosulfur compounds from the gas. Lab experiments have shown that a heterogeneous catalysis that forms H_2S is effective. However, since coke oven gas contains numerous different compounds in varying concentrations, it is crucial to perform tests with real process gas. For this purpose, a mobile and modular test plant containing a fixed-bed reactor was constructed and two measurement campaigns were conducted on the coke plant ArcelorMittal Bottrop GmbH in bypass with the real process.

The experiments with the test plant will be presented as well as the results that were used for modeling the main reaction, i.e. the heterogenous catalysis of organosulfur compounds. The experimental setup consists of three modules that perform different functions (Figure 1):

- Module 1: Gas-mixing (explosion-proof)
- Module 2: Heating and reaction
- Module 3: Cooling

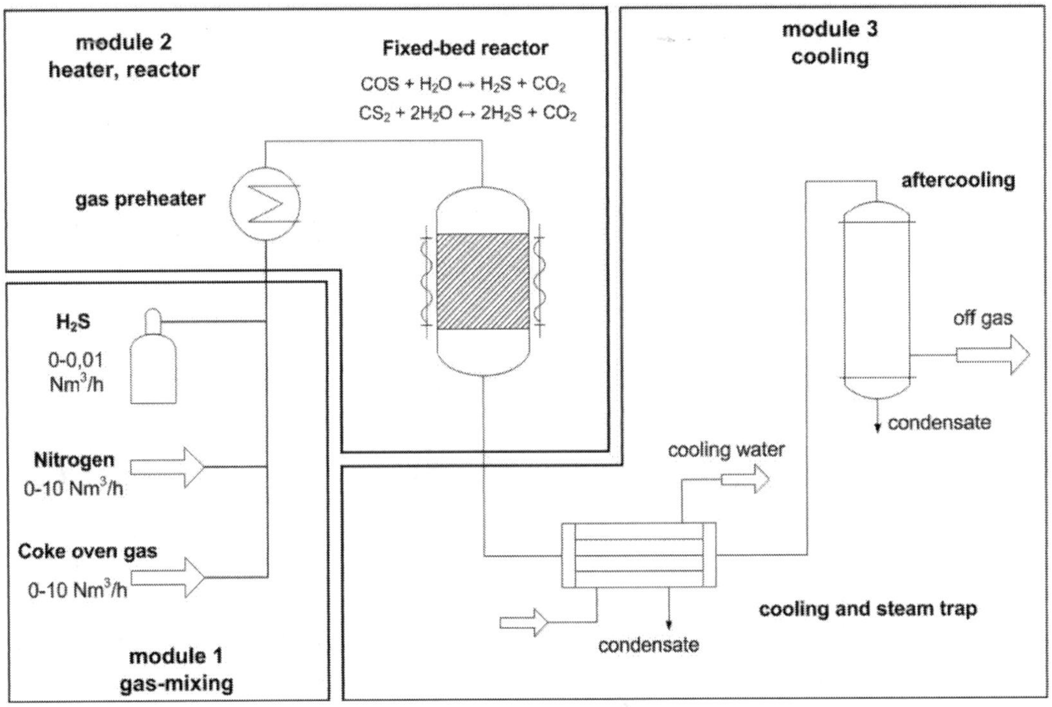

Figure 1: Set-up of the modular test plant.

During the first measurement campaign, that was conducted for 10 weeks, trend-setting results were gained. With the developed test plant it was possible to remove more than 90% of the organosulfur compounds. On this basis, a new series of experiments was planed and conducted. The focus of this second campaign was the detailed understanding of the conversion of the organosulfur compounds COS and CS_2 as a function of the parameters temperature and residence time. The gained data was used to find a suitable model for the reactions. The results are introduced here and will be compared with the previously gained lab experiments. The series of experiments demonstrate that the purification is successful and therefore coke oven gas can be used in alternative processes.

379166 A Novel Fiber Packed Reactor Under Laminar Regime for Biodiesel Production –a Preliminary Effort for Reactor Modeling

Abstract:

In addition to patents of fiber packed-bed fiber reactors (FBPR) [1], current experimental efforts at Tennessee Technological University have contributed to the development of novel technologies for fiber packed-bed fiber reactors. These fiber packed-bed fiber reactors take advantage of the unique features of the fibers that provide an enormous area of contact for two phases without the need of mixing [1]. For this reactor, to our best knowledge, there are no models yet that help predicting their performance. In this research, a collection of models (of increasing complexity) are being proposed and studied in order to predict the process yield and reactor behavior. The strategy is to first develop a relatively simple transport model (under laminar regime) with a relating simple associated kinetics to study the convective-diffusion transport process taking place inside the novel fiber-packed reactors. The novelty lies in the use of a continuum-based approach *(and up-scaling arguments)* with proper assumptions that will help produce a "*single-phase*" model in order to predict the conversion. The focus of the research is on a mathematical approach coupled with computational modeling and will focus on understanding how the different physical elements or components of transport processes and kinetics play a role in the production of biodiesel using either cylindrical or annular based geometry for the reactors. In this regard, limiting regimes will be studied in order to gain understanding of the (more complex) system behavior; both analytical and computational-based approaches will be used to illustrate the model's predictions. Preliminary results of this investigation using fundamental assumptions were obtained in Poiseuille flow by solving the differential equation and getting the asymptotic solutions. The research will lead to progressively the development of multi-scale models useful for the study of the kinetics, socioeconomic, environmental, and other aspects of biodiesel production.

References:

[1] Massingill. J. L.; Patel, P. N.; Guntupalli, M.; Garret, C.; Ji, C. High Efficiency Non-Dispersive Reactor for Two-Phase Reactions. *Organic Process Research & Development.* 12, 2008: 771-777.

381371 Design of Multi-Tubular Fixed-Bed Reactor for Gtl-FPSO Process Based on the CFD Simulation

Abstract:

World fuel consumption is ever increasing, together with the crude oil price on continuous rise. Recently, technology for liquid fuel production has been spotlighted in energy industry, and gas to liquid (GTL) process is one of alternative promising methods of petroleum fuel production technologies. For better use of the offshore platform, this research designs the GTL-FPSO process based on a new concept of combining GTL with floating production storage and offloading (FPSO). Laminating a multi-tubular type of compact fixed-bed reactor could be a first-choice solution. But it usually has problems of catalyst, equipment damage and low yield caused by hotspot and pressure drop.

This work is forcused on development of a spatio-temporal model to analyze the dynamic performance of a multi-tubular fixed-bed reactor (MTFBR), and the CFD simulation of the reactor interior is performed on a CFD tool. To prevent of hotspots, the cooling temperature was investigated different inlet temperature of shell side such as, 500, 520, 525, 530, 540 and 550 K. It was found that the optimal cooling temperature is 530 K. In order to reproduce the marine environment, the sloshing effect in this model was applied.

The optimum design of MTFBR for maximizing the heating and cooling efficiency was suggested by using the optimal cooling temperature and the results of sloshing effect simulation.

388973 Model-Based Design of Experiments to Achieve Kinetic Validation for Chemical-Looping Systems

Abstract:

Model-based design of experiments to achieve kinetic validation for chemical-looping systems

Lu Han, Zhiquan Zhou, George M. Bollas

Abstract

A model-based experimental design approach is applied to design experiments for model discrimination and parameter estimation of an ill-defined model structure. This concept is illustrated for a chemical-looping process, in which understanding of the intrinsic kinetics of the non-catalytic gas-solid reactions is crucial. Solid-state kinetic models are usually derived empirically and supported by advanced characterization techniques. These models can be classified according to their mechanistic basis, such as nucleation, geometrical contraction, diffusion, and reaction order. As these models vary in their degree of complexity, statistical approaches are used to discriminate the overall best-suited model that fits the experimental data. Recently, Zhou et al. used the corrected Akaike Information Criterion (AICc) and the F-test on twenty solid-state kinetic models for describing the reduction NiO by H_2 and oxidation of Ni by air [1]. All of the models were compared against experimental data from the literature and in-house experiments. However, due to subtle differences between the kinetic models, several cases presented by Zhou et al. appear to have multiple winning models. To address this issue, the approach employed in this work utilizes the rival models to design experiments that maximize the divergence of the model predictions. By inspection of the quality of fit, the choice of the winner model (i.e., model discrimination) can be made with improved statistical significance. The method for model discrimination is formulated as an optimal control problem [2]:

$$\boldsymbol{\varphi} = \arg\max_{\boldsymbol{\varphi}\in\boldsymbol{\Phi}} \sum_{k=1}^{n_{sp}} \sum_{l=1}^{N_M} \sum_{l'=l+1}^{N_M} T_{l,l'}\left(\boldsymbol{\varphi},\hat{\boldsymbol{\theta}},\hat{\boldsymbol{\theta}}',t_k\right)$$

$$T_{l,l'}\left(\varphi,\hat{\boldsymbol{\theta}},\hat{\boldsymbol{\theta}}',t_k\right)=\left[\hat{\mathbf{y}}_l\left(\boldsymbol{\varphi},\hat{\boldsymbol{\theta}},t_k\right)-\hat{\mathbf{y}}_{l'}\left(\boldsymbol{\varphi},\hat{\boldsymbol{\theta}}',t_k\right)\right]^T \times \mathbf{W}\left[\hat{\mathbf{y}}_l\left(\boldsymbol{\varphi},\hat{\boldsymbol{\theta}},t_k\right)-\hat{\mathbf{y}}_{l'}\left(\boldsymbol{\varphi},\hat{\boldsymbol{\theta}}',t_k\right)\right]$$

where is the vector of best available estimates of the model parameters, is the design vector, is the output trajectories, is the weighting vector, and

$\hat{\boldsymbol{\theta}}$

$\boldsymbol{\varphi}$

$\hat{\mathbf{y}}_l$

$\mathbf{W}$

t_k

is the discrete sampling time. This problem can be applied for practically any number of N_M rival models and it is not certain which of the models is the best.

Furthermore, it becomes important to decrease the size of the confidence intervals of each of the parameters in the best-suited model. Optimal experimental design is cast as an optimization problem of the control variables that maximizes the sensitivity of the output variables with respect to the model parameters. This is reflected in the Fisher information matrix, **F**:

$$\mathbf{F}=\int_0^{t_f}\left(\frac{\partial \mathbf{y}}{\partial \mathbf{p}}(t)\right)^T \mathbf{Q}\left(\frac{\partial \mathbf{y}}{\partial \mathbf{p}}(t)\right)dt$$

where **y** denotes the model outputs, **p** the set of unknown, parameters , and **Q** the inverse of the measurement error covariance matrix. The objective function is written as:

$\min_{\mathbf{u}(t)} \varphi(\mathbf{F})$,

Subject to: $\mathbf{f}[\mathbf{x},\mathbf{y},\mathbf{p},\mathbf{u},\mathbf{t}] = \mathbf{0}$, $\mathbf{x}(\mathbf{t}_0) = \mathbf{x}_0$

$\mathbf{h}[\mathbf{x},\mathbf{y},\mathbf{p},\mathbf{u},\mathbf{t}] = \mathbf{0}$

$\mathbf{g}[\mathbf{x},\mathbf{y},\mathbf{p},\mathbf{u},\mathbf{t}] \le \mathbf{0}$, $\mathbf{x}^L \le \mathbf{x} \le \mathbf{x}^U$, $\mathbf{u}^L \le \mathbf{u} \le \mathbf{u}^U$

where is a metric of the selected design criterion, **x** is the vector of state variables, **u** the vector of manipulated variables, **f** is the system of ordinary

φ

differential equations, **h** and **g** are equality and inequality algebraic constraints, and U and L are the upper and lower bounds for **x** and **u**. In this work, D-optimality criterion, aimed at maximizing the determinant of the information matrix, is used, as it maximizes the overall information while at the same time decreasing the degree of correlation between parameters [3].

The cases studies under investigation are performed in a bench-scale fixed-bed unit focusing on the reduction of NiO/γ-Al_2O_3-SiO_2 oxygen carrier by CH_4. In this configuration, over 20 redox cycles are typically conducted to analyze the stability of the oxygen carrier. Concerning only the reduction reactions with Ni, sets of parallel experiments are designed to maximize the statistical confidence in the Arrhenius rate constants. The set of manipulated variables include: reduction temperature, bed length, and CH_4 fraction. The experimental designs are evaluated in terms of the statistical quality of the model parameters fitted to the collected data. **Figure 2** shows the experimental results for two different temperatures and corresponding model predictions. The quality of fit is excellent and the estimated kinetic parameters (**Table 1**) are in good agreement with studies. A decrease in statistical uncertainty is achieved with this selection of experiments in comparison to performing a single experiment.

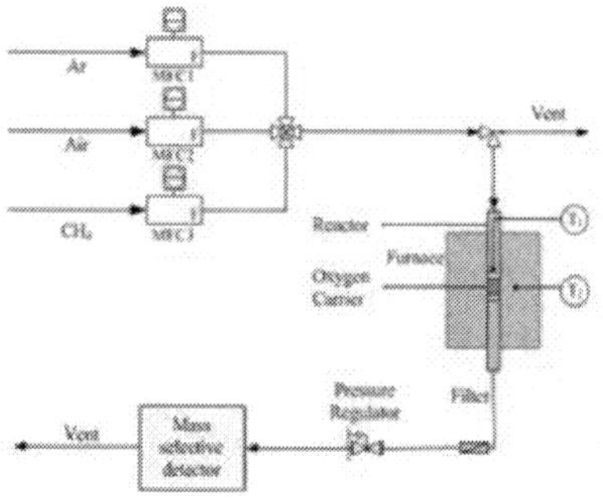

Figure 1: *Experimental set-up*

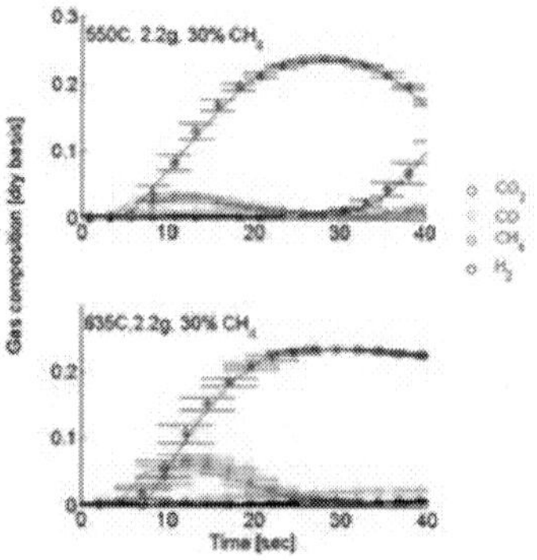

Figure 2: *Comparison of model predictions with experimental data*

Table 1: *Fitted kinetics from optimal experiments*

Reaction	Frequency factor [m/s]	Activation energy [kJ/mol]
CH_4+2NiOà2Ni+CO_2+H_2	1.7E-01	108
H_2+NiOàNi+H_2O	1.7E-03	56
CO+NiOàNi+CO_2	2.4E-03	36
CH_4+NiOàNi+2H_2+CO	6.0E-03	20

Acknowledgement: This material is based upon work supported by the National Science Foundation under Grant No. 1054718.

References

[1] Z. Zhou, L. Han, G.M. Bollas, Kinetics of NiO reduction and Ni oxidation at conditions relevant to chemical-looping combustion and reforming, Int. J. of Hydrogen Energy. 39 (2014) 8535–8556.

[2] S.P. Asprey, S. Macchietto, Statistical tools for optimal dynamic model building, Comput. Chem. Eng. 24 (2000) 1261–1267.

[3] G.E. Box, K.B. Wilson, On the Experimental Attainment of Optimum Conditions, J. R. Stat. Soc. Ser. B. 13 (1951) 1–45.

386239 Characterization of Tungsten and Zirconia Decorated Multi-Walled Carbon Nanotubes

Abstract:

Solid acid catalysts have found a wide range of applications, ranging from cracking of heavy polyols into smaller chains all the way to dehydration reactions that are useful in biomass conversion. The focus of this research is to develop strong solid acid catalysts supported on multi-walled carbon nanotubes (MWCNT) suitable for biomass processing applications. Previous work from our group has shown that zirconia supported on MWCNT is a hydrothermally stable catalyst and it is not altered when exposed to hot liquid water. Furthermore, this system has been sulfated to increase the catalytic acidity. However, it has been demonstrated that the sulfur species is not extraordinarily stable in aqueous media. Due to this fact, we have begun investigating the tungsten modified zirconia analogue that is expected to be more hydrothermally stable while maintaining comparable acidity. We plan a comparison between the sulfur and tungsten modified zirconia supported on MWCNT. The current work focuses on optimizing the synthesis of tungsten modified zirconia supported on MWCNT, investigating the particle size of the oxides when annealed at different temperatures with XRD and TEM, and conducting X-ray absorption spectroscopy on these materials. We are working towards kinetically characterizing the catalysts with a 2-butanol dehydration reaction. XRD and TEM indicate the zirconia particles are in the 2nm size range and the tungsten oxide particles are in the 6nm size range. XAS has shown shifts in the W L_3

the oxidation state of the tungsten. We expect to correlate average oxidation found in the X-ray absorption studies with increased activity in the probe reaction, enabling us to understand catalytic activity in terms of electronic structure.

Enantioselectivity of chiral metal nanostructures grown on chiral oxide surfaces

Simuck F. Yuk and Aravind Asthagiri

William G. Lowrie Department of Chemical and Biomolecular Engineering, The Ohio State University, Columbus, OH, 43210, USA

Email Address: yuk.4@osu.edu; asthagiri@chbmeng.ohio-state.edu

It is well known that biomolecules such as amino acids and proteins exist in homochiral form. Thus, enantiomer of a chiral drug can exhibit different physiological and toxicological properties compared to its mirror-image counterpart. Separating the racemic mixture is, however, not a simple task because enantiomers possess the same physical properties, rendering them indistinguishable for conventional separation processes. One of the recent techniques suggested for separating enantiomers is the use of naturally chiral metal surfaces. Due to the surface chirality imparted by step-kink edges on their surfaces, naturally chiral surfaces have been shown to be effective for enantioselective separation. Furthermore, enantioselectivity of chiral metal nanostructures can be optimized by the presence of heterogeneous electronic structure of underlying chiral oxide substrate.

Using first-principle calculations, we have examined the enantioselectivity of Pt nanostructures grown on the chiral $SrTiO_3$ substrate. In order to first analyze the electronic effects of oxide substrate on deposited metal, we investigated the CO adsorption on the Pt monolayers (MLs) supported on the $SrTiO_3$ (100). At the 1 ML of Pt, the presence of $SrTiO_3$ substrate greatly changes the CO adsorption energy and the geometry of deposited Pt compared to the bare Pt (100). Most importantly, the nature of deposited Pt is heavily dependent on its underlying oxide atoms. The hybridization between particular d orbitals of Pt and CO orbitals and the unique electronic structure of underlying oxide atoms is mainly responsible for these oxide-dependent behaviors of deposited Pt. The effects of oxide substrate significantly diminish after 2 ML of Pt, indicating the importance of deposited Pt thickness for maintaining any potential oxide effect.

As a next effort to understand the enantioselectivity of oxide-supported chiral metal structures, we studied the adsorption of FAM and methyl lactate on the $SrTiO_3$-supported Pt ML in comparison to the bare chiral metal surfaces. We found that the interaction between fluorine atom of FAM and metal surfaces is closely related to the enantioselectivity of chiral metal surfaces. Also, the presence of chiral oxide substrate significantly alters the adsorption energetic of chiral molecules on Pt MLs compared to the bare Pt surfaces. Further studies need to be conducted to properly characterize the electronic nature of oxide-supported chiral metal structures compared to the bare chiral metal structures.

375676 2D Transition Metal Chalcogenide Electrocatalysts

Abstract:

Two-dimensional transition metal chalcogenides (2D TMCs) have recently received great attention for electronic and catalytic applications. Here we report on organic solvothermal synthesis of 2D TMCs and their electrocatalytic applications for the hydrogen evolution reaction (HER). By thermal decomposition of molybdenum carbonyl and subsequent sulfurization with element sulfur, monolayer MoS_2 nanosheets obtained via preferential in-plane growth/attachment. A mild-temperature thermal treatment was developed to remove the organic ligands without changing the chemical composition or damaging the 2D nanostructure. Further high-temperature annealing was shown to induce stacking of the MoS_2 sheets but not inducing phase transition and lateral aggregation. The 2D MoS_2 produced by our approach are highly active for the HER, with the activity found to be further dependent on stacking of the nanosheets.

369199 V2O5-Nb/La-Al2O3 Catalysts for Oxidative Dehydrogenation of Ethane to Ethylene

Abstract:

The aim of this study is to investigate oxidative dehydrogenation (ODH) of ethane to ethylene in absence of gas phase oxygen using multimetallic V_2O_5-Nb/La-Al_2O_3 catalysts. In the catalyst formulation, La is used to modify γ-Al_2O_3, stabilizing bulk phase transformation of γ-Al_2O_3. A small amount of Nb is used as a promoter of the main active V_2O_5 species.

Catalysts were prepared by using an incipient wetness technique. V_2O_5 was loaded on La modfied γ-Al_2O_3 (260 m^2/g). Before V_2O_5 loading the support γ-Al_2O_3 was modified with La. The prepared catalysts were characterized using various physicochemical techniques in order to understand the different aspects of their performances including the reducibility, oxygen carrying capacity, stability and phase transformation under repeated redox cycles.

Temperature programmed reduction (TPR) evaluation shows that Nb-V_2O_5/La-Al_2O_3 catalyst is very stable over repeated reduction and oxidation cycles. In each cycle the amount of hydrogen consumption remains stable approximately at 80 cc/g catalyst. XRD analysis indicates that V_2O_5 on catalysts is an amorphous phase up to 10 % V loading. The pulse chemisorption show a good metal dispersion on the La modified support. SEM and EDX results reveal that La, Nb and V_2O_5 are present on the support surface as mono-vanadate and poly-vanadate in amorphous phase which further confirmed by XRD peak analysis.

The gas phase oxygen free ODH of ethane in the fluidized CREC Riser Simulator displays that approximately 30 % ethane conversion can be achieved in very short contact time reactions. The oxygen free (gas phase) ODH also found favorable to obtain high ethylene selectivity (upto 85 %). The Nb-V_2O_5/La-Al_2O_3 catalyst also shows stable performance in the repeated ODH-regeneration cycles which is consistent to cyclic TPR/TPO results.

Acknowledgement: We acknowledge the financial support by KACST research grant number ARP-30-252

388660 Pd-Impregnated Hyperbranched Aminosilica Materials As Hybrid Catalysts for the Selective Production of Olefins from Alkynes

Abstract:

The partial hydrogenation of alkynes to alkenes is an important class of chemical transformations with wide application in industry, for example in the elimination of alkynes from alkene feedstocks or the preparation of cis-olefins of high value. The Lindlar catalyst ($Pd/CaCO_3$ poisoned with $Pb(OAc)_2$) along with base additives (eg. quinoline) is commonly used in the semihydrogenation of alkynes or dienes. However undesirable lead compounds and other additives are required to achieve good selectivity, thus it is desirable to develop alternative, green catalysts for these catalytic transformations. In liquid phase hydrogenations, N or S donor ligands are often used as Pd modifiers to achieve improved selectivity. To this end, we sought to use hybrid silica-aminopolymer composite materials as supports for Pd centers in these reactions and a new aminopolymer-silica composite supported Pd catalyst was developed for the selective hydrogenation of alkynes to alkenes. Hyperbranched aminosilica (HAS) materials, which are prepared via the in-situ ring-opening polymerization of aziridine on porous silcia supports and were originally developed as carbon dioxide adsorbents,[1] were used as ligand/modifier-containing supports for the Pd species. The HAS solids were then impregnated with $Pd(OAc)_2$, followed by reduction to form Pd(0) nanoparticles wthin the aminopolymer matrix for use in catalytic tests. The resulting Pd-HAS and related materials were applied as catalysts using several different substrates, yielding good selectivity in the conversion of alkynes to alkenes in all cases, as well as excellent cis-alkene selectivity for disubstituted substrates. The catalysts were characterized in detail with multiple techniques to shed a light on the naure of the Pd species in the material. Recycle studies of the catalysts showed a negligible amount of palladium leaching during reaction, and the catalysis appeared to be occurring on the solid material.

Title:

Electrochemically Responsive Heterogeneous Catalysis for Controlling Reaction Kinetics

Authors:

Xianwen Mao[1], Wenda Tian[1], Jie Wu[2], Gregory C. Rutledge[1]*, and T. Alan Hatton[1]*

[1]Department of Chemical Engineering, [2]Department of Chemistry, Massachusetts Institute of Technology, 77 Massachusetts Avenue, Cambridge, MA 02139

Keywords:

heterogeneous catalysis, stimulus-responsive, electrochemical control, reaction kinetics

Abstract:

The ability to exert control over an entity or a process is arguably the ultimate demonstration of our understanding of that process, and enables the full exploitation of its potential. Chemists have recently begun to incorporate control elements into catalyst design in response to an increasing interest in responsive catalytic systems. Such systems enable new strategies for the modulation of reaction kinetics using various chemico-physical stimuli. The key to achieving stimuli-controlled catalysis is the development of a system in which the concentration or accessibility of the catalytic site in reaction media can be flexibly adjusted in response to external signals.

We report a conceptually new method to control reaction kinetics using an electrochemically responsive heterogeneous catalysis (ERHC) system consisting of a framework of conductive carbon fibers coated conformally with redox-switchable organometallic catalysts, prepared by a facile and controllable electrochemical deposition method. We show experimentally that the kinetics of a Michael addition reaction can be modulated continuously through changes in the applied voltage, while COMSOL simulations indicate that precise spatial and temporal control of reaction kinetics in a pack-bed-like flow reactor is possible. These factors suggest that under certain conditions the ERHC system could have distinct advantages over pH-, solvent-, or temperature-responsive catalytic systems based on soft materials.

Synthesis of stereostructured iron phosphate crystals as novel heterogeneous Fenton/ heroxidase like catalyst

Tongbao Zhang, Yangcheng Lu[*]*, Kai Wang and Guangsheng Luo*

The State Key Laboratory of Chemical Engineering, Department of Chemical Engineering, Tsinghua University, Beijing 100084, China

The synthesis of inorganic materials with controllable morphology and structure has gaining expansive research interest due to some fascinating structure and shape dependent properties with respect to applications, especially in catalysis. Iron phosphate compounds are a large family of inorganic materials with good catalytic activity, abundant resources and environmental friendliness. Most importantly, they can be endowed with variety of stereostructure and crystalline structure. These are of great importance for wide application and stable catalytic performance. However, the report on the synthesis of stereostructured iron phosphate crystals is limited and research on developing such material with unique structure as novel heterogeneous catalyst is highly desirable. Herein, two kinds of stereostructured iron phosphate crystals, (single-crystal dendritic iron hydroxyl phosphate $Fe_{1.18}\square_{0.82}(PO_4)(OH)_{0.54}(H_2O)_{0.46}$ (□ representsvacancy) and hierarchical iron hydrogen phosphate $Fe_3H_9(PO_4)_6(H_2O)_6$ consisted of nanosheets) were successfully synthesized through a facile one-step hydrothermal procedure. The products were systematically characterized in terms of morphology, crystal structure and composition. The formation mechanisms for the two unique stereostructured iron phosphate crystals were proposed as self-template process for the former and self-assembly process for the latter with informative evidence. Then, single-crystal dendritic iron hydroxyl phosphate was tested as a heterogeneous Fenton catalyst for the degradation of phenol, which presented numerous advantages including remarkable catalytic activity in a wide pH range (e.g., pH= 3-8), outstanding cycle performance (no observation of activity loss in five runs), and extremely low leaching iron concentration (below 0.5 mg L^{-1}). Moreover, hierarchical iron hydrogen phosphate crystals were first found to own peroxidase-like activity in pure chemically stable iron (III) salt crystals through colormetric reaction of 3,3',5,5'-tetramethylbenzidine (TMB) in the presence of

* Corresponding author, luyc@mail.tsinghua.edu.cn

H_2O_2. The property was further used for the detection of H_2O_2, which showed particular good reproducibility (RSD= 1.95% for 10 independent measurements), long-term stability (no activity loss after 10 cycles) and easily of recovery (by simple centrifugation).

Keywords: stereostructured morphology; iron phosphate; crystals; Fenton catalysis; peroxidase mimic; hydrothermal method

Methodology for virtual optimization of pore structure of catalyst support is described. The method combines 2D or 3D imaging of catalyst nanostructure, computer reconstruction of the structure and multi-scale mathematical modelling leading to computational evaluation of the pore size distribution, effective diffusivity and tortuosity in the reconstructed system. Experiments and modelling of the catalytic layer for CO and decane oxidation are compared.

Literature:

J. Kosek, F. Štěpánek, M. Marek, Modeling of transport and transformation processes in porous and multiphase bodies, *Adv. Chem. Eng.* 30 (2005) 137-203

V. Novák, P. Kočí, M. Marek, F. Štěpánek, P. Blanco-García, G. Jones, Multi-scale modelling and measurements of diffusion through porous catalytic coatings:an application to exhaust gas oxidation. *Catal. Today* 188(2012) 62-69

M. Dudák, V. Novák, P. Kočí, M. Marek, P. Blanco-García, G. Jones, Influence of coated $Pt/\gamma\text{-}Al_2O_3$ catalyst morphology on diffusivity and conversion of n-decane and CO, *Appl. Catal. B: Environ.* 150-151 (2014) 446-458

Novák V., Ortel E., Winter B., Butz B., Paul B., Kočí P., Marek M., Spiecker E., Kraehnert R. Prototyping of catalyst pore-systems by a combined synthetic, analytical and computational approach: Application to mesoporous TiO_2. *Chemical Engineering Journal* 248 (2014), 49-62.

Modification of supported metal catalysts with alkanethiol self-assembled monolayers

Rudy Kahsar, Dan Schwartz, and Will Medlin
Department of Chemical & Biological Engineering, University of Colorado, Boulder, CO, USA
karl.kahsar@colorado.edu
daniel.schwartz@colorado.edu
will.medlin@colorado.edu

To bias a supported metal catalyst to produce the less energetically favorable product, we deposit alkanethiol self-assembled monolayers (SAMs) on the surface of supported metal catalysts. These SAMs form a well-ordered structure on the metal catalyst surface, which provide a functional handle for tuning the near surface environment. Previous research has demonstrated the usefulness of these surface modifiers for enhancing the selectivity of a variety of hydrogenation reactions.

For example, our group has previously demonstrated the efficacy of SAMs by modifying Pd/Al_2O_3 for the selective hydrogenation of 1-epoxy-3-butene (EpB) [1,2]. The selectivity of this system was shown to be uniformly responsive to the presence of sulfur on the catalyst surface. An additional lever for tuning reaction selectivity was demonstrated for the hydrogenation of cinnamaldehyde, an α,β-unsaturated aldehyde, over a Pt/Al_2O_3 catalyst. Selective hydrogenation of cinnamaldehyde was performed using phenylated SAMs where interaction between the phenyl ring of the reactant (the large terminal end of the cinnamaldehyde molecule) and the SAM was used to direct hydrogenation of either the double bond or the aldehyde.

Here we will discuss ongoing work of how thiol SAM modifications to the near surface environment can be used to direct reactions. Control of the near surface environment of a catalyst system, as studied here, provides a unique, interesting, but not yet fully understood method for controlling the reaction environment.

References

1. Marshall, S., O'Brien, M., Oetter, B., & Corpuz, A. (2010). Controlled selectivity for palladium catalysts using self-assembled monolayers. Nature Materials, 9
2. Kahsar, K. R., Schwartz, D. K., & Medlin, J. W. (2012). Liquid- and vapor-phase hydrogenation of 1-epoxy-3-butene using self-assembled monolayer coated palladium and platinum catalysts. Applied Catalysis A: General 445-446

374576 Reaction Mechanistic Studies of NO Reduction By H2 on Transition Metal Surfaces

Abstract:

Nitric oxide (NO) is a major pollutant produced mainly during the combustion of fossil fuels. Though three-way catalysts (TWC) and selective catalytic reduction (SCR) have been widely applied to decrease NO_x emissions, the fundamental reaction mechanism is not well understood. Disadvantages associated with those methods also stimulate the development of other promising NO removal processes. In this study, we employ periodic, self-consistent density functional theory (DFT-GGA) calculations to investigate the reaction mechanism of NO reduction by H_2 on transition metal surfaces. In particular, we analyze the energetics of both direct NO dissociation and H-assisted pathways, elucidating the role of H in NO activation. Possible pathways leading to the formation of N_2, the most desirable product, and side-products N_2O and NH_3 are also proposed. Due to the strong binding of NO, we incorporate a high coverage of NO into our studies and show how NO spectator molecules affect the reaction mechanism. Comparisons among different transition metal surfaces are made to elucidate trends in catalytic activity and selectivity, as well as the structure sensitivity of NO reduction by H_2. Kinetic experiments and microkinetic modeling are used, in combination with first-principles analysis, to gain an improved understanding of the reaction network.

References

[1] R. M. Heck, R. J. Farrauto, S. T. Gulati, *Catalytic air pollution control: commercial technology*, 3rd ed., John Wiley & Sons, New Jessey, **2009**.

[2] J. L. C. Fajin, M. N. D. S. Cordeiro, J. R. B. Gomes, *Journal of Physical Chemistry C* **2009**, *113*, 8864.

[3] H. G. Stenger, J. S. Hepburn, *Energy & Fuels* **1987**, *1*, 412.

[4] T. P. Kobylinski, B. W. Taylor, *Journal of Catalysis* **1974**, *33*, 376.

374856 Si-Rich and Ti-Rich (Mn, Cu and Ce) Catalyst Formulations for the Selective Catalytic Reduction (SCR) of NOx with NH3: Surface Properties and Key Components in Relation with Activity of NOx Reduction

Abstract:

A series of M/TiO_2 and M/TiO_2-SiO_2 (with M = Mn, Cu and Ce) catalysts were prepared by adopting a wet-impregnation method and investigated for the selective catalytic reduction (SCR) of NO_x in the temperature range of 100-500 °C with excess (10 vol.%) oxygen. Our XPS depth profile analysis illustrate that the surface atomic ratio of Cu^{1+}/ Cu^{2+} greatly enhanced with increase in TiO_2 content in the TiO_2-SiO_2 support, and these results are consistent with the H_2-TPR results where the additional reduction peak evolved at 200 °C for copper loaded titania-rich (Cu/TiO_2) catalyst. The high activity of Cu-based TiO_2 formulations has been assigned to the enhancement in the formation of Cu^{1+} active sites, existence of surface Cu^{2+}, Cu^{1+} species and the increment of reduction potentials of surface copper species. The Ce^{3+}/Ce^{4+} and Ce^{3+}/Ce^{n+} atomic ratio (1.14 and 0.53, respectively) in Ce/TiO_2 catalyst calculated from deconvoluted XPS spectra are much higher than that of Ce/TiO_2-SiO_2 (1:1) and Ce/TiO_2-SiO_2 (3:1). The existence of higher Ce^{3+} surface species over CeO_2/TiO_2 illustrates the increment of surface oxygen vacancies and thus facilitates for the adsorption of oxygen species or activates reactants in the SCR reaction. When ceria was supported on pure TiO_2, the low-temperature reduction peak was broad and less defined, and the reducibility in the low temperature range was much less pronounced. On the other hand, the addition of ceria to titania with strong reciprocal interaction is generally perceived as a shift in the bulk reduction temperature to lower values, to about 500–650 °C. As bigger Ce^{4+} ions enter the lattice structure to proxy the Ti^{4+} ions with smaller ionic radii (248 and 215 pm, respectively), the lattice could become highly strained. The NO_x conversions and the apparent kinetic constant of the catalyst k_{ac}, over the Cu, Mn, Ce-loaded on different supports TiO_2 and TiO_2-SiO_2 (3:1 and 1:1) catalysts measured under steady-state conditions results demonstrated higher activity of the Ti-rich materials.

384646 Controlling Selectivity of Catalytic Reactions on Metal Nanoparticles through Direct Photoexcitation of Targeted Metal-Adsorbate Bonds

Abstract:

Controlling selectivity of chemical reactions on metal surfaces is of paramount importance for the design of efficient, environmentally friendly heterogeneous catalytic processes. Governing relationships between adsorption energies and activation barriers of competing chemical pathways on metal surfaces limit the potential to control selectivity when using thermal energy to drive reactions. It has recently been shown in a few cases that visible light irradiation of metal catalysts can overcome inherent limitations of thermal-driven catalysis, inducing new chemical pathways through electronically excited states. However these photon driven reactions typically occur through a substrate-mediated transfer of photon energy to adsorbate-metal bonds, where the initial photon absorption occurs in the bulk metal states, followed by charge transfer to adsorbates. Since only the bulk metal is involved in the initial photoexcitation step, the wavelength dependent quantum yields follow the absorption spectrum of the metal, and targeted activation of adsorbate-metal bonds (which is required to control the outcome of metal catalyzed reactions) cannot be controlled in a rational manner.

In this work, we show that catalytic processes on metal surfaces can be driven and controlled through direct, resonant photoexcitation of hybridized electronic states formed due to strong chemisorption of CO on Pt surfaces. As opposed to the substrate-mediated mechanism, where photons are absorbed in the metal bulk and energy is subsequently transferred to adsorbate-metal bonds, direct photoexcitation of adsorbate-metal bonds occurs at surface Pt atoms that are directly bound to adsorbates. Pt nanoparticle size dependent studies showed that activation of Pt-CO bonds by direct photoexcitation becomes the dominant mechanism driving chemistry, rather than the substrate-mediate mechanism, on nanoparticle catalysts <5 nm in diameter due to increased surface area to volume ratios of the particles.

By exploiting the resonant, adsorbate specific nature of electronic transitions formed between adsorbates and metal surfaces we show that targeted photoexcitation of catalytic 2.3 nm Pt nanoparticles with 450 nm photons enhances selectivity from 40 to 80% in the preferential oxidation of CO in an H_2 rich environment. It is expected that the development of insights into resonant photon induced electronic transitions between hybridized metal-adsorbate states should allow rational control of catalytic selectivity that cannot be achieved exclusively with thermal energy input.

Molecular Design of Advanced Supported ReO_x/Al_2O_3 Olefin Metathesis Catalysts

Soe Lwin and Israel E. Wachs
***Operando* Molecular Spectroscopy and Catalysis Laboratory**
Department of Chemical Engineering
Lehigh University, Bethlehem, PA 18015, USA

Olefin metathesis interconverts C=C bonds in hydrocarbons and can be tailored to produce an alkene of any desired length. Its versatility opens up a whole new series of reactions and results in numerous industrial applications that produce important petrochemicals, oleochemicals, polymers and specialty chemicals. Heterogeneous alumina and alumina-silica supported rhenium oxides (ReO_x) are extensively investigated due to their catalytic activity at near room temperature. Despite decades of catalysis research, the exact nature of catalytic active sites is not well understood because of the absence of detailed knowledge at molecular level and lack of modern characterization techniques in the past.

Extensive *in situ* and *operando* Raman, UV vis and IR studies, steady-state kinetics and temperature programmed surface reaction (TPSR) experiments with deuterium labeled olefins were undertaken for the first time to obtain unprecedented insights about the catalytic active sites (their anchoring sites, coordination and oxidation states) and surface kinetics to unravel the fundamental molecular structure-reactivity relationships. These new fundamental insights enabled the molecular design of advanced supported ReO_x/Al_2O_3 olefin metathesis catalysts.

390994 Copper Based Catalysts with Balanced Active Sites for Carbon-Oxygen Hydrogenolysis Reactions

Abstract:

Hydrogenolysis of carbon-oxygen (C–O) bonds (e.g., esters, ethers, furfural, and CO_2) has emerged as a versatile synthetic tool in organic methodology, as it could produce a variety of products (e.g., chemicals, fuels, and polymers).1 Copper-based catalysts have been intensively explored for hydrogenation reactions as the copper sites account for the selective hydrogenation of carbon-oxygen bonds and relatively inactive for the hydrogenolysis of carbon-carbon bonds. Despite extensive studies of C-O hydrogenolysis reactions, there is still a debate regarding the nature of active sites of Cu-based catalysts. Work on understanding the active sites of copper catalysts for hydrogenation of dimethyl oxalate (DMO),2 furfural, CO_2 indicted that both Cu^0 and Cu^+ species were crucial to the activity of Cu-based catalysts. However, researchers were not able to establish a relationship between activity and Cu^0/Cu^+ active species since the strong reductive H_2 and oxidising C=O make the Cu^0 and Cu^+ species unstable in the reaction.

We renctly present on efficient approaches for fabrication of a series of nanocatalysts and nanoreactors with stable and balanced Cu^0 and Cu^+ active species. We chose the hydrogenation of DMO (a highly exothermic and heat-sensitive system), and hydrogenation of CO_2(the symmetrical molecule with C=O bond) as probe reactions for a better understanding of properties of active sites and the structure-activity relationship.4 The results indicated a synergy between copper and the oxide components and the balanced surface Cu^0 and Cu^+ species can greatly improve the catalytic C-O hydrogenolysis performance of the catalysts.

In light of petroleum depletion and an increasing emission of greenhouse gases, there is an increasing interest in the use of renewable chemical feedstocks generated from lignocellulosic biomass. In particular, the conversion of lignocellulosic-derived furfural into 2-methylfuran is highly desirable, due to 2-methylfuran's viability as a chemical feedstock in a variety of industrially relevant processes. One promising conversion pathway is through the selective hydrodeoxygenation (HDO) of the furfural carbonyl group via a furfuryl alcohol intermediate. Recently, our group have shown a Ru/RuO_x catalyst to be a promising candidate for the selective HDO of furfural. However, the detailed reaction mechanism has yet to be established.

In this work, we seek to gain a mechanistic understanding of furfural HDO using plane-wave periodic slab density functional theory (DFT) calculations carried out using VASP and GPAW tools. Since experimentally it is established that it is a combination of metal and Lewis acid sites that is most effective in HDO, we model both Ru(0001) and $RuO_2(110)$ as possible constituents of the catalyst active site, and discuss their relative importance in the overall process. We examine three different chemistries: (1) C-H bond formation to hydrogenate the carbonyl group of furfural and weaken the C-O bond (Ru site); (2) C-O bond scission in the functional group through both direct and ring-activated pathways (Ru and RuO_2 sites); and (3) oxygen removal from the catalyst surface to maintain a sufficient number of vacant sites for high catalyst activity (Ru and RuO_2 sites). Using climbing image nudged elastic band (CI-NEB) DFT calculations, we estimate the highest barriers for the subsets 1 and 3 to be in the range of 1.0-1.2 eV. For the subset mechanism 2, we find that the direct pathway is preferred over the ring-activated pathway on Ru metallic sites, whereas on RuO_2 the ring-activated C-O bond hydrogenolysis is the most favorable. The mechanism is discussed in terms of the high oxygen affinity of Ru metal, and ways to improve the catalyst performance are proposed on the basis of electronic structure theory.

DFT and Microkinetic Modeling of *p*-cresol Hydrodeoxygenation on Pt(111)

Geun Ho Gu and Dionisios G. Vlachos

Depleting fossil-fuel resources combined with increasing energy demand and CO2 emissions drive the development of sustainable liquid fuels and chemicals. The only sustainable source of carbon is currently biomass, which can be converted to bio-oil through pyrolysis. However, the bio-oil contains high oxygen content that makes it incompatible with the petroleum infrastructure and unstable. The oxygen can be removed as water by hydrodeoxygenation (HDO) on a heterogeneous catalyst. The HDO of lignin portion of the bio-oil has been investigated experimentally using phenolic compounds as model molecules, but the reaction mechanism remains unknown due to the complexity of the reaction network.

To better understand the reaction network and identify the rate-limiting step, we employ density functional theory (DFT) combined with microkinetic modeling to simulate *p*-cresol, a model phenolic compound, on Pt(111). We show that the reaction network, without ring opening, consists of numerous reactions. We develop a scheme to prune the reaction network. With this reduced model, we are able to reproduce experimental results and demonstrate that *p*-cresol converts to toluene through a sequence of consecutive reactions: hydrogenations, dehydroxylation, and dehydrogenations. We find that a number of these reactions are partially-equilibrated and thus, the reaction network can considerably be simplified. The rate-determining step in the reaction network was identified to be the dehydroxylation reaction suggesting that acid catalyst can be utilized to decrease the dehydroxylation activation barrier and increase the overall activity. For this, we discuss the mechanism and reaction barriers on acid catalysts. Key reactions that control selectivity will also be discussed.

Reaction Path Analysis for CO Hydrogenation during Fischer Tropsch Synthesis: Role of Water

G. T. Kasun Kalhara Gunasooriya, Arghya Banerjee, Mark Saeys

Fischer-Tropsch (FT) synthesis has recently regained interest as a technology to convert a wide range of resources to long-chain hydrocarbon fuels. The detailed reaction mechanism, a complex combination of C-C bond formation, C-O scission and hydrogenation steps, remains intensely debated. In this study we compare different CO hydrogenation pathways, and analyze the experimentally observed kinetic role of water [1]. To accurately model the reactions under FT conditions, a realistic surface coverage needs to be taken into account. Two stable CO surface structures with coverages of 1/3 and 7/12 ML, respectively, have been observed on Co(0001) single crystals. Both the stability and the phase transition between the two stable phases are accurately described by DFT-revPBE-vdW-DF calculations [2]. The activation of C-O bonds can follow various reaction pathways. In addition to direct C-O dissociation, CO can be activated by insertion into growing hydrocarbon chains (CO insertion mechanism [3]), or through hydrogenation (hydrogen-assisted CO activation). Direct hydrogenation of the oxygen atom in CO and RCO however has a high barrier, 193 kJ/mol, while hydrogenation of the carbon atom leads to oxygenates. In addition to the direct hydrogenation pathway, a proton shuttling mechanism has been proposed [4]. In this mechanism, a water molecule acts as a proton shuttle and assists by bridging the distance between the surface hydrogen and the oxygen atom. However, the entropy cost to introduce a water molecule in the transition state increases the free energy barrier for this pathway to 162 kJ/mol under FT conditions on Co(0001). In this study we propose an alternative pathway where surface OH groups act as the source of hydrogen. The formation of OH groups from gas phase H_2O is slightly favorable under FT conditions with a free energy of -3 kJ/mol. Again, DFT- revPBE-vdW is necessary to describe the stability of surface OR groups in agreement with recent experiments [5]. The surface OH groups are found to be very active hydrogenating species, and the free energy barriers to hydrogenate the oxygen atoms in CO and in RCO are only 119 and 88 kJ/mol, respectively, opening a novel pathway to activate C-O bonds.

References

[1] Hibbitts, D.D., Loveless B.T., Neurock M., Iglesia E. *Angew. Chem.* **2013,** 52, 12273
[2] Gunasooriya, G.T.K.K., Saeys, M. in preparation
[3] Zhuo, M., Tan, K.F., Borgna, A., Saeys, M. *J. Phys. Chem. C* **2009,** *113,* 8357
[4] Zhao, Y.F., Yang, Y., Mims, C., Peden, C.H.F., Li, J., Mei, D. *J. Catal.* **2011**, 281, 199
[5] Lew, W., Crowe, M.C., Campbell, C.T., Carrasco, J., Michaelides, A. *J. Phys. Chem. C* **2011**, *115*, 23008

Single-Event Kinetic Model for Cracking and Isomerization of 1-Hexene on ZSM-5

T. von Aretin and O. Hinrichsen, Technische Universität München, Catalysis Research Center and Chemistry Department, D-85748 Garching b. München, Germany

Both ethene and propene offer a wide range of applications for the production of synthetic materials. They are mainly produced by steam cracking of naphtha or catalytic cracking of crude oil into petroleum fractions. Since the demand for propene is growing faster than for ethene, the production of low olefins on demand offers economic advantages. In this context the catalytic cracking of higher olefins on acid zeolites allows a process design with high propene yield. To obtain further mechanistic insights into the reaction pathways occurring during catalytic cracking of olefins, a kinetic model of the reaction is developed. Due to the large number of reaction pathways leading to an experimentally observed product distribution, the single event kinetic (SEK) modeling approach is applied [1].

Model Development:

The reactivity on the acid sites of the zeolite follows the well known carbenium ion chemistry [2]. Thus, the observed elementary reactions are (de)-protonations, alkyl shifts, hydride shifts, PCP-branchings, which proceed through a protonated cyclopropane (PCP) ring as the transition state, cracking reactions *via* beta scission as well as dimerization as the reverse reaction of cracking.

To incorporate this reactivity of olefins into a kinetic model, a reaction network including all possible elementary reactions on the acid catalyst is generated. In the reaction network every product has also been considered as a reactant, thus ensuring that all reaction possibilities are taken into account. A matrix notation, which ascribes a unique Boolean relation matrix to every olefin and carbenium ion, allows the computer aided generation of elementary reactions on the zeolite independent of the considered olefin. Using the concept of rate determining steps, these elementary reactions are combined to reaction pathways. Thereby, a reaction pathway comprises several elementary reactions leading from gas-phase educts *via* a rate determining step on an acid site to gas-phase product olefins.

To handle the large number of parameters necessary for the kinetic description of this reaction network, the SEK model is applied [3]. Using the single event approach the rate constant for every elementary reaction is divided into two parts. The first is an enthalpic part accounting for the different types of carbenium ions involved, e.g. secondary or tertiary ions. This procedure subdivides the elementary reactions into different reaction families, which are described by only one rate constant depending on the reactant and product carbenium ion. The second part of the rate constant accounts for the structure of the molecules and considers the entropy difference between reactant and transition state. Thus, multiple elementary reactions are comprehended by a limited amount of kinetic parameters.

Reaction pathways:

A test case with experimental data from literature [4] is used to show the suitability of the SEK model for the description of the complex reactivity observed in olefins cracking. The catalytic cracking of 1-hexene on ZSM-5 is considered, because a detailed product spectrum depending on conversion in the absence of deactivation is given [4]. In the literature, experimentally observed reactions include cracking and isomerization of hexenes as well as dimerization reactions forming dodecenes [4]. These dodecenes can undergo numerous isomerization and subsequent cracking reactions, leading to the observed product spectrum.

A reaction network for cracking and isomerization of 1-hexene describing these reaction pathways is generated with dodecenes being the largest considered olefins [4]. The influence of the catalyst on the reactivity is included into the model *via* steric constraints. Thus, compared to the free gas-phase chemistry reaction pathways are excluded from the reaction network. According to experimental observations [4] olefins with a high degree of branching or containing quaternary carbon are not observed due to pore size restrictions. Thus, the corresponding reaction pathways are excluded and not considered for calculating the reaction rates.

The rate constants for cracking and dimerization are not independent from each other since these two pathways are forward and reverse reactions. Thus, the single event rate constants for dimerization are expressed *via* the corresponding single event rate constants for cracking and an equilibrium condition to ensure microscopic reversibility of the reaction pathways.

Results:

Rate equations for every pathway and, based on these, the rates of formation for all olefins are set up according to the SEK model [1]. Since physisorption and protonation are regarded to be in quasi-equilibrium, the rate determining steps are methyl shift, PCP-branching, cracking and dimerization. Hereby, dimerization pathways *via* secondary educt and tertiary product carbenium ions are found to be excluded from the reaction network due to steric constraints in the pores of ZSM-5.

In total the kinetic model is based on 144 methyl shift, 784 PCP-branching, 141 cracking and 119 dimerization reactions with 593 different olefins being considered. Ratios of single event rate constants allowing to reproduce the experimental product distribution depending on conversion are estimated with a Levenberg-Marquardt algorithm [4].

The model predictions of 20 product olefins and the combined $C_7^=$ - $C_{12}^=$ olefins are in good agreement with the corresponding experimental mass fractions. Thus, the SEK modeling approach is capable of describing the complexity of the reaction pathways occurring during cracking and isomerization of 1-hexene on ZSM-5.

The advantage of applying the SEK modeling approach is the small number of parameters necessary to describe this complex reactivity. With quaternary carbon atoms being excluded from the reaction network, cracking reactions with tertiary product carbenium ions become impossible. The same holds for methyl shift reactions involving tertiary carbenium ions as well as for PCP-branching reactions between tertiary educt and product carbenium ions. Thus, the number of single event rate constants to be determined from the experimental data is reduced to six.

Insight into the ratios of reaction rates for different reaction pathways leading to the detailed experimental product distribution on ZSM-5 is obtained from the estimated single event rate constants. To describe the reaction rates of olefins on the zeolite, the concept of rate determining steps and quasi-equilibrated elementary reactions in the different reaction pathways can be used. Hereby, the complex product spectrum on ZSM-5 with experimental data points for 21 different species is reproduced by kinetic modeling based on the proposed reaction pathways. The agreement between model and experiment is exemplarily shown for two different olefins in figure 1.

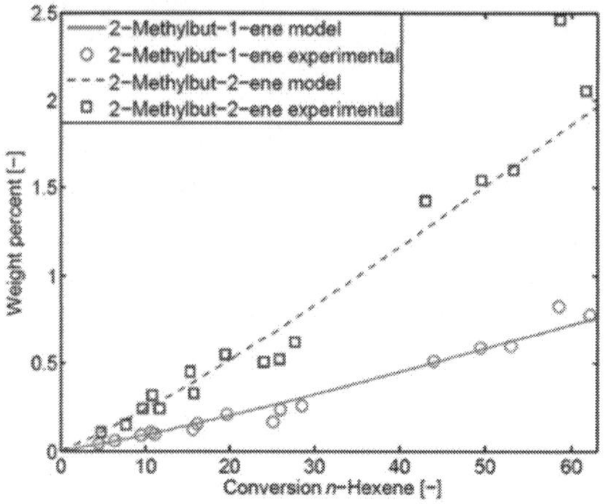

Figure 1: Comparison between experimental product distribution [4] and model predictions for 2-methylbutenes

Outlook:

The aim of further work on olefins cracking is to link catalyst design and product distribution. Therefore, catalyst descriptors like the acid strength of the zeolite shall be included into the kinetic model. In order to enhance the rate of reaction pathways leading to an increased propene yield, the effect of acid strength on the product distribution shall be investigated. By accounting for the temperature dependency of the reaction rate, the influence of experimental conditions on reaction pathways forming ethene or propene shall be studied. Thereby, the reactor model will be used for optimization purposes to increase the yield of lower olefins in the catalytic cracking on ZSM-5.

[1] Thybaut, J.W., Marin, G. B., Baron, G. V., Jacobs, P. A. and Martens, J. A., J. Catal. 202:324-339, 2001

[2] Brouwer, D.M., *Reactions of Alkylcarbeniumions in Relation to Isomerization and Cracking of Hydrocarbons,* Prins, R. and Schuit, G.C., Eds., *Sijthoff & Noordhoff, 1980, pp 137-160*

[3] Vynckier, E. and Froment, G.F., *Kinetic and Thermodynamic Lumping of Multicomponent Mixtures,* Astarita, G. and Sandler, S.I., Eds., *Elsevier, 1991, pp 131-161*

[4] Abbot, J. and Wojciechowski, B.W., Can. J. Chem. Eng. 63:451-461, 1985

364930 Experimental Study on Catalytic Pyrolysis Mechanism of n-Pentane over HZSM-22 Zeolite

Abstract:

Since 1950s, scientists have begun the catalytic pyrolysis research and paid attention to the application of catalytic pyrolysis in the heavy oil process. Some relative patents were sequentially published in many countries. Up to now, lots of reports about catalysts and technologies on catalytic pyrolysis have been studied and developed. Catalytic pyrolysis, which combines a catalytic cracking on catalyst and thermal pyrolysis reactions in catalyst void, has the advantages of both catalytic cracking and thermal pyrolysis. It can raise the yields of light olefins, expand the flexibility of products distribution, and simultaneously lower reaction temperature and decrease energy consumption for the whole system, thus it has broad application prospect. Zeolite-based catalyst have been introduced into the catalytic pyrolysis of hydrocarbons and it shows excellent selectivity to light olefins. Catalytic pyrolysis starts a new director to explore the cracking of hydrocarbons, however, the research about catalytic pyrolysis is still not enough, especially the lack of mechanism research.

In this research,the catalytic studies were performed on an homemade reactor with an internal diameter (ID) of 6 mm, packed with 200-400 mg of HZSM-22 (particle size in the 20-30 mesh range and diluted with SiC) at 600-900 °C and 1 atm. The catalyst bed length was fixed to be 200 mm long by adjusting the SiC loading. The reactor feed gas was a blend of n-pentane and nitrogen and the mole ratio was about 1:50. An Agilent Micro 3000GC equipped with three channels was used to collect chromatogram every 4 min.

This has permitted, thanks to design of experiments, to show the importance of temperature on n-paraffin catalytic pyrolysis mechanism. Through a series of n-pentane pyrolysis on HZSM-22 over a comprehensive temperature range, we find that the steady-state rate of n-pentane conversion has been related to catalytic activity of HZSM-22 and the reaction temperature. Low-temperature domain (below 700 °C) that thermal pyrolysis does not occur, HZSM-22 shows excellent catalytic activity for n-pentane. It is shown that the apparent activation energy of the overall n-pentane transformation decreases to about 115 $kJ \cdot mol^{-1}$ mol less than half of pure thermal pyrolysis. As temperature rising above 700 °C, emergence of thermal pyrolysis can accelerate the conversion of n-pentane and the rate is higher than the sum of pure catalytic cracking and thermal cracking at the same temperature. The catalytic/thermal cracking synergy in HZSM-22 has been studied. The results have revealed, for the first time, the detailed difference between pure catalytic cracking and the synergy effect of catalytic/thermal cracking which reduces above 1/3 the apparent activation energy. The activity of HZSM-22 nearly vanishes when the temperature exceeds 800 °C.

*Corresponding author. Email:gliu@tju.edu.cn

This work was supported by the National Natural Science Fund of China (U1232134).

Title: "Using Surface-Enhanced Raman Spectroscopy to Probe and Understand the Mechanism of Electron-mediated Chemical Reactions on Plasmonic Nanoparticle Catalysts"

An emerging field of heterogeneous catalysis involves the use of plasmonic metal nanoparticles as photocatalysts. These systems take advantage of the excitation of localized surface plasmon resonance (LSPR) in the presence of solar spectrum light to enhance rates of catalytic reactions on plasmonic metal nanoparticles. It has been proposed that these chemical transformations take place through an electron transfer from the plasmonic catalyst to the adsorbate, leading to an electron-driven activation of chemical transformations. While the general features of the process are reasonably well understood, there are certain aspects of the mechanism that are less clear. Among others, the nature of the interaction between the localized surface plasmon states and adsorbate electronic states is still under debate.

In these studies, probe molecules were used along with silver nanoparticle catalysts in a number of Raman spectroscopy experiments to provide further insights into the electron-driven mechanism of photo-chemical reaction on the surface of plasmonic metal nanoparticles. We will shed light on the nature of the interaction between localized surface plasmon states and the excited electronic states of the adsorbate and demonstrate how these interactions can lead to chemical transformations. We will discuss the implications of our studies on the chemical activity and selectivity of plasmon-mediated photo-chemical transformations on metal nanoparticles.

379004 Preparation of Uniform Nanostructures for Heterogeneous Catalysis By Fast Annealing of Monodisperse Metal Nanocrystals

Abstract:

Fundamental understanding of catalyst operation is key for the preparation of improved heterogeneous systems for several catalytic applications. In metal-supported systems, controlling the size, shape and composition of metal nanocrystals is pivotal for drawing structure-activity relationships that can help prepare systems with improved reaction rates, selectivity and stability. To this aim, nanotechnology tools and colloidal synthesis provide a wide variety of synthetic techniques for the preparation of tailored materials with desirable characteristics for catalysis. One of the main issues, however, are the changes to the nanostructures associated with calcination treatments to remove organic ligands and stabilize the overall structure or with the reaction conditions. Finding appropriate ways to activate the materials for catalytic reactions without altering the desired properties of the structures after their synthesis is therefore highly challenging but pivotal to fully exploit the potential of colloidal synthesis for the preparation of heterogeneous catalysts.

In this contribution, we show that uniform nanostructures can be prepared by starting from highly monodisperse metal nanocrystals and by dispersing them onto high-surface area supports. Key to preserving the uniformity of the metal particles is an activation procedure consisting of a fast annealing step that maintains the initial uniform structures in terms of size, shape and composition. Additionally, the activation step allows for an improved thermal stability of the systems by the creation of kinetically stabilized structures that are usually not obtained by conventional calcination treatments. We show how the stabilization of highly uniform metal nanocrystals for catalysis is helpful in understanding structure-activity relationships in heterogeneous systems.

The development of new and effective nano-catalysts requires an understanding and control of nanoparticles, their defining feature. The problem is two-fold; the outcome of a chemical reaction is governed by nanoparticle characteristics, which in turn are governed by nanoparticle synthesis conditions. Thus, a fundamental and empirical understanding of nanoparticle synthesis is integral to improved catalyst design. Nanomaterials are inherently difficult systems to study due to their small size. Study is further complicated by the poorly understood variables involved in their synthesis. Variability in the outcome of a nanoparticle synthesis is distributed amongst a large number of factors, ranging in scope from the type of synthesis method used to the specific types and amounts of chemicals used in the method. The overwhelming number of variables results in seemingly random fluctuations in outcome, despite the fact that each factor and factor interaction logically belongs to a specific outcome set.

Current research has been successful in identifying trends in response with respect to particular factors, but the complexities of the full design space are not yet well understood. There are many examples documented in the literature where the desired shape and size of nanoparticles was achieved through trial and error. Each of these trial and error efforts provide little to no information regarding the underlying curvature of the response surface

In this study statistical design of experiments was employed to elucidate the role of different design factors for one pot metal salt reduction, a widely used nanoparticle synthesis method. Using Cobalt as a case study, the full spectrum of possible numerical and categorical design factors were screened systematically using full and partial factorial experiments. Statistical analysis was used to discriminate between significant and insignificant experimental outcomes and guide subsequent experimental design. A reproducible response surface was developed by characterizing particle size and morphology using Transmission Electron Microscopy (TEM) and X-ray Diffraction (XRD).

Fischer-Tropsch Synthesis over Supported Nano-iron Catalysts Synthesized by Gas-Expanded Liquid Deposition Technique

Rui Xu, Pranav Vengsarkar, Christopher B. Roberts,

Department of Chemical Engineering, Auburn University, Auburn, AL

Conventional Fe-based low temperature Fischer-Tropsch synthesis (FTS) catalysts are commonly synthesized using impregnation or co-precipitation method with iron salt and promoters, followed by a high temperature calcination step. It is known that the traditional impregnation preparative methods have certain drawbacks, including slow diffusion of the metal salt into the pores of the support during impregnation because of the high viscosity of water, poor dispersion of metals, and inhomogeneous distribution of the active compounds.

In this study, a new catalyst was prepared using pre-synthesized iron oxide nanoparticles that were controllably deposited onto a support using a gas-expanded liquid precipitation technique (GXL). The effectiveness of this novel catalyst system was tested for FTS performance in a fixed bed tubular reactor. A GXL is a mixed solvent composed of a compressible gas like CO_2 or ethane dissolved in an organic solvent. Due to the expansion of the liquid-organic solvent upon gas dissolution, the viscosity, diffusivity and solubility are tunable through simple adjustments in the applied pressure. The lower viscosity, higher mass transfer coefficient, and diminished interfacial properties in the GXL enhance the diffusivity of the iron particles into the catalyst pores. Using the GXL technology, previously synthesized iron oxide nano-particles can be efficiently deposited onto the surface, and inside the pores, of a desired support material via CO_2 pressurization of a mixture of a support material, organic solvent, and dispersed iron oxide particles. This method allows us to generate a supported iron catalyst with a narrow size distribution of iron oxide nan-particles for the Fischer Tropsch synthesis.

This work reports the preparation and optimization of these new nano iron oxide catalysts and their catalytic performance in terms of suitable activity and product selectivity. These nano iron oxide catalysts were compared to conventional catalysts prepared by the incipient wetness methods using the same iron oxide loading. In addition, the influence of the type of support material employed, the iron oxide loading amount, as well as the surface areas of the support materials on the FT catalytic performance have been evaluated.

358210 Nanostructured Catalysts Prepared Using Atomic Layer Deposition

Abstract:

The size and composition of heterogeneous catalysts play important roles in achieving highly active, selective and stable catalytic systems. Atomic layer deposition (ALD) has been developed as a promising method to tailor size and composition of metal and metal oxide catalysts. Benefiting from self-limiting surface reactions, ALD enables conformal coatings of thin films and highly dispersed nanoparticles on high surface area supports. The size and composition of metal and metal oxides can be controlled precisely by tuning the ratio and sequence of ALD cycles of each substance. Nanostructured catalysts prepared by ALD have shown their advantages in several important catalytic systems, including fuel cells, alkane dehydrogenation, water splitting, and biomass conversion.

In the present study, the preparation of nanostructured Pt and Pd catalysts was characterized by *in situ* FT-IR, XPS, EXAFS, SAXS and STEM. The catalyst performance in the reactions involving carbon monoxide self-poisoning was studied and discussed. The nanostructured ALD catalysts showed promising properties that are resistant to coking formation and sintering. The selectivity of the reaction can be tuned by choosing proper combination of the catalysts and ALD oxide thin films.

375597 Synthesis and Electrocatalytic Application of Alloy Catalysts

Abstract:

Alloy nanoparticles (NPs) have attracted increasing interest due to their superior performance in various catalytic applications. For example, Pt alloys with transition metals (PtM with M = Fe, Co, Ni, etc.) have been found to be highly active for the oxygen reduction reaction in fuel cells. A lot of efforts have been dedicated to the synthesis of alloy catalysts, which are usually in the form of alloy NPs dispersed in high surface area carbon matrix. The approaches mostly include co-precipitation of metal salts in aqueous solution, impregnation, and electrodeposition. In spite of the successful preparation of various types of alloy catalysts, it yet remains challenging to synthesize monodisperse and homogeneous alloy catalysts with robust control over size and composition. Here we introduce our organic solvothermal approach towards high-quality alloy nanomaterials for electrocatalytic applications. Control over particle size, alloy composition and alloy homogeneity has been achieved by matching the different reduction and growth rates of the alloying elements, which has enabled systematic studies of the effects of these parameters on electrocatalysis. Our studies thus provide clues for the design and optimization of alloy nanomaterials for catalytic applications.

Continuous Synthesis of Platinum Nanoparticles in Bicontinuous Microemulsion in a Microfluidic Reactor

Barbara M. Kampa, Emanuela Negro, Michiel T. Kreutzer, Ger J. Koper, and J. Ruud van Ommen

Session: Science and Engineering of Catalyst Preparation

Noble metal nanoparticles with a size below 5 nm can serve as highly efficient and selective catalysts. Colloidal methods offer precision in controlling of size and shape of the nanoparticles (NPs). However, they typically deal with small amounts (up to 1g) of diluted solutions of NPs (<0.1 wt%). Concentrating the solutions and removing the ligands involves formation of large amounts of waste, making the process inefficient and unreliable with respect to the predicted NPs properties.

Microemulsions offer a facile route for the synthesis of nanoparticles. Water-in-oil microemulsions are composed of small water droplets dispersed in oil stabilized by surfactant. This approach gives monodisperse NPs but with very low yields. Increasing water content causes higher frequency of droplet merging which leads to aggregation of NPs. In order to make a dense dispersion and simultaneously preserve the size and shape control we use bicontinuous microemulsions (BMEs). BMEs are a special type of microemulsions with a high water content (up to 50 wt%), forming interconnected aqueous channels stabilized by large amounts of surfactant molecules dispersed in an oil phase. They have been successfully used as a template for very concentrated dispersions (up to 3 wt%) of highly monodisperse and extremely stable NPs due to the internal rigid structure which reduces particles agglomeration [1].

The objective of this study is to continuously synthesize dense, monodisperse nanoparticles dispersions. Here, for the first time, we combine the BMEs approach and segmented-flow microfluidic devices. Firstly, we reveal the reaction kinetics of the synthesis based on deconvolution of complex UV-Vis spectra containing information about the metallic precursor and nanoparticles. We made a kinetic model to gain insights into the reaction rate constants and mass transfer limitation in the system and how those affects the precursor conversion and the morphology of NPs. We will show how different variables influence the particle size (around 2 nm). Finally, we show the continuous synthesis of monodisperse platinum nanoparticles in BME containing up to 3 wt% of nanoparticles reaching high conversion levels. Combination of BMEs and plug-flow microreactors is generic, and can be applied for synthesizing of any other dense solutions of noble metal nanoparticles.

[1] Kowlgi, K., Lafont, U., Rappolt, M., Koper, G., Uniform metal nanoparticles produced at high yield in dense microemulsions, Journal of Colloid and Interface Science 372 (2011) 16-23.

Sulfur Poisoning of the NH_3-SCR Reaction over Cu-SAPO-34

William Epling*, Yong Liu, Li Zhang, Di Wang, and Yasser Jangjou
Department of Chemical and Biomolecular Engineering, University of Houston
*wsepling@central.uh.edu

Abstract

Sulfur is a common automotive catalyst poison and this also holds true for the newer metal-exchanged small pore zeolite selective catalytic reduction (SCR) catalysts. Sulfur poisoning modes likely involve formation of $(NH_4)HSO_4$, $(NH_4)_2SO_4$ and/or metal sulfates. With

1. literature evidence suggesting that SO_2 and SO_3 poisoning mechanisms of SCR catalysts differ[1-3], and
2. it being likely that the SCR catalyst will be exposed to SO_2, SO_3 (via oxidation of engine-out SO_2 on upstream diesel oxidation catalysts (DOCs) typically used in SCR catalyst aftertreatment systems) and H_2SO_4 (via reaction between SO_3 and H_2O);

it is critical to understand how sulfur species impact SCR performance.

In this study, we investigated the poisoning effects of the different sulfur species on the SCR reaction over a Cu-SAPO-34 catalyst. NH_3-SCR activity tests were carried out in a reactor specifically designed to evaluate monolith-supported emissions catalysts. The sample was exposed to SO_2, SO_3 or H_2SO_4. Performance regeneration was also evaluated via sample exposure to high temperatures (up to 735°C) in air. Surface species formed during exposure of the catalyst to sulfur, NO_x and NH_3 were characterized with in-situ Diffuse Reflectance Infrared Fourier Transform Spectroscopy (DRIFTS). Temperature programmed desorption (TPD) was also used to characterize the samples after exposure to NH_3 and NO_x in the presence of sulfur.

SCR reaction performance results show that all the sulfur species affected the NH_3-SCR activity with SO_2 having less of a negative impact than SO_3. SO_2 mainly affected low temperature activity, < 300°C, while SO_3 affected the activity of NH_3-SCR in the whole temperature range examined (100~600°C). With SO_2 exposure, there was also an increase in performance in the mid-temperature range. With high temperature exposure to air, the SO_2-poisoned catalyst activity could be recovered, but with other sulfur exposures activity was not re-attained. During temperature programmed desorption experiments after sulfur exposure, almost all the sulfur adsorbed with SO_2 exposure desorbed by ~300°C, explaining its limited impact on activity at higher temperature (>300°C). For SO_3 adsorption, significantly more sulfur adsorbed and desorption occurred over a much larger temperature range.

In-situ DRIFTS results of co-adsorbed SO_2 and NH_3 versus that of just NH_3 suggested a higher level of NH_4^+ formed. Furthermore, TPD results demonstrated that more NH_3 adsorbed in the presence of SO_2 and SO_3 and more sulfur adsorbed in the presence of NH_3. The extra NH_3 amount was in a 2:1 molar ratio with the extra sulfur that adsorbed, clearly demonstrating that ammonium sulfate formed[4]. The extra source of NH_3 and its decomposition onset at ~300°C led to the slightly improved performance in this temperature range, relative to the sample not exposed to SO_2.

In situ DRIFTS spectra obtained during NO_x adsorption show that NO_x was adsorbed mainly as bridging nitrate and bidentate nitrate species. However, the adsorbed bidentate nitrate disappeared in the presence sulfur. This bidentate nitrate has previously been reported as an important intermediate species for the NH_3-SCR reaction at low temperature[5]. This result demonstrates that sulfur adsorption also affected NO_x adsorbed and influenced the formation of the intermediate, affecting the low temperature activity accordingly.

Overall, two common modes of deactivation were observed. First ammonium sulfate formed, which could block zeolite pores thus limiting reactivity. This primarily affects low temperature performance since the ammonium sulfate decomposes at temperature higher than 300°C. Secondly, metal sulfates can also form, and these require more severe conditions, i.e. higher temperature, for decomposition. This second mode was more relevant under SO_3 exposure conditions.

REFERENCES

1. Y. Cheng, C. Montreuil, G. Cavataio, and C. Lambert, *SAE International* 2009-01-0898.
2. Y. Cheng, C. Lambert, D.H. Kim, J.H. Kwak, S.J. Cho, and C.H.F. Peden, *Catalysis Today* **151**(2010)266.
3. B. Ramachandran, R.G. Herman, S. Choi, H.G. Stenger, C.E. Lyman, and J.W. Sale, *Catalysis Today* **55**(2000)281.
4. L. Zhang, D. Wang, T. Liu, K. Kamasamudram, J. Li and W.S. Epling, *Applied Catalysis B: Environmental* **156-157**(2014)371.
5. D. Wang, L. Zhang, K. Kamasamudram and W.S. Epling, *ACS Catalysis* **3**(2013)871.

Deactivation Model for Co Fischer-Tropsch Catalysts

Kamyar Keyvanloo, William C. Hecker, and Calvin H. Bartholomew*

Department of Chemical Engineering, Brigham Young University, Provo, UT 84602

*hecker@byu.edu

One of the foremost technical challenges to the successful, economical practice of Fischer-Tropsch Synthesis (FTS), as with most catalytic processes, is catalyst life. Understanding the chemistry of deactivation pathways and modeling catalyst deactivation rates are imperatives to the design of more stable catalysts and the development of robust, realistic reactor models which facilitate FT reactor/process design and optimization. Carbon deposition on FT catalysts is not sufficiently understood well enough to enable its quantification. Our goal is to develop a deactivation rate model which predicts the deactivation of Co FT catalysts by carbon deposition as a function of reactant partial pressures and temperature.

The FTS reaction kinetics and deactivation kinetics can be linked by the activity of the catalyst. Activity (a) is a function of time, temperature, and concentrations and is defined as the CO depletion rate normalized by the initial rate. The rate of the main reaction (r_m) and deactivation rate (r_d) are defined as follows:

$$-r_m = k(T) f_1(T, C)\, a$$

$$-r_d = -da/dt = k_d\, f_2(T, C)(a - a_\infty)^d$$

where the constant a_∞ represents the asymptotic or steady state catalyst activity as time approaches infinity.

Five sets of deactivation data for a cobalt FT catalyst, each taken during 800 h runs, were obtained for reaction temperature of 230 °C, pressures of 20 bar, and H_2/CO ratios of 1-3. These data were found to be modeled well using concentration-dependent first order generalized power law expressions (GPLEs) which predict that activity approaches a non-zero asymptote. Deactivation rate was found to increase with both higher partial pressures of hydrogen and carbon monoxide, and partial pressure dependencies were determined.

Gas-Phase Conversion of 5-Methylfurfural to 2,5-Dimethylfuran Over Copper Chromite Catalyst Systems

Aaron Oberg and Dennis J. Miller
Department of Chemical Engineering and Materials Science
Michigan State University
East Lansing, Michigan 48824

Recently the production of furanics from biomass derived carbohydrates has gained significant attention. Among the furanics of interest is 2,5-dimethylfuran, which shows promise as a bio-fuel[1] in addition to being a precursor to the industrially important chemical intermediate p-xylene[2]. The focus of the current work is the development of alkali earth metal promoted copper chromite catalysts for the highly selective gas-phase conversion of 5-methylfurfural to 2,5-dimethylfuran. The copper chromite systems initially show high activity and selectivity, but deactivate continually over time. Deactivation in similar systems has previously been reported[3] in the literature and remains a primary technological barrier to the commercial utilization of bio-furanics. In this work, the rate of deactivation has been shown to be strongly related to variations in the 5-methylfurfural feed composition. Exploration of feed handling and purification methods to selectively remove catalyst poisons decreases the rate of deactivation. Removing feed impurities makes it possible to observe the intrinsic deactivation rate caused by the methyl furfural and/or reaction products. Catalyst characterization provides a picture of the copper chromite surface before and after reaction, and gives insight into the mode of catalyst deactivation.

1. Roman-Leshkov, Y., et al., *Production of dimethylfuran for liquid fuels from biomass-derived carbohydrates.* Nature, 2007. **447**(7147): p. 982-985.
2. Williams, C.L., et al., *Cycloaddition of Biomass-Derived Furans for Catalytic Production of Renewable p-Xylene.* ACS Catalysis, 2012. **2**(6): p. 935-939.
3. Liu, D., et al., *Deactivation mechanistic studies of copper chromite catalyst for selective hydrogenation of 2-furfuraldehyde.* Journal of Catalysis, 2013. **299**(0): p. 336-345.

360465 Effects of CO and NO on TiO2 Supported Pt, Pd, and Rh Nanoparticle Disintegration

Abstract:

Supported transition metal catalysts like Rh, Pd, and Pt are used in a broad range of economically and environmentally significant processes such as the conversion of toxic gases like NO and CO to less harmful species. Unfortunately these rare and expensive catalysts often suffer from deactivation. Thus there is a strong need to improve and understand these catalysts to increase their stability and efficiency. Due to its widespread catalytic importance, the focus of this study is to understand the effects of the presence of NO and CO on the stability of isolated Rh, Pd, and Pt nanoparticles supported on $TiO_2(110)$. Density functional theory and *ab initio* thermodynamics will be applied to understand these systems under a variety of experimental conditions. Formation and binding energies of metal adatoms and adatom-reactant complexes are computed as well as the influence of CO or NO adsorption on nanoparticle surface energetics. The disintegration free energy of supported Rh, Pd, or Pt particles under a large parameter space of reaction conditions and nanoparticle sizes is explored. In agreement with previous experiments, we compute that CO and NO both cause $Rh/TiO_2(110)$ to disintegrate at 300 K and 1 mbar via the formation of di-carbonyl or di-nitrosyl adatom-reactant complexes, and that NO is a more efficient reactant for particle redispersion than CO. On the other hand, we find that $TiO_2(110)$ supported Pt or Pd nanoparticles are more resistant to CO or NO induced disintegration via the formation of adatom complexes. Overall, this information could lead to strategies for facilitating the redispersion of aggregated nanoparticles.

Ultra-Dispersed Catalytic Hydroprocessing of Vacuum Gas Oil: Catalyst Recycle and Deactivation Study. **A. S. Carss** ascarss@ucalgary.ca and **P. Pereira-Almao** <ppereira@ucalgary.ca>

1.0. Introduction

The world will face many challenges with regards to meeting global energy demand in the ensuing years. In the wake of fossil fuels production growth, an era marked by management of greenhouse gas emissions and increasingly stringent environmental regulations is being entered. For example, the state of California has implemented a Low Carbon Fuel Standard that aims at reducing the carbon intensity of its fuels by 10% by the year 2020 (Jacobs Consultancy, 2009).

The province of Alberta (Canada) ranks third, next to the countries of Saudi Arabia and Venezuela, for their amount of proven global crude oil reserves. In 2012, Alberta produced 556,000 barrels per day (BPD) of conventional crude oil (API > 20) and 1,900,000 BPD of unconventional crude oil in the form of oil sands bitumen (API=8). The proven reserves of unconventional oil in Alberta are 169 billion barrels, and they represent 99% of total reserves (AB. Gov., 2013).

A multifaceted energy conundrum exists today. Liquid petroleum fuels consumption is increasing at a rate of 1.5% per annum combined with a demand shift in the energy pool away from fuel oil and towards diesel and distillates, unconventional crude oils are becoming increasingly prevalent, and meanwhile carbon emissions need to be reduced (EIA, 2013).

2.0. Background

It is evident that conversion of the "bottom of the barrel" will need to increase as energy demands change. The key to achieving this objective prudently and in an environmentally responsible fashion will be through process design improvements, new process design and new technologies such as those that incorporate state-of-the-art catalysts.

Conventional catalysts used for hydrocracking (HCK) and hydrotreating (HDT) of extra heavy oil fractions are porous particles dispersed on a solid support. Such catalysts suffer from deactivation problems especially due to: formation of coke around the catalyst particle, which causes blockage of the catalyst pores, and deposition of metals (nickel and vanadium) in and on the catalyst particle. These problems are exacerbated when heavy and extra heavy feedstocks are processed, and a service period of the catalyst of one year or less can typically be expected (Ramirez, 2007).

Due to the deactivation problems peculiar of conventional HCK and HDT catalysts, interest in dispersed catalytic upgrading is increasing. Dispersed catalysts are regarded as well suited to upgrade extra heavy charges that contain high amounts of metals and asphaltenes. Said catalysts are transition metal, non-porous particles that are freely suspended in a hydrocarbon matrix and are tailored to be as small as possible. Ultra dispersed (UD) catalysts are sub-micronic or even nanometric particles. UD catalysts specific surface area and ability to deliver activated hydrogen to a reacting hydrocarbon matrix is therefore maximized.

Dispersed catalysts have been used and proven to work successfully for HCK of heavy feedstocks in a once through basis. One of the major challenges, and one that affects the economic feasibility of UD catalytic processes, is the recycling and reuse of the spent catalyst during continuous operation. Additionally, the deactivation of UD catalysts when recycled is not known and needs to be investigated. This work deals with studying the recycling and deactivation of UD catalyst.

3.0. Experimental

In this study molybdenum based UD catalysts were prepared in a catalyst preparation pilot plant. The plant emulsified an aqueous solution of ammonium heptamolybdate in VGO and VGO/n-C_5 asphaltene solutions. The emulsion was subjected to a thermal treatment and UD particles were formed and suspended within the hydrocarbon matrix. $(NH_4)_2S$ was incorporated into the emulsion in order to pre-sulfide the catalyst.

The prepared UD catalyst feedstock was reacted in a separate pilot plant under low severity hydroprocessing (HDP) conditions. A range of processing temperatures, space-times and recycle ratio's were explored. Data was collected to calculate kinetic constants and activation energies of HCK and hydrodesulfurization (HDS) reactions. A simple global kinetic model was utilized as well as a 5-lump, lumped parameter model. The calculated kinetic parameters were vetted against literature reported values.

A diagram of the ultra dispersed catalyst-recycling unit (UDCRU) is presented in Figure 1. It consists of a feed, reaction and product separation zone. Following the hot separator there is a recycle pump, which when functioning in the recycle mode of operation, may return hot separator bottoms material containing catalyst and unconverted feed to the reaction zone for an additional pass through the reactor.

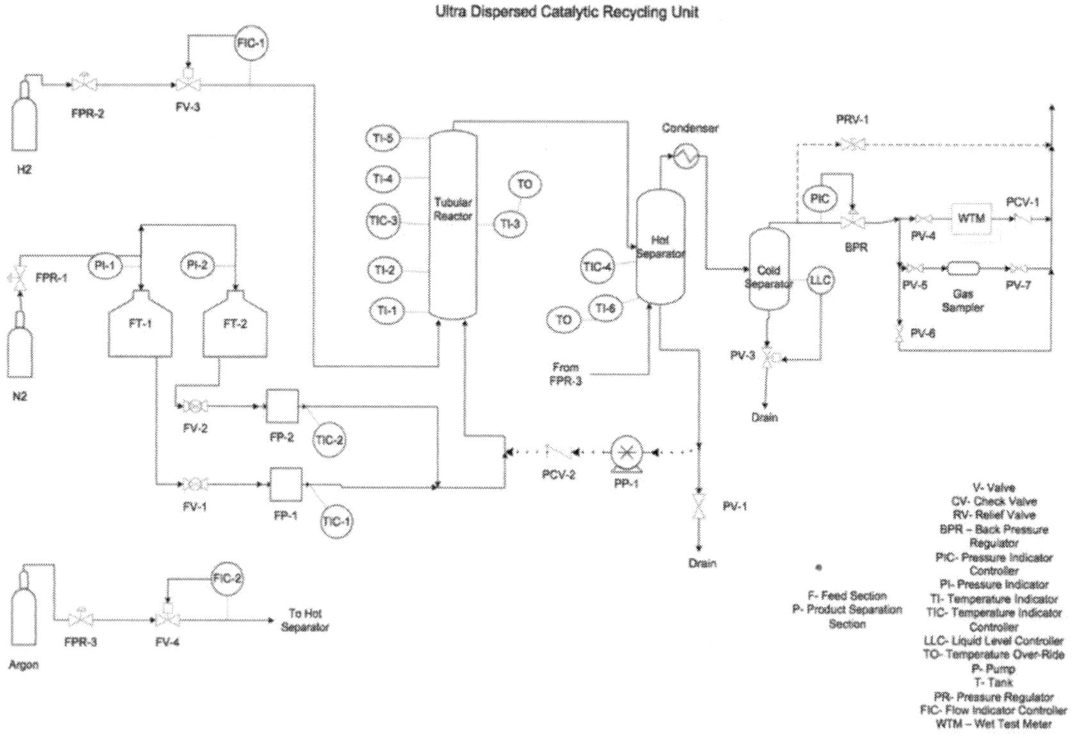

Figure. 1. Ultra-Dispersed Catalytic Recycling Unit Schematic

4.0. Achievements

A novel methodology for studying deactivation of UD catalyst under recycling mode of operation was developed and utilized in this work. A catalyst cutoff model was developed. In this model, at steady state in the recycling mode of operation, UD catalyst to the system was cutoff and replaced with virgin feedstock. In this way, the concentration of catalyst in the hot separator diminishes with time. An unsteady state component balance was solved for the hot separator and reactor concentration profile with time. The catalyst concentration evolution after catalyst cutoff for a recycle ratio of 1 is shown in Figure 2.

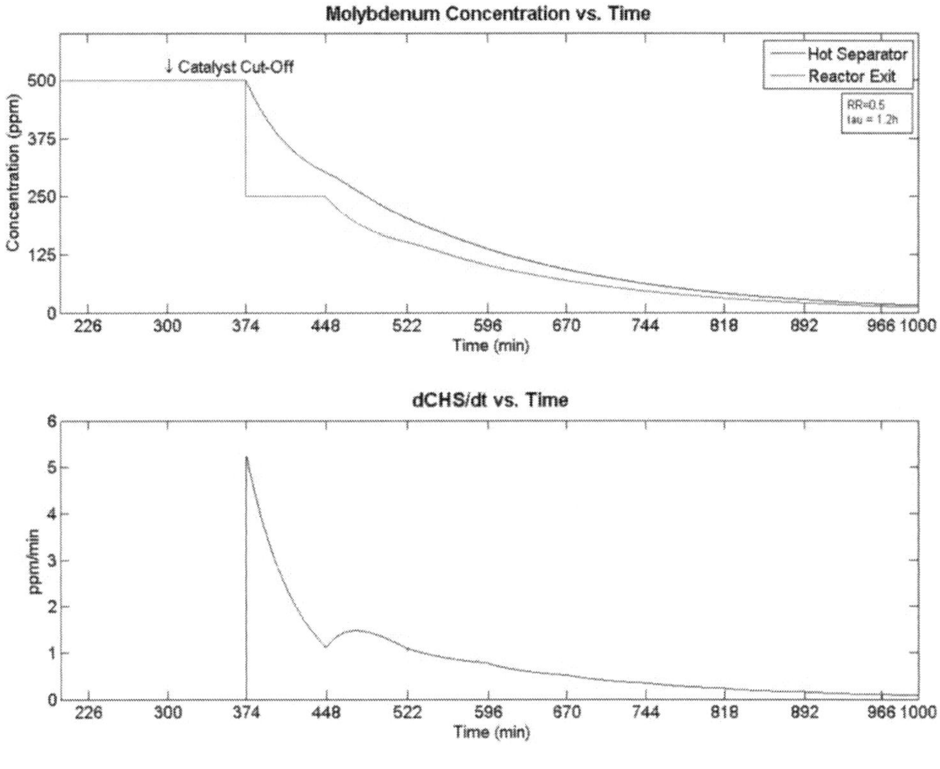

Figure. 2. Catalyst Concentration vs. Time for Catalyst Cutoff Experiments

The product quality of the hot separator bottoms material, such as the micro carbon residue can be related to the catalyst concentration present in the reacting media. Referring to figure 2, the following scenarios may occur: 1. The product quality may decline proportionally to the catalyst concentration, 2. The product quality may decline disproportionally faster than the catalyst concentration, or, 3. The product quality may decline disproportionally slower than the catalyst concentration. The three scenarios represent non-deactivating catalyst, deactivating catalyst, and catalyst present in higher than optimal concentration respectively.

It was found in this work that the theoretical unsteady state molybdenum component balance matched the experimental concentration data very well. Utilizing the catalyst deactivation model described, it was found in this work that the catalyst, after 3.3 passes through the reactor did not suffer from deactivation problems. Using fresh catalyst tie-in points, product quality data from the catalyst cutoff model was validated.

Dynamic light scattering (DLS) and Inductively Coupled Plasma Spectroscopy (ICP) were used to determine catalyst particle size and concentration in feedstock and hot separator bottoms products from reaction. As part of this work, a catalyst agglomeration study was carried out and the size of catalyst particles in the hot separator bottoms products was investigated. The experimental data indicated that the HDP reaction temperature was the variable that most strongly affected the hot separator bottoms particle size. Recycle ratio and space-time had a negligible effect on the catalyst particle size.

5.0. Conclusions and Recommendations

In this research UD catalyst was continuously recycled in a pilot plant, a catalyst cutoff deactivation model was developed and it was shown that UD catalyst was resistant to deactivation after several passes through the pilot plant reactor. The data obtained was used to solve for kinetic data and a catalyst agglomeration study was carried out.

Recommendations for future work were developed and include: experimenting with different feedstocks including vacuum residue, experimenting with different catalysts including bi and tri metallic catalysts, and changing the operating conditions of the plant. These recommendations will be presented and discussed.

6.0. References

AB. Gov. *Alberta Oil Sands Industry Quarterly Update Summer 2013,* obtained from http://albertacanada.com/files/albertacanada/AOSID_*Quarterly*_Update_Summer 2013.pdf. (accessed March 1, 2014), 2013.

EIA (U.S. Energy Information Administration), *International Energy Outlook 2013,* obtained from www.eia.gov/forecasts/ieo/more_highlights.cfm (accessed March 27th, 2014), 2013.

Jacobs Consultancy, *Life Cycle Assessment Comparison of North American and Imported Crudes*, obtained from http://www.eipa.alberta.ca/media/39640/life%20cycle%20analysis%20jacobs%20final%20report.pdf (accessed April 21, 2014), 2009.

Ramirez, J., Rana, M.S., and Ancheyta, J. *Characteristics of Heavy Oil Hydroprocessing Catalysts.* In: Ancheyta, J., and Speight, J.G., editors. Hydroprocessing of Heavy Oils and Residuas. Boca Raton (FL): Taylor and Francis Group, p. 121-190, 2007.

DYNAMICS OF CARBON DEPOSITION ON ALUMINA SUPPORTED NICKEL CATALYSTS DURING DRY REFORMING REACTION - AN IMPORTANT ROLE OF CARBON SPECIES ON THE NICKEL SURFACES –

Yuki Kitano, Naoya Mihara, Atsushi Okemoto, Keita Taniya, Yuichi Ichihashi, and Satoru Nishiyama
Department of Chemical Science and Engineering, Graduate School of Engineering, Kobe University, Rokkodai, Nada, Kobe 657-8501, Japan

Introduction

Methane dry reforming reaction ($CH_4 + CO_2 \rightarrow 2CO + 2H_2$) is one of the most promising processes for reduction of CO_2 emission accompanying H_2 production. As this reaction is a highly endothermic process, higher temperatures are required more than 873 K. These severer reaction conditions bring about heavy catalyst deactivation. The catalysts used for the reforming reactions are metal nanoparticles supported on porous materials, which can activate saturated aliphatic hydrocarbons at high temperatures. The reactant hydrocarbons are also unfavorably transformed to carbon species (coke) during the reactions and the surface of the active metal is covered by the deposited coke. In contrast to the steam reforming reaction, the reaction atmosphere of the dry reforming is significantly more reducing, because CO_2 is a weaker oxidizing reagent than H_2O. The coke deposition from hydrocarbon reactants would be more crucial during the dry reforming than the steam reforming. The coke deposition which brings about catalyst deactivation is usually considered to proceed via a side reaction pathway. In the methane dry reforming reaction over the supported Ni catalysts, the coke deposition process is suggested to be included in main reaction pathway [1]. Dehydrogenation of methane forming with carbon species on the nickel surfaces is a first step of the reaction, then the formed carbon species will react with the oxygen species which is generated by dissociative chemisorption of CO_2 on the nickel surfaces [1, 2]. In order to suppress the deactivation, the rate of coke deposition should be reduced. But in the dry reforming reaction, the suppression of the carbon deposition process will decrease the total reaction rate. The balance of the each elementary reaction step (the coke formation from methane and the surface reaction of the deposited coke and surface oxygen) would be very important to control the total process of the dry reforming reaction. The reactivity of the deposited coke plays an important role in the total process of the dry reforming.

In this paper, we will discuss about the detail of the reaction mechanism of the methane dry reforming and the behavior of carbon deposition. An effective procedure to optimize the activity and stability of the reaction will be proposed for the methane dry reforming reaction on the supported nickel catalysts.

Experimental

Catalyst Preparation

The alumina supported nickel catalysts, Ni/Al_2O_3, were prepared by a conventional impregnation method using one of the reference catalysts supplying from *The Catalysis Society of Japan*, ALO-6 and nickel nitrate in an aqueous solution. The loading of nickel was 15 wt% with respect to the support. After the impregnation, the catalyst precursors were dried at 373 K for 24 h and then calcined in flowing N_2 and O_2 mixture (N_2:O_2=80:20) at 1,073 K for 5 h. The calcined catalyst precursor was reduced in flowing H_2 and N_2 mixture (20 % of H_2) at 1,073 K just before the coke deposition experiment in the measurement apparatus.

Coke Deposition Measurement during the Dry Reforming

The coke deposition rate was measured in a DTA-TG apparatus (DTG60, Shimadzu, Japan) under the dry reforming condition, in the flowing gas mixture of CH_4, N_2, and/or CO_2. After the in situ activation of the catalyst at 1,073 K, the reactor was purged with N_2 and the furnace temperature was lowered to the corresponding deposition temperature in flowing N_2. The N_2 stream was switched to the gas mixture of CH_4, N_2, and/or CO_2 (100 mL/min, corresponding composition), then the coke deposition started. Weight change was measured by TG under atmospheric pressure. The charged catalyst sample weight was typically 40 mg and the flow rate of the stream was 100 mL/min (space velocity, SV=300,000 cm^3/h g-cat). The reference material for DTA-TG measurement was nonporous α-Al_2O_3.

Reactivity measurement of the deposited coke (Temperature Programmed Oxidation)

After the coke deposition experiment, the catalyst sample in the apparatus was cooled down to room temperature in N_2 flow. After the flow was re-switched to the gas mixture of N_2 and O_2 (N_2:O_2=80:20, 50 mL/min), the reactor start to heat with the constant rate of 4 K/min. The weight change, and exo- or endo-thermic signal were recorded by using the same apparatus for the coke deposition measurement. The successive coke deposition under the dry reforming condition and CO_2 oxidation experiments were also carried out at 1,073 K in the same apparatus.

Characterization of the deposited coke and deactivated catalysts

Characterization of the deactivated catalysts was carried out by using X-ray diffraction powder pattern and scanning electron microscopy. X-ray diffraction powder patterns were accumulated with

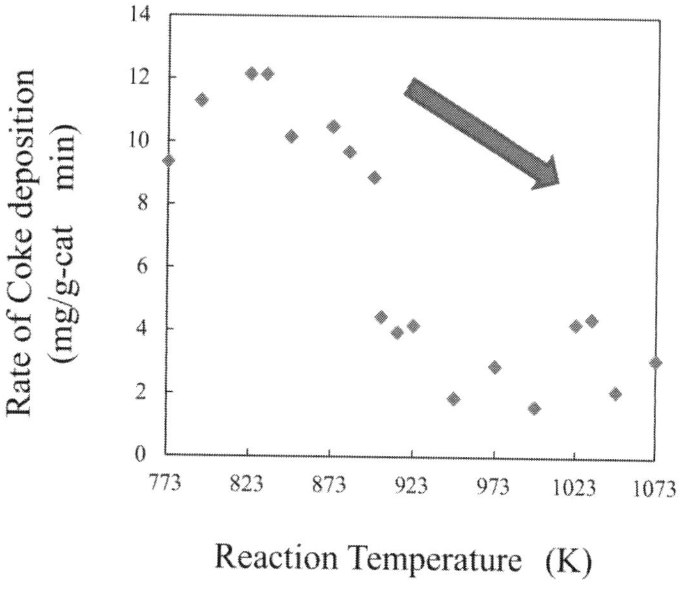

Figure 1. Temperature Dependence of the Rate of Coke Deposition on Ni/Al_2O_3
Catalyst weight: 40 mg
CH_4:CO_2:N_2=20:20:60 Reaction time: 5 min

Rint2100 X-ray diffractometer (Rigaku, Japan) to investigate the formation of graphite carbon and change of Ni oxidation state with Cu-Kα ray. Scanning electron micrographs of the deactivated catalysts were obtained by using a FE-SEM microscope, JSM-7500F (JEOL, Japan).

Results and Discussion

Coke deposition behavior during the dry reforming

Figure 1 shows the temperature dependence of the coke deposition rate over Ni/Al_2O_3 catalyst under a standard reaction condition. The rate of coke deposition did not indicate an Arrhenius type dependence. The rate at lower temperatures between 773 and 873 K was higher than that at between 873 and 1,073 K. As the coking process, decomposition of CH_4 ($CH_4 \rightarrow C + 2H_2$) is highly endo-thermic process, the low rate of coke deposition was not ascribed to equilibrium limitation. We investigated the stepwise dry reforming reaction as shown in **Fig. 2**. Methane was introduced to the catalyst first at 873 K, then the atmosphere was purged by N_2, then CO_2 was conducted at the corresponding temperatures. The coke deposition proceeded during coking step under CH_4 atmosphere (CH_4/N_2=20/80). After the flow was switched to CO_2 (CO_2/N_2=20/80), the deposited coke was gradually removed at 773 K. The removal rate of deposited coke increased with temperature as indicated in **Fig. 2**. It should be noted that some extent of deposited coke could not be removed at 773 K. It suggests that at least two kinds of coke

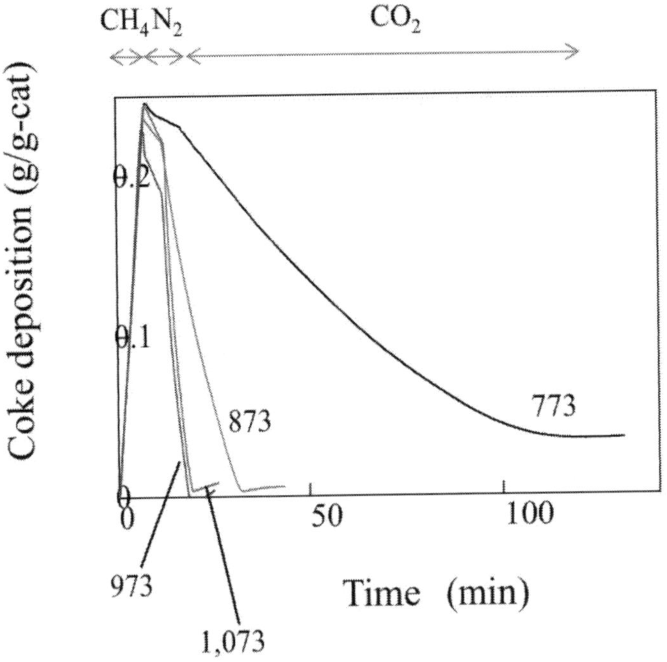

Figure 2. Oxidation Experiment of the Deposited Coke by CO_2 over Ni/Al_2O_3
Catalyst weight: 40 mg
CH_4 region CH_4:N_2=20:80 Temperature: 873 K
CO_2 region CO_2:N_2=20:80 Temperature: 773-1,073 K

should be considered on the catalysts. One was less reactive and could hardly react with CO_2 at 773 K. We also carried out the pulse experiments. Methane and CO_2 were successively introduced into the catalyst bed in a continuous flow reaction system. The produced gases were analyzed by a mass spectrometer. After the pulse of CH_4, H_2 formation was observed. The ratio of the evolved H_2 and the consumed CH_4 (molar ratio of H_2/CH_4) was ca. 2 [2]. These results indicate that CH_4 was reacted on the Ni surface of the catalyst and formed H_2 and carbon (coke) as shown in **Eq. 1**.

$CH_4 \rightarrow C + 2H_2$ **(Eq. 1)** (CH_4 decomposition)

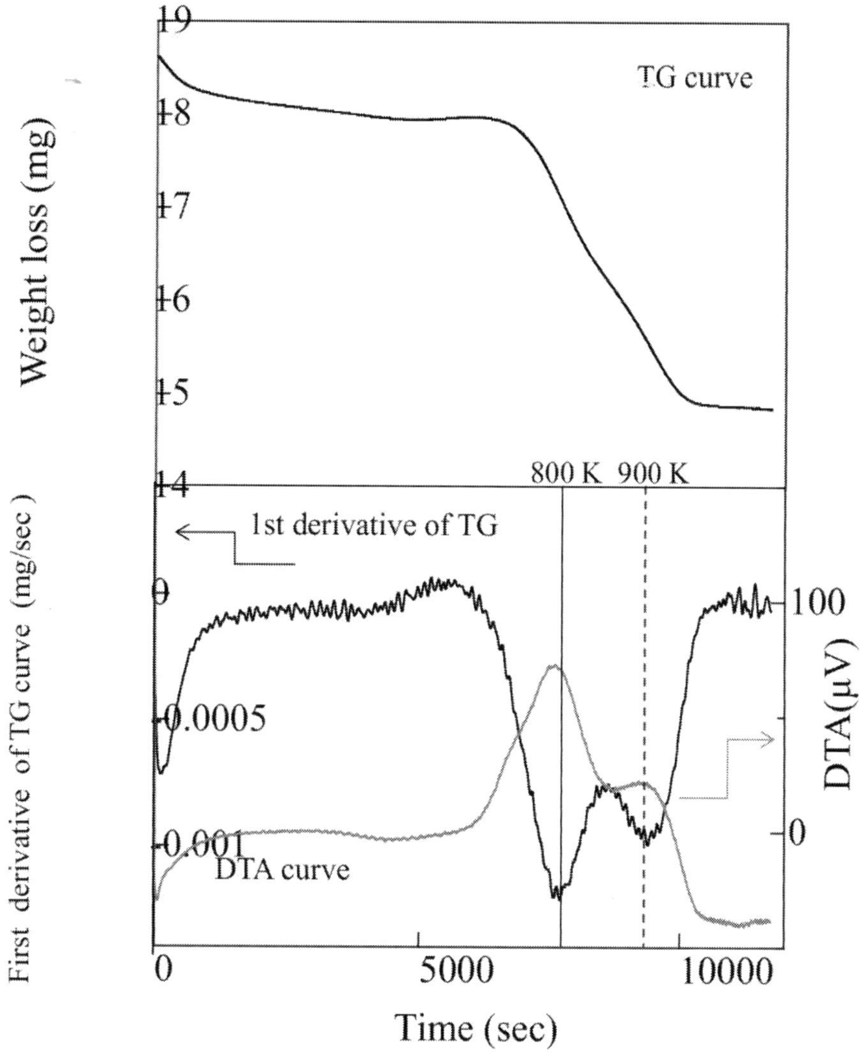

Figure 3. Temperature Programmed Oxidation of the Deposited Coke over Ni/Al_2O_3
Atmosphere O_2:N_2=20:80 Heating rate 4 K/min

After the successive pulse introduction of CO_2, significant amount of CO was formed. The ratio of the evolved CO and the reacted CO_2 (molar ratio of CO/CO_2) was also observed at 2. The second pulse of CO_2 brought about the following surface reaction as shown in **Eqs. 2 and 3**.

$CO_2 \rightarrow CO + O$ **(Eq. 2)** (dissociative adsorption of CO_2)

$C + O \rightarrow CO$ **(Eq. 3)** (surface reaction of adsorbed species)

After the pulse of CO_2, the color of the catalyst was changed from black to cobalt-blue. These results indicate that CO_2 was dissociatively adsorbed on the Ni surfaces and the surface Ni atoms were oxidized (or adsorbed oxygen on Ni). **Equation 1** indicates the coke formation process and **Eqs. 2 and 3** were coke removal processes. As shown in **Fig. 2**, the coke removal process was slower than the coke formation. At high temperatures, the rate of coke removal was much accelerated, as a results, the total of coke accumulation on the catalyst surface was considered to be lowered at the high temperatures.

Reactivity of the deposited coke during the dry reforming

The reactivity of the deposited coke was evaluated by temperature programmed oxidation (TPO) with flowing N_2 and O_2. **Figure 3** indicates TPO spectra with flowing air. The upper graph was indicates the change of the weight of the coke deposited catalyst (TG). The bottom graph also indicates

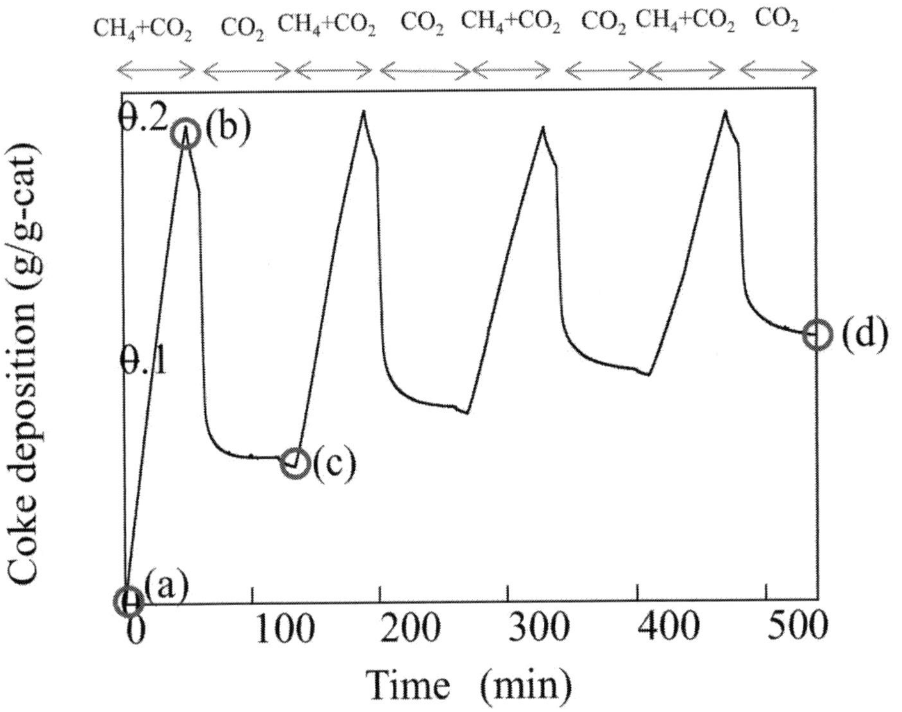

Figure 4. Successive Coke Deposition and CO_2 Oxidation over Ni/Al_2O_3
Catalyst weight: 40 mg Temperature: 1,073 K
$CH_4 + CO_2$ region $CH_4:CO_2:N_2$=20:20:60
CO_2 region $CO_2:N_2$=20:80

the DTA spectra (red line) and the first derivative curve (black line) of the TG. DTA curve and the first derivative curve were almost identical. These results indicate that at least two kind of coke with deferent reactivity are expected on the surface of the Ni catalysts during the dry reforming. The coke corresponding to the peak near 800 K was more readily oxidized than that near 900 K.

We have also carried out the successive coke deposition and CO_2 oxidation as shown in **Fig. 4**. The coke deposition under the dry reforming condition proceeded and the reactor was purged by N_2

flow, then the flow was switched by N_2 and CO_2. The deposited coke was partly removed by the reaction with CO_2 as shown in **Fig. 4**. The amount of unremovable deposited coke was gradually increased with the repeated sequence. It should be noted that the slope of curve during CO_2 oxidation process was decreased with the sequence. These results indicate that the oxidation of the deposited coke with CO_2 was deactivated, suggesting the catalyst deactivation during the dry reforming reaction by the coke deposition. The unremovable coke would bring about the severe catalyst deactivation. Four points indicating (a) to (d) in the figure were selected and the each catalyst was characterized by TPO, SEM and XRD.

Figure 5 shows the TPO spectra of the coke deposited catalysts at the points of (b), (c), and (d).

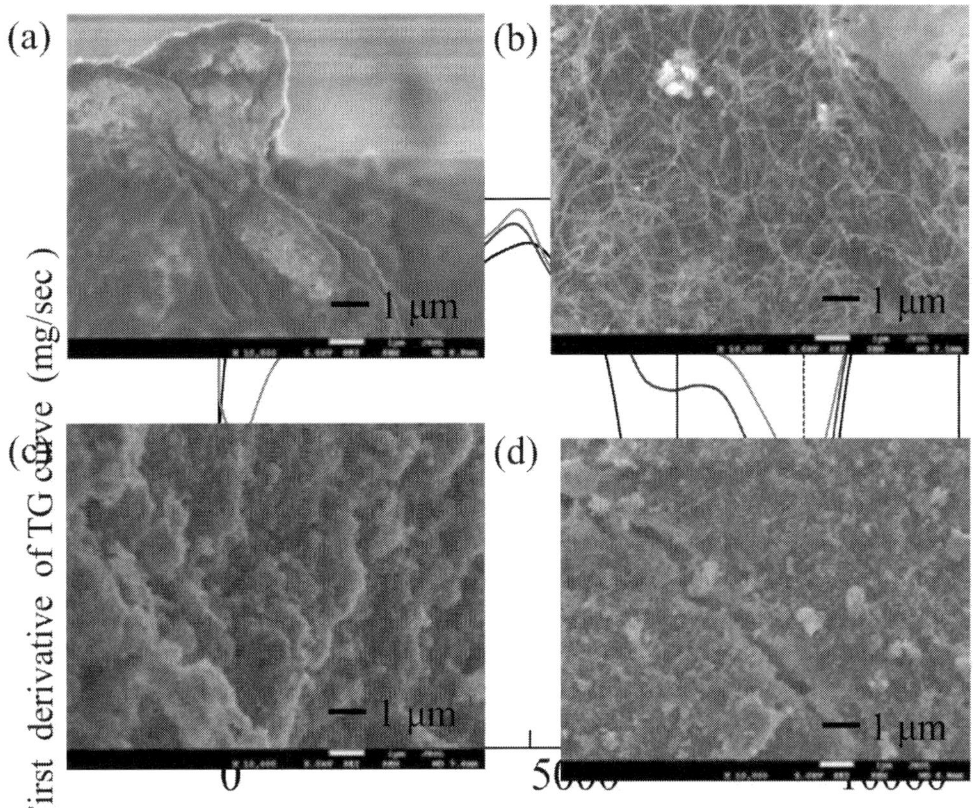

Figure 6. SEM Images of the Ni/Al_2O_3 Catalyst at the Points Indicated in Fig. 4

Magnification ×10,000

Figure 5. Temperature Programmed Oxidation of the Deposited Coke on Ni/Al_2O_3 at each point indicated in Fig. 4

Atmosphere O_2:N_2=20:80 Heating rate 4 K/min

The catalyst at point (a) was fresh and no deposited coke was observed. The catalyst at point (b), just after first coke deposition of the experiment, showed that two kinds of coke were observed at ca. 800 and 900 K. The catalyst at point (c) indicated that the peak at 900 K was major in the TPO spectrum. The catalyst at point (d) shows that the major peak was observed at ca. 900 K and that the peak intensity was increased with the sequence of the experiment. In XRD powder diffraction of the catalyst, the diffraction peak assignable to graphite carbon at 26.4 degree of 2θ was only observed in the catalyst at point (b). These results clearly indicated that the deposited coke bringing about deactivation was corresponding to the TPO peak at 900 K. It is also suggested that the peak at 800 K can be assigned to the graphite type carbon which is readily reacted with CO_2.

The surface morphology of the catalysts at the points of (a), (b), (c), and (d) was investigated by SEM observation as shown in **Fig. 6**. The micrograph (a) in **Fig. 6** shows the surface of the fresh catalyst. The surface of the fresh catalyst showed no specific structure. The micrograph of the catalyst at point (b) indicated significant amount of nanotube like structure, whisker carbon. After the CO_2 oxidation, no whisker structure was observed in the SEM image of the catalyst at point (c). Some irregular deposited structures were observed as shown in **Fig. 6 (c)**. The similar structure was also observed on the surface of the catalyst at point (d). These results strongly suggest that the whisker carbon is the coke forming during the dry reforming reaction, assigned the TPO peak at 800 K, and can be easily reacted with CO_2 accompanying with CO formation ($C + CO_2 \rightarrow 2CO$). The irregular structure which was observed in the SEM images (c) and (d) was an amorphous coke which can be assigned to the TPO peak at 900 K. The amorphous coke deposition would bring about the severe deactivation.

Dynamics of coke deposition and reaction mechanism of the dry reforming

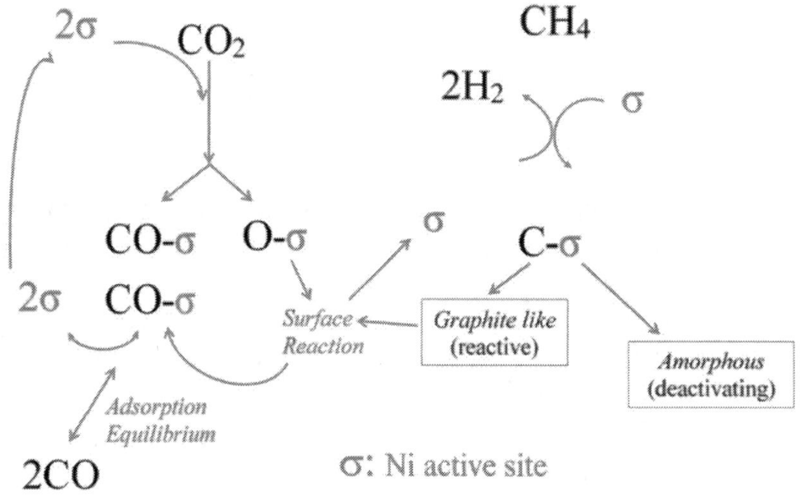

Scheme 1. Catalytic Cycle for Methane Dry Reforming Reaction Including Carbon Deposition Process over the Supported Ni Catalysts

The reaction mechanism proposed in this paper including coke formation process is summarized in **Scheme 1**. Methane is decomposed on the Ni surface and the two kinds of coke, the graphite type and amorphous coke, are formed. Carbon dioxide is dissociatively adsorbed with forming adsorbed CO and *reactive O atoms*. The graphite type coke and the reactive O atoms are readily reacted forming CO. The graphite type coke should be considered to be an *intermediate carbon* for the dry reforming reaction. The amorphous coke is less active and covers active sites on the Ni surface of Ni/Al_2O_3 catalysts.

In the CH_4 dry reforming reaction, the reaction atmosphere is markedly reducing and the coke deposition would be serious because of catalyst deactivation. In order to suppress the catalyst deactivation, the formation of the amorphous coke should be suppressed.

Conclusions

1) In the methane dry reforming reaction, at least two kinds of coke are formed over Ni/Al_2O_3 catalysts.

2) One is the graphite type coke and an intermediate transferring to CO by the surface reaction between the graphite type coke and O atoms which are supplied by the dissociative adsorption of CO_2.
3) The other amorphous coke is stable under the reaction condition and readily covers the Ni surface of the catalysts. This brings about the severe catalyst deactivation.
4) One of the most important points to design the reforming catalysts is to suppress the formation of amorphous coke without hindrance of the reactive graphite coke formation.

References

1. Ikeda, Y., T. Horie, K. Taniya, Y. Ichihashi, N. Ohmura, and S. Nishiyama (2011), "Process Intensification of Syngas Formation Reaction from Carbon Dioxide (Dry Reforming of Methane with Carbon Dioxide) by Kinetic Analysis," *Kagakukogaku Ronbunshu*, 2011, **37**, pp. 128-133.
2. Nakajima, A., K. Taniya, Y. Kitano, Y. Ichihashi, and S. Nishiyama, "Hydrogen Production by Methane Dry Reforming on Supported Nickel Catalysts – Enhanced Stability of the Reforming Process -," *Trans Tech Periodicals*, in press.

A pilot plant test to determine effectiveness factors of the commercial hydro-desulphurization (HDS) catalyst to produce ultra-low sulphur diesel (ULSD)

Nilesh Chandak,* *Adel Al Hamadi, Mohamed Yousef, Abdulhamid Mohamed, Kazuhiro Inamura, Mikael Berthod*
Takreer Research Centre, Abu Dhabi Oil Refining Company (TAKREER), P.O. Box: 3593, Abu Dhabi, U.A.E.
**Corresponding author. Email: nchandak@takreer.com, Tel # 971-2-6020903*

HIGHLIGHTS

- An ULSD production testing on pilot plant and applying it for industrial hydrodesulphurization process.
- Experimental data from pilot plant test to calculate hydrogen consumption & estimate catalyst deactivation rate
- Study the impact of change in feed & process conditions on the performance of commercial unit to produce ULSD.

ABSTRACT

A pilot plant test was carried out using a catalyst from the refinery hydrotreater unit to process gas oil blends to produce ultra-low sulphur diesel (ULSD) product. Effect of partial pressure variations have been studied for catalyst deactivation rate and hydrogen consumption. Test results shown that by increase of temperature, refinery unit can process such blend feed to reach 8 ppm sulphur product. However moderate increase in hydrogen consumption is seen which is due to higher aromatic content in the blend feed as compare to reference feed. Catalyst performance was evaluated at 55 / 42 / 32 bar hydrogen partial pressure (PPH_2) to determine catalyst deactivation rate and hydrogen consumption, targeting ULSD product. Decreases of PPH_2 from 52 bar to 32 bar resulted in reduction of H_2 consumption but also shows decrease of catalyst cycle length due to higher deactivation rate. Pilot plant test shows that by contriving computational methods and analysis techniques for hydrogen balance & catalyst deactivation rate from the pilot plant test data, it becomes possible to predict catalyst performance in commercial unit.

Keywords: Light Gas Oil, ULSD, HDS, H_2 Consumption, Catalyst Deactivation, HDT pilot plant test.

1. Introduction

Due to environmental concerns there is a continual thrust around the world to reduce the sulfur content of diesel fuel to ultra low levels (10 ppm) with the intention of lowering diesel engine's harmful exhaust emissions and improving air quality [1-6]. In line with this effort, Abu Dhabi government requires to supply the local market with Ultra Low Sulfur Diesel (ULSD) with 10 ppm sulfur by the end of 2014. Although the reduction of sulfur content in diesel fuel is beneficial from environment point of view, meeting the required stringent specifications represent a major and economic challenge for the petroleum refining industry [7-10]. Hence, several options were studied by refiners worldwide viz. the use of more active catalysts [11-19], the increase of reaction severity (e.g. increase of temperature and or/decrease of space velocity) [18-20], the change of reactor internals to improve the contacts efficiency between hydrogen and hydrocarbons or plainly to revamp the plant by installing more reactor in series to achieve deep hydrodesulphurization [5, 21-22].

Accordingly, as part of a green diesel project TAKREER (Abu Dhabi oil refining company) had studied the existing hydrotreater units in the refineries to process a blend of *LGO & HDT GO* to meet the demand of change in diesel product specification to less than 10 ppm sulphur (ULSD).

The refinery hydrotreating unit (Unit I) is designed to process LGO. The catalyst cycle length for this unit is 3 years minimum at full capacity. This unit was commissioned in 2011 and since then, it is producing on specification product. However, it has been consuming more hydrogen than estimated in the design basis.

Another hydrotreating unit (Unit II) is designed to produce HDT GO of 500 ppm sulphur content from SRGO feed. The refinery plan is to process HDT GO product of Unit II in Unit-I along with the current LGO.

This paper describes the pilot plant test that has been carried out for Unit-I catalyst when processing a blend made of 70 vol% *LGO* & 30 vol% *HDT GO* in order to achieve a sulphur content of 8 ppm in the product. The effects of reaction temperature, pressure and different feedstock on the hydrodesulphurization process are discussed to determine the following objectives of this study:

a. To show that refinery Unit-I with the current catalyst can produce 8 ppm sulphur products when processing a blend feed made of 70 vol% *LGO* & 30 vol% *HDT GO*
b. Determine the optimum operating conditions of the actual unit to produce ultra low sulphur content product of 8 ppm with this feed.
c. Study the impact of the process conditions on H_2 consumption and the catalyst cycle length.

2. Experimental

2.1 Hydrodesulphurization experiments

Hydrodesulphurization experiments were carried out according to refinery unit test conditions, with feed stock LGO (Density: 0.8263 g/cc, sulphur: 0.2393 wt%) and blend of 70 vol% *LGO* & 30 vol% *HDT GO* (Density: 0.8285 g/cc, sulphur: 0.2354 wt%) using a fresh catalyst ($CoMo/Al_2O_3$) in a pilot reactor at fixed operating conditions. Experiments were carried out at 52, 42 & 32 bar PPH_2, varying temperature from 312 °C to 332 °C, while maintaining LHSV at 1.0 $hr^{-1}$and H_2/Oil at 190 Nl/l during entire tests. Details of feed properties are mentioned in Table 1. Catalyst sample 100 cc was loaded to a continuous high-pressure reactor and sulphided by typical activation conditions, followed by the introduction of the feed at predetermined process conditions. Feed properties in Table 1 indicates that blend of 70 vol% LGO & 30 vol% HDT GO (*Blend Feed*) used during this experiment is heavier and have more aromatics, in comparison to 100% vol LGO (*Reference Feed*). For these reasons sulfur species presents in the *Blend Feed* will be more difficult to remove and, to reach the targeted 8 ppm in the products, an increase of the reactor bed temperature & H_2 consumption is expected. The pilot plant test was planned for 135 days to generate experimental data.

Table 1. Feed Properties

Sample Description		Test Method	Blend Feed : 70 vol% LGO & 30 vol% HDT GO	Reference Feed : 100 vol % *LGO*
Density @ 15°C	kg/l	ASTM D4052	0.8285	0.8263
Sulphur Total	% wt	ASTM D4294	0.2354	0.2393
Trace Nitrogen	ppm		14.0	3.6
Distillation D86		ASTM D86		
IBP	°C		174.3	214.5
5%	°C		234.3	242.5
50%	°C		272.3	272.7
90%	°C		305.9	295.5
95%	°C		322.8	301.3
FBP	°C		342.9	305.9
Total Aromatics	wt%	ASTM D6591/IP391	26.0	21.7

The catalyst loaded during this hydrodesulphurization experiments is recently-developed CoMo type commercial catalyst, used for the hydrodesulphurization and hydro-treatment of gas oil, diesel and kerosene fractions. Table 2 is the summary of catalyst properties obtained from catalyst manufacturer data sheet.

Table 2. Properties of the catalyst used in the study

Cobalt and molybdenum oxides on very high purity alumina		
Trilobe extrudates		
Diameter	1.6	mm
Apparent Bulk Density (ABD)	710	kg/m^3
Compact Bulk Density (CBD)	840	kg/m^3
Bulk crushing strength	1.5	MPa

2.2 Pilot Plant Configuration

The pilot plant used for this hydrodesulphurization catalyst testing is having following major sections.

- Feed Module (Gas Feed & Liquid Feed)
- Reactor Module
- Product Receiving Module
- Stripping and Atmospheric Column Module
- High pressure Scrubber and Gas Recycle Module
- Off Gas Scrubber Module

Figure 1 shows the pilot plant configuration & briefly the way it operates. This unit is designed to operate in various modes viz. five reactors in series or parallel mode of two & three reactors. For this experiment, pilot plant was configured for five reactors in series with the hydrogen flows in a once-through mode in the reactor, i.e. no recycle.

Reactor pressure and hydrogen partial pressure are the same since the 99.99% purity hydrogen is utilized in pilot units. Hydrogen at low pressure is boosted to high pressure using gas booster and is passed through calibrated mass flow controllers, which is preheated up to 100 °C before mixing with liquid into preheater zone of the reactor. The test feed liquid is weighted on accurate (1 g accuracy) weighing balance and preheated up to 100 °C, before mixing with hydrogen into preheater zone of the reactor. This reactant feed stream introduced into the isothermal fixed bed reactor, which is loaded with 100 cc catalyst in one reactor and operated in a down flow mode. The reactor has sensors for temperature and pressure. The temperature in the reactor is controlled across the fixed bed by using five different thermocouples inserted into the reactor thermowell which is in the center of the reactor. Reactor preheats, catalyst bed and post heat temperatures are maintained precisely (within ± 0.5 °C accuracy). Reactor intermediate sampling is done through automated sample loop. In the separation section, reactor outlet is passed through a high pressure separator (HPS), which is maintained at 60 °C temperature & same pressure as of reactor. The gas phase obtained in the top portion of the separator pass via knock out vessel to the acid gas scrubber. A pressure control valve in the off gas line controls the system pressure and off gas pass though the wet gas meter for volumetric measurement. Automated multiport valve in the off gas line allows diverting off gas sample to online refinery gas analyzer (RGA) during analysis period. Liquid level in the HPS is controlled precisely using vibrating paddle type level transmitter and further goes to low pressure separator (LPS) where pressure is controlled up to 3.5 bar and then drained into a product tank after stripping with nitrogen in the stripper column. Any trace of H_2S sample in the liquid product is removed during nitrogen stripping [23] and stripped sample is analyzed further for product sulphur, density, distillation and gas composition were analysed to assess the performance of the catalysts. This pilot plant has been used for testing of various hydrotreating & hydrocracking catalyst for its activity, selectivity, hydrogen consumption & deactivation rate to evaluate catalyst life, kinetic study data generation & detail reaction kinetic study [24-28].

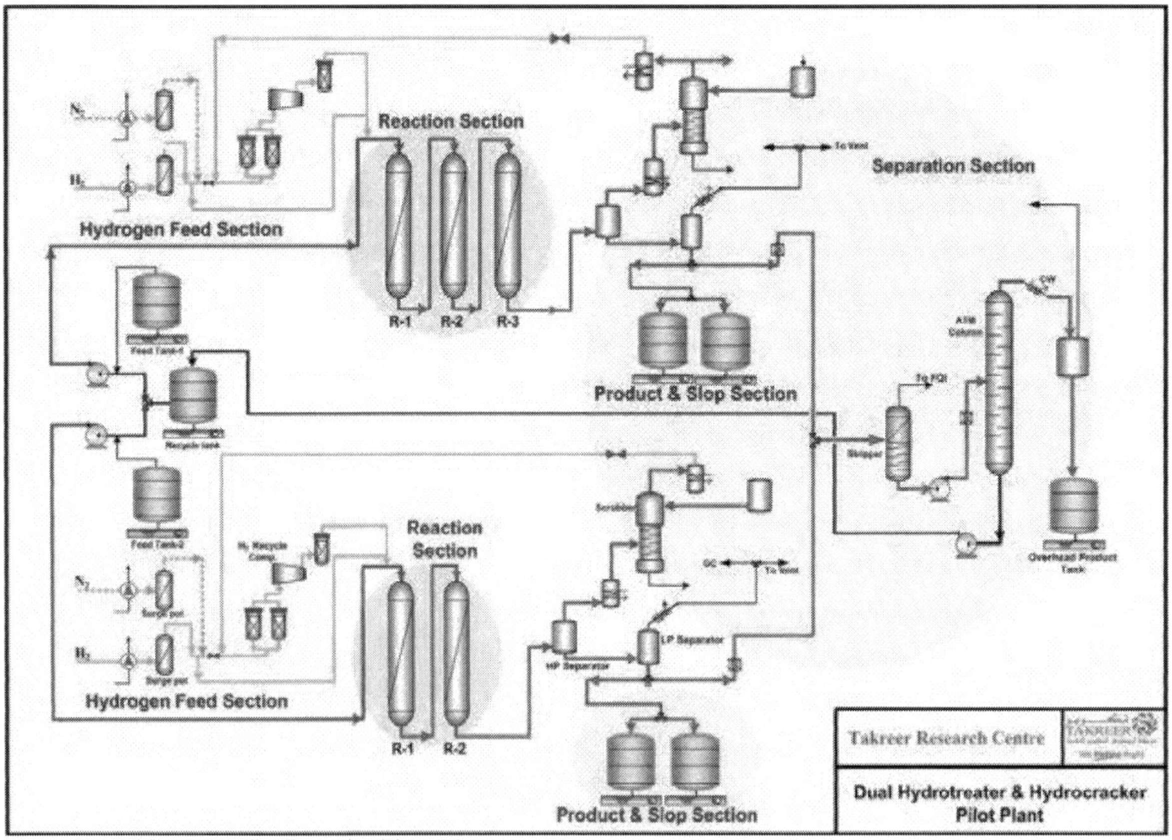

Figure 1. Pilot Plant Configuration Flow Diagram

2.3 Catalyst Sulphiding

For sulphiding this ULSD catalyst, Straight Run Light Gas Oil (SRGO) + Dimethyl Disulfide (DMDS) are used. DMDS quantity was blended to SRGO to adjust the sulphur content of the sulphiding feed to 2 wt%.

Two sulphiding steps at two different temperatures are mandatory to obtain a fully active catalyst:

- First step at 220°C (max outlet temperature 230 °C)
- Second step at high temperature, in the range 345-355°C

2.4 Catalyst Loading

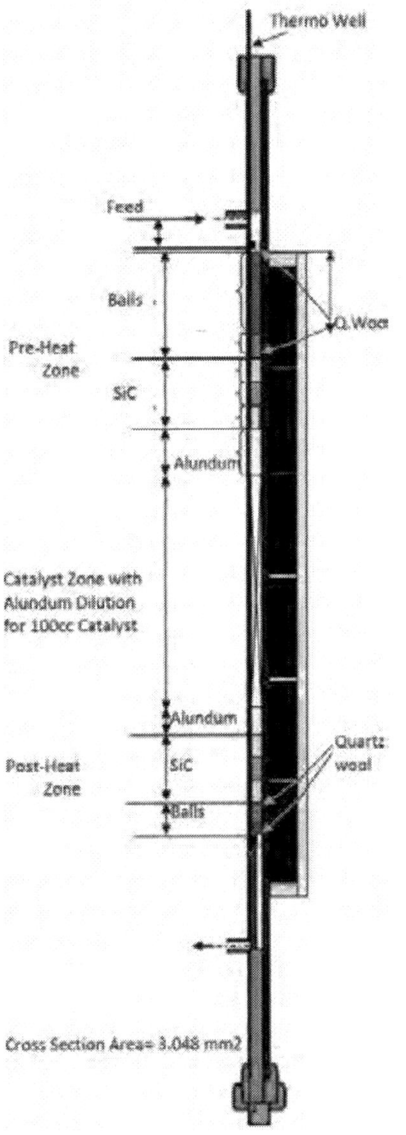

Figure 2. Catalyst Loading Diagram

The catalyst was loaded along with small size diluting particles for better distribution. To reduce pressure build up, plugging and wetting issues, Alundum (80 micron) was applied for diluting the catalyst bed. Different Silica Carbide (SiC) and ceramic balls sizes were applied for other sections of the reactor as supporting material. Both catalyst and diluents were loaded under vibration.

Figure 2 shows the catalyst packing configuration. For this ULSD catalyst packing into the pilot reactor tube, a 1:1 diluents ratio is applied for the whole catalyst bed. Catalyst loading amount has been determined on a mass basis. Catalyst target volume for loading has been derived based on catalyst Loss On Ignition (LOI, actual value) and Compact Bulk Density (CBD), dry base.

Measure the actual catalyst LOI after ramping from room temperature to 600 °C is calculated accordingly to:

$$\text{LOI} = \frac{(b-c)}{a}$$

where:
a = mass of the sample, grams
b = mass of the crucible + sample before ignition, grams
c = mass of the crucible + sample after ignition, grams
NOTE: The sample used for the determination of the loss on ignition can no longer be used for testing and should be discarded.

The weight of catalyst loaded in the pilot plant test is calculated according to:

$$\text{Wc} = \text{Vc}\frac{CBD}{1-LOI}$$

where:
Wc = Weight of catalyst loaded (as such), kg
Vc = target volume of catalyst to be loaded, kg
CBD= Compact Bulk Density (dry base), kg/m^3
LOI= loss on ignition after heating to 600 °C (wt fraction)

2.6 Pilot Plant Test Parameters

Hydrodesulphurization (HDS) experiments were carried out in the pilot plant by nearly 135 days to generate experimental data, as per the test plan parameters listed in Table 3.

Table 3. Test Plan Parameters

Test Point	Day Nr.	Feed	Estimated bed temp. (°C)	Product S cont. (ppm)	LHSV (h^{-1})	H_2 partial pressure (bar)	Reactor Inlet H_2/Oil (Nl/l)
Sulphiding		SRGO + DMDS					
1*	1-54	Ref. & blend Feed	315 to 332	6 to 12	1.0	32- 42 -52	190
2	55	Blend Feed	Adjusted (est: 322-326)	8	1.0	52	190
3	77	Blend Feed	Adjusted (est: 328-332)	8	1.0	32	190
4	106	Blend Feed	Adjusted (est: 324-328)	8	1.0	42	190

* First 54 days of experiment were spent in performing testing & stabilizing the unit. This data are not considered in this paper.

2.5 Analytical Methods

Laboratory analyses were performed on both gas & liquid products, during sulphiding and actual test conditions. Gas Compositions analyzed using online refinery gas analyzer (RGA) and liquid products were tested using trace sulfur & nitrogen analyzer, simulated distillation, X-Ray sulfur analyzer, TBP (True Boiling Point) distillation unit & other lab equipment.

Density of the sample has been obtained as per ASTM D4052 method. The liquid product sample were injected into Agilent 7890 SimDist Gas Chromatograph as per D2887 test method and obtained their distillation curve and correlated to D86 distillation. Actual D86 test also performed from initial boiling point (IBP) to final boiling point (FBP) and obtained distillation recoveries.

The total sulphur content of test feed was determined by the energy dispersive X-ray fluorescence spectroscopy technique using Oxford X Supreme 8000 analyzer, as per ASTM D4294 method (>100 ppm). Low sulphur concentration (sulphur calibration range 0-10 ppm) samples were determined using Antek low level sulphur analyzer 9000 series as per ASTM D5453 method. PAC Multitek analyzer was used to measure the total nitrogen content of some samples as per ATSM D4629 method.

Euro Vector EA3000 CHN analyzer was used for elemental analysis of petroleum liquids viz. C, H & N as per ASTM D5291 method.

Hydrogen analysis of various material balance (in liquid feed and liquid product sample) was carried out using Bruker MQ20 low resolution pulsed nuclear magnetic resonance (H-NMR) as per ASTM D7171 test method which is standard test method for hydrogen content of middle distillate petroleum products by H-NMR spectroscopy. Hydrogen analysis results from H-NMR analyzer were found to be having better repeatability than in CHN analysis.

Off gas analysis from pilot plant was carried out in the online Refinery Gas analyzer (Fast GC- RGA Agilent 7890A), as per UOP 539 method to determine the off gas composition, based on which molecular weight of off gas was considered for material balance and hydrogen consumption calculation.

3. Results and discussion

The test results (Table 4) have shown that it was possible to reach 8 ppm sulphur product using blend feed, with an increase of reactor temperature by 7°C at 52 bar PPH_2, 9°C at 42 bar PPH_2 and 14 °C at 32 PPH_2, while LHSV & H_2/Oil ratio remains same during entire experiment . Hence the blend feed will be more difficult feed to process in refinery unit at full capacity with a sulphur content target set at 8 ppm in comparison with the reference feed. The following Table 4 summarizes the results as compared to the test plan parameters:

From the catalyst pilot test parameters and laboratory analysis of the products at different operating conditions, kinetic parameters (reaction order, pre-exponential factors and activation energies), catalyst deactivation rate and hydrogen consumption were studied.

Table 4. Test points with actual temperature and sulphur concentration in products.

Test Point	Feed	Actual bed temp. (°C)	Product sulphur cont. (ppm)	LHSV (h^{-1})	H_2 partial pressure (bar)	Reactor inlet H_2/Oil (Nl/l)
Sulphiding	LGO + DMDS					
Point 1	Reference Feed	313	13	1.0	52	190
Point 1	**Reference Feed**	**318***	**8.0**	**1.0**	**52**	**190**
Point 2	Blend Feed	320	11	1.0	52	190
Point 2	Blend Feed	322	12	1.0	52	190
Point 2	**Blend Feed**	**325**	**7.8**	**1.0**	**52**	**190**
Point 3	Blend Feed	325	9.1	1.0	42	190
Point 3	**Blend Feed**	**327**	**8.1**	**1.0**	**42**	**190**
Point 4	Blend Feed	327	12	1.0	32	190
Point 4	Blend Feed	330	10	1.0	32	190
Point 4	**Blend Feed**	**332**	**8.0**	**1.0**	**32**	**190**

*calculated point, as reference point

3.1 Reaction Kinetics

Reaction Kinetics were calculated and utilized to predict the SOR temperature required to attain the target ULSD total sulfur. The reaction order for HDS was set 1.3 and the average activation energy was @ 30 kcal/mole based on our previous pilot plant experiment for LGO case using the same catalyst, both of which are in line with those reported for the typical ULSD reaction [26-27,29]. The Table 5 indicates different kinetics data calculation executed during pilot test. Kinetic rate factor calculation is based on Arrhenius law and corrected to calculate WABT for target product sulphur.

$$k = -(C^{1-n}-C_0^{1-n})/(1-n)*LHSV = A^* \exp^{(-E/RT)}$$

$$A= -(C^{1-n}-C_0^{1-n})/(1-n)*LHSV/\exp^{(-E/RT)}$$

C & Co= product & feed sulphur, n= reaction order, A= catalyst activity factor, E= activation energy, R= gas constant

k value is corrected for target sulphur and utilized to estimate WABT. Similar value of $k_{correct}$ (1.4617) for all three test point corresponds to constant target sulfur of 8 ppm and variation in calculated temperature for 8 ppm sulphur target is reflected. Experimental data shows that reaction rate increase with the increase of temperature even if PPH_2 is low which correlates reaction rate using reactor rate equation.

Table 5. Reaction Kinetics

Hydrogen Partial Pressure (PPH_2)	bar	52	42	32
WABT (Actual)	°C	325	325	330
WABT (Calculated @ 8ppm sulfur)	°C	324.8	326.1	332
LHSV	h^{-1}	1.0	1.0	1.0
Feed suphur	ppm	2354	2354	2354
Product sulfur	ppm	7.8	9.1	10
k_o		1.4753	1.3940	1.3460
Ao		-1.35*10^11	-1.28*10^11	-1.00*10^11

3.2 Deactivation Rate

Catalyst's deactivation is important criteria to be considered in its application in commercial unit and has been studied widely and deeply [30-32]. In the ULSD conditions, relatively rapid deactivations were reported and the main cause has been considered by the presence of the carbonaceous (coke) deposits on the catalyst [33-35]. As the coke deposits are supposed to occur under the circumstances in deficit of hydrogen, the deactivation rate is affected by PPH_2.

After the initial testing period to reach the 8 ppm sulphur product, test was continued at return point at 52 / 42 / 32 bar PPH_2 to estimate the catalyst deactivation rate, shown in figure 3. Here the temperature range studied was limited in the narrow range between 320°C and 335°C. In such a narrow temperature range, a linear deactivation rate is generally applicable. The results of deactivation rate study indicated very good stability of the catalyst at 52 bar PPH_2 with a rate of 0.1°C/month. When the PPH_2 is set at 32 bars, the catalyst starts to deactivate at a higher rate around 4°C/month. This increase in the deactivation rate when operating with a lower PPH_2 demands a higher reactor temperature. Nevertheless, 4.2°C / month is a very high deactivation rate even though in pilot plant the deactivation rate is often higher compared to commercial unit. Test shows with the blend feed, decreasing the pressure will reduce strongly the catalyst life cycle.

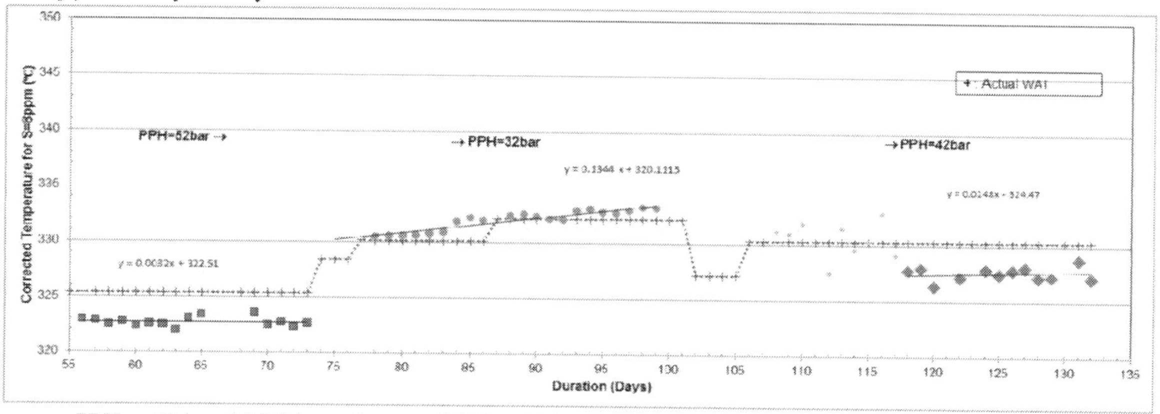

PPH_2 = 52 bar, 0.1 °C / month PPH_2 = 32 bar, 4.2 °C / month PPH_2 = 42 bar, 0.7 °C / month

Figure 3. Trend of actual and corrected reaction temperatures for ULSD at different H_2 pressure

3.3 Hydrogen Consumption

Hydrogen consumption has been estimated by performing a material balance (MB) for the selected test points accompanied by a series of analysis. During these selective test points, material balance accuracy was above 99% (i.e. error was less than 1%). MB correction principle using elemental balance was applied to obtain corrected hydrogen flow and product liquid flow to satisfy both Gross and Net material balance (MB) equations. Ideally, all major elements consisting of hydrocarbon, carbon, hydrogen, sulphur and nitrogen were balanced at in and out of reactor. Hydrogen analysis was obtained by low resolution pulsed nuclear magnetic resonance (H-NMR) using ASTM D7171 test method which is standard test method for hydrogen content of middle distillate petroleum products by H-NMR.

$$\Delta H = (H_{in} - H_{out}) \text{ but } \Delta H <<< H_{in} \text{ or } H_{out}$$

Gross material balance (Material balance in and out of the reactor)

Total feed to the reactor is equal to the total product and following balance is satisfied.
Combined feed gas (H_2 flow) + Feed oil = Combined product gas (Off gas) + Product oil, all in weight

Gross MB: (off gas + product oil) / (hydrogen flow + Feed oil)*100 = 100%

Net material balance (Material balance in and out of the process)

The output from the process, the total of the product gas and oil, should balance the input to the process, the total of makeup hydrogen and feed oil.

Net MB: Feed oil + make up hydrogen (hydrogen consumption) = Product gas + Product oil

And since the hydrogen solution loss can be neglected in pilot plants, make up hydrogen is assumed to be equal to the hydrogen consumption. Then dividing with the feed oil, above co-relation is transformed to:

100 + hydrogen consumption = gas yield + liquid yield

H_2 consumption obtained by optimizing both Gross and Net material balances

Accuracy of the gross material balance and the Net material balance is dependent on gas flow analyses and chemical analyses respectively, thus the latter is considered to be more accurate [36]. Main fluctuation for the former balance is derived from the small difference of mass flow rates of feed H_2, and that of the latter is from small error of liquid product yields. Using the solver in Microsoft excel, both Gross MB and Net MB were perfectly matched with the corrected H_2 feed flow rate and product liquid flow. As a result, the H_2 consumption by gas balance and element balance matched very well, and it was revealed that a typical deviation of H_2 consumption is ± 2 Nl/l by repeated measurements. Table 6 summarize the H_2 consumption & other relevant analyses data, at fixed LHSV=1.0 h^{-1} & H_2/Oil=190 Nl/l during this pilot test study at 52, 42 & 32 bar PPH_2.

Table 6. Hydrogen Consumption

Product Sample	Density kg/l @ 15 °C	H (wt%)	N (ppm)	Total Aromatics (wt%)	RI @ 20 °C	Color ASTM	Sim Dist (°C) (ASTM D2887) by D 86 Correlation 5%-50%-90%-95%	H_2 Cons. (Nl/l)
PPH_2= 52 bar T = 325 °C	0.8214	13.92	0.3	20.1	1.4558	<0.5	224-269-308-327	33 ± 2
PPH_2= 42 bar T = 330 °C	0.8219	13.90	0.3	21.1	1.4560	<0.5	228-269-317-341	28 ± 2
PPH_2= 32 bar T = 332 °C	0.8229	13.76	0.8	21.8	1.4572	<0.5	225-268-309-329	25 ± 2

There is a reduction of hydrogen consumption of 24% when the pressure is reduced by 20 bar. During previous pilot plant test for the same catalyst with the *Reference Feed LGO*, at 52 bar & 32 bar, H_2 consumption of 30 Nl/l and 17.5 Nl/l had been found respectively. With the *Blend Feed* used in this pilot plant test, its an increase of the hydrogen consumption. This can be explained by the aromatics contents (Table 7) which are higher in the *Blend Feed*. However the distillation data obtained by D 86 correlation from GC SimDist (ASTM D2887) test method of the product gasoil is surprisingly stable irrespective of change in pressure. Only 1 or 2 °C difference is seen between the different pressures.

The variation of hydrogen consumption can be explained by looking into the details of the aromatic content in the feed and the products (Table 7). Hydrogen consumption is mainly due to the chemical consumption in the various reactions occurring in the reactor. The most consuming reaction is the saturation of aromatics. For the hydro-desulphurization and for the same level of sulphur in the product, there is no change in its contribution to the overall hydrogen consumption whatever the pressure or the temperature. Due to the low level of Nitrogen in the feed, hydro-denitrogenation does not matter for hydrogen consumption. Mono aromatics increase in product indicates the stepwise conversion of aromatics from tri to di to mono aromatics and then to saturates.

While referring the two cases from Table 7, at 52 bar PPH_2 & 32 bar PPH_2, it is seen that total amount of aromatics have been reduced considerably at 52 bar pressure & also the di-aromatics have almost disappeared. While at 32 bar PPH_2, reduction in the total aromatic content has been less & moreover, almost 20% of initial di aromatics are still present. In this case, it is obvious that it needs less hydrogen to perform this redistribution. It explains why the consumption of hydrogen is lower at 32 bar PPH_2.

Table7. Feed Liquid & Product Liquid Aromatic Content Data

S.N.	Sample Description	Mono Aromatics wt %	Di Aromatics wt %	Tri Aromatics wt %	Total Aromatics wt %
1	Reference Feed	13.9	7.50	0.3	21.7
2	Blend Feed	18.3	7.52	0.14	26.0
3	Product 52 bar	19.3	0.78	0.00	20.1
4	Product 42 bar	19.5	1.57	0.00	21.1
5	Product 32 bar	20.4	1.29	0.08	21.8

4. Conclusion

The pilot plant testing of HDS catalyst using the blend feed, with in the design limits of operating conditions of refinery hydrotreater unit have shown that it is possible to reach 8 ppm ULSD product. This experiment shows that by contriving computational methods and analysis techniques for hydrogen balance & catalyst deactivation rate of pilot plant, it has become possible to calculate reasonable hydrogen consumption & catalyst deactivation rate using very small pilot plant test results to predict commercial unit performance. Compared with the reference feed, the temperature increase needed to reach this performance (using *Blend Feed*) is 7 °C at 52 bar PPH_2, 9 °C at 42 bar PPH_2 and 14 °C at 32 bar PPH_2. At 52 bar PPH_2, hydrogen consumption is higher (compared to the results at 32 bar PPH_2) @ 33 Nl/l and catalyst deactivation rate is just 0.1 °C / month. If pressure is reduced by 20 bar i.e. from 52 bar to 32 bar, chemical consumption of H_2 decreases significantly to 25 Nl/l but catalyst deactivation rate increases to 4.2 °C/month. Considering the high deactivation rate at 4.2°C / month, H_2 pressure at 32 bar operation is not practical. In case of lowering H_2 consumption, it is recommended to choose H_2 partial pressure at 42 bar operation where H_2 consumption is 28 Nl/l and deactivation rate is 0.7 °C / month.

Nomenclature

A	pre-exponential factor
HDS	Hydrodesulphurization
ABD	apparent bulk density (kg/m^3)
C & C0	product & feed sulphur concentration (ppm)
DMDS	dimethyl Disulfide
E	activation energy (kcal/mol)
FBP	final boiling point (°C)
GO	gas oil
H_2/Oil	hydrogen to liquid hydrocarbon flow ratio at the inlet of the reactor (Nl/l)
HDT	Hydrotreated
IBP	initial boiling point (°C)
k_0 & $k_{correct}$	raw & corrected reaction constant of HDS reactions
LGO	light gas oil
LHSV	liquid hourly space velocity (h^{-1})
LOI	loss on ignition
MB	material balance
n	reaction order
NMR	nuclear magnetic resonance
PPH_2	hydrogen partial pressure (bar)
RGA	refinery gas analyser
RI	refractive index
SRGO	straight run light gas oil
SOR	start of run temperature
T	absolute reactor temperature (K)
TBP	true boiling point
WABT	weighted average bed temperature (°C)

Acknowledgement:

Pilot Plant Operations, Laboratory Analysis & Catalysis Research sections at TRC are greatly acknowledge for their inputs in this work. The authors also acknowledge Idemitsu Kosan Co. Ltd. and the Japan Cooperation Center, Petroleum (JCCP) for their technical support.

References:

[1] S. F. Venner, Hydrocarbon Process 2000; 79; 51-60

[2] R. T. Yang, A. J. Hernandez-Maldonado, F.H. Yang, Science 2003; 301; 79-81

[3] S. Brunet, D. Mey, G. Perot, C. Bouchy, F. Diehl, App. Catal. A 2005; 278; 143-172

[4] N. Gupta, P.K. Roychoudhary, J.K. Deb, Appl. Microbiol. Biotechnol. 2005; 66; 356-366

[5] I.V. Babich, J.A. Moulijn, Fuel 2003; 82; 607-631

[6] C. Song, X. L. Ma, Appl. Catal, B 2003; 41; 207-238

[7] B.H Cooper, K.G. Knudsen, Practical advances in Petroleum Processing 2006; 10; 297

[8] M. Breysse, G, Djega-Mariadassou, S. Pessayre, C. Geantet, M. Vrinat, G. Perot, M. Lemaire, Catal. Today 2003; 84; 129

[9] C. Song, Catal, Today 2003; 86; 211-263

[10] S.G. Petrov, A.M. Donov, D.S. Stratiev, Int. Res. Publication, ISSN; 2002/03; 3; 1311

[11] H. Topsoe, B. Hineemann, J.K. Norskov, J.V. Lauritsen, F. Besenbacher, P.L. Hansen, Catal. Today 2005; 12; 107-108

[12] T. Fujikawa, Top. Catal, 2009, 52, 872

[13] M. Breysse, C. Geanter, P. Afanasiev, J. Blanchard, M. Vrinat, Catal. Today; 2008; 3; 130

[14] K. Segawa, K. Takahashi, S. Satoh, Catal. Today; 2000; 63; 123

[15] V. Costa, K. Marchand, M. Digne, C. Geanter, Catl. Today 2008; 69; 130

[16] N. Frizi, P. Blanchard, E. Payen, P. Baranek, C. Lancelot, M. Rabeiulleau, C. Dupuy, J. P. Dath, Catal. Today 2008, 32; 130

[17] S.T. Oyama, T. Gott, H. Zhao, Y.K Lee, Catal. Today 2009; 94; 143

[18] M. Andari, F. Abussedo, A. Stanislaus, H. Qabazard, Fuel, 1996, 75, 1664-1670

[19] K.G. Knudsen, B.H Cooper, H. Topsoe, Appl. Catal. A: Gen. 1999; 189; 205-215

[20] R.R. Bharvani, R.S/ Henderson, Hydrocarb. Process 2002; 81; 205-215

[21] Erik Resendiz, Jorge Ancheyta, Arnulfo Rosales–Quintero, Gustavo Marroquin, Fuel 2007; 86; 1247-1253

[22] Kemsley J. Chem Eng News 2003; 81(43):40

[23] R. Iwamoto, H. Koshika, N. Takahashi, K. Inamura, J. Japan Petroleum Institute 2005; 48; 319-323.

[24] H. Topsoe, B. S. Clausen, F. E Massaoth, in "Hydrotreating Catalysts Science and Technology", Fuel Science and Technology International, Springer-Verlab New York 1996.

[25] H. Toulhoat, P. Raybaud, "Catalysis by Transition Metal Sulphides", Edition Technip, 2013.

[26] A. Stanislaus, A. Marafi, M. S. Rana, Catal. Today 2010; 153; 1-68.

[27] J. Ancheyta, "Modeling of Process and Reaction for Upgrading of Heavy Petroleum", CRC Press, 2013

[28] C.L. Garcia, D. Hudebine, J.-M. Scweitzer, J.J. Verstraete, D. Ferre, Catal. Today 2010; 150; 279-299.

[29] Ikuo Saito et. al., Proceedings of 13th Saudi-Japanese Catalyst Symposium Dhahran, Saudi Arabia, December 2003.

[30] J. Oudar, H. Wise, "Deactivation and Poisoning of Catalysts", Marcel Dekker, Inc., 1985.

[31] E. Furimsky, F.E. Massoth, "Deactivation of Hydroprocessing Catalysts", Catal. Today 1999; 52; 381-495.

[32] L.E. Kallinikos, G.D. Bellos, N.G. Papayannakos, Fuel 2008, 87, 2444-2449

[33] B. Guichard, Chapter 2.5 "Life Cycle of an HDT Catalyst" in "Catalysis by Transition Metal Sulphides" Eds. H. Toulhoat, P. Raybaud, Edition Technip, 2013.

[34] P. Dufresne, Appl. Catalysis A 2007; 322; 67-75.

[35] N. Koizumi, Y. Urabe, K. Inamura, T. Itoh, M. Yamada, Catal. Today 2005; 106; 211-218.

[36] Claudia De la Paz-Zavala, Esteban Burgos-Vázquez, Javier Esteban Rodríguez-Rodríguez, Luis Felipe Ramírez-Verduzco, Fuel 2013; 110; 227-234

AICHE – Atlanta

Reactivity Descriptor in Solid Acid Catalysis: Predicting Turnover Frequencies for Propene Methylation in Zeotypes

Chuan-Ming Wang, Rasmus Y. Brogaard, Bert M. Weckhuysen, Jens K. Nørskov, and Felix Studt*

SUNCAT Center for Interface Science and Catalysis, SLAC National Accelerator Laboratory, 2575 Sand Hill Road, CA

* *studt@slac.stanford.edu*

Recent advances in the understanding of reactivity trends for chemistry at transition-metal surfaces have enabled in silico design of transition metal catalysts by employing a descriptor-based approach, establishing a correlation between a few well-defined properties of a material and its catalytic activity.[1] This theoretical work aims for a similar approach in solid acid catalysis, focusing on the reaction between propene and methanol catalyzed by Brønsted acidic zeotype catalysts. Experimentally, the ammonia heat of adsorption is often used as a measure of the strength of acid sites.[2] Using periodic DFT calculations, we show that this measure can be used to establish scaling relations for the energy of intermediates and transition states, effectively describing the reactivity of the acid site. This allows us to use microkinetic modeling to predict a quantitative relation between the ammonia heat of adsorption and the rate of propene methylation from first principles (see Figure 1).[3] We propose that this is the first step toward descriptor-based design of solid acid catalysts.

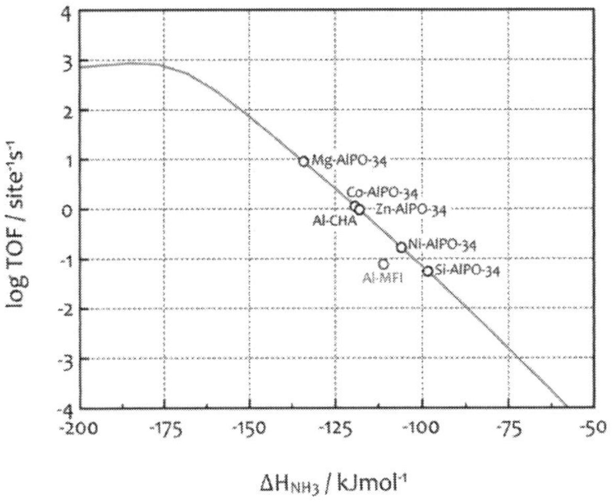

Figure 1. The relation between calculated $\Delta H_{ads}(NH_3)$ and the simulated TOF (see reaction conditions in the Computational Details section) for the methanol–propene reaction catalyzed by Me-CHA and Me-AlPO-34 materials. The calculated $\Delta H_{ads}(NH_3)$ values of selected catalysts are highlighted. The Al-MFI point represents the TOF extracted from the experiments.[4]

References

[1] J. K. Nørskov, F. Abild-Pedersen, F. Studt, T. Bligaard, *Proc. Natl. Aca. Sci. U. S. A.* **2011**, *108*, 937.
[2] E. Derouane, J. Védrine, R. R. Pinto, P. Borges, L. Costa, M. Lemos, F. Lemos, F. R. Ribeiro, *Catal. Rev.-Sci. Eng.* **2013**, *55*, 454.
[3] C. M. Wang, R. Y. Brogaard, B. M. Weckhuysen, J. K. Nørskov, F. Studt, *J. Phys. Chem. Lett.* **2014**, *5*, 1516.

Kinetic analysis of 4-nitrophenol hydrogenatioin catalyzed by bio-inspired shaped platinum nanoparticles

Xiaodi Li, Guozhu Li,* Wanting Zang, Li Wang, Xiangwen Zhang

Corresponding author. Tel & Fax:+86-22-27892340, Email: gzli@tju.edu.cn

Aqueous-phase hydrogenation of 4-nitrophenol (4-NP) was investigated in the presence of free-standing or supported Pt nanoparticles (NPs) of various shapes. Uniform cubic and *pseudo*-tetrahedral Pt NPs with sub-10-nm size were obtained by the direction of bio-inspired small molecules, 3-hydroxybutyric acid (3-HB) and tropic acid (TA), respectively. The catalytic activity was found to be strongly affected by the nanoparticle shape directed by the bio-inspired capping agent and the support of SBA-15. *Pseudo*-tetrahedral Pt NPs directed by TA have superior activity than the cubic ones with 3-HB. But when loaded on SBA-15, cubic Pt(3-HB) NPs show better performance than Pt(TA) NPs. The experimental data are fitted to the Langmuir-Hinshelwood (L-H) model involving a surface reaction controlling mechanism for 4-NP hydrogenation. We study the kinetic process of catalytic hydrogenation for 4-NP with computational simulation by the matrix laboratory to obtain the kinetic model most suitable for our system. This model follows a dual-site adsorption with molecular adsorption of substrate and dissociative adsorption of hydrogen. Results show that *pseudo*-tetrahedral Pt(TA) NPs have lower apparent activation energy than the cubic Pt(3-HB) NPs for 4-NP hydrogenation. But after loading these NPs onto SBA-15, apparent activation energy of Pt(3-HB)/SBA-15 decreases while that of Pt(TA)/SBA-15 increases.

376736 Effects of d-Band Shape on the Surface Reactivity of Transition-Metal Alloys

Abstract:

Transition-metal alloys represent an important class of materials that are used in many commercial catalytic processes. Rational design of alloy catalysts with optimized performance and minimal cost relies on a fundamental understanding of the electronic structure of an ensemble of surface atoms and its relationship with energetics of adsorbate-surface interactions. For that matter, identification of simple reactivity descriptors from complex electronic properties of materials becomes tremendously important. The d-band model, developed in the 90s', is the standard model in surface chemistry for understanding variations in chemisorption energies of adsorbates from one metal surface to another. In general, a metal site with higher d-band center exhibits stronger affinity to adsorbates. This simple design principle has proven to be extremely useful in search of optimal catalytic materials in many chemical and electrochemical reactions. However, outliers exist in many technological relevant alloy systems, i.e., increasing (decreasing) the d-band center of a metal site is not always associated with stronger (weaker) chemical bonding, which indicates the need for better electronic descriptors.

In this talk, I will start with a brief introduction to some conceptual aspects of the theory of surface chemisorption, followed by the discussion of alloy systems where the d-band center failed to describe the trend of surface reactivity. In particular, I will emphasize the effects of the shape of the d-band of a metal site, represented by higher moments of the d-band, on energetics of surface chemical bonding. The upper d-band edge, defined as the highest peak position of the Hilbert transform of the density of states projected onto d-orbitals of an active metal site, is identified as an electronic descriptor for the surface reactivity of transition metals and their alloys, regardless of variations in the d-band shape. The utilization of the upper d-band edge with scaling relations enables a considerable reduction of the parameter space in search of improved alloy catalysts and further extends our understanding of the relationship between the electronic structure and chemical reactivity of metal surfaces.

1. Xin, H., Vojvodic, A., Voss, J., Nørskov, J. K. & Abild-Pedersen, F. Effects of d-band shape on the surface reactivity of transition-metal alloys. *Phys. Rev. B* **89,** 115114 (2014).
2. Vojvodic, A., Nørskov, J. K. & Abild-Pedersen, F. Electronic Structure Effects in Transition Metal Surface Chemistry. *Top Catal* **57,** 25–32 (2014).

378995 Screening for Active Sites on Supported Metal-Oxide Catalysts Via Hybrid- Monte Carlo/Molecular Dynamics Simulations with the Reaxff Potential

Abstract:

The rational design of supported metal-oxide catalysts requires a detailed understanding of the catalyst surface structure as a function of temperature and pressure in the reactant gas phase. Generally, the gas phase can rearrange the metal/oxide lattice structure, forming unique surface, subsurface, or metal/oxide interfacial phases - each with catalytic behaviors that greatly vary from the parent system. Additionally, numerous phases or structures can co-exist on the catalyst surface, making it difficult to determine which phases are catalytically active, especially if the active site is attributed to a minority of the possible phases. Here, we demonstrate how ReaxFF-based Monte Carlo and molecular dynamics simulation techniques can be used in tandem to predict phase structure and stability as a function of temperature and pressure in the reactant phase above the catalyst. Using this method, we can generate an extensive set of catalyst structures that are representative of various stable geometries under the specified reaction conditions. These geometries then serve as appropriate models for subsequent kinetic studies assessing catalytic activity (either *via* quantum reaction barrier calculations or reactive molecular dynamics simulations), thus allowing geometries containing catalytically active sites to be identified. This is initially demonstrated by identifying under-coordinated sites on free Pd nanoclusters that are susceptible to Ar-induced surface reconstructions, in agreement with experimental findings. It is then extended to identifying unique sites on supported (TiO_2 and CeO_2) Pd clusters, which readily form oxide, hydride, and carbide phases impacting catalytic activity. Although demonstrated for Pd-based catalysts, this approach is readily extendable to other systems, providing a new screening tool for identifying active and selective sites under varying reaction conditions.

383940 Tuning Acid-Base Properties of Znxzryoz for Ethanol-to-Isobutene Conversion

Abstract:

With environment deterioration from global warming and the depletion of fossil resources, the fundament research on fuel production has been being shifted, focusing on the sustainable production of bio-fuels. As the most natural abundant renewable resource, biomass has been considered one of the most promising ones with the potential to replace petroleum in production of chemicals and liquid transportation fuels.[1-4] Isobutene is an important industrial intermediate, which has been widely used for the production of fuel additives (e.g. MTBE, ETBE and isooctane), butyl polymers, antioxidants, tert-butylamine, and tert-butanol et al.[3] Currently, isobutene is mainly produced by steam cracking of fossil resource. With the increasing demand/wide application of isobutene and the dwindling reserve of fossil fuel, it is highly demanded to explore sustainable techniques for isobutene production. We developed a new $Zn_xZr_yO_z$ catalyst with balanced acid-base pairs, on which a cascade reaction from bio-ethanol to isobutene has been achieved via acetone intermediate.[5-6] In this work, the effects of catalyst composition, reaction temperature, residence time, steam to carbon ratio, and ethanol molar fraction on isobutene yield have been studied. Isobutene yield as high as 79% was achieved with an ethanol concentration of 8.3% at 475 °C over $Zn_1Zr_8O_z$. Further durability and regeneration tests showed that the catalyst shows very slow deactivation via coke formation with isobutene yield maintained above 75% for more than 10 hours, and that the catalysts activity can be fully recovered after calcination in air at 550 °C for 2.5 hours. Characterizations such as XRD, TGA, Nitrogen sorption, and *in situ* FTIR techniques have been employed to correlate structure to the performance of the catalysts. Especially, the effect surface acid-base property on the reaction of reactant (e.g., ethanol) and possible intermediate (i.e., acetone) has been investigated and correlated.

References

1. Liu, C.; Wang, H.; Karim, A. M.; Sun, J.; Wang, Y., *Chem. Soc. Rev.*, DOI: 10.1039/c3cs60414d (2014).

2. Serrano-Ruiz, J. C.; Luque, R.; Sepulveda-Escribano, A., *Chem. Soc. Rev.* 40, 5266 (2011).

3. van Leeuwen, B.; van der Wulp, A.; Duijnstee, I.; van Maris, A.; Straathof, A., *Appl. Microbiol. Biotechnol.* 93, 1377 (2012).

4. Sun, J.; Wang, Y., *Acs Catal.* 4, 1078 (2014).

5. Sun, J.; Zhu, K.; Gao, F.; Wang, C.; Liu, J.; Peden, C. H. F.; Wang, Y., *J. Am. Chem. Soc.* 133, 11096 (2011).

6. Liu, C.; Sun, J.; Smith, C.; Wang, Y., *Appl. Catal., A* 467, 91 (2013).

390012 A Generalized Thermodynamic Activity and Reactivity Relation of Metal and Oxide Clusters and Their Periodic Reactivity Trend for Methanol Oxidative Dehydrogenation

Abstract:

Catalytic oxidation of alkanol and oxygen on transition metal or metal-oxide clusters leads to aldehydes, esters, syngas, or combustion products (CO_2, H_2O) with rates and selectivities depended largely on metal identity, coverages of surface intermediates, bulk chemical state, and operating pressures. We report, in this contribution, the mechanistic synergy and a generalized relation between the reactivity for methanol oxidation dehydrogenation and the thermodynamic quantity at Pt, Ru, Pd, Ag cluster surfaces with metal or oxide bulk, after rigorous elimination of transport artifacts by extensive dilution at intra-pellet and bed scales. First-order rate coefficients for CH_3OH-O_2catalysis on Pt, Ru, Pd, and Ag clusters are a single-valued function of the operating oxidant-to-reductant ratio, irrespective of the metal identity and bulk chemical state. We show that the oxidant-to-reductant ratio is a surrogate of the oxygen chemical potential through kinetic coupling of the oxygen and methanol activation steps for the case of irreversible oxygen activation. The ratio, as a result, dictates the identity and coverages of reactive intermediates at cluster surfaces as well as the chemical state of the cluster bulk. The single-value functional dependence of rate coefficients on oxidant-to-reductant ratio provides the direct, macroscopic relation between kinetic properties (rate coefficients) and oxygen chemical potential at cluster surfaces.

The rate coefficient values for methanol activation on Pt, Pd, and Ag cluster surfaces covered with reactive oxygen or on RuO_2 differ by five orders of magnitude in the decreasing reactivity order of Pt>Pd>Ru>Ag. These rate coefficients, when plotted against the oxygen binding energies, resemble a volcano type relation and contain the activation barrier and pre-exponential factor terms, which in the framework of transition state theory reflect the activation enthalpy and entropy required for the formation of O-H scission transition state of CH_3OH. The barriers on O* covered metal cluster surfaces (Ag: 77 kJ mol^{-1}; Pt: 84 kJ mol^{-1}; Pd 90 kJ mol^{-1}) increase as the O* binding strength on metal site increases, following a Brønsted-Evans-Polanyi type relation and then decrease on RuO_2 surfaces as the binding strength increases further. These reactivity trend with O* binding strength reflects that weakly bound O* are more effective as Brønsted base during the H abstraction step on oxygen covered Ag, Pt, and Pd surfaces and that there is a marked difference in active sites and O* reactivity between the RuO_2 and the O* covered metal surfaces. In addition to barriers, the reactivity difference among the metals is caused predominantly by the differences in the extent of CH_3O fragment stabilization at the CH_3OH activation transition state, as shown from the wide ranging activation entropy values (Pt=0±10, Pd=-18±10, Ag=-110±10, and RuO_2=-109±10 J mol^{-1} K^{-1}).

In summary, we report a generalized correlation between the first-order rate coefficient and the oxidant-to-reductant ratio and a direct connection of this ratio to the oxygen chemical potential at the metal or oxide cluster surfaces. The observed reactivity trend of Pt, Pt, and Ag metal and RuO_2 clusters reflects the reactivity of surface oxygen towards H abstraction and the extent of stabilization of the CH_3OH activation transition state.

Acknowledgement

We acknowledge Natural Sciences Council of Canada (NSERC) and Canadian Foundation for Innovation (CFI) for their financial supports.

References

[1] WF. Tu and Y-H. Chin, *J. Catal.* **313** (2014) 55-69.

[2] WF. Tu and Y-H. Chin, *Angew. Chem. Int. Edit.*, submitted.

Understanding Zeolites with the Tools of Surface Science: Using Scanning Tunneling Microscopy, X-ray Photoelectron Spectroscopy and Infrared Reflection Absorption Spectroscopy to Study a Two-Dimensional Zeolite

J.A. Boscoboinik

Center for Functional Nanomaterials, Brookhaven National Laboratory. Bldg. 735, Upton, NY 11973 (USA).

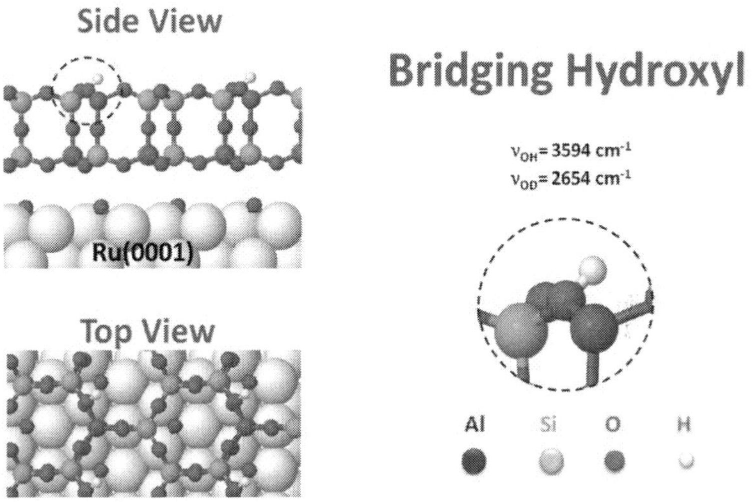

Although Surface Science has provided useful insights into a variety of materials of interest for catalysis, its contribution to the understanding of zeolites has been limited. This was mainly due to the lack of a well-defined surface that successfully mimics the properties of these important materials while allowing its analysis using the vast toolkit of surface science. Recently, a bilayer aluminosilicate structure that counts with bridging hydroxyl groups exposed on the surface was produced, [1] providing a suitable model system for surface science studies of zeolites. We will refer to this system as a "two-dimensional" zeolite. The hydroxyl groups present in the structure are very strong Brønsted acid sites and, in zeolite catalysts, they are the active sites for catalytic energy transformations, such as the cracking of crude oil and the methanol to gasoline conversion. This model system allows then to study the interaction of molecules involved in these catalytic processes and potentially contribute to the understanding of these chemical transformations. A description of the system itself as characterized in ultra-high vacuum (UHV) conditions will be given. [1,2] This will be followed by an analysis of the interaction of a series of probe molecules [3] in conditions ranging from UHV to near-ambient pressures. Both weak (CO and C2H4) and strong bases (NH3 and Pyridine) were used as probe molecules in this study. While strong bases readily protonate in contact with the bridging hydroxyl, weak bases bind weakly to the active site. The acidity of the bridging hydroxyl group was gauged by the red-shift in the O—H vibration when interacting with weak bases. The acidity measured in this way for the 2D-zeolite was similar to some of the most active 3D-zeolites, validating its use as a model system for further reactivity studies.

References

[1] J.A. Boscoboinik, X. Yu, B. Yang, F.D. Fischer, R. Wlodarczyk, M. Sierka, S. Shaikhutdinov, J. Sauer, H.-J. Freund, *Angew. Chem. Int. Ed.* 51 (**2012**) 24, 6005-6008. *Angew. Chem.* 124 (**2012**) 6107-6111.
[2] J.A. Boscoboinik, X. Yu, B. Yang, S. Shaikhutdinov, H.-J. Freund. *Micropor. Mesopor. Mater.* (**2013**) 165, 158-162.
[3] J.A. Boscoboinik, X. Yu, E. Emmez, B. Yang, S. Shaikhutdinov, F. Fischer, J. Sauer, H.-J. Freund. *J. Phys. Chem. C* (**2013**) 117, 13547-13556.

Research carried out in part at the Center for Functional Nanomaterials, Brookhaven National Laboratory, which is s upported by the U.S. Department of Energy, Office of Basic Energy Sciences, under Contract No. DE-AC02-98CH10886. The author acknowledges the Alexander von Humboldt Foundation for a fellowship, under which part of this research was carried out, during a stay at the Fritz Haber Institute of the Max-Planck Society in Berlin, Germany.

Mechanistic studies and modeling of the Vorbrüggen Reaction: using PAT and Reaction modeling to gain understanding of impurity formation

Authors: **Ben Cohen**, Greg Beutner, Michael Bultman, Junying Fan, Brenda Remy, Chris Sfouggatakis

Chemical Development, Bristol-Myers Squibb Co.,

Session: 20012 Reaction Engineering in Pharmaceuticals and Fine Chemicals

Abstract:

The Vorbrüggen Reaction is a popular method for the formation of nucleosides by treatment of a protected sugar acetate with a silylated heterocycle in the presence of a Lewis acid. Reaction performance is influenced by a wide range of parameters, including temperature, concentration, reagent charge rate, and stoichiometry of the reagents. Implementation of this chemistry during the development of a pharmaceutical candidate revealed the formation of undesired impurities which impacted both product quality and step yield. Due to the complexity of the system, the reaction mechanism was explored to enable reaction modeling, which led to a thorough understanding of the key reaction parameters.

An understanding of the reaction mechanism and formation of key impurities was developed through a combination of in situ FTIR monitoring, molecular modeling and kinetic modeling. The reaction mechanism rationalizes how changes in reaction conditions affected the rates of the desired reaction compared to impurity formation. The understanding of the reaction mechanism helped identify robust process conditions and supported the scale-up efforts to produce >2000 kg of product.

This presentation will focus on the application of PAT (FTIR) and reaction modeling (using Dynochem) to elucidate the reaction mechanism, and how this knowledge impacted process development.

374813 Catalysts for the Positional Isomerization of Long-Chain Olefins

Abstract:

Internal, long-chain olefins are primarily used in the synthesis of alkenyl succinic anhydrides (ASA), which are used as paper sizing agents. The purpose of this project is to compare catalysts for the selective double-bond isomerization of terminal, long-chain olefins – both strong acid or base catalysts and organometallic chain-walking catalysts. The application is a stringent test of a catalyst's selectivity because there are three possible classes of products: positional isomers, branched isomers and oligomers. Catalysts were evaluated by the positional isomerization of 1-hexadecene. Solid acid catalysts included sulfonated polystyrene/DVB, tungstated zirconia, acidic zeolites, and perfluorinated ion exchange resins supported on silica and alumina. The supported base catalyst is Na/Al_2O_3. Chain walking catalysts are based on $Fe(CO)_5$ and its derivatives. Catalysts were characterized by porosimetry, 1-propanamine TPD, and powder XRD. The solid catalysts were tested in packed bed reactors at temperatures up to 180°C. The reaction products were quantified by GC, GPC, HNMR, and UV-Vis. Lifetime studies were conducted on those solid acid and base catalysts with the highest selectivity to internal olefins. Amberlyst 35® (Dow), an "oversulfonated" poly(styrene-*co*-divinylbenzene) deactivated after 12 d but was regenerable with some loss of activity. SAC-13 (Nafion®/SiO_2, BASF) lasted over 10 d without deactivation. It is hypothesized that the reason for the disparity is the nature of the acid sites. An ensemble of closely-spaced acid sites are present in the sulfonated poly(styrenes), so the oligomers that are formed as byproducts cover more acid sites on Amberlyst 35 than SAC-13. SAC-13 also accumulates oligomers, but the individual acid sites are spaced further apart. The soluble $Fe(CO)_5$ catalyst was tested in a 4 L autoclave at 180°C by varying catalyst concentration (250-1000 ppm) and residence time (4-16 h). Our efforts in immobilizing derivatives of $Fe(CO)_5$ on solid supports are also described.

Development of an Immobilized Transamination Process for the Manufacture of API

Jonathan P. McMullen[1], Jon A. Jurica[1], Shane T. Grosser[1], Matthew D. Truppo[2]
[1]Chemical Process Development and Commercialization, Merck Manufacturing Division
[2]Department of Process Research, Merck Research Laboratories
Merck & Co., Inc., P.O. Box 2000, Rahway, New Jersey 07065

Within the pharmaceutical industry, a significant amount of effort is spent in the development of catalytic processes that produce enantiomerically pure chiral amines. Selective chemocatalytic approaches involve the need to screen numerous ligand/metal systems to identify catalysts that provide acceptable yield and enantiomeric excess, and typically require an isolation step to remove the undesired enantiomer. For manufacturing, chemocatalytic methods are further complicated by volatility in pricing of precious metals and the need to remove residual metals from the API to low levels. In biocatalytic processes, however, an enzyme catalyst can be more easily tailored to synthesize only the desired chirality. By immobilizing the enzyme to a solid support, additional process advantages can be realized. These improvements can include the reuse of the enzyme, the opportunity to operate in a wider array of solvent choices, and the ability to cost-effectively run the reaction at high catalyst loadings. Herein, we present the application and development of an immobilized transaminase process for the production of an active pharmaceutical ingredient.

An adsorption-based immobilization process that provides a highly active and selective enzyme was developed. Through immobilization, the enzyme is reused across multiple batches, significantly decreasing yearly enzyme requirements and the associated cost. Stability investigations demonstrated that enzyme activity is retained for over 400 hours at reaction conditions (caustic, organic solvent, 60 °C). This immobilized enzyme was used in a packed bed reactor to develop a transamination process that has line-of-sight to manufacturing. The transamination conditions were optimized to maximize product yield while satisfying constraints on cycle time and other operating considerations. Opportunities to recycle unreacted substrate through the packed bed reactor were also explored. Through this development, results were successfully scaled up from a ~10 mL reactor to a 1.2 L reactor to produce API at rate of ~1 kg/day.

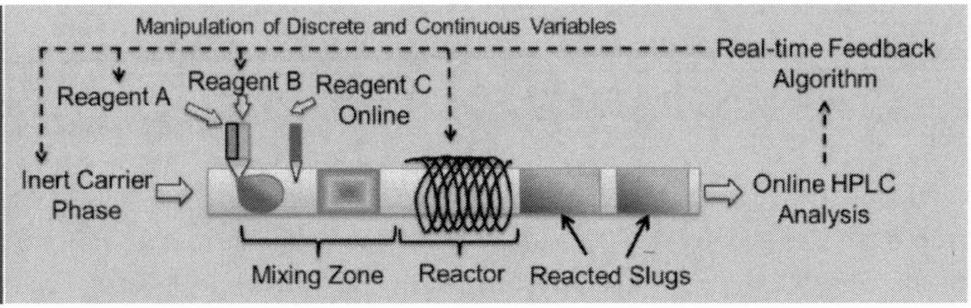

Simultaneous Screening and Reaction Optimization in an Automated Segmented-Flow Microreactor System

Brandon J. Reizman and Klavs F. Jensen

AIChE Annual Meeting: November 20, 2014

Simultaneous Screening/Optimization Overview

The Need for Optimization over Discrete and Continuous Variables

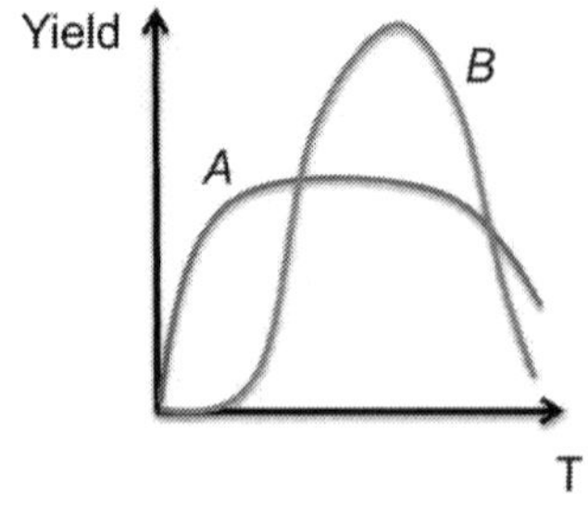

Reactions in Droplets: A Tool for Scalable Results

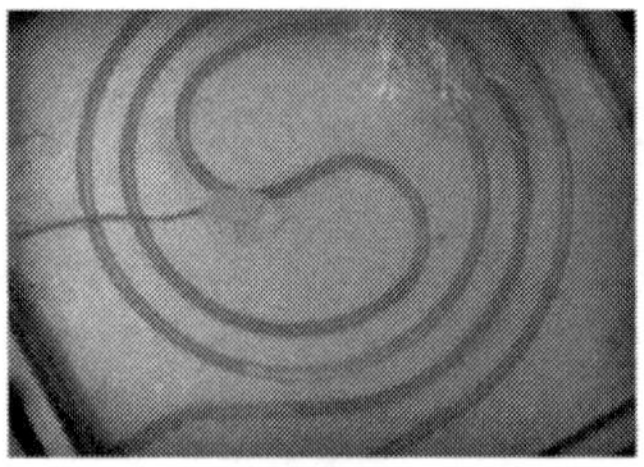

Implementation of a Real-time Optimization Algorithm

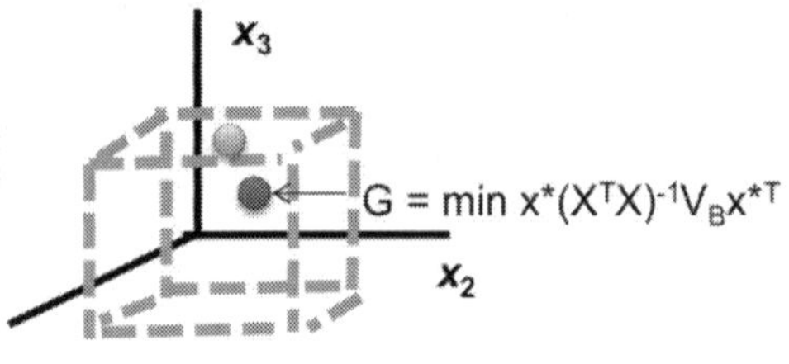

Application and Insight into Chemical Synthesis

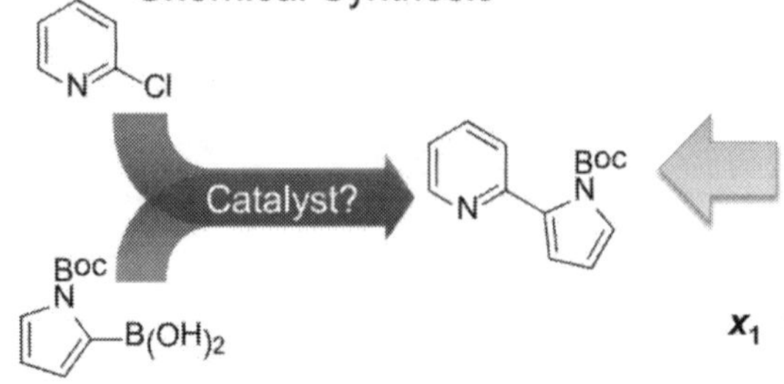

2

MIT NOVARTIS

Discovery and Development in Pharma are Challenged by Many Variables, Costly Experiments

- Consider a Suzuki-Miyaura cross-coupling reaction as an example

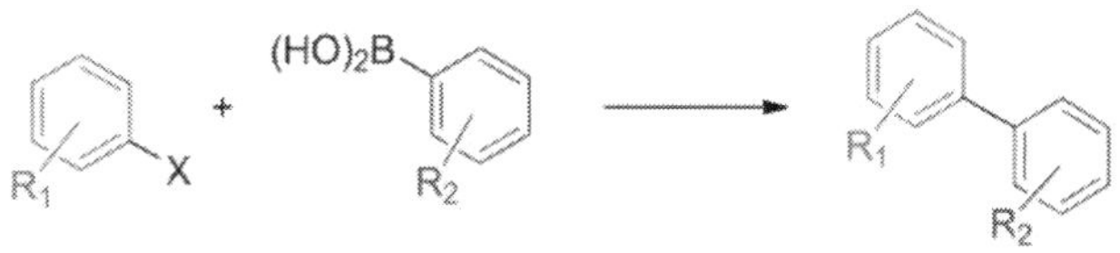

Variables to Consider

Continuous Variables

Temperature
Reaction Time
Catalyst Loading
Substrate Concentration
Base Concentration
Aqueous:Organic Ratio

Discrete Variables

Catalyst Source
Ligand
Solvent
Base
Phase Transfer Catalyst

3

Traditional Pharma Optimization Treats Discrete and Continuous Variables Separately

For Continuous Variables

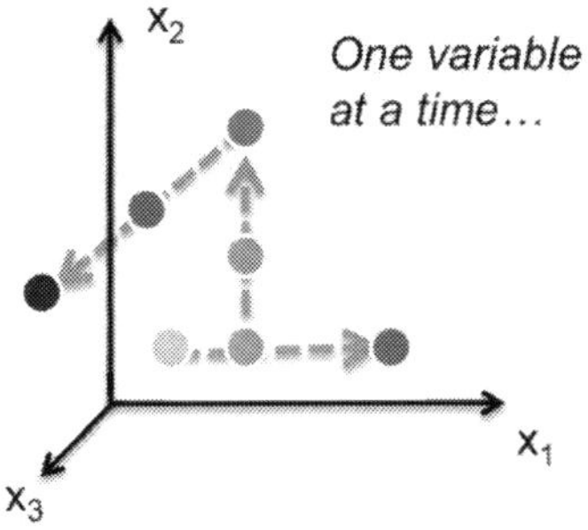

For Discrete Variables

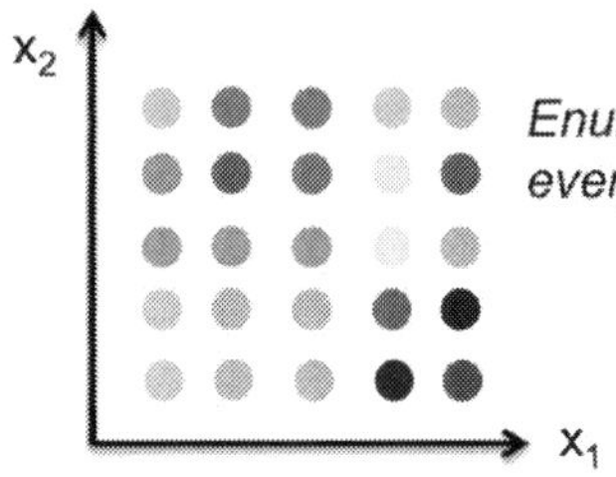

- But this is problematic when discrete and continuous variables interact

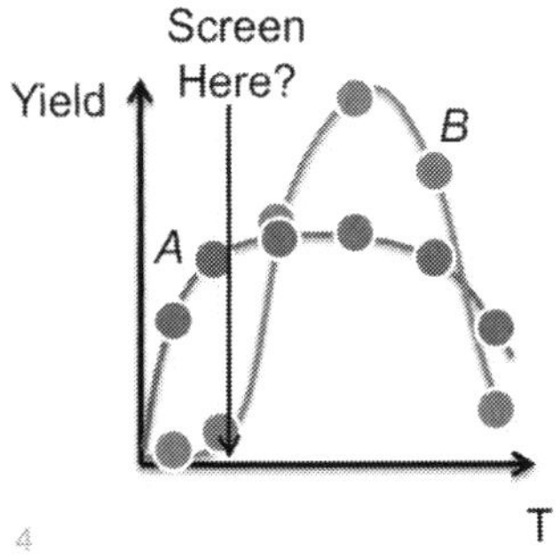

- With the one-variable-at-a-time approach, a screen at low T would miss identifying *B* as the better discrete variable
- With enumeration, we may waste many experiments resolving the maximum for *B*

4

With Microfluidic Slugs, We Can Screen Reagents while Manipulating Continuous Variables

- Reactions in slugs ensure good mixing and heat transfer for accurate reaction optimization and scalability

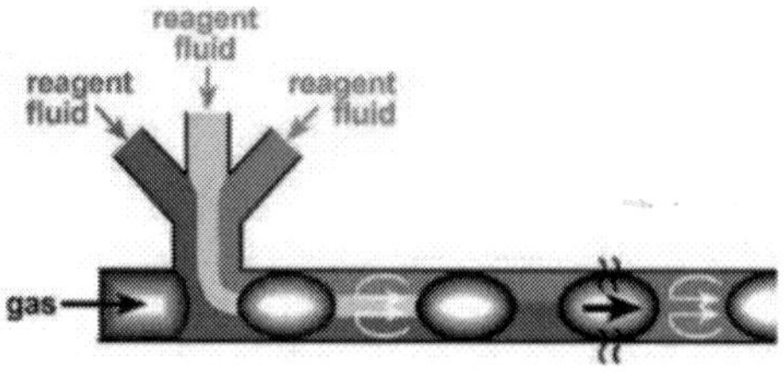

Song, H. *et al.*, *Angew Chemie Int Ed* **2006**, *45*, 7336.

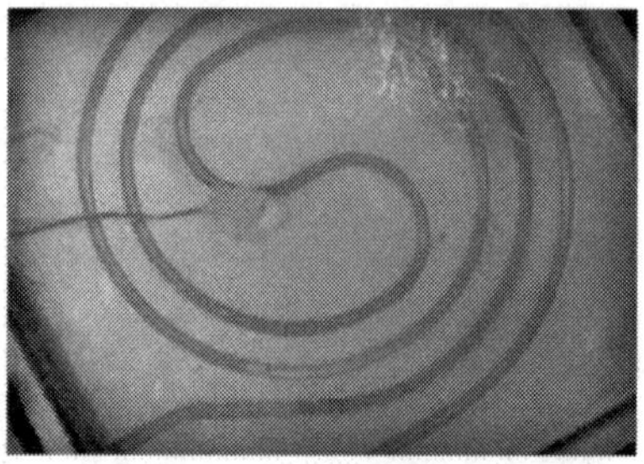

- Individual slug preparation allows reagents to be modulated

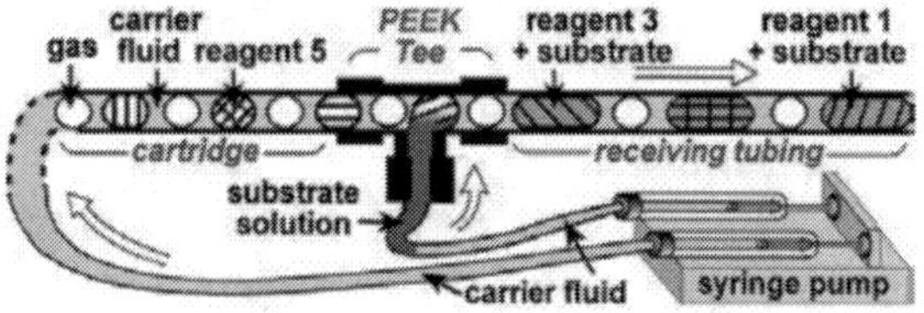

Hatakeyama ,T. *et al.*, *JACS* **2006**, *128*, 2518.

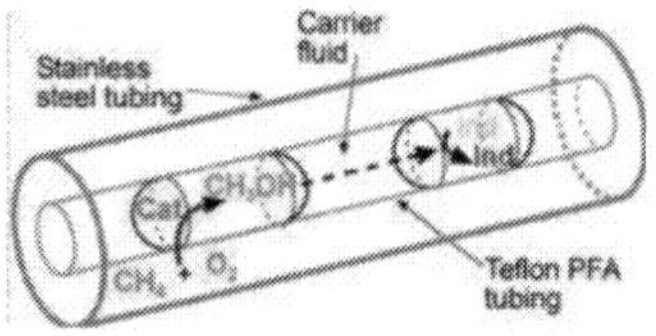

Kreutz ,J. *et al.*, *JACS* **2010**, *132*, 3128.

MIT NOVARTIS

Design of a Fully-Automated Screening Platform

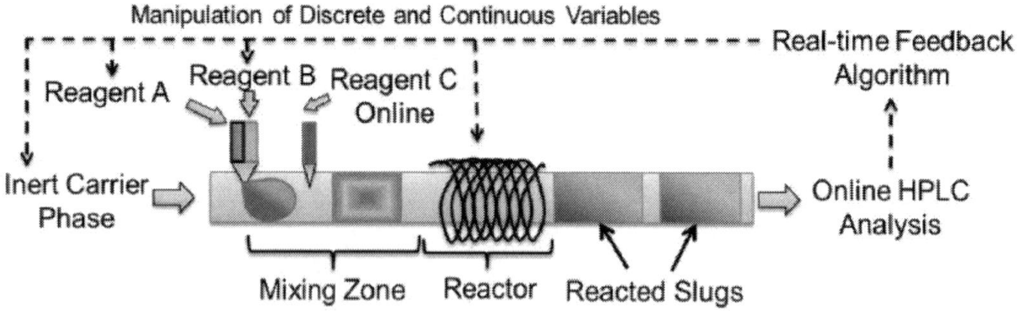

System Diagram

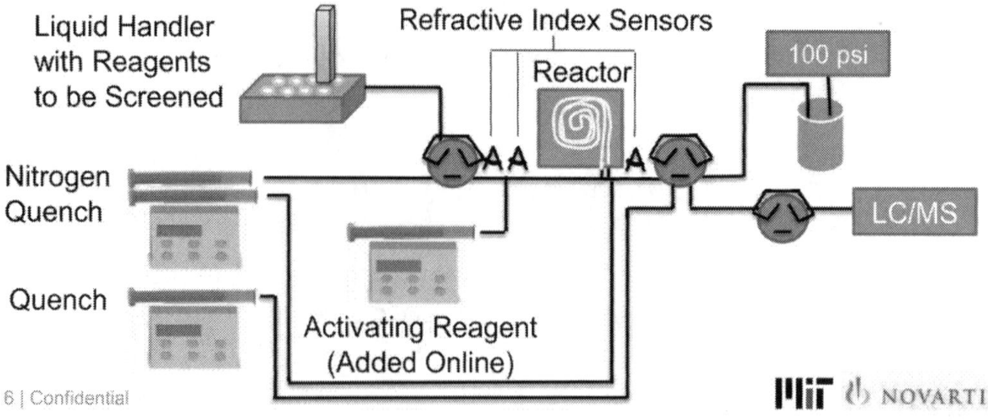

MIT NOVARTIS

A Response Surface Methodology Allows Efficient Optimization of Discrete and Continuous Variables

Initialize with Fractional Factorial Design for all Discrete Variables

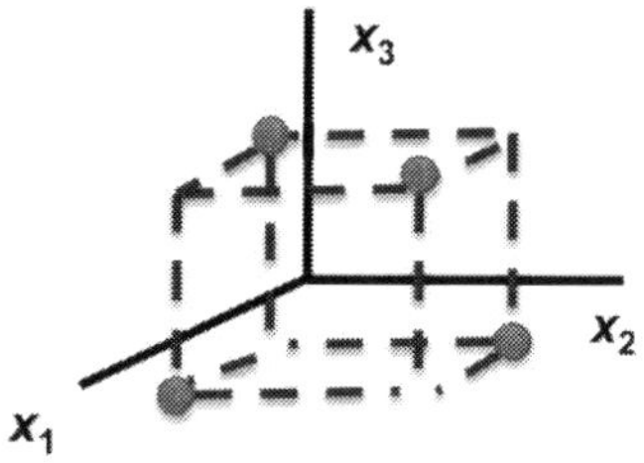

Optimize Linearized & Scaled Model. Run Fractional Fact Design around Optimum for each Discrete Variable

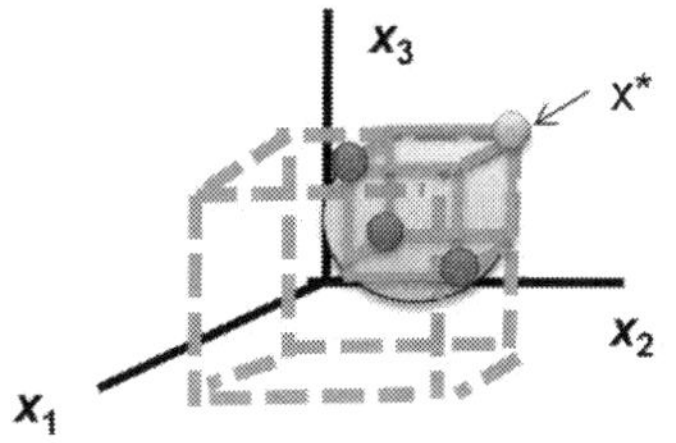

Optimize Scaled Quadratic Model. Choose New Experiments that Minimize Uncertainty in Optimum

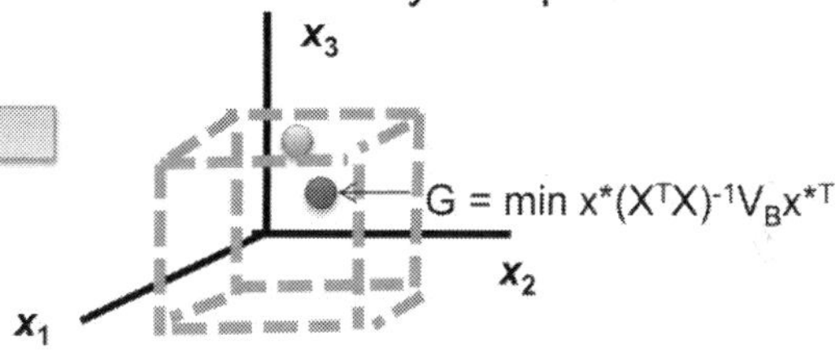

Conduct Quasi-Newton Search around Predicted Optimum

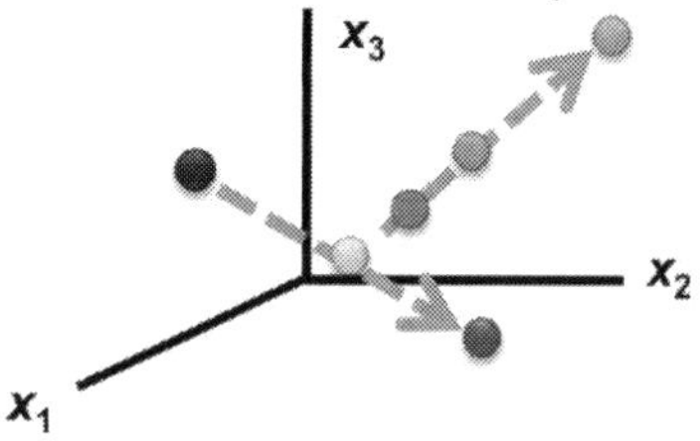

Atkinson, A. & Donev, A., *Optimum Experimental Designs* **1992**.
Box, G. *et al.*, *Statistics for Experimenters* **1978**.

Key Aspects of the Algorithm

- Response surfaces are updated iteratively
  - Surfaces for discrete variables share continuous variable interaction terms with other discrete variable surfaces
  - Each discrete variable is differentiated by a temperature term and an intercept—roughly corresponding to an activation energy and pre-exponential factor
- When the lower bound on the best possible discrete/continuous variable combination exceeds the maximum of another discrete variable, the "worse" discrete variable is fathomed

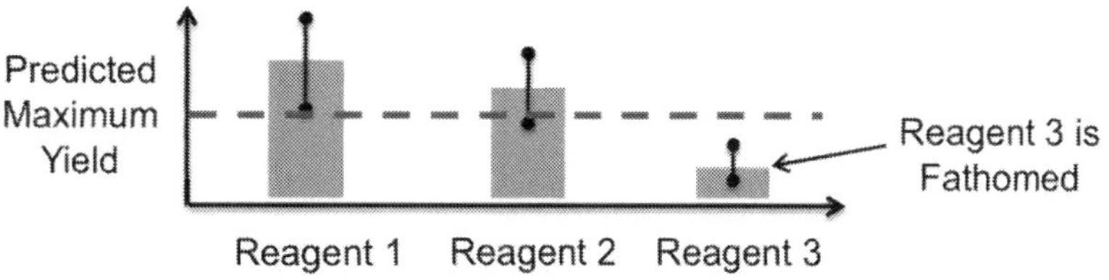

- A G-optimal search selects for the experiment(s) that minimize uncertainty in the optimum. This focuses the algorithm closer to a more likely "hit"

Example Application to Solvent Screening

- Selective alkylation of 1,2-diaminocyclohexane

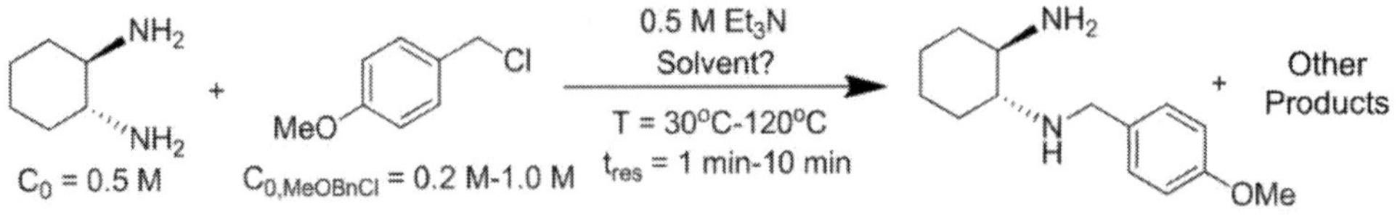

- Preliminary fractional factorial design results (4 experiments per solvent)

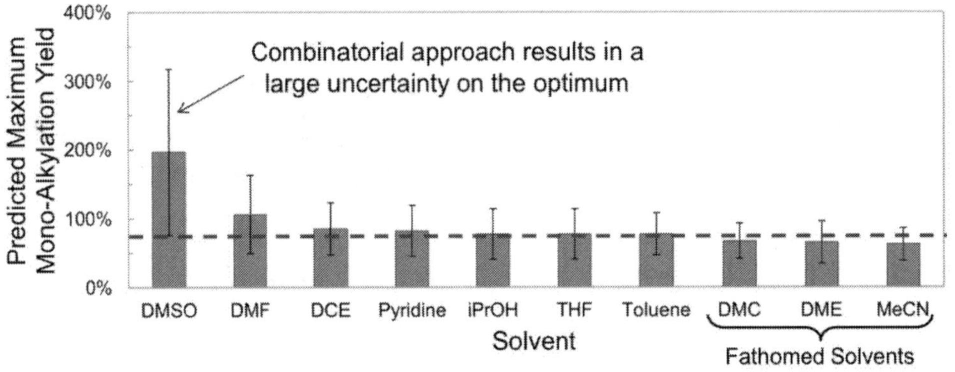

9

MIT NOVARTIS

Discrimination of Solvents by G-Optimality

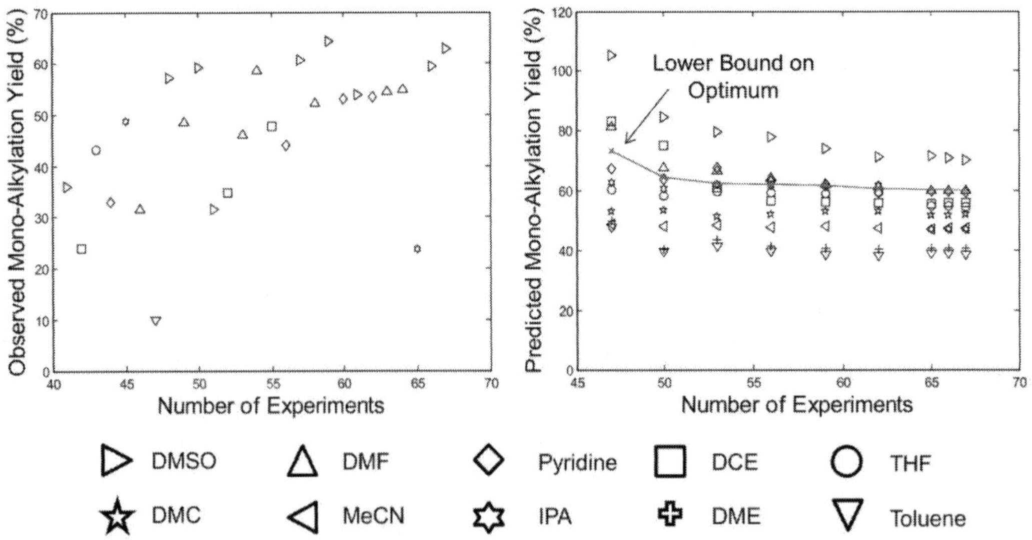

- A G-optimal search yields steady improvement in the observed optimum
- In 27 experiments:
  - Uncertainty reduced to 10%. Number of candidate solvents reduced to 1 (DMSO)

10

MIT NOVARTIS

Exploration of the Continuous Variable Space and Convergence to an Optimum

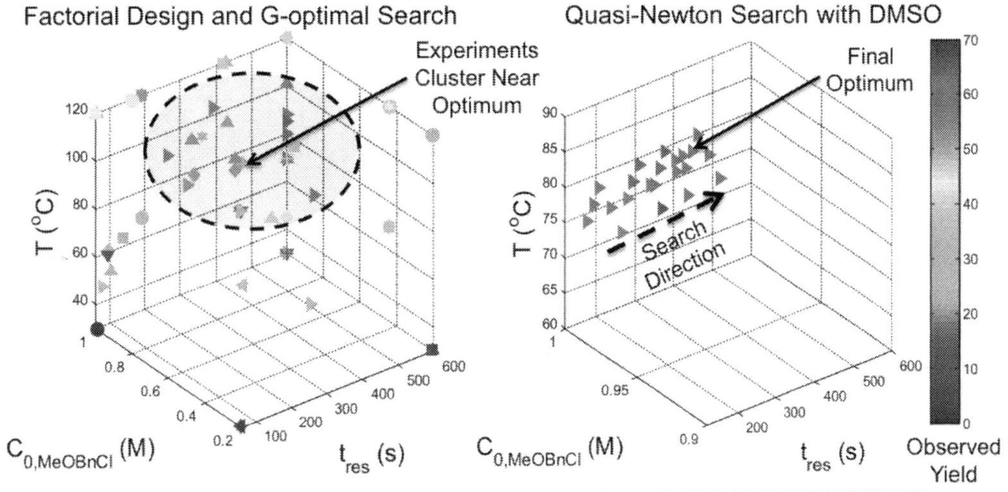

Scalability of Results		
Slug Screening System Optimum	HPLC	62% Yield
0.5 g Batch Synthesis with Chiral Amine	HPLC	61% Yield
0.5 g Batch Synthesis with Chiral Amine	Isolated	59% Yield

MIT NOVARTIS

What Do the Results Tell Us about Selectivity?

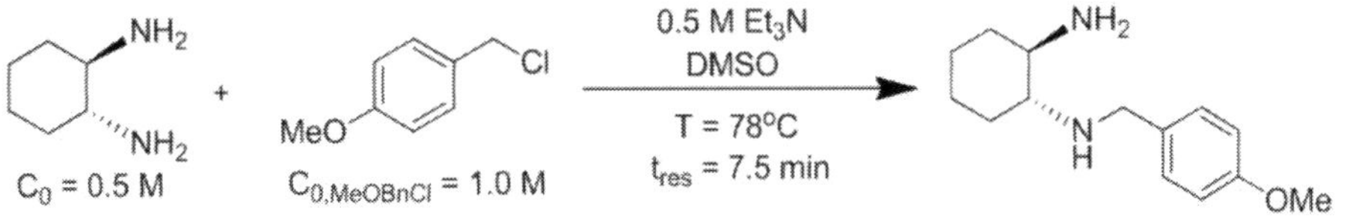

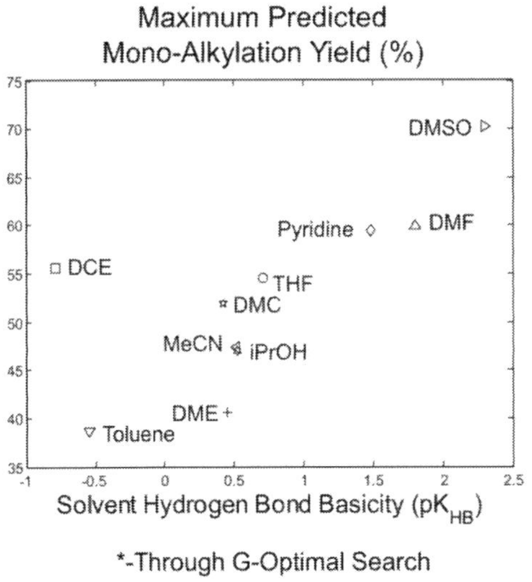

*-Through G-Optimal Search

- The fastest solvents are polar aprotic
  - Not surprising: the polar solvent stabilizes the charge on the TS

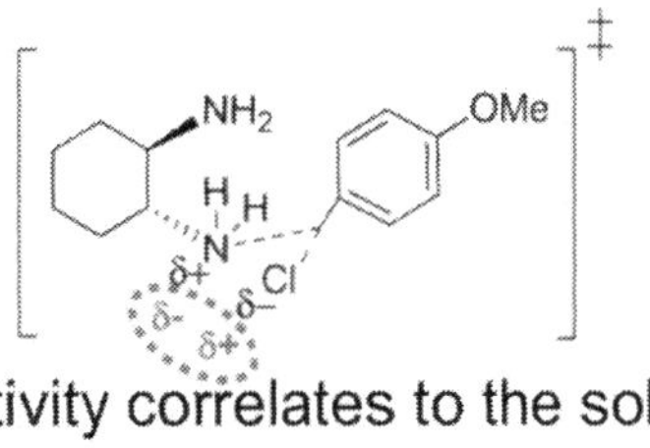

- Selectivity correlates to the solvent hydrogen bond basicity (pK_{HB})
  - Hypothesis: a good hydrogen bond donor counteracts the nucleophilicity of the mono-alkylated product

Laurence, C. & Gal, J.-F., *Lewis Basicity and Affinity Scales: Data and Measurement* **2009**
Lebleu, T. *et al.*, *Chem Commun* **2009**, *50*, 1836

MIT NOVARTIS

Simultaneous Catalyst Selection and Optimization

- Suzuki-Miyaura cross-coupling reaction for C-C bond formation

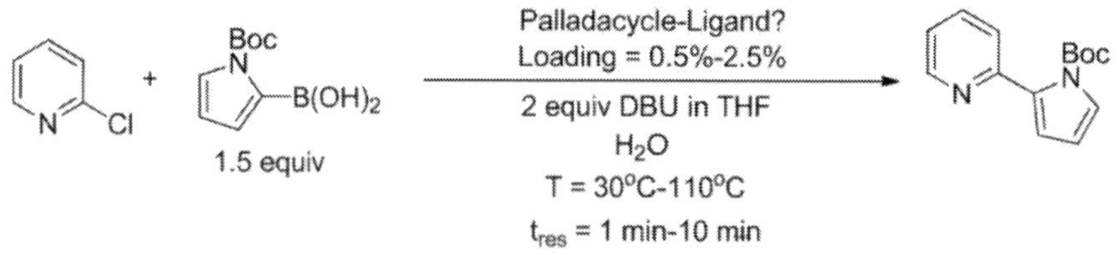

- Palladacycle-ligand combination key to high yield and low rate of protodeboronation

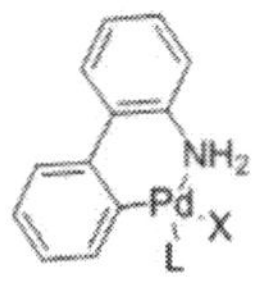

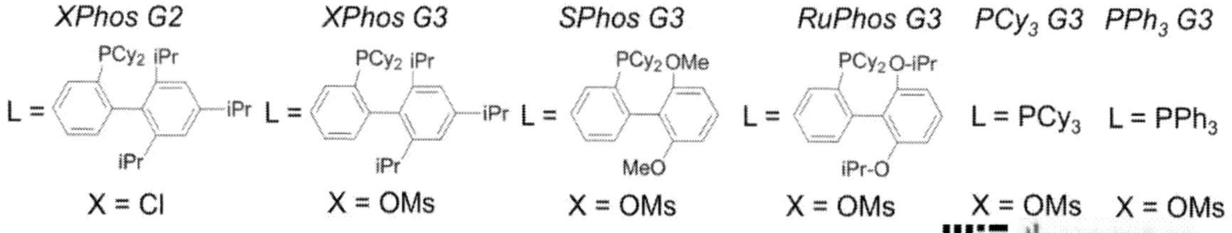

MIT NOVARTIS

Our System Rapidly Identifies the Best Catalysts

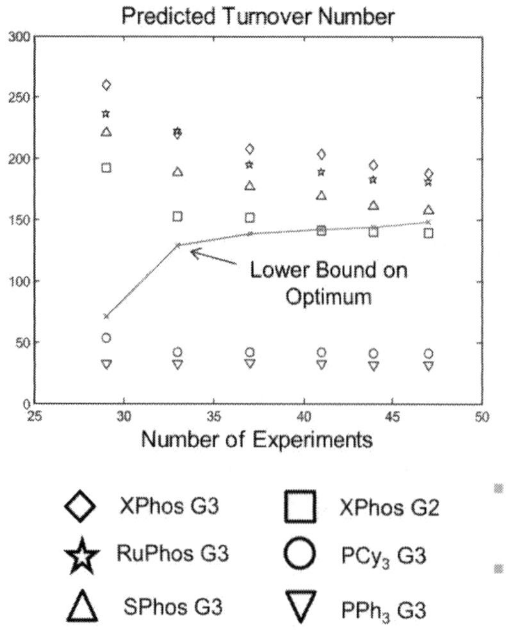

Observed Conditions of Optimal TON

Catalyst	t_{res} (s)	T (°C)	Loading (%)	Yield
SPhos G3	60	110	0.5	89%
XPhos G3	600	90	0.5	86%
RuPhos G3	600	89	0.5	86%
XPhos G2	600	110	0.5	75%
PCy_3 G3	600	110	0.5	27%
PPh_3 G3	600	110	0.5	15%

- Bulkier biarylphosphine ligands outperform smaller PCy_3 and PPh_3
- In the case of XPhos, OMs outperforms Cl as the palladacycle substituent
  - Agrees with batch results in THF/aq. K_3PO_4

Bruno, N. *et al.*, *Chem Sci* **2013**, *4*, 916.

MIT NOVARTIS

Conclusions and Future Work

- A continuous droplet screening system with feedback enables rapid optimization of reagents and continuous reaction variables
- Convergence is optimized with a G-optimal search, which fathoms discrete variables while minimizing uncertainty on the overall optimum
- Further investigation into screening and optimization results yields correlations among variables that can be probed in future studies
- Future work in this field will focus on merging multiple discrete variables with continuous variables, rapidly building reaction libraries, constructing reaction mechanisms, and elucidating reaction kinetics

15

MIT NOVARTIS

Acknowledgments

- System Development
  - Dr. Andrea Adamo
  - Dr. Baris Unal
  - Dr. Everett O'Neal
  - Dr. Patrick Heider
  - Michaelann Rodriguez
- Chemistry
  - Dr. Yiming Wang
  - Dr. Stephen Newman
  - Dr. Saurabh Shahane
  - Dr. Antony Fernandes
- Funding
  - Novartis Center for Continuous Manufacturing

16

MIT NOVARTIS

382936 Gold-Catalysed Ester Synthesis from Alcohols in a Continuous Flow System

Abstract:

Alcohols are a desirable starting point for many chemical synthesis routes as they tend to be stable and can be derived from natural products, therefore sustainably sourced. One such synthesis route is the formation of esters, which have many uses from solvents to food flavourings.

The formation of esters from alcohols and methanol is typically achieved in two steps: First by oxidation of the alcohol to a carboxylic acid, before esterification.[1] This process often requires the use of coupling agents and organic solvents, producing large quantities of by-products with further consumption of resources in the resultant work-up process. For these reasons has been a great deal of interest in the development of an alternative methodology, using oxygen[2]or hydrogen peroxide with a catalyst, giving water as the only by-product.

One promising group of catalysts for these oxidative esterification reactions are supported gold nanoparticles.[1-3]A major advantage of these catalysts are that they are heterogeneous, giving better catalyst-product separation, and so greater catalyst recovery and easier product purification. They are also suitable for incorporation into a continuous flow system as a packed bed.

A method was developed for gold-catalysed esterification/oxidation of alcohols in a flow reactor, to access the benefits delivered by a larger reaction space. The catalyst chosen was Au/TiO_2, with H_2O_2as the oxidant and a methanol/water solvent. Studies of the reaction kinetics of benzyl alcohol oxidation to methyl benzoate were carried out, with measurements of activation energies and an investigation into the origins of the selectivity between the methyl ester and benzoic acid, a side product.

References

1. F. Z. Su, J. Ni, H. Sun, Y. Cao, H. Y. He, K. N. Fan, *Chem. Eur. J.* **14**(2008) 7131.
2. I. S. Nielsen, E. Taarning, K. Egeblad, R. Madsen, C. H. Christensen, *Catal. Lett.* **116**(2007) 35
3. C. Marsden, E. Taarning, D. Hansen, L. Johansen, S. K. Klitgaard, K. Egeblad, C. H. Christensen, *Green Chem.* **10** (2008) 168

383151 Esterification of Citric Acid with Ethanol Using Homogeneous and Heterogeneous Catalysts

Abstract:

Citric acid esters are widely used as plasticizers for polymers with applications in the food, cosmetics, and health industries. Generally, production of alkyl citrates is carried out in batch reactors using homogeneous or heterogeneous acid catalysts. Operating temperature during reaction is constrained to be less than 150 °C to avoid citric acid polymerization or decomposition. The minimum acid to alcohol molar ratio is defined by the solubility of the acid, but in general large excess of alcohol is required.

Even though citric acid esters are well known, few literature reports present kinetic models to describe the process. In this regard, this work summarizes the experimental evaluation of a kinetic model for the esterification of citric acid with ethanol using both, Methane Sulfonic Acid and Amberlyst 70 as catalyst.

A Box-Behnken experimental design was developed to evaluate the kinetic models of reaction, varying reaction temperature (80 to 120 °C). Acid to alcohol feed ratio (1:7 to 1:15) and catalyst loading (0.5 to 1.5% w/w). Experiments were carried out in a 100 mL jacketed stirred SS reactor. Temperature was maintained by a thermal oil circulating bath. Reaction samples were withdrawn periodically and they were analyzed using HPLC, (previously calibrated). Reaction time was about 8 hours.

To reduce the amount of parameters to be fitted with the kinetic experiments, some independent chemical equilibrium experiments were carried out (80 to 120 °C). A set of 10 ml SS-hermetic tubes were loaded with a defined amount of the reactive mixture and the catalyst. These tubes were placed within an oven at given temperatures, and maintained in continuous agitation. After at least three days, tubes were sampled and the equilibrium concentrations were measured. Experimental equilibrium constants were obtained as the ratio of concentrations of products and reactants.

Regression of experimental data was performed in MatLabTM using a pseudo-homogeneous, and a heterogeneous model. Both agreed reasonably well with experimental kinetics ad can be used confidently for further process design.

Kinetic Study of Catalytic Hydrogenolysis for Pharmaceutical Applications

Hyun Tae Hwang, Rajamani Gounder and Arvind Varma

School of Chemical Engineering, Purdue University,
West Lafayette, IN 47907, USA

ABSTRACT

In multi-step organic syntheses, protection of functional groups such as amines, alcohols, and acids is frequently necessary to avoid undesired side reactions. For example, amines that are highly reactive under various reaction conditions can be protected by benzyl groups, which form toluene after removal by hydrogenolysis reactions. In the present study, the kinetic details of hydrogenolysis of N-(4-fluorophenyl)-benzenemethanamine, a compound subject to both debenzylation and deflourination, were investigated in a semi-batch reactor. The effects of various operating parameters (e.g., temperature, H_2 pressure, pH, solvent type) on different catalysts (e.g. Pd, Pt, Ru) were examined to identify factors that influence the selectivity between debenzylation and defluorination pathways.

387948 Preparation of Subnanometer Au Particles

Abstract:

It is of interest to prepare subnanometer Au particles as they possess unusual catalytic properties[1]. However, conventional preparation of colloidal Au particles usually resulted in particles of 5 nm or larger. We explored two different strategies to synthesize such particles. In the first method we used "designer siloxane polymers" to control the particle growth. The polymer possessed Si-H functionalities that could reduce the Au(I) precursor to Au(0) and amine groups that could protect the Au clusters from agglomeration. Various parameters such as steric constrain, temperature of nucleation and particle growth as well as the proximity of the reductant to the amine protective groups were examined to determine the critical factors that control the size of the Au nanoparticles. In the second method, nanocages were synthesized and the au particle size was constrained by the size of the cage interior. Characterizations of the Au particles were performed using TEM, DLS and Uv-visible.

Design of Ni double layer hydroxide based catalysts with long term deactivation stability for ethanol reforming

Allison Cross[1], Khachatur Manukyan[2], Armenuhi Yeghishyan[1], Sergei Rouvimov[3], and Eduardo E. Wolf[1*]
Department of Chemical and Biomolecular Engineering[1], Department of Physics[2], Department of Electrical Engineering[3], University of Notre Dame, Notre Dame, Indiana 46556 (USA)
*ewolf@nd.edu

Introduction and Background

In this research, we study the preparation and stability of the activity and selectivity of novel encapsulated, ultra-small (<5 nm) nickel nanoparticles as hydrogen generation catalysts via ethanol reforming for fuel cells. During recent years our group has investigated less expensive unsupported and supported transition metals (Ni, Cu, Fe) and their alloys in alcohol reforming[1-5] instead of noble metals, in an attempt to make on board hydrogen production for fuel cells economically viable. The catalysts studied previously show high activity and selectivity toward H_2 production at low temperature, but unfortunately undergo rapid deactivation, possibly due to several types of carbon deposition (including metal carbide formation) and sintering (Fig. 1, red curve)[6]. We studied how to rationally design a catalyst in which the metal is restricted from growth and inhibits carbon formation.

We found a new catalyst preparation method that fulfills the design objectives involving synthesis via a double layer hydroxide (DLH) used as a catalyst precursor. DLH based materials allow the active material and the support to be built in a framework together, which can protect the material from changing over time, leading to high catalytic stability. We studied the thermal treatment of the DLH precursors in a reductive atmosphere to prepare Ni-Al_2O_3 catalysts with a high surface area, tunable pore sizes, and resulting in Ni metal nanoparticles. We demonstrate that the design of DLH based catalysts result in highly active catalysts during ethanol reforming toward hydrogen generation, which exhibit orders of magnitude increase in stability towards deactivation with high conversion and selectivity lasting over 100 h.

We investigated the cause of the resistance to deactivation of the DLH catalysts by studying their structure and composition during reaction. A relationship is established between surface and bulk structure that confer stability to these catalysts and allow a molecular level understanding of the stability to deactivation found on the new DLH based catalysts.

Materials and Methods

A series of catalysts were prepared using a Ni-Al_2O_3 DLH template[7] varying the synthesis conditions. We investigated the catalysts under reaction to determine correlations between changes in the preparation resulting in differences in final product and their resistance to deactivation. The catalysts were characterized by activity-selectivity-time on stream (TOS) studies in a fixed bed reactor under a flowing stream of ethanol-N_2 as described previously[2,4,6]. The resulting effluent gas was analyzed by gas chromatography.

The various samples were characterized by BET (surface area, pore size distribution), XRD (crystal structure), with the atomic concentrations of the active catalysts determined through ICP-OES. The catalysts' structure throughout synthesis steps and the active sites of reaction were determined by FT-IR spectroscopy. Surface composition and oxidation state were characterized with UHV-XPS on the fresh and used catalysts to correlate activity with changes in surface structure. The morphology and dispersion of the catalysts were obtained via HRTEM.

Results and Discussion

Ni and Al are formed in layers between a layer of CO_3^{-2} through a mixture of metal nitrates, sodium carbonate, and sodium hydroxide. After precipitation of the solution, short-term thermal treatment of DLH under a reductive atmosphere results in the formation of a Ni-Al_2O_3 catalyst. Ultra-small metal nanoparticles with high active area are partially encapsulated by the support, securing their place in the Al framework. The as-prepared catalysts show very high activity-selectivity for hydrogen generation and orders of magnitude higher resistance to deactivation than previous catalysts (Fig. 1, black curve).

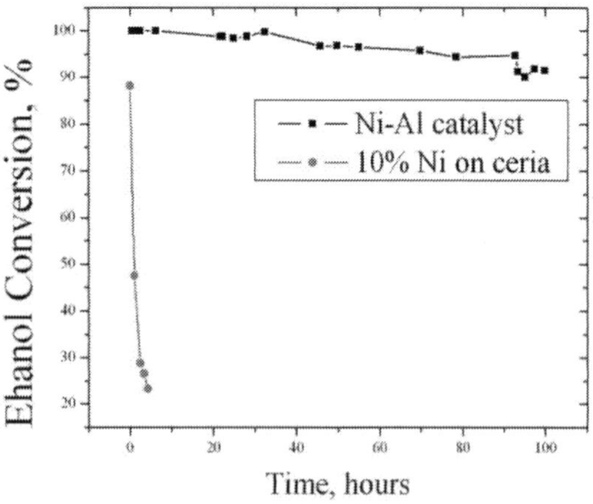

Fig 1. Stability tests of the Ni-Al DLH and previous Ni/ceria CS catalyst

The precursor Ni-Al DLH material was studied by FT-IR spectroscopy (Fig. 2). There is the presence of OH and H_2O bonding from the attached water at 1616 and 3467 cm^{-1} respectively, as well as a large peak at 1362 cm^{-1} of CO_3^{-2}, which is the separating layer that is removed during final processing of the catalyst. A cartoon of the structural development is shown in Fig. 3. The BET specific surface area of the precursor material is 140 m^2/g and consists of pores smaller than 20 nm in diameter, demonstrating a prime catalyst surface for high activity.

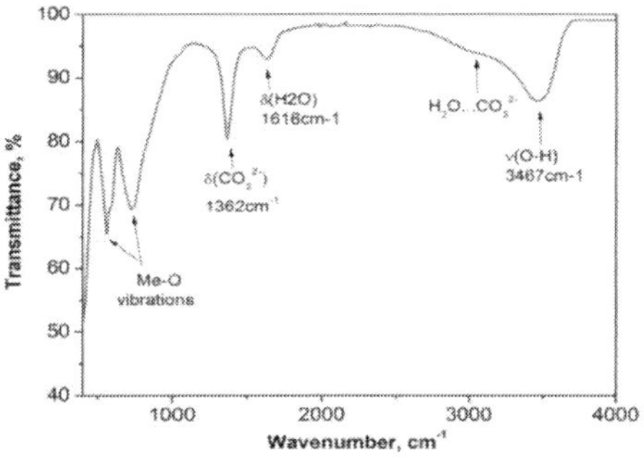

Fig. 2 FT-IR spectrum of Ni-Al DLH

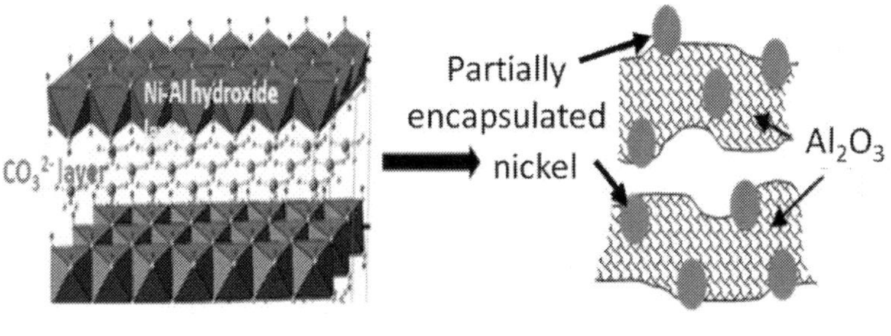

Fig. 3. Diagram of DLH precursor and cross section of reduced catalyst

TEM micrographs of the Ni-Al DLH after reduction using a slow temperature ramp under pure hydrogen to 500°C results in highly dispersed, Ni nanoparticles (Fig. 4). TEM images of used catalysts (not shown) show only little structural transformation even after 100 h of reaction, demonstrating a resistance of the catalyst to sintering.

Elemental XPS spectrum of the Ni-Al DLH catalyst and the Ni/ceria CS catalyst (Fig. 1, black and red curve, respectively) before and after the TOS studies show similar decreases in surface concentration of Ni (40%), but a difference result in the decrease of surface concentration of the support material. The concentration of Al in the Ni-Al DLH catalyst decreases by 19%, while in the Ni/ceria sample, Ce decreases by only 5% after reaction. The carbon formation on the Ni-Al DLH catalyst shows 106% increase of C, while on Ni/ceria the increase of C is 186%. Therefore, the structure of the Ni-Al DLH allows some protection from carbon formation, possibly by inhibiting carbon from preferentially forming on only the active material, causing deactivation like in the Ni/ceria catalyst. XPS also shows a small percentage of surface Ni (14%), while total Ni content is ~80%, supporting that the Ni is partially encapsulated by the support.

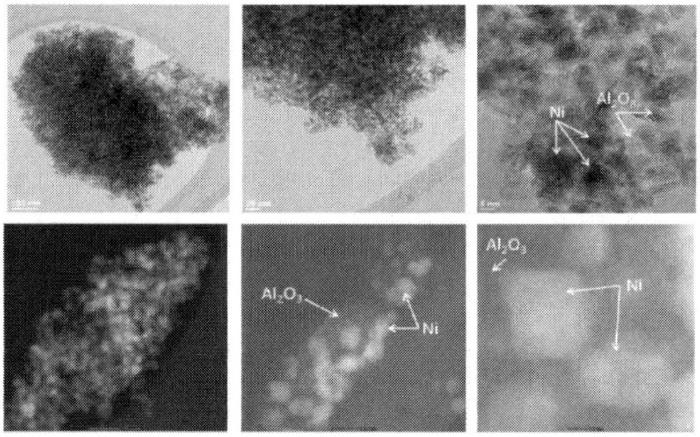

Fig. 4 TEM images of Ni-Al DLH reduced

Significance

Our expectation is that the understanding gained on the fundamental issues of catalysts' deactivation on the novel DLH based catalyst could extend the DLH preparation method to other catalysts for similar reforming reactions for hydrogen generation for fuel cell usage. Determining the correlation between the increased stability to deactivation found in our studies will allow a rational approach to design more stable catalysts, not only for the ethanol reforming reaction but other reactions in which deactivation by carbon formation and sintering are a problem. This can also be expanded to other active metals/supports and can allow a broad array of stable catalysts suitable for industrial environmental applications.

References

1. A. Kumar, A. S. Mukasyan, and E.E. Wolf, *"Impregnated layer combustion synthesis method for preparing a highly active and selective catalysts for hydrogen production from oxidative reforming of methanol"*, *Appl. Catal. A*, 372 (2010) 175-183.
2. A. Kumar, A. S. Mukasyan, and E. E. Wolf, *"Combustion synthesis of Ni, Fe and Cu multi-component catalysts for hydrogen production from ethanol reforming"*, *Appl. Catal. A*, 401 (2011) 20-28.
3. A. Kumar, A. S. Mukasyan, and E.E. Wolf , *"Modeling of Impregnated Layer Combustion Synthesis"* Paper accepted at the 21st ISCRE meeting, Philadelphia, June, 2010, Accepted in IECR, 2010.
4. A. Cross, A. Kumar, E. E. Wolf, and A. S. Mukasyan, *"Combustion synthesis of a nickel supported catalyst: effect of metal distribution on the activity during ethanol decomposition"*, *Ind. Eng. Chem. Res.*, (2012) 12004–12008.
5. K. Manukyan, A. Cross, S. Roslyakov, S. Rouvimov, A. S. Rogachev, E. E. Wolf, and A. S. Mukasyan, *"Solution combustion synthesis of nano-crystalline metallic materials: mechanistic studies"*, *J. Phys. Chem. C.*, 117 (2013) 24417-24427.
6. A. Cross, S. Zhang, J. Miller, F. Tao, A. S. Mukasyan, and E. E. Wolf. *"In-Situ XAS and AP-XPS of Ni-Cu Supported Catalysts under Ethanol Dehydrogenation Reaction" In preparation.*
7. L. He, Y. Huang, A. Wang, X. Wang, X. Chen, J. J. Degado, and T. Zhang. *"A noble-metal-free catalyst derived from Ni-Al hydrotalcite for hydrogen generation from $N_2H_4 \cdot H_2O$ decomposition"* *Angew. Chem. Int. Ed.*, 51 (2012) 6191-6194.

371832 Drying of Supported Catalysts in Packed Beds or on Belt Conveyors

Abstract:

Supported catalysts are used in many industrial processes and applications, ranging from petrochemical and catalytic converters to fuel cells. These catalysts have many advantages, such as a high surface area, a low amount of the often expensive active component (Pd, Pt, etc.) and high mechanical and thermal stability. Clearly, the catalyst design has a pronounced effect on the performance of a catalytic process. With respect to the distribution of the active component in the support materials, four main categories of metal profiles can be distinguished, i.e., uniform, egg-yolk, egg-shell and egg-white profiles. The choice of the desired metal profile is determined by the required activity and selectivity, and tailored for specific reactions and/or processes. Although the development and preparation of supported catalysts have been investigated for many years, many aspects of the various catalyst manufacturing steps are still not fully understood, and in industry the design of catalysts is predominated by trial and error experiments,

Belt dryers are used broadly in food, chemical and petro-chemical and catalyst industries. During the belt drying process, the wet materials are heated by hot air flowing from the bottom when they move forward on the belt conveyor. We have established a packed bed drying system, where the wet catalyst samples are dried in a bed by hot air flowing through the bed, to examine the belt drying process. During belt drying if the air flow doesn't move the catalyst particles on the belt, the variation of the catalyst properties, such as moisture content and metal distribution, as a function of location on the belt can be predicted by the variation of catalyst properties as a function of time in the packed bed. The packed drying process is much smaller and easier to control than the belt drying process. Packed bed experiments have been carried out in a Glatt GCPG-1 dryer and a mini-Glatt dryer. Drying profiles, such as temperature and relative humidity, are recorded during the process, and the sample metal distribution after drying is measured using micro-XRF technology. For all cases studied in this work, similar metal distribution is observed for samples taken from different layers in the batch.

We also developed a layering model to simulate the packed bed drying process. We found that at the beginning of drying an equilibrium state is reached between the wet catalyst samples and the humid drying air in upper layers so drying doesn't occur in upper layers. With further drying, the upper layers cannot hold the equilibrium state once the humidity in the drying air decreases to a certain point, and then the water evaporation starts in upper layers. It is very interesting to note that once the drying procedure passes the constant drying rate stage the falling rate stage is quite similar for different layers, leading to similar metal distribution among different layers after drying. This "self-reproducing" phenomenon indicates that during drying the metal distribution is mainly determined by the falling rate stage, where different layers share similar drying mechanisms. This will shed light on the scale-up of belt drying processes.

Currently we have only tested one heating zone system. Future work will focus on multiple heating zone systems, which have been used in many applications. Under this situation, the drying procedure could be quite different in different layers. This may lead to different metal distribution among the layers.

392058 The Synthesis of Noble Metal Bimetallic Catalysts Via Electrostatic Adsorption

Abstract:

Noble bimetallic catalysts are ubiquitous in academia and industry. A survey of the last decade of catalysis literature reveals over two hundred articles and a similar number of patents pertaining to bimetallic systems comprised of two noble metals, with predominant applications in air pollution abatement, petroleum refining, Fischer Tropsch synthesis, as well as fuel cell catalysts for both reactions occurring at the anodes and cathodes. More than 50% of the literature surveyed, however, employed dry impregnation to synthesize bimetallic catalysts. When using dry impregnation, the intimate interaction of the two cannot be well controlled and usually occurs only to a minimal extent. Furthermore, metal particles are often agglomerated and poorly dispersed.

In this talk we present several simple, rational, scalable methods to prepare supported bimetallic particles with high dispersion and intimate metal1-metal2 interaction. The points of zero charge of many pure noble metal oxides has been measured for the first time, as well as adsorption surveys of other noble metal complexes onto the bulk noble metal oxides. With this information we demonstrate how electrostatic adsorption can be used to synthesize well dispersed noble bimetallic nanoparticles that are either homogeneously alloyed or exist in core-shell morphologies.

Effects of Preparation Variables on an Alumina-Supported Fe/Cu/K Fischer-Tropsch Catalyst

Jonathon Horton[1], Kamyar Keyvanloo[1], Morris D. Argyle[1], William C. Hecker[1]

[1]Department of Chemical Engineering, Brigham Young University, Provo, UT 84602

Introduction

Fischer Tropsch Synthesis (FTS) is a catalyst-assisted conversion of syngas (CO + H_2) to hydrocarbons (C1–C50). Natural gas, one possible feedstock for the process, has recently had a historically low price. Additionally, the price of natural gas could remain low for the foreseeable future due to hydraulic fracturing. With this market condition and a relatively high price for crude oil, FTS has become an increasingly attractive process for production of liquid fuels. Of the two types of metals used for FTS catalysts in industry, iron (Fe) and cobalt (Co), Fe has some advantages over Co including lower cost, lower methane selectivity, and high water-gas shift activity [1]. Commercial Fe catalysts produced for FTS are unsupported because unsupported Fe catalysts have consistently been more active than supported Fe catalysts until recently [2–6]. Therefore, very little work on supported Fe catalysts is reported in the literature [2–4].

Despite their high activity and favorable selectivity, unsupported Fe catalysts are generally too mechanically weak to be used in slurry bubble column reactors (SBCR's), which are the most thermally efficient and economical reactors available [4]. The challenging conditions in SBCR's grind the weaker unsupported catalysts into fine powders, resulting in excessive catalyst loss and plugging the catalyst recovery system [4]. In contrast, supported Fe catalysts are far more attrition resistant [7]. The ideal Fe FTS catalyst would combine the superior activity of unsupported Fe catalysts with the improved strength of supported Fe.

The goal of this study was to gain deeper insights into supported Fe preparation methods as a foundation for future Fe FTS catalyst design. Although many preparation techniques for supported Fe FTS catalysts have been demonstrated [2–4], the reported data lack enough consistency to isolate the effects of specific preparation variables. This study examined five important variables in the preparation of supported Fe catalysts, systematically investigating each individual variable's effects by holding all other variables constant. A series of potassium (K) promoted Fe catalysts supported on alumina (Al_2O_3) were prepared to examine the effect of the following variables: (1) metal deposition method: aqueous incipient wetness deposition or non-aqueous slurry deposition, (2) Fe loading level, (3) K loading level, (4) timing of K and Fe deposition: sequential deposition or co-deposition, and (5) with or without direct K-promotion of the blank support.

Experimental

The six catalysts examined in this study (all Fe/Cu/K/Al_2O_3) are identified as K1–K6. The values of all five preparation variables studied are given for each catalyst in Table 1. Incipient wetness deposition involves using just enough solvent to dissolve and deposit the metal, while slurry deposition involves using excess solvent. Fe was loaded in 10 wt% steps (2 steps for 20 wt%, 4 steps for 40 wt%) with the copper (Cu) followed by drying at 80°C and calcination at 300°C. K was added (4K/100Fe or 8K/100Fe) in one step after all the Fe and Cu for sequential deposition, while for co-deposition, the K was added proportionally with each step of Fe and Cu.

Table 1. Preparation variable values for each of the six catalysts studied in this paper

	Variable 1		Variable 2	Variable 3	Variable 4	Variable 5
Catalyst	**Solvent**	**Deposition Method**	**Fe Loading[1]**	**K Loading**	**Deposition Timing**	**K on blank support**
K1	NA[2]	SD[4]	20	4K/100Fe	Sequential	None
K2	NA	SD	**40**	4K/100Fe	Sequential	None
K3	A[3]	IWD[5]	20	4K/100Fe	Sequential	None
K4	A	IWD	20	**8K/100Fe**	Sequential	None
K5	A	IWD	20	4K/100Fe	**Co-deposition.**	None
K6	A	IWD	20	4K/100Fe[6]	Sequential	**0.2 wt%**

[1]Catalyst wt% Fe. [2]NA = Non-aqueous (50% acetone, 50% isopropanol).
[3]A = Aqueous. [4]SD = Slurry Deposition. [5]IWD = Incipient Wetness Deposition
[6]Plus 0.2 wt% (of the support) K added directly onto the support

Results and Discussion

The effect of each preparation variable was observed by comparing catalyst performance. A few relevant observations for each of the examined preparation variables are given below:

1. Aqueous incipient wetness or non-aqueous slurry: K1 (non-aqueous) has 14% higher activity per gram catalyst and improved stability compared to K3 (aqueous).

2. Fe loading level: K1 (20 wt% Fe) has a 17% higher activity than K2 (40 wt% Fe).

3. K loading level: K3 (4K/100Fe) has a 20% higher activity than K4 (8K/100Fe). In addition, K4 did not show improved selectivity compared to K3.

4. Timing of deposition: K5 (co-deposition) has a 40% higher activity than K3 (sequential deposition).

5. With or without K-promotion of the blank support: K6 (direct K-promotion of the support) deactivates sharply in the first 200 hours on stream, while K3 (no K-promotion of the support) is far more stable.

The performance of the six catalysts of this study was also compared with some of the best catalysts reported in the literature. Two tested preparation methods in particular (non-aqueous slurry deposition and co-deposition) yielded supported Fe catalysts with better catalyst performance than previously thought possible. Figure 1 below shows these two supported iron catalysts (K1 and K5) are superior to the best supported Fe catalysts prior to this work and are even competitive with the best reported unsupported Fe FTS catalysts due to their very high activity (211–267 mmol ($CO + H_2$)/g_{cat}/MPa/h). K1 and K5 also exhibit high productivity (0.47–0.56 g_{HC}/g_{cat}/h), and reasonable stability. Their superior activity and productivity, coupled with the high strength of a supported catalyst, make these catalysts excellent candidates for use in industry for SBCR applications.

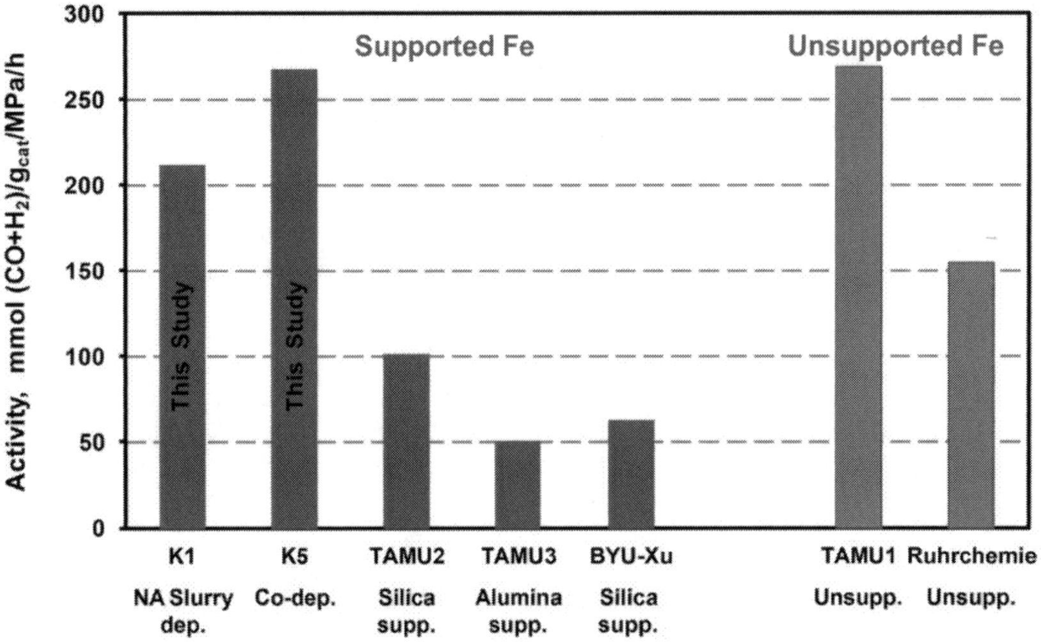

Fig. 1. FTS activity of two catalysts of this study compared with both unsupported and supported Fe catalysts from the literature

References:

1. Bartholomew, C.H. and R.J. Farrauto, *Fundamentals of Inductrial Catalytic processes*. Second ed. 2006: John Wiley & Sons.

2. Bukur, D.B. and C. Sivaraj, *Supported iron catalysts for slurry phase Fischer–Tropsch synthesis*. Applied Catalysis A: General, 2002. **231**(1-2): p. 201-214.

3. O'Brien, R.J., et al., *Activity, selectivity and attrition characteristics of supported iron Fischer–Tropsch catalysts*. Applied Catalysis A: General, 2000. **196**(2): p. 173-178.

4. Xu, J., et al., *Design, synthesis, and catalytic properties of silica-supported, Pt-promoted iron Fischer–Tropsch catalysts*. Topics in Catalysis, 2003. **26**(1-4): p. 55.

5. Bukur, D.B. and X. Lang, *Highly Active and Stable Iron Fischer–Tropsch Catalyst for Synthesis Gas Conversion to Liquid Fuels*. Ind. Eng. Chem. Res., 1999. **38**(9): p. 3270–3275.

6. Keyvanloo, K., et al., *Supported iron Fischer-Tropsch catalyst: superior activity and stability using a thermally stable silica-doped alumina support*. ACS Catalysis, 2014.

7. Anderson, R.B., *The Fischer-Tropsch Synthesis*. 1984, New York: Academic Press.

Controlled synthesis of bimetallic catalysts with uniform nanoparticle size and composition

Ana C. Alba-Rubio, Sikander H. Hakim, Thomas M. Gostanian, Canan Sener, Brandon O'Neill, and James A. Dumesic.

Department of Chemical and Biological Engineering, University of Wisconsin, Madison, Wisconsin 53706, United States.

The work to be presented will focus on a general strategy to synthesize supported bimetallic nanoparticles that possess well-defined particle size and composition. Traditional methods for bimetallic nanoparticle synthesis, such as impregnation, often fail to provide precise control over composition and structure, yielding a mixture of nanoparticles with varying particle size and composition. This heterogeneity complicates studies to link catalytic properties to structural and/or compositional changes in the catalyst.

The hydrogenolysis of C-O bonds is an important reaction that is essential for the deoxygenation of biomass to produce fuels and chemicals. Recent work [1] has demonstrated that catalysts containing a highly reducible metal, such as Rh, paired with an oxophilic additive, such as Re, provide high activity and selectivity for C-O hydrogenolysis reactions. Unfortunately, the dispersity of traditional RhRe catalyst in terms of size and composition, along with the bulk average data provided by X-ray absorption spectroscopy (XAS) [2], makes difficult the elucidation of the structure of the nanoparticles and the nature of the active sites. Additionally, this dispersity suggests that there is significant room for improvement in catalyst performance if bimetallic nanoparticles of the correct size and composition can be synthesized.

We will demonstrate a unique synthesis protocol to produce a series of RhRe bimetallic catalysts via controlled surface reactions. These catalysts show a sharp change in activity as a function of the composition, which is absent in catalysts synthesized by traditional methods.

[1] M. Chia, Y. J. Pagán-Torres, D. Hibbits, Q. Tan, H. N. Pham, A. K. Datye, M. Neurock, R. J. Davis, J. A. Dumesic. *JACS* **2011**, 133, 12675-12689.

[2] M. Chia, B. J. O'Neill, R. Alamillo, P. J. Dietrich, F. H. Ribeiro, J. T. Miller, J. A. Dumesic. *J. Catal.* **2013**, 308, 226-236.

376988 Synthesis of Graphene Oxide Via Sonication-Free Exfoliation of Graphite Oxide

Abstract:

Graphene oxide (GO) has attracted significant attention as catalysts or catalyst supports. Synthesis of GO is typically conducted by exfoliation of graphite oxide. Although efficient and high-yield exfoliation of graphite oxide can be achieved using sonication, the use of sonication causes complexity in the synthesis, which may limit the feasibility of large-scale manufacturing of GO. Furthermore, sonication fragments graphite oxide layers into relatively smaller sheet size. Thus, if the goal is to minimize fragmentation of nanosheets and prepare relatively large GO, a milder exfoliation is required. We have reported a scalable simple synthesis of GO via sonication-free exfoliation of graphite oxide. The method consists of repetitive cycles of rapid freezing of a solution containing graphite oxide and subsequent thawing of the resultant solid, which can efficiently exfoliate graphite oxide with approximately 80% GO yield. In this presentation, we will report the progress of the work and the application of the prepared GO to the synthesis of supported metal catalysts.

385435 Improved Photopolymerization Kinetics of Acrylate and Amide Monomers in the Presence of Bulky Organic Salts

Abstract:

As radical reactions are commonly used in the manufacture of polymers and polymer products, interest has increased in new methods of radical polymerization. Among these techniques is the process of photopolymerization in which the reaction proceeds through the production of radicals from a photoinitiator. Previously, ionic liquids (ILs) have been found to increase the polymerization rate of multiple vinyl monomers including styrene and methacrylates. In our study, coordinated ionic liquids were synthesized containing the anion bis(trifluoromethylsulfonyl)imide (Tf_2N) and Li^+ cations coordinated to the polar vinyl monomers methyl methacrylate, hydroxyethyl methacrylate, methyl acrylate, methacrylic acid, acrylic acid, and acrylamide. The resulting substances were photopolymerized and the monomer conversion measured using real-time Fourier transform infrared spectroscopy (FTIR). In addition, salt extraction tests were performed on the resulting polymers and the material properties were examined using differential scanning calorimetry (DSC).

pH-dependent Degradation Kinetics of Polylactic Acid

STEFANO LAZZARI, FABIO CODARI, DAVIDE MOSCATELLI[1], GIUSEPPE STORTI, MASSIMO MORBIDELLI

Institute for Chemical and Bioengineering, Department of Chemistry and Applied Biosciences, ETH Zurich, Valdimir-Prelog-Weg 1, CH-8093 Zurich (Switzerland)
[1] *Dipartimento di Chimica, Materiali e Ingegneria Chimica "Giulio Natta", Politecnico di Milano, I-20131 Milano (Italy)*

Abstract

Polylactic acid (PLA) is a biodegradable and biocompatible polymer. Due to its properties it raised significant interest in the biomedical field, where it is often employed in drug delivery applications. The knowledge of the degradation kinetics of PLA has a key role when designing such drug releasing devices, as it allows modulating their release profiles. Among the different parameters affecting the PLA degradation kinetics, pH is one of the most relevant ones. While previous studies clarified the pH-dependent degradation mechanisms occurring (acid-, base- and non-catalyzed), the values of the corresponding kinetic constants were actually provided only at specific pH values.[1-3]

In this frame, the aim of this work is to study the degradation of lactic acid oligomers as a function of pH in a wide range of values of interest for biomedical applications. In particular, the degradation of oligomers with chain length varying from 2 to 8 repeating units was followed at 37°C for pH values equal to 1, 3.5, 5.7 and 7.4. These experimental data, along with those previously reported at pH = 2[1], are used to develop a detailed kinetic scheme and evaluate the corresponding kinetic constants. The resulting deterministic, pH-dependent model is able to describe the experimental data with reasonable accuracy and an average error of about 5%. The following values of the kinetic constants have been estimated: $k_D^{\alpha} = 4.44 \times 10^{-3}$ $L^2mol^{-2}h^{-1}$, $k_D^{\beta} = 9.29 \times 10^{-4}$ $L^2mol^{-2}h^{-1}$ (acid-catalyzed), $k_{RH} = 1.56 \times 10^3$ $L^2mol^{-2}h^{-1}$, $k_{BB} = 2.88 \times 10^6$ $Lmol^{-1}h^{-1}$ (base-catalyzed) and $k_W = 2.49 \times 10^{-5}$ $Lmol^{-1}h^{-1}$ (non-catalyzed). Notably, these kinetic constants can be used to study also the degradation of PLA polymer chains, as the values have been found to be chain-length independent from the tetramer on. Taking advantage of the results of this kinetic analysis and accounting for the diffusion limitations typical of PLA biomedical devices (e.g. nano- and microparticles), a predictive modeling of the degradation behavior becomes possible for a wide range of pH conditions.

[1] F. Codari, S. Lazzari, M. Soos, G. Storti, M. Morbidelli, D. Moscatelli, *Polym Degrad Stabil* **2012**, *97*, 2460.
[2] S. J. de Jong, E. R. Arias, D. T. S. Rijkers, C. F. van Nostrum, J. J. Kettenes-van den Bosch, W. E. Hennink, *Polymer* **2001**, *42*, 2795.
[3] C. F. van Nostrum, T. F. J. Veldhuis, G. W. Bos, W. E. Hennink, *Polymer* **2004**, *45*, 6779.

Tandem Catalysis: A Prospective Method for Producing Ethylene/α-Olefin Copolymers from Ethylene Stock

Song Guo [1], Hong Fan [1], Bo-Geng Li [1,*] and Shiping Zhu [2]

[1]State Key Laboratory of Chemical Engineering, Zhejiang University

[2]Department of Chemical Engineering, McMaster University, Hamilton, Canada L8S 4L7

1

Introduction of polyethylene

Polyethylene (PE)
- Homopolymers
  HDPE LDPE
- Copolymers (propylene, 1-butene, 1-hexene, 1-octene. <20%)
  LLDPE, POE

Simplest Petroleum-Based Polymer

World production of polyolefins in 2010

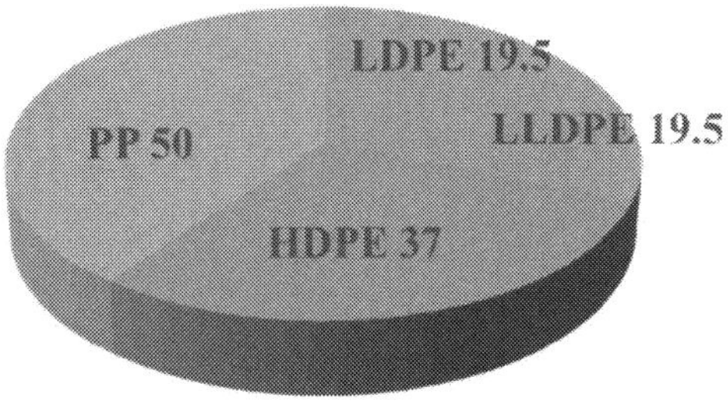

Unit: million tons

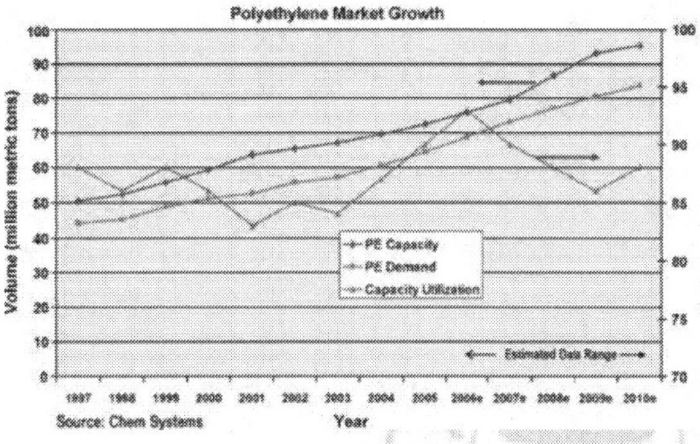

Figure 1. Market growth of polyethylene from 1997.

P. Chum, K. Swogger. Prog. Polym. Sci. 2008, 33, 797–819.

2

浙江大學 ZHEJIANG UNIVERSITY

Introduction of polyethylene

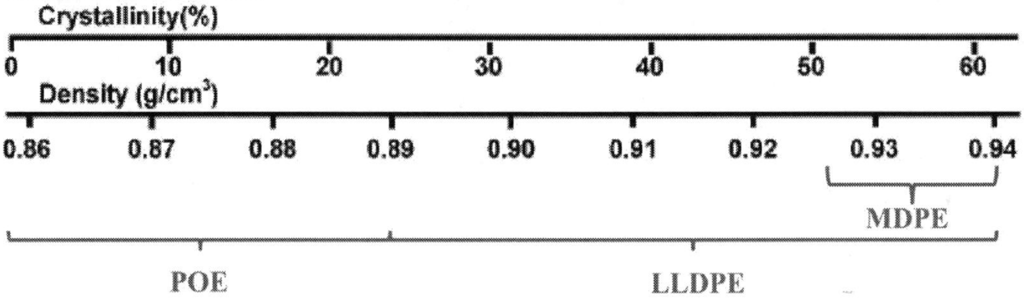

APPLICATIONS	film, sheet, fiber, foams, coatings, pipes
PROCESS	Gas Phase: UCC, BP Solution Phase: DuPont, Dow, DSM Slurry Phase: Philips, Solvay, Mistui
COMONOMER	1-butene, 1-hexene, 1-octene
PRODUCER Using 1-Hexene	ExxonMobil, Eastman, Equistar, Chevron Phillips, Dow, Basell, Innovene, Samsung Total

3

浙江大學 ZHEJIANG UNIVERSITY

Introduction of Tandem Catalysis for Preparation of PE

Tandem Catalysis

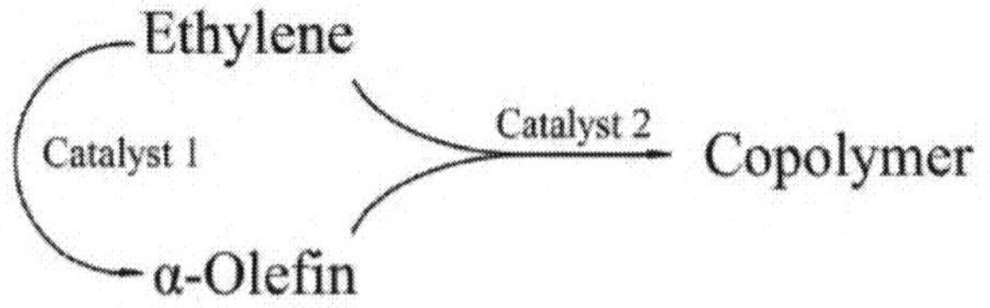

- **Using ethylene-only stock in single reactor**
- **Cost Save in comonomer production, transportation, and storage**

1980s-2000, ethylene oligomerization was mainly focused on dimerization.

SNS & PNP Catalysts for Ethylene Trimerization

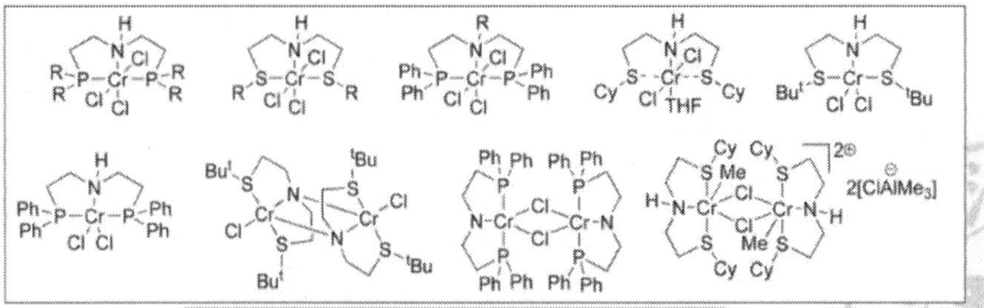

High activity: > 10^6 g/(mol Cr × h)

High Selectivity: > 98% for 1-hexene

D. S. McGuinness. *Chem. Rev.* 2011, 111, 2321–2341.

4

Linear low density polyethylene (LLDPE)

- T_m 120~125 °C, Comonomer content 5~10 wt%.
- Gas phase (1-C4, Unipol, UCC)
- Slurry phase (1-C4, Philips; 1-C6, Solvay)
- 1-C6 and 1-C8 better mechanical properties than 1-C4

Slurry Phase vs. Gas Phase

Cost in production	**Slurry Phase > Gas Phase, but comparable**
Cost in trademark switch	**Slurry Phase is lowest**

Preparation of EH LLDPE via Tandem Catalysts

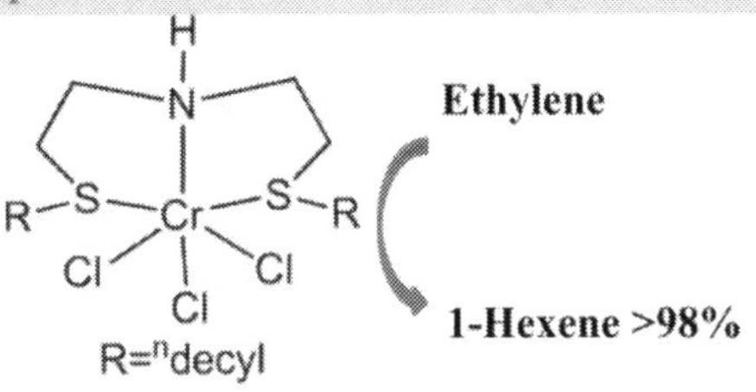

SNS-Cr

+

Supported Z-N catalyst

SLC-S donated
Mg: 6.66 wt%
Ti: 2.24 wt%
THF: 28.58 wt%
Average size: 12.40μm

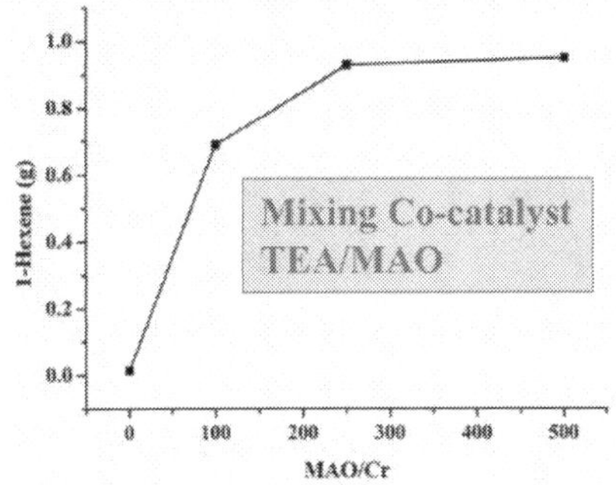

Figure 2. Ethylene trimerization results with different MAO/Cr ratios where TEA/Cr=200.

Advantages

- **Cost in MAO: low level**
- **PE byproduct: trace**
- **Optimal temperature of 70-90 °C: suitable for slurry phase**
- **Morphology of resulting polymer: spherical particle**

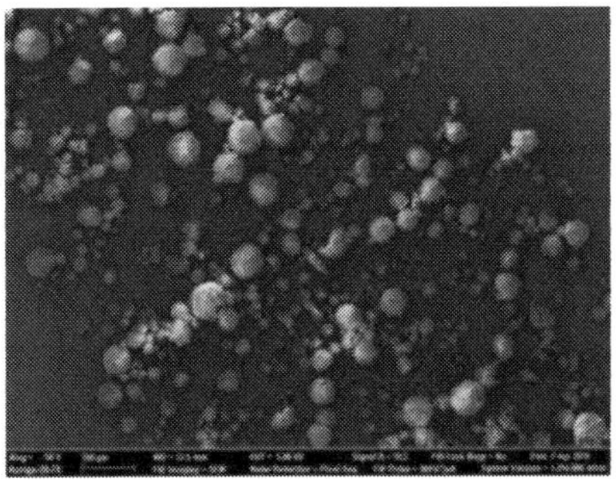

Ethylene homopolymer prepared with SLC-S.
T_m=131.8 °C, M_w=238 kg/mol, PDI=17.7

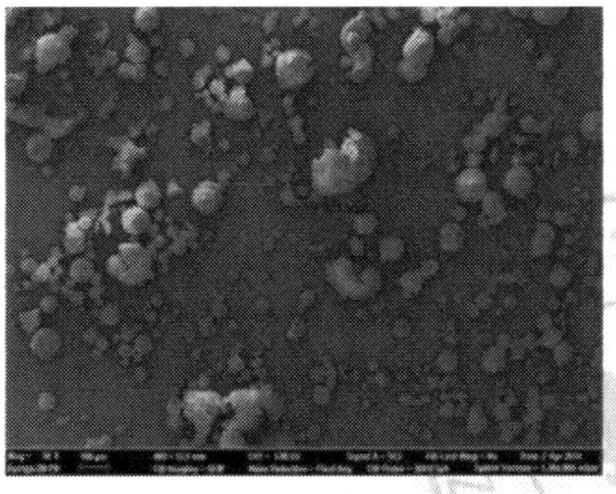

EH LLDPE prepared with SNS-Cr & SLC-S.
T_m=125.0 °C, M_w=149kg/mol, PDI=18.3

ZHEJIANG UNIVERSITY

Tandem Catalysis for Preparation of Polyethylene

High performance polyolefin elastomer (POE)

- High comonomer incorporation (>8 mol%)
- 1-C6 and 1-C8 better than 1-C4, but cost more
- Solution Process, such as INSITE™ of DOW

High Temperature is The Challenge for Tandem Catalyst System

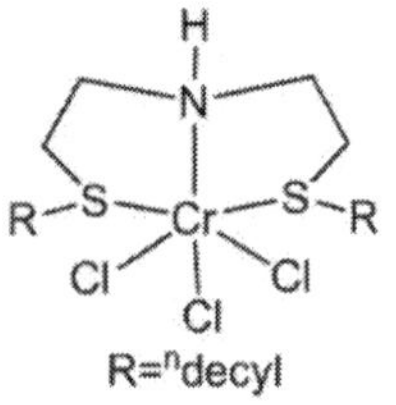

SNS-Cr

+

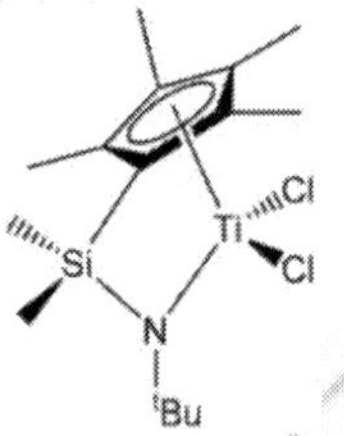

CGC

- High trimerization selectivity > 99%
- High 1-hexene selectivity > 99%
- Can be active over 110 °C

- High Copolymerization Capability
- Can be active over 140 °C
- Commercially used by Dow

7

ZHEJIANG UNIVERSITY

Synthesis of Ethylene-1-Hexene Copolymer via Tandem Catalysis System of SNS and CGC

75 °C & Atmospheric pressure

Range: Semi-crystallite LLDPE ~ Completely amorphous PE

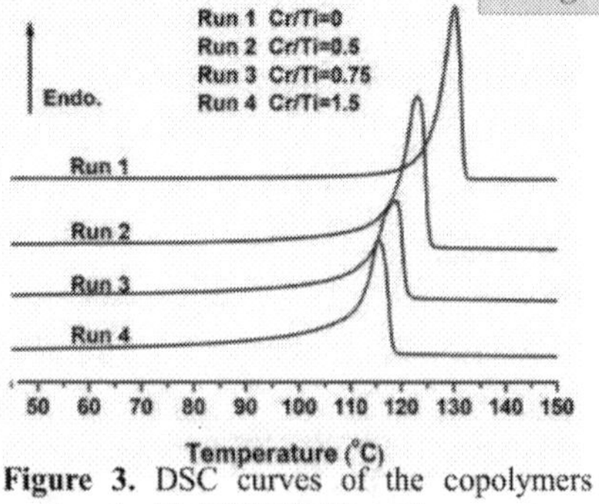

Figure 3. DSC curves of the copolymers produced with CGC-Ti=10 μmol.

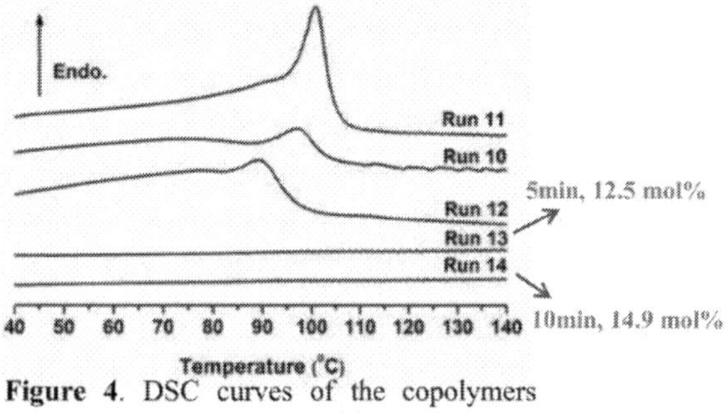

Figure 4. DSC curves of the copolymers produced with pre-trimerization.

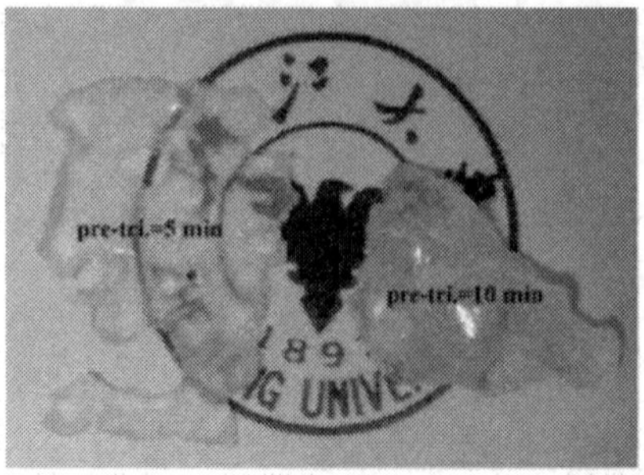

Figure 5. Photos of the products obtained in Run 13 and 14.

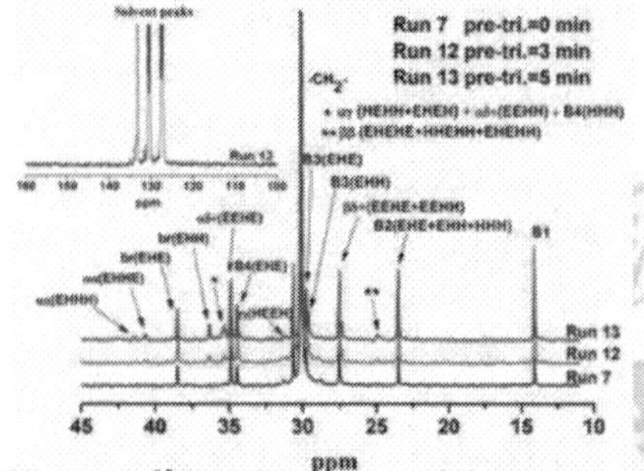

Figure 6. ^{13}C NMR spectra of the copolymers produced with different pre-trimerization times.

S. Guo, et al. ***Macromol. Chem. Phys.*** 2014, 215, 1661-1667.

8

High Temperature (110 °C) High Pressure (31 bar)

Table 1. Results of ethylene trimerization with SNS at high pressure[a)]

Run	T (°C)	1-C6 (g)	Productivity (10^6 g/mol Cr h)	1-C6 in C6 (%)	PE Product: Yield (g)	T_m (°C)	X_c[b)] (%)	Mw (kg/mol)	PDI
O1	75	50.5	6.05	99.8	0.05	125.6	40.2	539	18.3
O2	90	39.7	4.75	99.5	0.09	125.5	44.2	388	6.6
O3	110	28.3	3.39	99.3	0.04	125.8	49.0	284	8.6

[a)]Polymerization condition: SNS = 16.7μmol, Al/Cr = 1000, ethylene pressure = 31 bar, toluene volume = 250 ml, reaction time = 30 min; [b)]The heat of fusion of completely crystalline PE is 290 J/g.

Table 2. Ethylene polymerization with CGC and SNS at high pressure and high temperature[a)]

Run	CGC (μmol)	SNS (μmol)	P (bar)	Pre-tri.[b)] (min)	Productivity (10^6 g/mol Ti h)	F_2[c)] (mol %)	T_m (°C)	Xc (%)	M_w (kg/mol)	PDI
T1	5	10	31	0	3.94	1.5	101.0 110.4	34.3	124	3.0
T2	10	10	31	1	7.94	3.0	105.0	22.4	80.2	4.0
T3	10	20	31	5	6.48	3.1	103.8	21.9	64.2	2.5
T4	10	20	23	1	5.12	2.5	103.6	24.2	54.6	2.7

[a)]polymerization condition: copolymerization time = 30 min, reactor set temperature = 110 °C, 600 rpm, MAO = 15 mmol, 250 ml toluene; [b)]Trimerization time for SNS before introduction of CGC; [c)]Comonomer incorporation in copolymer.

S. Guo, et al. *Macromol. React. Eng.* 2014, DOI: 10.1002/mren.201400027.

Sample	T_m (°C)	Xc (%)	Mw (kg/mol)	PDI
CPOE	97.1	23.7	97.1	1.8

Comparable mechanical properties with commercial POE!

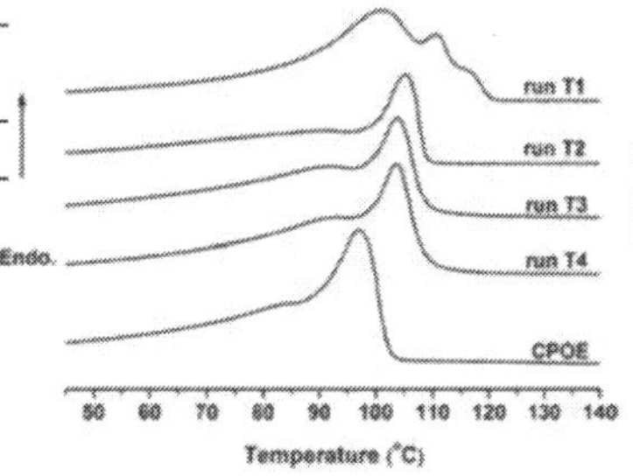

Figure 7. DSC curves of the samples prepared from tandem polymerization of SNS and CGC at high temperature and high ethylene pressure.

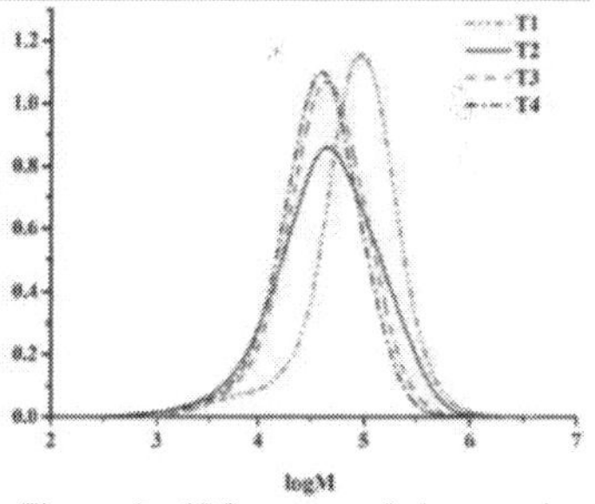

Figure 8. GPC curves of the samples prepared from tandem polymerization of SNS and CGC at high temperature and high ethylene pressure.

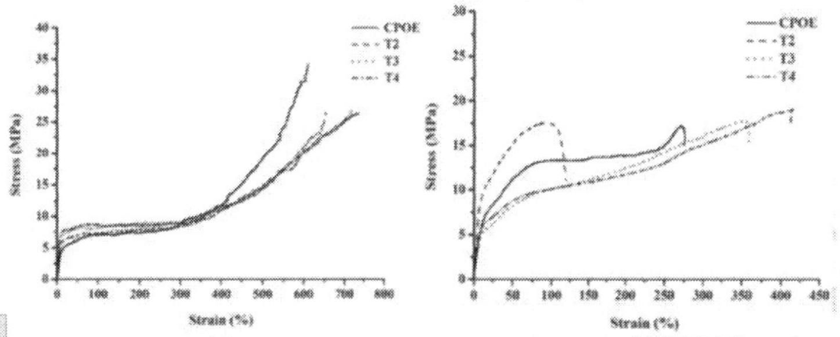

(a) (b)

Figure 9. Engineering stress-strain curves of Sample T2-T4 and CPOE. The tensile specimens were prepared by (a) compression molding and (b) injection molding.

Mechanism and Employing Moment Method to Build Kinetics Model

Trimerization mechanisms	
Active-site formation	$C_1 \xrightarrow{k_{a}} C_1^*$
Ethylene chelation	$C_1^* + 2M_A \xrightarrow{k_1} C_1M_AM_A^*$
Ethylene insertion	$C_1M_AM_A^* + M_A \longrightarrow C_1M_AM_AM_A^*$
1-Hexene elimination	$C_1M_AM_AM_A^* \longrightarrow M_B + C_1^*$
Deactivation	$C_1^* \xrightarrow{k_d} C_{d1}$

Copolymerization mechanisms	
Active-site formation	$C_2 \xrightarrow{k_{a2}} C_2^*$
Active-site initiation	$C_2^* + M_j \xrightarrow{k_{ij}} P_{1,j}^*, \quad j = A, B$
Chain Propagation	$P_{n,i}^* + M_j \xrightarrow{k_{pij}} P_{n+1,j}^*, \quad i = A, B \quad j = A, B$
Transfer to monomer	$P_{n,i}^* + M_B \xrightarrow{k_t} D_n + P_{1,B}^*$
β-hydrogen elimination	$P_{n,i}^* \xrightarrow{k_\beta} D_n + C_2^*$
Spontaneous deactivation	$P_{n,i}^* \xrightarrow{k_{d2}} D_n + C_{d2}$

Semi-Batch Experiments to Obtain Kinetics Parameters

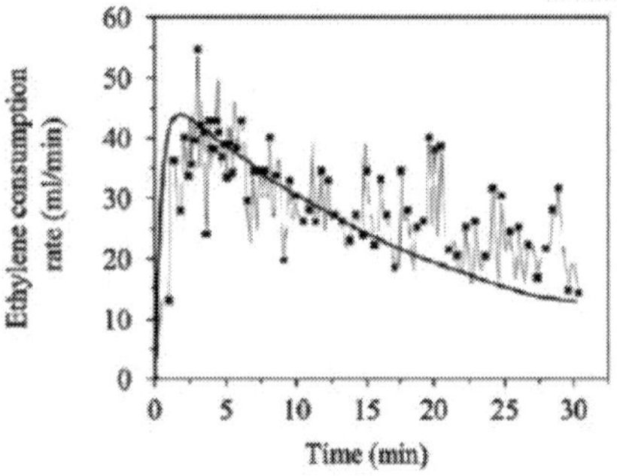

Trimerization

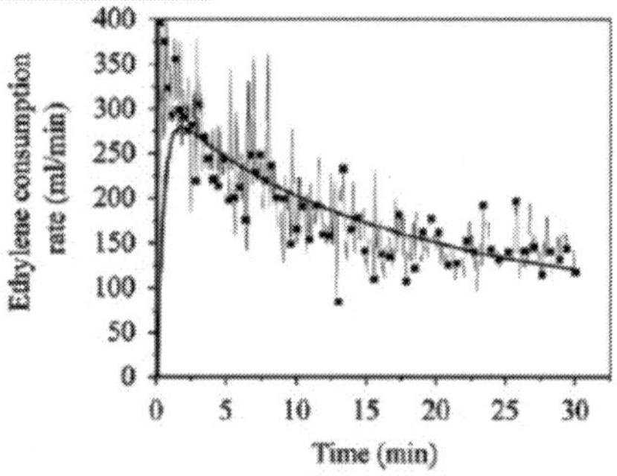

Tandem Polymerization

B -G Li, *et al. Chem. Eng. Sci.* 2008, *63*, 2057 – 2065.

Simulation in CSTR Steady state will be approached when switching process conditions

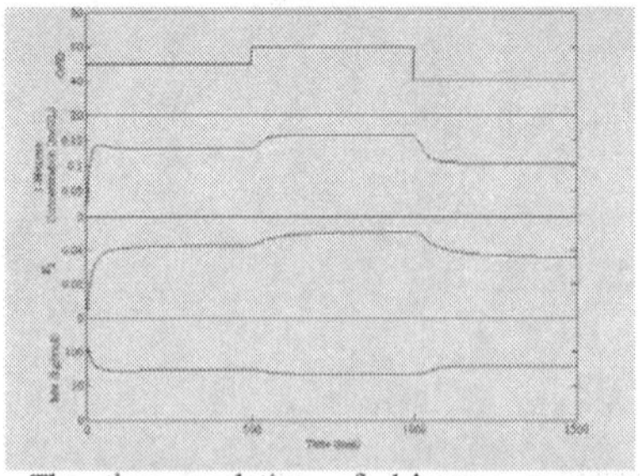

Figure 10. The time evolution of 1-hexene concentration, copolymer composition, and Mw for continuous copolymerization carried out with switching $[Cr]_{in}/[Zr]_{in}$.

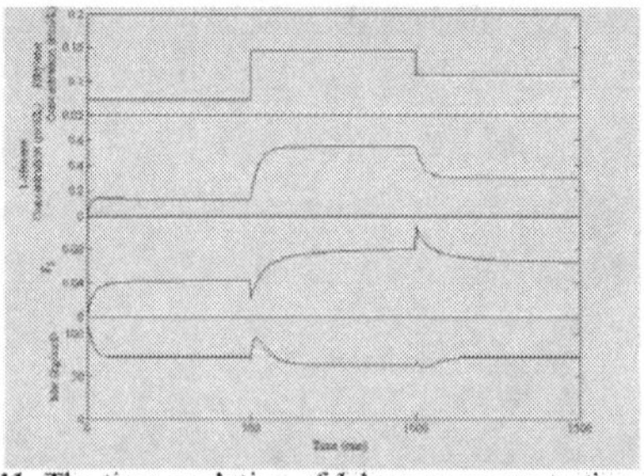

Figure 11. The time evolution of 1-hexene concentration, copolymer composition, and Mw with ethylene concentration switched from 0.0723 mol/L to 0.1446 mol/L, and to 0.1085 mol/L.

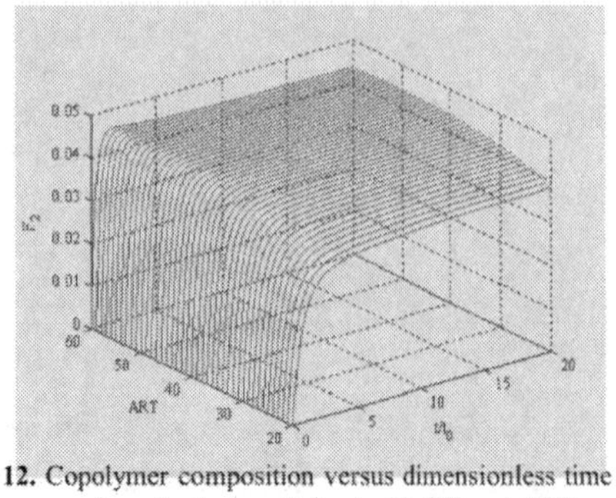

Figure 12. Copolymer composition versus dimensionless time for continuous copolymerization carried out with different ARTs.

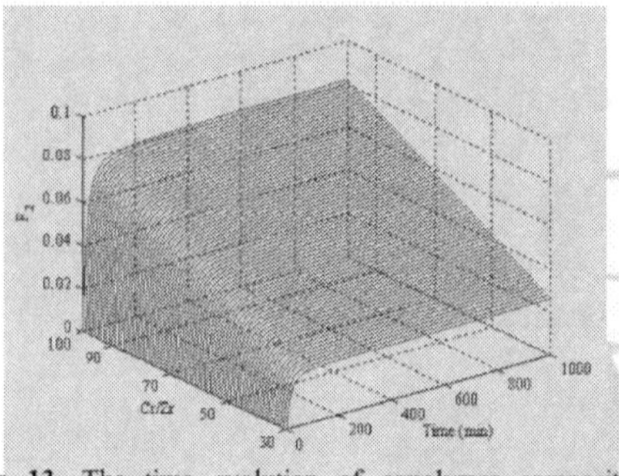

Figure 13. The time evolution of copolymer composition for continuous copolymerization carried out with different $[Cr]_{in}/[Zr]_{in}$.

浙江大學 ZHEJIANG UNIVERSITY

Summary

• The trial of preparing 1-hexene based LLDPE via tandem polymerization of SNS and supported Z-N catalyst appears promising.

• Metallocenes cooperate well with SNS complexes in the presence of MAO as co-catalyst, under 75 °C and atmospheric pressure.

• Ethylene-1-hexene copolymers with comonomer content up to 14.9 mol% was obtained by the tandem polymerization of SNS and CGC.

• Above 110°C, the resulting copolymers having comparable mechanical properties with commercial POE.

• The above results indicate the tandem polymerization carried out in CSTR under high-temperature high-pressure will be inspiring.

• According to the simulation result, higher Cr/Ti ratio and shorter residence time should be considered when applied in CSTR system under high-temperature high-pressure.

15

385176 Dispersed Microenvironments for Chemical Evolution

Abstract:

Biopolymers in current living organisms are usually synthesized with the help of enzymes and cell compartmentalization. Such mechanism was unlikely to exist in the prebiotic earth. Therefore, the first biopolymers for the evolution of early life forms should come from simpler systems with abiotic reaction mechanisms. Miller-Urey experiments have demonstrated the possibility of the formation of amino acids and hydroxy acids in the prebiotic world.[1] Those materials can be used as building blocks for biopolymers. However, even though the source of monomers is understood, the formation of polymers in the prebiotic ocean still faces a number of challenges. The high water concentration in the ocean can hamper the polycondensation reaction by shifting the equilibrium to the reactant side. This enhances the difficulty of forming long chain polymers and shortens their life time.

In this work, an oil/water biphasic reaction media is designed to solve this problem. It has been shown that some water immiscible organic solvents can exist in the prebiotic world.[2] Because of the lower water concentration in those oil solvents, they can facilitate the polymerization reaction by shifting the equilibrium to the product side and protect the synthesized polymers from hydrolysis by water. Also, the increased hydrophobicity of the growing polymer chain can further concentrate those polymers in the oil phase and speed up the evolutionary process. The oil phase can also be dispersed and stabilized by surfactants to form an emulsion. The dispersed oil phase has much higher interfacial area and can facilitate the absorption of building blocks from the prebiotic ocean.

The present work first focuses on the polymerization of lactic acid in the oil/water biphasic system. Polyesters have been proposed as one possible ancestor for polypeptides[3] and lactic acid is a significant product from the Miller-Urey experiments. Also, polylactic acid (PLA) is a biodegradable polymer that has been well characterized and can be used in drug delivery applications.

Two different abiotic reaction pathways have been studied in this work. The first one is the polycondensation reaction of lactic acid with dicyclohexylcarbodiimide as condensing agent and 4-dimethylaminopyridine as catalyst. It has proved to be an efficient condensing agent for the polymerization of lactic acid when organic solvent is used alone.[4] In our work, the interfacial reaction between the hydrophobic carbodiimide and hydrophilic lactic acid is able to produce long chain PLA. The second reaction pathway is starting from the ring-opening polymerization of lactide, the cyclic dimer of lactic acid. [5] Lactide has higher solubility in organic solvents than lactic acid. Therefore, the reaction can proceed directly in the oil phase. The ring-opening polymerization can be catalyzed by 4-dimethylaminopyridine which is a weak organic base and its distribution between oil and aqueous phase can be controlled by the pH of water. Polyester with high yield can be obtained and preserved in the oil phase under near neutral pH and room temperature.

The findings in this work provide another methodology to solve the mystery of the origin of life. The presence of oil phase successfully facilitates the polymerization and extends the life time of the synthesized polyester. The same biphasic reaction system can also be applied to the reaction and evolution of other biopolymers in the future.

Reference

1. Miller, S. L.; Urey, H. C., Organic compound synthesis on the primitive earth: Several questions about the origin of life have been answered, but much remains to be studied. *Science* **1959,** *130* (3370), 245-251.

2. Lasaga, A. C.; Holland, H. D.; Dwyer, M. J., Primordial oil slick. *Science* **1971,** *174* (4004), 53-55.

3. Orgel, L., Some consequences of the RNA world hypothesis. *Origins Of Life And Evolution Of The Biosphere* **2003,** *33* (2), 211-218.

4. Akutsu, F., Inoki, M., Uei, H., Sueyoshi, H., Kasashima, Y., Naruchi, K., Yamaguchi Y., Sunahara, M., Synthesis of poly(lactic acid) by direct polycondensation of lactic acid using 1,1'-carbonyldiimidazole, N,N,N',N'-tetramethylchloroformamidinium chloride and N,N'-dicyclohexylcarbodiimide as condensing agents. *Polym Journal* **1998,** *30*, 421.

5. Coulembier, O., Dubois, P., 4-dimethylaminopyridine-based organoactivation: From simple esterification to lactide ring-opening "Living" polymerization. *Journal of Polymer Science Part A: Polymer Chemistry* **2012,** *50* (9), 1672-1680.

Non-aqueous suspension polycondensation for the preparation of poly(*p*-phenylene terephthalamide)

Peijian Wang, Kai Wang, Jisong Zhang, Guangsheng Luo*

The State Key Laboratory of Chemical Engineering, Department of Chemical Engineering, Tsinghua University, Beijing, 100084, China

wpjqhu@gmail.com, 86-10-62797493

Outline

- Background
- Research Contents
- Conclusions

Background of poly(*p*-phenylene terephthalamide) (PPTA) 3

High performance properties :

- Tensile strength
- Corrosion resistance
- Roughness
- Weight and thermal stability

Wildly used in:

- Composite material
- Bulletproof material
- Rubber reinforced material
- Asbestos replacement material

Background of poly(*p*-phenylene terephthalamide) (PPTA) 4

Different methods for the synthesis of PPTA:

- Interfacial polycondensation method
- Vapor polycondensation method
- Direct polycondensation method
- Low temperature method

Reactant mixing

Isolation of oxygen and water

Rapid and strongly exothermic

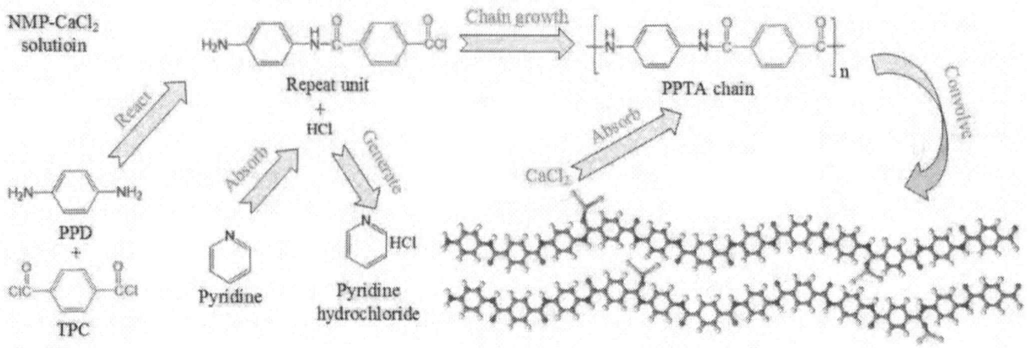

Experimental design

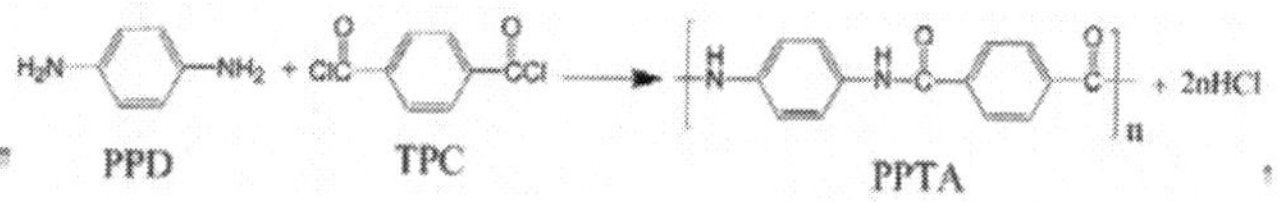

Non-aqueous suspension polycondensation method

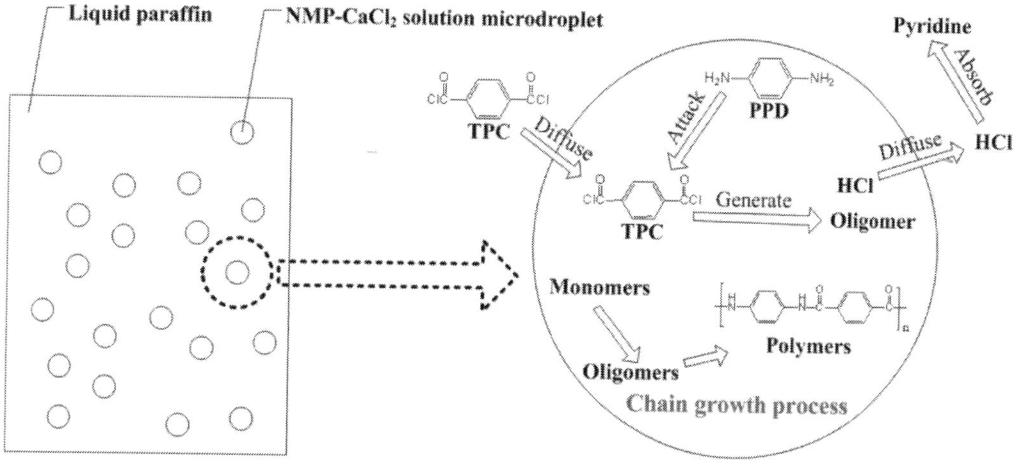

Surrounded by non-reactive solvent

Easy removal of HCl

Temperature and viscosity

Outline

6

- **Background Introduction**
- **Experimental section**
- **Conclusions**

Non-aqueous suspension

7

NMP microdroplets suspended in liquid paraffin under microscope

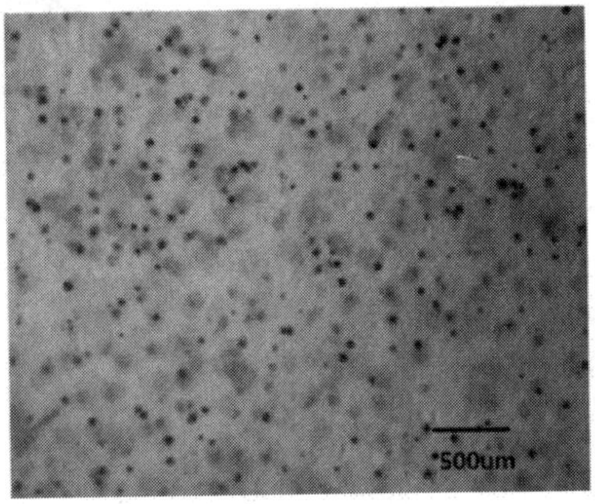

$$\overline{d} = k\frac{D_V \cdot R \cdot v_d \cdot \varepsilon}{D_S \cdot N \cdot v_m \cdot C_s}$$

NMP solution was dispersed in liquid paraffin under the strong shearing force from the emulsifying machine, forming a non-aqueous suspension with droplet size in dozens of micrometers.

Arshady, R., Colloid. Polym. Sci., 1992, 270(8): 717-732

Characterization

8

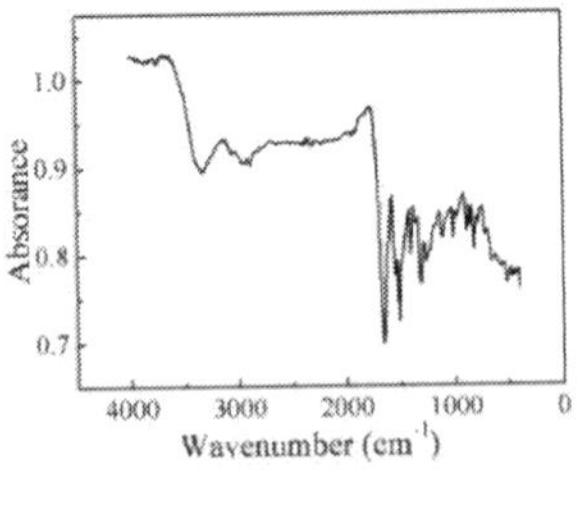

IR

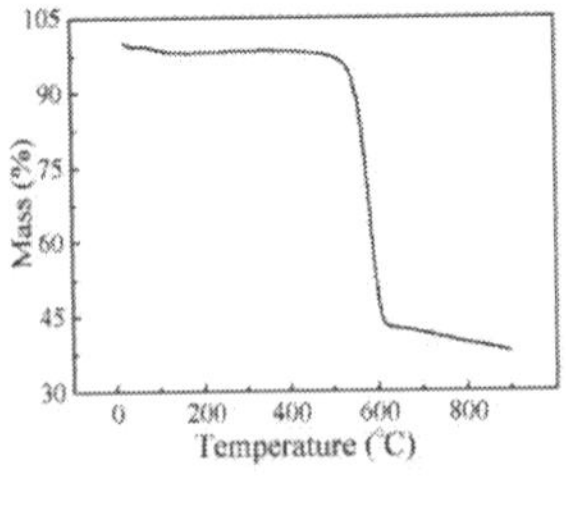

TGA

Sample No.	N	C	H	O
Theoretical	11.76	70.59	4.20	13.45
1	11.32	68.92	4.28	15.48
2	11.40	69.23	4.26	15.11
3	11.41	69.14	4.27	15.18

EA

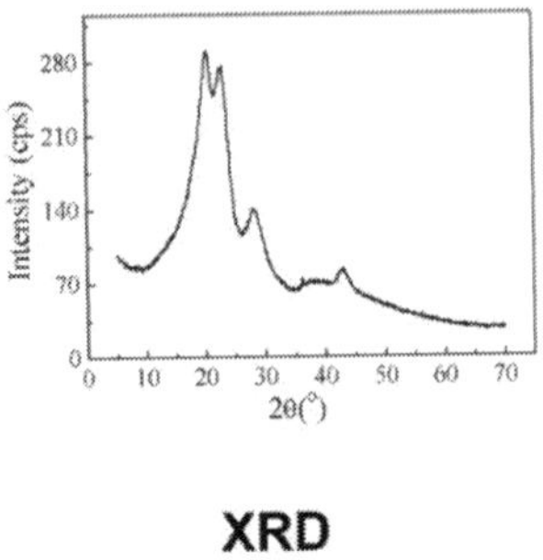

XRD

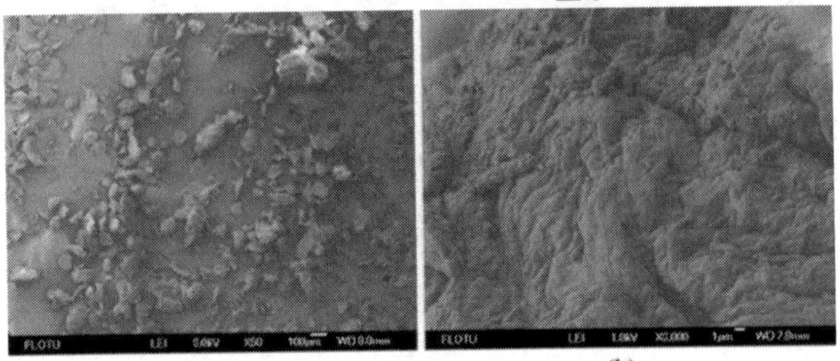

(a) (b)

SEM

Effects of different variables on the M_W of PPTA

9

Phase ratio

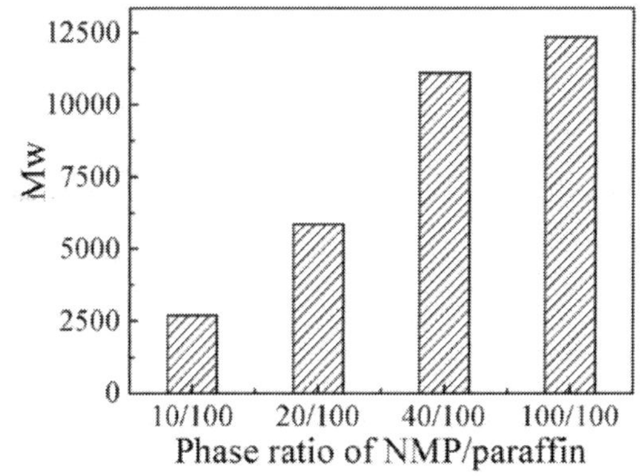

Reaction time

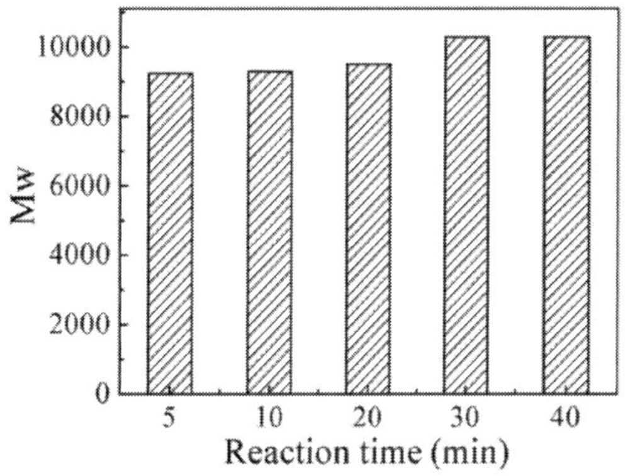

Small → Unbalance of monomer molar ratio

Large → Microdroplets coalesce quickly

Fast reaction

Effects of different variables on the M_W of PPTA

10

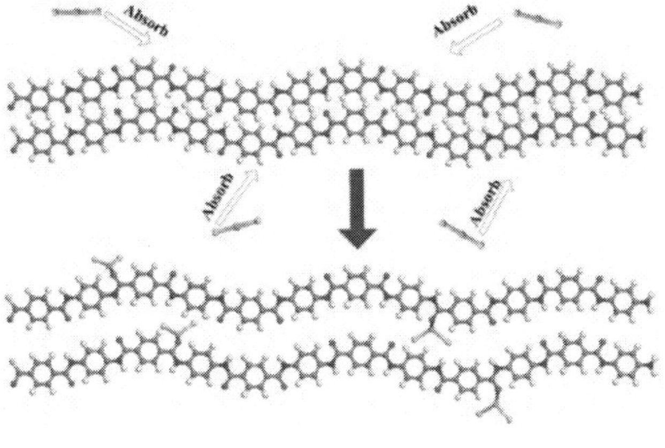

Complexation between calcium ions and carbonyl groups destroys the strong hydrogen bonding between PPTA chains

$CaCl_2$ concentratioin

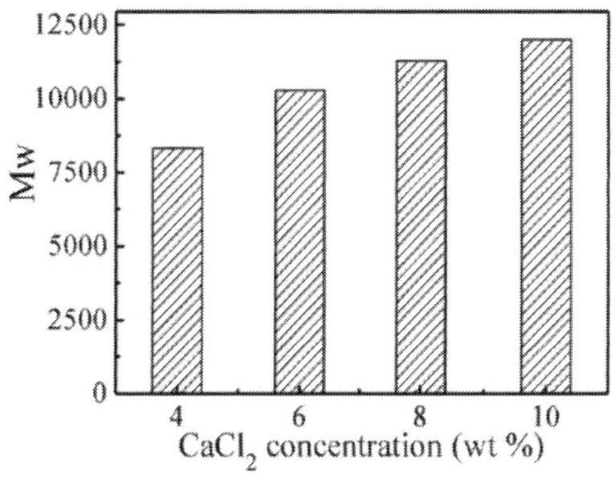

Effects of different variables on the M_W of PPTA

Monomer concentration

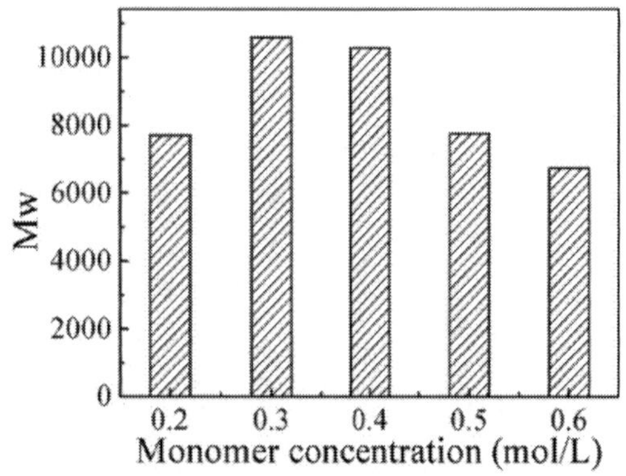

Medium monomer concentration range is better

Monomer molar ratio

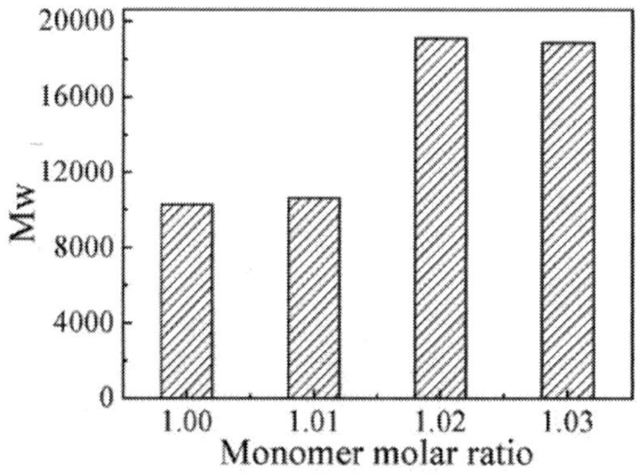

$$\overline{X}_n = \frac{1+r}{1+r-2rP}$$

About 2 percent more TPC

Effects of different variables on the M_W of PPTA

Temperature

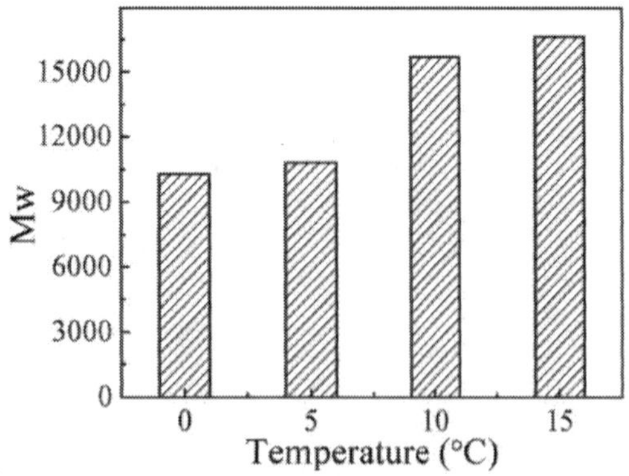

↑ Solubility of monomers and polymers

↓ Viscosity of the NMP phase

Stirring rate

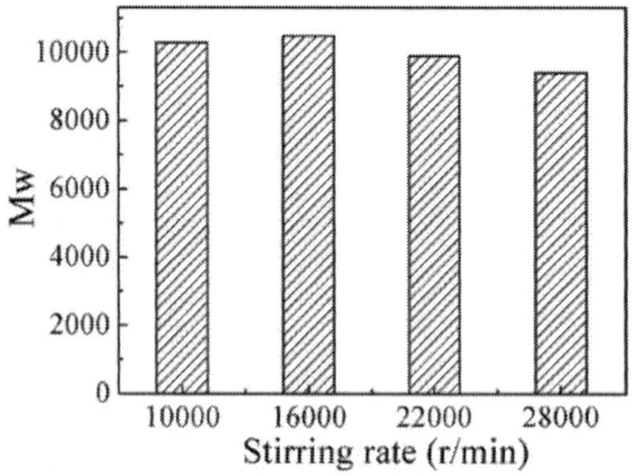

Overall effect is not huge

$$\overline{d} = k\frac{D_V \cdot R \cdot v_d \cdot \varepsilon}{D_S \cdot N \cdot v_m \cdot C_s}$$

Preparationof PPTA with a microreactor

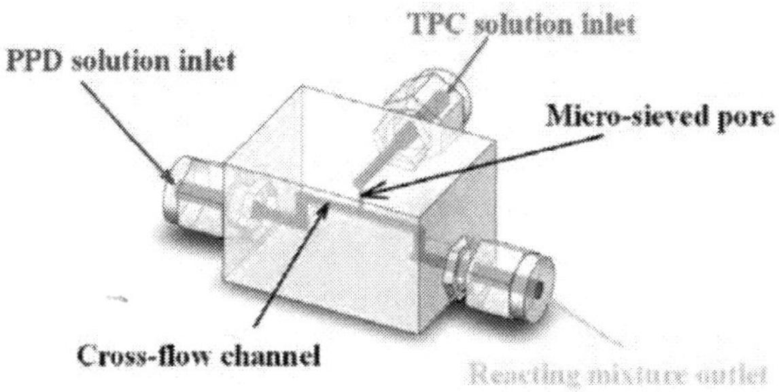

T-shaped microreactor

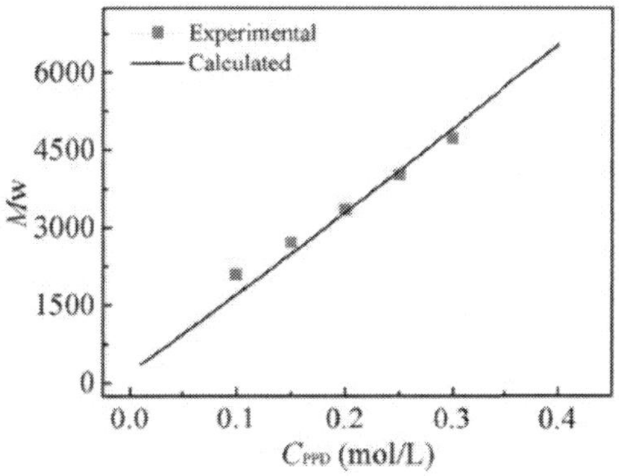

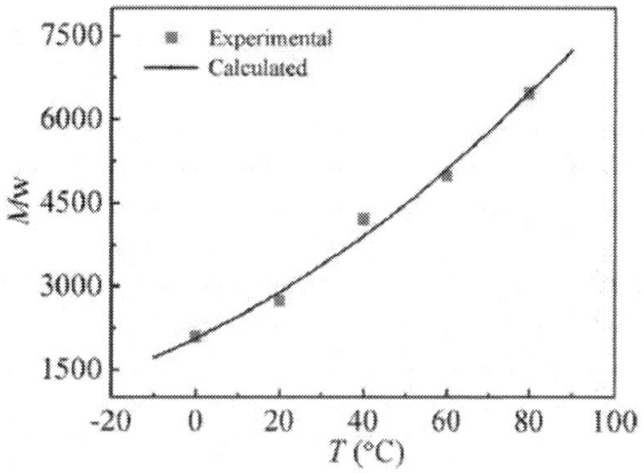

Outline

- **Background Introduction**
- **Research Contents**
- **Conclusions**

Conclusion

- The polycondensation between PPD and TPC was proposed in a non-aqueous suspension system to synthesize PPTA.
- Each of the NMP microdroplets suspended in paraffin liquid serves as a microreactor, in which the polycondensation takes places.
- Application of this new approach leads to less strict demand for removal of oxygen and water, easy removal of HCl, well-controlled temperature and low system viscosity
- The molecular weight can be adjusted easily by changing the following reaction parameters: phase ratio of NMP/paraffin, reaction time, $CaCl_2$ concentration, monomer concentration, monomer molar ratio, temperature and stirring speed.

Thank you for your listening!

Any question will be highly appreciated!

379180 Mathematical Model for the Bulk Polymerization of Styrene Using Multifunctional Initiators

Abstract:

A detailed mathematical model for bulk polymerization of styrene (St) using multifunctional initiators –both linear and cyclic- was developed. The model is based on a kinetic scheme that considers chemical and thermal initiation by sequential decomposition, propagation, transfer to monomer, diffusion-controlled termination by combination and re-initiation reactions and is used to predict the evolution of the chemical species and the molecular structure of the obtained polymeric species. The model calculates the evolution of total or average variables (e.g, total radicals, total polymer, molecular weight distributon moments) as well as the evolution of the detailed complete polymer molecular weight distribution, with polymer species characterized by chain length and number of undecomposed labile groups. The mathematical model was adjusted and validated using experimental data obtained from bulk polymerizations of styrene using various peroxide-type multifunctional initiators: diethyl ketone triperoxide (DEKTP, ctrifunctional cyclic), pinacolone diperoxide (DPP, bifunctional cyclic), 1,1-bis(tert-butylperoxy)cyclohexane (L331M80, bifunctional linear) and benzoyl peroxide (BPO, monofunctional linear) at reaction temperatures of 110-130 °C, where initiator decomposition is mostly sequential. The model is able to very adequately predict experimental polymerization rates and complete molecular weight distributions by adjusting the initiator decomposition constants. The model is used to theoretically evaluate initiator structure and functionality, as well as reaction conditions in a bulk St polymerization process. For a given initiator, the model predicts the existence of optimum temperature and initial initiator concentration resulting in high productivity and increased product quality, which was experimentally validated.

390814 Propylene Polymerization with Metallocene/Methylaluminoxane Catalysts: Mechanisms, Modeling and Simulation

Abstract:

AIChE-2014

Abstract

Polymers and co-polymers of propylene find extensive industrial and commercial applications like engineering plastics, elastomers and synthetic rubbers. Transition metal compounds are generally used as catalysts in polymerization of propylene, Ziegler Natta catalyst being the first successful kind among them. These catalysts are the derivatives of transition metals such as titanium ($TiCl_4$, $TiCl_3$), vanadium (VCl_4, $VOCl_3$), etc. and are generally employed with co-catalysts which are organometallic compounds, such as $Al(CH_3)_3$, $Al(C_2H_5)_3$, etc.

Ziegler-Natta catalysts are heterogeneous catalysts with multiple sites. Metallocene/methylaluminoxane complexes are the catalysts used next to Ziegler Natta catalysts in the polymer industry recently. Metallocene catalyzed propylene polymerization has recently attracted research interest since these catalysts allow the production of tailored macromolecules with properties those can be accurately designed. A broad spectrum of properties and applications of the polypropylene can be attained with metallocenes due to their single types of sites.

Two major types of metallocene catalysts are used in propylene polymerization –first type being bis-metallocene complexes of transition metals, (Cp_2MtCl_2) and the second type based on complexes containing one cyclopentadienyl group like [C_5H_4-$SiMe_2$-(tert-Bu)N]$TiCl_2$-MAO and [C_5Me_4-$SiMe_2$-(tert-Bu)N]$TiCl_2$-MAO, etc.
Homogeneity and single site characteristics of metallocenes enable to correlate metallocene structures with polymer properties such as molecular weight, stereochemical microstructure, behavior and mechanical properties. Understanding reaction mechanisms and kinetics of metallocene catalyzed polymerization reactions is very important to design/identify a catalyst for promising product properties or to modify the existing Ziegler-Natta based technology with metallocenes.

This article deals with the presents state of the art on the reaction mechanisms and kinetics involved in Ziegler-Natta and metallocene catalyzed propylene polymerization. Further, an approach of microscopic modeling and simulation of propylene polymerization through different metallocene catalyst systems is presented and discussed using a novel *Natural Logarithmic Differential Evolution* optimization algorithm.

References:

1. Kissin Y., In: Alkene Polymerization Reactions with Transition Metal Catalysts:Studies in surface science and catalysis, Elsevier publications, vol.173.

2. Ochoteco E., Vecino M., Montes M. and Cal J. C., Chemical Engineering Science 56: p. 4169–4179 (2001).

3. Song F., Hannant M. D., Cannon R. D. and Bochmann M., Macromol. Symp. 213: p. 173-185 (2004).

4. W. Kaminsky and Laban A., Applied Catalysis A: General 222: p. 47–612 (2001).

5. Zhang J., Wang X. and Jin G., Coordination Chemistry Reviews 250: p. 95–109 (2006).

367865 Simulation of Urethane Foaming Processes

Abstract:

Urethane formulations typically consist of two to five co-monomers and from four to eight additives including catalysts and surfactants. These thermoset polymers form in a process of rapidly changing physical and chemical properties. The complexity of this process has resulted in an industry that has been dominated by expertise embedded in people and formulation development that can accurately be described as an art. This creates a particularly difficult situation for advances in sustainability, especially use of soy-based polyols, because of how the momentum of proven formulations dominate the industry.

This paper is on the simulation of the complex urethane forming processes. A simulation program for urethane-forming reactions has been developed to assist in adopting soy-based resins in formulations. This paper describes Matlab program used to simulate urethane-forming reactions including the ability to simulate the reactions, viscosity, and gas formation in formulations containing up to three different polyols.

373748 Mechanical and Thermal Properties of Cross-Linked Phenolic Resins Using Molecular Dynamics

Abstract:

To gain insight into the design of materials, it is valuable to understand how the chemical make-up at the nano-scale can influence the thermal and mechanical bulk properties. An atomistic computational study allows us to manipulate the structural make-up of individual phenolic chains as well as generate various cross-linked (or cured) systems. In this study, molecular dynamics simulations of bulk phenolic systems were performed with the software LAMMPS. An all-atom force field was chosen to investigate how the strength and thermal conductivity of the phenolic material varies as a function of the degree of cross-linking and chemical make-up of the phenolic chains. Small-scale mechanical tests were performed to compute various moduli for the phenolic systems above and below the glass transition at varied degrees of cross-linking. The thermal conductivity was obtained using the Green-Kubo approach for the virgin phenolic system as well as the strained systems.

AUTHOR INDEX

AUTHOR INDEX

AUTHOR INDEX

AUTHOR INDEX

AUTHOR INDEX

AUTHOR INDEX

AUTHOR INDEX

AUTHOR INDEX

AUTHOR INDEX

AUTHOR INDEX

AUTHOR INDEX

AUTHOR INDEX

AUTHOR INDEX

AUTHOR INDEX

AUTHOR INDEX